Environmental Science

Environmental Science

Toward a Sustainable Future

Eighth Edition

Richard T. Wright

Gordon College

Bernard J. Nebel

Catonsville Community College

Upper Saddle River, New Jersey 07458

Library of Congress Cataloging-in-Publication Data

Wright, Richard T.
    Environmental science : toward a sustainable future / Richard T. Wright, Bernard J.
Nebel. — 8th ed.
        p. cm.
    Includes bibliographical references and index.
    ISBN 0-13-032538-4
        1. Environmental sciences. I. Nebel, Bernard J. II. Title.

GE105 .N42 2002
363.7 — dc21

2001021518

Executive Editor: Daniel Kaveney
Associate Editor: Amanda Griffith
Editor in Chief: Sheri L. Snavely
Marketing Manager: Martha McDonald
Production Editor: Shari Toron
Vice President of Production and Manufacturing: David W. Riccardi
Executive Managing Editor: Kathleen Schiaparelli
Assistant Managing Editor: Beth Sturla
Executive Marketing Manager: Jennifer Welchans
Assistant Managing Editor, Media: Alison Lorber
Media Production Support: Elizabeth Wright
Media Editor: Chris Rapp
Director, Creative Services: Paul Belfanti
Director of Design: Carol Anson
Art Director: Maureen Eide
Interior Design: Laura Garner
Cover Photo: Offshore oil platform–Ken Graham/Stone; Solar energy station–IFA Bilderteam/Leo De Wys;
    Modern Windmills–Library LTD/Leo De Wys; Windmill on farmstead–Greg Latza, PeopleScapes, Inc.
Managing Editor, Audio/Video Assets: Grace Hazeldine
Manufacturing Manager: Trudy Pisciotti
Manufacturing Buyer: Michael Bell
Photo Editor: Beth Boyd
Photo Researcher: Tobi Zausner
Photo Coordinator: Reynold Rieger
Illustrations: ArtWorks; MapQuest
Copy Editor: Brian Baker
Editorial Assistants: Margaret Ziegler/Nancy Bauer
Proofreader: Michael Rossa
Text Composition/Prepress: Karen Noferi

 ©2002, 2000, 1998, 1996, 1993, 1990, 1987, 1981 by Prentice-Hall, Inc.
Upper Saddle River, New Jersey 07458

Printed in the United States of America
ISBN: 0-13-032538-4 (College)
4567890

ISBN: 0-13-093654-5 (High School)
34567890

Prentice-Hall International (UK) Limited, London
Prentice-Hall of Australia Pty. Limited, Sydney
Prentice-Hall Canada Inc., Toronto
Prentice-Hall Hispanoamericana, S.A., Mexico
Prentice-Hall of India Private Limited, New Delhi
Prentice-Hall of Japan, Inc., Tokyo
Pearson Education Asia Pte. Ltd.
Editora Prentice-Hall do Brasil, Ltda., Rio de Janeiro

Total 10% Recycled Paper

All Post Consumer Waste

About the Authors

Richard T. Wright is Professor Emeritus of Biology at Gordon College in Massachusetts, where he taught environmental science for 28 years. He earned a B.A. from Rutgers University and a M.A. and Ph.D. in biology from Harvard University. For many years Wright received grant support from the National Science Foundation for his work in marine microbiology and, in 1981, he was a founding faculty member of Au Sable Institute of Environmental Studies in Michigan, where he also served as Academic Chairman for 11 years. He is a Fellow of the American Association for the Advancement of Science, and in 1996 was appointed a Fulbright Scholar to Kenya. He is a member of many environmental organizations, including the Nature Conservancy, Habitat for Humanity, the Union of Concerned Scientists, Massachusetts Audubon, and others, and is a supporting member of the Trustees of Reservations. Wright is involved full time in writing and speaking about the environment, and spends his spare time gardening, fishing, hiking, birding and enjoying his 7 grandchildren.

Bernard J. Nebel is Professor Emeritus of Biology at Catonsville Community College in Maryland. He earned his Bachelor of Arts from Earlham College and his Ph.D. from Duke University. Nebel was one of the first college professors to develop a comprehensive environmental science course and write a text for the subject. Nebel has recently developed an elementary (K-5) science curriculum designed to help children understand the world, their place in it, and their responsibility toward it. Nebel is a member of the American Association for the Advancement of Science, the American Institute of Biological Sciences, the American Solar Energy Society, and the National Association of Science Teachers. He strives to make a difference in the environment in his personal life; his urban backyard is a small ecosystem complex of pond, fruit trees, and garden that is supported by composted wastes. He is an active supporter of Freedom From Hunger, Habitat for Humanity, the World Wildlife Fund, Conservation International, and other environmental organizations.

Brief Contents

1   Introduction: Sustainability, Stewardship, and Sound Science   1

PART ONE   *Ecosystems and How They Work*   **23**

2   Ecosystems: What They Are   25
3   Ecosystems: How They Work   53
4   Ecosystems: Populations and Succession   81
5   Ecosystems and Evolutionary Change   107

PART TWO   *The Human Population*   **135**

6   The Human Population: Demographics   137
7   Issues in Population and Development   163

PART THREE   *Renewable Resources*   **187**

8   Soil and the Soil Ecosystem   189
9   Water: Hydrologic Cycle and Human Use   213
10   The Production and Distribution of Food   241
11   Wild Species: Biodiversity and Protection   263
12   Ecosystems as Resources   287

PART FOUR   *Energy*   **315**

13   Energy from Fossil Fuels   317
14   Nuclear Power: Promise and Problems   341
15   Renewable Energy   365

PART FIVE   *Pollution and Prevention*   **389**

16   Environmental Hazards and Human Health   391
17   Pests and Pest Control   415
18   Water Pollution and Its Prevention   439
19   Municipal Solid Waste: Disposal and Recovery   465
20   Hazardous Chemicals: Pollution and Prevention   485
21   The Atmosphere: Climate, Climate Change, and Ozone Depletion   509
22   Atmospheric Pollution   537

PART SIX   *Toward a Sustainable Future*   **567**

23   Economics, Public Policy, and the Environment   569
24   Sustainable Communities and Lifestyles   593

ABC Video Case Studies, Volume V   617
Appendix A—Environmental Organizations   621
Appendix B—Units of Measure   623
Appendix C—Some Basic Chemical Concepts   625
Bibliography and Additional Reading   631
Photo Credits   645
Glossary   647
Index   669

Essays

EARTH WATCH

Agenda 21	19
Biosphere 2	77
An Endangered Ecosystems Act?	87
What Is a Species?	120
Preserving Genes for Agriculture	130
Are We Living Longer?	139
An Integrated Approach to Alleviating the Conditions of Poverty	182
Water Purification	224
Return of the Gray Wolf	265
Nature's Corporations	298
Will Aquaculture Be Able to Fill the Gap?	302
Cogeneration: Industrial Common Sense	337
Radiation Phobia?	361
Economic Payoff of Solar Energy	375
The Ultimate Pest?	420
Monitoring for Sewage Contamination	444
The Algae from Hell	451
Regionalized Recycling	480
The Case of the Obee Road NPL Site	449
Portland Takes a Right Turn	557
The Clean Air Act Brings a Windfall	562
Green Fees and Taxes	584

ETHICS

What Is the Stewardship Ethic?	9
Can Ecosystems be Restored?	47
The Dilemma of Advocacy	85
Selection: Natural and Unnatural	113
The Dilemma of Immigration	148
Additional Incentives for Reducing Fertility	179
Erosion by Equation	195

The Lifeboat Ethic of Garret Hardin	259
Trading Wilderness for Energy in the Far North	333
Showdown in the New West	354
Transfer of Energy Technology to the Developing World	385
The Rights of Smokers?	403
DDT for Malaria Control: Hero or Villain?	419
Cleaning Up the Flow	459
Affluenza: Do You Have It?	481
Environmental Justice and Hazardous Waste	500
Stewardship of the Atmosphere	526
The Tangier Island Covenant	612

GLOBAL PERSPECTIVE

Light and Nutrients: The Controlling Factors in Marine Ecosystems	64
Fertility and Literacy	170
Three-Strata Forage System for Mountainous Drylands	209
The Death of the Aral Sea	228
The World Water Forum	237
World Food Summit	256
Biodiversity: Essential or Not?	281
An Unwelcome Globalization	404
Wasps 1, Mealybugs 0	428
Coping with UV Radiation	530
Mexico City: Life in a Gas Chamber	542
The World Trade Organization	589

CAREER LINK

Daniel S. Granz, EPA Environmental Engineer	502
Brian Hopper, Environmental Law Intern	613

Contents

Preface   *xvi*

1 Introduction: Sustainability, Stewardship, and Sound Science   1

1.1 The Global Environmental Picture   4
Population Growth   4
The Decline of Ecosystems   4
Global Atmospheric Changes   5
Loss of Biodiversity   6

1.2 Three Unifying Themes   6
Sustainability   6
Stewardship   8
Environmentalism   11
Sound Science   14
A New Commitment   18

Environment on the Web   18

Review Questions   21
Thinking Environmentally   21
Web References   21
Ethics: What Is the Stewardship Ethic?   9
Earth Watch: *Agenda 21*   19

PART ONE
Ecosystems and How They Work   23

2 Ecosystems: What They Are   25

2.1 What Are Ecosystems?   26

2.2 The Structure of Ecosystems   29
Trophic Categories   30
Trophic Relationships: Food Chains, Food Webs, and
Trophic Levels   35
Nonfeeding Relationships   37
Abiotic Factors   40

2.3 Global Biomes   42
The Role of Climate   42
Microclimate and Other Abiotic Factors   45
Biotic Factors   45
Physical Barriers   46

2.4 Implications for Humans   47
Three Revolutions   47

Environment on the Web   50

Review Questions   50
Thinking Environmentally   51
Web References   51
Ethics: Can Ecosystems Be Restored?   47

3 Ecosystems: How They Work   53

3.1 Matter, Energy, and Life   54
Matter in Living and Nonliving Systems   54
Energy Considerations   58
Energy Changes in Organisms and Ecosystems   61

3.2 Principles of Ecosystem Function   66
Energy Flow in Ecosystems   66
Biogeochemical Cycles   68

3.3 Implications for Humans   72
Sustainability   72
Value   74
Managing Ecosystems   75

Environment on the Web   78

Review Questions   79
Thinking Environmentally   79
Web References   79
Global Perspective: Light and Nutrients:
    The Controlling Factors in Marine Ecosystems   64
Earth Watch: Biosphere 2   77

4 Ecosystems: Populations and Succession   81

4.1 Population Dynamics   83
Population Growth Curves   83
Biotic Potential versus Environmental Resistance   84
Density Dependence and Critical Numbers   85

4.2 Mechanisms of Population Equilibrium   87
Predator–Prey Dynamics   87
Competition   90
Introduced Species   93

4.3 **Disturbance and Succession**   96
Ecological Succession   96
Disturbance and Resilience   98

4.4 **Implications for Humans**   102

Environment on the Web   104

Review Questions   105
Thinking Environmentally   105
Web References   105
Ethics: The Dilemma of Advocacy   85
Earth Watch: An Endangered Ecosystems Act?   87

5 Ecosystems and Evolutionary Change   107

5.1 **Selection by the Environment**   109
Change through Selective Breeding   109
Change through Natural Selection   110
Adaptations to the Environment   111

5.2 **Selection of Traits and Genes**   111
Genetic Variation and Gene Pools   114
Mutations: The Source of New Alleles   116

5.3 **Changes in Species and Ecosystems**   117
The Limits of Change   117
Speciation   118
Evolving Ecosystems?   122

5.4 **Plate Tectonics**   124
Tectonic Plates   124

5.5 **Evolution in Perspective**   127
The Fossil Record   127
Controversy over Evolution   127
Stewardship of Life   128

Environment on the Web   131

Review Questions   132
Thinking Environmentally   132
Making a Difference   132
Web References   133
Ethics: Selection: Natural and Unnatural   113
Earth Watch: What Is a Species?   120
Earth Watch: Preserving Genes for Agriculture   130

PART TWO
The Human Population   135

6 The Human Population: Demographics   137

6.1 **The Population Explosion and Its Cause**   138
The Explosion   138
Reasons for the Explosion   139

6.2 **Different Worlds**   140
Rich Nations and Poor Nations   141
Population Growth in Rich and Poor Nations   141
Different Populations Present Different Problems   143

6.3 **Environmental and Social Impacts of Growing Populations and Affluence**   144
The Growing Populations of Developing Countries   144
Effects of Increasing Affluence   150

6.4 **Dynamics of Population Growth**   150
Population Profiles   150
Population Projections   152
Population Momentum   155
The Demographic Transition   156

Environment on the Web   159

Review Questions   160
Thinking Environmentally   160
Web References   161
Earth Watch: Are We Living Longer?   139
Ethics: The Dilemma of Immigration   148

7 Issues in Population and Development   163

7.1 **Reassessing the Demographic Transition**   164
Factors Influencing Family Size   165
Conclusions   168

7.2 **Development**   168
Promoting the Development of Low-Income Countries   168
Past Successes and Failures of the World Bank   169
The Debt Crisis   172
World Bank Reform   173

7.3 **A New Direction for Development: Social Modernization**   174
Education   174
Improving Health   175
Family Planning   177
Enhancing Income   178
Putting It All Together   180

7.4 **The Cairo Conference**   181

Environment on the Web   184

Review Questions   184
Thinking Environmentally   185

Making a Difference	185
Web References	185
Global Perspective: Fertility and Literacy	170
Ethics: Additional Incentives for Reducing Fertility	179
Earth Watch: An Integrated Approach to	
Alleviating the Conditions of Poverty	182

PART THREE
Renewable Resources   *187*

8 Soil and the Soil Ecosystem   189

8.1 Plants and Soil	**191**
Soil Characteristics	191
Soil and Plants	194
Soil as an Ecosystem	196
8.2 Soil Degradation	**200**
Erosion and Desertification	200
Drylands	201
Causing and Correcting Erosion	202
Irrigation and Salinization	206
8.3 Addressing Soil Degradation	**207**
Public Policy and Soils	208
Helping Individual Landholders	209
Environment on the Web	**210**
Review Questions	211
Thinking Environmentally	211
Web References	211
Ethics: Erosion by Equation	195
Global Perspective: Three-Strata Forage System	
for Mountainous Drylands	209

9 Water: Hydrologic Cycle and Human Use   213

9.1 Water—A Vital Resource	**214**
9.2 The Hydrologic Cycle	**215**
Evaporation, Condensation, and Purification	215
Precipitation	217
Water over and through the Ground	219
Summary of the Hydrologic Cycle	220
9.3 Human Impacts on the Hydrologic Cycle	**221**
Changing the Surface of the Earth	221
Polluting the Water Cycle	222
Withdrawing Water Supplies	222
9.4 Sources and Uses of Fresh Water	**223**
9.5 Overdrawing Water Resources	**226**
Consequences of Overdrawing Surface Waters	226
Consequences of Overdrawing Groundwater	226
9.6 Obtaining More Water	**230**
9.7 Using Less Water	**231**
Irrigation	231
Municipal Systems	232
9.8 Desalting Sea Water	**233**
9.9 Storm Water	**233**
Mismanagement and Its Consequences	233
Improving Storm-water Management	235
9.10 Water Stewardship	**236**
Environment on the Web	**238**
Review Questions	238
Thinking Environmentally	239
Web References	239
Earth Watch: Water Purification	224
Global Perspective: The Death of the Aral Sea	228
Global Perspective: The World Water Forum	237

10 The Production and Distribution of Food   241

10.1 Crops and Animals: Major Patterns of	
Food Production	**242**
The Development of Modern Industrialized	
Agriculture	242
Subsistence Agriculture in the Developing World	245
Animal Farming and Its Consequences	246
Prospects for Increasing Food Production	248
The Promise of Biotechnology	249
10.2 Food Distribution and Trade	**252**
Patterns in Food Trade	252
Levels of Responsibility in Supplying Food	253
10.3 Hunger, Malnutrition, and Famine	**254**
Nutrition vs. Hunger	254
Extent and Consequences of Hunger	255
Root Cause of Hunger	256
Famine	257
Food Aid	258
Environment on the Web	**260**
Review Questions	261
Thinking Environmentally	261
Web References	261
Global Perspective: World Food Summit	256
Ethics: The Lifeboat Ethic of Garret Hardin	259

11 Wild Species: Biodiversity and Protection 263

11.1 Value of Wild Species 264
Biological Wealth 264
Two Kinds of Value 264
Sources for Agriculture, Forestry, Aquaculture, and Animal Husbandry 266
Sources for Medicine 267
Recreational, Aesthetic, and Scientific Value 267
Intrinsic Value 269

11.2 Saving Wild Species 269
Game Animals in the United States 269
The Endangered Species Act 271

11.3 Biodiversity 275
The Decline of Biodiversity 276
Reasons for the Decline 277
Consequences of Losing Biodiversity 280
International Steps to Protect Biodiversity 282
Stewardship Concerns 281

Environment on the Web 284

Review Questions 285
Thinking Environmentally 285
Web References 285
Earth Watch: Return of the Gray Wolf 265
Global Perspective: Biodiversity: Essential or Not? 281

12 Ecosystems as Resources 287

12.1 Biological Systems in a Global Perspective 288
Major Systems and Their Value 288
Ecosystems as Natural Resources 289

12.2 Conservation and Preservation 290
Patterns of Use of Natural Ecosystems 291

12.3 Biomes and Ecosystems under Pressure 296
Forest Biomes 296
Ocean Ecosystems 300

12.4 Public and Private Lands in the United States 307
National Parks and National Wildlife Refuges 307
National Forests 308
Private Land Trusts 310
Thoughts on Sustainability, Sound Science, and Stewardship 310

Environment on the Web 311

Review Questions 312
Thinking Environmentally 312
Making a Difference 312
Web References 313
Earth Watch: Nature's Corporations 298
Earth Watch: Will Aquaculture Be Able to Fill the Gap? 302

PART FOUR
Energy 315

13 Energy from Fossil Fuels 317

13.1 Energy Sources and Uses 318
Harnessing Energy Sources: An Overview 318
Electrical Power Production 320
Matching Sources to Uses 324

13.2 The Exploitation of Crude Oil 325
How Fossil Fuels Are Formed 325
Crude-Oil Reserves versus Production 325
Declining U.S. Reserves and Increasing Importation 327
The Oil Crisis of the 1970s 327
Adjusting to Higher Prices 328
Victims of Our Success 328
Problems of Growing U.S. Dependency on Foreign Oil 330

13.3 Other Fossil Fuels 333
Natural Gas 334
Coal 334
Oil Shales and Oil Sands 335

13.4 Sustainable Energy Options 335
Conservation 336
Development of Non-Fossil-Fuel Energy Sources 338

Environment on the Web 338

Review Questions 339
Thinking Environmentally 339
Web References 339
Ethics: Trading Wilderness for Energy in the Far North 333
Earth Watch: Cogeneration: Industrial Common Sense 337

14 Nuclear Power: Promise and Problems   341

14.1 Nuclear Power: Dream or Delusion?   342

14.2 How Nuclear Power Works   344
From Mass to Energy   344
Comparison of Nuclear Power with Coal Power   348

14.3 The Hazards and Costs of Nuclear Power 349
Radioactive Emissions   349
Radioactive Wastes   351
The Potential for Accidents   355
Safety and Nuclear Power   356
Economic Problems with Nuclear Power   357

14.4 More Advanced Reactors   358
Breeder Reactors   358
Fusion Reactors   359

14.5 The Future of Nuclear Power   360
Opposition   360
Rebirth of Nuclear Power?   361

Environment on the Web   362

Review Questions   363
Thinking Environmentally   363
Web References   363
Ethics: Showdown in the New West   354
Earth Watch: Radiation Phobia?   361

15 Renewable Energy   365

15.1 Principles of Solar Energy   367

15.2 Putting Solar Energy to Work   368
Solar Heating of Water   368
Solar Space Heating   369
Solar Production of Electricity   371
The Promise of Solar Energy   374
Solar Production of Hydrogen: The Fuel of the Future   375

15.3 Indirect Solar Energy   376
Hydropower   377
Wind Power   378
Biomass Energy   379

15.4 Additional Renewable Energy Options   381
Geothermal Energy   381
Tidal Power   382
Ocean Thermal Energy Conversion   382

15.5 Policy for a Sustainable Energy Future   382

Environment on the Web   386

Review Questions   386
Thinking Environmentally   387
Making a Difference   387
Web References   387
Earth Watch: Economic Payoff of Solar Energy   375
Ethics: Transfer of Energy Technology to the Developing World   385

PART FIVE
Pollution and Prevention   389

16 Environmental Hazards and Human Health   391

16.1 Links between Human Health and the Environment   392
The Picture of Health   392
Environmental Hazards   393
Cancer   399

16.2 Pathways of Risk   400
The Risks of Being Poor   400
The Cultural Risk of Smoking   401
Risk and Infectious Diseases   404
Toxic Risk Pathways   406

16.3 Risk Assessment   407
Risk Assessment by the EPA   407

Risk Management   409
Risk Perception   410

  Environment on the Web   412

  Review Questions   412
  Thinking Environmentally   413
  Web References   413
  Ethics: The Rights of Smokers?   403
  Global Perspective: An Unwelcome
    Globalization   404

17 Pests and Pest Control   415

17.1 The Need for Pest Control   416
Defining Pests   416
The Importance of Pest Control   416
Different Philosophies of Pest Control   416

17.2 Promises and Problems of the
Chemical Approach   417
Development of Chemical Pesticides and
  their Successes   417
Problems Stemming from Chemical
  Pesticide Use   419
Nonpersistent Pesticides: Are They the Answer?   424

17.3 Alternative Pest Control Methods   424
Cultural Control   425
Control by Natural Enemies   426
Genetic Control   429
Natural Chemical Control   430

17.4 Socioeconomic Issues in
Pest Management   431
Pressures to Use Pesticides   431
Integrated Pest Management   432
Organically Grown Food   433

17.5 Public Policy   434
FIFRA   434
FQPA of 1996   434
Pesticides in Developing Countries   435
New Policy Needs   436

  Environment on the Web   436

  Review Questions   437
  Thinking Environmentally   437
  Web References   437
  Ethics: DDT for Malaria Control: Hero or
    Villain?   419
  Earth Watch: The Ultimate Pest?   420
  Global Perspective: Wasps 1, Mealybugs 0   428

18 Water Pollution and
Its Prevention   439

18.1 Water Pollution   440
Pollution Essentials   440
Water Pollution: Sources, Types, Criteria   440

18.2 Eutrophication   447
Different Kinds of Aquatic Plants   447

The Impacts of Nutrient Enrichment   447
Combating Eutrophication   449

18.3 Sewage Management and Treatment   453
Development of Collection and Treatment Systems   453
The Pollutants in Raw Sewage   454
Removing the Pollutants from Sewage   454
Sludge Treatment   458
Alternative Treatment Systems   460

18.4 Public Policy   461

  Environment on the Web   462

  Review Questions   463
  Thinking Environmentally   463
  Web References   463
  Earth Watch: Monitoring for Sewage
    Contamination   444
  Earth Watch: The Algae from Hell   451
  Ethics: Cleaning Up the Flow   459

19 Municipal Solid Waste:
Disposal and Recovery   465

19.1 The Solid-Waste Problem   466
Disposing of Municipal Solid Waste   466
Landfills   467
Combustion: Waste to Energy   471
Costs of Municipal Solid Waste Disposal   472

19.2 Solutions   473
Source Reduction   473
The Recycling Solution   473
Composting   478

19.3 Public Policy and Waste Management 478
The Regulatory Perspective   478
Integrated Waste Management   478

  Environment on the Web   482

  Review Questions   482
  Thinking Environmentally   483
  Web References   483
  Earth Watch: Regionalized Recycling   480
  Ethics: Affluenza: Do You Have It?   481

20 Hazardous Chemicals: Pollution and Prevention   485

20.1 Toxicology and Chemical Hazards   486
Dose Response and Threshold   486
The Nature of Chemical Hazards: HAZMATs   487
Sources of Chemicals Entering the Environment   487
The Threat from Toxic Chemicals   488
Involvement with Food Chains   490

20.2 A History of Mismanagement   491
Methods of Land Disposal   491
Scope of the Mismanagement Problem   493

20.3 Cleaning Up the Mess   496
Assuring Safe Drinking Water   496
Groundwater Remediation   496
Superfund for Toxic Sites   496

20.4 Management of New Wastes   500
The Clean Air and Water Acts   501
The Resource Conservation and
  Recovery Act (RCRA)   501
Reduction of Accidents and Accidental Exposures   502

20.5 Looking toward the Future   504
Too Many or Too Few Regulations?   504
Pollution Avoidance for a Sustainable Society   505

Environment on the Web   506
Review Questions   507
Thinking Environmentally   507
Web References   507
Earth Watch: The Case of the Obee Road
  NPL Site   499
Ethics: Environmental Justice and
  Hazardous Waste   500
Career Link: Daniel S. Granz, EPA
  Environmental Engineer   502

21 The Atmosphere: Climate, Climate Change, and Ozone Depletion   509

21.1 Atmosphere and Weather   511
Atmospheric Structure   511
Weather   511

21.2 Climate   513
Climates in the Past   514
Ocean and Atmosphere   515

21.3 Global Climate Change   516
The Earth as a Greenhouse   516
Carbon Dioxide: Major Greenhouse Gas   518
Other Greenhouse Gases   519
Amount of Warming and Its Probable Effects   520
Coping with Global Warming   524

21.4 Depletion of the Ozone Layer   528
Radiation and Importance of the Shield   528
Formation and Breakdown of the Shield   529
Coming to Grips with Ozone Depletion   533

Environment on the Web   534
Review Questions   534
Thinking Environmentally   535
Web References   535
Ethics: Stewardship of the Atmosphere   526
Global Perspective: Coping with UV Radiation   530

22 Atmospheric Pollution   537

22.1 Air Pollution Essentials   538
Pollutants and Atmospheric Cleansing   538
The Appearance of Smogs   539

22.2 Major Air Pollutants and Their Impact   541
Major Pollutants   541
Adverse Effects of Air Pollution   542

22.3 Sources of Pollutants   546
Primary Pollutants   547
Secondary Pollutants   549

22.4 Acid Deposition   549
Acids and Bases   550
Extent and Potency of Acid Precipitation   552
Sources of Acid Deposition   552
Effects of Acid Deposition   553

22.5 Bringing Air Pollution under Control   555
Control Strategies   556
Coping with Acid Deposition   559

22.6 Taking Stock   561
Future Directions   561

Environment on the Web   564
Review Questions   564
Thinking Environmentally   565
Making a Difference   565
Web References   565
Global Perspective: Mexico City: Life in a
  Gas Chamber   542
Earth Watch: Portland Takes a Right Turn   557
Earth Watch: The Clean Air Act Brings a
  Windfall   562

PART SIX
Toward a Sustainable Future   *567*

23 Economics, Public Policy, and the Environment   569

23.1 Economics and Public Policy   570
The Need for Environmental Public Policy   570
Relationships between Economic Development and
  the Environment   570
Economic Systems   571

23.2 Resources and the Wealth of Nations   574
The Wealth of Nations   575
Shortcomings of the GNP   576
Resource Distribution   577

23.3 Pollution and Public Policy   578
Public Policy Development: The Policy Life Cycle   578
Economic Effects of Environmental Public Policy   580
Policy Options: Market or Regulatory?   583

23.4 Benefit–Cost Analysis   583
External and Internal Costs   583
The Costs of Environmental Regulations   584
The Benefits of Environmental Regulation   585
Cost-Effectiveness   586
Progress   587

23.5 Politics, the Public, and Public Policy   587
Politics and the Environment   587
Citizen Involvement   589

  Environment on the Web   590

  Review Questions   590
  Thinking Environmentally   591
  Web References   591
  Earth Watch: Green Fees and Taxes   584
  Global Perspective: The World Trade
    Organization   589

24 Sustainable Communities and Lifestyles   593

24.1 Urban Sprawl   594
The Origins of Urban Sprawl   594

Environmental Impacts of Urban Sprawl   598
Reining In Urban Sprawl: Smart Growth   600

24.2 Urban Blight   600
Economic and Ethnic Segregation   600
The Vicious Cycle of Urban Blight   602
Economic Exclusion of the Inner City   602
What Makes Cities Livable?   604

24.3 Moving toward Sustainable Communities   607
Sustainable Cities   607
Sustainable Communities   607
President's Council on Sustainable
  Development   608

24.4 Epilogue   610
Our Dilemma   610
Lifestyle Changes   611

  Environment on the Web   614

  Review Questions   614
  Thinking Environmentally   615
  Making a Difference   615
  Web References   615
  Ethics: The Tangier Island Covenant   612
  Career Link: Brian Hopper, Environmental
    Law Intern   613

ABC Video Case Studies, Volume V   617
Appendix A Environmental Organizations   621
Appendix B Units of Measure   623
Appendix C Some Basic Chemical Concepts   625

Bibliography and Additional Reading   *631*
Photo Credits   *645*
Glossary   *647*
Index   *669*

Preface

As we plunge into a new century and a new millennium, the environment is being called on to satisfy the growing needs of an expanding human population in the developing countries and increasing affluence in the developed countries. In many areas, we are already taking more from Earth's systems than they can provide in a sustainable fashion—and there are still billions of people who are not adequately housed, fed, or provided with health care or a paying job. Yet we must, as soon as possible, make a transition to a sustainable civilization, one in which a stable human population recognizes the finite limits of Earth's systems to produce resources and absorb wastes, and acts accordingly. This is hard to picture at present, but it is the only future that makes any sense. If we fail to achieve it by our deliberate actions, the natural world will impose it on us in highly undesirable ways.

Environmental science stands at the interface between humans and Earth and explores the interactions and relations between them. This relationship will need to be considered in virtually all future decision making. Our text considers a full spectrum of views and information in an effort to establish a solid base of understanding and a sustainable formula for the future.

You may already be informed about some of the issues we cover in the book, such as global warming, the extinction of species, air pollution, toxic wastes, overpopulation, recycling, and the destruction of tropical rain forests. However, what you have in your hands is a readable guide and up-to-date source of information that will help you to explore the issues in more depth. It will also help you to connect them to a framework of ideas and values that will equip you to become part of the solution to many of the environmental problems confronting us.

As the field of environmental science evolves and continues to change, so has this text. In this new edition, we hope to continue to lead the change in environmental science and have made every effort to address each of the following objectives:

- To write in a style that makes learning about environmental science both interesting to read and easy to understand, without overwhelming the student with details.
- To present well-established scientific principles and concepts that form the knowledge base for an understanding of our environment.
- To organize the text in a way that promotes sequential learning, yet allows individual chapters to stand on their own.

- To address all of the major environmental issues that confront our society and help to define the subject matter of environmental science.
- To present the latest information available by making full use of the resources of the World Wide Web.
- To give an assessment of options or progress in solving environmental problems.
- To support the text with excellent supplements for the instructor and the student that strongly enhance the teaching and learning processes.

Because we believe that learning how to live in the environment is one of the most important subjects in any student's educational experience, we have made every effort to put in your hands a book that will help the study of environmental science come alive.

A Guide to the Eighth Edition of Environmental Science

Overview

The **seventh edition,** published in 2000, involved the following major changes in the text, as the authorship shifted entirely to one author (Wright):

- Introduction of three central themes: sustainability, sound science, and stewardship.
- Reorganization into six parts instead of four.
- Addition of "Environment on the Web" essays at the end of each chapter.
- Two new chapters added (Chapter 16, "Environmental Hazards and Human Health," and Chapter 23, "Economics, Public Policy, and the Environment"), and two older chapters condensed into existing or new chapters.
- Completely updated and revised art program.
- Every chapter opening with a case study, to bring the subject matter into focus.

The **eighth edition** builds on the strengths of the seventh. The unifying themes of sustainability, sound science, and stewardship are retained and continue to provide important threads linking the different subjects and chapters of the text. In the eighth edition, I continue to provide a balance between pure science and the political, social, and historical perspectives of environmental affairs. I am also careful to reflect differences in interpretation of environmental concerns where they exist, while maintaining the standard of sound science for judging those concerns.

Most important, the eighth edition reflects the changing environmental scene in the United States, as well as in the rest of the world. Information from new books, journal articles, and Web-based reports from governmental and nongovernmental organizations has been incorporated into every chapter. New artwork has been introduced—51 new photos, 35 new diagrams, and seven new tables.

As in the seventh edition, each chapter opens with a case study or an illustrative story. Because of their relevance, a majority of these have been retained, while 10 have been replaced with new opening studies. Throughout, the high readability that the text has been known for has been maintained and strengthened. And the emphasis on science has continued, to the point where the text is more solidly grounded than ever in the physical and biological sciences as the bases for understanding every environmental issue.

Introduction

Chapter 1 provides an introduction to the rest of the book by discussing present global environmental concerns; **ecosystem decline** has been added as one of the most important of these. The chapter introduces the three themes that will provide the unifying threads throughout the book: **sustainability**—the practical goal that our interactions with the natural world should be working toward; **stewardship**—the ethical and moral framework that should inform our public and private actions (a **new ethics essay** explores this theme); and **sound science**—the basis for our understanding of how the world works and how human systems interact with it. I include in my coverage of sound science information on **the nature of science** and the **scientific method** in order to help students distinguish sound science from "junk" science (of which I give a prime example) as they encounter controversy over scientific information.

Part I. Ecosystems and How They Work

Part I (Chapters 2–5) explores natural ecosystems—what they are, how they function, how balances are maintained, and how they evolve and change. This examination, in addition to providing an appreciation of how the natural world functions, brings out **five basic sustainability principles** that keep ecosystems going. A new principle—**resilience**—is introduced and illustrated. The five principles serve as benchmarks to evaluate the sustainability of various courses of action presented in the rest of the text.

Part II. The Human Population

Part II (Chapters 6 and 7) first looks at the dynamics of the human population. The pressures on natural systems as a re-sult of the growth of that population are examined, with a focus on the **demographic transition**—the shift from high birth and death rates to low birth and death rates that has brought stable populations to the industrialized world. Then, the developing countries' difficulties through this transition are considered, and steps that are being taken on the part of the international community to address the needs of those countries are discussed. The eighth edition presents new population pyramids, the latest on **debt relief**, and the recent appraisal of the five-year anniversary of the International Conference on Population and Development, as well as completely updated population growth statistics.

Part III. Renewable Resources

Part III (Chapters 8–12) addresses the science of our natural resources of soil, water, and wildlife. Issues concerning the use of such resources in food production, forest growth, and fisheries management are examined in light of increasing population growth and increasing pressure on those resources: again, we all the while keep our eyes on sustainability. Some examples of issues receiving a **new emphasis** are (a) how erosion is measured and why the way it is measured is problematic, (b) dryland ecosystems and desertification, (c) the work of the World Water Forum, (d) the controversy surrounding **genetically modified food**, and (e) **restoration ecology** at work.

Part IV. Energy

Part IV (Chapters 13–15) presents the energy resources currently available and the consequences each can have on the environment. We learn how our past choices of energy sources to fuel the global economy have affected the environment on a global scale. The outlook regarding the impact on sustainability of the U.S. reliance on crude oil and the obvious prospects for renewable energy are presented in view of the most recent statistics and developments. Included is a discussion of the impact of **new standards** for appliances on energy conservation. Also examined is the option of nuclear power, despite the problems of its cost, the difficulty of nuclear waste storage and disposal, and the inherent dangers associated with nuclear power. Renewable energy is also discussed in light of its pros and cons; new information is presented on **fuel cells** and how they work.

Part V. Pollution and Prevention

Part V (Chapters 16–22) begins with a chapter on environmental health. The **precautionary principle** is introduced here and is discussed in connection with risk-based public policy. The text goes on to investigate the pollution of water, land, and air that results from human activities and our interactions with the environment that were discussed in earlier chapters. The coverage ranges from the use of pesticides to protect our crops, through sewage treatment and contamination of water, to municipal and hazardous wastes, and on to major atmospheric changes and more local and regional air pollution. Examples of some new issues introduced in this edition are (a) the controversy over **DDT for malaria control**, (b) the **dead zone** in the Gulf of Mexico, (c) the "**dirty dozen**" persistent organic pollutants (**POPs**), (d) the use of **MTBE** in gasoline, and (e) the **consequences of climate change**.

Part VI. Toward a Sustainable Future

Part VI (Chapters 23 and 24) directly addresses the relationship that exists among economics, public policy, and the environment, focusing especially on our present environmental concerns. A new box discusses the **World Trade Organization** as an environmental issue. The text then goes on to examine how inner cities have deteriorated as a result of migration to the suburbs and urban sprawl. Some communities are working toward renewal and sustainability with a plan called **Smart Growth**. Part VI closes with a look at personal involvement, lifestyles, and values as vital components of our efforts to enjoy a sustainable future.

Individual Text Elements

Essays: Environmental Science features four kinds of essays: "Earth Watch," "Ethics," "Global Perspective," and "Career Link." Lists of essays are found at the end of the outline for each chapter.

 "**Earth Watch**" essays provide further information that enhances the student's understanding of particular aspects of the topic being covered.

 "**Ethics**" essays focus on the fact that many environmental issues do not involve clear-cut rights or wrongs, but present ethical dilemmas.

 "**Global Perspective**" essays help the student appreciate the global nature and extent of the topic in question.

 "**Career Link**" essays present an individual who has chosen to work in some area of environmental concern. The essays discuss how the person got into his or her career.

Making a Difference: I believe that no amount of text-based learning about the environment truly becomes useful until students challenge themselves and those around them to begin making a difference. With this in mind, each of the six parts of the text concludes with a section that suggests courses of action that each student can take to bring about the needed changes to foster sustainability.

Chapter Opening: Each chapter begins with a set of "**Key Issues and Questions**"—succinct statements regarding key aspects of the issue being covered and questions inviting the student to explore those issues.

Chapter Outline: Chapter outlines may be found in the **Table of Contents**. Importantly, the text of each chapter is organized according to a logical outline of first-, second-, and third-order headings to assist student outlining, note taking, and learning.

Review Questions: Each chapter concludes with a set of "Review Questions" addressing each aspect of the topic covered. Of course, these questions may serve as learning objectives, as test items, or for review.

Thinking Environmentally: A set of questions, "**Thinking Environmentally**," is included at the end of each chapter. These questions invite the student to make connections between knowledge gleaned from the chapter and other areas of the environmental arena and to apply knowledge gained to specific environmental problems. The questions may be used also for testing or to focus class discussion.

Vocabulary: Each new term will be found in boldface type where it is first introduced and defined. All such items are found in the glossary at the end of the book.

Video Case Studies: Selected from the archives of ABC news, these timely and relevant video segments offer students an overview of a particular environmental issue or controversy. Case study material is found directly after the end of the text, but has direct application to particular chapters. A brief synopsis of each video, as well as a list of interesting questions, is included, in the hopes of stimulating healthy classroom debate and discussion of the various topics. Since videos from earlier volumes are also made available to instructors who adopt the eighth edition of Environmental Science, a list of these case studies is also provided.

Appendices: At many points in the text, reference is made to the work being done by various environmental organizations. A listing of major national environmental organizations is given in **Appendix A**. Most of these organizations and agencies have a home page on the Internet and can be located via the Web site that supports this text.

A conversion chart for various English and metric units is found in **Appendix B**.

For students who need some grounding in chemistry, a discussion of atoms, molecules, atomic bonding, and chemical reactions is provided in **Appendix C**.

Bibliography and Additional Reading: An updated listing of articles and books dealing with environmental topics follows the appendices, which are organized according to chapter, following a short list of general references. Also listed are virtually all of the newer references used in preparing this new edition of the text.

Glossary and Index: A comprehensive glossary provides definitions of virtually all of the special terms, treaties, legislation, and programs identified in the text in boldface type. The index gives page references for all of these terms and for thousands of other topics and issues dealt with in the text.

For the Instructor

Instructor's Resource Manual (0-13-091379-0)
By Nancy Ostiguy (Pennsylvania State University)
This thorough resource manual features a chapter outline, instructional goals, concepts and connections, a suggested lecture format, and answers to the chapter-opening "Key Issues and Questions," as well as creative discussion questions, activities, and labs.

Test Item File (0-13-091389-8)
By Steve Ailstock (Anne Arundel Community College) and Shari Snitovsky (Skyline College)
Contains over 1,800 test questions, including multiple-choice, short-answer, and essay questions.

Prentice Hall Custom Test
Available in formats for both Windows (0-13-091388-X) and Macintosh (0-13-091387-1) computers, and based on the powerful testing technology developed by Engineering Software Associates, Inc. (ESA), Prentice Hall Custom Test allows instructors to create and tailor exams to their own needs. With this on-line testing program, exams can be administered on-line and data can be automatically transferred for evaluation. A comprehensive desk reference guide is included along with on-line assistance.

Transparency Pack (013-091384-7)
A selection of 164 full-color transparencies of images from the text, as well as 87 black-and-white transparency masters.

Slides (013-091383-9)
A selection of the same 164 images presented on the transparencies, available in slide format.

The ABC News/Prentice Hall Video Library, Volume V (013-091386-3)
This unique video series contains segments from award-winning shows such as "*World News Tonight*," "*Nightline*," and "*Good Morning America*." Selected from the archives of ABC News, each video includes a written summary that ties the segment to particular sections of the text, making it easier to enhance your classroom presentation with timely and relevant video programs.

Environmental Science
Digital Image Gallery (0-13-091934-9).
This unique image bank contains all of the illustrations from the eighth edition of *Environmental Science*, as well as animations and videos in a digitized format for use in the classroom. The CD-ROM includes a navigational tool to allow instructors to browse easily through the images. The files are ideal for those professors who use PowerPoint® or a comparable presentation software for their classes or for professors who create text-specific Web sites for their students.

For the Student

Study Guide (013-091391-x)
By Clark Adams (Texas A & M University)
This study guide helps students identify the important points from the text and then provides them with review exercises, study questions, self-check exercises, and vocabulary review. In addition, the author has included PowerPoint® presentations for student review and for the professor to go over in class.

Environmental Science World Wide Web Home Page
http://www.prenhall.com/wright. This unique tool is designed to launch student exploration of environmental science resources on the World Wide Web. The home page is regularly updated and linked specifically to chapters in the text. In addition to providing a juried guide to many interesting Web-based resources, the site features review exercises (from which the students receive immediate feedback), updates of environmental issues by region, a guide to environmental careers, and a guide to help students learn how to start making a positive difference for Earth's environment.

Science on the Internet: A Student's Guide
By Andrew T. Stull.
The perfect guide to help students take advantage of the Environmental Science Web site. This unique resource gives clear steps for accessing the regularly updated Environmental Science resource area, as well as an overview of general navigation and research strategies.

Course Management
Prentice Hall is proud to partner with many of the leading course management system providers on the market today. These partnerships enable us to combine our market-leading on-line content with the powerful course management tools Blackboard and WebCT, as well as with our proprietary course management system, CourseCompass. Visit our demo site, www.prenhall.com/demo, for more information, or contact your local Prentice Hall representative, who can provide a live demonstration of these exciting tools.

Reviewers

I offer my sincere thanks to those who reviewed the seventh edition and previous editions of this text. Their comments, suggestions, and constructive criticisms have all been carefully considered and in many instances have led to significant improvements in the text. I thank the following people:

M. Stephen Ailstock
Anne Arundel Community College
Darren Divine
University of Nevada, Las Vegas
David Gardner
Owens Community College

Ray Grizzle
University of New Hampshire
Kathleen Keating
Cook-Rutgers University
David Liscio
Endicott College
Stephen R. Overmann
Southeast Missouri State University
Max R. Terman
Tabor College
Richard E. Terry
Brigham Young University

Reviewers of Previous Editions

John Blachley
College of the Desert
Robert H. Blodgett
Austin Community College
Norm Dronen
Texas A & M University
William P. Hayes
Catholic University
Robert Kistler
Bethel College
Alberto L. Mancinelli
Columbia University
Kenneth E. Mantai
State University of New York, Fredonia
Nancy Ostiguy
Pennsylvania State University
Julia D. Schroeder
John A. Logan College
Morris L. Sotonoff
Chicago State University
Christy N. Stather
Illinois State University
Phillip L. Watson
Ferris State University
Narayanaswamy Bharathan
Northern State University, Aberdeen
Roger G. Bland
Central Michigan University
Jack L. Butler
University of South Dakota
Ann S. Causey
Auburn University
Robert W. Christopherson
American River College
Lynnette Danzl-Tauer
Rock Valley College
Phil Evans
East Carolina University, Pitt Community College
Gian Gupta
University of Maryland, Eastern Shore
John P. Harley
Eastern Kentucky University

Vern Harnapp
University of Akron
Stanley Hedeen
Xavier University
Clyde W. Hibbs
Ball State University
John C. Jahoda
Bridgewater State College
Karolyn Johnston
California State University, Chico
Guy R. Lanza
East Tennessee State University
John Mathwig
College of Lake County
Richard J. McCloskey
Boise State University
SuEarl McReynolds
San Jacinto College, Central
Eric Pallant
Allegheny College
David J. Parrish
Virginia Polytechnic Institute
Carol Skinner
Edinboro University of Pennsylvania

Acknowledgments

Although the content and accuracy of this text are the responsibility of the authors, it would never have seen the light of day without the dedicated work of many other people. I want to express my heartfelt thanks to all those at Prentice Hall who have contributed to the book in so many ways.

A special thanks goes to my executive editor, Daniel Kaveney, for his encouragement and attention to the oversight of this entire project. Shari Toron was my production editor, keeping me focused on the details of transcribing manuscripts into a published product, and doing it so cheerfully. Thanks also go to Tobi Zausner, photo researcher, who scoured the Web and other media sources for the pictures I was looking for. In addition, I thank Clark Adams for writing the study guide, Nancy Ostiguy for doing a fine job with the instructor's guide, Steve Ailstock and Shari Snitovsky for their good work on the test bank, and Isobel Heathcote (of the University of Guelph) for her fine work on the "Environment on the Web" essays, and for crafting and keeping up the book's home page on the World Wide Web.

Ten years ago, Prentice Hall editor David Brake asked me if I would be interested in helping Bernard Nebel write the fourth edition of his environmental science text. Because of my longtime concern about environmental issues and my interest in writing, I accepted the offer. As the years passed, my commitment to environmental stewardship and deep concerns about our society's interactions with the environment have led me to direct more and more of my energy and ability to writing and speaking about environmental issues.

As I have accepted more of the responsibility for writing this text, I have realized what an amazing job Bernie did in producing the first three editions alone while also teaching full time. He did it because he was frustrated with existing environmental science texts and was convinced he could produce a more readable and effective book—and he did! Bernie Nebel and I share very similar philosophical and educational values and have enjoyed collaborating over the years. Although I alone have been responsible for the seventh and eighth editions, I am deeply indebted to Bernie for his wonderful sense of organization and beautiful and clear prose that still form an important part of the book. Both of us have offered this book in its successive editions as our contribution to the students who are now entering this new century, in the hope that they will join us in helping to bring about the environmental revolution that must come—hopefully sooner than later.

I wish to offer some very personal thanks to my wife, Ann, who has been with me since the beginning of my work in biology and has provided the emotional base and companionship without which I would be far less of a person and a biologist. Her love and patience have sustained me in immeasurable ways. Finally, I offer my gratefulness to God, who is the author of the amazing Creation I love so much. I count it a privilege to be involved in promoting the care of His Creation.

Richard T. Wright

Introduction: Sustainability, Stewardship, and Sound Science

Key Issues and Questions

1. History is a saga of rises and falls of civilizations. What are the factors that brought about the collapse of the Easter Island civilization in the South Pacific? Are there any parallels to the present?

2. There is cause for concern about the general condition of the global environment today. What are four global trends that are of particular concern?

3. *Sustainability* is the practical goal toward which we should be working. What is meant by sustainability? For a sustainable society, what are the principal prerequisites?

4. *Sustainable development* is now a broadly accepted ideal. How do different disciplines address sustainable development?

5. *Stewardship* represents the ethical and moral framework that should inform our public and private actions. How can stewardship be applied to the natural world and to the concern for justice?

6. The modern environmental movement has achieved much in recent years. How did this movement start, and what have been some recent reactions against it?

7. Sound science is the basis for understanding how the world works and how human systems interact with it. What is the essence of science and the scientific method?

8. Science occurs within a community of scientists. How does this community function to prevent the occurrence of poor, or "junk," science?

9. What are some indications that show hope for a new commitment to environmental concerns and sustainability?

Imagine that you are assigned the task of traveling throughout the world to document the range of human interactions with the environment. Armed with a camera, you start your trip in South America. Boating the Amazon River through Peru's rain forest, you might photograph a village along the shore where there are a few small houses. Constructed of poles cut from the forest, lashed together with vines, and covered with a thatch of braided palm leaves, these primitive dwellings provide simple, but adequate, shelter in a climate where temperatures range from 75° to 85°F (24° to 29°C) year-round.

People living here have no running water or sewers, no electricity or telephones, and the closest shops or markets are many miles downriver. The rain forest and the river provide all basic needs: fish, game, fruits, and even medicines for those who learn which plants to use. People here have lived this way for centuries, but their way of life is gradually changing as modern civilization makes its inroads, as forests are cleared for pasture or logging, or as mining and oil exploration bring technologies far upriver.

Your next stop is Tanzania, where you might observe a much harsher life: Women and girls in rural villages may have to walk miles across a denuded, eroded landscape each day to collect the water—often polluted—that they will use for drinking, cooking, and washing. Similar treks of increasing length must be taken to collect the firewood for cooking (Fig. 1–1). Food is mostly coarse grains, such as sorghum and maize, raised in small landholdings by the women, and the amounts are often limited.

Both of these pictures are from the **developing countries**, nowadays often referred to as the **South**, because the majority of the countries below the equator are developing countries. They are pictures of people closely tied to their environment. You return home to the United States and observe that here and in other **industrialized countries** (the **North**), people appear to be more detached from their environment. People live in well-built homes surrounded by manicured lawns. They adjust the indoor temperature to their liking, and any amount of safe hot or cold water is available with the turning of a faucet handle. Travelers go

◀ Primitive people and their dwellings along a river in a South American rain forest.

▲ FIGURE 1–1 *Impact on people's singular source of fuel as a result of landscape depletion.* For one-third of the world population, the only source of fuel for cooking is firewood. Many women in less developed countries must spend several hours each day gathering wood. Treks become longer and more difficult as the landscape is increasingly denuded.

nearly anywhere they wish in the air-conditioned comfort of private cars. Their greatest concern about food is that they not overeat. News and entertainment from around the world are displayed on several television sets in each home, and most families own computers with global connections. There is an appearance of a civilization that is detached from its environment (Fig. 1–2). This is an illusion. Human societies, whether they appear to be closely tied to their environment or not, are absolutely dependent on the natural environment, and their impact on the environment is crucial for their continued success.

To illustrate this point, we can travel to Rapa Nui (formerly Easter Island), a remote spot of land in the South Pacific. Here we find giant stone heads (Maoi) standing as sentinels with their backs toward the sea (Fig. 1–3). These statues are evidence of a once sophisticated civilization. Yet, Easter Island natives encountered by 18thcentury explorers were living at a primitive level, scratching out a meager existence on a desolate island. When asked about the great stone heads, the natives could say only, "Our legends say that our ancestors made them, but we don't know how or why." The past culture and civilization of the island were not sustained.

Working from the legends that natives told, and conducting excavations for evidence, archaeologists have

▶ FIGURE 1–2 *Living in industrialized countries.* Life in industrialized countries appears to be detached from the environment.

▲ **FIGURE 1–3** *Easter Island.* The great stone heads and other artifacts found on Easter Island provide evidence of a once prosperous culture. The present barren, eroded landscape indicates that the civilization collapsed as a result of overexploitation of forest and soil resources. Is the story of Easter Island a parable for modern civilization?

pieced together the following chronology of events: The original inhabitants of Easter Island were Polynesians who probably arrived on the island as part of a deliberate colonization mission sometime between 400 and 800 A.D. The evidence from pollen grains found in the soil and in artifacts shows that these early arrivals found an island abundantly forested with a wide variety of trees, including palms, conifers, and mahogany. As their population grew and flourished, they cut trees for agriculture, for structural materials, and to move the huge stone heads from the quarries to their erection sites. By 1600, all the trees were gone. Without plant roots, the cleared land failed to hold water, and the soil washed into the sea, killing the fish and shellfish near the shore. The eroded soil baked hard and dry after rains, offering little support for agriculture.

As the forest was depleted and soil and water resources were degraded, the work necessary for existence became harder and the rewards fewer. The gap between the ruling and worker classes widened, apparently becoming intolerable. In 1678, there was a sudden revolt of the workers. In the great war that ensued, virtually the entire ruling class was killed. Still, the situation worsened. Anarchy broke out among the workers, who splintered into groups and continued to fight among themselves. Starvation and disease be-

came epidemic. Without any trees, no one could escape the island by boat. A population that had numbered about 8,000 at the time of the revolt was down to a few hundred people by the mid-1800s. Many have puzzled over the reasons why the Easter Islanders, who could walk around their 166-km^2 island in a day, were unable to foresee the consequences of their practices.

The lesson of Easter Island is all too clear: When a society fails to care for the environment that sustains it, when its population increases beyond the capacity of the land and water to provide adequate food for all, and when the disparity between *haves* and *have-nots* widens into a gulf of social injustice, the result is disaster. The civilization collapses. History is replete with the ruins of other civilizations, such as the Mayans, Greeks, Incas, and Romans, that failed to recognize the constraints of their environment.

Are there some parallels between Easter Island and the start of the 21st century? Is there evidence that we are making some of the same mistakes made by the Easter Islanders? Like Easter Island, Earth has only limited resources to support human societies and their demands. There is no escape. The social unrest on Easter Island may parallel the tensions between the industrialized and the developing countries. Is there some possibility that we will so damage our environment that our civilization will begin to decline? Are we already in trouble? Can we take corrective action before it is too late?

These are important questions, and to answer them will take the rest of this text. The task before us—both in the text and as a 21st century civilization—can be laid out in four steps:

- To understand how the natural world works
- To understand how human systems are interacting with natural systems
- To accurately assess the status and trends of crucial natural systems
- To promote and follow a long-term, sustainable relationship with the natural world

1.1 The Global Environmental Picture

The arrival of the new millennium and a new century became the occasion for taking stock in many areas of human concern, including, in particular, the global environment. The picture of the state of our planet that emerged from a number of surveys was troubling. Four global trends are of particular concern: (a) population growth and increasing consumption per person, (b) a decline of vital life-support ecosystems, (c) global atmospheric changes, and (d) a loss of biodiversity. Each of these issues will be explored in greater depth in later chapters.

Population Growth

The world's human population, over 6.1 billion persons in 2001, has grown by 2 billion in just the last 25 years. It is continuing to grow, adding nearly 78 million persons per year. Even though the growth rate is gradually slowing, the world population at 2050 could be 8.9 billion, according to the most recent projections from the U.N. Population Division (Fig. 1–4). Each person creates a certain demand on Earth's resources, and the demand tends to increase with greater affluence. Compare, for example, the resources required to support a typical American or Canadian lifestyle with those required to support indigenous people living along the banks of the Peruvian Amazon.

The 3 billion persons added to the human population by 2050 will all have to be fed, clothed, housed, and supported by gainful employment. Virtually all of the increase will be in the developing countries, where birthrates are still much higher than death rates. In these same countries, 1.2 billion experience severe poverty, lacking sufficient income to meet their basic needs for food, clothing, and shelter. Over 800 million—one out of every five—remain undernourished. At the same time, global economic production continues to rise, tripling since 1980. In spite of that growth, the gap between the average incomes of the wealthiest 20% and the poorest 20% in the world is growing wider. In 1990 it was 60 to 1, and in 1999 it widened to 74 to 1. Stabilizing population growth in the developing countries is the most important prerequisite for closing the economic gap between those nations and the industrialized countries.

The Decline of Ecosystems

Vital resources are stressed by the dual demands of increasing population and increasing consumption per person. Around the world we see groundwater supplies depleted, agricultural soils degraded, oceans overfished, and forests cut faster than they can regrow. A recent report from the United Nations entitled *Pilot Analysis of Global Ecosystems*, or PAGE, addressed these trends as it examined the status of the five major ecosystems which deliver the goods and services that support human life and the economy: coastal/marine systems, freshwater systems, agricultural

▶ **FIGURE 1–4 *World population in the last 2000 years.*** World population started a rapid growth phase in the early 1800s and has grown sixfold in the last 200 years. It is growing by nearly 78 million people per year. (See Chapter 6.) Future projections are based on assumptions of a continued decline in birthrates.

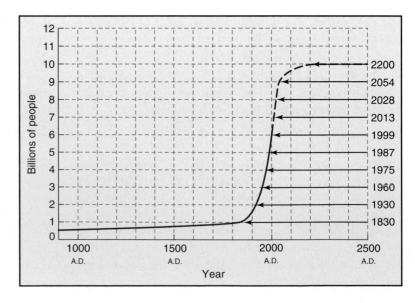

lands, grasslands, and forests. PAGE represents the first step in an ongoing project designed to measure the impact of human actions on these vital ecosystems. To quote the report's summary, "nearly every measure we use to assess the health of ecosystems tells us we are drawing on them more than ever and degrading them at an accelerating pace."

One of the most disturbing findings of PAGE is that human activities are now beginning to significantly affect the natural chemical cycles—water, carbon, nitrogen, phosphorus—on which all ecosystems depend. In spite of the critical importance of these cycles to human affairs, we still lack much of the knowledge needed to assess present conditions and to move toward deliberate and wise management of natural ecosystems. And what makes the situation worse is that, in light of our absolute dependence on the goods and services supplied by these natural ecosystems, few enterprises could be more important than the pursuit of that knowledge.

Global Atmospheric Changes

Historically, pollution has been a relatively local problem, affecting a given river, lake, or bay, or the air in a city. Today, scientists are analyzing pollution on a global scale. A case in point is the danger of global climate change due to carbon dioxide (CO_2), an unavoidable by-product of burning fossil fuels—crude oil, coal, and natural gas. Because of the large amount of fossil fuels currently being burned, CO_2 levels in the atmosphere have grown from about 280 parts per million (ppm), or 0.028%, in 1900 to over 375 ppm as we enter the first decade of the new century. Indeed, the level of atmospheric CO_2 is increasing at 0.4% per year and is expected to double during the next century.

Carbon dioxide is a natural component of the lower atmosphere, along with nitrogen and oxygen. It is required by plants for photosynthesis and is important to the Earth–atmosphere energy budget. Carbon dioxide is transparent to incoming light from the Sun, but absorbs infrared (heat) energy radiated from Earth's surface, thus delaying its loss to space. This process warms the lower atmosphere in a phenomenon known as the *greenhouse effect*. Although the concentration of CO_2 is a small percentage of the atmosphere, even slight increases in the volume of the gas affects temperatures. Figure 1–5 graphs air temperatures from 1880 to the present and illustrates the clear warming trend. Referring to this trend, the latest report of the Intergovernmental Panel on Climate Change (IPCC), released in 2000, stated that anthropogenic greenhouse gases have "contributed substantially to the observed warming over the last 50 years."

Concern about global climate change led representatives of 166 nations to meet in Kyoto, Japan, in December of 1997 to negotiate a treaty to reduce emissions of carbon

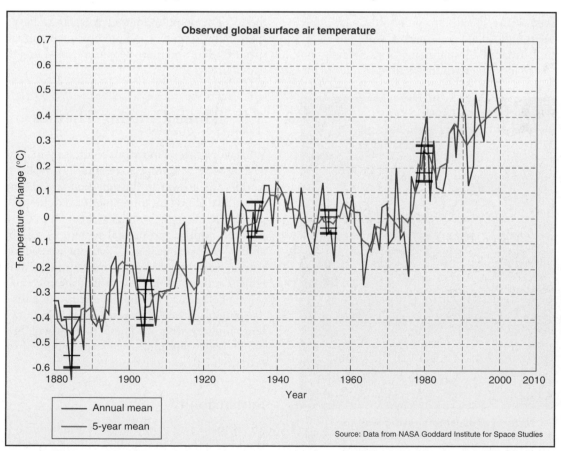

▲ **FIGURE 1–5** *Annual mean global surface atmospheric temperatures.* The baseline, or zero point, is the 1880–1999 long-term average temperature. The warming trend since 1970 is conspicuous. (*Source:* National Climatic Data Center, NOAA).

dioxide and other greenhouse gases. Most of the industrialized nations agreed to reduce emissions to below 1990 levels, to be achieved by the year 2010. Still, if the treaty is ratified and adhered to by all parties, greenhouse gases will continue to rise. At issue are the conflicting concerns between the short-term economic impacts of reducing the use of fossil fuels and the long-term consequences of climate change for the planet and all its inhabitants. Yet, stabilizing the atmospheric concentration of CO_2 is essential to stabilizing the global climate itself, and this cannot be accomplished without serious reductions in the current rate of use of fossil fuels.

Loss of Biodiversity

The rapidly growing human population, along with increasing consumption, is accelerating the conversion of forests, grasslands, and wetlands to agriculture and urban development (Fig. 1–6). The inevitable result is the loss of most of the wild plants and animals that occupy those natural habitats. If the species involved have no populations at other locations, they are doomed to extinction by such **habitat alteration**. **Pollution** also degrades habitats—particularly aquatic and marine habitats—destroying the species they support. Further, hundreds of species of mammals, reptiles, amphibians, fish, birds, and butterflies, as well as innumerable plants, are exploited for their commercial value. Even where species are protected by law, they continue to be hunted, killed, and marketed illegally.

Thus, Earth is rapidly losing many of its species—no one knows how many. The term that is used to refer to the total diversity of living things—plants, animals, and microbes—that inhabit the planet is **biodiversity**. About 1.75 million species have been described and classified, but scientists estimate that at least 13 million remain unidentified. Because so many species remain unidentified, the exact number of species becoming extinct can only be estimated.

Why is losing biodiversity so critical? For one thing, all domestic plants and animals used in agriculture are derived from wild species, and we still rely on introducing genes from wild species into our domestic species to keep them vigorous and capable of adapting to different conditions. For another, some 80% of the human population depends on traditional medicines, which in turn are highly dependent on biodiversity. Many modern prescription drugs were originally derived from plants, even though only a few percent of plants have been thoroughly studied from this medicinal viewpoint. Biodiversity is the mainstay of agricultural crops and of medicines. The loss of biodiversity can only curtail development in these areas. Biodiversity is also a critical factor in maintaining the stability of natural systems and enabling them to recover after disturbances such as fires or volcanic eruptions.

There are also aesthetic and moral arguments for maintaining biodiversity. Forty or 50 years from now, do you want to show your grandchildren pictures of animals such as rhinoceroses, tigers, pandas, and orangutans and have to say, "These animals don't exist anymore. We killed them"? The question for society is, do we have a moral responsibility to protect and preserve such animals and other species? More and more people are answering this question with a firm yes.

1.2 Three Unifying Themes

What will it take to move our civilization in the direction of a long-term sustainable relationship with the natural world? The answer to this question is not simple, but we would like to present three interrelated themes that are applicable to changing or giving direction to the interactions between human and natural systems (Fig. 1–7): **sustainability**—the practical goal that our interactions with the natural world should be working toward; **stewardship**—the ethical and moral framework that informs our public and private actions; and **sound science**—the basis for our understanding of how the world works and how human systems interact with it. These themes will be applied to public policy and individual responsibility throughout the text; we briefly explore them here.

Sustainability

To say that a system or process is sustainable is to say that it can be continued indefinitely without depleting any of the material or energy resources required to keep it running. The term was first applied to the idea of **sustainable yields**

▲ **FIGURE 1–6 *Natural ecosystems giving way to development.*** Continuing growth requires a massive reorganization and exploitation of natural resources, bringing record levels of degradation of natural ecosystems. Here we see the stripping away of a mature forest in the northeastern United States.

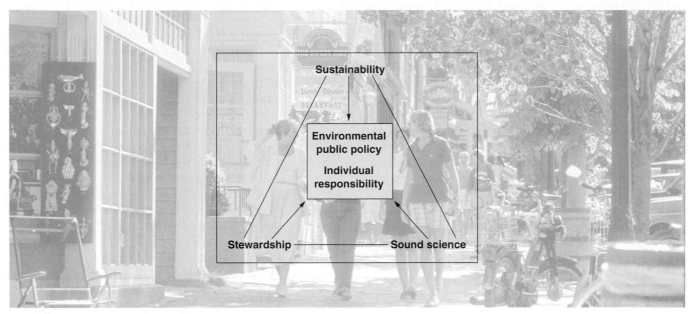

▲ **FIGURE 1–7** *Three unifying themes.* Sustainability, stewardship, and sound science represent three vital concepts that must be embraced by our society and employed in the development of environmental public policy and private environmental concern.

in human endeavors such as forestry and fisheries. Trees, fish, and other biological species normally grow and reproduce at rates faster than that required just to keep their populations stable. This built-in capacity allows every species to increase or replace a population following some natural disaster.

Thus, it is possible to harvest a certain percentage of trees or fish every year without depleting the forest or reducing the fish population below a certain base number. As long as the number harvested stays within the capacity of the population to grow and replace itself, the practice can be continued indefinitely. The harvest then represents a sustainable yield. It becomes unsustainable only when the cutting of trees or catching of fish exceeds the capacity for their present population to reproduce and grow. The concept of a sustainable yield can also be applied to freshwater supplies, soils, and the ability of natural systems to absorb pollutants without being damaged.

The notion of sustainability can be extended to include ecosystems. *Sustainable ecosystems* are entire natural systems that persist over time by recycling nutrients and maintaining a diversity of species in balance and by using the Sun as a source of sustainable energy. We will see how these systems can model sustainability for us in the next several chapters.

Applying the concept to human systems, we can picture a *sustainable society* as a society that is in balance with the natural world, continuing generation after generation, neither depleting its resource base by exceeding sustainable yields nor producing pollutants in excess of nature's capacity to absorb them. Many primitive societies were sustainable in this sense for thousands of years, but in time, their populations expanded and required them to convert to agriculture in order to meet their food needs.

When we apply the concept of sustainability to modern societies, we immediately recognize that many of our interactions with the environment are *not sustainable*, as seen in the aforementioned global trends. Although population growth in the industrialized countries has almost halted, these countries are using energy and other resources at unsustainable rates, producing pollutants that are accumulating in the atmosphere, water, and land. In contrast, developing countries are experiencing a continued rapid population growth, yet are often unable to meet the needs of many of their people in spite of heavy exploitation of their natural resources. Based on the expectations of continued economic growth and progress, the crux of the problem is modern society's inexperience with sustainability. No civilization on Earth has ever done it. How do we resolve this dilemma? One answer is the concept of **sustainable development**.

Sustainable Development. Sustainable development is a term that was first brought into common use by the World Commission on Environment and Development, a group appointed by the United Nations. The commission made sustainable development the theme of its final report, *Our Common Future*, published in 1987. The report defined the term as a form of development or progress that "meets the needs of the present without compromising the ability of future generations to meet their own needs." The concept arose in the context of a debate between the *environmental* and *developmental* concerns of different groups of countries. **Development** refers to the continued improvement of living standards by economic growth, usually in the developing countries of the South. Both groups of countries (North and South) have embraced the concept of sustainable development, although the industrialized countries are usually more concerned about environmental sustainability, while the developing countries are more concerned about economic development.

The concept is now so well entrenched in international circles that it has become almost an article of faith. It sounds comforting, so people want to believe it is possible, and it appears to incorporate some ideals that are sorely needed, such as **equity**—whereby the needs of the present are actually met and where future generations are seen as equally deserving as those living now. However, sustainable development means different things to different people, and this divergence is well illustrated by the viewpoints of three important disciplines traditionally concerned with the processes involved. **Economists** are concerned mainly with growth, efficiency, and maximum use of resources. **Sociologists** focus on human needs and on concepts like equity, empowerment, social cohesion, and cultural identity. **Ecologists** show their greatest concern for preserving the integrity of natural systems, for living within the carrying capacity of the environment, and for dealing effectively with pollution. Yet, it can be argued that *sustainable solutions* will be found only where the concerns of these three groups intersect, as illustrated in Fig. 1–8.

There are many dimensions to sustainable development—environmental, social, economic, political—and it is clear that no societies today have achieved anything resembling it. Nevertheless, as with justice, equality, and freedom, it is important to uphold sustainable development as an ideal, a goal toward which all human societies need to be moving, even if we have not achieved it completely anywhere.

The transition to a sustainable civilization is hard to picture at present. We are talking about a stable human population that recognizes the finite limits of Earth's systems to produce resources and absorb wastes and that acts

accordingly. However, if we fail to achieve sustainability by our deliberate actions, the natural world will impose it on us in highly undesirable ways (e.g., through famine, disease, and deprivation). To achieve sustainability will require a special level of dedication and commitment to care for the natural world and to act with justice and equity toward one another. In their recent book, *The Local Politics of Global Sustainability*, authors Thomas Prugh, Robert Costanza, and Herman Daly argue that *communities* will be the main arenas for achieving sustainability. They point out that the process must be a political one and that the present political and economic systems will likely only continue to move us in the wrong direction. They propose a political evolution to "strong democracy," where citizens become much more informed and involved in decision-making at all levels. As The Class of 2000 Report on environmental education states, "Sustainability requires that society itself, within and among nations, becomes a steward of the planet." We now turn to stewardship as a concept that captures much of what is needed in the realm of ethics and values in order to achieve sustainability.

Stewardship

Stewardship is a concept that emerged from the institution of slavery. A steward was a slave put in charge of the master's household, responsible for maintaining the welfare of the people and the property of the owner. Since a steward did not own the property himself, the steward's ethic involved a faithful caring for something on behalf of someone else.

Applying this concept to our current scene, we see that stewards are those who care for something—from the natural world or from human culture—that is not theirs and that they will pass on to the next generation. Modern-day stewardship, therefore, is an ethic that provides a guide to actions taken to benefit the natural world and other people. Stewardly care is clearly compatible with the goal of sustainability; it is just as clearly different from the goal itself, however, in that stewardship deals more directly with *how* sustainability is to be achieved—what values and ethical considerations must be foremost as different choices are weighed. (See "Ethics," p. 9.)

How is stewardship accomplished? As one example, it is common for foresters, hunters, or fishers—who began as exploiters—to become aware of the vulnerability and beauty of the resources they are using and instead to turn exploitation into responsible, stewardly care for the natural systems in which they are working. Indeed, sometimes stewardship leads people into battle to stop the destruction of the environment or to stop the pollution that is degrading human neighborhoods and health. Examples of this kind of stewardly action are Lois Gibbs and other homeowners at Love Canal, who drew attention to the chemical wastes buried in their neighborhood; Rachel Carson, who, in her book *Silent Spring*, alerted the public about pesticides; Rodolfo Montiel, who organized fellow peasants to

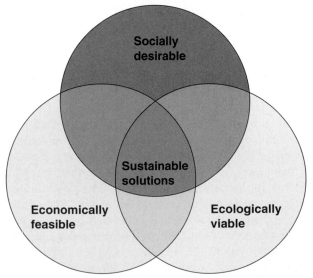

▲ **FIGURE 1–8** *Sustainable solutions.* The concerns of sociologists, economists, and ecologists must intersect in order to achieve sustainable solutions in a society.

ETHICS

WHAT IS THE STEWARDSHIP ETHIC?

Ethics is about the *good*—those values and virtues that we should encourage—and about the *right*—our moral duties as we face practical problems. Obviously, ethics is a "normative" discipline—it tells us what we ought to do. Some things are right, and some things are wrong. How do we know what we ought to do? In other words, how does an ethic work? A fully developed ethic has four ingredients:

1. *Cases*. These refer to specific acts and ask whether a particular act is morally justified. The answer must be based on moral rules.

2. *Moral rules*. These are general guidelines that can be applied to various areas of concern—for example, the rules that govern how we should treat endangered species.

3. *Moral principles*. Moral rules are based on more general principles, which are the broadest ethical concepts. They are considered to be valid in all cases. An example is the principle of distributive justice, which states that all persons should be treated equitably.

4. *Bases*. Ethical principles are justified by reference to some philosophical or theological basis. This is the foundation for an ethical system.

As we have defined it, a stewardship ethic is concerned with right and wrong in reference to taking care of the natural world and the people in it. Since a steward is someone who cares for the natural world on behalf of others, we might ask, To whom is the steward responsible? Many would say, to present and future generations of people who depend on the natural world as their life-support system. For people with religious convictions, stewardship stems from a belief that the world and everything in it be-

longs to a higher being; thus, they are stewards on behalf of God. For others, stewardship becomes a matter of concern that stems from a deep understanding and love of the natural world and the necessary limitations on human use of that world.

Is there a well-established stewardship ethic? Insofar as human interests are concerned, ethical principles and rules are fairly well established, even if they are often violated by public and private acts. However, there is no firmly established ethic that deals with care for natural lands and creatures for their own sake. Most of our ethic concerning natural things really deals with how those things serve human purposes; that is, our current ethic is highly anthropocentric. For example, a basic ethical principle from the U.N. Declaration on Human Rights and the Environment is as follows: "All persons have the right to a secure, healthy, and ecologically sound environment."

Many organizations have developed stewardship principles for their own guidance, and these can sometimes serve as broader guidelines for society, although they usually address very specific areas of concern and are generally anthropocentric. For example, the U.S. Forest Service has published a "Land and Service Ethic," which is supposed to guide Forest Service personnel as they carry out their activities. The Forest Service's Land Ethic principle is to "promote the sustainability of ecosystems by ensuring their health, diversity, and productivity." The agency goes on to explain, "Through ecosystem sustainability, present and future generations will reap the benefits that healthy, diverse, and productive ecosystems provide." (The Forest Service: Ethics and Course to the Future. USDA Forest Service, Washington, DC, October 1994.)

A stewardship ethic is concerned with the following issues, among others:

1. How to define the common good in cases where conflicting needs emerge, such as the economic need to extract resources from natural systems versus the need to maintain those systems in a healthy state.

2. How to balance the needs of present generations versus those of future generations.

3. How to preserve species when doing so clearly means limiting some of the property rights commonly enjoyed by people and organizations.

4. How to encourage people and governments to exercise compassion and care for others who suffer profoundly from a lack of access to the basic human needs of food, shelter, health, and gainful work.

5. How to promote virtues that are conducive to stewardly care of the natural world, such as benevolence, accountability, and responsibility. (There are many others.)

6. How to limit consumption while at the same time allowing people the freedom to choose their lifestyles? In other words, how to balance individual rights and responsibilities with those of the community with regard to the environment?

These are some of the questions that a stewardship ethic must grapple with in order to be truly helpful in accomplishing our goal of providing the "ethical and moral framework that informs our public and private actions," as we have stated it. The work of establishing a well-recognized, broadly applied stewardship ethic is still to be done.

block loggers from cutting virgin forests near his Mexican village and, as a result, was tortured and imprisoned on trumped-up charges; and Ken Saro-Wiwa, who was executed by the Nigerian dictatorship for defending the Ogoni people of Nigeria when oil spills and toxic wastes were devastating their farming and fishing (Fig. 1–9).

Justice and Equity. The stewardship ethic is concerned not only for the care of the natural world, but also for the establishment of just relationships among humans. This concern

for justice has been applied to the United States in what is called the *environmental justice movement*. The major problem addressed by the movement is **environmental racism**—the placement of waste sites and other hazardous facilities in towns and neighborhoods in which most of the residents are nonwhite. People of color are seizing the initiative to correct these wrongs, creating citizen groups and watchdog agencies to bring effective action and to monitor progress. An example of this can be found in the neighborhoods of Piney Woods and Alton Park, suburbs of Chattanooga, Tennessee,

▲ FIGURE 1–9 *Stewardship at work.* Ken Saro-Wiwa was a Nigerian writer who lost his life defending the Ogoni people against government-backed interests that were degrading their environment. He is an example of unusual environmental stewardship at work.

populated largely by African-Americans. In those neighborhoods, there are 42 known hazardous waste sites, 12 of which have been listed as Superfund sites. Cancer and asthma rates there were unusually high, so people organized the group Stop TOxic Pollution (STOP) to draw attention to the toxic waste and health problems and, eventually, to acquire a Technical Assistance Grant from the Environmental Protection Agency (EPA) to help fence off the worst sites.

The flip side of this problem is seen when wealthier, more politically active, and often predominantly white communities receive a disproportionately greater share of facilities such as new roads, public buildings, and water and sewer projects.

Justice is especially crucial for the developing world, where unjust relationships often leave people without land, with inadequate food, and in poor health. Abject poverty is the condition of at least 1.2 billion people, whose poverty is often brought on by injustices within societies where

wealthy elites maintain political power and, through corruption and nepotism, steal money and create corporations that receive preferential treatment. The names Marcos (of the Philippines), Mobutu (Zaire), and Sukarno (Indonesia) are reminders of the capacity of some leaders to divert billions of dollars from a country's resources for their private gain while ignoring the needs of the desperately poor and powerless majorities in their countries (Fig. 1–10). These leaders plundered their countries and enriched their families until they were finally driven out of office.

Some of the poverty of the developing countries can be attributed to unjust economic practices of the wealthy industrialized countries. The history of European colonialism demonstrates how a dominant culture can enrich itself by seizing power in an undeveloped region and exploiting the peoples and resources. Although colonialism is now dead, its legacy is still seen in the dependency fostered in so many of the former colonies and the poorly developed technological, educational, and social conditions there. Other sources of the continued poverty of many of the developing countries are current patterns of international trade and the problem of international debt. By imposing restrictive tariffs and import quotas, industrialized countries have maintained inequities that discriminate against the developing countries. The United Nations reports that the rate of protection against exports from developing countries is significantly higher than the rate against exports from industrial countries. Although these barriers are falling, they are still in place for many manufactured goods coming from the developing world. Such barriers deprive people in developing countries of jobs and money that would go far to improve their living conditions.

One form of trade the developing countries can do without is the international trade in toxic wastes. Some countries or local jurisdictions in search of ready cash have found a source in the form of toxic waste shipments, which invariably come from more developed countries and are usually illegal (Fig. 1–11). The *Basel Action Network* is a nongovernmental organization that seeks to prevent such

(a)

(b)

(c)

▲ FIGURE 1–10 *Leaders who stole from their people.* (a) Mobutu Sese Seko, former president of Zaire; (b) Sukarno, former president of Indonesia; and (c) Ferdinand Marcos, former president of the Philippines. These national leaders diverted billions of dollars from the people of their countries to enrich themselves and their families.

▲ **FIGURE 1–11** *International toxic waste.* The *Wan He* is carrying toxic waste from U.S. military bases in Japan. It was refused entry in the United States and Canada. Greenpeace protesters are shown on board in Yokohama, Japan, where the ship was docked pending a final decision on its destination.

trade when it is not in conformity with the Basel Convention, an international agreement that places a ban on most international toxic waste trade. The Network is vigilant in publicizing and coordinating legal challenges to incidents involving toxic waste shipments and has made it clear that this trade is a matter of international environmental justice.

The international debt problem is complex; debtor nations are most commonly those in the process of development, and the borrowed funds were intended to speed up that process. Unfortunately, much of the borrowed money went to the purchase of arms, to ill-planned projects, or to corruption. One example is the Bura Irrigation project in Kenya, which cost over $100 million, displaced thousands of people, and was then never implemented. Today, some 41 low-income countries are basically unable to repay their debts (amounting to approximately $207 billion), according to the World Bank. These countries must continue to make scheduled repayments that are often greater than the amount they spend on basic education and health care. One can argue that the leaders of those countries should not

have borrowed the funds, but such arguments do nothing to help the poor, who are most affected by the problem. Although the issue of international justice is a difficult one, it is clearly part of the mission of stewardly action to address it.

The degradation of the natural world and concerns about environmental justice have led many to become involved in efforts to resolve these problems; such people are called environmentalists, and the worldview they embrace has become known as environmentalism.

Environmentalism

A Brief History. Although what we now term the environmental movement began less than 50 years ago, it had its roots in the late 19th century, when some people realized that the unique, wild aspects of the United States were disappearing. The 1890 U.S. census demonstrated the "closing of the frontier," an event that was noted with some sadness. No longer could any area of the country be classified as totally uninhabited. Around that time, several groups devoted to conservation formed: the National Audubon Society, the National Wildlife Federation, and the Sierra Club, founded in California by naturalist John Muir, who helped popularize the idea of wilderness. To counter the wide-scale misuse of natural resources, President Theodore Roosevelt promoted the conservation of public lands, setting aside some key sites as national parks. The national environmental consciousness was stirring.

Technological achievements following World War I, however, eventually helped create an environmental crisis—the Dust Bowl of the 1930s, when wheat failed to hold soil in place and the topsoil eroded (Fig. 1–12). During the Great Depression (1930–1936), conservation provided a means both of restoring the land and providing work for the unemployed. Many trails, erosion-control projects, and other improvements in national parks and forests were originally installed by the Civilian Conservation Corps (CCC), which played a major role in pulling the country out of the depression. The two decades following World War II

◀ **FIGURE 1–12** *The Dust Bowl.* In the 1930s, modern farming methods combined with an extended drought led to the loss of enormous amounts of topsoil due to wind erosion.

(1945–1965) were full of optimism. The United States and its allies had won the "greatest of all wars," in part because of technology. The tremendous productive capacity built up during the war, together with new developments ranging from rocket science to computers and from pesticides to antibiotics, could be redirected to peacetime applications. Except for the tensions of the Cold War during those years and some apprehension about atomic energy, it seemed as if nothing but opportunity and prosperity lay ahead.

Although economic expansion enabled most families to have a home, a car, and other possessions, certain problems became obvious. The air in and around cities was becoming murky and irritating to people's eyes and respiratory systems. Streetlights from St. Louis to Pittsburgh were left on during the day because of the smoke from coal-burning industries. In winter, even freshly fallen snow soon turned gray as the soot fell from city chimneys. Rivers and beaches were increasingly fouled with raw sewage, garbage, and chemical wastes. Conspicuous declines occurred in many bird populations (including our national symbol, the bald eagle), aquatic species, and other animals.

It was easy to identify some culprits: belching industrial smokestacks, open burning dumps, and municipal and industrial sewers, discharging raw sewage and chemical wastes into waterways. The decline of the bald eagle and other bird populations was traced to the accumulation in their bodies of DDT, the long-lasting pesticide that had been used in large amounts since the 1940s. In short, it was clear that we were seriously contaminating our environment.

The Modern Environmental Movement. In 1962 biologist Rachel Carson wrote *Silent Spring,* presenting her scenario of a future with no songbirds and with other dire consequences if pollution of the environment with DDT and other pesticides continued. Carson's voice was soon joined by others, many of whom formed organizations to focus and amplify the voices of thousands more in demanding a cleaner environment. This was the beginning of the modern **environmental movement**, in which a newly militant citizenry demanded the curtailment of pollution, the cleanup of polluted environments, and the protection of pristine areas. It is significant that the environmental movement began as a grassroots initiative, and it maintains its momentum and force today only by continuing to command public interest and support.

Members of the environmental movement of the 1960s joined forces with older organizations, such as the National Audubon Society and the National Parks Conservation Association, that had a considerable history of dedication to preserving wildlife and natural habitats. There is an obvious overlap between the goal of reducing pollution and that of protecting wildlife. Wildlife cannot survive in polluted surroundings, and preserving an uncontaminated environment presupposes that the plant and animal species within that environment will be protected. Therefore, established wildlife-preservation organizations and their members became significant players in the blossoming environmental movement, along with newly formed organi-

zations such as the Environmental Defense Fund, the Natural Resources Defense Council, Greenpeace, and Zero Population Growth. This dual focus continues today. As the broad category of "environmentalists" includes everyone concerned with reducing pollution or protecting wildlife, it is no wonder that at least 70% of the population identify themselves as environmentalists.

By almost any measure, the environmental movement has been successful. Pressured by a concerned citizenry, Congress created the Environmental Protection Agency (EPA) in 1970 and passed numerous laws promoting pollution control and wildlife protection, including the National Environmental Policy Act of 1969, the Clean Air Act of 1970, the Clean Water Act of 1972, the Marine Mammals Protection Act of 1972, the Endangered Species Act of 1973, and the Safe Drinking Water Act of 1974. You will read more about these laws and their effects in subsequent chapters. In the area of pollution abatement alone, our society has spent hundreds of billions of dollars, both on installing pollution control devices and on redesigning procedures and products so that less pollution is created. The pollution control devices on our cars are an example. Governments have spent additional billions on upgrading sewage treatment, on refuse disposal, and on other measures to reduce pollution. As a result of these expenditures, the air in most cities and the water in numerous lakes and rivers are cleaner now than they were in the late 1960s. Without a doubt, they are immensely cleaner than they would be had the environmental movement not come into existence.

Environmentalism Acquires Its Critics. In the early stages of the environmental movement, the sources of the problems were specific and visible. That made it easy to identify certain polluters as the "bad guys" and take the side of the environmental movement. Likewise, the solutions seemed relatively straightforward: Install waste treatment and pollution control equipment and ban the use of DDT, substituting safer pesticides. At first, many industries conformed to the reforms, but as regulatory demands increased, a number of the more polluting industries, such as mining and utilities, pressed for a halt in regulation, and some began demanding a significant *deregulation.* They believed that environmental regulation was harmful to the economy, and they found a political ally in President Ronald Reagan, elected in 1980. Soon, Reagan forces allied themselves with some western opponents of public-lands regulation and others committed to reducing the role of federal government, whereupon they initiated an attempt to dismantle the environmental regulatory framework of government. Fortunately, this attempt largely failed, due in large part to a public backlash against Reagan's new policies. Nevertheless, the EPA and the Department of the Interior were embroiled in political controversy for a number of years.

In the 1990s, environmental protection once again was back on track with the election of President William Clinton and his appointment of strong environmental protection supporters to head the federal agencies. However, the opponents of federal land management and widespread pollution regulations began a new and more sophisticated attack against environmentalism, the *wise-use movement*. Briefly, this movement brought together at the grassroots level many disparate groups whose interests were affected by environmental regulations. Eventually, the movement reached effective political strength when the 104th Congress was elected in 1994.

The bitterness of the conflict between environmentalists and those with special business interests is illustrated by the case of the spotted owl that inhabits old-growth forests of the Northwest (Fig. 1–13). Saving the spotted owl from extinction required putting large tracts of old-growth forests off limits to logging. Indeed, the central interest of environmentalists is to save some of these old-growth forests in their natural state. However, curtailing logging of the remaining old-growth forests was portrayed as threatening the jobs of many loggers, the business of some logging companies, and even the existence of whole communities whose economies were based solely on logging. Many loggers were hostile toward environmentalists who would effectively trade their jobs for "some bird no one ever heard of before the controversy started." Similarly, many landowners became hostile toward environmentalists as they found themselves prevented from developing their property because it was home to an endangered species or was subject to some other environmental restriction. Landowners also became hostile toward the government in general, since the government was responsible for enforcing the regulations.

Opposing the wise-use movement on many fronts have been nongovernmental organizations such as the Sierra Club, the National Audubon Society, and Greenpeace. These grassroots organizations regard themselves as the caretakers of the public interest, and in addition to opposing wise-use, they do not hesitate to criticize government actions that are perceived as harmful to the environment. They continue to mobilize support for environmental causes and are regarded as the frontline manifestations of environmentalism.

Environmentalism to Stewardship. At this point, we might ask, What does the future hold for environmentalism? Along with most other professionals in the environmental arena, we feel that our future depends on redefining the role of environmentalism. Most important is that the *ethic of stewardship* be broadly communicated and practiced by whole societies—individuals, communities, corporations, and nations—as we move toward the goal of sustainability. But how can this be accomplished? The President's Council on Sustainable Development addressed this concern in its initial report, *Sustainable America: A New Consensus:*

> The answer is multifaceted, but it starts with understanding the dynamics at work in the environment and the connection among environmental protection, economic prosperity, and social equity and well-being. It depends on the processes by which individuals, institutions, and government at all levels can work together toward protecting and restoring the country's inherited natural resource base. Education, information, and communication are all important for developing a stewardship ethic. Also important is the widespread understanding that people, bonded by a shared purpose, can work together to make sustainable development a reality.

◀ **FIGURE 1–13** *Spotted owl and old-growth forests of the Pacific Northwest.* Listing of the spotted owl as a threatened species led to battles between environmentalists, who wanted to preserve the old-growth forests, and the logging interests, who wanted to cut them.

In its report, the Council identified a number of *policy recommendations* to accomplish the *stewardship of natural resources*. Each recommendation is further developed in the report through the proposal of specific actions:

- Enhance, restore, and sustain the health, productivity, and biodiversity of terrestrial and aquatic ecosystems through cooperative efforts to use the best ecological, social, and economic information to manage natural resources.
- Create and promote incentives to stimulate and support the appropriate involvement of corporations, property owners, resource users, and government at all levels in the individual and collective pursuit of stewardship of natural resources.
- Manage and protect agricultural resources to maintain and enhance long-term productivity, profitability, human health, and environmental quality.
- Establish a structured process involving a representative group of stakeholders to facilitate public and private efforts to define and achieve the national goal of sustainable management of forests by the year 2000.
- Restore habitats and eliminate overfishing to rebuild and sustain depleted wild stocks of fish in U.S. waters.
- Create voluntary partnerships among private landowners at the local and regional levels to foster environmentally responsible management and protect biological diversity, with government agencies providing incentives, support, and information.

It should be evident that these recommendations are directed at both the present and the future; where they are implemented, they represent actions taken that will lead in the direction of sustainability.

As we have suggested, sustainability encompasses the goals that we should work toward, while stewardship defines an ethic that should guide our practices. Still missing from our picture, however, is the crucial information that tells us how the natural world works and what is happening to it as a result of human activities. This is the domain of sound science and the scientist.

Sound Science

Many environmental issues are embroiled in controversies that are so polarized that no middle ground seems possible. On the one hand are persons who argue from apparently sound facts and proven theories. On the other are persons who disagree and present opposing theories to interpret the facts. Both groups may have motives for arguing their case that are not at all apparent to the public. In the face of such controversy, many people are understandably left confused. It is our objective to give a brief overview of the nature of science and the scientific method so that you can evaluate for yourself the two sides of such controversies.

Science and the Scientific Method. In its essence, science is simply a way of gaining knowledge; the way is called the **scientific method**. The term *science* further refers to all the knowledge gained through that method. We employ the term *sound science* to distinguish legitimate science from what has been called *junk science*, information that is presented as valid science, but that does not conform to the rigors of the methods and practice of legitimate science. What is the scientific method?

First, the scientific method rests on four basic *assumptions* that most of us accept without argument. The first assumption is that what we perceive with our basic five senses represents an objective reality; that is, our perceptions are not some kind of mirage or dream. The second assumption is that this "objective reality" functions according to certain basic principles and natural laws that remain consistent through time and space.

The third assumption, which follows directly from the second, is that every result has a cause, and every event, in turn, will cause other events. In other words, we assume that events do not occur without reason and that there is an explainable cause behind every happening. The fourth and final assumption is that through our powers of observation, manipulation, and reason, we can discover and understand the basic principles and natural laws by which the universe functions.

Although assumptions, by definition, are premises that cannot be proved, the fact is that the assumptions underlying science have served us well and are borne out by everyday experience. For example, we suffer severe consequences if we do not accept our perception of fire as real. Similarly, our experience confirms that gravity is a predictable force acting throughout the universe and that it is not subject to unpredictable change. And the same can be said for any number of other phenomena that we observe. If our car fails to start, we know that there is a logical reason, and we call a mechanic to fix it.

Thus, whether we are conscious of the fact or not, we all accept the basic assumptions of science in the conduct and understanding of our everyday lives. Scientists and scientific investigations focused on the natural world extend the boundaries of everyday experience, deepen our understanding of cause–effect relationships, and enable us to have a greater appreciation of the principles and natural laws that seem to determine the behavior of all things, from the outcome of a chemical reaction to the functioning of the biosphere.

Observation. In previous schooling, you may have learned that the scientific method consists of the following sequence: question, hypothesis, test (experiment), and theory (Fig. 1 14). This sequence is an oversimplification in that it fails to describe the unique thought processes often involved in the four steps it comprises and it omits what is really the most fundamental aspect of science.

The foundation of all science and scientific discovery is **observation** (seeing, hearing, smelling, tasting, feeling).

Indeed, many branches of science, such as natural history (where and how various plants and animals live, mate, reproduce, etc.), astronomy, anthropology, and evolutionary biology, are based almost entirely on observation, because experimentation is either inappropriate or impossible. For example, experimentation is obviously counterproductive if you want to discover what plants and animals do under completely natural conditions, and it simply is impossible to conduct experiments on stars or past events.

Likewise, many of the data and conclusions of other sciences, such as zoology, botany, geology, comparative anatomy, and taxonomy (the classification of organisms), are based on nothing more (or less) than the careful observing and chronicling of things and events by persons taking the pains and time to do so. Even experimentation, as we will discuss shortly, is conducted to gain another window of observation. Simply put, in all science, careful observation is the keystone.

So how can we be sure that observations are accurate? Well, as a matter of fact, not every reported observation is accurate, for reasons ranging from honest misperceptions to calculated mischief. Therefore, an important aspect of science, and a trait of scientists, is to be skeptical of any new report until it is confirmed or verified. Such confirmation usually involves other investigators' repeating and checking out the observations of the first investigator and validating (or invalidating) their accuracy. As observations are confirmed by more and more investigators, they gain the status of factual data. In

other words, **facts** are things or events that have been confirmed by more than one observer and remain open to be reconfirmed by additional people. Things or events that do not allow such confirmation—UFOs, for example—remain in the realm of speculation from a scientific standpoint.

Various observations by themselves, like the pieces of a puzzle, may be put together into a larger picture—a model of how a system works. To give a simple example, we observe that water evaporates and leads to moist air and that water from moist air condenses on a cool surface. We also observe clouds and precipitation. Putting these observations together logically, we derive the concept of the hydrologic cycle. Water evaporates and then condenses as air is cooled, condensation forms clouds, and precipitation follows. Water thus makes a cycle from the surface of Earth into the atmosphere and back to Earth (Fig. 9–3). Note how this simple example incorporates the four assumptions described previously: There is an objective reality, it operates according to principles, every result has a cause, and we can discover the principles according to which reality operates. Note also how the example broadens our everyday experiences of the evaporation of water and falling of rain into an understanding of a cycle involving both.

Thus, the essence of science and the scientific method may be seen as a process of making observations and logically integrating those observations into a model of how the world works. To be sure, many areas of science get more complex and difficult to comprehend. Still, the basic process—constructing a logically coherent picture of causes and effects from basic observations—is the same.

Where, then, does experimentation, that additional hallmark of science, fit in?

Experimentation. Experimentation is basically setting up situations to make more systematic observations regarding causes and effects. For example, the number of chemical reactions one can readily observe in nature is limited. However, in the laboratory, it is possible to purify elements or compounds, mix them together in desired combinations and proportions, and carefully observe and measure how they react (or fail to react). From the way chemicals react, chemists have constructed a coherent cause–effect picture, the **atomic theory**, and they have determined the attributes of each element. Similarly, biologists put plants or animals into specific situations in which they can carefully observe and measure their responses to particular conditions or treatments (Fig. 1–15). (Again, note that experimentation is necessarily limited to things that lend themselves to artificial manipulation. In many cases, such as stars, geological events, atmospheric events, and events that have occurred in the past, one has only observations to work with in the construction of the broader picture.)

Some experimentation may be more or less spontaneous and random—the childhood inclination to "do this and see what happens." Sometimes, valuable information may be obtained in this way if careful accounts are kept so that one has an accurate record of causes and effects. How-

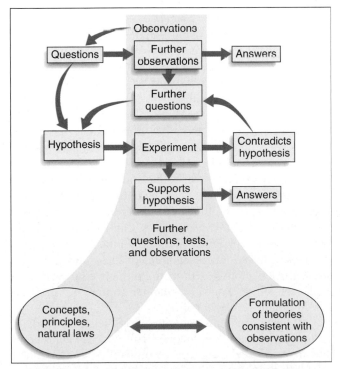

▲ **FIGURE 1–14** *Steps of the scientific method.* The process leading to theory formation and the postulation of natural laws and concepts is a continual interplay between observations, experimentation, hypothesis formulation, and further refinement.

▲ **FIGURE 1–15** *Experiment demonstrating the importance of mycorrhizae.* Mycorrhizae is a root–fungus symbiosis that greatly enhances the ability of the plant to absorb nutrients. The seedlings shown were grown under identical conditions, except that the plant on the left lacked mycorrhizal fungi, and the plant on the right was grown with them.

ever, to solve a particular problem (i.e., What is the cause of *this* event?), a more systematic line of experimentation is used. This is where the sequence consisting of question, hypothesis, test (experiment), and theory comes in. Let's consider the observation of the die-off of submerged aquatic vegetation in Chesapeake Bay. (See Chapter 18.)

The question is, What has caused the die-off? The next step is to make educated guesses as to the cause. Each such educated guess is a **hypothesis**. Each hypothesis is then tested by making further field observations or conducting experiments to determine whether the hypothesis really accounts for the observed effect. Historically speaking, in this particular case, the hypothesis or presumed cause first proposed was industrial wastes. However, this hypothesis was disproved by observations revealing that the die-off occurred in locations where no industrial wastes were present. A second hypothesis, that the die-off was due to herbicides used on farmlands, was rejected as the primary cause by laboratory testing and field measurement showing that herbicide levels were generally too low to cause damage. A third hypothesis, that a reduced amount of light, due to increased turbidity, was reaching the water, stood up to tests whereby photosynthesis was measured at various light intensities. Additional observations concerning the causes of the increased turbidity filled in the picture and brought what started out as a hypothesis to the status of the proven cause, which we might express as a *theory* or *concept*.

Theories. We have already noted how various specific observations (e.g., evaporation, precipitation, etc.) may

fit together to give a logically coherent cause–effect picture of a broader phenomenon (i.e., the hydrologic cycle). At first, the broader picture is termed a hypothesis because it is really a tentative explanation as to how various observations are related. It becomes a **theory** only after much testing and confirmation showing that it is logically consistent with all observations (Fig. 1–14). (Of course, a hypothesis may be *dis*confirmed, too, if it is inconsistent with an observation.)

Through cause–effect reasoning, theories will generally suggest or predict certain events. If an event predicted by a theory is indeed observed, the observation provides strong evidence for the truth of the theory. Predictions require experiments, testing, further data gathering, and more observation. When theories reach a state of providing a logically consistent framework for all relevant observations (facts), and when they can be used to reliably predict outcomes, they represent a valid interpretation of reality. For example, we have never seen atoms directly (or, up until recently, even indirectly), but innumerable observations and experiments are coherently explainable by the theory that all gases, liquids, and solids consist of various combinations of only slightly more than a hundred kinds of atoms. Hence, we fully accept the atomic theory of matter.

Sometimes people will argue that because a theory is not proven fact, one theory is as good as another. That notion is false! One theory may have overwhelming supporting evidence, whereas much evidence may contradict another theory. In evaluating a theory, we must ask, What is the supporting evidence? Is there more evidence, or less, supporting an alternative theory?

Natural Laws and Concepts. Like a theory, the second assumption underlying science—that the universe functions according to certain basic principles and natural laws that remain consistent through time and space—cannot be proved with absolute certainty. Still, every observation and test that scientists have performed up to the present have borne it out. All our observations, whether direct or through experimentation, demonstrate that matter and energy behave neither randomly nor even inconsistently, but in precise and predictable ways. We refer to these principles by which we can define and precisely predict the behavior of matter and energy as **natural laws**. Examples are the *law of gravity*, the *law of conservation of matter*, and the various *laws of thermodynamics*. Our technological success in space exploration and many other fields is in no small part due to our recognition of these principles and precise calculations based on them. Conversely, trying to make something work in a manner contrary to a natural law invariably results in failure.

In many situations, the outcome of scientific work results in well-established theories that must be expressed in the mathematical language of probability and statistics. Those theories are probabilistic in nature. This is especially true of biological phenomena such as predator–prey relationships or the effects of pesticides. Accordingly, here it is appropriate to speak of *concepts* rather than laws. **Con-**

cepts are perfectly valid explanations of data gathered from the natural world, and they can also be predictive. They model the way we believe the natural world works and enable us to make qualified predictions of future outcomes. Thus, we might say that, on the basis of our understanding of the effects of DDT (a potent pesticide) on mosquitoes, and on the basis of our observation that mosquitos can, and do, develop resistance to DDT, continued spraying of the local salt marshes with DDT to eradicate unwelcome mosquitoes is *likely* to result in the development of a more resistant mosquito population in the future (not to mention the other effects of DDT on fish and bird populations). Here, the *concept* of pest resistance is used to modify public policy.

The Role of Instruments in Science. Complex instrumentation is another hallmark of science, one that often gives it an aura of mystery. Yet, regardless of complexity, all scientific **instruments** perform one of three basic functions. First, they may extend our powers of observation, as do telescopes, microscopes, X-ray machines, and CAT scans, for example (Fig. 1–16). Second, instruments are used to quantify observations; that is, they enable us to measure exact quantities. For example, we may feel cold, but a thermometer enables us to measure and quantify exactly how cold it is. Comparisons, communication, and verification of different observations and events would be impossible if it were not for such measurement and quantification. Third, instruments such as growth chambers and robots help us achieve conditions and perform manipulations required to make certain observations or perform certain experiments.

All instruments used in science are themselves subjected to testing and verification to be sure that they are giving as accurate a representation of reality as we can attain with their use.

Scientific Controversies. With the scientific method capable of coming to objective, well-established conclusions, why does so much controversy still prevail? There are four main reasons. First, we are continually confronted by new observations—the hole in the ozone layer, for instance, or the dieback of certain forests. It takes some time before all the hypotheses regarding the cause of what we have observed can be adequately tested. During this time, there may be honest disagreement as to which hypothesis is most likely. Such controversies are gradually settled by further observations and testing, but the process leads into the second reason for continuing controversy, namely, that certain phenomena, such as the hole in the ozone layer or the loss of forests, do not lend themselves to simple tests or experiments. Therefore, it is difficult and time consuming to prove the causative role of one factor or to rule out the involvement of another. Gradually, different lines of evidence come to support one hypothesis and exclude another and enable the issue to be resolved.

When is there enough evidence to say unequivocally that one hypothesis is right and another wrong? Deciding that there is enough evidence to be convincing involves subjective judgment. The biases or vested interests of a person may affect the amount of information that person requires to be convinced. The Tobacco Institute, a lobbying association for the tobacco industry, provides a prime example. For years, it promoted the point that the connection between smoking and illness had not been proved and that more studies were necessary. By harping on the absence of absolute proof (a scientific impossibility anyway, as we have seen) and simply ignoring the overwhelming amount of evidence supporting the connection between smoking and illness, the tobacco lobby succeeded in keeping the issue controversial and thereby delayed regulatory restrictions on smoking. Thus, the third reason for controversy is that there are many vested interests that wish to maintain and promote disagreement, because they stand to profit by doing so. The need to keep a watch for this kind of behavior in evaluating the two sides of a controversy is crucial and self-evident.

The fourth reason for controversy, which may be seen as a generalization of the third, is that subjective value judgments, as well as subjective judgments of facts, may be involved. This is particularly true in environmental science, because the discipline deals with the human response to environmental issues. For example, there is virtually no controversy regarding nuclear power, as long as it is considered at the purely scientific level of physics. However, when

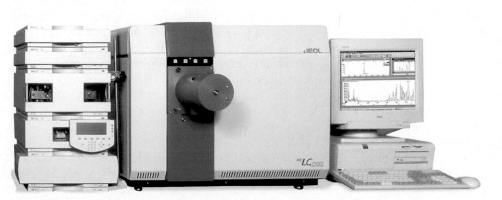

◀ **FIGURE 1–16** *Mass Spectrometer.* This machine, the JEOL LCmate, greatly extends our ability to make chemical analyses. The mass spectrometer produces charged particles (ions) from a chemical substance to be analyzed and then determines the masses and relative abundances of the ions present.

it comes to the environmental level of deciding whether to promote the further use of nuclear energy to generate electrical power, controversy arises, and it stems from the fact that different people have different subjective feelings regarding the relative risks and benefits involved.

The Scientific Community. We have seen that the scientific method is a way of gaining understanding which starts with basic observations that are developed into a logically coherent picture of broader phenomena through proper cause–effect reasoning. The value of this approach is seen in the scientific and technological progress it has made possible. However, what is often left out of the usual presentation of the scientific method is the fact that science and its outcomes take place in the context of a **scientific community** and a larger society. There is no single authoritative source that makes judgments on the soundness of scientific theories. Instead, it is the collective body of scientists working in a given field who, because of their competence and experience, establish what is sound science and what is not. They do so by communicating their findings to each other and to the public as they publish their work in peer-reviewed journals. The process of *peer review* is crucial; in it, experts in a given field review the analyses and results of their colleagues' work. Careful scrutiny is given to grant proposals and published research, with the objective of rooting out poor or sloppy "science" and affirming work that is clearly meritorious.

As we have just learned, controversy often arises in the context of environmental science, proof that science is a thoroughly human enterprise. Some controversy is the inevitable outcome of the scientific process itself, but much of it is attributable to less noble causes, such as junk science.

Junk Science. **Junk science** is information that is presented as scientifically valid, but that does not conform to the rigors imposed by the scientific community. According to the Union of Concerned Scientists (UCS), junk science can take many forms: the presentation of selective results (picking and choosing only those results which agree with one's preconceived ideas), politically motivated distortions of scientifically sound papers, or the publication of poor work in quasi-scientific, unreviewed journals and books. Quite often, junk science is generated by special-interest groups trying to influence the public debate about science-related concerns. Information presented by the Tobacco Institute, for example, is clearly suspect. Unfortunately, the media and the public may not be aware of the true nature of the information—whether it is sound or junk science—and will often give equal credibility to opposing views on an issue. The result is that the junk science is given equal respectability with the sound science in the public arena. The UCS is so concerned about this problem that it has launched the **Sound Science Initiative**, which is intended to "help scientists build public awareness about global environmental issues. The goal is to present accurate credible information to the media, the public, and policymakers, and to counter misinformation about environmental science."

As an example of this problem, in March, 1998, tens of thousands of U.S. scientists received a mailing from a nationally prominent scientist, inviting readers to sign a petition urging U.S. lawmakers to reject the 1997 Kyoto climate treaty. The mailing also contained an article that looked like—but wasn't—a reprint from the prestigious Proceedings of the National Academy of Sciences. The article presented the benefits of rising levels of CO_2 in the atmosphere and concluded that predictions of global warming were in error. The lead author of the article, a chemist, had never published research in the field of atmospheric science. The National Academy of Sciences took exception to the mailing and made it clear that the position presented in it was just the reverse of the Academy's views on the subject. One atmospheric scientist stated, "The mailing is clearly designed to be deceptive by giving people the impression that the article, which is full of half-truths, is a reprint and has passed peer review."

Evaluating Science. Whether we are scientists or laypersons, we can use facets of the scientific method to judge the relative validity of alternative viewpoints and also to develop our own capacity for logical reasoning. The basic questions to ask are the following:

- What are the observations (facts) underlying the conclusion (theory)?
- Can the observations be satisfactorily verified?
- Do the conclusions follow logically from the observations?
- Does the conclusion account for all of the observations? (If the conclusion is logically inconsistent with any observations, it must be judged as questionable at best.)
- Is the conclusion supported by the community of scientists with the greatest competence to judge the work? If not, it is highly suspect.

In sum, *sound science* is absolutely essential to the project of forging a sustainable relationship with the natural world. Our planet is dominated by human beings—our activities have reached such an intensity and such a scale that we are now one of the major forces affecting nature. Our influence on the natural ecosystems that support most of the world economy and process our wastes is strong and widespread, and we need to know how to manage the planet so as to maintain a sustainable relationship with it. It is obvious that the information gathered by scientists needs to be accurate, credible, and communicated clearly to decision makers and the public.

A New Commitment

People in all walks of life—scientists, sociologists, workers and executives, economists, government leaders, and clergy, as well as traditional environmentalists—are recognizing that "business as usual" is not sustainable. Many global

EARTH WATCH

AGENDA 21

Agenda 21 ("21" refers to the 21st century) is the official document signed by world leaders representing 98% of the global population at the United Nations Earth Summit in Rio de Janeiro, Brazil, in June of 1992. The following is an excerpt from an abridged version edited by Daniel Sitarz:[1]

> *Agenda 21* is, first and foremost, a document of hope.... [I]t is the principal global plan to confront and overcome the economic and ecological problems of the late twentieth century. It provides a comprehensive blueprint for humanity to use to forge its way into the next century by proceeding more gently upon the Earth....
>
> Humanity is at a crossroads of enormous consequence. Never before has civilization faced an array of problems as critical as the ones now faced. As forbidding and portentous as it may sound, what is at stake is nothing less than the global survival of humankind.
>
> ...Where once nature seemed forever the dominant force on Earth, evidence is rapidly accumulating that human influence over nature has reached a point where natural forces may soon be overwhelmed. Only very recently have the citizens of Earth begun to appreciate the depth of the potential danger of the human impact on our planet.... Scientists around the world, in every country on Earth, are documenting the hazard of ignoring our dependence upon the natural world.... For the first time in history, humanity must face the risk of unintentionally destroying the foundation of life on Earth.... To prevent such a collapse is an awesome challenge for the global community....
>
> *Agenda 21* is not a static document. It is a plan of action. It is meant to be a hands-on instrument to guide the development of the Earth in a sustainable manner.... It is based on the premise that sustainable development of the Earth is not simply an option: it is a requirement—a requirement increasingly imposed by the limits of nature to absorb the punishment which humanity has inflicted upon it. *Agenda 21* is also based on the premise that sustainable development of the Earth is entirely feasible.
>
> The bold goal of *Agenda 21* is to halt and reverse the environmental damage to our planet and to promote environmentally sound and sustainable development in all countries on Earth. It is a blueprint for action in all areas relating to the sustainable development of our planet into the 21st-century.... It includes concrete measures and incentives to reduce the environmental impact of the industrialized nations, revitalize development in developing nations, eliminate poverty world-wide and stabilize the level of human population.
>
> *Agenda 21* provides a myriad of opportunities. Suggestions are furnished to develop new industries, pioneer innovative technologies, evolve fresh techniques and institute novel trade arrangements.
>
> Various meetings involving leaders of both governments and nongovernmental organizations are continuing to take place around the world to develop, refine, and implement strategies to confront the world's environmental problems. In particular, the UN Commission on Sustainable Development is charged with the responsibility of overseeing *Agenda 21*. "Gradually it is being understood that the issues of poverty, population growth, industrial development, depletion of natural resources and destruction of the environment are all very closely interrelated."

[1]Daniel Sitarz, *Agenda 21* (Boulder, CO: Earth Press, 1993), pp. 1–5.

---

trends are on a collision course with the fundamental systems that maintain our planet as a tolerable place to live. A finite planet cannot go on adding almost 78 million additional persons annually, nor can we tolerate the current degradation of ecosystems, atmospheric changes, losses of species, and depletion of water resources without arriving at a point where resources are no longer adequate to support the human population and where, consequently, civil order breaks down. As one observer put it, "If we don't change direction, we will end up where we are heading."

The news is not all bad, however. Food production has improved the nutrition of millions in the developing world, and the percentage of individuals who are undernourished has declined from 35% to 20% over the past 30 years. Population growth rates continue to decline in many of the developing countries. A rising tide of environmental awareness in the industrialized countries has led to the establishment of policies, laws, and treaties that have improved the protection of natural resources and significantly reduced the pollution load. It is clear that environmental degradation can be slowed down and reversed, that people can be freed from hunger and poverty, and that peoples' behavior toward the environment can be transformed from an exploitative mode to a conserving one. Numerous caring people are beginning to play an important role in changing society's treatment of Earth. For example, many are now adding their voices to the literally hundreds of traditional environmental and professional organizations devoted to controlling pollution and protecting wildlife. People in business have formed the Business Council for Sustainable Development, economists have formed the International Society for Ecological Economics, religious leaders have formed the National Religious Partnership for the Environment, and philosophers are speaking out for a new ethic of "caring for Creation."

Sustainable development was the primary focus of the 1992 world summit meeting of leaders and representatives from 180 nations—the United Nations Conference on Environment and Development (UNCED)—held in Rio de Janeiro, Brazil. The major outcome of this conference, a "blueprint" intended to guide development in sustainable directions into and through the 21st century, was published in book form as *Agenda 21*. (See "Earth Watch" essay, this page.) Following the conference, major regions, nations, and communities adopted the concepts promoted in the document and began to develop their own versions of how sustainable development would be achieved in their various jurisdictions. For example, **Baltic 21** is the program for 11 countries surrounding the Baltic Sea. It addresses economic, social, and environmental objectives—the three sectors shown in Figure 1–8—as they apply to establishing sustainability in different sectors: agriculture, energy, fisheries, forests, industry, tourism, and transportation.

Five years later, in June 1997, the U.N. General Assembly held a special session to review the progress made since UNCED. Sadly, many observers concluded that there were more instances of failure than progress, citing a continued deterioration in the global environment, a further decline in aid to the developing countries, and a failure to set targets for reducing CO_2 emissions. The conference did, however, recognize some achievements and served to revitalize the UNCED commitments to sustainable development.

Thus, we are seeing a continuing concern on the part of many individuals and groups that appreciate the problems jeopardizing sustainability. Together, they are working to bring about corrective measures. In this light, the textbook you are holding is our own "best effort" to contribute to the cause from our perspective as scientists and educators. Given the vastness of what is now the environmental arena, we are the first to say that this text is far from exhaustive, nor should it be taken as the last word. Of course, viewpoints are subject to change as human experience and understanding increase. Our basic premise, however, is that sound environmental public policy in the long run must be based on sustainability, stewardship, and sound science.

In the first part of the text (Chapters 2 through 5), we describe the basics of how natural systems function and perpetuate themselves, and we discuss their limits in terms of adaptability to changing conditions. In addition to giving us some appreciation for natural systems, this study will reveal certain basic principles underlying sustainability. We contend that, to achieve sustainability, we must incorporate these principles into the functioning of our own society. Subsequent chapters will address population and development, resources, energy and land, and pollution in its various forms. In each case, we will attempt to give you a deeper understanding of just what the problems are, how far we have come, and a view of the path ahead toward sustainability.

ENVIRONMENT ON THE WEB

THE MYTH OF OBJECTIVE SCIENCE

The image of the white-coated scientist is a familiar and reassuring one. Scientific evidence now provides the foundation for many important public debates, including those in environmental management. But how accurate is the image of the scientist as an objective seeker of truth?

Scientific analysis is not just a straightforward process of observation and reporting but, rather, a complex series of personal decisions, value judgments, and guesses, influenced by the scientist's unique combination of personal experiences, fears, hopes, desires, and values. Like the rest of us, scientists worry about their careers, their families, and their finances. Ultimately, these human qualities influence the ways that individuals see and interpret scientific information. For this reason, it is not uncommon for two scientists to examine the same data set but reach very different conclusions.

Beth Savan, in her book *Science Under Siege*, cites the example of Stephen Jay Gould, a professor of geology at Harvard University. Gould reanalyzed data compiled by Samuel Morton, a 19th century physician, on the physical and intellectual differences among human races. Gould's analysis showed that Morton had consciously or unconsciously manipulated his data to arrive at the conclusion—widely held when Morton was alive—that white people are a superior race. Yet in his own analysis, Gould misread one of Morton's figures, leading him to underestimate racial differences in the data and thus to arrive at a conclusion more in keeping with his own preconceptions—that the differences among races are small.

This example demonstrates another feature of scientific analysis—that we tend to favor familiar, widely accepted views, while demanding a higher standard of proof for new ideas. Sometimes these biases can create obstacles to sound decision making. For example, a group of Western scientists planned to conserve Peary caribou in the High Arctic by protecting females and juveniles but allowing some hunting of adult males. Inuit hunters, knowledgeable about the social structure of caribou herds, warned that this practice would instead speed the decline of the population. Subsequent monitoring has confirmed the validity of the Inuit position.

Human emotions and values underlie most of the environmental disputes of this century. Divergent scientific analyses are often seen in the development of environmental standards. Environmental managers can reveal these subjective influences and make them explicit in decision making by including a wide range of viewpoints in their analysis and by recognizing and, where possible, compensating for their own unique values and biases.

Web Explorations

The Environment on the Web activity for this essay describes some of the debate—much of it driven by the human cultural and economic context of the decision—surrounding U.S. EPA's recent reevaluation of the standard for the cancer-causing agent dioxin. Go to the Environment on the Web activity (select Chapter 1 at **http://www.prenhall.com/wright**) and learn for yourself:

1. how human biases may influence applications of the scientific method;

2. how science can be used to persuade and influence political and social decisions; and

3. how individual interpretations of scientific data can vary greatly.

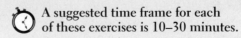 **A suggested time frame for each of these exercises is 10–30 minutes.**

REVIEW QUESTIONS

1. Describe several ways that people in different parts of the world live and interact with their environments.
2. What factors brought about the collapse of the Easter Island civilization? Draw parallels and counterparallels between the current global situation and the prelude to the collapse of the Easter Island civilization.
3. Cite four global trends which indicate that we are "still losing the environmental war."
4. List some indications that things are not all bad and that progress toward sustainability is possible.
5. Define sustainability and a sustainable society.
6. Define and give features of sustainable development.
7. Describe the origins of the stewardship ethic and its modern usage.
8. What are the concerns of the environmental justice movement?

9. How does justice become an issue between the industrialized countries and the developing countries?
10. How and when did environmentalism arise?
11. What are the successes of the environmental movement?
12. What are the grounds and concerns of present critics of environmentalism?
13. Explain the role of observation, experimentation, and theory formation in the operation of scientific work.
14. Give several reasons for the existence of scientific controversies.
15. Describe the major distinction between sound science and junk science.
16. What are the new directions, new focus, and new adherents of modern environmental concern?

THINKING ENVIRONMENTALLY

1. Have a class debate between environmental critics and people representing environmentalists. Characterize the two sides in terms of long-term vs. short-term viewpoints, personal interests vs. the interests of society in general, and a local vs. global perspective.
2. Some people say that the concept of sustainable development either is an oxymoron or represents going back to some kind of primitive living. Present an argument demonstrating that neither is the case, but that sustainable development is the only course that will allow the continued advancement of civilization.

3. List all the prerequisites for a sustainable society that really works. Why is it necessary that people from all walks of life be involved? State the roles (in general) that each needs to play in achieving a sustainable society.
4. How can you contribute to a sustainable society?
5. Set up a debate between proponents and opponents of smoking. Use the concepts of sound science and junk science to conduct the debate, and show how the two kinds of science will lead to different conclusions.

WEB REFERENCES

On-line resources for this chapter can be found on the World Wide Web at: **http://www.prenhall.com/wright**. (Click on Chapter 1 on the Chapter Selector.)

Chapter 2
Ecosystems: What They Are

Chapter 3
Ecosystems: How They Work

Chapter 4
Ecosystems: Populations and Succession

Chapter 5
Ecosystems and Evolutionary Change

PART ONE

Ecosystems and How They Work

Tropical rain forests—humid, warm, dense, full of unusual plants and animals. These are the impressions that our senses bring to us when we take the first few steps into a rain forest. However, our senses will not tell us that the tropical rain forests are home to a greater diversity of living things than anywhere else on Earth, that they are the result of millions of years of adaptive evolution, or that they are the site of storage of more carbon than is in the entire atmosphere. Nor can we sense how energy flows through these forests or how nitrogen and phosphorus are cycled and recycled. This information comes from the work of many scientists who have been asking the basic questions of how such natural systems function, how they came to be, and how they relate to the rest of the world.

How important is the information that comes from these scientists? In this first part of the text, we hope to convince you that information about how the natural world works is absolutely crucial to the human enterprise of living on our planet. Ecosystems—the name we give to natural units like the tropical rain forest—are not just the backdrop for human activities; they are the basic context of life on Earth, including human life. They are self-sustaining systems, and if we accept the notion that the human enterprise should also be self-sustaining (and currently is not), then perhaps our study of ecosystems can bring us some lessons on how to construct a sustainable society. Environmental science begins by understanding how ecosystems work.

◄ **Tropical rain forest.**

23

Ecosystems: What They Are

Key Issues and Questions

1. Natural communities are organized into units we call ecosystems. What are ecosystems? How are they organized into larger units?

2. The organisms in every ecosystem can be assigned to trophic categories. What are the categories? How do they function together in food webs to make a sustainable system?

3. Many nontrophic relationships also exist in ecosystems. What role do mutualism and competition play in ecosystems?

4. Environmental factors can be categorized into conditions and resources. How do these categories act as limiting factors in the distribution of different species in ecosystems?

5. Precipitation and temperature are the predominant determinants of climate. How do these factors interact to produce the different biomes around the globe?

6. Revolutionary changes have occurred in human culture that have greatly altered the relationship between humans and the environment. How have the Neolithic and Industrial Revolutions affected the natural environment, and what is meant by the Environmental Revolution?

Barrier islands are found along the east and Gulf Coasts of the United States. Plum Island, in Massachusetts, is a nine-mile-long barrier island in the northern part of the state. (See opposite page.) The southern two-thirds of the island is part of the Parker River National Wildlife Refuge, while the northern third is occupied by a host of summer cottages and year-round houses. On the refuge part of the island, sand dunes emerge at the upper edge of the ocean beach and extend back several hundred meters. Occupying the sand dunes is a mosaic of vegetation. The primary dunes, closest to the ocean, are colonized by beach grass (*Ammophila brevigulata*). Behind these dunes, a heathlike low growth occurs, dominated by false heather (*Hudsonia tomentosa*). Farther back, a shrub community can be found, with poison ivy and bayberry as dominants. Even farther back, we encounter the maritime forest, in which we find pitch pine, poplar, wild cherry, and a few other deciduous trees. Here

and there is a depression that reaches the water table, occupied by cranberry plants. Animals on the barrier island—mice, rabbits, deer, skunk, red foxes—are seldom seen by visitors, but leave their tracks on the sand. A diverse bird community can be found here, drawing birders who often find rarities like the peregrine falcon. Life for the plants—especially close to the ocean—is harsh: The sand holds little water or nutrients, salt spray from the ocean stresses leaf surfaces, and wind either blows loose sand away from plants' roots or piles it up around the plants and buries them. Yet the barrier island community is a functioning system that has endured for centuries.

On the northern third of the island, scores of streets were laid out in the early 1900s, and hundreds of small cottages were built. The dunes were either removed or built upon. Now, the only vegetation is a few beach grass plants that lie between the beach and the first houses. There are no dunes to speak of, no shrubs or maritime forest, no deer

◀ *A barrier island.* Plum Island in northern Massachusetts is a nine-mile-long sand island separating the ocean from a bay, a salt marsh, and the upland. The inset shows the northern, developed end of the island.

or foxes. It is not hard to imagine the entire island like this; only the presence of the protected refuge prevents further development. The contrast between these "communities" is unforgettable.

From time to time, the "environment" tests this barrier island, as it tests all barrier islands, with storms. On the southern section, the dunes absorb the powerful waves that pound the beach and wash upward toward the land. After a storm, the beach grass and primary dunes are intact and the back dunes untouched. The dune communities continue to thrive, forming a true barrier to the ocean that protects the fragile salt marshes and land behind the island. However, on the northern section, storms often wash beach sand by the houses and onto the streets, and occasionally a house is swept into the ocean. No one visits the northern Plum Island dunes to view birds or animals; in fact, there are no dunes, birds, or animals. In comparison, thousands of visitors enjoy the Parker River National Wildlife Reserve for its wildlife and beauty each year.

This situation is a microcosm of our interaction with the natural environment in general. Where we have managed to save the natural plants and animals occupying the land, they repay us by providing many goods and services. Where we have replaced or damaged them, we lose those benefits and often suffer the consequences. Can we learn to interact with the natural world and conduct our own endeavors in ways that are sustainable—that is, ways which meet the needs of the present without compromising the ability of future generations to meet their own needs?

The study of natural ecosystems—those groupings of plants, animals, and other organisms that we think of as unspoiled forests, grasslands, coral reefs, ponds, and barrier islands—can serve several objectives.

First, our agriculture and technology notwithstanding, we remain dependent on the natural world and its biodiversity for such basic needs as clean air, clean water, a suitable climate for growing food, and a host of other things. Our study of natural ecosystems will help us understand the relationships between the environment and living things and between the natural world and ourselves. From this, we can more clearly understand the impact of humans on the natural world and the consequences that may have. We can then use our knowledge to improve our management of ecosystems.

Second, natural ecosystems are models of sustainability. The individual kinds of trees and other plants, animals, and microbes making up a forest, for example, are known to have propagated themselves over many tens of thousands, or even millions, of years. Our premise is that bringing out the basic principles that foster the perpetuation or sustainability of natural ecosystems will provide insights into the ways in which we need to direct human development in order to achieve a sustainable future.

Finally, it is our hope that the study of natural ecosystems will lead you to a greater appreciation of the amazing diversity and beauty of our planetary home and a heightened desire to care for it as stewards. Thus, our study begins with an examination of natural ecosystems.

2.1 What Are Ecosystems?

The grouping or assemblage of plants, animals, and microbes we observe when we study a natural forest, a grassland, a pond, a coral reef, or some other undisturbed area is referred to as the area's **biota** (*bio*, living) or **biotic community**. The plant portion of the biotic community includes all vegetation, from large trees down through microscopic algae. Likewise, the animal portion includes everything from large mammals, birds, reptiles, and amphibians through earthworms, tiny insects, and mites. Microbes encompass a large array of microscopic bacteria, fungi, and protozoans. Thus, one may speak of the biotic community as comprising a plant community, an animal community, and a microbial community.

The particular kind of biotic community that we witness in a given area is, in large part, determined by **abiotic** (nonliving, chemical, and physical) factors, such as the amount of water or moisture present, the temperature, the salinity, and the type of soil in the area. These abiotic factors both support and limit the particular community. For example, a relative lack of available moisture prevents the growth of most species of plants, but supports certain species, such as cacti; we recognize such areas as deserts. Land with plenty of available moisture and a suitable temperature supports forests. Obviously, the presence of water is the major factor that sustains aquatic communities.

The first step in investigating a biotic community may be simply to catalogue all the *species* present. **Species** are the different kinds of plants, animals, and microbes in the community. Each species includes all those individuals which have a strong similarity in appearance to one another and which are distinct in appearance from other such groups (robins vs. redwing blackbirds, for example). Similarity in appearance suggests a close genetic relationship. Indeed, the *biological definition* of a species is the entirety of a population that can interbreed and produce fertile offspring, whereas members of different species generally do not interbreed, or if they do, no fertile offspring are produced. Breeding is often impractical or impossible to observe, however, so for purposes of identification, the aspect of appearance usually suffices.

In cataloguing the species of a community, one observes that each species is represented by a certain **population**—that is, by a certain number of individuals that make up the interbreeding, reproducing group. The distinction between *population* and *species* is that the term *population* is used to refer only to those individuals of a certain species that live within a given area, whereas the term *species* is all inclusive—it refers to all the individuals of a certain kind, even though they may exist in different populations in widely separated areas.

Continuing our study, we might want to identify the biotic community in order to understand how it fits into the landscape or how it differs from other biotic communities. The approach we would most likely take is to work with the plant community: Vegetation is readily measured and is a strong indicator of the environmental conditions of a site. The most basic kind of plant community is the **association**, defined as a plant community with a definite composition, uniform habitat characteristics, and uniform plant growth. In recent years, scientists have constructed the U.S. National Vegetation Classification System and have classified over 4,100 plant associations to facilitate community identification. (See "Earth Watch," p. 35.)

Along with the wonderment caused by the incredible variety of species and communities, it is impressive that the species within a community depend on and support one another. Most evident is the fact that certain animals will not be present unless particular plants that provide their necessary food and shelter are also present. Thus, the plant community supports (or limits by its absence) the animal community. In addition, every plant and animal species is adapted to cope with the abiotic factors of the region. For example, every species that lives in temperate regions is adapted in one way or another to survive the winter season, which includes a period of freezing temperatures (Fig. 2–1). We shall explore these interactions among organisms and their environments later. For now, the point is that the populations of different species within a biotic community are constantly interacting with each other and the abiotic environment.

This brings us to the concept of an *ecosystem*, which joins together the biotic community *and* the abiotic conditions that it lives in. The ecosystem concept includes considerations of the ways populations interact with each other and the abiotic environment to reproduce and perpetuate the entire grouping. In one sentence, an **ecosystem** is a grouping of plants, animals, and microbes occupying an explicit unit of space and interacting with each other and their environment. For study purposes, an ecosystem may be taken to be any more or less distinctive biotic community living in a certain environment. Thus, a forest, a grassland, a wetland, a marsh, a pond, a barrier island, and a coral reef, each with its respective species in a particular environment, can be studied as distinct ecosystems.

Since no organism can live apart from its environment or from interacting with other species, ecosystems are the functional units of sustainable life on Earth. The study of ecosystems and the interactions that occur among organisms and between organisms and their environment is the

▲ **FIGURE 2–1** *Winter in the forest.* Many trees and other plants of temperate forests are so adapted to the winter season that they actually require a period of freezing temperature in order to recommence growth in the spring.

science of **ecology**, and the investigators who conduct such studies are called **ecologists**.

While it is convenient to divide the living world into different ecosystems, any investigation soon reveals that there are seldom distinct boundaries between ecosystems, and they are never totally isolated from one another. Many species will occupy (and thus be a part of) two or more ecosystems at the same time. Or they may move from one ecosystem to another at different times, as in the case of migrating birds. In passing from one ecosystem to another, one may observe only a gradual decrease in the populations of one biotic community and an increase in the populations representing another. Thus, one ecosystem may grade into the next through a transitional region, known as an **ecotone**, that shares many of the species and characteristics of the two adjacent ecosystems (Fig. 2–2).

The ecotone between adjacent systems may also include *unique* conditions that support distinctive plant and animal species. Consider, for example, the marshy area that often occurs between the open water of a lake and dry land (Fig. 2–3). Ecotones may be studied as distinct ecosystems in their own right.

Furthermore, what happens in one ecosystem will definitely affect other ecosystems. For this reason, ecologists have begun using the concept of **landscapes**, defined as a group of interacting ecosystems. Thus, a barrier island, a saltwater bay, and the salt marsh behind it could be considered a landscape. Landscape ecology is then the science that studies the interactions between ecosystems.

Similar or related ecosystems or landscapes are often grouped together to form major kinds of ecosystems called **biomes**. Tropical rain forests, grasslands, and deserts are

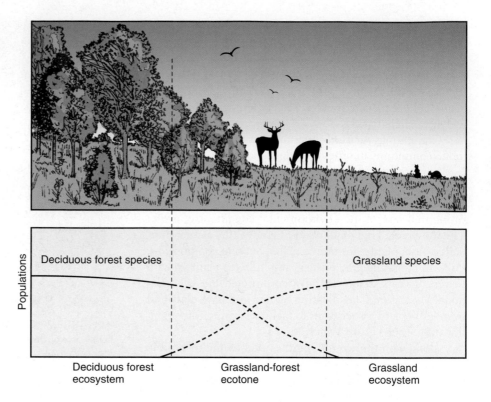

▶ **FIGURE 2–2 *Ecotones on land.*** Ecosystems are not isolated from one another. One ecosystem blends into the next through a transitional region—an ecotone—that contains many species common to the two adjacent systems.

examples. While more extensive than an ecosystem in its breadth and complexity, a biome is still basically a distinct type of biotic community supported and limited by certain abiotic environmental factors. As with ecosystems, there are generally no sharp boundaries between biomes, but one grades into the next through transitional regions.

Likewise, there are major categories of aquatic and wetland ecosystems that are determined primarily by the depth, salinity, and permanence of water. Among these ecosystems are lakes, marshes, streams, rivers, estuaries,

bays, and ocean systems. As units of study, these aquatic systems may be viewed as ecosystems, as parts of landscapes, or as major biomelike features such as seas or oceans. (The biome category has been reserved exclusively for terrestrial systems.) Table 2–1 presents the six major aquatic systems and their primary characteristics.

Regardless of how we choose to divide (or group) and name different ecosystems, it is important to recognize that they all remain interconnected and interdependent. Terrestrial biomes are connected by the flow of rivers between them

▶ **FIGURE 2–3 *Terrestrial-to-aquatic-system ecotone.*** An ecotone may create a unique habitat that harbors specialized species not found in either of the ecosystems bordering it. Typically, cattails, reeds, and lily pads grow here, along with several species of frogs and turtles as well as egrets and herons.

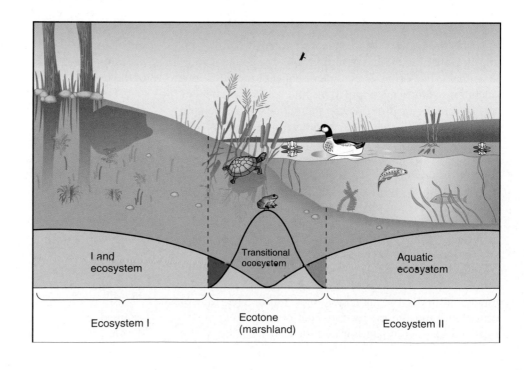

TABLE 2–1	Major Aquatic Systems			
Aquatic Systems	Major Environmental Parameters	Dominant Vegetation	Dominant Animal Life	Distribution
Lakes and Ponds (freshwater)	Bodies of standing water; low concentration of dissolved solids; seasonal vertical stratification of water	Rooted and floating plants, phytoplankton	Zooplankton, fish, insect larvae, ducks, geese, herons	Physical depressions in the landscape where precipitation and groundwater accumulate
Streams and Rivers (freshwater)	Flowing water; low level of dissolved solids; high level of dissolved oxygen, often turbid	Attached algae, rooted plants	Insect larvae, fish, amphibians, otters, raccoons, wading birds, ducks, geese, swans	Landscapes where precipitation and groundwater flow by gravity toward oceans or lakes
Inland Wetlands (freshwater)	Standing water, at times seasonally dry; thick organic sediments; high nutrients	Marshes: grasses, reeds, cattails. Swamps: water-tolerant trees. Bogs: sphagnum moss, low shrubs	Amphibians, snakes, numerous invertebrates, wading birds, ducks, geese, alligators, turtles	Shallow depressions, poorly drained, often occupy sites of lakes and ponds that have filled in
Estuaries	Variable salinity; tides create two-way currents, often rich in nutrients, turbid	Phytoplankton in water column, rooted grasses like salt-marsh grass, mangrove swamps in tropics with salt-tolerant trees and shrubs	Zooplankton, rich shellfish, worms, crustaceans, fish, wading birds, sandpipers, ducks, geese	Coastal regions where rivers meet the ocean; may form bays behind sandy barrier islands
Coastal Ocean (saltwater)	Tidal currents promote mixing; nutrients high	Phytoplankton, large benthic algae, turtle grass, symbiotic algae in corals	Zooplankton, rich bottom fauna of worms, shellfish, crustaceans, echinoderms; coral colonies, jellyfish, fish, turtles, gulls, terns, ducks, sea lions, seals, dolphins, penguins, whales	From coastline outward over continental shelf; coral reefs abundant in tropics
Open Ocean	Great depths (to 11,000 meters); all but upper 200 m dark and cold; poor in nutrients except in upwelling regions	Exclusively phytoplankton	Diverse zooplankton and fish adapted to different depths; seabirds, whales, tuna, sharks, squid, flying fish	Covering 70% of Earth, from edge of continental shelf outward

and by migrating animals. Sediments and nutrients washing from the land may nourish or pollute the ocean. Seabirds and mammals connect the oceans with the land, and all biomes share a common atmosphere and water cycle.

Therefore, all the species on Earth, along with all their environments, can be seen as one vast ecosystem, often called the **biosphere**. Although the separate local ecosystems are the individual units of sustainability, they are all interconnected to form the biosphere. The concept is analogous to the idea that the cells of our bodies are the units of living systems but are all interconnected to form the whole body. Carrying the analogy further, we may ask, to what degree can individual ecosystems be degraded or destroyed before the entire biosphere is affected? And to what degree can basic global parameters, such as the atmosphere and

the temperature, be altered before all ecosystems on Earth are affected? Let us now begin a more intensive study of ecosystems to discover how they are structured.

The key terms introduced in this section are summarized in Table 2–2.

2.2 The Structure of Ecosystems

Structure refers to parts and the way they fit together to make a whole system. There are two key aspects to every ecosystem: the biota, or biotic community, and the abiotic environmental factors. The way different categories of organisms fit together is referred to as the **biotic structure**, and the major feeding relationships between organisms

TABLE 2–2	Important Terms
Species	All the members of a specific kind of plant, animal, or microbe; a kind given by similarity of appearance or capacity for interbreeding and producing fertile offspring.
Population	All the members of a particular species occupying a given area.
Association	A plant community with a definite composition, uniform habitat characteristics, and uniform plant growth.
Biotic community	All the populations of different plants, animals, and microbes occupying a given area.
Abiotic factors	All the factors of the physical environment: moisture, temperature, light, wind, pH, soil type, salinity, etc.
Ecosystem	The biotic community together with the abiotic factors; all the interactions among the members of the biotic community, and between the biotic community and the abiotic factors, within an explicit unit of space.
Landscape	A group of interacting ecosystems in a particular area.
Biome	A grouping of all the ecosystems of a similar type (e.g., tropical forests or grasslands).
Biosphere	All species and physical factors on Earth functioning as one unified ecosystem.

constitute the **trophic structure** (*trophic*, feeding). All ecosystems have the same three basic categories of organisms that interact in the same ways.

Trophic Categories

The major categories of organisms are (1) *producers*, (2) *consumers*, and (3) *detritus feeders* and *decomposers*. Together, these groups produce food, pass it along food chains, and return the starting materials to the abiotic parts of the environment.

Producers. Producers are organisms that capture energy from the Sun or from chemical reactions to convert carbon dioxide (CO_2) to organic matter. Most producers are green plants, which use light energy to convert CO_2 and water to a sugar called glucose and then release oxygen as a by-product. This chemical conversion, which is driven by light energy, is called **photosynthesis**. Plants are able to

manufacture all the complex molecules that make up their bodies from the glucose produced in photosynthesis, plus a few additional *mineral nutrients* such as nitrogen, phosphorus, potassium, and sulfur, which they absorb from the soil or from water (Fig. 2–4).

Plants use a variety of molecules to capture light energy for photosynthesis, but the most predominant of these is **chlorophyll**, a green pigment. Hence, plants that carry on photosynthesis are easily identified by their green color. In some plants, additional red or brown photosynthetic pigments (in red and brown algae, for example) may overshadow the green. Producers range in diversity from microscopic photosynthetic bacteria and single-celled algae through medium-sized plants such as grass, daisies, and cacti, to gigantic trees. Every major ecosystem, both aquatic and terrestrial, has its particular producers carrying on photosynthesis.

The term **organic** is used to refer to all those materials that make up the bodies of living *organ*isms—molecules such as proteins, fats or lipids, and carbohydrates. Likewise, materials that are specific products of living organisms, such as dead leaves, leather, sugar, and wood, are considered *organic*. On the other hand, materials and chemicals in air, water, rocks, and minerals, which exist apart from the activity of living organisms, are considered **inorganic**

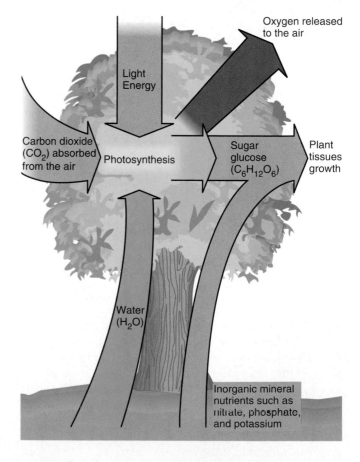

▲ **FIGURE 2–4** *Green plant photosynthesis.* The producers in all major ecosystems are green plants.

(Fig. 2–5). Interestingly, there are bacteria that are able to use the energy in some inorganic chemicals to form organic matter from CO_2 and water. This process is called **chemosynthesis**, and these organisms clearly are also producers.

The key feature of *organic* materials and molecules is that they are in large part constructed from bonded carbon and hydrogen atoms, a structure that is not found among *inorganic* materials. This carbon–hydrogen structure has its origins in the process of photosynthesis, in which hydrogen atoms taken from water molecules and carbon atoms taken from carbon dioxide are joined together to form organic compounds (Fig. 2–4). Green plants use light as the energy source to produce all the complex organic molecules their bodies need from the simple inorganic chemicals present in the environment. As this conversion from inorganic to organic occurs, some of the energy from light is stored in the organic compounds.

All organisms in the ecosystem *other than the producers* feed on organic matter as their source of energy. These organisms include not only all animals, but also **fungi** (mushrooms, molds, and similar organisms), most bacteria, and even a few higher plants that do not have chlorophyll and that therefore cannot carry on photosynthesis.

Thus, green plants, which carry on photosynthesis, are absolutely essential to every ecosystem (with the exception of a few ecosystems, such as deep-sea hydrothermal vents, which depend on chemosynthesis). The photosynthesis and growth of green plants constitute the production of organic matter, which sustains all other organisms in the ecosystem.

Indeed, all organisms in the biosphere can be divided into two categories—*autotrophs* and *heterotrophs*—on the basis of whether they do or do not produce the organic compounds they need to survive and grow. Those organisms, such as green plants and chemosynthetic bacteria, which produce their own organic material from inorganic constituents in the environment through the use of an external energy source, are **autotrophs** (*auto*, self; *troph*, feeding). The most important and common autotrophs by far are green plants, which use chlorophyll to capture light energy for photosynthesis. All other organisms, which must *consume* organic material to obtain energy, are **heterotrophs** (*hetero*, other). Heterotrophs may be divided into numerous subcategories, the two major ones being **consumers** (which eat living prey) and **detritus feeders** and **decomposers**, both of which feed on dead organisms or their products.

Consumers. Consumers encompass a wide variety of organisms ranging in size from microscopic bacteria to blue whales. Among consumers are such diverse groups as protozoans, worms, fish and shellfish, insects, reptiles, amphibians, birds, and mammals (including humans).

For the purpose of understanding ecosystem structure, consumers are divided into various subgroups according to their food source. Animals, as large as elephants or as small as mites, that feed directly on producers are called **primary consumers** or **herbivores** (*herb*, grass).

Animals that feed on primary consumers are called **secondary consumers**. Thus, elk, which feed on vegetation, are primary consumers, whereas wolves, because they feed on elk, are secondary consumers (Fig. 2–6). There may also be third, fourth, or even higher levels of consumers, and certain animals may occupy more than one position on the consumer scale. For instance, humans are

Inorganic

Organic

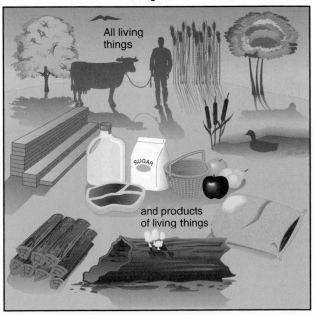

▲ **FIGURE 2–5** *Organic and inorganic.* Water and the simple molecules found in air, rocks, and soils are *inorganic*. The complex molecules that make up plant and animal tissues are *organic*.

► **FIGURE 2-6** *Secondary consumers.* Gray wolves have brought down an elk.

primary consumers when they eat vegetables, secondary consumers when they eat beef, and third-level consumers when they eat fish that feed on smaller fish that feed on algae. Secondary and higher order consumers are also called **carnivores** (*carni*, meat). Consumers that feed on both plants and animals are called **omnivores** (*omni*, all).

In any relationship in which one organism feeds on another, the organism that does the feeding is called the **predator**; the organism that is fed on is called the **prey**. Together, the two are said to have a **predator–prey** relationship. Predation thus ranges from the classic predator–prey interactions of carnivores and herbivores to herbivores feeding on plants and parasites feeding on their hosts. In fact, **parasites** are another important category of consumers. Parasites are organisms—either plants or animals—that become intimately associated with their "prey" and feed on it over an extended period of time, typically

without killing it, but sometimes weakening it so that it becomes more prone to being killed by predators or adverse conditions. The plant or animal that is fed upon is called the **host**; thus, we speak of a **host–parasite** relationship.

A tremendous variety of organisms may be parasitic. Various worms are well-known examples, but certain protozoans, insects, and even mammals (vampire bats) and plants (dodder) (Fig. 2–7a) are also parasites. Many serious plant diseases and some animal diseases (such as athlete's foot) are caused by parasitic fungi. Indeed, virtually every major group of organisms has at least some members that are parasitic. Parasites may live inside or outside their hosts, as the examples shown in Fig. 2–7 illustrate.

In medical circles, a distinction is generally made between, on the one hand, bacteria and viruses that cause disease (known as **pathogens**), and, on the other, parasites, which are usually larger organisms. Ecologically, however,

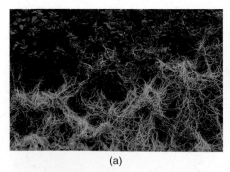

(a)

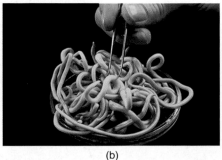

(b)

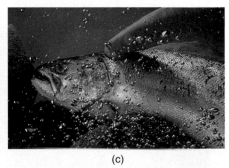

(c)

▲ **FIGURE 2-7** *Diversity of parasites.* Nearly every major biological group of organisms has at least some members that are parasitic on others. Shown here is (a) dodder, a plant parasite that has no leaves or chlorophyll. The orange "strings" are the dodder stems, which suck sap from the host plant. (b) Nematode worms (*Ascaris lumbricoides*), the largest of the human parasites, reach a length of 14 inches (35 cm). (c) Lamprey attached to a salmon. Lampreys are parasitic on fish.

there is no real distinction. Bacteria are foreign organisms, and viruses are organism-like entities feeding on, and multiplying in, their hosts over a period of time and doing the same damage as do other parasites. Therefore, disease-causing bacteria and viruses can be considered highly specialized parasites. Representative examples of producers and consumers, and the feeding relationships among them, are shown in Fig. 2–8.

Detritus Feeders and Decomposers.   Dead plant material, such as fallen leaves, branches and trunks of dead trees, dead grass, the fecal wastes of animals, and dead animal bodies, is called **detritus**. Many organisms are specialized to feed on detritus, and we refer to such consumers as detritus feeders, or *detritivores*. Some examples are earthworms, millipedes, fiddler crabs, termites, ants, and wood beetles. As with regular consumers, one can identify *primary* **detritus feeders** (those which feed directly on detritus), *secondary* **detritus feeders** (those which feed on primary detritus feeders), and so on.

An extremely important group of primary detritus feeders is the decomposers, namely, fungi and bacteria. Much of the detritus in an ecosystem—particularly dead

leaves and the wood of dead trees or branches—does not appear to be eaten as such, but rots away. Rotting is the result of the metabolic activity of fungi and bacteria. These organisms secrete digestive enzymes that cause the breakdown of wood, for example, into simple sugars that the fungi or bacteria then absorb for their nourishment. Thus, the rotting we observe is really the result of material being consumed by fungi and bacteria. Even though fungi and bacteria are called decomposers because of their unique behavior, we group them with detritus feeders because their function in the ecosystem is the same. In turn, such secondary detritus feeders as protozoans, mites, insects, and worms (Fig. 2–9) feed upon decomposers. When a fungus or other decomposer dies, its body becomes part of the detritus and the source of energy and nutrients for still more detritus feeders and decomposers.

In sum, despite the apparent diversity of ecosystems, they all have a similar *biotic structure*. They all can be described in terms of (1) autotrophs, or producers, which produce organic matter that becomes the source of energy and nutrients for (2) heterotrophs, which are various categories of consumers and detritus feeders and decomposers (Fig. 2–10).

◀ **FIGURE 2-8** *Trophic relationships among producers and consumers.*

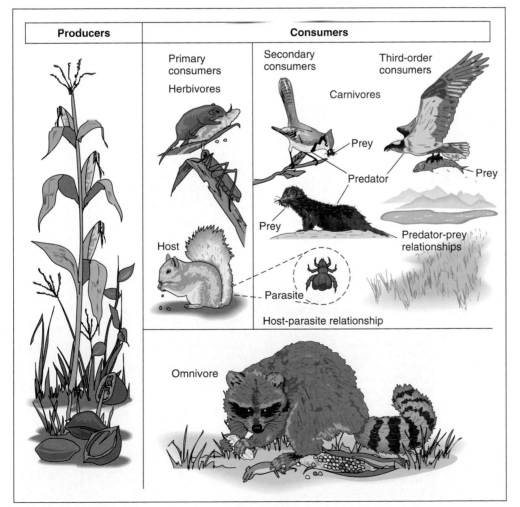

Producers	Consumers		
	Primary consumers — Herbivores	Secondary consumers — Carnivores	Third-order consumers

Prey

Predator

Prey

Prey

Predator-prey relationships

Host

Parasite

Host-parasite relationship

Omnivore

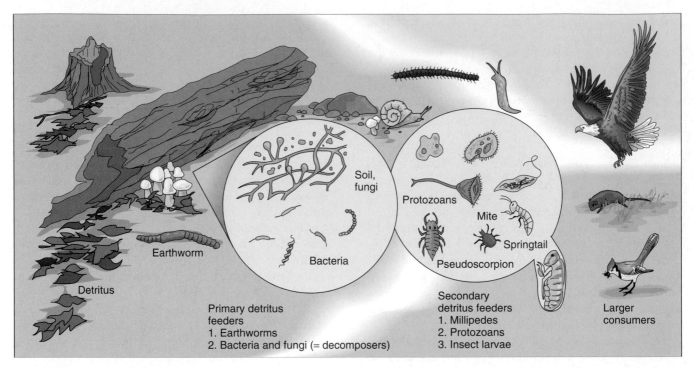

▲ **FIGURE 2–9** *Detritus food web.* The feeding (trophic) relationships among primary detritus feeders, secondary detritus feeders, and consumers.

▼ **FIGURE 2–10** *Trophic categories.* A summary of how living organisms are ecologically categorized according to feeding attributes.

Autotrophs Make their own organic matter from inorganic nutrients and an environmental energy source	**Heterotrophs** Must feed on organic matter for energy	
Producers	**Consumers**	**Detritus feeders and decomposers:** organisms that feed on dead organic material
Photosynthetic green plants: use chlorophyll to absorb light energy	Primary consumers/herbivores: animals that feed exclusively on plants	Decomposers: fungi and bacteria that cause rotting
Photosynthetic bacteria: use purple pigment to absorb light energy	Omnivores: animals that feed on both plants and animals	Primary detritus feeders: organisms that feed directly on detritus
Chemosynthetic bacteria: use high-energy inorganic chemicals such as hydrogen sulfide	Secondary consumers/ carnivores: animals that feed on primary consumers	Secondary and higher orders of detritus feeders: feed on primary detritus feeders
	Higher orders of consumers/carnivores: animals that feed on other carnivores	
	Parasites: plants or animals that become associated with another plant or animal and feed on it over an extended period of time	

EARTH WATCH

TAKING STOCK

As important as ecosystems are to human affairs, it might be assumed that we have a good inventory of them. However, not only do we not know *what* we have, but we also don't know much about the *goods and services* specific ecosystems can produce, nor do we know how they *respond to changes* brought on by human activities. No one doubts that we are going to have to manage natural ecosystems more intensely and deliberately if we want to enjoy their benefits in the future. The question is, How can you manage something you do not even know you have, let alone understand how we are affecting it?

This is a problem that has risen to the top of the agendas of a number of agencies and organizations. One such effort, prepared by U.S. Geological Survey scientists, is entitled *The Status and Trends of the Nation's Biological Resources*. Painting in broad strokes, this work provides a historical perspective on ecosystems and biodiversity and then documents what we know about how they are changing in response to human activities. A trend such as the enormous decline in native tallgrass prairie habitats, the effect of introducing 25 nonnative fish species to the Great Lakes, and the impact of 90 introduced plant species on the Hawaiian Islands are presented as serious threats to our natural heritage and, indeed, our welfare.

A more ambitious and, arguably, valuable work that is fully available on the Internet (URL: http://consci.tnc.org/library/pubs/class/index.html) is entitled *Terrestrial Vegetation of the United States*. Prepared by ecologists with The Nature Conservancy and the Natural Heritage Network, this two-volume publication is designed to function as an on-line resource, with constant updating. The title really does not do the work justice; it is no less than a standardized methodology and detailed inventory of land vegetation across the United States. By collaborating with various governmental agencies and the Ecological Society of America, the authors have presented a classification system that is meant to be a standard for all future work on land vegetation in the nation: the United States National Vegetation Classification (USNVC) system.

How does the USNVC work? First, it is based on vegetation because vegetation is such a measurable, vital, and sensitive indicator of an ecosystem's health. Once the plant community is known, the other ecosystem components can often be more readily assessed. Second, it is based on both **physiognomy** (the general structure or overall appearance of a community) and **floristics** (the actual plant species present). Both of these are quite measurable, and, ulti-

mately, it is hoped, quantifiable, although the methodology is not presently quantified. Third, it presents a hierarchical taxonomic scheme for classification with seven levels, which start with the class (e.g., forests) and end at the level of the **association** (a unique combination of plants on a given site). Unique associations are described (ideally, by trained ecologists) by their dominant plant forms. The USNVC includes over 4,100 associations at present.

A typical description of an association would start with the name of the association [e.g., *Pinus strobus* (eastern white pine)–*Tsuga canadensis* (eastern hemlock) Great Lakes forest]. Then the hierarchical classification would follow, and finally, for some of the associations, a detailed description is provided that could serve as the basis for management and comparison with other associations. Much work remains to be done to make the USNVC broadly useful; many more associations need to be identified, and most of the descriptions are lacking. Nevertheless, The Nature Conservancy and the Natural Heritage Network have performed a valuable service. The USNVC is more than a start; it is a solid contribution to taking stock of our ecosystems—the vital systems that provide us with so many essential goods and services.

Trophic Relationships: Food Chains, Food Webs, and Trophic Levels

In describing the trophic structure of ecosystems, we can identify innumerable pathways wherein one organism is eaten by a second, which is eaten by a third, and so on. Each such pathway is called a **food chain**. While it is interesting to trace these pathways, it is important to recognize that food chains seldom exist as isolated entities. A herbivore population feeds on several kinds of plants and is preyed upon by several secondary consumers, or omnivores. Consequently, virtually all food chains are interconnected and form a complex *web* of feeding relationships—the **food web**.

Despite the number of theoretical food chains and the complexity of food webs, there is a simple overall pattern: They all basically lead through a series of steps or levels—namely, from producers to primary consumers (or primary detritus feeders) to secondary consumers (or secondary de-

tritus feeders), and so on. These *feeding levels* are called **trophic levels**. All producers belong to the first trophic level; all primary consumers (in other words, all herbivores) belong to the second trophic level; organisms feeding on these herbivores belong to the third level, and so forth.

Whether we visualize the biotic structure of an ecosystem in terms of food chains, food webs, or trophic levels, we should see, through each feeding step, that there is a fundamental movement of the chemical nutrients and stored energy they contain from one organism or level to the next. These movements of energy and nutrients will be described in more detail later. A diagrammatic comparison of a food chain, a food web, and trophic levels is shown in Fig. 2–11a. A marine food web is shown in Fig. 2–11b.

How many trophic levels are there? Usually, no more than three or four in any ecosystem. This answer comes from straightforward observations. The **biomass**, or total combined (net dry) weight, (often, per unit area or volume) of all the organisms at each trophic level can be estimated

Third
trophic
level:
all
primary
carnivores

Second
trophic
level:
all
herbivores

First
trophic
level:
all
producers

(a)

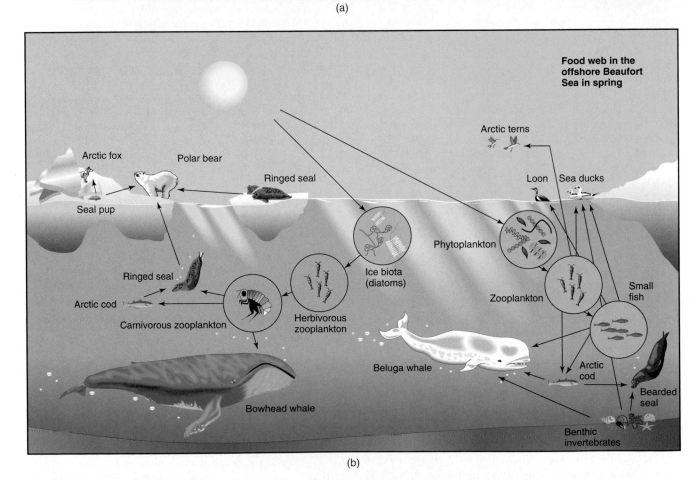

Food web in the offshore Beaufort Sea in spring

Arctic terns

Arctic fox   Polar bear

Ringed seal

Seal pup

Loon   Sea ducks

Phytoplankton

Ringed seal

Arctic cod

Carnivorous zooplankton

Herbivorous zooplankton

Ice biota (diatoms)

Zooplankton

Small fish

Arctic cod

Beluga whale

Bowhead whale

Bearded seal

Benthic invertebrates

(b)

▲ **FIGURE 2–11   *Food webs.*** (a) Specific pathways, such as that from nuts to squirrels to foxes (shown by green arrows), are referred to as *food chains*. A *food web* refers to the collection of all food chains, which are invariably interconnected (all arrows). Trophic levels, indicated by shading at the left, stresses the general pattern that food always flows from producers to herbivores to carnivores. (b) A marine food web.

by collecting (or trapping) and weighing suitable samples. In terrestrial ecosystems, the biomass is about 90–99% less at each higher trophic level. Thus, if the biomass of producers in a grassland is 10 tons (20,000 lb) per acre, the biomass of herbivores will be about 2,000 pounds and that of carnivores about 200 pounds. Clearly, you can't go through very many trophic levels before the biomass approaches zero. Depicting this graphically gives rise to what is commonly called a **biomass pyramid** (Fig. 2–12).

The biomass decreases so much at each trophic level for two reasons: First, much of the food that is consumed by a heterotroph is not converted to the body tissues of the heterotroph; rather, it is broken down so that the stored energy it contains can be released and used by the heterotroph. Second, much of the biomass—especially at the producer level—is never eaten by herbivores and goes directly to the decomposers. Hence, there is an inevitable loss of biomass with the movement to higher trophic levels. It is obvious that all heterotrophs depend on the continual input of fresh organic matter produced by the autotrophs (green plants). Without such input, the heterotrophs would run out of food and starve.

As organic matter is broken down, its chemical elements are released back to the environment, where in the inorganic state, they may be reabsorbed by autotrophs (producers). Thus, there is a continuous cycle of nutrients from the environment through organisms and back to the environment. The spent energy, on the other hand, is lost as heat given off from bodies (Fig. 2–13). These concepts of recycling chemical nutrients and the flow of energy will be discussed in greater detail in Chapter 3.

In sum, all food chains, food webs, and trophic levels *must start with producers*, and producers must have suitable environmental conditions to support their growth. Populations of all heterotrophs, including humans, are ultimately limited by what plants produce, in accordance with the concept of the biomass pyramid. Should any factor cause the productive capacity of green plants to be diminished, all other organisms at higher trophic levels will be diminished accordingly.

Nonfeeding Relationships

Mutually Supportive Relationships. The overall structure of ecosystems is characterized by feeding relationships, as we have just seen. We generally think that in any feeding relationship, one species benefits and the other is harmed to a greater or lesser extent. However, many relationships provide a mutual benefit to both species. This phenomenon is called **mutualism**. A common example is the relationship between flowers and insects: The insects benefit by obtaining nectar from the flowers, and the

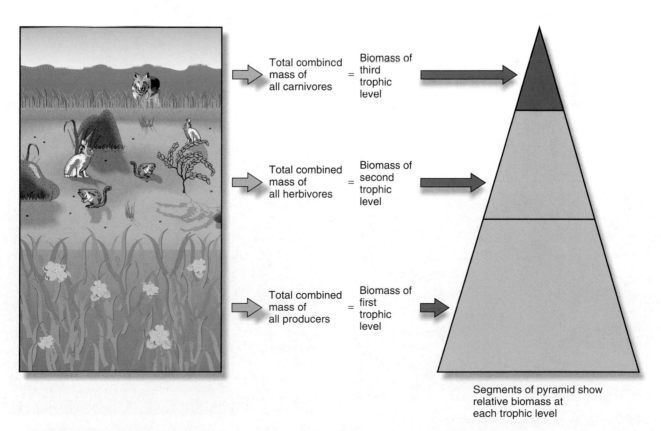

▲ **FIGURE 2–12** *Biomass pyramid.* A graphic representation of the biomass (the total combined mass of organisms) at successive trophic levels has the form of a pyramid.

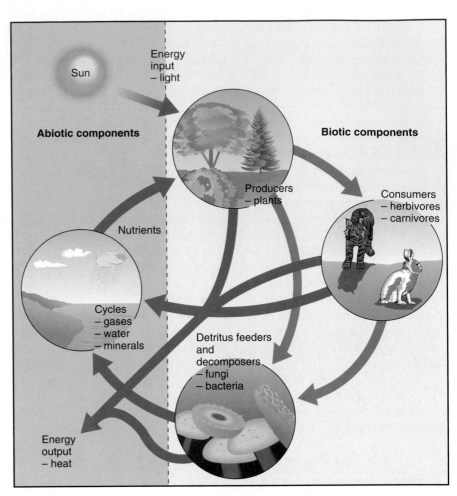

▶ FIGURE 2–13 **Nutrient cycles and
energy flow.**   The movement of nutrients (blue
arrows), energy (red arrows), and both (brown
arrows) through the ecosystem. Nutrients follow a
cycle, being used over and over. Light energy
absorbed by producers is released and lost as
heat energy as it is "spent." (From p. 594 in
GEOSYSTEMS 2/e by Robert W. Christopherson.
Copyright © 1994 by Prentice Hall, Inc. Reprinted
by permission of Pearson Education, Inc. Upper
Saddle River, NJ 07458)

plants benefit by being pollinated in the process (Fig.
2–14). Another example is observed in tropical seas:
Clown fish are immune to the toxin in the tentacles of sea
anemones, which the anemones use to immobilize their
prey. Thus, the clown fish are able to feed on detritus
around the anemones, at the same time receiving protec-

tion against would-be predators that are *not* immune. The
anemones benefit by being cleaned.

In some cases, the mutualistic relationship has become so close that the species involved are no longer capable of living alone. A classic example is the group of plants
known as *lichens* (Fig. 2–15). Lichens are actually com-

▲ FIGURE 2–14 **A mutualistic relationship.**   Insects like
this honeybee are attracted to flowers by their nectar, a good food
source for the bees. As they move from plant to plant, the bees pollinate
the flowers, enabling them to set fertile seed for the next generation.

▲ FIGURE 2–15 **Lichens.**   The crusty-appearing "plants"
commonly seen growing on rocks or the bark of trees are actually
comprised of a fungus and an alga growing in a symbiotic relationship.

prised of two organisms: a fungus and an alga. The fungus provides protection for the alga, enabling it to survive in dry habitats where it could not live by itself, and the alga, which is a producer, provides food for the fungus, which is a heterotroph. These two species living together in close union are said to have a *symbiotic* relationship. However, **symbiosis** by itself simply refers to the fact of "living together" in close union (*sym*, together; *bio*, living); it does not specify a mutual benefit or harm. Therefore, symbiotic relationships may include parasitic relationships as well as mutualistic relationships.

While not categorized as mutualistic, countless relationships in an ecosystem may be seen as aiding its overall sustainability. For example, plant detritus provides most of the food for decomposers and soil-dwelling detritus feeders such as earthworms. Thus, these organisms benefit from plants, but the plants also benefit because the activity of the organisms is instrumental in releasing nutrients from the detritus and in returning them to the soil, where they can be reused by the plants. In another example, insect-eating birds benefit from vegetation by finding nesting materials and places among trees, while the plant community benefits because the birds feed on and reduce the populations of many herbivorous insects. Even in predator–prey relationships, some mutual advantage may exist. The killing of individual prey that are weak or diseased may benefit the population at large by keeping it healthy. Predators and parasites may also prevent herbivore populations from becoming so abundant that they overgraze their environment.

Competitive Relationships. Given the concept of food webs, it might seem that species of animals would be in a great "free-for-all" competition with each other. In fact, fierce competition rarely occurs, because each species tends to be specialized and adapted to its own *habitat* or *niche*.

Habitat refers to the kind of place—defined by the plant community and the physical environment—where a species is biologically adapted to live. For example, a deciduous forest, a swamp, and an open, grassy field are types of habitats. Different types of forests (e.g., coniferous vs. deciduous) provide markedly different habitats and support different species of wildlife.

Even when different species occupy the same habitat, competition may be slight or nonexistent because each species has its own *niche*. An animal's **ecological niche** refers to what the animal feeds on, where it feeds, when it feeds, where it finds shelter, how it responds to abiotic factors, and where it nests. Seeming competitors can coexist in the same habitat, but have separate niches. Competition is minimized because potential competitors are using different resources. For example, woodpeckers, which feed on insects in deadwood, are not in competition with birds that feed on seeds. Bats and swallows both feed on flying insects, but they are not in competition, because bats feed on night-flying insects and swallows feed during the day. Sometimes the "resource" can be the space used by different species as they forage for food, as in the case of five species of warblers that coexist in the spruce forests of Maine (Fig. 2–16). This is a well-known case of what is

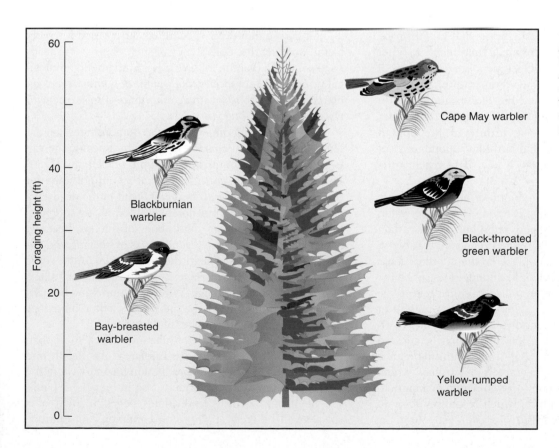

◀ **FIGURE 2–16**
Resource partitioning.
Five species of North American warblers reduce the competition among themselves by feeding at different levels and on different parts of trees.

Foraging height (ft)

Cape May warbler

Blackburnian warbler

Black-throated green warbler

Bay-breasted warbler

Yellow-rumped warbler

called *resource partitioning*. It is assumed that by adapting to each other's presence over time, these species avoid competition and all benefit. (See Chapter 5.)

Depending on how a set of resources is "divided up" among species, there may be overlap between the niches of the species. All green plants require water, nutrients, and light, and where they are growing in the same location, one species may eliminate others through competition. (Hence, maintaining flowers and vegetables against the advance of weeds is a constant struggle.) However, different plant species are also adapted and specialized to particular conditions. Thus, each species is able to hold its own against competition where conditions are well suited to it. The same concepts hold true for species in aquatic and marine ecosystems.

If two species compete directly in many respects, as sometimes occurs when a species is introduced from another continent, one of the two generally perishes in the competition. This is the *competitive exclusion principle*. For example, the introduction of the European rabbit to Australia has led to the decline and disappearance of several small marsupial animal species, due to direct competition for food and burrows.

Abiotic Factors

We now turn to the *abiotic* side of the ecosystem. As noted before, the environment involves the interplay of many physical and chemical factors that different species respond to, or **abiotic factors**. It is helpful to make a distinction between two types of abiotic factors: **conditions** and **resources**. *Conditions* are abiotic factors that vary in space and time, but are not used up or made unavailable to other species. Such factors include temperature (extremes of heat and cold, as well as average temperature), wind, pH (acidity), salinity (saltiness), and fire. For example, within aquatic systems, the key conditions are salinity (freshwater vs. saltwater), temperature, the texture of the bottom (rocky vs. silty), the depth and turbidity (cloudiness) of water (determining how much, if any, light reaches the bottom), and currents.

Resources are any factors—biotic or abiotic—that are consumed by organisms. Abiotic resources include water, chemical nutrients (like nitrogen, phosphorus, and carbon dioxide), light (for plants), and oxygen. The concept also includes spatial needs, such as a place on the intertidal rocks or a hole in a tree. Resources, unlike conditions, can be the objects of competition between individuals or species.

The degree to which each abiotic factor is present or absent and high or low profoundly affects the ability of organisms to survive. However, each species may be affected differently by each factor. We shall find that this difference in response to environmental factors determines which species may or may not occupy a given region or a particular area within a region. In turn, which organisms do or do not survive determines the nature of a given ecosystem.

Optimum, Zones of Stress, and Limits of Tolerance. In any study of ecology, a primary observation is that *different species thrive under different environmental regimes*. This principle applies to all living things, both plants and animals. Some survive where it is very wet, others where it is relatively dry. Some thrive in warmth, others in cooler situations. Some tolerate freezing, while others do not. Some require bright sun; others do best in shade. Aquatic systems are divided into freshwater and saltwater regimes, each with its respective fish and other organisms.

Laboratory experiments clearly bear out the fact that different species are best adapted to different factors. Organisms can be grown under controlled conditions in which one factor is varied while other factors are held constant. Such experiments demonstrate that for every factor, there is an **optimum**, a certain level at which the organisms do best. At higher or lower levels the organisms do less well, and at further extremes they may not be able to survive at all. This concept is shown graphically in Fig. 2–17. Temperature is shown as the variable in the figure, but the idea pertains to any abiotic factor that might be tested.

The point at which the best response occurs is called the optimum, but since this often occurs over a range of several degrees, it is common to speak of an *optimal range*. The entire span that allows any growth at all is called the **range of tolerance**. The points at the high and low ends of the range of tolerance are called the **limits of tolerance**. Between the optimal range and the high or low limit of tolerance are **zones of stress**. That is, as the factor is raised or lowered from the optimal range, the organisms experience increasing stress, until, at either limit of tolerance, they cannot survive.

Of course, not every species has been tested for every factor; however, the consistency of such observations leads us to conclude that the following is a fundamental biological principle: *Every species (both plant and animal) has an optimum range, zones of stress, and limits of tolerance with respect to every abiotic factor.*

This line of experimentation also demonstrates that different species vary in characteristics with respect to the values at which the optimum and the limits of tolerance occur. For instance, what may be an optimal amount of water for one species may stress a second and result in the death of a third. Some plants cannot tolerate any freezing temperatures, others can tolerate slight, but not intense, freezing, and some actually require several weeks of freezing temperatures in order to complete their life cycles. Also, some species have a very broad range of tolerance, whereas others have a much narrower range. While optimums and limits of tolerance may differ from one species to another, there may be great overlap in their ranges of tolerance.

The concept of a range of tolerance affects more than just the growth of individuals: Insofar as the health and vigor of individuals affect reproduction and survival of the next generation, the population is also influenced. That is, the population density (individuals per unit area) of a species will be greatest where all conditions are optimal,

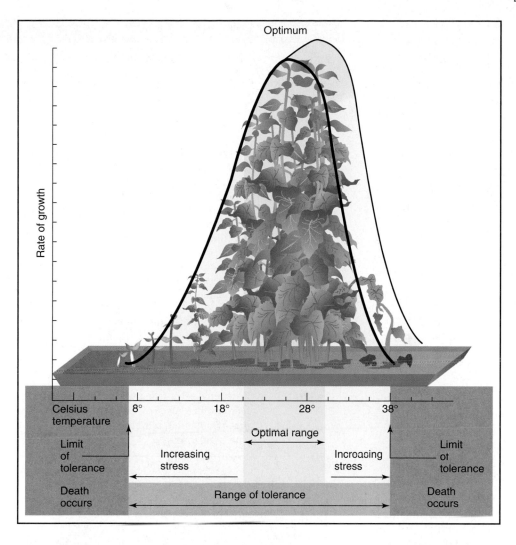

Optimum

Rate of growth

Celsius temperature   8°   18°   28°   38°

Optimal range

Limit of tolerance

Increasing stress

Increasing stress

Limit of tolerance

Death occurs

Range of tolerance

Death occurs

◀ **FIGURE 2–17** *Survival curve.* For every factor influencing growth, reproduction, and survival, there is an optimum level. Above and below the optimum, stress increases, until survival becomes impossible at the limits of tolerance. The total range between the high and low limits is the range of tolerance.

and it will decrease as any one or more conditions depart from the optimum. Different ranges of tolerance for different factors make an important contribution to the identity of an ecological niche for a given species.

Law of Limiting Factors. In 1840, Justus von Liebig studied the effects of chemical nutrients on plant growth. He observed that restricting any one of the many different nutrients at any given time had the same effect: It limited growth. A factor that limits growth is called a **limiting factor**. *Any one factor* being outside the optimal range will cause stress and limit the growth, reproduction, or even the survival of a population. This observation is referred to as the **law of limiting factors**, or Liebig's law of minimums.

The limiting factor may be a problem of "too much," as well as a problem of "too little." For example, plants may be stressed or killed not only by underwatering or underfertilizing, but also by overwatering or overfertilizing, which are common pitfalls for beginning gardeners. Note also that the limiting factor may change from one time to another. For instance, in a single growing season, temperature may be limiting in the early spring, nutrients may be limiting later, and then water may be limiting if a drought

occurs. Also, if one limiting factor is corrected, growth will increase only until another factor comes into play. Of course, the organism's genetic potential is an ultimate limiting factor: A daisy will never grow to be the height of a tree, nor a mouse to the bulk of an elephant, regardless of optimal environmental factors.

Observations made since Liebig's time show that his law has a much broader application: Growth may be limited not only by abiotic factors, but also by biotic factors. Thus, the limiting factor for a population may be competition or predation from another species. This is certainly the case with our agricultural crops, where it is a constant struggle to keep them from being limited or even eliminated by weeds and "pests."

Finally, while one factor may be determined to be limiting at a given time, several factors outside the optimum may combine to cause additional stress or even death. In particular, pollutants may act in a way that causes organisms to become more vulnerable to disease or drought. Such cases are examples of **synergistic effects**, or **synergisms**, which are defined as two or more factors interacting in a way that causes an effect much greater than one would anticipate from the effects of each of the two acting separately.

2.3 Global Biomes

We can now use the concepts of optimums and limiting factors to gain a better understanding of why different regions or even localized areas may have distinct biotic communities, creating a variety of ecosystems, landscapes, and biomes.

The Role of Climate

The **climate** of a given region is a description of the average temperature and precipitation—the weather—that may be expected on each day throughout the entire year. (See Chapter 21.) Climates in different parts of the world vary widely. In general, equatorial regions are continuously warm, with high rainfall and no discernible seasons. Above and below the equator, temperatures become increasingly seasonal (characterized by warm or hot summers and cool or cold winters); the farther we go toward the poles, the longer and colder the winters become, until at the poles it is perpetually winter. Likewise, colder temperatures are found at higher elevations, so that there are even snow-capped mountains on or near the equator.

Annual precipitation in any area also may vary greatly, from virtually zero to well over 100 inches (250 cm) per year. Precipitation may be evenly distributed throughout the year or occur in certain months, dividing the year into wet and dry seasons.

Different temperature and rainfall conditions may occur in almost any combination, yielding a wide variety of climates. In turn, a given climate will support only those species that find the temperature and precipitation levels optimal or at least within their ranges of tolerance. As indicated in Fig. 2–17, population densities will be greatest where conditions are optimal and will decrease as any condition departs from the optimum. A species will be excluded from a region (or local areas) where any condition is beyond its limit of tolerance. How will this affect the biotic community?

To illustrate, let us consider six major types of biomes and their global distribution. Table 2–3 describes these terrestrial biomes and their major characteristics, and Figure 2–18 shows the distribution and variations of the biomes as they occur globally. Within the temperate zone (between 30° and 50° of latitude) the amount of rainfall is the key limiting factor. The **temperate deciduous forest biome** is found where annual precipitation is 30–80 inches (75–200 cm). Where rainfall tapers off or is highly seasonal (10–60 inches, or 25–150 cm, per year), **grassland biomes** are found, and regions receiving an average of less than 10 inches (25 cm) per year are occupied by a **desert biome** (Table 2–3).

The effect of temperature, the other dominant parameter of climate, is largely superimposed on that of rainfall. That is, 30 inches (75 cm) or more of rainfall per year will usually support a forest, but temperature will determine the *kind* of forest. For example, broad-leafed

TABLE 2–3   Major Terrestrial Biomes

Biome	Climate and Soils
Deserts	Very dry; hot days and cold nights; rainfall less than 10 in./yr; soils thin and porous
Grasslands	Seasonal rainfall, 10 to 60 in./yr; fires frequent; soils rich and often deep
Tropical Rain Forests	Nonseasonal; annual average temperature 28°C; rainfall frequent and heavy, average over 95 in./yr; soils thin and poor in nutrients
Temperate Forests	Seasonal; temperature below freezing in winter, summers warm, humid; rainfall from 30–80 in./yr; soils well developed
Coniferous Forests	Seasonal; winters long and cold; precipitation light in winter, heavier in summer; soils acidic, much humus and litter
Tundra	Bitter cold except for an 8- to 10-week growing season with long days and moderate temperatures; precipitation low, soils thin and underlain with permafrost

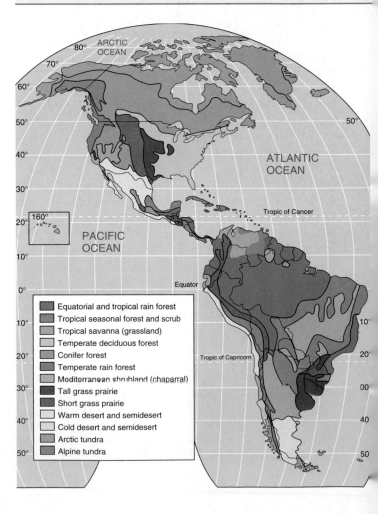

Equatorial and tropical rain forest
Tropical seasonal forest and scrub
Tropical savanna (grassland)
Temperate deciduous forest
Conifer forest
Temperate rain forest
Mediterranean shrubland (chaparral)
Tall grass prairie
Short grass prairie
Warm desert and semidesert
Cold desert and semidesert
Arctic tundra
Alpine tundra

Dominant Vegetation	Dominant Animal Life	Geographic Distribution
Widely scattered thorny bushes and shrubs, cacti	Rodents, lizards, snakes, numerous insects, owls, hawks, small birds	N. and S.W. Africa; parts of Middle East and Asia; S.W. United States, northern Mexico
Grass species, from tall grasses in areas with higher rainfall to short grasses where drier; bushes and woodlands in some areas	Large grazing mammals: bison, goats; wild horses; kangaroos; antelopes, rhinos, warthogs, prairie dogs, coyotes, jackals, lions, hyenas; termites important	Central North America, central Asia, subequatorial Africa and South America, much of southern India, northern Australia
High diversity of broad-leafed evergreen trees, dense canopy, abundant epiphytes and vines; little understory	Enormous biodiversity; exotic colorful insects, amphibians, birds, snakes; monkeys, small mammals, tigers, jaguars	Northern South America, Central America, and western central Africa, islands in Indian and Pacific Oceans, S.E. Asia
Broad-leafed deciduous trees, some conifers; shrubby undergrowth, ferns, lichens, mosses	Squirrels, raccoons, opossums, skunk, deer, foxes, black bears; snakes, amphibians, rich soil microbiota, birds	Western and central Europe, eastern Asia, eastern North America
Coniferous trees (spruce, fir, pine, hemlock), some deciduous trees (birch, maple); poor understory	Large herbivores such as mule deer, moose, elk; mice, hares, squirrels; lynx, bears, foxes, fishers, marten; important nesting area for neotropical birds	Northern portions of North America, Europe, Asia, extending southward at high elevations
Low-growing sedges, dwarf shrubs, lichens, mosses, and grasses	Year-round: lemmings, arctic hares, arctic foxes, lynx, caribou, musk ox; summers: abundant insects, many migrant shorebirds, geese, and ducks	North of the coniferous forest in northern hemisphere, extending southward at elevations above the coniferous forest

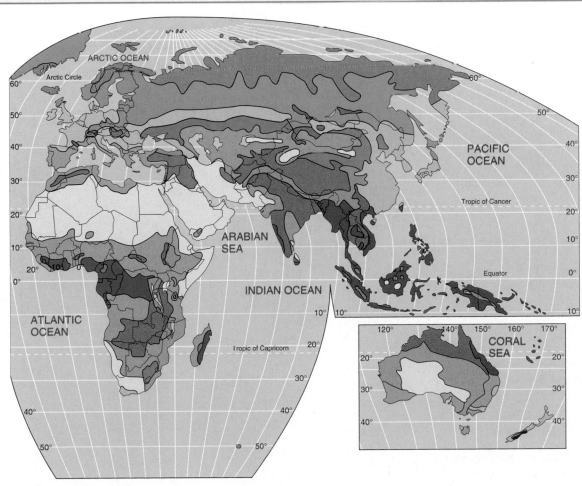

▼ FIGURE 2–18
World distribution of the major terrestrial biomes.
(Figure 20-4 from GEOSYSTEMS Introduction to Physical Geography 4/e by Robert W. Christopherson. Copyright © 2000 by Prentice Hall, Inc. Reprinted by permission of Pearson Education, Inc. Upper Saddle River, NJ 07458)

evergreen species, which are extremely vigorous and fast growing, but cannot tolerate freezing temperatures, predominate in the **tropical rain forest**. By dropping their leaves and becoming dormant each autumn, deciduous trees are well adapted to freezing temperatures. Therefore, wherever rainfall is sufficient, deciduous forests predominate in temperate latitudes. Most deciduous trees, however, cannot tolerate the extremely harsh winters and short summers that occur at higher latitudes and higher elevations. Therefore, northern regions and high elevations are occupied by the **coniferous forest biome**, as conifers are better adapted to those conditions.

Temperature by itself limits forests only when it becomes low enough to cause **permafrost** (permanently frozen subsoil). Permafrost prevents the growth of trees, because roots cannot penetrate deeply enough to provide adequate support. However, a number of grasses, clovers, and other small flowering plants can grow in the topsoil above permafrost. Consequently, where permafrost sets in, the coniferous forest biome gives way to the **tundra biome** (Table 2–3). Of course, at still colder temperatures, the tundra gives way to permanent snow and ice cover.

The same relationship of rainfall effects being primary and temperature effects secondary applies in deserts. Any region receiving less than about 10 inches (25 cm) of rain per year will be a desert, but the plant and animal species found in hot deserts are different from those found in cold deserts.

Temperature also exerts considerable influence on an ecosystem by its effect on the rate of evaporation of water. Higher temperatures effectively reduce the amount of available water, because more is lost through evaporation. As a result, the transitions from deserts to grasslands and from grasslands to forests are found at higher precipitation levels in hot regions than in cold regions.

A summary of the relationship between biomes and temperature and rainfall conditions is given in Fig. 2–19.

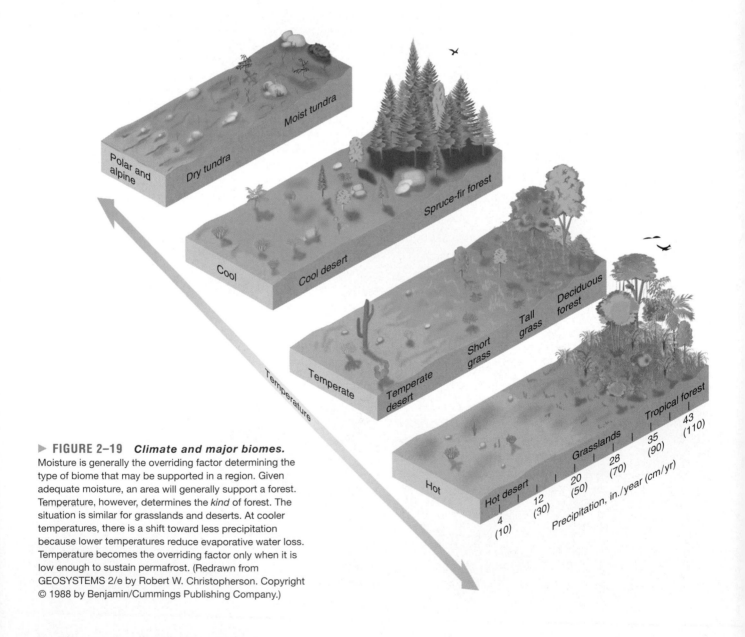

▶ **FIGURE 2–19** *Climate and major biomes.*
Moisture is generally the overriding factor determining the type of biome that may be supported in a region. Given adequate moisture, an area will generally support a forest. Temperature, however, determines the *kind* of forest. The situation is similar for grasslands and deserts. At cooler temperatures, there is a shift toward less precipitation because lower temperatures reduce evaporative water loss. Temperature becomes the overriding factor only when it is low enough to sustain permafrost. (Redrawn from GEOSYSTEMS 2/e by Robert W. Christopherson. Copyright © 1988 by Benjamin/Cummings Publishing Company.)

The average temperature for a region varies with both latitude and altitude, as shown in Fig. 2–20.

Microclimate and Other Abiotic Factors

A specific site may have temperature and moisture conditions that are significantly different from the overall climate of the region in which it is located, which is necessarily an average. For example, a south-facing slope, which receives more direct sunlight, will be relatively warmer and hence also drier than a north-facing slope (Fig. 2–21). Similarly, the temperature range in a sheltered ravine will be narrower than that in a more exposed location, and so on. The conditions found in a specific localized area are referred to as the **microclimate** of that location. In the same way that different climates determine the major biome of the region, different microclimates result in variations of the biotic community within the biome.

Soil type and topography may also contribute to the diversity found in a biome, because these two factors affect the availability of moisture. For example, in the eastern United States, oaks and hickories generally predominate on rocky, sandy soils and on hilltops, which retain little moisture, whereas beeches and maples are found on richer soils, which hold more moisture, and red maples and cedars inhabit low, swampy areas. In the transitional region between desert and grassland [10–20 inches (25–50 cm) of rainfall per year], a soil with good water-holding capacity will support grass, but a sandy soil with little ability to hold water will support only desert species.

In certain cases, an abiotic factor other than rainfall or temperature may be the primary limiting factor. For example, the strip of land adjacent to a coast frequently receives a salty spray from the ocean, a factor that relatively few plants can tolerate. Consequently, an association of salt-tolerant plants frequently occupies this strip, as on the barrier island. Relative acidity or alkalinity (pH) may also have an overriding effect on a plant or animal community.

Biotic Factors

Limiting factors may also be biotic—that is, caused by other species. Grasses thrive when rainfall is more than 30 inches (75 cm). However, when the rainfall is great enough to support trees, increased shade may limit grasses. Thus, the factor that limits grasses from taking over high-rainfall regions is biotic: overwhelming competition from taller species. The distribution of plants may also be limited by the presence of certain herbivores; elephants, for example,

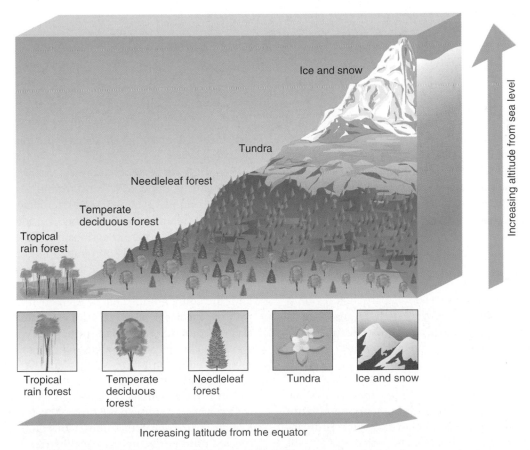

▲ **FIGURE 2–20** *Effects of latitude and altitude.* Decreasing temperatures that result in the biome shifts noted in Fig. 2–20 occur both with increasing latitude (distance from the equator) and increasing altitude. (From p. 595 in GEOSYSTEMS 2/e by Robert W. Christopherson. Copyright © 1996 by Prentice Hall, Inc. Reprinted by permission of Pearson Education, Inc. Upper Saddle River, NJ 07458)

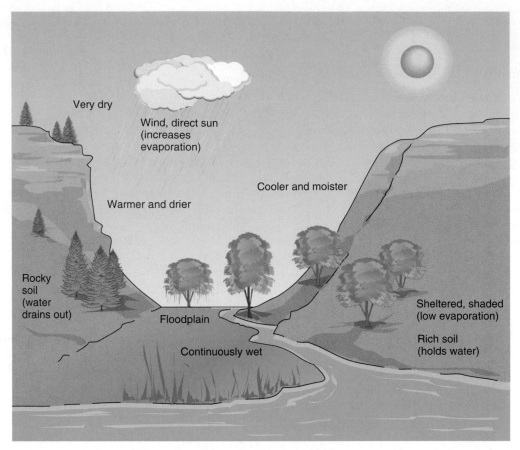

Very dry

Wind, direct sun
(increases
evaporation)

Cooler and moister

Warmer and drier

Rocky
soil
(water
drains out)

Floodplain

Sheltered, shaded
(low evaporation)

Rich soil
(holds water)

Continuously wet

▲ **FIGURE 2–21** *Microclimates.* Abiotic factors such as terrain, wind, and type of soil create different microclimates by influencing temperature and moisture in localized areas.

are notorious destroyers of woodlands, and their presence leads predominantly to grasslands.

The concept of limiting factors also applies to animals. As with plants, the limiting factor may be abiotic—cold temperatures or lack of open water, for instance—but it is more frequently biotic, in the form of the absence of a plant community that would otherwise provide suitable food or habitat.

Physical Barriers

A final factor that may limit species to a particular region is the existence of a physical barrier, such as an ocean, a desert, or a mountain range, that species are unable to cross. Thus, species making up the communities on separate continents or remote islands are usually quite different despite living in similar climates.

When such barriers are overcome—for example, by humans transporting a species from one continent to another—the introduced species may make a successful "invasion." Examples of the impact of imported species will be explored further in Chapter 4. Note also that humans

erect barriers—dams, roadways, cities, and farms—that may block the normal movement of populations and cause their demise.

In sum, the biosphere consists of a great variety of environments, both aquatic and terrestrial. In each environment we find plant, animal, and microbial species that are adapted to all the abiotic factors. In addition, they are adapted to each other, in various feeding and nonfeeding relationships. Each environment supports a more or less unique grouping of organisms interacting with each other and with the environment in a way that perpetuates or sustains the entire group. That is, each environment, together with the species it supports, is an ecosystem. Every ecosystem is tied to others through species that migrate from one system to another and through exchanges of air, water, and minerals common to the whole planet. At the same time, each species—and, as a result, each ecosystem—is kept within certain bounds by limiting factors. The spread of each species is at some point limited by its not being able to tolerate particular conditions, to compete with other species, or to cross some physical barrier. The distribution of species is always due to one or more limiting factors.

2.4 Implications for Humans

So far, we have been looking at ecosystems as natural functioning systems, unaffected by humans. In reality, the impact of humans must be taken into consideration because we have become such a dominant presence on Earth. We have replaced many natural systems with agriculture and urban and suburban developments; we make heavy use of most of the remaining "natural" systems, for wood, food, and other commercial products; and the by-products of our economic activities have polluted and degraded ecosystems everywhere. Our involvement with natural ecosystems is so pervasive that if we want to continue to enjoy their goods and services, we must learn how to *manage* natural ecosystems to keep them healthy and productive. A brief look into the past will help us understand the changes that have oc-

curred to make humans such a dominant part of the landscape and will better equip us to deal with the future.

Three Revolutions

Neolithic Revolution. Natural ecosystems have existed and perpetuated themselves on Earth for hundreds of millions of years, while humans are relative newcomers on the scene. Evidence gained through archaeology and anthropology shows that hominid ancestry goes back several million years. Evidence also indicates that several different hominid species ascended the evolutionary pathway from our primate ancestors to our present-day human species, which emerged about 100,000 years ago.

Early hominids survived in small tribes as hunter–gatherers, catching wildlife and gathering seeds, nuts, roots,

---

ETHICS

CAN ECOSYSTEMS BE RESTORED?

The human capacity for destroying ecosystems is well established. To some degree, however, we also have the capacity to restore them. In many cases, restoration simply involves stopping the abuse. For example, it has been found that water quality improves and fish and shellfish gradually return to previously polluted lakes, rivers, and bays after pollution is curtailed. Similarly, forests may gradually return to areas that have been cleared. Humans can speed up the process by seeding, planting seedling trees, and reintroducing populations of fish and animals that have been eliminated.

In some cases, however, specific ecosystems have been eliminated or disturbed to such an extent that they require the efforts of a new breed of scientist, the *restoration ecologist.* Two types of ecosystem that have suffered most in this regard are the prairie and wetland ecosystems. The potential for restoration of any ecosystem rests on three assumptions: (a) Abiotic factors must have remained unaltered or can be returned to their original state, (b) viable populations of the species formerly inhabiting the ecosystem must still exist, and (c) the ecosystem must not have been upset by the introduction of one or more foreign species that cannot be eliminated and that may preclude the sur-

vival of reintroduced native species. If these conditions are met, revival efforts have the potential to restore the ecosystem to some semblance of its former state.

For example, let us assume that the Natural Lands Trust has acquired land in the Great Plains and wishes to restore the prairie that once flourished there. The problems are many: Lack of grazing and regular fires have led to much woody vegetation, exotic species abound in the region and can continuously rain seeds on the experimental prairie, and there may be no remnants of the original prairie grasses and herbs on the site. How do we proceed?

First, we take an *inventory* of what is on the site and, from historical records, what used to be present. Then we develop a hypothetical *model* of the desired ecosystem in structural and functional terms. We set *goals* for the restoration efforts by defining the desired future condition of the site. Next, we design an *implementation plan* that can convert the goals into specific actions. For example, it may be necessary to remove all herbaceous vegetation with herbicides and plow and plant the land with native grasses. Then, to maintain the prairie, it may be necessary to conduct regular burnings of it or, in the case of larger landholdings, to introduce

bison. Finally, we should monitor the results of, and frequently make midcourse corrections to, the implementation plan.

Why should we be involved in restoring ecosystems? Restoration ecologists Steven Apfelbaum and Kim Chapman cite several compelling reasons in a recent piece on ecological restoration.[1] First, we should do so for aesthetic reasons: Natural ecosystems are often beautiful, and the restoration of something beautiful and pleasing to the eye is a worthy project that can be uplifting to many people. Second, we should do so for the benefit of human use: The ecosystem services of a restored wetland, for example, can be enjoyed by present and future generations. Finally, we should do so for the benefit of the species and ecosystems themselves: Nature has value and a right to continued existence. It can be argued that people should act to preserve and restore ecosystems and species in order to preserve that right. Do you find these reasons compelling?

[1] "Ecological Restoration: A Practical Approach," Ch. 15 in *Ecosystem Management: Applications for Sustainable Forest and Wildlife Resources,* ed. by Mark S. Boyce and Alan Haney (New Haven, CT: Yale University Press, 1997).

berries, and other plant foods (Fig. 2–22). Settlements were never large and were of relatively short duration because, as one area was "picked over," the tribe was forced to move on. As hunter–gatherers, hominids were much like other omnivorous consumers in natural ecosystems. Populations could not expand beyond the sizes that natural food sources supported, and deaths from predators, disease, and famine were common. In time, however, the hunter–gatherer culture was successful enough to foster population increases, and the pressures of rising populations led to the next stage.

About 10,000 years ago, a highly significant change in human culture occurred: Humans in the Middle East began to develop animal husbandry and agriculture—the domestication of wild species. Animal husbandry and agriculture are processes of taking particular animal and plant species out of the wild, clearing space, and providing other conditions to grow them preferentially. Plants are protected from competitors (weeds) and other would-be consumers, and additional nutrients (fertilizer) and water may be provided. Animals are protected from predators and are given food for optimal growth.

The development of agriculture provided a more abundant and reliable food supply, but it created a turning point in human history for other reasons as well. Because of this profound effect, it is referred to as a *revolution*—specifically, the **Neolithic Revolution**. Conducting agriculture does not just allow, but *requires*, permanent (or at least long-term) settlements and the specialization of labor. Some members of the settlement specialize in tending crops and producing food, freeing others to specialize in other endeavors. With this specialization of labor in permanent settlements, there is more incentive and potential for technological development: better tools, better dwellings, and better means of transporting water and other materials. Trade with other settlements begins, and thus commerce is born. Also, living in settlements permits better care and protection for everyone; therefore, the number of early deaths is reduced. This reduced mortality rate, coupled with more reliable food production, supports population growth, which in turn supports (and is supported by) expanding agriculture. In short, modern civilization had its origins in the invention of agriculture about 10,000 years ago.

Industrial Revolution.   For another 9,000-plus years, the human population increased and spread throughout the Earth. Agriculture and natural ecosystems provided the support for the growth of a civilization and culture that increased in knowledge and mastery over the natural world. With the birth of modern science and technology in the 17th and 18th centuries, the human population—by 1800 almost a billion strong—was on the threshold of another revolution: the **Industrial Revolution** (Fig. 2–23). It is fair to say that this revolution created the modern world, with its worldwide commerce, factories, large cities, and pollution, on a massive scale. The Industrial Revolution and its technological marvels were energized by fossil fuels—first coal and then oil and gas. Pollution and exploitation took on new dimensions as the industrial world turned to the extraction of raw materials from all over the world (hence the desire for colonies). In time, every part of Earth was affect-

▲ **FIGURE 2–22 *Hunter–gatherer culture.***   Before the advent of agriculture, all human societies had to forage for their food as these bushmen from Namibia are doing.

ed by this revolution and continues to be even today. As a result, we are now living in a time of continued population growth and economic expansion, with all of the environmental problems we outlined in Chapter 1 and will be dealing with in the rest of the text.

In this historical progression, we see natural systems being displaced by the **human system**, a term we will use to refer to our total system, including animal husbandry, agriculture, and all human developments. It must be noted, however, that the human system still depends heavily on goods and services provided by natural ecosystems. If the human system functioned as a true ecosystem, it would recognize this dependence and would also establish sustainable practices in its own right.

Indeed, the human system does have some features in common with natural ecosystems; the series of trophic levels from crop producers to human consumers is an example. But, in other respects, it is far off the mark—failing to break down and recycle its "detritus" (e.g., trash and chemical wastes) and other by-products and suffering the consequences of pollution is one example. Remaining highly dependent on fossil fuels and thereby suffering the buildup of carbon dioxide in the atmosphere is another. Another revolution is needed.

Environmental Revolution.   In Chapter 1, we suggested that a "business as usual" approach in human affairs will not work and that a new commitment is necessary, one that leads to a society characterized by stewardship as its leading ethic and sustainability as its goal. Some observers have referred to this shift as the **Environmental Revolution**, recognizing that all of the changes that will be required to move the human system from its present state to one that is sustainable are indeed revolutionary. We shall be exploring those changes in succeeding chapters.

Revolutions suggest overthrowing something, and indeed, what is involved is an overthrow of reigning attitudes toward our economy and the environment. This does not have to be a violent revolution; it could take place so peacefully that it would take a future generation to look back and realize that a major revolution had taken place. Yet it appears that the options are limited as we look to the future: We can choose to undergo the changes necessary to achieve sustainability by planning properly and learning as we go, or we can ignore the signs of unsustainability and move toward a gradual disintegration of biological systems and the global atmosphere. At some point, the environmental revolution will be thrust upon us by the inability of the environment to support an irresponsible human population. At the end of a chapter entitled "Challenges of the New Century" in Worldwatch Institute's book *State of the World 2000* (p. 21), Institute President Lester R. Brown makes this point very clear: "There is no middle path. The challenge is either to build an economy that is sustainable or to stay with our unsustainable economy until it declines. It is not a goal that can be compromised. One way or another, the choice will be made by our generation, but it will affect life on Earth for all generations to come."

▲ **FIGURE 2–23** *Industrial Revolution.*   The Industrial Revolution began in England in the 1800s. Coal was the energy source, and economic growth and pollution were the consequences.

ENVIRONMENT ON THE WEB

CHANGING TROPHIC STRUCTURE IN ESTUARIES AFFECTED BY HUMAN ACTIVITIES

Estuaries—the zones where rivers empty into the ocean—are among the most productive ecosystems on Earth, with the most complex food webs. The combination of terrestrial habitat, fresh water and saltwater, the tidal flux of nutrients and other materials, and the flushing action of ocean currents create a system in which many habitat types can coexist. This diversity of habitat in turn allows the coexistence of a wide range of species that would not normally be found together.

Estuaries are also a focus for impacts from human activities, like dam building, in upstream areas. Such practices can fundamentally alter downstream ecosystems through habitat degradation, blockage of migratory routes, and increased pollution levels.

Think, for example, about the food web of the harp seal. Although it is not strictly speaking an estuarine species, the harp seal feeds on a wide range of organisms at several trophic levels, some of them estuarine in adult or juvenile forms. (A migratory species, the harp seal spends time both in near-shore waters and in offshore waters and thus is tied to different marine ecosystems.) This food web is complex, and perhaps because of its complexity is able to withstand minor stresses imposed by fluctuating water-quality conditions, temperature, and flows. However, when major estuarine flow and sedimentation changes occur, as would be the case when a dam is built in a river that flows to the coast, habitat alteration may be so significant that some species no longer thrive and may die out locally. In particular, as the reduction of sediment loads to downstream estuaries decline, so do the number of organisms that depend on it for habitat through some or all of their life cycles. These can include sediment-dwelling worms and other invertebrates, but also higher level consumers, such as fish, which require a particular type of substrate for effective spawning and rearing of young.

Even modest changes in the number and abundance of sediment-dwelling polychaete worms, for example, can have a "snowball" effect, reducing the number of flatfish and sand lance (a small fish). As species at higher trophic levels fail to find enough prey, their own numbers will decline, while competing species, such as capelin and redfish, flourish. The net result is a simplified food web, now dependent on a few "keystone" species. And although the harp seal itself, which feeds on a variety of prey, may be largely unaffected in such a system, the food web overall is now much more vulnerable to the effects of other temperature, climatic, or chemical stressors.

Web Explorations

The Environment on the Web activity for this essay describes how human activities like dam building can influence the ecology of rivers and estuaries. Go to the Environment on the Web activity (select Chapter 2 at **http://www.prenhall.com/wright**) and learn for yourself:

1. how infrared images can reveal the nature and extent of habitat alteration in estuaries;
2. about human influences on the development of hydrologic systems in Everglades National Park, Florida; and
3. how human activities affect food webs and aquatic habitat.

🕐 **A suggested time frame for each of these exercises is 10–30 minutes.**

REVIEW QUESTIONS

1. What is the difference between the biotic community and the abiotic environmental factors of an ecosystem?
2. Define and compare the terms *species*, *population*, *association*, and *ecosystem*.
3. In relation to the definition of an ecosystem, what are an ecotone, landscape, biome, and biosphere?
4. Identify and describe the biotic and the abiotic components of the biome of the region in which you live.
5. Name and describe the roles of the three main trophic categories that make up the biotic structure of every ecosystem. Give examples of organisms from each category.
6. How do the terms *organic* and *inorganic* relate to the biotic and abiotic components of an ecosystem?
7. Name and describe the attributes of the two categories into which all organisms can be divided.
8. Give four categories of consumers in an ecosystem and the role that each plays.
9. State the similarities and differences between detritus feeders and decomposers in terms of what they do, how they do it, and the kinds of organisms that occupy each category.
10. Differentiate between the concepts of *food chain*, *food web*, and *trophic levels*.
11. Relate the concept of the biomass pyramid to the fact that all heterotrophs are dependent upon autotrophic production.
12. Describe three nonfeeding relationships that exist between organisms.
13. How is competition among different species of an ecosystem reduced?
14. Differentiate between the two types of abiotic factors. What is the effect on a population when any abiotic factor shifts from the optimum to the limit of tolerance and beyond? What things in addition to abiotic factors may act as limiting factors?

15. Describe how differences in climate cause the Earth to be distributed into six major biomes.

16. What are three situations that might cause microclimates to develop within an ecosystem?

17. What is significant about each of the following revolutions: Neolithic, Industrial, and Environmental?

THINKING ENVIRONMENTALLY

1. From local, national, and international news, compile a list of the many ways humans are altering abiotic and biotic factors on a local, regional, and global scale. Analyze ways that local changes may affect ecosystems on larger levels and ways that global changes may affect local levels.

2. Write a scenario of what would happen to an ecosystem or to the human system in the event of one of the following: (a) All producers are killed through a loss of fertility of the soil or through toxic contamination. (b) All parasites are eliminated. (c) Decomposers and detritus feeders are eliminated. Support all of your statements with reasons drawn from your understanding of the way ecosystems function.

3. Consider the various kinds of relationships humans have with other species, both natural and domestic. Give examples of relationships that (a) benefit humans, but harm other species, (b) benefit both humans and other species, and (c) benefit other species, but harm humans. Give examples in which the relationship may be changing—for instance, from exploitation to protection. Discuss the ethical issues involved in changing relationships.

4. Explore how the human system can be modified into a sustainable ecosystem in balance with (i.e., preserving) other natural ecosystems without losing the benefits of modern civilization.

WEB REFERENCES

On-line resources for this chapter can be found on the World Wide Web at: **http://www.prenhall.com/wright**. (Click on Chapter 2 on the Chapter Selector.)

Ecosystems: How They Work

Key Issues and Questions

1. All the elements that make up living things come from the environment. What are these key elements? Where is each found?

2. All chemical reactions taking place in living things involve energy. What are the different forms of energy, and what are the laws that govern energy exchanges?

3. Photosynthesis and cell respiration are the two fundamental biological processes. What matter and energy changes occur in these two processes? Relate them to the dynamics of ecosystems.

4. Detritivores and decomposers promote the breakdown of organic matter. What are they, and how important is their role?

5. The flow of energy is one of the vital processes that occurs in all ecosystems. Describe how energy flows in terms of trophic levels. How efficient are the transfers of energy?

6. The recycling of elements is another vital functional process occurring in all ecosystems. Describe the biogeochemical cycles for carbon, phosphorus, and nitrogen. How have humans influenced these three cycles?

7. The goods and services produced by natural ecosystems are essential to human survival. What is their overall value, and of what significance is it to measure this value?

Pictured in the opening photo, the Serengeti is a vast tropical savanna ecosystem of 25,000 km^2 in northern Tanzania and southern Kenya. The rain falls bimodally: Short rains normally occur in November and December, long rains during March through May. A rainfall gradient passes from the drier southeastern plains (50 cm/yr) to the wet northwest in Kenya (120 cm/yr). The southeastern plains are a treeless grassland, and as the volcanic soils thicken to the north and west, the grasses gradually shift to woodlands—mixed woods and grassy patches. Large herbivores (wildebeest, zebra, and Thomson's gazelle) dominate the ecosystem. Herds of over 1.5 million animals are common on the plains during the rainy seasons. Just below the surface of the land, hordes of termites in extensive galleries process organic detritus as fast as it appears and restore nutrients to the soil.

As the rain fails, the large herds move to the woodlands in the north, where forage is available year-round. There the animals face heavier predation from lions and hyenas, which are more tied to the woodlands because of their need for cover for raising young and for successful stalking. This migration, in which the animals travel 200 or more kilometers and back again each year, is energetically costly. Why do they do it if, as it seems, there is sufficient forage in the woodlands? The answer is still being investigated, but evidence points to two possible factors: (1) The vegetation in the plains is high in phosphorus, which the herbivores need for successful growth and lactation. Their annual presence there maintains the high phosphorus content as their wastes are broken down and nutrients are returned to the soils. (2) The presence of high numbers of predators in the woodlands may force the flocks to migrate to the grasslands, where they are less vulnerable, especially when giving birth to their young.

In this ecosystem known as the Serengeti, it is clear that producers, herbivores, carnivores, and scavengers or detritus feeders interact in a sustainable set of relationships. In this chapter, we explore how ecosystems like the Serengeti work at the fundamental level of chemicals and energy. We will also gain some understanding of how

◀ **The Serengeti.** A large herd of wildebeest and zebra graze on the lush grasses of the plains.

TABLE 3–1 **Principles of Ecosystem Sustainability**

For sustainability,

- Ecosystems use sunlight as their source of energy.
- Ecosystems dispose of wastes and replenish nutrients by recycling all elements.
- The size of consumer populations in ecosystems is maintained such that overgrazing and other forms of overuse do not occur.
- Ecosystems show resilience when subject to disturbance.
- Ecosystems depend on biodiversity.

natural ecosystems sustain human life, look at an estimate of what ecosystem goods and services are worth to humankind, and consider how best to manage such vital resources. Our examination of how ecosystems work will reveal five underlying principles that enable natural systems to be sustainable and will provide insight into the pathways we must take to make our human system sustainable (Table 3–1). Two principles of sustainability will be discussed in this chapter, and the remaining three will be covered in Chapter 4.

3.1 Matter, Energy, and Life

The basic building blocks of all **matter** (all gases, liquids, and solids in both living and nonliving systems) are **atoms**. Only 94 different kinds of atoms occur in nature, and these are known as the naturally occurring **elements**. In addition, physicists have created 13 more in the laboratory, but they are unstable and break down into simpler elements. (See Table C–1, p. 626.)

How can these relatively few building blocks make up the innumerable materials of our world, including the tissues of living things? Like blocks, the elements can be put together to build a great variety of things. Also like blocks, nature's materials can be taken apart into their separate constituent atoms, and the atoms can then be reassembled into different materials. All chemical reactions, whether they occur in a test tube, in the environment, or inside living things, and whether they occur very slowly or very fast, involve rearrangements of atoms to form different kinds of matter.

Atoms do not change during the disassembly and reassembly of different materials. A carbon atom, for instance, will always remain a carbon atom. Furthermore, atoms are neither created nor destroyed during chemical reactions. This constancy of atoms is regarded as a fundamental natural law, the *law of conservation of matter*.

On the chemical level, then, the cycle of growth, reproduction, death, and decay of organisms can be seen as a continuous process of taking various atoms from the environment, assembling them into living organisms (growth) and then disassembling them (decay) and repeating the process. Driving the cycle is the irresistible, genetically predisposed urge of living things to grow and reproduce.

Which atoms make up living organisms? Where are those atoms found in the environment? How do they become part of living organisms? We answer these questions next.

Matter in Living and Nonliving Systems

A more detailed discussion of atoms—how they differ from one another, how they bond to form various gases, liquids, and solids, and how we use chemical formulas to describe different chemicals—is given in Appendix C (p. 000). Studying that appendix now may give you a better comprehension of the material we are about to cover. At the very least, the definitions of two terms are essential: *molecule* and *compound*.

A **molecule** consists of two or more atoms bonded together in a specific way. The properties of a material are dependent on the specific way in which atoms are bonded to form molecules, as well as on the atoms themselves. Similarly, a **compound** consists of two or more *different kinds* of atoms bonded together. Note the distinction that a molecule may consist of two or more of the *same kind*, as well as different kinds, of atoms bonded together. In contrast, in a compound, at least two different kinds of atoms are always involved. For example, the fundamental units of oxygen gas, which consists of two oxygen atoms bonded together, are molecules, but not a compound. Water, on the other hand, can be referred to as either a molecule or a compound, since the fundamental units are two hydrogen atoms bonded to an oxygen atom. Some further distinctions are given in Appendix C.

The key elements in living systems (and their chemical symbols) are carbon (C), hydrogen (H), oxygen (O), nitrogen (N), phosphorus (P), and sulfur (S). You can remember them by the acronym N-CHOPS. These six elements are the essential ones in the organic molecules that make up the tissues of plants, animals, and microbes.

We have said that growth and decay can be seen as a process of atoms moving from the environment into living things and returning to the environment. It is useful to think of the environment as three open abiotic (nonliving) systems, or "spheres," occupied by living things—the biosphere (Fig. 3–1). The **lithosphere** is Earth's crust, made up of rocks and minerals. The **hydrosphere** is water in all of its compartments: oceans, rivers, ice, and groundwater. The **atmosphere** is the thin layer of gases separating Earth from outer space. Matter is constantly being exchanged within and among these four spheres. By looking at the chemical nature of the spheres, we shall see where our six key elements and others occur in the environment (Table 3–2).

The lower atmosphere is a mixture of molecules of three important gases—oxygen (O_2), nitrogen (N_2), and carbon dioxide (CO_2)—along with water vapor and trace amounts of several other gases that have no immediate biological importance (Fig. 3–2). The gases in the atmosphere are normally stable, but under some circumstances they react chemically to form new compounds (for example, the production of ozone from oxygen in the upper atmosphere, described in Chapter 21).

While the atmosphere is a major source of *carbon* and *oxygen* for all organisms (and a source of *nitrogen* for a few organisms), the source of the key element *hydrogen* is the hydrosphere. Each molecule of water consists of two hydrogen atoms bonded to an oxygen atom, as indicated by the formula for water: H_2O. A weak attraction between water molecules is known as *hydrogen bonding*. At temperatures below freezing, hydrogen bonding holds the molecules in position with respect to one another, and the result is a solid (ice or snow). At temperatures above freezing, but below vaporization (evaporation), hydrogen

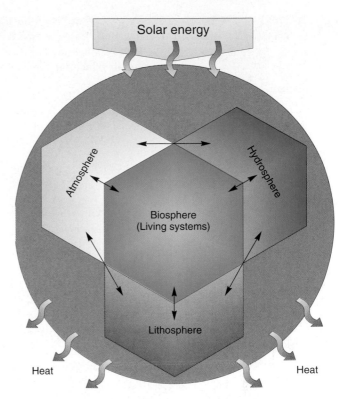

▲ **FIGURE 3–1** *The four spheres of Earth's environment.* The biosphere is all of life on earth, and it depends on, and interacts with, the atmosphere (air), the hydrosphere (water), and the lithosphere (soil and rocks).

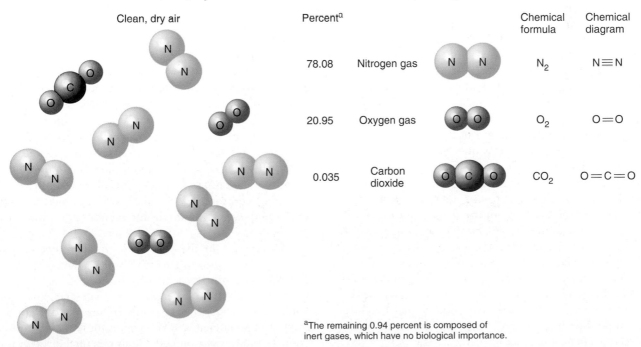

Clean, dry air is a mixture of molecules of three important gases.

	Percent[a]		Chemical formula	Chemical diagram
	78.08	Nitrogen gas	N_2	N≡N
	20.95	Oxygen gas	O_2	O=O
	0.035	Carbon dioxide	CO_2	O=C=O

[a]The remaining 0.94 percent is composed of inert gases, which have no biological importance.

▲ **FIGURE 3–2** *The major gases of clean, dry air.* From a biological point of view, the three most important gases of the lower atmosphere are nitrogen, oxygen, and carbon dioxide.

TABLE 3–2   Elements Found in Living Organisms and the Locations of Those Elements in the Environment[b]

Element (Kind of Atom)	Biologically Important Molecule or Ion in Which the Element Occurs[a]			Location in the Environment[b]		
	Symbol	Name	Formula	Atmosphere	Hydrosphere	Lithosphere
Carbon	C	Carbon dioxide	CO_2	X	X	$X(CO_3^-)$
Hydrogen	H	Water	H_2O	X	(Water itself)	
Atomic oxygen (required in respiration)	O	Oxygen gas	O_2	X	X	
Molecular oxygen (released in photosynthesis)	O_2	Water	H_2O		(Water itself)	
Nitrogen	N	Nitrogen gas	N_2	X	X	Via fixation
		Ammonium ion	NH_4^+		X	X
		Nitrate ion	NO_3^-		X	X
Sulfur	S	Sulfate ion	SO_4^{2-}		X	X
Phosphorus	P	Phosphate ion	PO_4^{3-}		X	X
Potassium	K	Potassium ion	K^+		X	X
Calcium	Ca	Calcium ion	Ca^{2+}		X	X
Magnesium	Mg	Magnesium ion	Mg^{2+}		X	X
Trace Elements[c]						
Iron	Fe	Iron ion	Fe^{2+}, Fe^{3+}		X	X
Manganese	Mn	Manganese ion	Mn^{2+}		X	X
Boron	B	Boron ion	B^{3+}		X	X
Zinc	Zn	Zinc ion	Zn^{2+}		X	X
Copper	Cu	Copper ion	Cu^{2+}		X	X
Molybdenum	Mo	Molybdenum ion	Mo^{2+}		X	X
Chlorine	Cl	Chloride ion	Cl^-		X	X

NOTE: These elements are found in *all* living organisms—plants, animals, and microbes. Some organisms require certain elements in addition to the ones listed. For example, humans require sodium and iodine.
[a]A molecule is a chemical unit of two or more atoms bonded together. An ion is a single atom or group of bonded atoms that has acquired a positive or negative charge as indicated.
[b]"X" means element exists in indicated "sphere."
[c]Only small or trace amounts of these elements are required.

bonding still holds the molecules close, but allows them to move around one another, producing the liquid state. Vaporization occurs as hydrogen bonds break and water molecules move into the air independently. With a lowering of temperature, all these changes in state go in the reverse direction (Fig. 3–3). We reemphasize, that regardless of the changes in state, the water molecules themselves retain their basic structure of two hydrogen atoms bonded to an oxygen atom. It is only the *relationship* between the molecules that changes.

All the other elements required by living organisms, as well as the 72 or so elements that are not required by them, are found in the lithosphere, in the form of rock and soil minerals. A **mineral** is any hard, crystalline, inorganic material of a given chemical composition. Most rocks are made up of relatively small crystals of two or more minerals, and soil generally consists of particles of many different minerals. Each mineral is made up of dense clusters of two or more kinds of atoms bonded together by an attraction between positive and negative charges on the atoms, as explained in Appendix C and Fig. 3–4.

Air, water, and minerals interact with each other in a simple, but significant, manner. Gases from the air and ions (charged atoms) from minerals may dissolve in water. Therefore, natural water is inevitably a *solution* containing variable amounts of dissolved gases and minerals. This solution is constantly subject to change, as any dissolved substances in it may be removed by various processes, or additional materials may dissolve in it. Molecules of water enter the air by evaporation and leave it by means of condensation and precipitation. (See the hydrologic cycle, Chapter 9.) Thus, the amount of moisture in the air is constantly fluctuating. Wind may carry a certain amount of dust or mineral particles, and this amount is also changing constantly, since the particles gradually settle out from the air. The various interactions are summarized in Fig. 3–5.

In contrast to the relatively simple molecules that occur in the environment (for example, CO_2, H_2O, and N_2), the key atoms in living organisms (C, H, O, N, P, S) bond into very large, complex molecules known as proteins, carbohydrates (sugars and starches), lipids (fatty substances), and nucleic acids. Some of these molecules may contain millions of atoms, and their potential diversity is infinite. Indeed, the diversity of living things is a reflection of the diversity of such molecules.

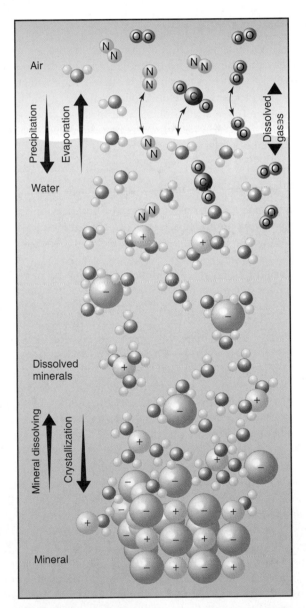

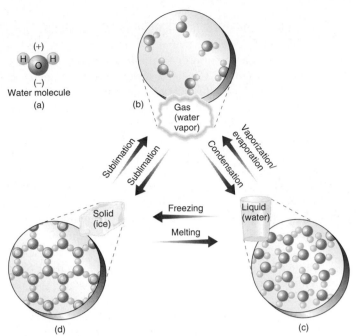

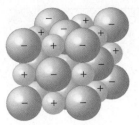

(a) Water molecule

(b) Gas (water vapor)

Sublimation

Condensation

Vaporization/evaporation

Freezing

Melting

Solid (ice)

Liquid (water)

(d)

(c)

▲ **FIGURE 3–3** *Water and its three states.* (a) Water consists of molecules, each of which is formed by two hydrogen atoms bonded to an oxygen atom (H₂O). (b) In water vapor, the molecules are separate and independent. (c) In liquid water, the weak attraction between water molecules known as hydrogen bonding gives the water its liquid property. (d) At freezing temperatures, hydrogen bonding holds the molecules firmly, giving the solid state—ice.

▲ **FIGURE 3–4** *Minerals.* Minerals (hard crystalline compounds) are composed of dense clusters of atoms of two or more elements. The atoms of most elements gain or lose one or more electrons, becoming negative (−) or positive (+) ions. The ions are held together by an attraction between positive and negative charges.

The molecules that make up the tissues of living things are constructed mainly from carbon atoms bonded together into chains with hydrogen atoms attached. Oxygen, nitrogen, phosphorus, and sulfur may be present also, but the key common denominator is carbon–carbon and carbon–hydrogen bonds (Fig. 3–6). Recall (Chapter 1) that material making up the tissues of living organisms is referred to as *organic*. Hence, the carbon-based molecules that make up the tissues of living organisms are called **organic molecules**. (Don't miss the connection between the words *organic* and *organism*.) **Inorganic**, then, refers to molecules or compounds with neither carbon–carbon nor carbon–hydrogen bonds.

Causing some confusion is the fact that all plastics and countless other human-made compounds are based on carbon–carbon bonding and are, chemically speaking, organic compounds. We resolve this confusion by referring to the compounds making up living organisms as **natural organic compounds** and the human-made ones as **synthetic organic compounds**.

In conclusion, the elements essential to life (C, H, O, and so on) are present in the atmosphere, hydrosphere, or lithosphere in relatively simple molecules. In living *organisms* of the biosphere, on the other hand, they are *organized* into highly complex *organic* compounds. These organic compounds in turn make up the various parts of cells, which in their turn make up the tissues and organs of the

▲ **FIGURE 3–5** *Interrelationship among air, water, and minerals.* Minerals and gases dissolve in water, forming solutions. Water evaporates into air, causing humidity. These processes are all reversible: Minerals in solution recrystallize, and water vapor in the air condenses to form liquid water.

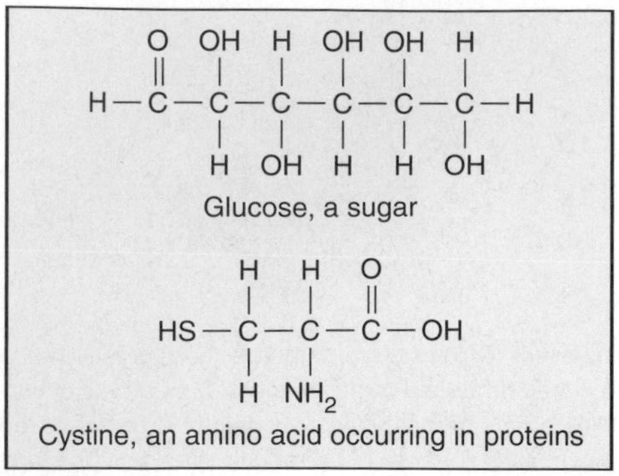

▲ **FIGURE 3–6** *Organic molecules.* The organic molecules that make up living organisms are larger and more complex than the inorganic molecules found in the environment. Glucose and cystine show this relative complexity.

body (Fig. 3–7). Growth and reproduction, then, may be seen as using the atoms from simple molecules in the environment to construct the complex organic molecules of an organism. Decomposition and decay may be seen as the reverse process. We shall look at each of these processes in more detail later in the chapter; first, however, we must consider another factor: *energy*.

Energy Considerations

In addition to rearranging atoms, chemical reactions involve the absorption or release of energy. To grasp this concept, let us examine the distinction between matter and energy.

Matter and Energy. The universe is made up of *matter* and *energy*. A more technical definition of **matter** than the one given earlier in this chapter is anything that occupies space and has mass—that is, anything that can be weighed when gravity is present. This definition obviously covers all solids, liquids, and gases, and living as well as nonliving things.

Atoms are made up of protons, neutrons, and electrons, which in turn are made of still smaller particles. Since atoms are the basic units of all elements and remain unchanged during chemical reactions, it is practical to consider them as the basic units of matter.

Light, heat, movement, and *electricity*, on the other hand, do not have mass, nor do they occupy space. (Note that *heat*, as used here, refers not to a hot object, but to the heat energy we can feel radiating from the hot object.) These are the common forms of energy that we are familiar with. What do the various forms of energy have in common? They *affect* matter, causing changes in its *position* or its *state*. For example, the release of energy in an explosion causes things to go flying, a change in position. And heating water causes it to boil and change to steam, a change in state. On a molecular level, changes in state may be seen as movements of atoms or molecules. For instance, the degree of heat energy contained in the substance is actually a measure of the relative vibrational motion of the atoms and

molecules of the substance. Therefore, we can define **energy** as the ability to move matter.

Energy is commonly divided into two major categories: *kinetic* and *potential* (Fig. 3–8). **Kinetic energy** is energy in action or motion. Light, heat energy, physical motion, and electrical current are all forms of kinetic energy. **Potential energy** is energy in storage. A substance or system with potential energy has the capacity, or *potential*, to release one or more forms of kinetic energy. A stretched rubber band, for example, has potential energy; it can send a paper clip flying. Numerous chemicals, such as gasoline and other fuels, release kinetic energy—heat energy, light, and movement—when ignited. The potential energy contained in such chemicals and fuels is called **chemical energy**.

Energy may be changed from one form to another in innumerable ways. How many examples can you think of in addition to those shown in Fig. 3–9? Besides seeing that potential energy may be converted to kinetic energy, it is especially important to recognize that kinetic energy may be converted to potential energy. (Consider, for example, charging a battery or pumping water into a high-elevation reservoir.) We shall see shortly that photosynthesis does just that.

Because energy does not have mass or occupy space, it cannot be measured in units of weight or volume, but it can be measured in other kinds of units. One of the most common units is the **calorie**, which is defined as the amount of heat required to raise the temperature of 1 gram (1 milliliter) of water 1 degree Celsius. Since this is a very small unit, it is frequently more convenient to speak in terms of kilocalories (1 kilocalorie = 1,000 calories), the amount of heat required to raise 1 liter (1,000 milliliters) of water 1 degree Celsius. (Kilocalories are sometimes denoted as "Calories" with a capital "C." Food Calories, which are a measure of the energy in given foods, are actually kilocalories.) Many forms of chemical energy can be measured in calories by converting a substance to heat energy in a device called the calorimeter and measuring the heat in terms of a rise in temperature. Temperature is a measurement of the molecular motion in a substance caused by the kinetic energy present in it.

We defined energy as the ability to move matter. Conversely, no movement of matter can occur *without* the absorption or release of energy. Indeed, no *change* in matter—from a few atoms coming together or apart in a chemical reaction to a major volcanic eruption—can be separated from respective changes in energy.

Energy Laws: Laws of Thermodynamics. Knowing that energy can be converted from one form to another has led numerous would-be inventors over the years to try to build machines or devices that would produce more energy than they consumed. A common idea is to use the output from a generator to drive a motor that, in turn, drives the generator to keep the cycle going and yields additional power in the bargain. Unfortunately, all such devices have one feature in common: They don't work. When all the inputs and outputs

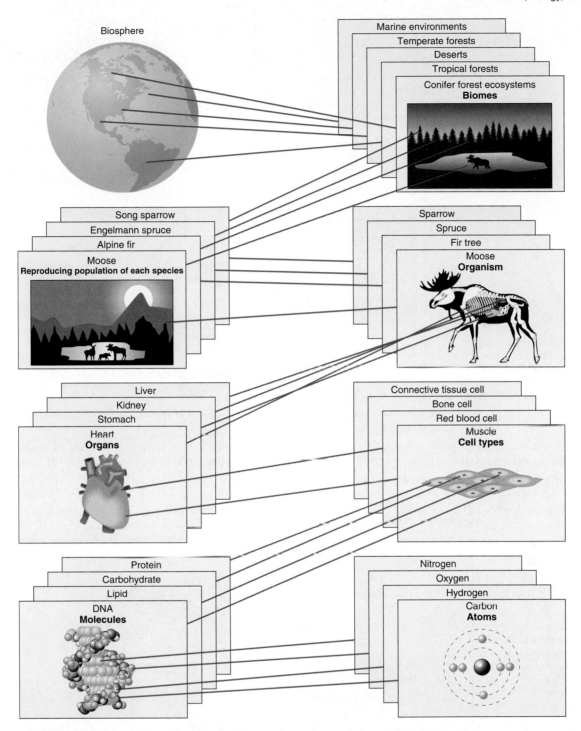

▲ **FIGURE 3–7** *Life as a hierarchy of organization of matter.* In the inorganic sphere, elements are arranged simply in molecules of the air, water, and minerals. In living organisms, they are arranged in complex organic molecules, which in turn make up cells that constitute tissues, organs, and, thus, the whole organism. Levels of organization continue up through populations, species, ecosystems, and, finally, the whole biosphere.

of energy are carefully measured, they are found to be *equal*. There is no net gain or loss in total energy. This observation is now accepted as a fundamental natural law, **the law of conservation of energy**, also called the **first law of thermodynamics**: *Energy is neither created nor destroyed, but may be converted from one form to another.* The law is also commonly stated as "You can't get something for nothing."

Fanciful "energy generators" fail for two reasons: First, in every energy conversion, a portion of the energy is converted to heat energy (thermal infrared). Second, there is no way of trapping and recycling heat energy without expending even more energy in doing so. Consequently, in the absence of energy inputs, any and every system will sooner or later come to a stop as its energy is converted to heat and

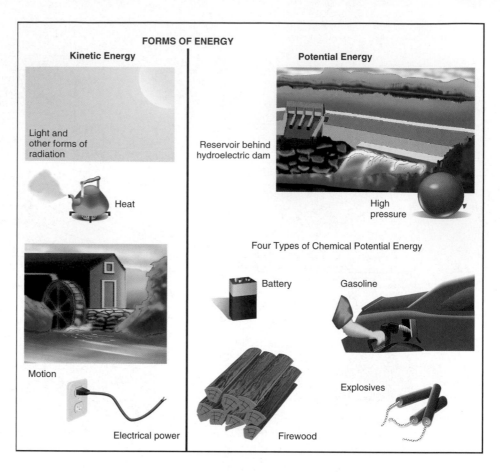

FORMS OF ENERGY

Kinetic Energy

Light and other forms of radiation

Heat

Motion

Electrical power

Potential Energy

Reservoir behind hydroelectric dam

High pressure

Four Types of Chemical Potential Energy

Battery

Gasoline

Explosives

Firewood

▶ **FIGURE 3–8** *Forms of energy.* Energy is distinct from matter in that it neither has mass nor occupies space. It has the ability to act on matter, though, changing the position or the state of the matter. Kinetic energy is energy in one of its active forms. Potential energy is the potential that systems or materials have to release kinetic energy.

lost. This is now accepted as another natural law, the **second law of thermodynamics**. Basically, the second law says that *in any energy conversion, some of the usable energy is always lost.* So, not only can you not get something for nothing (the first law), but you can't even break even.

A principle that underlies the loss of usable energy to heat is the principle of increasing *entropy*. **Entropy** refers to the degree of disorder in a system: Increasing entropy means increasing disorder. The principle is that without energy inputs, everything goes in one direction only: toward

▶ **FIGURE 3–9** *Energy conversions.* Any form of energy except heat energy can spontaneously transform into any other form. Heat is a form of energy that flows from one system or object to another because the two are at different temperatures; therefore, heat can spontaneously transfer only to something cooler.

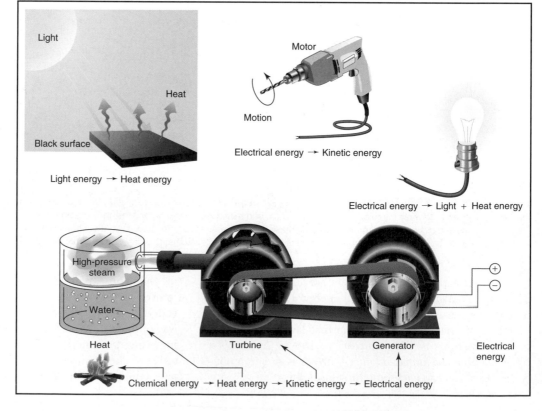

Light

Heat

Black surface

Light energy → Heat energy

Motor

Motion

Electrical energy → Kinetic energy

Electrical energy → Light + Heat energy

High-pressure steam

Water

Heat

Turbine

Generator

Electrical energy

Heat

Chemical energy → Heat energy → Kinetic energy → Electrical energy

increasing entropy. This principle of ever-increasing entropy is most readily apparent in the fact that all human-made things tend to deteriorate. Students often like to speak of the increasing disorder of their dormitory rooms as the semester wears on as an example of entropy.

The conversion of energy and the loss of usable energy to heat are both aspects of increasing entropy. Heat energy is the result of the random vibrational motion of atoms and molecules. Thus, it is the lowest (most disordered) form of energy, and its spontaneous flow to cooler surroundings is a way for that disorder to spread. Therefore, the second law of thermodynamics is nowadays more generally stated as follows: *Systems will go spontaneously in one direction only—toward increasing entropy*. The second law also says that systems will go spontaneously only toward *lower* potential energy, a direction that releases heat from the systems (Fig. 3–10).

Very important in the statement of the second law is the word *spontaneously*. It is possible to pump water uphill, charge a battery, stretch a rubber band, compress air, or otherwise increase the potential energy of a system; however, inherent in such words as *pump*, *charge*, *stretch*, and *compress* is the fact that energy is being put into the system. In contrast, flow in the opposite direction, which releases energy, occurs spontaneously (Fig. 3–11a).

The conclusion is that whenever you see something gaining potential energy, you should realize that that energy is being obtained from somewhere else (the first law). Moreover, the amount of energy lost from that "somewhere else" is greater than the amount gained (the second law). Let us now relate these concepts of matter and energy to organic molecules, organisms, ecosystems, and the biosphere.

Energy Changes in Organisms and Ecosystems

All organic molecules, which make up the tissues of living organisms, contain *high potential energy*. This is evident from the simple fact that they burn: The heat and light of the flame are the potential energy being released as kinetic energy. On the other hand, try as you might, you will not be able to get energy by burning inorganic molecules, such as carbon dioxide, water, or mineral compounds that occur in nature. Indeed, many of these materials are used as fire extinguishers. This extreme nonflammability is evidence that such materials have very *low potential energy*. Thus, the production of organic material from inorganic material involves a *gain* in potential energy. Conversely, the breakdown of organic matter involves a *release* of energy.

In this relationship between the formation and breakdown of organic matter on the one hand and the gain and release of energy on the other, we can see the energy dynamics of ecosystems. *Producers* (green plants) play the role of making high-potential-energy organic molecules for their bodies from low-potential-energy raw materials in the environment—namely, carbon dioxide, water, and a few dissolved compounds of nitrogen, phosphorus, and other elements. Such "uphill" conversion is made possible by the light energy absorbed by chlorophyll. On the other hand, all *consumers*, *detritus feeders*, and *decomposers* obtain their energy requirement for movement and other body functions from feeding on and breaking down organic matter made by producers (Fig. 3–11b). Let us now look at this *energy flow* in somewhat more detail for each category of organisms.

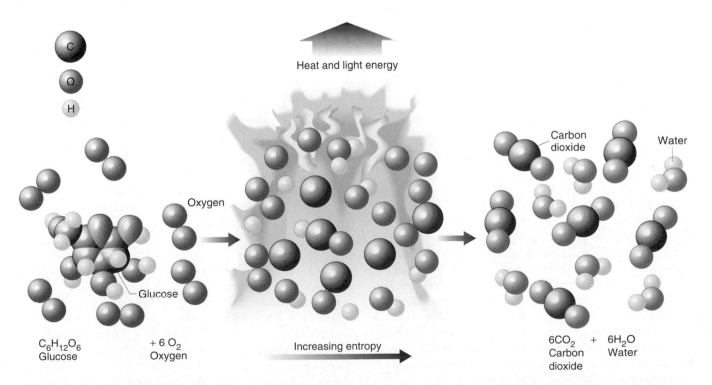

$C_6H_{12}O_6$ Glucose $+ 6 O_2$ Oxygen   Increasing entropy $6CO_2$ Carbon dioxide $+$ $6H_2O$ Water

▲ **FIGURE 3–10** *Entropy.* Systems go spontaneously only in the direction of increasing entropy. When glucose, the building-block molecule of wood, is burned, heat is released, and the atoms become more and more disordered. Both of these phenomena are aspects of increasing entropy.

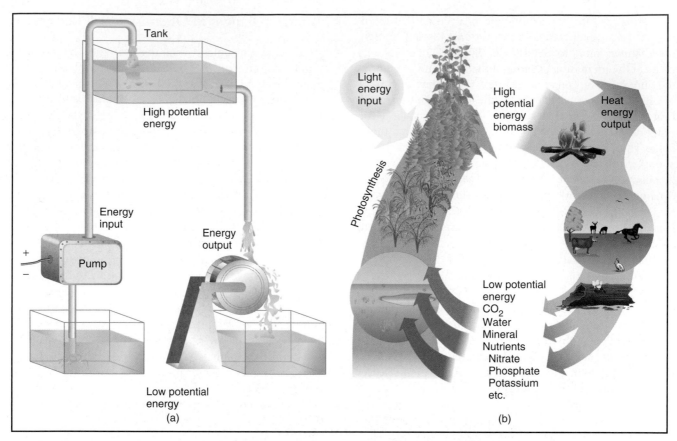

▲ FIGURE 3–11 *Storage and release of potential energy.* (a) A simple physical example of the storage and release of potential energy. (b) The same principle of storage and release of potential energy shown in ecosystems.

Producers. Recall from Chapter 2 that producers are green plants, which use light energy in the process of *photosynthesis* to make sugar (glucose, stored chemical energy) from carbon dioxide and water and release oxygen gas as a by-product. The process is described by the following formula:

Photosynthesis
(An energy-demanding process)

$$6\,CO_2 + 12\,H_2O \longrightarrow C_6H_{12}O_6 + 6\,O_2 + 6\,H_2O$$

carbon	water	light	glucose	oxygen	water
dioxide		energy		(gas)	
(gas)		input			

(low potential energy)   (high potential energy)

The kinetic energy of light is absorbed by chlorophyll in the cells of the plant and used to remove the hydrogen atoms from water (H_2O) molecules. The hydrogen atoms join carbon atoms coming from carbon dioxide as the carbons join a chain to begin forming a glucose molecule. After the removal of hydrogen from water, the oxygen atoms that remain combine with each other to form oxygen gas, which is released into the air. Water is seen on both sides of the equation because 12 molecules are consumed and 6 molecules are newly formed during photosynthesis.

The key energy steps in photosynthesis are removing the hydrogen from water molecules and joining carbon

atoms together to form the high-potential-energy carbon–carbon and carbon–hydrogen bonds of glucose in place of the low-potential-energy bonds in water and carbon dioxide molecules. But the laws of thermodynamics are not violated in this process. Careful measurements show that the rate of photosynthesis (which determines the amount of glucose formed) is proportional to the intensity of light, and, at most, 2 calories of sugar are formed for each 100 calories of light energy falling on the plant. Thus, plants are not particularly efficient "machines" in performing this conversion of light energy to chemical energy.

The glucose produced in photosynthesis plays three roles in the plant. First, either by itself or along with nitrogen, phosphorus, sulfur, and other mineral nutrients absorbed by the plant's roots, glucose is the raw material used for making all the other organic molecules (proteins, carbohydrates, and so on) that make up the stem, roots, leaves, flowers, and fruit of the plant. Second, the synthesis of all these organic molecules requires additional energy, as does the plant's absorption of nutrients from the soil and certain other functions. This energy is provided when the plant breaks down a portion of the glucose to release its stored energy in a process called *cell respiration*, which will be discussed shortly. Third, a portion of the glucose produced may be stored for future use. For storage, the glucose is generally converted to starch, as in potatoes, or to oils, as in seeds. These conversions are summarized in Fig. 3–12.

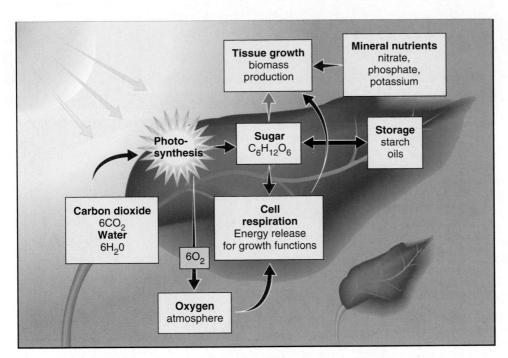

◀ FIGURE 3–12 *Producers as chemical factories.* Using light energy, producers make glucose from carbon dioxide and water, releasing oxygen as a by-product. Breaking down some of the glucose to provide additional chemical energy, they combine the remaining glucose with certain nutrients from the soil to form other complex organic molecules that the plant then uses for growth.

As the plants in ecosystems convert sunlight into new organic matter, they are "setting the table" for the components of other living ecosystems—the herbivores, carnivores, and decomposers we met in the food webs in the previous chapter. Since they are creating *new* organic matter for the ecosystem, we refer to them as *primary* producers. Given suitable conditions and resources, the producers of an ecosystem will maintain their photosynthetic activity over time in the process we call *primary production*. The total amount of photosynthetic activity in producers is called *gross primary production*; subtracting the energy consumed by the plants themselves yields the *net primary production*. Thus, net primary production refers to the *rate* at which new organic matter is made available to consumers in an ecosystem. In the Serengeti, the grasses in the plains have to produce at the rate of 560 kg dry weight per km^2 per day in order to keep up with the rate at which they are being grazed.

Consumers. Obviously, consumers need energy to move about and to perform such bodily functions as pumping blood. In addition, consumers need energy to synthesize all the molecules required for growth, maintenance, and repair of the body. Where does this energy come from? It comes from the breakdown of organic molecules of food (or of the body's own tissues if food is not available). About 60–90 percent of the food that we and other consumers eat and digest acts as "fuel" to provide energy.

First, the starches, fats, and proteins that you eat are digested in the stomach or intestine, which means that they are broken down into simpler molecules—starches into sugar (glucose), for example. These simpler molecules are then absorbed from the intestine into the bloodstream and transported to individual cells of the body.

Inside each cell, organic molecules may be broken down through a process called **cell respiration** to release the energy required for the work done by that cell. Most commonly, cell respiration involves the breakdown of glucose, and the overall chemical equation is basically the reverse of that for photosynthesis:

Cell respiration
(An energy-releasing process)

$$C_6H_{12}O_6 + 6\,O_2 \longrightarrow 6\,CO_2 + 6\,H_2O$$

| Glucose | oxygen | energy released | carbon dioxide | water |

(high potential energy)   (low potential energy)

Again, the key point of cell respiration is to release the potential energy contained in organic molecules to perform the activities of the organism. However, other aspects of the chemistry are also significant. Note that *oxygen* is released in photosynthesis, but in cell respiration it is *used* to complete the breakdown of glucose to carbon dioxide and water. Oxygen is absorbed through the lungs with every inhalation (or through the gills, in the case of fish) and is transported to all the body's cells via the circulatory system. Carbon dioxide, which is formed as a waste product, moves from the cells into the circulatory system and is eliminated through the lungs (or gills) with every exhalation.

In keeping with the laws of thermodynamics, the energy conversions involved in the body's use of the potential energy from glucose to do work are not 100% efficient. Considerable waste heat is produced, and this is the source of *body heat*. This heat output can be measured in cold-blooded animals and in plants, as well as in warm-blooded animals. It is more noticeable in warm-blooded animals only because they produce extra heat, via cell respiration, to maintain their warm body temperature.

The basis of weight gain or loss should become evident here. Organic matter is broken down in cell respiration only

GLOBAL PERSPECTIVE

LIGHT AND NUTRIENTS: THE CONTROLLING FACTORS IN MARINE ECOSYSTEMS

Even though running on solar energy and recycling nutrients are basic principles of sustainability, light and nutrients are limiting factors in marine ecosystems. First, light is diminished as water depth increases, because even clear water absorbs light. The layer of water from the surface down to the greatest depth at which there is adequate light for photosynthesis is known as the **euphotic zone**. Below the euphotic zone, by definition, photosynthesis does not occur. In clear water, the euphotic zone may be as deep as 600 feet (200 m), but in very turbid (cloudy) water, it may be a matter of only a few centimeters. In coastal waters, where the euphotic zone extends to the bottom, the bottom may support abundant plant life—that is, aquatic vegetation attached to or rooted in sediments. If the euphotic zone does not extend to the bottom, however, the bottom will be barren of plant life.

That the euphotic zone does not extend to the bottom does not preclude an ecosystem from existing in it. Phytoplankton—algae and photosynthetic bacteria that grow as single cells or in small groups of cells—can maintain themselves close to the surface in the euphotic zone. Phytoplankton support a diverse food web, from the zooplankton (small crustaceans, protozoans) that feed on them to many species of fish and sea mammals (whales and porpoises) at the higher trophic levels.

Also, an entire ecosystem operates in the cold, dark depths below the euphotic layer nourished by detritus raining down from above and, closer to the ocean floor, by vents and fissures that produce mineral-rich water and warmth.

In a phytoplankton-based system, nutrients dissolved in the water become critically important. If the water contains too few dis-

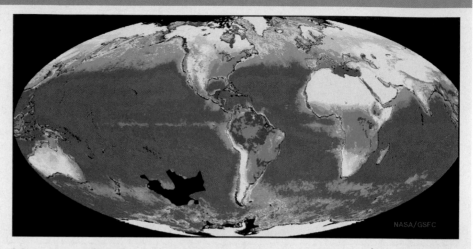

Magenta—mid oceans: lowest productivity (0.1 mg chlorophyll/m^3 or less).
Red/orange—along coasts: highest productivity (10 mg chlorophyll/m^3 or more).

solved nutrients such as phosphorus or nitrogen compounds, the growth of phytoplankton and, hence, the rest of the ecosystem will be limited. If the bottom receives light, it may support vegetation despite nutrient-poor water, because such vegetation draws nutrients from the bottom materials. Indeed, in some estuaries, nutrient-rich water is counterproductive to bottom vegetation, because the dissolved nutrients support the growth of phytoplankton, which makes the water turbid and shades out the bottom vegetation.

Let us put these concepts together to understand particular marine environments. The most productive areas of the ocean—the areas supporting the most abundant marine life of all sorts—are mostly found mostly within 200 miles (300 km) of shorelines. The reason is that either the bottom is within the euphotic zone and thus supports abundant vegetation, or nutrients washing in from the land support an abundant primary production of phytoplankton.

In the open ocean, there is less and less marine life as one moves farther from shore. Indeed, marine biologists speak of most of the open ocean as being a "biological desert." The scarcity of life occurs both because the bottom is well below the euphotic zone and because the water is poor in nutrients. The nutrients carried to the bottom with the settling detritus are released into solution by decomposers, thus enriching the bottom water with nutrients. This nutrient-rich bottom water is carried along by deep-running ocean currents. Where the currents hit underwater mountains or continental rims, the nutrient-rich water is brought to the surface. Phytoplankton flourish in these areas of **upwelling** (rising) nutrient-rich water and support a rich diversity of fish and marine mammals.

As a result, the world's oceans are far from being uniformly stocked with fish. By far the richest marine fishing areas are continental shelves and regions of upwelling, as shown on the accompanying map.

---

as it is needed to meet the energy demands of the body; this is why your breathing rate, the outer reflection of cell respiration, varies with changes in your level of exercise and activity. If you consume more calories from food than your body needs, the excess may be converted to fat and stored, and the result is a gain in weight. Conversely, the principle of dieting is to eat less and exercise more, to create an energy demand that exceeds the amount of energy contained in the food being consumed. This imbalance forces the body

to break down its own tissues to make up the difference, and the result is a weight loss. Of course, carried to an extreme, this imbalance leads to *starvation* and death when the body runs out of anything expendable to break down for its energy needs.

The overall reaction for cell respiration is the same as that for simply burning glucose. Thus, it is not uncommon to speak of "burning" our food for energy. Such a breakdown of molecules is also called **oxidation**. The distinction between burning and cell respiration is that in cell respiration the oxidation takes place in about 20 small steps, so that the energy is released in small "packets" suitable for driving the functions of each cell. If all the energy from glucose molecules were released in a single "bang," as occurs in burning, it would be like heating and lighting a room with large firecrackers—energy, yes, but hardly useful energy.

The Fate of Food. Whereas 60–90% of the food that consumers eat, digest, and absorb is oxidized for energy, the remaining 10–40%, which is converted to the body tissues of the consumer, is no less important. This is the fraction that enables the body to grow, as well as to maintain and repair itself. A portion of what is ingested by consumers is not digested, but simply passes through the digestive system and out as fecal wastes. For consumers that eat plants, such waste is largely the material of plant cell walls, or **cellulose**. We often refer to it as *fiber, bulk,* or *roughage.* Some fiber is a necessary part of the diet in order for the intestines to have something to push through them so that they can keep clean and open. Waste products can also include compounds of nitrogen, phosphorus, and any other elements present, in addition to the usual carbon dioxide and water. These by-products are excreted in the urine (or as similar waste in other kinds of animals) and returned to the environment, where they may be reabsorbed by plants. Here you can see the movement of elements in a cycle between the environment and living organisms. We expand on these cycles shortly.

In sum, organic material (food) eaten by any consumer follows one of three pathways: (1) More than 60% of what is digested and absorbed is oxidized to provide energy, and waste products are released back to the environment; (2) the remainder of what is digested and absorbed goes into body growth, maintenance and repair, or storage (fat); and (3) the portion that is not digested or absorbed passes out as fecal waste (Fig. 3–13). Recognize, then, that in an ecosystem, it is only that portion of the food which becomes body tissue of the consumer that becomes food for the next organism in the food chain. This process is often referred to as *secondary* production, and like primary production, it also can be expressed as a rate (amount of growth of the consumer or consumer trophic level) over time.

Detritus Feeders and Decomposers—the Detritivores. Recall that detritus is mostly dead leaves, the woody parts of plants, and animal fecal wastes. Hence, it is largely cellulose. Nevertheless, it is still organic and high in potential energy for those organisms that can digest it—namely, the decomposers we learned about in Chapter 2. Beyond having this ability to digest cellulose, decomposers (various species of fungi and bacteria, as well as a few other microbes) act as any other consumer, using the cellulose as a source of both energy and nutrients. Termites and some other detritus feeders can digest woody material by virtue of maintaining decomposer microorganisms in their guts in a mutualistic symbiotic relationship. The termite (a detritus feeder) provides a cozy home for the microbes (decomposers) and takes in the cellulose, which the microbes digest for both their own and the termites' benefit.

Most decomposers make use of cell respiration. Thus, the detritus is broken down into carbon dioxide, water, and mineral nutrients. Likewise, there is a release of waste heat, which you may observe as the "steaming" of a manure or compost pile on a cold day. The release of nutrients by decomposers is highly important to the primary producers, as it is the major source of nutrients in most ecosystems.

Some decomposers (certain bacteria and yeasts) can meet their energy needs through the partial breakdown of glucose that can occur in the absence of oxygen. This modified form of cell respiration is called **fermentation**. It

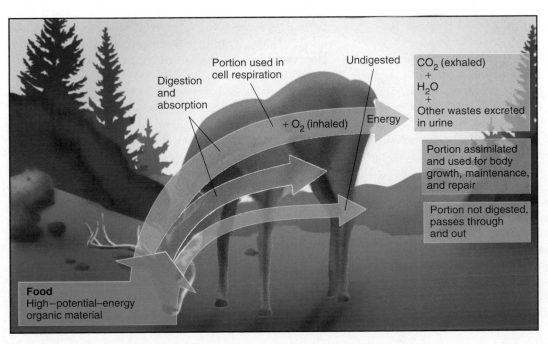

Digestion and absorption

Portion used in cell respiration

$+ O_2$ (inhaled)

Energy

Undigested

CO_2 (exhaled) + H_2O + Other wastes excreted in urine

Portion assimilated and used for body growth, maintenance, and repair

Portion not digested, passes through and out

Food
High–potential–energy organic material

◄ **FIGURE 3–13**
Consumers. Only a small portion of the food ingested by a consumer is assimilated into body growth, maintenance, and repair. A larger amount is used in cell respiration to provide energy; waste products are carbon dioxide, water, and various mineral nutrients. A third portion is not digested and becomes fecal waste.

results in such end products as ethyl alcohol (C_2H_6O), methane gas (CH_4), and acetic acid ($C_2H_4O_2$). The commercial production of these compounds is achieved by growing the particular organism on suitable organic matter in a vessel without oxygen. In nature, **anaerobic**, or *oxygen-free*, environments commonly exist in the sediments of lakes, marshes, or swamps and in the guts of animals, where oxygen does not penetrate readily. Methane gas is commonly produced in such locations. A number of large grazing animals, including cattle, maintain fermenting bacteria in their digestive systems in a mutualistic, symbiotic relationship similar to that just described for termites. Both cattle and termites produce methane as a result.

For simplicity, our orientation in this chapter is directed toward terrestrial ecosystems. It is important to realize that exactly the same processes occur in aquatic ecosystems. As aquatic plants and algae absorb dissolved carbon dioxide and mineral nutrients from the water, they employ photosynthesis to produce food and dissolved oxygen that sustain consumers and other heterotrophs. Likewise, aquatic heterotrophs return carbon dioxide and mineral nutrients to the aquatic environment. Of course, aquatic and terrestrial systems are never entirely isolated from one another, and exchanges between them go on all the time.

3.2   Principles of Ecosystem Function

The preceding examination of how the different biotic components of ecosystems function reveals that two common processes underlie them all: (a) the *flow of energy*, using sunlight as the basic energy source, and (b) the *recycling of nutrients*. In turn, the two proccesses reveal basic principles underlying the sustainability of ecosystems. Let us examine these processes further at the ecosystem level.

Energy Flow in Ecosystems

Primary Production.   In most ecosystems, sunlight, or solar energy, is the initial source of energy absorbed by producers through the process of photosynthesis. (The only exceptions are ecosystems near the ocean floor or in dark caves, where the producers are chemosynthetic bacteria.) As we saw, the process of primary production is capable of capturing at best only about 2% of incoming solar energy. Even though this seems like a small fraction, it is enough to fuel all of life on Earth—estimated at some 115 billion tons of organic matter per year as net production! In a given ecosystem, the actual biomass of primary producers at any given time is referred to as the *standing crop biomass*. Both biomass and primary production vary greatly in different ecosystems. For example, a forested ecosystem maintains a very large biomass in comparison with a tropical grassland, yet the rate of primary production could be higher in the grassland, where animals continually graze newly produced organic matter.

It is instructive to examine the productivity of different types of ecosystem (e.g., terrestrial biomes and aquatic ecosystems) to evaluate their contribution to global productivity and to investigate why some are more productive than others. Figure 3–14 presents (a) the average net primary productivity, (b) the percentage of different ecosystems over Earth's surface, and subsequently, (c) the percentage of global net primary productivity attributed to 19 of the most important types of ecosystem. Some key relationships between abiotic factors and specific ecosystems can be seen in the data. Tropical rain forests are both highly productive and contribute considerably to global productivity; they cover a large area of the land and are characterized by ideal climatic conditions for photosynthesis—warm temperature and abundant rainfall. The open oceans, because they cover 65% of Earth's surface, account for a large portion of global productivity, yet the actual rate of production is low enough to refer to them as veritable biological deserts. Although light, temperature, and water are abundant, primary production in the oceans is limited by the scarcity of nutrients—a good lesson in the significance of limiting factors (Chapter 2). The seasonal effects of differences in latitude can also be seen by comparing productivity in tropical, temperate, and boreal (coniferous) forests.

Energy Flow and Efficiency.   As primary producers are consumed by herbivores, energy is transferred from the one to the other. Recall that we refer to each of these components as a trophic level. Thus, we can examine energy flow in an ecosystem by considering how the energy moves from one trophic level to another. Figure 3–15 presents a scheme of energy flow through three trophic levels of a grazing food web. Notice that at each trophic level, some energy goes into growth (production), some is converted to heat (respiration), and some is given off as waste or is not consumed. It is quite clear that as energy flows from one trophic level, only a small fraction is actually passed on to the next trophic level. This is a consequence of three things: (1) Much of the preceding trophic level is standing biomass and is not consumed; (2) much of what is consumed is used for energy; and (3) some of what is consumed is undigested and passes through the organism.

As Fig. 3–15 shows, a very large proportion of the primary-producer trophic level is not consumed in the grazing food web. As this material dies (leaves drop, grasses wither and die, etc.), it is joined by the fecal wastes and dead bodies of higher trophic levels and represents the starting point for a separate food web, the detritus food web, pictured earlier in Fig. 2–9. It is often the case that the majority of energy in an ecosystem flows through the detritus food web.

Because of the losses of energy that occur when it is transferred at each trophic level, it is clear that each successive trophic level will capture only a fraction of the energy that entered the previous trophic level and usually will be represented by a much smaller biomass. Calculations show that the efficiency of transfer in a number of ecosystems ranges from 5 to 20%, with 10% being the average. Thus,

▶ **FIGURE 3–14** *Productivity of different ecosystems.* (a) The annual net primary productivity of different ecosystems; (b) the percentage of different ecosystems over Earth's surface area; (c) percentage of global net primary productivity.

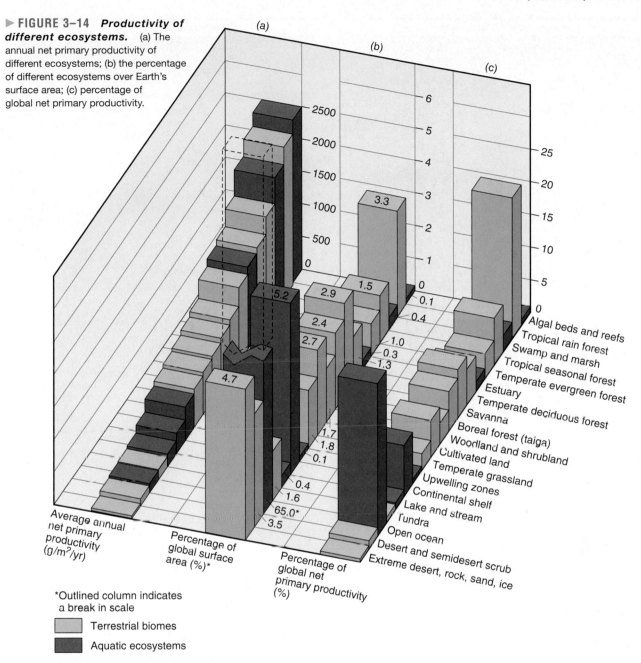

there is an estimated 90% *loss* of energy as it moves from one trophic level to the next. This loss gets quite critical at increasingly higher trophic levels and is the reason carnivores are much less abundant than herbivores; carnivores that eat other carnivores are even less abundant, and so forth. This helps to explain why, in any given ecosystem, there are usually only three to five trophic levels; there simply isn't enough energy left to pass along to "supercarnivores."

What happens to all the energy entering ecosystems? High potential energy in living plants, synthesized by using high-energy solar radiation, is either passed along to the next trophic level or degraded into the lowest and most disordered form of energy—heat—as it undergoes decomposition. Eventually, all of the energy escapes as heat energy. The laws of thermodynamics require that no energy actually be lost, but because so many energy conversions are tak-

ing place in ecosystem trophic activities, entropy is increased and all the energy is degraded to a form unavailable to do further work. The ultimate result is that *energy flows in a one-way direction through ecosystems*; it is not recycled, so it must be resupplied by sunlight.

Running on Solar Energy. We have seen that no system can run without an input of energy, and living systems are no exception. For all major ecosystems, both terrestrial and aquatic, the initial source of energy is *sunlight*. Using sunlight as the basic energy source is fundamental to sustainability for two reasons: It is both *nonpolluting* and *nondepletable*.

Nonpolluting. Light from the Sun is a form of pure energy; it contains no substance that can pollute the environment. All the matter and pollution involved in the pro-

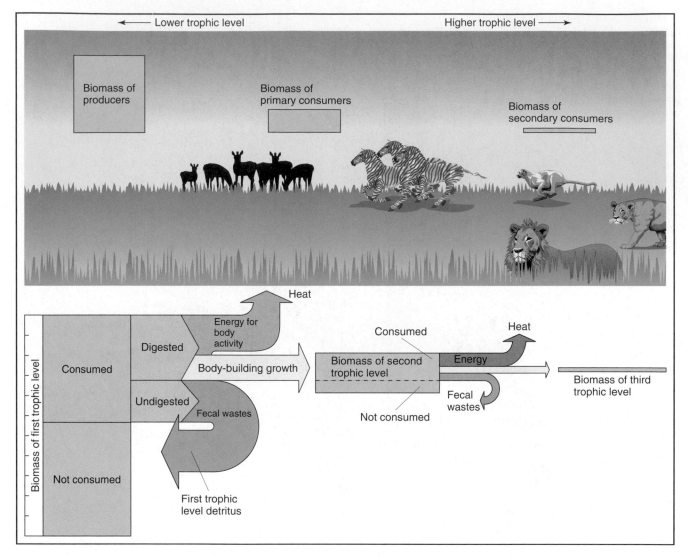

▲ **FIGURE 3–15** *Energy flow through trophic levels in a grazing food web.* Each trophic level is represented as biomass boxes, and the pathways taken by the energy flow are indicated with arrows.

duction of light energy are conveniently left behind on the Sun some 93 million miles (150 million kilometers) away in space.

Nondepletable. The Sun's energy output is constant. How much or how little of this energy is used on Earth will not influence, much less deplete, the Sun's output. For all practical purposes, the Sun is an everlasting source of energy. True, astronomers tell us that the Sun will burn out in another 3–5 billion years, but we need to put this figure in perspective. One thousand is only 0.0001% of a billion. Thus, even the passing of millennia is hardly noticeable on this time scale.

Hence, we uncover the **first basic principle of ecosystem sustainability**:

For sustainability, ecosystems use sunlight as their source of energy

Energy flow is one of the two fundamental processes that make ecosystems work. The second process is the recycling of nutrients in what are known as biogeochemical cycles.

Biogeochemical Cycles

Looking at the various inputs and outputs of producers, consumers, detritus feeders, and decomposers, you should be impressed by how they fit together. The products and by-products of each group are the food or essential nutrients for the other. Specifically, the organic material and oxygen produced by green plants are the food and oxygen required by consumers and other heterotrophs. In turn, the carbon dioxide and other wastes generated when heterotrophs break down their food are exactly the nutrients needed by green plants. Such recycling is fundamental, for two reasons: (a) It prevents the accumulation of wastes that would cause problems; and (b) it guarantees that the ecosystem will not run out of essential elements. Thus, we encounter the **second basic principle of ecosystem sustainability**:

For sustainability, ecosystems dispose of wastes and replenish nutrients by recycling all elements.

If we consider again the natural law of conservation of matter, which says that atoms cannot be created, destroyed, or changed, we can see that recycling is the only possible way to maintain a dynamic system, and the biosphere has mastered the process to a remarkable degree. We can see this even more clearly by focusing on the pathways of three key elements heavily affected by human activities: carbon, phosphorus, and nitrogen. Because these pathways all lead in circles and involve biological, geological, and chemical processes, they are known as biogeochemical cycles. (Recall that energy is not recycled.)

The Carbon Cycle.   For descriptive purposes, it is convenient to start the carbon cycle (Fig. 3–16) with the "reservoir" of carbon dioxide (CO_2) molecules present in the air and bicarbonate (HCO_3^-) molecules present in water. Through photosynthesis and further metabolism, carbon atoms from CO_2 become the carbon atoms of the organic molecules making up a plant's body. The carbon atoms then move into food webs and become part of the tissues of all the other organisms in the ecosystem. At any point, a given atom may be respired and returned to the atmosphere; in aquatic systems, it will be returned to the inorganic carbonate in solution. Processes other than trophic transfer are significant. The figure indicates two in particular: (1) geological sedimentation and limestone formation of carbon in the oceans, which removes carbon from solution, and (2) weathering

and combustion of fossil-fuel carbon that was laid down millions of years ago by biological systems. Notice how all four "spheres" are involved in this cycle.

By calculating the total amount of carbon dioxide in the atmosphere and the amount of primary production (photosynthesis) occurring in the biosphere, scientists have concluded that about a third of the total atmospheric carbon dioxide is taken up in photosynthesis in a year, but an equal amount is returned to the atmosphere through cell respiration. This means that, on the average, a carbon atom makes a cycle from the atmosphere through one or more living things and back to the atmosphere every three years.

Our intrusion into this cycle is significant. As we will see later, we are diverting or canceling out a large proportion (40%) of terrestrial primary production to support human enterprises. And through our burning of fossil fuels and destruction of forests, we have increased the atmospheric carbon dioxide by 33% over preindustrial levels, a topic we explore in Chapter 21.

The Phosphorus Cycle.   The phosphorus cycle is representative of the cycles for all the mineral nutrients—those elements which have their origin in the rock and soil minerals of the lithosphere. (See Table 3–2.) We focus on phosphorus because its shortage tends to be a limiting factor in a number of ecosystems and its excess can be a serious stimulant for unwanted algal growth in freshwater systems.

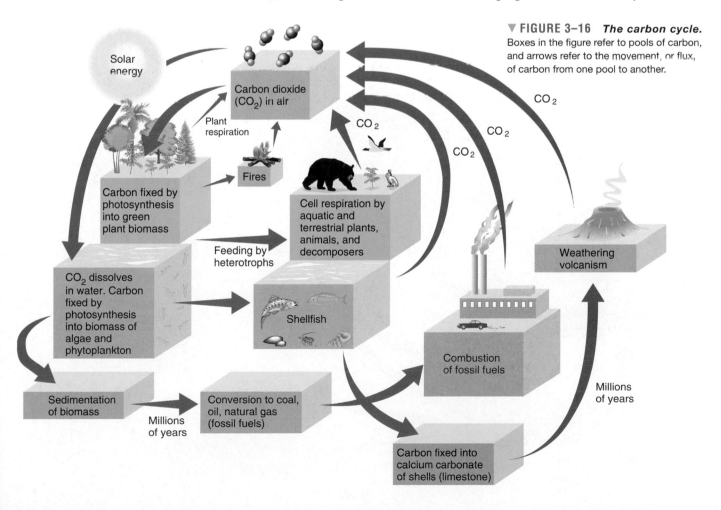

▼ **FIGURE 3–16** *The carbon cycle.*
Boxes in the figure refer to pools of carbon, and arrows refer to the movement, or flux, of carbon from one pool to another.

The phosphorus cycle is illustrated in Fig. 3–17, again set up as a set of pools and fluxes to indicate key processes. Phosphorus exists in various rock and soil minerals as the inorganic ion *phosphate* (PO_4^{3-}). As rock gradually breaks down, phosphate and other ions are released. PO_4^{3-} dissolves in water, but does not enter the air. Plants absorb PO_4^{3-} from the soil or from a water solution, and when the phosphate is bonded into organic compounds by the plant, it is referred to as **organic phosphate**. Moving through food chains, organic phosphate is transferred from producers to the rest of the ecosystem. As with carbon, at each step there is a high likelihood that the organic compounds containing phosphate will be broken down in cell respiration, releasing PO_4^{3-} in urine or other waste material. The phosphate may then be reabsorbed by plants to start another cycle.

Phosphorus enters into complex chemical reactions with other substances that are not shown in this simplified version of the cycle. For example, PO_4^{3-} forms insoluble chemical precipitates with a number of cations (positively charged ions), such as iron (Fe^{3+}), aluminum (Al^{3+}), and calcium (Ca^{3+}). If these cations are in sufficiently high concentration in soil or aquatic systems, the phosphorus can be tied up in chemical precipitates and rendered largely unavailable to plants. The precipitated phosphorus can slowly release PO_4^{3-} as plants withdraw naturally occurring PO_4^{3-} from soil, water, or sediments.

There is an important difference between the carbon cycle and the phosphorus cycle. No matter where CO_2 is released, it will mix into and maintain the concentration of CO_2 in the atmosphere. Phosphorus, however, which does not have a gas phase, is recycled only if the wastes containing it are deposited in the *ecosystem from which it came*. The same holds true for other mineral nutrients. Of course, in natural ecosystems, wastes (urine, detritus) are deposited in the same area, so that recycling occurs efficiently. As mentioned earlier, the rich growth of grasses on the plains of the Serengeti is traced to phosphorus brought and maintained there by the dense herbivore herds. Humans have been extremely prone to interrupt this cycle, however.

A very serious case of disruption of the phosphorus cycle is the cutting of tropical rain forests. This type of ecosystem is supported by a virtually 100% efficient recycling of nutrients. There are little or no reserves of nutrients in the soil. When the forest is cut and burned, the nutrients that were stored in the region's organisms and detritus are readily washed away by the heavy rains, and the land is thus rendered unproductive. Another human effect on the cycle is that much phosphate from agricultural croplands makes its way into waterways—either directly, in runoff from the croplands, or indirectly, in sewage effluents. When humans use manure, compost (rotted plant wastes), or sewage sludge on crops, lawns, or gardens, the foregoing natural cycle is duplicated. But in too many cases it is not, and some of the applied chemical fertilizers end up leaching (being carried by water seepage) into waterways. Calculations indicate that leaching and erosion have almost tripled the amount of phosphorus

▼ **FIGURE 3–17** *The phosphorus cycle.*

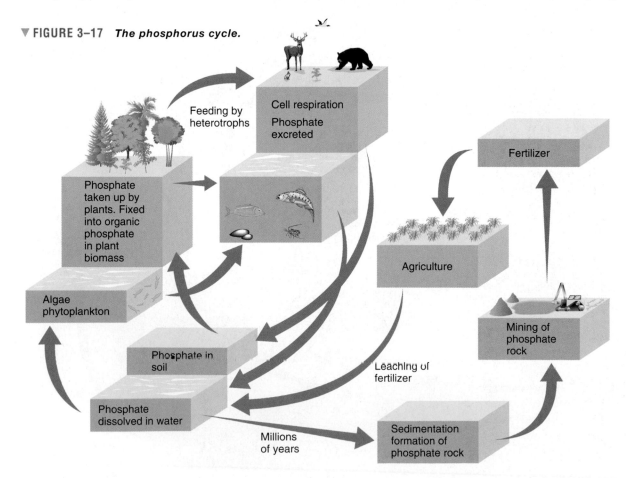

annually entering surface waters (originally 8 million metric tons per year to the present 22 million metric tons per year). Because there is essentially no return of phosphorus from water to soil, this addition results in overfertilization of bodies of water, which in turn leads to a severe pollution problem known as eutrophication. (See Chapter 18.)

The Nitrogen Cycle.   The nitrogen cycle (Fig. 3–18) is unique; it has aspects of both the carbon cycle and the phosphorus cycle. It is also unique in that many of the steps of the cycle are performed by bacteria in soils, water, and sediments. Like phosphorus, nitrogen is in high demand by plants in both aquatic and terrestrial systems and is a major limiting factor in many. The figure shows some of the details of the cycle.

The main reservoir of nitrogen is the air, which is about 78% nitrogen gas (N_2). Plants and animals cannot use nitrogen gas directly from the air. Instead, the nitrogen must be in mineral form, such as ammonium ions (NH_4^+) or nitrate ions (NO_3^-). We can begin with the uptake of nitrates by green plants, which incorporate the nitrogen contained therein into essential organic compounds such as proteins and nucleic acids. The nitrogen then follows the classic energy-flow pattern from producers to herbivores to carnivores and, finally, to decomposers (referred to as heterotrophs in the figure). At various points, nitrogen wastes are released, primarily in the form of ammonium compounds. A group of soil bacteria, the nitrifying bacteria, converts the ammonium to nitrate through oxidation, which yields energy to the bacteria. (This is a chemosynthetic process.) At this point, the nitrate is once again available for uptake by green plants—an internal ecosystem cycle within the larger cycle. Note that in most ecosystems, the supply of nitrate or ammonium nitrogen is quite limited, yet there is an abundance of nitrogen gas—if it can be accessed.

A number of bacteria and cyanobacteria (chlorophyll-containing bacteria, formerly referred to as blue-green algae) can convert nitrogen gas to the ammonium form, a process called biological **nitrogen fixation**. In terrestrial ecosystems, the most important among these nitrogen-fixing organisms is a bacterium in the genus *Rhizobium*, which lives in nodules on the roots of legumes, the plant family that includes peas and beans (Fig. 3–19). (This is another example of mutualistic symbiosis: The legume provides the bacterium with a place to live and with food (sugar) and gains a source of nitrogen in return.) From the legumes, nitrogen enters the food web. Many ecosystems, then, are "fertilized" by nitrogen-fixing organisms; legumes, with their symbiotic bacteria, are by far the most important. The legume family includes a huge diversity of plants, ranging from clovers (common in grasslands) to desert shrubs and many trees. Every major terrestrial ecosystem, from tropical

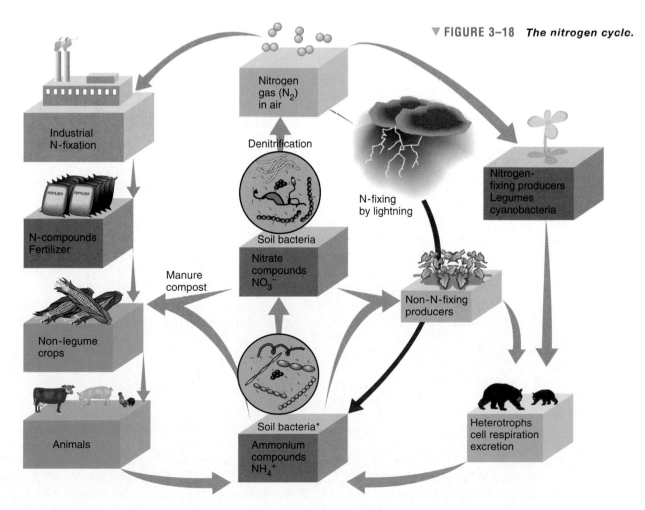

▼ **FIGURE 3–18**   *The nitrogen cycle.*

▲ **FIGURE 3–19** *Nitrogen fixation.* Bacteria in root nodules of legumes convert nitrogen gas in the atmosphere to forms that can be used by plants.

rain forest to desert and tundra, has its representative legume species, and legumes are generally the first plants to recolonize a burned-over area. Without them, all production would be sharply impaired because of a lack of available nitrogen—precluding the formation of proteins, nucleic acids, and other building blocks of life. The nitrogen cycle in aquatic ecosystems is similar. There, cyanobacteria are the most significant nitrogen fixers.

Three other important processes also "fix" nitrogen. One is the conversion of nitrogen gas to the ammonium form by discharges of lightning in a process known as *atmospheric nitrogen fixation*; the ammonium then comes down with rainfall. The second is the *industrial fixation* of nitrogen in the manufacture of fertilizer, and the third is a consequence of the *combustion of fossil fuels*, during which nitrogen from coal and oil is oxidized; some nitrogen gas is also oxidized during high-temperature combustion. Both of these processes lead to nitrogen oxides in the atmosphere, which are converted to nitric acid and then brought down to Earth as acid precipitation.

One final microbial process must be introduced: **denitrification**. This process occurs in soils and sediments, where oxygen is unavailable for normal bacterial decomposition. A number of microbes can take nitrate (which is highly oxidized) and use it as a substitute for oxygen. In so doing, the nitrogen is reduced (it gains electrons) to nitrogen gas and released back into the atmosphere; much organic matter is decomposed in this manner. Farmers seek to avoid denitrification, as it results in a loss of soil fertility. Toward that end, they plow as early as possible in the spring to restore oxygen to the soil. In sewage treatment systems, denitrification is a desirable process and is promoted in order to remove nitrogen from the wastewater before it is released to the environment (Chapter 18).

Human involvement in the nitrogen cycle is significant and a major cause for concern. Many agricultural crops are legumes (peas, beans, soybeans, alfalfa), and because they draw nitrogen from the air, they increase the normal rate of nitrogen fixation on land. Crops that are nonleguminous (corn, wheat, potatoes, cotton, etc.) are heavily fertilized with nitrogen derived from industrial fixation. And fossil-fuel combustion fixes nitrogen from the air. All told, these processes are estimated to add some 140 billion metric tons of nitrogen to terrestrial ecosystems annually. This is approximately the same amount of nitrogen fixation as that which occurs naturally. In effect, we are doubling the rate at which nitrogen is moved from the atmosphere to the land.

The consequences of this doubling for natural ecosystems are serious. Acid deposition has destroyed thousands of lakes and ponds and caused extensive damage to forests (Chapter 22). The surplus nitrogen has led to "nitrogen saturation" of many natural areas, wherein the nitrogen can no longer be incorporated into living matter and is released into the soil. There it leaches cations like calcium and magnesium from the soil and leads to mineral deficiencies in trees and other vegetation. Washed into surface waters, the nitrogen makes its way to estuaries and coastal oceans, where it promotes rich "blooms" of algae. Some of these algae are toxic to fish and shellfish; then, when the algal blooms die, they sink to deeper water or sediments, where they reduce the oxygen supply and kill bottom-dwelling organisms like crabs, oysters, and clams. These are just the observable effects of nitrogen enrichment; there may be other effects that have not yet been documented, such as a loss of biodiversity by encouraging luxuriant growth of a few dominant species.

While we have focused on the cycles of carbon, phosphorus, and nitrogen, it should be evident that cycles exist for oxygen, hydrogen, and all the other elements that play a role in living things. Also, while the routes taken by distinct elements may differ, it should be evident that all of the cycles are going on simultaneously and that all come together in the tissues of living things. And as the elements move in these cyclical patterns in ecosystems, energy flows in from the Sun and through the living members of the ecosystems. The links between these two fundamental processes of ecosystem function are shown in Fig. 3–20.

3.3 Implications for Humans

Sustainability

We have said that a major part of our purpose in studying natural ecosystems lies in the fact that they are models of sustainability. We have also said that if we can elucidate the principles that underlie their sustainability, we may be able to apply those principles toward our own efforts to achieve a sustainable society. Let us look at the first principle of sustainability: *For sustainability, ecosystems use sunlight as their source of energy.* How are we making use of solar energy?

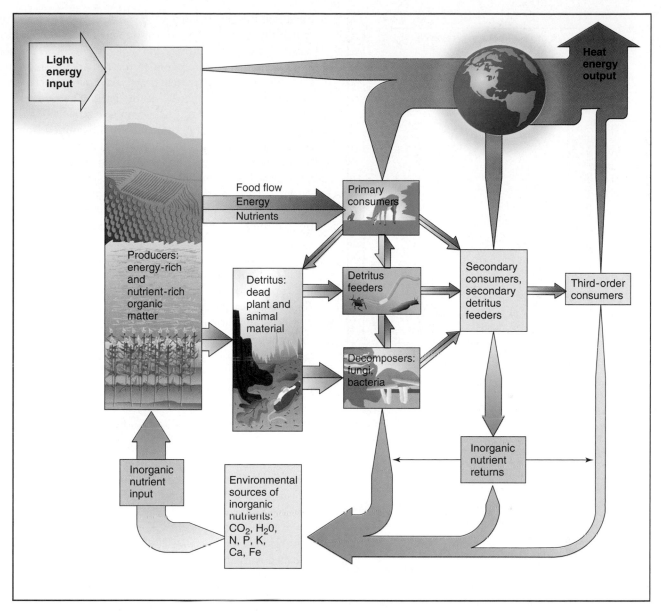

▲ **FIGURE 3–20** *Nutrient recycling and energy flow through an ecosystem.* Arranging organisms by feeding relationships and depicting the energy and nutrient inputs and outputs of each relationship show a continuous recycling of nutrients (blue) in the ecosystem, a continuous flow of energy through it (red), and a decrease in biomass (thickness of arrows).

Significance of Energy Flow. The human system makes heavy use of the energy flowing through natural and agricultural ecosystems. Agricultural systems provide most of our food. Grasslands provide animals for labor, meat, wool, leather, and milk. Forest biomes provide us with 3.3 billion cubic meters of wood annually for fuel, building material, and paper. Finally, some 15% of the world's energy consumption is derived directly from plant material.

The fact is, we have converted almost 11% of Earth's land area from forest and grassland biomes to agricultural systems. Calculations indicate that, because of our use of primary production in terrestrial systems, both natural and cultivated, we are diverting 27% of primary production to support human enterprises. And because of our conversion of many natural and agricultural lands to urban and suburban housing, highways, dumps, factories, and the like, we

cancel out an additional 13% of potential primary production. Thus, we divert 40% of the land's primary production to support human needs. In so doing, we have become the dominant biological force on Earth.

One practical application of our insights into energy flow involves the human fondness for meat. A trend that parallels increasing affluence in every country observed thus far is increasing meat consumption. Because of the principles involved in the biomass pyramid, it takes about 10 pounds of grain to grow a pound of meat [more for beef, less for chicken (Fig. 3–21)]. Therefore, for every increase in meat consumption, there is a tenfold increase in the demand on plant production and, consequently, on the land, fertilizer, pesticides, energy used, and pollution produced. Of course, the reverse is also true: Dropping down a trophic level alleviates the demand proportionally. The implication

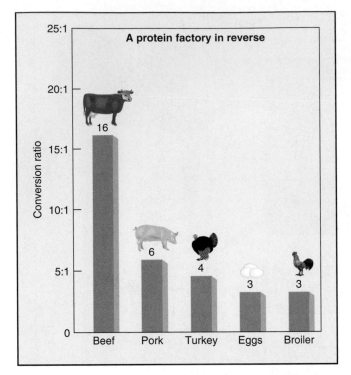

▲ FIGURE 3–21 *Conversion of grain to protein.* To obtain 1 pound of meat, poultry, or eggs, farmers must invest the number of pounds of grains and soy feeds shown. To get 1 pound of beef requires an expenditure of 16 pounds of feed. Said another way, the grain consumed to support one person eating meat could support 16 persons eating the grains directly.

of this is profound when you consider that half the cultivated acreage in the United States (roughly the size of Texas and Oklahoma) produces animal feed!

Another Energy Source.   In addition to running on solar energy, which is nonpolluting and nondepletable, we have constructed a human system that is heavily dependent on fossil fuels—coal, natural gas, and crude oil. Crude oil is the base for refinement of all liquid fuels: gasoline, diesel fuel, fuel oil, and so on. Even in the production of food, which is fundamentally supported by sunlight and photosynthesis, it is estimated that we use about 10 calories of fossil fuel for every calorie of food consumed. This additional energy is used in the course of preparing fields, fertilizing the plants, controlling pests, harvesting the food, and processing, preserving, transporting and, finally, cooking it.

The most pressing problem in connection with consuming these fuels is the limited capacity of the biosphere to absorb the waste by-products produced from burning them. Air pollution problems, including urban smog, acid rain, and the potential for global climate change, are the result of these by-products. Also, problems stemming from the depletion of fuels, particularly crude oil, are on the horizon. So you see why most people concerned about sustainability are also solar-energy advocates. Solar energy is extremely abundant. Just as important, we already have the technology to obtain much more of our energy needs from sunlight and the forces it causes, such as wind. (See Chapter 15.)

Sustainability and Nutrient Cycling.   Our second sustainability principle is very simple: *For sustainability, ecosystems dispose of wastes and replenish nutrients by recycling all elements.* In contrast to the remarkable recycling seen in natural ecosystems, we have constructed our human system, in large part, on the basis of a *one-directional flow* of elements (Fig. 3–22). We have already noted that the fertilizer–nutrient phosphate, which is mined from deposits, ends up going into waterways via land runoff and effluents from sewage treatment. The same one-way flow can be seen with such metals as aluminum, mercury, lead, and cadmium, which are the "nutrients" of our industry. At one end, they are mined from the Earth; at the other, they end up in dumps and landfills as items containing them are discarded. Is it any wonder that there are depletion problems at one end and pollution problems at the other?

The Earth has vast deposits of most minerals; however, the capacity of ecosystems (even the whole biosphere) to absorb wastes without being disturbed is comparatively limited. This limitation is aggravated by the fact that many of the products we use are nonbiodegradable. Accordingly, can you see the rationale for expanding the concept of recycling to include not just paper, bottles, and cans, but everything from sewage to industrial wastes as well?

Value

How much are natural ecosystems worth to us? Answering this question requires that we introduce a concept more fully developed in Chapter 23: **natural capital**. Use of the term "capital" suggests that there is wealth involved, which can produce income. One component of natural capital is the natural assets of a country in the form of nonrenewable minerals and renewable land, timber, rangelands, fisheries, and the like—the so-called natural resources of a country, which are exploited for meeting human needs and generating economic profit. The second component, however, is the services performed by natural systems that benefit human life and enterprises. This natural capital is represented in natural ecosystems, which provide *goods* (like lumber, fiber, and food) and vital *services* (such as waste assimilation and nutrient cycling). In a first-ever attempt of its kind, a team of 13 natural scientists and economists[1] recently collaborated to produce a report entitled "The Value of the World's Ecosystem Services and Natural Capital." Their reason for making such an effort was that the goods and services provided by natural ecosystems are not easily seen in the market (meaning the market economy that normally allows us to place value on things) and in fact may not be in the market at all. Thus, things like clean air to breathe, soil formation, the breakdown of pollutants, and the like never pass through the market economy. People are often not even aware of their importance. Because of this, these things are undervalued or not valued at all.

The team identified 17 major ecosystem goods and services that provide vital functions we depend on. The team

---

[1] Robert Costanza et al., "The value of the world's ecosystem services and natural capital," *Nature* 387 (1997): 253–260.

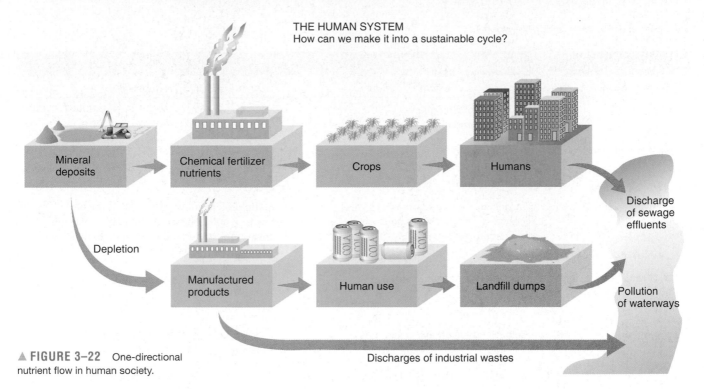

▲ **FIGURE 3–22** One-directional nutrient flow in human society.

also identified the ecosystem functions that actually carry out the vital human support and gave examples of each (Table 3–3), making the point that it is useless to consider human welfare without these ecosystem services, so in one sense, their value as a whole is infinite. However, one can calculate the **incremental value** of each type of service— how changes in the quantity or quality of various types of services may influence human welfare. For example, removing a given forest will affect the ability of the forest to provide lumber in the future, as well as perform other services, such as soil formation and promotion of the hydrologic cycle. This value can be calculated in economic terms. So, by calculating incremental values, making many approximations, and collecting information from other researchers who have worked on individual processes, the research team produced a table of the annual global value of ecosystem services performed. So calculated, the total value to human welfare of a year's services amounts to $33 trillion, and that is considered a conservative estimate! Note, though, that it is far more than the $18 trillion for the global gross national product calculated for the world economy.

The team pointed out that the real power of their analysis lies in its use for making local decisions. Thus, the value of a wetlands cannot be represented just by the amount of soybeans that could be grown on the land if it were drained; wetlands provide other vital ecosystem services, and these should be added to the value of the soybeans in calculating costs and benefits of a given proposed change in land use. The bottom line of their analysis is, in their words, "that ecosystem services provide an important portion of the total contribution to human welfare on this planet." For this reason, we need to give the natural capital stock (the ecosystems and populations in them, including the lakes and wetlands) adequate weight in public policy decisions involving changes to them. Because these services are outside the market and uncertain, they are too often ignored or under-

valued, and the net result is human changes to natural systems whose social costs far outweigh their benefits.

This remarkable analysis is a good example of *sound science* being employed to generate vital information for public policy. Our great dependence on natural systems should lead, then, to proper management of those systems.

Managing Ecosystems

Practically everything we do affects natural ecosystems. It is obvious that ecosystems must be healthy in order to provide the goods and services we deem so valuable. By using ecosystems for these goods and services, we actively "manage" them. Management simply may be a consequence of reckless exploitation (poor management!), or it may be intentional, as in the case of silviculture practice in our national forests. Ecosystems have continued on in time past without us, and if we were to pass from the scene, they would undoubtedly continue into the future. Thus, it is in our own interest to manage ecosystems well.

The **goal** of ecosystem management should be *managing ecosystems so as to assure their sustainability*, where sustainability is defined as maintaining ecosystems such that they continue to provide the same level of goods and services indefinitely into the future. This, in very simple terms, means preserving the productivity and maintaining the biodiversity of ecosystems.

How can such a goal best be accomplished? The Ecological Society of America's Committee on Ecosystem Management has identified the following necessary elements: (1) clear operational goals, (2) sound ecological models and understanding, (3) an understanding of complexity and interconnectedness, (4) recognition of the dynamic character of ecosystems, (5) attention to context and scale, (6) acknowledgment of ignorance and uncertainty, (7) commitment to adaptability and accountability, and

TABLE 3–3   Ecosystem Services and Functions

Ecosystem Service	Ecosystem Functions	Examples
Gas regulation	Regulation of atmospheric chemical composition	CO_2–O_2 balance, O_3 for UVB protection, and SO_x levels
Climate regulation	Regulation of global temperature, precipitation, and other biologically mediated climatic processes at global or local levels	Greenhouse gas regulation, dimethylsulfoxide production affecting cloud formation
Disturbance regulation	Capacitance, damping, and integrity of ecosystem response to environmental fluctuations	Storm protection, flood control, drought recovery, and other aspects of habitat response to environmental variability controlled mainly by vegetation structure
Water regulation	Regulation of hydrological flows	Provisioning of water for agricultural (such as irrigation) or industrial (such as milling) processes or transportation
Water supply	Storage and retention of water	Provisioning of water by watersheds, reservoirs, and aquifers
Erosion control and sediment retention	Retention of soil within an ecosystem	Prevention of loss of soil by wind, runoff, or other removal processes; storage of silt in lakes and wetlands
Soil formation	Soil-formation processes	Weathering of rock and the accumulation of organic material
Nutrient cycling	Storage, internal cycling, processing, and acquisition of nutrients	Nitrogen fixation, N, P, and other elemental or nutrient cycles
Waste treatment	Recovery of mobile nutrients and removal or breakdown of excess nutrients and compounds	Waste treatment, pollution control, detoxification
Pollination	Movement of floral gametes	Provisioning of pollinators for the reproduction of plant populations
Biological control	Trophic-dynamic regulations of populations	Keystone predator control of prey species, reduction of herbivory by top predators
Refugia	Habitat for resident and transient populations	Nurseries, habitat for migratory species, regional habitats for locally harvested species, or overwintering grounds
Food production	That portion of primary production extractable as food	Production of fish, game, crops, nuts, and fruits by hunting, gathering, subsistence farming, or fishing
Raw materials	That portion of primary production extractable as raw materials	The production of lumber, fuel, or fodder
Genetic resources	Sources of unique biological materials and products	Medicine, products for materials science, genes for resistance to plant pathogens and crop pests, ornamental species (pets and horticultural varieties of plants)
Recreation	Provision of opportunities for recreational activities	Ecotourism, sport fishing, and other outdoor recreational activities
Cultural	Provision of opportunities for noncommercial uses	Aesthetic, artistic, educational, spiritual, or scientific values of ecosystems

Source: Reprinted, by permission, from Robert Costanza, Ralph d'Arge, Rudolf de Groot, Stephen Farber, Monica Grasso, Bruce Hannon, Karin Limburg, Shahid Naeem, Robert V. O'Neill, Jose Paruelo, Robert G. Raskin, Paul Sutton, and Marjan van den Belt, "The value of the world's ecosystem services and natural capital," *Nature* 387 (1997): 253–260.

EARTH WATCH

BIOSPHERE 2

The proof of a theory lies in testing it. If the biosphere functions as we have described—running on solar energy and recycling all the elements from the environment through living organisms and back to the environment—then it might be possible to create an artificial biosphere that functions similarly.

The largest such experiment to date is Biosphere 2, constructed in Arizona 30 miles north of Tucson. Biosphere 2 was developed entirely with private venture capital ($200 million), with a view toward gaining information and experience that might be used in creating permanent space stations on the Moon or other planets or in long-distance space travel. In addition, it was hoped that Biosphere 2 would yield information that would further our understanding of our own biosphere—Biosphere 1.

As originally constructed, Biosphere 2 was a supersealed "greenhouse" enclosing an area of 2.5 acres (1 ha). Entry and exit were through a double air lock. Different environmental conditions within the containment supported several ecosystems. Accordingly, there were areas of tropical rain forest, savanna, desert, fresh- and saltwater marshes, and a miniocean complete with a coral reef, each stocked with respective species—over 4,000 in all. An agricultural area and living quarters for a crew of up to 10 "Biospherians" completed the arrangements.

All water, air, and nutrient recycling took place within the structure; wastes, including human and animal excrement, were treated and recycled to support the growth of plants. A crew of four men and four women began a two-year mission in September 1991. In addition to monitoring and collecting data on the natural systems, their main occupation was engaging in intensive organic agriculture to produce plant foods both for themselves and for feeding a few goats and chickens, which produced eggs, milk, and a little meat. The living quarters included the comforts and conveniences of modern living, but all communication with the outside world was via electronics. The crew's environment, including its plants and animals, was totally sealed from that of the outside world.

Not everything went perfectly; in fact, quite a few things went wrong. At one point additional oxygen had to be introduced, because oxygen was absorbed by the huge concrete structure that supports the greenhouse. The amount of nitrous oxide rose to dangerous levels and had to be controlled in order to preserve the health of the inhabitants. Nearly all the birds and animals thought to be able to tolerate the enclosure died off, with the exception of cockroaches and ants.

At the end of their two-year experimental sojourn, the "Biospherians" emerged somewhat thinner, but all in good health. Grave doubts were cast on the scientific value of the experiment, however, because of the necessary interventions and the loss of so many species. Perhaps the greatest value of the experiment was to demonstrate that even $200 million could not reproduce a functioning, self-contained ecosystem capable of supporting just eight people.

Biosphere 2 received new life when Columbia University's Lamont–Doherty Earth Observatory agreed to take it over in 1996. The facility has become an educational center for undergraduates and classroom teachers. Its ecosystems, with the capability of controlling many environmental parameters, are now being used for major research projects. Next to Earth, it is the world's largest greenhouse—a suitable place to explore the greenhouse effect. In particular, researchers are investigating the effects of increased levels of atmospheric CO_2 on the coral reef ecosystem and on the tropical rain forest. Early results indicate that such increases will greatly affect these systems—negatively in the case of the coral

reefs and only slightly positively in that of the rain forests. Highly satisfied with the results of its first three years of management, Columbia will spend $50 million in the next decade to transform Biosphere 2 and its environs into a major "west campus."

Some writers have suggested that if we trash our own planet, we may end up living in Biosphere 2–like enclosures, although the cost of escaping a polluted environment in that manner would be prohibitive. An important lesson to be learned from Biosphere 2 is that we should appreciate the operational complexity of Biosphere 1. If we fail to maintain our natural biosphere, there will be no alternative for survival.

---

(8) acknowledgment of humans as ecosystem components. It is highly significant that, since 1992, the U.S. Forest Service, the U.S. Bureau of Land Management, the U.S. Fish and Wildlife Service, the U.S. National Park Service, and the U.S. Environmental Protection Agency have all officially adopted ecosystem management as their management paradigm. This is, in the words of Jerry Franklin[2] a leading forester, "a paradigm shift of massive proportions."

It should be evident that there is no way to accomplish this kind of ecosystem management without the involvement of sound science. Every one of these agencies recognizes the need to use the best available science as a guiding policy.

[2]p. 21, Chapter 2: Ecosystem Management: An Overview by Jerry Franklin. IN: Ecosystem Management: Applications for Sustainable Forest and Wildlife Resources, Mark S. Boyce and Alan Haney. Yale Univ. Press, New Haven, 1997

This commitment of our federal agencies is in fact a commitment to stewardship, an acceptance of responsibility "for protecting the integrity of natural resources and their underlying ecosystems, and, in so doing, safeguarding the interests of future generations," as the President's Council on Sustainable Development[3] put it. However, there are many pitfalls on the way to these good intentions, and we will explore the recent record of our federal agencies in subsequent chapters as we look at how our natural resources are in fact being managed. For the moment, let us be glad that there is now clear recognition, at the highest levels of resource management, of the goal of sustainability and the ethic of stewardship as guiding principles.

There is no doubt that, on a more global scale, the future growth of population and rising consumption levels will severely challenge the goal of sustainable ecosystem management. If we consider a *50-year perspective* and the expected increase of at least 3 billion more people by that time, the needs and demands of this expanded human population will create unprecedented pressures on the ability of Earth's systems to provide goods and services. Estimates based on cur-

[3]See Bibliography, p. 631 (8e), General References, for President's Council on Sustainable Development.

rent trends indicate that impacts on the nitrogen and phosphorus cycles alone will cause global fertilization of two to three times the amounts of these nutrients presently added to the land and water. Demands on agriculture will likely require at least 10% more agricultural land; irrigated land area is expected to double, placing great stress on an already strained hydrologic cycle. Such growing demands cannot be met without facing serious trade-offs between different goods and services. More agricultural land means more food, but less forest, and therefore less of the important services performed by forests. More water for irrigation means more diversion of rivers and less that is available for domestic use and for sustaining the riverine ecosystems. The point here is that such projections are not necessarily predictions; by looking ahead, we may choose other alternatives if we clearly understand the consequences of simply continuing present practices. This is the challenge to ecosystem management.

We turn our attention in the next chapter to the question of how ecosystems are balanced and how they change over time. As we do so, we will have a chance to examine the consequences of such ecosystem behavior for management purposes and will also encounter the remaining principles of sustainability.

ENVIRONMENT ON THE WEB

HUMAN IMPACTS ON THE BIOGEOCHEMICAL CYCLES

Biogeochemical cycles describe the movement and transformation of chemical substances through the global environment. Humans impact these cycles in many ways. Sometimes the change arises because of the ways that humans have changed the face of the planet, through the building of cities and widespread agriculture. Other changes arise because of human additions to or removal of substances from the environment. In a sense, biogeochemical cycles provide simple conceptual models that help us understand the sources, pathways, and sinks of natural materials—and the implications of human intervention in those cycles. The following examples illustrate this point.

Sometimes humans impact biogeochemical cycles by changing the rates at which materials move from one stage to another. An example of this can be found in the carbon cycle. In undisturbed ecosystems, large quantities of carbon are tied up in biological tissues and are released again into the atmosphere only when the organism dies and its tissues decay. However, when human activities affect the way the system operates—for instance, through the cutting or the burning of forests—the rate at which organisms die and carbon is released into the atmosphere increases greatly.

Another example comes from the phosphorus cycle. In nature, phosphorus enters lakes and streams gradually as phosphorus is released from rock weathering and decay of biological tissue. Rock weathering in particular is a very slow process. Human activities such as phosphate extraction for fertilizer manufacture greatly increase the rate at which mineral phosphorus becomes available for biological processes. In agricultural systems, clearance of the natural vegetative cover and planting of simpler plant ecosystems contributes to the smoothness of the landscape and increases the

rate at which water and dissolved nutrients like phosphorus can move through and over soils to reach waterways.

In other cases, human activities actually add new sources to a biogeochemical cycle. The best example of this can be found in the nitrogen cycle, where human manufacture of fertilizer creates new sources of some nitrogen compounds. These fertilizers are not naturally occurring substances (as in the case of phosphate fertilizers) but rather are new products formed from raw materials containing nitrogen. So although these processes do not create new nitrogen, they do fundamentally change the proportions of nitrogen compounds in the cycle.

Web Explorations

The Environment on the Web activity for this essay describes how people use biogeochemical cycles to understand the impact of human activities on ecosystems. Go to the Environment on the Web activity (select Chapter 3 at **http://www.prenhall.com/wright**) and learn for yourself

1. about your own "carbon budget;"

2. how you can use simple knowledge of the nitrogen cycle to identify sources of nitrate pollution; and

3. about the use of the phosphorus cycle in the management of cottage development around lakes.

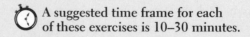 A suggested time frame for each of these exercises is 10–30 minutes.

REVIEW QUESTIONS

1. What are the six key elements in living organisms, and where does each occur in the atmosphere, hydrosphere, and lithosphere?

2. What is the "common denominator" that distinguishes between organic and inorganic molecules?

3. In one sentence, define matter and energy, and demonstrate how they are related. What are the three categories of matter?

4. Give five examples of potential energy. In each case, how can the potential energy be converted into kinetic energy?

5. State the two energy laws. How do they relate to entropy?

6. What is the chemical equation for photosynthesis? Examine the origin and destination of each molecule referred to in the equation. Do the same for cell respiration.

7. Food ingested by a consumer follows three different pathways. Describe each pathway in terms of what happens to the food involved and what products and by-products are produced in each case.

8. Compare and contrast the decomposers with other consumers in terms of matter and energy changes that they perform.

9. What factors affect the rate and amount of primary production?

10. What three factors account for decreasing biomass at higher trophic levels? That is, how do those factors account for the food pyramid?

11. What are the two principles of ecosystem sustainability stated in this chapter?

12. Describe the biogeochemical cycle for carbon as it moves into and through organisms and back to the environment. Do the same for phosphorus and nitrogen.

13. What is natural capital? What value has been assigned to the goods and services provided annually by natural capital?

14. What is the goal of ecosystem management? How do future projections make this goal difficult?

THINKING ENVIRONMENTALLY

1. Describe the consumption of fuel by a car in terms of the laws of conservation of matter and energy. That is, what are the inputs and outputs of matter and energy? (*Note:* Gasoline is an organic carbon–hydrogen compound.)

2. Relate your level of exercise—breathing hard and "working up an appetite"—to cell respiration in your body. What materials are consumed, and what products and by-products are produced?

3. Using your knowledge of photosynthesis and cell respiration, create an illustration showing the hydrogen cycle and the oxygen cycle.

4. Tundra and desert ecosystems support a much smaller biomass of animals than do tropical rain forests. Give two reasons for this fact.

5. Evaluate the sustainability of parts of the human system, such as transportation, manufacturing, agriculture, and waste disposal, by relating them to the two principles of sustainability set forth in this chapter. How can such things be modified to make them more sustainable?

WEB REFERENCES

On-line resources for this chapter can be found on the World Wide Web at: **http://www.prenhall.com/wright**. (Click on Chapter 3 within the Chapter Selector.)

Ecosystems: Populations and Succession

Key Issues and Questions

1. Population growth is the result of a balance between biotic potential and environmental resistance. What do these terms mean, and what are the basic patterns of growth of natural populations?

2. Density-dependent mechanisms help to regulate natural populations. Explain density dependence and density independence, and apply the concept of critical number.

3. Predators, parasites, and grazers are important in controlling populations. Explain how these organisms can operate in a density-dependent manner.

4. Competition between species can be an important check on populations. How is competition between plant species minimized by different adaptations and a balanced herbivory?

5. The introduction of a foreign species frequently has disruptive ecological results. Why is this the case?

6. Natural ecosystems may undergo gradual succession until they reach a climax—that is, a more stable state of ongoing adaptation. What is meant by "succession," and what factors are responsible for it?

7. Fire is a major form of disturbance to terrestrial ecosystems. Of what significance are disturbances, and what role can they play in maintaining high levels of biodiversity?

8. Although ecosystems are dynamic and viewed as nonequilibrium systems, they may still function normally even when disturbed. What is meant by ecosystem resilience?

9. What is adaptive ecosystem management, and what are some its features?

I n May and June of 1988, following an extended drought, lightning started a large number of fires in Yellowstone National Park, a 2.2-million-acre gem in the U.S. National Park System. The fires burned slowly for a while, and then, fanned by high winds behind a series of cold fronts with no rain, they broke out and swept across hundreds of thousands of acres. Prior to the 1988 fires, the Park Service policy on fires was to allow them to burn their course unless they were near human habitations. This itself was a reversal of the Smokey Bear policy of earlier years of suppressing and preventing all possible forest fires. The Yellowstone fires touched off a great political controversy over Park Service policy and led to fruitless efforts to put the fires out. In spite of the largest fire-fighting effort in U.S. history, the fires were finally put out by a snowfall, leading to a standing joke at the time: "How do you put out the Yellowstone fires? Pour a hundred million dollars on it and wait for it to snow."

The fires burned nearly 10% of the park area, leaving behind a "patchy" landscape, with heavily or lightly burned areas interspersed among areas untouched by the fires. However, within two weeks, grasses and herbaceous vegetation began sprouting from the ashes, and a year later, abundant vegetation covered the burned areas (Fig. 4–1a). Herbivores such as bison and elk fed on the lush new growth. Ten years later, lodgepole pines several feet tall carpeted much of the ground, competing with aspen sprouts (Fig. 4–1b). It is expected that some 25 years after the fires, the diversity of plants and animals in the burned areas will have completely recovered; in the meantime, the total diversity of the park has been unaffected.

◄ **Yellowstone National Park.** This scene illustrates the patchy landscape of the park as it appeared shortly after the devastating fires of 1988.

(a) (b)

▲ **FIGURE 4–1** *Recovery from fire.* In 1988, a fire swept through Yellowstone National Park. (a) One year after the fire, ground vegetation was well established. (b) Eight years after the fires, lodgepole pines were replacing the herbaceous vegetation.

It is hard to imagine a more devastating disaster to a forest ecosystem than a roaring fire. Yet, in the long run, as events at Yellowstone proved, the total landscape was recovering, populations of herbivores and predators reacted positively to the disturbance, and biodiversity was, if anything, enhanced. Many populations of plants revealed remarkable adaptations to fire, flourishing far more than in the unburned areas. The natural landscape demonstrated the capacity to recover from a profound disturbance that was looked on at the time as a disaster.

In our previous two chapters, our focus has been on the structure and function of ecosystems—viewing them as units of sustainability in the natural landscape. We considered the environmental conditions and resources that act as limiting factors in the broad distribution of ecosystems on Earth. We saw how important ecosystems are in human affairs, we sought to assign a value to the goods and services we derive from them, and we maintained that such importance should be reflected in the way in which we manage them, for manage them we must. In addition, we are recommending that ecosystems serve as models of sustainability, and we are in the process of deriving principles of sustainability extracted from our understanding of how ecosystems work.

In this chapter, we take a closer look at the populations of different species that make up the living community of ecosystems. We address the very important questions of how it is that populations are sustained over time, what kinds of interactions between populations we can expect in ecosystems, and how important these interactions are for ecosystems. In the process, we encounter questions of balance, or equilibrium, in ecosystems: What does it mean for an ecosystem to be in balance? Are populations always in a state of equilibrium? What happens when a major disturbance like fire interrupts the "balance" that we observe in an ecosystem? Can ecosystems and the natural populations in them change over time and yet maintain the important processes that make them sustainable units?

Far from retaining the notion that nature is stable and will be in balance if we leave it alone, we will see that an ecosystem is a dynamic system in which changes are constantly occurring. The notion of a balanced ecosystem, then, must be carefully defined. Our objective in this chapter is to examine more of the basic mechanisms that underlie the sustainability of all ecosystems, including human systems and the rest of the biosphere. We do so by first considering the perspective of populations and how they are controlled; then we examine ways in which changes occur over time in ecosystems and the reasons for those changes.

4.1 Population Dynamics

Each species in an ecosystem exists as a population—that is to say, a reproducing group. Studies indicate that over a long period of time, the populations of many species in an ecosystem tend to remain more or less constant in size and geographic distribution. In short, on average, deaths equal births; otherwise the population would shrink or grow accordingly. We speak of a balance between births and deaths as **population equilibrium**. Yet we know that populations are capable of growth under the right conditions. How can we relate growth to equilibrium?

Population Growth Curves

When the size of a population is plotted over time, two basic kinds of curve can be seen: *S*-curves and *J*-curves (Fig. 4–2). For example, suppose some abnormally severe years have reduced a population to a low level, but then conditions return to normal. Once normal conditions return, the population may increase exponentially for a time, but then either of two things may occur: (1) Natural mechanisms come into play and cause the population to level off and continue in a dynamic equilibrium. This pattern is known as the *S*-curve. (2) In the absence of natural enemies, the population keeps growing until it exhausts essential resources—usually food—and then dies off precipitously due to starvation and, perhaps, diseases related to malnutrition. This pattern is known as a *J*-curve.

An important characteristic of *J*-curve growth is the rapidity with which a population can go from modest levels to the peak and then crash. Suppose, for example, that an insect population is doubling (hence doubling the amount it eats) *each week*. Suppose also that it has taken the population eight weeks to devour one-half of a crop. How long will it take to devour the second half? The answer is *1 week!* In any doubling sequence, the last doubling necessarily includes one half of the total. You may test this for yourself by taking a sheet of graph paper and blackening first one square, then two, then four, then eight, and so on, doubling the number each time. You will find that regardless of the size of the paper or the size of the squares, the last turn will involve blackening half the paper. If you double the size of the paper, how many more turns will that provide?

What follows the *J*-curve crash? Any one of three scenarios may unfold. First, if the ecosystem has not been too seriously damaged, the producers may recover, allowing a recovery of the herbivore population, and the *J*-curve may be repeated. This scenario is seen in periodic outbreaks of certain pest insects, even in natural ecosystems. Second, after the initial *J*, natural mechanisms may come into the picture as the ecosystem recovers and bring the population into an *S*-balance. There is evidence that such a balance has been established in the eastern United States for the introduced gypsy moth (Fig. 4–3). Stands of oak trees that were devastated by the initial invasion of gypsy moths a few years ago have recovered, and the insect is remaining at low levels. In the third scenario, damage to the ecosystem may be so severe that recovery does not occur, but small

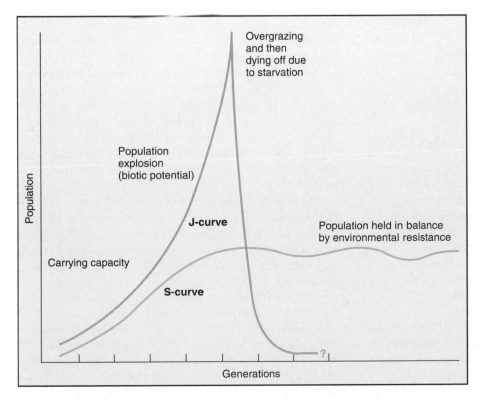

◀ **FIGURE 4–2 *Two types of growth curves.*** The *J*-curve (blue) demonstrates population growth under optimal conditions, with no restraints. The *S*-curve (green) shows a population at equilibrium.

▲ **FIGURE 4–3** *Gypsy moth caterpillar.* The gypsy moth, an introduced species that has often caused massive defoliation of trees, now seems to have been brought under natural control in forests.

surviving populations eke out an existence in a badly degraded environment.

The outstanding feature of natural ecosystems—ecosystems that are more or less undisturbed by human activities—is that they are made up of populations that are usually in the dynamic equilibrium represented by S-curves. J-curves come about when there are unusual disturbances, such as the introduction of a foreign species, the elimination of a predator, or the sudden alteration of a habitat. A J-curve is a picture of nonsustainability—a population growing out of control. Such increases are only temporary in animal populations; there is an inevitable dying off as resources are exhausted. Nevertheless, under the right conditions, population increase is always a possibility for species.

Biotic Potential versus Environmental Resistance

We refer to the ability of populations to increase as the **biotic potential**, the number of offspring (live births, eggs laid, or seeds or spores set in plants) that a species may produce under ideal conditions. Looking at different species, you can readily see that their biotic potential varies tremendously, averaging from less than one birth per year in certain mammals and birds to many millions per year in the case of many plants, invertebrates, and fish. However, to have any effect on the size of subsequent generations, the young must survive and reproduce in turn. Survival through the early growth stages to become part of the breeding population is called **recruitment**.

Examining differences in biotic potential and recruitment, we observe two different **reproductive strategies** in the natural world. The first strategy is to produce massive numbers of young, but then leave survival to the whims of nature. This strategy often results in very low recruitment.

Thus, despite a high biotic potential, a population's increase may be nil because recruitment is so low. (Note that "low recruitment" is a euphemism for high mortality of the young.) The second strategy is to have a much lower reproductive rate, but then care for and protect the young to enhance recruitment.

Additional factors that influence population growth and geographic distribution are the ability of animals to migrate, or of seeds to disperse, to similar habitats in other regions; the ability to adapt to and invade new habitats; defense mechanisms; and resistance to adverse conditions and disease.

Taking into account all these factors for population growth we find that every species has the capacity to increase its population when conditions are favorable. Furthermore, the growth of a population under absolutely ideal conditions will be *exponential*. For example, a pair of rabbits producing 20 offspring, 10 of which are female, may grow by a factor of 10 each generation: 10, 100, 1,000, 10,000 (10^1, 10^2, 10^3, 10^4), and so on. Such a series is called an **exponential increase**. When this occurs in a population, it is commonly called a **population explosion**. A basic feature of an exponential increase is that the numbers increase faster and faster as the population doubles and redoubles, with each doubling occurring in the same amount of time. If we plot numbers over time during an exponential increase, the pattern produced is our familiar J-curve (Fig. 4–2).

Population explosions are seldom seen in natural ecosystems, however, because biotic and abiotic factors tend to decrease population. Among the biotic factors are predators, parasites, competitors, and lack of food. Among the abiotic factors are unusual temperatures, moisture, light, salinity, pH, lack of nutrients, and fire. The combination of all the abiotic and biotic factors that may limit population increase is referred to as **environmental resistance**.

You may already foresee the result of the interplay between the factors promoting population growth and those leading to population decline. Conditions are always changing. When they are favorable, populations increase. When they are unfavorable, populations decrease. If we plot numbers over time in this case, the pattern is an S-curve (Fig. 4–2).

In general, the reproductive ability of a species, or its biotic potential, remains fairly constant, because that ability is part of the genetic endowment of the species. What varies substantially is recruitment. It is in the early stages of growth that individuals (plants or animals) are most vulnerable to predation, disease, lack of food (or nutrients) or water, and other adverse conditions. Consequently, environmental resistance effectively reduces recruitment. Of course, some adults also perish, particularly the old or weak. If recruitment is at the **replacement level**—that is, just enough to replace these adults—then the population will remain constant, at equilibrium. If recruitment is not sufficient to replace losses in the breeding population, then the population will decline.

ETHICS

THE DILEMMA OF ADVOCACY

What should a scientist do who has uncovered knowledge that points very clearly to the need for a change in some policy or practice involving the environment? If she published her results in a respected journal, presented them at scientific meetings, and taught them to her students, she would have met all of the traditional expectations of her university and her scientific field. The trouble is, doing so would likely have absolutely no impact on the problem she has uncovered.

An alternative course of action might be not only to do all of these things, but also to become an advocate for the policy implications of her research. This could involve contacting Congress, writing letters, working with the news media, and, in general, not being content until her message is heard by those who are in a position to do something about it and corrective actions are taken. Increasingly, some scientists are becoming advocates as they work on serious problems like the loss of biodiversity, declin-

ing ecosystems, and changing biogeochemical cycles and see how things might be different. These scientists are taking a number of risks: (1) They may get criticism from some of their colleagues who feel that it is right to do the research and point to the problems, but not to cross a line that separates science from social activism. (2) Their actions may blur the distinction between environmental scientists and environmental activists, and thus they may be perceived by policymakers and the public as just another group with an agenda to push. (3) They may be tempted to step out beyond the immediate implications of their work and make projections or predictions of disaster that never come true, thus harming the credibility of sound science.

More and more leading scientists, however, are calling for a closer engagement of science with societal problems. The issues, they argue, are so important that to ignore them is to renege on the traditional social contract between science and the public:

The public supports scientific research in exchange for the knowledge science uncovers and the expectation that scientists will produce something useful from their work. Thus, the argument goes, in view of the serious nature of our impacts on the environment, more scientists should be working on issues such as the depletion of resources, endangered species, climate change, and the like. And when they do, they have a deep responsibility to see that their findings are clearly communicated to the public and to policymakers and resource managers. However, they should do so carefully, making sure that their listeners know when they are speaking as a scientist and when they are speaking as a private citizen taking a stand because of their values. Some scientists are getting used to wearing two hats: one as a scientist and one as an activist. Others, however, warn of a "creeping advocacy syndrome" that threatens to ruin the objectivity that is so important to sound science. What do you think?

However, there is a definite upper limit to the population of any particular animal that an ecosystem can support. This limit is known as the *carrying capacity* (Fig. 4–2). More precisely, **carrying capacity** is defined as the maximum population of an animal that a given habitat will support without the habitat being degraded over the long term—in other words, a sustainable system. If a population greatly exceeds the habitat's carrying capacity, it will undergo a *J*-curve crash, as Fig. 4–2 shows. Because conditions within habitats also undergo change from year to year, the habitat's carrying capacity will vary accordingly.

In certain situations, environmental resistance may affect reproduction, as well as causing mortality directly. For example, the loss of suitable habitat often prevents animals from breeding. Also, certain pollutants adversely affect reproduction. However, we can still view these situations as environmental resistance that either blocks a population's growth or causes its decline.

In sum, whether a population grows, remains stable, or decreases is the result of an *interplay between its biotic potential and environmental resistance* (Fig. 4–4). In general, a population's biotic potential remains constant; it is changes in environmental resistance that allow populations to increase or cause them to decrease. For example, a number of favorable years (low environmental resistance) will

allow a population to increase; then a drought or other unfavorable conditions may cause it to die back, and the cycle may be repeated.

We emphasize that population balance is a **dynamic balance**, which implies that additions (births) and subtractions (deaths) are occurring continually and the population may fluctuate around a median (Fig. 4–2). Some populations fluctuate very little; others fluctuate widely, but as long as decreased populations restore their numbers and the ecosystem's carrying capacity is not exceeded, the population is considered to be at an equilibrium. Still, the questions remain: What maintains the equilibrium within a certain range? What prevents a population from "exploding" or, conversely, becoming extinct?

Density Dependence and Critical Numbers

In general, the size of a population remains within a certain range because most factors of environmental resistance are *density dependent*. That is, as **population density** (the number of individuals per unit area) increases, environmental resistance becomes more intense and causes such an increase in mortality that population growth ceases or declines. Conversely, as population density decreases, environmental resistance is generally less-

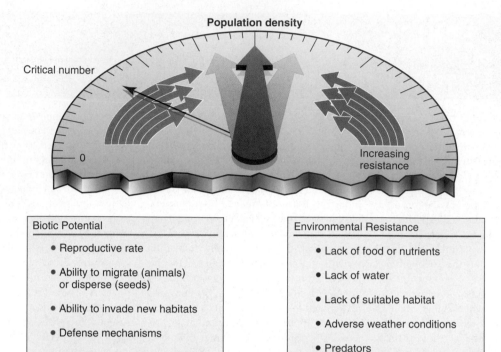

Population density

Critical number

0

Increasing resistance

Biotic Potential	Environmental Resistance
• Reproductive rate	• Lack of food or nutrients
• Ability to migrate (animals) or disperse (seeds)	• Lack of water
• Ability to invade new habitats	• Lack of suitable habitat
• Defense mechanisms	• Adverse weather conditions
• Ability to cope with adverse conditions	• Predators
	• Disease
	• Parasites
	• Competitors

▲ **FIGURE 4–4** *Biotic potential and environmental resistance.* A stable population in nature is the result of the interaction between factors tending to increase population (biotic potential) and factors tending to decrease population (environmental resistance).

ened, allowing the population to recover. This balancing act will become clearer as we discuss specific mechanisms of maintaining a population at equilibrium in the next section. Factors in the environment that cause mortality can also be *density independent*. That is, their effect is independent of the density of the population. A sudden deep freeze in spring, for example, can cause high mortality in early germinating plants, independently of their density. Still, although density-independent factors can be important sources of mortality, they are not responsive to population density and thus are not involved in maintaining population equilibria.

There are no guarantees that a population *will* recover from low numbers. Extinctions can and do occur in nature. For example, where are the dinosaurs? The survival and recovery of a population depends on a certain minimum population base, which is referred to as the population's **critical number**. You can see the idea of critical number at work in terms of a herd of deer, a pack of wolves, a flock of birds, or a school of fish. Often, the group is necessary to provide protection and support for its members. In some cases, the critical number is larger than a single pack or flock, because interactions between groups may be necessary as well. In any case, if a population is depleted below the critical number needed to provide such supporting interactions, the sur-

viving members actually become more vulnerable, breeding fails, and extinction is virtually inevitable. Thus, we must recognize that the density-dependent decline and recovery of a population occurs well above the critical number for the population.

As we will see subsequently, human activities are responsible for the decline and even extinction of many plants and animals (Chapter 11). Why is this? It is simply that human impacts, such as altering habitats, pollution, hunting, and other forms of exploitation are not density dependent; they can even intensify as populations decline toward extinction. Concern for these declining populations eventually led to the Endangered Species Act, which calls for the recovery of two categories of species. Species whose populations are declining rapidly are classified as **threatened**. If the population is near what scientists believe to be its critical number, the species may be classified as **endangered**. These definitions, when officially assigned by the U.S. Fish and Wildlife Service, set into motion a number of actions aimed at stemming the negative impacts on, and providing protection for and even artificial breeding of, the species in question. (See Chapter 11.) Nongovernmental organizations, such as the World Wildlife Fund and the National Wildlife Federation, are also playing a great role in protecting threatened and endangered species.

EARTH WATCH

AN ENDANGERED ECOSYSTEMS ACT?

The Endangered Species Act is a formal recognition of the importance of saving species that are in serious trouble. The Act has enjoyed some successes (Chapter 11), but is considered seriously flawed by many environmentalists because it does not directly address the major reason species become endangered, namely, the loss of crucial ecosystems that are their habitats. Thus, a level of diversity higher than biodiversity is also important to maintain: ecosystem diversity.

Ecosystem diversity posits that there should be a variety of ecosystems in a given area, be it local, national, or global. Protecting many different types of ecosystem consequently means protecting many endangered species and preventing the bulk of other species from becoming endangered. But maintaining the goods and services provided to us by ecosystems requires going beyond preserving basic *types* of ecosystems to protecting landscapes and regions of ecosystems large enough to sustain the functions we value. Thus, we are talking about the qualitative as well as quantitative preservation of ecosystems.

Is there a need for this kind of preservation in the United States? Wildlife ecologists Reed Noss and Michael Scott performed an extensive review of ecosystem declines and presented their results in terms of critically endangered, endangered, and threatened ecosystems of the United States.[1] They considered critically endangered ecosystems to be those which have suffered greater than a 98% decline in the area formerly occupied. Twenty-eight ecosystems fit this description, including such ecosystems as native grasslands in California, spruce–fir forest in the southern Appalachians, and tallgrass prairie east of the Missouri River. In addition, over 40 ecosystems were considered endangered (an 85–98% decline), including all tallgrass prairies, coastal redwood forests, and large streams and rivers in all major regions of the United States. Many more made the category of threatened ecosystems (a 70–84% decline).

What has happened to these ecosystems? In large measure, they have been altered considerably by urban development, agriculture, exploitation of their resources, and pollution. Even though these transformations have served many legitimate human needs, Noss and Scott maintain that much could have been done to reduce the damage. Further, there are pressures—largely economic—to continue the pattern of transformation for many of the endangered ecosystems. In response to such real and continuing losses, many scientists and environmentalists have called for a national Endangered Ecosystems Act.

Such an act would involve taking steps similar to those taken to protect endangered species—that is, taking an inventory of the status of ecosystems to identify those in trouble, protecting endangered ecosystems from further activities that would damage them, establishing recovery plans for many of the most critical ecosystems, and promoting research and monitoring to maintain surveillance and to gain knowledge for better management and protection. What do you think about this suggestion? Moreover, what could be done to make it less controversial than the Endangered Species Act?

[1]Reed F. Noss and J. Michael Scott, "Ecosystem Protection and Restoration: The Core of Ecosystem Management," Ch. 12 in Ecosystem Management: Applications for Sustainable Forest and Wildlife Resources, eds. Mark S. Boyce and Alan Haney (New Haven, CT: Yale University Press, 1997.) See also Robert L. Peters and Reed F. Noss, "America's Endangered Ecosystems," Defenders Magazine, Fall 1995.

4.2 Mechanisms of Population Equilibrium

With our general understanding of population equilibrium as a dynamic interplay between biotic potential and environmental resistance, we now turn our attention to some specific kinds of population interactions. (Keep in mind, however, that in the natural world, a population is subjected to the total array of all the biotic and abiotic environmental factors around it. Single factors are seldom totally responsible for the regulation of a given population; rather, regulation results from many factors acting together.)

Predator–Prey Dynamics

Predation on Animals.   An important mechanism of population control is regulation of a population by a predator—that is, a *predator–prey* relationship. A well-documented example is the interaction between wolves and moose on Isle Royale, a 45-mile-long island in Lake Superior that is now a national park.

During a hard winter early in this century, a small group of moose crossed the ice to the island and stayed. Their population grew considerably in the absence of predators. Then, in 1949, a pack of wolves also managed to reach the island. Nine years later, in 1958, wildlife biologists began carefully tracking the populations of the two species (Fig. 4–5). As seen in the figure, a rise in the moose population is followed by a rise in the wolf population, and a decline in the moose population is followed by a decline in the wolf population. The data can be interpreted as follows: A paucity of wolves represents low environmental resistance for the moose, so the moose population increases. Then, the abundance of moose represents optimal conditions (low environmental resistance) for the wolves, so the wolf population increases. The growing wolf population means higher predation on the moose, so the moose population falls. The decline in the moose population is followed by a decline in the wolf population, because now there are fewer prey. We can see how this cycle can be repeated indefinitely, providing a dynamic balance between the populations of moose and wolves.

The most recent data from Isle Royale show another cycle in progress, but the wolf population has not increased

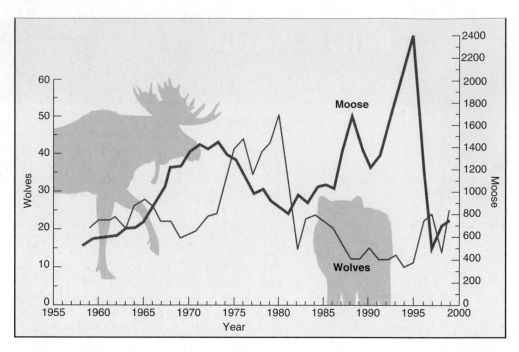

▶ **FIGURE 4–5**
Predator–prey relationship.
Wolves and moose populations on
Isle Royale from 1955 to 1996.

as rapidly as expected. Apparently, the wolves suffered a decline in the 1980s due to a canine virus introduced by human visitors and their dogs. Also, the most recent dramatic fall in the moose population cannot be attributed entirely to predation by the small number of wolves. Deep snow and an increase in ticks are thought to have caused additional mortality. The sharp decline of moose is thought to be responsible for keeping the wolf population low. In both predator and prey species, other factors may come into play to influence the observed fluctuations in population densities. For example, the shortage of vegetation that occurs as a herbivore population grows may stress the animals, especially the old, sick, and young, and make them more vulnerable to predators, parasites, and disease. It has been observed, for instance, that wolves are incapable of bringing down an adult moose in good physical condition. The animals they kill are the young and those weakened by another factor—which is not the case when the predators are human hunters.

The observation that predators are often incapable of killing individuals of their prey that are mature and in good physical condition is extremely significant. This is what prevents predators from eliminating their prey. As the prey population is culled down to those healthy individuals that can escape attack, the predator population will necessarily decline, unless it can switch to other prey. Meanwhile, the survivors of the prey population are healthy and can readily reproduce the next generation.

Parasites.   Much more abundant and equally important as predators in population control is a huge diversity of *parasitic organisms*. Recall from Chapter 2 that these organisms range from tapeworms, which may be a foot or more in length, to microscopic disease-causing bacteria, viruses, protozoans, and fungi. All species of plants and animals (in-

cluding the predators), and even microbes themselves, may be infected with parasites.

In terms of population impacts, parasitic organisms often act in the same way as predators do—in a density-dependent manner. As the population density of the host organism increases, parasites and their *vectors* (agents that carry the parasites from one host to another), such as disease-carrying insects, have little trouble finding new hosts, and infection rates increase, causing higher mortality. Conversely, when the population density of the host is low, the transfer of infection is less efficient, and there is a great reduction in levels of infection, a condition that allows the host population to recover.

You can readily see how a parasite can work in conjunction with a predator to control a given herbivore population: Parasitic infection breaks out in a dense population. Individuals weakened by infection are more easily removed by predators, leaving a smaller, but healthier, population.

The wide swings observed in the populations of moose and wolves on Isle Royale are seen to occur in very simple ecosystems involving relatively few species. However, most food webs are more complicated than that, and a population of any given organism is affected by a number of predators and parasites simultaneously. As a result, the population density of a species can be thought of more broadly as a consequence of the relationships the species has with its food sources and all of its **natural enemies**. Relationships between a prey population and its natural enemies are generally much more stable and less prone to wide fluctuations than when only a single predator or parasite is involved, because different predators or parasites come into play at different population densities. Also, when the preferred prey is at a low density, the population of the predator may be supported by switching to something else. Thus, the lag time between an increase in the prey population and that of the predator is diminished. These factors have a great damping effect on the rise and fall of the prey population.

Plant–Herbivore Dynamics. As we have just seen, herbivore populations are commonly held in check by various natural enemies. Since we have defined predation (in Chapter 2) as any situation wherein one kind of organism feeds on another, herbivores represent the natural enemies of plants. Just as too many wolves can bring the moose population dangerously low, too many herbivores can do the same to their plant food.

In a grazing situation, it is readily apparent that if the herbivores eat the plants faster than they can grow, sooner or later the plants will be depleted, and the animals will suffer. This situation is known as **overgrazing**. The best way to appreciate the potential for herbivore overgrazing is to observe what occurs when the herbivore's natural enemies are *not* present. A classic example with good documentation is the case of reindeer on St. Matthew Island, a 128-square-mile island in the Bering Sea midway between Alaska and Russia. In 1944 a herd of 29 reindeer (5 males and 24 females) was introduced onto the island, where they had no predators. From these 29 animals, the herd multiplied some two-hundredfold over the next 19 years. Early in the cycle, the animals were observed to be healthy and well nourished, as supporting vegetation was abundant. However, by 1963, when the size of the herd had reached an estimated 6,000, the animals were obviously malnourished. Lichens, an important winter food source, had been virtually eliminated and replaced by unpalatable sedges and grasses. During the winter of 1963–64, this factor, combined with harsh weather, resulted in death by starvation of nearly the entire herd; there were only 42 surviving animals in 1966 (Fig. 4–6)

The lesson is that no population can escape ultimate limitation by environmental resistance. But the form of environmental resistance and the consequences may differ. If a population is not held in check, it may explode, over-graze, and then crash as a result of starvation. It is also crucial to note that the consequence is not just to the herbivore in question. One or more types of vegetation may be eliminated and replaced by other forms or not replaced at all, leaving behind a seriously degraded ecosystem. Other herbivores that were dependent on the original vegetation, and secondary and higher levels of consumers dependent on them, also are eliminated as food chains are severed. For example, innumerable extinctions have occurred among the unique flora and fauna of islands because goats were introduced by sailors to create a convenient food supply for return trips.

Eliminating predators or other natural enemies upsets basic plant–herbivore relationships in the same way as introducing an animal without natural enemies does. Examples of this type of folly abound as well. For example, in much of the United States, deer populations were originally controlled by wolves, mountain lions, and bear, most of which were killed because they were felt to be a threat to livestock and even humans. Now, if it were not for human hunting in place of these natural predators, deer populations in most areas would increase to the point of overgrazing. Indeed, drastic population increases do occur where hunting is prevented. As another example, sea urchins can be a problem in coastal marine systems of eastern Canada as a result of heavy exploitation of lobsters, which prey on the urchins. Sea urchin populations increase and graze down the seaweeds of the subtidal rocky coasts, creating large areas of bare rock incapable of supporting the complex subtidal community ordinarily found there.

A second factor influencing plant–herbivore balance is that large herbivores—bison in the American West or elephants in Africa, for example—were originally able to roam vast regions. As forage was reduced in one area, a herd would simply migrate before overgrazing could occur, as with the large Serengeti herbivore populations (Chapter 3). Because humans have fenced such regions for agriculture

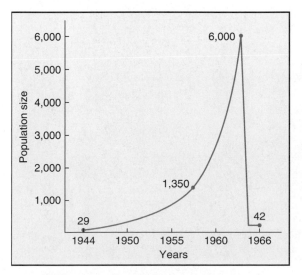

▲ **FIGURE 4–6** *Plant–herbivore interaction.* In 1944, a population of 29 reindeer was introduced onto St. Matthew Island, where they increased exponentially to about 6,000 and then died off due to overgrazing.

and cattle ranching, however, wild herbivores are increasingly confined to areas such as parks and reserves. Overgrazing in these presumably "protected" areas has become an increasing danger.

We see, then, that in undisturbed ecosystems, herbivore populations are held in check by a number of important interactions—especially by predators and parasites. As a consequence, herbivores rarely increase to populations numerous enough to overgraze their food. The result is that primary producers (plants) are able to maintain a substantial standing biomass and sustain their production of the organic matter that is so important to the entire ecosystem. We can readily see that this is another feature that is fundamental to sustainability. Hence, here is the **third basic principle of ecosystem sustainability**:

> *The size of consumer populations in ecosystems is maintained such that overgrazing and other forms of overuse do not occur.*

Competition

Competition between Plant Species. In Chapter 2, we discussed the significance of competition between species and introduced the concept of the ecological niche. If two species are competing for some scarce resource, we say that their niches *overlap*. One outcome of such competition may be the elimination of one of the species, called *competitive exclusion*. However, competitive exclusion requires a stable and homogeneous environment and may take decades or longer. Consider the situation for plants. A natural ecosystem may contain hundreds or even thousands of species of green plants, all competing for nutrients, water, and light. What prevents one plant species from driving out others in competition—again, an event that may occur in the case of introduced species? What role could such **interspecific competition** (competition between members of different species) play in maintaining population equilibria?

First, in Chapter 2 we observed that because of differences in topography, soil type, and so on, the landscape is far from uniform, even within a single ecosystem. Actually, it is composed of numerous microclimates or microhabitats. That is to say, the specific conditions of moisture, temperature, light, and so on differ from location to location. Thus, the adaptation of a species to specific conditions enables it to thrive and overcome its competitors in one location, but not in another. An example is the distribution of trees along streams and rivers. In the Great Plains states, trees grow only along waterways, because elsewhere the environment is too dry. This creates what are called **riparian** woodlands (Fig. 4–7). In the eastern United States, sycamore and red maple, which can thrive in water-saturated soil, grow along riverbanks. Oaks, which require well-drained soil, occupy higher elevations. In the West, white alder, willow, and cottonwoods can survive in water-saturated soils.

A second factor affecting the competition between plant species is the fact that a single species generally cannot utilize all of the resources in a given area. Therefore,

▲ **FIGURE 4–7** *Riparian woodlands.* Competition between plant communities is often maintained by differing amounts of available moisture. Riparian woodlands are shown growing only along a river in this prairie region.

any resources that remain may be gathered by other species having different adaptations (Chapter 2). For example, grasslands contain both grasses that have a fibrous root system and plants that have taproots (Fig. 4–8). These different root systems enable the plants to coexist because they get their water and nutrients from different layers of the soil. Also, trees in a forest, while competing with each

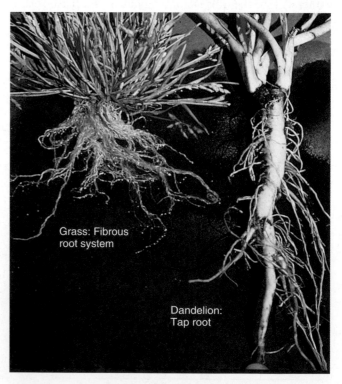

Grass: Fibrous root system

Dandelion: Tap root

▲ **FIGURE 4–8** *Coexistence in plants.* Plants with fibrous roots may coexist with plants having taproots because each is drawing water and nutrients from a different part of the soil.

other for light in the canopy (the "layer" of treetops), leave lots of space near the ground, and this space may be occupied by plants (ferns and mosses, for example) that can tolerate the reduced light intensity. Another adaptation is the plethora of spring wildflowers that inhabit temperate deciduous forests. Sprouting from perennial roots or bulbs in the early part of that season, these plants take advantage of the light that can reach the forest floor before the trees grow leaves.

Another option is a form of mutualistic symbiosis. For example, in warm, humid climates, the branches of trees are often covered with **epiphytes**, or air plants (Fig. 4–9). Such plants are not parasitic. There is some evidence that the epiphytes help to gather the minute amounts of nutrients that come with rainfall and make them accessible to the tree on which the epiphytes are located.

A third and very important factor in multiple-plant balance is called *balanced herbivory*. It is easiest to understand if we start from the point of view of a **monoculture**— the growth of a single species over a wide area, a practice commonly followed in agriculture and forestry for economic efficiency.

Experience shows that monocultures are exceedingly vulnerable to insects, fungal diseases, or other pests, while diverse ecosystems are much more resistant to them. To understand why this should be, consider the following: First, insects, fungal diseases, and other parasites are, for the most part, **host specific**. That is, they will attack only one species and, possibly, its close relatives. They are unable to attack species unrelated to their specific host. Second, such organisms have an enormous biotic potential. An individual often produces thousands of offspring—even millions, in the case of fungal spores—and they have a generation time of only a few days or weeks.

Now, a monoculture may be seen as a continuous, lush food supply for its particular host-specific attacker, a situation highly conducive to supporting a population explosion. Indeed, the pest population may explode so fast that its natural enemies cannot keep up with it even if they are present. Only virtual elimination of the monoculture halts the multiplication of the pest—a scenario not unlike that of the reindeer described earlier. [It is for this reason that many farmers and forest managers use chemical pest controls despite recognizing that they may present certain environmental hazards.(See Chapter 17.)]

On the other hand, in a diverse ecosystem—one consisting of a mixture of many different species of plants growing together—the host-specific attacker has trouble reaching its next host. With this limitation, most of the pest's offspring may perish, and the surviving population may be held in check by its natural enemies.

To conclude, visualize a monoculture developing in a natural situation. Its being largely wiped out by an outbreak of its host-specific pest or herbivore would leave space that might be invaded by another plant species, which in turn might be largely wiped out by an outbreak of *its* pest, leaving space that might be occupied by a

▲ **FIGURE 4–9** *Epiphytes in the Amazon region.* The plants (bromeliads) growing on the tree branches are epiphytes. They are not parasitic, but perch on the branches of trees to gain access to light.

third plant species, and so on. The end result of this process would be a diversified plant community, with each species held down in density by its specific herbivore(s) and the herbivores held in check by their natural enemies (Fig. 4–10a). Thus, a **balanced herbivory** may be defined as a balance among competing plant populations being maintained by herbivores feeding on the respective populations.

A prime example of a balanced herbivory is seen in the tropical rain forests of the Amazon River basin in Brazil and Peru (Fig. 4–10b). A single acre may contain a hundred or more species of trees, but often no more than a single individual of each. The next individual of the same species may be as much as a half mile away. Evidence that this diversity is maintained by a balanced herbivory is seen in that attempts to create plantations of single species—rubber plantations in South America, for example—met with failure because outbreaks of various pests proved uncontrollable in the monoculture situation. Yet rubber trees thrive in the natural rain forests of South America.

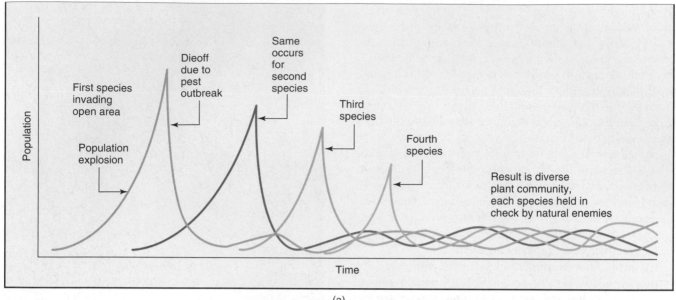

(a)

▶ **FIGURE 4–10 *Balanced herbivory.*** (a) A plant species may experience a population explosion as it invades an open area. If it dies back and is held down by a herbivore, space is opened up for a second invader, which may experience the same fate, etc. The result is a plant community of many species, each held in check by its specific herbivore. (b) View from canopy walkway overlooking the Amazonian rain forest near Iquitos, Peru. Nearly every tree in view is a different species.

(b)

Territoriality.   In discussing predator–prey balances, we said that in lean times the carnivore population—the wolf, for instance—had to switch to other prey or starve. Actually, another factor is often involved in the control of carnivore and some herbivore populations of a great number of species, ranging from fish to birds and mammals: *territoriality*. **Territoriality** refers to individuals or groups such as a pack of wolves defending a territory against the encroachment of others of the *same species*. This is **intraspecific competition**—competition between members of the same species. For example, the males of many species of songbirds claim a territory at the time of nesting. Their song has the function of warning other males to keep away (Fig. 4–11). The males of many carnivorous mammals, including dogs, "stake out" a territory by marking it with urine, the smell of which warns others to stay away. If others encroach, there may be a fight, but in most species a large part of the battle is intimidation—an actual fight rarely results in death.

In territoriality, what is really being protected by the defender or sought after by the "invader" is the claim to an area from which adequate food resources can be obtained in order to rear a brood successfully. Hence, the territory is defended only against others that would cause the most direct competition for those resources—usually, members of the same species. As a consequence of territoriality, some members of the population are able to gain access to sufficient food resources to rear a well-fed next generation. Thus, a healthy population of the species survives. If, instead, there were an even rationing of inadequate resources to all the members, with all of them trying to raise broods, the entire population would become malnourished and might perish. Through territoriality, breeding is restricted to only those individuals capable of claiming and defending territory, and thus, population growth is curtailed.

Individuals unable to claim a territory are in large part the young of the previous generation(s). Some may hang out

▲ **FIGURE 4–11** *Territoriality.* A red-winged blackbird is announcing its claim on a territory.

on the fringes and seize their opportunity as they become more mature and older members with territories weaken. Some fall prey to one or another factors of environmental resistance as they are continually chased out of one territory after another. Finally, some may be driven to disperse. Of course, it is an open question whether such migration will lead them to another region where they can successfully breed or whether it will lead them to perish in conditions beyond their limit of tolerance along the way. In any case, territoriality may be seen as a powerful force behind the dispersal, as well as the stabilization, of populations.

Introduced Species

Another way of seeing how important predation and competition are in maintaining population balances is to observe what happens when species from foreign ecosystems are introduced. Over the past 500 years, and especially now that there is a vast global commerce, thousands of species of plants, animals, and microbes have been accidentally or de-liberately introduced to new continents and islands. A recent attempt to estimate the economic losses due to such introductions in the United States calculated the total cost in excess of $138 billion per year. A couple of examples suffice to make the point:

In 1859, rabbits were introduced to Australia from England to be used for sport shooting. The Australian environment proved favorable to the rabbits and contained no carnivore or other natural enemies capable of controlling them. The result was that the rabbit population exploded and devastated vast areas of rangeland by overgrazing (Fig. 4–12). The devastation was extremely damaging to both native kangaroos and ranchers' sheep. It was finally brought under control by the introduction of a disease-causing virus in rabbits. Over time, however, the rabbits adapted to the virus and began to repeat their explosive growth. After the recent release of a different virus, rabbits have undergone a 95% decline in some regions, and kangaroos and rare plants are thriving once again.

Prior to 1900, the dominant tree in the eastern deciduous forests of the United States was the American chestnut, which was highly valued for both its high-quality wood and its prolific production of chestnuts, eaten by wildlife as well as people. In 1904, however, a fungal disease called the chestnut blight was accidentally introduced when some Chinese chestnut trees carrying the disease were planted in New York. The fungus spread through the forests, killing nearly every American chestnut tree by 1950. Although oaks filled in where the chestnuts died, the ecological and commercial loss was incalculable. There is hope, however: Researchers recently crossbred the American and Chinese chestnut, resulting in a hybrid that is 94% native and resistant to the blight. It will be several decades, though, before enough of the hybrids are available for sale to nurseries and wider distribution.

Most of the important insect pests in croplands and forests—Japanese beetles, fire ants, and gypsy moths, for example—are species introduced from other ecosystems. Do-

◀ **FIGURE 4–12** *Rabbits overgrazing in Australia.* On one side of a rabbitproof fence, there is lush pasture; on the other side, the land is barren.

mestic cats introduced into island ecosystems have often proved to be effective predators and have exterminated many species of wildlife unique to the islands. They are also responsible for greatly diminished songbird populations in urban and suburban areas, including parks (Chapter 11). Because of their voracious appetites, goats introduced onto islands have been devastating to both native plants and animals dependent on that vegetation.

Such unfortunate introductions are not all in the past. Indeed, with expanding world trade and travel, the problem is increasing. In the 1980s, the zebra mussel and the closely related quagga mussel were introduced into the Great Lakes with the discharge of ballast water from European ships (Fig. 4–13a). The mussels are now spreading through the Mississippi River basin and may go much farther, causing untold ecological and commercial damage as they displace native mussel species and clog water-intake pipes. Of course, such problems may be exported as well as imported. In 1982, several species of jellyfishlike animals known as ctenophores were similarly transported from the east coast of the United States to the Black Sea and the Sea of Azov in Eastern Europe. The ctenophores have cost Black Sea fisheries an estimated $250 million and have totally shut down those in the Sea of Azov, for they both kill larval fish directly and deprive larger fish of food (Fig. 4–13b).

Plant species have also been moved all over the world, often by horticulturists. In some cases, they have raised havoc. For example, in 1884, the water hyacinth, a plant originally from South and Central America, was introduced into Florida as an ornamental flower. It soon escaped into waterways, where it had little competition and few natural enemies. It eventually proliferated to the extent of making navigation difficult or impossible on some waters, covering lakes and ponds with up to 200 tons of hyacinths per acre (Fig. 4–14a). With a combination of herbicides, harvesting, and biological control by several insect species, Florida's water hyacinth problem is now under "maintenance control"; the hyacinth is still present, but constant surveillance

keeps it from proliferating. Sadly, this same species has now spread throughout the major rivers and lakes of Africa, where it is presently causing enormous economic and biological damage.

Kudzu, a vigorous vine introduced from Japan in 1876, was widely planted on farms throughout the southeastern United States, with the idea of using it for cattle fodder and also for erosion control. From wherever it is planted, however, kudzu invades and climbs over adjacent forests. The species now occupies 7 million acres of the deep South (Fig. 4–14b). Because it is a threat to forest growth, considerable efforts are now being exerted by the Forest Service to contain it.

Spotted knapweed (Fig. 4–14c), which was probably unwittingly introduced into this country with alfalfa seed imported from Europe in the 1920s, has spread over millions of acres of rangelands in the northwestern United States and southwestern Canada. Bitter tasting and virtually inedible, knapweed is endangering wildlife such as elk and rendering lands worthless for grazing by domestic cattle as it displaces native plants and grasses. Knapweed has also been found to promote greater soil erosion where it is abundant. To date, control efforts have been unable to check the spread of this invader.

Wetlands throughout temperate regions of the United States and Canada, already reduced by over 50% by development pressures, are now being further degraded by the invasion of purple loosestrife, which was introduced from Europe in the 1800s as an ornamental and medicinal plant (Fig. 4–14d). By outcompeting and eliminating native wetland vegetation and being inedible itself, purple loosestrife is threatening many wildlife species, including waterfowl. Control of such weeds by the introduction of plant-eating insects is being investigated. However, it is hard to find an insect (or other organism) that will control the target weed but not attack desired species. (See Chapter 17.)

The ecological lesson to be learned from our experience of unfortunate introductions is twofold. First, we

(a)   (b)

▲ **FIGURE 4–13** *Introduced species.* (a) Zebra mussels introduced from Europe are now proliferating throughout the Great Lakes and the Mississippi valley. (b) Ctenophores originating on the South Atlantic Coast of the United States and introduced into Europe have destroyed fishing in the Black Sea and the Sea of Azov.

should emphasize again that the regulation of populations is a matter of complex interactions among the members of the biotic community. Second, and just as important, the relationships are specific to the organisms in each particular ecosystem. Therefore, when we transport a species over a physical barrier from one ecosystem to another, it is unlikely to fit into the framework of relationships in the new biotic community. In most cases, it finds the environmental resistance of the new system too severe and dies out. No harm is then done: The system that received the "invader" continues unaltered. In some instances, the introduced species simply joins the native flora or fauna and does no measurable harm (as in the case of the ring-necked pheasant, a valuable game bird). In the worst case, the transported species finds physical conditions and a food supply that are hospitable, together with an insufficient number of natural enemies to stop its population growth. Then its population explodes, and it drives out native species by outcompeting them for space, food, or other resources if not by predation. Such disruptions may be caused by any category of organism—plant, herbivore, carnivore, or parasite—large or small.

You may be wondering why the interactions among species in different ecosystems should differ. It is because ecosystems on different continents or remote islands have been isolated by physical barriers for millions of years. Consequently, the species within each system have developed adaptations to other species *within their own ecosystem, and these are independent of those that have developed in other ecosystems*. We shall address the question of how species adapt to and develop such balances with each other in Chapter 5.

The seemingly obvious solution to the takeover by a "foreign invader" is to introduce a natural enemy. Indeed, this approach has been used in a number of cases; rabbit control in Australia is one. Others are discussed in Chapter 17 in connection with the biological control of pests. However, such control is more easily said than done. Recall that control is achieved as a result of the interplay among *all* the factors of environmental resistance, which often include several natural enemies, as well as *all* the abiotic factors. Thus, a single natural enemy that will control a pest simply may not exist. In fact, there is no guaran-

(a)

(b)

(c)

(d)

▲ FIGURE 4–14 *Introduced plant species.*   (a) Water hyacinth overgrowing waterways. (b) Kudzu overgrowing forests. (c) Spotted knapweed taking over rangeland in the northwestern United States and southwestern Canada. (d) Purple loosestrife, a native of Eurasia, now is abundant in wetlands of the northeastern United States and Canada.

tee that the natural enemy, when introduced into the new ecosystem, will focus its attention on the target pest. Control of the rabbit population in Australia was initially attempted by introducing foxes. However, the foxes soon learned that they could catch other Australian wildlife more easily than rabbits and thus went their own way. In short, to prevent doing more harm than good, a great amount of research needs to be done before a natural enemy is introduced.

4.3 Disturbance and Succession

We have just considered a variety of biotic mechanisms that contribute to the regulation of species in ecosystems. *Equilibrium theory* suggests that ecosystems are stable environments in which species are able to interact constantly in well-balanced predator–prey and competitive relationships. Thus, *biotic interactions* determine the structure of living communities within ecosystems. This approach has led to the popular idea of "the balance of nature"—that natural systems maintain a delicate balance over time which lends great stability to them. However, in our previous analysis of ecosystems, we have largely ignored the dimension of *time*. Are species and ecosystems at equilibrium all the time? If the environment were entirely stable and uniform, we might expect the answer to be yes. But this is not the case. So how do ecosystems respond to disturbances?

To answer this question, we must examine the possibility that time brings disturbances that affect the sustainability of ecosystems.. Indeed, many studies have shown that ecosystems are in fact very patchy environments: Conditions and resources within them vary in temperature, moisture, exposure to sunlight, soil conditions, and the like. The distribution of species is similarly patchy, reflecting the patchiness in conditions. In a forest, for example, it is common to find species of trees varying independently in space, such that it is difficult to predict what species will tend to be found associated together in a given stand at a given time. But if neither conditions nor distributions of species are uniform even within ecosystems, then perhaps the concept of a stable equilibrium should be questioned. Indeed, the patchiness of ecosystems has caused many ecologists to think of them as *nonequilibrium systems* that seldom exhibit the characteristics of a true equilibrium. This controversy is well illustrated in the phenomenon known as ecological succession.

Ecological Succession

There are situations in nature wherein, over the course of years, we observe one biotic community gradually giving way to a second, the second perhaps to a third, and even the third to a fourth. This phenomenon of *transition* from one biotic community to another is called **ecological**, or **natural, succession**. Succession occurs because the physical environment may be gradually modified by the growth of the

biotic community itself, such that the area becomes more favorable to another group of species and less favorable to the present occupants.

The succession of species does not go on indefinitely, however: A stage of development is reached during which there appears to be a dynamic balance between all species and the physical environment. This final state is referred to as a **climax ecosystem**, an assemblage of species continuing on in space and time. The major biomes discussed up to this point in the text are climax ecosystems. Of course, it remains important to keep in mind that all balances are relative to the *current* biotic community and the *existing* climatic conditions. Therefore, even climax systems are subject to change if climatic conditions change or if new species are introduced or old ones removed. Nevertheless, natural succession may be seen as a progression toward a relatively more stable climax, one that no longer changes over time. Sometimes, there may be several "final" stages, or a polyclimax condition, with adjoining ecosystems in the same environment at different stages. The following are three classic examples:

Primary Succession. If the area has not been occupied previously, the process of initial invasion and then progression from one biotic community to the next is termed "primary succession." An example is the gradual invasion of a bare rock surface by what eventually becomes a climax forest ecosystem. Bare rock is an inhospitable environment. It has few places for seeds to lodge and germinate, and if they do, the seedlings are killed by lack of water or by exposure to wind and sun on the rock surface. However, certain species of moss are uniquely adapted to this environment. Their tiny spores, specialized cells that function reproductively, can lodge and germinate in minute cracks, and moss can withstand severe drying simply by becoming dormant. With each bit of moisture, moss grows and gradually forms a mat that acts as a sieve, catching and holding soil particles as they are broken from the rock or as they blow or wash by. Thus, a layer of soil, held in place by the moss, gradually accumulates (Fig. 4–15). The mat of moss and soil provides a suitable place for seeds of larger plants to lodge, and the greater amount of water held by the mat supports their germination and growth. The larger plants in turn collect and build additional soil, and eventually there is enough soil to support shrubs and trees. In the process, the fallen leaves and other litter from the larger plants smother and eliminate the moss and most of the smaller plants that initiated the process. Thus, there is a gradual succession from moss through small plants and, finally, to trees that form a climax forest ecosystem. The nature of the climax ecosystem, of course, differs according to the prevailing abiotic factors of the region, giving us the biomes typical of different climatic regions, as described in Chapter 2.

Because bare rock substrate can be exposed by glaciation, earthquakes, landslides, and volcanic eruptions, there are always places for primary succession to start anew.

Secondary Succession. When an area has been cleared by fire or by humans and then left alone, the surrounding

▲ **FIGURE 4–15** *Primary succession on bare rock.* Moss invades bare rock and acts as a collector, accumulating a layer of soil sufficient for additional plants to become established.

ecosystem may gradually reinvade the area—not at once, but through a series of distinct stages termed *secondary succession*. The major difference between primary and secondary succession is that secondary succession starts with the preexisting soil substrate. Therefore, the early, prolonged stages of soil building are bypassed. Still, as you can readily observe, a clear area has a microclimate quite the opposite from the cool, moist, shaded conditions beneath a forest canopy. Those plant species which propagate themselves in the microclimate of the forest floor cannot tolerate the harsh conditions of the clearing. Hence, the process of reinvasion necessarily begins with different species and proceeds accordingly. The steps leading from abandoned agricultural fields in the eastern United States back to deciduous forests provide a classic example of secondary succession (Fig. 4–16a).

On an abandoned agricultural field, crabgrass is predominant among the initial invaders. Crabgrass is particularly well adapted to invading bare soil. Its seeds germinate in the spring, and it grows and spreads rapidly by means of runners; moreover, it is exceptionally resistant to drought. In spite of its vigor on bare soil, crabgrass is easily shaded out by taller plants. Consequently, taller weeds and grasses, which take a year or more to develop, eventually take over

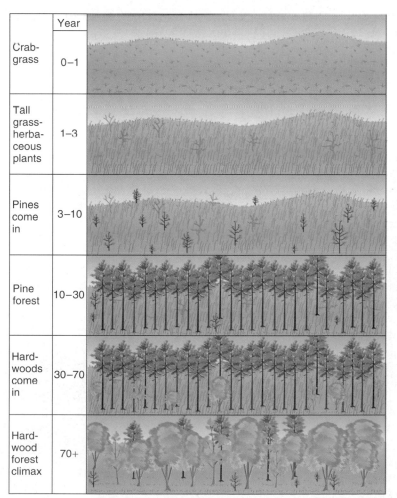

	Year	
Crab-grass	0–1	
Tall grass-herba-ceous plants	1–3	
Pines come in	3–10	
Pine forest	10–30	
Hard-woods come in	30–70	
Hard-wood forest climax	70+	

(a)

(b)

◄ **FIGURE 4–16** *Secondary succession.* (a) Reinvasion of an agricultural field by a forest ecosystem occurs in the stages shown. (b) Hardwoods (species of oak) growing up underneath and displacing pines in eastern Maryland.

from the crabgrass. Next, young pine trees, which are well adapted to thrive in the direct sunlight and heat of open fields, gradually develop and shade out the smaller, sun-loving weeds and grasses, eventually forming a pine forest. But pine trees also shade out their own seedlings, which need bright sun to grow. Thus, the seedlings of deciduous trees, not pines, develop in the cool shade beneath the pine trees (Fig. 4–16b). Consequently, as the pines die off (their life span is limited), they are replaced by oaks, hickories, beeches, maples, and other species of hardwoods that characterize eastern deciduous forests. The seedlings of the latter continue to flourish beneath the cover of their parents, providing a stable balance—the climax deciduous forest ecosystem.

We emphasize again that secondary succession requires a suitable soil base to start. If this base is lost by erosion, even the initial plants cannot get a start. Therefore, the bare subsoil may continue to erode, leaving a situation that more closely resembles the much slower primary succession process.

Aquatic Succession.   Another example of natural succession is seen as lakes or ponds are gradually filled and taken over by the surrounding terrestrial ecosystem. This process occurs because a certain quantity of soil particles inevitably erodes from the land and settles out in ponds or lakes, gradually filling them. Aquatic vegetation produces detritus that also contributes to the filling process. As the buildup occurs, terrestrial species can advance, and aquatic species move farther out into the lake. In short, the shoreline gradually advances toward the center of the lake until, finally, the lake disappears altogether (Fig. 4–17). Here, the climax community may be a bog or a forest.

Climate Change.   Pollen records inform us that the climate itself has been changing over time spans of hundreds and thousands of years, as glaciation and changes in the Earth's orbit have occurred (Chapter 21). In response to past climate changes, the forests of the temperate zone have shifted from coniferous to deciduous, with many changes in the dominant species of trees. In this light, a so-called climax ecosystem is perhaps only a figment of our limited perspective. Note, however, that these changes have taken place over many hundreds or thousands of years. What happens if the climate changes more rapidly? Can forests adapt to a warming that takes place over decades?

Currently, one of our most serious environmental concerns is global climate change. Because of our burning of fossil fuels, we are increasing the carbon dioxide level of the atmosphere to the point where, by the mid-21st century, it will have doubled since the beginning of the Industrial Revolution. Since carbon dioxide is a powerful greenhouse gas (Chapter 21), a general warming of the atmosphere is expected, with many other effects on the Earth's climate. Numerous ecosystems will be affected by the changes, and it is questionable whether they can maintain their functional sustainability in the face of such rapid changes. This threat of disturbance on an enormous scale is a major reason the nations of the world have begun to take steps to reduce emissions of greenhouse gases.

Disturbance and Resilience

In order for natural succession to occur, the spores and seeds of the various invading plants and the breeding populations of the various invading animals must already be present in the vicinity. Ecological succession is not a matter of new species developing or even old species adapting to new conditions; it is a matter of populations of existing species taking advantage of a new area as conditions become favorable. Where do these early-stage species come from if their usual fate is to be replaced by late-stage or climax species?

▼ **FIGURE 4–17** *Aquatic succession.*   (a) Ponds and lakes are gradually filled and invaded by the surrounding land ecosystem. (b) In this photograph, taken in Banff National Park in the Canadian Rockies, you can visualize the lake that used to exist in the low-level area. It is now filled with sediment and covered by scrub willow. Spruce and fir forest is gradually encroaching.

(a)

The answer is a key to the nonequilibrium theory: They come from other surrounding ecosystems in early stages of succession. Similarly, the late-stage species are recruited from ecosystems in later stages of succession. Thus, in any given landscape, all stages of succession are likely to be represented in the ecosystems. The reason for this is that disturbances are constantly creating gaps or patches in the landscape. It will be readily seen that when a variety of successional stages is present in a landscape, as opposed to one single climax stage, a greater diversity of species can be expected; in other words, biodiversity is enhanced by disturbance! And consequently, natural succession is affected.

Of course, if certain species have been eliminated, natural succession will be blocked or modified. For example, beginning with the early colonization of Iceland by Norsemen in the 11th and 12th centuries, the forests of that country were cut for fuel, a process that accelerated with further European colonization in the 18th and 19th centuries. By 1850 not a tree was left standing, and Iceland remained a barren, tundralike habitat, because there was no remaining source of seeds to foster natural regeneration (Fig. 4–18). Tree seedlings are now being imported and planted in Iceland in the hope that a natural succession may be reestablished.

Fire and Succession. Fire is an abiotic factor that has particular relevance to succession. It is a major form of disturbance common to terrestrial ecosystems. About 75 years ago, forest managers interpreted the potential destructiveness of fire to mean that all fire is bad and embarked on fire-prevention programs that eliminated fires from many areas. Unexpectedly, fire prevention did not preserve all ecosystems in their existing state. In pine forests of the southeastern United States, for instance, economically worthless scrub oaks and other broad-leafed species began to displace the more valuable pines. Grasslands were gradually taken over by scrubby, woody species that hindered grazing. Pine forests of the western United States that were once clear and open became cluttered with the trunks and branches of trees that had died in the normal aging process. This deadwood became the breeding ground for wood-boring insects that proceeded to attack live trees. In California, the regeneration of redwood seedlings began to be blocked by the proliferation of broad-leafed species.

Scientists now recognize that fire, which is often started by lightning, is a natural and important abiotic factor. As with all abiotic factors, different species have different degrees of tolerance to fire. In particular, grasses and pines have their growing buds located deep among the leaves or needles, where they are protected from fire; broad-leafed species, such as oaks, have their buds exposed; hence, they are sensitive to damage from fire. Consequently, where these species exist in competition, periodic fires are instrumental in maintaining a balance in favor of pines, grasses, or redwood trees. In relatively dry ecosystems, where natural decomposition is slow, fire may also play a role in releasing nutrients from dead organic matter. Some plant species actually depend on fire. The cones of lodgepole pine, for example, will not release their seeds until they have been scorched by fire (Fig. 4–1b).

Ecosystems that depend on the recurrence of fire to maintain the existing balance are now referred to as **fire climax ecosystems**. The category includes various grasslands and pine forests. Fire is now increasingly used as a tool in

(b)

▲ **FIGURE 4–18** *Iceland.* Forests that originally covered much of this island nation were totally stripped for fuel in the 18th and 19th centuries. With natural succession impossible, Iceland has remained barren and tundralike, as seen here.

▲ **FIGURE 4–19** *Ground fire.* Periodic ground fires are necessary to preserve the balance of pine forests. Such fires remove excessive fuel and kill competing species.

the management of such ecosystems. In pine forests, if ground fires occur every few years, relatively little deadwood accumulates. With only small amounts of fuel, fires usually just burn along the ground, harming neither pines nor wildlife significantly (Fig. 4–19). In forests where fire has not occurred for many decades, however, so much deadwood has accumulated that if a fire does break out, it will almost certainly become a crown fire. That is, so much heat is generated that entire living trees are ignited and destroyed. This long-term lack of fire was a major factor in the fires in Yellowstone National Park in the summer of 1988.

Crown fires do occur in nature, since not every area is burned on a regular basis, and exceedingly dry conditions can make a forest vulnerable to crown fires even when large amounts of deadwood are not present. Thus, humans have only promoted the potential for crown fires by fire prevention programs. Even crown fires, however, serve to clear the deadwood and sickly trees that provide a breeding ground for pests, to release nutrients, and to provide for a fresh ecological start. Burned areas soon become productive meadows as secondary succession starts anew. We have seen this in Yellowstone National Park. Thus, periodic crown fires create a patchwork of meadows and forests at different stages of succession that lead to a more varied, healthier habitat which supports a greater diversity of wildlife than does a uniform, aging conifer forest (Fig. 4–1).

The devastating fire that spread from brushland in Oakland, California, and destroyed some 1,000 homes in the fall of 1991 is another example of poor fire management. The grasslands on the hills east of San Francisco are a fire climax ecosystem. Preventing fire in this area for many years produced a superabundance of woody shrubs, not the least of which was eucalyptus, which had been introduced and which has a particularly flammable wood. Homeowners created additional fuel by planting a lush, forested landscape, unnatural for this dry-summer climatic

regime. The result, finally, was a fire that burned almost 1,000 homes before it was controlled. Of course, as in any human-managed enterprise, fire management can go awry, as it did in the spring of 2000 in Los Alamos, New Mexico (Fig. 4–20).

Nonequilibrium Systems.   In sum, the concept most important to recognize is that disturbances such as fires, floods, windstorms, and droughts are important in structuring ecosystems. The disturbances remove organisms, reduce populations, and create opportunities for other species to colonize the ecosystem. Much of the patchiness observed in natural landscapes is evidence of periodic disturbances. Some ecosystems are more stable than others and experience disturbances very infrequently. Biotic relationships then become important in maintaining the stability of the ecosystem. Thus, the sustainability of ecosystems is dependent upon equilibria among populations of the species in the biotic community, and it is also dependent upon existing relationships between the biotic community and abiotic factors of the environment, such as disturbances. We have seen that the natural biotic community itself may induce changes in abiotic factors that in turn result in changes in the biotic community (succession). Given this dynamic succession, it should be clear that if any one or more physical factors of the environment are shifted, portions of the biotic community may again be pushed into a state of flux in which certain species that are stressed by the new conditions die out and other species that are better suited to those conditions thrive and become more abundant. This is the essence of the nonequilibrium theory of ecosystem structure.

Although our focus in this chapter has been primarily on populations and their interactions, we must not lose

◄ FIGURE 4–20 *Fire management gone wrong.* In May 2000, National Park Service personnel started a prescribed fire at the Bandelier National Monument, New Mexico. High winds increased and spread the fire to Los Alamos, destroying 235 homes.

sight of the importance of ecosystem functions such as the trophic interactions of energy flow and the efficient cycling of nutrients that we examined in Chapter 3. How does the nonequilibrium theory of ecosystem structure relate to these functions? Actually, disturbances and shifting biotic relationships not only may have little detrimental effect on an ecosystem, but may actually contribute to its ongoing functioning. Ecologists have referred to this condition as **resilience**; a resilient ecosystem is one that maintains its normal functioning—its integrity—even through a disturbance. For example, fire certainly can appear to be a highly destructive disturbance to a forested landscape. Yet, as we have seen, fire releases nutrients that nourish a new crop of plants, and in time the burned area is repopulated with trees and is indistinguishable from the surrounding area. The processes of replenishment of nutrients, dispersion by surrounding plants and animals, rapid regrowth of plant cover, and succession to a forest can all be thought of as **resilience mechanisms** (Fig. 4–21). Resilience is clearly important to maintaining the sustainability of ecosystems, and so we encounter the **fourth sustainability principle**:

Ecosystems show resilience when subject to disturbance.

Ecosystem resilience has its limits, however. If forests are removed from a landscape by human intervention and the area is prevented from reforestation by overgrazing, so that the soil is eroded away, a degraded state is reached which shows little of the original ecosystem's functioning (Fig. 4–21). As in Iceland (Fig. 4–18), some disturbances may be so profound that they can overcome normal re-

silience mechanisms and create an entirely new and far less useful (in terms of ecosystem goods and services) ecosystem. This new ecosystem can also have its own "pathological" resilience mechanisms, which may resist any restoration to the original state. A recently published example of this situation is seen in lakes, which show a resilience that maintains all of the desirable qualities that lakes have, unless the lake is subjected to pollution; when that happens, the degraded lake has a new set of factors which act to maintain its degraded state.[1]

Biodiversity. Throughout this chapter, you may have noted a great importance on maintaining a diversity of species—that is, biodiversity. To reiterate some of the major points, the most stable population equilibria are achieved by a diversity of natural enemies. Simple systems, especially monocultures, are inherently unstable. Most or all succession depends on a preservation of biodiversity, and succession underlies the ability of an ecosystem to recover from damage (e.g., from fire or blight, as in the case of the American chestnut) and its ability to accommodate to disturbances. Thus, we begin to see another basic principle of sustainability, the fifth. Very simply, the **fifth principle of ecosystem sustainability** may be stated as

Ecosystems depend on biodiversity.

[1] Stephen R. Carpenter and Kathryn L. Cottingham, "Resilience and Restoration of Lakes," *Conservation Ecology* [on-line]1(1):2 (1997). URL: **http://www.consecol.org/vol1/iss1/art2**.

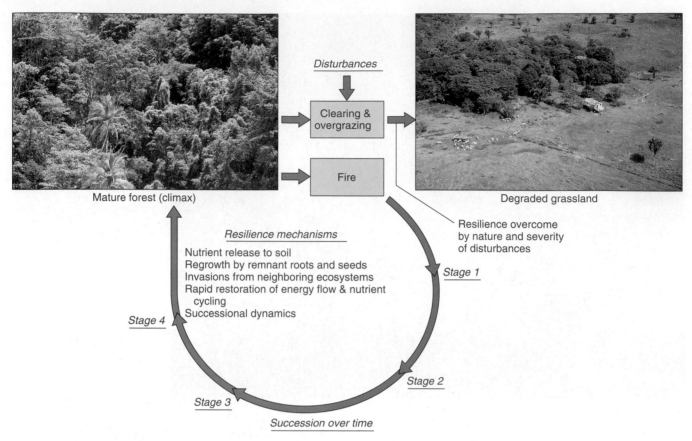

Mature forest (climax)

Degraded grassland

Disturbances

Clearing & overgrazing

Fire

Resilience overcome by nature and severity of disturbances

Stage 1

Resilience mechanisms

Nutrient release to soil
Regrowth by remnant roots and seeds
Invasions from neighboring ecosystems
Rapid restoration of energy flow & nutrient cycling
Successional dynamics

Stage 4

Stage 2

Stage 3

Succession over time

▲ **FIGURE 4–21** *Resilience in ecosystems.* Disturbances in ecosystems are usually ameliorated by a number of resilience mechanisms that result in the restoration of normal ecosystem function in a short time.

In Chapter 5, we will find that biodiversity is also required in order to sustain a much longer term, evolutionary adaptation of species to changing conditions that are inevitable, even without human interference.

Our five principles of ecosystem sustainability are listed in Table 3–1 for review purposes.

4.4 Implications for Humans

How can we use the information presented in this chapter to create a sustainable future? The answer to this question has two aspects. One is protecting or managing the natural environment to maintain the goods and services vital to the human economy and to life support, as well as maintaining the beauty, interest, biodiversity, and other intrinsic values of the natural world. The second aspect is establishing a balance between our own species and the rest of the biosphere. The two aspects are interrelated.

Adaptive Management. With regard to the first aspect, as we stated in Chapter 3, ecosystem management is the paradigm now adopted by all federal agencies with responsibilities for natural resource management. This paradigm recognizes that some degree of management is necessary to maintain normal structure and function where there are human demands on ecosystems—in other words, to assure

the sustainability of those ecosystems. In more or less pristine ecosystems, simple *protection* from adverse human impacts and avoiding the introduction of foreign species may be adequate. Even if disturbances occur—and they will—the natural resilience of ecosystems will likely ensure that they persist over time and provide the goods and services so important to human affairs.

In most cases, however, some degree of active management will be necessary. For this to be successful, managers will be well advised to pay careful attention to the principles of ecosystem science, as briefly introduced in Chapters 2–4. (This is *sound science* in action.) For example, managers should expect that ecosystems will change over time; that individual species and interactions between species will have important impacts on ecological processes; that different sites and regions are unique, with unique organisms and abiotic factors; and that disturbances are important and frequent events which can have a profound impact on the ecosytem. Based on these principles, a set of practical guidelines has been proposed by the Ecological Society of America's Committee on Land Use. Here are several of the Society's suggestions for land managers:

- Plan for long-term change and unexpected events
- Avoid land uses that deplete natural resources
- Minimize the introduction and spread of alien species

- Implement land-use and management practices compatible with the natural potential of the area
- Examine impacts of local decisions in a regional context
- Retain large connected areas that contain critical habitats

Because of the existence of uncertainties wherever the natural world is concerned, management should above all else be **adaptive**. That is to say, practices and policies should be seen as tentative and experimental, with a readiness to change in the face of necessity. This is especially important in light of our understanding that wherever humans intervene, changes will occur; in addition, as time goes on, surprises will be inevitable, and new uncertainties will appear.

For example, if an ecosystem has suffered disturbances from human interference, restorative measures may become necessary. These measures involve a number of options, depending on the damage. *Restoration* may mean the painstaking reintroduction of a predator or parasite, or the hunting of a species that is multiplying beyond the carrying capacity of the system. Depending on the nature of the disturbance, more may be needed. For instance, how do you restore some semblance of a natural ecosystem after an area has been totally altered by such activities as development, agriculture, or mining? In such cases, restoration begins with the creation of the desired physical environment and the introduction of appropriate plants to support the desired animals. An example is the replacement of soil onto land stripped by surface mining, followed by introducing desirable cover plants and trees.

One of the more difficult problems for adaptive management is preventing the entry of alien species. The most common source of these invaders is global trade, which is increasing exponentially. Most insects and plant pathogens come in as stowaways in containers or on horticultural stock, but even imported plants themselves can become a problem, as we have seen. International treaties exist to govern such trade by providing for inspection and quarantine of trade goods, but the measures taken under the treaties are not strong enough to afford effective control against the invaders. Indeed, the World Trade Organization may rule that a nation's attempt to restrict the importation of a species is an unlawful trade barrier instead of an effort to exclude pests. The most prudent approach to this problem is to assume new entries to be "guilty until proven innocent," rather than the other way around, as is the current practice. It is far more cost effective to prevent invasions than to deal with them once the invaders have arrived and become established. With the continued increase in global trade, there is every likelihood of an increase in economic and ecological damage worldwide due to the spread of alien species.

The Pressures of Population.   In the long run, efforts to protect natural ecosystems will be overwhelmed and thwarted by increasing pressures from the human system if current trends of population growth and exploitation of natural ecosystems continue. Note that a graph of the human population over the last 200 years (see Fig. 1–4) has a distinct similarity to the upsweeping portion of the *J*-curve for the reindeer population on St. Matthew Island (Fig. 4–6). We can also draw parallels in that humans have overcome, for the most part, those factors which generally keep natural populations in balance—disease, parasites, and predators. Therefore, we may observe that the human population is behaving much like any population in the absence of natural enemies.

But what is the urgency? In discussing the reindeer population on St. Matthew Island, we noted the rapidity with which an exponentially growing population can deplete its resource base. Half of the original resource base is consumed in the last doubling. This is the same as saying that half of the original resource base is needed just to accomplish the last doubling, and then the crash will occur anyhow, because no resources remain to sustain the high population, much less to allow it to grow further. Such was the case with many ancient civilizations that perished (Fig. 4–22).

We need to ask the question, What is the carrying capacity of our planet for human beings? That is, how many of us can the Earth support sustainably? Many experts have considered and debated this question. Unfortunately, there

▲ **FIGURE 4–22** *An unsustainable civilization.*   On the Yucatán Peninsula of Mexico, the ruins of Chichen-Itza are only one example of civilizations that arose, prospered, and then passed into oblivion due to a failure to maintain a suitable balance between their human population and environmental resources.

is no uniform conclusion, because humans are capable of such diverse behavior. However, we observe that major environmental parameters, such as forests, fisheries, soils, water resources, and the atmosphere, are in a state of decline under the pressures of the present population of 6.2 billion. Therefore, it may be that we are already over the Earth's carrying capacity. What will an additional 3 billion do to natural systems? It is certain that the next few decades will be the most crucial in terms of achieving sustainability. Indeed, it can be argued that there is no more important human enterprise than coming to grips with the realities of our environmental limits. But this cannot be expected to happen automatically through "business as usual." It will require using our intelligence and wisdom to understand and appreciate ecological principles and using our technological skills to bring our human system into compliance with those principles.

In relation to the concepts presented in this chapter, anything that we are doing, or that we may do, toward pre-serving or reestablishing natural equilibria among biota and protecting natural biota from destructive human impacts may be seen as progress toward sustainability. Likewise, protecting ecosystems, to say nothing of the total biosphere, from unnatural changes in abiotic factors such as pollution or atmospheric alterations is equally important. We must work toward stabilizing the human population in morally acceptable ways and husbanding resources such that they remain available to future generations. A concern for posterity is one of the basic elements of the stewardship ethic.

Chapters 6 and 7 will focus on the problems of stabilizing the human population, and later chapters will treat husbanding resources. Before proceeding with those issues, however, we will examine in Chapter 5 the question of how species become adapted to one another and their abiotic environment in such a way as to form balanced, sustainable ecosystems—the perspective of evolutionary change.

ENVIRONMENT ON THE WEB

FIRE IN PROTECTED ECOSYSTEMS: FRIEND OR FOE?

In 1994, the Smokey Bear campaign—"Only *you* can prevent forest fires!"—celebrated its fiftieth anniversary, testimony to the long fight humans have waged against the natural force of fire. The battle has been successful: Forest fires are now suppressed throughout North America, saving millions of dollars in damages every year. But what is the cost to ecosystems of removing a natural cycle of disturbance and succession?

Ecologists now know that many ecosystems have evolved in the presence of fire and need periodic fire to develop normally. In ecosystems like the Indiana Dunes National Lakeshore, maintenance of the natural parklike oak savannah depends on periodic fire for the preservation of its characteristic open habitat. Also, species such as jack pine need the heat of an intense fire to burn off the pitch that protects seeds, permitting normal germination. When fire is suppressed, jack pine populations decline, and species that depend on jack pine habitat (such as the endangered Kirtland's Warbler) are jeopardized. Yet fire, however natural, also has real costs in managed ecosystems. Historically, national parks have suppressed fire in an attempt to protect valued aesthetic characteristics—and associated tourism revenues. This chapter describes the 1988 fire in Yellowstone Park, one of the most visited parks in the U.S. National Park System. In a controversial policy decision, park managers allowed the fire to burn freely, destroying thousands of acres of forest. Was this the right decision?

Evidence from other ecosystems shows that letting the fire burn was probably the right thing to do. In nature, forests are subjected to occasional high winds and lightning-induced fire, forces that create openings in the canopy of trees above. Opportunistic species colonize these spaces, creating a mosaic of species of different ages, shapes, and reproductive strategies. Periodic fires are usually moderate in intensity and are confined to dry litter and understory plants, with relatively little damage to older trees. In a sense, regular fires remove accumulated "fuel," reducing the potential for catastrophic fire in a future blaze. Fire also provides a mechanism for the remobilization of nutrients stored in plant and animal tissue. When fire is suppressed, however, the result may be major changes in habitat structure and species richness. In commercially valuable forests, research suggests that fire suppression results in more, but smaller, trees. And because dense vegetation is more vulnerable to drought and insect pests, the system may be less able to withstand normal stresses than a fire-controlled forest. Letting the fire burn may therefore be a case of "short-term pain, long-term gain."

Web Explorations

The Environment on the Web activity for this essay compares clearance by natural forces with clearcutting and observes the impacts these different forces have on an ecosystem. Go to the Environment on the Web activity (select Chapter 4 at **http://www.prenhall. com/wright**) and learn for yourself.

1. about trends in human control of fires in Florida;
2. how natural clearance by fire or wind differs from clearcutting; and
3. about the importance of the scale of disturbance in its impact on an ecosystem.

⏱ **A suggested time frame for each of these exercises is 10–30 minutes.**

REVIEW QUESTIONS

1. What are the two fundamental kinds of population growth curves? What are the causes and consequences of each?
2. Define biotic potential and environmental resistance, and give factors of each. Which generally remains constant, and which controls a population's size?
3. Differentiate between the terms *critical number* and *carrying capacity*.
4. Describe the predator–prey relationship between the moose and wolves of Isle Royale. What other factors influence these two populations?
5. What is the third basic principle of ecosystem sustainability? How does it relate to the current U.S. white-tailed deer population?
6. State the three main ways that enable different plant species to coexist in the same region.
7. What is a balanced herbivory, and how do competition and predation accomplish it?
8. What is meant by territoriality, and how does it control certain populations in nature?
9. What problems arise when a species is introduced from a foreign ecosystem? What is the basic explanation as to why these problems occur?
10. Define the terms *ecological succession* and *climax ecosystem*. How do disturbances allow for ecological succession?
11. What role may fire play in ecological succession, and how may fire be used in the management of certain ecosystems?
12. What is the nonequilibrium theory of ecosystem stability? What is the evidence for it?
13. What is meant by ecosystem resilience? How is this accomplished, and what can cause it to fail?
14. What are the fourth and fifth principles of ecosystem sustainability?
15. Give five guidelines for ecosystem managers. What is adaptive management?

THINKING ENVIRONMENTALLY

1. Describe, in terms of biotic potential and environmental resistance, how the human population is affecting natural ecosystems.
2. Give two density-dependent factors of environmental resistance that would act on a field mouse population. Give two density-independent factors.
3. Consider the plants, animals, and other organisms present in a natural area near you, and then do the following: (a) Think of how the area has undergone ecological succession. (b) Analyze the population-balancing mechanisms that are operating among the various organisms. (c) Choose one species and predict what will happen to it if two or three other species native to the area are removed and, again, if two or three foreign species are introduced into the area.
4. Evaluate, in terms of supporting sustainable ecosystems, such practices as legal hunting, controlling pests with chemical sprays, using or preventing fires, and poaching endangered species.
5. Make an argument, pro or con, regarding the sustainability of the human system in terms of concepts you have learned in this chapter. What new directions, if any, do humans need to take to achieve sustainability?

WEB REFERENCES

On-line resources for this chapter are on the World Wide Web at: **http://www.prenhall.com/wright**. (Click on Chapter 4 within the Chapter Selector.)

Ecosystems and Evolutionary Change

Key Issues and Questions

1. We have already identified resilience and biodiversity as two important properties of ecosystems essential to their sustainability. How have these properties come into existence?

2. The characteristics of a population can be modified by differential reproduction. Give an example of this in terms of selective breeding and in terms of natural selection.

3. Mutations and differential reproduction lead to inevitable changes in the gene pool of a species. Explain how this occurs.

4. In the face of environmental changes, some species will survive, whereas others will become extinct. What attributes influence the survival of a species?

5. Natural selection can lead to the development of new species. How does this occur? How is speciation related to biodiversity?

6. Ecosystems change over time and vary greatly in different parts of the world, yet ecosystem resilience is maintained. Explain these facts in terms of evolution at the species level.

7. The past history of Earth has involved the movement of entire continents. How does the theory of plate tectonics explain the movement of continental land masses?

8. Evolution is the most widely accepted explanation for origins, but it is also controversial. What are the reasons for the controversy?

9. Loss of biodiversity undercuts the ability of species, ecosystems, and agriculture to adapt to changing conditions. Why is this the case?

Some 600 miles (970 km) off the west coast of South America, the Galápagos Islands are home to a unique array of plants and animals, many found nowhere else on Earth. The 13 larger islands and numerous smaller ones were formed by volcanic activity; the island archipelago is situated on a hot spot where three tectonic plates join and volcanic magma is moving vertically to the surface. The islands are spawned by the volcanic forces and gradually move their way eastward (at a rate of 7 cm per year!). The oldest island, Española, is about 3 million years old. On the islands, dry lowland ecosystems, punctuated with tree cactuses and thornbushes, cling to the harsh volcanic soils. In contrast, the rocky coastline is abundant with marine life, nourished by cold upwelling currents rich in nutrients. Inland, mountains occupied by fertile, humid forests rise to several thousand feet. Among the unique animals of these ecosystems are giant tortoises, marine iguanas, and a group of finch species often referred to as "Darwin's finches," so-called because they were first discovered by Charles Darwin in 1835 during his famous voyage on HMS *Beagle*.

The islands are a showcase of evolution; after Darwin returned home to England, he began to examine the many specimens he had collected and realized how unique they were. Different islands had different races of giant tortoises and different species of finches, and the islands in general had a great many life-forms found nowhere else. His thinking developed gradually into the grand thesis introduced to the world in 1859 as "evolution by natural selection." Many other scientists have followed up on his work and have fine-tuned evolutionary theory to the point where it is today: the most thorough and convincing concept of how living things have developed over time to their present tremendous biodiversity.

◀ *Galápagos Islands.* Found nowhere else in the world, marine iguanas bask on the rocks in the Galápagos Islands.

In Chapter 4, we examined the factors that contribute to the sustainability of ecosystems. In particular, we encountered predation and competition as mechanisms that keep populations in equilibrium, and we also considered what happens when ecosystems are subject to disturbances. We observed that ecosystems demonstrate remarkable resilience in the face of disturbances like fire, maintaining the important ecosystem functions that sustain life even as the species present change over time. We also observed that a moderate amount of disturbance could make a positive contribution to the biodiversity of an ecosystem. In particular,

we saw that the introduction of a species from a foreign ecosystem often created a major disturbance in the relative abundance of native species. The effects of these disturbances led us to conclude that balances among populations are not "automatic," but apparently develop as species live together over many millennia.

The foregoing observations imply an important concept: that species are gradually "molded" to cope with other members of the biotic community and with the abiotic environment (Fig. 5–1). How does such "molding," or **adaptation**, of a species occur? Further, the fossil record reveals a

▼ **FIGURE 5–1** *Adaptation of certain species.* Modifications of body shape and color that allow species to blend into the background and thus provide protection from predation are among the most amazing adaptations. Shown are the (a) Thomisus spider, (b) spanworm, (c) leaf katydid, and (d) giant Australian stick insect.

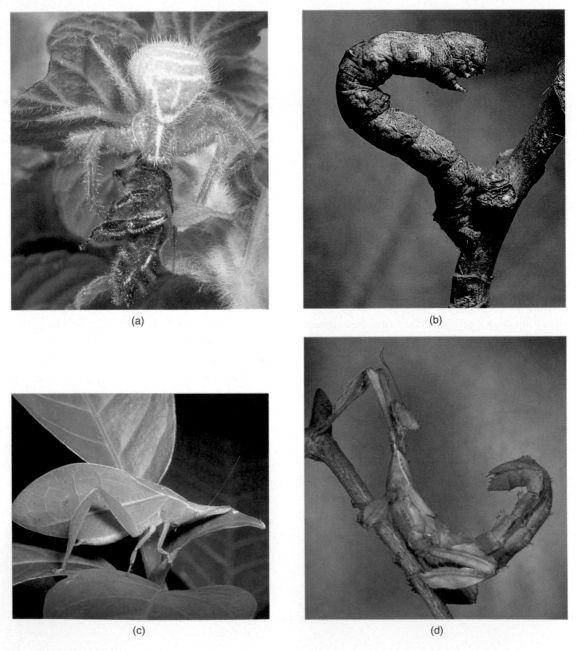

(a)

(b)

(c)

(d)

remarkable degree of change in species over time, suggesting that the extinction and the creation of species are common events in the history of life. How do new species arise, and what leads to extinctions? Also, if the generation of new species and the development of balances are natural processes, can we assume that nature will take care of the disturbances caused by humans and nullify the consequences? Or can humans understand these processes sufficiently to manipulate them toward a desired end?

Our objective in this chapter is to examine the processes by which species change and adapt to their abiotic environments and to other members of their biotic community and by which new species are created. By understanding these processes, we will understand the source of ecosystem resilience—our fourth principle of sustainability. We will also see the source of biodiversity—our fifth principle of sustainability.

5.1 Selection by the Environment

First, consider our modern understanding that all the inherited characteristics of an individual organism (plant, animal, or microbe) are expressions of its genetic (DNA) makeup. Second, we are familiar with the fact that any two individuals (except for identical twins, which will be considered later) can be identified or distinguished on the basis of their DNA, as well as on the basis of their physical appearance. In other words, the differences between individuals that we see on the physical level correspond to measurable differences on the genetic level. We speak of these genetic differences among individuals as **genetic variation** in the population. We find genetic variation in the populations of *all* species.

Third, imagine that we take a DNA sample from every individual of a species. This would give us what is referred to as the **gene pool** of the species. Although we have not sampled every individual, we have sampled enough to find that the DNA gene pool of one species differs from that of another, reflecting the degree of physical differences between the species.

As the members of a species reproduce, the genetic makeup of the species is passed from one generation to the next. Can the gene pool of the species change in the process? It can be mathematically demonstrated that if breeding among all the members of a species were totally random, and if each member produced the same number of offspring, the gene pool would remain constant in its genetic makeup over time. However, such constancy is not likely. Invariably, we expect some individuals to reproduce more than others, a phenomenon called **differential reproduction**.

As a result of differential reproduction, the particular genes carried by the prolific individuals are reproduced and passed on, and genes uniquely carried by individuals that reproduce little or not at all become increasingly rare or nonexistent in the population. In this way, a gene pool can change over generations. Differential reproduction will lead to a gradual modification in the gene pool as some genes become increasingly common in the population and other genes diminish in frequency. A change in the gene pool of a species over the course of generations is the essence of **biological evolution**.

The changes in the genetic characteristics of populations that can occur through differential reproduction are most dramatically illustrated through *selective breeding* of domestic plants and animals.

Change through Selective Breeding

Virtually all agricultural plants and domestic breeds of animals have been derived from wild populations through the process of **selective breeding**. The primary technique behind all such breeding is for the breeders to envision the traits they would like to achieve in a given species: a dog with short, squat legs that is able to wiggle into animal burrows to aid hunters, for example. The breeders then examine the existing population of dogs and *select* those individuals that show the sought-after trait a little more than other members of the population. The selected individuals are then bred. The offspring tend to be like the parents, but some offspring may express the particular trait *more* than the parents do, and others express it *less*. Those offspring which show the trait most are *selected* to be the parents for the next generation, whereas the other offspring are prevented from breeding. Repeating this process of selection and breeding over many generations gradually yields the traits desired by the breeder, as illustrated in Fig. 5–2. In this example, the desired outcome is seen in the dachshund.

The fact that all the different breeds of dogs, from Great Danes to Chihuahuas, are derived from the same original species of wolf (Fig. 5–3) demonstrates that a gene pool of a species may be molded in any number of different ways through selective breeding. Note, however, that all breeds of dogs are still considered the same species, because they can interbreed and produce fertile offspring. (Chihuahuas and Great Danes, though, might have trouble doing so!)

The development or enhancement of desired traits through selective breeding has been practiced since antiquity, many centuries before there was any knowledge about genes or genetics. Purely on the basis of visible characteristics, breeders selected or excluded which animals or plants to breed, and it worked. Only now do we recognize that the visible characteristics are actually the manifestation of certain genes. In selecting particular individuals that express a certain trait more than others, the breeder is in fact selecting to increase the proportion of certain genes in the gene

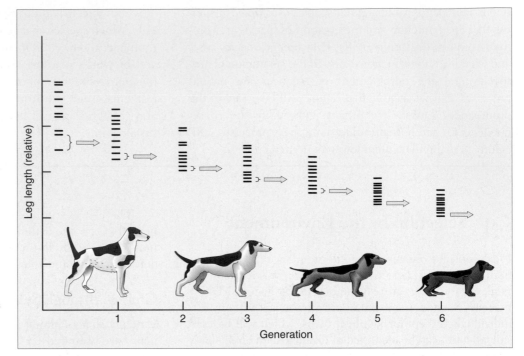

▶ **FIGURE 5–2** *Selective breeding of short-legged dogs (dachshund).* In each generation, the pups with the shortest legs are selected as parents of the next generation. The lines represent the variation in leg length for a given generation. As this process is repeated over many generations, the desired feature (short legs) is gradually developed.

pool and to decrease others. Because the process is based on the breeder's preference and the breed is carefully maintained distinct from other breeds, we refer to it as **artificial selection**.

Change through Natural Selection

A similar type of selection occurs in nature. We note that in nature most offspring do not survive; rather, they fall victim to various factors of environmental resistance, as described in Chapter 4. These factors—predators, parasites, drought—can now be seen as **selective pressures**. That is, each factor plays a role in determining which individuals survive and reproduce and which are eliminated. For ex-

ample, if a predator is present, prey animals having traits that protect them or allow them to escape from their enemies (such as coloration that blends in with the background) tend to survive and reproduce (Fig. 5–1), and those without such traits tend to become the predator's dinner. Any individual with a gene that slows it down or makes it conspicuous will tend to be eaten. Thus, predators may be seen as a selective force favoring the reproduction of genes that enhance the prey's ability to escape or protect itself and causing the elimination of any genes handicapping those functions. The need for food can also be seen as a selective pressure acting on the predator, enhancing those characteristics benefiting predation—such as keen eyesight and sharp claws.

▼ **FIGURE 5–3** *Variation in a gene pool.* (a) The ancestor of dogs, the wolf. (b) Different breeds of dogs whose genes were present in the original wolf gene pool.

(a)

(b)

Every factor of environmental resistance is a selective pressure resulting in the survival and reproduction of those individuals with a genetic endowment that enables them to cope with their surroundings. In nature, there is undeniably a constant selection and, consequently, a modification of a species' gene pool toward features that enhance survival and reproduction *within the existing biotic community and environment*. Since the process occurs naturally, it is known as **natural selection**.

These concepts were first presented by Charles Darwin in his famous book, *The Origin of Species by Natural Selection* (1859). The phrase "survival of the fittest," often identified with Darwin's work, means the survival of those individuals having traits that best enable them to cope with the biotic and abiotic factors of their environment. The modification of the gene pool that occurs through natural selection over the course of many generations is the sum and substance of *biological evolution*. Darwin deserves great credit for constructing his theory purely from his own observations, without any knowledge of genes or genetics—information that wasn't discovered until several decades later. Today, our modern understanding of DNA, mutations, and genetics fully supports Darwin's theory of evolution by natural selection. Indeed, our knowledge of reproductive and genetic technologies has opened up possibilities for the human race with many moral and ethical implications. (See the "Ethics" essay, p. 113.)

Adaptations to the Environment

The key distinction between the genetic changes that can be brought about through artificial selection by humans and those brought about through natural selection lies in the end point. For human breeders, the end point toward which selection is carried on is some image in their minds—a better tasting tomato or a fatter pig. In natural selection, the criteria for selection in every generation are simply survival and reproduction.

Under the selective pressures exerted by the factors of environmental resistance, the gene pool of each population is continually tested such that the population remains well adapted for survival and reproduction in the particular biotic community and environment in which it exists. Indeed, virtually all traits of any organism can be looked at in terms of features that adapt the organism for survival and reproduction, or, in Darwinian terms, **fitness**. Essentially all characteristics of organisms can be grouped as follows:

- Adaptations for coping with climatic and other abiotic factors.
- Adaptations for obtaining food and water in the case of animals and for obtaining nutrients, energy, and water in the case of plants.
- Adaptations for escaping from or protecting against predation and for resistance to disease-causing or parasitic organisms.

- Adaptations for finding or attracting mates in animal populations and for pollinating and setting seed in plant populations.
- Adaptations for migrating in the case of animals and for dispersing seeds in the case of plants.

The fundamental question about any trait is, Does it facilitate survival and reproduction of the organism? If the answer is yes, the trait will be maintained through natural selection. Consequently, various organisms have evolved different traits to accomplish the same function. For example, the ability to run fast, to fly, or to burrow, and protective features such as quills, thorns, and an obnoxious smell or taste, all support the function of reducing predation and are seen in various organisms (Fig. 5–4).

You may have observed that in our discussion we speak of survival and reproduction together. It is obvious that an organism will not reproduce if it fails to survive. However, it should be equally obvious that survival by itself is not enough. In terms of passing its genes on to future generations, an individual that lives to a "ripe old age" without ever reproducing is no different from one that died in infancy. Any effect on the genetics of future generations is accomplished only through reproduction. Furthermore, the genetic effect of an individual on the next generation is directly proportional to the number of offspring it produces that survive and reproduce in turn. Thus, fitness always involves both survival and reproduction.

5.2 Selection of Traits and Genes

Selection invariably takes place at the level of the *individual organism*—that is, it is the individual organism that survives and reproduces or else ends up as somebody's dinner. Yet, by differential reproduction—weeding out certain individuals while others reproduce—it is the *population* that gradually becomes adapted to its particular physical environment and the other species it interacts with. Because the genetic makeup of the individual often is the determining factor in its success or failure to survive and reproduce, differential reproduction leads to a change in the gene pool of the population. In this section, we explore more fully how characteristics are determined by genes, how genes are passed from one generation to the next, and how they may be altered. We begin with a brief consideration of the first issue.

Describe an organism in every detail, and you are describing its *traits*. A **trait** is any particular characteristic of the individual. Such characteristics include all features of physical appearance, metabolism, aptitude, and behavior. Body height, shape of the nose, eye color, body build, lung capacity, and size of the appendix are examples of *physical* traits. *Metabolic* traits include such things as sensitivity to allergens, digestive capacity, tolerance to heat or cold, and resistance to disease. Traits pertaining to *aptitude* refer to natural abilities, such as running, swimming, jumping, and,

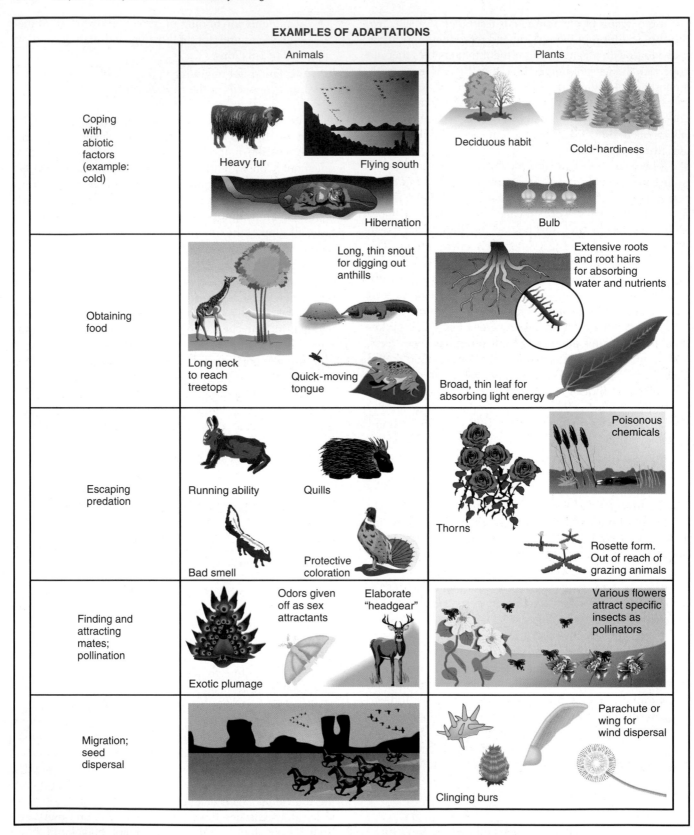

EXAMPLES OF ADAPTATIONS

▲ **FIGURE 5-4** *Adaptation for survival and reproduction.*   The five general features listed at the left are essential for the continuation of every species. Each such feature is found in each species in terms of particular adaptations that enable the species to survive or reproduce. Across the various species, a multitude of adaptations will accomplish the same function. Thus, a tremendous diversity of species exists, each adapted in its own special way.

ETHICS

SELECTION: NATURAL AND UNNATURAL

The fact that the gene pool of a species changes only through differential reproduction has tremendous practical, ethical, and moral implications for humans. First, we must restate a thoroughly discredited, but still commonly held, notion called the *inheritance of acquired characteristics*. This is the *false* notion that characteristics acquired by parents through training, learning, or accident will somehow be inherited by the offspring. Numerous experiments and common experience show that such acquired characteristics are not inherited, because they absolutely do *not* affect the DNA of the parent. The offspring will inherit the same genes, no matter what kind of skills the parents have developed.

A special aspect of this false notion concerns genetic disorders. Many congenital or genetic diseases, defined as conditions that are caused by certain mutant genes, can be held in check with medicines or treatments. Certain forms of diabetes, sickle-cell anemia, and hemophilia are examples. Medicines do not correct the defective gene, however. Thus, any offspring of those who are affected may inherit the disorder and require the same treatment.

You may have heard the statement "The human race is evolving toward bigger brains because we are required to use them," or "The human race is evolving toward smaller legs because we don't use them." These statements are basically expressions of the discredited notion of the inheritance of acquired characteristics. The only way such events would come to pass is if short-legged people consistently had more children than people with normal-size legs. The way to respond to such concerns regarding the future evolution of the human race is to reflect upon the following question: Are people with [name a characteristic, such as big brains or short legs] consistently reproducing more than aver-

age? If the answer is no, then there is no basis for projecting the change.

In short, the only way the genetics of a population will change from one generation to the next is through differential reproduction. The population will become increasingly dominated by the characteristics of those who reproduce the most, and a smaller and smaller proportion will bear the genetic traits of those who reproduce the least.

Let us speculate about what might be done with the human species if we could selectively breed humans. The study of improving the genetic future of humans is known as *eugenics* and has a cadre of serious advocates, but implementing any eugenics program collides with serious moral and ethical issues. Consider, for example, the ethics of one group dictating how many children another group should or should not produce. Who is to make the decisions, and how are they to be implemented? In China, families are allowed only one or two children—a direct effort to slow population growth. Traditionally, a male child is valued more than a female. If more males reach adulthood, will there be enough females for marriage? Take a moment and assess the outcome of this situation.

Yet, through suitable education, information, and understanding, some people may choose to practice a form of eugenics on a personal level. Prospective parents may obtain counseling regarding the genetic basis of any defect they may have and the probability of its being passed on to an offspring. They can then choose whether or not to bear their own children. Also, it is now possible to perform DNA analyses on a fetus to determine the presence or absence of a number of genetic defects. Parents can then make the decision whether to have the fetus aborted. Of course, this assumes that having an abortion is an acceptable option, a moral dilemma in itself.

A practice already widely used in animal breeding is to collect sperm samples from certain males—a prize racehorse or bull, for example. The sperm is kept frozen and then is used to artificially inseminate females whenever and wherever desired. Thus, the genes of the prize animal are passed on to a much larger portion of the next generation than would be possible through natural breeding. How do you feel about applying this practice to human breeding? Artificially inseminating women with sperm from an unknown man is already commonly practiced. Some have seriously proposed that the sperm of Nobel laureate men should be made available (at a high price!) to women so that they can give birth to highly intelligent children.

Advances in biotechnology have made possible some procedures only dreamed of a decade ago. Transgenic animals—animals bearing the genes from different kinds of organisms—are now commonly produced. Sheep bearing the human gene for the blood-clotting protein factor IX are now bred. The clotting factor is obtained in their milk and then purified and marketed. In 1997, Dolly captured the public eye—a sheep cloned from the nucleus of an adult sheep's mammary gland. Since then, cattle and pigs have also been cloned, and there seem to be no technical obstacles to cloning humans.

Through advances in genetic engineering, it may be possible in the relatively near future to manipulate and "correct" genes in the developing embryo. To what degree should such manipulation be conducted? Who should make the decision, and who will pay the costs?

In short, technologies for identifying, manipulating, and propagating particular genes are becoming increasingly available. More and more, we will have to face the ethical dilemmas involved in choosing to use or not to use these technologies.

in humans, additional things such as talent in mathematics or music. *Behavioral* traits are the instinctive ways organisms act, such as a spider spinning a web, a bird flying south for the winter, and sexual behavior.

Many traits may be modified or developed more fully by experience, learning, or training; examples are physical ability, hunting ability, tolerance to cold, and avoidance of

danger. However, the underlying basis of traits is *hereditary*, or *genetic*. That is to say, each individual (plant, animal, or microbe) is born with a certain potential for each trait as a result of the transfer of genetic material (genes) from the parents to the offspring in the process of reproduction.

Since the 1940s, research has made it abundantly clear that, regardless of species, the genetic material responsible for all traits is the chemical known as deoxyribonucleic acid (DNA), which resides in the cells of every organism (save some viruses that have a related chemical, RNA). DNA is a long, chainlike molecule arranged on structures called **chromosomes**, which are found in the nucleus of each cell. The sequence of four subunits (nucleotides) along the length of a strand of DNA provides a "code" analogous to the way the dots and dashes of the Morse code provide a code for various letters. The genetic code in DNA is translated into the production of proteins through the cell's metabolism.

Some of the proteins specified by DNA, called *structural proteins*, determine not only the physical structure of the body—whether the body is that of a mouse, an elephant, a human, or a turnip—but also such details as whisker length, nose shape, and hair texture (straight or curly). Another set of proteins, called *enzymatic proteins* or *enzymes*, catalyzes or speeds up all the chemical reactions that go on in the body. Other proteins may serve as hormones or control the production of hormones that in turn regulate overall body functions such as growth, development, metabolism, and sex; these are hormonal proteins. In short, DNA specifies proteins, and the proteins determine everything else (Fig. 5–5). One **gene** can be thought of as a segment of DNA that codes for one particular protein. The physical structure and functioning of every individual organism are the result of the coordinated interaction of many thousands of genes. The number of genes in humans is estimated to be between 30,000 and 40,000, arranged on 46 chromosomes.

For every individual, this interaction involves *two sets of genes* on chromosomes—one set received from the male parent by way of a sperm cell and the other received from the female parent by way of an egg cell. In that way, each individual manifests characteristics of both parents and also has certain unique traits stemming from the new combinations of genes. The genetic makeup of the new individual is determined when the sperm enters the egg and the two sets of genes combine in the process of fertilization. The fertilized egg divides and redivides. As the cells continue to divide and grow, they gradually develop through various stages into an embryo and so on. During each cell division, all the genes are replicated (copied) exactly, and each new cell resulting from the division receives a complete copy of both sets of genes. Humans, for example, have two sets of 23 chromosomes—one set from each parent. The chromosomes from each parent are quite similar, with the exception of the X and Y chromosomes—the sex chromosomes. Males receive an X chromosome from their mother and a Y chromosome from their father. Females receive two X chromosomes, one from each parent. Since a male has one X and one Y, and donates either one or the other via the sperm, the father's genetic contribution determines the sex of the child.

Genetic Variation and Gene Pools

It is easy to see that there is *variation* among individuals of the same species. The human species is a good example. We come in a wide range of sizes, shapes, and colors; our bodies have different chemical abilities, such as the ability to digest certain foods; we have different tolerances to various conditions; and we have widely different abilities apart from any training or teaching we may receive. Similar variation exists among the individuals of all other species: dogs, elephants, oak trees, mushrooms, flies, and on and on. For example, the only reason we tend to think of all houseflies as identical is that we do not examine them closely. When we do, it is not hard to find differences in size, color, or any other trait we choose to focus on.

These variations imply that there are variations in the genes controlling each particular aspect of the individual, and indeed, there are. Variations of a single trait such as eye color result from different forms of a particular gene for eye color. Different forms of a gene are called **alleles**. Thus, we

▶ **FIGURE 5–5** *DNA, genes, and traits.* Segments along the DNA's length code for the synthesis of specific proteins, which in turn determine and control physical structure, metabolism, and other hereditary attributes of the organism, as indicated.

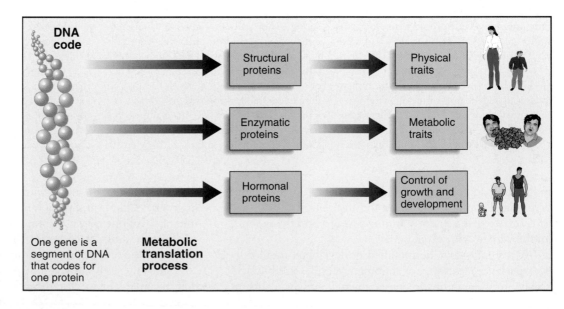

say that there are two major alleles for eye color: blue and brown. Other, less common alleles for eye color exist as well, giving rise to other colors sometimes seen. Similarly, differences in natural hair color, height, blood type, and so on all fundamentally result from different versions or alleles of the particular genes in question. Likewise, hereditary diseases such as sickle-cell anemia, hemophilia, cystic fibrosis, and Huntington's disease are the result of alleles of particular genes that fail to function as the normal gene should.

Since each individual has two sets of chromosomes—one from the father and one from the mother—each individual carries two alleles for each gene. The two alleles for any one gene may be different from each other, or they may be identical. How they interact determines the trait that the individual shows. You are probably familiar with the ideas of dominance and recessiveness as they apply to eye color. The allele for brown eye color dominates over that for blue. Hence, brown eyes result from either the presence of two alleles for brown eyes or the presence of just one allele for brown and one for blue. Blue eyes, the recessive condition, result only when the allele for brown is absent—that is, when both alleles are for blue. By the same token, can you see how a person may carry an allele for a genetic disease without showing any symptoms of that disease?

Understanding that each individual has two alleles (which may be the same or different) for each characteristic is important, because, in the formation of an egg or sperm cell, there is a segregation process such that the egg or sperm receives *one and only one* allele for each characteristic. Furthermore, if the alleles are different, there is a 50–50 chance that either of the two will end up in any given egg or sperm. Said another way, half the sex cells will carry one of the alleles, and half will carry the other. Then, as egg and sperm unite in fertilization, the alleles recombine so that, again, there will be two for each gene, but one will have come from each parent. Can you understand from this how two brown-eyed parents may have a blue-eyed child? (See Fig. 5–6.)

Such examples show how a particular characteristic may be determined by the alleles of just one gene. These cases provide a basic understanding of the process involved. However, many characteristics, such as overall body stature, may be influenced by numerous genes, each with different alleles playing a greater or lesser role in the final manifestation. Thus, one may have any of the gradations between the extremes of the particular characteristic, as well as the extremes themselves. Other exceptions that you may be familiar with in humans are the so-called sex-linked

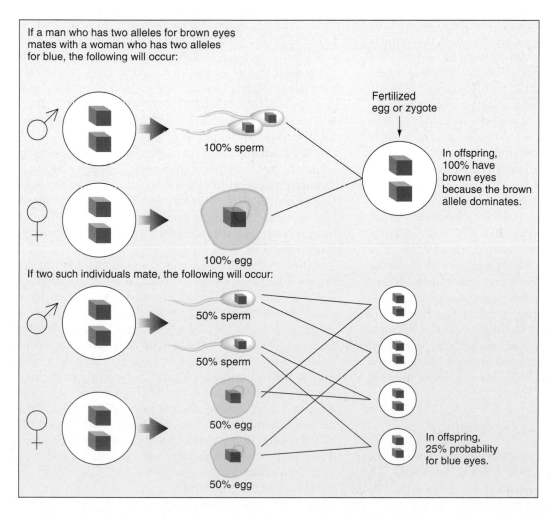

If a man who has two alleles for brown eyes mates with a woman who has two alleles for blue, the following will occur:

100% sperm

Fertilized egg or zygote

100% egg

In offspring, 100% have brown eyes because the brown allele dominates.

If two such individuals mate, the following will occur:

50% sperm

50% sperm

50% egg

50% egg

In offspring, 25% probability for blue eyes.

◀ FIGURE 5–6
Inheritance of eye color. The gene for eye color has two common alleles: one for brown eyes and one for blue eyes. Matings with different combinations of these alleles are shown.

characteristics, such as color blindness. The alleles for these characteristics are located on the X chromosome and do not have a counterpart on the male Y chromosome. Males carry only one X chromosome (inherited from the mother), and characteristics on the X chromosome will be automatically expressed. Also, keep in mind that an individual is always the sum total of all its many thousands of traits, each being determined by the appropriate gene or combination of genes.

In sum, the different combinations of alleles in each individual are responsible for the genetic variation among the individuals of all species.

Let us now relate this idea of different alleles to the concepts of DNA and the gene pool of a species discussed earlier. Each different allele is a slightly different piece of DNA coding for a slightly different protein that causes some variation of the trait in question. The total of all the alleles in the population represents the gene pool of the population. Note again that some alleles will be very widespread in the population, while other alleles may be rare, meaning that only a few individuals carry or have the particular traits associated with them.

Thus, we should try to visualize the reproduction of a species as far more complicated than popping out additional copies in cookie-cutter fashion. We should visualize the reproduction of a population over generations as a constant recombining of all the alleles within the population's gene pool. With so many genes and alleles involved, it is virtually certain that each individual will be genetically unique (that is, have a unique combination of alleles). Obvious exceptions exist, of course—namely, identical twins and *clones*.

Identical twins arise in the following way: Instead of the fertilized egg developing into a single embryo, the two cells formed by the first division separate and each goes on to form an embryo and, hence, two individuals. We have noted that in regular cell division, all the DNA (the set of genes) is replicated exactly. Therefore, the two cells derived from the same fertilized egg and, subsequently, the two individuals who develop from that egg are genetically identical.

Clones, which are populations of genetically identical individuals, are propagated by asexual reproduction. The most common example is the propagation of plants by means of cuttings or grafts. Such propagation involves only regular cell divisions in which all genes are replicated exactly. Therefore, each "offspring" is an exact genetic copy of the "parent." For example, all McIntosh apples are the same because all McIntosh apple trees are a clone propagated from cuttings from the original tree. But how did that original McIntosh variety occur? We address this question in the next section.

Mutations: The Source of New Alleles

We have noted that the forces of environmental resistance acting on a population serve to "weed out" individuals with any traits that handicap any aspect of survival or reproduc-

tion. Thus, those alleles and combinations of alleles responsible for such traits are removed from the population, while those which enhance survival and reproduction become more abundant. If this weeding out of less desirable, and reproduction of more desirable, traits were the only process at work, however, we should expect a limit to be reached as the most advantageous combination of alleles is attained.

Experience gained through selective breeding shows that such limits do not occur. Breeders find that, regardless of how far the development of a given trait is pursued, they still observe a variation that provides the basis for further enhancement of the trait in question. Thus, chickens having tail feathers 15 feet (5 meters) long have been bred, and still the breeding for longer feathers continues.

The potential for genetic change seems to have no boundaries, except insofar as it may lead to a nonfunctional organism. For example, breeders have found that they can produce a dachshund with legs so short that it is unable to walk.

Modern knowledge of DNA allows us to understand this potential for unlimited genetic change. In the course of cell division and reproduction, the DNA molecules are generally copied exactly, but occasionally mistakes occur. A **mutation** is any change in the DNA molecule; it could be as simple as one nucleotide exchanging for another in the code of a gene. Such a change may cause the protein coded by the changed segment to be altered. In turn, the trait determined by that protein will be modified in one way or another. In this manner, mutations introduce new alleles into the population and, hence, new modifications of traits. If a mutation involves a large change in the DNA, one or more proteins may be so altered that the organism fails to function and consequently dies. Mutations that result in death are termed **lethal mutations**.

Very important is the fact that *mutations are random events*. So far, there is no known way, either in nature or through human technology, of causing a specific mutation to cope with a particular selective pressure. Indeed, like randomly connecting wires in an engine, most mutations result in modifications that are harmful; only rarely is one beneficial.

How does nature cope with the problem of sorting out the good mutations from the bad? Again, it acts through the process of natural selection. Any individual having a mutation resulting in a handicap will usually perish before it can reproduce. Thus, the harmful allele is eliminated from the gene pool. The individual with the rare mutation (allele) that enhances survival, however, is likely to produce offspring. As these offspring, which inherit the new trait, reproduce in turn, the population will comprise more and more individuals that carry the new allele. Mutations that neither benefit nor harm their owners are called **neutral**. Such neutral alleles continue to be reproduced in the same proportions in the population and manifest themselves as the variations we see among the individuals of the population.

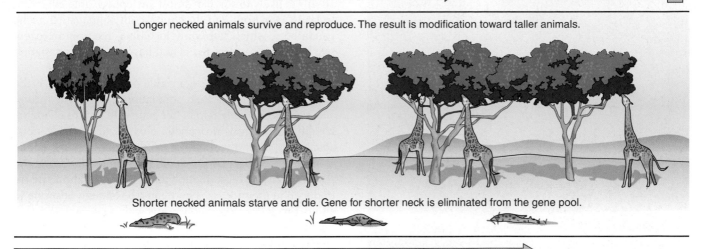

Selective pressure toward a longer necked, taller animal

Selective pressure for taller animal stops

Longer necked animals survive and reproduce. The result is modification toward taller animals.

Shorter necked animals starve and die. Gene for shorter neck is eliminated from the gene pool.

Many generations

(a)

5.3 Changes in Species and Ecosystems

Understanding the process of how the gene pool of a population may be changed through mutation, breeding, and natural selection enables us to comprehend how species evolve. We need only recognize, as did Darwin, that members of the same population are invariably in competition with one another for survival and that more offspring are produced than can survive. Individuals with any characteristic that provides an advantage over their cohorts stand a better chance of survival and reproduction. Thus, the alleles for the beneficial characteristic are differentially reproduced, while nonadaptive alleles are eliminated from the population. For example, if a population takes to browsing on leaves overhead as its main source of food, it will be the individuals with the longest necks that get the most food and produce the most offspring; shorter individuals will perish. As such natural selection occurs over many generations, the end result, as you may guess in this case, will be the giraffe (Fig. 5–7). Similarly, a population that takes to feeding on ants and termites will, over generations, be modified to be more and more proficient in this regard, and we will have the giant anteater of Brazil (Fig. 5–8).

The Limits of Change

When facing a new selective pressure—such as a different climatic condition or a new species invading the ecosystem—species have three, and only three, alternatives:

1. Adaptation. The population of survivors may gradually adapt to the new condition through natural selection.

(b)

▲ **FIGURE 5–7** *Selective pressure.* (a) The process of natural selection favoring giraffes with longer necks. (b) The Masai giraffe in its native habitat.

2. Migration. Surviving populations may migrate and find an area where conditions are suitable to them.

3. Extinction. Failing the first two possibilities, extinction is inevitable.

▲ FIGURE 5–8   *Adaptation of giant anteater.*   The major herbivores (in terms of consumption of biomass) on the savannas of Brazil are termites that build large earthen mounds as nests. The giant anteater, actually a termite eater, has evolved huge forelimbs and claws that it uses to break open the mounds and a head and mouth structure adapted to feeding on the termites.

Migration and extinction need no further explanation. The critical question is, What factors determine whether a species will be able to adapt to new conditions, as opposed to its failing and becoming extinct?

We can reason out the answer to this question if we return to the basic concept that adaptation occurs by selective pressures eliminating all those individuals that cannot tolerate the new condition. *For adaptation to occur, there must be some survivors*—individuals with traits (alleles) that enable them to survive and reproduce under the new conditions. Furthermore, there must be enough survivors to maintain a viable breeding population. If a breeding population is maintained, one can expect the process of natural selection to lead to increased adaptation over successive generations. Of course, if a viable population is not maintained at any stage, extinction is assured. For example, the California condor population had declined to about 20 birds by the early 1980s. To save the species from extinction, wildlife biologists captured the remaining wild condors and initiated a breeding program. The population is now over 150 individuals, and about 50 birds are now living in the wild. However, this is still not a viable breeding population.

There are four key variables among species that will affect whether or not a viable population of individuals is likely to survive new conditions: (1) geographical distribution, (2) specialization to a given habitat or food supply, (3) genetic variation within the gene pool of the species, and (4) reproductive rate relative to the rate of environmental change.

Consider how these factors work. It is unlikely that any change will affect all locations uniformly or equally. Therefore, a species such as the housefly—which is present over most of Earth, does not have requirements for a specialized habitat or food supply, and has a high degree of genetic variation—is likely to survive almost any conceivable alteration in Earth's environment. In contrast, an animal such as the panda bear, which lives only in China, requires an extensive habitat of bamboo to eat and has little genetic variation in its population.

Beyond the survival of a viable population, there is the question of how fast that population can evolve further adaptations to enable it to better cope with new or changing conditions. There is no genetic change, hence no genetic adaptation, over the lifetime of the *individual*. Genetic change occurs only as the genes are "shuffled" during the process of sexual reproduction and as new mutations are introduced in the process. There must be enough offspring to afford the loss of those offspring with maladaptive traits. Again, we see that species such as houseflies, which have a generation time measured in a few weeks and which produce several hundred offspring in each generation, are in a position to evolve much more rapidly than large animals.

Of course, the rate at which changes in the environment occur is an important consideration. If the environmental changes are very slow, populations of slowly reproducing species may even be able to adapt. When changes in environmental conditions occur more rapidly, however, more and more species will be eliminated in the contest to adapt accordingly. These factors are summarized in Fig. 5–9.

Speciation

The infusion of new variations from mutations and the pressures of natural selection serve to adapt a species to the biotic community and the environment *in which it exists*. In this process of adaptation, the final "product"—giraffe, anteater, redwood tree—may be so different from the population that started the process that it is considered a different species. This is one aspect of the process of **speciation**.

The same process also may result in two or more species developing from one. There are only two prerequisites. The first is that the original population separate into smaller populations that do not interbreed with one another. This **reproductive isolation** is significant, because if the subpopulations continue to interbreed, all the alleles will continue to mix through the entire population, keeping it as one species. The second prerequisite is that separated subpopulations be exposed to different selective pressures. As the separated populations adapt to these different selective pressures, they may gradually become so different as to be considered different species and thus be unable to interbreed with one another, even if they later come together again.

For example, consider our present arctic and gray foxes. It is assumed that some thousands of years ago a northern subpopulation became separated from the main population, perhaps as a result of the effects of glaciation in North America. In the Arctic, selective pressures favor individuals that have heavier fur, a shorter tail, legs, ears, and

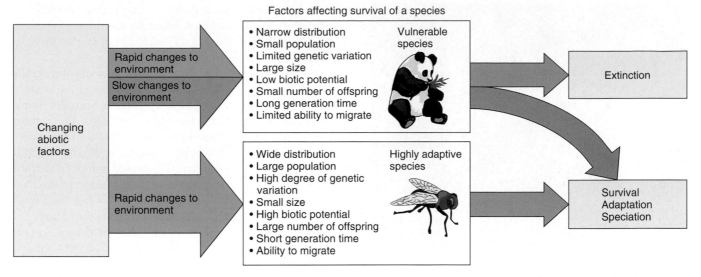

Factors affecting survival of a species

▲ FIGURE 5–9 *Vulnerability of different organisms to environmental changes.* A summary of factors supporting the survival and adaptation of species, as opposed to their extinction.

nose (all of which help conserve body heat); and a white color (which helps animals hide in snow). In the southern regions, selective pressures regarding adaptation to temperature and background color are the reverse. Mutations creating alleles adaptive for the Arctic would actually be harmful in southern animals, because in warmer climates animals need a thinner coat to dissipate excessive body heat, and a white coat would make the animals more conspicuous. The geographic isolation of the two subpopulations over many generations made possible the fixation of these sets of traits into two species (Fig. 5–10).

Within the plant kingdom, tundra plants provide striking examples of speciation. They have a short, dense, mosslike vegetative structure, yet, as the similarity of their flowers reveals, they are closely related to taller plants at lower elevations (Fig. 5–11). It is not hard to visualize ice crystals driven by strong winds being the selective pressure propelling tundra plants toward their compact form. Such crystals are common at high elevations in the Arctic and destroy any plant taller and less dense.

A renowned example of speciation is one first observed by Darwin himself. It is the 14 species of finches living on

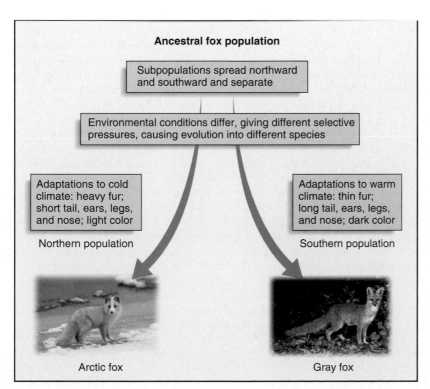

Arctic fox

Gray fox

◀ FIGURE 5–10 *Evolution of new species.*
A population spread over a broad area may face various selective pressures. If the population splits so that interbreeding among the subpopulations does not occur, the different selective pressures may result in the subpopulations evolving into new species, as shown here for the arctic fox and the gray fox.

▲ **FIGURE 5–11** *Adaptation in tundra plants.* Frequently subjected to the harsh abrasive force of ice and sand particles driven by high winds, taller vegetation is destroyed, giving way to tundra plants that have low, dense, mosslike habits.

the Galápagos Islands. Darwin was struck by the differences between finches on different islands. By similarities in their overall body structure, Darwin could see that all were clear-

ly members of the finch family. He wondered whether the differences in size and beak structure could have appeared after the subpopulations were isolated from one another on separate islands (Fig. 5–12).

Although Darwin himself did not work out the details of how the different Galápagos finches evolved, it is likely that at some time in the past a few finches from the South American mainland were blown westward by a strong storm and became the first terrestrial birds to inhabit the relatively new (10,000-year-old) volcanic islands. As the initial population grew and members faced increased competition from one another, groups dispersed to nearby islands, where they were separated from the main population. These subpopulations encountered different selective pressures and became specialized for feeding on different things (cactus fruit, insects). In time, when the changed populations dispersed back to their original islands, they were different enough from the parent species to be distinguishable as new species, so interbreeding among them did not occur. Since the process of speciation is a gradual change occurring over many generations, you might ask, At what stage in the process do the separated populations become different enough to be considered different species? Indeed, this presents a problem for those who name species. (See the "Earth Watch" essay, this page.)

EARTH WATCH

WHAT IS A SPECIES?

A species was originally, and to a considerable extent still is, defined simply as a distinct kind of organism. The problem is, What constitutes a "distinct kind"? For example, two populations of tree frogs, one in New England and one in Georgia, appear to be distinct enough to be classified as different species. In the intervening states, however, we find tree frog populations that, when all put together, create a smooth transition of variations from the New England to the Georgia populations. Should all these frogs be classified as one large species with much variation, as two species, or as many very similar species? This problem exists with numerous kinds of plants and animals, and there is no clear-cut answer.

Taxonomists, the term for those who make a profession of classifying organisms, fall into two philosophical camps: splitters and lumpers. Splitters recommend dividing groups, such as the frogs just mentioned, into multiple species; lumpers recommend putting them all into one species.

In an attempt to remedy the situation, the definition of *species* was amended to in-

clude the aspect of interbreeding. The current definition also recognizes the ecological and geographical properties of the species. Thus, we have the *biological species concept*, which states that a species is a group of interbreeding natural populations that does not breed with other groups, because of behavioral or physiological barriers. If interbreeding occurs among individuals and the offspring are fertile, then they are of one species, regardless of how different they may appear. Conversely, if interbreeding does not occur between two groups, they should be considered separate species. Or if two different populations interbreed, but produce sterile offspring, as in the case of horses and donkeys producing mules, which are sterile, the two populations are likewise different species.

This definition may help in an intellectual sense, but often does not help in a practical sense, mainly because interbreeding is often impractical or impossible to test or observe. Similar-looking animals may not interbreed simply because they occupy different geographical ranges. Or, although

natural interbreeding may not occur between two populations of similar animals (which would indicate that they were separate species), when members of the two populations are placed together *in captivity*, they may interbreed readily. This situation is even more problematic in plants that do not readily cross-pollinate in nature, but that can be crossed by artificial pollination. The offspring from populations that generally do not interbreed in nature are termed **hybrids**; many species form hybrids that never go on to breed or affect the parent species' gene pools.

So what is a species, then? We must conclude that the concept of a species as a particular kind of organism distinct from all others is a hypothetical construct of the human mind. (We like to categorize things and put them in discrete pigeonholes.) However, if species are undergoing a process of separation and change, then examples such as those described are what we should expect to observe. For the sake of discussion, though, we still need to use the term *species* to refer to kinds of organisms.

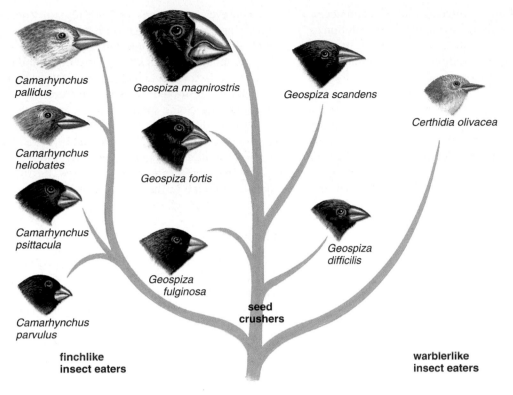

Camarhynchus pallidus

Geospiza magnirostris

Geospiza scandens

Certhidia olivacea

Camarhynchus heliobates

Geospiza fortis

Camarhynchus psittacula

Geospiza difficilis

Geospiza fulginosa

seed crushers

Camarhynchus parvulus

finchlike insect eaters

warblerlike insect eaters

ancestral South American finch

(a)

(b)

(c)

▲ **FIGURE 5–12** *Some of Darwin's finches.* (a) The similarities among these birds attest to their common ancestor. Selective pressures to feed on different foods have caused modification and speciation in adapting subpopulations. (b) The insect-eating *Certhidia olivacea.* (c) The large seed-crushing beak of *Geospiza magnirostris.* Notice the difference in beak between (b) and (c) (From p. 310 in BIOLOGY: Life on Earth 4/e by Teresa Audesirk & Gerald Audesirk. Copyright © 1994 by Prentice Hall, Inc. Reprinted by permission of Pearson Education, Inc., Upper Saddle River, NJ 07458.).

In sum, the key feature of our understanding of speciation is that new species are not formed from scratch; they are formed only by the gradual modification of existing species. In addition, the gene pool of a species may be molded in many different directions when different populations within the species are isolated and subjected to different selective pressures, as illustrated by Darwin's finches. This concept is entirely consistent with the fact that in nature we generally find groups of closely related species rather than distinct species with no close relatives. (Recall

the five species of warblers in Fig. 2–16). It is also consistent with, and explains the fact that, all living things can be classified into a few relatively major groups. All the members of these major groups can be seen as variations on the same theme and, hence, derived from a common ancestor.

Although it is difficult to imagine, the present array of plants, animals, and microbes, in all of its diversity, is likely to have originated through speciation over long periods of time and in every geographic area on Earth. This, then, is the source of our current biodiversity.

How Rapid Is Speciation?   Since natural selection necessarily acts on each generation, Darwin pictured evolution as a slow, but continuous, modification of species to become ever better adapted to their surroundings, a thesis referred to as *gradualism*. However, a more detailed picture of the fossil record, developed since Darwin's time, shows that the evolution of many species, plotted against time, has been a stepwise process, with long periods of relative stability (called *stasis*) and quite short periods of rapid change. This is the theory of *punctuated equilibrium* (Fig. 5–13). Instead of natural selection working by the accumulation of small changes over long periods of time (millions of years), the fossil record indicates that species might undergo changes so abruptly—on the order of thousands rather than millions of years—that the intermediate forms would be relatively rare and would not likely be found. After this theory was proposed (by Gould and Eldridge in 1972), a great deal of debate ensued. It is fair to say that both gradual changes and punctuations may be found in the fossil record, depending on the species examined, and that both processes are important in the workings of evolution.

We have seen that ecosystems tend to develop sustainable dynamic balances among species. As this occurs, selective pressures on species within a system reach an equilibrium. After a species becomes well adapted for survival and reproduction in its particular ecosystem, there is no great selective pressure for further adaptation, especially since specialized adaptations have their disadvantages as well as advantages. For example, a giraffe with a neck longer than necessary to reach the treetops has no advantage over other giraffes and may in fact suffer a disadvantage: It is harder for that giraffe to bend over for a drink of water than it is for its shorter necked companions. Thus, selective pressures tend to preserve the status quo—*as long as conditions remain constant.*

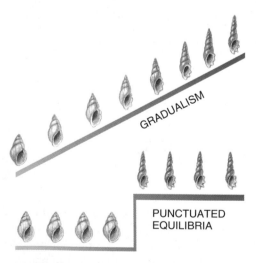

▲ **FIGURE 5–13** *Punctuated Equilibrium.* Has the process of speciation proceeded by gradual changes over long periods of time or by long periods of no change punctuated by rapid bursts of speciation?

However, conditions do not remain constant. The coming and going of ice ages with associated changes in sea level, the uplifting of mountain ranges, and massive volcanic eruptions are well known. As climate and other environmental factors are changed by such events, existing balances are disrupted; selective pressures for new conditions intensify. Some species will not be able to cope with the new conditions, and they will become extinct. Populations of surviving species, however, may evolve and speciate relatively rapidly (geologically speaking) under the intensity of new selective pressures and, perhaps, the availability of vacant niches left by those species which died out.

Evolving Ecosystems?

To understand the concept of change through natural selection, we have focused on how a species may be modified in a particular way by a specific selective pressure. However, each species is always living and reproducing within the context of an entire ecosystem, and therefore, it is simultaneously adapting to a host of biotic and abiotic factors present in the ecosystem. By the same token, all the species in the ecosystem are *simultaneously adapting to each other!* Does this mean, then, that there is a process of natural selection acting at the ecosystem level as well as at the species level?

This is a question that has been debated for many years. Most evolutionary scientists believe that natural selection operates only on individuals, eliminating or favoring them and the genes that they bear. The species is affected by this process when a significant number of individuals are subjected to similar selection, and thus the gene pool of the species is changed. While we do not think natural selection acts at the ecosystem level, we know that ecosystems do change over time as a result of the impacts of biotic and abiotic factors. (Recall the process of succession in Chapter 4.) Hence, an ecosystem is affected by the species found within it, and the basic characteristics of any ecosystem are determined by the interactions of the species and the abiotic factors within it.

For example, the Serengeti region of East Africa is occupied by populations of large herbivores, some of which feed on grasses and herbs (wildebeest, zebra, Cape buffalo), and some of which feed on bushes and trees (giraffe, elephant) (Fig. 5–14). Elephants are notorious for pulling down trees to feed on the bark and leaves, and their presence in the Serengeti normally maintains a healthy mixture of forested areas and grasslands. If elephant populations are removed—as they have been recently by ivory poachers—whole regions in the Serengeti will undergo a change from grasslands to woodlands. Conversely, if elephants become too numerous, as they have in some of Kenya's national parks, their presence will cause the decline of forests and the appearance of pure grasslands within a few years.

Thus, if relationships within an ecosystem are affected by the removal or introduction of a dominant organism or by a change in climatic conditions, the entire ecosystem may develop into one that bears little resemblance to its

◀ **FIGURE 5–14** *Dominant organism in an ecosystem.* In some parts of the Serengeti in East Africa, elephants are numerous enough to be major ecosystem engineers. In time, their activities can change a woodlands ecosystem to a grasslands ecosystem. Here, one elephant is seen feeding on a tree.

precursor. We do not then say that the ecosystem has evolved; rather, it has developed into a different *kind* of ecosystem because of changing physical conditions or biota. It still has all of the functional components of an ecosystem: producers, herbivores, carnivores, decomposers, and so on. But it now has a different cast of characters—a different biotic community—adapted to the new set of conditions. They may have come by various means of dispersal, possibly from similar ecosystems not far distant. Over time, species in these changed ecosystems or biomes (forests, grasslands, tundra) continue to evolve in ways that adapt them to particular abiotic factors (such as temperature, moisture, and light) and to particular roles in the entire system as a result of selective pressures brought on by the presence of other organisms. And these adaptations to abiotic factors or to other organisms contribute to the remarkable resilience of ecosystems. Indeed, such adaptations are the primary source of ecosystem resilience.

But why have species evolved so differently on different continents and remote islands, even though the climates and other environmental factors in these regions may be very similar? This is explained by the fact that natural selection can modify only preexisting species. Thus, a landmass will bear species that are modifications of the species that were there when the landmass became isolated from other landmasses or that managed to disperse to the isolated landmass. For example, when Australia broke away from a central landmass some 65 million years ago (via shifting plates of Earth's crust; see Section 5.4), marsupials (pouched animals) were present, but there were no placental mammals. Therefore, the animal life on Australia today is marked by many modifications (species) of marsupials—in particular, kangaroos (Fig. 5–15)—while other continents have modifications of the placental mammals.

Likewise, giant tortoises are the major herbivores on the Galápagos Islands (Fig. 5–16), a reflection of the fact that ancestral turtles dispersed to the islands, probably on floating logs, after the islands arose from volcanic action, and other herbivores did not. Similarly, before humans brought in cattle, the grasslands of southern Brazil had essentially no large herbivores at all; termites largely occupied the primary consumer niche, and giant anteaters, which actually eat termites (Fig. 5–8) were the major "carnivore."

▲ **FIGURE 5–15** *Marsupials in Australia.* In Australia, the marsupials present 65 million years ago evolved into the prime grazing animals of today. Nowadays, kangaroos and other marsupials dominate the grasslands.

▲ **FIGURE 5–16**　**The Galápagos tortoise.**　On the Galápagos Islands, mammals were not present, but turtles were. Therefore, tortoises evolved as the prime grazing animal.

the past—*plate tectonics*. This phenomenon has had a decisive influence on the course of evolution.

5.4　Plate Tectonics

Many scientists are working on understanding the causes behind shifting weather patterns and long-term changes in climate so that we can predict what the future climate of any given region will be as a result of either human or natural causes. To date, the ability to make exact predictions is still elusive. However, everyone agrees that over the long term, change is inevitable. One slow, but spectacular, global change is that continents are continually on the move toward different relative positions on Earth; their movement helps us to understand earthquakes and volcanic activity. Let us look at the theory behind this movement of continental landmasses, known as **plate tectonics**.

Tectonic Plates

The interior of Earth is molten rock kept hot by the radioactive decay of unstable isotopes remaining from the time when the solar system was formed, about 5 billion years ago. Earth's crust, which includes the bottoms of oceans as well as the continents, is a relatively thin layer (ranging from 10 to 250 kilometers) that can be visualized as huge slabs of rock floating on an elastic layer beneath, much like crackers floating next to each other in a bowl of soup. (An elastic layer of rock consists of rock that flows under heat and pressure.) These slabs of rock are called **tectonic plates**. Some 14 major plates and a few minor ones make up the Earth's crust (Fig. 5–17).

Thus, although ecosystems do not evolve in the Darwinian sense, they do undergo changes as a result of the evolution of species found in them. Different landmasses will therefore reflect unique ecosystems because of their long isolation and the unique dispersal of certain organisms to them. It is clear, then, why species introduced from one region to another by humans may bring about major disruptions: because the species already present in the region have not evolved adaptations to the intruder.

Let us now turn our attention to one of the most important changes in global conditions that has occurred in

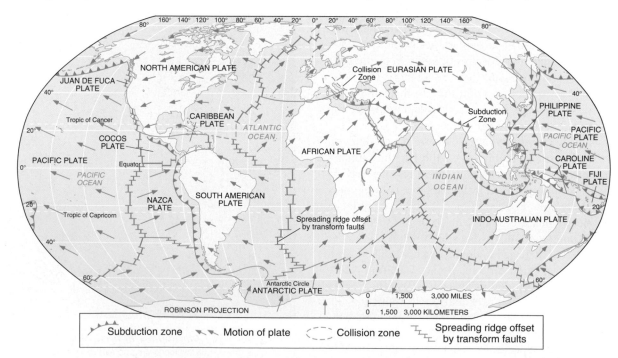

▲ **FIGURE 5–17**　**Earth's major tectonic plates.**　The 14 major tectonic plates making up Earth's crust and their directions of movement. The arrows in the figure indicate 20 million years of movement. (From GEOSYSTEMS: Introduction to Physical Geography 3/e by Robert W. Christopherson. Copyright © 1997 by Prentice Hall, Inc. Reprinted by permission of Pearson Education Inc., Upper Saddle River, NJ 07458.)

Tectonic plates are not stationary. The evidence indicates that about 225 million years ago all the continents were positioned as one major continent, which we now call *Pangaea* (Fig. 5–18a). Within Earth's semimolten interior, hot material rises toward the surface and spreads out at some locations, while cooler material sinks toward the interior at other locations. Riding atop these convection currents, the plates move slowly, but inexorably, with respect to one another, as crackers might move if the soup below were gently stirred. The spreading process of the past 225 million years has brought the continents to their present positions and accounts for the other interactions between tectonic plates (Fig. 5–18b). The average rate of a plate's movement is about 6 centimeters per year, but over 100 million years this adds up to almost 8,000 kilometers in the fastest moving segments. Movement of the crust itself, by contrast, is *not*

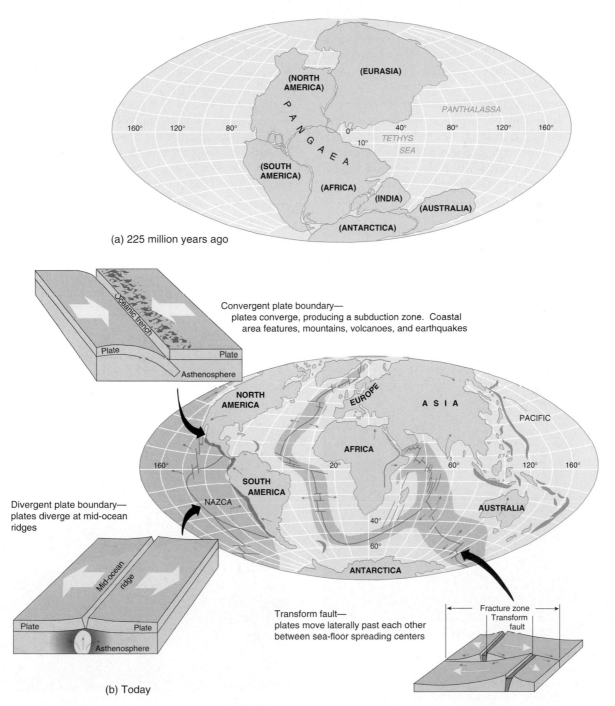

(a) 225 million years ago

Convergent plate boundary—
plates converge, producing a subduction zone. Coastal area features, mountains, volcanoes, and earthquakes

Divergent plate boundary—
plates diverge at mid-ocean ridges

Transform fault—
plates move laterally past each other between sea-floor spreading centers

Fracture zone
Transform fault

(b) Today

▲ **FIGURE 5–18** *Drifting continents.*   (a) Similarities in types of rock, the distribution of fossil species, and other lines of evidence indicate that 225 million years ago all the present continents were formed into one huge landmass that we now call Pangaea. (b) Slow, but steady, movement of the tectonic plates over the intervening time caused the breakup of Pangaea and brought the continents to their present positions. (From GEOSYSTEMS: Introduction to Physical Geography 3/e by Robert W. Christopherson. Copyright © 1997 by Prentice Hall, Inc. Reprinted by permission of Pearson Education Inc., Upper Saddle River, NJ 07458.)

gradual, for the plate boundaries are locked by friction and hence are regions of major disturbances.

You can understand plate tectonics by pressing your thumb firmly on a tabletop and simultaneously sliding it across the surface. You will find that your thumb moves in jerks as it sticks, jumps, sticks, and jumps. Analogously, a plate boundary may not move for decades and then will snap suddenly during dramatic events—namely, earthquakes and volcanic eruptions—as pressures from movement gradually build until a break occurs. The earthquake or volcanic eruption releases the pressures and is then followed by a quiescent period during which pressures gradually build again, eventually causing another break or eruption. So you can see how misleading an average rate of motion can be.

Adjacent tectonic plates may move with respect to each other in three basic ways. First, where rising convection currents of molten material reach the surface, *divergent boundaries* are formed as plates are forced apart and the gap between them is filled with molten material that solidifies. Such regions are presently seen as mid-ocean ridges and rift valleys, and there is considerable volcanic activity in these regions.

In the second kind of interaction, two plates may gradually slide past each other, creating *transform boundaries*. An example of this sliding can be seen in the United States with the San Andreas fault in California, which marks the boundary between the Pacific plate, which is moving in a northwesterly direction, and the North American plate, which forms the bulk of the United States (Fig. 5–19).

Along this fault, there is a major earthquake every 50 to 100 years as the pressures built up by the movement of 1–2 centimeters per year are suddenly released in a jump of 1–2 meters. The last two major slippages on the central California stretch of the San Andreas fault were the Loma Prieta earthquake of 1989 and the Los Angeles earthquake of 1994. These two earthquakes caused over $20 billion in damage and took 113 lives, and neither was close to the magnitude of the great San Francisco earthquake of 1906. California residents live with the knowledge that the "big one" could happen anytime.

In the third kind of interaction, plates move toward each other, forming *convergent boundaries*. *Subduction* occurs when an *oceanic* plate slides under a *continental* plate. This type of movement is currently taking place in the northwestern United States, where part of the Pacific plate is sliding under Oregon and Washington. There are periodic earthquakes because the sliding is not smooth, but in this case another dramatic event also occurs: As the edge of the plate that is sliding under is forced down beneath the continental crust, it melts and periodically erupts to the surface. Over geologic time, this movement has produced a volcanic mountain chain from northern California through Oregon and Washington. The most recent major eruption in the Cascade Mountains was that of Mount Saint Helens in 1980 (Fig. 5–20).

▲ **FIGURE 5–19   *San Andreas Fault.*** The narrow valleylike scar running from the top to the bottom of the picture is the San Andreas fault as it crosses the Carrizo Plain about 300 miles south of San Francisco. Every 50 to 100 years, mounting pressures of plate movement cause the fault to rupture in a major earthquake, and the left side moves away from the viewer a distance of 3 to 6 feet relative to the right side. Hills on either side of the fault are pressure ridges formed as a result of many movements.

▲ **FIGURE 5–20   *Volcanic eruption.*** Volcanic eruptions, such as that of Mount Saint Helens in southern Washington state on May 18, 1980, are a result of tectonic movement in which one plate off the coast is sliding under the continental plate.

In the interaction described in the previous paragraph, two continental plates converge, resulting in a *continental-plate collision*. Again, this produces periodic earthquakes, but the gradual result is the crumpling and uplifting of the plates into mountain ranges similar to the way the hoods of colliding cars crumple. An example of mountains caused by this type of movement is the Himalayas, which resulted from the plate that bears India moving northward and crashing into the Eurasian plate.

In addition to the periodic catastrophic destruction that may be caused in localized regions by earthquakes and volcanic eruptions, tectonic movement may gradually lead to major shifts in climate in three ways. First, as continents gradually move to different positions on the globe, their climates change accordingly. Second, the movement of continents alters the direction and flow of ocean currents, which in turn have an effect on climate. Third, the uplifting of mountains alters the movement of air currents, which also affect climate. (For example, see the rain shadow effect in Fig. 9–6).

The fact that volcanic eruptions and earthquakes continue to occur is evidence that tectonic plates are continuing to move today as they have over millions of years. In fact, volcanoes and earthquakes mark the boundaries between the plates and have provided some of the best evidence for the plate tectonic theory. Because we find every region on Earth occupied by well-adapted living organisms, it is clear that they have been able to evolve, speciate, and adapt to these geological changes. Indeed, we can understand the present-day distribution of plants and animals only in the light of continental drift and plate tectonics. (Recall our earlier remarks about Australian marsupials.) To get another perspective on the past, let us briefly examine the evolution of species as it is revealed in the fossil record.

5.5 Evolution in Perspective

The Fossil Record

The fossil record provides us with one of the strongest arguments for the theory of evolution. An overview of the major events in the evolution of species, as it is revealed in the fossil record, is shown in Fig. 5–21. One way to appreciate the relative periods involved in the different episodes of evolution is to make the entire geological history of Earth, 4.6 billion (4,600 million) years, equivalent to 1 calendar year. With that stipulation, each day on our calendar represents about 12.6 million years; each hour is thus 525,000 years. At first, the planet was entirely molten, and it took several hundred million years—till the end of February on our calendar—for the surface to cool sufficiently to allow life. By the end of March (3.5 billion years before the present [BP]), bacterialike organisms are in evidence. The first photosynthesis began in cyanobacteria at about this time, beginning the accumulation of atmospheric oxygen gas. Nothing other than bacteria existed for another four months (1.5 billion years). Then, in late July (2 billion years BP), the first more complex cells typical of today's higher life forms are seen. Still, these cells seem to have done little more than reproduce as single cells or filaments of cells for nearly another four months (1.5 billion years).

During the third week of November (600 million to 500 million years BP), an episode known as the **Cambrian explosion** occurs in which there appear the representatives of all the major groups of organisms, including the first primitive forerunners of vertebrates. The last few days of November and the first few of December (436 million to 367 million years BP) are marked by the development of the primitive vertebrates into many species of fish and the evolution of land plants and land arthropods (organisms with an external skeleton). The development of amphibians, reptiles (leading to dinosaurs), and insects follows in close order over the second week of December (350 million to 250 million years BP). During the third week of December (250 million to 150 million years BP), reptiles become the dominant animals of Earth; also, the first primitive mammals and birds are seen. The dinosaurs dominate until December 26 (65 million years BP), when the Earth is apparently struck by an asteroid causing their rapid demise, but mammals survive and proliferate.

It is not until about 3 P.M. on December 31 (4.4 million years BP) that hominid creatures with an upright body posture appear—the australopithecines. Even then, hominids and early humans exist as hunter–gatherers, much like other animals in the ecosystem, for most of the 4.4 million years till the present. Humans develop only in just the last 8 hours, civilization since the advent of agriculture occurs in the last 2 minutes, and progress since the Industrial Revolution occupies only the last 2 seconds of the year. Thus, relatively speaking, humans have barely opened their eyes on the evolutionary scene.

Controversy over Evolution

Evolution by natural selection, the process we have been describing, is the most widely accepted explanation for the origins of the plants, animals, and microbes populating Earth. The evidence for it is abundant and strong. Nevertheless, evolution is also one of the most controversial scientific theories ever presented. Part of the controversy over evolution stems from uncertainties about its mechanisms. As we have seen, although Darwin postulated speciation by gradual changes accumulating over long periods of time, the fossil record supports the more recent view of punctuated equilibrium. The debate over gradualism versus more abrupt processes continues to this day. The debate centers on **macroevolutionary** changes—those changes involved in the appearance of major new features and new classes of organisms. Trying to deduce what happened over millions of years of evolution is like trying to solve a crime—we can see that it happened, but we have only a limited amount of evidence testifying to the sequence of events that brought it about.

THE EMERGENCE OF LIFE

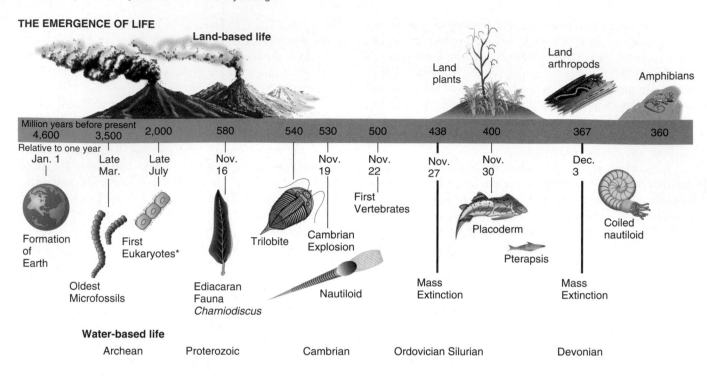

▲ **FIGURE 5–21** *Geological time scale.*   Contrasting the geological time scale with a single year gives an appreciation of the relative amount of time taken for various evolutionary stages. Note that two-thirds of the time is taken in the development of cells; then the pace quickens. (*Eukaryotes are cells that possess a membrane-enclosed nucleus that contains chromosomes of DNA and proteins.)

Some of the evidence for evolution involves present-day processes and is less controversial. This is **microevolution**— evolution within a species, or speciation itself. Speciation can be observed at various stages in many species. Natural selection by diverse components of environmental resistance can be evaluated for many organisms. Also, the DNA of living organisms is a powerful argument for evolution; closely related species show very minor differences in their DNA code sequences, while more distantly related species show greater differences. Indeed, the DNA of a species can be thought of as a capsule view of its evolutionary past. The base sequences of the DNA code of many groups of species are now yielding new information about the relationships between closely and more distantly related species.

In spite of the strong scientific support for evolution, the theory continues to generate opposition from people who object to it on grounds that it conflicts with their religious beliefs. There are, however, many people in the scientific and religious communities who have found that the controversy can be resolved without sacrificing either scientific integrity or religious faith. This view holds that the scientific and religious explanations of origins are complementary, each dealing with questions that are appropriate for their domains. Although this is an interesting topic, a thorough treatment is beyond the scope of the text.

Whereas the controversy over evolution is focused on what has happened in the past, it could be argued that a more important question facing us is what is happening now and what will happen in the future as life continues to

develop over time. As we continue to alter the surface of the Earth, move species from continent to continent, degrade and eliminate entire ecosystems, and drive a rising number of species to extinction, our choice is not whether we will change the future course of evolution. *We are already doing so.* The real alternative is whether we will simply chronicle the loss of more and more species as we pursue "business as usual" or whether we will choose to adopt a stewardship perspective, understanding that we have a responsibility to care for life in all of its diversity.

Stewardship of Life

The fossil record contains evidence of events that caused massive extinctions of many species (Fig. 5–21). Some people are drawn to speculate when the next such catastrophe may strike. In fact, *it may already be happening.* Humans are currently causing great changes in the biosphere, ranging from direct impacts on local ecosystems to altering the global atmosphere. As a result, the rate of extinction of species has shown a drastic increase. And if projections of global warming come to pass (Chapter 21), the climatic changes we cause in the next 100 years will be greater than all the climatic changes that have occurred over the past 10,000 years. That is, they will be occurring at a rate 100 times faster! Such changes may rival any of the natural geological changes that have occurred in the past. The fact that the geological record shows that biodiversity can recover from catastrophes might make us

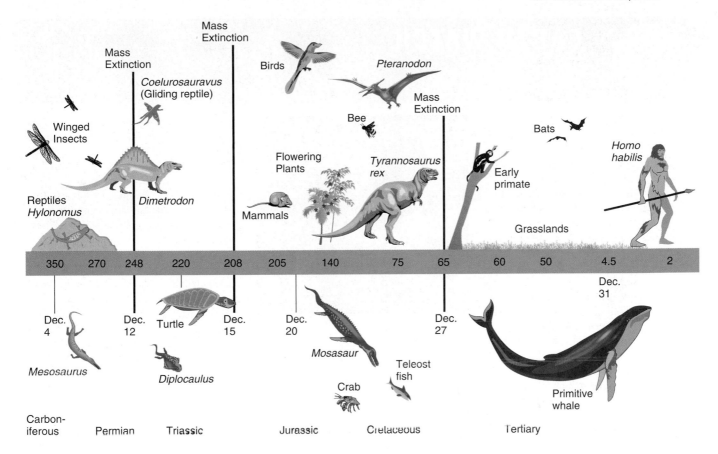

think that it doesn't matter what humans do, because nature will eventually repair the system and restore those balances. But the flaw in this line of thinking is that the natural restoration, while "relatively rapid," is still on the order of thousands or millions of years. The species and ecosystems that reemerge may be totally different from those that perish. Therefore, it is of more than academic interest to direct our efforts towards preventing extinctions and preserving biodiversity.

The information set forth in this chapter enables us to comprehend the significance of the fifth principle of ecosystem sustainability, presented in Chapter 4: *For sustainability, ecosystems depend on biodiversity.* First, notice that biodiversity is really an expression of the gene pools, and the genetic variation within those gene pools, of all the species that inhabit the planet. We have seen that the ability of a species to adapt to new conditions depends, to a large extent, upon the genetic variation within its gene pool. Therefore, as we reduce the size of surviving populations—which we are doing with countless large mammals, birds, and other species—we inevitably reduce the genetic variation within their gene pools and thereby undercut their potential for future adaptation. Small surviving populations are increasingly vulnerable. Even if a population is brought back from the brink of extinction and restored in numbers, it will still carry the legacy of the genetic uniformity of that small surviving population for many generations. Therefore, it will remain vulnerable. For example, an outbreak of a disease may eliminate an entire population because the

variations that would have conferred resistance to disease on some members may have been lost. Of course, when a species becomes extinct, its gene pool and everything it might have given rise to are lost forever.

This connection between the lack of genetic variation and the inability to adapt to new conditions has particular relevance in agriculture. Around the world, a major trend of modern agriculture, particularly in the last 50 years, has been to replace local varieties of crop plants and animal breeds with a single high-yielding variety or breed. This practice has increased production and has been a major factor in the world's ability to support a growing human population. (See the discussion of the Green Revolution in Chapter 10.) However, as discussed in Chapter 4, such genetically uniform monocultures are extremely vulnerable to outbreaks of pests and disease. Therefore, not only are we creating a monocultural system with dubious stability and sustainability, but with the loss of biodiversity represented by all the local varieties and breeds, future options for adapting agricultural species to changing conditions will be sharply curtailed. (See the "Earth Watch" essay, p. 130.)

Some argue that biotechnology, or genetic engineering—the transferring of genes from one species to another—may be our saving grace. But the potential of biotechnology depends on a source of those genes we wish to transfer. We are still a long way from being able to synthesize a gene for a specific characteristic, much less a new species. Currently, naturally occurring species are the sources of the genes that biotechnology depends on. In short, the loss of biodiversity

EARTH WATCH

PRESERVING GENES FOR AGRICULTURE

There are two basic approaches to agriculture: You can select and grow plants suited to the environmental conditions, or you can change the conditions to suit the plants. Having limited ability to control the conditions, premodern cultures selected and cultivated plants that had the ability and tenacity to reproduce even in unfavorable conditions, without the benefits of irrigation, pesticides, or fertilizer.

Along the Missouri River in North Dakota, Native Americans grew numerous species of corn and squash. On the dry, wind-shifting sands of the Colorado Plateau, Hopi farmers tended sunflowers. In the dry lands of the Southwest, Indians grew hundreds of species of beans. By sending roots as much as six feet into the soil to find moisture, these beans had a high yield despite scorching 115°F (46°C) heat and almost no rain. In Peru, each valley grew its particular variety of potato or other tuber. Thus, ancient people survived for millennia by growing remarkably adapted crops. These crops are often referred to as heirloom vegetables by horticulturists, because they were grown only in one particular locality and their seeds were passed down through the generations.

Modern agriculture, however, has taken the approach of modifying environmental conditions to fit the plant and then modifying the plant to fit the modified environment. It has focused on using relatively few high-yielding varieties and obtaining

Worker at the Seed Saver's Exchange in Decorah, Iowa, inserts dried beans into moisture-proof packages for storage. Such seed banks play an important role in preserving the biodiversity of agricultural species.

maximum yields through the intensive use of pesticides, fertilizer, and irrigation. Agriculturists are well aware of the vulnerability of this monocultural system. In 1980, for example, 20% of the U.S. corn crop was lost to a fungal disease because the disease was able to spread so rapidly through the monoculture. Resistance to disease and pests and the vigor of modern varieties can be improved through the infusion of genes from the old heirloom varieties, but only so long as those varieties continue to survive. (Recall that new genes cannot be "invented.")

Recently, however, scientists have mastered the ability to bioengineer organisms through recombinant DNA technology. The emerging field of biotechnology promises to break new ground in the development of new varieties of plants. (See Chapter 10.) Yet, the genes used to introduce new traits into existing plants still need a natural source.

In the shift to modern agriculture, many of the old varieties and species were abandoned, and many have been lost forever. Of all the food plants that were grown by Native Americans at the time of Columbus, roughly 75% no longer exist, and more are at risk. The loss of these species and the alleles they contain threatens to limit the future potential of modern agriculture. Aware of this problem, a worldwide network of "seed banks"—institutions devoted to the collection and preservation of seeds—has been established under the auspices of the World Bank's Consultative Group on International Agricultural Research. Also, a number of nongovernmental organizations, such as Seed Savers Exchange of Decorah, Iowa, Native Seed/SEARCH of Phoenix, Arizona, and the National Gardening Association, operate seed exchanges whereby members can obtain, grow, and keep alive heirloom varieties. The viability of future agriculture may depend on the efforts of those who are creating and maintaining these repositories.

will handicap the potential of biotechnology as well as the traditional methods of breeding.

An additional argument for preserving biodiversity that may have even more immediate importance is that about a quarter of all prescription drugs marketed in the United States have active ingredients derived from plants found in the wild. Thus, an additional bonanza of wonder drugs likely awaits discovery, since 98% of tropical flora is still to be screened for medicinal properties. If this flora is destroyed before such screening occurs, we may lose medicines of incalculable value.

Still, many people remain unconcerned about losing future values when there is economic gain in the short-term exploitation of ecosystems. Early in the environmental movement, it was frequently stated that, since all living things were intricately connected in food webs, the extinction of certain species might cause the collapse of entire

ecosystems. But many species have become extinct, and ecosystems are obviously still here, so it is argued that there is nothing wrong with continued exploitation of ecosystems.

However, does the fact that ecosystems have tolerated some losses of species without undue disturbance indicate that they can tolerate further, sustained losses? How many examples can you think of in which a little damage or injury is tolerable, but too much is disastrous? Perhaps it is more appropriate to view the loss of any species, even if it seems insignificant, the way canaries were viewed by miners in the early days of coal mining (Fig. 5–22): as sentinels, due to their increased sensitivity to poisonous gases. The death of the canaries served as an early warning of "bad air" in the mine. Unlike the miners, however, we cannot stay out of the "mine."

In any event, these arguments miss the most fundamental point: What kind of world do we ultimately wish to create and leave to our children and grandchildren? The reins of future evolution are in our hands—like it or not, we are now the stewards of life on Earth. Concerns regarding sustainability aside, do we wish to lead the process of evolution to an impoverished Earth that consists almost entirely of cities, suburbs, and the agricultural systems devoted to supporting them? Or would we rather see natural ecosystems with a richness of biodiversity flourish and evolve naturally beside human culture? Probably few, if any, of us really desire the former outcome over the latter. The problem is that most of us have simply assumed that nature would always be there and would take care of itself despite human activities. In the last few decades, the choice has become clear: We need to make a conscious, concerted effort to avoid exterminating a major portion of the biodiversity on the planet. Otherwise, we will be taking the former pathway by default, as the present trend of mounting loss of biodiversity attests.

On the basis of this foundation, we now turn to the actual "mechanics" of how we may achieve the goal of sustainable development. We begin in the next chapter with a study of human population.

▲ **FIGURE 5–22** *The canary in the cage.* Coal miners used to carry canaries into their mines. The canaries were especially sensitive to poisonous gases that might be present. Is the current loss of species a signal to us that our environment is in danger of irreparable damage?

ENVIRONMENT ON THE WEB

GENETIC DIVERSITY AND MEASURING SPECIES CHANGE

Genetic diversity within a species has been compared to a baseball team. If you only have a catcher and some outfielders, you do not have enough diversity to make a viable team. And if your "pool" of players doesn't contain the necessary components—for instance, good pitchers—it is impossible to create a winning team. Similarly, gene pools must be diverse enough to support a self-sustaining population. If diversity is low, the population of a species runs the risk of declining rapidly or becoming extinct. For this reason, environmental managers now track genetic diversity carefully in managed populations and constantly evaluate the significance of change for the viability of the species overall. One way that environmental managers track genetic diversity is through population genetics. Population genetics is a powerful tool used to define a species and its characteristics. Using "marker" genes, geneticists can quickly identify specific DNA sequences that are unique to a particular population. In this way, scientists can track the genetic origins of a given population—and determine when significant change has occurred locally and in the species overall.

This technique has been especially important in aquaculture, where genetically engineered populations of native species such as salmon or carp may be "farmed" in large tanks or in-ocean enclosures. Occasionally, farmed fish escape and interbreed with native populations, producing offspring with genetics that reflect both parent populations. Although native and farmed individuals may appear identical, they can differ in their growth potential and other important characteristics, such as resistance to disease. Because wild populations provide a valuable reservoir of genetic diversity, it is important that they be conserved, both for their inherent value and as a rich gene pool that can be used to augment farmed populations.

Genetic analysis is also helpful in determining the potential for extinction of a population or species. For instance, anecdotal evidence from Acadia National Park, Maine, suggested that the native beaver population was eradicated a hundred years ago. Beavers are still observed in the park, although recent studies reveal a dramatic decline in numbers—perhaps as much as 60% in the last 15 years. Genetic analysis has demonstrated that the beavers now living in the park are descended from two or three pairs introduced from the Mid-Atlantic states. The low genetic diversity in this population—a result of the small gene pool of ancestors—may be a factor in the current population decline, as well as an indication of the beaver's increased potential for extinction.

Web Explorations

The Environment on the Web activity for this essay describes the use of genetic analysis in recent efforts to preserve endangered species such as the Florida panther and the California condor. Go to the Environment on the Web activity (select Chapter 5 at **http://www.prenhall.com/wright**) and learn for yourself:

1. how genetic analysis techniques are used to identify and monitor endangered species;
2. how captive propagation techniques are being applied to help preserve and replenish genetic diversity; and
3. why a rich genetic diversity is vital to a healthy population.

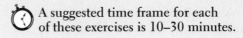 **A suggested time frame for each of these exercises is 10–30 minutes.**

REVIEW QUESTIONS

1. Where are the Galápagos Islands, and why are they important?
2. What is meant by the gene pool of a species? By differential reproduction?
3. What is selective pressure, and how does it relate to differential reproduction in nature?
4. What are the similarities and differences between selective breeding and natural selection in terms of the process and the outcome?
5. What five groupings of characteristics of organisms categorize their different kinds of adaptations for survival and reproduction?
6. What is the chemical structure of genes, and what is the relationship of genes to alleles and to observable traits or characteristics?
7. What is responsible for most of the immediate variation among individuals of a species? What is the role of mutations?

8. What factors determine whether a species will adapt to a change or whether the change will render it extinct?
9. How can natural selection lead to the development of new species? What is the process called?
10. How may natural selection lead to the development of resilience in ecosystems?
11. What factors are responsible for the similarities and differences among ecosystems?
12. What is plate tectonics, and how does it help to explain the past movement of continents? How does tectonic movement affect present-day conditions?
13. What does the fossil record tell us about the history of life? How recently have humans appeared on the scene?
14. What are two important reasons for the controversy over evolution? How can the controversy be resolved?
15. What are the implications of diminishing biodiversity for the future of agriculture? For the future survival of wildlife?

THINKING ENVIRONMENTALLY

1. Select a particular species and describe how its various traits support its survival and reproduction. Describe how these traits may have developed by, and are maintained through, natural selection.
2. How do strains of disease-causing organisms become resistant to medicines used against them?
3. Why is it biologically impossible for a "Ninja turtle" to arise from a single mutation?
4. Select a particular ecosystem and describe how adaptations found in its species facilitate resilience in the ecosystem.

5. Take a survey of the opinions of your class regarding evolution: how many support it, how many are skeptical about it, and how many oppose it. Find out the reasons for the class' opinions.
6. Speculate as to the future course of evolution, including the evolution of humans, if substantial biodiversity is lost.
7. Discuss how we humans should carry out our stewardship responsibilities in terms of each of the five basic principles of ecosystem sustainability.

MAKING A DIFFERENCE

PART ONE: Chapters 2, 3, 4, 5

1. Find out where and how your school, college, community, or city gets its water and power, how it disposes of sewage and refuse, and what is the trend of land development and preservation around the area.
2. Find out whether there is an environmental organization or club on your campus or in your community. Join if there is, or create one if there isn't.
3. Learn the common names of trees and other plants, birds, and mammals found in your area. This knowledge is a prerequisite to understanding and describing an actual ecosystem. Then, pass your knowledge on to children, who are often curious about the natural world.

4. Take courses offered by colleges or local environmental organizations that give you a greater appreciation for the natural world; amaze and influence your friends by becoming a local expert in some taxonomic group, such as birds, wildflowers, or butterflies.
5. Consider small animals, birds, butterflies, and other wildlife that might be present in your yard, school, or campus grounds were it not for human-created limiting factors. Investigate and begin a project to create natural habitats that will attract and support additional wildlife.
6. Create a list of the things you do and the things that you use and throw away. Write a brief evaluation of

each in terms of the principles of sustainability. Consider how you might change these habits to move toward a more sustainable option. Begin by making one such change.

7. Rather than using chemical pesticides and fertilizers to maintain your lawn, which is an "unnatural" monoculture, introduce clover and other low-growing flowering plants that will create a sustainable balance with only mowing.

8. Read local papers, make contact with environmental organizations, and become aware of efforts to protect natural areas locally. Support these efforts to protect such areas locally, regionally, and globally.

9. Select a current environmental issue related to biodiversity or endangered species, and write your congresspersons to express your concern and ask for their support for legislation that effectively addresses the issue.

10. Support and join efforts and organizations, such as the World Wildlife Fund and Conservation International, that are devoted to protecting endangered and threatened species.

11. Use the World Wide Web to investigate an environmental issue, and create an annotated bibliography of Websites that address the issue. Post your bibliography on an appropriate listserver that others can access.

12. Continue your education toward a profession that is important for environmental concerns and make your lifework one that helps rather than hinders the environmental revolution.

13. Use your citizenship to support environmentally sound practices and policies by voting in your local, state, and national elections for candidates who are clearly environmentally aware.

WEB REFERENCES

On-line resources for this chapter are on the World Wide Web at: **http://www.prenhall.com/wright**. (Click on Chapter 5 within the Chapter Selector.)

Chapter 6
The Human Population: Demographics

Chapter 7
Issues in Population and Development

PART 2

The Human Population

From a population perspective, the 20th century was remarkable; we started the century with less than 2 billion people, and ended it with 6 billion and counting. The United Nations Population Division suggests that our numbers may reach 9 billion by the middle of the century—a 50% increase in 50 years. Virtually all of the increase will be in the developing countries, which are already densely populated and straining to meet the needs of their people for food, water, health, shelter, and employment.

In the industrialized world populations have almost stopped growing. Yet the developed countries, with only 20% of the world's people, account for two-thirds of the world's energy use and create pressures on water, forests, and fisheries that are leading to the depletion of these renewable resources. When we picture a world of 9 billion, with a greater proportion of people straining to improve their economic standards and catch up to the developed world, we may be staring into the face of disaster.

In Chapter 6, we first look at the dynamics of the human population and focus especially on the demographic transition—the shift from high birth and death rates to low birth and death rates that has brought stable populations to the industrialized world. We then consider, in Chapter 7, what needs to be done to bring the developing countries through this transition. It is appropriate that we take up population concerns before other significant issues like resource use and pollution, because population affects every environmental issue, and we have no chance of achieving sustainability in these other areas until we achieve population stability.

The Human Population: Demographics

The Mombasa Highway leads out from the center of Nairobi, Kenya, and is lined with factories for several miles. Outside every factory gate every morning is a crowd of people—mostly men—waiting for the gates to open. Dressed in tattered clothes and worn-out shoes, they are the "casual workers," hired for periods of up to three months, but never under any employment contract. Many are there just for the hope that they will be hired for the day. One such worker—we will call him Charles—has worked at one of these factories for almost a year. He is paid 100 Kenya shillings a day, about $1.70. He walks 10 km to work from his home in one of the shantytowns on the edge of the city. He can't afford transportation and can only occasionally buy lunch, which is a bowl of githeri—a 10-shilling dish of corn and beans sold in the kiosks that cluster around every factory. Charles has no work contract, nor do any of his coworkers. Virtually all of the factories pay workers 100 shillings a day; working six days a week for 10 hours a day, they earn, at most, $50 a month. Any worker who becomes

a union representative is fired. Any worker who complains about an injury on the job is fired. For every job, there are scores of applicants, so no one dares attract attention. They work in dusty, dangerous conditions. Cement-factory workers have no masks to prevent them from breathing the dust. Metalworkers lack protective gear and often lose limbs or their lives due to the nature of the jobs, which are often monotonous, making the workers lose their concentration. There are 14,000 such factories in cities like Nairobi, Mombasa, Kisumu, Nakuru, and Eldoret, in Kenya.

Scenes like the Mombasa highway can be found around most cities in the developing world. The incredible growth of cities is one of the realities of the population explosion in that world. It is no secret why: The rural countryside provides only subsistence farming and livestock herding for most of its people—and with population growth rates as high as 3% per year, there are continually more people than there is land to till and cattle to herd. It is the tragic story of *Cry the Beloved Country* (a classic novel by Alan Paton, set

◄ **Casual workers in Kenya.** High population growth and slow economic development combine to produce a large unemployed and underemployed workforce. Competition for the scarce, low-paying factory jobs leaves many men like these without employment.

in South Africa), multiplied by hundreds of thousands. So the young people and the men migrate to the cities to find work. Such is the case in Nairobi. With one of the more rapidly expanding populations in the world, Kenya has an enormous unemployed and underemployed workforce.

The global human population has recently undergone an explosion. It has more than tripled in the last century, and while the rate of growth is slowing down, the increase in absolute numbers continues to be great. Remarkable changes in technology and substantial improvements in living standards have accompanied this growth. It comes as no surprise that the great growth in the human population and the accompanying demands for resources have affected the natural environment and will continue to do so. Much of this text is dedicated to documenting concerns over the use of resources and over pollution stemming from the human system in all of its extensions.

What are the implications of these trends for sustainability? Another way to put this question is to ask, What is the carrying capacity of Earth for humans? How many people can Earth support, and at what quality of life? Some observers are convinced that there are already too many people and that the global environment is being stressed beyond its limits even now. No one argues about the fact that the human population cannot continue to increase indefinitely. There is broad consensus that continued population increase makes solving other problems more difficult, and thus, there are many calls for bringing human population growth to a halt. The focus is especially on the developing world, where 99% of world population growth is occurring.

Yet, specific efforts to reduce population growth often collide with ethical and moral values, creating controversy. Moreover, not everyone agrees that the population issue is serious. The major counterarguments are as follows:

- Some claim that population growth is beneficial, in that more people provide more ideas, creativity, and work. This belief is supported by the fact that the greatest technological advances and improvements in living standards have occurred in parallel with the population explosion of the past 100 years.

- Because of their religious convictions, some take the position that any artificial interference in the reproductive process (including the use of sex education, contraceptives, and, especially, abortions) is immoral. Therefore, for them, any thought of altering the course of population growth, except by abstinence, is not debatable.

- Some argue that population growth is not the issue as much as consumption is, at least for the present. What needs to be achieved more than reducing growth in numbers, they say, is adopting measures that will reduce consumption.

- Others take the position that population growth will level off by itself well within the capacity to support the Earth's population. This view is supported by the facts that we have been able to expand agricultural production even faster than population growth and that the average number of births per woman is decreasing.

- Still others are disturbed by the fact that population programs often seem to have the trappings of social engineering—the rich trying to hold down the poor or minorities by preventing them from having children.

In spite of these objections, efforts to provide help to couples in the form of family planning have been mounted in almost every country; these efforts and other socioeconomic factors brought about a remarkable decline in reproduction rates during the second half of the 20th century, and there is every expectation that population growth rates will continue to slow down into the present century.

Our objective in this chapter is to gain an understanding of the dynamics of population growth and its social and environmental consequences. We affirm that a continually growing population is unsustainable, and so our focus in this and the next chapter will be population stability and what it will take to get there.

6.1 The Population Explosion and Its Cause

Considering all the thousands of years of human history, the explosion of the global human population is a recent and unique event—a phenomenon of just the past 100 years (Fig. 6–1). Let us look more closely at this event and why it occurred.

The Explosion

From the dawn of human history until the beginning of the 1800s, population increased slowly and variably with periodic setbacks. It was roughly 1830 before world population reached the 1 billion mark. But by 1930, just 100 years later, the population had doubled to 2 billion. Barely 30 years later, in 1960, it reached 3 billion, and in only 15 more years, by 1975, it had climbed to 4 billion. Thus, the population doubled in just 45 years, from 1930 to 1975. Then, 12 years later, in 1987, it crossed the 5 billion mark! In 1999, world population passed 6 billion, and it is currently growing at the rate of nearly 80 million people per year. This rate is equivalent to fitting into the world every year the combined populations of New York, Los Angeles, Chicago, Philadelphia, Detroit, Dallas, Boston, and 10 other U.S. metropolitan areas.

On the basis of current trends (which assume a continued decline in fertility rates), the U.N. Population Division

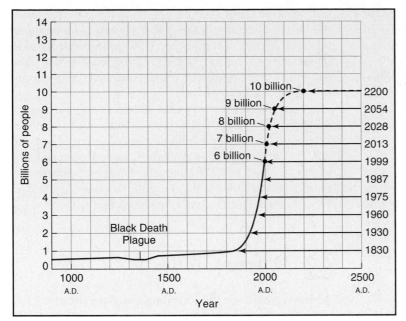

◀ **FIGURE 6–1** *The world population explosion.* For most of human history, population grew slowly, but in modern times it has suddenly "exploded." (*Sources:* Basic plot from Joseph A. McFalls, Jr., "Population: A Lively Introduction," *Population Bulletin* 46, no. 2 [1991]:4; updated from U.N. Population Division, median projection)

projects that world population will pass the 7 billion mark in 2013, the 8 billion mark in 2028, and the 9 billion mark in 2054, before population finally levels off at around 10 billion by the end of the 22nd century (Fig. 6–1).

Reasons for the Explosion

The main reason for the slow and fluctuating population growth prior to the early 1800s was the prevalence of diseases that were often fatal, such as smallpox, diphtheria, measles, and scarlet fever. These diseases hit infants and children particularly hard. It was not uncommon for a woman who had seven or eight live births to have only one or two children reach adulthood. In addition, epidemics of diseases such as the black plague of the 14th century, typhus, and cholera would eliminate large numbers of adults. Famines also were not unusual.

Biologically speaking, prior to the 1800s, the population was essentially in a dynamic balance with natural enemies—mainly diseases—and other aspects of environmental resistance. High reproductive rates were largely balanced by high mortality, especially among infants and children. With high birth and death rates, the population growth rate was low in these preindustrial societies.

In the 1800s, Louis Pasteur and others made the major discovery that diseases were caused by infectious agents (now identified as various bacteria, viruses, and parasites) and that these were transmitted via water and food, insects, and other vermin. With these discoveries came major improvements in sanitation and personal hygiene. Then, techniques of providing protection by means of vaccinations came into play. Later, in the 1930s, the discovery of penicillin, the first in a long line of antibiotics, resulted in cures for otherwise often-fatal diseases such as pneumonia. Im-

EARTH WATCH

ARE WE LIVING LONGER?

It is commonly said that with the introduction of modern medical technology and disease control, average longevity increased from about 40 to 75 years of age. While mathematically correct, this statement may be misleading. Many people take it to mean that nearly everyone used to die around age 40 and now everyone lives to around 75. However, the years between 35 and 45 are generally the healthiest period of human life. With or without modern medicine, a relatively small portion of the population dies in this age range. The "average longevity of 40" before modern techniques of disease control was chiefly a function of a large fraction of the population's dying in childhood, thus counterbalancing another fraction of people that lived into their sixties and beyond.

Through disease control, most of those who once would have died in childhood now live past 60. Extending the life span of this group raises the average age of death of the population. But the basic life span of the human species has changed little, if at all, as a result of modern medicine. All modern medicine has done is to increase the proportion of people who get to or near the maximum age.

provements in nutrition began to be significant as well. In short, better sanitation, medicine, and nutrition brought about spectacular reductions in mortality, especially among infants and children, while birthrates remained high.

From the biological point of view, the human population entered into exponential growth, as does any natural population on being freed from natural enemies and other environmental restraints. (Note that the rapidly declining death rates must be compared with the birthrates, which remained high.)

In the last few decades, the average *fertility rate*—that is, the average number of babies born to a woman over her lifetime—has declined, resulting in a decreasing *rate of growth* of population. (Table 6–1 provides a list of definitions of technical terms used in this chapter.) Still, as the number of people of reproductive age has expanded with the increasing numbers of children growing up, the human race continued until 1990 to add absolute numbers faster than at any other time in history (Fig. 6–2). Extrapolating the trend of lower fertility rates leads to the U.N. Population Division's projection that the global human population will reach 8.9 billion by 2050, but will continue to increase, leveling off at around 10 billion well into the 22nd century. (Note that this projection is the U.N.'s *medium* scenario; other scenarios are based on different fertility assumptions. U.N. projections to 2150 range from 3.2 billion to 24.8 billion, making the point that long-range population projections are largely exercises in mathematics.)

The projected leveling off at around 10 billion raises the question of whether Earth can sustain such numbers. Where are the additional billions of people going to live, and how are they going to be fed, clothed, housed, educated, and otherwise cared for? Will enough energy and material resources be available for them to fulfill a satisfying lifestyle? What will the natural environment look like by then?

6.2 Different Worlds

To begin to answer the foregoing questions, we must first recognize the tremendous disparities among nations. In fact, people in wealthy and poor countries live almost in separate worlds, isolated by radically different economic and demographic conditions.

TABLE 6–1 Demographic Terms Used in This Chapter	
Term	**Definition**
Growth Rate (annual rate of increase)	The rate of growth of a population, as a percentage. Multiplied by the existing population, this rate gives the net yearly increase for a population.
Total Fertility Rate	The average number of children each woman has over her lifetime, expressed as a yearly rate based on fertility occurring during a particular year.
Replacement-level Fertility	A fertility rate that will just replace a woman and her partner, theoretically 2.0, but adjusted slightly higher because of mortality and failure to reproduce.
Infant Mortality	Infant deaths per thousand live births.
Population Profile (age structure)	A bar graph plotting numbers of males and females for successive ages in the population, starting with youngest at the bottom.
Population Momentum	The tendency of a population to continue growing even after replacement-level fertility has been reached, due to continued reproduction by already existing age groups.
Crude Birth Rate	The number of live births per thousand in a population in a given year.
Crude Death Rate	The number of deaths per thousand in a population in a given year.
Doubling Time	The time it takes for a population increasing at a given growth rate to double in size.
Epidemiologic Transition	The shift from high death rates to low death rates in a population as a result of modern medical and sanitary developments.
Fertility Transition	The decline of birthrates from high to low levels in a population.
Demographic Transition	The tendency of a population to shift from high birth and death rates to low birth and death rates as a result of the epidemiologic and fertility transitions. The consequence is a population that grows very slowly, if at all.

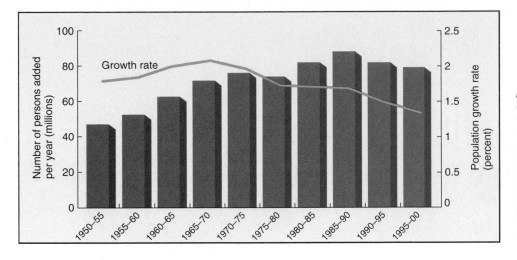

◀ **FIGURE 6–2** *World population growth rate and absolute growth.* Declining fertility rates in the last three decades have resulted in a decreasing rate of population growth. However, absolute numbers are still adding 80 million per year. [*Source*: Data from Shiro Horiuchi, "World Population Growth Rate," *Population Today* (June 1993): 7 and June/July 1996): 1; updated from U.S. Census Bureau, International Data Base.]

Rich Nations and Poor Nations

The World Bank, an arm of the United Nations, divides the countries of the world into three main economic categories according to average per capita gross national product. (Fig. 6–3):

1. **High-income, highly developed, industrialized countries.** This group includes the United States, Canada, Japan, Australia, New Zealand, the countries of western Europe and Scandinavia, Singapore, Taiwan, Israel, and several Arab states. (1998 gross national product per capita, $9,361 or above; avg. of $25,510.)

2. **Middle-income, moderately developed, countries.** These are mainly the countries of Latin America (Mexico, Central America, and South America), northern and southern Africa, some of eastern Asia, eastern Europe, and countries of the former U.S.S.R. (1998 gross national product per capita ranges from $761–$9,360; avg. of $2,950.)

3. **Low-income countries.** This group comprises the countries of eastern, western, and central Africa, India and other countries of central Asia, and China. (1998 gross national product per capita, less than $760; avg. of $520.)

The high-income nations are commonly referred to as **developed countries**, whereas the middle- and low-income countries are often grouped together and referred to as **developing countries**. The terms *more developed countries* (MDCs), *less developed countries* (LDCs), and *Third World countries* are being phased out, although you may still hear them used. (The Second World was the former Communist bloc, which no longer exists. Therefore, referring to the developing countries as Third World countries is obsolete.)

The disparity in distribution of wealth among the countries of the world is mind boggling. The highly developed countries hold just 20% of the world's population, yet they control about 80% of the world's wealth. Thus, developing countries, which have 80% of the world's population, have only about 20% of the world's wealth. Of course, the distribution of wealth *within* each country is also disproportionate. Between 10% and 15% of the people in more developed countries are recognized as poor (unable to afford adequate food, shelter, or clothing), while about 60% of those in developing countries are.

The disparity of wealth is difficult to understand just by looking at monetary figures. Therefore, Allen Durning of the WorldWatch Institute describes the relative wealth for different peoples of the world in terms of their access to different kinds of food, drink, and transportation, as is shown in Table 6–2. It is sobering that although some people in developed nations such as the United States may feel poor, they are still well within the rich category in terms of representative living standards for the world. The people in the lower middle and poor categories of consumption, predominantly in low- and middle-income countries, live along a bare "subsistence margin," with little beyond the minimum requirements for survival. Roughly half of these—somewhat over a billion people—live in a condition of "absolute poverty," defined by Robert McNamara, former president of the World Bank, as "a condition of life so limited by malnutrition, illiteracy, disease, squalid surroundings, high infant mortality, and low life expectancy as to be beneath any reasonable definition of human decency." Yet, it is among the poorest people that population growth is usually the highest.

Population Growth in Rich and Poor Nations

The population growth shown in Figs. 6–1 and 6–2 is for the world as a whole. If we look at population growth in developed versus developing countries, we find a discrepancy that parallels the great differences in wealth of these two groups of countries. The developed world, with a population of 1.18 billion in mid-2001, is growing at a rate of 0.1% per year. The remaining countries, whose mid-2001 population was 4.97 billion, are increasing at a rate of 1.7% per year. Consequently, *99% of world population growth is occurring in the developing countries.* Why is this so?

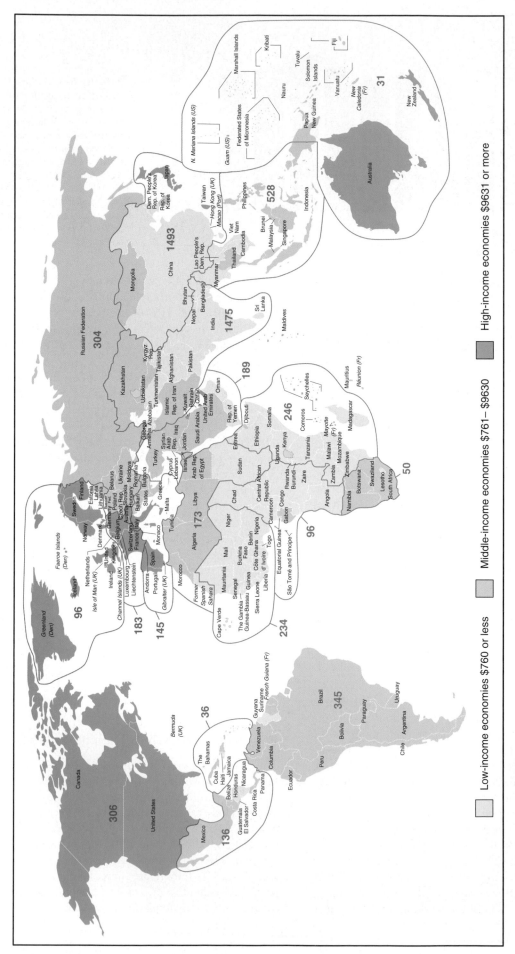

▲ FIGURE 6–3 *Major economic divisions of the world.* Nations of the world are grouped according to gross national product per capita. The population of various regions (millions) is also shown by magenta lines and numbers. [*Sources:* Reprinted, by permission of the publisher, from the World Development Report, (New York: Oxford University Press, Inc., 1994). Copyright © 1994 by the International Bank for Reconstruction and Development/The World Bank. Population and per capita gross national product data from the Population Reference Bureau, *World Population Sheet.* Updated from the World Bank, *World Development Report*, 1999–2000.]

Low-income economies $760 or less

Middle-income economies $761–$9630

High-income economies $9631 or more

TABLE 6–2	World Consumption Classes		
Category of consumption	Rich (1.2 billion)	Middle (3.6 billion)	Poor (1.2 billion)
Diet	Meat, packaged food, soft drinks	Grain, clean water	Insufficient grain, unsafe water
Transport	Private cars	Bicycles, buses	Walking
Materials	Throwaways	Durables	Local biomass

Source: Data from Alan Durning, *How Much Is Enough* (Washington, DC: Worldwatch Inst., 1992), p. 27; updated by Population Reference Bureau, Inc., 2000.

TABLE 6–3	Population Data for Selected Countries	
Country	Total Fertility Rate	Doubling Time of Population (Years)
World	2.9	51
Developing Countries		
Average (excluding China)	3.7	36
Egypt	3.3	35
Kenya	4.7	33
Madagascar	6.0	24
India	3.3	39
Iraq	5.7	25
Vietnam	2.5	48
Haiti	4.7	40
Brazil	2.4	45
Mexico	2.7	36
Developed Countries		
Average	1.5	809
United States	2.1	120
Canada	1.5	178
Japan	1.3	462
Denmark	1.7	472
Germany	1.3	—
Italy	1.2	—
Spain	1.2	6,931

Source: Data from *2000 World Population Data Sheet* (Washington, DC: Population Reference Bureau, 2000).

Note: Dash indicates doubling time cannot be calculated, because growth is negative.

In the absence of high mortality, the major determining factor for population growth is **total fertility rate**—the average number of children each woman has over her lifetime. It follows that a total fertility rate of 2.0 will give a stable population, since two children per woman will just replace the parents when they eventually die. Fertility rates greater than 2.0 will give a growing population, because each generation is replaced by a larger one. And barring immigration, a total fertility rate less than 2.0 will lead to a declining population, because each generation will eventually be replaced by a smaller one. Given that infant and childhood mortality are not in fact zero, and that some women do not reproduce, **replacement-level fertility**—the fertility rate that will just replace the population of parents—is slightly greater: 2.1 for developed countries and higher for developing countries, which have higher infant and childhood mortality.

For reasons we shall discuss later, total fertility rates in developed countries have declined over the past several decades to the point where they now average 1.5. The one major exception is the United States, with a total fertility rate of 2.1 (2000). In developing countries, on the other hand, although fertility rates have come down considerably in recent years, they still average 3.2, with some as high as 6 or more, rates that will cause the populations of those countries to double in just 20 to 40 years (Table 6–3). Thus, the populations of poor countries will continue growing, while the populations of more developed countries will stabilize or even decline. As a consequence, the percentage of the world's population living in developing countries—already 80%—is expected to climb steadily to over 90% by 2075 (Fig. 6–4). However, this is not to say that only the developing countries have a population problem.

Different Populations Present Different Problems

Today, increasing numbers of people put increasing demands on the environment, both through demands for resources, including food, energy, and water, and through the production of wastes. However, it should also be clear that the demand each individual makes on the environment depends on how much and what that individual consumes. Each additional thing purchased represents a certain additional demand on resources for its production, as well as additional wastes produced in the course of its production, use, and, finally, disposal. Therefore, negative effects on the environment also increase dramatically as consumption increases.

For example, it is estimated that, because of differences in consumption, the average American places at least 20 times the demand on Earth's resources, including its ability to absorb pollutants, as does the average person in Bangladesh, a poor Asian country. Major world pollution problems, including the depletion of the ozone layer, the impacts of global climate change, and the accumulation of toxic wastes in the environment, are largely the consequence of the high consumption associated with affluent lifestyles. For instance, it is people who drive cars and heat and cook with fossil fuels that contribute most significantly to rising levels of carbon dioxide in the atmosphere. Likewise, much of the global deforestation and loss of biodiversity is a consequence of consumer demands in developed countries.

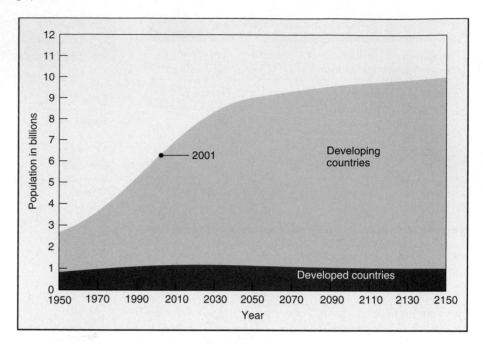

▶ **FIGURE 6–4** *Population increase in developed and developing countries.* Because of higher populations and higher birthrates, developing countries represent a larger and larger share of the world's population.

Environmental impacts of affluent lifestyles may be moderated to a large extent by a concern for environmental stewardship. For example, suitable attention to wildlife conservation, pollution control, energy conservation and efficiency, and recycling may offset, to some extent, the negative impact of a consumer lifestyle. One might make a good argument that a life devoted to conservation or other aspects of environmental stewardship might entirely offset the negatives and have a highly positive effect overall. The relationship between these factors may be expressed as

$$\text{Environmental impact} \propto \frac{\text{Population size} \times \text{Affluence of lifestyle}}{\text{Stewardship}}.$$

This proportionality should be read as "Environmental impact is proportional to the population multiplied by the affluence of the population's lifestyle, moderated by the stewardly action of the population." Note that it is not strictly a mathematical proportionality, because meaningful numbers cannot be assigned to any of the factors except population.

You may hear debates among people, one side arguing that population growth is the main problem, another claiming that our highly consumption-oriented lifestyle is the main problem, and a third maintaining that it is our inattention to stewardship that is the chief shortcoming. Our contention in this text is that in order to reach sustainability, all *three* areas must be addressed: Stabilize population, decrease consumption, and increase stewardly care. For the present, we wish to keep focusing on the factors and consequences of population growth, principally in developing countries.

6.3 Environmental and Social Impacts of Growing Populations and Affluence

Both growing populations and increasing affluence have numerous environmental and social implications for the whole world.

The Growing Populations of Developing Countries

Prior to the Industrial Revolution, most of the human population survived through subsistence agriculture. That is, families lived on the land and produced enough food for their own consumption and perhaps enough extra to barter for other essentials. Natural forests provided firewood, structural materials for housing, and wild game for meat. With a small, stable population, this system was basically sustainable. As the older generation passed away, the land and natural systems could still support the next generation. Indeed, many cultures sustained themselves in this way over thousands of years, and a few areas of Latin America, Africa, and Asia maintain the tradition today.

After World War II, modern medicines—chiefly vaccines and antibiotics—were introduced into developing nations, whereupon death rates plummeted and population growth increased. What are the impacts of rapid growth on a population that is largely engaged in subsistence agriculture? Five basic alternatives are possible, all of which are being played out to various degrees by people in these societies:

1. Subdivide farms among the children of the next generation, or intensify cultivation of existing land to increase production per unit area.
2. Open up new land to farm.
3. Move to cities and seek employment.
4. Engage in illicit activities for income.
5. Emigrate to other countries, legally or illegally.

In addition, rapid population growth especially affects women and children. Let us look at each of the preceding alternatives and their consequences in a little more detail.

Subdividing Farms and Intensifying Cultivation. Over wide areas of Asia and Africa, plots of land have been divided and redivided to the point that the U.N. Food and Agriculture Organization (FAO) estimates that over a billion rural people live in households that have too little land to meet even their own meager needs for food and fuel, much less to produce extra for income to barter.

Adding to the problem is the quest for water and firewood. Some 3 billion people, or 60% of the world's population, do not have gas, electric, or even kerosene stoves; hence, they depend on firewood in the preparation of their daily food. As a consequence, local woodlands are being cut faster than they can regenerate. Much of East Africa, Nepal, and Tibet, as well as many slopes of the Andes in South America, have been deforested as a result. Already, women in many developing countries spend a large portion of each day on increasingly long treks to gather firewood and water (Fig. 6–5). As many as 3 billion poor people face acute shortages of firewood. Worse, in the long term, as the trees are removed, soil erosion occurs (examined in Chapter 8).

With regard to the intensification of cultivation, the introduction of more highly productive varieties of basic food grains has had a dramatic positive effect in supporting the growing population, but is not without some concerns, as will be described in Chapter 10. However, the intensification of cultivation also means simply working the land harder. For example, traditional subsistence farming in Africa involved rotating cultivation among three plots. In that way, after being cultivated for one year, the soil in each plot had two years to regenerate. With pressures for increasing productivity, plots have been put into continuous production with no time off. The results have been a deterioration of the soil, decreased productivity (ironically), and erosion.

In addition, the increasing intensity of grazing is damaging the land (desertification; see Chapter 8.) Given the countertrends of rapidly increasing population and deterioration of land from overcultivation, food production per capita in Africa, for example, is currently on a downward course.

Opening Up New Lands for Agriculture. With respect to opening up new lands for agriculture, first consider that there really is no such thing as "new land." Instead, it is always a matter of converting the land of natural ecosystems to agricultural production, which means increasing pressure on

▲ **FIGURE 6–5** *Gathering firewood.* Like the Nepalese woman seen here, some 60% of the world's population still depends on gathering firewood for cooking and other fuel needs. The resulting deforestation is causing both ecological and human tragedy.

wildlife and an unavoidable loss of biodiversity. Even then, converted land may not be well suited for agriculture. (Most good agricultural land is already in production.) For example, it is estimated that two-thirds of the tropical deforestation that is occurring in Brazil and Central America is for the purpose of increasing agricultural production (Fig. 6–6). Much of this deforestation is done by poor, young people who are seeking an opportunity to get ahead, but are unskilled and untrained in the unique requirements of maintaining tropical soils. Consequently, beyond the loss of biodiversity, we have the additional problem that between a third and a half of the cleared land becomes unproductive within three to five years, again leaving the people in absolute poverty. Some "new lands" include steep slopes, which suffer horrendous erosion when plowed, and areas with minimal rainfall, which turn to deserts under cultivation.

Migration to Cities. Faced with the poverty and hardship of the countryside, many hundreds of millions of people in developing nations continue to migrate to cities in search of employment and a better life. The result is that a number of cities of the developing world are now among the world's largest and are still growing rapidly (Fig. 6–7a). Opportunities in many cities have not expanded fast enough to handle the influx. People are forced to live in sprawling, wretched squatter settlements and slums that do not even provide adequate water and sewers, much less other services (Fig. 6–7b). It is not a coincidence that three epidemics of

▲ **FIGURE 6–6** **Deforestation in the tropics.** Millions of acres of rain forest in Central and South America are being cut down each year to make room for agriculture, as shown in this photograph from Peru.

cholera have occurred in the developing world in recent years. Cholera is caused by a bacterium spread via sewage that gets into drinking water. The disease causes extreme vomiting and diarrhea, which result in great loss of body fluids; it is frequently fatal if not properly treated.

Worse, these cities often do not provide the jobs people seek. Indeed, the high numbers of rural immigrants in the cities dilute the value of the one thing they have to sell—their labor. As we saw earlier, a common wage for a day's unskilled work is often equivalent to no more than a dollar or two, not enough for food, much less housing, clothing, and other amenities. Thousands, including many children, make their living by scavenging in dumps to find items they can salvage, repair, and sell. Many survive by begging—or worse.

Illicit Activities.   Anyone who doesn't have a way to grow sufficient food must gain enough income to buy it—and sometimes, desperate people break the law to do this. Of course, it is difficult to draw the line between the need and the greed that also draws people into illicit activities. However, it is undeniable that the shortage of adequately paying employment exacerbates the problem. Besides the rampant petty thievery and corruption that pervade many developing countries, income is also obtained from the following illegal activities.

Illicit Drugs.   A small peasant farmer with too little land to make a living growing food can make a decent income growing the various crops from which illicit drugs are made. Of course, those who synthesize and sell the drugs make even greater profits.

Poaching of Wildlife.   A person in a tropical developing country can make a considerable income hunting or trapping various fish, birds, reptiles, and other animals and selling them into the "pet" trade. Illegal trade involving endangered species has grown steadily over the years; in black-market activity, it is second only to drugs. Since few of the animals collected survive, and fewer yet are put into situations in which they will breed, poaching is a major factor in pushing many species toward extinction.

Given the poverty in developing countries, can you see why these activities persist despite efforts to stop them? Of course, it shouldn't escape your notice that the affluent consumer at the end of the chain provides the basic incentive for such activities.

Emigration and Immigration.   Facing the stresses and limited opportunities in their own countries, many people of developing countries see emigration to developed countries as their best hope for a brighter future. Historically, the New World was colonized by the overflow population from

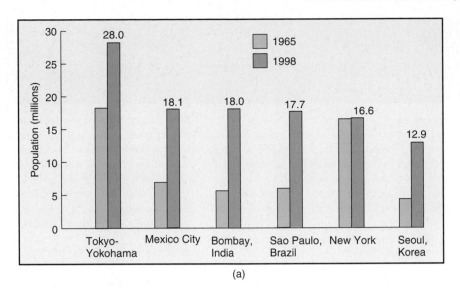

(a)

(b)

◀ **FIGURE 6-7** *Growing cities.* (a) Growth of some major world metropolitan areas. Since 1965, cities in the developing world have grown phenomenally, and a number of them are now among the world's largest. (b) Slums on the outskirts of São Paulo, Brazil, where 32% of the city's population lives. (*Source:* Data for [a] from *Christian Science Monitor* [Dec. 1990]: 12; updated from U.S. Census Bureau.)

Europe, which experienced its population explosion early in the Industrial Revolution, and certainly, immigrants have contributed immensely to the development and economic growth of the United States, Canada, and other countries. However, it isn't feasible for the United States or any other developed nation simply to open its doors to all who would flee from poverty and lack of opportunity in the developing world.

The Worldwatch Institute estimates that there are already over 23 million environmental refugees—people living outside their homeland because they can't make a living there—and the number is growing rapidly. The gates for these immigrants are rapidly closing, however. Concerned over already high unemployment and strains on state welfare systems, the European Union countries are taking steps to severely limit immigration from the developing world.

The one developed country that accepts more immigrants than all others combined is the United States. (See the "Ethics" essay, p. 148.) The current (2001) population of the United States—283 million—is growing at about 2.5 million per year. Over a third of this growth is new immigrants.

Impoverished Women and Children. The hardships and deprivation of poverty fall most heavily on women and children. Men are more free to roam and pick up whatever work is available, and they may keep their wages for themselves. Some men take no responsibility at all for the women they make pregnant, much less the children that are produced. Even many married men, under the stress of poverty, abandon wives and children. Few developing countries have a

ETHICS

THE DILEMMA OF IMMIGRATION

For people trapped by poverty or lack of opportunity in their homeland, emigration to another country has always seemed a way to achieve a better life. As the New World opened up, many millions of people recognized this dream by emigrating to the United States and other countries. The United States and Canada are countries composed largely of immigrants and their descendants. Until 1875, all immigration into the United States was legal; all who could manage to arrive could stay and become citizens. This openness was inscribed on the Statue of Liberty, which reads, in part, "Give me your tired, your poor, your huddled masses yearning to breathe free, the wretched refuse of your teeming shore....Send these, the homeless, tempest-tossed to me."

Emigration from the Old World created a flood of migrants immigrating to the New World. This emigration relieved population pressures in European countries and aided in the development of the New World. It is apparent, however, that a totally open policy toward immigration would be untenable today. The United States, with its current population of 283 million, is no longer a vast, open land awaiting development; yet hundreds of millions of people would immigrate if they could. How much immigration should be permitted, and should some groups be favored over others? For example, in 1882, the U.S. Congress passed the Chinese Exclusion Act, which barred the immigration of Chinese laborers but not Chinese teachers, diplomats, students, merchants, or tourists. This act remained in effect until 1943, when China and the United States became allies in World War II. However, the current immigration policy still makes it easier for trained people to gain citizenship and relatively difficult for untrained people to do so. This policy created what is commonly referred to as a "brain drain." Brainpower, many pointed out, is the "export" that developing nations can least afford.

Under the current immigration laws, the United States now accepts almost 900,000 new immigrants per year, a number larger than we have received at any time since the 1920s and larger than is accepted by all other countries combined. Further, a lottery system was adopted for selection. At present, immigration accounts for about 33% of U.S. population growth, which is presently 2.4 million per year. The remainder of the population growth, called the natural increase, is the excess of the number of births over the number of deaths. If the fertility rate remains low, immigration will account for a growing proportion of population growth.

The preceding deals, of course, with legal immigration; illegal immigration is another matter. Hundreds of thousands, unable to gain access through legal channels, seek ways to enter the country illegally. The United States maintains an active border patrol, especially along the border with Mexico, that several thousand people try to cross each night. Most are caught and returned, but an undetermined number—estimated at 200,000 to 300,000 per year—slip through. The Illegal Immigration Reform Act of 1996 addressed this problem by strengthening the border patrol and stepping up efforts to locate and deport illegal aliens, actions that are supported by most observers.

In 1991 and 1992, Haitian people made headlines as they fled their country in crowded, unseaworthy boats and headed for the United States. These people presented a new issue—political asylum. U.S. immigration policy admits people to the United States who are fleeing their country because of political oppression. Although political oppression was certainly present in Haiti, the evidence—as reflected in a Supreme Court decision—indicated that most of the boat people were *economic* refugees. That is, they were simply seeking a better life. They were deported back to Haiti, and subsequently, the United States intervened in Haiti in an attempt to restore democracy and economic progress to that impoverished country. More recently, the Haitian Refugee Immigration Fairness Act of 1998 allows Haitians living in the United States since 1995 to remain if returning to Haiti would impose a hardship on the foreigner or his or her family.

A recent report by the National Research Council examined the economic impacts of our immigration policies and concluded that legal immigration basically benefits the U.S. economy and has little negative impact on native-born Americans. A few areas of the country where immigrants are especially numerous (e.g., California, Texas, and Florida) experience some challenges in assimilating new immigrants, but the overall impact is positive, according to the Council.

In its report to the nation, the President's Council on Sustainable Development addressed the question of immigration in the context of achieving a sustainable society, one in which population growth has come to a halt. The Council decided not to take a definite stand on immigration policy, but instead deferred to a coming report from the U.S. Commission on Immigration Reform, established by Congress in 1990 and directed to make its report at the end of 1997. The Commission recommended cutting back immigration to a core level of 550,000 per year and allowing 150,000 additional visas annually for spouses and minor children of legal permanent residents. This policy is to be phased out when the existing backlog is eliminated—judged to take up to eight years. Interestingly, the Commission recommended discontinuing the immigration of unskilled laborers, while continuing to welcome more highly skilled immigrants (though under quotas). The Commission has now disbanded, and it is up to Congress to take up the report and act on it—or not. (To date, it is *not.*) Note that the commission's recommendations, if adopted, will result in only a small reduction in the current rate of legal immigration. It appears that immigration will continue to be a large and increasing part of growth in the United States well into the future.

As population pressures in developing countries continue to mount, the questions of how many immigrants to accept, from what countries, and where to draw the line regarding asylum seem certain to become more and more pressing. In addition to compassion, the social, economic, and environmental consequences—both national and global—of the alternatives must be weighed in making the final decision. Where do you stand?

welfare system that will provide care in such situations. Too often, the women cannot cope; children are abandoned, and women turn to begging, stealing, or prostitution.

And what happens to their children? If they survive at all, it is by begging, scrounging through garbage, stealing, and finding shelter in any hole or crevice they can find (Fig. 6–8). The problem is great: Nearly every sizable city in the developing world has thousands of these "stray" children, on the order of 20 million in all by some estimates, and their numbers are growing. Forced child labor, child prostitution, and selling children for adoption are additional problems that exist in no small measure. One can speculate as to the kinds of adults these children become as they grow up. At the very least, all of the factors tend to lock the poor into the vicious cycle of illiteracy and squalid conditions that defines absolute poverty.

A summary of the consequences of rapid population growth in developing countries is given in Fig. 6–9. It is clear that population growth, poverty, and environmental degradation are not separate issues. They are very much interrelated.

▲ **FIGURE 6–8** *Foraging in trash.* In cities of the developing world, many poor people, including mothers and children, subsist only by scrounging through refuse for bits of food and items they can resell.

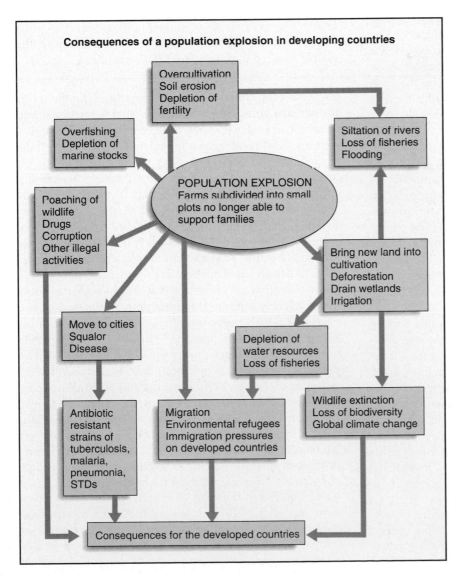

Consequences of a population explosion in developing countries

- Overcultivation
  Soil erosion
  Depletion of fertility
- Overfishing
  Depletion of marine stocks
- Siltation of rivers
  Loss of fisheries
  Flooding

POPULATION EXPLOSION
Farms subdivided into small plots no longer able to support families

- Poaching of wildlife
  Drugs
  Corruption
  Other illegal activities
- Bring new land into cultivation
  Deforestation
  Drain wetlands
  Irrigation
- Move to cities
  Squalor
  Disease
- Depletion of water resources
  Loss of fisheries
- Wildlife extinction
  Loss of biodiversity
  Global climate change
- Antibiotic resistant strains of tuberculosis, malaria, pneumonia, STDs
- Migration
  Environmental refugees
  Immigration pressures on developed countries

Consequences for the developed countries

◀ **FIGURE 6–9** *Consequences of exploding populations in the developing world.* The diagram shows the numerous connections between unchecked population growth and social and environmental problems.

Effects of Increasing Affluence

Increasing the average wealth of a population affects the environment both positively and negatively. An affluent country certainly can and does provide such things as safe drinking water, sanitary sewage systems and sewage treatment, and the collection and disposal of refuse. Thus, many forms of pollution decrease, and the environment we live in improves with increasing affluence. In addition, if we can afford gas and electricity, we are not destroying our parks and woodlands for firewood. In short, we are able to afford conservation and management, better agricultural practices, and pollution control and thereby improve our environment.

But affluence has its negative aspects as well. For example, by using such large quantities of fossil fuel (coal, oil, and natural gas) to drive our cars, heat and cool our homes, and generate electricity, the United States is responsible for a large share of the production of carbon dioxide. Specifically, with about 4.5% of the world's population, the United States is responsible for about 25% of the emissions of carbon dioxide that may be changing global climate. Similarly, emissions of chlorofluorocarbons (CFCs) that degrade the ozone layer, emissions of chemicals that cause acid rain, emissions of hazardous chemicals, and the production of nuclear wastes are all largely the by-products of affluent societies.

Economic factors place further demands on the environment. The world's wealthy 20% are responsible for 86% of all private consumption and 80% of world trade. As a consequence, 11 of the 15 major world fisheries are either fully exploited or overexploited, and old-growth forests in southern South America are being clear-cut and turned into chips to make fax paper. Oil spills are a "by-product" of our appetite for energy. Tropical forests are harvested to satisfy the desires of the affluent for exotic wood furnishings. We have already mentioned the pressure being put on endangered species from people willing to pay exorbitant prices for exotic "pets" or for "medicines" made from animal parts. As increasing numbers of people strive for and achieve greater affluence, it seems more than likely that these and similar pressures will mount.

One way of generalizing the effect of affluence is to say that it enables humans to clean up their immediate environment by disposing of their wastes to more distant locations. It also allows them to obtain resources from more distant locations such that they do not see or feel the impacts of obtaining those resources. Thus, in many respects, the affluent isolate themselves and may become totally unaware of the environmental stresses they cause with their consumption-oriented lifestyles. On the other hand, affluence also provides people with opportunities to exercise lifestyle choices that are consistent with the concerns for stewardship and sustainability.

With this picture of population growth and its impacts in view, we turn now to some additional dimensions of population growth to provide a more thorough understanding of population issues.

6.4 Dynamics of Population Growth

In considering population growth, we consider more than just the increase in numbers, which is simply births minus deaths; we also consider how the numbers of births ultimately affect the entire population over the **longevity**, or lifetimes, of the individuals.

Population Profiles

A **population profile** is a bar graph showing the number or proportion of people (males and females separately) at each age for a given population. The data are collected through a census of the entire population, a process in which each household is asked to fill out a questionnaire concerning the status of each of its members. Various estimates are made for those who do not maintain regular households. In the United States and most other countries, a detailed census is taken every 10 years. Between censuses, the population profile may be adjusted by using data regarding births, deaths, immigration, and, of course, the aging of the population. The field of collecting, compiling, and presenting information about populations is called **demography**; the people engaged in this work are **demographers**.

Profile for the United States. A population profile shows the **age structure** of the population—that is, the proportion of people in each age group at a given date. It is a snapshot of the population at a given time. Population profiles of the United States for 1990 and 2000 are shown in Figs. 6–10a and b. Leaving out the complication of emigration and immigration for the moment, recognize that each bar in the profile started out as a *cohort* of babies at a given point in the past, and that cohort has only been diminished by deaths as it has aged. In developed countries such as the United States, the proportion of people who die before age 60 is relatively small. Therefore, the population profile below age 60 is an "echo" of past events insofar as those events affected birthrates. For example, in Fig. 6–10a, you can observe that smaller numbers of people were born between 1931 and 1935 (ages 55–59 in 1990). This is a reflection of lower birthrates during the Great Depression. The dramatic increase in people born in 1946 and in following years (ages 30–44 in 1990) is a reflection of returning veterans and others starting families and choosing to have relatively large numbers of children following World War II—the "baby boom." The general drop in numbers of people born from 1961 to 1976 (ages 10 to 29 in 1990) is a reflection of sharply declining fertility rates, with people choosing to have significantly fewer children—the "baby bust." The rise in numbers of people born in more recent years (ages 0–9 in 1990) is termed the "baby boom echo," the large baby-boom generation producing a similarly large number of children, even though the actual total fertility rate remains near 2.0. Of course, the tapering off of numbers

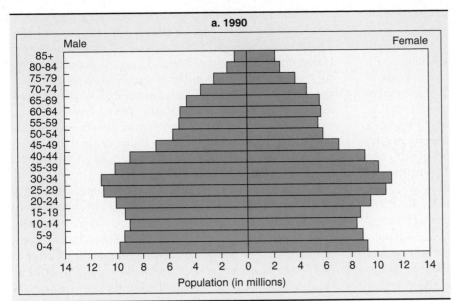

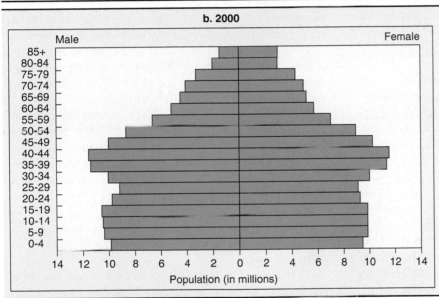

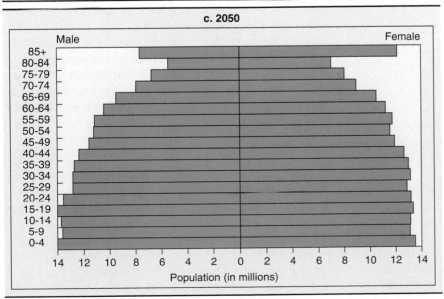

◄ **FIGURE 6-10** *Population profiles of the United States.* The age structure of the U.S. population in (a) 1990, (b) 2000, and (c) projected to 2050. (*Source*: U.S. Census Bureau, International Data Base.)

of people beyond age 60 is a function of the increasing number of deaths that occur with aging.

More than a view of the past, however, a population profile provides governments and businesses with a means of realistic planning for future demand for various goods and services, ranging from elementary schools to retirement homes. Consumer demands are largely age specific; that is, what children need and want differs from what teenagers, or young adults, or older adults, or, finally, people entering retirement want and need. Using a population profile, one can literally see the projected populations of particular age cohorts and plan to expand or retrench accordingly. A number of industries expanded and then contracted as the baby-boom generation moved through a particular age range, and this phenomenon is not yet past. In sequence, schools, then colleges and universities, and then the job market were affected by the large influx of "baby boomers." As the large baby-boom generation, now in middle age, moves up the population profile, any business or profession in the area of providing goods or services to seniors is looking forward to a period of growth. For example, look for a dramatic increase in the construction of retirement homes and long-term care facilities in the near future.

Much has been made of the future demand for Social Security outlays to retirees, especially as the baby boomers retire. This is sometimes wrongly pictured as an impending disaster. The current condition of Social Security is robust, with annual cash flow surpluses of $150 billion. These surpluses are invested in Treasury bonds, and the income from them is used to pay benefits to existing retirees. However, as the number of retirees rises, the combined income from reserves and ongoing Social Security payments of workers will eventually (in 35 years) be insufficient to pay 100% of the promised benefits. This deficit will develop slowly, and there is every reason to believe that Congress will address it long before it is expected to occur. Even if Congress does nothing, revenues will continue to cover approximately 73% of benefits.

Population Projections

Most significantly, we can use a population profile and certain other data to estimate the future overall growth of a population.

The Forecasting Technique. The technique of estimating future population growth (or decline) is one of estimating numbers of future births and deaths. Simple subtraction of deaths from births gives the absolute change in the size of the population.

Estimating Births. Using the population profile, you can see the numbers of young people moving into reproductive ages, and from other statistics, you know what percentage of women at each age have babies. From there, it is simply a process of multiplying the number of women coming into each age by the percent who have babies at that age, then totaling, and then adding that total number of babies as a new bar at the bottom of the profile.

Estimating Deaths. From statistics regarding deaths, you know the percentage of people who die at each age. These are the actuarial tables used by insurance companies to determine life insurance rates. Using the population profile, you can see the number of people moving into each age. Particularly significant are the numbers of people moving into older ages, for which death rates are highest. Again, multiplying the number of people moving into each age by the percentage of people who die at that age and then totaling gives the overall deaths for the year, and all the bars on the profile are adjusted accordingly.

Repeating this process over and over allows demographers to project births and deaths and, by subtraction, changes in population as far into the future as is desired. You can see the need for many calculations, but computers make the process relatively straightforward. Of course, such projections become increasingly subject to error as they are extended further into the future, because one has only current statistics and trends to work from, and many things may occur to either increase or decrease longevity. Just as troublesome in making long-term projections are changes in fertility rates, which are basically a function of how many babies women choose to have at any given age. As we noted at the beginning of the chapter, the projection that the world population will level off at around 10 billion is based on the assumption that fertility rates will continue their gradually declining trend. However, fertility rates tend to rise and fall, for reasons that are not always well understood or predictable. We have already seen how fertility rates in the United States rose after World War II, producing the baby-boom generation, and then how the rates declined sharply after 1961. Also imponderable, of course, are conditions that may drastically increase death rates, something we wish to make every effort to avoid.

Therefore, different projections are made mainly on the basis of different assumptions regarding future fertility rates. The United Nations gives three different projections of future world population (Fig. 6–11). The medium-fertility scenario assumes that replacement-level fertility will be reached by 2050; the high-fertility scenario assumes a total fertility rate of 2.5 by 2050 and the continuation of that rate; the low-fertility scenario assumes that the total fertility rate will reach 1.5 by 2050 and will be maintained. Note how each fertility assumption generates profoundly different world populations.

Population Projections for Developed Countries. The 2000 population profile of Italy, a developed country in southern Europe, shown in Fig. 6–12a, reflects the fact that Italian women have had a low fertility rate for some time. If we assume that the 2000 total fertility rate of 1.2 remains constant for the next 25 years, we obtain the profile presented in Fig. 6–12b. The profile shows a dramatic increase in the number of older people and a great reduction in the number of children and young people.

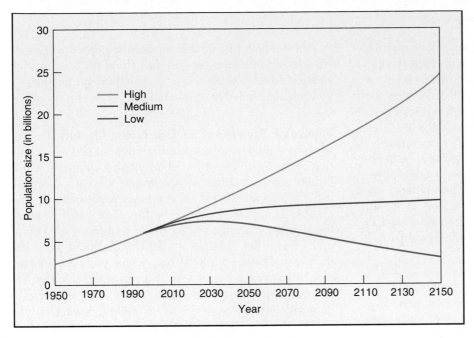

◄ **FIGURE 6–11** *Projected world population according to three different fertility scenarios.* U.N. projections of the future world population, using different total fertility rates. (*Source*: Population Division of the Dept. of Economic and Social Affairs at the United Nations Secretariat, *World Population Projections to 2150, 1998 Revision* (New York: United Nations, 1999).

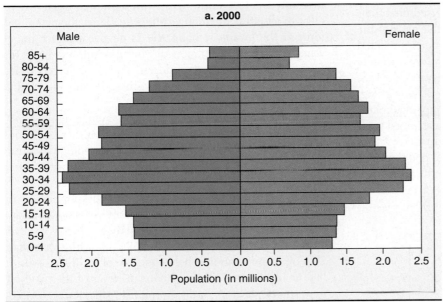

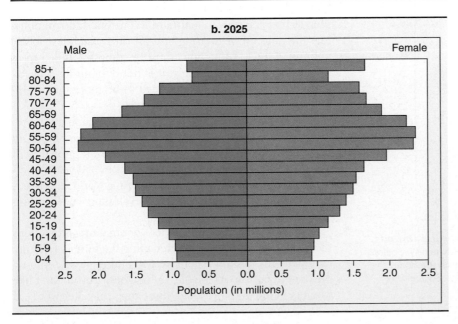

◄ **FIGURE 6–12** *Projecting future populations: developed country.* A population profile representative of a highly developed country, Italy, (a) in 2000 and (b) projected to 2025. Note how larger numbers of persons are moving into older age groups and the number of children is diminishing. (*Source*: U.S. Census Bureau, International Data Base.)

For the next 25 years, Italy's population will be **graying**, a term used to indicate that the proportion of elderly people is increasing. In fact, a net population decrease of 6% is expected to occur. What opportunities and risks does the changing population profile imply for Italy? If you were an adviser to the Italian government, what would your advice be for the short term? For the longer term? Unless a smaller population is the goal, our advice here might well be to encourage and to provide incentives for Italian couples to bear more children. We might also advocate allowing more immigration. But allowing the declining numbers of Italian people to be replaced by an immigrant population requires one to reflect on the effect that would have on Italian culture, religion, etc.

The very low fertility rate and prospective declining population seen in Italy are typical of an increasing number of highly developed nations. Therefore, the preceding analysis and questions can be applied to any of them. The one exception is the United States: In contrast to other developed countries, the U.S. fertility rate reversed directions in the late 1980s and started back up. On the basis of the lower fertility rate, the U.S. population had been projected to stabilize at between 290 and 300 million toward the middle of the next century. With a higher fertility rate of 2.1, the U.S. population is projected to be 404 million by 2050 (see Fig. 6–10c for a profile) and to continue growing indefinitely (Fig. 6–13). For this projection, immigration is assumed to remain constant at current levels—880,000 per year.

These projections for the United States show what a profound effect on population differences in the total fertility rate and high immigration make when they are extrapolated 50 or more years forward. In light of the concerns for sustainable development, what do you think the population policy of the United States should be?

Population Projections for Developing Countries.   Developing countries are in a vastly different situation from that of developed countries. In developing countries, while fertility rates are generally declining, they are still well above replacement level. (The average is currently 3.2, excluding China.) Because of even higher past fertility rates, the population profiles of developing countries have a pyramidal shape. For example, the 2000 population profile for the western Asian country of Iraq, which has a fertility rate of 5.7, is shown in Fig. 6–14a. Even assuming that this fertility rate will gradually decline to 2.7 by 2025, the population will increase from 23 to 40 million, giving us the profile shown in Fig. 6–14b. Note that even the high infant mortality rate (currently 127 per thousand) comes nowhere close to offsetting the high fertility rate. Thus, the pyramidal form of the profile remains the same because, for many years, the rising generation of young adults produces an even larger generation of children. The pyramid gets wider and wider, until the projected declining fertility rate begins to take effect.

While highly developed countries are facing the problems of a graying population, the high fertility rates in developing countries maintain an exceedingly young population. An "ideal" population structure with equal numbers of persons in each age group and a longevity of 75 years would have one-fifth, or 20%, of the population in each 15-year age group. By comparison, in many developing countries, 40%–50% of the population is below 15 years of age. In contrast, in most developed countries, less than 20% of the population is below the age of 15 (Table 6–4).

Consider what all this means in terms of the need for new schools, housing units, hospitals, roads, sewage collection and treatment facilities, telephones, etc. If a country such as Iraq is simply to maintain its current standard of living, the amount of housing and all other facilities (to say nothing of food production) must be almost doubled in as little as 25 years. Accordingly, it is not difficult for a developing country's efforts to get ahead to be more than nullified by its own population growth.

A comparison of present and projected population profiles for developed and developing countries is shown in Fig. 6–15. The figure shows that while little growth will occur in developed countries over the next 50 years, enormous growth is in store for the developing world, and this is assuming that fertility rates in the developing world continue their current downward trend!

Again, these or any other population projections should not be confused with predicting the future. They are intended only to show where we will end up if we continue the present course. As the old saying goes, "If you don't like where you are going, change direction." In other words, if

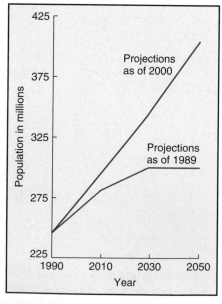

▲ **FIGURE 6–13** *Population projections for the United States.* Projections shift drastically with changes in fertility. Contrast the 1988 projection, based on a fertility rate of 1.8, with the 2000 projection, based on the increased fertility rate of 2.1. (*Source:* U.S. Census Bureau, International Data Base.)

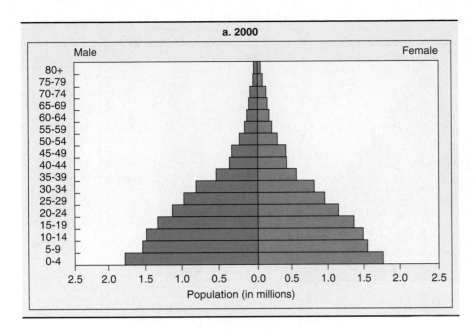

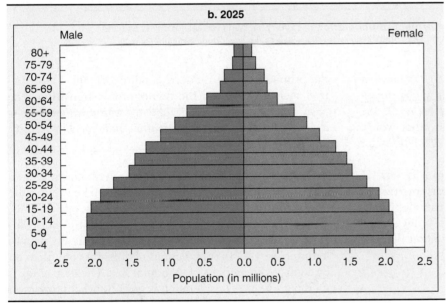

◀ **FIGURE 6–14** *Projecting future populations: developing country.* (a) The 2000 population profile of Iraq, a developing country. (b) Projection for the year 2025, based on the assumption that the total fertility rate of 5.7 will decline to 2.7 by 2025. (*Source*: U.S. Census Bureau, International Data Base.)

we feel that the projected population growth is not desirable, we can try to bring fertility rates down faster. However, even bringing the fertility rates of developing countries down to 2.0 will not stop their growth immediately. This is because of a phenomenon known as *population momentum*, which we examine next.

Population Momentum

Countries with a pyramid-shaped population profile, such as Iraq, will continue to grow for 50–60 years, even after the total fertility rate is reduced to the replacement level. This phenomenon, called ***population momentum***, occurs because such a small portion of the population is in the upper age groups, where most deaths occur, and many children are entering their reproductive years. Even if these rising

TABLE 6–4 Populations by Age Group

Region or Country	Percent of Population in Specific Age Groups		
	<15	15 to 65	>65
sub-Saharan Africa	44	53	3
Latin America	33	62	5
Asia	32	62	6
Iraq	43	54	3
Europe	18	68	14
Germany	16	68	16
China	25	68	7
United States	21	66	13

Source: Data from *2000 World Population Data Sheet* (Washington, DC: Population Reference Bureau, 2000).

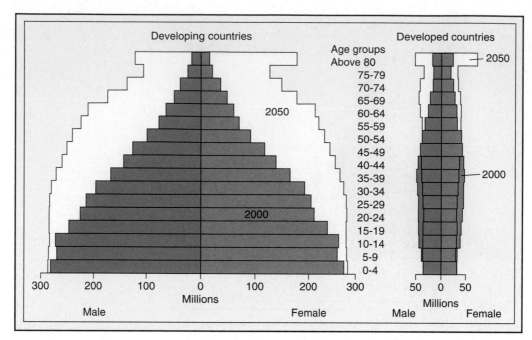

▲ **FIGURE 6–15** *Comparing projected populations.* 2000 population profiles for developed and developing countries, projected to the year 2050. (*Source*: U.S. Census Bureau, International Data Base.)

generations have only two children per woman, the number of births will far exceed the number of deaths. The imbalance will continue until the current children reach the Iraqi limits of longevity—50 to 60 years. In other words, only a population at or below replacement-level fertility for many decades will achieve a stable population.

To acknowledge population momentum is not, of course, to say that efforts to stabilize population are fruitless. It is, however, to say that population growth cannot be halted quickly; the earlier the fertility rates are reduced, the greater is the likelihood of averting a crisis point decades into the future.

The Demographic Transition

The concept of a stable, nongrowing global human population based on people freely choosing to have smaller families is possible because it is already the fact in developed countries. If we can understand the factors that have brought this about in developed countries, then perhaps we can make those factors operative in developing countries.

Early demographers observed that the modernization of a nation brings about more than just a lower death rate resulting from better health care; a decline in fertility rate also occurs as people choose to limit the size of their families. Thus, as economic development occurs, human societies move from a primitive population stability, in which high birthrates are offset by high infant and childhood mortality, to a modern population stability, in which low infant and childhood mortality are balanced by low birthrates. This gradual shift in birth and death rates from

the primitive to the modern condition in the industrialized societies is called the **demographic transition**. *The basic premise of the demographic transition is that there is a causal link between modernization and a decline in birth and death rates.*

Birth Rates and Death Rates. To understand the demographic transition, we need to introduce two new terms: the **crude birth rate (CBR)** and **crude death rate (CDR)**. These terms are defined as the number of births or deaths per 1,000 of the population per year. By giving the data in terms of "per 1,000," populations of different countries can be compared regardless of their total size. The term *crude* is used because no consideration is given to what proportion of the population is old or young, male or female. Subtracting the CDR from the CBR gives the increase (or decrease) per 1,000 per year. Dividing this result by 10 then puts it in terms of "per 100," or percent. Mathematically,

CBR		CDR		Natural increase
Number of *births* per 1,000 per year	−	Number of *deaths* per 1,000 per year	=	(or decrease) in population per 1,000 per year ÷ 10 = %

Of course, a stable population is achieved if, and only if, the CBR and CDR are equal.

The **doubling time**, or the number of years it will take a population growing at a constant percent per year to double, is calculated by dividing the percentage rate of growth into 70. (The 70 has nothing to do with population; it is derived from an equation for population growth.) The CBR,

TABLE 6–5 Crude Birth and Death Rates for Selected Countries

Country or Region	Crude Birth Rate	Crude Death Rate	Annual Rate of Increase (%)	Doubling Time (Years)
World	22	9	1.4	51
Developing Nations				
Average (excluding China)	29	9	1.9	36
Egypt	26	6	2.0	35
Kenya	35	14	2.1	33
Madagascar	44	14	2.9	24
India	27	9	1.8	39
Iraq	38	10	2.8	25
Vietnam	20	6	1.4	48
Haiti	33	16	1.7	40
Brazil	21	6	1.5	45
Mexico	24	4	2.0	36
Developed Nations				
Average	11	10	0.1	809
United States	15	9	0.6	120
Canada	11	7	0.4	178
Japan	9	8	0.2	462
Denmark	12	11	0.1	472
Germany	9	10	-0.1	–
Italy	9	10	-0.1	–
Spain	9	9	0.0	6,931

Source: Data from *2000 World Population Data Sheet* (Washington, DC: Population Reference Bureau, 2000).
Note: Dash indicates doubling time cannot be calculated, because growth is negative.

the CDR, and the doubling time of various countries are shown in Table 6–5.

Epidemiologic Transition. Throughout most of human history, crude death rates were high—40 or more per thousand for most societies. As we have seen, by the middle of the 19th century, the epidemics and other social conditions responsible for high death rates began to recede, and death rates in Europe and North America underwent a decline. It is significant that this decline was gradual in the now-developed countries, lasting for many decades and finally stabilizing at a CDR of about 11 per thousand. At present, cancer and cardiovascular disease and other degenerative diseases account for most mortality, and many people survive to old age. This pattern of change in mortality factors has been called the *epidemiologic* transition (Fig. 6–16) and represents one element of the demographic transition. (Epidemiology is the study of diseases in human societies.)

Fertility Transition. Another pattern of change over time can be seen in crude birthrates. Again, in the now-developed countries, birth rates have undergone a decline from a high of 40 to more than 50 per thousand down to 8 to 12 per thousand—a *fertility* transition. As Fig. 6–16 shows, this does not

happen at the same time as the epidemiologic transition; it is delayed by decades or more. Since net growth is the difference between the CBR and the CDR, the time during which these two patterns are out of phase is a time of rapid population growth. The developed countries underwent such a time during the 19th and early 20th centuries, and one result of this was the massive emigration from the Old World to the less populated New World.

Phases of the Demographic Transition. The demographic transition is typically presented as occurring in four phases, as shown in Fig. 6–16. **Phase I** is the primitive stability resulting from a high CBR being offset by an equally high CDR. **Phase II** is marked by a declining CDR—the epidemiologic transition. Because fertility and, hence, the CBR remain high during Phase II, this is a phase of accelerating population growth. **Phase III** is a phase of declining CBR resulting from a declining fertility rate; population growth is still significant. Finally, **Phase IV** is reached, in which modern stability is achieved by a continuing low CDR, but an equally low CBR.

Basically, developed countries have completed the demographic transition. Developing countries, on the other hand, are still in Phases II and III. Death rates have declined markedly, and fertility and birth rates are declining,

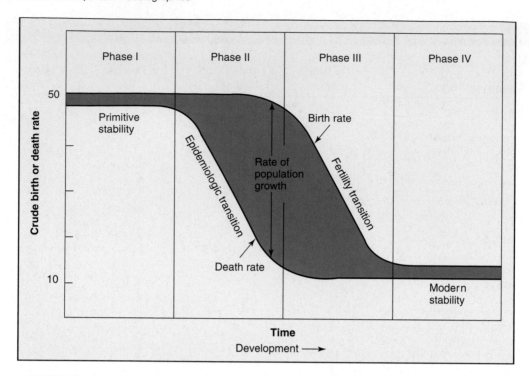

▲ **FIGURE 6–16**   ***The demographic transition.***   The epidemiologic transition and the fertility transition combined to produce the demographic transition in the developed countries over many decades. (Adapted from Joseph A. McFalls, Jr., "Population: A Lively Introduction," *Population Bulletin* 53, no. 3 [1998]: 39. Courtesy of Population Reference Bureau, 2000.)

but remain considerably above replacement levels. Therefore, populations in developing countries are still growing rapidly. Using the concept of the demographic transition, we can plot the major regions of the world in terms of current birth and death rates (Fig. 6–17). We can draw a dividing line through this plot; nations to the right of the line

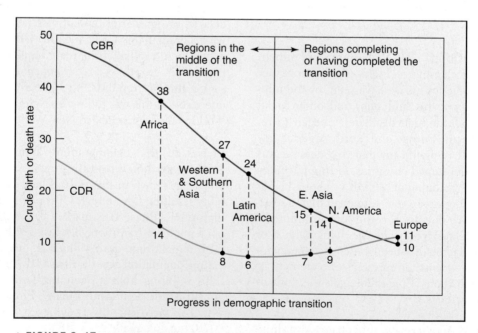

▲ **FIGURE 6–17**   ***World regions in the process of demographic transition.***   Crude birth rates and crude death rates are shown for major regions of the world. A dividing line separates countries at or well along in the demographic transition from those still in the middle of the transition. (*Source*: Data from *2000 World Population Data Sheet* [Washington, DC: Population Reference Bureau, 2000].)

are on a fast track to completing the demographic transition or are already there. Nations to the left of the line (about half the world) are in the middle of the demographic transition, most in the fourth decade of rapid population increase. They appear to be trapped there, with serious consequences.

It is on the basis of the demographic transition that some argue that we not worry about population; it will stabilize by itself as developing countries reach Phase IV. Therefore, the argument goes, we need only focus on free world trade and other factors that will speed economic growth in developing countries. But there are major flaws in this argument. First, it must be emphasized that the demographic transition occurred in the developed countries over a period of many decades; modernization did not happen overnight. Second, many of the most populous developing countries are still very far behind the developed countries economically (Fig. 6–3) and are making very slow progress toward modernization. If they must modernize before population growth comes under control, their population growth and its demands for resources and services will undercut the very economic growth that is so necessary—a catch-22 with profound consequences. Third, we emphasize again that present stresses on the biosphere and loss of biodiversity are largely a consequence of the consumption-oriented lifestyles of the current 1.2 billion people in developed nations. Recognizing that severe stresses are being caused by the lifestyles of 1.2 billion people makes any notion of a world with 10 billion people living with the same lifestyle utterly absurd. Finally, and most important, the demographic transition really shows only a *correlation* between development and changing birth and death rates; it does not prove that development is necessary for the demographic transition to occur. Other factors may be much more significant.

In Chapter 7, we will investigate the factors that influence birth and death rates and explore the relative contributions of economic development and family planning for bringing about the fertility transition.

ENVIRONMENT ON THE WEB

TRACKING THE DEMOGRAPHIC TRANSITION

This chapter describes how a demographic transition is likely to occur as a population (or economy) modernizes. When we read about population projections for this or that country, it is easy to think of ourselves as somehow removed from the system. In fact, we are all a part of some demographic trend, whether rising or falling, and our own actions will affect the size and composition of future populations. Like us, people in other countries are faced with lifestyle decisions—for instance, about education, work, and family size. These decisions are influenced by our personal histories and the patterns we have observed in our own families and the society around us.

For example, in more industrialized societies, a larger proportion of the population stays in school longer, is engaged in high paying, complex work, and defers marriage and childbearing. With fewer childbearing years, the individual has fewer children than someone with less education and a menial job—or no job at all. Conversely, people in less developed societies may not go to school at all, are less likely to work at complex jobs, and tend to marry and bear children earlier. In these societies, fertility rates are generally higher than in developed nations.

Therefore, it is possible to observe some general patterns in birth and death rates as societies modernize and to be able to predict these demographics for management purposes. Yet, in order to predict demographic change, we must understand the human behaviors that cause it. We must also understand that it is much more difficult to predict how these behaviors may or may not change over time. A key challenge in predicting demographics relates to uncertainties in birth rates. While medical advancements, and thus changes in death rates, can occur over a matter of years, birth rates may take generations to respond to societal change. People like to have children, especially if they have come from a large family or a culture that values children highly. Even when a society has made good progress on industrialization, it can be hard to encourage a move to a smaller family size.

Web Explorations

The Environment on the Web activity for this essay describes the challenges involved in predicting and managing demographic change and the human behaviors that underlie these factors. Go to the Environment on the Web activity (select Chapter 6 at **http://www.prenhall.com/wright**) and learn for yourself:

1. about the relationship between population and birth rate;
2. about optimistic and pessimistic models for population growth;
3. how population pyramids can help forecast health care and educational needs; and
4. how decision makers can use population growth models for long-term resource management planning.

A suggested time frame for each of these exercises is 10–30 minutes.

REVIEW QUESTIONS

1. How has the global human population changed from early times to 1800? From 1800 until the present? What is projected over the next 50 years?

2. How is the world divided in terms of relative per capita incomes? Fertility rates? Population growth rates?

3. What two factors are multiplied to give total environmental impact? Are developed nations exempt from environmental impact? Why not? What third factor affects environmental impact, and how does it do so?

4. What are the social and environmental consequences of rapid population growth in developing countries? How may developed countries be affected?

5. What information is given by a population profile? How is the information presented?

6. How are numbers of future births and deaths predicted from a population profile?

7. How do the population profiles, fertility rates, and population projections of developed countries differ from those of developing countries? How might future population goals of developed and developing countries contrast?

8. What is meant by population momentum, and what is its cause?

9. Define the crude birth rate (CBR) and crude death rate (CDR). Describe how these rates are used to calculate the percent rate of growth and the doubling time of a population.

10. What is meant by the demographic transition? Relate the epidemiologic transition and the fertility transition, two elements of the demographic transition, to its four phases.

11. How do developed and developing nations differ regarding their current positions in the demographic transition?

THINKING ENVIRONMENTALLY

1. Make a cause-and-effect "map" showing the many social and environmental consequences linked to unabated population growth. Include crossovers between the developed and developing worlds.

2. It has been proposed that excess human populations be accommodated by building orbiting space stations that would house about 10,000 persons each. Each station would be able to produce its own food. How many space stations would be required to accommodate the world's projected population growth over the next 10 years? What kind of population policy would have to be enforced on the space stations? Are space stations a logical solution to the population problem?

3. Starting with a hypothetical population of 14,000 people and an even age distribution (1,000 in each five-year age group from 1–5 to 65–70), assume that this population initially has a total fertility rate of 2 and an average longevity of 70 years. Project how the population will change over the next 60 years under each of the following conditions:

a. Total fertility rate and longevity remain constant.
b. Total fertility rate changes to 4; longevity remains constant.
c. Total fertility rate changes to 1; longevity remains constant.
d. Total fertility rate remains at 2; longevity increases to 100.
e. Total fertility rate remains at 2; longevity decreases to 50.

4. From the 2000 crude birth and crude death rates, calculate the rate of population growth and the population doubling time for each of the following countries:

	CBR	CDR
Algeria	29	6
Ethiopia	45	21
Argentina	19	8
Iran	21	6
Russia	9	14
France	13	9
Germany	9	10

WEB REFERENCES

On-line resources for this chapter are on the World Wide Web at: **http://www.prenhall.com/wright**. (Click on Chapter 6 within the Chapter Selector.)

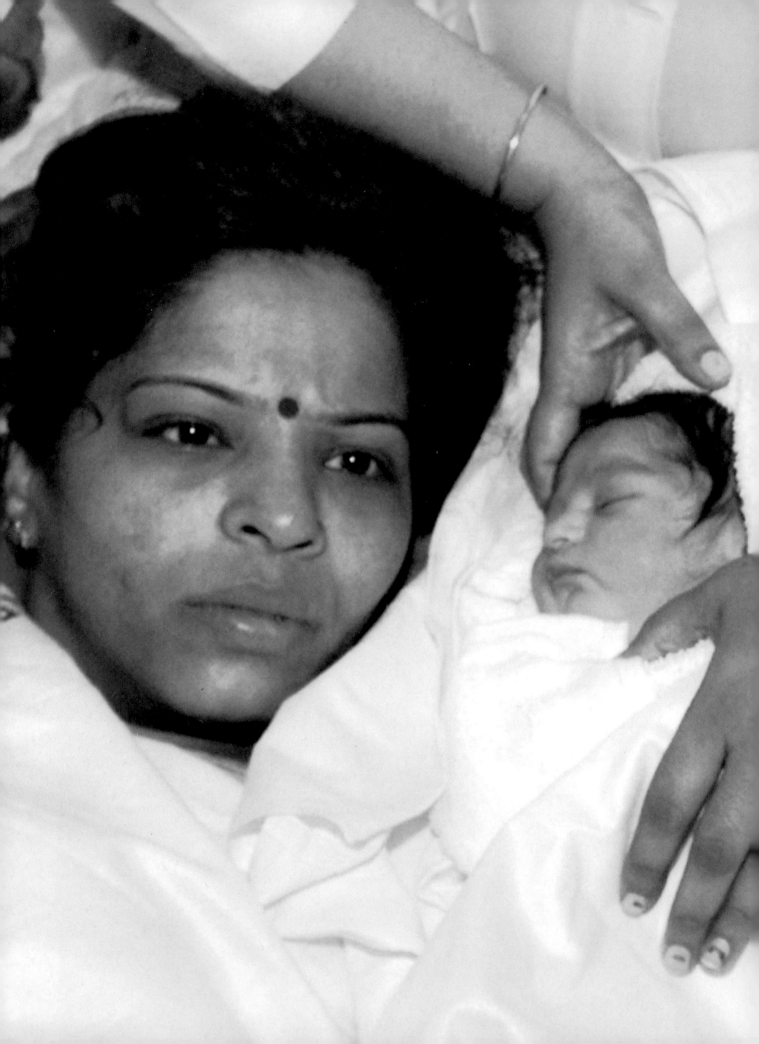

Issues in Population and Development

1. Many regions of the developing world are stuck in the middle of the demographic transition. Must they modernize before the fertility of their populations declines, or must they bring fertility down before they can modernize?

2. The factors that influence fertility rates are more specific than development in general. What are the factors that actually influence the number of children desired?

3. The industrialized countries established the World Bank as a means of encouraging development in the poor countries. How has this agency carried out its role? In particular, what is the debt crisis and how is it being addressed by the World Bank and the developed countries?

4. The shift from high to low fertility in developing countries is being accomplished by social modernization. How is this different from industrialization, and on what five areas must development efforts focus for social modernization to be successful?

5. In 1994, world leaders met in Cairo, Egypt, at the U.N. Conference on Population and Development. What was the significance of this meeting? What are the agreed-upon strategies for addressing the problems of poverty, excessive population, and environmental degradation?

O n May 11, 2000, Astha Arora, a baby girl born in Safdarjang Hospital in New Delhi, became India's one billionth person. With Astha, whose name means "faith" in Hindi, India joined China in the exclusive club of population billionaires. In the 52 years since India gained its independence, its population has tripled; it is now growing at a rate of 15.5 million per year and is expected to surpass China's by midcentury. In spite of many decades of family-planning programs, India's population has continued its relentless growth, canceling out most of the country's gains in food production, health care, and literacy. More than half of India's children are undernourished, a third of the population lives below the poverty line, and half of its adults remain illiterate. With a per capita gross national product (GNP) of $420 per year, India has a very long way to go to achieve the kind of development that characterizes the low population growth of developed countries. It might be tempting to consider this a hopeless situa-

tion if it were not for one region of India where things are quite different: Kerala.

Kerala is the southernmost state of India, with a population of 33 million occupying an area of 39.9 thousand square miles. This makes Kerala the second most densely populated state in India and close to the most densely populated region in the world (Fig. 7–1). Situated only 10° north of the equator, Kerala is tropical, with lush plantations of coffee, tea, rubber, and spices in the highlands, and rice, coconut, sugarcane, tapioca, ginger, and bananas typically grown in the lowlands. With the Arabian Sea on the coastline, fishing is an important industry. Kerala is very much like the rest of India in some ways: It is crowded, per capita income is low—less than $400 per year—and food intake is around 2,200 calories per day, considered adequate, but on the low end. Here the comparison ends, however.

The people of Kerala have a life expectancy of 71 years, vs. 61 for all of India. Infant mortality in the state is 17 per

◀ **Astha Arora,** India's billionth person, with her mother, Anjana. 42,000 babies are born every day in India.

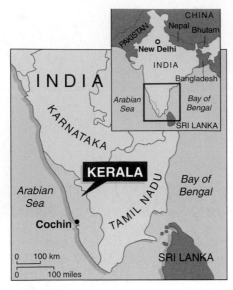

▲ **FIGURE 7–1** Kerala state, in southwestern India, has achieved remarkable progress towards a stable population.

thousand, vs. 72 per thousand for India. The fertility rate is 1.8 (below replacement level), vs. 3.3 for India. Literacy is over 95%, almost all villages in the state have access to school and modern health services, and women have achieved high offices in the land and are as well educated as the men. Even though Kerala must be considered a poor region by every economic measurement, its people are well on the way to achieving a stable population.

What is different about Kerala is a strong public policy commitment to health development and education. Land distribution is relatively equitable, food distribution is efficient, and the old caste system of India has all but disappeared. These investments in social policy have paid off. Kerala's total fertility rate dropped from 3.7 to 1.8 in just two decades; in the last few years, Kerala has been making twice the progress of India in per capita income growth. In short, Kerala is an example of the possibility of bringing a developing region from the midphase of the demographic transition to the threshold of its completion without the thorough economic development that characterized the industrial countries as they underwent their demographic transition.

In this chapter, we follow up on the demographic information presented in Chapter 6. We begin with a reassessment of the demographic transition, examining the factors that promote the movement of a country through the transition. Next, we examine various policies and programs that have been used in the past to improve development and lower fertility. We then look at evidence for a new direction in development, and finally, we examine the international efforts being made to address population issues in the developing world.

7.1 Reassessing the Demographic Transition

At the end of Chapter 6, we presented evidence that many regions of the developing world are stuck in the middle phases of the demographic transition (Fig. 6–17), resulting in continuing rapid population growth. The countries in these regions would all like to experience the economic development that has brought South Korea, Indonesia, Malaysia, Brazil, and others into the middle- and even high-income nation groups. If they did so, it is likely that they would then move through the demographic transition. Yet there are great disparities in economic growth among developing countries; some 100 countries have been experiencing economic stagnation or decline and remain mired in their poverty. These same countries have the most rapid growth in population. The **key question** is; Must the developing countries bring down population growth before they can grow economically, or must they make serious progress toward modernization before their population growth declines? This question has been debated for some time.

In 1798, Thomas Malthus, a British economist, pointed out that populations tend to grow exponentially, but there are ultimate limits to the expansion of agriculture. Thus, at the very beginning of the population explosion, Malthus foresaw a world headed toward calamity if something was not done to control population. But Malthus could foresee neither the tremendous expansion of agriculture that would come with the Industrial Revolution nor the demographic transition—the fact that fertility rates would decline with the Industrial Revolution.

Therefore, from early on, there were two basic, conflicting schools of thought regarding the growth of population: (1) We need to concentrate on population policies and family-planning technologies to bring down birthrates. (2) If we concentrate on development, population growth will slow down "automatically," as it did in the developed countries. These two schools of thought were reflected at two previous U.N. population conferences, the first in Bucharest, Romania, in 1974, and the second in Mexico City in 1984. At the Bucharest conference, the United States was a strong advocate of population control through family planning, while the developing nations argued that "development was the best contraceptive." Their resistance to family planning was also bolstered by feelings that the developed world's promotion of population control was another form of economic imperialism or even genocide.

At the second conference in Mexico City, the sides were somewhat reversed. Developing countries facing real problems of excessive population growth were asking for more assistance with family planning, whereas the United

States, under pressure from "right-to-life" advocates, took the position that development was the answer and terminated all contributions to international family-planning efforts, a policy that remained in effect until 1993. The other developed countries, however, remained convinced that family planning was essential and continued to support international efforts to aid the developing countries in their attempts to implement policies designed to bring fertility rates down.

A third conference, the International Conference on Population and Development (ICPD), was held in Cairo in 1994. Poverty, population growth, and development were clearly linked at this conference, and an important focus was also placed on resources and environmental degradation. The ideological split between developed and developing countries was a thing of the past, but the conference and its Program of Action (detailed later in the chapter) left to the individual countries the implementation of the goals to bring down population growth and bring about sustainable development. As we have already seen, there is still no clear consensus on how to do this.

To put the debate in perspective, it is instructive to examine the factors that influence people to have more or fewer children. Obviously, no one makes a decision whether to have a child on the basis of his or her country's average GNP (considered a good measure of the country's level of development) or fertility rate. However, a plot of the fertility rate against GNP per capita does show a weak correlation (Fig. 7–2). There is no question that fertility rates in the developed countries have declined with development. Therefore, we need to determine the specific aspects of development that influence childbearing.

Factors Influencing Family Size

Many students in developed countries find it difficult to understand why poor women in developing countries have large numbers of children. It is obvious from our perspective that more children spread a family's income more thinly and handicap efforts to get ahead economically. "Why," many ask, "do poor people behave so irrationally?"

What we fail to recognize is that the poor in developing countries live in a very different sociocultural situation. When we understand that situation, we find that their choices for larger families are quite logical. Numerous studies and surveys reveal the following as primary reasons that the poor in developing countries desire large numbers of children:

1. **Security in one's old age.** A traditional custom and expectation in developing countries is that old people will be cared for by their children. Social Security, welfare, Medicare, retirement, and nursing homes are all relatively new developments, found in high-income nations alone. Such things are not available to the poor of developing countries. Therefore, a primary reason given by poor women in developing nations for desiring many children is "to assure my care in old age."

2. **Infant and childhood mortality.** Closely coupled with the desire for security in one's old age is the experience of high infant and childhood mortality. According to the Population Reference Bureau, 20,000 children below one year of age die every day in the developing world. This high infant mortality is the most profound indicator of the conditions of squalor and poverty in which people are living; it is unacceptable on any moral, ethical, or religious

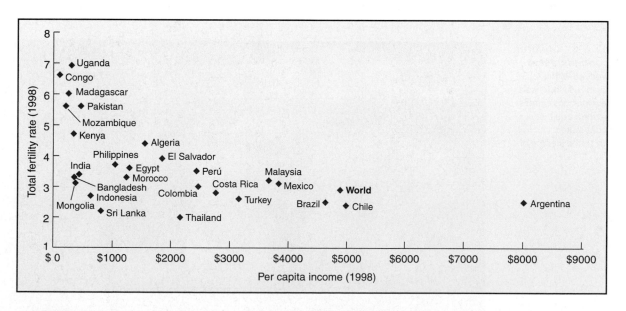

▲ **FIGURE 7–2** *Fertility rate and income for selected developing countries.* There is a weak correlation between income and lower total fertility. Factors that affect fertility more directly are health care, education for women, and the availability of contraceptive information and services. (*Source:* Data from *1998 World Population Data Sheet* and *2000 World Population Data Sheet* [Washington, DC: Population Reference Bureau, 1998, 2000].)

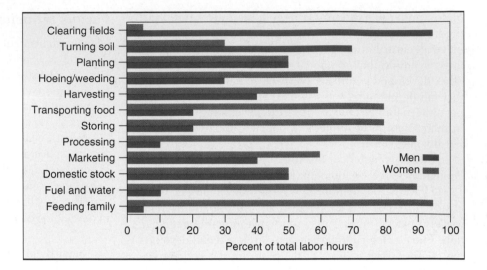

▶ **FIGURE 7–3** *Gender-related work in developing countries.* In developing countries—especially in Africa—women do most of the work relating to care and maintenance of the family, including heavy farming tasks. (*Source*: Data from Jodi Jacobson, "Gender Bias: Roadblock to Sustainable Development," *Worldwatch Paper* 110 [1992].)

ground. Nor does it serve to stabilize population, for the following reason: The common and often personal experience of children dying leads people to desire additional children as an "insurance policy" for security in their old age. Therefore, high infant mortality actually leads to higher fertility rates. It is only when there is a very high likelihood that children will survive (low childhood mortality) that people feel secure in having just two children or even one child.

3. **Helping hands**. A third reason given by women of developing nations for desiring many children is "to help me with my work." In the subsistence-agriculture societies of the developing world, it has been, and remains, traditional for women to do most of the work relating to the direct care and support of the family. Clearing fields and turning the soil in preparation for planting is done largely by men, but all of the rest of the work—

from planting, weeding, and harvesting to going to the market and gathering firewood and water—falls to the women (Fig. 7–3). A child as young as 5 can begin to help with many of these chores, and 12-year-olds can do an adult's work (Fig. 7–4). In short, children are seen as an economic asset. As resources become scarcer and more time is required to fetch water, collect firewood, and tend crops, the asset of having many little hands becomes even greater. It is only in an urban setting in a developed nation that opportunities for children to contribute to the economic welfare of the family become extremely limited, that the costs of feeding, clothing, and educating children are prolonged, and that the economic burden of children is acutely felt.

4. **Importance of education**. The importance given by a society to education is closely allied to whether chil-

▶ **FIGURE 7–4** *Children as an economic asset.* Children working with adults in the fields in Bali, Indonesia. In most developing countries, children perform adult work and thus contribute significantly to the income of the family.

dren are seen as an economic asset or a liability. If it is felt that children do not need to be educated, then it is easy to cast them in the role of simply being the "many little hands" to help with the chores of everyday survival—the more, the better. If, on the other hand, it is required that children go to school, they not only are removed from the labor force, but also need additional economic support for suitable clothing, school supplies, and school fees. In short, the requirement that children be educated changes their position from being seen as an economic asset to being viewed as an economic liability and influences fertility rates. Again, in traditional, subsistence-agriculture societies, education has been deemed unnecessary, and this remains the case for many children in the developing world, especially girls. Thus, the continued high fertility of parents is supported by a belief that educating girls is unnecessary. Of course, education does much more than keep children from working; it opens up any number of additional opportunities that also affect fertility rates. (See the "Global Perspective" essay, p. 170.)

5. **Status of women: opportunities for women's education and careers.** The traditional social structure in many developing countries still discourages and, in many cases, bars women from obtaining higher education, owning businesses, owning land, and pursuing many careers. Such discrimination against women forces them into doing what only they can do: bear children. Worse, in many of these countries, the male's respect for a woman is often proportional to the number of children she bears. Breaking down such barriers of discrimination so that girls are educated and gain status outside the context of raising children has probably contributed more than anything toward the very low fertility rates seen in developed countries today. Indeed, studies show that women with just an eighth-grade education have, on average, only half the number of children of their uneducated counterparts.

6. **Availability of contraceptives.** There can be no doubt regarding the importance of contraceptives in achieving a lower fertility rate. Studies show a strong correlation between lower fertility rates and the percentage of couples using contraception (Fig. 7–5): Each 12% increase in contraceptive use translates into one fewer child. In the developed world, we take the availability of contraceptives almost for granted. Perhaps the most profound finding in surveys of women in the developing world is that large numbers state that they want to delay having their next child or that they do not want any more children. Yet many of these women are not using contraceptives. Women in rural areas report that contraceptives are frequently not available. Women in cities also have trouble getting them in spite of free clinics, because the clinics may be too far away or crowded or they may run out of contraceptives. Providing contraceptives to women is a major facet of family planning.

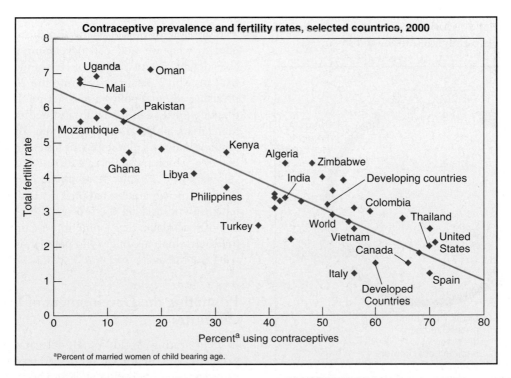

▲ FIGURE 7–5 *Prevalence of contraception and fertility rates.* More than any other single factor, lower fertility rates are correlated with the percentage of the population using contraceptives. (*Source*: Data from *2000 World Population Data Sheet* [Washington, DC: Population Reference Bureau, 2000].)

Students frequently raise the point that religious beliefs play a role in determining family size. To some extent, this may be true, but it becomes less of a factor as educational and other barriers to women are broken down. For example, Italy and Spain, predominantly Catholic countries, have some of the lowest fertility rates in the world. Obviously, many Catholics are not taking the Pope's admonitions seriously.

Conclusions

Reflecting on the foregoing items, we can see that the factors supporting large families are common to preindustrialized, agrarian societies, while those conducive to raising small families (or no children) generally appear with industrialization and development. The latter factors include the relatively high cost of raising children, the existence of pensions and a Social Security system, the existence of opportunities for women to join the workforce, free access to inexpensive contraceptives, adequate health care, wide educational opportunities and high educational achievement, and a higher age at marriage. Fertility rates in developing countries remain high not because people in those countries are behaving irrationally, but because the sociocultural climate in which they live favors high fertility and because contraceptives often are not available. Furthermore, we should be able to understand how poverty, environmental degradation, and high fertility drive one another in a vicious cycle (Fig. 7–6). Increasing population density leads to a greater depletion of rural community resources like firewood, water, and land, which encourages couples to have more children to help gather resources and so on.

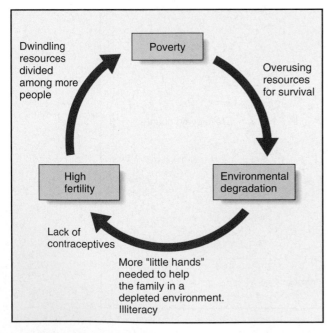

▲ **FIGURE 7–6** *The poverty cycle.* Poverty, environmental degradation, and high fertility rates become locked in a self-perpetuating vicious cycle.

One may ask how the now industrialized nations came through the demographic transition without getting caught in the poverty–population trap. Two points are significant in this regard. First, the improvements in disease control that lowered death rates occurred gradually through the 1800s and early 1900s. Industrialization, which introduced factors that lowered fertility, occurred over the same period. Therefore, there was never a huge discrepancy between birth and death rates (Fig. 7–7a). Second, and perhaps even more significant, surplus population from European nations could, and did, readily emigrate to the United States, Canada, Latin America, New Zealand, and Australia.

In contrast, modern medicine was introduced to the developing world relatively suddenly, bringing about a precipitous decline in death rates, while the fertility-lowering effects of development have been slow in coming (Fig. 7–7b). All these observations point to the fact that it is not economic development by itself that leads to declining fertility rates. Rather, fertility rates decline insofar as development provides (1) security in one's old age apart from the ministrations of children, (2) lower infant and childhood mortality, (3) universal education for children, (4) opportunities for higher education and careers for women, and (5) unrestricted access to contraceptives.

We turn our attention now to efforts that have been made to bring about development in the countries that most need it.

7.2 Development

The absence of development and the impoverished societies in what we now call developing countries are to a considerable extent a legacy of 18th- and 19th century colonialism, which persisted well into the 20th century. Thus, reversing colonial policy and fostering the development of these countries can be amply justified on all grounds: humanitarian, political, and economic. By providing better jobs and incomes, development fosters improved standards of living, which in turn create expanded markets for the developed world. Finally, development fosters trade, cooperation, and peace among nations. If fertility rates decline to replacement levels in the process, then, from the perspective of sustainable development, we can look forward to a future in which a stable world population lives in a state of relative affluence.

Promoting the Development of Low-income Countries

In 1944, during World War II, delegates from around the world met in Bretton Woods, New Hampshire, and conceived a vision of development for poor countries. They established the International Bank for Reconstruction and Development (or, as it is more commonly known now, the **World Bank**). The World Bank functions as a special

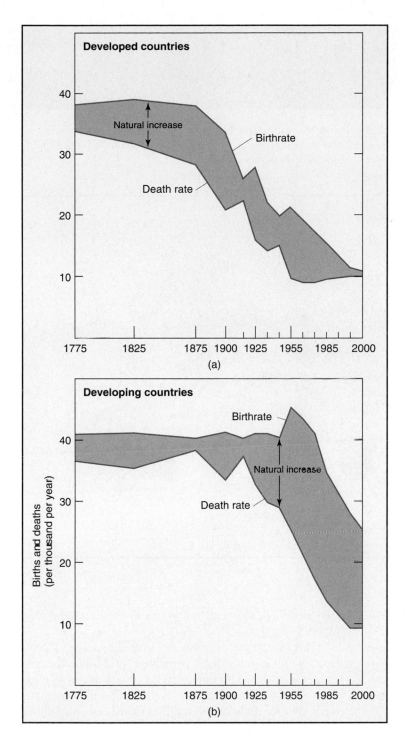

◀ **FIGURE 7-7** *Demographic transition in developed and developing countries.* (a) In developed countries, the decrease in birthrates proceeded soon after, and along with, the decrease in death rates, so very rapid population growth never occurred. (b) In developing countries, both birth and death rates remained high until the mid-1900s. Then a precipitous decline in death rates was caused by the sudden introduction of modern medicine, whereas birthrates remained high, causing very rapid population growth. (Redrawn with permission of Population Reference Bureau, 1998.)

agency within the United Nations. With deposits from governments and commercial banks in the developed world, the World Bank lends money to developing nations for a variety of projects at interest rates somewhat below the going market rates. In effect, the World Bank helps governments of developing countries (the bank loans only to governments) borrow large sums of money for projects they otherwise could not afford.

Annual loans from the World Bank have climbed steadily, from $1.3 billion in 1949 to $29 billion in 1999. With the power to approve or disapprove loans, and through the amount of money it lends, the World Bank has been the major single agency providing aid to developing countries for the past 50 years. How does the World Bank's record look?

Past Successes and Failures of the World Bank

Many developing countries have made remarkable *economic* progress. The gross national products of some countries have increased as much as fivefold, bringing them from low- to medium-income status, and some medium-income

GLOBAL PERSPECTIVE

FERTILITY AND LITERACY

Illiteracy, particularly among women, is one of the prime indicators of poverty and high fertility. The accompanying map depicts countries according to female illiteracy rate and total fertility rate. The strong correlation is evident. A conclusion which may be drawn is that improving women's literacy will have an impact on lowering fertility.

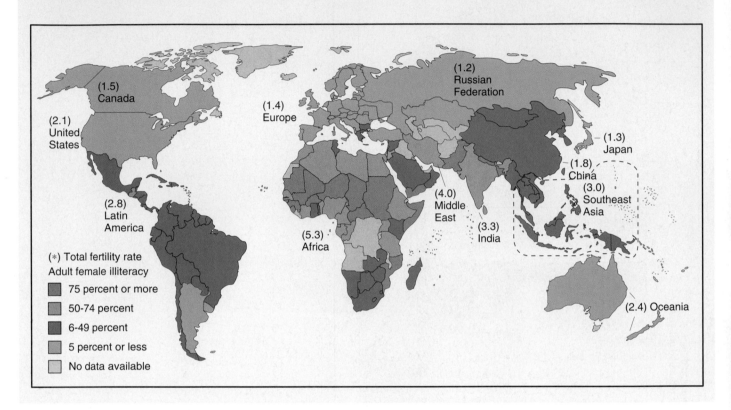

nations have achieved high-income status. Although the world economy is still strongly dominated by the industrialized countries, the developing countries have become more and more involved in what is now an integrated global economy. Foreign investment is playing a large role in this development; in the last decade, foreign investment in developing countries increased fourfold.

In addition, great strides have been made in *social* progress. Efforts from other branches of the United Nations, such as the World Health Organization (WHO), Food and Agricultural Organization (FAO), U.N. Educational, Scientific, and Cultural Organization (UNESCO), and United Nations Children's Fund (UNICEF), have augmented the work of various government programs. Private charitable organizations have also played a large role. Literacy rates, the percentage of the population with access to clean drinking water and sanitary sewers, and other social indicators of development, generally speaking, have improved (Table 7–1). Further, in keeping with the concept of the demographic transition, the fertility rates of most developing countries

have declined, although not yet to the replacement level (Table 7–2).

However, these successes are dulled by some sobering facts: A fifth of the world's population (1.2 billion people) lives on less than $1 a day, even more lack access to clean water (1.5 billion) and sanitary facilities (2.4 billion), an estimated 790 million are malnourished, environmental degradation is rampant, and fertility rates remain high in the poorest countries. Also, the gap between the rich and poor countries is growing: The difference between the average income of the richest 20% and the poorest 20% increased from 30 to 1 in 1960 to 74 to 1 in 1999. In short, the vicious cycle of high fertility, poverty, and environmental degradation persists, presenting the international community with an enormous challenge. In 1997, representatives from the United Nations, the World Bank, and the Organization for Economic Cooperation and Development (OECD) met to formulate a set of goals for international development that would primarily address poverty and its various aspects (Table 7–3). These goals, adopted by all U.N.

TABLE 7–1   Improvement in Indicators of Development, 1970–1995

	Low-income Countries		Upper- and Middle-income Countries	
	1970	1995	1970	1995
Population with access to				
Safe drinking water (percent)	22	51	59	84
Sanitary sewers (percent)	29	38	61	79
Age group enrolled in education:[1]				
Primary education (percent of total)	55	105[2]	94	99
Percent females per 100 males	61	88	95	99
Secondary education (percent of total)	13	27	32	70
Percent females per 100 males	44	83	95	100
Infant mortality, Per 1,000 live births	114	69	70	35
Life expectancy at birth, Years	47	63	62	69

[1]Data from 1993.
[2]Adults and children enrolled.
Source: From the WORLD DEVELOPMENT REPORT. Copyright © 1994 by the International Bank for Reconstruction and Development/The World Bank. By permission of Oxford University Press..

countries, are intended to encourage accountability as the various international agencies and individual countries partner to address the needs of the poorest in the developing countries. A U.N. report on progress toward the goals, published in 2000 and titled *A Better World for All*, shows that even though some progress has been made, the goals will not be reached unless efforts are stepped up in addressing all seven of the goals.

It is not accurate either to credit the World Bank for all the progress made or to blame it for all areas in which progress has been lacking. However, critics point to many examples wherein the bank's projects have actually exacerbated the cycle of poverty and environmental decline. For example, the World Bank loaned India nearly a billion dollars to create a huge electric-generating facility consisting of

five coal-burning power plants at Singraali and to develop open-pit coal mines to support the plants. However, the increased power does little for the poor, who cannot afford electrical hookups. Worse, the project displaced over 200,000 rural poor people who had farmed the fertile soil of the region for generations, and it moved them to a much less fertile area without allowing them any say in the matter and giving them little, if any, compensation. In addition, the project has caused extensive air and water pollution. Hydroelectric dams in a number of countries have similarly displaced people and exacerbated their poverty.

Nowhere are the failures and environmental destructiveness of large-scale projects more evident than in agriculture.

TABLE 7–2   Decline in Total Fertility Rate

	1981	1985	1990	1995	2000
Africa	6.4	6.3	6.2	5.8	5.3
Latin America and Caribbean	4.4	4.2	3.5	3.1	2.8
Asia (excluding China)	5.5	4.6	4.1	3.5	3.3
China	2.3	2.1	2.3	1.9	1.8
Developed countries	2.0	2.0	2.0	1.6	1.5

Source: Data from *2000 World Population Data Sheet* (Washington, DC: Population Reference Bureau, 2000).

TABLE 7–3   International Development Goals

1. Reduce the proportion of people living in extreme poverty by half between 1990 and 2015.
2. Enroll all children in primary school by 2015.
3. Make progress toward gender equality and empowering women by eliminating gender disparities in primary and secondary education by 2005.
4. Reduce infant and child mortality rates by two-thirds between 1990 and 2015.
5. Reduce maternal mortality ratios by three-quarters between 1990 and 2015.
6. Provide access for all who need reproductive health services by 2015.
7. Implement national strategies for sustainable development by 2005 so as to reverse the loss of environmental resources by 2015.

Sources: United Nations, World Bank, and Organization for Economic Cooperation and Development.

▲ FIGURE 7–8 *Cattle Ranch in Eastern Brazil.* Large World Bank loans went to clear rain forest and convert the land into rangeland, as seen here.

The World Bank funneled $1.5 billion into Latin America from 1963 to 1985 for clearing millions of acres of tropical forests. Most of the cleared land was given over to large cattle operations for producing beef for export (Fig. 7-8). However, no type of agriculture requires less labor per acre than ranching. Spreads of more than a million acres are run by millionaire cattle barons checking their herds by aircraft. Meanwhile, the poor were pushed into more marginal lands or into cities, as described in Chapter 6. Ironically, because the soil of cleared tropical forests is so poor, some of the ranches have already been abandoned, and most of the remaining ones are only marginally profitable. In other countries, projects have emphasized growing cash crops for export, fostering huge mechanized plantations while leaving the poor likewise marginalized.

The Debt Crisis

Another consequence of promoting development through World Bank loans is similar to enticing people to buy on credit. Borrowers become overwhelmed by interest payments. Theoretically, development projects were intended to generate additional revenues that would be sufficient for the recipient to pay back the loan with interest. However, a number of things have gone wrong with this theory, such as corruption, mismanagement, and, perhaps, honest miscalculations, not the least of which are the responsibility of the recipient countries. In their eagerness to obtain a loan for a billion-dollar project, government officials often overestimate the virtues of, and expected revenues from, the project.

In any case, far from paying off loans, developing countries as a group have become increasingly indebted. Their total debt reached $2.5 trillion in 1998 (Fig. 7–9a). Of course, interest obligations climb accordingly, and any failure to pay interest gets added to debt, increasing the interest owed—the typical credit-debt trap. The developing countries are now expected to pay $300 billion a year (an average of 17% of their earnings from exports) in interest (Fig. 7–9b), although many countries have fallen far behind in their debt service payments.

The debt situation continues to be an economic, social, and ecological disaster for many developing countries. In order to keep up even partial interest payments, poor countries have done one or more of the following:

1. Focused agriculture on large-scale growing of cash crops for export. This has occurred at the expense of peasant farmers growing food. Thus, hunger and malnutrition have increased, and so has poverty, as peasants have been pushed from the land.

2. Adopted austerity measures. Government expenditures have been drastically reduced so that income can go to pay interest. But what is cut? Usually, it is funds for schools, health clinics, police protection in poor areas, building and maintenance of roads in rural areas, and other goods and services that benefit not only the poor, but the country as a whole. For example, in 1996, Zambia (a small East African country) spent $312 million on servicing its debt, which is $2\frac{1}{2}$ times the amount the country spent on health and education.

3. Invited the rapid exploitation of natural resources (e.g., logging of forests and extraction of minerals) for quick cash. With the emphasis on quick cash, few, if any, environmental restrictions are imposed. Thus, the debt

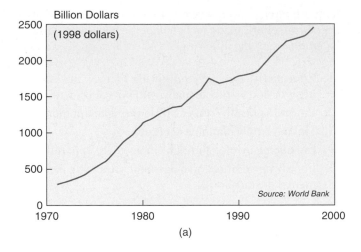

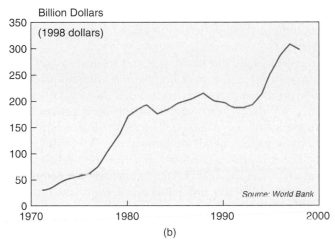

▲ **FIGURE 7–9** *External debt of developing countries.*
(a) The total debt has risen to $2½ trillion, a fivefold increase in a 25-year period. (b) Servicing the developing world's debt requires expenditures of $300 billion a year, a heavy burden to countries already straining to meet their needs for social services. (*Source*: Vital Signs 2000 [Washington DC, Worldwatch Institute, 2000].)

crisis has meant disaster for the environment. And ironically, this forced exploitation of material resources has placed such oversupplies of commodities on the market that prices have been severely depressed, so that actual earnings are low.

In essence, these measures are all examples of liquidating natural capital assets to raise cash for short-term needs. Clearly, they do not represent sustainability. Also, it is evident that the brunt of the measures falls on the poor. Thus, many observers point out that it is the poor, who gained nothing from the development, who are now being required to pay for it. As is typical with the credit trap, many countries have paid back in interest many times what they originally borrowed, yet the debt remains. Is there any point—humanitarian, ecological, or economic—to keeping such debts in place, especially when it is clear that they can never be paid off?

In sum, the concept of fostering the development of poor countries through massive loans for large-scale projects—whatever the advantages in enhancing gross national products—has not broken the cycle of excess population, poverty, and environmental degradation. For decades, policymakers at the World Bank saw development in terms of the hallmarks of the industrialized world: *energy*—in particular, electrification through centralized plants; *transportation*—especially building roads and cars; and the *mechanization of agriculture*. Thus, about three-fourths of the money loaned by the World Bank has been for projects in these three areas. Moreover, the policymakers at the Bank were mainly economists and engineers, and the traditional training in these two professions often involves viewing Earth as yielding endless resources and seeing humans as able to manipulate anything to their benefit. Thus, such professionals, lacking any ecological training, were notoriously slow to appreciate the environmental ramifications of the projects they proposed.

World Bank Reform

It is fair to say that the World Bank has undergone major reform in recent years in regard to how it does business with the developing world. The Bank now has a well-organized and influential Environment Department that provides "intellectual and practical leadership and support in fulfilling the World Bank's … environmental and social agenda," as the department's mission statement reads. The Environment Department has established policies and procedures aimed at ensuring that World Bank–financed projects are not environmentally damaging. The department is also engaged in helping developing countries to strengthen their own environmental institutions and policies. Most recently, the Environment Department has turned its emphasis toward "incorporating environmental and social values into the everyday operation of the major sectors in which the Bank invests," according to Director Robert Watson.

The difficulty of this task can be seen in a project in central Africa, where the Bank recently approved a loan of $193 million to promote the development of a crude oil industry in Chad and a pipeline through Cameroon for export. The profits—expected to be $2 billion in oil revenues in 25 years—could go toward financing health, education, and other important human services. Critics fear, however, that the money will be misused and that the pipeline will open up forests to illegal logging, as well as holding the potential for oil spills. The bank has agreed that the project has a certain risk, but has instituted safeguards such as depositing Chad's share of revenues into overseas accounts under the control of a multinational group that will dole out the money for approved purposes. The bank has also performed major environmental studies to ensure that the pipeline will have a minimal impact on the environment.

Debt Relief. The World Bank is addressing the problems of poverty more directly through two new initiatives: the Consultative Group to Assist the Poorest (CGAP) and the

Heavily Indebted Poor Country (HIPC) initiative. The first of these is designed to increase access to financial services for very poor households through what is called "microfinancing" (to be discussed shortly). The HIPC initiative addresses the debt problem of 41 of the poorest developing countries—amounting to $207 billion in 1998—by providing direct debt relief to a level deemed sustainable. To qualify, the countries have to demonstrate a track record of successfully carrying out economic and social reforms that lead to greater stability. By 2000, 16 countries had qualified for debt relief totaling $30 billion. In addition to these initiatives, the bank has produced a major document laying out its "Poverty Reduction Strategy," designed to focus attention on all of the bank's efforts to address poverty in the developing world.

Some of the debt of the HIPCs is bilateral, consisting of low-interest loans or export credit debts (to promote exports of the lending country). Recently, the G7 group of industrialized nations (the United States, Japan, Germany, France, Britain, Canada, and Italy) promised to cancel such bilateral loans to the HIPCs. Thus, Germany will write off about $5 billion, Britain $3.3 billion, France $7 billion, and the United States about $6 billion. The total debt relief for the HIPCs, from the G7 countries and the World Bank, amounts to some $90 billion. In most cases, it is tied to the requirement that the debtor countries apply the benefits they will receive to relieving poverty in their lands. Much of this renewed emphasis on debt relief can be traced to *Jubilee 2000*, a worldwide coalition of people and organizations—in particular, religious groups, since Jubilee is a biblical principle of debt cancellation every 50 years—concerned about poverty. Through demonstrations, petitions, and lobbying, Jubilee 2000 has focused world attention on the problem of debt relief and can take much of the credit for the recent G7 debt cancellations.

Undoubtedly, the work of the World Bank and other international agencies is sorely needed to provide assistance for economic development in the developing countries. Such work represents a vital element in the response of the industrialized world to the needs of the developing world. Meeting the objectives of Agenda 21 for sustainable development (see the "Earth Watch" essay, Chapter 1, p. 19) will require continued large-scale funding from these sources. However, the developing countries themselves must also address the root causes of poverty, and this usually means public policy changes at more local levels. To illustrate, we return to our assessment of the relative roles played by development and family planning in an attempt to resolve this decades-long debate.

7.3 A New Direction for Development: Social Modernization

As the case of Kerala demonstrates, demographic experts are recognizing that the shift from high to low fertility rates in the poorer developing countries does not require the eco-

nomic trappings of a developed country. Instead, what is needed are efforts made on behalf of the poor, with particular emphasis on the following:

1. Education—especially improving literacy and educating girls and women equally with boys and men.
2. Improving health—especially lowering infant mortality.
3. Making family planning accessible.
4. Enhancing income through employment opportunities.
5. Improving resource management (reversing environmental degradation).

In all of these areas, the focus should be on women, because they not only bear the children, but are also the primary providers of nutrition, child care, hygiene, and early education. In short, it is women who are most relevant in determining the numbers and welfare of subsequent generations.

Fortunately, the world is by no means starting from scratch in any of the preceding areas. Numerous programs, both private and government funded, have been in existence for many years, and a wealth of experience and knowledge has been gained. The result of success in these efforts is called **social modernization**. We shall look at each area in somewhat more detail; it should be evident that efforts aimed at social modernization are completely consistent with the international development goals set forth in Table 7–3.

Education

The education we are speaking of in this context is not college or graduate school, nor even advanced high school. It is basic literacy—learning to read, write, and do simple calculations (Fig. 7–10). Illiteracy rates among poor women in developing countries are commonly between 50 and 70%, in part because the education of women is not considered important and in part because the population explosion has overwhelmed school systems and transportation systems. Providing basic literacy will empower people to glean information from pamphlets on everything from treating diarrhea to conditioning soils with compost to baking bread. Consider how much more efficiently information can be distributed with written materials as opposed to one-on-one oral instruction, which is required if people cannot read. An educated populace is thus an important component of the wealth of a nation.

Investing in the education of children represents a key element of the public policy options of a developing country, one that returns great dividends. For example, Pakistan and South Korea both had similar incomes and population growth rates (2.6%) in 1960, but very different school enrollments—94% in Korea vs. 30% in Pakistan. Within 25 years, Korea's economic growth was three times that of Pakistan's, and now the rate of population growth

▲ **FIGURE 7–10** *Education in developing countries.* Providing education in developing countries can be quite cost effective, since any open space can suffice as a classroom and few materials are required. This is a class in Bombay, India.

in Korea has declined to 0.6% per year, while Pakistan's is still 2.4% per year. Kerala is another case in point. The lowest net enrollment of children in primary schools (about 40%) is in sub-Saharan Africa, where 45 million children are out of school. This region has made no progress in enrollment ratios since 1980, a clear consequence of rapid population growth.

Improving Health

Like education, the health care required most by poor communities in the developing world is not high-tech bypass surgery or chemotherapy; rather, it is the basics of good nutrition and hygiene—steps such as boiling water to avoid the spread of disease and properly treating infections and common ailments such as diarrhea. (In developing countries, diarrhea is a major killer of young children, but is easily treated by giving suitable liquids, a technique called oral rehydration therapy.) The discrepancy in infant mortality rates (infant deaths per 1,000 live births) between the developed countries (8/1,000) and developing countries (63/1,000) speaks for itself.

Health care in the developing world must emphasize pre- and postnatal care of the mother, as well as that of the children. Many governmental, charitable, and religious organizations are involved in providing basic health care, and when this is extended to rural countrysides in the form of

clinics, it is one of the most effective ways of delivering family-planning information and contraceptives to women. Maternal mortality is over 1,000 per 100,000 live births in a number of developing countries where maternal health care is almost nonexistent.

AIDS. One of the greatest challenges to health care in the developing countries is the human immunodeficiency virus (HIV), which leads to the sexually transmitted disease called acquired immune deficiency syndrome (AIDS). Unfortunately, the global epidemic of AIDS is most severe in many of the poorest developing countries, which are least able to cope with the consequences (Fig. 7–11). More than 90% of all HIV-infected people (35 million) live in the developing countries, and most of these people are not aware that they are infected—thus guaranteeing that the epidemic will continue. In sub-Saharan Africa the virus is spread by heterosexual contact, resulting in high incidences of female infection, often transmitted to their children. The incidence of infected adults is over 20% in some countries (Zimbabwe, Botswana, and South Africa).

The impact of this epidemic is horrendous for the developing world. Mortality rates are climbing in many countries. Life expectancy in Botswana has declined from 61 years in the late 1980s to 44 years at present. The long-term impact of AIDS can be seen in the projected

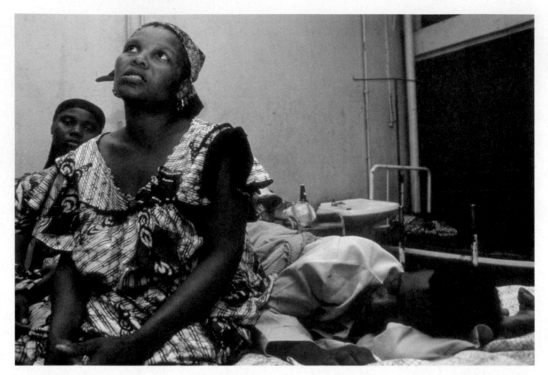

▲ **FIGURE 7–11** *AIDS.* Four members of a Central African family are victims of AIDS. The disease is often acquired from other family members who are unaware that they are infected.

population age structure of Botswana for 2020 (Fig. 7–12). In many countries, AIDS is tearing apart the very structure of the society. At least 860,000 elementary school students in sub-Saharan Africa have lost teachers to the epidemic; there may be as many as 40 million AIDS orphans in the developing world by 2010. Already inadequate health care systems are being swamped by the victims. Because there is no cure and the disease is invari-

▶ **FIGURE 7–12** *Effect of AIDS on the future population structure of Botswana.* The AIDS epidemic will sharply reduce the total population, especially in the number of children born and surviving. Similar effects may be expected in many of the sub-Saharan African countries. (Source: Data from "US Census Bureau World Population Profile 2000" as reprinted with "AIDS in a New Millennium" by Bernard Schwartlander, et al, SCIENCE, 289 (July 7, 2000). Copyright © 2000 by American Assn. for the Advancement of Science. Reprinted by permission.)

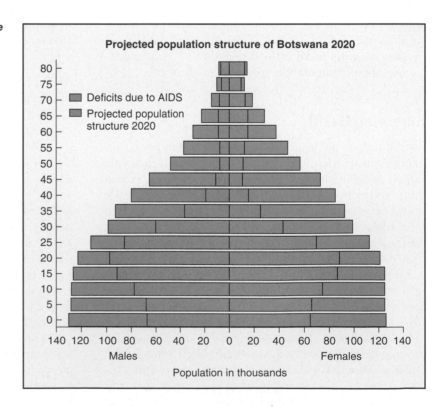

ably fatal, the only hope is to convince people to change their sexual behavior—to practice "safe sex," limit their partners, avoid prostitution, and delay sexual activity until marriage. These things can be done, as witness Uganda, where significant declines in HIV infection have been registered in recent years, especially among pregnant women. Again, education and literacy are important components of combating this social problem. The Ugandan government has mounted a national AIDS control program that includes a major education campaign.

Family Planning

For those who can pay, information on contraceptives, related materials, and treatments are readily available from private doctors and health-care institutions. The poor, however, must depend on family-planning agencies, which are supported by a combination of private donations, government funding, and small amounts the clients may be able to afford. The stated policy of family-planning agencies (or private services) is, as the name implies, to enable people to plan their own family size—that is, to have children only if and when they want them. In addition to helping people avoid unwanted pregnancies, planning often involves determining and overcoming fertility problems for those couples who are having reproductive difficulties. More specifically, family-planning services include the following:

- Counseling and education for singles, couples, and groups regarding the reproductive process, the hazards of sexually transmitted diseases (AIDS, in particular), and the benefits and risks of various contraceptive techniques.
- Counseling and education on achieving the best possible pre- and postnatal health for mother and child. The emphasis is on good nutrition, sanitation, and hygiene.
- Counseling and education to avoid high-risk pregnancies. Pregnancies that occur when a woman is too young or too old and pregnancies that follow too closely on a previous pregnancy are considered high risk; they seriously jeopardize the health and even the life of the mother. Any existing children are also at risk if the mother suffers injury or death.
- Provision of contraceptive materials or treatments after people have been properly instructed about all alternatives.

The vigorous promotion and provision of contraceptives has proved all by itself to have a decided effect in lowering fertility rates, as seen in Fig. 7–5. Those countries which have implemented effective family-planning programs have experienced the most rapid decline in fertility. For example, Thailand initiated a vigorous family-planning program as part of a national population policy in 1971. Population growth subsequently declined from 3.1% per year to the present 1.0% per year, and the Thai economy has shown one of the most rapid rates of increase over the years. Encouraging and implementing family planning is the first and most important step a country can take to improve its chances to develop economically.

Unmet Need.   A society's fertility is determined not only by the choices made by couples to have children, but also by births that are unplanned. Women who are not currently using contraception, but who want to postpone or prevent childbearing, are said to have an unmet need. Thus, both the spacing of children and the limiting of family size are involved. One major goal of family-planning policies and agencies is to ensure that all those who want or need reproductive health services actually have access to them. This goal is far from achieved in many parts of the developing world. The unmet need for family planning ranges from 7% to almost 50% of women in the different developing countries and involves a total of 120 million women. Where women's unmet need is minimal, fertility rates are invariably lower. Countries desiring to bring down their fertility are aware that meeting the unmet need for family-planning services is perhaps their most crucial objective. Indeed, much international aid is targeted at helping these countries provide reproductive health services to their families.

Abortion.   Unfortunately, to many people, family planning conjures up images of the abortion clinic. Abortions, by definition, are terminations of unwanted pregnancies. Nearly everyone agrees that an abortion is the least desirable way to avoid having an unwanted child—especially in view of the other alternatives that are available. As we will see, the document resulting from the most recent (Cairo) population conference explicitly states that abortions should *never* be used as a means of family planning. Therefore, it is particularly important to observe that the primary functions of family planning are education and the provision of services directed at avoiding unwanted or high-risk pregnancies. If family-planning services were universally available and people availed themselves of them, there would be far fewer unwanted pregnancies and, hence, fewer abortions. As it is, 46 million women have abortions each year, almost half of which are illegal. The figure represents one in every five pregnancies, an indication of the continuing high occurrence of unplanned pregnancies. This disturbing recourse to abortions can be seen as a consequence of the lack of family-planning education and services. All studies show that cutbacks in family-planning services result in *more* unwanted pregnancies and *more* demand for abortions, not *less*. Indeed, following a cutoff of family-planning monies by the Reagan administration in 1984, abortions increased an estimated 70,000 a year.

Furthermore, if abortions are not legal in a particular country, they tend to be provided illegally, or women will attempt to induce them on their own. Both of these practices have tremendous adverse health consequences. It is estimated that 500,000 women die annually from pregnancy-related problems. Many of these deaths result from illegal and self-induced abortions.

Planned Parenthood, which operates clinics throughout the world, is probably the best known family-planning agency. Another significant player is the U.N. Population Fund (UNFPA), which provides financial and technical assistance to developing countries at their request. The emphasis of UNFPA is on combining family-planning services with maternal and child health care and expanding the delivery of such services in rural and marginal urban areas. Support for this U.N. agency and other family-planning agencies has become a political football in the United States, where the Clinton administration—through the Agency for International Development—tried to continue funding for these agencies and many Republican members of the House opposed it. The argument, used since the Reagan years, is that these agencies promote abortions or, in the case of China, condone government efforts that have forced couples to have no more than one child. (See the "Ethics" essay, p. 179.) One far-reaching consequence of this battle is that for years Congress continued to reject bills authorizing payment of our commitment to the United Nations, placing us in arrears for over $1 billion. (The United Nations puts the figure at $1.52 billion.) This means that UNICEF and all the other U.N. agencies are receiving less than they need to carry out their mission, with tragic consequences for the developing world. As of late 2000, the U.S. and the United Nations reached an agreement whereby the U.S. would pay $926 million of its arrears, and the United Nations would accept that amount as full payment and also institute a number of reforms called for by the U.S. Congress.

Enhancing Income

The bottom line of any economic system is the exchange of goods and services. At its simplest level, this entails a barter economy in which people agree on direct exchanges of certain goods or services. Barter economies are still widespread in the developing world.

The introduction of a cash economy facilitates the exchange of a wider variety of goods and services, and everyone may prosper, as they have a wider market for what they can provide and a wider choice of what they can get in return. In a poor community, the ironic twist is that everyone may have the potential to provide certain goods or services and may want other things in return, but there is no money to get the system off and running. In a going economy, people who wish to start a new business venture generally begin by obtaining a bank loan to "set up shop." However, the poor are considered high credit risks, they may want a smaller loan than what a commercial bank wants to deal with, and many are women who are denied credit because of gender discrimination alone. For these three reasons, poor communities are handicapped in getting start-up capital. The situation is changing, however.

In 1976, Muhammad Yunus, an economics professor in Bangladesh, conceived and created a new kind of bank (now known as the Grameen Bank) that would engage in **microlending** to the poor. As the name implies, microloans are small—they average just $67—and they are short term, usually just four to six months. Nevertheless, they provide such basic things as seed and fertilizer for a peasant farmer to start growing tomatoes, some pans for a baker to start baking bread, a supply of yarn for a weaver, some tools for an auto mechanic, and so on.

Yunus secured his loans by having the recipients form **credit associations**, groups of several people who agreed to be responsible for each other's loans. With this arrangement, the Grameen Bank experienced an exceptional rate of payback: Less than 3% of the people defaulted on their loans. Loans from the Grameen Bank have had outstanding results when applied to small-scale agriculture. In a rural area of Bangladesh, small loans, along with horticultural advice, are now enabling peasant farmers to raise tomatoes and other vegetables for sale to the cities. These people have doubled their incomes in three years.

Microlending has been found to have the greatest social benefits when focused on women, because, as Yunus observed, "When women borrow, the beneficiaries are the children and the household. In the case of a man, too often the beneficiaries are himself and his friends." Note that the credit associations also create another level of cooperation and mutual support within the community, particularly when the loans are directed toward women.

The unqualified success of microlending in stimulating economic activity and enhancing the incomes of people within poor communities has been so remarkable that the concept has been adopted with various modifications by a considerable number of private organizations dedicated to alleviating hunger and poverty—Oxfam and Freedom from Hunger among them. Freedom from Hunger, which has projects in nine countries around the world, combines its lending with "problem-solving education" in a program called Credit with Education. (See the "Earth Watch" essay, p. 182.)

Recently, the U.S. Agency for International Development (AID) has entered the picture by providing grants to other organizations that wish to do microlending. Even the World Bank is expressing the need to be more sensitive to local communities in its lending practices and is making $200 million available for microlending. A major development in this arena has been the work of the Mi-

ETHICS

ADDITIONAL INCENTIVES FOR REDUCING FERTILITY

What are our options as we foresee the limits of Earth's carrying capacity for the human species? What are the ethical and moral implications of each option?

1. We can argue, as some optimists do, that the problems do not really exist—that technology will always have a solution to make life better and better for more and more people. Does global environmental evidence support this position, or is adopting such a position tantamount to blinding oneself to reality? Is it simply delaying making tough choices? Is it creating an even greater crisis for the future, as more people will have to support themselves with ever more depleted or degraded resources?

2. Perhaps a disease, a famine, or a natural disaster—or even war—will come along and take care of the situation for us in "nature's way." Again, this argument seems to lack an appreciation for the magnitude of the numbers involved. For example, the current annual world death toll from AIDS, tragic as it is, is about 2.7 million—less than 4% of what would actually stabilize population. Therefore, looking for a "death solution" is looking for untold human suffering. It is doubtful that anyone would escape the ramifications of that suffering. The entire human endeavor has been to try to avoid such calamities. Are we going to change that approach?

3. Some consider forced sterilization or abortion as a solution to the population problem. This option immediately implies that some person or group is deciding who is going to reproduce and who is not. The backlash from that kind of endeavor in the few cases where it has been tried has been so severe that the agencies or governments which have tried it have been summarily thrown out of power.

4. The most acceptable option—the one that has been the theme of this chapter—is to create an economic and social climate in which people, of their own volition, will de-

sire to have fewer children. Along with this climate, society will provide the means (family planning) to enable people to exercise that choice. This option, of course, has been realized "unconsciously" in developed countries; still, we are confronted by dilemmas in attempting to apply it to developing countries. For example, how far can one go in manipulating the social or economic environment before choice becomes coercion?

With its current population of 1.3 billion (a fifth of the world's people), China provides the most comprehensive example of a country that offers extensive economic incentives and disincentives geared toward reducing population growth. Some years ago, China's leaders recognized that unless population growth was stemmed, the country would be unable to live within the limits of its resources. Because of inevitable population momentum, the leaders felt that the country could not even afford a total fertility rate of 2.0; they set a goal of a one-child family, and to achieve that goal, they instituted an elaborate array of incentives and deterrents. The prime *incentives* were as follows:

- Paid leave to women who have fertility-related operations—namely, sterilization or abortion procedures.
- A monthly subsidy to one-child families.
- Job priority for only children.
- Additional food rations for only children.
- Housing preferences for single-child families.
- Preferential medical care to parents whose only child is a girl. (There is a strong preference for sons in China, and parents generally wish to have children until at least one son is born.)

Penalties for an excessive number of children in China included the following:

- Repayment to the government of bonuses received for the first child if a second is born.
- Payment of a tax for having a second child.
- Payment of higher prices for food for a second child.
- Maternity leave and paid medical expenses only for the first child.

Besides improving economic opportunities, these incentives and deterrents have helped China achieve a precipitous drop in its fertility rate, from about 4.5 in the mid-1970s to an official government rate of 1.8 in 2000 (although recent work by Chinese demographers have pegged the actual fertility rate at 2.1). In practice, the one-child policy has been pursued only in urban areas, some 29% of the population; people in rural areas and minorities (who are exempted from the policy) have a higher fertility rate. Despite these efforts, the population of China is still growing because of momentum. (A large percentage of the population is still at or below reproductive age.) Recently China has begun to phase out its one-child policy, concerned about a future in which there will not be enough adult children to help care for aging parents. One disturbing consequence of China's policy is the skewed ratio of males to females: Males are preferred in Chinese families, and if only one child is permitted, there is a tendency on the part of some parents to make sure that their one child is a male, through such extreme measure as abortion and even infanticide of female children.

Is China's population policy morally just or unjust? Is it possible that some other countries will have to resort to a China-like policy in the near future as they face severe limits on resources? What about India? At what point are such policies necessary? What do you think?

crocredit Summit Campaign, an outgrowth of the 1995 U.N. Fourth Conference on Women in Beijing. An update from this campaign indicates that by the end of 1999, some 1,065 institutions had granted loans to 13.8 million "poorest clients," of which 80% were women. The ultimate goal of the campaign is to reach 100 million families with credit for self-employment and other business services by 2005!

Improving Resource Management

The world's poor, almost by definition, are dependent on local resources, particularly water, soil for growing food, and forests for firewood. We have amply described how the pressures of excessive population are degrading these resources. However, a considerable part of the problem lies in poor utilization of resources—failing to replant trees and prevent erosion of the soil, for example. Conversely, a major factor in enhancing income is providing the technical skills and information necessary to manage those basic resources more effectively (Fig. 7–13). This in itself can greatly slow or even reverse the tide of environmental degradation. Allowing the environment to become impoverished can only lead to further impoverishment of the people.

Putting It All Together

Each of the five components just described both depends on and supports the other components. For example, better health and nutrition support better economic productivity, better economic productivity supports obtaining a better education, and a better education leads to a delay in marriage and the desire to have fewer children. The availability of family-planning services is essential to realizing the desire of parents to have fewer children. In short, all the components work together in harmony to alleviate the conditions of poverty, reverse environmental degradation, and stabilize population (Fig. 7–14). Conversely, the lack of any component—especially family-planning services—will undercut the ability to achieve all the other components.

It is not too hard to imagine putting these components together without undue expense. Most important, doing so involves the local people's uplifting themselves. For example, Zimbabwe is training 5,000 women to be preventive health-care workers in their own communities. Already, a number of developing countries have itinerant "nurses" who make the rounds of rural villages, treating illnesses and injuries, giving advice regarding better health maintenance, and dispensing contraceptive supplies. Another one or two persons traveling with the nurses might conduct classes to improve literacy, provide technical skills, and dispense microloans.

The importance of putting the five components together into a single "package" may seem self-evident. However, probably the biggest reason that U.N., governmental, and charitable programs have not achieved greater success is because they have often focused on only a single component in their approach. This brings us back to the significance of the Cairo Conference.

▲ **FIGURE 7–13** *Improving resource management*   A major step in enhancing incomes and protecting the environment is to encourage better resource management. Here, local people are learning the skills to raise tree seedlings that will later be transplanted in a reforestation project. This is part of the Greenbelt Movement in Kenya.

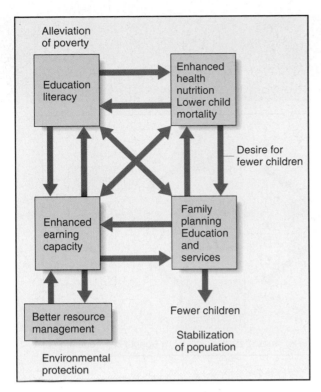

▲ **FIGURE 7–14** *Social development.* Five main aspects of enhancing the well-being of the poor are mutually supporting and dependent on one another as illustrated.

7.4 The Cairo Conference

In September 1994, some 15,000 leaders and representatives from 179 nations and nearly 1,000 nongovernmental organizations (NGOs) met in Cairo, Egypt, for the Third U.N. Conference on Population and Development. Before the delegates was a draft document of a "Program of Action" to address the world's persistent problems of poverty and population. In her keynote address to the conference, Norway's then Prime Minister Gro Harlem Brundtland, stated,

> Population growth is one of the most serious obstacles to world prosperity and sustainable development....
>
> We may soon be facing new famine, mass migration, destabilization, and even armed struggle as people compete for ever more scarce land and water resources....
>
> Today's newborns will be facing the ultimate collapse of vital resource bases.... Only when people have the right to take part in the shaping of society by participating in democratic political processes will changes be politically sustainable. Only then can we fulfill the hopes and aspirations of generations yet unborn.

Some objections were raised to the draft document by the Vatican and by Muslim countries regarding the wording of certain sections alluding to birth control methods and abortions. However, differences were resolved, and in the end, all 179 nations—large and small, rich and poor—signed the final document, committing themselves to achieve basic goals set forth therein by the year 2015.

The historic significance of this event is that for the first time in history, the political, religious, and scientific communities of the world reached a consensus on the population issue. There is clearly worldwide agreement that the intertwined issues of excessive population, poverty, and environmental degradation are jeopardizing the future of the planet. If they are not mitigated over the next 20 years, the world is more than likely to see a future of unprecedented biologic and human impoverishment.

The decades since World War II must be recognized as a time when a substantial portion of world leaders held views to the effect that technological progress could always support more people and prevent environmental degradation or that, with more general development, the population problem would take care of itself. They were also decades in which we were gaining knowledge and experience regarding the effectiveness of different kinds of programs. Therefore, it is perhaps understandable why the world has not reached a consensus before now.

Recognizing this diversity of opinion makes the consensus reached at the Cairo Conference in 1994 all the more significant. Effectively, *all nations agreed that population is an issue of crisis proportions that must be confronted forthrightly.* This sentiment was summed up in the words of Lewis Preston, then president of the World Bank: "Putting it bluntly, if we do not deal with rapid population growth, we will not reduce poverty—and development will not be sustainable." Hearing these words emanate from the World Bank, which traditionally had been a proponent of the "development will take care of it" view, reflects a fundamental change in attitudes on the part of numerous leaders and organizations.

The document that nations have signed and thus given their commitment to achieve is technically known as the *1994 International Conference on Population and Development Program of Action* (1994 ICPD Program of Action). It sets forth various principles and goals to be achieved by 2015.

Importantly, the goals of the 20-year Program of Action are not cast simply in terms of reducing fertility or population growth per se. Instead, the 1994 ICPD document asks that "interrelationships between population, resources, the environment and development should be fully recognized, properly managed and brought into a harmonious, dynamic balance." Thus, the goals are set in terms of creating an economic, social, and cultural environment in which *all people*, regardless of race, gender, or age, can share equitably in a state of well-being. In this context of providing opportunities such that people can improve their quality of life, it is assumed that they will choose to have fewer children as a result and that population growth will level off to the medium projection shown in Fig. 6–11. This means that no more than 7.5 billion

EARTH WATCH

AN INTEGRATED APPROACH TO ALLEVIATING THE CONDITIONS OF POVERTY

Freedom from Hunger is a nongovernmental organization that is pioneering an integrated approach toward improving the lives of the poor. Freedom from Hunger's **Credit with Education** program provides opportunities for extremely poor women to invest in their own small businesses and to save for emergency needs. These financial services are linked to education for better health, nutrition, and family planning—always respecting local beliefs and culture.

Women interested in receiving loans come together to form credit associations, composed of about 20–30 members, the great majority of whom are very poor. The members guarantee repayment of each other's loans, so they must agree that each woman is capable of making a sufficient profit from her proposed income-earning activity. After training the new members to manage their own association within specified rules, Freedom from Hunger makes a four- to six-month loan to the credit association. The members then break the large loan into small loans averaging $78 per individual. Women invest in activities in which they are already skilled and need no technical assistance, such as baking and selling food, raising chickens, operating a small shop, and making or buying and selling clothing.

Initially attracted by the offer of credit, the women of the association become engaged in weekly "learning sessions" to discuss and plan how to provide more and better food for their children. In addition, learning about family planning—with nat-

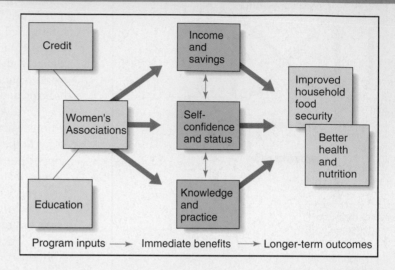

Program inputs ⟶ Immediate benefits ⟶ Longer-term outcomes

ural or artificial methods that prevent or postpone conception (rather than with postconception methods)—is critically important, because multiple, closely spaced births often lead to maternal and child malnutrition, poor health, and even death.

Freedom from Hunger has found that poor women who enter the Credit with Education program are good credit risks; their repayment rate currently stands at 97% over three or more years. The impact of Credit with Education, however, goes beyond augmented incomes and knowledge gained: The women increase their confidence and self-esteem, and they go on to provide more active leadership in their communities.

Freedom from Hunger's desire is to have its program adopted by others. The or-

ganization is actively enlisting the direct support of banks and other financial and nonfinancial institutions in each country and training them to deliver Credit with Education. Thus, the program is designed to become self-financing and therefore can be expanded to reach very large numbers of women and families in numerous countries. Started in 1989, Credit with Education programs now exist in nine nations: Bolivia, Burkina Faso, Ghana, Guinea, Honduras, Mali, Philippines, Togo, and Uganda. As of 1999, over 112,000 women have been helped by the program.

From Freedom from Hunger, Davis, CA 95617

persons will have to share the world by 2020 and 8.9 billion by the year 2050.

By signing the Program of Action, the governments of 179 nations have indicated their commitment to achieving, over the next 20 years, a host of objectives, including the following, in seven different categories:

General

1. Implementing strategies for income generation and employment, especially for the rural poor and those living within or on the edge of fragile ecosystems. Strategies should include maintaining and enhancing the productivity of natural resources.

Empowerment of Women

2. Eliminating gender discrimination in hiring, wages, benefits, training, and job security, with a view toward eradicating gender-biased disparities in income.

3. Changing customs and laws, where necessary, such that women can buy, hold, and sell property and land, obtain credit, and negotiate contracts equally with men.

4. Promoting the full involvement of women in community life, with speaking and voting rights equal to those of men.

Family

5. Promoting the full involvement of men in family life and creating policies to ensure men's responsibility to, and financial support for, their children and families.

6. Assuring that programs are created and administered in ways that encourage keeping family members together.

Reproductive and Basic Health

7. Making access to basic health care and health maintenance central strategies for reducing mortality and morbidity, especially among infants, children, and childbearing women.

8. Ensuring community participation in planning health policy.

9. Strengthening education and communication regarding health and nutrition.

10. Making reproductive health counseling and information accessible to all.

11. Placing special emphasis on controlling the spread of AIDS.

12. Integrating basic health and reproductive health services.

Education

13. Ensuring complete access to primary school education for both girls and boys.

14. Removing any gender-based barriers that prevent girls from going on to reach their full potential in education.

15. Sensitizing parents to the value of education for girls.

Migration

16. Addressing the root causes of migration into cities and emigration to other areas.

International Cooperation

17. Transferring technology from the developed countries to the developing countries such that the need for contraceptives and other basic items can be met by local production.

18. Setting aside 0.7% of the GNP of each member country of the developed world for the achievement of the Program of Action's objectives throughout the globe. The developed countries are specifically called upon to cooperate in achieving the last two objectives.

What is to prevent the 1994 ICPD Program of Action document from simply being forgotten? It is significant that the program does not spell out exactly how the objectives should be met, and it does not have any power of enforcement. Each nation is left to decide on its own how it will meet the objectives. National governments and multilateral agencies like the World Bank and UNFPA will implement actions. NGOs like the Population Institute and a number of women's groups are expected to play the role of watchdog, becoming involved at the grassroots level.

A review and appraisal of the ICPD took place during 1999, five years after the original Cairo conference. The process involved several preliminary forums and roundtables held in different parts of the world, culminating in a Special Session of the U.N. General Assembly. The review demonstrated that the Program of Action is indeed being implemented in many developing countries. Changes in laws and programs have occurred: National population policies in countries like India have switched their focus from strict family planning and targeted numbers to providing a wide range of reproductive health services, delivered to local levels; many African countries have outlawed female genital mutilation; the right to contraceptives has been established in many countries; and in Bangladesh, government policy has encouraged the fuller participation of girls in school through the secondary level. Further, population and development have been more fully integrated in many nations, the use of family-planning methods is on the rise, and mortality in most countries has continued to fall. However, there are some serious exceptions: Women and girls continue to face discrimination in many societies; the HIV/AIDS epidemic has led to a rise in mortality in numerous countries; infectious diseases continue to take an unacceptable toll in sickness and death; and millions of couples still lack access to reproductive health information and services. At the review, a new set of benchmark indicators was adopted, giving some quantitative measures for future progress. These were for the most part identical to the international development goals shown in Table 7–3.

Funding remains a problem. The developing countries have agreed to cover two-thirds of the costs of realizing the objectives of the program and have fulfilled 70% of their commitment to date. Unfortunately, the industrialized donor countries have not kept their promises to fund the rest—providing only about 35% of their one-third share of the costs. Donor contributions were $2 billion in 1995 and have been declining ever since. If the 0.7%-of-GNP recommendation made at the ICPD conference were followed, the United States, with by far the world's largest GNP (over $8 trillion in 1998), would be a major donor. According to Population Action, the United States contributed $638 million in 1996, only one third of what its contribution should be to meet the 2000 funding goals of the program. It is significant—and perhaps shameful for the nation—that two U.S. billionaires—Ted Turner and Bill Gates—have stepped into the gap with multimillion-dollar donations to support U.N. population activities.

Regardless of what happens in the United States, it is clear that the international community is taking the ICPD Program of Action seriously and is, on the whole, engaged in implementing its recommendations. This is good news for everyone—even us—as it means, in the end, greater world security and a much better chance of achieving sustainability in our interactions with the environment that supports us.

ENVIRONMENT ON THE WEB

WOMEN AS THE "KEY TO DEVELOPMENT"

In 1994, the Cairo Conference helped to focus the world's attention on the need to satisfy women's rights and needs (education, reproductive health care, economic and political equality) as part of attaining the world's goals of development and population stabilization. Indeed, women, as bearers of children, have a central role to play in demographic change. However, rather than focus on women solely as the targets of increasing birth control policies, the conference saw women as a central factor in development. Some authors have gone so far as to call women the "key to development."

In India, there is a long tradition of empowerment of women through community groups. Sometimes these groups are initiated by aid organizations such as Oxfam or the U.S. Agency for International Development (U.S. AID). In other cases, a single individual within the community can begin activities that eventually draw in many women. Often, a woman is attracted to a community group because of its social opportunities—especially the chance to get away from the home and the burden of routine work, and to interact with other women with similar interests. Once a participant, the woman may feel able to share painful experiences of infant death, family violence, or simple poverty with the others. Sometimes these women have few other opportunities for support and advice, and the community group becomes a highly valued part of their lives. Some travel dozens of kilometers to attend a meeting, often bringing with them their children and friends.

Oxfam, through its Community Aid Abroad (CAA) program, sponsors a number of women's groups in India. Its RUCHI (Rural and Urban Center for Human Interest) program began as an attempt to halt the rapid deforestation in the northern Indian state of Himachel Pradesh. Women had found that adequate fuel for cremation was rarely available, and excessive tree harvesting had also encouraged erosion, which resulted in landslide risks and poor productivity on hillside agricultural fields. Working through women, RUCHI and similar programs have made tremendous progress on the environmental problems they set out to solve, while raising women's awareness of issues and options for themselves and their communities.

Chaitanya, another CAA program in India, began as a simple community nonprofit savings-and-loan organization, encouraging women to save a few rupees a month (ten to twenty cents). As the program developed, participants came to realize that the more they can save—even if only a rupee or two—the more they will earn in interest. Women have begun to plan their savings strategically and to look for other ways of earning income, such as through cottage crafts.

Through women's advocacy groups, women are becoming more autonomous and confident. They are taking control of land and household improvements as well as making economic and political decisions. Also, as they gain control over their futures, more of these women will be equipped to make sound decisions about their reproductive health, family size, and timing. As their knowledge and confidence grow, so does their awareness of their own abilities. Whereas most used to spend all of their time around home and children, they are now traveling to distant meetings, making community decisions, and building the capacity for a brighter future.

Web Explorations

The Environment on the Web activity for this essay describes some of the ways in which governments have intervened in women's reproductive health. Go to the Environment on the Web activity (select Chapter 7 at **http://www.prenhall.com/wright**) and learn for yourself:

1. how education can affect a woman's childbearing and child-rearing behavior;
2. about the availability of birth control in U.S. schools;
3. about family planning programs; and
4. about the ethics of forced birth control.

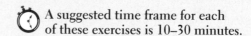 **A suggested time frame for each of these exercises is 10–30 minutes.**

REVIEW QUESTIONS

1. What have been the two basic schools of thought regarding population growth?
2. Discuss the six specific factors that influence the number of children a couple desires.
3. Describe how poverty, environmental degradation, and high fertility rates drive one another in a vicious cycle.
4. What has been the major agency and mechanism for promoting development in poor nations over the past 45 years? Discuss its past successes and failures.
5. What is meant by the debt crisis of the developing world? What is being done in the form of debt relief to help resolve this crisis?
6. How has the World Bank undergone major reform in recent years?
7. What are the five interdependent components that must be addressed to bring about social modernization?
8. What are the key aspects of family planning, and why is family planning of critical importance to all other aspects of development?

9. What is meant by "unmet need?"

10. What is microlending? How does it work?

11. How can each of the following be addressed in a cost-effective way: education; reduction of infant mortality; income enhancement; environmental degradation?

12. What was the significance of the 1994 Cairo Conference? What are the agreed-upon strategies for addressing the problems of poverty, excessive population, and environmental degradation?

THINKING ENVIRONMENTALLY

1. Is the world population below, at, or above the optimum? Defend your answer by pointing out things that may improve and things that may worsen by increasing population.

2. Suppose you are the head of an island nation with a poor, growing population, and the natural resources of the island are being degraded. What kinds of programs would you initiate, and what help would you ask for to try to provide a better, sustainable future for your nation's people?

3. List and discuss the benefits and harms of writing off debts owed by developing nations.

4. Describe what you think will be the long-term results (for the people, the society, and the environment) of limiting family-planning services in developing nations.

5. What priority do you give to the cyclic factors—population, poverty, and environmental decline? Give reasons pro and con as to your ranking. What proportion of national security expenditures would you allocate toward addressing the population–poverty issue?

MAKING A DIFFERENCE

PART TWO: Chapters 6 and 7

1. Think carefully about your own reproduction. What concerns will you and your mate weigh as you plan your family?

2. Become involved in the abortion debate, pro or con. Whether the United States supports international family planning, whether abortions remain legal in your state, and other issues will be determined by votes cast by you or your representatives.

3. Become involved in, and support, programs promoting effective sex education and responsible sexual behavior. Consider the advantages of abstinence and monogamy as ways to avoid needing an abortion or contracting a sexually transmitted disease.

4. Share your knowledge and energy by joining the Peace Corps or another organization engaged in providing appropriate technology to a developing country.

5. Encourage sustainability by buying products that originate from appropriate technology in the developing world.

WEB REFERENCES

On-line resources for this chapter are on the World Wide Web at: **http://www.prenhall.com/wright**. (Click on Chapter 7 within the Chapter Selector.)

Chapter 8
Soil and the Soil Ecosystem

Chapter 9
Water: Hydrologic Cycle and Human Use

Chapter 10
The Production and Distribution of Food

Chapter 11
Wild Species: Biodiversity and Protection

Chapter 12
Ecosystems as Resources

PART THREE

Renewable Resources

A mountainside on the island of Bali is terraced for rice cultivation; tallgrass prairies of Illinois and Iowa are plowed under to raise corn; tropical rain forests in Brazil are converted to pasture for cattle; desert in Israel is irrigated to raise vegetables. Forests are harvested for wood and paper pulp; coastal oceans are fished; grasslands are grazed by sheep. We get food and fiber from a host of natural ecosystems, some thoroughly managed and some not. All of these activities depend on water, nutrients, and sunlight—the basis for productivity in natural systems. All are renewable resources—that is, they are replenished as energy flows and water and nutrients cycle in ways that have long sustained life on Earth. These same systems provide us with priceless services, like soil formation, waste degradation, pest control, climate modification, flood control, and many others.

At the same time, we embrace the beauty of wild ecosystems and turn to them for enjoyment and rest. Can we have it both ways? Can we hope to preserve nature while we also make use of its goods and services, especially in light of continuing population growth and, therefore, continuing pressure on all resources? In this part of the text, it is our task to dig more deeply into the science of soil, water, food production, forest growth, and fisheries. We will examine all of these renewable resources and ways of managing them as stewards, while again keeping our eyes on sustainability.

◀ *Delaware River Valley, New Jersey and Pennsylvania.*

Soil and the Soil Ecosystem

Key Issues and Questions

1. Certain properties of soils are described by soil scientists for purposes of understanding how they were formed. What is a typical soil profile, and how can soil texture be differentiated?

2. Soils are classified in a complex taxonomy. What are four common groups of soils important for agriculture and forestry?

3. The soil environment must provide plants with water, nutrients, and air for the roots. What are the key attributes of the soil that bear on its being able (or not being able) to provide these things?

4. A dynamic interaction between mineral particles, detritus, and organisms in the soil is most important in developing the soil's key attributes. Describe this dynamic interaction and how the soil develops these attributes.

5. Soil degradation is devastating to the future productivity of a landscape. What are the major practices leading to erosion and desertification, and what are some ways to combat these harmful processes?

6. Certain irrigation practices may be nonsustainable. Why? Describe the problems that may develop from irrigation.

7. To bring about better stewardship of soil resources, soil conservation must be promoted at the level of public policy. Describe some encouraging changes that are taking place in this regard in the United States.

8. Individual landholders and herders hold the key to sustainable soil stewardship. What are some of the steps being taken to promote sustainable agriculture in the developing world?

Twenty-five years ago, Gabino López attended classes on sustainable agriculture in San Martín Jilotepeque, Guatemala, and then returned to his village to apply what he had learned to his own small farm. He planted vegetation barriers along the contours of his land to stop erosion and applied cow manure to enrich the soil. His maize harvest increased from 0.8 to 1.3 tons/hectare (1,050 lb/acre). Gabino's harvest jumped to 3.4 tons/hectare when he plowed under his crop residues, as opposed to when he burned them or rotated crops. Aware of his success, his friends asked him for advice, and before long, Gabino López was an international consultant in *ecological agriculture*, ranging from Mexico to Ecuador and even India. He and his fellow farmers employ a number of practices that follow the "five golden rules of the humid tropics": (1) keeping the soil covered, (2) using minimal or even no tillage, (3) using mulch to provide nutrients for crops, (4) maximizing biomass production, and (5) maximizing biodiversity. As a result, the maize harvest has averaged 4.3 tons/hectare

in a number of local villages. Gabino teaches a form of agriculture adapted to the needs of village farmers—a form based on a process of agricultural development that involves villagers experimenting and teaching others what they learn (Fig. 8–1). It is an ongoing process, and it is leading to sustainable agriculture for people in serious need of all the food they can raise.

In 1997, in Komsomolsky, southeastern Russia, Alexandra Bakhtayeva told a *Boston Globe* reporter that she has to dig herself out every day from the sand blowing in from surrounding dry land that was once prime grazing land. The windblown sand is the signal that the area has undergone desertification—conversion from stable, rolling grasslands to drifting sand. The desert area is growing, expanding hundreds of thousands of acres annually in this Russian southland (Fig. 8–2). What used to be soil is gone and is replaced by sand, and the local environmental officials lack the financial support to reverse the process. It is a human-made desert, the result of failed Communist agricultural policies

◀ *Reaping the harvest in Iowa.*

▶ FIGURE 8–1 *Farming in Guatemala.* Maize is a mainstay of nutrition for the people of Guatemala. It is possible to dramatically improve maize production with proper treatment of the soil.

and, earlier than that, the removal of the local Kalmyks by Stalin as he forced collectivization of the rural areas of the Soviet Union. (Before their removal by Stalin, the Kalmyks very carefully avoided plowing the land and, instead, rotated the land being grazed by their cattle and sheep. Shepherds even wore flat-bottomed shoes to avoid breaking up the top-soil, and their fat-tailed sheep were a breed with flat-bottomed hooves.) The combination of plowing the thin soil to attempt to raise crops and grazing millions of Caucasian sheep with sharp hooves gradually broke down the soil and gave it to the winds. Now the sand has already buried 25 towns, with

more on the way; nearly half of the territory of Kalmykia is severely desertified. Without soil, the region is quickly be-coming a wasteland where a few people scratch out a living protected from the blowing sand by tall fences.

Ninety percent of the world's food comes from land-based agricultural systems, and the percentage is growing as the ocean's fish and natural ecosystems are increasingly de-pleted. Protecting and nurturing agricultural soils, which are the cornerstone of food production, must be a central feature of sustainability. Yet, it is a feature that has been overlooked repeatedly in the past. The story of Easter Island in Chapter 1 is just one of many. In his book, *A Green His-*

▶ FIGURE 8–2 *Desertification in Kalmykia.* Poor agricultural policy and practice and brutal repression of indigenous people led to wide-scale conversion of the fragile landscape to sandy desert.

TABLE 8–1 Global Degradation of Land in Crops, Pasture, Forest, and Woodlands

Degradation Category	Amount of Land Affected (Millions of Hectares)	Lost Production (Percent)
Total land	8,735	
Not degraded	6,770	0
Degraded	1,965	
Lightly	650	5
Moderately	904	18
Heavily	411	50

tory of the World,[1] Clive Ponting documents how the fall of the ancient Greek, Roman, and Mayan empires was more the result of a decline in agricultural sustenance due to soil erosion than of outside forces.

As the world's population moves beyond 6 billion into the 21st century, croplands and grazing lands are being increasingly pressed to yield more crops and other products. Yet, according to a U.N. Environment Program report, poor agricultural practices of the past 50 years have led to the degradation of 22% of land used for crops, grazing, or forestry—almost 2 billion hectares, or 4.8 billion acres. (See Table 8–1.) Throughout the world, agricultural soils have been (and continue to be) degraded by erosion, the buildup of salts, and

[1] Ponting, Clive. 1991. A Green History of the World. Penguin Books. New York, NY.

other problems that can only undercut future productivity. In the United States, another kind of loss has also occurred: The loss of farmland to development averaged some 819 thousand acres per year from 1982 to 1992 and then began to accelerate to 1.23 million acres per year from 1992 to 1997.

Our objective in this chapter is to provide an understanding of the attributes of soil required to support good plant growth, how these attributes may deteriorate under various practices, and what is necessary to maintain a productive soil. We will then look at what is being done nationally and internationally to rescue this most vital resource and to establish practices that are sustainable. We begin by considering some characteristics of soil and then examining the relationships between plants and soil that sustain productivity.

8.1 Plants and Soil

Soil can make the difference between harvesting a luxuriant crop or abandoning the field to a few meager weeds. A rich soil that supports a luxuriant crop is so much more than the dirt you might get out of any hole in the ground. Indeed, agriculturists cringe when anyone refers to soil as dirt. You have already learned (Chapters 2 and 3) that various detritus feeders and decomposers feed upon detritus in an ecosystem. Nutrients from the detritus are thus released and reabsorbed by producers, fostering the recycling of nutrients. As we focus on a productive soil, we will see that the detritus feeders and decomposers constitute a biotic community of organisms that not only facilitates the transfer of nutrients but also creates a soil environment that is most favorable to the growth of roots. In short, a productive topsoil involves dynamic interactions among the organisms, detritus, and mineral particles of the soil (Fig. 8–3).

Soil Characteristics

Most soils are hundreds of years old and change very slowly, if at all, under most circumstances. Soil science is an integrative science that is at the heart of agricultural and

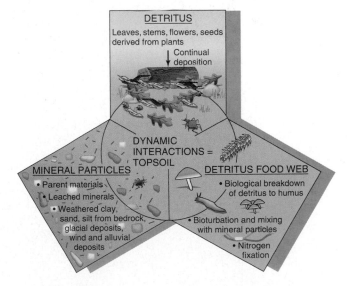

▲ **FIGURE 8–3** *Topsoil formation.* Soil production involves a dynamic interaction among mineral particles, detritus, and members of the detritus food web.

forestry practice. The results of many years of study of soils have provided a system of classification of soil profiles and soil structure and a taxonomy of soil types from all over the world—as well as a large body of scientific literature investigating relevant topics. We shall examine just enough of this information to enable us to open the door to soil science and, hopefully, stimulate you to follow up on your own in this fascinating field.

Soil Profiles. The processes of soil formation create a vertical gradient of layers that are often (but not always) quite distinct, as you may know if you have ever dug up a small tree from a forest. These horizontal layers are called **horizons**, and a vertical slice through the different horizons is called the **soil profile** (Fig. 8–4). The profile tells us a great deal about the factors that interact in the formation of a soil. The topmost layer, the **O horizon**, consists of dead organic matter (detritus) deposited by plants—leaves, stems, fruits, seeds, etc. Thus, the O horizon is high in organic content and is the primary energy source for the soil ecosystem. Toward the bottom of the horizon, the processes of decomposition are well advanced, and the original materials may be unrecognizable. At this point, the material is dark and is called **humus**.

The next layer is the **A horizon**; it is a mixture of mineral soil from below and humus from above and is familiarly termed **topsoil**. Fine roots from the overlying vegetation cover permeate this layer. The A horizon is usually dark because of the humus present and may be shallow or thick, depending on the overlying ecosystem. In many soils, the next layer is the **E horizon**; E stands for **eluviation**, which refers to the process of leaching of many minerals because of the downward movement of water. This layer is often more pale in color than the two layers above it.

The next layer downward is the **B horizon**, which is characterized by the deposition of minerals that have leached from the A and E horizons above and thus is often

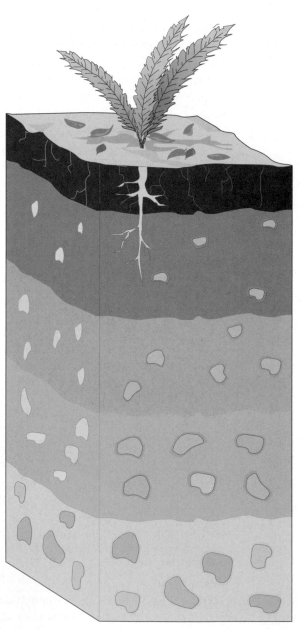

◀ **FIGURE 8–4** *Soil profile.* Major horizons from the surface to the parent material in an idealized soil profile.

O Horizon: Humus. (surface litter, decomposing plant matter)

A Horizon: Topsoil. (mixed humus and leached mineral soil)

E Horizon: Zone of leaching. (less humus, minerals resistant to leaching)

B Horizon: Subsoil. (accumulation of leached minerals like iron and aluminum oxides)

C Horizon: Weathered parent material. (partly broken-down minerals)

high in iron, aluminum, calcium, and other minerals. This layer is frequently referred to as the **subsoil**, is often high in clay content, and is reddish or yellow in color. Below the B horizon is the **C horizon**, which is the parent mineral material originally occupying the site, representing weathered rock, glacial deposits, or volcanic ash—usually revealing the geological process that created the landscape. This layer is affected little by the biological and chemical processes that go on in the overlying layers.

Because there are innumerable soils across the diverse landscapes of the continents, soil profiles will differ in the thickness and the contents of the layers. However, all soils exist in layers and can be characterized by their texture.

Soil Texture. As rock weathers, it breaks down into smaller and smaller fragments. Below the size of small stones, we classify these fragments, called soil separates, as sand, silt, and clay. Sand is made up of particles from 2.0 to 0.02 mm in size, silt particles range from 0.02 down to 0.002 mm, and clay is anything finer than 0.002 mm. You are familiar with sand as individual rock particles, and you are probably also familiar with the finer particles called silt, but it may surprise you to note that clay is actually made up of microscopic particles. Consider washing clay in water: The water immediately takes on a cloudy or muddy appearance because the clay particles are suspended in it. The moldable, "gooey" quality of clay appears when just enough water is added for the particles to slide about one another on a film of water, but still cling together. On drying further, the clay particles adhere in hard clods.

The sand, silt, and clay particles constitute the mineral portion of soil. **Soil texture** refers to the relative proportions of each in a given soil. If one predominates, we speak of a sandy, silty, or clayey soil. A proportion that is commonly found consists of roughly 40% sand, 40% silt, and 20% clay. A soil with such proportions is called a **loam**. You can determine the texture of a given soil by shaking a small amount of it with water in a large test tube to separate the particles and then allowing them to settle. Because particles settle according to their weight, sand particles settle first, silt second, and clay last. The proportion of each can then be seen. Soil scientists classify soil texture with the aid of a triangle that shows the relative proportions of sand, silt, and clay in a given soil (Fig. 8–5).

Three basic considerations determine how several important properties of the soil are influenced by its texture: infiltration, nutrient- and water-holding capacities, and aeration. 1. Larger particles have larger spaces separating them than smaller particles have. (Visualize packing softballs versus golf balls in containers.) 2. Smaller particles have more surface area relative to their volume than larger particles have. (Visualize cutting a block in half again and again. Each time you cut it, you create two new surfaces, but the total volume of the block remains the same.) 3. Nutrient ions and water molecules tend to cling to surfaces. (When you drain a nongreasy surface, it remains wet.)

Soil attributes, as they vary with particle size (sand, silt, clay), are shown in Table 8–2. Do you see how they logically correspond to the items in the list?

Soil texture also affects **workability**, the ease with which a soil can be cultivated. This fact has an important bearing on agriculture. Clayey soils are very difficult to work, because,

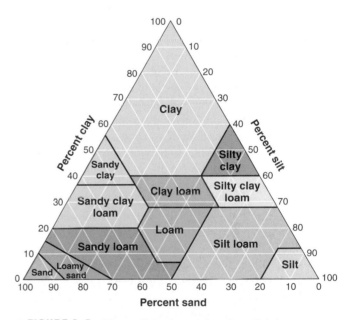

▲ **FIGURE 8–5** **The soil texture triangle.** Relative proportions of sand, clay, and silt are represented on each axis. Major soil classes are indicated on the triangle. For clay, read across horizontally; for silt, read diagonally downward; for sand, read diagonally upwards to the left. Any soil should total 100% if read properly.

TABLE 8–2	Relationship between Soil Texture and Soil Properties				
Soil Texture	Water Infiltration	Water-holding Capacity	Nutrient-holding Capacity	Aeration	Workability
Sand	Good	Poor	Poor	Good	Good
Silt	Medium	Medium	Medium	Medium	Medium
Clay	Poor	Good	Good	Poor	Poor
Loam	Medium	Medium	Medium	Medium	Medium

with even modest changes in moisture content, they go from being too sticky and muddy to being too hard and even bricklike. Sandy soils are very easy to work, because they become neither muddy when wet nor hard and bricklike when dry.

Soil Classes. Soils come in an almost infinite variety of vertical structures and textures. To give some order to this diversity, soil scientists have created a taxonomy of soils. It works much like the biological taxonomy with which you may be familiar. The most inclusive group in the taxonomy is the soil *order*; if you are classifying a soil, you find the order first and then work your way downward through the taxonomic categories (*suborders, groups, subgroups, families*) until you come to the soil *class* that best corresponds to the soil in question. There are literally hundreds of soil classes. For our purposes, we will briefly examine the characteristics of 4 of the major soil orders most important for agriculture, animal husbandry, and forestry. (A total of 11 major orders has been established.)

Mollisols: Mollisols are very fertile, dark soils found in temperate grassland biomes. They are the world's best agricultural soils and are encountered in the midwestern United States, across temperate Ukraine, Russia, and Mongolia, and in the pampas in Argentina. They have a deep A horizon and are rich in humus and minerals; precipitation is insufficient to leach the minerals downward.

Oxisols: These are soils of the tropical and subtropical rain forests. They have a layer of iron and aluminum oxides in the B horizon and have little O horizon, due to rapid decomposition of plant matter. Most of the minerals are in the living plant matter, so oxisols are of limited fertility for agriculture. If the forests are cut, a few years of crop growth can be obtained, but in time the intense rainfall leaches the minerals downward, forming a hardpan, which resists further cultivation.

Alfisols: Alfisols are widespread, moderately weathered forest soils. Although not deep, they have well-developed O, A, E, and B horizons. They are typical of the moist, temperate forest biome and are suitable for agriculture if supplemented with organic matter or mineral fertilizers to maintain soil fertility.

Aridisols: These are very widespread soils of drylands and deserts. The paucity of vegetation and precipitation leaves aridisols relatively unstructured vertically. They are thin and light colored and, in some regions, may support enough vegetation for rangeland animal husbandry. Irrigation used on these soils usually leads to salinization, as high evaporation rates draw salts to surface horizons, where they accumulate to toxic levels.

It is clear that different soils vary significantly in their properties. What are the characteristics of soils that enable them to support plant growth, especially agricultural crops?

Soil and Plants

For their best growth, plants need a root environment that supplies optimal amounts of mineral nutrients, water, and air (oxygen). The pH (relative acidity) and salinity (salt concentration) of the soil are also critically important. **Soil fertility**, the soil's ability to support plant growth, often refers specifically to the presence of proper amounts of nutrients. But the soil's ability to meet all the other needs of plants is also involved in soil fertility. Farmers speak of a given soil's ability to support plant growth as the *tilth* of the soil.

Mineral Nutrients and Nutrient-holding Capacity. Mineral nutrients—phosphate (PO_4^{3-}), potassium (K^+), calcium (Ca^{2+}), and other ions—are present in various rocks along with nonnutrient elements. Minerals initially become available to roots through the gradual chemical and physical breakdown of rock—the processes collectively referred to as **weathering**. But weathering is much too slow to support anything approaching normal plant growth. The nutrients that support plant growth in natural ecosystems are supplied mostly through the breakdown and release (recycling) of nutrients from detritus. (See Fig. 3–20.)

But as we have seen, nutrients may literally be washed from the soil as water moves through it, a process called **leaching**. Leaching not only lessens soil fertility; it also contributes to pollution when materials removed from the soil enter waterways. Consequently, the soil's capacity to bind and hold nutrient ions until they are absorbed by roots is just as important as the initial supply of those ions. This property is referred to as the soil's **nutrient-holding capacity** or its **ion-exchange capacity**.

In agricultural systems, there is an unavoidable removal of nutrients from the soil with each crop, because nutrients absorbed by plants are contained in the harvested material. Therefore, agricultural systems require inputs of nutrients to replace those removed with the harvest. Nutrients are replenished with applications of **fertilizer**, material that contains one or more of the necessary nutrients. Fertilizer may be organic or inorganic. **Organic fertilizer** includes plant or animal wastes or both; manure and compost (rotted organic material) are two examples. **Inorganic fertilizers** are chemical formulations of required nutrients without any organic matter included. Inorganic fertilizers are much more prone to leaching than are organic fertilizers.

Water and Water-holding Capacity. A plant's need for water is self-evident. What is less obvious is the constant stream of water being absorbed by the roots, passing up through the plant, and exiting as water vapor through microscopic pores in the leaves—a process called **transpiration** (Fig. 8–6). The pores, called *stomata* (singular, *stoma*), are essential to permit the entry of carbon dioxide and the exit of oxygen in photosynthesis; however, the plant's loss of water through the stomata is dramatic. A field of corn, for example, transpires an equivalent of a field-sized layer of water 17 inches (43 cm) deep in a single growing season. Inadequate water results in wilting, a condition that conserves water, but also shuts off photosynthesis by closing the stomata and preventing gas

ETHICS

EROSION BY EQUATION

By any name or measure, soil erosion is a highly undesirable process. It is a highly visible one, too: ugly gashes can be seen in fields and slopes that are not well vegetated (Fig. 8–14), streams and rivers run brown with sediment after a heavy rain, deltas form where rivers empty into larger bodies of water, and reservoirs fill up with sediment over time. This chapter makes the point that eroded soil represents lost agricultural productivity, and the process can certainly lead to desertification if it is unchecked. Erosion is also a natural geological process, wherein land surfaces undergo weathering and sediments form over geological time, leading to the laying down of thick layers of sediment that form sedimentary rock. Thus, some erosion is bound to occur no matter what farming methods are practiced.

The National Resources Inventory reports erosion rates on an annual basis. For 1997, this amounted to about 1.9 billion tons of eroded material. Much erosion is judged to be "excessive," originating from over 110 million acres of cropland deemed vulnerable. A serious question is, How is erosion measured, especially if measurements are taken for the entire United States? The bottom-line answer is, mostly via map-reading, aerial photography, and the judicious use of a "universal soil loss equation" and a "wind erosion equation." Necessarily, these are estimates rather than measurements. The actual rates are not measured and, indeed, are probably not measurable for the whole country.

What happens when someone actually measures erosion and compares it with estimated rates? UCLA geographer Stanley W. Trimble asked this question in a study of the Coon Creek Basin in Wisconsin, published in 1999.[1] He had the good fortune of having available a previous historical study of the basin by the Soil Conservation Service that left many benchmarks in the sediments of the basin and a huge amount of archived data. Trimble took soil profiles all around the basin, dated the layers of sediment, and was able to work back to the original surface of organic soil of the prairie, which dated to about 1850, when the basin was first farmed. He found rates of erosion that were high in the late 19th century, went even higher in the 1920s and 1930s, and then began to decline as farmers began to adopt conservation practices. Most remarkably, the rates from the 1970s to the 1990s were only 6% of their peak rate. Although the nationally measured erosion rates showed a slight decline during the 1970s to the 1990s, compared with earlier rates, the low rates found by what can be called "ground truth" in Trimble's work do not come close to the much higher estimates. Trimble's work has come under some criticism, but his response is that "the burden of proof is on those who have been making these pronouncements about big erosion numbers.... They owe us physical evidence. For one big basin, I've measured the sediment and I'm saying, I don't see it." At the very least, this kind of work suggests that more such actual measurements should be made and that the published erosion estimates should be taken with, well, a grain of silt. What do you think?

[1]From "Decreased Rates of Alluvial Sediment Storage in the Coon Creek Basin, Wisconsin, 1975-1999" by Stanley Trimble in SCIENCE, 285, August 20, 1999. Copyright © by American Assn. for the Advancement of Science. Reprinted by permission of the Amerian Assn. for the Advancement of Science.

WATER TRANSPORT BY TRANSPIRATION

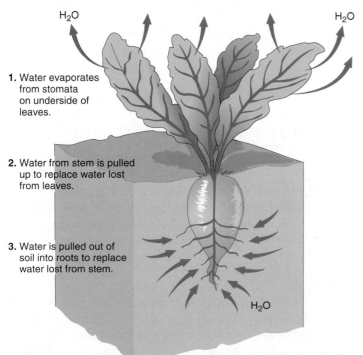

H_2O

H_2O

1. Water evaporates from stomata on underside of leaves.

2. Water from stem is pulled up to replace water lost from leaves.

3. Water is pulled out of soil into roots to replace water lost from stem.

H_2O

◄ FIGURE 8–6 *Transpiration.* When water evaporates from the leaves of a plant, a vacuum is created that pulls the water up through the plant tissues. The roots draw water from the soil to replenish the evaporated water. The sun's energy drives transpiration via the initial evaporation process.

exchange. Of course, if the wilted condition is too severe or too prolonged, the plants die.

Water is resupplied to the soil naturally by rainfall or artificially by **irrigation**. Three attributes of the soil are significant in this respect. First is the soil's ability to allow water to **infiltrate**, or soak in. If water runs off the ground surface, it won't be useful. Worse, it may cause erosion, which we shall discuss shortly.

Second is the soil's ability to hold water after it infiltrates, a property simply called **water-holding capacity**. Poor water-holding capacity implies that most of the infiltrating water percolates on down below the reach of roots—not far in the case of seedlings and small plants—again becoming useless. What is desired is a good water-holding capacity—the ability to hold a large amount of water like a sponge—providing a reservoir on which plants can draw between rains. If the soil does not have such water-holding capacity, the plants will have to depend on frequent rains or irrigation or suffer the consequences of drought. Sandy soils are notorious for their poor water-holding capacity.

The third critical attribute is **evaporative water loss** from the soil surface. It is plain that such evaporation will deplete the soil's water reservoir without serving the needs of plants. The O horizon functions well to reduce evaporative water loss by covering the soil. These aspects of the soil–water relationship are summarized in Fig. 8–7.

Aeration.   Novice gardeners commonly kill plants by overwatering, or "drowning," them. Roots need to "breathe." Basically, they are living organs and need a constant supply of oxygen for energy via metabolism. Land plants depend on the soil being loose and porous enough to allow the diffusion of oxygen into, and carbon dioxide out of, the soil, a property called **soil aeration**. Overwatering fills the air spaces in the soil, preventing adequate aeration. So does

compaction, or packing of the soil, such as occurs with excessive foot or vehicular traffic. Compaction also reduces infiltration and increases runoff. Again, soil texture strongly influences this property, as indicated in Table 8–2.

Relative Acidity (pH).   The term **pH** refers to the acidity or alkalinity (basicity) of any solution. A solution that is neither acidic nor alkaline is said to be neutral and has a pH of 7. The pH scale, which runs from 1 to 14, is discussed more fully in Chapter 22, on acid deposition. For now, it is important to know that most plants (as well as animals) require a pH near neutral, and most natural environments provide this pH.

Salt and Water Uptake.   A buildup of salts in the soil makes it impossible for the roots to take in water. Indeed, if salt levels in the soil get high enough, water can be drawn *out* of the plant, resulting in dehydration and death. This is why highly salted soils are virtual deserts supporting no life at all. We shall see the importance of the problem later when we study how irrigation may lead to the accumulation of salts in soil (salinization).

Soil as an Ecosystem

Summarizing the preceding section, we see that, to support a good crop, the soil must (1) have a good supply of nutrients and a good nutrient-holding capacity; (2) allow infiltration, have a good water-holding capacity, and resist evaporative water loss; (3) have a porous structure that permits good aeration; (4) have a pH near neutral; and (5) have a low salt content. Moreover, these attributes must be sustained. How does the natural soil ecosystem—that dynamic interaction of minerals, detritus, and soil organisms—provide and sustain these attributes?

A soil's mineral attributes—in particular, its texture—are crucial to its ability to support plant growth. Which is the best soil? Recall the principle of limiting factors. The poorest attribute is the limiting factor; the poor water-holding capacity of sandy soil, for example, may preclude agriculture altogether because the soil dries out so quickly. The best texture proves to be silt or loam, because limiting factors are moderated in these two types of soil. But the good qualities are also moderated, so this "best" is really only "medium." The organic parts of the soil ecosystem—the detritus and soil organisms—are necessary to optimize all attributes.

Detritus, Soil Organisms, Humus, and Topsoil.   The accumulation of dead leaves, roots, and other detritus on and in the soil supports a complex food web, including numerous species of bacteria, fungi, protozoans, mites, insects, millipedes, spiders, centipedes, earthworms, snails, slugs, moles, and other burrowing animals (Fig. 8–8). The most numerous and important organisms are the smallest—the bacteria. With the use of fluorescent staining (a process that makes bacteria glow when viewed in a fluorescence microscope), literally millions of bacteria can be seen and counted in a gram of soil. As these organisms feed, the bulk of the

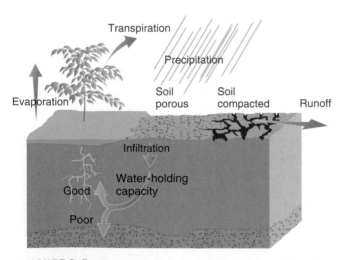

▲ **FIGURE 8–7** *Plant–soil–water relationships.* Water lost from the plant by transpiration must be replaced from a reservoir of water held in the soil. In addition to the amount and frequency of precipitation, the size of this reservoir depends on the soil's ability to allow water to infiltrate, hold water, and minimize direct evaporation.

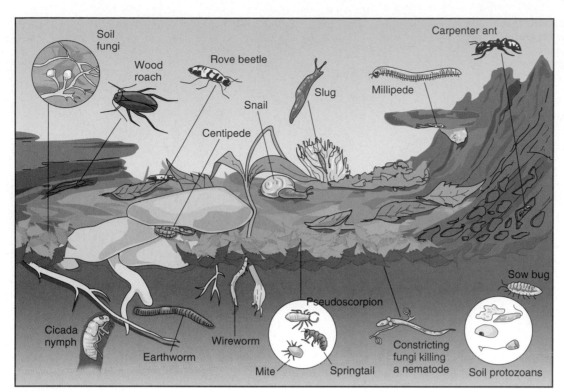

◄ **FIGURE 8–8** *Soil as a detritus-based ecosystem.* A host of organisms, major examples of which are shown here, feed on detritus and burrow through the soil, forming a humus-rich topsoil with a loose, clumpy structure.

detritus is consumed through their cell respiration, and carbon dioxide, water, and mineral nutrients are released as by-products (as described in Chapter 3). However, each organism leaves a certain portion undigested—that is, a certain portion resists breakdown by the organism's digestive enzymes. This residue of partly decomposed organic matter is **humus**, found in high concentrations at the bottom of the O horizon. A familiar example is the black or dark brown spongy material remaining in a dead log after the center has rotted out (Fig. 8–9). **Composting** is the process of fostering the decay of organic wastes under more or less controlled conditions, and the resulting compost is the same as humus. (Composting as a means of processing and recycling various wastes is discussed further in Chapter 19.)

As animals feed on detritus on or in the soil, they often ingest mineral soil particles as well. For example, it is estimated that as much as 15 tons per acre (37 tons per hectare) of soil pass through earthworms each year in the course of their feeding. As the mineral particles go through the worm's gut, they actually become "glued" together by the indigestible humus compounds. Thus, earthworm excrements—or **castings**, as they are called—are relatively stable "clumps" of inorganic particles plus humus. The burrowing activity of organisms keeps the clumps loose. This loose, clumpy characteristic, which you may experience as softness under foot as you walk through a woods, is referred to as **soil structure** (Fig. 8–10). Whereas soil texture describes the *size* of soil particles, soil structure refers to their *arrangement*. You can readily visualize what this characteristic means as regards infiltration, aeration, and workability. In addition, humus has a phenomenal capacity

for holding both water and nutrients, as much as one hundred fold greater than clay, on the basis of weight. The clumpy, loose, humus-rich soil that is ideal for supporting plant growth is the topsoil, represented in a soil profile as the A horizon (Fig. 8–4).

▲ **FIGURE 8–9** *Formation of humus.* As seen in this rotted log, humus is the residue of organic matter that remains after the bulk of organic material has been decomposed by fungi and microorganisms.

▶ **FIGURE 8–10** *Humus and the development of soil structure.* On the left is a humus-poor sample of loam. Note that it is a relatively uniform, dense "clod." On the right is a sample of the same loam, but rich in humus. Note that it has a very loose structure, composed of numerous aggregates of various sizes.

If plants are grown on adjacent plots, one of which has had all its topsoil removed, the results are striking: The yield from plants grown on subsoil is only 10% to 15% of that from plants grown on topsoil. In other words, a loss of all the topsoil would result in an 85% to 90% decline in productivity. The development of topsoil from subsoil or parent material is a process that takes hundreds of years or more. Figure 8–11 shows a site where the topsoil was removed 12 years before; although a lichen cover has developed on the exposed gravel, this is only the beginning of soil formation on that site.

There are other interactions between plants and soil biota. A very significant one is the symbiotic relationship between the roots of some plants and certain fungi called **mycorrhizae**. Drawing some nourishment from the roots, mycorrhizae penetrate the detritus, absorb nutrients, and

▲ **FIGURE 8–11** *The results of removal of topsoil.* This "soil" is nothing but gravel with the topsoil removed. After 12 years, only lichens are able to grow on it.

transfer them directly to the plant. Thus, there is no loss of nutrients to leaching. Another important relationship is the role of certain soil bacteria in adding nitrogen to the soil, as discussed in Chapter 3. Not all soil organisms are beneficial to plants. For example, *nematodes*, small worms that feed on living roots, are highly destructive to some agricultural crops. In a flourishing soil ecosystem, however, nematode populations may be controlled by other soil organisms, such as a fungus that forms little snares to catch and feed on the nematodes (Fig. 8–12).

Soil Enrichment or Mineralization. Coming back to the context of the total ecosystem, we can now see how the aboveground portion and the soil portion of the ecosystem support each other. The bulk of the detritus, which supports the soil organisms, is from green-plant producers, so green plants support the soil organisms. But then, by feeding on detritus, the soil organisms create the chemical and physical soil environment that is most beneficial to the growth of producers.

Green plants protect the soil and consequently themselves in two other important ways as well: The cover of living plants and detritus (1) protects the soil from erosion and (2) reduces evaporative water loss. Thus, you can see the desirability of maintaining an organic mulch around those of your garden vegetables which do not maintain a complete cover themselves.

Unfortunately, the mutually supportive relationship between plants and soil can be broken all too easily. The maintenance of topsoil depends on additions of detritus sufficient to balance losses. Without continual additions of detritus, it should be obvious that soil organisms will starve, and their benefit in keeping the soil loose will be lost. However, additional consequences occur. Although resistant to digestion, humus does decompose at the rate of about 2% to 5% of its

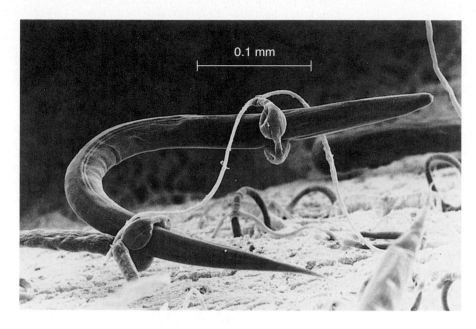

◄ **FIGURE 8–12** *Predatory fungus.* Soil nematode (roundworm), a root parasite, captured by the constricting rings of the predatory fungus *Arthrobotrys anchonia.*

volume per year, depending on conditions. As the soil's humus content declines, the clumpy aggregate structure created by soil particles "glued" together with the humus breaks down. Water- and nutrient-holding capacities, infiltration, and aeration decline accordingly. This loss of humus and the consequent collapse of topsoil is referred to as **mineralization** of the soil, because what is left is just the gritty mineral content—sand, silt, and clay—devoid of humus.

Thus, topsoil must be seen as a dynamic balance between, on the one hand, detritus additions and humus-forming processes and, on the other, the breakdown and loss of detritus and humus (Fig. 8–13). If additions of detritus are not sufficient, there will be a gradual deterioration of the soil. Conversely, you can see how mineralized soils can potentially be "rejuvenated" through generous additions of compost or other organic matter.

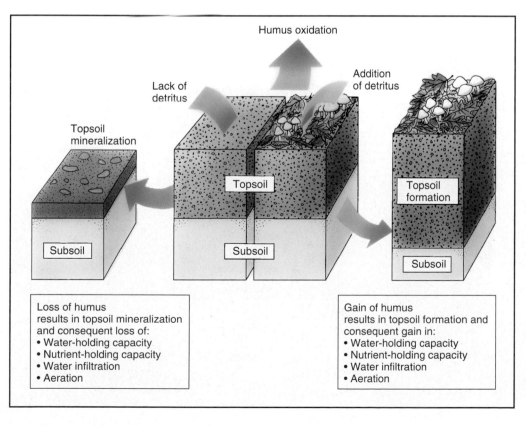

◄ **FIGURE 8–13** *The importance of humus to topsoil.* Topsoil must be recognized as a dynamic balance between, on the one hand, detritus additions and humus-forming processes and, on the other, the breakdown and loss of detritus and humus. If additions of detritus are not sufficient, the soil will gradually deteriorate.

8.2 Soil Degradation

In natural ecosystems, the demand for detritus is no problem: There is always a turnover of plant material. However, when humans come on the scene and cut forests, graze livestock, or grow crops, the soil system is at the mercy of our management or mismanagement. Contrary to knowledge and common sense, so much land is still mismanaged that the sustainability of current levels of food production in many areas is in doubt. Almost 40% of the world's agricultural land—2 billion hectares—have been degraded. Let us examine the key factors involved in this loss.

Erosion and Desertification

How is topsoil lost? The most pervasive and damaging force is **erosion**, the process of soil and humus particles being picked up and carried away by water or wind. Erosion follows any time soil is bared and exposed to the elements. The removal may be slow and subtle, as when soil is gradually blown away by wind, or it may be dramatic, as when gullies are washed out in a single storm.

In natural terrestrial ecosystems other than deserts, a vegetative cover protects against erosion. The energy of falling raindrops is dissipated against the vegetation, and the water infiltrates gently into the loose topsoil without disturbing its structure. With good infiltration, runoff is minimal. Any runoff that does occur is slowed as the water moves through the vegetative or litter mat, so the water has too little energy to pick up soil particles. Grass is particularly good for erosion control, because when runoff volume and velocity increase, well-anchored grass simply "lies down," forming a smooth mat over which the

water can flow without disturbing the soil underneath. Similarly, vegetation slows the velocity of wind and holds soil particles.

When soil is left bare and unprotected, however, it is highly subject to erosion. Water erosion starts with what is called **splash erosion** as the impact of falling raindrops breaks up the clumpy structure of topsoil. The dislodged particles wash into spaces between other aggregates, clogging the pores and thereby decreasing infiltration and aeration. The decreased infiltration results in more water running off and carrying away the fine particles from the surface. This is called **sheet erosion**. As further runoff occurs, the water converges into rivulets and streams, which have greater volume, velocity, and energy, and hence greater capacity to pick up and remove soil. The result is the erosion of gullies, or **gully erosion** (Fig. 8–14). Once started, erosion can readily turn into a vicious cycle if it is not controlled. Eroding soil is less able to support the regrowth of vegetation and is exposed to further erosion, rendering it even less able to support vegetation, and so on.

Another very important and devastating feature of wind and water erosion is that both types always involve the *differential* removal of soil particles. The lighter particles of humus and clay are the first to be carried away, while rocks, stones, and coarse sand remain behind. Consequently, as erosion removes the finer materials, the remaining soil becomes progressively coarser—sandy, stony, and rocky. Such coarse soils frequently reflect past or ongoing erosion. Did you ever wonder why deserts are full of sand? The sand is what remains after the finer, lighter clay and silt particles have blown away. Often in deserts, the removal of fine material by wind has left a thin surface layer of stones and coarse sand called a **desert pavement**, which protects the underlying soil against further erosion. Can you see how

▶ **FIGURE 8–14**
Erosion. Severe erosion due to poor farming practices, near Bosencheve, Mexico.

damaging this surface layer with vehicular traffic allows another episode of erosion to commence (Fig. 8–15)?

Recall that clay and humus are the most important components of soil as regards both nutrient- and water-holding capacity. As clay and humus are removed, nutrients are removed as well, because they are bound to these particles. The loss of water-holding capacity is even more serious, however. Regions that have sparse rainfall or long, dry seasons support grass, scrub trees, or crops only insofar as soils have good water-holding and nutrient-holding capacity. As these soil properties are diminished by the erosion of topsoil, such areas become deserts, both ecologically and from the standpoint of production (Fig. 8–16). Indeed, the term **desertification** is used to denote this process. Note that "desertification" doesn't mean "advancing deserts"; rather, the term refers to *the formation and expansion of degraded areas of soil and vegetation cover in arid, semiarid, and seasonally dry areas, caused by climatic variations and human activities.* These lands are traditionally called drylands.

Drylands

Dryland ecosystems cover over one-third of the land area of Earth; they are defined by precipitation, not temperature. Beyond the relatively uninhabited deserts, much of the land receives only 10 to 30 inches (25 to 75 cm) of rainfall a year, a minimal amount to support rangeland or nonirrigated cropland. Droughts are common features of the climate in drylands, sometimes lasting for years. These lands, occupying some 5.2 billion hectares, are home to more than a billion people, and are found on every continent, except Antarctica. According to the United Nations, some 70% of these dryland types of rangelands and nonirrigated croplands have been adversely affected by erosion and desertifi-

cation, directly affecting the quality of life of some of the poorest people in the world.

Recognizing the severity of this problem, the United Nations has established the **Convention to Combat Desertification (UNCCD)**, which was signed and officially ratified by over 100 nations in late 1996. The first Conference of the Parties (COP) was held in October 1997 in Rome. Since then, COPs have been held annually, the latest in December 2000. These conferences and other regular UNCCD meetings have been concerned with such issues as how to fund projects to reverse land degradation, "bottom-up" programs that enable local communities to help themselves, and the gathering and dissemination of "traditional knowledge" on effective drylands agricultural practices. Under the aegis of the UNCCD, affected countries have been developing National Action Programs, with assistance and funding to carry out the programs provided from donor countries and international agencies like the World Bank and U.N. agencies.

The focus of the 1999 COP-3 was countries in sub-Saharan Africa, which are deemed to be at the greatest risk. COP-4 dealt with the other regions affected by desertification. The activities of the UNCCD and affected countries have made it clear that desertification, once started, is not easy to reverse. With diminished productivity, the soil is left unprotected. Further erosion takes place, causing a further reduction of productivity, and the vicious cycle goes on and on until it ends in a desert landscape that supports virtually no growth at all. The human costs are seen in severe food insecurity and subsequent population movements in efforts to survive. Even though variations in climate often play a role in these processes, it is human agency that is the greatest threat to the health of dryland ecosystems. Toward this end, let us take a closer look at practices that lead to erosion and desertification.

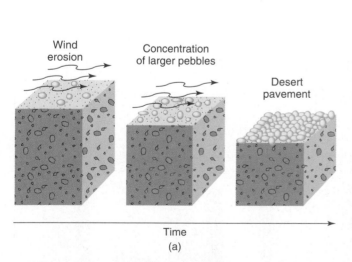

(a)

(b)

▲ **FIGURE 8–15** *Formation of desert pavement.* (a) As wind erosion removes the finer particles, the larger grains and stones are concentrated on the surface. (b) The result is desert pavement, which protects the underlying soil from further erosion. Traffic such as off-road vehicles break up the desert pavement and initiate further erosion.

▲ **FIGURE 8–16** *Desertification.* Cultivation without suitable protection against erosion allows finer components of the soil to be blown away, leaving the soil increasingly coarse and stony and diminishing its water-holding capacity. Persistence gradually renders the area an ecological desert, as seen in this photo, taken in east Africa.

Causing and Correcting Erosion

The major practices that expose soil to erosion and lead to desertification are (1) overcultivation, (2) overgrazing, and (3) deforestation (Fig. 8–17).

Overcultivation. Traditionally, the first step in growing crops has been (and, to a large extent, still is) plowing to control weeds. The drawback is that the soil is then exposed to wind and water erosion. Further, it may remain bare for a considerable time before the newly planted crop forms a complete cover; after harvest, much of the soil again may be left exposed to erosion. Runoff and erosion are particularly severe on slopes, but wind erosion may extract a heavy toll on any terrain with minimal rainfall.

It is ironic that plowing is frequently deemed necessary to "loosen" the soil to improve aeration and infiltration, for all too often the effect is just the reverse. Splash erosion destroys the soil's aggregate structure and seals the surface so that aeration and infiltration are decreased. The weight of tractors used in plowing may add to the compaction. In addition, plowing accelerates the oxidation of humus and evaporative water loss.

Despite the harmful impacts inherent in cultivation, systems of crop rotations—a cash crop such as corn every third year, with hay and clover (which fixes nitrogen as well as adding organic matter) in between—have proved sustainable. However, as food or economic demands cause the abandonment of rotations, degradation and erosion exceed regenerative processes, and the result is a gradual decline in soil quality, or desertification. This is the essence of overcultivation.

A technique that permits continuous cropping, yet minimizes soil erosion is **no-till agriculture**, which is now routinely practiced over much of the United States. The field is first sprayed with herbicide to kill weeds; then a planting apparatus pulled behind a tractor accomplishes several operations at once. A steel disk cuts a furrow through

▶ **FIGURE 8–17** *Main causes of dryland soil degradation.* Deforestation, overgrazing, and overcultivation result in the degradation of soils in every region of the world.

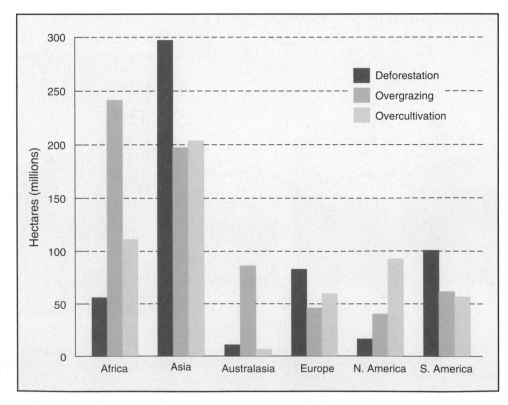

the mulch of dead weeds, drops seed and fertilizer into the furrow, and then closes it (Fig. 8–18). At harvest, the process is repeated, and the waste from the previous crop becomes the detritus and mulch cover for the next. Thus, the soil is never left exposed, erosion and evaporative water loss are reduced, and there is enough detritus, including roots from the previous crop, to maintain the topsoil. Of course, the drawbacks are the use of chemical herbicides that may have unrecognized side effects, and the use of pest control, which may become more problematic as the pests overwinter in the crop residues left on the surface.

Another aspect of overcultivation involves the use of inorganic versus organic fertilizer. There is no question that optimal amounts of required nutrients can be efficiently provided by the suitable application of inorganic chemical fertilizer. The failing of chemical fertilizer is in its lack of organic matter to support soil organisms and build soil structure. Under intensive cultivation—a cash crop every year—nutrient content may be kept high with inorganic fertilizer, but mineralization and thus desertification proceed in any case. Then, with the soil's loss of nutrient-holding capacity, applied inorganic fertilizer is prone to simply leach into waterways, causing pollution.

This is not to say that chemical fertilizers do not have a valuable place in agriculture. The exclusive use of organic material may not provide enough of certain nutrients required to support optimal plant growth. What is required is for growers to understand the different roles played by organic material and inorganic nutrients and then to use each as necessary.

Regardless of cropping procedures and the use of fertilizer, a number of other techniques are widely employed to reduce erosion. Among the most conspicuous are **contour strip cropping** and the establishment of **shelterbelts** (Fig. 8–19). The U.S. Natural Resource Conservation Service (NRCS), formerly and still widely known as the Soil Conservation Service (SCS), was established in response to the Dust Bowl tragedy of the early 1930s. Through a nationwide network of regional offices, the NRCS provides information to farmers or other interested persons regarding soil or water conservation practices. (The regional offices can be found through county government telephone listings. Federal extension service offices will test soil samples, provide an analysis thereof, and make recommendations.) The NRCS performs a regular inventory of erosion losses in the United States. (See the Ethics essay, p. 195.) Estimates indicate that soil loss from croplands has lessened in recent years: 1.9 billion tons in 1997 vs. 2.2 billion tons in 1992. This is likely a consequence of improved conservation practices put in place, such as windbreaks, grassed waterways, and field border strips.

Overgrazing. Grasslands that receive too little rainfall to support cultivated crops have traditionally been used for grazing livestock. Likewise, forested slopes too steep for cropping are commonly cleared and put into grass for grazing. Unfortunately, such lands are too often overgrazed. As grass production fails to keep up with consumption, the land becomes barren; wind and water erosion follows, and desertification results. According to the World Resources Institute, overgrazing during the past 40 years has degraded 679 million hectares.

Overgrazing is not a new problem. In the United States in the 1800s, the American buffalo (bison) was slaughtered to starve out the Native Americans and stock the rangelands

◀ **FIGURE 8–18** *Apparatus for no-till planting.* The "wheel" (a) opens a furrow in the soil, fertilizer is dispensed from the tank (b), seeds are dropped in from the hopper (c), and the furrow is closed by the trailing wheels (d). Weeds are controlled with herbicides. Thus, all operations of field preparation, planting, and cultivation are accomplished without disturbing the protective mulch layer on the surface of the soil.

(a)

(b)

▶ **FIGURE 8–19** *Contour farming and shelterbelts.* (a) Cultivation up and down a slope encourages water to run down furrows and may lead to severe erosion. The problem is reduced by plowing and cultivating along the contours at a right angle to the slope. In this photograph, strip cropping is also being utilized: The light green bands are one type of crop, the dark bands another. (b) Shelterbelts—belts of trees around farm fields—break the wind and protect the soil from erosion.

with cattle. Overgrazing was rampant, leading to desertification and encroachment by hardy desert plants such as sagebrush, mesquite, and juniper, which are not palatable to cattle. Western rangelands now produce less than 50% of the livestock forage they produced before the advent of commercial grazing. Yet, according to a U.S. General Accounting Office (GAO) study, 20% of the rangelands remain overstocked.

The broader ecological impact of overgrazing should not go unnoticed either. The World Resources Institute reports, "Overgrazing has profoundly upset the dynamics of many range ecosystems, reducing biodiversity and altering the feeding and breeding patterns of birds, small mammals, reptiles, and insects." Further, about one-third

of the endangered species in the United States are in jeopardy because of overgrazing or other practices associated with raising cattle, such as predator-control programs and the suppression of fire. Particularly hard hit are the wooded zones along streams and rivers that are trampled by cattle seeking water. The resulting water pollution by sediments and cattle wastes is high on the list of factors making fish species the fastest disappearing wildlife group in the United States.

Overgrazing occurs in many cases because the rangelands are "public lands" not owned by the people who own the animals. Where this is the case, herders who choose to withdraw their livestock from the range sacrifice income, while others continue to overgraze the range. Therefore, the incentive is for all to keep grazing despite realizing that the range is being overgrazed. This problem, known as the "tragedy of the commons," is discussed further in Chapter 12.

In a variation of the tragedy of the commons, the U.S. Bureau of Land Management (BLM) leases huge tracts of government- (i.e., taxpayer)-owned lands for grazing rights at a nominal fee of $1.35 per animal unit (one cow and calf pair or five sheep) per month, about one-fourth of what they would pay a private owner for the same rights. The fee structure was established by the Taylor Grazing Act of 1934, which effectively prohibited any reduction in grazing levels and kept the fees well below market value. The income to the government ($14 million in 1999) is far below the costs of administering the program, and the amount of livestock permitted by the BLM is considered too high by nearly all range experts. Attempts in Congress or the Department of the Interior (which houses the BLM) to raise grazing fees are consistently met with opposition from western congressmen, who threaten to cut the BLM budget or to hand over the rangelands to the western states. A "reform" bill—the Forage Improvement Act— was under consideration, but is opposed by the secretaries of the Interior and Agriculture because it would raise the grazing fees only slightly, while greatly loosening up control over the grazing process by administrators. The bill died in the Senate.

The obvious solution lies in better management. Since rangelands are now producing at only a 50% level or less, their restoration could benefit both wildlife and cattle production. The NRCS has a project that provides information and support to enable ranchers who own their lands to burn unwanted woody plants, reseed the land with perennial grass varieties that hold water, and manage cattle so that herds are moved to a new location before overgrazing occurs. Early results from this project are encouraging.

Deforestation. Forest ecosystems are extremely efficient systems both for holding and recycling nutrients and for absorbing and holding water, because they maintain and protect a very porous, humus-rich topsoil. Investigators at Hubbard Brook Forest in New Hampshire found that converting a hillside from forest to grass resulted in doubling the amount of runoff, and the leaching of nutrients increased manyfold.

Much worse is what occurs if the forest is simply cut and soil is left exposed. Pounding raindrops quickly seal the soil; the topsoil becomes saturated with water and slides off the slope in a muddy mass into waterways, leaving barren subsoil that continues to erode.

The problem is particularly acute with the cutting of tropical rain forests. The soils of equatorial regions have been subjected to heavy rains and leaching for millennia. Parent materials are already maximally weathered, and any free nutrients have long since been leached away. Conse-

quently, tropical soils (oxisols) are notoriously lacking in nutrients. When the forests are cleared, the thin layer of humus with nutrients readily washes away. Only the nutrient-poor clayey subsoil, which is very poor for agriculture, is left.

Forests continue to be cleared at an alarming pace— about 27.9 million acres (11.3 million hectares) per year, most of which is in the developing countries, according to a 1999 report from the U.N. Food and Agriculture Organization (FAO). Over the past decade, an area three times the size of France has been lost, and much more has been degraded by fragmentation (partial clearing and thinning). About 80% of the deforestation is for agricultural purposes despite the poor soil, and much of agriculture is directed toward the production of cash crops or conversion to grass to grow beef. Very little is managed with a clear aim of sustainable food production.

In all, less than 2% of tropical forests are under any kind of management plan for the protection or harvesting of forest products on a sustainable basis, although nongovernmental organizations such as Conservation International are active in this regard. Also, the FAO has developed several bodies designed to address and coordinate issues of management of the tropical forests—the Committee on Forestry and six regional forestry commissions provide a forum for member countries to discuss and develop policies that will lead to sustainability.

The Other End of the Erosion Problem. In addition to soil degradation and the loss of productivity resulting from erosion, we must consider the altered pathway of water and where the **sediments**, as the eroding soil is called, go. Water that is unable to infiltrate flows over the surface immediately into streams and rivers, overfilling them and causing flooding. Eroding sediments are likewise carried into streams and rivers, where they clog channels and exacerbate flooding, fill reservoirs, kill fish, and generally upset the ecosystems of streams, rivers, bays, and estuaries. Around the world, coral reefs are dying because of sediments and other pollutants carried with them. Indeed, excess sediments and nutrients resulting from erosion are recognized as the number-one pollution problem of surface waters in many regions of the world. Meanwhile, groundwater resources are also depleted as the rainfall runs off rather than refilling the reservoir of water held in the soil or recharging groundwater.

Summary. Lands suffering from or prone to soil degradation cover much of the globe (Fig. 8–20). In order to save this soil, the foundation of all terrestrial productivity, we must save the natural ecosystems that build, maintain, and nurture that soil or at least provide mechanisms that duplicate those ecosystems. Erosion is a particularly insidious phenomenon, because the first 20% to 30% of the topsoil may be lost with only marginal declines in productivity, a loss that may be compensated for by additional fertilizer and a favorable distribution of rains. As the loss of topsoil continues, however, the decrease in

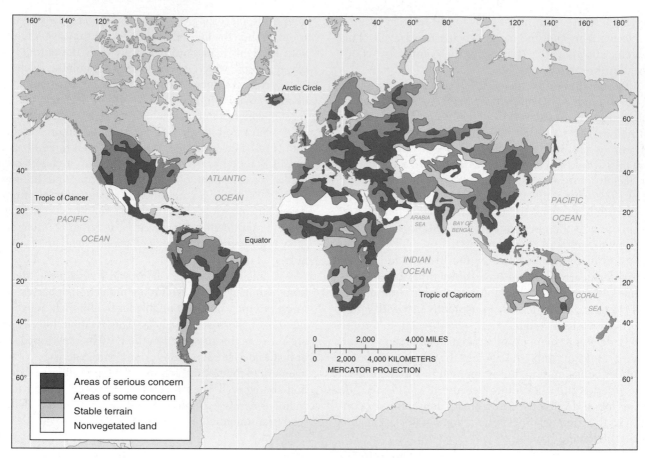

▲ FIGURE 8–20   *Areas subject to soil degradation.*   Throughout the world, overcultivation, overgrazing, and deforestation are causing soil degradation in vast areas.

productivity and increase in vulnerability to drought will also continue.

Irrigation and Salinization

Irrigation—supplying water to croplands by artificial means—has dramatically increased crop production in regions that typically receive inadequate rainfall. Traditionally, water has been diverted from rivers through canals and flooded through furrows in fields, a technique known as **flood irrigation** (Fig. 8–21). In recent years, **center-pivot irrigation**, a procedure in which water is pumped from a central well through a gigantic sprinkler that slowly pivots itself around the well, has become much more popular. (See Fig. 9–16b.) In the United States, the Bureau of Reclamation is the major governmental agency involved with ongoing supplies of irrigation water to regions of the west. More than 600 dams and reservoirs were constructed by the Bureau since its inception in 1902, and it now provides irrigation water for 10 million acres (4 million hectares) of farmland. Total 1997 irrigated land acreage in the United States was 62.1 million acres (25 million hectares), according to the Natural Resources Conservation Service.

Worldwide, irrigated acreage has increased dramatically in the last few decades; the International Food Policy Research Institute estimates that total irrigated land now amounts to 657 million acres (266 million hectares), a 70% increase over the past 30 years. New irrigation projects continue to be built, but the expansion in acreage is now being offset in large part by another ominous trend—*salinization*. **Salinization** is the accumulation of salts in and on the soil to the point where plant growth is suppressed (Fig. 8–22). Salinization occurs because even the freshest irrigation water contains at least 200 to 500 parts per million (0.02 to 0.05%) of dissolved salts. Adding water to dryland soils also dissolves the high concentrations of soluble minerals often present in those soils. As the applied water leaves by evaporation or transpiration, the salts remain behind and gradually accumulate as a precipitate. Because it happens in drylands, salinization is considered a form of desertification, since it renders the land less productive or even useless. Worldwide, an estimated 10 million acres (4 million hectares) of agricultural land is lost yearly to salinization and waterlogging. In the United States, the problem is especially acute in the lower Colorado River basin and in the San Joaquin Valley of California, areas in which a total of 400,000 acres (160,000 hectares) have been rendered unproductive, representing an economic loss of more than $30 million per year. Adding to the problems, water supplies are fast being depleted by withdrawals for irrigation. This and related problems will be discussed further in Chapter 9.

◄ FIGURE 8–21 *Flood irrigation.* The traditional method of irrigation is to flood furrows between rows with water from an irrigation canal. This method is extremely wasteful, because most water either evaporates or percolates beyond the root zone or the water table rises to waterlog the soil and prevent crop growth.

Salinization can be avoided, and even reversed, if sufficient water is applied to leach the salts down through the soil. However, unless there is suitable drainage, the soil will become a waterlogged quagmire in addition to being salinized. Drainage pipe may be installed 10–13 feet (3–4 meters) under the surface, but only at great expense, and then attention must be paid to where the salt-laden water is drained. The Kesterson National Wildlife Refuge in California was destroyed by pollution from irrigation drainage from selenium-enriched soils. The area was classified as a toxic-waste site after only three years of receiving the salty drainage water. Problems of both salinization and depletion of water resources have been exacerbated by the government's building of dam and irrigation projects and providing water to farmers at far below cost, practices that encourage the use of excessive water. On the positive side, the Department of the Interior has established the National

Irrigation Water Quality Program, designed to identify and address the irrigation-related water-quality problems associated with the Interior's water projects in the West.

8.3 Addressing Soil Degradation

Thus far in this chapter, we have examined the nature of soils and how they support the growth of plants. We have seen the crucial role that soils play in maintaining agricultural production. Then we examined the factors that lead to erosion and desertification—in particular, overcultivation, overgrazing, and deforestation. Along the way, we mentioned some agencies and organizations working to preserve and protect the soils. There is broad consensus that soils are vital to human societies and that they should be preserved. This is encouraging, and so

◄ FIGURE 8–22 *Salinization.* Millions of acres of irrigated land are now worthless because of the accumulation of salts left behind as water evaporates.

long as this consensus translates into action, we will be moving in the direction of sustainability.

Soil conservation must be practiced at two levels. The most important is the level of the *individual landholder*. As our opening story of Gabino López clearly showed, those working on the land are best situated to put into practice the conservation strategies that not only enhance their soils, but also bring better harvests and improve their way of life. A great deal of traditional knowledge exists with landholders, who routinely practice soil conservation techniques. This is stewardship in action. The other level is the level of *public policy*. In particular, farm policies play a crucial role in determining how soils are conserved. Policies governing the grazing of public lands and forestry practices also lead to either good or poor stewardship of the soils that underlie the resources being exploited. Our second opening story, of Komsomolsky, introduced us to agricultural and human-resource policies that led to a soil-related disaster on an epic scale. The two levels must work in harmony to bring about effective, sustainable stewardship of soil resources. Let us examine briefly what is being done in these two areas, beginning with public policy.

Public Policy and Soils

We have seen that, despite the crucial importance of saving agricultural land, forests, and rangeland, soil degradation in the form of erosion, desertification, and salinization continues to occur at disturbing rates. The losses are on a collision course with sustainability. Who is responsible for making changes?

Farm policy in the United States has traditionally focused on a single goal: increasing production. Success in attaining this goal cannot be denied; continuing high yields and surpluses speak for themselves. With the focus on increasing production, however, various practices having questionable sustainability have been allowed and even promoted through government subsidies. The National Research Council summed it up in a 1989 report titled *Alternative Agriculture*:

> As a whole, federal policies work against environmentally benign practices and the adoption of alternative [sustainable] agricultural systems, particularly those involving crop rotations, certain soil conservation practices, reductions in pesticide use, and increased use of biological and cultural means of pest control. These policies have generally made a plentiful food supply a higher priority than protection of the resource base.

We have seen how similar subsidies to ranchers for grazing on federal lands undercut the objectives of soil conservation for those lands.

In comparison, the goals of sustainable agriculture are to (1) maintain a productive topsoil, (2) keep food safe and wholesome, (3) reduce the use of chemical fertilizers and pesticides, and, last but far from least, (4) keep farms economically viable. Professional agriculturists are becoming increasingly aware of the shortfalls of modern farming and are at work developing alternatives. Many options actually mimic practices of the past: contouring, crop rotation, terraces, smaller equipment, and a reduced amount of chemicals or no chemicals at all. In 1988, the U.S. Department of Agriculture started the **Low Input Sustainable Agriculture (LISA)** program, which provides funding for "alternative" methods. A small, but growing, number of farmers are experimenting with and developing alternative systems on their own. Also, agricultural colleges and universities across the country are revising their curricula to emphasize sustainable agriculture. For example, the University of California at Davis has established the Sustainable Agriculture Research and Education Program, whose mission is to provide "leadership and support for scientific research and education to encourage farmers, farmworkers, and consumers in California to produce, distribute, process and consume food and fiber in a manner that is economically viable, sustains natural resources and biodiversity, and enhances the quality of life in the state's diverse communities for present and future generations."

Other winds of change have been blowing across the U.S. farm scene as well. In 1985, Congress passed the **Conservation Reserve Program,** under whose authority 40 million acres (16 million hectares) of highly erodible cropland were established as a "conservation reserve" of forest and grass. Farmers are paid about $50 per acre ($125 per hectare) per year for land placed in the reserve. As of 1997, 32.7 million acres (13.2 million hectares) had been placed in the reserve, saving an estimated 700 million tons of topsoil per year from erosion—but it is also costing taxpayers over a billion dollars per year for the "deficiency payments," as they are called. Under another bill, the **Food Security Act of 1985,** farmers are required to develop and implement soil-conservation programs in order to remain eligible for price supports and other benefits provided by the government.

The **Federal Agricultural Improvement and Reform Act (FAIR) of 1996** (also referred to as the "Freedom to Farm" Act) brought even more reforms. Subsidies and controls over many farm commodities were reduced or eliminated, giving farmers greater flexibility in deciding what to plant, but also forcing them to rely more heavily on the market to guide their decisions (and for their income!). The Conservation Reserve Program was reauthorized and capped at 36 million acres, and incentives were provided to forego cultivating highly erodible land, land adjacent to waterways, and land susceptible to flooding. Two new programs were introduced: the Environmental Quality Incentive Program (EQUIP), to assist farmers in dealing with environmental and conservation improvements on their farms, and the Natural Resources Conservation Foundation, a nonprofit, private organization whose mission is to promote and fund innovative solutions to conservation problems through partnerships. Unfortunately, "Freedom to Farm" has also become "Freedom to Fail," as farmers experienced sharply falling livestock and grain prices in the late 1990s. The Congress had to pump $8.7 billion in 1999 alone into a farm-aid package in order to establish a safety net to prevent wide-scale economic disaster in the farming industry.

Public policies aside, farmers should have incentives to nurture the soil on lands under cultivation. The more farmers incorporate conservation practices in their operations, the more productive their farms will become in the long run. The same applies to those who care for livestock that graze open rangelands. Most observers agree that the most significant obstacle to establishing sustainable soil conservation is a lack of knowledge about what such conservation can do. In particular, this applies to the developing countries, where agricultural policies often are poorly developed and the majority of landholders have little access to agricultural extension services.

Helping Individual Landholders

It is the individual landholders and herders who hold the key to sustainable soil stewardship. They must be convinced that what they do will work, that it is affordable, and that in the long run, their own well-being will improve. Frequently, this requires taking a number of small, realistic steps. It may also require help in the way of microlending (see Chapter 7), sound advice (often best coming from their own countrymen and countrywomen), or simply encouragement to experi-

ment (See the Global Perspective essay, this page). Toward this end, the U.N. Convention to Combat Desertification is a large step in the right direction. To be sure, reforms are often needed in agricultural policies in the affected countries so that the landholders and herders are encouraged and enabled to initiate changes in their practices. Development specialists agree that the last thing small farmers in the developing countries need is a transfer of the high-technology agricultural practices employed in the industrialized countries. Most of the farmers that need to be reached are subsistence farmers working on small plots of marginal land.

Various U.N. agencies and nongovernmental organizations (NGOs) are working in the developing world to foster sustainable agriculture for subsistence farmers. To give an idea of these efforts, let us look into one organization implemented by the FAO: the **Farmer-centred Agriculture Resource Management (FARM)** Program. FARM is a cooperative venture of eight Asian countries: China, India, Indonesia, Nepal, Philippines, Sri Lanka, Thailand, and Vietnam. The mission of the organization is to "support improved sustainable agricultural resource management and the attainment of household security through innovative approaches in rainfed areas of Asia."

The major thrust of FARM is to find successful cases of sustainable agriculture used in the member countries and then to create the conditions for replicating these successes.

GLOBAL PERSPECTIVE

THREE-STRATA FORAGE SYSTEM FOR MOUNTAINOUS DRYLANDS

The Hindu Kush Himalayas (HKH) is a mountain chain straddling eight Asian countries from China and Myanmar in the east to Afghanistan and Pakistan in the west. The world's highest mountain region, it is also the most populous. It occupies 1.65 million square miles (4.29 million km^2) and is home to more than 155 million people. Farming, forestry, and livestock sustain most of the people in the region; the land is fragile and is now threatened with an imbalance between population and available productive land. Rapid population growth and intensifying agriculture have led to, and are still producing, great increases in erosion and landslides as people clear forested slopes in a desperate attempt to meet their basic needs for fuelwood, timber, animal fodder and cultivation. Aware of this intensifying problem, the International Centre for Integrated Mountain Development has promoted studies on how such a system can be brought into sustainable use. What is

needed for the HKH region is an approach that will rehabilitate degraded lands, provide a sustainable source of fodder for animals, and also furnish fuelwood for domestic use. The *three-strata forage system* (TSFS) can accomplish all of these aims, and has been shown to work in other dryland environments.

The three strata are (1) grass and legumes, (2) shrubs, and (3) fodder trees. Given the right combination of plants, this vegetation will prevent soil erosion, yield both fodder and human food and medicines, and create a pleasant environment where some greenery will persist all year. Such a system has worked well in the dry regions of Bali and is in use by almost all of the farmers there. In the HKH region, researcher Shaheena Hafeez has recommended a number of plants (grasses, legumes, shrubs, and trees) to use in a TSFS application for the mountainous terrain. One of the most interesting varieties is a tall shrub commonly found throughout the region—sea

buckthorn (*Hippophae* spp.). This plant can root and thrive in poor soils, because it is a nitrogen fixer. Its root system is highly developed and holds the soil well. It supplies palatable forage for animals all year and produces up to 1,300 pounds of berries per acre. The berries are a rich source of vitamins and can be made into jams, juices, and beverages. Oil from the berries is valued for its medicinal properties. A six-year-old sea buckthorn forest can provide over 7 tons of fuelwood per acre on a sustainable basis.

The three-strata forage system is only one of a number of strategies needed for meeting what are called the "silent emergencies" of the remote HKH region (as opposed to the "loud emergencies" of global climate change, loss of biodiversity, etc.): poverty, energy and water shortages, fuelwood and fodder scarcity, and poor access to sanitation, health, and education. Yet, it is an important strategy, because it promises to provide most of the basic goods and services essential to life in these arid mountain lands.

This is done by building information systems describing the effective technologies (both traditional and modern), communicating this information to professionals and policy-makers in the various countries, and training farmers and development workers in these proven technologies. Funding comes from a variety of sources: the member countries, private donors, NGOs, and international donors like the U.N. Development Program (UNDP) and the World Bank. FARM has identified one or two field sites in each member country—small villages located in fragile ecosystems characterized by significant poverty. Site Working Groups (SWGs) composed of local farmers, community organizations, and government representatives are established at each site. As the SWGs develop solutions to problems, their work is communicated at regional and national levels by other FARM teams, with the objective of developing national farm strategies that incorporate the solutions found.

Under the FARM Program, four "farmer field schools" teaching integrated soil management have been estab-lished. A typical school program involves local farmers and begins with an examination of basic soil characteristics and typical soil-management problems that occur in the area, usually conducted by a qualified soil specialist. Then the farmers select several solutions to the problems and test the solutions in their own fields. Many of the ideas they test originate in traditions not well known by others in the group. At the end of a growing season, the results are measured and evaluated. The farmers make all the decisions and do all the fieldwork. The program is still in its youth, but with its emphasis on the empowerment of local people, use of traditional knowledge blended with modern agricultural science, and strategy of communicating results at regional and national levels, it promises to deliver just what sustainable development needs. To quote James Gustave Speth, administrator of the UNDP, "Development will bring food security only if it is people centered, if it is environmentally sound, if it is participatory, and if it builds local and national capacity of self reliance."

E N V I R O N M E N T   O N   T H E   W E B

COMMUNITY OUTREACH AS KEY TO SOIL CONSERVATION

As any farmer will tell you, no two farms are alike. Each has its unique configuration of soil type and slope, crop or livestock production, farm buildings, equipment, and people. Programs to preserve and enhance the farm environment must therefore be tailored to the specific needs of a given operation and the people who run it. This chapter describes a number of regulatory initiatives that encourage soil conservation. Most of them share an emphasis on farmer-centered planning, because experience has shown that farmer cooperation, often based on site-specific initiatives, is central to the success of soil conservation programs.

Since the dust bowl years of the 1930s, when soil erosion and drought were at their worst levels in decades, governments around the world recognized that farmer education is key to land stewardship. In the intervening 70 years, many governments developed ambitious agricultural "extension" programs to provide conservation information to farmers. These programs are usually staffed by agricultural professionals—people who understand the technological and economic issues facing farmers and can discuss options for a particular farm operation.

However, securing farmer cooperation for these programs is much more difficult than it might appear. Farmers are understandably nervous about any action that might adversely affect farm income. Those whose families have worked a farm for generations may be skeptical that governments have much to teach them about the land they know so well.

Surveys show that farmers prefer to get their information from other farmers. In fact, farmers rank other farmers first, government experts next, and university researchers last in terms of the value of their information—and the trust to be placed in it. As a result, over the past decade, farming organizations have begun to assume more of the responsibility for community outreach at a grassroots level. This approach has proved a powerful one. In Canada, Ontario's highly successful Environmental Farm Plan (EFP) program, for example, was started by Ontario farming organizations and has involved farmers at every stage of development. It involves voluntary preparation of documents that highlight a farm's environmental strengths and weaknesses and propose a realistic action plan to correct identified problems. More than 20% of Ontario farmers now participate in the EFP process.

The power of this farmer-to-farmer outreach can perhaps be traced to the strong sense of heritage most farm families feel. Learning from other farmers how to protect and improve their land is one way farmers can protect that heritage for their family's future.

Web Explorations

The Environment on the Web activity for this essay describes the role of farmer-to-farmer outreach in soil conservation. Go to the Environment on the Web activity (select Chapter 8 at **http://www.prenhall.com/wright**) and learn for yourself:

1. how the Universal Soil Loss Equation is used to test the effectiveness of alternative soil conservation strategies;

2. about social, economic, and other factors that influence farmer behavior; and

3. about farmer-to-farmer outreach in the promotion of soil conservation activities.

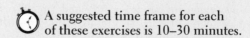 **A suggested time frame for each of these exercises is 10–30 minutes.**

REVIEW QUESTIONS

1. What are the different layers, or horizons, that compose the soil profile? What makes each one distinct from the others?

2. Name and describe the properties of the three main soil textures. Do the same for the four major soil orders listed in this chapter.

3. What are the main things that plant roots must obtain from the soil? Name and describe a process (natural or unnatural) that can keep plants from obtaining the amounts of each of these things they need for survival.

4. Describe, in terms of both content and physical structure, the soil environment that will best fulfill the needs of plants.

5. State and define the relationship between soil structure, humus, detritus, and soil organisms. Describe the role each of these factors plays in creating a soil environment that best supports the needs of plants.

6. What is meant by mineralization? What causes it? What are its consequences for soil?

7. How does a vegetative cover protect and nurture the soil in a way that supports a plant's growth?

8. What is the impact of water and wind on bare soil? Define and describe the process of erosion in detail.

9. What is meant by desertification? Describe how the process of erosion leads to loss of water-holding capacity and, hence, to desertification.

10. What are drylands? Where are they found, and how important are they for human habitation and agriculture?

11. What are the three major cultural practices that expose soil to the weather?

12. What is meant by salinization, and what are its consequences? How does salinization result from irrigation?

13. What are the two levels at which soil conservation must be practiced? What is being done at each level?

THINKING ENVIRONMENTALLY

1. Why is soil considered to be a detritus-based ecosystem? Describe how the aboveground portion and the belowground portion of an ecosystem act as two interrelated and interdependent ecosystems.

2. Why do human societies, past and present, seem to place so little value in maintaining topsoil? What do such attitudes bode for sustainability? What might be done to change attitudes toward soil and farming?

3. Evaluate the following argument: Erosion is always with us; mountains are formed and then erode, and rivers erode canyons. You can't ever eliminate soil erosion, so it is foolish to ask farmers to do so.

4. Suppose you were going into farming. Describe the type of operation you would create, and give a rationale for each measure you would take in terms of sustainability.

5. With the aim of ending hunger in developing countries, would it be best to emphasize the development of large-scale industrialized agriculture as seen in the United States, or should small-scale diversified agriculture be encouraged? Give arguments for your recommendation.

WEB REFERENCES

On-line resources for this chapter are on the World Wide Web at: **http://www.prenhall.com/wright**. (Click on Chapter 8 within the Chapter Selector.)

Water: Hydrologic Cycle and Human Use

Key Issues and Questions

1. All water on Earth is constantly recycled, repurified, and reused. How does the hydrologic cycle accomplish these processes?
2. Humans have three major impacts on the hydrologic cycle. What are they, and what are their effects?
3. All the water humans use must come out of the hydrologic cycle. What are the major uses, points of withdrawal, and limitations and consequences of overdrawing water?
4. Historically, humans have addressed water problems by obtaining more water. To what degree is this not a viable option for the future?
5. Humans can reduce their demand for water in numerous ways. How can demand be reduced in agriculture, industry, and domestic use?
6. Urbanization seals surfaces with pavement, increasing runoff from storms and decreasing concentration times. Discuss related problems caused by paving over soil. How should the ideas you examine influence development?
7. There is a potential for all parties getting together to work out compromises for water usage between agriculture, cities, and natural ecosystems. What are some policy options that would encourage discussion?

Mono Lake is a 63-square-mile (163-km²) saline lake in east-central California. The lake has no outlet, but is fed by freshwater tributaries from snowmelt off the Sierra Nevada. Like other saline lakes around the world, Mono Lake supported numerous species of wildlife, including huge flocks of aquatic birds that nested there and fed on the unique species of brine shrimp in the lake.

Beginning in 1941, however, much of the freshwater inflow was diverted to support water-hungry Los Angeles, 350 miles to the south. Soon Mono Lake was losing more water to evaporation than it was receiving, and gradually it began to shrink. As the lake shrank, the salt in its water became more concentrated, threatening to turn the lake into a sterile, hypersaline system. The streams feeding the lake lost much of their water and failed to support once abundant trout populations. In the meantime, the city of Los Angeles continued to grow, and it became increasingly dependent on imported water.

In 1978, concerned about what was happening to the lake, a group of students formed the Mono Lake Committee, which brought legal action under the Public Trust Doctrine, an old law originally intended to protect navigation. The idea that destruction of the integrity of an ecosystem constituted a violation of the public trust was a new concept that had to be tried in the courts, and it was bitterly opposed by the city of Los Angeles. Nevertheless, the citizens of the Mono Lake Committee persevered through many appeals and won the final decision in September 1994. The State Water Resources Control Board ruled that Los Angeles must reduce diversions of water sufficiently to allow Mono Lake, some 40 feet below its original level, to recover by at least 15 feet (5 m) above the 1995 level. The lake is now on its way back up, having gained 10 feet in height since the Water Board's decision. Recovery of the entire 15 feet could take 20 years, but at least a major ecological collapse of wildlife populations supported by the lake and the feeder streams will probably have been averted.

◀ **Mono Lake.** With its freshwater input diverted to Los Angeles, Mono Lake was drying up, leaving unique sculptured forms of tufa (precipitated calcium carbonate) and threatening ecological collapse.

And Los Angeles? The decision forces L.A. to initially reduce its takings from the Mono Lake supply by 80%, which amounts to a 12% cut in the city's total water supply. When the lake reaches its prescribed height, L.A. will be able to divert about one-third of its historical take of water. This controversy and frequent droughts have had the effect of forcing Angelenos to develop water-saving habits and additional conservation and recycling schemes, which have been more than adequate to make up for the loss. In recent years, Mono Lake has been joined by the Aral Sea, the Colorado River, the Nile, the Jordan, the Danube, and countless other bodies of water that have become the focus of national and international battles over access to fresh water.

9.1   Water—a Vital Resource

Water is absolutely fundamental to life as we know it. It is difficult even to imagine a form of life that might exist without water. Happily, Earth is virtually flooded with water; a total volume of some 325 million cubic miles (1.4 billion cubic kilometers) covers 71% of Earth's surface. Yet, in many locations, it is still difficult to obtain desired amounts of water of suitable purity.

All major terrestrial biota, ecosystems, and humans depend on **fresh water**, water that has a salt content of less than 0.1% (1000 ppm). Over 97% of Earth's water is the salt water of oceans and seas. Then, of the 2.5% that is fresh water, two-thirds is bound up in the polar ice caps and glaciers. Only 0.77% of all water is found in lakes, wetlands, rivers, groundwater, biota, soil, and the atmosphere (Fig. 9–1). Yet, evaporation from the seas and precipitation continually resupply that small percentage through the solar-powered hydrologic cycle, as we shall describe in detail shortly. Thus, fresh water is a continually renewable resource.

As we know, precipitation patterns around the globe are far from even. Regions with abundant precipitation support lush forest ecosystems; other regions have hardly any rainfall and are deserts as a result. Thus, we can visualize different volumes of flow through different natural regions (over 1 million gallons of water per acre per year in a temperate forest region, 2,500 gallons or less per acre per year in a desert region). Human societies must draw on the same water for drinking, irrigating crops, and supplying industries.

In high-rainfall regions, there is plenty of water for both human demands and natural biota. However, in dryer regions with growing human populations, conflicts are increasingly arising between human needs and those of the natural ecosystems (Fig. 9–2). Around the world, countless ecosystems are under stress or already dead because of diversions of water for human uses. Moreover, within the human arena, there is growing contention between agricultural, urban, and industrial demands and between countries that share a common water source.

Hydrologists (experts on water) estimate that severe constraints are placed on food production, economic development, and the protection of natural ecosystems when available water drops below about 1,000 cubic meters per person per year (725 gallons per day). So defined, water shortages loom in at least 26 countries. In many other countries, seasonal shortages of water occur. Water shortages increase conflicts and public health problems, reduce food production, and endanger the environment. As populations grow, more countries will be joining the list. And as shortages become more severe, already contentious relations between nations sharing common supplies will become more so. Furthermore, dried-up and polluted water sources created an estimated 25 million refugees in 1998. Finally, a considerable number of countries and regions are now satisfying their water needs only by withdrawing groundwater faster than it is replenished, thereby depleting their supply for future generations. Obviously, these trends are not sustainable.

A sustainable future will depend on learning stewardship of water resources. There are abundant opportunities for stewardly management in this arena. Our objective in this chapter is threefold: (1) to understand the natural water cycle, its capacities, and its limitations, (2) to understand how we are overdrawing certain water resources and to envisage the consequences thereof, and (3) to understand how water must be managed if we are to achieve sustainable supplies.

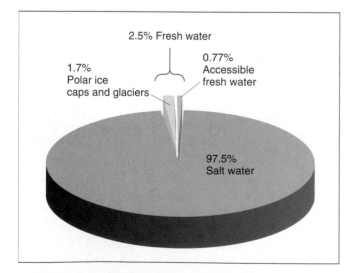

▲ **FIGURE 9–1   Earth's water.**   The Earth has an abundance of water, but terrestrial ecosystems, humans, and agriculture depend on accessible fresh water, which constitutes only 0.77% of the total.

(a) (b)

◄ **FIGURE 9–2**
Differences in availability. Human conditions range from (a) being able to luxuriate in water to (b) having to walk long distances to obtain enough simply to survive. In many cases, however, the abundance of water is illusory, since water supplies are being overdrawn.

9.2 The Hydrologic Cycle

Earth's **water cycle**, also called the **hydrologic cycle**, is represented in Fig. 9–3. The basic cycle consists of water rising to the atmosphere through either evaporation or transpiration and returning to the land and oceans through condensation and precipitation. Modern references to the hydrologic cycle distinguish between water vapor and liquid water as *green water* and *blue water*, respectively. These two forms of water are vitally linked in the hydrologic cycle, as we shall see.

Evaporation, Condensation, and Purification

As we discussed in Chapter 3, a weak attraction known as hydrogen bonding tends to hold water molecules (H_2O) together. Below 32°F (0°C), the kinetic energy of the molecules is so low that the hydrogen bonding is enough to hold the molecules in place with respect to one another, and the result is ice. At temperatures above freezing, but below boiling, the kinetic energy of the molecules is such that hydrogen bonds keep breaking and re-forming with different molecules. The result is liquid water. As the water molecules absorb energy from sunlight or an artificial source, the kinetic energy they gain may be enough to allow them to break away from other water molecules entirely and enter the atmosphere. This is the process we know as **evaporation**, and the water molecules are said to be in the gaseous state.

We speak of water molecules in the air as **water vapor**; the amount of water vapor in the air is the **humidity**. Hu-

midity is generally measured as **relative humidity**, the amount of water vapor as a percentage of what the air can hold *at a particular temperature*. For example, a relative humidity of 60% means that the air contains 60% of the maximum amount of water vapor it could hold at a given temperature. The amount of water vapor air can hold increases with rising temperature and decreases with falling temperature. Consequently, relative humidity will decrease as air warms and increase as air cools, quite apart from any change in the actual amount of water vapor present. The key point is that, when warm, moist air is cooled, and its relative humidity rises until it reaches 100%, further cooling causes the excess vapor to *condense*, because the air can no longer hold as much vapor (Fig. 9–4).

Condensation simply is water molecules rejoining by hydrogen bonding to form liquid water or ice. If the droplets form in the atmosphere, the result is fog and clouds. (Fog is just a very low cloud.) If the droplets of condensing vapor form on the cool surfaces of vegetation, the result is dew. Condensation is greatly facilitated by the presence of *aerosols* in the atmosphere. The **aerosols**—microscopic liquid or solid particles originating from land and water surfaces—provide surfaces that attract water vapor and promote the formation of droplets of moisture. Most aerosols originate naturally, from sources like volcanoes, wind-stirred dust and soil, and sea salts. Some, however, are formed by pollutants released into the atmosphere; for example, sulfur dioxide from burning coal forms sulfate aerosols.

One very important aspect of evaporation and condensation is that these processes result in natural *water purification*.

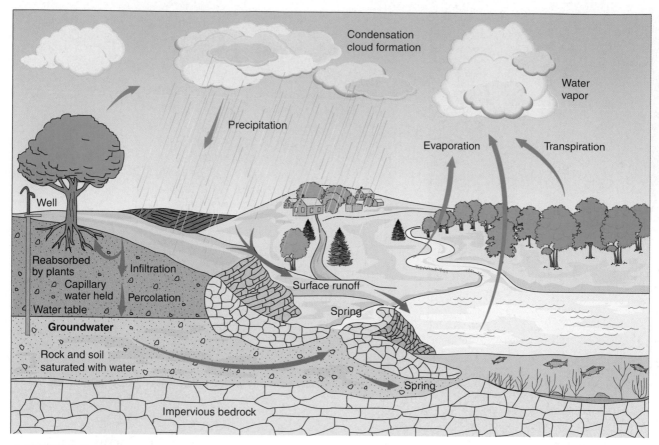

▲ FIGURE 9–3   *The hydrologic cycle.*   The Earth's fresh waters are replenished as water vapor enters the atmosphere by evaporation or transpiration from vegetation, leaving salts and other impurities behind. Note that as precipitation hits the ground, three additional pathways are possible. Colored arrows (blue and green) distinguish between blue and green water flows. (See text.)

▶ FIGURE 9–4   *Condensation.*   The amount of water vapor that air can hold increases and decreases with corresponding changes in temperature. Therefore, as warm, moist air is cooled, the amount of water it can hold decreases. Cooling air beyond the point where the relative humidity (RH) reaches 100% forces excess moisture to condense, forming clouds. Further cooling and condensation results in precipitation.

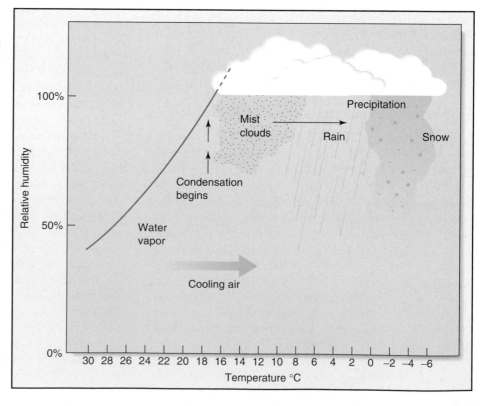

When water evaporates, only the water molecules leave the surface; the salts and other solids in solution remain behind. (We noted this phenomenon when we discussed the problem of salinization in Chapter 8.) The condensed water is thus purified water—except as it picks up pollutants and other aerosols like sea salts in the air. (The most chemically pure water for use in laboratories is obtained by distillation, a process of boiling water and recondensing the vapor.) Thus, evaporation and condensation of water vapor are the source of all natural fresh water on Earth. Fresh water from precipitation falling on the land gradually makes its way through aquifers, streams, rivers, and lakes to oceans or seas. In the process, salts from the land are constantly flushed toward locations where the only exit is by evaporation, and the salts accumulate at those points. Oceans are the prime example, but there are notable inland salt seas or lakes, such as the Great Salt Lake in Utah. Salinization of irrigated croplands is a noteworthy human-made example.

Precipitation

Recall that during evaporation air readily picks up water vapor from any wet or moist surface. Similarly, as transpiration occurs, air picks up water vapor from vegetation (a process called "green water flow"). Warm air rises because it is less dense than the cooler air above, and as it encounters lower atmospheric pressure, it gradually cools as it expands—a process known as *adiabatic cooling*. When the relative humidity reaches 100% and cooling continues, condensation occurs and clouds form. With intensifying condensation, water droplets become large enough to fall as precipitation. *Adiabatic warming* occurs as air descends and is compressed by the higher air pressure in the lower atmosphere.

The distribution of precipitation over Earth, which ranges from near zero in some areas to more than 100 inches (2.5 m) per year in others, basically depends on patterns of rising or falling air currents. As air rises, cooling and condensation occur, and precipitation results. As air descends, it tends to become warmer, causing an increase in evaporation and dryness. A rain-causing event that you see in almost every television weather report is the movement of a cold front. As a cold front moves into an area, the warm, moist air already in the area is forced upward because the cold air of the advancing front is denser. The rising warm air cools, causing condensation and precipitation along the leading edge of the cold front.

Two factors may cause more or less continuously rising or falling air currents over particular regions, with major effects on precipitation. First are global *convection currents*. Solar heating of Earth is most intense over and near the equator, where rays of sunlight are almost perpendicular to Earth's surface. As the air at the equator is heated, it expands, rises, and cools; condensation and precipitation occur. The constant intense heat in these equatorial areas ensures that this process is repeated often, thus causing high amounts of rainfall, which, along with continuous warmth, in turn supports tropical rain forests.

Rising air over the equator is just half of the convection current, however: The air must come down again. Pushed from beneath by more rising air, it literally "spills over" to the north and south of the equator and descends over subtropical regions (25° to 35° north and south of the equator), resulting in subtropical deserts. The Sahara of Africa is the prime example. The two halves of the system composed of the rising and falling air make up a **Hadley cell** (Figs. 9–5a and b). Because of Earth's rotation, winds are deflected from the strictly vertical and horizontal paths indicated by a Hadley cell and tend to flow in easterly and westerly directions—the *trade winds* (Fig. 9–5c), which blow almost continuously from the same direction.

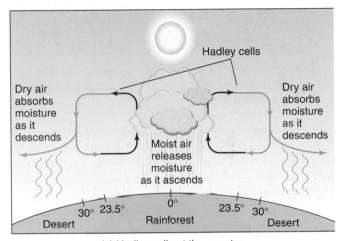

(a) Hadley cells at the equator

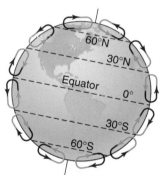

(b) Global air flow patterns

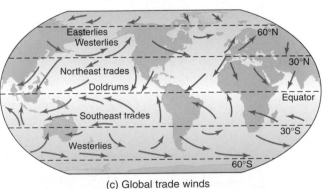

(c) Global trade winds

▲ **FIGURE 9–5** *Global air circulation.* (a) The two Hadley cells at the equator. (b) The six Hadley cells on either side of the equator, indicating general vertical airflow patterns. (c) Global trade-wind patterns as a result of Earth's rotation.

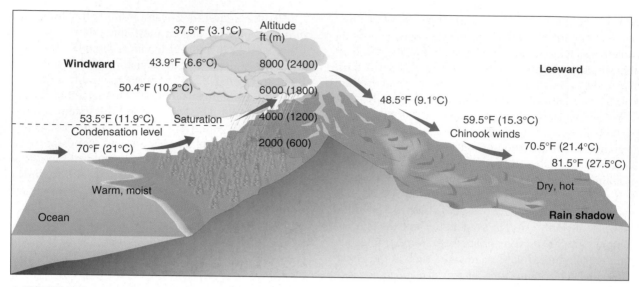

▲ **FIGURE 9–6** *Rain shadow.*   Moisture-laden air in a trade wind cools as it rises over a mountain range, resulting in high precipitation on the windward slopes. Desert conditions arise on the leeward side as the descending air warms and tends to evaporate water from the soil.

The second situation that causes continually rising and falling air occurs where trade winds hit mountain ranges. As the moisture-laden air in the trade winds encounters a mountain range, the air is deflected upward, causing cooling and high precipitation on the windward side of the range. As the air crosses the range and descends on the other side, it becomes warmer and increases its capacity to pick up moisture. Hence, deserts occur on the leeward sides of mountain ranges. The dry region downwind of a mountain range is referred to as a **rain shadow** (Fig. 9–6). The severest deserts in the world are caused by the rain-shadow effect. For example, the westerly trade winds, full of

▶ **FIGURE 9–7** *Global precipitation.*   Note the high rainfall in equatorial regions and the regions of low rainfall to the north and south.

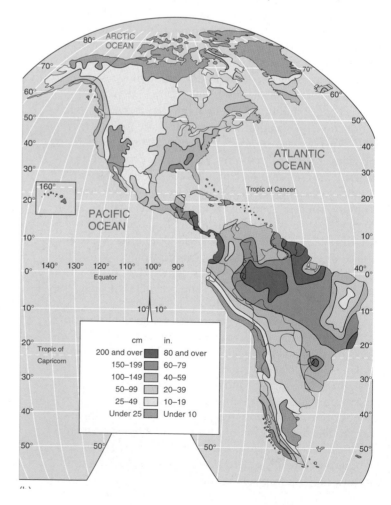

moisture from the Pacific Ocean, strike the Sierra Nevada in California. As the winds rise over the mountains, large amounts of water precipitate out, supporting the lush forests on the western slopes. Immediately east of the southern Sierra Nevada lies Death Valley, a result of the rain shadow.

Figure 9–7 shows general precipitation patterns for the land; the general atmospheric circulation (Fig. 9–5) and the effects of mountain ranges (Fig. 9–6) combine to give great variation in the precipitation reaching the land. This in turn is a major determinant of the ecosystems found in a given region.

Water over and through the Ground

As precipitation hits the ground, it may follow either of two pathways: If it soaks into the ground, it is called **infiltration**; if it runs off the surface, it is termed **runoff** (*blue water* flow). We speak of the amount that soaks in compared with the amount that runs off as the **infiltration–runoff ratio**.

Runoff flows over the surface of the ground into streams and rivers, which make their way to the ocean or to inland seas. All the land area that contributes water to a particular stream or river is referred to as the **watershed** for that stream or river. All ponds, lakes, streams, rivers, and other waters on the surface of Earth are referred to as **surface waters**.

Water that infiltrates has two more alternatives (Fig. 9–3). The water may be held in the soil, in an amount that depends on the water-holding capacity of the soil, as was discussed in Chapter 8. This water, called **capillary water**, returns to the atmosphere either by way of evaporation from the soil or by transpiration through plants (*green water* flow). The combination of evaporation and transpiration is referred to as **evapotranspiration**.

The second alternative is **percolation** (*blue water* flow). Infiltrating water that is not held in the soil is called **gravitational water** because it trickles, or percolates, down through pores or cracks under the pull of gravity. Sooner or later, however, gravitational water comes to an impervious layer of rock or dense clay. There it accumulates, completely filling all the spaces above the impervious layer. This accumulated water is called **groundwater**, and its upper surface is the **water table** (Fig. 9–3). Gravitational water becomes groundwater as it reaches the water table in the same way that rainwater is defined as lake water as it hits the surface of a lake. Wells must be dug to below the water table; then groundwater, which is free to move, seeps into the well and fills it to the level of the water table.

Groundwater will seep laterally as it seeks its lowest level. Where a highway has been cut through rock layers, you can frequently observe groundwater seeping out. Layers

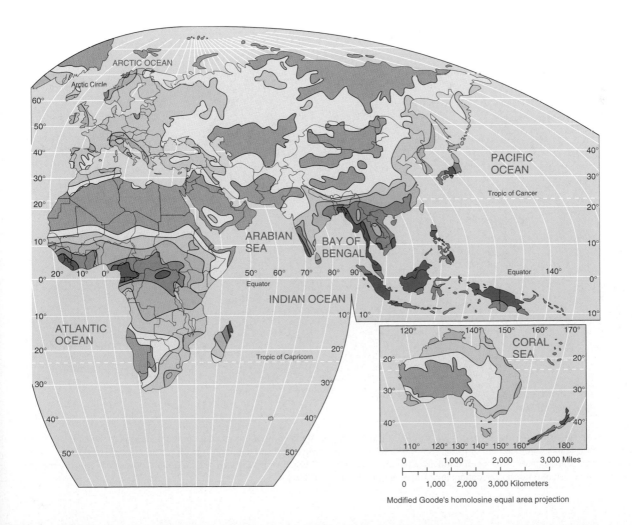

Modified Goode's homolosine equal area projection

of porous material through which groundwater moves are called **aquifers**. It is often difficult to determine the location of an aquifer. Many times, layers of porous rock are found between layers of impervious material, and the entire formation may be folded or fractured in various ways. Thus, groundwater in aquifers may be found at various depths between layers of impervious rock. Also, the **recharge area**—the area where water enters an aquifer—may be many miles away from where the water leaves the aquifer.

As water percolates through the soil, debris and bacteria from the surface are generally filtered out. However, water may dissolve and leach certain minerals. Underground caverns, for example, are the result of the gradual leaching away of limestone (calcium carbonate). In most natural situations, the minerals that leach into groundwater are not harmful. Indeed, calcium from limestone is considered beneficial to health. Thus, groundwater is generally high-quality fresh water that is safe for drinking. A few exceptions occur in which the groundwater leaches minerals containing sulfide, arsenic, or other poisonous elements that make the water unsafe to drink.

Drawn by gravity, groundwater may move through aquifers until it finds some opening to the surface. We observe such natural exits as springs or seeps. In a **seep**, water flows out over a relatively wide area; in a **spring**, water exits the ground as a significant flow from a relatively small opening. Since springs and seeps feed streams, lakes, and rivers, groundwater joins and becomes part of surface water. A

spring will flow, however, only if it is lower than the water table. Whenever the water table drops below the level of the spring, the spring will dry up.

Summary of the Hydrologic Cycle

The hydrologic cycle consists of four physical processes: evaporation, condensation, precipitation, and gravitational flow. There are three principal "loops" in the cycle: (1) In the *evapotranspiration loop* (consisting of *green water*), the water evaporates and is returned by precipitation; on land, this water, the main source for natural ecosystems and rain-fed agriculture, is held as capillary water and then returns to the atmosphere by way of evapotranspiration. (2) In the *surface runoff loop* (containing *blue water*), the water runs across the ground surface and becomes part of the surface water system. (3) In the *groundwater loop* (also containing blue water), the water infiltrates, percolates down to join the groundwater, and then moves through aquifers, finally exiting through springs, seeps, or wells, where it rejoins the surface water. The last two loops are the usual focus for human water resource management. In the hydrologic cycle, there are substantial exchanges of water between the land, the atmosphere, and the oceans. Estimates of these exchanges provide a quantitative overview of the cycle (Fig. 9–8) and establish a basis for calculating the human appropriation of water therein.

Terms commonly used to describe water are given in Table 9–1.

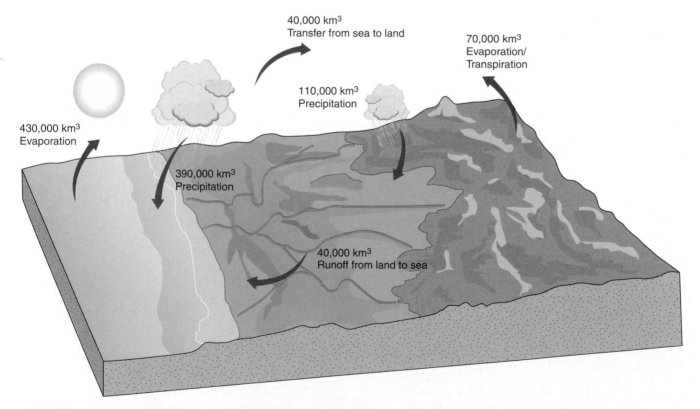

▲ **FIGURE 9–8   *Water balance in the hydrologic cycle.***   The data show (1) the contribution of water from the oceans to the land via evaporation and then precipitation, (2) movement of water from the land to the oceans via runoff and seepage, and (3) the net balance of water movement between terrestrial and oceanic regions of Earth. (Figure 4.2 from EARTH SCIENCE 8e by Tarbuck and Lutgens. Copyright © 1997 by the authors. Reprinted by permission of Pearson Education, Inc., Upper Saddle River, NJ 07458.)

TABLE 9–1	Terms Commonly Used to Describe Water
Term	**Definition**
Water quantity	The amount of water available to meet desired demands.
Water quality	The degree to which water is pure enough to fulfill the requirements of various uses.
Fresh water	Water having a salt concentration below 0.1%. As a result of purification by evaporation, all forms of precipitation are fresh water, as are lakes, rivers, groundwater, and other bodies of water that have a throughflow of water from precipitation.
Salt water	Water, typical of oceans and seas, that contains at least 3% salt (30 parts salt per 1,000 parts water).
Brackish water	A mixture of fresh and salt water, typically found where rivers enter the ocean.
Hard water	Water that contains minerals, especially calcium or magnesium, that cause soap to precipitate, producing a scum, curd, or scale in boilers.
Soft water	Water that is relatively free of minerals.
Polluted water	Water that contains one or more impurities, making the water unsuitable for a desired use.
Purified water	Water that has had pollutants removed or is rendered harmless.
Storm water	Water from precipitation that runs off of land surfaces in surges.
Green water	Water in the soil and in organisms that eventually ends up as water vapor—the main source of water for natural ecosystems and rain-fed agriculture.
Blue water	Renewable surface water runoff and groundwater recharge—the focus of management and the main source of water for human withdrawals.

9.3 Human Impacts on the Hydrologic Cycle

A large share of the environmental problems we face today stem from direct or indirect impacts on the water cycle. These impacts can be classified into three categories: (1) changing Earth's surface, (2) pollution, and (3) withdrawals for human use.

Changing the Surface of Earth

In earlier chapters, we were concerned by the direct loss of forests and other ecosystems to various human enterprises, a loss that diminishes the goods and services these systems have been providing. However, the effects of human-caused changes on the water cycle may be even more profound. In most natural ecosystems, there is relatively little runoff; rather, precipitation is intercepted by vegetation and infiltrates into porous topsoil, as we have seen in Chapter 8. From there, water provides the lifeblood of the natural and human-created ecosystems that sustain us with essential goods and services. The evapotranspiration that takes place in these ecosystems—the green water flow—not only sustains the ecosystems, but also recycles the water to local rainfall. Some of the water that infiltrates percolates down to recharge the groundwater reservoir. Then its gradual release through springs and seeps maintains the flow of streams and rivers at relatively uniform rates. The reservoir of groundwater may be sufficient to maintain a flow even during a prolonged drought. In addition, dirt, detritus, and microorganisms are filtered out as the water percolates through soil and porous rock, resulting in groundwater that is drinkable in most cases.

Similarly, streams and rivers fed by springs contain high-quality water.

As forests are cleared or land is overgrazed, the pathway of the water cycle is shifted from infiltration and groundwater recharge to runoff, and the water runs into streams or rivers almost immediately. This sudden influx of water into waterways not only is likely to cause a flood, but also brings along sediments and other pollutants from surface erosion.

Of course, floods are not unknown in nature. However, in many parts of the world, the frequency and severity of flooding are increasing—not because precipitation is greater, but because both deforestation and cultivation cause erosion and reduce infiltration. For example, extreme flooding in Bangladesh (a very flat country only a few feet above sea level) is now common because Himalayan foothills in India and Nepal have been deforested. Due to sediment deposited from upriver erosion, the Ganges river basin has risen 15–22 feet (5–7 m) in recent years! The 1998 flood inundated 60% of the country and caused an estimated 918 deaths (Fig. 9–9).

Increased runoff necessarily means less infiltration and therefore less evapotranspiration and groundwater recharge. Lowered evapotranspiration means less moisture for local rainfall. Groundwater may not be sufficient to keep springs flowing during dry periods. Dry, barren, and lifeless stream beds are typical of deforested regions—a tragedy for both the ecosystems and the humans dependent on the flow. Wetlands function to store and release water in a manner similar to the groundwater reservoir. Therefore, the destruction of wetlands has the same impact as deforestation: Flooding is exacerbated, and waterways are polluted during wet periods and dry up during droughts.

Urban and suburban development provides an extreme case of altering Earth's surface, replacing porous soil with

▲ FIGURE 9–9 *Flooding in Bangladesh.*   A flooded street in Dhaka, the capital of Bangladesh. Floods in 1998 in India and Bangladesh claimed more than 900 lives and displaced millions of people.

asphalt. This problem and remedial measures will be discussed later in the chapter.

Polluting the Water Cycle

Clearly, the water cycle involves the entire biosphere. Therefore, wherever wastes are put, they are inevitably introduced into the water cycle. Any smoke or fumes exhausted or evaporated into the air will come back down as contaminated precipitation. Acid rain (see Chapter 22) is a case in point. Chemicals that we use on the soil, such as fertilizers, pesticides, and road salt, may either leach into groundwater or be carried into streams by runoff. The same is true of any oil, grease, or other materials we drop or spill on the ground. Any waste we bury in the ground (landfills) may eventually leach into groundwater (Fig. 9–10). (Note that modern landfills are constructed so as to minimize this problem; see Chapter 19.) Of course, all the water we use in the course of washing or flushing away wastes directly adds the pollutants into surface waters, unless the water is treated (Chapter 18).

Withdrawing Water Supplies

Finally are the many problems centering around withdrawals of water for human use, not the least of which is having insufficient water to sustain human needs. We shall expand on this issue in the sections that follow.

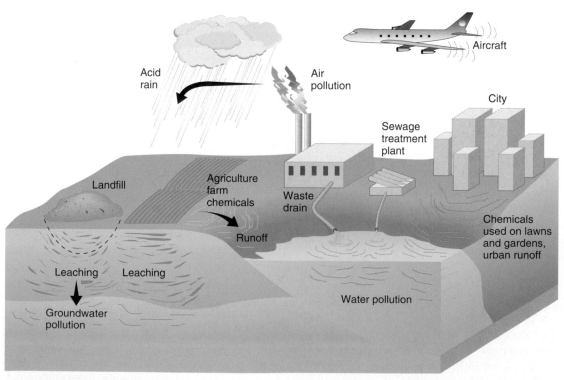

▲ FIGURE 9–10 *Pollution of the hydrologic cycle.*   Human activities introduce pollution into the water cycle at numerous points as shown.

9.4 Sources and Uses of Fresh Water

Human concerns regarding water can be divided into two categories: *quantitative* and *qualitative*. **Quantitative** refers to issues such as whether we will have enough water to meet our needs and what the impacts are of diverting water from one point of the cycle to another. **Qualitative** refers to issues such as whether the water will be of sufficient purity so as not to harm human or environmental health. In the remainder of the chapter, we shall focus primarily on the quantitative aspects, although you should bear in mind that quality is an ever-present concern. In Chapter 18, we shall focus on the qualitative aspects—that is, pollution.

The major uses of fresh water are given in Table 9–2. Most of the water used in homes and industries is for washing and flushing away undesired materials, and the water used in electric power production is for taking away waste heat and increasing the efficiency of the process. Such uses are termed **nonconsumptive** because the water, though now contaminated with the wastes, remains available to humans for the same or other uses if its quality is adequate or if it can be treated to remove undesired materials. In contrast, irrigation is a **consumptive** use, because the applied water does not return to the water resource. It can only percolate into the ground or return to the atmosphere through evapotranspiration. Of course, in either case, the water does reenter the overall water cycle, but is gone from human control.

Worldwide, the largest use of water is for irrigation (70%); second is for industry (20%) and third for direct human consumption (10%). These percentages vary greatly from one region to another, depending on natural precipitation and the degree of development of the region (Fig. 9–11).

Humans take fresh water from whatever source they can. In some cases, this means capturing precipitation directly in a rain barrel under a downspout. The major

TABLE 9–2 U.S. Demands on Fresh Water	
Use	Gallons (liters) used per person per day
Consumptive	
Irrigation and other agricultural use	700 (2,800)
Nonconsumptive	
Electric power production	600 (2,400)
Industrial use	370 (1,500)
Residential use	100 (400)

sources of fresh water, however, are surface waters (rivers and lakes) and groundwater. In the United States, about half of domestic water comes from each of these sources. Before municipal water supplies were made available, each family dipped into a local stream or shallow well for its own use. This method is still employed to a considerable extent, and women in many developing countries walk long distances each day to fetch water. Because surface waters and shallow wells often receive runoff, they frequently are polluted with various wastes, including animal excrement and human sewage likely to contain pathogens (disease-causing organisms). Yet, unsafe as it is, this polluted water is the only water available for an estimated 1.4 billion poor people in less developed countries (Fig. 9–12). It is commonly consumed without treatment, but not without consequences: According to the World Health Organization's (WHO's) *World Health Report 1999*, contaminated water is responsible for 80% of the diseases in the developing world and for the deaths of 3–4 million people per year, including over 2 million young children who die from simple diarrhea.

In the developed countries, also, the major freshwater sources are rivers and lakes, but methods for collection,

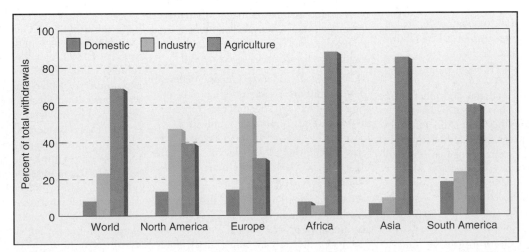

▲ **FIGURE 9–11** *Human usage of water.* The percentage used in each category varies with climate and relative development of the country. A less developed region with a dry climate (e.g., Africa) uses most of its water for irrigation, whereas an industrialized region (e.g., Europe) requires the largest percentage for industry.

▲ **FIGURE 9–12** *Water in the developing world.* In many villages and cities of the developing world, people withdraw water from rivers, streams, and ponds. Such sources are often contaminated with pathogens and other pollutants. This woman in Burkina Faso is collecting water from a pond used by people and animals.

treatment, and distribution are more sophisticated. Dams are built across rivers to create reservoirs, which hold water in times of excess flow and can be drawn down at times of lower flow. In addition, dams and reservoirs may provide for power generation, recreation, and flood control. Water for municipal use is piped from the reservoir to a treatment plant, where it is treated to kill pathogens and remove undesirable materials, as shown in Fig. 9–13. After treatment, the water is distributed through the water system to homes, schools, and industries. Wastewater, collected by the sewage system, is carried to a sewage-treatment plant, where it is treated before being discharged into a natural waterway. (See Chapter 18.) Often, wastewater is discharged into the same river from which it was withdrawn, but farther downstream.

Whenever possible, both water and sewage systems are laid out so that gravity maintains the flow through the system. This arrangement minimizes pumping costs and increases reliability. On major rivers, such as the Mississippi, water is reused many times. Each city along the river takes water, treats it, uses it, and then returns it to the river. In developing nations, the wastewater frequently is discharged with minimal or no treatment. Thus, as the water moves downstream, each city has a higher load of pollutants to

EARTH WATCH

WATER PURIFICATION

Polluted water is defined as water that contains one or more materials that make it unsuitable for a given use. Water purification is any method that will remove one or more of the materials. Several methods may be used in combination to obtain water that is sufficiently pure for a given use. The following are methods that are commonly used in water purification:

1. *Settling.* Soil particles and other solid material carried by flowing water may be removed by holding the water still and allowing the solids to settle. Clarified water is removed from the top. Settling may be aided by the addition of alum (aluminum sulfate). The +3 charge on aluminum ions pulls clay and other particles, which are negatively charged, into clumps that settle more readily than do the individual particles.

2. *Filtration.* Filtration is the passage of water through a porous medium. Any materials larger than the pores will be filtered out. A bed of sand is often used for this purpose.

3. *Adsorption.* Passing water through an adsorbing material that binds and holds other materials on its surface will remove certain pollutants. Activated carbon is a material commonly used in this way to remove organic contaminants from water or air.

4. *Biological oxidation.* Organic material (detritus and organisms) is fed upon by detritus feeders and decomposers, broken down in cell respiration, and thus removed. The passage of water through systems supporting the growth of such organisms removes organic material. (See Chapter 18.)

5. *Distillation.* Distillation is the evaporation and condensation of water. All materials present in the water before the evaporation step remain behind in the holding tank and are therefore not present when the water vapor is condensed.

6. *Disinfection.* Water is treated with chlorine or other agents that kill disease-causing organisms.

The natural water cycle includes all of these purification methods except disinfection. Sitting in lakes, ponds, or the oceans, water is subject to settling. As it percolates through soil or porous rock, it is filtered. Soil and humus are good chemical adsorbents. As water flows down streams and rivers, detritus is removed by biological oxidation. As water evaporates and condenses, it is distilled.

Thus, numerous sources of fresh water might be safe to drink were it not for human pollution. The most serious threat to human health is contamination with disease-causing organisms and parasites, which come from the excrement of humans and their domestic animals. In human settlements, you can see how these organisms get into water and are passed on to people before any of the natural purification processes begin working.

contend with than the previous city had, and ecosystems at the end of the line may be severely affected by the pollution. Pollutants include industrial wastes as well as pollutants from households, since industries and residences generally utilize the same water and sewer systems.

Reservoirs created by dams on rivers are also major sources of water for irrigation. In this case, no treatment is required. As noted earlier, croplands are the end point for this water, except insofar as it reenters the water cycle through evaporation and infiltration into groundwater.

Both to augment surface water supplies and to obtain water of higher quality, in the past few decades there has been an increasing trend of drilling wells and tapping groundwater. Since 1950, hundreds of cities have drilled huge wells for municipal supplies, and millions of wells have been drilled for individual households in suburbs

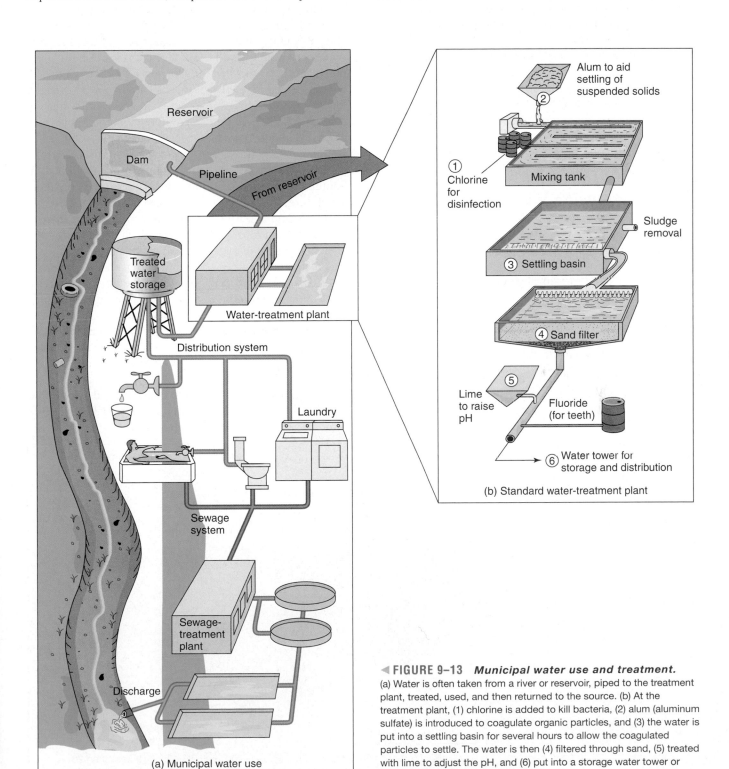

◀ **FIGURE 9–13** *Municipal water use and treatment.*
(a) Water is often taken from a river or reservoir, piped to the treatment plant, treated, used, and then returned to the source. (b) At the treatment plant, (1) chlorine is added to kill bacteria, (2) alum (aluminum sulfate) is introduced to coagulate organic particles, and (3) the water is put into a settling basin for several hours to allow the coagulated particles to settle. The water is then (4) filtered through sand, (5) treated with lime to adjust the pH, and (6) put into a storage water tower or reservoir for distribution to your home.

beyond municipal supply systems. In addition, farmers have turned to the center-pivot irrigation system. (See Fig. 9–16c.) The use of such wells has risen tremendously in the last 25 years, increasing agricultural production, as noted in Chapter 10, but also consuming huge amounts of groundwater. One system may use as much as 10,000 gallons (40,000 liters) per minute. From your understanding of the water cycle, you can see that both surface water and groundwater are replenished through the cycle. Therefore, in theory at least, such water represents a sustainable or renewable (self-replenishing) resource, but it is not inexhaustible. If humans attempt to extract amounts that exceed those of natural flow, shortages will occur. Furthermore, as water is diverted from its natural pathway for human use, ecological consequences ensue. Using and diverting water in volumes that lead to shortages or other undesirable consequences are spoken of as *overdrawing*, or *overdraft of*, water resources. Let's turn our attention now to those consequences.

9.5 Overdrawing Water Resources

Because the consequences of overdrawing surface waters differ somewhat from those of overdrawing groundwater, we shall consider these two categories separately. However, because all water is tied together in the same overall cycle, you should be aware of the similarities and interconnections between the two discussions.

Consequences of Overdrawing Surface Waters

Inevitable Shortages.   There are wet years and dry years, and surface-water flows vary accordingly. On the average of once every 20 years, surface-water flow may drop to only 30% of its annual average. Therefore, the rule of thumb is that no more than 30% of a river's average flow can be taken out each year without risking a shortfall every 20 years. This rule has not always been heeded, however. In some river systems in the United States, for instance, water demand has grown to 100% (and even *more than* 100%!) of average flow, making water shortages of increasing length and severity inevitable (Fig. 9–14). Southern California is a case in point. Because of a seven-year drought, much of that part of the state entered the spring of 1992 with reservoirs down to 20%–40% of capacity. If the drought had continued another year, there would have been a severe crisis, but fortunately, rains came and mitigated the situation—a narrow escape. Of course, people blamed the water shortage on the drought, but was the drought the real cause? Southern California has a desert climate, and prolonged droughts are not abnormal. Their likelihood must be taken into account in long-term planning for sustainability. Currently, 85% of the water supply goes to agriculture for irrigation. In all likelihood, as Southern California continues to increase in population, agriculture will gradually be squeezed out of the region.

The Colorado River has been providing a large share of Southern California's water supply; in fact, California has been drawing more than its allotment for a number of years, and recently the other states bordering on the Colorado River or through which the river flows have been pressuring the federal government to limit California's use. The Colorado is so heavily used that it sometimes fails to reach the Gulf of California. Around the world, in numerous instances tensions are mounting as different nations share common rivers and, hence, compete for the same water.

Ecological Effects.   When a river is dammed and its flow is diverted to cities or croplands, the waterway below the di-

▶ **FIGURE 9–14 *Water shortage in the United States.*** Droughts occur on an average of every 20 years and may reduce normal water flows by 70%. Therefore, no more than 30% of the average surface-water flow can be counted on to be continuously available. By the year 2000, large areas of the United States were above the 30% level, making severe, recurring water shortages inevitable.

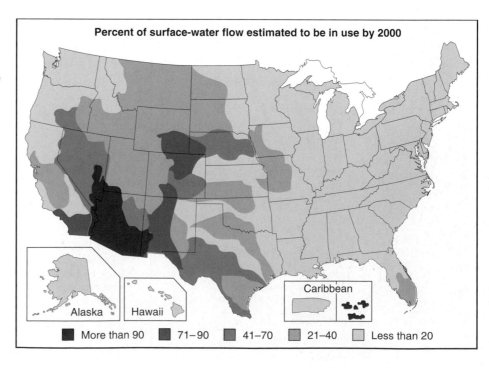

Percent of surface-water flow estimated to be in use by 2000

Alaska   Hawaii   Caribbean

■ More than 90   ■ 71–90   ■ 41–70   ■ 21–40   □ Less than 20

version is deprived of that much water. The impact on fish and other aquatic organisms is obvious, but the ecological ramifications go far beyond the river. Wildlife that depends on the water or on food chains involving aquatic organisms—that is to say, virtually all wildlife—are also adversely affected. Wetlands along many rivers, no longer nourished by occasional overflows, dry up, resulting in a tremendous die-off of waterfowl and other wildlife that depended on those habitats (Fig. 9–15). Fish such as salmon, which swim from the ocean far upriver to spawn, are seriously affected by the reduced water level and have problems getting around the dam, even one equipped with fish ladders (a stepwise series of pools on the side of a dam, where the water flows in small falls that fish can negotiate). If the fish do get upriver, the hatchlings have similar problems getting back to the ocean.

The problems extend to **estuaries**, bays in which fresh water from a river mixes with seawater. Estuaries are among the most productive ecosystems on Earth; they are rich breeding grounds for many species of fish, shellfish, and waterfowl. As a river's flow is diverted to other locations, less fresh water enters and flushes the estuary. Consequently, the salt concentration increases, profoundly upsetting the estuary's ecology. The San Francisco Bay is a prime example. Over 60% of the fresh water that once flowed from rivers into the bay has been diverted for irrigation in the Central Valley and for municipal use in Southern California. Without the freshwater flows, salt water from the Pacific has intruded into the bay, with devastating consequences: Chinook salmon have become all but nonexistent; sturgeon, Dungeness crab, and striped bass populations are greatly reduced; and water quality for drinking and irrigation in the region is at risk. New standards will require fresh-

water flows to be restored to the delta (the convergence of rivers leading into the bay), but central and Southern California will have to give up some water.

The problem is not limited to the United States. The southeastern end of the Mediterranean Sea was formerly flushed by water from the Nile River. Because this water is now held back and diverted for irrigation by the Aswan High Dam in Egypt, that part of the Mediterranean is suffering ecological consequences.

The world's most dramatic example of water mismanagement is the Aral Sea, an inland sea in south-central Russia. (See the "Global Perspective" essay, p. 228.)

Consequences of Overdrawing Groundwater

To augment supplies of high-quality fresh water, humans have increasingly turned to groundwater; indeed, groundwater is the largest reservoir of fresh water now available to us. In tapping groundwater, we are tapping a large, but not unlimited, natural reservoir. Its sustainability ultimately depends on balancing withdrawals with rates of recharge.

In some dry regions, the groundwater found is actually water that accumulated millennia ago, when the climate in the region was wetter. Current rates of recharge are nil. The practice of tapping such pockets of ancient water is frequently spoken of as "mining fossil water," to emphasize that the resource will ultimately be depleted regardless of rates of withdrawal.

Falling Water Tables and Depletion. Rates of groundwater recharge aside, the simplest indication that groundwater withdrawals are exceeding recharge is a falling water table, a situation that is common throughout the world.

(a)

(b)

▲ **FIGURE 9–15** *Ecological effects of diversion* (a) Ducks and geese at Klamath, Oregon. Vast flocks of such waterfowl are supported by wetlands. (b) What happens to these birds when wetlands dry up because of water diversion or a falling water table?

GLOBAL PERSPECTIVE

THE DEATH OF THE ARAL SEA

In the 1930s, economic planners sitting in thick-walled stone buildings in Moscow set in motion a chain of events that has almost killed an entire sea. In order to create vast fields of cotton in the drylands of Soviet central Asia, the planners had long irrigation canals dug, fed by the waters of two rivers, the Amu Darya and the Syr Darya, that flow into the inland Aral Sea. In the statistics of the central planners, the project was a huge success. The cotton harvests grew until the Soviet Union became the world's second-largest cotton exporter, after China. But the statistics did not show the effect of diverting most of the river's flow away from the Aral Sea. By 1989, the sea was receiving only one-eighth the amount of water it received in 1960. Its water level had dropped by 47 feet (16 meters), and its volume had shrunk by two-thirds. Once the size of North America's Lake Huron, the Aral Sea had diminished in area by 44%.

The shores of the sea receded, leaving fishing villages tens of miles from the shore. A new desert was created around the sea, with salt strewn in massive dust storms across a vast area. At the same time, Moscow's demands for cotton cultivation were met with the saturation use of pesticides, fertilizers, and herbicides that flowed into rivers and canals. The population around the sea, deprived of clean drinking water and living on poisoned soil, experienced rising rates of disease and infant mortality. "The problem of the Aral Sea is very simple," said Igor Zonn, a specialist on the subject: "The Aral Sea will be dead, not soon and not completely, but it will be dead."

Dr. Zonn and fellow Russian scientist Nikita Glazorsky, head of the Institute of Geology and until recently deputy environment minister of Russia, have been working for a long time to try to save the Aral Sea. The Russian scientists say a technical solution for the sea is already well formulated. Much of the water is now wasted because of evaporation and drainage out of unlined irrigation canals and primitive irrigation technology.

At least half the (120-km^3) flow of the two rivers could be saved by rebuilding the canals and introducing new irrigation systems. That would be enough to begin stabilizing the Aral Sea at its present level, though not enough to restore the waters to the 1960 level. At the same time, there must be a program of health care and a concerted effort to shift the economy away from cotton cultivation, they say.

But such measures require resources and a political will that is not present in today's Russia. The breakup of the Soviet Union and its replacement by a loose commonwealth have given the Russian government an opportunity to dump the problem into the laps of the five former Soviet central Asian states that form the Aral Sea's water basin. This is a prime source of anger for central Asians who see the cotton monoculture imposed by Moscow as a classic example of colonial-style exploitation.

"We are left face to face with the Aral Sea problem," Kazakhstan president Nursultan Nazarbayev said. "Meanwhile, 97% of the cotton of central Asia and Kazakhstan is taken out to the European part of the

Commonwealth of Independent States, where the employment of 10 million workers depends on cotton use."

Russia has an ethical responsibility to help, agrees Dr. Glazorsky. "All of us live in this system, and we must solve this problem together."[1]

(Eight years later, the Aral Sea continues to shrink; it has now lost 80% of its original volume, and the water level is down more than 53 feet. The cultivation of cotton is shrinking rapidly as fields rapidly accumulate salt; the drinking water is contaminated with high salt concentrations. Toxic dust storms continue to plague the health of people in the lake basin, the fishing industry has completely collapsed, and over 100,000 people have moved away from the area. The international community is responding to the area's needs: The World Bank and several other international agencies will spend $600 million through 2002 in efforts to provide pure drinking water and upgrade the irrigation and drainage systems throughout the region. There is no hope for restoring the Aral Sea, however; it would take stopping all irrigation from the two rivers for half a century just to increase the sea's area to twice its present size, and the surrounding country's economies would be completely destroyed. Most likely, the Aral Sea will break up into several smaller lakes, one or two of which might regain a fishery if the river is allowed to fill them directly.)

[1] By Daniel Sneider. Reprinted by permission from *The Christian Science Monitor*; ©1992 The Christian Science Publishing Society. All rights reserved.

Since irrigation consumes far and away the largest amount of fresh water, depleting water resources will ultimately have its most significant impact on crop production. A prime example is the Great Plains region of Texas, Oklahoma, New Mexico, Colorado, Kansas, Wyoming, and Nebraska. Within this arid region, the Ogallala aquifer supplies irrigation water to 10.4 million acres (4.2 million hectares), one-fifth of the irrigated land in the United States (Fig. 9–16). However, the withdrawal rate has been about 23 million acre-feet (28 billion cubic meters) per year, two orders of magnitude higher than the recharge rate. Water tables have dropped 100–200 ft (30–60 meters) and are lowering at 6 ft per year. Irrigated farming has already come to

a halt in some sections, and it is predicted that over the next 20 years another 3 million acres (1.2 million hectares) in this region will be abandoned or converted to dryland farming (ranching and production of forage crops) because of water depletion.

Although running out of water is the obvious eventual conclusion of overdrawing groundwater, falling water tables have other consequences. Let us now examine some of them.

Diminishing Surface Water.   Surface waters are also affected by falling water tables. In various wetlands, for instance, the water table is essentially at or slightly above the ground surface. Dropping water tables result in such wet-

(a)

(b)

(c)

▶ **FIGURE 9–16** *Exploitation of an aquifer.*
(a) Pumping up water from the Ogallala aquifer has made this arid region of the United States into some of the most productive farmland in the country. (b) Water is applied by means of center-pivot irrigation, in which the water is pumped from a central well to a self-powered boom that rotates around the well, spraying water as it goes. (c) Aerial photograph shows the extent of center-pivot irrigation throughout the region. Groundwater depletion will bring an end to this kind of farming.

lands drying up, with the ecological results described earlier. Further, as water tables drop, springs and seeps dry up, diminishing even streams and rivers to the point of dryness. Thus, excessive groundwater removal leads to the same effects as diversion of surface water.

Land Subsidence. Over the ages, groundwater has leached cavities in Earth. Where these spaces are filled with water, the water helps support the overlying rock and soil, but as the water table drops, this support is lost. Then there may be a gradual settling of the land, a phenomenon known as **land subsidence**. The rate of sinking may be 6–12 inches (10–15 cm) per year. In some areas of the San Joaquin Valley in California, land has settled as much as 29 feet (9 m) because of groundwater removal. Land subsidence causes building foundations, roadways, and water and sewer lines to crack. In coastal areas, subsidence causes flooding unless levees are built for protection. For ex-

ample, where a 4,000-square-mile (10,000-km²) area in the Houston–Galveston Bay region of Texas is gradually sinking because of groundwater removal, coastal properties are being abandoned as they are gradually being inundated by the sea. Land subsidence is also a serious problem in New Orleans, sections of Arizona, Mexico City, and many other places throughout the world.

Another kind of land subsidence, a **sinkhole**, may develop suddenly and dramatically (Fig. 9–17). A sinkhole results when an underground cavern, drained of its supporting groundwater, suddenly collapses. Sinkholes may be at least 300 feet (91 m) across and as much as 150 feet deep. The formation of sinkholes is particularly severe in the southeastern United States, where groundwater has leached numerous passageways and caverns through ancient beds of underlying limestone. An estimated 4,000 sinkholes have formed in Alabama alone, some of which have "consumed" buildings, livestock, and sections of highways.

▲ FIGURE 9–17 **Sinkhole.** The removal of groundwater may drain an underground cavern until the roof, no longer supported by water pressure, collapses. The result is the sudden development of a sinkhole, such as this one, which consumed a home in Frostproof, Florida, July 12, 1991.

Saltwater Intrusion. Another problem resulting from dropping water tables is **saltwater intrusion**. In coastal regions, springs of outflowing groundwater may lie under the ocean. As long as a high water table maintains a sufficient head of pressure in the aquifer, fresh water will flow into the ocean. Thus, wells near the ocean yield fresh water (Fig. 9–18a). However, a lowering of the water table or a rapid rate of groundwater removal may reduce the pressure in the aquifer, permitting salt water to flow back into the aquifer and hence into wells (Fig. 9–18b). Saltwater intrusion is a problem at many locations along U.S. coasts.

9.6 Obtaining More Water

Despite the obvious and growing negative impacts of overdrawing water resources, expanding populations create an ever-increasing demand for additional water for both irrigation and municipal use. In the United States, more than 75,000 dams tame wild rivers, provide irrigation water, and harness energy. The Snake River, once home to great salmon runs, now courses from dam to dam—four have been built—and as a consequence, the salmon have all but disappeared.

▶ FIGURE 9–18 **Saltwater intrusion.** (a) Where aquifers open into the ocean, fresh water is maintained in the aquifer by the head of fresh water inland. (b) Excessive removal of water may reduce the pressure, so that salt water moves into the aquifer.

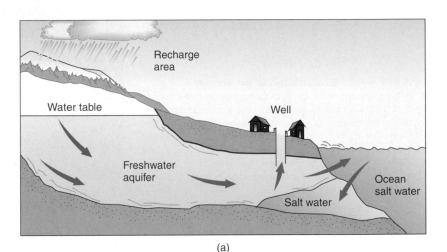

(a)

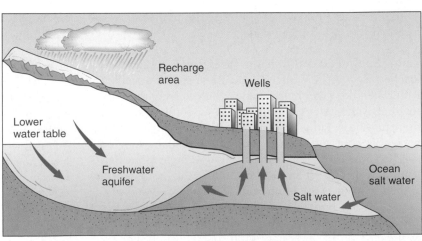

(b)

Increasingly, however, people are recognizing the inevitable trade-offs that occur with such projects and are considering them to be unacceptable. Intense counterlobbying by various environmental organizations has, in recent years, led to the rejection of a number of proposals to erect dams. And even existing dams are being challenged in several states; some 465 have already been dismantled, with others contemplated for the same fate. Although the Snake River dams provide power for 300,000 homes and irrigate 35,000 acres of farmland, serious consideration has been given to breaching the dams in order to restore the salmon runs and protect other endangered fish species.

Protection has been accorded to some rivers with the passage of the Wild and Scenic Rivers Act of 1968, which protects rivers designated as "wild and scenic" from damming and other harmful operations. Some 10,900 miles of rivers have been protected, but more than 60,000 miles qualify for protection under the act. These rivers designated as wild and scenic are in many ways equivalent to national parks, and thus, like national parks, they need pubic supporters and defenders, such as the organization American Rivers. The Clinton administration found other ways to protect rivers; recently, a 51-mile undammed stretch of the Columbia River, the Hanford Reach, was protected as a national monument, countering pressures to dam and dredge the river.

Bitter controversies between environmentalists and dam-building interests continue and are not limited to the industrialized world. An example currently unfolding is the construction of the Three Gorges Dam across a scenic stretch of the Yangtze River in China. The project is the centerpiece of the Chinese government's effort to industrialize and join the modern age. When completed in 2009, the dam will be the largest in the world, generating 18,000 MW of electricity; it is also expected to provide control over disastrous flooding of a river that has taken over 300,000 lives in the 20th century. More than 1.2 million people—including entire cities, farms, homes, and factories—will be displaced and relocated to make way for the 370-mile-long (600 km) reservoir. Critics point to the enormous human, ecological, and aesthetic costs of the dam and claim that alternative sources of electric power are cheaper and more readily available. An international campaign to stop the dam has prevented many funding agencies, like the World Bank, from getting involved, but Chinese officials have effectively stifled internal criticism and are proceeding with work on the dam.

A lesson might be taken from the Aswan High Dam, which was constructed across the Nile River in the 1970s to bring Egypt into the modern industrial age. The loss of fisheries and productive wetlands below the dam, the loss of land flooded by the reservoir above, and the unbridled population growth that ensued have largely canceled out any gains: The people in general are as bad off as they were before the dam was built.

In conclusion, it does not appear that a sustainable future can be found in large-scale water exploitation and diversion projects. Fortunately, there are alternatives, based on the conservation and recycling of water.

9.7 Using Less Water

In the past, water has been treated as an inexhaustible resource that can be taken for granted. This viewpoint has led to the extravagant and wasteful use of water. A developing-nation family living where water must be carried several miles from a well finds that one gallon per day per person is sufficient to provide for all essential needs, including cooking and washing. Yet, a typical household in America consumes an average of 106 gallons (400 L) per day per person. If all indirect uses are added (including irrigation), this increases to a per capita use of 1,350 gallons (5,100 L). Similarly, a peasant farmer may irrigate by carefully ladling water onto each plant with a dipper, while typical modern irrigation floods the whole field. This is not to say that we in the developed world should take up developing-world practices, but it does suggest that water consumption can be reduced without people suffering hardship. It is through such cutbacks that we have an opportunity to meet our needs without undercutting needs of future generations or natural ecosystems—the essence of sustainable development.

Let's consider some specific measures that are being implemented to reduce water demands.

Irrigation

Where irrigation water is applied by traditional flood or center-pivot systems, 30%–50% is wasted in evaporation, percolation, or runoff. Several strategies have been employed recently to cut down this waste. One is the *surge flow* method, in which microprocessor control allows the periodic release of water, in contrast to the continuous-flood method; the practice can cut water use in half. Another is the *drip irrigation* system, a network of plastic pipes with pinholes that literally drip water at the base of each plant (Fig. 9–19). With this system, less than 5% of the water is wasted, although such systems are costly. They have the added benefit of retarding salinization. (See Chapter 8.) Although drip irrigation is spreading worldwide, especially in arid lands like Israel and Australia, 97% of the irrigation in the United States and 99% throughout the world is still done by traditional flood or center-pivot methods.

The reason for the low changeover is the cost to farmers, about $1,000 per acre to install a drip system. In comparison, water for irrigation is heavily subsidized by the government; the farmer pays next to nothing for it. Therefore, it makes financial sense to use the cheapest system for distributing water, even if it is wasteful. Calculations of construction costs and energy subsidies to provide irrigation water to farmers indicate an annual subsidy of $4.4 billion for the 11 million acres of irrigated land in the western United States, an average of $400 an acre. Reducing this subsidy can greatly encourage water conservation through

▲ **FIGURE 9–19**   *Drip irrigation.*   Irrigation consumes the most water. Drip irrigation offers a conservative method of applying water.

the use of more efficient irrigation technologies. For example, the Broadview Water District in California has instituted a tiered pricing structure according to which farmers are charged their customary fee for the first 90% of their average water use and much higher prices for the remaining 10%. Given this incentive, farmers have conserved from 9% to 31% on water use, depending on the crop, and crop yields were not affected at all.

Municipal Systems

The water consumption of 106 gallons (400 liters) per day per person in modern homes is used mostly for washing and flushing away wastes: flushing toilets (3–5 gallons per flush), taking showers (2–3 gallons per minute), doing laundry (20–30 gallons per wash), and so on. Watering lawns, filling swimming pools, and other indirect consumption add to this use.

Water conservation has long been promoted as a "save the environment" measure, though without much effect. Now numerous cities are facing the stark reality that it will be extremely expensive and in many cases impossible to increase supplies by the traditional means of building more reservoirs or drilling more wells. The only practical alternative, they are discovering, is to take real steps toward reducing water consumption and wastage. A considerable number of cities have programs whereby leaky faucets will be repaired and low-flow shower heads and water-displacement devices in toilets will be installed free of charge. Phoenix is paying homeowners to replace their lawns with **xeriscaping**—landscaping with desert species that require no additional watering—and the business is thriving. Many cities and towns ban certain uses of water when droughts reduce the available supply. Brown lawns

and dusty automobiles are the result, signs that a city's water supply is threatened.

In 1997, the last phase of a regulation authorized by the 1992 National Energy Act took effect, and it became illegal to sell 3.5-gallon commodes. In their place is the new wonder of the flush world—the 1.6-gallon toilet (Fig. 9–20). When the new toilets first came into use, homeowners found that they could no longer depend on an easy flush; in fact, plumbers would usually offer a free plunger with installation of the early 1.6-gallon models. Newer versions, however, work perfectly well, and save 10 gallons or more a day per person. The 50 million low-flow toilets now in place in the United States save an estimated 600 million gallons of water a day. New York City is providing significant rebates to people who replace their old toilets with the new models, and Los Angeles is offering the low-flow toilets free as part of its efforts to restore Mono Lake.

Also, *gray-water* recycling systems are being adopted in some water-short areas. **Gray water**, the slightly dirtied water from sinks, showers, bathtubs, and laundry tubs, is collected in a holding tank and used for such things as flushing toilets, watering lawns, and washing cars. Going further, a number of cities are using treated wastewater (sewage water) for irrigation, both to conserve water and to abate the pollution of receiving waters. (See Chapter 18.) If the idea of reusing sewage water turns you off, recall that *all* water is recycled by nature. There is hardly a molecule of water you drink that has not moved through organisms—including humans—numerous times. A number of communities are already treating their wastewater to such a degree that its quality surpasses what many cities take in from lakes and rivers.

▲ **FIGURE 9–20**   *The 1.6-gallon commode.*   Now required for all new installations, this device saves 10 gallons of water or more a day per person.

9.8 Desalting Seawater

The world's oceans are an inexhaustible source of water, not only because they are vast, but because any water removed from them will ultimately flow back in. Traditionally, seawater has not been used, since it is costly to remove the salt and pump it uphill to where it will be used and adequate sources of natural fresh water are available. However, with increasing water shortages and most of the world's population living near coasts, there is a growing trend toward **desalination** (desalting) of seawater. More than 7,500 desalination plants already exist, primarily in Saudi Arabia, Israel, and other countries of the Middle East.

Two technologies are commonly used for desalination: microfiltration, or reverse osmosis, and distillation. Small desalination plants generally employ microfiltration, in which great pressure forces seawater through a membrane filter fine enough to remove the salt. Large plants, particularly those which can draw from a source of waste heat (e.g., from electrical power plants), generally use distillation (evaporation and recondensation of vapor). Efficiency is gained by using the heat given off by condensing water to heat the incoming water. Even where waste heat is used, however, the costs of building and maintaining the plant, which is subject to corrosion from seawater, are considerable. Under the best of circumstances, the production of desalinized water costs about $3 per 1,000 gallons (4,000 L). This is three to six times what most city dwellers in the United States currently pay, but it is still not a high price to pay for drinking water. Still, the cost might cause some people to cut back on watering lawns and to implement other conservation measures, but most people in the United States could afford desalinized water without undue strain.

Irrigation, however, commonly consumes 500,000 gallons (1.9 million L) or more per acre to produce one crop. Farmers currently pay as little as 2 cents per 1,000 gallons ($10 per acre of crop) for water. Thus, it is cost effective to irrigate crops that bring in only a few hundred dollars per acre. However, paying $3 per 1,000 gallons of desalinized water would raise the cost to $1,500 per acre of crop. Then there would be the additional pumping costs to get the water from the coast to farmlands. As these costs are added up, to say nothing of future constraints on energy use (see Chapter 13), a future of irrigating croplands with desalinized seawater seems out of the question.

9.9 Storm Water

Earlier, we examined the general aspects of how changing the surface of the soil alters the infiltration–runoff ratio and, in turn, the rest of the hydrologic cycle. This problem is particularly acute wherever development results in exchanging porous, water-receiving topsoil for asphalt pavements and rooftops that shed nearly 100% of the water falling on them. Even the soil in lawns is much more compacted than in a natural ecosystem and sheds a high percentage of rainwater.

On the other side of the infiltration–runoff ratio, decreased infiltration also exacerbates saltwater intrusion, land subsidence, and other problems related to falling water tables. The traditional practice was, and in many cases still is, to channel storm runoff down storm drains, which lead to the nearest convenient off-site location to discharge the water, usually the side of a valley or a natural streambed. The ramifications of this practice are many.

Mismanagement and Its Consequences

The increase in stream flow during rains because of the sudden influx of runoff is well documented, as is the decrease in flow between rains as springs have gone dry because of reduced groundwater recharge. Most streams in urban and suburban areas that used to flow quietly year-round and that supported a diverse biota are now little more than open storm drains alternating between surges of water when it rains and dry, dead beds when it doesn't (Fig. 9–21). Indeed, many such streams have been covered and simply incorporated into the storm-drain system. In urban areas, great numbers of streams and their tributaries no longer exist, except as underground storm drains.

The surges from increasing runoff have a number of other effects as well.

Flooding. The increased potential for flooding is self-evident. Countless communities, many of them expensive, new suburban developments, have experienced flooding with increasing frequency and severity as development has led to paving more and more of the upstream watershed.

Stream-bank Erosion. The erosion resulting from poorly placed storm-drain outlets can be horrendous (Fig. 9–22). The effects continue downstream. Even short of flooding, the surges of water greatly accelerate erosion of stream banks, undercutting trees and causing them to topple into the stream. As a result, water is diverted against and over the banks, causing further erosion. While finer soil particles (clay and silt) are carried on to finally settle out in lakes and bays, coarser materials (sand, stones, and rocks) eroded from gullies below poorly placed storm drains and stream banks are deposited at the bottom of the stream channel itself. This filling of the channel causes the water to eat away even more at the sides, exacerbating the erosion and rendering the stream channel wider and shallower. Indeed, a stream channel may become completely filled, such that the water is diverted and simply floods the valley floor (Fig. 9–23). Then the topsoil erodes and many trees die because of waterlogged soil. Gradually, a narrow, tree-lined stream is converted into a broad wash of sand and gravel.

Increased Pollution. Consider all the materials used, carelessly dropped or spilled on the ground, or, even worse,

(a)

(b)

▲ **FIGURE 9–21** *Effect of development on streamflow.*   Before development, this stream maintained a continuous, generally modest flow of water throughout the year. Now, after development of the upstream watershed, the stream is dry most of the time. (a) During rains, there are high surges of water because of increased runoff. (b) A few hours later the stream is dry again because of the depletion of groundwater.

▲ **FIGURE 9–22** *Erosion gully.*   The gully seen in this photograph is the result of erosion caused by water exiting from a storm drain, just out of the picture in the background, funneling the runoff from a parking lot onto the hillside. The soil and stones eroded from such gullies add sediments to, and clog the channels of, waterways below. (See Fig. 9–23.)

▲ **FIGURE 9–23** *Erosion deposits.*   Sediment from upstream erosion fills and clogs stream channels, causing flooding and further erosion of the stream valley as the water is forced to find new pathways. In this photo, you can see how the stream channel has been nearly filled with fine gravel sediment, which eroded from upstream areas.

disposed of down storm drains. All may wash away with runoff directly into adjacent natural waterways. Indeed, urban runoff is now recognized as a major nonpoint source of pollution for many rivers and estuaries (Chapter 18). The following are the major categories of contaminants:

- Nutrients from lawn and garden fertilizer
- Insecticides and herbicides used on lawns and gardens
- Bacteria from fecal wastes of pets
- Road salt and other chemicals from surface treatments or spills
- Grime and toxic chemicals from settled vehicle exhaust and other air pollution
- Oil and grease picked up from road surfaces or disposed of in storm drains
- Trash and litter carelessly discarded on the ground

How many other things can you think of?

Channelization.   What is to be done about the problems arising from increasing runoff from storms? The old approach to stream-bank erosion and flooding was **channelization**. The streambed was dredged, straightened, or gently curved and lined with rock or concrete to prevent erosion of its banks (Fig. 9–24). To be sure, a channelized stream carries both water and sediment efficiently and reduces flooding and erosion in the area of the channel. However, as you may observe, channeling obliterates any semblance of a natural stream ecosystem. The channelized stream is simply an open storm drain; indeed, it may be covered and made into a storm drain proper, as indicated previously.

The real defeat of channelization, however, came from recognizing that channelized sections simply served to transfer water farther downstream and create flooding there. The flooding on the Mississippi River in 1993 is an extreme example. Millions of acres of upstream tributaries were channelized to drain wetlands that normally would have held the water, and extensive levees were constructed downriver to protect floodplains. Heavy spring and summer rains, together with the channelization and levees, led to a disastrous flood that swept over the levees, engulfed more than 9.3 million acres, and took 50 lives. More than 70,000 people were made homeless by the flood.

A major impetus behind preserving wetlands, besides seeking to protect their natural ecology, is to maintain their capacity to store storm water. In contrast, developing wetlands invariably translates into increasing runoff, lowering water tables, and degrading water quality. Understanding these relationships is what impels environmentalists to preserve wetlands, an effort that is threatened by "takings legislation"—legislation which basically says that a landowner is free to develop wetlands or any other ecotype unless he or she is compensated for not developing it.

Improving Storm-water Management

Most states now require storm-water management to be a part of any new development. The management technique most commonly used is to construct a *storm-water retention reservoir* at the low extremity of the site being developed. The **storm-water retention reservoir** simply is a pond that receives and holds runoff from the site during storms. From the pond, water may gradually infiltrate into the soil, or it may trickle out slowly through a standpipe mounted in the

◀ **FIGURE 9–24** *A channelized stream.* To reduce flooding and stream-bank erosion, many streams and rivers have been channelized; that is, their course has been straightened and their banks lined with concrete or stone.

pond. Thus, the pond plays a role imitating groundwater storage, and it may also create a pocket of natural wetland habitat supporting wildlife (Fig. 9–25). Large storm-water retention–flood control reservoirs may serve, in addition, as recreational areas for boating, fishing, and so on.

Techniques of storm-water management on small sites include such things as wells and trenches filled with rock that receive water and allow it to percolate, terraces that receive and hold water on their steps and allow it to infiltrate, and rooftops and parking lots designed to pond water and let it trickle away slowly. Of course, storm water collected in any sort of reservoir can act as a source of water for watering lawns, washing cars, and other nondrinking purposes.

9.10 Water Stewardship

Our study of the hydrologic cycle shows us that nature provides a finite flow of water through each region of Earth. When humans come on the scene, this flow of water is inevitably divided between the needs of the existing natural ecosystems and the agricultural, industrial, and domestic needs of humans, with human needs usually being met first. In fact, recent calculations indicate that humans now use 26% of total terrestrial evapotranspiration and 54% of the accessible precipitation runoff. We are major players in the hydrologic cycle, and as we have seen, many facets of our use of water are unsustainable.

We have also seen many opportunities for conserving and recycling water. The question is, How can we get ourselves and others to implement these measures, short of social or ecological disaster? Laws, litigation, and the courts have been, and are being, used to settle disputes, but we can see that this is probably the most costly and time-consuming method; besides, parties are left embittered and antagonistic, and wins are questionable. For example, we saw that the recovery of Mono Lake will be long and only partial, despite the win by environmentalists.

We might at least imagine a better way. With suitable leadership, the contending parties might be brought together and provided with information regarding such questions as the following: How much water may be removed from a given source without disrupting natural ecosystems? If that amount is less than desired, what are the conservation or recycling measures available to reduce needs for irrigation and cities? Is irrigating crops really essential? (In a number of cases, cities are buying farmlands simply for the water rights.) How can revenues be raised and costs be shared to implement such measures? In short, how can we best balance our direct needs for water with the broader objectives of Earth stewardship and creating a sustainable future?

The report of the President's Council on Sustainable Development addresses these concerns with some helpful policy recommendations:

- Executive orders should be issued by the president and state governors directing federal agencies to promote voluntary multistakeholder collaborative approaches toward managing and restoring natural resources.
- Public and private leaders, community institutions, nongovernmental organizations, and individual citizens should take collective responsibility for practicing environmental stewardship.
- The federal government should play a more active role in building a consensus on difficult issues and identifying actions that would allow stakeholders to work together toward common goals.
- Government agencies, conservation groups, and the private sector should expand the use of ecosystem approaches to natural-resource management by using

▶ **FIGURE 9–25** *Storm-water management.* Rather than letting excessive runoff from developed areas cause flooding and other environmental damage, runoff can be funneled into a retention pond, as shown here. Then it can drain away slowly, maintaining natural streamflow, or recharge groundwater. The retention pond may be designed to retain a certain amount of water and thus create a pocket wildlife habitat in an otherwise urban or suburban setting.

collaborative partnerships, developing compatible information databases, and carrying out appropriate incentives for responsible stewardship.

As a case study of how its recommendations could be accomplished, the council reported on the protection of coastal wetlands in Louisiana. Early in the process of addressing erosion of the wetlands, the Louisiana Coastal Wetlands Interfaith Stewardship Group began getting involved with state, business, and environmental agencies. The religious community helped keep up front the message of environmental stewardship and social justice for the coastal people as different solutions to the problem were being considered. As a result of the collaboration of all these groups, the people of Louisiana voted in 1989 to establish the Louisiana Wetlands Conservation and Restoration Trust Fund. Subsequently, $1.5 billion in funds was appropriated by Congress to help restore Louisiana's wetlands.

Thus, the choice is either to find similar sustainable solutions that meet the economic needs of people and the ecological needs of the land that supports us or to continue disputing, in which case all parties ultimately stand to lose.

On the international front, the World Commission on Water for the 21st century sponsors the World Water Forum, which convened in 2000 and meets every three years. (See Global Perspective essay, this page.) The findings of the forum relate directly to solving the great water-related needs of Earth, especially the developing countries. With a large proportion of the developing world already experiencing shortages of clean water or water for agriculture, the population increases and pressure to develop their resources that are certain to come in the next few decades will undoubtedly subject hundreds of millions to increased water stress. The problem is not that Earth contains too little fresh water; rather, it is that we have not yet learned to manage well the water that our planet provides. To quote the World Water Council's Vision report: "There is a water crisis, but it is a crisis of management. We have threatened our water resources with bad institutions, bad governance, bad incentives, and bad allocations of resources. In all this, we have a choice. We can continue with business as usual, and widen and deepen the crisis tomorrow. Or we can launch a movement to move from vision to action—by making water everybody's business."

GLOBAL PERSPECTIVE

THE WORLD WATER FORUM

Established in 1998, the World Commission on Water for the 21st century was given the responsibility for developing a long-term vision for water for that century. The Commission, co-sponsored by the World Bank, the FAO and UNEP of the United Nations, and other major organizations, delivered its vision report at the Second World Water Forum in The Hague, the Netherlands, March 17–22, 2000. More than 4,500 from all over the world attended the meetings. Actually, two reports were delivered, both available on the Internet:[2] the Commission Report: A Water Secure World and a Vision Report: Making Water Everybody's Business (which was the document the commission drew on for most of its background information).

Both reports documented a world water crisis, focusing primarily on the unsafe water and inadequate sanitation half of the world's population—primarily the poor—have no choice but to contend with. Although progress has been made in providing safe drinking water and irrigation water

to a greatly expanded human population over the last century, major problems remain. Seven world water challenges were explored in the two reports: water scarcity (especially linked to heavy uses of irrigation), inaccessibility, the deterioration of water quality, world peace and security, awareness by decision makers and the public, a decline in financial resources for water allocation, and the fragmentation of water management. The reports documented steps that should be taken (the vision part) to meet these challenges, such as limiting the growth rate of irrigation for agriculture, increasing the productivity (efficiency) of water, increasing water storage, promoting cooperation in the use of international water basins, valuing ecosystem functions that involve water, and mobilizing financial resources to deal with future uses of water.

Several strategies, explained in the documents' "Framework for Action," are needed to accomplish these goals: involving all stakeholders in integrated management, moving toward full-cost pricing of all water

services, increasing public funding for research and innovation in the public interest, and massively increasing investments in water. The commission's documents are clearly visionary, and criticism came from a group of nongovernmental organizations (NGOs) attending the forum. In effect, they claimed that the forum's vision was, so to speak, watered down. In their words, "We need reform of the governance of water based on the skills, experience and legitimacy of local people and communities, on recognition of the primacy of human needs and rights, and on sound understanding of ecosystems and river basin management. We need targets and timetables for improvement." The critics called for a thorough review of the commission's framework, to "ensure that it addresses the delicate balance between meeting the world's needs for safe drinking water, food, sanitation and energy, while conserving the environment."

[2] At http://watervision.org/.

ENVIRONMENT ON THE WEB

WATER CONSERVATION: HOW FAR CAN WE GO?

Arid regions have long recognized the need for careful use of precious water resources. In recent years, even moist temperate zones have come to realize the value of water conservation. As climatologists debate the likelihood of global warming, water managers are confronting the need for careful planning for water supplies into the 21st century. In Southern Canada, for instance, water managers are facing the possibility that river levels may drop as much as 1 m under worst-case climate projections. In the United Kingdom, water companies are analyzing expected changes in the distribution of precipitation and associated impacts on water supply. Faced with the potential for shortages, water managers must either develop new water sources or find ways to encourage water users to conserve.

In many parts of North America, domestic water use now exceeds 600 L per person per day, yet in dry African nations people consume only 8–20 L per person per day. Clearly, North American water usage is often wasteful compared to other water-poor nations. But how much reduction is feasible as a resource conservation measure?

Associations such as the American Water Works Association and individual municipalities suggest that 25% of domestic water use could be saved through simple measures such as water-saving toilets, shower heads, and leak controls, without noticeable impact on current lifestyles. Some estimates place potential savings as high as 50%. But despite its obvious advantages for resource protection, water conservation has been a hard sell in many municipalities, especially those in water-rich areas. And some uses, such as firefighting and electric power generation, require immense volumes of water that cannot easily be reduced without jeopardiz-

ing the benefits of such water use. Industrial uses of water can also be massive, particularly in older operations.

One of the simplest and most effective measures available for water conservation is water metering. In a metered system, a small device measures usage, and the consumer is billed for each unit of water used. So-called "increasing block" pricing, in which the consumer pays a proportionately higher rate with higher use, is a particularly effective way to encourage water efficiency. Most new residential developments in North America have water metering, but in some areas of the world, such as the United Kingdom, metering has met with strong resistance from politicians and the public alike. To be effective, water-conservation strategies must therefore be developed in the context of local conditions—including the needs and priorities of the users themselves.

Web Explorations

The Environment on the Web activity for this essay describes some of the debate around residential water conservation and the tools for achieving efficient water use. Go to the Environment on the Web activity (select Chapter 9 at **http://www.prenhall.com/wright**) and learn for yourself:

1. about the debate surrounding residential water conservation;
2. what tools are available to achieve efficient water use; and
3. how your local water use compares to that of neighboring areas.

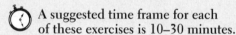 **A suggested time frame for each of these exercises is 10–30 minutes.**

REVIEW QUESTIONS

1. How does water change its state with changes in energy (heat)?

2. What are the two processes that result in natural water purification? State the difference between them. Distinguish between green water and blue water.

3. Describe how a Hadley cell works, and explain how Earth's rotation creates the trade winds.

4. Why do different regions receive different amounts of precipitation?

5. Define precipitation, infiltration, runoff, capillary water, transpiration, evapotranspiration, percolation, gravitational water, groundwater, the water table, an aquifer, a recharge area, a seep, and a spring.

6. Use the terms defined in Question 5 to give a full description of the hydrologic cycle, including each of its three "loops." What is the water quality (purity) at

different points of the cycle? Explain the reasons for the differences.

7. How does changing Earth's surface, for example, by deforestation change the pathway of water? How does it affect streams and rivers? Humans? Natural ecology?

8. Describe how groundwater and wetlands act as natural reservoirs to moderate the flow of streams and rivers in wet periods and dry periods. What impact will the development of wetlands have on water quality and quantity?

9. What are the three major categories of water use? From what point(s) in the cycle is water generally withdrawn? How is it withdrawn and distributed?

10. What are the indicators of overdrawing water supplies? What are the human, ecological, and environmental consequences of overdrawing surface waters? Of overdrawing groundwater?

11. Describe how water demands might be reduced in agriculture, industry, and households.

12. What is meant by storm-water runoff, and what causes it? What are the consequences of storm-water runoff for groundwater? For water quality in streams and rivers? For streamflow during rains? For streamflow between rains? For the natural environment and ecology?

13. Describe techniques now being implemented to manage storm-water runoff.

14. Are new policies called for to achieve sustainable water supplies? How is the government addressing this need? What policies would you support or promote?

THINKING ENVIRONMENTALLY

1. Pretend you are a water molecule, and describe your travels through the many places you have been and might go in the future as you make your way around the hydrologic cycle time after time. Include travels through organisms.

2. Commercial interests wish to create a large new development on what is presently wetlands. Have a debate between those representing the commercial interests and environmentalists who will bring out the environmental and economic costs of development. Work toward negotiating a compromise.

3. Describe how many of your everyday activities, including your demands for food and other materials, add pollution to the water cycle or alter it in other ways. How can you be more stewardly with your water consumption?

4. Describe the natural system that maintains uniform streamflow despite fluctuations in weather. How do humans upset this regulation? What are the consequences?

5. An increasing number of people are moving to the arid southwestern United States despite the fact that water supplies are already being overdrawn. If you were the governor of one of the states in this region, what policies would you advocate regarding the situation?

6. Commercial interests wish to develop a golf course on presently forested land next to a reservoir used for city water. Describe the impacts this development might have on water quality in the reservoir.

7. Divide the class into five groups: environmentalists, farmers, city officials responsible for domestic water, industrialists, and developers. The actual supply of water will support only 50% of prospective demands. Negotiate a compromise that will honor the interests of all parties.

WEB REFERENCES

On-line resources for this chapter are on the World Wide Web at: **http://www.prenhall.com/wright**. (Click on Chapter 9 within the Chapter Selector.)

PRODUCT 229

229???30877?

The Production and Distribution of Food

Key Issues and Questions

1. In industrialized societies, an agricultural revolution has radically affected the practice of farming and its environmental impact. What is industrialized agriculture, how did it develop, and what are its environmental costs?
2. The agricultural revolution has been transferred to the developing world in a process called the Green Revolution. What are the origins and impacts of the Green Revolution?
3. In most of the developing world, agriculture is still practiced in traditional ways, called subsistence agriculture. How important is subsistence agriculture?
4. Continued population growth puts pressure on agricultural practice to keep producing more food. What are the prospects for increasing food production?
5. Genetically modified crops are being developed at a breathtaking pace. What are the promises and problems associated with this technology?
6. World trade in foodstuffs is lively and important. What are the global patterns of food trade, and what are their consequences?
7. Hunger and malnutrition still plague human societies. What is the extent of hunger, malnutrition, and undernutrition in the world?
8. Famines continue to occur. What are the causes of famine, and which geographical areas are affected?
9. Food aid is distributed to countries all over the world. Is food aid necessary? How is aid distributed?

Half the people in the world eat rice daily, most of them in the developing countries. Although an excellent food for energy, rice is not a sufficient source of vitamins and other nutrients; hence, it must be supplemented with other foodstuffs for a balanced diet. Unfortunately, millions cannot afford other foods and depend heavily on rice for their nutrition. The result is two of the most common nutritional deficiencies, with tragic consequences: vitamin A deficiency, which leads to blindness and immune-system failures, and iron deficiency, ultimately causing anemia and also immune problems. Over a million children die each year and 350,000 children are blinded yearly, due to a lack of vitamin A alone. If somehow rice could be supplemented with iron and vitamin A, countless millions of lives could be saved.

Biochemist Ingo Potrykus, working at the Swiss Federal Institute of Technology in Zurich, believed that it could be done and in the early 1990s began a program of research that would lead to what many now believe may be the most remarkable accomplishment of plant biotechnology—golden rice (Fig. 10–1). Over a period of seven years, using the techniques of gene splicing, Potrykus, colleague Peter Beyer of the University of Freiburg in Germany, and other coworkers incorporated into a rice genome seven genes that would enable the plant to synthesize beta carotene (which the body can use, in turn, to synthesize vitamin A) and iron in the body of the grain. (Both beta carotene and iron are normally present in the leaves of the plant, but when the rice is milled, only the grain is saved.) Just 300 grams (2/3 lb) of golden rice can provide a full daily complement of vitamin A and iron. The miracle rice will not be ready for distribution to developing-world farmers for several years, but efforts are being made to ensure that it will be provided without cost when it is ready.

However, golden rice may meet with a less-than-universal welcome, because it is a genetically modified organism, containing genes from daffodils and beans. Such organisms are called by many "Frankenfoods"—foods that

◄ **Frankenfood protest.** Antibiotechnology protestors demonstrating in March, 2000, at the BIO 2000 Conference, where scientists met to review developments in bioengineering.

▲ **FIGURE 10–1** *Golden rice.* The yellow rice kernels contain beta carotene, inserted into them with biotechnology to enable the rice to help meet nutritional needs for vitamin A.

incorporate foreign genes. The international debate over these foods has injected a new set of fears into the food production and distribution arena (see chapter-opening photo), and the debate is far from settled. Yet, as we shall see, if society is to meet the challenge of feeding another 3 or 4 billion people in this century, and doing so with better

nutrition, genetic engineering technology is likely to play the key role.

Throughout the world, three decades of rapid population growth have left hundreds of millions dependent on food aid, while one in five in the developing world remains undernourished. As the world population continues its relentless rise (Fig. 6–1), no resource is more vital than food. Can the world's farmers and herders produce food fast enough to keep up with population growth? After all, our planet holds only a finite amount of the resources needed for food production: suitable land, water, energy, and fertilizer.

By many measures, human societies have done very well at putting food on the table. More people are being fed than ever before, with more nutritional food. In the last 25 years, world food production has more than doubled, rising even more rapidly than population. A lively world trade in foodstuffs forms the bulk of economic production for many nations. The number of hungry people has declined by 40 million in the past decade. Optimists would say that, although regions of chronic hunger and malnutrition—and occasional famines—exist, these are the exceptions to an otherwise remarkable accomplishment. Those viewing the situation through rose-colored glasses are convinced that when the world population levels off (at 10 billion?), enough food will still be produced to sustain it. Others question whether that is at all possible.

This chapter examines food-production systems, problems surrounding the lack of food, and the global effort being made to reduce hunger and malnutrition.

10.1 Crops and Animals: Major Patterns of Food Production

Some 10,000 years ago, the Neolithic Revolution saw the introduction of agriculture and animal husbandry, which probably did more than anything else to foster the development of human civilization. Virtually all of the major crop plants and domestic animals were established in the first thousand years of agriculture. Between 1450 and 1700, world exploration and discovery led to an exchange of foods that greatly influenced agriculture and nutrition in the Western world. From the New World came potatoes, maize (corn), beans, squash, tomatoes, pineapples, and cocoa. Rice came from the Orient. In exchange, Europeans brought to the New World wheat, onions, sugar cane, and a host of domestic animals—horses, pigs, cattle, sheep, and goats. With the advancing capabilities of science and technology by the 1800s, the stage was set for a remarkable change in agriculture.

The Development of Modern Industrialized Agriculture

Until 150 years ago, the majority of people in the United States lived and worked on small farms. Human and animal labor turned former forests and grasslands into systems that produced enough food to supply a robust and growing nation (Fig. 10–2a). Farmers used traditional approaches to combat pests and soil erosion: Crops were rotated regularly, many different crops were grown, and animal wastes were returned to the soil. The land was good, and farming was efficient enough to allow a substantial segment of the population to leave the farm and join the growing ranks of merchants and workers living in cities and towns. Then, in the mid-1800s, the Industrial Revolution came to the United States, and it had a major impact on farming.

The Transformation of Traditional Agriculture. The Industrial Revolution contributed to a revolution in agriculture so profound that today less than 3% of the U.S.

(a)

(b)

▲ **FIGURE 10–2** *Traditional vs. modern farming.* (a) Traditional farming practiced in Arkansas. For hundreds of years, traditional practices on American farms supported the growing U.S. population. (b) Modern agricultural practice, illustrated by a combine harvesting wheat in America's Midwest.

workforce produces enough food for all the nation's needs, plus a substantial amount for trade on world markets (Fig. 10–2b). Indeed, this revolution has achieved such gains in production that the United States has had to formulate policies to cope with surpluses of many crops.

Virtually every industrialized nation has experienced this agricultural revolution. The pattern of developments in U.S. agriculture could just as well describe that of France, Australia, or Japan. Crop production has been raised to new heights, doubling or tripling yields per acre (Fig. 10–3). However, there are environmental costs: According to many agricultural experts, expanding production has reached, or even exceeded, sustainable limits. Let us examine the components of the agricultural revolution.

Machinery. The shift from animal labor to machinery has created a dependency on fossil-fuel energy that adds significantly to the energy demands of the industrial societies (Chapter 13). As the price of oil rises, it will certainly have an impact on the costs of food production and begin to be reflected in the market. And as we have seen, the continued use of farm machines for plowing, planting, and harvesting causes some soil compaction (Chapter 8).

Land under Cultivation. Before 1960, much of the increased production in the United States came from bringing new land into production. Since then, attempts have been made to increase the land used to raise grain, but the new land areas were not well suited for agriculture and have been abandoned because erosion or depletion of water resources has rendered them no longer productive. Current farm policy—the **Conservation Reserve Program**—reimburses farmers for "retiring" erosion-prone land and planting it to produce trees or grasses. The law allows up to

36 million acres to be placed in this program; land in the program currently totals 33.5 million acres.

Essentially all of the good cropland in the United States is now under cultivation or held in short-term reserve. Worldwide, on a *per capita* basis, cropland is on the decline as population continues to rise. Any significant expansion in cropland will come at the expense of forests and wetlands, which are both economically important and ecologically fragile.

Fertilizers and Pesticides. When fertilizers were first employed, 15 to 20 additional tons of grain were gained

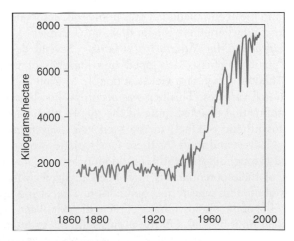

▲ **FIGURE 10–3** *U.S. corn yield.* This graph demonstrates three phenomena: the long-term rise in yields, the recent leveling off in yields, and the effects of droughts in 1970, 1973, 1980, 1983, and 1988 (8,000 kg/hectare = 3.5 tons/acre). (Source: Data from *State of the World* 1998 by Lester Brown et al. Copyright © 1998. Reprinted by permission of Worldwatch Institute.)

from each ton of fertilizer used. Between 1950 and 1990, worldwide fertilizer use rose tenfold. Now, however, farmers in developed countries are applying near-optimal levels of fertilizers, and the gain is less than 2 tons of grain per additional ton of fertilizer applied. When levels of fertilizer are too high, plants become more vulnerable to attack by insects and other pests; when fertilizer is washed away, the result is groundwater and surface-water pollution (Chapter 18). The worldwide use of fertilizer resumed its rise in 1995 following a six-year decline traced to economic conditions in the countries of the former Soviet Union, but, since 1997, has been holding steady at around 135 million tons. Most of the current increase is in "underfertilized" countries like China, India, and Brazil.

Chemical pesticides have provided significant control over insect and plant pests, but the pests have become resistant to most of the pesticides as a result of natural selection. Also, efforts to reduce the use of pesticides have begun because of side effects to human and environmental health. Many of the chemicals in use have not been adequately tested in terms of human concerns and their potential for causing genetic defects. As we shall see in Chapter 17, progress is being made toward developing natural means of control that will be environmentally safe, but these new methods will be unlikely to increase yields.

Irrigation. Worldwide, irrigated acreage increased about 2.6 times from 1950 to 1980 and, by 1999, represented 17% of all cropland—some 255 million hectares—and produced 40% of the world's food. Irrigation is still expanding, but at a much slower pace because of limits on water resources. (Irrigation is by far the greatest consumer of water.) More ominous, much current irrigation is not sustainable, since groundwater resources are being depleted. In addition, production is being adversely affected on as much as one-third of the world's irrigated land, because of waterlogging and the accumulation of salts in the soil, consequences of irrigating where there is poor drainage (Chapter 8).

High-yielding Varieties of Plants. Several decades ago, plant geneticists developed new varieties of wheat, corn, and rice that gave yields of double to triple those of traditional varieties. This feat was accomplished by selecting strains that diverted more of the plant's photosynthate (photosynthetic product) to the seed and away from the stems, leaves, and roots. As these new varieties were introduced throughout the world, production soared. However, most of their potential has now been realized, with the seeds of modern wheat, rice, and corn now receiving more than 50% of the photosynthate, close to the calculated physiological limit of 60%. On the minus side, the widespread use of genetically identical crops has given rise to major pest damage, as pests have become adapted to the new varieties and resistant to pesticides.

The Green Revolution. The same technologies that gave rise to the agricultural revolution in the industrialized countries were eventually introduced into the developing world. There they gave birth to the remarkable increases in crop production called the **Green Revolution**.

In 1943, the Rockefeller Foundation sent agricultural expert Norman Borlaug and three other U.S. agricultural scientists to Mexico, with the objective of exporting U.S. agricultural technology to a less developed nation that had serious food problems. Their aim was to improve the traditional crops grown in Mexico, especially wheat. Mexican wheat was well adapted to the subtropical climate, but it gave low yields and responded to fertilization by growing very tall stalks that were easily blown over. Using wheat from other areas of the world, Borlaug and his coworkers bred a dwarf hybrid with a large head and a thick stalk; the hybrid did well in warm weather when provided with fertilizer and sufficient water (Fig. 10–4). The program was highly successful: By the 1960s, Mexico had closed the gap between food production and food needs, wheat production had tripled, and Mexican wheat appeared on the export market.

Research workers with the Consultative Group on International Agricultural Research (CGIAR) extended the work done in Mexico, introducing both high-yielding wheat and high-yielding rice to other developing countries. To cite one success, Borlaug induced India to import hybrid Mexican wheat seed in the mid-1960s, and in six years, India's wheat production tripled. Within a few years, many of the world's most populous countries turned the corner from being grain importers to achieving stability and, in some cases, even becoming grain exporters. Thus, while world population was increasing at its highest rate (2% per year), rice and wheat production underwent increases in yields of 4% or more per year. This Green Revolution has probably done more than any other single scientific or other achievement to prevent hunger and malnutrition. Borlaug was awarded the Nobel peace prize in 1970 in recognition of his contribution.

The high-yielding grains are now cultivated throughout the world and have become the basis of food production in China, Latin America, the Middle East, southern Asia, and, of course, the industrialized nations. Because the technology raises yields without requiring new agricultural lands, the Green Revolution has also held back a significant amount of deforestation in the developing world. But as remarkable as it was, the Green Revolution is not a panacea for all of the world's food-population difficulties, for the following reasons:

1. Most of its potential has been realized; many of the most populous countries are reaching a plateau in their grain production and in acreage planted to high-yielding varieties, while their populations continue to increase.

2. More water is required to raise high-yielding grains; thus, these grains are highly dependent on irrigation. Water shortages have begun to occur as a result of this dependence (Chapter 9). Also, such grains require constant inputs of fertilizers, pesticides, and energy-

(a)

(b)

▲ **FIGURE 10–4** *Traditional vs. high-yielding wheat.* Comparison of (a) an old variety of wheat, shown growing in Rwanda, with (b) a new, high-yielding variety of dwarf wheat growing in Mexico.

using mechanized labor, all of which are often in short supply in developing countries.

3. Because it is patterned after agriculture in the developed world, Green Revolution agriculture tends to benefit larger landholders. More food is raised by a smaller farm workforce, causing many farm laborers and small landholders to become displaced and migrate to the cities, joining the ranks of the unemployed.

4. The most important African food crops (sorghum, millet, and yams) are not commonly used in the developed world, so they have not benefited from Green Revolution technology. The fact is, the Green Revolution has had little impact on a large part of the developing world where another kind of agriculture—subsistence agriculture—is practiced.

Subsistence Agriculture in the Developing World

In most of the developing world, plants and animals continue to be raised for food by *subsistence farmers*, using traditional agricultural methods. These farmers represent the great majority of rural populations. **Subsistence farmers** live on small parcels of land that provide them with the food for their households and, it is hoped, a small cash crop. From the point of view of the modern world, such farmers are very poor, although many do not consider themselves to be so. Like past agricultural practice in the United States, subsistence farming is labor intensive and lacks practically all of the inputs of industrialized agriculture. Also, it is often practiced on marginally productive land (Fig. 10–5).

Typically, a family owns a small parcel of land for growing food and maintains a few goats, chickens, or cows. Such a family is making the best use of very limited resources, and very often the people are adapted well enough to the prevailing social and environmental conditions to provide a

▼ **FIGURE 10–5** *Subsistence farming.* Mexican farmer plowing marginally productive land with a team of oxen. Subsistence farming feeds more than 1.4 billion people in the developing world.

livelihood for a household. An important fact to remember, however, is that subsistence agriculture is practiced in regions experiencing the most rapid population growth, even though that kind of agriculture is best suited for low population densities. An estimated 1.4 billion people in Latin America, Asia, and Africa—over one-third of the people there—are sustained by subsistence agriculture.

The pressures of population and the diversion of better land to industrialized agriculture often lead to practices that are at best nonsustainable and at worst ecologically devastating. In many regions in developing countries, woodlands and forests are cleared for agriculture or removed for firewood and animal fodder, forcing the gatherers to travel farther and farther from their homes and leaving the soil susceptible to erosion. The ensuing scarcity of firewood leads the residents to burn animal dung for cooking and heat, thus diverting nutrients from the land. Erosion-prone land suited only to growing grass or trees is planted to produce annual crops. Good land is forced to produce multiple crops instead of being left fallow to recover nutrients. Growing populations compel the continued subdivision of land, which diminishes the land's ability to support each household. All these factors tend to increase the poverty that is characteristic of populations supported by subsistence agriculture, and, in a relentless cycle (see Fig. 7–6), the added poverty in turn puts increased pressures on the land to produce food and income.

Because subsistence agricultural practice varies with the local climate and with local knowledge, it is difficult to draw sweeping generalities. In some areas, subsistence agriculture involves shifting cultivation within tropical forests—often called slash-and-burn agriculture. Research has shown that such a practice can be sustainable (Fig. 10–6). The cultivators create highly diverse ecosystems in which the cleared land supports a few years of crops and gradually shifts into agroforestry—a system of tree plantations with different ground crops employed as the trees grow.

In other areas, subsistence farmers are showing remarkable success in adapting to the changing needs of local societies as they are forced to support expanding populations on the same land. For example, in Kenya, land in the Machakos region was seriously degraded during the 1930s. Now the land has recovered and supports a population six times larger, as the 1.5 million people there have diversified their agriculture and practiced soil and water conservation.

Animal Farming and Its Consequences

Raising livestock—sheep, goats, cattle, buffalo, and poultry—has many parallels to raising crops (Table 10–1), and there are also direct connections between the two. Fully one-fourth of the world's croplands are used to feed domestic animals; in the United States alone, 70% of the grain crop goes to animals—on half the cultivated acreage. The care, feeding, and "harvesting" of the estimated 15 billion domestic animals constitute one of the most important economic activities on the planet. The primary force driving this livestock economy is the large number of the world's

▶ **FIGURE 10–6** *Slash-and-burn agriculture.* Countryside in Mindanao, Philippine Islands, showing the results of sustainable slash-and-burn agriculture. Cultivated fields are interspersed with trees and natural areas in a diverse ecosystem.

TABLE 10–1	Parallels between Plant and Animal Farming	
	Plant	**Animal**
Major products	Grains, fruits, and vegetables for food	Meat, dairy products, and eggs for food
Other important products	Oils, fabrics, rubber, specialty crops (spices, nuts, etc.)	Labor, leather, wool, manure, lanolin
Modern practices	Industrialized agriculture on former grasslands and forests	Ranching, dairy farming, and stall-feeding
Traditional practices	Subsistence agriculture on marginally productive lands	Pastoral herding on nonagricultural lands
Current global land use	3.7 billion acres (1.5 billion ha, or 11% of land surface)	8.4 billion acres (3.4 billion ha, or 26% of land surface)

people who enjoy eating meat and dairy products—primarily, most of the developed world and growing numbers of people in less developed nations. As with crop farming, however, there are two patterns: In the developed world and on ranches in the developing world, livestock is raised in large herds and often under factorylike conditions; in rural societies in the developing world, livestock is raised on family farms or by pastoralists who are subsistence farmers.

Industrial-style animal farming can affect the environment in a host of nonsustainable ways. Since so much of the plant crop is fed to animals, all of the problems of industrialized agriculture apply to animal farming. In addition, rangelands are susceptible to overgrazing, either because of mismanagement of prime grazing land or because the land used for grazing is marginal dry grasslands, used in that manner because the better lands have been converted to producing crops. Much of the western rangeland in the United States is public land leased at subsidized fees that easily lead to overgrazing (Chapter 8). Another serious problem is the management of animal manure. In developing countries, manure is a precious resource that is used to renew the fertility of the soil, build shelters, and provide fuel. In developed countries, it is a wasted resource. Close to 1.3 billion tons of animal waste is produced each year in the United States, some of which leaks into surface waters and contributes to die-offs of fish, contamination with pathogens, and a proliferation of algae. With factory farms on the sharp increase, wastes from manure and meat processing are either bypassing or overwhelming the often inadequate treatment systems available and polluting the nation's waterways. The EPA asserts that animal-based agriculture is the most widespread source of pollution in the nation's rivers.

In Latin America, more than 49 million acres (20 million ha) of tropical rain forests have been converted to cattle pasture. Even though most of this land is best suited for growing rain forest trees, some of it could support a rural population of subsistence farmers producing a diversity of crops. Instead, it is held by relatively few ranchers who own huge spreads. According to the Intergovernmental Panel on Climate Change, deforestation and other changes in land use in the tropics release an estimated 1.6 billion tons of carbon to the atmosphere annually, contributing a significant amount of carbon dioxide to the greenhouse effect.

Also, because their digestive process is anaerobic, through belching and flatulence, cows and other ruminant animals annually eliminate some 100 million tons of methane, another greenhouse gas. Anaerobic decomposition of manure leads to an additional 30 million tons of methane per year. All this methane released by livestock makes up about 3% of the gases causing global warming (Chapter 21).

Even though their animals also contribute to the methane problem, it is callous to fault the subsistence farmers whose domestic animals enhance their diet and improve their quality of life. In fact, one of the most important kinds of sustainable development aid brought to rural families is the gift of a cow or a few goats or rabbits, as is carried out by Heifer Project International. Working in 115 countries, the project has distributed large animals, beehives, fowl, and fish fingerlings to families. The organization works with grassroots groups of local people who oversee a given project, for which there must be a genuine need, which must improve the environment, and for which training must be effective and the people committed to passing on the gift in the form of offspring of livestock to other needy people.

The lives of millions in the developing world are tied very directly to the animals they raise, and their impact on the environment is often sustainable. Animals that are well managed can enhance the soil and enable rural farmers to maintain a balanced farm ecosystem. In sum, animal farming is far more likely to be sustainable in the context of rural farms and pastoral herding than in the beef ranches and hog pens of the developed world, where the pressing need is to address problems of pollution, overgrazing, and deforestation.

Prospects for Increasing Food Production

On a world basis, grain harvests (grains are the staple food for most people) have declined 1% since the record harvest of 1.88 billion tons in 1997; *per capita* grain production has been slowly declining since 1984 (Fig. 10–7). Nonetheless, there is no shortage of food in the world as a whole: Food production appears to be keeping up with demand. Indeed, the production of meat has risen dramatically in recent years, indicative of dietary shifts to higher protein consumption as living standards continue to rise. The greatest concern continues to be about the future, as demand rises. How will we manage a 40% increase in food production needed over the next two decades?

One suggestion is to bring more land into cultivation. Where this is done, however, it means the loss of grasslands, forests, or wetlands and the ecosystem goods and services they provide. Between 1990 and 1995, there was a net loss of 65 million hectares of forests in the developing countries, primarily to agriculture. This loss will undoubtedly continue as population increases in the developing countries. Yet most of the prime arable land in the world is now in cultivation, and the newly converted land is likely to be marginal land that is highly subject to erosion and soil degradation. Indeed, the problem of land degradation is already with us, as we saw in Chapter 8. Cutting-edge satellite technology has been used to analyze land use, and the results indicate that almost 40% of agricultural lands are seriously affected by soil degradation. Most of these lands are in the developing countries, where the need for increased food production is greatest. These opposing trends of degrading agricultural lands and opening up new lands are likely to offset each other, in the opinion of agricultural experts. However, more lands probably will be placed under irrigation, putting yet more pressure on groundwater supplies; again, water tables are dropping (see Chapter 9), and it is likely that very shortly the new irrigated lands will be offset by abandoned irrigated lands.

Thus, we appear to have only two prospects for increasing food production: (1) Continue to increase crop yields and (2) begin growing food crops on land that is now used for feedstock crops or cash crops. As we have seen, a dramatic rise in grain yields in the developing countries was the major accomplishment of the Green Revolution. Surprisingly, grain yields have continued to rise in some of the developed countries. In France, wheat yields have quadrupled since 1950, reaching over 7 tons per hectare. Rice yields in Japan have risen 67% in the same period. Indeed, the genetic potential exists for yields as high as 14 tons per hectare or higher for wheat. Can we expect yields to continue to increase up to their genetic potential?

Actually, great differences in yields of grain between regions (Table 10–2) have little to do with the genetic strains used. Egypt and Mexico, for example, achieve higher yields than the United States because most of their wheat is irrigated, while U.S. wheat is rain fed and grown on former grasslands. Australian wheat yields are even lower, as a consequence of the country's sparse rainfall. Once the agricultural land is planted with high-yielding strains and fertilized to the maximum, other factors—soil, rainfall, and available sunlight—place limits on productivity. These environmental limits are a reminder that agricultural sustainability is highly dependent on soil and water conservation (Chapter 8) and the weather. Variations in climate, especially those associated with El Niño and La Niña (Chapter 21), affect harvests locally and globally, injecting significant instability in food production that makes future planning extremely chancy. Another imponderable in future food production is global climate change: As the predicted warming of the 21st century occurs, it is impossible to predict how rainfall patterns will change.

In the developing world, sub-Saharan Africa has the greatest need for increasing crop yields. With much of its land arid and populations continuing to grow at exponential rates, the region continues to fall behind in per capita food production and is becoming increasingly dependent on

▶ FIGURE 10–7 *Per capita food production.* World per capita grain production, 1950–1999. (*Source:* Vital Signs, 2000, Worldwatch Institute, 2000.)

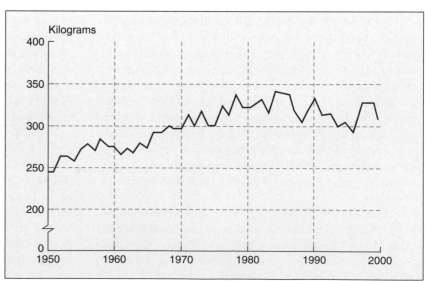

TABLE 10–2 Annual Wheat Yield per Hectare in Various Countries	
United Kingdom	8.1
France	7.2
Egypt	6.4
Mexico	4.8
China	4.0
Poland	3.5
United States	2.9
Canada	2.6
India	2.6
Argentina	2.5
Ukraine	2.3
Pakistan	2.2
Australia	1.8
Russian Federation	1.4
Kazakhstan	1.3

Source: FAOSTAT Agricultural Data, FAO, 1999.

but is a complex undertaking, for it involves such issues as land reform and maintaining a balance of trade.

In the face of these constraints, Gordon Conway, president of the Rockefeller Foundation, has called for a new Green Revolution, what he has called a **Doubly Green Revolution**, "a revolution that is even more productive than the first Green Revolution and even more 'green' in terms of conserving natural resources and the environment. During the next three decades, it must aim to repeat the successes of the Green Revolution on a global scale in many diverse localities and be equitable, sustainable, and environmentally friendly." Conway believes that this new revolution is possible, in large part because of the emergence of biotechnology and the kinds of genetic manipulations we saw in the development of golden rice.

The Promise and Problems of Biotechnology

The revolutionary potential of genetic engineering has made it possible to crossbreed genetically different plants and to incorporate desired traits into crop lines and animals—so-called transgenic breeds. It has also led to the cloning of domestic animals like cows and goats. Without a doubt, this technology can help the developing world to produce more food. Genetically modified organisms (GMOs) are under intense research and development, with new products announced almost daily. Still, in spite of the obvious potential, serious objections continue to arise to the development and use of GMOs. We need to look at both sides of this controversy.

food imports. Yields would have to increase 2.5% annually just to keep up with population growth. Yet, the potential is there to make great progress with some very basic steps. With the aid of an extension service promoted by former President Jimmy Carter, Ethiopia harvested record crops in 1995–96, showing a 32% increase in production and a 15% increase in yield in one year. The key to this growth was the use of a simple fertilizer providing nitrogen and phosphorus to the nutrient-starved soils. Poor agricultural support and a lack of capital to purchase fertilizer and new seeds are preventing many other African countries from making similar improvements.

Now, what of the possibility of switching from the production of feed grain and cash crops to food for people? Fully 70% of domestic grain in the United States is used to feed livestock; the percentages drop over other regions of the world, in proportion to the economic level of the region. Sub-Saharan Africa and India, for example, use only 2% of their grain for livestock feed. Feed grain can be considered a buffer against world hunger; if the food supply becomes critical, it might force more of the world's people to eat lower on the food chain (less meat, more grain). As we have seen, however, the trend is in exactly the opposite direction. Converting land use from cash crops to food crops is possible,

The Promise. The first genetically altered products to be marketed were (1) the Flavr Savr™, a tomato that can be vine ripened and subsequently brought to market and kept fresh much longer than locally produced "ordinary" tomatoes; (2) cotton plants with built-in resistance to insects that comes from genes taken from a bacterium (so-called Bt, standing for *Bacillus thuringiensis*; the name of the bacterium) and (3) numerous crop plants resistant to the herbicide Roundup®, allowing farmers to employ no-till techniques. More recently, *genomics*—as the technique has come to be known—has developed sorghum (an important African crop) resistant to a parasitic plant known as witchweed, which infests many crops in Africa; corn, potatoes, and cotton resistant to insects; rice resistant to bacterial blight disease; and salmon that grow very rapidly, to name a few organisms. In just five years since bioengineered seeds were commercially available, half of soybeans, half of the cotton, and one-third of the corn acreage in the United States are planted transgenic breeds; worldwide, more than 75 million acres (30 million hectares) are now planted with bioengineered crops. To date, most of the new crops have been developed for use in the industrialized countries.

Biotech crop research that can benefit the developing countries is proceeding at a rapid pace, often in those very countries. Among other aims, the objectives are: (1) to engender resistance to diseases and pests that attack important tropical plants, (2) to produce a tolerance to environmental conditions, such as drought and a high salt level, that stress most plants, (3) to improve the nutritional value of commonly eaten crops (as has been done to produce golden rice), and (4) to incorporate vaccines against major human diseases into the cells of commonly eaten plants like bananas. Thus, products under development include incorporating resistance to viruses into cassava, sweet potatoes, melons, and papaya (Fig. 10–8), protein enhancement into corn and soybeans, resistance to fungus into bananas, antidiarrheal vaccine into bananas and tomatoes, and tolerance to drought into sorghum and corn. The potential for *transgenic* crops and animals seems almost unlimited.

Among the important environmental benefits of bio-engineered crops are reductions in the use of pesticides, since the crops are already resistant to pests; less erosion, because no-till cropping is facilitated by the use of herbicide-resistant crops; and less environmental damage associated with bringing more land into production, because of the enhanced productivity of existing agricultural lands.

The Problems.   Concerns about the new technology involve three considerations: environmental problems, food safety, and access to the new techniques. A major environmental concern focuses on the pest-resistant properties of the transgenic crops. With such a broad exposure to the toxin or some other resistance incorporated into the plant, it is possible that the pest will develop its own resistance to the toxin more rapidly and thus render it ineffective as an independent pesticide. Of course, the transgenic crop then loses its advantage. Another concern is the ecological impact of the crops. For example, pollen from Bt corn disperses in the wind and can spread to adjacent natural areas where beneficial insects may pick it up and be killed by the toxin. This effect has been shown to occur in laboratory tests with monarch butterflies, but as yet has not been tested extensively in the field. Genes for resistance to herbicides or for tolerance of drought and other environmental conditions also can spread by pollen, to wild relatives of crop plants, possibly creating new or "super" weeds.

Food safety issues center around the fact that transgenic crops contain proteins from different organisms and might trigger an unexpected allergic response in people who consume the food. Tests have shown that a Brazil nut gene incorporated into soybeans was able to express a protein in the soybeans that induced an allergic response in individuals already allergic to Brazil nuts. Also, antibiotic-resistance genes are often incorporated into transgenic organisms in order to provide a way to trace cells that have been transformed. The transgenic product could then convey the antibiotic resistance to pathogens in human systems by a separate process of gene transfer, or the product could prevent the antibiotic from being effectively used in an individual who has consumed the transgenic food. Another concern relates to the possibility that a plant might produce new, toxic substances in its tissues in response to the presence of the foreign genes. Critics also point out that the possibility of unknown and unanticipated harm is always present.

The problems concerning access to the new technologies relate to the developing world. So far, almost all GMOs have been developed by large agricultural–industrial firms with profit as the primary motive. Farmers are forbidden by contract from simply propagating the seeds themselves and

▼ **FIGURE 10–8** *Papaya plants in Hawaii.* The healthy plants have been genetically altered to prevent ring spot, a viral disease of papayas; the plants on the right are traditional ones. The developers of a papaya that is resistant to a deadly virus, from left, Dr. Richard Manshardt, Dr. Dennis Gonsalves, and Maureen Fitch, a Hawaiian graduate student.

must purchase seeds annually. Farmers in the developing countries are far less able to afford the higher costs of the new seeds, which must be paid up front each year. Aware of the potential for farmers simply to begin to produce their own seeds, researchers developed what is called the *terminator technology*, a transgenic technique that renders any seeds from the crops sterile. Faced with a firestorm of criticism, the large biotech corporation Monsanto backed down on a decision to adapt terminator technology to its transgenic crops. Still the problem of cost remains, and because most farmers in developing countries are poor, the major biotech firms are not likely to develop products aimed at addressing their unique crop needs. Fortunately, some noncommercial and donor-funded laboratories are taking aim at this problem, as we have seen with the development of golden rice.

All of these concerns and, perhaps beyond them, the general fear of what unknown technologies can do have resulted in a huge controversy over the spreading use of GMOs by farmers and the foods developed from such organisms. Activists around the world have rallied to protest the development of GMOs and the "Frankenfoods" derived from them. The protest has gone as far as vandalizing agricultural test plots in the United States and Europe. The protests have been strongest in Europe, where concerns over food have been exacerbated by other recent food scares. Far less concern is found in the United States. Observers believe that the fact that U.S. companies have been in the forefront of this technology might have something to do with the European response. In the light of the controversy, major food-producing companies have begun to pull back from the use of GMOs in foodstuffs (Frito Lay is telling its suppliers not to use genetically altered corn, for example, and Gerber will not put any such crops in baby foods), suggesting that we may begin to see a reduction in the use of this technology in common food crops like corn and soybeans. Unfortunately, farmers are caught in the middle and are scrambling to maintain a balance between conventional and genetically modified versions of their crops in order to satisfy an unpredictable market.

Policies. Biotechnology and its application to food crops do not exist in a regulatory vacuum. The Environmental Protection Agency (EPA), the U.S. Department of Agriculture (USDA), and the Food and Drug Administration (FDA) all have had regulatory oversight of different elements of the application of biotechnology to food crops. Concern over their oversight led to a report from the National Academy of Science's National Research Council, a prestigious and influential body. The report, titled *Genetically Modified Pest-Protected Plants: Science and Regulation*, published in April 2000, concluded that transgenic crops have been adequately tested for environmental and health effects, but that the three agencies involved need to do a better job of coordinating their work and sharing their information more with the public. Overall, the report en-

dorsed the science, technology, and regulation, stating, "The committee is not aware of any evidence that foods on the market are unsafe to eat as a result of genetic modification." At the same time, the report called for more research on the environmental and safety issues.

On the international level, the U.N. Convention on Biodiversity sponsored a key conference in Montreal, Canada, in January 2000, to deal with trade in GMOs. After heated negotiations, the conference reached an agreement, called the **Cartagena Protocol**. Surprisingly, the agreement was welcomed by governments, the private sector, and environmental groups alike. At issue was a fundamental philosophy about how technologies, particularly new ones, should be regulated. Those skeptical of the claims of biotech companies about the safety of their products wanted *proof that the GMOs were safe* before they were allowed into their countries. Those advocating the use of GMOs wanted to see the *suspected dangers proven* before GMOs could be denied access to overseas markets.

The Cartagena Protocol states that the "lack of scientific certainty due to insufficient relevant scientific information and knowledge…shall not prevent [a country] from taking a decision" on the import of LMOs, or living modified organisms, as the Protocol calls GMOs. The Protocol puts the right to deny entry of any GMOs in the hands of the importing country, but its decision must be based on sound science (involving an assessment of the risks involved) and the broad sharing of information about the products. Thus, the agreement makes operational for the first time in international environmental affairs the so-called **precautionary principle**, which states that *where there are threats of serious or irreversible damage, lack of scientific certainty should not be used as a reason for failing to take measures to prevent potential damage.* As another plank in the Protocol, shipments that contain food commodities made with GMOs must be labeled in wording which says that they *may* contain GMOs; this wording was a compromise that enabled opposing parties to reach a consensus on the entire Protocol. As with other agreements from U.N. conventions, participating countries must ratify the Protocol, which will become effective when 50 countries have signed on.

Clearly, a regulatory framework is being erected around the use of this new technology. The public response is still highly negative in Europe and positive in the United States, although activists have been quite vocal on both sides of the Atlantic and are starting to be heard in the developing countries. However, the future almost certainly will have to include major advances in food production from biotechnology, and many of these advances are likely to benefit the developing countries. A new plant revolution is underway, and it is unlikely to be stopped. Former President Jimmy Carter put it well: "Responsible biotechnology is not the enemy; starvation is." If food production is to keep pace with population growth, biotechnological advances will be essential, in the view of most observers.

10.2 Food Distribution and Trade

For centuries, the general rule for basic foodstuffs—grains, vegetables, meat, and dairy products—was *self-sufficiency*. Whenever climate, blight (as in the 19th century Irish potato famine), or war interrupted the agricultural production of a nation or region, the inevitable result was famine and death, sometimes on the scale of millions. Once colonies were established in the New World, timber, furs, tobacco, fish, and, later, sugar, coffee, cotton, and other raw materials began to flow back to the Old World. In turn, the Old World exported manufactured goods, which helped transform the colonies into societies much like the European ones that had given birth to them. With the Industrial Revolution, trade between nations intensified, and soon it became economically feasible to ship basic foodstuffs from one part of the world to another. In time, a lively and important world trade in foodstuffs arose, and as it did, the need for self-sufficiency in food diminished. Like other sectors of the economy, food has become globalized.

Patterns in Food Trade

Today, agricultural production systems do much more than supply a country's internal food needs. For some nations (such as the United States and Canada), the capacity to produce more basic foodstuffs than the home population needs represents an extremely important entry into the international market. And for many countries (especially those of the developing world), special commodities such as coffee, fruit, sugar, spices, cocoa, and nuts provide their only significant export product. This trade clearly helps the exporter, and it allows importing nations to use foods that they are not able to raise. Given the realities of a market economy, the exchange works well only as long as the importing nation can pay cash for the food. Cash is earned by exporting raw materials, fuel, manufactured goods, or special commodities. In this way, for example, Japan imports $128 billion worth of food, livestock feed, and other raw materials, but more than makes up for it by exporting $421 billion in manufactured goods (cars, electronic equipment, and so on) annually.

The most important foodstuff on the world market is grain: wheat, rice, corn, barley, rye, and sorghum. To be sure, much of this is feed grain imported by some high- and middle-income countries to satisfy the rising demand for animal protein. It is instructive to examine the pattern of global trade in grain over the past six decades (Table 10–3). In 1935, only Western Europe was importing grain; Asia, Africa, and Latin America were self-sufficient. By 1950, new patterns were emerging, and today the trade in grains—as well as other basic foodstuffs—represents a development with great economic and political implications. As the table shows, North America has become the major source of exportable grains—the world's "breadbasket" or, in another sense, the world's "meat market." Over the past 40 years, Asia, Latin America, and Africa have shown an increasing dependence on imported grain. These three regions also have in common 40 years of continued rapid population growth. For example, Mexico, birthplace of the Green Revolution, now imports 12 million metric tons of grain per year; population growth has eaten up all the gains of the Green Revolution. Although most of the food needs of these regions are met by internal production (Mexico, for example, produced 29 million tons of grain in 1998), the trend toward greater dependency is an ominous signal.

At no time in recent history has the world grain supply run out; on the average, carryover stocks (the amount in storage as a new harvest begins—in 1998, 293 million tons) are enough to supply more than a year's worth of interna-

TABLE 10–3 World Grain Trade since 1935								
	Amount Exported or Imported (Million Metric Tons)[1]							
Region	1935	1950	1960	1970	1980	1990	1995	1998
North America	5	23	39	56	131	123	108	94
Latin America	9	1	0	4	−10	−12	−4	−13
Western Europe	−24	−22	−25	−30	−16	28	26	18
Eastern Europe and Former U.S.S.R.	5	0	0	0	−46	−38	−5	5
Africa	1	0	−2	−5	−15	−31	−31	−39
Asia	2	−6	−17	−37	−63	−82	−78	−73
Australia and New Zealand	3	3	6	12	19	15	13	20

[1]A minus sign in front of a figure indicates a net import; no sign indicates a net export.
(*Source*: FAOSTAT Food Balance Sheets, Food and Agricultural Organization.)

tional trade and aid. In other words, enough food is produced to satisfy the world market—enough to feed millions of animals and keep a healthy surplus in storage. Why, then, are there people in every nation who are hungry and malnourished? Shouldn't every nation make an all-out effort to provide food for its people? And if it is unable to, shouldn't the rest of the world assume some of the responsibility for providing food to a hungry nation? Where does the responsibility lie for meeting the need for this most basic resource?

Levels of Responsibility in Supplying Food

The Bread for the World's Hunger Institute defines **food security** as "assured access for every person to enough nutritious food to sustain an active and healthy life." To begin to answer the question of who is responsible for food security, it is helpful to examine Fig. 10–9. The figure displays *three major levels of responsibility for food security*: the **family**, the **nation**, and the **global community**. At each level, the players are part of a market economy as well as a sociopolitical system. It is important to note that in a market economy, food flows in the direction of *economic demand*. *Need* is not taken into consideration. To illustrate, in the event that there are hungry cats and hungry children, the food will go to the cats if the owners of the cats have money and the children's parents don't. In other words, the cash economy, following the rules of the market, provides the *opportunity* to purchase food, but not the food itself. Where the economic

status of the player (a destitute breadwinner, a poor country) is very low, the next level may be willing and able to provide the needed purchasing power or the food (but maybe not).

In the United States, this help is described as the "safety net" and is represented by a variety of welfare measures, such as the Food Stamp Program and the Supplemental Security Income program. Recently, the safety net was radically transformed by the Personal Responsibility and Work Opportunity Reconciliation Act of 1996. The new law is intended to move welfare recipients from public assistance to independence by requiring that they join the workforce or perform "workfare" and by limiting the duration during which they are publicly supported. Eligibility for food stamps for adults also has work requirements. Critics of the new law state that it represents an unprecedented and unjust attack on the poor, one that has resulted in pushing many families and children into severe poverty—in effect, pulling the safety net out from under them. Indeed, over the past four years, food stamp enrollment has declined 28% (8 million people). Other measures indicate that the decline in the poverty rate is much lower than 28%.

The most important level of responsibility is the family. The *goal* at this level is to meet the nutritional needs of everyone in the family to an extent that grants freedom from hunger and malnutrition. For an individual, there are four legitimate options for attaining food security: Purchase the food, raise the food, gather it from natural

Family	Country	Global community

Goal: Personal and family food security
Policies:
—*Employment security*
—*Adequate land or livestock*
—*Good health and nutrition*
—*Adequate housing*
—*Effective family planning*
—*Access to food*

Goal: Self-sufficiency in food and nutrition
Policies:
—*Just land distribution*
—*Support of sustainable agriculture*
—*Effective family planning*
—*Promotion of market economy*
—*Avoidance of militarization*
—*Effective safety net*

Goal: Sustainable food and nutrition for all
countries
Policies:
—*Food aid for famine relief*
—*Appropriate technology in development aid*
—*Aid for sustainable agricultural
development*
—*Debt relief*
—*Fair trade*
—*Disarmament*
—*Family planning assistance*

▲ **FIGURE 10–9** *Responsibility for food security.*   Goals and strategies for meeting food needs at three levels of responsibility: the family, the country, and the globe.

ecosystems, or have it provided by someone, usually in the context of the family.

In the event of economic or agricultural failure at the family level, the fourth option implies that there is an effective safety net—that, at some level, national, state, or local, there exist policies whose objective is to meet the food security needs of all individuals in the society. Thus, an appropriate goal at the national level would be *self-sufficiency in food*—enough food to satisfy the nutritional needs of all of a country's people, thereby providing *an effective safety net*. The nation can either produce all the food its people needs or buy it on the world market. This goal implies policies to eliminate chronic hunger and malnutrition in the society. Many nations, though, are not self-sufficient in food, and those which cannot afford to buy all they need must turn to the global community for *food aid*. (See section 10.3.)

There also are some less obvious factors to consider when we are talking about global food needs. One of the most serious is the debt crisis in developing countries, addressed in Chapter 7. Another factor in meeting global nutrition needs is the trade imbalance between the industrial and developing countries. The developing countries typically export commodities like cash crops, mineral ores, and petroleum and import more sophisticated manufactured products such as aircraft, computers, machinery, and the like. Prices for the latter have risen, while those for the commodities from developing countries often decline. (Making matters worse is **tariff escalation**, wherein processed products from developing countries are heavily taxed.) To some extent, the increasing use of labor in the developing countries for electronic assembly and clothing manufacture has offset this imbalance. Indeed, the globalization of markets has begun working to the advantage of the developing countries, which have a surplus of labor and low labor costs and which have thus been the recipients of much outsourcing of manufacture by industrial corporations. What the poor need is jobs, and even if the pay is far below what workers in industrial countries receive, it is often enough to help workers in developing countries improve their living conditions and their diets.

Consequently, being part of the global market system is a major opportunity for many developing countries, but there is often a downside to the process: Workers are exploited in many developing countries, by both domestic and foreign-owned businesses. Labor laws are often ignored, workers frequently live in squalor, and in some countries, children are drafted into the workforce. According to the International Labor Organization, some 120 million children ages 5 to 14 are fully employed in the developing countries. To their credit, most industrial countries are trying to reduce this scourge by prohibiting the importation of goods manufactured with child labor.

Another consequence of the trade imbalance is the tendency of the developing countries to "mine" their natural resources for the external market. Thus, lumber, fish prod-

ucts, minerals, and other resources are exploited in order to generate a cash flow. Often, such exploitation destroys the environment.

Appropriate global food security goals for the wealthy nations are listed in Fig. 10–9. These are policies that promote self-sufficiency in food production and sustainable relationships between the poorer nations and their environments, as well as addressing the trade imbalance and human-exploitation problems.

10.3 Hunger, Malnutrition, and Famine

At a U.N. World Food Conference in 1974, delegates from all nations subscribed to the objective "that within a decade no child will go to bed hungry, that no family will fear for its next day's bread, and that no human being's future and capacities will be stunted by malnutrition." In 1996, 22 years later, the United Nations held the World Food Summit to address *continuing* hunger, malnutrition, and famine in the world. (See the "Global Perspective" essay, p. 256.) And in 1999, the FAO published an update showing that over 800 million are still suffering from hunger. In light of the evident availability of enough food to feed everyone, and in view of the global market in foodstuffs, why is the 1974 U.N. objective still unfulfilled on such a large scale?

Nutrition vs. Hunger

Hunger is the general term referring to a lack of basic food required for energy and for meeting nutritional needs such that the individual is unable to lead a normal, healthy life. **Malnutrition** is the lack of essential nutrients such as amino acids, vitamins, and minerals, and **undernourishment** is the lack of adequate food energy (usually measured in Calories). What is adequate nutrition, and how can individuals be sure that what they eat provides it? Recently, the U.S. Department of Agriculture developed the **Food Guide Pyramid** to help Americans achieve better nutrition (Fig. 10–10). The problem of overnutrition is widespread: Almost 23% of the U.S. population is clinically obese, and at least 1.2 billion people worldwide overeat and are overweight. To address this and other nutritional problems, the USDA established six food groups and arranged them in a pyramid in order to indicate the relative proportions of each that should be consumed.

Following these guidelines facilitates weight control and cuts down on one's chances of developing diabetes, heart disease, high blood pressure, and some cancers, typical problems of the overnourished. However, these disorders are not very common in most parts of the developing world. There, the major nutritional problems are a lack of proteins and some vitamins (malnutrition), especially in children, and a lack of Calories for food energy (undernourishment) for all ages.

Extent and Consequences of Hunger

Accurate, reliable figures on the worldwide extent of hunger are unavailable, mainly because few governments make any effort to document such figures. On the basis of surveys of available food in the society, with accessibility to food factored in, the FAO estimates that 790 million people in the developing world and an additional 34 million in the developed world are underfed and undernourished. These numbers may be an underestimate. Using normal weight as a measure of nutrition, a recent study showed that 49% of adults and 53% of children in India are underweight; by contrast, the FAO estimated only 21% as being chronically undernourished. The World Health Organization estimates that half the people in the world, 3 billion, suffer from some form of malnutrition, and a survey of studies from other U.N. agencies indicates that at least 1.2 billion are undernourished. Whichever numbers we accept, it is fair to say that hunger and malnutrition still take a terrible toll on human life and productivity.

Almost two-thirds of the undernourished (526 million) live in Asia and the Pacific. The regions most seriously affected are India and Bangladesh, Southeast Asia, and sub-Saharan Africa. Almost half of the people in central, eastern, and southern Africa are undernourished. It is important to remember that hunger has faces and personalities. Listen to these quotes from the World Bank's *Voices of the Poor*: from Senegal, "Your hunger is never satisfied, your thirst is never quenched; you can never sleep until you are no longer tired"; from a 10-year-old child from Gabon, "When I leave for school in the mornings, I don't have any breakfast. At noon there is no lunch, in the evening I get a little supper, and that is not enough. So when I see another child eating, I watch him, and if he doesn't give me something, I think I'm going to die of hunger"; and from Ukraine, "In the evenings, eat sweet potatoes, sleep; in the mornings, eat sweet potatoes, work; at lunch, go without."

The consequences of malnutrition and undernutrition vary. The effects are greatest in children and next greatest in women. Hunger can prevent normal growth in children, leaving them thin, stunted, and often mentally and physically impaired (Fig. 10–11), a condition known as protein-energy malnutrition. The U.N. Subcommittee on Nutrition has identified malnutrition and hunger in children of southern Asia as the world's most serious nutritional problem, involving at least 108 million children who are underweight because of lack of food. Research has shown that undernutrition in early childhood can seriously limit one's intellectual development throughout life. Women in the developing world are especially vulnerable because of widespread patterns of low status and the performance of heavy labor.

Sickness and death are companions of hunger. Because poor nutrition lowers a person's resistance to disease, measles, malaria, and diarrheal diseases become common and are major causes of death in the malnourished and undernourished. According to the Bread for the World Institute on Hunger and Development, "Almost 40,000 children under five die each day from malnutrition and infection . . . . The number of deaths is the same as if one hundred jumbo

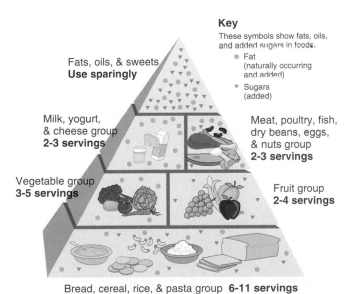

▲ **FIGURE 10–10**   *The Food Guide Pyramid.*   A guide produced by the U.S. Department of Agriculture to help people evaluate their food needs and food intake so as to provide adequate nutrition and keep their weight under control. Suggested numbers of servings are given as a range, because energy requirements vary for people, depending on their size, age, and level of activity. Thus, older adults and many women should use the minimum number of servings in order to cut down on calories (they might need only 1,600 calories per day), while teenage boys and active men might need the maximum number of servings to get the 2,800 or so calories they need daily.

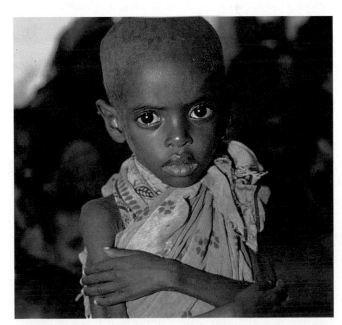

▲ **FIGURE 10–11**   *Serious malnourishment.*   A child in the Denunay Feeding Center, Somalia, operated by World Vision International. Famines and food shortages in Somalia were largely a consequence of civil conflict, not drought.

GLOBAL PERSPECTIVE

WORLD FOOD SUMMIT

Under the leadership of the Food and Agricultural Organization (FAO) of the United Nations, representatives and heads of state from 100 countries met from November 13–17, 1996. Called the World Food Summit, the meeting was the result of two years of planning and negotiations. The objective was to bring about a renewed commitment around the world to eradicate hunger and malnutrition and to promote conditions leading to food security for individuals, families, and countries everywhere. The need for this high-level attention was emphasized by new data from the FAO indicating that 840 million people, or 18% of the population of the developing world, are malnourished or hungry. In light of continuing increases in population, rising costs of grain, and declining per capita grain production, the summit was a unique opportunity for world leaders to take a new look at the meaning of sustainability.

At the start of the meeting, the delegates adopted, by acclamation, two major documents: the Rome Declaration on World Food Security and the World Food Summit Plan of Action. These documents,

in the words of the FAO, "set forth a seven-point plan stipulating concrete, political actions to ensure:

1. conditions conducive to food security;
2. the right of access to food by all;
3. sustainable increases in food production;
4. trade's contribution to food security;
5. emergency relief when and where needed;
6. the required investments to accomplish the plan;
7. concerted efforts to achieve results by countries and organizations."

One concrete objective of the plan was a 50% reduction in the number of hungry people by the year 2015. As a measure of success, this objective was criticized by some participants as being too timid, since it assumed that 420 million people would still be malnourished or hungry. The plan emphasizes the responsibilities of countries (especially developing countries) to enact reforms that would promote food security—policies that (1) do not discriminate against agriculture or small farmers, (2) open trade,

(3) foster greater efforts to bring down population growth, and (4) encourage investing in the country's infrastructure.

The World Food Summit mandated a new program, the Food Insecurity and Vulnerability Information and Mapping Systems (FIVIMS), which produced its first report in 1999: *The State of Food Insecurity in the World.* Although the total number of hungry in the world has declined by 40 million, only 37 countries achieved reductions (totaling 100 million), while the rest of the developing world saw increases in hungry people (totaling 60 million). At the current rate of reduction of 8 million a year, there will still be over 600 million hungry by 2015, indicating that the objective of 50% reduction will not be attained. The FAO estimates that reaching that goal would cost an additional $6 billion a year; other studies indicate that the cost would be somewhere between $2.6 and $4 billion. Although this sounds like a lot of money, consider that the world spent more than $500 billion to fix the Y2K computer problem. It has been estimated that hunger causes a loss in human productivity of $16 billion a year—the cost of *not* ending world hunger.

jets, each loaded with 400 infants and young children, crashed to Earth each day—one every 14 minutes." Hunger is often a seasonal phenomenon in rural areas that are supported by subsistence agriculture, as people are forced to ration their stored food in order to survive until the beginning of the next harvest. Anyone who travels in the developing world cannot help but notice the fact that few people in the rural areas look well fed; most are thin and spare from a lifetime of hard work and limited access to food.

Root Cause of Hunger

In the view of most observers, *the root cause of hunger is poverty.* Our planet produces enough food for everyone alive today. However, hungry and malnourished people lack either the money to buy food or adequate land to raise their own.

Inadequate food is only one of many consequences of poverty. Alan Durning of the Worldwatch Institute defines **absolute poverty** as "the lack of sufficient income in cash or kind to meet the most basic biological needs for food, clothing and shelter"; sufficient income is measured as income equal to or more than $1 a day. On the basis of this definition, 1.2 billion people live in absolute poverty today. Most

of these people reside in rural villages, are illiterate, spend half or more of their income on food, and represent races, tribes, or religions that suffer discrimination. They are powerless to do anything about their plight, and quite often the society in which they live is content to keep them that way. Millions are trapped in a cycle of poverty that leads to the degradation of resources and perpetuates the poverty—all made worse by the high fertility of these rural populations.

Hunger and poverty do not always go from bad to worse. A number of Asian countries, including China, Indonesia, and Thailand, significantly reduced the extent of poverty and hunger in them during the 1980s. Deliberate public policies and social services have greatly improved the welfare of millions in China, where food security is a matter of high national priority. Indonesia has benefited from oil exports and Green Revolution technology, and it continues to put a major emphasis on rural development and the country's social infrastructure. Clearly, it is possible for societies to address the needs of the hungry poor with appropriate public policies and to make progress in reducing the extent of absolute poverty and hunger. On the other hand, the most severe kind of hunger—famine—is found in societies that are regressing into disorder and chaos, and it is in those societies that international responsibility comes most sharply into focus.

Famine

By definition, a **famine** is a severe shortage of food accompanied by a significant increase in the death rate. Famine is a clear signal that a society is either unable or unwilling to distribute food to all segments of its population. Two factors have been immediate causes of famines in recent years: *drought* and *warfare*.

Drought is blamed for famines that occurred in 1968–74 and again in 1984–85 in the Sahel region of Africa (Fig. 10–12). The Sahel is a broad belt south of the Sahara desert, occupied by 60 million people who practice subsistence agriculture or tend cattle, sheep, and goats. (Such people are called *pastoralists.*) Although the region normally has enough rainfall to support dry grasslands or savanna ecosystems, the rainfall is seasonal, undependable, and prone to failure. Making matters worse, population increases in the region have led to unsound agricultural practices and overgrazing by the expanding herds of livestock. Beginning in 1965, the region experienced 20 years of subnormal rainfall, with tragic results. Crops withered, forage for livestock declined, watering places dried up, and livestock died. Both farmers and pastoralists began abandoning their land and migrating toward urban centers, where they were often herded into refugee camps. Unsanitary conditions in the camps and the already weakened condition of the refugees led to the spread of infectious diseases such as dysentery and cholera, and many thousands died before effective aid could be organized. The 1984–85 Sahelian famine is thought to have been responsible for 750,000 deaths; the number would have been in the millions, except for aid extended by Africans and numerous international agencies.

In 1999 and 2000, drought returned to the eastern Sahel, bringing on massive food shortages and famine. At least 10 million in Ethiopia were facing famine as food relief from the World Food Program and the U.S. Agency for International Development was being organized. Eritrea, Kenya, Uganda, and Tanzania have also seen droughts in the past year and are dealing with serious food shortages. The Famine Early Warning System Network, a U.S.-funded partnership of a number of agencies involved in satellite operations, has been alerting the world to conditions in the Horn of Africa, where almost 20 million are facing serious food shortages. Food aid to this region is being mobilized and monitored, with the network continually measuring rainfall and agricultural conditions in the region. The network issues regular bulletins and special reports on the World Wide Web, giving governments and relief agencies accurate and timely assessments of the status of food security in African countries.

Famines in which the causative factor was not drought, but war, threatened several African nations in the 1990s: Ethiopia, Eritrea, Somalia, Rwanda, Sudan, Mozambique, Angola, and Congo (Fig. 10–12). Devastating and prolonged

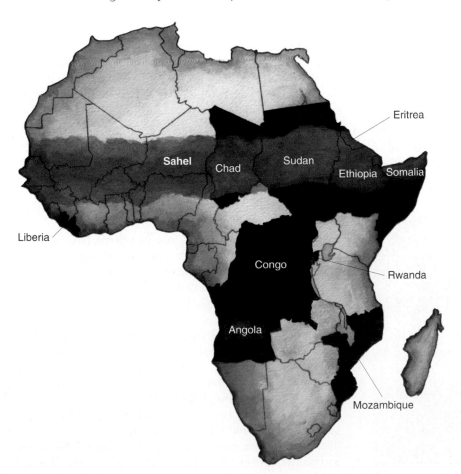

◄ **FIGURE 10–12** *The geography of famine.* Famines have occurred repeatedly in sub-Saharan Africa, especially in the Sahel (a band of dry grasslands that stretches across the continent). The map shows the countries where civil wars and droughts recently brought on serious famines.

civil warfare put millions of Africans at risk of famine. The civil wars disrupted the farmers' normal planting and harvesting and forced the displacement of millions from their homes and food sources. Governments in power maintained control over food and relief supplies; relief agencies operated under dangerous conditions and frequently experienced casualties. In some areas, the problem was made worse by persistent droughts. In Mozambique alone, 900,000 people died from direct military action or from indirect effects of the war there.

Clearly, famines from drought and war are preventable. India, Brazil, Kenya, and southern Africa have coped with droughts in recent years by mobilizing effective relief in the form of food, clothing, and medical assistance. Indeed, the drought of the early 1990s accelerated the peace process in Mozambique and helped change the political landscape in southern Africa. Cooperation between South Africa and the 10 nations of the Southern Africa Development Community to prevent famine lowered barriers between South Africa and its neighbors prior to the coming of democracy to that troubled nation.

Two countries recently affected by serious famines are Sudan and North Korea. In Sudan, civil war and drought combined to put almost 1 million people in peril, triggering a massive airdrop by the World Food Program to provide almost 10,000 tons of food a month to the southern parts of Sudan (Fig. 10–13). Many humanitarian aid organizations, such as Doctors Without Borders and World Vision, are doing what they can to help, but the situation will not get better until there is peace in the region. Another set of circumstances has brought famine to North Korea. The primary cause has been the failure of centrally planned agricultural policies in a country that is one of the last bastions of communism. Floods and droughts alternatively have made conditions worse, to the point where millions are judged to be close to starvation and as many as two mil-

lion have already perished. For years, North Korea refused to admit that there was a problem, but recently the country has permitted the United States to deliver several hundred thousand tons of food and medical assistance.

These two recent cases open the question of the function and effectiveness of food aid.

Food Aid

The World Food Program of the United Nations coordinates global food aid, receiving most of its donations from the United States, Canada, Australia, and the countries of the European Union. In addition, many nations conduct their own programs of bilateral food aid to needy countries. Although international food aid was cut in half between 1993 and 1996 (to 6.2 million tons), it has been on the increase since then and reached 13.2 million tons in 1999 (Fig. 10–14). Much of the increase in 1999 was the result of a major shortfall in grain production in the Russian Federation and a consequent extension of food aid primarily from the United States (admittedly, to boost U.S. farm exports, which had been falling).

Food aid is distributed to countries all over the world, not just where famines are threatened. This raises the question, What is the proper role of food aid? Clearly, aid is vital in saving lives where famines occur. But what about the people who suffer from chronic hunger and malnutrition? The basic question is, When should food be given to those in need, instead of being distributed according to market economics?

Numerous humanitarian campaigns to end world hunger have been mounted in the last 50 years. The United States and Canada have been world leaders in giving away food (which is first purchased from farmers and therefore represents a subsidy). As mentioned earlier, a number of serious famines have been moderated or averted by these efforts. As virtuous as such efforts seem on the surface, however, routinely supplying food aid in an attempt to alleviate chronic hunger in developing countries may be the worst thing to do.

The problem is, people will not pay more than they must for food. Therefore, free or very cheap foreign food undercuts the local market. In effect, local farmers must compete economically with free or low-cost imported food. When they cannot earn a profit, they stop producing and eventually enter the ranks of the poor. The cycle continues, with people who sell goods to the farmer also suffering when the farmer loses buying power. In the long run, the entire local economy deteriorates. Hence, the donation of food, while well intended, often aggravates the very conditions that it is meant to alleviate. Meanwhile, population pressures continue to build and the magnitude of the problem increases. (See the "Ethics" essay, p. 259.)

Some food aid is strictly humanitarian: Bangladesh receives over a million tons per year, about half as much as it purchases on the market. The African continent receives 2–3 million tons per year. All the signs indicate that this fig-

▼ FIGURE 10–13   *Food Aid.*   Women receive aid from the World Food Program in Sudan.

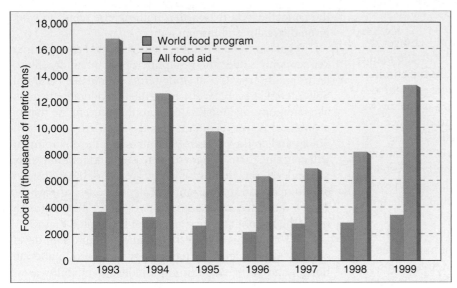

◀ **FIGURE 10–14** *Global food aid.*
Global food aid from 1993 to 1999. Blue indicates food aid delivered by the World Food Program; food aid from all sources is indicated in red. (*Source:* World Food Program statistics, UNFAO.)

ure will increase, especially in light of the recent drought. A significant amount of politically expedient food aid is flowing to Eastern Europe and the countries of the former Soviet Union, largely to bolster their fragile economies.

Food aid will undoubtedly continue to be an international responsibility. It is at best a buffer against famine, and it will probably continue to be awarded to some countries for political reasons. As part of the solution to the chronic hunger and malnutrition among the poor, however, simple food aid is clearly counterproductive. Much more good will be accomplished by a restructuring of the economic arrangements between rich and poor nations and by the extension of development aid directed toward fostering self-sufficiency in food and sustainable interactions with the environment.

ETHICS

THE LIFEBOAT ETHIC OF GARRET HARDIN

Biologist Garret Hardin has published several provocative essays addressing the worldwide food-vs.-population issue. Here we give you an opportunity to respond to Hardin's thinking.

We begin with the concept of carrying capacity: the number of a species that can be supported indefinitely without degrading the environment. For human societies, the carrying capacity means the ability to meet food needs over the long term—that is, sustainably. If ecology had a decalogue, Hardin says, the first commandment would be "Thou shalt not transgress the carrying capacity." A look at the world scene reveals that numerous countries are pressing against the limits of, or have exceeded, their carrying capacity. This, says Hardin, is their problem, not ours, and he uses a lifeboat metaphor to show why.

Picture a number of lifeboats in the sea after a ship has sunk—some crowded with people and some in which people are riding in relatively uncrowded luxury. Each lifeboat has a limited capacity. The people in the crowded boats are continually falling overboard, leaving the people on the uncrowded boats with the problem of whether to take them on board or not. Imagine an uncrowded boat with 50 on board and room for 10 more, with 100 people treading in water and begging to be taken on board. There are several options: (1) Assume that all people have an equal right to survival, and take everyone on board. This, says Hardin, would lead to catastrophe for all. (2) Admit only 10, filling all the space on the boat. Two problems with this option are that you lose your margin of safety and that you must find a basis on which to discriminate among all the people in the water. (3) Admit no more to the boat. This preserves your margin of safety and guarantees the long-term survival of the people on your boat; it is the rational solution to the lifeboat problem.

The metaphor, of course, is to be applied to the problem of food aid. Some people would argue that if less grain went to feeding animals, there would be more available to feed the hungry in poor countries. Or, since there are often agricultural surpluses in the rich nations, we could use those surpluses to feed the hungry. The real problem, then, is a problem of food distribution, not of the quantity produced. We should give food to the hungry wherever they are.

These arguments, says Hardin, are foolish. In giving away food, we would only be encouraging the "population escalator": A population growing rapidly reaches the limits of its food capacity and is supplied with food from abroad, encouraging still further growth and necessitating still further food aid, and so on. Our hearts, says Hardin, tell us to send food, but our heads should tell us not to. We only postpone the day of reckoning, and in the end, the amount of suffering will be greater. Overpopulated, food-poor countries have transgressed the first commandment of ecology. If we want to help, we should direct our aid toward limiting population growth, according to Hardin. What do you think? (Sample the opinion of those to whom you speak.)

Closing Thoughts on Hunger

In 1980, a Presidential Commission on World Hunger delivered its report to President Jimmy Carter, with the major recommendation "that the United States make the elimination of hunger the primary focus of its relations with the developing world." The thrust of the report was that if we respect human dignity and have a sense of social justice, we must agree that hunger is an affront to both. The right to food must be considered a basic human right. It follows, then, that we as a nation have a moral obligation to respond to world hunger, the report concluded. Today the question is, Has that moral obligation made its way into public policy in the ensuing years?

By now it should be clear that, although food is our most vital resource, we do not treat it as a commons (free to all that need it). Indeed, the production and distribution of food constitute one of the most important economic enterprises on Earth. Alleviating hunger, as we have seen, is primarily a matter of addressing the absolute poverty that afflicts one of every five people on Earth. To treat food as a commons would be to treat wealth as a commons. Even though this is one of the tenets of socialist systems, it has proved to be a completely unworkable one. We are part of the world economy now, and it is a market economy—for the perfectly good reason that nothing else seems to work, given the realities of human nature.

What has not been done, however, is to bring this market economy under the discipline of sustainability. Short-term profit crowds out long-term sustainable restraint. Self-interest at every level—from the individual to the global community—generates decisions that prevent the sharing of political power, economic goods, knowledge, and technology and, in the process, guarantees that the environment will continue to be degraded. Why isn't land reform carried out in developing countries? Why are the rich nations content to maintain the current imbalance between prices for commodities and for manufactured goods? Why is hunger not at the top of the policy agenda of the United States?

Our understanding of the situation is quite well developed. No new science or technology is needed to alleviate hunger and at the same time promote sustainability as we grow our food. The solutions lie in the realm of political and social action at all levels of responsibility. Given the current groundswell of concern about the environment and the disappearance of the Cold War between capitalism and communism, there may never be a better time to turn things around and take more seriously our responsibilities as stewards of the planet and as our brothers' keepers.

ENVIRONMENT ON THE WEB

GETTING THE FOOD WHERE IT'S NEEDED

It is sometimes suggested that there is enough food produced in the world to feed all of Earth's population, yet the real difficulty is to get the food where it is most needed. Indeed, close examination of recent famine situations reveals that food aid is often hindered by one of three factors: (1) poor infrastructure, such as roads and railways, for food distribution; (2) political instability or war, causing disruptions in the food distribution system; or (3) corruption within the system, so that the most powerful, rather than the poorest, individuals receive the food. In warring nations, food aid is sometimes misdirected to feed soldiers of the regime in power.

The recent famine in the Democratic People's Republic of Korea (North Korea) illustrates these points. Devastating floods in 1995 displaced more than half a million North Koreans and left more than 2.5 million critically short of food. Subsequent droughts and the Asian economic crisis have deepened the problem. Now, more than 6 million North Koreans rely on food aid. Though North Korea has experienced chronic food shortages for many years, Catherine Bertini, head of the U.N. World Food Programme, says that now "there is almost no food in the country." Western governments and relief organizations have been quick to respond, shipping more than 800,000 tons of food to the country since January 1995. Analysts believe, however, that the true need may be more than a million tons a year, just to meet minimum survival needs.

The roots of this famine lie in North Korea's unstable agricultural economy; the weak state of its agricultural policies seems to indicate that more than food aid will be needed to provide a permanent solution to the problem. A stable agricultural economy will also require political stability, particularly with neighboring South Korea. Currently, South Koreans fear that their neighbors to the north will attack out of desperation for food. The "cold war" context of the conflict between North and South Korea further complicates the distribution of food aid—and increases the risk that food will be misdirected to serve political ends.

In recent years, North Korea has strictly limited the number of international aid monitors and restricted the movement of those currently in the country. Some aid agencies are therefore calling for food contributions to be contingent on adequate needs assessment and monitoring of distribution systems to avoid the problems of persistent famine experienced in Congo, Zaire, and Bosnia in recent years.

Web Explorations

The Environment on the Web activity for this essay discusses the ethics of food aid and the need for systems that prevent famine. Go to the Environment on the Web activity (select Chapter 10 at http://www.prenhall.com/wright) and learn for yourself:

1. about the ethics of food aid;
2. about international rights and responsibilities in the management of famine; and
3. about the underlying causes of famine.

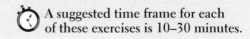 A suggested time frame for each of these exercises is 10–30 minutes.

REVIEW QUESTIONS

1. Describe the five components of the agricultural revolution. What are the environmental costs of each?

2. What is the Green Revolution? What have been its limitations?

3. Describe subsistence agriculture, and discuss its relationship to sustainability.

4. How do sustainable animal farming and industrial-style animal farming affect the environment on different scales?

5. What are our major prospects for increasing food production in the future?

6. What are the promise and problems associated with using biotechnology in food production?

7. What is the Cartagena Protocol, and how does it embody the precautionary principle?

8. Trace the patterns in grain trade between different world regions over the last 60 years. (See Table 10–3, p. 252.)

9. Describe the three levels of responsibility for meeting food needs. At each level, list several ways food security can be improved.

10. Define hunger, malnutrition, and undernourishment. What is the extent of these problems in the world?

11. How are hunger and poverty related?

12. Discuss the causes of famine, and name the geographical areas most threatened by it.

13. Why does food aid often aggravate poverty and hunger?

THINKING ENVIRONMENTALLY

1. Imagine that you have been sent as a Peace Corps volunteer to a poor African nation experiencing widespread hunger. How would you begin to address the needs of the people of that nation?

2. Record your food intake over a two- to three-day period. Analyze the nutritional value of your diet. Which nutrients are lacking? Which are in excess? What changes in your diet would reconcile these differences?

3. Of the following methods for increasing food production, which do you feel are viable options? Why?

   clearing forests
   increasing yield on farmland already in production
   irrigating arid lands
   developing transgenic crops with biotechnology
   aquaculture

WEB REFERENCES

On-line resources for this chapter are on the World Wide Web at: **http://www.prenhall.com/wright**. (Click on Chapter 10 within the Chapter Selector.)

Wild Species: Biodiversity and Protection

Key Issues and Questions

1. Preserving wild species may require that we find a way to show that they have value. How can we establish the value of natural species?
2. Wild species and their habitats are threatened and endangered by human activities, a matter of concern for many people. How does public policy in the United States protect endangered species?
3. Many naturalists claim that we are losing much of the biodiversity that has enriched Earth for millions of years. What is biodiversity, and what is the current extent of it?
4. It seems certain that humans are the cause of the decline in biodiversity. What human enterprises in particular are responsible for this decline?
5. Much of the loss in wild species and biodiversity is occurring outside the United States. How has the international community acted to protect biodiversity?

Puffins are roly-poly little seabirds that live in cold coastal waters on both sides of the North American continent. The coast of Maine once enjoyed flourishing puffin colonies on a number of offshore islands, but from colonial times the birds were hunted for their eggs, meat, and feathers, until, by 1900, only one very small colony remained. Protective laws were passed, but the islands once inhabited by the puffins began to host large colonies of predatory gulls; the puffins never returned. National Audubon Society biologist Steven Kress was teaching ornithology to summer visitors on the Maine coast in the early 1970s and began wondering whether it might be possible to bring the puffins back to Maine. Against the advice of many biologists, Kress launched a program to accomplish his dream—the Puffin Project, as it was called. His plan was to obtain puffins from Newfoundland, where they were still numerous, and install them on Eastern Egg Rock, a five-acre island seven miles off the Maine coast. This was no easy task. First the predatory gulls had to be removed,

with the approval of the Fish and Wildlife Service. Then, starting with four chicks in 1973, Kress and a team of volunteers and interns installed the chicks in burrows and fed them daily until they were ready to go to sea. (Puffins spend most of their time at sea.) Each year, more chicks were brought to the island. Kress' team had to convince the puffins that the island was a thriving colony, so they installed numerous painted wooden puffin decoys, played tape-recorded puffin noises, and installed bogus eggs until the first few birds began to return. Finally, in 1981, four pairs of puffins set up housekeeping on the island and raised the first Maine-bred chicks. The island now has a thriving colony of 25 pairs.

The techniques used by Kress and his coworkers were innovative, and his success has blossomed into an enterprise with an annual budget of $400,000, now called the Seabird Restoration Project of the Audubon Society. Using his methods, Kress has presided over a number of reintroductions of seabirds, including petrels in the Galápagos

◄ **The Atlantic puffin.** Dedicated researchers are helping to return puffins to the Northeastern United States.

Islands and, recently, Laysan albatrosses in Hawaii. The Fish and Wildlife Service has availed itself of Kress' methods to introduce seabirds to other islands where they once bred; many budding biologists got their life's inspiration from working with Kress. Today, children young and old can cruise out in special touring boats from nearby Boothbay Harbor to see puffins buzzing around Eastern Egg Rock and a number of other islands to which they have also been restored. Kress says, "Sometimes it's not enough to pass pro-tective laws or buy land to create new sanctuaries. Every case is different, so first you try to understand the species' needs and then you go in to re-create communities. We can either document the loss of a population or design techniques to try to save it."

In this chapter, we give attention to wild species. We consider their importance to humans, examine the perspective of biodiversity, and look at the public policies that seek to protect wild species.

11.1 Value of Wild Species

In Chapter 3, we saw that ecosystems and the wild species living in them are of enormous value to humankind, providing goods and services that are conservatively estimated to be worth $33 trillion a year. We also saw that if we wish to maintain the sustainability of these ecosystems, we must engage in their active management; in particular, we must preserve their productivity and maintain their biodiversity. Thus, it seems only sensible to put a high priority on protecting wild species and the ecosystems in which they live. Yet, views diverge widely on the kind of protection we might afford to wild species. Some want them protected so as to provide recreational hunting. Others feel strongly that we should not engage at all in hunting for sport. Many have a deeper concern for the ecological importance of wild species, and they view the ongoing loss of biodiversity as an impending tragedy. On the other hand, many in the developing world depend on collecting or killing wild species in order to eat or make money; their personal survival is at stake. These are all attitudes involving values we place on wild species. Can these differing values be reconciled in a way that still leads to sustainable management of these important natural resources?

Biological Wealth

About 1.75 million species of plants, animals, and microbes have been examined, named, and classified, but it is common knowledge that many species have not yet been found or catalogued. On the basis of ratios of known to unknown species in different groups, Australian taxonomist Nigel Stork believes that 13 million would be a good working figure, and in fact, it compares well with the working number of 13.6 million estimated by the 1995 U.N. Environmental Program (UNEP) survey, "Global Biodiversity Assessment."

These natural species of living things, collectively referred to as biota, are responsible for the structure and maintenance of all ecosystems. They and the ecosystems they form represent a kind of wealth—the **biological wealth**—that sustains human life and economic activity with goods and services. It is as if the natural world were an enormous bank account of capital assets, capable of paying vital, life-sustaining dividends indefinitely, but, like a bank account, only as long as the capital is maintained. The **biota** found in each country represents a major component of the country's wealth, and we have referred (Chapter 3) to the stocks of biota in a country as its **natural capital**. (See Chapter 23 for an economic view of the concept.) More broadly, this richness of living species is referred to as Earth's **biodiversity**.

Humans have always been dependent on Earth's biological wealth for food and materials. Some 10,000 years ago, our ancestors began learning to select certain plant and animal species from the natural biota and to propagate them, and the natural world has never been the same. Over time, forests, savannas, and plains were converted to fields and pastures as the human population grew and human culture flourished. In the process, many living species were exploited to extinction, and others disappeared as their habitats underwent development. At least 500 plant and animal species have become extinct in the United States alone, and thousands more are at risk. We have been drawing down our biological capital, with unknown consequences.

Now that we live in cities and suburbs in the industrialized world and get all our food from supermarkets, our connections to nature seem remote. That is an illusion, however: Our interactions with the natural world may have changed, but we are still dependent on biological wealth. And millions of our neighbors in the developing world are not so insulated from the natural world; their dependence is much more immediate, as they draw sustenance and income directly from forests, grasslands, and fisheries. Because of overwhelming economic pressures, these people, too, are often engaged in unsustainable practices by drawing down their biological capital, with consequences that are obvious and grave. A root source of our problem is the way we regard and value wild nature.

Two Kinds of Value

It was not so long ago that hunters on horseback would ride out to the vast herds of bison roaming the North American prairies and shoot them by the thousands, often taking only the tongues for markets back in the east. The passenger pigeons that darkened the skies in huge flocks were ruthlessly killed at their roosts to fill a lively demand for their meat, until the species was gone (Fig. 11–1). Plume hunters decimated egrets and other shorebirds to satisfy the demands of fashion in the late 1800s. Appalled at this wanton destruction, 19th century naturalists called for an end to the slaughter, and the U.S. public began to be sensitized to the losses. People began to see natural species as worthy of preservation, and naturalists began to look for ways to justify their calls to conserve nature. Thus, there was an emerging sense that species should not be hunted to extinction. But why? Were these early conservationists just concerned that there might not be any animals left to hunt or trees left to chop down? Their problem then, and ours now, is to establish the thesis that wild species have some *value* that makes it essential that they be preserved. If we can identify that value, then we will be able to justify the sometimes costly action we must take to preserve wild species.

Philosophers who have addressed this problem inform us that two kinds of value should be considered. The first is **instrumental value**. A species or individual organism has instrumental value if its existence or use benefits some other entity. This kind of value is usually *anthropocentric*; that is, the beneficiaries are human beings. Clearly, many species of plants and animals have instrumental value to humans and will tend to be preserved (or conserved, as we would say) so that we can continue to enjoy the benefits we derive from them.

▲ FIGURE 11–1 *Extinct passenger pigeon.* Clouds of passenger pigeons darkened American skies during the 18th and 19th centuries, but relentless hunting extinguished the species in the early 20th century.

EARTH WATCH

RETURN OF THE GRAY WOLF

Alaska, with 5,000 to 7,000 gray wolves, is the only state in which the wolf is not endangered and therefore not protected by the Endangered Species Act (ESA). Prior to the act, wolves had been exterminated from all but two of the lower 48 states. Currently, gray-wolf populations are on the increase; more than 3,200 of the animals are now found in nine states. Symptomatic of a profound shift in public attitude toward wolves, instead of states poisoning and trapping wolves and paying bounties for them, Defenders of Wildlife pays ranchers for any losses they suffer from wolf predation. (In the last 10 years, Defenders has compensated ranchers $121,800 for such losses.)

In a highly publicized program, the U.S. Fish and Wildlife Service began releasing captured Canadian wolves in Wyoming's Yellowstone National Park and in central Idaho. The 14 wolves released in Yellowstone in 1995 demonstrated their approval of their new home by producing nine pups, successfully preying on the abundant elk and bison in the park, aggressively harassing the numerous coyotes in the park, and rewarding some 6,000 visitors with views of their activities. Encouraged by their success, the wildlife biologists added 17 more wolves to Yellowstone in 1996 and have halted the reintroductions. As of mid-2000, the Yellowstone wolf population numbered 118, signaling the wolves' successful adaptation to the national park. This is bad news for the coyote population, which wolves have reduced by 50% in some areas as they regain their status as top dogs in the park. A subspecies, the Mexican gray wolf, recently was reintroduced to the Southwest, where some 22 wolves are hanging on in an area with a heavy presence of range cattle. Serious proposals have been made to reintroduce the wolf to the Northeast—in particular, northern Maine, where suitable habitat and prey are believed to be excellent. The USFWS announced plans to downgrade the status of all Great Lakes wolf populations from endangered to threatened, on the basis of their successful recovery under the protection of the ESA.

Research biologist David Mech, probably the leading authority on wolves, believes that the key to continuing the wolf's comeback is to stop short of complete control. The conflicting interests of protectionists on the one hand and farmers and ranchers on the other will have to be worked out with a careful program of regulation of wolf populations—just as many other large animals, such as bears, cougars, and coyotes, are kept under control by regulations and hunting quotas. When the wolf is removed from the endangered list, in the words of David Mech, it "can be accepted as a regular member of our environment, rather than as a special saint or sinner, [and] this will go a long way toward ensuring that the howl of the wolf will always be heard throughout the wild areas of the northern world."

The second kind of value we must consider is **intrinsic value**. We assign intrinsic value to something when we agree that it has value *for its own sake*; that is, it does not have to be useful to us to possess value. How do we know that something has intrinsic value? That is a philosophical question, and it comes down to a matter of moral reasoning. People often disagree about intrinsic value, as illustrated by the animal-rights controversy. In this regard, some people argue that animals should have legal standing to be protected from being used for food, fur and hides, or experimentation. On the other side, although most people agree that cruelty to animals is not right, they do not believe that any animals should be given rights at all like those afforded to humans.

As we study the problem of loss of species, we find that some claim that no species on Earth except *Homo sapiens* has any intrinsic value. However, if no other species are held to have instrinsic value, then it is difficult to justify preserving many that are apparently insignificant or very local in distribution. Still, in spite of the problems of establishing intrinsic value for species, support is growing in favor of preserving species that not only may be useless to humans, but also may never be seen by anyone except a few naturalists or systematists—biologists who are experts on classifying organisms.

The value of natural species can be categorized into four areas, which we examine in this chapter:

- sources for agriculture, forestry, aquaculture, and animal husbandry.
- sources for medicines and pharmaceuticals.
- recreational, aesthetic, and scientific value.
- intrinsic value.

The first three categories reflect mostly instrumental value. In the case of aesthetic and scientific value, it could be argued that these sometimes represent intrinsic value.

Sources for Agriculture, Forestry, Aquaculture, and Animal Husbandry

Since most of our food comes from agriculture, we tend to believe that it is independent of natural biota. This is not true. Recall that in nature, both plants and animals are continuously subjected to the rigors of natural selection. Only the fittest survive. Consequently, wild populations have numerous traits for resistance to parasites, competitiveness, tolerance to adverse conditions, and other aspects of vigor.

In contrast, populations grown for many generations under the "pampered" conditions of agriculture tend to lose these traits, because they are selected for production, not vigor. For example, a high-producing plant that lacks resistance to drought is irrigated, and the resistance to drought is ignored. Also, in the process of breeding plants for maximum production, virtually all genetic variation is eliminated. Indeed, the cultivated population is commonly called a cultivar (for *culti*vated *vari*ety), indicating that it is a highly

selected strain of the original species, with a *minimum* of genetic variation. When provided with optimal water and fertilizer, cultivars do give outstanding production under the specific climatic conditions to which they are adapted. With their minimum genetic variation, however, they have virtually no capacity to adapt to any other conditions.

To maintain vigor in cultivars and to adapt them to various climatic conditions, plant breeders comb wild populations of related species for the desired traits. When found, these traits are introduced into the cultivar through cross-breeding or, more recently, biotechnology (Chapter 10). For example, in the 1970s, the U.S. corn crop was saved from blight by genes from a wild strain of maize. The point is, this trait came from a related *wild* population—that is, from natural biota. If natural biota with wild populations are lost, the options for continued improvements in food plants will be greatly reduced.

Also, the potential for developing *new* agricultural cultivars will be lost. From the hundreds of thousands of plant species existing in nature, humans have used perhaps 7,000 in all, and modern agriculture has tended to focus on only about 30. Of these, three species—wheat, maize (corn), and rice—fulfill about 50% of global food demands. This limited diversity in agriculture makes it ill suited to production under many environmental conditions. For example, we tend to think of arid regions as being unproductive without irrigation. However, many wild species belonging to the bean family produce abundantly under dry conditions. Scientists estimate that 30,000 plant species with edible parts could be brought into cultivation. Many of these could increase production in environments that are less than ideal. For example, consider the winged bean, native to New Guinea (Fig. 11–2). This plant is a veritable supermarket, with every part edible: pods, flowers, stems, roots, and leaves. Recently introduced to many developing countries, it has already made a significant contribution to improving nutrition. Loss of biological diversity undercuts similar future opportunities.

▼ **FIGURE 11–2** *The winged bean.* A climbing legume with edible pods, seeds, leaves, and roots, this tropical species is an example of the great potential of wild species for human use.

Another area in which biodiversity has instrumental value to humans is pest control. In Chapter 17, we will discuss the tremendous and invaluable opportunities to control pests by introducing natural enemies and increasing genetic resistance. Natural enemies and genes for increasing resistance can come only from natural biota. Destroying natural biota may destroy such opportunities.

Since we select species from nature for animal husbandry, forestry, and aquaculture, essentially all the same arguments can be made in connection with those important enterprises.

To return to our concept of biological wealth, we can look at natural biota as a bank in which the gene pools of all the species involved are deposited. As long as natural biota are preserved, we have a rich endowment of genes in the bank that we can draw upon as needed. Thus, natural biota are frequently referred to as a **genetic bank**. Depleting this bank cannot help but deplete our future.

Sources for Medicine

Earth's genetic bank also serves medicine. For thousands of years, the indigenous people of the island of Madagascar used an obscure plant, the rosy periwinkle, in their folk medicine (Fig. 11–3). If this plant, which grows only on Madagascar, had become extinct before 1960, hardly anyone outside Madagascar would have cared. In the 1960s, however, scientists extracted two chemicals called vincristine and vinblastine, with medicinal properties, from the plant. These chemicals have revolutionized the treatment of childhood leukemia and Hodgkin's disease. Before their discovery, leukemia was almost always fatal in children; today, with vincristine treatment, there is a 95% chance of remission. These two drugs now represent a $100-million-a-year industry.

The story of the rosy periwinkle is just one of hundreds. The venom from a Brazilian pit viper (a poisonous snake) led to the development of the drug Capoten, used to control high blood pressure. Paclitaxel (trade name Taxol), an extract from the bark of the Pacific yew, has proved to be valuable for treating ovarian, breast, and small-cell cancers. For a time, this use threatened to decimate the Pacific yew; six trees were required to treat one patient for a year. Now the substance is extracted from the leaves of the English yew tree, a horticultural plant that is easy to maintain. Stories like these have created a new appreciation for the field of *ethnobotany*, the study of the relationships between plants and people. To date, some 3,000 plants have been identified as having anticancer properties. Drug companies are now financing field studies of the medicinal use of plants by indigenous peoples and are even funding the creation of parks and reserves to promote the preservation of natural ecosystems that are home to both the people and the plants.

According to the World Health Organization, 80% of the world's people depend on non-Western medicine that in turn depends on natural products. It is a fact that 25% of pharmaceuticals in the United States contain ingredients originally derived from native plants, representing $8 bil-

▲ **FIGURE 11–3** *The rosy periwinkle.* This native to Madagascar is a source of two anticancer agents that are highly successful in treating childhood leukemia and Hodgkin's disease.

lion of annual revenue for drug companies and better health and longevity for countless people. Table 11–1 shows a number of well-established drugs that were discovered as a result of analyzing the chemical properties of plants used by traditional healers. It is likely that the search for such chemicals has barely scratched the surface.

Recreational, Aesthetic, and Scientific Value

The species in natural ecosystems also provide the foundation for numerous recreational and aesthetic interests, ranging from sportfishing and hunting to hiking, camping, bird-watching, photography, and so on (Fig. 11–4). Interests may range from casual aesthetic enjoyment to serious scientific study. Virtually all our knowledge and understanding of evolution and ecology have come from studying wild species and the ecosystems in which they live. Pleasure and satisfaction may even be indirect. For instance, one may never see a whale, but knowing that whales and similar exciting animals exist provides a certain aesthetic pleasure. The great popularity of nature films attests to this. Further, knowing that the Earth and its biosphere continue to support and maintain such wildlife provides a sense of well-being.

Recreational and aesthetic values constitute a very important source of support for maintaining wild species. Recreational and aesthetic activities support commercial interests. **Ecotourism**—whereby tourists visit a place in order to observe wild species or unique ecological sites—represents the largest foreign-exchange-generating enterprise in many developing countries. As the amount of leisure time available to people increases, more and more money is spent on recreation. Since some percentage of these recreational dollars will be spent on activities related to the

TABLE 11–1   Modern Drugs from Traditional Medicines

Drug	Medical Use	Source	Common Name
Aspirin	Reduces pain and inflammation	*Filipendula ulmaria*	Queen of the meadow
Codeine	Eases pain; suppresses coughing	*Papaver somniferum*	Opium poppy
Ipecac	Induces vomiting	*Psychotria ipecacuanha*	Ipecac
Pilocarpine	Reduces pressure in the eye	*Pilocarpus jaborandi*	Jaborandi plant
Pseudoephedrine	Reduces nasal congestion	*Ephedra sinica*	*Ma-huang* shrub
Quinine	Combats malaria	*Cinchona pubescens*	Cinchona tree
Reserpine	Lowers blood pressure	*Rauwolfia serpentina*	Rauwolfia
Scopolamine	Eases motion sickness	*Datura stramonium*	Jimsonweed
Theophylline	Opens bronchial passages	*Camellia sinensis*	Tea
Tubocurarine	Relaxes muscles during surgery	*Chondrodendron tomentosum*	Curare vine
Vinblastine	Combats Hodgkin's disease	Catharanthus roseus	Rosy periwinkle

natural environment, any degradation of that environment affects commercial interests. These activities involve a great number of people and represent a huge economic enterprise. To cite one example, 84% of Canadians take part in some form of wildlife-related recreation, spending an estimated $9.4 billion yearly. In the most recent national survey, it was found that some 91 million Americans spent $101 billion on wildlife-related recreation during 1996.

▲ **FIGURE 11–4** *Recreational, aesthetic, and scientific uses.* Natural biota provide numerous values, a few of which are depicted here.

Very likely, the broadest public support for preserving wild species and habitats is traceable to the aesthetic and recreational enjoyment people derive from them.

Intrinsic Value

The usefulness (instrumental value) of many wild species is apparent. But what about those other species which have no obvious value to anyone—probably the majority of plant and animal species, many of which are rare or inconspicuous in the environment? Some observers believe that the most important strategy for preserving all wild species is to emphasize the *intrinsic* value of species, rather than the unknown or uncertain ecological and economic instrumental values. Thus, we should recognize that the extinction of a species per se is an irretrievable loss of something of value.

Environmental ethicists argue that long-established existence carries with it a right to continued existence. They hold that humans have no right to terminate a species that has existed for thousands or millions of years and that represents a unique set of biological characteristics. Some support this view by arguing that there is value in every living thing and that one kind of living thing (e.g., humans) has no greater value than any other. This argument, however, can lead to some difficulties, such as having to defend the rights of pathogens and parasites. A more common viewpoint held by ethicists is that because humans have the ability to make moral judgments, they also have a special responsibility toward the natural world, and that responsibility includes concern for other species. It should be pointed out that until recently, Western philosophers argued that only humans were worthy of ethical consideration; the Western philosophical tradition has been strongly anthropocentric.

Many ethicists find their basis for intrinsic value in religion. For example, Old Testament writings express God's concern for wild species when He created them. Jewish and Christian scholars alike maintain that by declaring His creation good (see Genesis 1) and giving it His blessing, God was saying that all wild things have intrinsic value and therefore deserve moral consideration. They are created for Him, not for humans alone, and we ought to regard the creation with respect and stewardly care. The Islamic Quran (Koran) proclaims that the environment is the creation of Allah and should be protected because it praises the Creator. Many Native American tribes have a strong environmental ethic; for example, the Lakota hold that humans and other life-forms should interact like members of a large, healthy family. Clearly, ethical concern for wild species underlies many religious traditions and represents a potentially powerful force for preserving biodiversity.

Thus, we see that even if species have no demonstrable use to humans, it can still be argued that they have a right to continue to exist. Only rarely (as in the case of parasites and pathogens) can we claim that there is any moral justification for driving other species to extinction.

11.2 Saving Wild Species

When we place a value on wild species, we are saying that we want to have them around for some reason. One compelling reason is that they provide recreation and food for people who hunt them.

Game Animals in the United States

Game animals are those traditionally hunted for sport, meat, or pelts. In the early days of the United States, there were no restrictions on hunting, and a number of species were hunted to extinction (great auk, heath hen, passenger pigeon) or near extinction (bison, wild turkey). As game animals became scarce in the face of increasing pressure from hunting, regulations were enacted. State governments, backed up by the federal government, enacted laws establishing hunting seasons and bag limits and hired wardens to enforce them. Some species were given complete protection in order to allow populations to build up to numbers that would once again allow hunting.

One success story is the wild turkey (Fig. 11–5). A favorite game species, this bird was hunted to the brink of extinction, but was making a slow comeback by the 1930s as a result of hunting restrictions. At that time, there was a total population of about 30,000 individuals in a few scattered states. After World War II, state and federal programs addressed the need for protecting turkey habitats. The birds were reintroduced into areas they had once inhabited, and hunting quotas were strictly limited. The turkey is now found in 49 states, and its population has risen to a total of 4.5 million as a result of these measures.

Using hunting and trapping fees as a source of revenue, state wildlife managers enhance the habitats supporting important game species and provide special areas for hunting. They monitor game populations and adjust seasons and bag limits accordingly. Another source of revenue for wildlife conservation comes from the excise taxes placed on hunt-

▼ **FIGURE 11–5** *Wild turkey.* Restoration efforts have successfully reestablished the wild turkey throughout the United States, such that hunting within carefully controlled limits now takes place.

ing, fishing, shooting, and boating equipment. In 1999, this source provided $434 million to state fish and wildlife conservation programs. Game preserves, parks, and other areas where hunting is prohibited are maintained to protect habitats as well as certain breeding populations.

In spite of what seems on the surface to be a destruction of wildlife, hunting has many positive aspects. Besides the fees hunters pay, many hunters belong to organizations dedicated to the game they are interested in hunting. Organizations like Ducks Unlimited, The National Wild Turkey Federation, and Pheasants Forever raise funds that are used for the restoration and maintenance of natural ecosystems vital for the game they are interested in hunting. For example, Ducks Unlimited recently announced that 8 million acres of vital wetland habitat for wildfowl have been conserved in North America. The organization's activities have resulted in stemming the tide of loss that has cut North America's wetlands by more than 50% over the years. The activities are in great measure responsible for the remarkable fall flights of ducks migrating from northern breeding grounds to wintering areas in the southern United States, Mexico, and points south—a record 105 million ducks in 1999.

Defenders of hunting and trapping point out that their prey are often animals that lack natural predators and would increase to the point of destroying their own habitat. This is often the case with larger animals, like deer and elk. However, many members of the nonhunting public object to the killing of wildlife, and some groups, such as the Humane Society of the United States and People for the Ethical Treatment of Animals (PETA), actively campaign to limit or end hunting and trapping. Some practices that are regarded as especially cruel, such as the use of leghold steel traps, have been banned in several states as a result of ballot initiatives.

Common game animals, such as deer, rabbits, doves, and squirrels, are well adapted to rural and suburban environments. Thus adapted and protected from overhunting, viable populations of these animals are being maintained. Some predatory animals are also on the increase. However, serious problems have emerged:

1. The number of animals killed on roadways now far exceeds the number killed by hunters. As rural areas are developed, increasing numbers of animals found on roadways are a serious hazard to motorists. With 3.9 million miles of roadways and 200 million vehicles in the United States, over a million animals a day are killed in collisions. Over 200 human fatalities occur as a consequence of these collisions. In Maine and New Hampshire, more than 800 moose are killed annually in accidents that often also kill motorists because of the size of the animals. Overpasses and tunnels are increasingly being built to provide wildlife with safe corridors (Fig. 11–6).

2. Many nuisance animals are thriving in highly urbanized areas, creating various health hazards. Opos-

sums, skunks, and deer are attracted to urban areas by opportunities for food; unsecured garbage cans and pet food left outside will ensure visits from raccoons. In 1998, the Center for Disease Control reported 7,962 cases of rabies in the United States, 72% of which occurred in raccoons and skunks. A number of European countries have eradicated rabies in wild animals by spreading bait laced with rabies vaccine; several states are now employing this strategy, which shows great promise.

3. Some game animals have no predators except hunters and tend to reach population densities that push them into suburban habitats, where they cannot be effectively hunted. The white-tailed deer, for example, has become a pest to gardeners and fruit nurseries; it also poses a public health risk because it is often infested with ticks that carry Lyme disease. Many public parks, in which hunting is not feasible, are home to high densities of deer. For example, Sharon Woods, a one-square-mile park near Columbus, Ohio, had a deer population of 500 and became an ecological disaster. Deer browsed virtually every tree from six feet down, stripped bark from many, and removed the understory. In such areas, municipalities have resorted to hiring sharpshooters to cull the deer herds. In wilder areas, rising deer populations have reduced populations of many endangered or threatened plant species by their grazing.

4. In recent years, suburbanites have been increasingly attacked by cougars or mountain lions as urbanization encroaches on the wild (Fig. 11–7). California and Oregon have cougar populations numbering in the thousands as a result of diminished hunting, with resulting increases in contact between people and these large predators.

5. Coyotes, which roamed only in the Midwest and western states, are now found in every state and are increas-

▲ FIGURE 11–6   *Bear underpass.*   To reduce black-bear roadkill, Florida has established this bear corridor under State Route 46 to connect important bear habitats transected by the highway. Twelve other species use the corridor.

ing in numbers. A highly adaptable predator, coyotes will eat almost anything. In the Adirondacks, they eat mostly deer; in suburban neighborhoods, they often dine on cats, small dogs, and human garbage. (One particularly adaptive pair had a den on the median strip of Boston's beltway, Route 128, where they lived off roadkill.) Very difficult to control, the 35-lb animals occasionally attack small children.

6. Suburban parks, lawns, college campuses, and golf courses have become home to exploding flocks of Canada geese. These large birds are, ecologically, grazing herbivores, able to consume large quantities of grass and defecating as often as every eight minutes. Protected by wildlife laws, the geese present a challenge to efforts to control them. Few predators are able to tackle these large birds, which originated from geese captured and bred to act as decoys to wild migrating flocks. One imaginative solution used around golf courses and colleges is to "hire" border collies (and their trainers), which are taught to harass the geese and keep them from comfortably grazing the lush grass they relish.

One highly controversial effort for controlling unwanted animals is carried out by the **Animal Damage Control** (ADC) program, an agency of the U.S. Department of Agriculture. (In 1998, the agency changed its name to **Wildlife Services** in an effort to soften its image.) This agency responds to requests from livestock owners, farmers, and others concerned with human health and safety to remove nuisance animals and birds. "Removal" virtually always means killing, and the ADC routinely uses poisons, traps, and other devices to kill up to 1 million animals and birds yearly. In 1998, for example, the ADC killed 78,000 coyotes, 31,000 beavers, 2,100 bobcats, 5,000 feral hogs, and

7,100 raccoons, among other animals. In spite of attempts by conservation organizations to limit the budgets and activities of the ADC, it is still drawing $30 million to $40 million of taxpayers' money and enjoys strong support from Congress. Interestingly, this is the same Congress that for several years has been unable to agree to renew the major piece of legislation that provides protection to wild animals—the Endangered Species Act.

The Endangered Species Act

In colonial days, huge flocks of snowy egrets inhabited coastal wetlands and marshes of the southeastern United States. In the 1800s, when fashion dictated fancy hats adorned with feathers, egrets and other birds were hunted for their plumage. By the late 1800s, egrets were almost extinct. In 1886, the newly formed Audubon Society began a press campaign to shame "feather wearers" and end the "terrible folly." The campaign caught on, and gradually attitudes changed and new laws followed.

Florida and Texas were first to pass laws protecting plumed birds. Then, in 1900, Congress passed the **Lacey Act**, forbidding interstate commerce to deal in illegally killed wildlife, making it more difficult for hunters to sell their kill. Since then, numerous wildlife refuges have been established to protect the birds' breeding habitats. With millions of people visiting these refuges and seeing the birds in their natural locales, attitudes have changed significantly. Today, the thought of hunting these birds would be abhorrent to most of us, even if official protection were removed. Thus protected, egret populations were able to recover substantially. In the meantime, the Lacey Act has become the most important piece of legislation protecting wildlife from illegal killing or smuggling. In 1999, for example, Fish and Wildlife Service special agents worked on more than 1,500

◀ **FIGURE 11–7** *Cougar on the roof.* As suburbs encroach on cougar habitats in the western United States, attacks on humans have become more numerous. This particular cougar on the roof of a home in a California suburb has been shot with a tranquilizer dart.

violations of the act, and in 1997, the prosecution of over 134 Lacey Act wildlife crimes resulted in 198 years of jail time and $183,000 in fines.

Congress took another major step when it passed a series of acts to protect endangered species. The most comprehensive and recent of these acts is the **Endangered Species Act (ESA)** of 1973 (reauthorized in 1988). As we learned in Chapter 4, an **endangered species** is a species that has been reduced to the point where it is in imminent danger of becoming extinct if protection is not provided (Fig. 11–8). The act also provides for the protection of **threatened species**, which are judged to be in jeopardy, but

(a) Male Pine Barrens Tree Frog

(b) Silversword

(c) Red-cockaded Woodpecker

(d) Karner Blue Butterfly

(e) Swamp Pink

(f) Whooping Cranes

(g) Devil's Hole Pupfish

(h) White Oryx

(i) Manatee

▲ **FIGURE 11–8 *Endangered species.*** Some examples of endangered species—species whose populations in nature have dropped so low that they are in imminent danger of becoming extinct unless protection is provided.

not on the brink of extinction. When a species is officially recognized as being either endangered or threatened, the law specifies substantial fines for killing, trapping, uprooting (plants), or engaging in commerce in the species or its parts. The legislation forbidding commerce includes wildlife threatened with extinction anywhere in the world.

The ESA requires the U.S. Fish and Wildlife Service (USFWS), under the Department of the Interior, to draft recovery plans for protected species. Habitats must be mapped and a program for the preservation and management of critical habitats must be designed, such that the species can rebuild its population. A 1995 Supreme Court decision made it clear that federal authority to conserve critical habitats extended to privately held hands. As Table 11–2 indicates, 1,233 U.S. species are currently listed for protection under the act, and recovery plans are in place for 923 of them.

Some critics of the ESA believe that the act does not go far enough. A major shortcoming is that protection is not provided until a species is officially listed as endangered or threatened by the USFWS and a recovery plan is established. Species usually will not make the list until their populations have become dangerously low. In the last two years, the USFWS has been working intensely on listing species and developing recovery plans for them. (There are 94 new listings, 4 delistings, and 144 more recovery plans.) One of the species recently removed from the list, and an amazing recovery story, is the American peregrine falcon (Fig. 11–9). The bald eagle also is scheduled to be removed from the list soon.

Both the peregrine falcon and the bald eagle were driven to extremely low numbers because of the use of DDT as a pesticide from the 1940s through the 1960s. Carried up to

▲ **FIGURE 11–9** *The peregrine falcon.* The Endangered Species Act has worked well in bringing this magnificent falcon to the point where it can be dropped from the list.

TABLE 11–2 Federal Listings of Threatened and Endangered U.S. Plant and Animal Species, 2000

Category	Endangered	Threatened	Total Listings*	Species with Recovery Plans†
Mammals	63	9	340	47
Birds	78	15	274	76
Reptiles	14	22	115	30
Amphibians	10	8	27	12
Fishes	69	44	124	90
Snails	20	11	32	20
Clams	61	8	71	45
Crustaceans	18	3	21	12
Insects	30	9	43	28
Arachnids	6	0	6	5
Plants (flowering, conifers, ferns, others)	593	142	738	558
Total	962	271	1,791	923

Total endangered U.S. species:	962 (369 animals, 593 plants)
Total threatened U.S. species:	271 (129 animals, 142 plants)
Total listed U.S. species:	1,233 (498 animals, 735 plants)
Total listed non-U.S. species	558 (555 animals, 3 plants)

Source: Department of the Interior, U.S. Fish and Wildlife Service, Division of Endangered Species, Sept. 14, 2000.
*Total listings refer to both U.S. and non-U.S. species.
†Recovery plans pertain only to U.S.-listed species.

these predators through the food chain, DDT caused a serious thinning of the birds' eggshells that led to nesting failures in the two species and in numerous other predatory birds (Chapter 17). By 1975, a survey indicated that there were only 324 pairs of nesting peregrines in North America. DDT use was banned in both the United States and Canada in the early 1970s, and the stage was set for recovery of the bird. Working with several nonprofit captive-breeding institutions such as the Peregrine Fund, the USFWS sponsored efforts that resulted in the release of some 6,000 captive-bred young falcons in 34 states over a period of 23 years. There are now about 1,600 known breeding pairs in the United States and Canada, well above the targeted recovery population of 631 pairs.

A few species have gained exceptional public attention, and heroic efforts have been mounted to save them. Efforts to save the whooping crane, for example, include virtually full-time monitoring and protection of the single remaining flock, which for years numbered only in the teens. The migratory wild flock is increasing steadily and has reached a population of 188 individuals from a low of 14 cranes in 1939. This flock migrates from wintering grounds on the Texas Gulf Coast to Wood Buffalo National Park in Canada; the migration is a hazardous affair, and accidents along the way take a toll on the birds. A nonmigratory flock is taking hold in central Florida, using eggs from captive-breeding flocks. Over 70 birds now make up the flock, and in 2000, 15 pairs formed bonds and three pairs laid eggs. Two chicks hatched, but did not survive to fledge; however, the adult cranes appear to be on the way to a stable population, and the successful hatching is viewed as a very good sign. The whooping crane recovery program has been so successful that other countries with endangered crane species have adopted many of the strategies developed by the USFWS.

Some critics of the ESA claim that it goes too far. The most famous case in point is the northern spotted owl. This species became a focal point in the battle to save some of the remaining old-growth forests of the Pacific Northwest. The owl is found only in these forests and had dwindled to a population between 6,000 and 8,000. In June 1990, the USFWS listed the owl as a threatened species. Years of bitter controversy followed, pitting environmentalists against the timber industry, until a compromise (the **Northwest Forest Plan**) was worked out in 1994. The plan, employing the ecosystem management approach now adopted by the Forest Service (Chapters 3 and 4), set aside 7.4 million acres of federal land in California, Oregon, and Washington, where logging is prohibited in stands of trees older than 80 years.

The latest controversy in the Pacific Northwest centers around salmon species, which have plummeted from runs of 16 million down to less than 1 million. Recovery rules issued in the summer of 2000 generated enormous concern because they involve such a broad range of economic activities in the region; the regulations govern storm sewer runoff, building construction, farm irrigation, and timber harvests. Because migrating fish runs involve thousands of square miles and hundreds of thousands of people, the $3 billion that has been spent in the last 15 years on salmon recovery will seem a pittance compared with the costs of restoring the wild salmon runs. However, there is broad political support for the regulations, at least at this point in the restoration efforts.

The ESA was scheduled for reauthorization in 1992, but Congress, fearing a veto from President Bush, decided to defer consideration of the bill until after the coming presidential election. With the election of Bill Clinton, environmentalists were optimistic that the ESA would get swift consideration—and were disappointed when it did not. With the shift of Congress to Republican control in 1994, the ESA has been in continuous trouble and has been operating on year-to-year extensions. Opponents such as development, lumber, and mineral interests claim that the act is costly, is inefficient, and favors the rights of owls and flowers over private-property owners and legitimate business interests. No fewer than 50 bills involving the ESA were introduced in the 106th Congress, but not many were given any chance of passage.

In recent years, the USFWS has made progress in bringing together landowners and regulators by making increasing use of a procedure authorized in a 1982 amendment to the ESA—the **Habitat Conservation Plan** (HCP). The USFWS can issue a permit for activities that potentially harm a protected species if the landowners take special steps to protect the species over the critical part of the land, so as to minimize any harm done. There are now more than 300 such plans involving 200 endangered or threatened species over an area of 20 million acres. As an example, the Plum Creek Timber Company and the USFWS have agreed on a plan covering 170,000 acres of land in the northern Cascade Mountains of Washington, home to spotted owls and several other endangered or threatened species. A small amount of this land is old-growth forest, which the company has agreed to leave alone. In addition, the company has consented to allow extrawide forested buffers to remain around streams and to employ selective logging to reserve the more mature trees. In return, the company will gradually log its mature forests over a 30-year period. The one aspect of the HCP that is objectionable to many is that the plan will be in effect for 50 years and will guarantee the timber company that no new restrictions (a "no surprises" clause) will be imposed even if new evidence that certain species are becoming threatened or endangered surfaces. A recent study of the HCP indicates that, except for a few very bad exceptions, the amendment makes good use of sound science and moves us in the right direction, although monitoring of results is still inadequate.

In the final analysis, the ESA is a formal recognition of the importance of preserving wild species, regardless of whether they are economically important. Species listed as endangered or threatened have legal rights to protection under the law. The act is something of a last resort for wild species, but it embodies an encouraging attitude toward nature that has now become public policy. Actions taken under auspices of the act have demonstrated a unique commitment to preserve wild species. The reauthorization of a strong ESA will be a major signal of our continuing commitment to consider the value of wild species on grounds other than economic or political.

We turn now to the broader topic of biodiversity, or the sum of richness of species and habitats on Earth, and examine reasons for its decline and strategies for its preservation.

11.3 Biodiversity

We saw in Chapter 5 that, over time, natural selection leads to speciation—the creation of new species—as well as extinction—the disappearance of species. Over geological time, the net balance of these processes has favored the gradual accumulation of more and more species—in other words, biodiversity. The concept of biodiversity is often extended to the genetic diversity within species, as well as the diversity of ecosystems and habitats. Its main focus, however, is the species.

How much biodiversity is there? Estimates range widely, and the fact is, no one knows. Today, the only two certainties are that 1.75 million species have been described and many more than that exist. Most people are completely unaware of the great diversity of species within any given taxonomic category. Groups especially rich in species are the flowering plants (270,000 species) and the insects (950,000 species), but even less diverse groups, such as birds or ferns, are rich with species that are unknown to most people (Table 11–3). Taxonomists are aware that their work in finding and describing new species is incomplete. Groups that are conspicuous or commercially important, such as birds, mammals, fish, and trees, are much more fully explored and described than, say, insects, very small invertebrates like mites and soil nematodes, fungi, and bacteria. Trained researchers who can identify major groups of organisms are few, and

TABLE 11–3 Known and Estimated Species on Earth

Taxonomic group	Number of Known Species	High	Estimated Numbers Low	Working Number	Accuracy
Viruses	4,000	1,000,000	50,000	400,000	Very poor
Bacteria	4,000	3,000,000	50,000	1,000,000	Very poor
Fungi	72,000	2,700,000	200,000	1,500,000	Moderate
Protozoa	40,000	200,000	60,000	200,000	Very poor
Algae	40,000	1,000,000	150,000	400,000	Very poor
Plants	270,000	500,000	300,000	320,000	Good
Nematodes	25,000	1,000,000	100,000	400,000	Poor
Arthropods:					
Crustaceans	40,000	200,000	75,000	150,000	Moderate
Arachnids	75,000	1,000,000	300,000	750,000	Moderate
Insects	950,000	100,000,000	2,000,000	8,000,000	Moderate
Mollusks	70,000	200,000	100,000	200,000	Moderate
Chordates	45,000	55,000	50,000	50,000	Good
Others	115,000	800,000	200,000	250,000	Moderate
Total	1,750,000	111,655,000	3,635,000	13,620,000	Very poor

Source: United Nations Environment Programme, "Global Biodiversity Assessment" (Cambridge, U.K.: Cambridge University Press, 1995): 118, Table 3.1–1.

the task of fully exploring the diversity of life will require a major sustained effort in systematic biology.

Estimates of the number of species on Earth today are based on recent work in the tropical rain forests, which hold more living species than all other habitats combined. Costa Rica, less than half the size of New York State, is estimated to contain at least 5% of all living species. The upper boundary of the estimate keeps rising as taxonomists explore the rain forests more and more. Some (enthusiastic?) estimates place the high end at 112 million species. Whatever the number, the planet's biodiversity represents an amazing and diverse storehouse of biological wealth.

The Decline of Biodiversity

Known and Unknown Losses. The biota of the United States is as well known as any. At least 500 species native to the United States are known to have become extinct since the early days of colonization. Over 100 of these are vertebrates. A 1997 inventory of 20,439 wild plant and animal species in the 50 states, carried out by a team of biologists from The Nature Conservancy and the National Heritage Network, concluded that almost one-third (31.9%) are vulnerable, imperiled, or already extinct (Fig. 11–10). Mussels, crayfish, fishes, and amphibians—all species dependent on freshwater habitats—are most at risk. Flowering plants are also of great concern, with one-third of their numbers in trouble.

Across North America, general population declines of wild species are also occurring. Commercial landings of many species of fish are down; puzzling reductions in amphibian populations are occurring in North America and other parts of the world; and many North American songbird species, such as the Kentucky warbler, wood thrush, and scarlet tanager, have been declining and have disappeared entirely from some local regions. These birds are Neotropical migrants; that is, they winter in the neotropics of Central and South America and breed in temperate North America. Almost 100 species of these migrants have shown a drop in numbers over the past 30 years.

Worldwide, the loss of biodiversity is even more disturbing. At least 484 animal species and 654 plant species have become extinct since 1600. Most of the known extinctions of the past several hundred years have occurred on oceanic islands, where small landmasses limit the size of populations and human intrusions are most severe. The known causes of animal extinctions, recently assessed by the World Conservation Union, are shown in Fig. 11–11. The "Global Biodiversity Assessment," commissioned by UNEP to provide information for the Convention on Biological Diversity, estimates that some 5,400 known species of animals and 26,000 plant species are in danger of becoming extinct.

Biodiversity is richest in the tropics—a richness that is almost unimaginable. Biologist E. O. Wilson identified 43 species of ants on a single tree in a Peruvian rain forest, a level of diversity equal to the entire ant fauna of the British Isles. Other scientists found 300 species of trees in a single 2.5-acre (1-ha) plot and as many as 10,000 species of insects on a single tree in Peru. Assuming the existence of 2 million species in the tropical forests (a conservative estimate) and a clearance rate of 1.8% per year for those forests, Wilson calculated that tropical deforestation is responsible for the loss of 4,000 species a year. Other scientists have projected that as many as 17,000 species are lost per year, or between 500,000 and 2 million species over the last half of the 20th century! Not all scientists agree with these calculations, however, as the estimates are based on many assumptions. The most important conclusion we can draw is that many species are in decline and some are becoming extinct. No one knows how many, but the loss is real, and it represents a continuing depletion of the biological wealth of our planet.

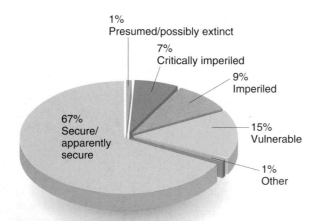

▲ **FIGURE 11–10** *The state of U.S. species.* Fully one-third of over 20,000 species of plants and animals surveyed were found by biologists to be at risk of extinction. (*Source:* Data from The Nature Conservancy.)

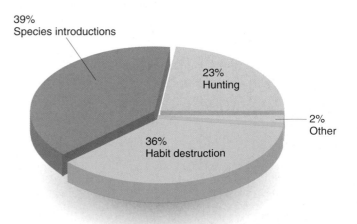

▲ **FIGURE 11–11** *Known causes of animal extinctions since 1600.* The introduction of foreign species, the destruction of habitats, and hunting have brought about the majority of known animal extinctions since 1600. (*Source:* Data from World Conservation Monitoring Centre.)

Reasons for the Decline

Physical Alteration of Habitats. Although multiple causes are usually the rule in losses of biodiversity, one of the greatest sources of loss is the physical alteration of habitats through the processes of conversion, fragmentation, and simplification. Habitat destruction has already been responsible for 36% of known extinctions (Fig. 11–11).

Conversion. Natural areas are converted to farms, housing subdivisions, shopping malls, marinas, and industrial centers. When, for example, a forest is cleared, it is not just the trees that are destroyed; every other plant and animal that occupies the destroyed ecosystem, either permanently or temporarily (e.g., migrating birds), also suffers. The idea that this wildlife simply will move "next door" and continue to live in an undisturbed section is erroneous. Any loss of natural habitat can result in only one thing: a proportional reduction in all populations that require that habitat. Thus, the decline in songbird populations cited earlier has been traced to a combination of the loss of winter forest habitat in Central and South America and the increasing fragmentation of summer forest habitat in North America.

Fragmentation. Natural landscapes generally have large patches of habitat that are well connected to other, similar patches. Human-dominated landscapes, however, consist of a mosaic of different land uses, resulting in small, often geometrically configured patches that frequently contrast highly with neighboring patches (Fig. 11–12). For the continued survival of any natural population, the number of individuals must never fall below a *critical number*, and that requires a certain minimum area. This minimum area must be large enough to compensate for years of adverse weather. That is, more area will be required during a dry year than during a normal year. If development reduces the habitat to a point where it cannot support the critical number during an adverse year, the entire population will perish. Similarly, development (such as a highway) that fragments a territory and prevents migration between the two fragments will cause a population to perish if neither area can adequately support the critical number. Also, reducing the size of a habitat creates a greater proportion of edges, a situation that favors some species, but may be detrimental to others. For example, the Kirtland's warbler is an endangered species that is highly dependent on large patches of second-growth jack pines. The species is endangered because its habitat has been greatly fragmented, creating edges that favor the brown-headed cowbird, a nest parasite that can invade the forest and lay its eggs in the nest of the rare warbler.

Simplification. Human use of habitats often simplifies them. We might, for example, remove fallen logs and dead trees from woodlands for firewood, thus diminishing an important microhabitat on which several species depend. When a forest is managed for the production of a few or one species of tree, tree diversity obviously declines, and with it so does the diversity of a cluster of plant and animal species dependent on the less favored trees. Streams are sometimes "channelized"—their beds are cleared of fallen trees and riffles, and sometimes the stream is straightened out by dredging. Such alterations inevitably bring on a loss of diversity of fish and invertebrates that live in the stream.

The Population Factor. Past losses in biodiversity can be attributed to the expansion of the human population over the globe. Continuing human population growth will bring on continued alteration of natural ecosystems and the inevitable loss of more wild species. The losses will be greatest

◀ **FIGURE 11–12**
Fragmentation.
Development usually leads to the breaking up of natural areas, resulting in a mosaic of habitats that may not support local populations of species.

where biodiversity is greatest and human population growth is highest—in the developing world. Africa and Asia have lost almost two-thirds of their original natural habitat. People's desire for a better way of life, the desperate poverty of rural populations, and the global market for timber and other natural resources are powerful forces that will continue to draw down biological wealth on those continents.

In East Africa, where human population growth has been explosive for several decades, the conversion of savanna and woodlands to cultivation or intensive grazing by goats and cattle has driven most of the African elephant population into the existing wildlife reserves, causing a great reduction in their numbers. The other large African mammals have experienced similar reductions, as the needs of the rural population inevitably conflict with those of the large wild animals of East Africa. Similar losses in large-mammal populations (wolves, bison, elk, bear, cougar, etc.) occurred as North America was gradually transformed into a great agricultural and industrial continent.

One key to holding down the loss in biodiversity lies in bringing human population growth down. If the human population increases to 10 billion, as some demographers believe that it will, the consequences for the natural world are frightening.

Pollution.   Another major factor causing loss of biodiversity is pollution, which can directly kill many kinds of plants and animals, seriously reducing their populations. For example, pollutants from the Mississippi River have created a 7,700-square-mile "dead zone" in the Gulf of Mexico where oxygen completely disappears from depths below 20 meters every summer. (See Chapter 18) Shrimp, fish, crabs, and other commercially valuable sea life are either killed or forced to migrate away from the huge area along the Mississippi and Louisiana coastline. Every oil spill kills seabirds and, often, sea mammals. In January 2000, a massive cyanide spill into the Tisza River (a tributary of the Danube) from a Romanian gold mine wiped out life in the river for 250 miles downriver , and immediately put 15,000 fishermen out of work (Fig. 11–13). Recent declines in populations of frogs, turtles, alligators, and fish have been tentatively traced to the widespread presence of chemicals known as endocrine disrupters—commonly used chlorine compounds and plastics, as well as some pesticides. Because of their molecular structure, these compounds may mimic the effects of some hormones and thus disrupt the normal course of embryological development or sexual activity of animals.

Pollution destroys or alters habitats, with consequences just as severe as those caused by deliberate conversions. Acid deposition and air pollution cause forests to die; sediments and nutrients kill species in lakes, rivers, and bays; DDT devastates wild-bird populations; and the depletion of the ozone layer increases the impact of ultraviolet light on wild species. The list is endless. Some scientists project that global warming from the greenhouse effect may be the

greatest catastrophe to hit natural biota in 65 million years. (The fossil record reveals that a massive extinction of plants and animals occurred at that time, probably caused by a major asteroid hitting Earth.) Scientists speculate that the rate of climatic warming will far outpace the ability of most species of plants to migrate northward, thus trapping them in inhospitable climates. Every species that dies out will doubtlessly take others with it. If forest trees are unable to migrate, neither can the rest of the wildlife that depend on them for food and habitat.

Most of the global pollution problems can be traced to the industrialized world, where energy-generating and other technologies continue to pour pollutants into the air and water at increasing rates. For this reason, it is important not to point the finger of blame primarily at the developing world, where population growth is such a problem. Global climate change and pollution from toxic substances are the legacy of the already developed nations.

Exotic Species.   An exotic species is a species introduced into an area from somewhere else, often a different continent. Because the species is not native to the new area, it is often unsuccessful in establishing a viable population and quietly disappears. This is the fate of many pet birds, rep-

▲ **FIGURE 11–13 Cyanide spill.** Cyanide from a Romanian gold mine spilled into the Tisza River in January 2000, killing fish and other aquatic life for 250 miles downriver and into the Danube.

tiles, and fish that escape or are deliberately released away from their native habitats. Occasionally, however, an introduced species finds the new environment very much to its liking. As we will see in Chapter 17, most of the insect pests and plant parasites that plague agricultural production were accidentally introduced. Exotic species are major agents in driving native species to extinction and are responsible for an estimated 39% of all extinctions of animals since 1600. Chapter 4 documents a number of introduced species and their ecological impact.

The transplantation of species by humans has occurred throughout history, to the point where most people are not aware of the distinction between native and exotic species living in their lands. The European colonists brought literally hundreds of weeds and forage plants to the Americas, so that now most of the common field, lawn, and roadside plant species in eastern North America are exotics. Fully one-third of the plants in Massachusetts are alien or introduced—about 900 species! Among the animals, the most notorious have been the house mouse and the Norway rat; others include the wild boar, donkey, horse, nutria, and even the red fox, which was brought to provide better fox hunting for the early colonial equestrians. One of the most destructive exotics is the house cat. The 60 million domestic and feral cats in the United States are highly efficient at catching small mammals and birds. One recent study of cat predation in Wisconsin indicated that cats kill 19 million songbirds and 140 million game birds each year.

Deliberate introductions sanctioned by the U.S. Natural Resources Conservation Service (in the interest of "reclaiming" eroded or degraded lands) have brought us kudzu and other runaway plants such as autumn olive, multiflora rose, and Amur honeysuckle. Many other exotic plants are introduced as horticultural desirables. One notorious example is purple loosestrife, which infests wetland habitats and replaces many plants valuable to wildlife. (Figure 4–14 illustrates purple loosestrife, kudzu, and other exotic plants.) Another is Oriental bittersweet, an aggressive vine that chokes and shades small shrubs and trees. Often, it smothers young trees and prevents the forest from regrowing. In Oregon and California, the introduced star thistle invades and ruins pastureland, grows wildly, and is a fire hazard.

We might think that introduced species would add to the biodiversity of a region, but many introductions have exactly the opposite effect. Foreign plants crowd out native plants. New animal species are often successful predators that eliminate native species not adapted to their presence. For example, the brown tree snake was introduced to the island of Guam as a stowaway on cargo ships during World War II (Fig. 11–14). This snake is mildly poisonous, grows to a length of 13 feet, and eats almost anything smaller than itself. In a little over 40 years, the snakes have eliminated most of the island's birds; nine of Guam's 12 native species of birds have become extinct, and populations of the remaining birds are so decimated that survivors are rarely seen or heard. Efforts at controlling the snakes have proved fu-

▲ **FIGURE 11–14** *The brown tree snake.* This snake, accidentally introduced to Guam, has decimated bird life on the island. Often lacking in natural predators, islands are especially vulnerable to the harmful effects of exotic species.

tile. The animals have no natural predators on the island, and the snake population has reached such a high density that snakes have invaded houses in search of prey (such as puppies and cats). At present, wildlife officials are concentrating their efforts on preventing the snakes from spreading to other islands in the Marianas chain.

Introduced exotic species may also drive out native species by competing with them for resources, as documented in Chapter 4. Brazilian pepper, an ornamental shrub resembling holly, has become the most widespread exotic in Florida. It is an aggressive colonizer, taking over many native plant communities and fundamentally changing ecosystems like the Everglades (Fig. 11–15). The plant now occupies more than 700,000 acres in central and southern Florida; a task force recently concluded that "the uncontrolled expansion of Brazilian pepper constitutes one of the most serious ecological threats to the biological integrity of Florida's natural systems." (Brazilian Pepper Management Plan for Florida, 1997)

Overuse.   It should be obvious that killing whales, fish, or trees faster than they can reproduce will lead to the ultimate extinction of the species. In spite of that, overuse is another major assault against wild species, responsible for 23% of recent extinctions. Overuse is driven by a combination of economic greed, ignorance, and desperation. The plight of birds in Europe is a good example. It is said, with some justification, that a line can be drawn across the continent north of which people watch birds and south of which they eat them. Some 700,000 "protected" birds are shot in Greece each year, Malta accounts for about 3 million a year, and Italy holds a shameful record of an astronomical

(a)

(b)

▲ **FIGURE 11–15** *Brazilian pepper bush.* (a) A typical everglades landscape, with wooded hammocks surrounded by marshes; (b) a virtual monoculture of Brazilian pepper covers this everglades marsh area, obliterating the native biota.

50 million birds killed and eaten each year! Most of these are small songbirds.

Another prominent form of overuse is the trafficking in wildlife and products derived from wild species. Much of this "trade" is illegal—worldwide, it is estimated to be at least $10 billion a year. It flourishes because some consumers are willing to pay exorbitant prices for such things as furniture made from tropical hardwoods like teak, exotic pets, furs from wild animals, traditional medicines from animal parts, and innumerable other "luxuries," including polar-bear rugs, baskets made from elephant and rhinoceros feet, ivory-handled knives, and reptile-skin shoes and handbags. For example, some Indonesian and South American parrots sell for up to $10,000 in the United States, a panda-skin rug can bring in $25,000, and the gallbladder of the North American black bear, which is valued as folk medicine in Korea, can fetch as much as $2,000. Such prices create a powerful economic incentive to exploit the species involved.

The long-term prospect of extinction does not curtail the activities of exploiters, because the prospect of a huge immediate profit outweighs it. Even when the species is protected, the economic incentive is such that poaching and black-market trade continue. Tiger and rhinoceros populations in the wild have declined drastically in the last two decades (a 90% decline for rhinos) and are on the brink of extinction. Driving this decline is the widespread, but unfounded, belief in Far Eastern countries that parts from these animals have medicinal or aphrodisiac properties. The World Wildlife Fund reported that some 66 Sumatran tigers were killed recently, representing 20% of the Sumatran tiger population.

The USFWS is the agency that has been given jurisdiction over the illegal trade in wildlife in the United States. Its Special Operations force of 235 agents actively investigates cases of wildlife smuggling and poaching. Recently, an agent "purchased" an African leopard skin and a frozen still-born tiger cub for $1,500 on eBay and nabbed the seller as a result. The agents now monitor a number of Internet online auction sites, where collectors and traffickers often do business in endangered species and their parts.

It is easy to place the blame on the economic greed of the persons doing the killing. However, equal blame must be shared by the consumers who offer the "reward." It is their ignorance or insensitivity to the fact that their money is fostering the extinction of invaluable wildlife that is fueling the situation.

Particularly severe is the growing fad for exotic "pets": fish, reptiles, birds, and houseplants. In many cases, these plants and animals are taken from the wild; often, only a small fraction even survives the transition. When they are removed from the natural breeding populations, they are no better than dead, in terms of maintaining the species. At least 40 of the world's 330 species of parrots face extinction because of this fad. Numerous species of tropical fish, birds, reptiles, and plants are thus headed toward extinction because of exploitation by the pet trade.

Poor management represents another form of overuse leading to a loss of biodiversity. Forests and woodlands are overcut for firewood, grasslands are overgrazed, game species are overhunted, fisheries are overexploited, and croplands are overcultivated. These practices not only deplete the resource in question, but often set into motion a cycle of erosion and desertification, with effects far beyond the exploited area. The problem is explored in depth in Chapter 12.

Consequences of Losing Biodiversity

What will happen as more and more rare or unknown species pass from the Earth because of our activities? A few people will mourn their passing, but as one observer put it, the Sun will continue to come up every morning, and the newspaper will appear on our doorstep. Currently, we seem to be getting away with it, as evidenced by the fact that many species have already become extinct as a consequence of human activities, and life goes on. However, we really do not know what we are losing when we lose species. Some ecologists have likened the loss of biodiversity to an airplane flight during which we continually pull out rivets as the plane cruises along. How many rivets can we pull out before disaster occurs? Is this a wise activity? So far, we have gotten away with driving species to extinction, but the natural world is certainly less beautiful and more monotonous as a result.

Can ecosystems lose some species and still remain productive? Research evidence indicates that it is the dominant plants and animals that determine major ecosystem processes like energy flow and nutrient cycling. (See "Global Perspective," this page.) Thus, simplifying ecosystems by driving rarer species to extinction is not necessarily expected to lead to ecosystem decline. We must be careful not to justify preserving wild species and biodiversity with predictions of disasters. However, it is possible to lose what ecologists call **keystone species**—species whose role is absolutely vital for the survival of many other species in an ecosystem. For example, through felling trees and building dams, beavers significantly alter forest and stream ecology and in-crease ecosystem diversity in a region. Or perhaps the keystone species in an ecosystem is a predator that keeps herbivore populations under control. Sometimes, keystone species are the largest animals in the ecosystem, and because of their size, they have the greatest demands for unspoiled habitat. Thus, wolves, elephants, tigers, moose, and other large animals are considered "umbrella" species for whole ecosystems. If they can survive, there will likely still be enough undisturbed habitat for the other living species.

As we have seen, it is also possible to introduce species (e.g., purple loosestrife) that can become new dominants in ecosystems; they have an impact on both biodiversity and functionality as they crowd out existing species, and the results are very likely to be undesirable.

Biodiversity will continue to decline as long as we continue to remove and constrict the natural habitats in which wild species live. This loss is bound to be costly, because natural ecosystems provide vital services to human societies. Recreational, aesthetic, and commercial losses will be inevitable. The loss of wild species, therefore, will bring certain and unwelcome consequences, because it is linked directly to the degradation or disappearance of ecosystems (our focus in Chapter 12).

GLOBAL PERSPECTIVE

BIODIVERSITY: ESSENTIAL OR NOT?

A fierce debate is raging within the ecological academy, focused on a very important question that has serious political and economic ramifications: Will the loss of species to an ecosystem bring it closer to collapse? A corollary question asks just what the role of biodiversity is in promoting normal functioning like productivity in an ecosystem. On one side of this controversy are some of the leading ecologists in the United States (especially David Tilman) and the United Kingdom (John Lawton), who have published experimental work indicating that biodiversity is essential to the healthy functioning of an ecosystem. For example, one line of experimental work—the Ecotron Experiment—was carried out in the United Kingdom. Researchers added various plants, insects, and worms to replicate plots in enclosed chambers. They found that the more species they added to a plot, the more biomass the plot produced, thus the greater the productivity. Other work carried out in Minnesota showed that plots with a greater diversity of species were more resistant to drought. More important, Tilman and Lawton were among the coauthors of "Biodiversity and Ecosystem Functioning: Maintaining Natural Life Support Processes," an important *Issues in Ecology* paper published by the Ecological Society of America in 1999 and widely distributed on the Internet and to members of Congress. Citing their experimental work, the authors concluded, "Given its importance to human welfare, the maintenance of ecosystem functioning should be included as an integral part of national and international policies designed to conserve local and global biodiversity."

But wait a minute, say Michael Huston of Oak Ridge National Laboratory in Tennessee, Phil Grime of the University of Sheffield in the United Kingdom, and others. There is a good chance that these experimental studies and the conclusions drawn from them were "politically manipulated." Not only that, but in some important cases the experiments were flawed and therefore are inconclusive. Years ago, theoretical ecologist Robert May concluded on the basis of mathematical modeling that biodiversity has no consistent effect on the functioning of an ecosystem. Rather, what is more important, the critics say, is how individual organisms respond to environmental change. Citing experimental evidence in support of this view, these researchers note that if particular species are present, they will often dominate ecosystem functioning because of their superior adaptations to a given environment; the presence of many other species can be irrelevant. This is not to say that biodiversity is unimportant; in fact, greater biodiversity will make the presence of key species more likely in an ecosystem. Thus, biodiversity is like an insurance policy against environmental variability. But to say that biodiversity is the key to ecosystem functioning and then to transfer this conclusion to policy recommendations goes beyond what the science supports, according to the critics. They believe that other arguments should be used to support biodiversity as an important property of natural systems and that to tie support of biodiversity to normal ecosystem functioning, as Tilman and others do, is to erect a "house of cards" that could collapse in the face of new experimental evidence.

International Steps to Protect Biodiversity

Convention on Trade in Endangered Species. Endangered species outside the United States are only tangentially addressed by the ESA. However, under the leadership of the United States, the Convention on Trade in Endangered Species of Wild Fauna and Flora, or CITES, was established in the early 1970s. CITES is not specifically a device to protect rare species; instead, it is an international agreement (signed by 118 nations) that focuses on trade in wildlife and wildlife parts. The treaty recognizes three levels of vulnerability of species, the highest being species threatened with extinction. Covering some 30,000 species, restrictive trade permits and agreements between exporting and importing countries are applied to those species, sometimes with the result of a complete ban on trade if the CITES nations agree. Every two or three years, the signatory countries meet at a Conference of Parties (COP). The latest (the 11th) COP was held in Nairobi, Kenya, in April 2000.

Perhaps the best known act of CITES was to ban the international trade in ivory in 1990 in order to stop the rapid decline of the African elephant (from 2.5 million animals in 1950 to about 600,000 today). In response to requests from African nations, the 10th COP, in 1997, voted to lift the ban to allow a one-time ivory sale for three African countries (Namibia, Botswana, and Zimbabwe) that had demonstrated the ability to maintain sustainable populations of their elephants. As many predicted, the lifting of the ban led to a resumption in poaching: Over 400 elephants were killed by gangs in Zimbabwe alone. As a result, the 11th COP refused to allow any new ivory sales, in spite of a proposal from the three countries to expand their trade in ivory. Understandably, the three southern African countries were unhappy about the results, because the one-time sale netted $5 million for their cash-strapped economies.

Convention on Biological Diversity. Although CITES provides for some protection of species that might be involved in international trade, it is inadequate to address broader issues pertaining to the loss of biodiversity. People in industrialized nations have expressed a growing concern about the decline of well-loved species such as pandas, elephants, and rhinos, the rapid rate of deforestation in tropical rain forests and the feared loss of many species as a consequence, and inroads into lands occupied by indigenous peoples. It is clear that the main target of their concern is the plight of biological diversity in the developing countries, where most biodiversity is found. On the other hand, people in the developing countries regard their forests and animals as part of their natural resources; millions of rural individuals depend on access to these resources, and many governments look to their development as the key to more rapid economic growth. Thus, a problem arises wherein some of the world wants to see the protection of resources that are not in its territory, but which it values nevertheless.

With the support of UNEP, an ad hoc working group proposed an international treaty that would resolve these differences and form the basis of action to conserve biological diversity worldwide. After several years of negotiation, the **Convention on Biological Diversity** (CBD) was drafted and became one of the pillars of the 1992 Earth Summit in Rio de Janeiro. The Biodiversity Treaty, as the convention is called, was ratified in December 1993 and is now in force. The treaty establishes a Conference of the Parties as the agency that will provide oversight and report on its task during periodic meetings. Five COP meetings have been held, the latest in Nairobi in May 2000. One of the decisions reached at that meeting was to adopt "the ecosystem approach" in managing land and water resources. Many of the principles of ecosystem management discussed in Chapters 3 and 4 can be seen in the list of principles adopted at this latest COP.

The preamble sets forth basic guidelines for the Biodiversity Treaty: a concern for the intrinsic value of biodiversity, its significance for human welfare, the sovereignty of a nation over its biodiversity, and the nation's obligations to protect and conserve biodiversity.

The treaty requires the following from the signatory countries:

- Adopt specific national biodiversity action plans and strategies.
- Establish a system of protected areas and ecosystems within the country.
- Establish policies that provide incentives to promote the sustainable use of biological resources.
- Restore habitats that have been degraded.
- Protect threatened species.
- Respect and preserve the knowledge and practices of indigenous peoples.
- Respect the ownership of genetic resources by countries, and share the technologies developed from those resources.

Concern about access to genetic resources and funding mechanisms led President Bush to refuse to sign the treaty in Rio in 1992. President Clinton signed the treaty in June 1993, but in order to become binding, the treaty must be ratified by a two-thirds majority vote of the U.S. Senate. Continued objections led the Republican-dominated Senate to shelve ratification, leaving both the treaty and U.S. involvement in it in a state of limbo. Many of the provisions of the treaty are being carried out by different U.S. agencies; for example, the National Biological Service is working on several of the treaty's provisions. The United States continues to send delegations as observers to the COPs and participates in other CBD meetings, such as the recent meeting on genetically modified organisms held in Montreal. (See Chapter 10.)

Critical Ecosystem Partnership Fund. A new player on the international scene emerged in August 2000: the Critical Ecosystem Partnership Fund. Jointly sponsored by the

World Bank, Conservation International, and the Global Environment Facility, the fund provides $150 million for conservation activities in developing countries. The aim is to protect biodiversity "hot spots," 25 regions in which 60% of the biodiversity has been located in just 1.4% of Earth's land surface (Fig. 11–16). The fund will provide money to protect and manage these hot spots. Since 1989, some $400 million has already gone to protecting some of them, much of the money coming from the John D. and Catherine T. MacArthur Foundation.

Stewardship Concerns

What we do about wild species and biodiversity reflects on our *wisdom* and our *values*. *Wisdom* dictates that we take steps to protect the biological wealth that sustains so many of our needs and economic activities. The scientific team that put together the "U.N. Global Biodiversity Assessment" focused on four themes in its recommendations:

1. *Reforming policies that often lead to declines in biodiversity.* For example, many governments subsidize the exploitation of natural resources and agricultural activities that consume natural habitats. These policies are often counterproductive, leading to the decline of important natural goods and services, which tend to be undervalued economically because they do not enter the market.

2. *Addressing the needs of people who live adjacent to or in high-biodiversity areas or whose livelihood is derived from exploiting wild species.* Although protecting biodiversity benefits entire societies, the people who are closest to the protected areas are key to the success of efforts to maintain sustainability of natural resources. These people must be given ownership rights or communal-use rights to natural resources; they must be involved in the protection and management of wildlife resources. Experience has shown that when local people benefit from protection of wildlife, they will more readily take up stewardship responsibility for the wildlife. Recently, community-based natural resource management (CBNRM) emerged as a strategy for wildlife-rich African countries to preserve their biodiversity.

3. *Practicing conservation at the landscape level.* Too often, we tend to believe that a given refuge or park will protect the biodiversity of a region, when, in reality, many wild species require larger contiguous habitats or a variety of habitats. For example, to address the needs of migratory birds, summer and winter ranges and flyways between them must be considered. Buffer zones should be created between major population centers and wildlife preserves. This requires cooperation between governments, private-property owners, and corporations.

4. *Promoting more research on biodiversity.* We know far too little about how to manage diverse ecosystems and landscapes; much remains to be known about the species within the boundaries of different countries. For example, Costa Rica is engaged in a scientific inventory of its biodiversity, with an eye especially toward the development of new pharmaceutical or horticultural products. To that end in the United States, the

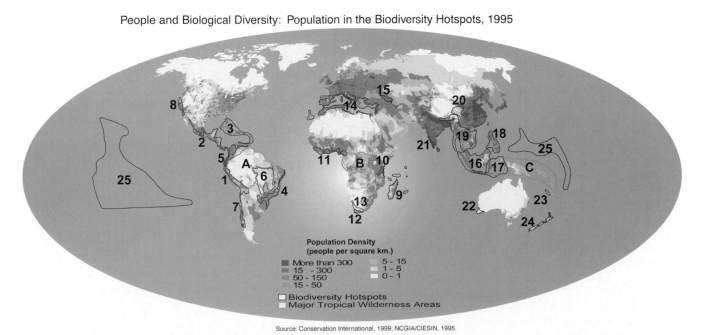

People and Biological Diversity: Population in the Biodiversity Hotspots, 1995

Population Density (people per square km.)
- More than 300
- 15 - 300
- 50 - 150
- 15 - 50
- 5 - 15
- 1 - 5
- 0 - 1
- ☐ Biodiversity Hotspots
- ☐ Major Tropical Wilderness Areas

Source: Conservation International, 1999; NCGIA/CIESIN, 1995.

▲ **FIGURE 11–16** *Biodiversity Hot Spots.* Some 25 regions in which 60% of the biodiversity has been located in just 1.4% of Earth's land surface.

National Biological Service of the U.S. Department of the Interior coordinates research like taxonomic studies designed to inventory the nation's biological resources, one of the goals of the CBD.

Our *values* also are demonstrated in our approach to wild species. Do we consider the natural world simply as grist for our increasing consumption? Or do we treat nature as something toward which we have a responsibility of stewardship and are challenged to manage sustainably and pass on to our children and their children? Because different people answer these questions differently and for different reasons, some people feel a need for alternative sets of values, other than or in addition to simply survival or utilitarianism, to guide our course into the future. What each individual turns to will depend on his or her beliefs, needs, education, and religious or ethnic background. For instance, some environmentalists have embraced Deep Ecol-

ogy, a movement that reminds humans that we are a part of the Earth and not separate from it, thus placing nature at the forefront of human awareness. For others, religious faith forms the basis for their interaction with nature. Significantly, Harvard University's Center for the Study of World Religions has held a series of conferences exploring each of the major world faiths and its ecological views. The American Academy of Arts and Sciences, the United Nations, and the American Museum of Natural History have held interdisciplinary gatherings on faith and the environment.

Further strategies to address the loss of biodiversity must focus on preserving the natural ecosystems that sustain wild species. In the next chapter, we turn our attention to those ecosystems—how we depend on them, what we are doing to them, and what we must do to resist the forces that seem to be leading to a future in which all that is left of wild nature is what we have managed to protect in parks, preserves, and zoos.

ENVIRONMENT ON THE WEB

MEASURING BIODIVERSITY

Ecologists have long argued about the best ways to track the diversity and the integrity—soundness—of ecosystems. The debate centers on whether we should measure ecosystem structure or function, or both. This is like trying to decide whether you will measure the adequacy of your house by counting the number of rooms, or by assessing the amount of energy the house uses annually—both measurements provide useful insight.

The traditional way of measuring structural diversity is simply to count the number of species present and the number of individuals in each. A system with many species and few individuals in each is considered diverse relative to one with few species, each with many individuals. Such an approach, however, ignores the complexity of the abiotic environment and overemphasizes the importance of common species.

More recent approaches combine measurements of several factors that together are thought to reflect ecosystem integrity. For example, in recent years, Canada and the United States have been grappling with the challenge of measuring biodiversity in the Great Lakes basin, as part of their joint responsibility for managing the health of that shared ecosystem. Several key points have emerged in their discussions. First, because ecosystems are organized in a hierarchical—or layered—fashion (for example, a food chain), it may be necessary to measure biodiversity at several levels. This layering makes it possible to observe good species richness and primary productivity at one level, for instance among soil microorganisms, but poorer structure and function at a larger scale, such as in woody plants. Since the behavior of the ecosystem depends on the integrity of its component parts, ignoring any one layer can lead to a skewed measurement of diversity.

Second, the "right" measurement depends on the question being asked. If, on the one hand, the question is "What species are present?" then traditional species counts are a useful tool. Conversely, if the question is "How genetically diverse are these species?" it may be necessary to map the genetic "fingerprint" of

each population. However, if you ask "How well is the ecosystem functioning?" or "How well is the ecosystem able to withstand stress?" then more integrative measurements, such as net primary productivity (the ecosystem's ability to produce new biomass), community composition (e.g., relative proportions of trees, shrubs, and herbaceous species), or even gross soil erosion provide insight into how the ecosystem works. Hence, several structural and functional measurements may be needed to develop an adequate assessment of a system's biodiversity.

The fifth principle of ecosystem sustainability—*For sustainability, biodiversity is maintained*—implicitly denotes that more biodiversity is better than less biodiversity and that biodiversity can be measured. However, sound science doesn't always give us clear answers about what level of biodiversity exists in a given area. Regardless, measuring the biodiversity of a system is a form of developing a baseline condition against which to evaluate future trends. It is a form of understanding the cause-and-effect relationships that are becoming more and more essential to us in our stewardship of Earth's systems.

Web Explorations

The Environment on the Web activity for this essay describes the challenges involved in transboundary management of ecosystem diversity. Go to the Environment on the Web activity (select Chapter 11 at http:// www.prenhall.com/wright) and learn for yourself:

1. about priorities for conservation of biological diversity;
2. about global patterns of biodiversity; and
3. how biodiversity protection can sometimes conflict with other transboundary issues.

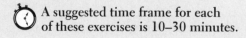 **A suggested time frame for each of these exercises is 10–30 minutes.**

REVIEW QUESTIONS

1. Define biological wealth and apply the concept to human use of that wealth.

2. Define instrumental value and intrinsic value as they relate to determining the worth of natural species.

3. What are the four categories into which human dependence on natural species can be divided?

4. What means are used to preserve game species, and what are some problems emerging from the adaptations of many game species to the humanized environment?

5. How does the Endangered Species Act (ESA) preserve threatened and endangered species in the United States? What are the shortcomings of, and controversies surrounding, the act? Illustrate.

6. What is biodiversity? What are the best current estimates of its extent?

7. Describe both the known and estimated losses in biodiversity.

8. How do habitat conversion, fragmentation, and simplification affect biodiversity?

9. How do population, pollution, exotic species, and overuse affect biodiversity?

10. What are the possible consequences of extinctions of species?

11. What is CITES, and what are its limitations? Give three requirements of the Convention on Biological Diversity.

12. Describe ways in which our wisdom and values are reflected through our actions as stewards of biodiversity.

THINKING ENVIRONMENTALLY

1. Choose an endangered or threatened animal or plant species, and research what is currently being done to preserve it. What dangers is it subject to? Write a protection plan for the species.

2. You have been given the task of maintaining or increasing biodiversity on a small island. What steps will you take?

3. Some people argue that each individual animal has an intrinsic right to survival. Should this right extend to

plants and microorganisms? What about the *Anopheles* mosquito, which transmits malaria? Tigers that kill people in India? Bacteria that cause typhoid fever? Defend your position.

4. Monies for enforcing the Endangered Species Act are limited. How should we decide which species are the most important to save?

5. Is it right to use animals for teaching and research? Justify your answer.

WEB REFERENCES

On-line resources for this chapter are on the World Wide Web at: **http://www.prenhall.com/wright**. (Click on Chapter 11 within the Chapter Selector.)

Ecosystems as Resources

Key Issues and Questions

1. Natural ecosystems perform a number of "natural services" that are vital to human interests. What are these services?
2. As natural resources, ecosystems are put to both consumptive and productive uses. What are the distinctions between these two kinds of use?
3. The exploitation and overuse of natural systems is pervasive and nonsustainable. How do the concepts of the maximum sustained yield and the tragedy of the commons help us to understand the problems that arise?
4. In many parts of the world, forests are being cleared. Why is this happening, and what are its immediate consequences?
5. Temperate and tropical forests can be managed sustainably. What are some patterns of forest management, and what are the components of sustainable forestry?
6. Tropical forests are disappearing rapidly. What are some promising trends that can slow the pace of deforestation?
7. Overfishing is occurring in 11 out of 15 of the world's major fishing areas. Why is this happening, and how can the declines be reversed?
8. Public lands occupy 40% of the land area of the United States and represent a vital national treasure. What are some problems in the management of the federal lands, and how are the lands being managed for sustainability?

The Pacific halibut is a large flatfish (up to 8 ft in length and 800 lb in weight) found across the north Pacific in bottom waters of the continental shelf. It is a prized table fish and has supported an active commercial fishery since the late 1800s. (See the opposite page.) The eastern Pacific fishery has yielded annual catches as high as 50,000 tons and has been heavily regulated since the 1920s by the International Pacific Halibut Commission. Typically, the commission would set an annual **total allowable catch** (TAC), based on its assessment of stocks, and then the fishing fleet would fish until the TAC was reached. Soon this took on all the aspects of a fishing derby, with increasingly shorter seasons until, in 1990, the season lasted only six days. Almost all of the fish caught in that short time had to be frozen, which lowers the quality of the fish and the price received by the fishers. The intensive fishing led to lost and damaged gear and endangered the crew and vessels as they fished in spite of adverse weather conditions.

Canada was the first country to move away from this fishing derby approach, in 1991, followed by the United States in 1995. Both countries adopted a management strategy called the **individual quota** (IQ) system, whereby owners of fishing vessels would be allocated a percentage of the annual TAC based on the size of their vessel and their recent fishing performance. Then they could decide when to fish within a season that lasted as long as 8 months. Entry into the fishery was strictly limited. The results of this shift were highly successful. The fishers saw an increase in their income, as they could sell their fish in the more lucrative fresh-fish market. They could fish in better weather, and they lost less fishing gear. And to everyone's surprise, fish stocks improved so well that, from 1995 to 1999, the TAC increased by 55%. The change in management strategy took place with the full involvement of the fishing community, and even though the fishery is now more thoroughly regulated than ever (fishers must report when they intend to go out, fish catches are carefully monitored, and observers are present on many vessels), the fishery is healthy and the fishers are better off.

The Canadian Pacific halibut story is unusual; more often, natural ecosystems and the species in them are in a

◀ **Halibut fishing.** Fishermen harvest Pacific halibut on board the *Mar del Norte,* off Kodiak Island, Alaska.

losing battle. Yet there are many notable exceptions—places around the world where people are successfully managing natural resources in sustainable ways.

As we traced the loss of biodiversity in Chapter 11, we observed that the key to saving wild species is to preserve their habitats—the ecosystems in which they are found. Populations of wild species continue to decline, however, because ecosystems are also in decline throughout the world. Beyond the loss of species is the degradation or loss of the natural goods and services ecosystems provide.

In this chapter, we look at major ecosystems that sustain human life and economy, examine the gains achieved and losses incurred as we convert these ecosystems to other uses, and consider in particular those ecosystems that are in the deepest trouble. To begin, we will examine the approaches to natural-resource use that lead to trouble and contrast them with sustainable approaches.

12.1 Biological Systems in a Global Perspective

Major Systems and Their Value

As we learned earlier, Earth is occupied by ecosystems that vary greatly in the composition of their species, but that exhibit common functions, such as energy flow and the cycling of matter. The major types of terrestrial and aquatic ecosystems—the biomes—reflect the response of biotas to different climatic conditions. We can divide Earth's land area into several broad categories for the purposes of this chapter (Table 12–1): **forests and woodlands, grasslands and savannas, croplands, wetlands,** and **desert lands and tundra** (see Fig. 2–18). We can separate the oceanic ecosystems into **coastal ocean and bays, coral reefs,** and **open ocean**. These eight systems provide all of our food; much of our fuel; wood for lumber and paper; leather, furs, and raw materials for fabrics; oils and alcohols; and much more. The world economy is directly dependent on the exploitation of the **natural goods** we extract from ecosystems.

Ecosystems also perform a number of valuable **natural services** as they process energy and circulate matter in the normal course of their functioning (Fig. 12–1). Some of the services are general and pertain to essentially all ecosystems; others are more specific. Some of the most important natural services performed by ecosystems are as follows (a more complete list can be found in Table 3–3):

1. **Modification of climate**. Because plants absorb considerable amounts of solar radiation and release water vapor through transpiration, ecosystems moderate temperature and help to maintain an even climate.
2. **Maintenance of the hydrologic cycle**. Plants absorb water from soils and release it through transpiration, returning the water to the atmosphere. Forests, grasslands, and wetlands maintain a favorable distribution and even flow of water, absorbing it when it is abundant and releasing it gradually. In doing so, they prevent much flooding.
3. **Erosion control and soil building**. Plant cover and ground litter absorb the potentially destructive impact of rainfall and prevent the breakup of soil; plant roots bind the soil. Forests are especially crucial in preventing erosion on hilly terrain. All of the plants, small animals, and microorganisms in terrestrial systems contribute to the formation of the soil.

TABLE 12–1 Natural Ecosystems on the Earth's Surface			
Ecosystem	Area (million km²)	(million mi²)	Percent of Total Land Area
Forests and woodlands	47.3	18.2	31.7
Grasslands and savannas	30.2	11.6	20.2
Croplands	16.1	6.2	10.7
Wetlands	5.2	2.0	3.5
Desert lands and tundra	51.0	19.6	34.0
Total land area	149.8	57.6	
Coastal ocean and bays	21.8	8.4	
Coral reefs	0.52	0.2	
Open ocean	426.4	164.0	
Total ocean area	448.72	172.6	

▲ **FIGURE 12–1** *Ecosystem services.* Natural ecosystems perform invaluable natural services if they are treated well. On the left is an aerial view of Baranof Island, Alaska, showing a salmon-spawning stream, meadows, and a coniferous forest. At the right is a view of Kin-Buc landfill in New Jersey, a Superfund site where hazardous waste disposal has damaged vegetation and interrupted the normal functioning of the area biota.

4. **Maintenance of the oxygen and nitrogen cycles**. Photosynthesis in green plants continually regenerates oxygen, and microbes maintain the soil's fertility through nitrogen fixation. They also prevent the buildup of potentially harmful nitrogenous compounds (for example, ammonium and nitrite).

5. **Waste treatment**. Wetlands are particularly significant in stabilizing sediments, absorbing excess nutrients, and thus preventing eutrophication. In all ecosystems, microbes transform many toxic organic and inorganic chemicals into harmless products.

6. **Pest management**. Natural ecosystems contain a diverse array of insect predators. Even small patches of natural habitat, such as hedgerows and woodlots, contribute to the control of pests in adjacent agricultural lands.

7. **Carbon storage and maintenance of the carbon cycle**. The global carbon cycle is maintained by energy flow within natural ecosystems. Over 500 billion metric tons of carbon are stored in the standing biomass of forests, more than is found in the entire atmosphere. Much more is present in the organic matter of soils.

In their normal functioning, all natural and altered ecosystems perform some or all of the preceding natural services, free of charge. They also do it repeatedly, year after year. We tend to take these services for granted until the ecosystems, and thus the services, are lost. For example, deforestation in India is largely responsible for the massive siltation and flooding that cause repeated human tragedy and suffering in Bangladesh. Loss of wetlands, which provide sediment and nutrient control, is a major factor in the eutrophication of Chesapeake Bay (Chapter 18). The removal of trees for firewood in Africa has resulted in extensive erosion of the soil and lowered the productivity of arable lands. And centuries of forest removal and overgrazing in the Mediterranean basin have produced a climate that is much hotter and drier than it once was.

As we saw in Chapter 3, a team of natural scientists and economists has calculated that the total value of a year's global ecosystem goods and services is $33 trillion! To give an example of ecosystem value, scientists have calculated that it would cost more than $100,000 a year to artificially duplicate the water purification and fish propagation capacity provided by a single acre of natural tidal wetland (Fig. 12–2). Even if such sums were available to carry out these processes artificially, the energy expenditures involved (producing and burning the requisite fuels) would probably lead to a net increase in pollution rather than a decrease. Thus, there is no real way we can compensate for the losses of natural services that are incurred as natural ecosystems are degraded or destroyed. We simply suffer the consequent deterioration in environmental quality.

Ecosystems as Natural Resources

If the natural services performed by ecosystems are so valuable, why are we still draining wetlands and removing forests? The answer is revealing: *A natural area will receive protection only if the value a society assigns to its natural functions is higher than the value the society assigns to exploiting its natural resources.* When we refer to natural ecosystems and the biota in them as **natural resources**, we are deliberately placing them in an economic setting, and it becomes easy to lose sight of their ecological importance. A resource is expected to produce something of economic value for its owner. Thus, even though an acre of wetland

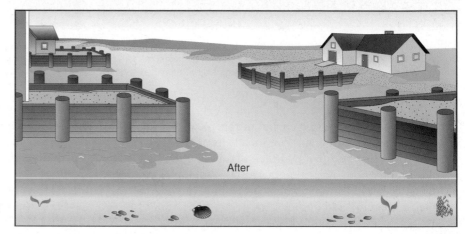

▶ **FIGURE 12–2** *Wetland services.*
Tidal wetlands provide a number of valuable services, worth more than $100,000 a year for just one acre. Especially important are water purification and fish propagation. These services are lost when wetlands are bulkheaded and converted for vacation homes.

provides valuable services for Chesapeake Bay, the owner of that acre will benefit economically by draining or filling it in and selling it to a builder. Its conversion removes the acre as a functional wetland and thus represents a *loss* in natural services, but an economic *gain* to the owner.

Some natural ecosystems are maintained in a natural or seminatural state because that is how they provide the greatest economic value for their owners. For example, most of the state of Maine is owned by private corporations that harvest the timber for lumber and for paper manufacturing. However, if Maine experienced a population explosion and corporations could sell their land to developers for more money than could be gained from harvesting timber, forested lands would quickly become lots for houses.

Finally, some natural ecosystems either are publicly owned (state and federal lands) or cannot be owned (ocean ecosystems). These ecosystems are still considered natural resources and may be subject to economically motivated exploitation. Obviously, *sustainable* exploitation of such systems will maintain the natural services they perform. Fortunately for us, ecosystems not only can sustain continued exploitation, but can also recover in time from very destructive uses that have degraded them and removed their natural biota.

As population continues its relentless rise, pressures on ecosystems to provide resources will increase: More food,

more wood, more water, and more fisheries products will be needed. Maintaining sustainable exploitation of natural resources will become more difficult, in view of the already intense pressure being put on those resources. Trade-offs among goods and services will become more commonplace; to increase its agricultural land, a country may clear its forests, thus losing the timber, flood control, and other services provided by the forests.

12.2 Conservation and Preservation

An ecosystem's remarkable natural capacity to regenerate makes it (and its biota) a **renewable resource**. In other words, an ecosystem has the capacity to replenish itself through reproduction despite certain quantities of organisms being taken from it, and this renewal can go on indefinitely. Recall from Chapter 4 that every species has the biotic potential to increase its numbers and that, in a balanced population, the excess numbers fall prey to parasites, predators, and other factors of environmental resistance. It is difficult to find fault with activities that effectively put some of this excess population to human use. The trouble occurs when users (hunters, fishers, loggers, and so on) take more than the excess and deplete the breeding populations, threatening or causing extinction of the species.

Conservation of natural biotas, then, does not—or at least should not—imply *no* use by humans whatsoever, although this may sometimes be temporarily expedient in a management program to allow a certain species to recover its numbers. Rather, the aim of conservation is to *manage or regulate use* so that it does not exceed the capacity of the species or system to renew itself. Clearly, conservation is capable of being carried out sustainably, and when sustainability is adopted in principle, conservation has a well-defined goal.

Preservation is often confused with conservation. The objective of the preservation of species and ecosystems is to *ensure their continuity, regardless of their potential utility.* Effective preservation often precludes making use of the species or ecosystems in question. For example, it is not possible to maintain old-growth (virgin) forests and at the same time harvest the trees. Thus, a second-growth forest can be *conserved* (trees can be cut, but at a rate that allows recovery of the forest), but an old-growth forest must be *preserved* (it must not be cut down at all).

There are times when conservation and preservation come into conflict. The Muriqui monkey of Brazil was once thought to require virgin forests, leading to a concern for protecting such forests for the sake of the species (Fig. 12–3). Recent research, however, has shown that the monkeys actually do better in second-growth forests, which support a greater range of vegetation on which the monkeys feed. Indefinite preservation of the virgin forests would lead to a decline in the population of the Muriqui monkey, a seriously endangered species. Thus, *conservation* of the forests is essential for *preservation* of this New World monkey.

Patterns of Use of Natural Ecosystems

Before we begin to examine the specific biomes and ecosystems that are under the highest pressure from human use, we should look at a few general patterns of human use that will help us understand what is happening in these natural systems and why it is happening. We should also keep in mind the three themes that have been serving to provide guidelines to "a long-term relationship with the natural world," as stated in Chapter 1: What policies and practices will lead to *sustainable* resource use? How can we enlist *sound science* in the cause of understanding how natural systems work and what is happening to natural resources? and How can we encourage attitudes of *stewardship* on the part of resource users and managers so that they will look beyond their own use and practices to the needs of other people, both those living now and those of future generations?

Consumptive Use vs. Productive Use. People in more rural areas make use of natural lands or aquatic systems that are in close proximity to them. Where people harvest natural resources in order to provide for their needs for food,

▲ **FIGURE 12–3** *Muriqui monkey.* Protection of this species of monkey in Brazil involves allowing some forest cutting in order to provide second-growth forests.

shelter, tools, fuel, and clothing, we refer to the practice as **consumptive use.** This kind of exploitation does not appear in the calculations of the market economy of a country. People may barter or sell such goods in exchange for other goods or services in very local markets, but most consumptive use involves family members engaged in hunting and gathering to meet their own needs. Thus, people are hunting for game (Fig. 12–4), fishing, or gathering fruits and nuts in order to meet their food needs; or they are gathering natural products like firewood, forage for animals, or wood and palm leaves for constructing shelters, or natural products for traditional medicines.

For example, wild game, or "bush meat," is harvested in many parts of Africa and provides a large proportion of the protein needs of people—in Congo, as high as 75% of a person's protein intake. Without access to these resources, people's living standard would decline and their very existence could be threatened. This kind of dependence is most commonly associated with the developing world, but may also exist in the more rural areas of the developed world, where people may be equally dependent on wood for fuel or wild game for food; their lives would be seriously affected if they were denied access to these "free" resources.

Productive use refers to the exploitation of ecosystem resources for economic gain. Thus, products such as timber and fish are harvested and sold for national or international

► **FIGURE 12–4** *Consumptive use.* A hunter from the Asmat tribe, in Irian Jaya, Indonesia, brings home game. Wild game is an important food source for millions in many parts of the world.

markets (Fig. 12–5). This is an enormously important source of revenue and employment for people in every country. The commercial trade in wood products for pulp, lumber, and fuel amounts to $142 billion per year; nonwood forest products like rattan, cork, or Brazil nuts are worth an additional $11 billion per year. Another productive use is the collection of wild species of plants and animals for cultivation or domestication. The wild species may provide the initial breeding stock for commercial plantations or ranches, or they may be used as sources of genes to be introduced into existing crop plants or animals to improve their resistance or to add other desirable genetic traits. As we saw in Chapter 11, wild species continue to provide sources of new medicines.

▲ **FIGURE 12–5** *Productive use.* Loggers load poplar logs to be transported to a lumber mill, where they will be made into wafer board for the building industry.

Consumptive and productive use of natural ecosystem products is strongly influenced by the political and legal institutions of a country or region. For example, private ownership clearly restricts access to natural areas, whereas community ownership is permissive of use by members of the community. State ownership of natural resources implies regulated use; here, public policy is the key to access these resources. In all of these arrangements—private, communal, or state—there is the potential for sustainable use of the resources, or for unsustainable use leading to the loss of the resources as well as the services provided by the natural ecosystem being exploited. Therefore, the perspective of stewardship is vital. Do the resource managers and harvesters see themselves as caretakers, as well as exploiters, of the natural resource? Are public policies that encourage stewardship and sustainability in place? Are sound scientific principles employed in management strategies? Keep these questions in mind as we next examine two patterns that are commonly seen where people are exploiting natural resources.

Maximum Sustainable Yield. The central question in managing a renewable natural resource is, How much continual use can be sustained without undercutting the capacity of the species or system to renew itself? The term employed to describe this amount of use is **maximum sustainable yield (MSY)** *the highest possible rate of use that the system can match with its own rate of replacement or maintenance.* MSY applies to more than just the preservation of natural biotas: It is also the central question in maintaining parks, air quality, water quality and quantity, soils, and, indeed, the entire biosphere. *Use* can refer to the cutting of timber, hunting, fishing, the number of park visitations, the discharge of pollutants into air or water, and so on. Natural systems can withstand a certain amount of use (or abuse, in terms of pollution) and still remain viable. However, a point exists at which increasing use begins to destroy regenerative capacity. Just short of that point is the MSY.

An important consideration in understanding MSY is the **carrying capacity** of the ecosystem—the maximum population the ecosystem can support on a sustainable basis. If a population is well below the carrying capacity of the ecosystem (Fig. 12–6a), then allowing that population to grow will increase the number of reproductive individuals and, thus, the yield that can be harvested. However, as the population approaches the carrying capacity of the ecosystem, new individuals must compete with older individuals for food and living space. As a result, recruitment may fall drastically (Fig. 12–6b). When a population is at or near the ecosystem's carrying capacity, production—and, hence, sustainable yield—can be increased by thinning the population so that competition is reduced and optimal growth and reproductive rates are achieved. Thus, the MSY cannot be obtained with a population that is at the carrying capacity. Theoretically, the **optimal population** is just half the population at the carrying capacity (Fig. 12–6).

The matter is further complicated by the fact that the carrying capacity and, therefore, the optimal population are not constant. They may vary from year to year as the weather fluctuates. Replacement may also vary from year to year, because some years are particularly favorable to reproduction and recruitment, while others are not. Of course, human impacts, such as pollution and other forms of altering habitats, adversely affect reproductive rates, recruitment, carrying capacity, and, consequently, sustainable yields. For these reasons alone, managing natural populations to achieve the MSY is fraught with difficulties. Furthermore, accurate estimates of the size of the population and the recruitment rate must be made continually, and these data are hard to come by for many species.

The conventional approach has been to use the estimated MSY to set a fixed quota—in fishery management, the **total allowable catch**. If the data on population and recruitment are not accurate, it is easy to overestimate the TAC, especially when there are economic pressures to maintain a high harvest quota. In the face of such uncertainties, and also in response to repeated situations of overuse and depletion of resources, resource managers have been turning to the **precautionary approach**: Where there is uncertainty, resource managers must favor the protection of the living resource. This means setting the exploitation limits well enough below the MSY to allow for uncertainties. Of course, those using the resource ordinarily want to push the limits higher, and the result is frequent conflicts between users and managers. We will return to this shortly when we look at the problems of fisheries.

Tragedy of the Commons. Where a resource is owned by many people in common or by no one, it is known as a **common pool resource**, or a **commons**. Examples of natural-resource commons are many: federal grasslands where private ranchers graze their livestock; coastal and open-ocean fisheries used by commercial fishers; groundwater drawn for private estates and farms; nationally owned woodlands and forests harvested for fuel in the developing world; and the atmosphere, which is polluted by private industry and traffic.

The exploitation of such common pool resources presents some serious problems and can lead to the eventual ruin of the resource—a phenomenon called the *tragedy of the commons*, after biologist Garrett Hardin's classic essay (1968) by that title. Sustainability requires that common

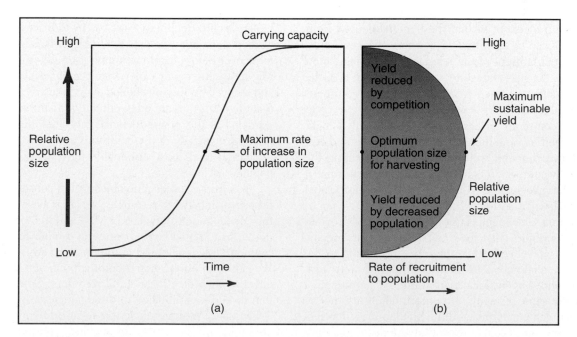

▲ FIGURE 12–6 *Maximum sustainable yield.* The maximum sustainable yield occurs when the population is at the optimal level. (the rate of increase in population is at a maximum). (a) The logistic curve of population size in relation to carrying capacity. (b) Recruitment plotted against population size, showing the effects of competition and decreased population levels.

pool resources be maintained for the benefit of future users, not just present ones.

As described by Hardin, the original "commons" were areas of pastureland in England provided free by the king to anyone who wished to graze cattle. In the parable, herders were quick to realize that whoever grazed the most cattle stood to benefit the most. Even if they realized that a particular commons was being overgrazed, those who withdrew their cattle simply sacrificed personal profits, while others went on using that commons. One herder's loss became another's gain, and the commons was overgrazed in any case. Consequently, herders would add to their herds until the commons was totally destroyed. They were locked into a system that led to their ruin.

Hardin's parable applies to a limited but significant set of problems in which there is open access to the commons, but where there is no regulating authority (or there is one, but it is ineffective) and no functioning community. Exploitation of the commons then becomes a free-for-all in which profit is the only motive. Coastal and offshore fisheries have consistently demonstrated the reality of the tragedy of the commons, as stocks of desirable fish have declined all over the world. Obviously, the tragedy can be avoided by limiting freedom of access.

One arrangement that can mitigate the tragedy is *private ownership.* When a renewable natural resource is privately owned, access to it is restricted, and, in theory, it will be exploited in a manner that guarantees a continuing harvest for its owner(s). This theory does not hold, however, when an owner maximizes immediate income and then moves on. Most owners, though, will be in for the long run and manage the resources more responsibly.

Where private ownership is unworkable, the alternative is to *regulate access to the commons.* Regulation should allow for (1) protection, so that the benefits derived from the commons can be sustained, (2) fairness in access rights, and (3) mutual consent of the regulated. Such regulation can reside in the state, but it does not have to. In fact, the most sustainable approach to maintaining the commons may be local community control, wherein the power to manage the commons resides in those who directly benefit most from their use and there are strong social ties and customs that can function well over time in protecting the commons. By contrast, in many situations, state control of a commons has accelerated its ruin and led to an associated social breakdown and impoverishment of people.

Harvesting soft-shelled clams in New England provides a modern example (Fig. 12–7). The clam flats in any coastal town are effectively a commons belonging to the town; commercial access to them is limited to town residents and policed by the local clam warden. Clamming is a way of life for some residents, but unfortunately, it is not always a dependable enterprise. Although clammers may recognize that the clams are being overharvested, any clammer who curtails his or her own take diminishes personal income, while competitors do not. The clammer's loss becomes their gain, and the clams are depleted in any case.

▲ FIGURE 12–7   *Exploiting a commons.*   Clam digger harvesting the soft-shelled clam, *Mya arenaria,* from a coastal Massachusetts clam flat.

Thus, while the size and number of the clams diminish, indicating overharvesting, digging continues, because the clammers have to make a living. Higher prices brought on by the shortage of clams continue to make the digging profitable despite declining harvests. In most cases, the clam warden steps in and closes a number of flats to allow them to recover before local clamming is totally ruined. Coastal Massachusetts towns experience regular cycles of scarcity and recovery in the soft-shelled clam industry, an indication that the commons are often difficult to maintain. The clam warden has the power to make the difference between a full-blown tragedy of the commons and a system in which regulation leads to a sustainable harvest—not an easy or popular task!

In sum, to achieve the objectives of conservation in industries using living resources, we must be aware of both the concept and limitations of MSY and the social and economic factors that cause overuse and other forms of environmental degradation that diminish the sustainable yield. We can then establish and enforce public policies that are effective in protecting natural resources. Some principles that should be embodied in these policies are presented in Table 12–2. Where policies based on principles like these are in place, it is possible for people to continue to put natural resources to sustainable consumptive and productive uses. The vast harvest we take from the natural world each year is proof that sustainable use not only is possible, but is

TABLE 12–2 Principles That Should Be Incorporated into Public Policies to Protect Natural Resources

1. Natural resources cannot be treated as open commons. Biotas should be put under an authority that is responsible for their sustainability and that can regulate their use.
2. Sound science should be employed to accurately assess the health of the resource and to set sustainable limits.
3. To accommodate uncertainty, the precautionary approach should be used in setting limits for exploitation.
4. Regulations should be enforced.
5. Economic incentives that promote the violation of regulations should be eliminated.
6. Subsidies that support exploitation of the resource should be removed.
7. Suitable habitats for the resource should be preserved and protected from pollution.
8. The sustenance needs of people living close to the resource should be met.

actually occurring in many resources and locations around the world. However, in some situations, exploitation and degradation have gone too far, and the resilience of an ecosystem to disturbance has been overcome. In those cases, natural services and uses can be restored only if the habitats are deliberately restored. In recent years, the need for the restoration of damaged ecosystems has become clear, and a new subdiscipline with that objective has appeared—**restoration ecology**.

Restoration. A great increase in restoration activity has occurred during the last 30 years, spurred on by federal and state programs and the growing science of restoration ecology. The intent of ecosystem restoration is simple: Repair the damage so that normal functioning returns and native flora and fauna are once again present. Because of the complexity of natural ecosystems, however, the task of restoration is often difficult in practice. Frequently, soils have been disturbed, pollutants have accumulated, important species have disappeared, and other—often exotic—species have achieved dominance. For these reasons, a thorough knowledge of ecosystem and species ecology is essential to the success of restoration efforts. The ecological problems that can be ameliorated by restoration include those resulting from soil erosion, surface strip mining, draining wetlands, coastal damage, agricultural use, deforestation, overgrazing, desertification, and eutrophication of lakes. In Chapter 2, we described the process involved in restoring a prairie ecosystem. Every case is different, as in the coastal restoration taking place in Baytown, Texas.

Baytown is on the Texas coast, 20 miles from Houston. In the 1940s and 1950s, a subdivision called Brownwood—nearly 400 homes—was built on a 500-acre peninsula surrounded by Galveston Bay that was occupied by dunes, marshes, woodlands, and ponds. Never high above the water, the peninsula had experienced occasional flooding from high tides and hurricanes. However, hurricane **Carla** in 1961 devastated the Texas coast, flooding Brownwood and putting a stop to any further development. Then, in the 1970s and 1980s, Brownwood and adjacent coastal lands sank 10 to 15 feet because oil and chemical facilities, as well as domestic water suppliers, were pumping out groundwater faster than it could be replenished. Bay tides and storms

repeatedly flooded Brownwood, and the city built pumping stations and an elevated road around the development in 1973. Two months after the road was completed, a tropical storm poured water over it, and Brownwood became a huge swimming pool. Following Hurricane **Alicia** in 1983, residents abandoned Brownwood forever. In time, the subdivision became an eyesore, with submerged streets, rotting houses, uprooted trees, and trash littering the whole area.

Faced with this problem, the city of Baytown decided that it might be possible to restore some semblance of what was once there and create a public park and wildlife sanctuary from the ruins, but a lot of problems had to be solved before that could happen. In time, the city bought up the properties from the absent owners, and a combination of federal funds, local bond issues, and a consortium of 200 industrial companies came up with the money to transform the eyesore into a functioning natural ecosystem again. The houses, utilities, and most of the roads were removed, the land was graded, channels were created to link with Galveston Bay, freshwater ponds were created on remaining upland areas, and by 1995 plant life began to return. The Baytown Nature Center (Fig. 12–8) is now a functioning sanctuary, with new residents like alligators, deer, armadillos, bobcats, raccoons, and foxes, as well as so many birds

▲ **FIGURE 12–8** *The Baytown Nature Center.* Formerly a housing development, the area is now a functioning sanctuary with new residents.

that it is now a stop on the Great Texas Birding Trail, drawing birders from all over the country.

As the values of natural ecosystems become more recognized, efforts to restore damaged or lost ones will become increasingly important. John Berger, executive director of Restoring the Earth (a private organization dedicated to the task of restoration), states that "the business of restoration is likely to become a multibillion dollar global enterprise" because of its capacity to ameliorate many of our worst environmental problems. We turn our attention now to a number of still functioning biomes and ecosystems in trouble because of their great economic importance.

12.3 Biomes and Ecosystems under Pressure

The biomes and ecosystems of the world provide us with priceless services and economically valuable goods. Although human activities affect virtually all biomes and ecosystems, some are under more pressure than others. We look first at the forest biomes, indisputably the most important of all types of biome in terms of their economic significance and potential for active human management.

Forest Biomes

Forests are the normal ecosystems in regions with year-round rainfall adequate to sustain tree growth. They are also the most productive systems the land can support, and they are self-sustaining. Forests and woodlands (ecosystems with mixed trees and grasses) perform a number of vital natural services: They conserve biodiversity, moderate regional climates, prevent erosion, store carbon and nutrients, and provide recreational opportunities (among other things). They afford us a number of vital goods: lumber, the raw material for making paper, fodder for domestic animals, fruit and nuts, and fuel for cooking and heating. The total monetary value assigned to global forest goods and services amounts to some $4.7 trillion a year. Of this, $153 billion is related directly to commercial exploitation of the forests. In spite of this value, the major threat to the world's forests is not simply their *exploitation*, but rather, their *total destruction*.

The U.N. Food and Agricultural Organization (FAO) releases a global assessment of forest resources every two years and compares the data with previous information in order to evaluate deforestation trends. The agency's 1999 findings indicate that, worldwide, slightly more than half of the area originally covered by forests and woodlands is still forested, but only one-fifth of the originally forested area represents intact forest ecosystems, or what are called **frontier forests.** Deforestation in the developed countries has been halted, and forest cover is slowly expanding. But because of deforestation in the developing world, there was a net loss of 56.3 million hectares (140 million acres) between 1990 and 1995 (Fig. 12–9).

Why are forests being cleared when they could be managed for wood production on a sustainable basis? The answer is simple: Even though forests are highly productive systems, it has always been difficult for humans to exploit them for food. Most of the energy in forests goes to detritus and decomposer food webs, not to the grazing food web (Chapter 3). In contrast, grasslands have short food chains in which herbaceous growth supports large herbivores that can yield meat and other animal products. Alter-

▶ **FIGURE 12–9   *Changes in forest area.*** Forest area of different regions of the world in 1995 compared with 1980. Forests are on the increase in the developed world, but are decreasing in the developing world. (Source: From *State of the World's Forests* 1999. Copyright © 1999 by Food and Agriculture Organization of the United Nantions. Reprinted by permission.)

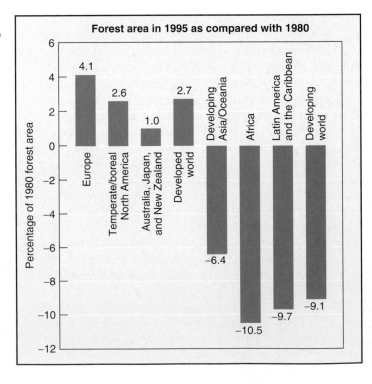

natively, natural grasses can be replaced with cultivated plants and used directly. Thus, the forests have always been an obstacle to conventional animal husbandry and agriculture. The first task the early European colonists faced when they came to the Western Hemisphere was to clear the forests so they could raise crops. Once the forests were cleared, continual grazing or plowing effectively prevented regrowth of the trees.

Clearing a forest has immediate consequences for the land and its people:

1. The overall productivity of the area is reduced.
2. The standing stock of nutrients and biomass, once stored in the trees and leaf litter, is enormously reduced.
3. Biodiversity is greatly diminished.
4. The soil is more prone to erosion and drying.
5. The hydrologic cycle is changed, as water drains off the land instead of being released by transpiration through the leaves of trees or percolating into groundwater.
6. A major carbon dioxide sink (removal of CO_2 from the air) is lost.
7. The land no longer yields either wood for fuel and building or nonwood forest products.
8. People dependent on harvesting forest products lose their livelihood.

Unless forests are cleared and converted to other uses, they can yield a harvest of wood for fuel, paper, and building. Approximately 3.4 billion cubic feet of wood are harvested annually from the world's forests, half for fuel and half for wood and paper (Fig. 12–10). It is unreasonable to expect the countries of the world—especially the developing countries—to forgo making use of their forests. However-er, forest-management practices vary greatly in the impact on the forest itself and on surrounding ecosystems.

Types of Forest Management. The term given to the practice of forest management, usually with the objective of producing a specific crop (hardwood, pulp, softwood, wood chips, etc.), is **silviculture**. Several practices are employed, with quite different impacts on the total forest ecosystem involved. Since trees take from 25 to 100 years to mature to harvestable size, the normal process of management is a *rotation*, which is a cycle of decisions about a particular stand of trees, from its early growth to the point of harvest. Two options are open to the forester: *even-aged management* and *uneven-aged management*. With the former, trees of a fairly uniform age are managed until the point of harvest, cut down, and then replanted, with the objective of continuing the cycle in a dependable sequence.

Typically, fast-growing, but economically valuable, trees are favored (others are culled from the stand); in the Pacific Northwest, Douglas fir is frequently managed this way. Harvesting is often accomplished by *clear-cutting*—removing an entire stand at one time. This management strategy has been under severe criticism, because it creates a fragmented landscape with serious impacts on biodiversity and adjacent ecosystems; also, a clear-cut leaves an ugly, unnatural-looking site for many years. (See the "Earth Watch" essay, p. 298.) Large timber companies often use it because it is efficient at the time of harvest and it does not involve much active management and planning.

Uneven-aged management of a stand of trees can result in a more diverse forest and lends itself to different harvesting strategies. For example, *selective cutting* can be employed, wherein some mature trees are removed in small groups or singly, leaving behind a forest that continues to maintain a

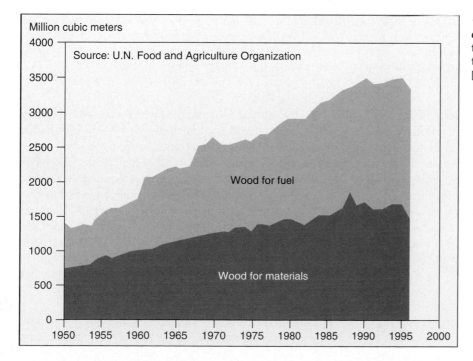

◀ FIGURE 12–10 *World wood consumption.* World wood consumption for fuel, building material, and paper. (*Source:* Data from Lester Brown, et al., *Vital Signs, 1999* [Washington, DC: Worldwatch Institute, 1999].)

EARTH WATCH

NATURE'S CORPORATIONS

The spotted-owl controversy pitted jobs against the preservation of the old-growth forests of the West. Timber interests maintained that preservation would result in 20,000 lost jobs. In their view, the controversy came down to this: We should continue to cut the old-growth forests for the sake of keeping loggers employed. This view not only is shortsighted; it is illogical.

General Motors and other major businesses facing hard times lay off thousands of workers, and no one questions their need or right to do so. Clearly, the survival of General Motors is more fundamental than keeping all of the company's workers employed. That is, it is assumed that those laid off will find other employment. Thus, according to conventional wisdom, corporate America must survive if the economy is to have a chance of recovering from economic bad times.

In a real sense, ecosystems are the corporations that sustain the economy of the biosphere. (See the inset.) Forested ecosystems can be viewed as *nature's corporations*, providing lumber and firewood for the economy. Coastal fisheries provide seafood for millions. If we want these ecosystems to survive and recover, we have to tighten our belts and withdraw some of the workforce engaged in exploiting them. The maintenance of these systems is obviously more important than some jobs: Why can't laid-off loggers seek other employment the way laid-off autoworkers, steelworkers, or computer engineers do? Why should a natural ecosystem be "bankrupted" just to maintain the *tempo-*

Forest ecosystems can be viewed as nature's corporations, providing lumber and firewood for the economy.

rary employment of a few (temporary because the loggers will be out of a job in a few years, when the old-growth forests are finally all cut)?

With the extent of the logging cutbacks due to the now-established protection of endangered species, the economy in the Pacific northwest is picking up. Oregon's unemployment rate of 5% is at its lowest in 25 years. Although some 15,000 wood-related jobs have been lost, there has been a net gain in employment due to rapid growth in high-technology industries and federal retraining aid to many towns and workers. Industries locate in the region because of the

impact of the "second paycheck," which refers to the high quality of life that attracts workers, even at lower pay scales. The same high quality of life—the forests, streams, scenic vistas, hunting, and fishing—requires preservation of the ecosystems that were being logged over.

Clearly, it is shortsighted to assume that ecosystems are there just to provide jobs and that the jobs are more important than the ecosystems. When the old-growth forests are gone, we shall have lost more than just loggers' jobs; we shall have lost a priceless heritage and a major part of the natural world that provides us with vital services.

diversity of biota and normal ecosystem functions. Replanting is usually unnecessary, as the remaining trees provide seeds. Another strategy is *shelter-wood cutting*, which involves cutting the mature trees in groups over a period of, say, 10 or 20 years, such that at any time there are enough trees both to provide seeds and to give shelter to growing seedlings. Like selective cutting, this method takes more active management and skill, but unlike clear-cutting, it leaves a functional ecosystem standing. In effect, it should lead to a sustainable forest, an objective that the entire wood and paper industry is currently focusing on—at least in principle.

Sustainable Forestry.   In forestry, sustainability must be carefully defined. A common objective of forestry management is **sustained yield**, which means that the production of

wood is the primary goal and the forest is managed to continuously harvest wood without being destroyed. **Sustainable forest management**, on the other hand, means that forests are managed as ecosystems and that maintaining the biodiversity and function of the ecosystem is the primary objective. Thus, the forests are managed and trees are harvested in ways that will preserve most or all of the vital ecological services they perform. It is encouraging that the American Forest and Paper Association, whose membership of privately owned forest and paper companies controls 90% of the industrial forestland in the United States, has adopted the *Sustainable Forestry Initiative*, consisting of the following principles:

1. **Sustainable forestry**—to meet the needs of the present without compromising the ability of future generations

to meet their own needs by practicing a land steward-ship ethic that integrates managing the reforestation, growing, nurturing, and harvesting of trees for useful products with the conservation of soil, air and water quality, wildlife and fish habitats, and aesthetics.

2. **Responsible practices**—to use in its own forests, and to promote among other forest landowners, sustainable forestry practices that are economically and environ-mentally responsible.

3. **Forest health and productivity**—to protect forests from wildfire, pests, diseases, and other damaging agents in order to maintain and improve long-term for-est health and productivity.

4. **Protecting special sites**—to manage its forests and lands that are of special significance (e.g., biologically, geologically, or historically) in a manner that takes into account their unique qualities.

5. **Continuous improvement**—to continuously improve the practice of forest management and to monitor, measure, and report the performance of the associa-tion's members in achieving their commitment to sus-tainable forestry.

Later in the chapter, we will address federal forest-management policy in the United States when we address the use of public lands in general.

Tropical Forests. Because of continued deforestation, tropi-cal forests are of greatest concern. They are the habitat for mil-lions of plant and animal species, a vast number of which are still unidentified. Climatologists reason that these forests are also crucial in maintaining Earth's climate, serving as a major sink for carbon and restraining the buildup of global carbon dioxide. Yet, tropical forests continue to be removed at a rapid rate. Between 1960 and 1990, 20% of the tropical forests (over

445 million hectares, or 1.1 billion acres)—equivalent to two-fifths of the land area of the United States—were converted to other uses. From 1990 to 1995, an additional 65 million hectares of forest in developing countries was lost, mostly in the tropics. Although this is a disturbing rate, it is lower than the rate of deforestation of the previous decade. However, the losses continue, as reports of deforestation come from various parts of the tropics: Tanzania is losing up to 500,000 hectares of forest annually; Indonesia's annual loss is at least 1.5 million hectares; 2.6 million hectares of Congo Basin forests are de-stroyed annually; Cambodia's forests are being logged so rapidly that they will be gone by 2003; and in 1999, the Brazil-ian Amazon lost 1.7 million hectares to illegal logging and agricultural clearing.

Typically, an area to be cleared is cut and then allowed to dry for a few weeks before being burned. The unmistakable sig-nal of tropical deforestation is the plumes of smoke from burn-ing vegetation that can often be detected from satellites or space shuttles. Although this process of burning has been occurring for many years, it reached unusual proportions in 1997, "the year the earth caught fire," in the view of the World Wildlife Fund. Indonesian fires destroyed 25 million acres (10 million hectares) of forest and were so intense and numerous that tens of millions of people across Southeast Asia were exposed to in-tense smoke for weeks. The Indonesian fires alone were respon-sible for contributing 30% of all global carbon emissions in 1997. Fires burned large areas in Brazil, Papua New Guinea, and Colombia. Most of the fires were set by people wanting to clear the forest for agriculture or to establish plantations: often, the fires burned out of control because the forests were already unusually dry—the result of El Niño's impact on global climate (Fig. 12–11). Although rainfall has returned to normal, new rounds of fires burn in Indonesia yearly, set by plantation com-panies intent on agricultural development.

Deforestation is being caused by a number of factors, all of which come down to the fact that the countries

◀ **FIGURE 12–11** *1997—"The year the earth caught fire."* The morning haze is thick in Metapang, Borneo (Indonesia), as a result of fires burning the dried vegetation. Similar fires, fed by drought and land clearing, burned out of control in Southeast Asia, blanketing huge areas of land and sea with heavy smoke.

involved are in need of greater economic development and have rapid population growth. The 1999 FAO study concluded that the current major causes of deforestation are expanding subsistence agriculture and large resettlement programs in Asia and Latin America. Many governments in the developing world are promoting deforestation by encouraging the colonization of forested lands. For example, Indonesia has embarked on a program of resettlement and intends to convert 20% of its remaining forests to agricultural production—an area of 1 million hectares to be converted to rice paddies. Squatters are moving into national parks in Guatemala's northern Petén region, logging and clearing forests to plant cornfields. The clearing of forests for subsistence agriculture is especially intense in Africa and Asia, clearly a function of population growth in those areas. With no end to population growth in sight for these regions, clearing is expected to account for increasing deforestation in the coming decades as people with no other options do what they have to do to attain food security.

There is no doubt that the forests of the developing world are an important resource that can generate much needed revenue. Developing-world countries account for 20% of the international trade in forest products, about $28 billion. Often, concessions are sold to multinational logging companies, which reap high profits by exploiting the economic desperation of the developing countries and harvest the timber with little regard for regenerating the forest. More than 12 million hectares of Amazon forest were put under concession to Asian timber companies in 1996 alone. More important is the consumptive use of forests by millions of people living in or on their edges. Owning few private possessions and often little land, they are absolutely dependent on foraging in and extracting goods from the forests. Among the encouraging trends in forest management in the developing world—trends which indicate that those in power are paying more attention to the needs of indigenous people and the importance of forest goods and services other than just wood—are the following:

1. *Sustainable logging.* Several such strategies are employed: Cutting the vines before removing a desired tree often saves many surrounding smaller trees; planning a route for removing desired trees can minimize the damage done by dragging and by building roads; strip clear-cutting a 30- to 40-meter swath of trees creates gaps that can quickly be filled by seeding in from surrounding trees.

2. *Plantations of trees for wood or other products (cacao, rubber, etc.).* These are becoming increasingly common. Although much biodiversity is lost, plantations can continue to recycle nutrients, hold soil, and recycle water. Plantations are much better than typical agricultural uses in preserving the natural functions of forests, and they can be managed sustainably.

3. *Extractive reserves that yield nontimber goods.* Among these goods are latex, nuts, fibers, and fruits. Recent calculations show that some forests are worth much more as extractive reserves than as a source of timber.

4. *Preserving forests as part of a national heritage and putting them to use as tourist attractions.* This practice can often generate much more income than logging can.

5. *Putting forests under the control of indigenous villagers.* The villagers can then collectively use the forest products in traditional ways. Given tenure over the land, the villagers tend to exercise stewardship over their forests in a way that is sustainable. Where this practice has been implemented, the forests have fared better than where they have been placed under state control.

6. *Remote sensing.* A new $1.4 billion Brazilian project (SIVAM) is poised to monitor the Amazon with radar and satellite imagery. Existing satellites provide rapid information on the rate of deforestation.

One of the documents signed by heads of state at the Rio de Janeiro Earth Summit of 1992 was the **Statement on Forest Principles**. Treaty negotiations building up to the summit proved too contentious to lead to any binding agreements. The more developed countries showed great concern about the deforestation of tropical forests, whereas the less developed countries made it clear that they regarded the forests as resources for exploitation. In the end, the signatories agreed to a statement of 17 nonbinding principles that stressed sustainable management. The less developed countries insisted on their sovereign right to develop their forest resources, and all of the signatories agreed to consider revising the principles in the future, to assure continuing cooperation on issues pertaining to the forests.

A recent initiative directed toward the certification of wood products is the creation of the *Forest Stewardship Council.* This alliance of nongovernmental organizations like the National Wildlife Federation, industry representatives, and forest scientists has developed into a major international organization with the mission of promoting sustainable forestry by certifying forest products for the consumer market. By 2000, over 22 million hectares in 32 countries had been certified.

All of these developments suggest that global awareness of the importance of the tropical forests has reached the level of serious concern. In some cases, this concern is being translated into action that is slowing the high rate of deforestation experienced in the 1980s. In others, the outcome is uncertain. The demand for tropical wood has not diminished, and the economic rewards of exploitation will undoubtedly continue to promote unsustainable uses of the forests in many tropical countries.

Ocean Ecosystems

Marine Fisheries.   Fisheries provide employment for at least 200 million people and account for one-fifth of the total human consumption of protein. (The term **fishery** refers either to a limited marine area or to a group of fish or shellfish species being exploited.) For years, the oceans beyond a 12-mile limit were considered international commons. By the end of the 1960s, however, numerous regions of the sea were being seriously depleted of many species by overfishing on the part of international fleets equipped with factory ships and modern fish-finding technology. In the mid-1970s, as a result of agreements forged at a series of U.N. Conferences on the Law of the Sea, nations extended their limits of jurisdiction to 200 miles offshore. The United States accomplished this with the Magnuson Act of 1976. Since many prime fishing grounds are located between 12 and 200 miles from shore, this action effectively removed the fish from the international commons and placed them under the authority of particular nations. As a result, some fishing areas recovered, while nationally based fishing fleets expanded to exploit the fisheries.

Currently, the tragedy of the commons is playing itself out in fishing waters everywhere. Overfishing is occurring in 11 of the world's 15 major fishing areas, offering little hope that there will be any sustained increase in the harvest. The world fishing fleet consists of more than 3 million vessels, many equipped with fish-finding sonar and under-

water television cameras and aided by spotter planes and helicopters. Because of heavy government subsidies, the fleet is able to land $70 billion worth of fish at a cost of $124 billion. It is estimated that the current fleet now has 50% more capacity for catching fish than is necessary.

The total recorded harvest from marine fisheries and fish farming has increased remarkably since 1950, when it was just 20 million metric tons (Fig. 12–12). By 1998, it had reached 117 million metric tons. Aquaculture accounted for 31 million tons in 1998, or one-fourth of world fish supplies that year. (See "Earth Watch" essay, p. 302.) Eighty-six million metric tons of the total of 117 million metric tons was the freshwater and marine fisheries catch, to which must be added some 20 million tons of bycatch (fish caught and discarded). However, many species and areas continue to be overfished, and when this happens, fishers turn to new areas and formerly less desirable species. For example, 73% of the rise during the 1980s is attributed to an increased catch of just five species not previously exploited; many other species yielded decreasing harvests. If 18 of the important declining species were restored to their former productivity, the catch could increase by at least 10 million metric tons. Therefore,

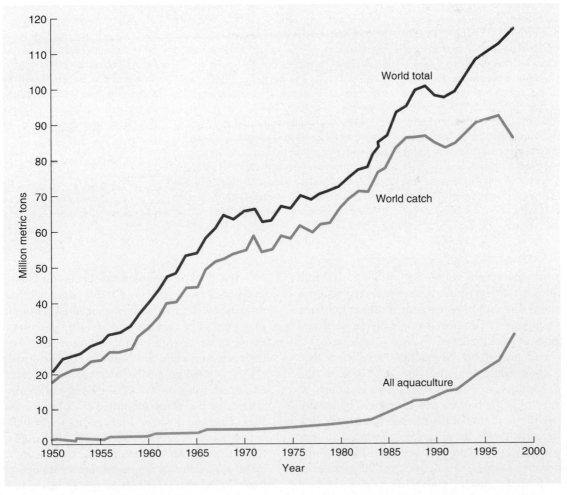

▲ FIGURE 12–12 *The global fish harvest.* The global fish catch plus fish farming equals the world total. Shown are data for 1950–98. (*Source:* Data from Lester Brown, et al., *Vital Signs, 2000* [Washington, DC: Worldwatch Institute, 2000].)

EARTH WATCH

WILL AQUACULTURE BE ABLE TO FILL THE GAP?

Modern aquaculture, which is the farming of aquatic organisms, includes species ranging from seaweeds to bivalve molluscs to freshwater and marine fish. Globally, aquaculture production has increased dramatically in the past 10 years, particularly the farming of marine and estuarine (coastal) species. (See Fig. 12–12.) To some extent, this production is offsetting the continuing declines in some marine fisheries, but coastal aquaculture is not without its problems and critics. For example, shrimp farming in many tropical areas has come under criticism for destroying mangrove habitats. In this case, ironically, shrimp aquaculture is being blamed, at least in part, for accelerating the decline of fisheries for some species that depend on the mangroves for a part of their life cycle.

Coastal aquaculture sites are typically in estuaries and lagoons or in ocean waters very close to the shoreline. These are the waters of the coastal zone that are the most used by humans for many other purposes: sportfishing, boating, waterskiing, etc. The nearshore coastal waters are also the most likely to be affected by sewage effluents and land runoff carrying many kinds of pollutants. The result in many cases is conflict among the various users, and the long-term outlook for coastal aquaculture is that nearshore waters are becoming less and less available for farming. There does not seem to be much argument about the need for ex-

panded farming of coastal species to offset the declines in many marine fisheries, but the industry appears be facing more problems than ever before. So what can be done to make aquaculture more sustainable in the long term?

One approach is open-ocean aquaculture, which is the farming of coastal species in ocean waters several kilometers from the shoreline. This is not a new practice in some areas, but it is in North America. And even in Europe and other areas where it has been long practiced, total production levels are far lower than they could be. In 1996, the first international conference on open-ocean aquaculture was held in Portland, Maine. Scientists, educators, businesspeople, government regulators, and aquaculture practitioners attended. The participants came from both coasts of the United States, Canada, and nine other countries. The conference marked the beginning of concerted, international efforts to expand traditional aquaculture practices into the open ocean. Time will tell whether such efforts meet with success, but there are many reasons for optimism.

For example, the most polluted areas in many coastal zones are the estuaries and embayments very close to shore. Hence, open-ocean waters are relatively clean. High water quality is important for aquaculture, but it is especially critical for species such as bivalve molluscs, which concentrate pathogens and

other pollutants that can be passed on to human consumers. Also, new satellite-based remote-sensing efforts are revealing nutrient-rich upwelling zones in ocean waters heretofore not known to display such conditions. In some areas in the Gulf of Maine, phytoplankton concentrations can exceed the levels thought only to occur in enriched estuarine waters. This means that open-ocean waters may be able to support a high productivity of farmed species like the blue mussel. In addition, open-ocean waters are generally not nearly as intensively used for other human activities, so the potential for conflicts among users is reduced. Finally, the nearshore aquaculture of many species has a longstanding tradition in many areas. Knowledge gained from this aquaculture will supply the base upon which open-ocean aquaculture practices are developed.

The problems facing aquaculture today are not unusual; aquaculture is yet another complex practice that consists of competing demands on our environment. The problems can be solved only by approaches that in the long term are truly sustainable. In this case, it means taking into account the needs and demands of many interest groups.

(Contributed by Ray Grizzle, Research Associate Professor of Zoology at the Jackson Estuarine Laboratory, University of New Hampshire.)

scientists reason, the oceans could yield well over 100 million metric tons (considerably above present levels) on a sustained basis. But this could happen only if fishing for many species was temporarily eliminated or greatly reduced and the major fisheries were managed on a sustainable basis.

Events on Georges Bank, Massachusetts, which is New England's richest fishing ground, are indicative of the trends. Cod, haddock, and flounder (called groundfish, as they feed on or near the bottom and are caught by bottom trawling) were the mainstay of the fishing industry for centuries. In the early 1960s, these species amounted to two-thirds of the fish population on the bank. After 1976, fishing on Georges Bank doubled in intensity, resulting in a decline in the desirable species and a rise in the so-called rough species—dogfish (sharks) and skates—which now constitute the bulk of the fish population on the bank.

The decline of the prized species on Georges Bank reflects poor management of the fishery. The Magnuson Act established eight regional management councils, made up of government officials and industry representatives. The councils are responsible for setting management plans for their regions, while the National Marine Fisheries Service (NMFS) provides advice and scientifically based information on assessing stocks and also has the authority to reject elements of the plans submitted to it by the councils. The New England Fishery Management Council (NEFMC) began its task by setting TAC quotas, but fishers claimed that the TACs were set too low and successfully argued for an indirect approach that employed mesh with openings of a size that allowed smaller fish to escape. In less than 10 years, the number of boats fishing Georges Bank doubled, with disastrous results, as indicated by the cod landings (Fig. 12–13).

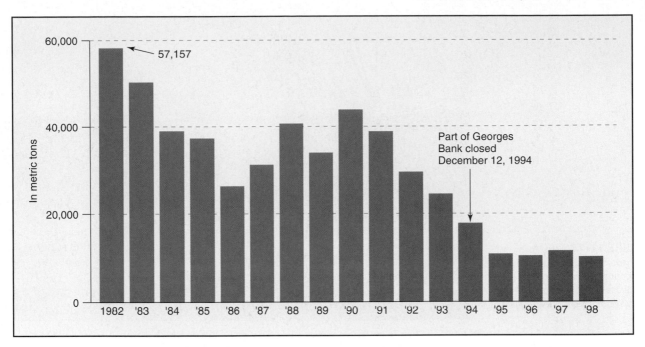

▲ FIGURE 12–13 *Fishery in distress.* Data show cod landings from Georges Bank, 1982–98. (*Source*: Northeast Fisheries Science Center, Woods Hole Laboratory, 2000.)

With collapse of the fishery imminent, the council took a more drastic approach: In 1997, all vessels were restricted to 50% of their normal days at sea; almost 13,000 km² (about one-third of the fishing area) was totally closed to fishing, and target TACs were set at levels designed to reduce fishing mortality to 15% of the stock per year. (In 1998, mortality was still 22%.) In addition, no new vessels were allowed to enter the fishery, and the Department of Commerce began a buyout program that pays the fishers to scrap their boats and surrender their fishing licenses. The buyout has resulted in the retirement of 79 vessels, or 20% of the fleet. In the meantime, many of the vessels turned northward to the Gulf of Maine, where regulations were less stringent. The result was predictable: Cod landings from the Gulf of Maine are at an all-time low, with the fishery now in danger of total collapse. Back on Georges Bank, although the yield is still very low, there are encouraging signs that the measures taken have begun to have their desired effect: Haddock and flounder stocks have rebounded very well and are close to a sustainable level.

What does it take to restore a fishery such as Georges Bank? The Canadian answer to an even worse problem on the Grand Banks off Newfoundland was to close the fishery indefinitely in 1992. In this once-richest fishing grounds in the world, the cod were decimated by Canadian fishers, who were encouraged by their government to exploit the fishery and were given little warning of its imminent collapse. By 1991, the cod population had dropped to one-hundredth of its former size. The closing cost the jobs of 35,000 fishers. So far the cod population has not rebounded, and cod are at their lowest levels in the century. The good news is that recovery is possible: A recent study of 128 heavily exploited fish stocks indicated that the stocks were ready to rebound once fishing pressure was reduced. Norwegian cod landings were dropping rapidly in the mid-1980s, and the government immediately cut quotas by 50%. Since 1990, the catch has tripled, and Norwegian cod is now supplying the North American market demand created by the decline in the Canadian and New England fisheries.

The basic problem of the fisheries is obvious to all: There are too many boats, rigged with high technology that gives the fish little chance to escape, chasing too few fish. Equipped with sophisticated electronic equipment that can find fish in the depths and with trawlers with 200-foot-wide openings that scoop up everything on and close to the bottom, the fishing industry has become overefficient (Fig. 12–14). The answer is a better system of management, according to which the fishing fleet is reduced in size and remaining fishers are better able to make a living without overfishing. What is needed is a limited system of access, such as the individual quota (IQ) system in place for managing the Pacific halibut fishery. The system effectively gives fishers the equivalent of property rights in the fishery. The rights are transferable and allow the fishers to determine how and when they will harvest their catch. Coupled with accurate population assessments by fishery scientists, the system has produced an economically stable and environmentally sustainable Pacific halibut fishery, and there is no reason that it should not be applied to fisheries everywhere.

In 1996, the Magnuson Act was reauthorized by the Sustainable Fisheries Act. The latter act also requires that depleted fisheries be rebuilt and that fishing be maintained

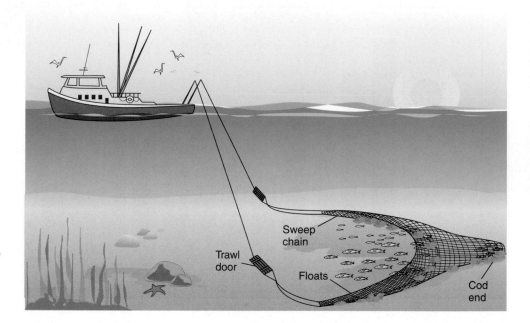

▶ **FIGURE 12–14** *Bottom trawling.* Because of the way the ocean bottom is degraded, this method, used to harvest groundfish, has been compared to clear-cutting forests.

Labels in figure: Sweep chain, Trawl door, Floats, Cod end

at biologically sustainable levels. The structure of the regional councils was retained, but the councils were given clear mandates to employ, in their fishery management plans, a number of means to see that the fish stocks are rebuilt and maintained, including IQs, buyouts of fishing vessels, and the requirement that scientific information be employed in setting yields. The bill also requires steps to be taken to assess and minimize the bycatch. The management councils are in the process of grappling with these new regulations. For example, in 2000 the NEFMC estimated that a 67% reduction in fishing mortality will be required to rebuild all the species of groundfish on Georges Bank and the Gulf of Maine. Accordingly, the organization has set TACs for 11 species at 36% lower than the 1998 landings, as a start towards meeting the requirements of the Sustainable Fisheries Act.

Many other issues pertaining to fisheries are of concern: banning shark finning, a practice in which just the fins of sharks are taken to satisfy the demand for shark-fin soup; the certification of "dolphin-safe" tuna (to avoid the practice of fishing that sets nets on dolphin schools because tuna are usually feeding below them); shrimping in the Gulf of Mexico that kills endangered sea turtle species; declining stocks of swordfish and tuna even as fishing intensity increases; and aquaculture demands placed on ocean fish because of the use of fish meal as feed for pen-raised fish. Another "fishery" of great concern to the international community is the great whales.

International Whaling. Whales, found in the open ocean as well as in coastal waters, were heavily depleted by overexploitation that was finally halted in the late 1980s. Whales once were harvested for their oil, but are now prized for their meat, considered a delicacy in Japan and a few other countries. In 1974, the International Whaling Commission (IWC), an organization of nations with whaling interests,

decided to regulate whaling according to the principle of maximum sustainable yield. Whenever a species of whale dropped below the optimal population for such a yield, the IWC instituted a ban on hunting that species in order to allow the population to recover. At that time, three species (the right whale, bowhead whale, and blue whale) were at very low levels and were immediately protected. Because of difficulties in obtaining reliable data on and enforcing catch limits, the IWC took more drastic action by placing a moratorium on the harvesting of all whales beginning in 1986. In spite of the moratorium, some limited whaling by Japan and Norway continues, as well as harvesting by indigenous people in Canada, Alaska, and Greenland. The World Wildlife Fund estimates that more than 18,000 whales have been harvested since the moratorium.

The International Union for Conservation of Nature publishes the status of whale populations in its *Red Data Book.* Table 12–3 lists the status of 10 whale species of commercial interest. Estimates of the numbers remaining range from 300 northern right whales to approximately 935,000 minke whales. Although reliable data are still hard to acquire, many of the whale species appear to be recovering. The bowhead whale, for example, recently was upgraded from endangered to vulnerable. In the time between the IWC moratorium and the present, however, the basis of the ethical controversy over whaling has shifted from conservation to animal rights. Many people worldwide simply believe that it is wrong to kill and eat such large and unique mammals. People from whaling nations counter with the argument that their culture includes eating whales, just as other cultures include eating cows or turkeys.

There has been heavy pressure from two members of the IWC—Japan and Norway—to reopen whaling. The interests of these countries focus on the minke whale. In 1993, Norway resumed whaling for minkes, citing the country's rights to refuse to accept specific IWC rulings. Norway based its deci-

TABLE 12–3 Remaining Numbers of Large Whales

Endangered Species	Vulnerable Species	Insufficient Knowledge
Blue whale (5,000)	Bowhead whale (7,800)	Minke whale (935,000)
Northern right whale (300)	Fin whale (110,000)	Sperm whale (400,000)
	Humpback whale (10,000)	
	Sei whale (54,000)	
	Southern right whale (2,900)	
	Gray whale (21,000)	

Sources: International Union for Conservation of Nature, *Red Data Book* (status in 1996), and International Whaling Commission (population estimates for 1998).

sion on information submitted by the Scientific Committee of the IWC, which judged the minke population to be numerous enough to absorb a sustainable exploitation. However, environmental scientists maintain that the data on whale populations is far too uncertain to support such a judgment. Under continued pressure from antiwhaling nations (the United States, France, Australia, and others) the IWC has thus far refused to set a whaling quota on the minke. Currently, Japan and Norway each take up to 500 minke whales per year, the Japanese for "scientific purposes" and the Norwegians for human consumption. It is no secret that the Japanese "scientific" harvest ends up in the markets and commands a healthy price; the industry generates $27–36 million in revenue a year. In 2000, in spite of pleas from many nations to refrain from the expanded hunt, a Japanese whaling fleet took 43 Bryde's whales and 10 sperm whales in addition to its usual harvest of minkes. In response, President Clinton denied Japan access to fishing rights in U.S. waters and threatened unspecified trade sanctions. Japanese officials announced their intention to continue the "research" whaling, claiming that they needed the in-

formation obtained from the hunts in order to assess the impact of whales on the livelihood of Japanese fishers.

The IWC is caught in the middle of this controversy. At a recent meeting, the commission adopted a draft version of the Revised Management Procedure (RMP) for setting quotas that could lead to a resumption of whaling. In subsequent annual meetings, however, IWC member countries have refused to authorize quotas on any whales until effective monitoring systems are put in place. At its latest meeting in Adelaide, Australia, in 2000, the IWC agreed to complete the rules for the RMP by its 2001 meeting, raising the possibility that limited whaling sanctioned by the IWC might resume. IWC officials believe that it would be advisable to bring existing whaling under international oversight.

One reason for the rising interest in protecting whales is the opportunity many people have had to observe them firsthand. Indeed, whale watching has become an important tourist enterprise in coastal areas (Fig. 12–15). Stellwagen Bank, within easy reach of boats from Boston, Cape Ann, and Cape Cod, Massachusetts, has become the center

◀ **FIGURE 12–15 *Whale watching.*** A humpback whale entertains a boatload of whale watchers in the Gulf of Maine off the Massachusetts coast. Whale watching has replaced whaling in New England waters that were once famous as whaling centers.

of a whale-watching industry estimated to generate $17 million annually. From spring through fall, scores of boats venture offshore daily to watch the whales that congregate over the bank. Many of the humpback whales seem to enjoy entertaining the visitors and often frolic alongside the boats for hours. The International Fund for Animal Welfare reported recently that whale-watching now flourishes in 87 countries and generates more than $1 billion in business annually.

Besides having its obvious aesthetic and entertainment values, whale watching is of scientific value. Whale-watching tour boats usually carry a biologist along who identifies the whale species and interprets the experience for the visitors. The biologists are often associated with groups such as the Cetacean Research Unit and the Whale Conservation Institute, which have studied the whales of Stellwagen Bank since 1979 and have published many papers on the humpback whale. Largely in response to its importance for whales, Stellwagen Bank was designated a National Marine Sanctuary in 1993.

Coral Reefs and Mangroves. In a band from 30° north to 30° south of the equator, coral reefs occupy shallow coastal areas and form atolls, or islands, in the ocean. Coral reefs are among the most diverse and biologically productive ecosystems in the world (Fig. 12–16). The reef-building coral animals live in a symbiotic relationship with photosynthetic

algae called **zooxanthellae** and therefore are found only in water shallower than 75 meters. The corals are important in building and protecting the land shoreward of the reefs and are spectacular attractions for tourists enjoying the warm, shallow waters. In addition, because the reefs attract a great variety of fish and shellfish, they are important sources of food for local people. In recent years, coral reefs have been the subject of concern as they have shown signs of deterioration in areas close to human population centers.

The 1997–98 El Niño (see Chapter 21) brought record-high sea-surface temperatures to many tropical coasts; the intensity and frequency of these events is thought to be one consequence of global warming. Extensive *coral bleaching* (wherein the coral animals lose their symbiotic algae) resulted in the death of 15% of the world's reef corals during the summer of 1998, according to Reef Check, a global affiliation of divers and marine scientists. Some coral bleaching is an annual phenomenon related to the normal high temperatures and light intensities of summer, but the most recent incidents are unprecedented in scope. Belize, which has the largest barrier reef in the northern hemisphere, sweltered under sustained temperatures of 31.5°C (89°F), killing off most of the coral. The most recent survey by Reef Check indicates that many of the devastated reefs are recovering more rapidly than expected, as young corals have begun again to colonize the dead reefs. Researchers point out, however, that repeated El Niño episodes (forecasted in global-warming scenarios) could permanently wipe out coral reefs over vast areas of the shallow tropical oceans, which would be an enormous loss of biodiversity and economic value.

Coral reefs are also being damaged by eutrophication of coastal waters. It has been shown that when nutrients—especially phosphorus—are carried to the reefs by currents in the vicinity of major developed coastlines, they encourage the growth of macroscopic algae and other submerged vegetation. The coral then becomes shaded, the coral animals become starved for oxygen, and the reefs become brittle and stunted. Wastes from resort areas are frequently the sources of the nutrients that are polluting the water. (Recent work indicates that the algae thrive on the corals because herbivorous turtles and fish that normally graze on the corals have been removed.) Not long ago, Bermuda's reefs, the key to a $9-million-a-year tourist industry, received protection from algal overgrowth by the establishment of no-fishing zones.

Another source of damage can be traced to poverty and greed. Islanders and coastal people in the tropics often scour the reefs for fish, shellfish, and other edible sea life. Although much of their use of the reefs is for food, some exploitation is driven by the lucrative trade in tropical fish. Local residents sometimes use dynamite and cyanide to flush the fish out of hiding in the coral, a practice that is not only destructive to the coral, but also dangerous to the perpetrators. "Fishing" with cyanide is a growing (illegal) practice that has been encouraged by exporters in the Indo-Pacific region. Diners in Hong Kong pay handsome sums for live reef fish, which are kept in aquaria and then are cooked and consumed on the spot. This

▲ **FIGURE 12–16** *Coral reefs.* Coral reefs like this one on the coast of Belize were devastated by high sea-surface temperatures during the 1997–98 El Niño.

trade in Southeast Asia is worth $1.2 billion, with a large percentage of these fish being captured by means of cyanide poisoning. Aquarium owners in Europe and North America admire the colorful imported tropical fish, unaware of how they were caught. The technique is destructive: The intention is to stun the fish with cyanide squirted into coral crevices, but the fish may die in the process. Meanwhile, the deadly poison damages the coral, often severely. Then divers often use crowbars to pry apart corals sheltering stunned fish. Halting this practice will take a concerted effort on the part of the Asian countries involved to police the reefs and work toward retraining the cyanide fishers to use nondestructive fishing methods; it will also require monitoring on the part of importing countries and a process of certification for cyanide-free imports.

Often, just inland of the coral reefs is a fringe of mangrove trees. These trees have the unique ability to take root and grow in shallow marine sediments. There, they protect the coasts from damage due to storms and erosion and form a rich refuge and nursery for many marine fish. Mangroves are under assault from three exploitative activities: coastal development, logging, and shrimp aquaculture. Between 1983 and the present, half of the world's 45 million acres of mangroves were cut down—from 40% (e.g., in Cameroon and Indonesia) to nearly 80% (e.g., in Bangladesh and Philippines). Logging companies removed mangroves for paper pulp and chipboard; logging and development are believed to be responsible for most of the mangrove deforestation. Another destructive practice, however, is the development of ponds for raising shrimp. In this practice, the mangrove forests are cleared and the land is bulldozed into shallow ponds, where black tiger prawns are fed with processed food. The aquaculture of prawns is now a multibillion-dollar industry, satisfying the great demand for shrimp in the developed countries. Unfortunately, in many cases, the ponds become polluted and no longer support shrimp; new ponds are then created by clearing more mangroves. The regular flushing of the ponds required to remove excess nutrients and wastes often degrades adjacent coastal environments. The massive removal of mangroves has brought on the destabilization of entire coastal areas, causing erosion, siltation of sea grasses and coral reefs, and the ruin of local fisheries. Developing nations that once considered the mangroves as useless swampland are now beginning to recognize the natural services they performed; local and international pressure to stop the destruction of mangroves is growing.

Let us now turn in our final section to the connection between public policy and natural-resource management as it is reflected in public and private lands in the United States.

12.4 Public and Private Lands in the United States

As we saw in Chapter 11, to save wild species, we must protect their habitats. The time has passed when we could justify losses of ecosystems on the grounds that there were suitable substitute habitats just over the hill. With the rising human population, industrial expansion, and pressure to convert natural resources to economic gain, however, there will always be reasons to exploit natural ecosystems. The last resort for many species and ecosystems is protection by law in the form of national parks, wildlife refuges, and reserves. Worldwide, some 6,930 areas, representing 4.8% of the national land area on the planet, have received this kind of protection. Three hundred areas are designated more restrictive *biosphere reserves*. Yet in the developing world, many of these are "paper parks," where exploitation continues and protection is given only lip service.

The United States is unique among the countries of the world in having set aside a major proportion of its landmass for public ownership. Nearly 40% of the country's land is publicly owned and is managed by state and federal agencies for a variety of purposes. The distribution of public lands, shown in Fig. 12–17, is greatly skewed toward Alaska and other Western states, a consequence of historical settlement and land-distribution policies. Nonetheless, although most of the East and Midwest are in private hands, natural ecosystems still function on many of their lands.

Land given the greatest protection (preservation) is designated **wilderness**. Authorized by the Wilderness Act of 1964, it includes over 104 million acres at 466 locations, almost 4% of the land area of the United States. The act provides for the permanent protection of these undeveloped and unexploited areas so that natural ecological processes can operate freely. Permanent structures, roads, motor vehicles, and other mechanized transport are prohibited. Timber harvesting is excluded. Some livestock grazing and mineral development are allowed where such use existed previously; hiking and other similar activities are also allowed. Proposals by several presidents to increase wilderness protection to 5 million acres of national parklands have received no action by Congress for over 30 years. Opposition to these proposals has been mounted by various recreational organizations, which fear that the move would lock up lands they are accustomed to enjoying.

National Parks and National Wildlife Refuges

The **national parks** (administered by the National Park Service) and **national wildlife refuges** (administered by the Fish and Wildlife Service) provide the next level of protection to 172 million acres. Here, the intent is to protect areas of great scenic or unique ecological significance, protect important wildlife species, and provide public access to view these scenic wonders. The dual goals of protection and providing public access often conflict with each other, because the parks and refuges are extremely popular, drawing so many visitors (over 280 million visits a year) that protection can be threatened by those who want to see and experience the natural sites. At one of the most popular parks—Zion National Park in Utah—auto traffic has been replaced by shuttle vehicles because of the problems of parking and habitat destruction accompanying the use of

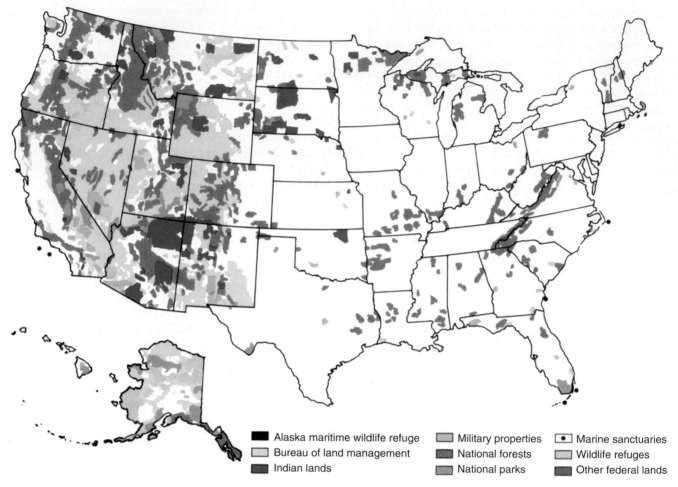

Alaska maritime wildlife refuge
Bureau of land management
Indian lands
Military properties
National forests
National parks
Marine sanctuaries
Wildlife refuges
Other federal lands

▲ FIGURE 12–17   *Distribution of federal lands in the United States.*   Because the East and Midwest were settled first, federally owned lands are concentrated in the West and Alaska. (Council on Environmental Quality, *Environmental Trends*, 1989.)

motor vehicles within the parks. Other parks will soon follow suit. All of the federal agencies involved are trying to cope with the rising tide of recreational vehicles like snowmobiles and off-road vehicles (ORV) used on most federal lands. The Blue Ribbon Coalition, a lobbying group representing ORV owners and manufacturers, fiercely defends keeping federal lands accessible to its clients.

Increasingly, agencies, environmental groups, and private individuals are working together to manage natural sites as part of larger ecosystems. For example, the Greater Yellowstone Coalition has been formed to conserve the larger ecosystem that surrounds Yellowstone National Park (Fig. 12–18). Yellowstone and Grand Teton National Parks form the core of the ecosystem, to which seven national forests, three wildlife refuges, and private lands are added to form the greater ecosystem of 18 million acres (7.2 million hectares). The coalition acts to restrain forces threatening the ecosystem; it has challenged the Forest Service's logging activities, addressed residential sprawl and road building on the private lands, and acted to protect crucial grizzly bear habitat and to curtail the senseless shooting of bison that wander off the national parklands in search of winter grazing.

This cooperative approach is important for the continued maintenance of biodiversity, since so much of the nation's natural lands remain outside of protected areas. Cooperation may also help restrict development up to the borders of the parks and refuges, avoiding a situation in which the refuge is a fairly small natural island in a sea of developed landscape.

National Forests

There is no doubt that forests in the United States represent an enormously important natural resource, providing habitat for countless wild species, as well as supplying natural services and products. U.S. forests range over 740 million acres, of which about two-thirds are managed for commercial timber harvest. Almost three-fourths of the managed commercial forestland is in the East and privately owned; the remainder is mainly in the West and is administered by a number of governmental agencies, primarily the National Forest Service and the Bureau of Land Management (BLM).

Deforestation is no longer a problem in the United States. Although we have cut all but 5% of the forests that were here

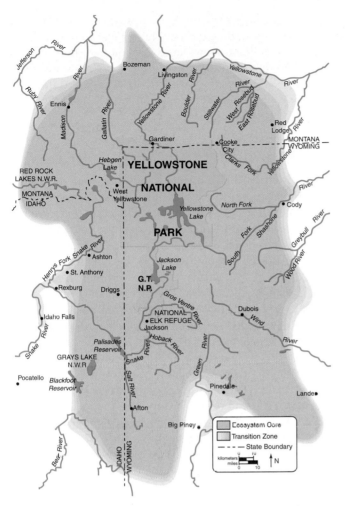

▲ **FIGURE 12–18** *Greater Yellowstone ecosystem.* The national park is the center of a much larger ecosystem, which has received attention from the Greater Yellowstone Coalition.

some second-growth forests have aged to the point where they look almost as they did before their first cutting, except for the absence of the American chestnut.

The Forest Service is responsible for managing 192 million acres (78 million hectares) of national forests. BLM land, some 270 million acres (109 million hectares), is a great mixture of prairies, deserts, forests, mountains, wetlands, and tundra. Some 6% of federal timber production comes from BLM land, the remainder from National Forest lands.

The Forest Service's management principle in the 1950s and 1960s was *multiple use*, which meant a combination of extraction of resources (grazing, logging, and mining), recreation, and the protection of watersheds and wildlife. Although the intent was to achieve a balance among these uses, in fact, multiple use emphasized the extractive uses; it was output oriented and served to justify the ongoing exploitation of public lands by private, often favored, interest groups (ranchers, miners, and the timber industry).

To harvest timber, government foresters select tracts of forest they judge ready for harvesting and then lease the tracts to private companies, which log and sell the timber. The timber harvest from national forests has been a controversial issue. Historic data on timber sales (Fig. 12–19) reflect major shifts in logging: a rapid rise following World War II to satisfy the housing boom underway; a drop reflecting rising environmental concern in the 1970s; increased logging during the Reagan–Bush years; and a steep drop with the Clinton administration as a result of renewed environmental concern.

A forestry-management strategy was introduced in the late 1980s called **New Forestry**. This practice of forestry is directed more toward protecting the ecological health and diversity of forests than toward producing a maximum harvest of logs. Thus, New Forestry involves cutting trees less frequently (at 350-year intervals instead of every 60 to 80 years); leaving wider buffer zones along streams, to reduce erosion and protect fish habitats; leaving dead logs and debris in the forests, to replenish the soil; and protecting broad landscapes—up to 1 million acres—across private and public

when the first colonists arrived, second-growth forests have regenerated wherever forestlands have been protected from conversion to croplands and lots for houses. There are more trees in the United States today than there were in 1920. In the East,

◀ **FIGURE 12–19** *National forest timber sales.* Timber sales over the years reflect changes in resource needs and political philosophy.

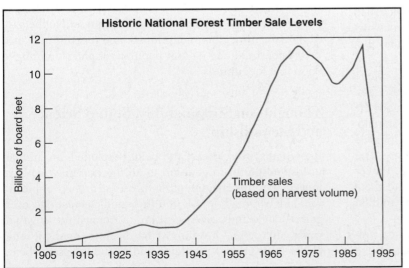

boundaries, while involving private landowners in management decisions. The Forest Service began adopting some of these management principles in the early 1990s, and they formed the core of what is now the official management paradigm of the Forest Service—**ecosystem management**. As we noted in Chapter 3, this paradigm has been adopted by all federal agencies managing public lands. It is a major philosophical shift from the orientation toward maximizing timber profits to the long-term stewardship of forest ecosystems. According to forestry scientist Jerry Franklin, ecosystem management comprises several principles:

- It takes an integrated view of terrestrial and aquatic ecosystems. In particular, watersheds and riparian stands of trees are nurtured so as to minimize the impact of logging on streams and rivers.
- It integrates ecological concepts at a variety of spatial scales. Leaving dead logs and soil litter in place, called the ecosystem's "biological legacy," enhances the biodiversity of small organisms. At larger scales, corridors for animal migration are preserved between stands of forests, and fragmentation is minimized.
- It incorporates the perspectives of landscape ecology so that the range of possible landscapes in an ecosystem is recognized and preserved.
- It is an evolving paradigm. Managers are given the opportunity to learn from experiments and to employ new knowledge from forestry and landscape science in their practice.
- It incorporates the human element. The goods and services of forested ecosystems are recognized, and, through active involvement in monitoring and management, local people are included as important elements in stewardship of the resources.

Above all, the paradigm incorporates the objective of ecological sustainability. The Clinton administration's Forest Service chief, Mike Dombeck, embraced the new paradigm, and one of his first acts was to declare a moratorium on building new logging roads. The Forest Service had come under fire for its road building, because of the destructive effects the existing roads had on forests and streams. Before any new roads are built, the damage from old roads is to be addressed, and this is a substantial undertaking. There are some 380,000 miles of forest roads, and their poor maintenance contributes to soil erosion, landslides, and the degradation of streams and rivers. In response to the moratorium, the timber industry and many members of Congress from states with national forests raised a hue and cry. The problem they perceived was that without new roads, timber harvests in roadless areas would come to a halt and there would be no ORV access to these areas. Some 43 million acres of forestland are affected by the moratorium.

Other controversies over the use of federal lands continue to fester. Recently, the Clinton administration issued a host of grazing reforms aimed at misuse of federal lands by the sheep and cattle industry: Grazing fees were raised, the density of cattle on many lands was lowered, livestock was excluded from sensitive river corridors, and, in some areas, grazing was completely banned. In May 2000, the U.S. Supreme Court upheld the administration's action in a suit brought by livestock groups.

Private Land Trusts

With so much land in private hands, the future of natural ecosystems in the United States is always uncertain. This is particularly true of land having special appeal for recreation and aesthetic enjoyment, because there the potential for development is great. Often, landowners and townspeople want to protect natural areas from development, but are wary of turning the land over to a governmental authority. One very creative option is the **private land trust**, a nonprofit organization that will accept either outright gifts of land or **easements**—arrangements in which the landowner gives up development rights into the future, but retains ownership. The land trust may also purchase land to protect it from development. The land trust movement is growing exponentially: There were 429 land trusts in the United States in 1980 and over 1,220 in 2000, as reported by the Land Trust Alliance, an umbrella organization in Washington serving local and regional land trusts. The latter protect over 4.7 million acres (1.9 million hectares) of land; the Nature Conservancy, a large international land trust agency, has helped protect an additional 11 million acres (4.4 million hectares) in the United States and 60 million acres (24 million hectares) in other countries.

Land trusts are proving to be a vital link in the preservation of ecosystems. The oldest trust is the Trustees of Reservations in Massachusetts, founded in 1891 and now guardian of 34,000 acres throughout the state. The Trustees, as the organization is often called, has received some ecologically prime and scenic land through the years and now maintains many of its properties for public use that is compatible with preservation. The land trusts are serving the common desires of landowners and rural dwellers to preserve the sense of place that links the present to the past. At the same time, the undeveloped land remains in its natural state, sustaining natural populations and promising to do so into the future.

Thoughts on Sustainability, Sound Science, and Stewardship

Ecosystems everywhere are being exploited for human needs and profit. In addition to all the examples we have looked at in this chapter, other areas that are in trouble are wetlands drained for agriculture and recreation, overgrazed rangelands, rivers that are overdrawn for irrigation water, and more. Our purpose has been to highlight some of the most critical problems and to indicate steps being taken to correct them. It is certain that greater pressures will be put on natural ecosystems as the human population continues to rise. These pressures must be met with increasingly effective protective measures if we want to continue to enjoy the fruits of our various ecosystems. The problems are most difficult in the developing world, where poverty forces people to take from nature in order to survive. If natural areas are to be preserved, the needs of people must be met in ways that do not involve destroying ecosystems. People must be provided with alternatives to exploitation, a situation requiring both wise leadership and effective international aid.

Once again, we maintain that sustainability must be the goal of all our interactions with natural systems. As nations formulate policies for dealing with the environment in sustainable ways, they will need better information on what is happening and what will be the consequences of those policies. Sound scientific information is essential, and in many areas there are large gaps in our knowledge. The Pilot Analysis of Global Ecosystems (PAGE) has made a good start toward closing those gaps (Chapter 1). To build

on this analysis, in 2001 the United Nations, through its Development, Environment, and FAO programs, together with the World Bank, will launch the **Millennium Ecosystem Assessment**, a four-year, $20 million effort. The objective of this effort is to find out what is happening to the world's ecosystems and to help countries both deal with ecosystems that are in trouble and meet the goals of the numerous international environmental conventions now in place. If the project is carried out properly, it will yield the most accessible and rigorous data set on world ecosystems ever assembled.

U.N. Secretary-General Kofi Annan presented his millennium report to the General Assembly in April 2000 with a stern warning that far too little concern was being given to the sustainability of our planet: "If I could sum it up in one sentence, I should say that we are plundering our children's heritage to pay for our present unsustainable practices....We must preserve our forests, fisheries, and the diversity of living species, all of which are close to collapsing under the pressure of human consumption and destruction." "In short," Annan stated, "We need a new ethic of stewardship."

ENVIRONMENT ON THE WEB

CERTIFICATION OF SUSTAINABLE FOREST MANAGEMENT

Agenda 21 of the U.N. Convention on Biodiversity Agreement (discussed in Chapter 11), signed at Rio de Janeiro in 1992, recognized the special value of forested ecosystems in maintaining Earth's biosphere. However, consumption by individuals has created the product demand that drives forest destruction. How do consumers, who are often far removed from the sources of the products they use, and industries that produce wood products, contribute to sustainable forest management?

The answer may lie in certification programs that reward sustainability. New programs now encourage forest-product companies to manage their lands sustainably. As one example, the Canadian Standards Association (CSA), Canada's central standard-setting body, has recently developed standards for the management of Canadian forests. This far-reaching initiative, which grew out of nationwide public and industry consultation, establishes a voluntary program to help responsible forest organizations move toward sustainable forest management.

The program combines performance and management system requirements that incorporate both national sustainable forest management criteria and recognition of local conditions. This means that each participating industry must demonstrate that it has adequate in-house management systems, such as record keeping and management oversight, *and* that it achieves specific performance objectives in the forests it manages. The CSA standards, which are administered through the Canadian Sustainable Forestry Certification Coalition, are based on broad sustainability goals. In order to put them into practice, though, it is necessary to translate goals into indicators of sustainable use and then into specific performance targets. For example, a broad goal of biological diversity conservation might translate to a local goal of a healthy woodpecker population,

which could be realized by increasing suitable habitat. The local performance target might therefore be to increase the number of standing dead trees in the target area. To demonstrate progress on this indicator, a company would have to prove that it had increased the number of standing dead trees within a specified time period.

The Sustainable Forestry Certification program calls for an independent third-party audit of both the management system and actual environmental performance. Companies that can demonstrate acceptable performance are awarded a certificate of registration that can be used in advertising or negotiations with regulatory agencies, thus allowing consumers to identify and be informed about eco-friendly products.

Web Explorations

The Environment on the Web activity for this essay discusses the development of specific forest management criteria from broader sustainability goals. Go to the Environment on the Web activity (select Chapter 12 at **http://www.prenhall.com/wright**) and learn for yourself:

1. about what drives local initiatives for sustainable forest management;

2. about forestry certification programs and the availability of eco-friendly wood products; and

3. about the interpretation of criteria for sustainable forest management.

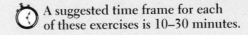 **A suggested time frame for each of these exercises is 10–30 minutes.**

REVIEW QUESTIONS

1. What were the problems in the Pacific halibut fishery, and how were they resolved?
2. Describe seven natural services that all ecosystems perform.
3. Under what conditions will a natural area receive protection?
4. Compare and contrast the terms *conservation* and *preservation*.
5. Differentiate between consumptive use and productive use. Give examples.
6. What does *maximum sustainable yield* mean? What factors complicate its application?
7. What is the tragedy of the commons? Give an example of a common pool resource and how it can be mistreated. How can such resources be protected?
8. When are restoration efforts needed? What is the objective of restoration?
9. What is the basic reason for clearing forests, and what are six consequences of that clearing?
10. Compare and contrast the two types of forest management. What are the five principles of the Sustainable Forestry Initiative?

11. What factors are responsible for deforestation of the tropics? Name six trends in developing world forest management that emphasize both people's needs and sustainability.
12. What is the global pattern of exploitation of fisheries?
13. Explain the objectives of the Magnuson Act and the Sustainable Fisheries Act.
14. How does the New England fishery exhibit the worst problems of exploitation?
15. What two countries are pressuring the IWC to reopen commercial whaling, and what is their rationale for resuming the killing?
16. How are coral reefs and mangroves being threatened, and how is this destruction linked to other environmental problems?
17. Contrast the different levels of ecosystem protection given to public lands in the United States.
18. Describe the progression of the management of our national forests during the last half century. What are the principles of ecosystem management?
19. How do land trusts work, and what roles do they play in preserving natural lands?

THINKING ENVIRONMENTALLY

1. Identify and study an area in your community where the destruction or degradation of natural land or wetland is an issue. What, if anything, is being done to protect this land? Propose a program for restoration of the degraded land.
2. What incentives and assistance could the United States offer Brazil or African countries to keep their tropical rain forests from further harm due to development?
3. Imagine that you are the clam warden for a New England town. What policies would you put in place to prevent a tragedy of the commons from occurring?

4. What natural-resource commons do you share with those around you? Are you exploiting or conserving these commons?
5. Does our national park system encompass enough area, or should it be expanded? Either way, defend your answer. If you believe the system should be expanded, propose some new areas to be added.
6. Research the preservation efforts and effectiveness of one of the following conservation groups: the Sierra Club, the Appalachian Mountain Club, The Nature Conservancy, and the Audubon Society.

MAKING A DIFFERENCE

PART THREE: Chapters 8, 9, 10, 11, 12

1. Demonstrate your concern for the hungry and homeless in your area by getting involved in local soup kitchens and by collecting goods for the needy as organized by Second Harvest.
2. Stay informed about food- and hunger-related issues; join Bread for the World, a nationwide organization

that seeks justice for the world's hungry by lobbying our nation's decision makers. Write your congressperson asking for his or her support of legislation that benefits the hungry.

3. Produce some of your own food by planting a vegetable garden and raising animals. Practice sustainable agriculture: Avoid the use of herbicides and pesticides, compost kitchen wastes, and yard wastes, and build up

a humus-rich soil. Any space—even a rooftop—with six hours of sunlight is enough.

4. Contact your county soil-conservation or farm-extension agent, and ask for help in communicating to others the work being done to prevent the erosion and runoff of farm chemicals.

5. Investigate the following issues pertaining to your community or city: Where does the water come from and how is it treated? Are water supplies adequate, or are they being overdrawn? How is storm water managed in your area? What policies or actions are being considered to meet future needs?

6. Observe where storm water from your own home and yard goes. (Does it mostly run off or infiltrate?) Install a system—perhaps just a barrel at a downspout—to capture and use storm water for watering your lawn or garden and for washing your car.

7. Examine your own eating habits, and opt for a nutritionally balanced diet that includes less fat and red meat and more fruit and vegetables. Use the USDA food pyramid to select your diet.

8. Do not buy any item that might come from an endangered species (for example, turtle shell products, seashells and corals, animal skins, and many species of cactus).

9. If you hunt or fish, observe all pertinent laws and kill only what you can use; alternatively, become a photographer of nature—your pictures can be a source of enjoyment to many.

10. When you visit parks and other natural areas, stay on the designated trails. Avoid making new trails that might lead to erosion, and never disturb nesting birds and wild animals raising their young.

11. Purchase food products that are derived from the extractive use of tropical forests rather than the degradative use thereof (for example, "Rainforest Crunch" ice cream and Brazil nuts, as opposed to tropical wood products).

12. Avoid high-impact activities that cause the most environmental damage, such as activities involving jet skis, off-road vehicles, and snowmobiles.

WEB REFERENCES

On-line resources for this chapter are on the World Wide Web at: **http://www.prenhall.com/wright**. (Click on Chapter 12 within the Chapter Selector.)

Chapter 13
Energy from Fossil Fuels

Chapter 14
Nuclear Power: Promise and Problems

Chapter 15
Renewable Energy

PART 4

Energy

The Alaska oil pipeline snakes its way down through tundra and forest to the port of Valdez, an intrusion into a beautiful and fragile wilderness that is symbolic of the dilemma facing our society. The industrial countries are so deeply committed to fossil fuel and nuclear energy that we cannot imagine an economy apart from these energy sources. Developing countries are following the same example, hoping to enhance the well-being of their people. Oil prices were at their lowest in decades in 1998, but rose sharply in 2000. A recent analysis of the world's oil reserves suggests that oil production will peak within 10 years and then begin a decline that will drive prices up to new heights. Will we invade wildlife refuges and fragile coastal waters to suck up every drop of oil from the ground? Looming in the background is the specter of global climate change, and the struggle of dealing with this threat that is tied so closely to the use of fossil fuels.

Many, including the power industry, the fossil fuel industry, automobile manufacturers, and some highly placed politicians, deny that any threat exists. Recently though, they have undergone a significant change in attitude. For example, the new chairman of Ford Motor Company warns his fellow automakers not to dismiss global warming, Royal Dutch/Shell will invest more than half a billion dollars in renewable energy technology, and British Petroleum will cut its greenhouse gas emissions well below the amount called for in the Kyoto Protocol. But what about other forms of energy? Nuclear energy is an option that does not generate greenhouse gases and is already well entrenched (but slipping) worldwide. Is renewable energy—solar cells, wind turbines, and biomass combustion—the energy of the future? If so, what form will it take? How can we develop an alternative energy source to fuel our vehicles? These are some of the vital questions we address in Part Four.

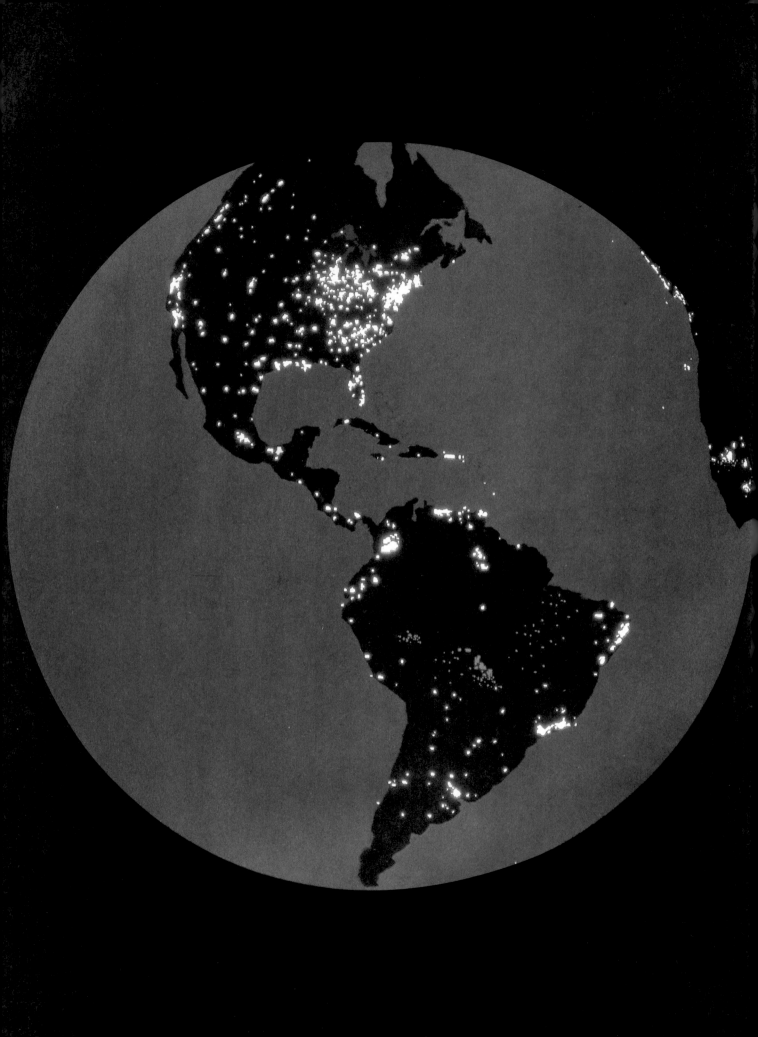

Energy from Fossil Fuels

Key Issues and Questions

1. Modern civilization depends greatly on a number of energy sources. How have the three primary fossil fuels been harnessed in recent history?

2. Much energy use is devoted to generating electrical power. How are fossil fuels coupled with electrical power, and what are the major environmental impacts of the processes involved?

3. What are the major categories of primary energy use and the energy sources that match those uses? How does an analysis of end-use energy demand help us to design energy supplies?

4. Crude oil is an essential energy source for transportation. What are the reserves of this resource, and how are they estimated?

5. The United States now imports more than 50% of the crude oil it uses. What led to the U.S. dependency on foreign oil? What are the problems of acquiring and shipping the quantities of oil the United States needs?

6. Oil shortages and oil gluts have rocked the world economy in recent years. What are the reasons for these violent shifts, and what are their long-term consequences?

7. Coal and natural gas are also important fossil fuels in the United States. How much of a reserve of these resources is there in the world and how is that reserve presently being put to use?

8. Sustainable energy options are clearly desirable for the development of a sustainable future. What are some options that will meet our energy needs with the least economic and environmental consequences?

The *Sea Empress*, a 147,000-ton tanker, left the Firth of Forth in Scotland in early February 1996, laden with 131,000 tons of crude oil bound for the Texaco Refinery in Milford Haven, Wales. The ship approached the entrance to the harbor shortly after 8 P.M. on February 15, at which time it ran aground and immediately began leaking oil (Fig. 13–1). After pulling off of the rocks, the ship lost steerage and ran aground again, still spilling oil. Tugs managed to pull it away from the rocks, but by that time the ship was listing and severely damaged. Subsequently, heavy weather set in and battered the salvage vessels attempting to recover oil and stabilize the ship; two days later the *Sea Empress* was swept back onto the rocks, damaging her again and releasing even more oil. By the time the tugs were finally able to tow the ship to harbor and the remaining oil was off-loaded, some 72,000 tons of crude oil had been spilled into St. George's Channel. The oil washed over 200 kilometers of the Welsh shoreline and hundreds of square kilometers of coastal ocean waters.

Thousands of seabirds died in the days following the spill and for several months afterward. Some 7,000 dead and dying birds were recovered, and wildlife personnel estimated that an additional 9,000 had perished. Unknown numbers of seals and porpoises died. A $20 million fishing industry was devastated, putting some 700 fishers out of work for months. An enormous and costly cleanup was initiated, and, finally, the coastline was freed of large oil accumulations within a few months.

Two years later, the ban on fishing off of southwest Wales was lifted, and the seabird populations were reestablished on their nesting islands. A final assessment of the spill's environmental impact indicated that the cleanup was quite successful and that no long-term impacts could be detected. The spill could have been much worse if the weather had been calmer; the heavy seas and offshore winds dispersed the majority of the oil, keeping it from being deposited on the fragile intertidal shoreline. On the other hand, if the *Sea Empress* had been a double-hulled vessel, it might not have ruptured at all.

◀ **Western Hemisphere at night.** Satellite photos demonstrate the delivery of electricity to cities, towns, and villages throughout the hemisphere. In the United States, demand is growing faster than supply, threatening power disruptions. (White patches are city lights, and the red spots in South America and Africa show large-scale forest burning.)

▶ **FIGURE 13–1** *The Sea Empress oil spill.* On February 15, 1996, the tanker *Sea Empress* ran aground on the coast of Wales, spilling 72,000 tons of crude oil and killing thousands of seabirds, fish, and porpoises.

The *Sea Empress* spill is just one of many that have occurred in recent years, and it certainly will not be the last. Oil for the global economy is moved from the oil fields and ports and transported by oil tankers to refineries around the world. Millions of gallons are in transit at any given time, and it is inevitable that oil spills will occur. The environmental consequences of oil spills are chalked up as one of the many costs of doing business as we mine the earth for fossil fuels and employ them in generating energy to run our economic enterprises. In future chapters we will encounter some of the other costs, such as air pollution and its impact on human and environmental health, and rising greenhouse gases leading to global climate change. Our objective in this chapter is to gain an overall understanding of the sources of energy that support our transportation, homes, and industries and of their relative sustainability. We will consider the consequences of our dependence on fossil fuels and begin to look at options for the future.

13.1 Energy Sources and Uses

Try to imagine what life would be like without all the energy we use. There would be no transportation beyond walking, riding on horseback, or traveling on boats powered by paddles or the wind. There would be no home appliances, air conditioners, lighting, or communications equipment. There would be little manufacturing beyond what could be done by hand, on looms, or on potters' wheels. Farming would depend on hand labor and on draft animals such as horses and oxen. In short, anything beyond a primitive existence is irrevocably tied to harnessing energy sources. Thus, it behooves us to look at this matter in a little more detail.

Harnessing Energy Sources: An Overview

Throughout human history, the advance of technological civilization has been tied to the development of energy

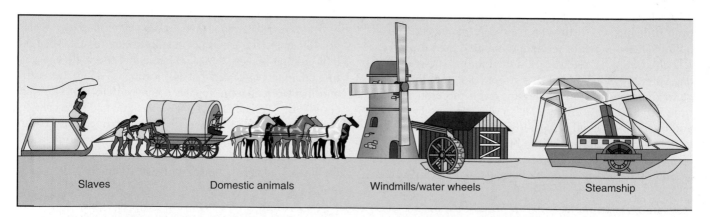

Slaves   Domestic animals   Windmills/water wheels   Steamship

▲ **FIGURE 13–2** *Energy and time.* The development of energy sources has in large part supported the development of civilization.

sources (Fig. 13–2). In early times (and even today in less developed regions), the major energy source was muscle power. Some people lived in relative luxury by exploiting the labor of others—slaves, indentured servants, and minimally paid workers. Human labor was supplemented by the use of domestic animals for agriculture and transportation, by water- or wind power for milling grain, and by the Sun.

By the early 1700s, inventors had already designed many kinds of machinery. The limiting factor was a continual power source to run them. The breakthrough that launched the Industrial Revolution was the development of the steam engine in the late 1700s. In a steam engine, water is boiled in a closed vessel to produce high-pressure steam, which pushes a piston back and forth in a cylinder. Through its connection to a crankshaft, the piston turns the drive wheel of the machinery. Steam engines rapidly became the power source for steamships, steam shovels, steam tractors, steam locomotives, and stationary engines to run sawmills, textile mills, and virtually all other industrial plants.

At first, the major fuel for steam engines was firewood. Then, as demands for energy increased and firewood around industrial centers became scarce, coal was substituted. By the end of the 1800s, coal had become the dominant fuel, and it remained so into the 1940s. In addition to being used as fuel for steam engines, coal was widely used for heating, cooking, and industrial processes. In the 1920s, coal provided 80% of all energy used in the United States and similar percentages in other industrialized countries.

Although coal and steam engines powered the Industrial Revolution, which greatly improved life for most people, those two sources of energy had many drawbacks. The smoke and fumes from the numerous coal fires made air pollution in cities far worse than anything seen today. Writers of the time recorded that often they could not see as far as a city block away because of the smoke. Coal is also notoriously hazardous to mine and dirty to handle, and burning coal results in large quantities of ashes that must be removed. As for steam engines, because of the size and bulk of the boiler, the engines were heavy and awkward to operate (Fig. 13–3). Frequently, the fire had to be started several

hours before the engine was put into operation in order to heat the boiler sufficiently.

In the late 1800s, the simultaneous development of three technologies—the internal combustion engine, oil-well drilling, and the refinement of crude oil into gasoline and other liquid fuels—combined to provide an alternative to steam power. The replacement of coal-fired steam engines and furnaces with petroleum-fueled engines and oil furnaces was an immense step forward in convenience. Also, the air quality improved greatly as cities were gradually rid of the smoke and soot from burning coal. (It was only in the 1960s, with the tremendous proliferation of cars, that pollution from gasoline engines became a problem.) Further, the gasoline internal-combustion engine provided a valuable power-to-weight advantage that allowed rapid advances in technology. A 100-horsepower gasoline engine weighs only a tiny fraction of what a 100-horsepower steam engine and its boiler weigh, and jet engines have an even greater power-to-weight ratio. Automobiles and other forms of transportation would be cumbersome, to say the least, without this power-to-weight advantage, and airplanes would be impossible.

Development throughout the world, so far as it has occurred since the 1940s, has been largely predicated on technologies that consume gasoline and other fuels refined from crude oil. The replacement of one energy technology with another is a very gradual process, however, because it is more cost effective to use existing machinery until it wears out before replacing it. It was not until 1951 that crude oil became the dominant energy source for the United States. Since then, it has become the mainstay not only for the United States, but also for most other countries, both developed and developing. **Crude oil** currently provides about 39% of total U.S. energy consumption. The story is similar throughout the rest of the world, although the timing of events differs. Earlier, we noted that many poor regions of the world remain dependent on firewood. In the countries of Eastern Europe and in China, coal is still the dominant fuel. In the United States, **coal** is still a major source of energy, providing about 22% of total energy fuel input.

▶ **FIGURE 13–3** *Steam power.* Steam-driven tractors marked the beginning of the industrialization of agriculture. Today, tractors half the size do 10 times the work and are much easier to operate.

Natural gas, the third primary fossil fuel, is found in association with oil or during the search for oil. Largely methane, which produces only carbon dioxide and water as it combusts, natural gas burns more cleanly than coal or oil; thus, in terms of pollution, it is a more desirable fuel. Despite the obvious fuel potential of natural gas, at first there was no practical way to transport it from wells to consumers. Any gas released from oil fields was (and in many parts of the world still is) flared—that is, vented and burned in the atmosphere, a tremendous waste of valuable fuel. Gradually, the United States constructed a network of pipelines connecting wells with consumers. With the completion of these pipelines, the use of natural gas for heating, cooking, and industrial processes escalated rapidly because of that fuel's cleanliness, convenience (no storage bins or tanks are required on the premises), and relatively low cost. Currently, **natural gas** provides about 23% of U.S. energy demand.

Thus, three fossil fuels—crude oil, coal, and natural gas—provide 84% of U.S. energy fuel consumption. The remaining 16% is provided by nuclear power, waterpower, and renewables. The changing pattern of energy sources in the United States is shown in Fig. 13–4. The picture is similar for other developed countries of the world, although the percentages differ somewhat, depending on the energy resources of the country.

Electrical Power Production

A considerable portion of total energy used is electrical power, called a **secondary energy source** because it depends on a **primary energy source** (coal or waterpower, for instance) to turn a generator. Electricity now defines our modern technological society, powering appliances and computers in our homes, lighting up the night (see chapter opening photo), and enabling the global Internet and countless other vital processes in business and industry to operate.

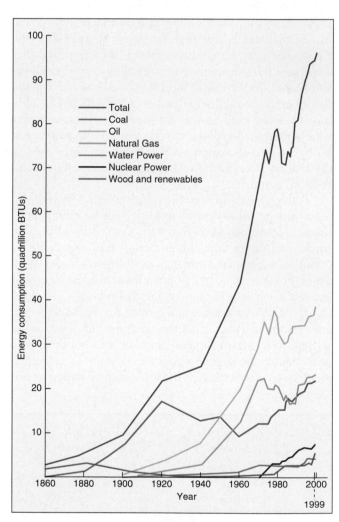

▲ **FIGURE 13–4** *Energy consumption in the United States.* Total consumption and major primary sources are shown here, from 1860 to 1999. Note how the mix of primary sources has changed over the years and how the total amount of energy consumed has continued to grow. Note also the skyrocketing increase in the use of oil after World War II (1945) as private cars and car-dependent commuting became common. (*Source:* Data from Energy Information Administration, U.S. Department of Energy, *Annual Energy Review 1999*, September 2000.)

Electric generators were invented in the 19th century. In 1831, the English scientist Michael Faraday discovered that passing a coil of wire through a magnetic field causes a flow of electrons—an electrical current—in the wire. An electric generator is basically a coil of wire that rotates in a magnetic field or that remains stationary while a magnetic field is rotated around it (Fig. 13–5). This process seems like cost-free energy until we consider the first and second laws of thermodynamics (Chapter 3). These laws remind us that we can't have something for nothing; in fact, we can't even break even. As the current flows in the wire, it creates a new magnetic field that is in opposition to the first and thus resists the flow of current. For example, in the process of converting the chemical energy in coal to electrical energy, much of the energy is lost as heat. Some is lost in transmitting the electricity over wires from the generating plants to the end users. In the end, it takes three units of primary energy to create one unit of electrical energy that actually is put to use. This energy-losing proposition is justified only because electrical power is so useful and, now, indispensable.

In the most widely used technique for generating electrical power, a primary energy source is employed to boil water, creating high-pressure steam that drives a **turbine**—a sophisticated "paddle wheel"—coupled to the generator (Fig. 13–6a). The combined turbine and generator are called a **turbogenerator**. Any primary energy source can be harnessed for boiling the water; coal, oil, and nuclear energy are most commonly used at present, but burning refuse, solar energy, and geothermal energy (heat from Earth's interior) may be more widely used in the future.

In addition to steam-driven turbines, gas and water (hydro-) turbines are also used. In a gas turbogenerator, the high-pressure gases produced by the combustion of a fuel (usually natural gas) drive the turbine directly (Fig. 13–6b). For hydroelectric power, water under high pressure—at the base of a dam or at the bottom of a pipe from the top of a waterfall—is used to drive a hydroturbogenerator (Fig. 13–6c). Wind turbines are also coming into use. (See Fig. 15–20.) A utility company supplying electricity to its customers will employ a mix of power sources, depending on the availability of each and the demand cycle.

Fluctuations in Demand. Where does your electricity come from? The answer to this question is, "It depends." Most utility companies are linked together in what are called *pools*. In New England, for example, ISO New England ties together some 400 electric power-generating plants through 8,100 miles of transmission lines. The utility is responsible for balancing electricity supply and demand, regardless of daily or seasonal fluctuations. The generating capacity of the ISO New England utilities is 24,800 megawatts (MW). A variety of energy sources are responsible for this capacity, as shown in Table 13–1. The greatest demands on electrical power in New England are in the summer months, when air-conditioning is at its highest use.

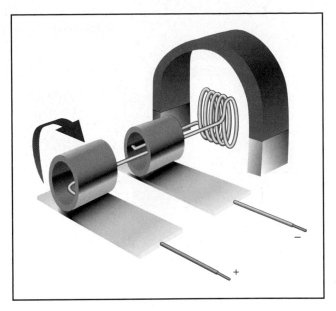

▲ **FIGURE 13–5** *Principle of an electric generator.* Rotating a coil of wire in a magnetic field induces a flow of electricity in the wire.

The peak load (highest demand) for 2000 was 21,900 MW, during a hot day in June. A power pool must accommodate daily and weekly electricity demand, as well as seasonal fluctuations, and will employ different mixes of energy sources, depending on the demand cycle.

The typical pattern of daily and weekly demand is shown in Fig. 13–7. **Baseload** represents the constant supply of power provided by the large coal-burning and nuclear plants with capacities as high as 1,000 MW. As demand rises during the day, the utility draws on additional plants that can be turned on and off—the intermediate and peak-load power sources that represent the utility's **reserve capacity.** These are the gas turbines and diesel plants, or the use of pumped storage hydroelectric power (water is pumped to a reservoir using off-peak power and is allowed to flow through turbines when needed). Occasionally, a deficiency in available power will prompt a brownout (a reduction in voltage) or blackout (a total loss of power), disrupting whole regions temporarily. Such events are most likely during times of peak power demand and may be precipitated by a sudden loss of power by the shutdown of a base-load plant.

In our current thriving economy, demand for electricity has been rising faster than supply, resulting in reserve capacity declining from 25% in the 1980s to its current level of 15%. Summer heat waves represent the greatest source of sudden pulses in demand, and as the climate continues to grow warmer in many parts of the United States, more and more regional utilities have been pushed to the edge of their ability to satisfy summer electricity demand. Power brownouts and blackouts are a serious threat to an economy that is now highly dependent on computers and other electronic devices. The power

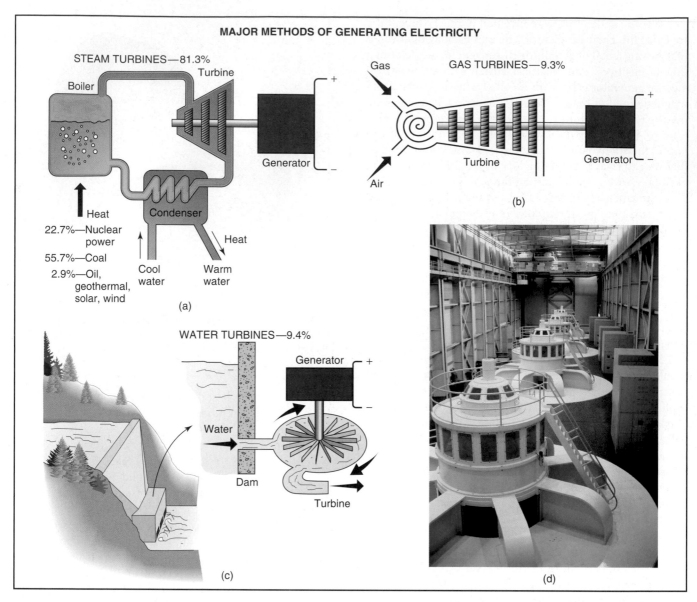

MAJOR METHODS OF GENERATING ELECTRICITY

STEAM TURBINES—81.3%

22.7%—Nuclear power
55.7%—Coal
2.9%—Oil, geothermal, solar, wind

(a)

GAS TURBINES—9.3%

(b)

WATER TURBINES—9.4%

(c)

(d)

▲ **FIGURE 13–6** *Electrical power generation.* Electricity is produced commercially by driving generators with (a) steam turbines, (b) gas turbines, and (c) water turbines. The percentage of the U.S. electricity supply derived from each source (1999) is indicated. Small amounts of power are now also coming from solar wind energy, solar thermal energy, and photovoltaics (solar cells). (d) Array of generators in a hydroelectric plant; each is attached by a shaft to a turbine propelled by water.

TABLE 13–1 Energy Sources for Generating Electrical Power (% of total)			
Source	New England*	United States	World
Nuclear power	24.1	22.7	17.7
Natural gas	15.7	9.3	14.8
Coal	13.4	55.7	38.4
Oil	18.9	2.8	9.3
Hydro	4.8	9.4	18.4
Other (wood, refuse, renewables)	7.4	<1	1.4

*Does not include an additional 15.7% from national energy grid.

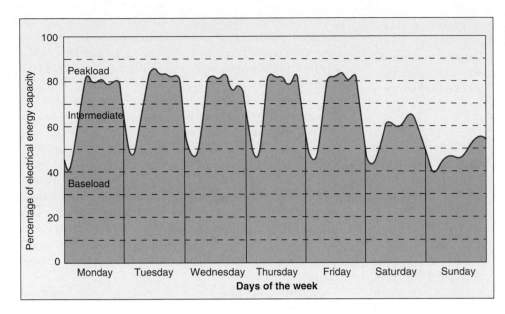

◀ FIGURE 13–7 **Weekly electrical demand cycle.** Electrical demand fluctuates daily, weekly, and seasonally. A variety of generators must be employed to meet base, intermediate, and peak electricity needs.

industry's response to this threat is to continue building more small gas-fired turbines that can be easily turned on at times of peak demand.

Clean Energy? Electrical power is often promoted as being the ultimate clean, nonpolluting energy source. This is true at the point of its use: Using electricity creates essentially no pollution. Hidden, however, is the fact that pollution is simply transferred from one part of the environment to another: Switching to "clean" electric heat implies more demand for electricity, which must be generated from coal, hydropower, nuclear energy, or alternative energy sources. Coal-burning power plants are the major source of acid deposition and contribute substantially to other air-pollution problems (Chapter 22). If hydroelectric power is used, creating a dam and reservoir involves displacing people, farmland, and wildlife and disrupts the migration of fish, such as salmon. Finally, nuclear power, discussed in Chapter 14, is a much-feared technology because of the potential for accidents and the problem of disposal of the nuclear waste products. Additional adverse effects of using electrical energy apply to other parts of the fuel cycle, such as mining coal or processing uranium ore.

The problem of transferring pollution from one place to another becomes even more pronounced when we consider efficiency: The thermal production of electricity has an *efficiency of only 30–40%*. The 60–70% loss is accounted for partly by some heat energy from the firebox going up the chimney and partly by a large amount of heat energy remaining in the spent steam as it comes out the end of the turbine. These unavoidable energy losses are a consequence of the need to maintain a high heat differential between the incoming steam and the receiving turbine, so as to maximize efficiency. Heat energy will go only toward a cooler place, so it cannot be recycled into the turbine. The most common practice is simply to dissipate it into the environment by means of a con-

denser. In fact, the most conspicuous features of coal-burning or nuclear power plants are the huge cooling towers used for that purpose (Fig. 13–8).

As an alternative to using cooling towers, water from a river, lake, or ocean can be passed over the condensing system (Fig. 13–6a). Thus, the waste heat energy is transferred to the body of water. All the small planktonic organisms, both plant and animal, drawn through the condensing system with the water are cooked, and the warm water added back to the waterway may have deleterious effects on the aquatic ecosystem. Therefore, waste heat energy discharged into natural waterways is referred to as **thermal pollution**.

▲ FIGURE 13–8 **Cooling towers.** These structures, the most conspicuous feature of both coal-fired and nuclear power plants, are necessary to cool and recondense the steam from the turbines before returning it to the boiler. Air is drawn in at the bottom of the tower, over the condensing apparatus, and out the top by natural convection. In some systems, cooling may be aided by the evaporation of water that condenses in a plume of vapor as the moist air exits the top of the tower, as seen in this photo.

Electricity is a form of energy that is indispensable to modern societies as they are currently structured. Worldwide, the electrical power industry generates revenues valued at more than $800 billion a year and, at the same time, generates a disturbing array of environmental problems.

Matching Sources to Uses

In addressing questions of whether energy resources are sufficient and where we can get additional energy, we need to consider more than just the source of energy, because some forms of energy lend themselves well to one use, but not to others. For example, the U.S. transportation system of highways, cars, and trucks is virtually 100% dependent on liquid fuels. Practically all other machines used for moving, such as tractors, airplanes, locomotives, and bulldozers, are likewise dependent on liquid fuels. Reserves of crude oil are needed to support this transportation system. The nuclear industry has promoted nuclear power as a solution to the energy shortage. However, nuclear power will do little to mitigate our demand for crude oil, because it is suitable only for boiling water to drive turbogenerators, which will do essentially nothing for our trans-

portation system. Although this situation might change with more widespread use of electric vehicles, their practicality for long distances still requires a technological breakthrough in the form of a low-cost, lightweight battery that will store large amounts of power.

Primary energy use is commonly divided into four categories: (1) transportation, (2) industrial processes, (3) commercial and residential uses (heating, cooling, lighting, and appliances), and (4) the generation of electrical power, which is used in categories (2) and (3). The major pathways from the primary energy sources to the various end uses are shown in Fig. 13–9. Observe again that transportation, which represents 27% of energy use, is totally dependent on petroleum, whereas nuclear power, coal, and waterpower are largely limited to the production of electrical power. Natural gas and oil are more versatile energy sources.

Figure 13–9 also shows the proportion of consumed energy that goes directly to waste heat energy rather than for its intended purpose. Some waste is inevitable, as dictated by the second law of thermodynamics, but the current losses are much greater than necessary. The efficiency of ener-

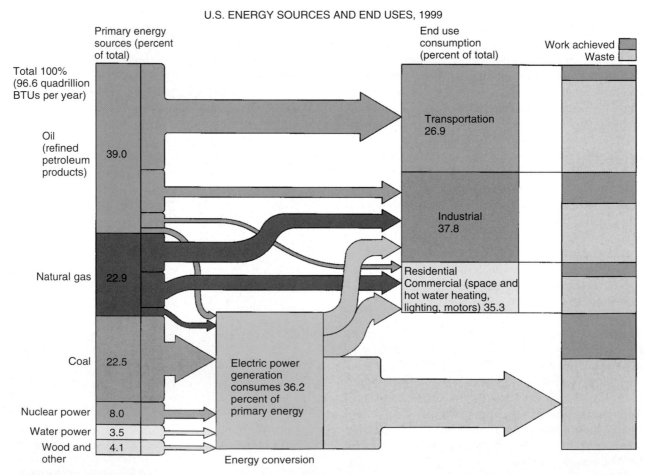

U.S. ENERGY SOURCES AND END USES, 1999

▲ **FIGURE 13–9** *Pathways from primary energy sources to end uses.* Only major pathways are shown. Note that end uses are connected to primary sources in specific ways. Note also the large percentage of energy that is wasted as a large portion of the energy consumed is converted to heat and lost. (*Source:* Data from Energy Information Administration, U.S. Department of Energy, *Annual Energy Review 1999*, September 2000.)

gy use can be at least doubled for some uses: We can make cars that get twice as many miles per gallon and appliances that consume half as much power, for example. For uses like electric power generation, however, somewhat lower increases in efficiency can be expected. Increasing the efficiency of energy use would have the same effect as increasing the energy supply, at far less expense. Such increases in efficiency are the essence of energy conservation and are discussed in more detail later in the chapter.

In addressing our energy future, we need to consider more than gross supplies and demands: We must also examine how particular primary sources can be matched to particular end uses. Further, we need to consider the most cost-effective ways of balancing supplies and demands—by increasing supplies or by decreasing consumption through *conservation*, *efficiency*, and *demand management*.

13.2 The Exploitation of Crude Oil

The United States has abundant reserves of coal, adequate reserves of natural gas (at least for the time being), and the potential for expansion of nuclear power. Therefore, no shortages loom in these areas now. (Note that we are setting aside the environmental concerns regarding coal and nuclear power for the moment.) Petroleum to fuel our transportation system, however, is another matter. U.S. crude oil supplies fall far short of meeting demand; we are now dependent on foreign sources for over 50% of our crude oil, and this dependency is steadily growing as our reserves diminish. Of course, foreign sources come at increasing costs: trade imbalances, necessary military actions, pollution of the oceans, economic disruptions, and coastal oil spills. Let us examine the situation, starting with where crude oil comes from and how it is exploited.

How Fossil Fuels Are Formed

The reason crude oil, coal, and natural gas are called *fossil fuels* is that all three are derived from the remains of living organisms (Fig. 13–10). In the Paleozoic and Mesozoic geological eras, 100 to 500 million years ago, freshwater swamps and shallow seas, which supported abundant vegetation and phytoplankton, covered vast areas of the globe. Anaerobic conditions in the lowest layers of such bodies of water impeded the respiration of decomposers and hence the breakdown of detritus. As a result, massive quantities of dead organic matter accumulated. Over millions of years, this organic matter was gradually buried under layers of sediment and converted by pressure and heat to coal, crude oil, and natural gas. Which fuel was formed depended on the nature of the organic material buried, the specific environmental conditions, and the time involved. Just as anaerobic treatment (digestion) of sewage sludge yields methane gas and a residual organic sludge, a similar anaerobic process

occurred with the buried vegetation. Natural gas is the trapped methane, and crude oil represents the residual sludge. Coal is highly compressed organic matter—mostly leafy material from swamp vegetation—that decomposed relatively little.

Additional fossil fuels may still be forming through natural processes to this day. However, they cannot be considered renewable resources: We are using fossil fuels far faster than they ever formed. To accumulate the amount of organic matter that the world now consumes each day, 1,000 years are needed. Recognizing how fossil fuels were formed, we can easily see that these resources are finite; sooner or later we will run out. The pertinent question is, When?

Crude-Oil Reserves vs. Production

The total amount of crude oil remaining represents Earth's oil resources. Finding and exploiting these resources is the problem. The science of geology provides information about the probable locations and extents of ancient shallow seas. On the basis of their knowledge and their field experience, geologists make educated guesses as to where oil or natural gas may be located and how much may be found. These educated guesses are the world's **estimated reserves**. Of course, such estimates may be far off the mark: There is no way to determine whether estimated reserves actually exist, except by the next step, exploratory drilling.

If exploratory drilling strikes oil, further drilling is conducted to determine the extent and depth of the **oil field**. From that information, a fairly accurate estimate can be made of how much oil can be economically obtained from the field. That amount then becomes **proven reserves**. Amounts are given in barrels of oil (1 barrel = 42 gallons), and the content of each reserve field is expressed in probabilities. Thus, a 50% probability, or P50, of 700 million barrels for a field means that 700 million barrels is just as likely as not to come from the field. Proven reserve estimates can range from P05 to P90; obviously, a P90 estimate is the more reliable figure, but oil producers are often vague about probabilities and might prefer to use a P05 or a P10 estimate in order to leave the impression of a large reserve for political or economic reasons. Also, proven reserves hinge on the economics of extraction. Thus, such reserves may actually increase or decrease with the price of oil, because higher prices justify exploiting resources that would not be worth extracting for lower prices. The final step, the withdrawal of oil or gas from the field, is called **production** in the oil business. Of course, *production*, as used here, is a euphemism. "It is production," says ecologist Barry Commoner, "only in the sense that a boy robbing his mother's pantry can be said to be producing the jam supply." In reality, this step is *extraction* from Earth.

Production from a given field cannot proceed at a constant rate, because crude oil is a viscous fluid held in spaces in sedimentary rock as water is held in a sponge. Initially,

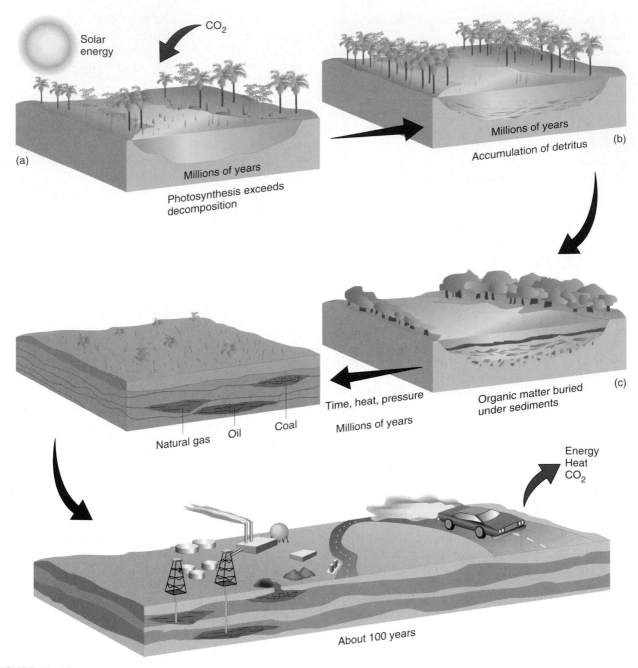

▲ FIGURE 13–10

Energy flow through fossil fuels. Coal, oil, and natural gas are derived from biomass that was produced many millions of years ago. Deposits are finite and, since formative processes require millions of years, they are nonrenewable.

the field may be under so much pressure that the first penetration produces a gusher. Gushers, however, are generally short lived, and the oil then seeps slowly from the sedimentary rock into a well from which it is withdrawn by pumping. Ordinarily, only 25% of the oil in an oil field can be removed by conventional pumping, or **primary recovery**. Further removal, to 50 or 60% or more of the oil in the field, is often possible, but it is more costly, as it involves manipulating pressure in the oil reservoir by injecting brine, steam, or other substances—so-called **secondary or tertiary recovery**—that force the oil into the wells.

Obviously, economics—the price of a barrel of oil—comes into play. No oil company will spend more money for extracting oil than it expects to make selling the oil. At $10 per barrel of crude oil, the price in the late 1990s, it is economical to extract no more than 25–35% of the oil in a given field. Thus, an increase in price makes more reserves available. In the 1970s and early 1980s, increasing oil prices justified reopening old oil fields and created an economic boom in Texas and Louisiana. Then an economic bust occurred in the late 1980s, as a drop in oil prices caused the fields to be shut down again.

Declining U.S. Reserves and Increasing Importation

A consideration of the relationships among production, price, and the exploitation of reserves will help us understand the U.S. energy dilemma. A brief history is in order. Up to 1970, the United States was largely oil independent. That is to say, exploration was leading to increasing proven reserves, such that production could basically keep pace with growing consumption (Fig. 13–11). But 1970 marked a significant turning point: New discoveries and hence reserve additions fell short, and production turned down, while consumption continued its rapid growth. This downturn was a complete vindication of the predictions of the late geologist M. King Hubbart, who proposed that oil exploitation in a region would follow a bell-shaped curve. Hubbart observed the pattern of U.S. exploitation and predicted that U.S. production would peak between 1965 and 1970. At that point, about half of the available oil will have been withdrawn, and production would decline gradually as reserves were exploited. Figure 13–11 demonstrates a "Hubbart" curve in progress for U.S. oil production.

To fill the energy gap between rising consumption and falling production, the United States became increasingly dependent on imported oil, primarily from the Arab countries of the Middle East. European countries and Japan did likewise. Since imported oil cost only about $2.30 per barrel ($8.30 in 1996 dollars) in the early 1970s and Middle Eastern reserves were more than adequate to meet the demand, this course seemed to present few problems.

The Oil Crisis of the 1970s

Thus, in the early 1970s, the United States and most other highly industrialized nations were becoming more and more dependent on oil imports. A group of predominantly

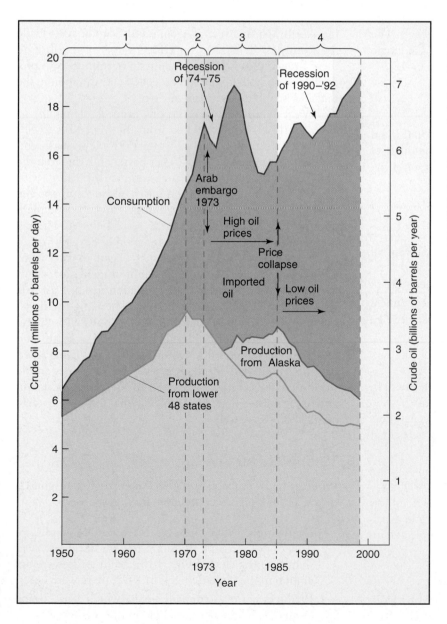

◄ **FIGURE 13–11** *Oil production and consumption in the United States.* Four stages can be seen: (1) Up to 1970, the discovery of new reserves allowed production to parallel increasing consumption. (2) From 1970 to 1973, a lack of new oil discoveries caused production to turn downward, while consumption continued to climb, resulting in a vast increase in oil imports. (3) In the late 1970s to early 1980s, high oil prices promoted both lowered consumption and increased production—which included bringing the Alaskan oil field into production. (4) From 1986, when oil prices declined sharply, until the present, consumption has again increased while production has resumed its decline, making us increasingly dependent on foreign oil. (*Source:* Data from Energy Information Administration, U.S. Department of Energy, *Annual Energy Review 1999*, September 2000.)

Arab countries known as the Organization of Petroleum Exporting Countries (**OPEC**) formed a cartel and agreed to restrain production and raise prices. Imported oil began to cost more in the early 1970s, and then, in conjunction with an Arab–Israeli war in 1973, OPEC initiated a boycott of oil sales to countries like the United States that gave military and economic support to Israel. The effect was almost instantaneous, because we depend on a fairly continuous flow from wells to points of consumption. (Compared with total use, little oil was held in storage.) Spot shortages occurred, and because we are so dependent on our autos, the spot shortages quickly escalated into widespread panic. We were willing to pay almost anything to have oil shipments resumed—and pay we did. OPEC resumed shipments at a price of $10.50 a barrel ($26 in 1996 dollars), almost four times the previous price.

Then, by continuing to limit oil production all through the 1970s, OPEC was able to keep supplies tight enough to force prices higher and higher. In the early 1980s, a barrel of oil delivered to the United States cost $36 ($58 in 1996 dollars), and the United States was paying accordingly. The purchase of foreign oil became, and remains, the single largest factor in our balance-of-trade deficit (Fig. 13–12).

Adjusting to Higher Prices

In response to the higher prices, the United States and other industrialized countries quickly made a number of adjustments during the 1970s. Following are the most significant changes made by the United States:

To increase domestic production of crude oil,
- Exploratory drilling was stepped up.
- The Alaska pipeline was constructed. (The Alaskan oil field had actually been discovered in 1968, but it remained out of production because of its remoteness.)
- Fields that had been closed down as uneconomical when oil was only $2.30 a barrel were reopened (e.g., in Texas).

To decrease consumption of crude oil,
- Standards were set for automobile fuel efficiency. In 1973, cars averaged 13 miles per gallon (mpg). The government mandated stepped increases for new cars to average 27.5 mpg by 1985. Speed limits were set at 55 mph for all U.S. highways.
- Other conservation goals were promoted for such things as insulation in buildings and efficiency of appliances.
- The development of alternative energy sources was begun. The government supported research and development efforts and gave tax breaks to people for installing alternative energy systems.

To protect against another OPEC boycott,
- A strategic oil reserve was created. As of 1998, the United States has "stockpiled" 574 million barrels of oil (equivalent to 58 days' worth of imports) in underground caverns in Louisiana.

It is noteworthy that, with the exception of the stockpile, all these steps are economically appealing only when the price of oil is high. If the price is low, the economic returns or savings may not justify the necessary investments (in spite of the fact that the steps are environmentally desirable).

The results of the new actions were not immediate, nor could quick returns have been expected. For example, it takes at least three to five years to design a new car model and start production. Then another six to eight years pass before enough new models are sold and old ones retired to affect the overall highway "fleet."

As we entered the 1980s, however, the efforts were definitely having an impact. The United States remained substantially dependent on foreign oil, but consumption was headed down, and production, up after the Alaska pipeline was completed, was holding its own. (See Fig. 13–11.) Elsewhere in the world, significant discoveries in Mexico and in the North Sea (between England and Scandinavia) made the world less dependent on OPEC oil. As a result of all these factors, plus OPEC's inability to restrain its own production, world oil production exceeded consumption in the mid-1980s, and there was an oil "glut."

What happens when supply exceeds demand? In 1986, world oil prices crashed from the high $20s to $14 per barrel. From that time until late 1999, oil prices fluctuated in the range of $10 to $20 per barrel and, due to inflation, were almost as low as they were in the early 1970s (Fig. 13–12). The effect of the lower oil prices was to the apparent benefit of consumers, industry, and oil-dependent developing nations, but it hardly solved the underlying problem. Indeed, it set the stage for another developing crisis: the steep rise in oil prices that began in late 1999. OPEC was concerned about the continuing oil glut and its low prices and decided to cut production just as East Asia was coming out of a serious recession. The rapid rise in demand brought on a shortfall in the oil supply, with the result that the price of crude oil rose to $37 a barrel and remained in the $30s throughout 2000. Amid rising international concern about the impact of higher prices on the world economy, OPEC promised to increase production and has indicated that it will aim at a long-term price range of $22–$28 per barrel.

Victims of Our Success

Whereas high oil prices stimulated constructive responses, the collapse in oil prices undercut those responses:

- Exploration, which becomes more costly as wells are drilled deeper and in more remote locations, was sharply curtailed. In 1982, 3,105 drilling rigs were operating in the United States; in 1992, there were 660.
- Production from older oil fields, which was costing around $10–$15 per barrel to pump, was terminated.

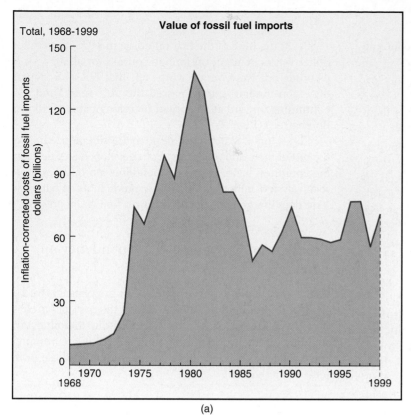

(a)

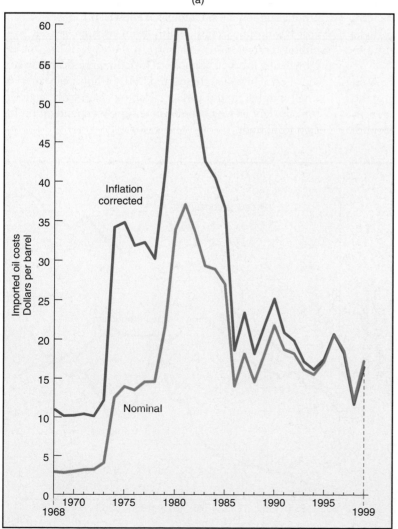

(b)

◄ **FIGURE 13–12** *The cost of fossil fuel imports.* The inflation-corrected cost (a) fluctuates sharply with the delivered price (b) of foreign oil (data from 1968–1999).
(*Source*: Energy Information Administration, U.S. Department of Energy, *Annual Energy Review 1999*, September 2000.)

The resulting drop in production caused economic devastation in Texas and Louisiana.

• Conservation efforts and incentives were abandoned. Government standards for automobile fuel efficiency were frozen at 27.5 mpg with no intention to increase them further. Speed limits were set to 65 and even 75 mph.

• Tax incentives and other subsidies for the development or installation of alternative energy sources were terminated, destroying many new businesses engaged in solar and wind energy. Grants for research and development of alternative energy sources (except nuclear power) were cut sharply.

• The need for conservation and for development of alternative ways to provide transportation seems to have largely passed from the mind of the public.

U.S. production of crude oil continues to drop as proven reserves are drawn down. The declining production now includes Alaska, which reached its peak production in 1988. At the same time, consumption of crude oil is rising again, because both the total number of cars on the highways and the average number of miles per year that each car is driven are increasing. Adding to that consumption, consumers are buying vehicles that are less fuel efficient. Minivans, four-wheel-drive sport-utility vehicles, light trucks, and other less fuel-efficient models now constitute 50% of the new-car market.

Altogether, following considerable improvement in our energy situation in the late 1970s and early 1980s, the United States has reverted to old patterns. Our dependence on foreign oil has grown steadily since the mid-1980s. At the time of the first oil crisis, in 1973, the United States was dependent on foreign sources for about 35% of its crude oil. Now we are worse off: In 1993, U.S. dependency on foreign sources passed the 50% mark, and it is continuing upward at an annual increase of about 140 million barrels. (Fig. 13–13).

The United States is not alone in this situation. Much of the rest of the world, including both developed and developing countries, has a growing dependency on oil from those relatively few nations with surplus production. What problems does this growing dependency on foreign oil present?

Problems of Growing U.S. Dependency on Foreign Oil

As the U.S. dependency on foreign oil grows again, we are faced with problems at three levels: (1) the costs of purchasing oil, (2) the risk of supply disruptions due to political instability in the Middle East, and (3) ultimate resource limitations in any case. Let's consider each of these problems in more detail.

Costs of Purchase. The yearly cost to the United States of purchasing foreign oil and gas is shown in Fig. 13–12. The cost has dropped considerably from its high of over $130 billion (corrected for inflation) in the early 1980s, the decline being brought about both by falling prices and by conservation. However, the current $70 billion per year is no small number, and it is now climbing again with no downturn in sight as we become increasingly dependent on foreign fossil fuels.

▶ **FIGURE 13–13** *Consumption, domestic production, and imports of petroleum products.* After the oil crises of the 1970s, the United States became less reliant on foreign oil. But since the mid-1980s, consumption of foreign oil has for the most part been rising steadily.
(*Source*: Data from Energy Information Administration, U.S. Department of Energy, *Annual Energy Review 1999*, September 2000.)

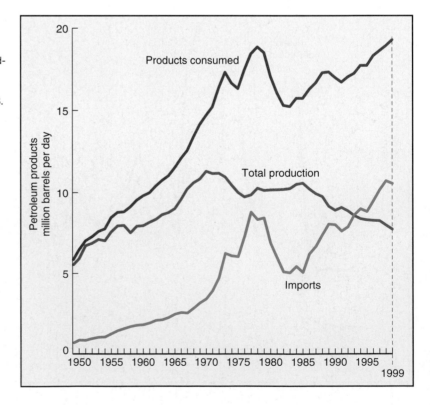

This cost for crude oil represents about 26% of our current balance-of-trade deficit ($271 billion in 1999), which has significant consequences for our economy. The price we pay at the pump is basically the same whether the oil is produced here or abroad. However, if the oil is produced at home, the money we pay for it circulates within our own economy, providing jobs and producing other goods and services. When we buy oil from a foreign country, we are effectively providing "foreign aid" to that country and that much of a drain on our own economy.

Of course, the ecological costs of oil spills such as the *Sea Empress* spill and the estimated $15 billion *Exxon Valdez* incident and of other forms of pollution from the refining and consumption of fuels must not go unnoticed—although estimating total damage in dollars is sometimes difficult. There are also military costs, as we will describe in the next section.

Risk of Supply Disruptions. The Middle East is a politically unstable, unpredictable region of the world. As noted earlier, it was the unexpected Arab boycott that plunged us into the first oil crisis in 1973. Recognizing this political instability, the United States maintains a military capacity to ensure our access to Middle Eastern oil. This military capacity was tested in the fall of 1990 when Saddam Hussein of Iraq invaded Kuwait, then producing 6 million barrels per day. The Persian Gulf War of early 1991 prevented Iraq from advancing further. Nevertheless, Kuwait's oil production was knocked out for more than a year after the war, because the Iraqis dynamited and set fire to nearly all of Kuwait's almost 700 wells before leaving (Fig. 13–14).

Military costs associated with maintaining access to Middle Eastern oil can be looked at as the U.S. government's subsidizing our oil consumption. Our military presence there costs the United States $50 billion per year. On the basis of importing about 884 million barrels a year from Persian Gulf sources (in 1999), the dollar amount of this "subsidy" comes to about $56 per barrel. In other words, when military "support services" are added to the $35 market price (2000), we are actually paying about $91 per barrel!

Maintaining access to Middle Eastern oil is also a major reason for our efforts toward negotiating peace agreements between various parties in that region. Even with all this effort, however, continuing risks are present. The rise of Islamic radicals, who threaten to take over governments of Middle Eastern countries, may pose a great danger.

It is significant that the United States has turned more and more to oil suppliers in the Western Hemisphere; recently, Canada replaced Saudi Arabia as the number-one exporter of oil to the United States, with about 14% of the market. Venezuela and Mexico are also major sources of oil. The Western Hemisphere sources now provide about 43% of our imports. The political stability and proximity of these sources are clear reasons for the United States to turn to them and away from the Middle East. However, we will continue to depend on OPEC and the Persian Gulf for a substantial share of our oil imports, for the present and even more so in the future.

Resource Limitations. No responsible geologist questions the fact that U.S. crude oil production is decreasing because of diminishing domestic reserves. Nor do geologists hold out hope for major new finds. The dim hope for new U.S. finds derives from what is called the "Easter-egg hypothesis": As your searching turns up fewer and fewer eggs, you draw the logical conclusion that nearly all the eggs have been found. In terms of oil, North America is already the most intensively explored of any continental landmass. The last major find was the Alaskan oil field in 1968 (see "Ethics" essay, p. 333). Discoveries since then have been small in comparison, with increasing numbers of dry holes in between. Indeed, most of the "new" oil in the United States is coming from innovative computer mapping of geological structures and horizontal drilling technology that enables oil to be identified and withdrawn from small isolated pockets that were previously missed within old fields. Yet, an increasing percentage of our needs must continue to be met by imported oil.

How much oil is still available for future generations? To put this question in perspective, it is instructive to explore what has already been used and the current world rate of use. About 800 billion barrels (BB) of oil has already been used, almost all in the 20th century. Currently, the nations of the world are using about 75 million barrels per day, or some 27 BBs a year. Rising demand by the developing

▲ **FIGURE 13–14** *The Gulf War of 1991.* Even military might may not guarantee uninterrupted access to world oil supplies. Although we were the winner of the Persian Gulf War of 1991, Iraq's forces succeeded in setting fire to nearly 700 of Kuwait's oil wells before their retreat. Six million barrels per day were thus taken out of production for nearly a year.

TABLE 13–2   World Proven Reserves and Annual Use of Fossil Fuels

Region	Petroleum (billion barrels)*	Reserves Natural Gas (trillion ft³)*	Coal (billion tons)*
North America	75.7	291.4	285.2
South and Central America	89.5	219.1	23.8
Western Europe	18.9	161.5	99.7
Eastern Europe and Former U.S.S.R	58.9	1,999.4	288.4
Middle East	673.6	1,749.5	0.2
Africa	75.4	361.1	67.7
Far East and Oceania	43	359.6	322.3
Total	1,033.2	5,141.6	1,087.2
	Use Petroleum	Natural Gas	Coal
World use, 1998*	26.6 billion barrels	83 trillion ft³	5.04 billion tons
World use, 1998 (quads Btus)	152 quads	85 quads	89 quads

* The fuels may be compared by calculating their energy content in British thermal units (Btus). One billion barrels of petroleum yields ca. 5.7×10^{15} Btus, or 5.7 quadrillion (quads) Btus. One trillion cubic feet of natural gas yields ca. 1.02 quadrillion Btus, and 1 billion tons of coal yields ca. 17.65 quadrillion Btus.
Source: Data from *Oil and Gas Journal* and Energy Information Administration, U.S. Department of Energy, 1999.

countries and the continued rise in oil demand by the developed countries point to an increase of about 2% per year, so that demand will reach 40 BBs a year by 2020. How long can the existing oil reserves provide that much oil?

As with many issues, it depends on whom you believe. Two oil geologists, Colin Campbell and Jean Laherrère, recently provided an analysis that puts the world's proven reserves at about 850 BBs. This is less than the 1,020–1,160

BBs claimed by some oil industry reports (Table 13-2); Campbell and Laherrère derived their estimate from the P50 values of *proven* reserves and discounted some reports of reserves that were inflated for political reasons. On the basis of their estimates and the known amount already used, the two authors present a Hubbart curve indicating that peak oil production will occur sometime in this decade (Fig. 13–15). When that happens, production will begin to

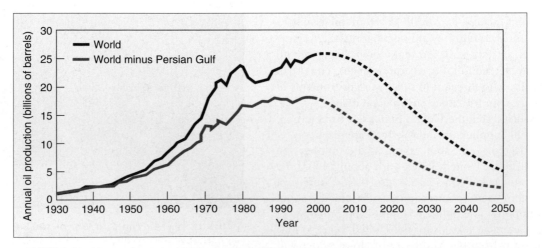

▲ FIGURE 13–15   *Hubbart curves of oil production.*   Oil production in a region follows a bell-shaped curve. The evidence indicates that world oil production will peak sometime in this decade. (*Source*: Data from Richard A. Kerr, "The Next Oil Crisis Looms Large—and Perhaps Close," *Science* 281 [1998]:1128–1131. Reprinted by permission of American Assn. for the Advancement of Science.)

ETHICS

TRADING WILDERNESS FOR ENERGY IN THE FAR NORTH

The search for new energy sources to satisfy the enormous energy appetite of the U.S. economy has turned northward in recent years. The discovery and development of the huge Prudhoe Bay oil field in Alaska and the building of the 800-mile (1,230-km) Alaska pipeline have dramatized the stark contrasts between wilderness and modern technology. Scenes of caribou and grizzly bears are shown with a background of oil rigs and the pipeline carrying oil south to Valdez.

East of Prudhoe Bay is the 19 million-acre Arctic National Wildlife Refuge (ANWR), the largest single undeveloped area in the United States. ANWR is home to 130,000 caribou and several hundred native people whose subsistence lifestyle is tied to the caribou. In addition, ANWR is the summer breeding ground for innumerable birds and other wildlife. Unfortunately, it is also believed to be on top of another large oil field that industrial interests are anxious to exploit.

ANWR has become a major political battleground. On one side, energy interests and many political interests see values in terms of energy supplies for a thriving economy and national security; the oil would also keep the Alaska pipeline in business. On the other side, native people, environmentalists, and other politicians see values in terms of the beauty and biodiversity of the northern wilderness, as well as the rights of native peoples. Those favoring development claim that it will be done in an environmentally sound way and the native people will be compensated. Those favoring preservation are not satisfied with good intentions. They point to the *Exxon Valdez* disaster, in which 11 million gallons of crude oil were spilled in the pristine waters of Prince Edward Sound, despite all safeguards. They point out that present oil development in Alaska has introduced much of the worst of American culture to native peoples; a sense of lost identity is apparent as their traditional culture is swamped by consumer goods and pop culture. Environmentalists also question development on the basis of sustainability: The oil reserve underlying ANWR could be as high as 11.6 billion barrels, but more likely around 5.7, according to the U.S. Geological Survey; this is about nine months' supply at the current rate of use in the United States. Are there ways to have adequate power without purchasing it at the expense of wilderness?

Current law forbids oil and gas leasing in ANWR; it would take congressional action to change the law. The refuge was declared off limits to oil exploration by the Clinton administration, but, of course, administrations change. Some of the pressure was taken off when part of the National Petroleum Reserve, an area of almost 4 million acres just west of Prudhoe Bay, was opened for exploration in the summer of 1998. This area was set aside 75 years ago for oil exploration by President Warren G. Harding and is thought to contain smaller amounts of oil than ANWR, but nevertheless is worth exploiting. The area is also environmentally sensitive, a famous breeding ground for many migratory waterfowl and home to caribou, moose, musk oxen, and polar bears.

How do we balance our profligate need for energy, jobs, and economic growth against caribou, musk oxen, and polar bears? How do we weigh the traditions and cultures of native peoples against the impacts of our own?

---

taper off in spite of rising demand, and the world will enter a new era of rising oil prices. Even if Campbell and Laherrère are wrong and the oil industry estimates are right, it would only put off the peak by a decade at most. To quote the authors, "The world is not running out of oil—at least not yet. What our society does face, and soon, is the end of abundant and cheap oil on which all industrial nations depend."

The U.S. Geological Survey (USGS) released a 2000 assessment of world oil and gas in the form of "*undiscovered* reserves"—reserves from fields that have not yet been found, but that the USGS believes exist, on the basis of extensive geological studies. Thus, the Geological Survey believes that there are undiscovered reserves of 732 BBs of oil remaining from unknown fields, and an additional 688 BBs in known fields that could be unaccounted for by conventional analyses. Campbell believes that the USGS estimates are overly optimistic, based as they are on oil yet to be found (not *proven* reserves!). And even if much of this oil is found, it will at best move peak oil production back a few years or a decade, given the rising consumption rates. Clearly, the 21st century will become known as an era of declining oil use, just as the 20th century has seen continuously rising use.

As oil becomes more scarce, the OPEC nations will once again achieve dominance of the world oil market. The Middle East possesses 65% of the world's proven reserves, and eventually the developed world—in particular, the United States—will have to rely on that region for its oil imports (Table 13–2). It would seem the path of wisdom to begin to prepare for that day by reducing our dependency on foreign oil now. We can become more independent in three ways: (1) We can use the other fossil-fuel resources we have to make fuel for vehicles; (2) we can reduce our energy demand through conservation and energy efficiency; and (3) we can develop alternatives to fossil fuels, many of which are ready now. The best answer probably lies in a suitable mix of all three choices.

13.3 Other Fossil Fuels

Unlike crude oil, considerable reserves of natural gas and comparatively vast reserves of coal and oil sands are still available in the United States and elsewhere in the world. Natural gas is gradually coming into use as a fuel for vehicles, and there is some potential in this respect for coal and oil sands.

Natural Gas

In the United States, natural gas has been rising as a fossil fuel of choice—so much so, that 16% of our annual use now comes from imports, almost all by pipeline from Canada. Although most of its use is for space heating and cooking, natural gas is increasingly being employed for electrical power generation. Like oil, its costs are subject to market fluctuations in supply and demand, and they often rise and fall with the price of oil. (Between 1999 and 2000, gas prices doubled.) At the current rate of use, proven reserves in the United States hold only an eight-year supply of gas. However, this is somewhat deceiving, because natural gas is continually emitted from oil and gas-bearing geological deposits. The expected long-term recovery of natural gas in the United States is about six times greater than the proven reserves, which puts the supply closer to 50 years.

Worldwide, the estimated natural gas resource base is even more plentiful (Table 13–2). Almost four times as much equivalent gas (energy equivalency) is likely to be available as oil. Much of it, however, is relatively inaccessible. Gas must either be pumped through a pipeline (a process limited by the length of the line) or be subjected to high pressures so that it remains a liquid at room temperature. Nevertheless, liquefied natural gas is increasing in production and use.

Natural gas can also meet some of the fuel needs for transportation. With the installation of a tank for compressed gas in the trunk and some modifications of the engine fuel-intake system—costing about a thousand dollars—a car will run perfectly well on natural gas, although trunk space is a problem. Such cars are in widespread use in Buenos Aires, where service stations are equipped with compressed gas to refill the tank. Natural gas is a clean-burning fuel; hydrocarbon emissions are nil. Indeed, natural gas is coming into use in the United States as a vehicle fuel and is already being used by many buses and a number of private and government car fleets.

With the aid of a chemical process, natural gas can be converted to a hydrocarbon that is liquid at room temperature and pressure—basically, a synthetic oil. Researchers are hotly pursuing ways to convert natural gas to synthetic oil more efficiently. Presently, this process produces a fuel that is only about 10% more costly than oil. Already, some diesel fuel in California is partly derived from the conversion process. Estimates indicate that enough natural gas is available to produce 500 billion barrels of synthetic oil.

These two applications of natural gas to the transportation sector—as a fuel and as a synthetic oil—promise to extend the oil phase of our economy, but they will not be a sustainable solution, because the natural-gas reserves are also limited. However, what about coal, the most abundant fossil fuel?

Coal

In the United States, 56% of our electricity comes from coal-fired power plants, and current reserves are calculated to be 274 billion tons. At the current rate of production (1.1 billion tons/yr), we have 250 years of supply. Unlike oil and

natural gas, we produce more coal each year than we use, and we export about 5% of coal production annually.

Coal can be obtained by two methods: surface mining and underground mining. The mining work can be hazardous, and injuries and fatalities are not uncommon. Both methods have substantial environmental impacts. In underground mining, at least 50% of the coal must be left in place to support the roof of the mine. Land subsidence and underground fires often occur in conjunction with underground mines. In the more common surface, or strip, mining, gigantic power shovels turn aside the rock and soil above the coal seam and then remove the coal (Fig. 13–16). Obviously, this method results in total ecological destruction. Although federal regulations require that such areas be reclaimed—that is, graded and replanted—it takes many decades before an ecosystem resembling the original one can develop. In arid areas of the West, water limits may prevent the ecosystem from ever being reestablished. Consequently, surface-mined areas may be turned into permanent deserts. Furthermore, erosion and acid leaching from the disturbed ground have numerous adverse effects on waterways and groundwater in the surrounding area.

Ninety percent of the coal the United States produces is used to generate electricity. Once it is mined, coal is transported to large power plants. A typical 1,000-MW plant burns 8,000 tons of coal a day—a mile-long train's worth of

▲ **FIGURE 13–16** **Strip mining of coal.** It is economical to exploit most coal deposits only by strip mining, whereby gigantic power shovels remove as much as 100 feet (32 m) of rock and soil from the surface to get at the coal seam.

coal every day. Combustion of this much coal releases 20,000 tons of carbon dioxide, and 800 tons of sulfur dioxide (most of which is removed if the power plant is well provided with scrubbers—see Chapter 22). Then there is the waste: 800 tons of fly ash and 800 tons of boiler residue ash daily, requiring a special landfill.

Coal may be chemically converted to a liquid or gas fuel. Such coal-derived fuels are referred to as **synthetic fuels,** or **synfuels.** With both government and corporate support, a great deal of research went into upscaling these chemical processes to commercial production in the late 1970s and early 1980s. The projects were all abandoned, however, because they proved too expensive, at least for the present. The production of coal-based synthetic fuels would be profitable only if oil prices rose to $60 to $70 per barrel.

Thus, mining and burning coal involves substantial hazards and pollutants. The thought of doubling or tripling this impact to run vehicles on coal-derived fuel makes a mockery of the country's progress in environmental protection. In addition, converting coal to liquid fuel creates by-products that pollute the environment and that would be difficult and expensive to control.

Oil Shales and Oil Sands

The United States has extensive deposits of oil shales in Colorado, Utah, and Wyoming. **Oil shale** is a fine sedimentary rock containing a mixture of solid, waxlike hydrocarbons called *kerogen*. When shale is heated to about 1,100°F (600°C), the kerogen releases hydrocarbon vapors that can be recondensed to form a black, viscous substance similar to crude oil, which can then be refined into gasoline and other petroleum products. However, about a ton of oil shale yields little more than half a barrel of oil. The mining, transportation, and disposal of wastes necessitated by an operation producing, for instance, a million barrels a day (just 5% of U.S. demand) would be a Herculean task, to say nothing of its environmental impact. Perhaps to our environmental advantage, oil shale, like oil from coal, has proved economically impractical for the present. Also, the production of oil from oil shale requires large amounts of water, something in short supply in the West.

Oil sands are a sedimentary material containing bitumen, an extremely viscous, tarlike hydrocarbon. When oil sands are heated, the bitumen can be "melted out" and refined in the same way as crude oil. Northern Alberta, Canada, has the world's largest oil-sand deposits (300 billion barrels), and Canada is commercially exploiting them, because the cost is competitive with current oil prices. Suncor Energy of Canada sells 34 million barrels of this unconventional oil a year and is constructing facilities to expand to 80 million barrels a year by 2002. The United States has smaller, poorer oil-sand deposits, mostly in Utah, that might produce oil for $50–$60 per barrel.

Worldwide, the combination of oil shales and oil sands has been estimated to hold several trillion barrels of oil.

Large-scale commercial development of these resources will not occur until oil prices rise dramatically—something that may be underway already. Both sources involve substantial environmental impacts.

13.4 Sustainable Energy Options

Fossil-fuel use has been rising at rates of 2–3% per year, with no end in sight. Yet, at the same time that we are rethinking our energy future on the basis of availability of resources and managing our demands, we need to consider environmental factors looming on the horizon, primarily global climate change. Note again that all fossil fuels are carbon-based compounds and none can be burned without releasing carbon dioxide (CO_2) as a by-product. Of the top three fossil fuels—coal, oil, and natural gas—coal produces the most, and natural gas produces the least, CO_2 (Table 13–3). Because CO_2 is a greenhouse gas, the increasing atmospheric concentration of CO_2 is likely to bring about major changes in the global climate. (See Chapter 21.) This threat of climate change—and mounting evidence indicates that it is already occurring—is likely to force a curtailment of fossil-fuel consumption long before limited resources do. Therefore, fossil-fuel contributions to global warming are probably the most compelling reason to reduce energy demand through conservation and to examine non-fossil-fuel alternatives. On the basis of carbon emissions per capita (Fig. 13-17), the United States could make the greatest contribution to solving the problem.

Also, in rethinking our energy strategies, we need to start with the concept that energy has no real meaning or value by itself. Its only worth is in the work it can do. Thus, what we really want is not energy, but comfortable homes, transportation to where we need to go, manufactured goods, computers that work, and so forth. Therefore, many energy analysts are saying that we should stop thinking in terms of where we can get additional supplies of the old fuels to keep things running as they are. Instead, we should be thinking in terms of how we can satisfy these needs with minimum expenditure of energy and the least environmental impact.

TABLE 13–3 CO_2 Emissions per Unit Energy for Fossil Fuels (Natural Gas = 100%)	
Natural Gas	100%
Gasoline	134%
Crude Oil	138%
Coal	178%

Source: Data from Energy Information Administration, U.S. Department of Energy, 1999.

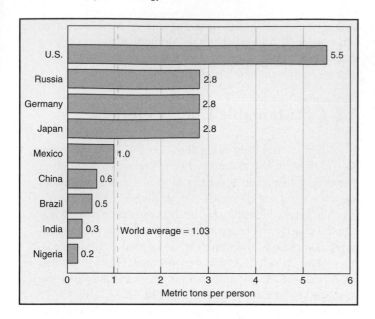

▲ **FIGURE 13–17** *Annual carbon emissions per capita from burning fossil fuels, selected countries, 1998.* (*Source:* Data from Energy Information Agency, 1999; Population Reference Bureau, 1998.)

Conservation

Imagine the discovery of a new oil field having a production potential of at least 6 million barrels a day—three times the capacity of the Alaskan field even at its height. Furthermore, assume that this new-found field is inexhaustible and that its exploitation will not adversely affect the environment. Of course, such an oil field sounds like a dream, but this picture is effectively what can be achieved through fossil-fuel conservation.

As noted earlier in the chapter, this "conservation reserve" has already been tapped to some extent, especially as the average fuel efficiency of cars was doubled from 13 to 27.5 mpg. Thus, drivers of those autos are satisfying their transportation needs with only half the earlier expenditure of energy and dollars. This and other conservation measures are already saving about $100 billion per year in oil imports. In short, advances in conservation have been extremely cost effective as well as environmentally benign.

A number of energy experts believe that most of the "conservation reserve" remains to be tapped. Cars are currently available that get at least double the current average of 27.5 mpg, and many manufacturers have prototypes that get from 70 to 100 mpg. The General Motors Precept and the Ford Prodigy, unveiled in 2000, both get nearly 80 mpg; they are "concept cars" developed under a federally sponsored program, Partnership for a New Generation of Vehicles. To guarantee high sales, however, most automakers are waiting for government mandates, consumer demand, or fuel shortages before putting such models into production. Of course, carpooling with just one person in addition to the driver effectively doubles the mpg.

A technique known as **cogeneration** (see the "Earth Watch" essay, p. 337) is another way of gaining large amounts of energy from the conservation reserve. And in recent years, a new technology has been increasingly employed to generate electricity: the **combined-cycle natural-gas unit**. Two turbines are employed, the first burning natural gas in a conventional gas turbine (Fig. 13–6b), and the second in a steam turbine (Fig. 13–6a) that runs on the excess heat from the gas turbine. The use of this technology boosts the efficiency of conversion of fuel energy to electrical energy from around 30% to as high as 50%. The cost of combined-cycle systems is half that of conventional coal-burning plants, and the pollution from such systems is much lower. A large proportion of new units being built and planned for the future is combined-cycle systems.

Other major changes are under way for the entire electric power industry, as an increasing number of states adopt **deregulation,** which requires the utilities to maintain distribution services (transmission lines, telephone poles, etc.), but to divest themselves of their power-generating facilities. Thus, consumers will be able to choose energy suppliers as the utilities lose their monopolies on electrical power supply. In Massachusetts, deregulation has spawned a number of new power plants, all natural-gas fired. Consumers can, for example, opt to purchase "green" energy from AllEnergy, a company developing power from landfill methane (see Chapter 19) and other renewable sources. The final outcome of deregulation is still uncertain, but energy conservation and renewable energy appear to be much more achievable as a consequence.

Many conservation measures have been put in place and are reducing energy costs. An example of how important such measures are can be seen in the standards for refrigerators and freezers, which consume energy in every home. New standards put in place (accompanied by changes in technology) have reduced the energy demand per unit to one-quarter of what it was in 1974. The net impact is remarkable: By 2001, homeowners will have saved $10 billion, and 40 1,000-watt power plants and 32 million tons of carbon emissions per year will have been avoided.

New building codes require improved insulation and double-pane windows for homes and buildings. Some options are underutilized: Substitution of fluorescent lights, which are between 20% and 25% efficient (Fig. 13–18), for conventional incandescent lightbulbs, which are only 5% efficient, would cut back on our electricity demands for lighting by 75–80%. Energy-conserving lightbulbs and appliances are more expensive than conventional models, but calculations show that they more than pay for themselves in energy savings. Interestingly, some utility companies are subsidizing the installation of fluorescent lightbulbs and energy-efficient appliances for consumers as a cost-effective alternative to building additional power plants.

Energy savings are coming from an unexpected source: the Internet. On-line shopping, working from home computers (telecommuting), business-to-business Web sites, and

EARTH WATCH

COGENERATION: INDUSTRIAL COMMON SENSE

In many situations, a single energy source can be put to use to produce both electrical and heat energy. This process, called cogeneration, can be employed to conserve energy in two fundamental ways: (1) A power plant whose primary function is to generate electricity also can be used to heat surrounding homes and buildings. As one example, the Consolidated Edison electric utility has been using steam from waste heat in New York City for years. (2) A facility whose primary function is to produce heat or mechanical energy can also generate electricity, which is used by the facility and can also be sold to the local utility. The system can work for all sorts of fuels and heat engines. This is by far the most common application of cogeneration.

The arrangement makes perfect sense. Every building and factory requires both electricity and heat. Traditionally, these two forms of energy come from separate sources—the electricity from power lines brought in from some central power station and the heat from an on-site heating system, usually fueled by natural gas or oil. Ordinary power plants are,

at best, 30% efficient in their use of energy. Combining the two sources in cogeneration can provide an overall efficiency of fuel use as high as 80%. Besides providing savings on fuel, the cogenerating systems have at least 50% lower capital costs per kilowatt generated than do conventional utilities. This cost savings makes cogeneration quite competitive with utility-produced power.

The Santa Barbara (California) County Hospital recently installed a cogeneration system based on natural gas as the fuel source. Natural gas drives a turbine with the capacity to generate 7 MW of electricity, much of which is used by the hospital. Boilers use the waste heat from the turbine to produce high-pressure steam, which provides all of the hospital's heating and cooling needs. Any steam not required by the hospital is directed back to the turbine to increase its power. Electrical power not used by the hospital is sold to the local utility. The system has the added advantage of protecting the hospital against wide-scale blackouts or brownouts that may occur in the region's centralized utility system.

The enactment of the Public Utility Regulatory Policies Act (PURPA) in 1978 was essential for stimulating the use of cogeneration. Under this act, a facility with cogenerative capacity is entitled to use fuels not usually available to electricity-generating plants (for example, natural gas), must produce above-standard efficiency of fuel use, and is entitled to sell cogenerated electricity to a utility at a fair price. After many challenges by the utilities (which considered the procedures an interference with their business), the Supreme Court decided in favor of PURPA, and the legislation is now firmly established.

Cogeneration is becoming increasingly popular in U.S. industries as a major strategy for cutting energy costs. Given the likelihood of major price increases for petroleum and of reductions in CO_2 emissions, pressures favoring energy conservation will increase, and cogeneration will continue to grow. Its impact can be seen in the electric power sector of the economy: In 1999, 13% of total generation of electricity came from nonutility power production such as cogeneration.

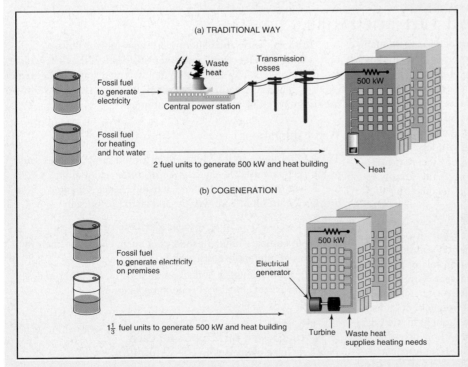

Cogeneration. (a) The traditional method of providing the energy needed for buildings. (b) In cogeneration, the heat lost from an on-site power-generating system provides heating needs.

▲ FIGURE 13–18 *Energy-efficient light bulbs.* Replacing standard incandescent light bulbs with screw-in fluorescent bulbs shown here can cut the energy demand for lighting by 75–80%. The higher cost of the fluorescent bulbs is more than offset by greater efficiency and much longer lifetime of the bulb.

e-mailing—all are reducing energy costs substantially. With fewer catalogs and much less retail and warehouse space needed, there will be savings on paper costs, fuel for driving, and construction and heating costs for businesses, to name a few. A study by the Center for Energy and Climate Solu-

tions ("The Internet Economy and Global Warming: A Scenario of the Impact of E-Commerce on Energy and the Environment") published in 2000 suggests that the Internet will eventually bring about a 4.5 billion-square-foot reduction in commercial spaces, an energy savings worth 21 power plants per year and, through paper savings alone, a reduction of 10 million tons of CO_2 per year. Essentially, the American economy will be able to continue to grow without corresponding growth in energy consumption.

Development of Non-Fossil-Fuel Energy Sources

Conservation of energy is extremely important, but remember that we are still talking about just *reducing* our use of fossil fuels, not eliminating it. Lying before us are two major pathways for developing non-fossil-fuel energy alternatives: to pursue nuclear power and to promote solar-energy applications. Of course, nuclear power has been in use for 40 years, and various solar-energy systems have also been developed. In other words, both nuclear and solar alternatives are at a stage where they could be expanded to provide further energy needs if suitable technological solutions and public acceptance (in the case of nuclear power) or pressure and government support (in that of solar energy) were provided. The next chapter focuses on the nuclear option, and Chapter 15 focuses on solar options. We address public policy needs at the end of Chapter 15.

ENVIRONMENT ON THE WEB

THE HIDDEN COSTS OF FOSSIL FUEL UTILIZATION

Fossil fuel use incurs real costs for society. Some of these costs, such as those related to exploration and the construction of generating plants, are "direct" and have a monetary value passed on to the consumer through electricity bills. However, there are also many hidden or "indirect" costs of fossil fuel use, often related to the environmental or human health impacts of fossil fuel use. These include the tangible or intangible costs of additional health care, diminished quality of life (including shorter lives), loss of wildlife habitat, and loss of aesthetic value (such as loss of a scenic vista).

Because these costs are externalities—not paid directly by consumers or producers—and often difficult to quantify, they are usually excluded from energy pricing. Yet they are nevertheless real costs, with real implications for society. The National Academy of Sciences estimates, for instance, that air pollution causes billions of dollars a year in environmental and structural damage. The health care burden of air pollution related to fossil fuel combustion adds another $20 billion a year. Greenhouse gas emissions arising from fossil fuel use have also been linked to climate change. Although the potential impacts of global warming are still poorly understood, they are thought to include effects on crop yields, water supply, redistribution of populations, and the impacts of altered weather patterns on hurricanes, typhoons, and similar phenomena. All of these are likely to carry a cost to society.

Although some jurisdictions are now moving toward "full-cost pricing" of electricity and other utilities to include a wider range of costs, such as anticipated maintenance and replacement of costly generating equipment, the full hidden costs of fossil fuel use still may not be apparent to consumers.

Web Explorations

The Environment on the Web activity for this essay explores some of the methods used to put a price on some externalities. Go to the Environment on the Web activity (select Chapter 13 at http://www.prenhall.com/wright) and learn for yourself:

1. about other externalities associated with fossil fuel use;
2. how to compare different sources of energy based on their tangible and intangible costs; and
3. about the advantages and disadvantages of retaining and upgrading existing fossil fuel generating facilities.

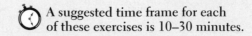 **A suggested time frame for each of these exercises is 10–30 minutes.**

REVIEW QUESTIONS

1. How have the fuels to power homes, industry, and transportation changed from the beginning of the Industrial Revolution to the present?
2. What are the three primary fossil fuels, and what percentage does each contribute to the U.S. energy supply?
3. Electricity is a secondary energy source. How is it generated, and how efficient is its generation?
4. Name the four major categories of primary energy use in an industrial society, and match the energy sources that correspond to each.
5. What is the distinction between estimated reserves and proven reserves of crude oil, and what factors cause the amounts of each to change?
6. How does the price of oil influence the amount produced?
7. What were the trends in oil consumption, in the discovery of new reserves, and in U.S. production before 1970? After 1970?
8. What did the United States do in the early 1970s to resolve the disparity between oil production and consumption? What events caused the sudden oil shortages of the mid-1970s and then the return to abundant, but more expensive, supplies?
9. What responses were made by the United States and other industrialized nations to the "oil crisis" of the 1970s, and what were the impacts of those responses on production and consumption?
10. Why did an oil glut occur in the mid-1980s, and what were its consequences in terms of oil prices and the continuing pursuit of the responses listed in Question 9?
11. What have been the directions of U.S. production, consumption, and importation of crude oil since the mid-1980s?
12. What are the liabilities and risks of our growing dependence on foreign oil? How do they relate to the Persian Gulf War of 1991?
13. What alternative fossil fuels might the United States exploit to supplement declining reserves of crude oil? What are the economic and environmental advantages and disadvantages of each?
14. What phenomenon may restrict the consumption of fossil fuels before resources are depleted?
15. To what degree has energy conservation served to mitigate energy dependency? What are some prime examples of energy conservation? To what degree may conservation alleviate future energy shortages and the cost of developing alternative sources of energy?

THINKING ENVIRONMENTALLY

1. Statistics show that less developed countries use far less energy per capita than developed countries. Explain why this is so.
2. Between 1980 and 1999, gasoline prices in the United States declined considerably relative to inflation. Predict what they will do in the next 10 years, and give a rationale for your prediction.
3. Suppose your region is facing power shortages (brownouts). How would you propose solving the problem? Defend your proposed solution on both economic and environmental grounds.
4. List all the environmental impacts that occur during the "life span" of gasoline—that is, from exploratory drilling to the consumption of gasoline in your car.

WEB REFERENCES

On-line resources for this chapter are on the World Wide Web at: **http://www.prenhall.com/wright**. (Click on Chapter 13 within the Chapter Selector.)

Nuclear Power: Promise and Problems

Key Issues and Questions

1. Nuclear power currently generates 20% of U.S. electricity and 7% of U.S. energy overall, yet it is controversial. What are the history and current status of nuclear power in the United States?

2. The objective of nuclear power technology is to use a controlled nuclear reaction to drive a generator. How does a nuclear power plant work?

3. In order to understand the hazards of nuclear power, we must understand something about radioactive materials. What are radioactive materials, and what hazards do they pose?

4. The disposal of nuclear waste is one of the two major issues environmentalists are concerned about. What are the problems surrounding the disposal of nuclear waste?

5. The other major issue of great concern is the possibility of accidents in nuclear plants. Of what significance are the Three Mile Island and Chernobyl nuclear accidents?

6. Fusion-based energy has long been considered a candidate for a pollution-free energy source of the future. What is the prognosis for fusion-based energy?

7. In light of current energy choices, nuclear power still appears to be an attractive option. What future is there for nuclear power?

Tokaimura, a village 120 km northeast of Tokyo, is home to several nuclear installations, including a nuclear power plant, the Japan Atomic Energy Research Institute, and a fuel-reprocessing plant (Fig. 14–1). On September 30, 1999, three workers in the reprocessing plant were preparing enriched uranium for a research reactor. They mixed nitric acid with 35 lb of highly enriched uranium, adding the chemicals to a tank in buckets instead of using the plant's mechanized equipment. The amount added was well above the approved levels, and the mixture reached a critical mass and began a nuclear-fission chain reaction. The reaction was not intense enough to produce an explosion, but it emitted high levels of gamma rays and neutrons, triggering alarms that sent the three workers running out of the building. It took an emergency response team almost 24 hours to shut down the reaction. By that time, enough gamma radiation had escaped to expose 439 townspeople and workers to varying levels of radiation. The three workers received very high doses, and two of the work-

ers subsequently died of radiation poisoning. The third worker received a lower dose and survived the accident. Fortunately, only trace amounts of radioactive gases were emitted, and the impact of the accident was limited to a few hundred meters of the plant. However, the town was partially evacuated as a safety precaution until the extent of the accident became clear.

An investigation revealed that workers frequently violated regulations, and the company, JCO, had illegally revised a government-approved manual to allow workers to mix uranium fuels with the use of buckets. The Japanese government has revoked the license of JCO, and a criminal investigation is under way. Nevertheless, with its violent introduction into the nuclear age via the bombing of Hiroshima and Nagasaki, Japan is determined to increase its nuclear power capacity. Nuclear power now supplies 36% of electrical power in Japan, and the goal is to reach 43% by 2010. The accident at Tokaimura however, has created a groundswell of public protest in Japan against any expan-

◀ *Indian Point Nuclear Power Station in Buchanan, New York.*

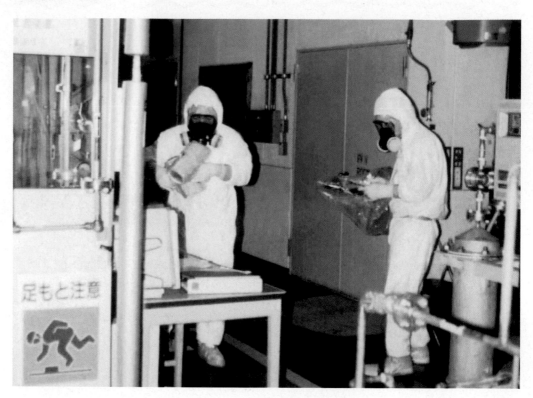

▶ FIGURE 14–1   *Nuclear accident at Tokaimura, Japan.*   Weeks after the accident, workers wearing protective gear test for remaining radiation in a technical room at the uranium-processing plant.

sion of nuclear power and has embarrassed the Japanese government and shaken the country's nuclear industry.

The controversy brought on by Tokaimura is only one of many in the troubled history of nuclear power. In this chapter, we look into that history through the twin perspectives of science and public policy as we sort out the many concerns about nuclear power.

14.1 Nuclear Power: Dream or Delusion?

As the use of fossil fuels continues to rise, the threat of global climate change gathers momentum. Also, the Gulf War of 1991 provided graphic evidence of how developed nations dependent on overseas sources of crude oil are willing to use military force to protect those sources. After many years of low prices, OPEC cut oil production in late 1999, and the price of oil shot upwards, causing economic hardship in numerous countries. In light of these serious energy-related problems, it seems prudent to do everything possible to develop non-fossil-fuel sources of energy. Nuclear power is an alternative that does not contribute to global warming, and there is sufficient uranium to fuel nuclear reactors well into the 21st century, with the possibility of extending the nuclear fuel supply indefinitely through reprocessing technologies. Thirty-one nations now have nuclear power plants either in place or under construction, and in some of these countries, nuclear power generates more electricity than any other source. Is nuclear energy the key to the future of global energy—the best route to a sustainable relationship with our environment?

For years, geologists have recognized that fossil fuels would not last forever. Sooner or later, other energy sources would be needed. It was anticipated that nuclear power could produce electricity in such large amounts and so cheaply that we would phase into an economy in which electricity would take over virtually all functions, including the generation of other fuels, at nominal costs. Following World War II, determined to show the world that the power of the atom could benefit humankind, the U.S. government embarked on a course to lead the world into the "Nuclear Age."

Thus, the government moved into the research, development, and promotion of commercial nuclear power plants (along with the continuing development of nuclear weaponry). Utilizing this research, companies such as General Electric and Westinghouse constructed nuclear power plants that were ordered and paid for by utility companies, with assurances from the federal government, via the Price–Anderson Act of 1957, that these corporations and utilities would be exempt from any liabilities incurred therefrom. The Nuclear Regulatory Commission (NRC), formerly the Atomic Energy Commission, an agency in the Department of Energy (DOE), set and enforced safety standards for the operation and maintenance of these plants, as it does today. In the 1960s and early 1970s, utility companies moved ahead with plans for numerous nuclear power plants (Fig. 14–2). By 1975, 53 plants were operating in the United

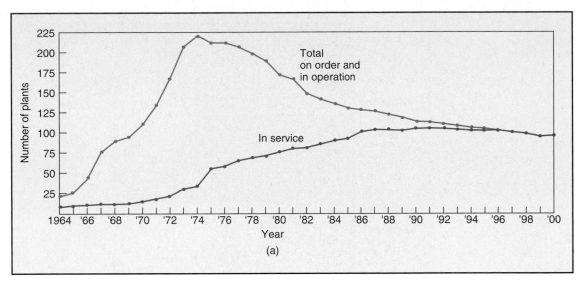

▲ **FIGURE 14–2** *Nuclear power in the United States.* Since the early 1970s, when orders for plants reached a peak, few utilities called for new plants, and many canceled earlier orders. Nevertheless, the number of plants in service increased steadily as plants under construction were completed. The number of operating plants peaked at 110 and is now declining. (*Source*: Data from U.S. Department of Energy.)

States, producing about 9% of the nation's electricity, and another 170 plants were in various stages of planning or construction. Officials estimated that by 1990 several hundred plants would be on line and by the turn of the century as many as a thousand would be operating. A number of other industrialized countries got in step with their own programs, and some less developed nations were going nuclear by purchasing plants from industrialized nations.

Since 1975, however, the picture has changed dramatically. Utilities stopped ordering nuclear plants, and numerous existing orders were canceled. The last order for a plant in the United States that was not subsequently canceled was placed in 1974. In some cases, construction was terminated even after billions of dollars had already been invested. Most striking, the Shoreham Nuclear Power Plant on Long Island, New York, after being completed and licensed at a cost of $5.5 billion, was turned over to the state in the summer of 1989, to be dismantled after generating electricity for only 32 hours (Fig. 14–3). The reason was that citizens and the state deemed that there was no way to evacuate people from the area should an accident occur. Similarly, just a few weeks earlier, California citizens voted to shut down the Rancho Seco Nuclear Power Plant, located near Sacramento, after a 15-year history of troubled operation.

At the end of 2000, 104 nuclear power plants were operating in the United States. When the Watts Bar nuclear plant in Tennessee came on line in February 1996, no more plants were left under construction. The 104 plants existing generate 20% of U.S. electrical power. Thus, U.S. nuclear power, which leads the world, peaked at 112 plants and is headed downward as more and more plants are decommissioned. Including plants in the United States, the world has a total of 433 operating nuclear plants, with an additional 37 under construction. In general, the story abroad is similar to that in the United States: drastic downward revisions, if not total moratoriums, on the construction of new plants. Nuclear power generates about 17% of the world's electricity (23% in those countries with nuclear units), but dependence on nuclear energy varies greatly among the 31 countries operating nuclear power plants (Fig. 14–4). Of the major industrial nations, only France and Japan remain fully committed to pushing forward with nuclear programs. Indeed, France now produces 75% of its electricity with nuclear power and has plans to go to more than 80%.

After the catastrophic accident at Chernobyl in April 1986, it is not hard to see why nuclear power is being

▲ **FIGURE 14–3** *The Shoreham Nuclear Plant on Long Island, New York.* With little electricity produced, this plant was closed because of concerns about whether surrounding areas could be evacuated in case of an accident. The plant is now being dismantled.

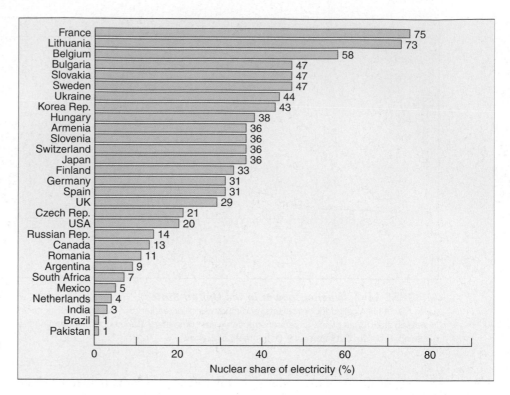

▶ **FIGURE 14–4** *Nuclear share of electrical power generation.* In 1999, those countries lacking fossil-fuel reserves tended to be the most eager to employ nuclear power. (*Source:* Data from International Atomic Energy Agency.)

rethought. Yet, the demand for electricity is more robust than ever, and the other means of generating electricity have their own problems. Coal-powered electricity generates more greenhouse gases than any other form of energy. Oil supplies are more limited, and oil is vital to transportation and home heating. Hydroelectric power is already heavily developed. Solar-power technology is still lagging. If the technological and economic issues can be resolved, can nuclear power supply future energy needs? In order to react intelligently to this issue, we need a clear understanding of what nuclear power is and what are its pros and cons.

14.2 How Nuclear Power Works

The objective of nuclear power technology is to control nuclear reactions so that energy is released gradually as heat. As with plants powered by fossil fuels, the heat energy produced by a nuclear plant is used to boil water and produce steam, which then drives conventional turbogenerators. Nuclear power plants are base-load plants (see Chapter 13); they are always operating unless they are being refueled, and they are large plants, generating up to 1,400 megawatts.

From Mass to Energy

The release of nuclear energy is completely different from the burning of fuels or any other chemical reactions we have discussed. To begin with, materials involved in chemical reactions remain unchanged at the atomic level, although the visible forms of these materials undergo great transformation as atoms are rearranged to form different compounds. Nuclear energy, however, involves changes at the atomic level through one of two basic processes: fission and fusion. In **fission**, a large atom of one element is split to produce two smaller atoms of different elements (Fig. 14–5a). In **fusion**, two small atoms combine to form a larger atom of a different element (Fig. 14–5b). In both fission and fusion, the mass of the product(s) is less than the mass of the starting material, and the lost mass is converted to energy in accordance with the law of mass–energy equivalence ($E = mc^2$), first described by Albert Einstein. The amount of energy released by this mass-to-energy conversion is tremendous: The sudden fission or fusion of a mere 1 kg of material, for example, releases the devastatingly explosive energy of a nuclear bomb.

The Fuel for Nuclear Power Plants. All current nuclear power plants employ the fission (splitting) of uranium-235. The element uranium occurs naturally in various minerals in Earth's crust. It exists in two primary forms, or isotopes: uranium-238 (^{238}U) and uranium-235 (^{235}U). The **isotopes** of a given element contain different numbers of neutrons, but the same number of protons and electrons.

The number 238 or 235 that accompanies the chemical name or symbol for uranium is called the **mass number** of the element. It is the sum of the number of neutrons and the number of protons in the nucleus of the atom. ^{238}U contains 92 protons and 146 neutrons, while ^{235}U contains 92 protons and 143 neutrons. However, since, by definition, all atoms of any given element must contain the same number of protons, variations in mass number represent variations in numbers of neutrons. Therefore, whereas ^{238}U contains three more neutrons than ^{235}U, both isotopes contain the 92 protons that *define* the element uranium. Although all isotopes of a given element behave the same chemically, their other characteristics may differ

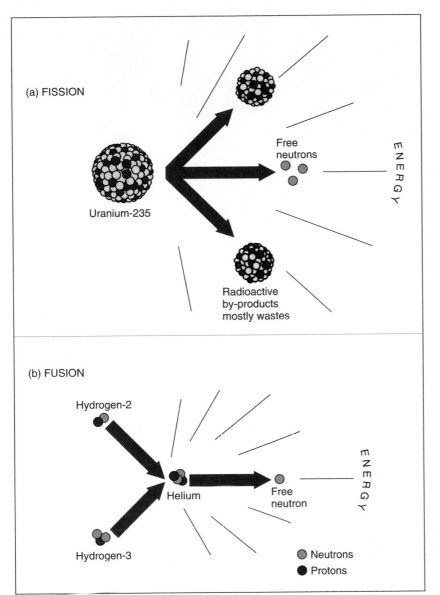

(a) FISSION

Uranium-235

Free neutrons

Radioactive by-products mostly wastes

ENERGY

(b) FUSION

Hydrogen-2

Hydrogen-3

Helium

Free neutron

ENERGY

● Neutrons
● Protons

◄ **FIGURE 14–5** *Nuclear fission and fusion.*
Nuclear energy is released from either (a) fission, the splitting of certain large atoms into smaller atoms, or (b) fusion, the fusing together of small atoms to form a larger atom. In both cases, some of the mass of the starting atom(s) is converted to energy.

profoundly. In the case of uranium, ^{235}U atoms will readily undergo fission, but ^{238}U atoms will not. Like uranium, many other elements exist in more than one isotopic form.

It takes a neutron hitting the nucleus at just the right speed to cause ^{235}U to undergo fission. Since ^{235}U is an unstable isotope, a small, but predictable, number of the ^{235}U atoms in any sample of the element undergo radioactive decay and release neutrons, among other things. If one released neutron moving at just the right speed hits another ^{235}U atom, the latter becomes ^{236}U, which is highly unstable and undergoes fission immediately into lighter atoms (**fission products**). The fission reaction gives off several more neutrons and releases a great deal of energy (Fig. 14–5a). Ordinarily, these neutrons are traveling too fast to cause fission, but if they are slowed down and then strike another ^{235}U atom, they can cause fission to occur again. In this way, more neutrons and more energy are released, with the potential to repeat the process. A domino effect, known as a chain reaction, may thus occur (Fig. 14–6a).

To make nuclear "fuel," uranium ore is mined, purified into uranium dioxide (UO_2), and enriched. Since 99.3% of all uranium found in nature is ^{238}U—only 0.7% is ^{235}U—**enrichment** involves separating ^{235}U from ^{238}U to produce a material containing a higher concentration of ^{235}U. Since ^{238}U and ^{235}U are chemically identical, enrichment is based on their slight difference in mass. The technical difficulty of enrichment is the major hurdle that prevents less developed countries from advancing their own nuclear capabilities.

When ^{235}U is *highly* enriched, the spontaneous fission of an atom can trigger a chain reaction. In nuclear weapons, small masses of virtually pure ^{235}U or other fissionable material are forced together so that the two or three neutrons from a spontaneous fission cause two or three more atoms to undergo fission; each of these in turn triggers two or three more fissions, and so on. The whole mass undergoes fission in a fraction of a second, releasing all the energy in one huge explosion (Fig. 14–6b).

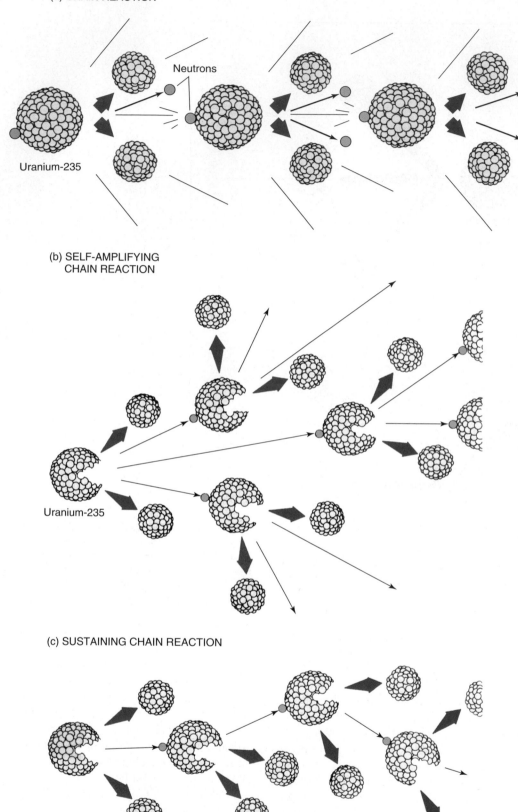

▲ FIGURE 14–6 *Fission reactions.* (a) A simple chain reaction. When a ^{235}U atom fissions, it releases two or three high-energy neutrons, in addition to energy and split "halves." If another ^{235}U atom is struck by a high-energy neutron, it fissions, and the process is repeated, causing a chain reaction. (b) A self-amplifying chain reaction leading to a nuclear explosion. Since two or three high-energy neutrons are produced by each fission, each may cause the fission of two or three additional atoms. (c) In a sustaining chain reaction, the extra neutrons are absorbed in control rods, so that amplification does not occur.

The Nuclear Reactor. A nuclear reactor for a power plant is designed to sustain a continuous chain reaction (Fig. 14–6c), but not allow it to amplify into a nuclear explosion. Control is achieved by enriching the uranium to only 4% ^{235}U and 96% ^{238}U. This modest enrichment will not support the amplification of a chain reaction into a nuclear explosion.

A chain reaction can be achieved in a nuclear reactor only if a sufficient mass of enriched uranium is arranged in a suitable geometric pattern and is surrounded by a material called a *moderator*. The **moderator** slows down the neutrons that produce fission, so that they are traveling at the right speed to trigger another fission. In slowing down the neutrons, the moderator gains heat. In nuclear plants in the United States, the moderator is very pure water, and the reactors are called light-water reactors (LWRs). [The term "light water" denotes ordinary water (H_2O), as opposed to the substance known as heavy water, or deuterium (D_2O).] Other moderators employed in reactors of different design are graphite and deuterium.

To achieve the geometric pattern necessary for fission, the enriched uranium dioxide is made into pellets that are loaded into long metal tubes. The loaded tubes are called **fuel elements** or **fuel rods** (Fig. 14–7a). Many fuel rods are placed close together to form a reactor core inside a strong reactor vessel that holds the water, which serves as both moderator and heat-exchange fluid (coolant) (Fig. 14–7b). Over time, daughter products that also absorb neutrons

(a)

Control rods

Fuel elements

(b)

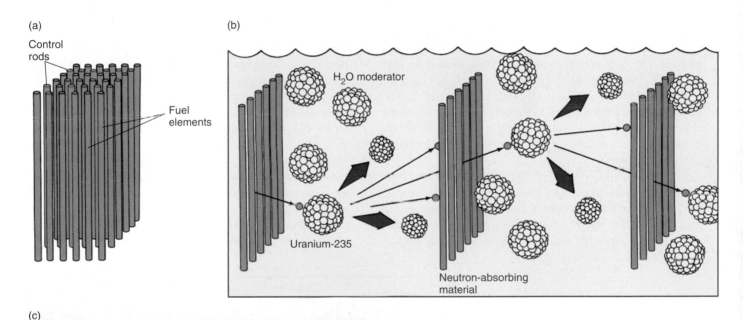

H$_2$O moderator

Uranium-235

Neutron-absorbing material

(c)

◀ **FIGURE 14–7** *A nuclear reactor.* (a) In the core of a nuclear reactor, a large mass of uranium is created by placing uranium in adjacent tubes, called fuel elements. The rate of the chain reaction is moderated by inserting or removing rods of neutron-absorbing material (control rods) between the fuel elements. (b) The fuel and rods are surrounded by the moderator fluid, pure water. (c) Technicians ready the core housing to receive uranium fuel elements in this nuclear reactor.

accumulate in the fuel rods and slow down the rate of fission and heat production. The highly radioactive spent-fuel elements are then removed and replaced with new ones.

The chain reaction in the reactor core is controlled by rods of neutron-absorbing material, referred to as **control rods**, inserted between the fuel elements. The chain reaction is started and controlled by withdrawing and inserting the control rods as necessary. Hence, the nuclear reactor is simply the assembly of fuel elements, moderator–coolant, and movable control rods (Fig. 14–7). As the control rods are removed and a chain reaction is initiated, the fuel rods and the moderator become intensely hot.

The Nuclear Power Plant. In a nuclear power plant, heat from the reactor is used to boil water to provide steam for driving conventional turbogenerators. One way to boil the water is to circulate it through the reactor. In most power plants in the United States, however, a double loop is employed. The moderator–coolant water is heated to over 600°F by circulating it through the reactor, but it does not boil, because the system is under very high pressure (2,100 psi). The superheated water is circulated through a heat exchanger, boiling other unpressurized water that flows past the heat exchanger tubes. This produces the steam used to drive the turbogenerator.

The double-loop design of this primary system and secondary system isolates hazardous materials in the reactor from the rest of the power plant. However, it has one serious drawback: If the reaction vessel should break, the sudden loss of water from around the reactor, called a "loss-of-coolant accident" (LOCA), could result in the core's overheating. The sudden loss of the moderator–coolant water would cause fission to cease, since the moderator would no longer be present. However, even though fission stops, the fuel core can still overheat because 7% of the reactor heat comes from radioactive decay in the newly formed fission products. In time, the uncontrolled decay would release enough heat energy to melt the materials in the core, a situation called a **meltdown**. Then, the molten material falling into the remaining water could cause a steam explosion. To guard against all this, backup cooling systems keep the reactor immersed in water should leaks occur, and the entire assembly is housed in a thick concrete containment building (Fig. 14–8).

Comparison of Nuclear Power with Coal Power

Because nuclear plants are base-load plants that provide the foundation for meeting the daily and weekly electrical demand cycle, they must be replaced with other base-load plants. With the recent cancellations and continuing shutdowns of nuclear power plants, the sole option is replacing such large plants with coal-burning plants. The United States has abundant coal reserves, but is burning coal the course of action we wish to pursue? Nuclear power has some decided environmental advantages over coal-fired power (Fig. 14–9). Comparing a 1,000-megawatt nuclear plant with a coal-fired plant of the same capacity, each operating for one year, we find the following characteristics:

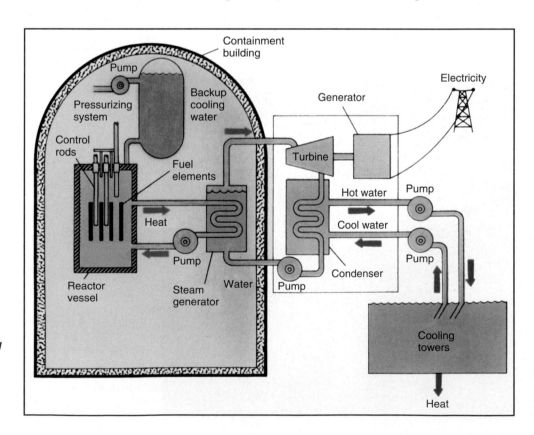

▶ **FIGURE 14–8** *Pressurized nuclear power plant.* The double-loop design isolates the pressurized water from the steam-generating loop that drives the turbogenerator.

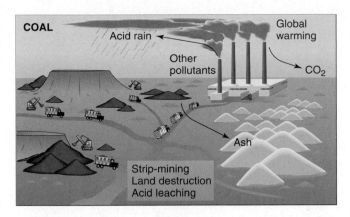

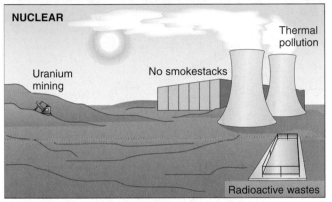

▲ **FIGURE 14–9** *Nuclear power vs. coal-fired power.* Both methods of generating electricity have their environmental advantages and disadvantages. The nuclear power option assumes perfect containment of radioactivity and the availability of some method for waste storage and disposal.

* *Fuel needed.* The coal plant consumes about 3 million tons of coal. If this amount is obtained by strip mining, some environmental destruction and acid leaching will result. If the coal comes from deep mines, there will be human costs in the form of accidental deaths and impaired health. The nuclear plant requires about 30 tons of enriched uranium obtained from mining 75,000 tons of ore, with much less harm to humans and the environment. The fission of about 1 pound (0.5 kg) of uranium fuel releases energy equivalent to burning 50 tons of coal. Thus, one fueling of the reactor with about 60 tons is sufficient to run the nuclear power plant for as long as two years.

* *Carbon dioxide emissions.* The coal plant emits over 7 million tons of carbon dioxide into the atmosphere, contributing to global climate change. The nuclear plant emits none.

* *Sulfur dioxide and other emissions.* The coal plant emits over 300,000 tons of sulfur dioxide and other acid-forming pollutants, which must be captured by precipitators and scrubbers. The coal plant also releases low levels of many radioactive chemicals found naturally in the coal. The nuclear power plant produces no acid-forming pollutants, but may release low levels of radioactive waste gases.

* *Solid wastes.* The coal plant produces about 600,000 tons of ash requiring land disposal. The nuclear plant produces about 250 tons of highly radioactive wastes requiring safe storage and ultimate safe disposal. (The handling of these radioactive wastes remains an unresolved enigma.)

* *Accidents.* A worst-case accident in the coal plant could result in fatalities to workers and a destructive fire, a situation common to many industries. Accidents in a nuclear plant can range from minor emissions of radioactivity to catastrophic releases that can lead to widespread radiation sickness, scores of human deaths, untold numbers of cancers, and widespread, long-lasting environmental contamination.

Of course, it is the handling of radioactive wastes and releases and the potential for accidents that have led the public to reject nuclear power. But can't we overcome these problems? Or are other energy options less problematic and less costly than nuclear power?

14.3 The Hazards and Costs of Nuclear Power

Assessing the hazards of nuclear power necessitates our understanding radioactive substances and their danger.

Radioactive Emissions

When uranium or any other element undergoes fission, the split "halves" are atoms of lighter elements: iodine, cesium, strontium, cobalt, or any of some 30 other elements. These newly formed atoms—called the direct products of the fission—are generally unstable isotopes of their respective elements. Unstable isotopes (called **radioisotopes**) become stable by spontaneously ejecting subatomic particles (alpha particles, beta particles, and neutrons), high-energy radiation (gamma rays), or both. Radioactivity is measured in **curies**; one gram of pure radium-226 gives off one curie per second, which is approximately 37 billion spontaneous disintegrations into particles and radiation. The particles and radiation are collectively referred to as **radioactive emissions**. Any materials in and around the reactor may also be converted to unstable isotopes and become radioactive by absorbing neutrons from the fission process (Fig. 14–10). These indirect products of fission, along with the direct products, are the **radioactive wastes** of nuclear power. (Radioactive fallout from nuclear explosions also consists of direct and indirect fission products.)

Biological Effects. A major concern regarding nuclear power is that large numbers of the public may be exposed to low levels of radiation, thus elevating their risk of cancer and

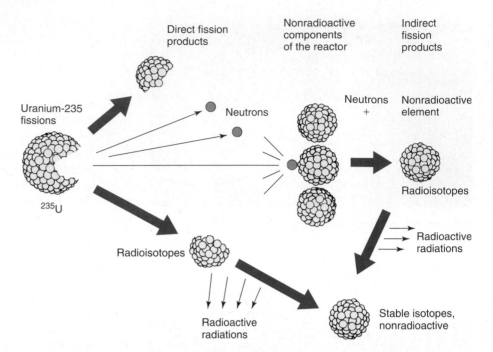

▶ **FIGURE 14–10** *Radioactive wastes and radioactive emissions.* Nuclear fission results in the production of numerous unstable isotopes, designated radioactive waste. These isotopes give off potentially damaging radiation until they regain a stable structure.

other disorders. Is this a valid concern? Radioactive emissions can penetrate biological tissue; their ability to do damage is measured in units called sieverts (Sv; 1 sievert = 100 rem, an older unit of measurement). The emissions leave no visible mark, nor are they felt, but they are capable of dislodging electrons from molecules or atoms they strike. Left behind are ions (charged particles): therefore, the emissions are called ionizing radiation. The process of ionization may involve breaking chemical bonds or changing the structure of molecules in ways that impair their normal functions. In high doses, radiation may cause enough damage to prevent cell division. Thus, in medical applications, radiation can be focused on a cancerous tumor to destroy it. However, if the whole body is exposed to such levels of radiation (over 1 sievert is considered a high dose), a generalized blockage of cell division occurs that prevents the normal replacement or repair of blood, skin, and other tissues. This result is called radiation sickness and may lead to death a few days or months after exposure. The two stricken Japanese workers in the Tokaimura accident received from 10 to 17 sieverts of radiation, ultimately a lethal dose. Very high levels of radiation may totally destroy cells, causing immediate death.

In lower doses, radiation may damage DNA, the genetic material inside the cell. Cells with damaged (mutated) DNA may then begin dividing and growing out of control, forming malignant tumors or leukemia. If the damaged DNA is in an egg or a sperm cell, the result may be birth defects (mutations) in offspring. The effects of exposure to radiation may go unseen until many years after the event; ten to 40 years is typical. Other effects include a weakening of the immune system, mental retardation, and the development of cataracts.

Health effects are directly related to the level of exposure. There is broad agreement that doses between 100 and 500 millisieverts (10 and 50 rem; 1 millisievert = a thousandth of a sievert) result in an increased risk of developing cancer. Evidence for this hypothesis comes from studies of patients with various illnesses who were exposed to high levels of X-rays in the 1930s, when the potential harm of such radiation was not realized. People in these groups subsequently developed higher-than-normal rates of cancer and leukemia. Many scientists believe that no dose is without some harm; others point to the ability of living cells to repair small amounts of damage to DNA and believe that there is a threshold of radiation below which no biological effects occur. If there is such a threshold, it is virtually impossible to demonstrate, because of the many different causes of cancer, the long time it takes for cancer to develop, and the variety of sources of radiation to which people are exposed. Federal agencies employ the concept of no threshold and a direct relationship between the amount of radiation and the incidence of cancer; federal standards are set at 1.7 mSv/yr (170 mrem/yr) as the maximum exposure permitted for the general populace per year, except for medical X-rays.

Sources of Radiation. Nuclear power is by no means the only source of radiation: There is also normal background radiation from radioactive materials, such as the uranium and radon gas that occur naturally in the Earth's crust, and cosmic rays from outer space. For most people, background radiation is the major source of radiation exposure. In addition, we deliberately expose ourselves to radiation from medical and dental X-rays, by far the largest source of human-induced exposure and, for the average person, equal to one-fifth the exposure from background sources. The average person in the United States receives a dose of about 3.6 millisieverts (mSv) per year (Table 14–1). Thus, the argument becomes one of relative hazards: Does or will the radiation from nuclear power significantly raise radiation levels and elevate the risk of developing cancer?

TABLE 14–1 Relative Doses from Radiation Sources

Source	Dose
All sources (avg.)	3.6 mSv/year
Cosmic radiation at sea level*	0.26 mSv/year
Terrestrial radiation (elements in soil)	0.26 mSv/year
Radon in average house	2 mSv/year
X-rays and nuclear medicine (avg.)	0.50 mSv/year
Consumer products	0.11 mSv/year
Natural radioactivity in the body	0.39 mSv/year
Mammogram	0.30 mSv
Chest X-ray	0.10 mSv
Gastrointestinal series of X-rays	14 mSv
Continued fallout from nuclear testing	0.01 mSv/year
Living near a nuclear power station (assuming perfect containment)	<0.01 mSv/year
Federal standards for the general populace	1.7 mSv/year

*Increases 0.005 mSv for every 100 ft of elevation.
Source: Data from U.S. Nuclear Regulatory Commission.

During normal operation of a nuclear power plant, the direct fission products remain within the fuel elements, and the indirect products are maintained within the containment building that houses the reactor. No routine dis-charges of radioactive materials (other than routine stack emissions) into the environment occur. Even very close to an operating nuclear power plant, radiation levels from the plant are lower than normal background levels. A radiation detector will pick up more radiation from the ground or concrete on a basement floor than it will when held within 150 yards of a nuclear power plant. Careful measurements have shown that public exposure to radiation from normal operations of a power plant is less than 1% of natural back-ground radiation.

The main concern about nuclear power, therefore, should not focus on a plant's normal operation; the real problems are about the storage and disposal of radioactive wastes and the potential for accidents.

Radioactive Wastes

To understand the problems surrounding nuclear waste dis-posal, we must understand the concept of **radioactive decay**. As unstable isotopes eject particles and radiation, they become stable and cease to be radioactive. This process is known as radioactive decay. As long as radioactive materials are kept isolated from humans and other organ-isms, the decay proceeds harmlessly.

The rate of radioactive decay is such that half of the starting amount of a given isotope will decay in a certain pe-riod of time. In the next equal period, half of the remainder (half of a half, which equals one-fourth of the original) de-cays, and so on, as shown in Fig. 14–11. The time for half of the amount of a radioactive isotope to decay is known as the isotope's **half-life**. The half-life of an isotope is always the same, regardless of the starting amount.

Each particular radioactive isotope has a characteristic half-life. The half-lives of various isotopes range from a frac-tion of a second to many thousands of years. Uranium

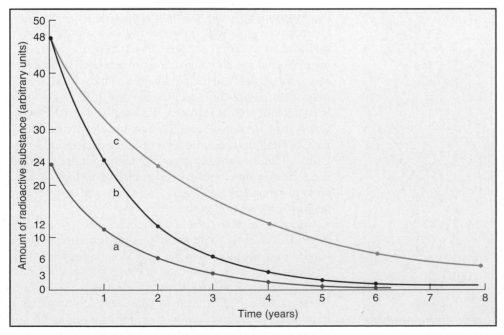

◀ **FIGURE 14–11** *Radioactive decay.* Regardless of the starting amount, one-half of a radioactive substance decays during each successive half-life. (a) A substance with a half-life of one year, starting with 24 units; (b) the same substance starting with 48 units; (c) a substance with a half-life of two years. (Note that the decay of a radioisotope never equals 100%; radioactivity is reduced by only half in each half-life, and there is always an undecayed portion remaining. However, it is generally held that radiation is reduced to insignificant levels after 10 half-lives.)

fissioning results in a heterogeneous mixture of radioisotopes, the most common of which are listed in Table 14–2. Some of this material—in particular, the remaining ^{235}U and plutonium (^{239}Pu) that has been created by the neutron bombardment of ^{238}U—can be recovered and recycled for use as nuclear fuel. This operation is called reprocessing; U.S. policy prohibits the use of reprocessed fuel, because of concerns over plutonium and atomic weapons.

Disposal of Radioactive Wastes.   The development of, and commitment to, nuclear power went ahead without ever fully addressing the issue of ultimate long-term containment. Proponents of nuclear power generally assumed that the long-lived wastes could be solidified, placed in sealed containers, and buried in deep, stable rock formations (geologic burial) as the need for such containment became necessary. However, this has not yet happened. Thus, the current problem of nuclear waste disposal is twofold:

- Short-term containment, to allow the radioactive decay of short-lived isotopes (Table 14–2). In 10 years, fission wastes lose more than 97% of their radioactivity. Wastes can be handled much more easily and safely after this loss occurs.

TABLE 14–2   Common Radioactive Isotopes Resulting from Uranium Fission and Their Half-Lives

Short-Lived Fission Products	Half-Life (days)
Strontium-89	50.5
Yttrium-91	58.5
Zirconium-95	64.0
Niobium-95	35.0
Molybdenum-99	2.8
Ruthenium-103	39.4
Iodine-131	8.1
Xenon-133	5.3
Barium-140	12.8
Cerium-141	32.5
Praseodymium-143	13.6
Neodymium-147	11.1

Long-Lived Fission Products	Half-Life (years)
Krypton-85	10.7
Strontium-90	28.0
Ruthenium-106	1.0
Cesium-137	30.0
Cerium-144	0.8
Promethium-147	2.6

Additional Product of Neutron Bombardment	Half-Life (years)
Plutonium-239	24,000

- Long-term containment (EPA recommends a 10,000-year minimum, and the National Research Council opted for 100,000 years), to provide protection from the long-lived isotopes. Government standards require isolation for 20 half-lives. (Plutonium has a half-life of 24,000 years.)

For short-term containment, spent fuel is first stored in deep swimming pool–like tanks on the sites of nuclear power plants. The water in these tanks dissipates waste heat (which is still generated to some degree) and acts as a shield against radiation. The storage pools can typically accommodate 10 to 20 years of spent fuel. However, the capacity of storage pools at U.S. nuclear plants reached 30% in 2000 and will be at 90% by 2012. After a few years of decay, the spent fuel may be placed in air-cooled dry casks in order to save space. The casks are engineered to resist floods, tornadoes, and extremes of temperature. Currently, 10 nuclear plants have turned to dry storage, with the prospect of many more as pool storage capacity is exhausted.

In the meantime, the wastes from the world's commercial reactors have been accumulating at a rate of 10,000 tons a year, reaching 180,000 tons at the end of 2000, all of which is stored on-site at the power plants. Of these wastes, 41,000 tons are in the United States. Furthermore, because of neutron bombardment of the reactor walls, all nuclear power plants will eventually add to the stockpile of radioactive wastes. Scientists estimate that dismantling a decommissioned power plant will generate more nuclear waste than the plant produced during its active life!

Military Radioactive Wastes.   Some of the worst failures in handling wastes have occurred at military facilities in the United States and in the former Soviet Union, in connection with the manufacture of nuclear weapons. Liquid high-level wastes stored at many U.S. facilities have leaked into the environment and contaminated wildlife, sediments, groundwater, and soil. Activities at these sites have been shrouded in secrecy; only recently have documents revealing past accidents and radioactive releases been declassified and made available to the public. Deliberate releases of uranium dust, xenon-133, iodine-131, and tritium gas have been documented at Hanford, Washington; Fernald, Ohio; Oak Ridge, Tennessee; and Savannah River, South Carolina. Cleaning up these sites is now the responsibility of the DOE, which has already spent $50 billion on this problem and estimates that the final cost could run as high as $250 billion, probably the largest environmental project the country will ever undertake.

Russian military weapons facilities have been even more irresponsible. The worst is a giant complex called Chelyabinsk-65, located in the Ural Mountains. For at least 20 years, nuclear wastes were discharged into the Techa River and then into Lake Karachay. At least 1,000 cases of leukemia have been traced to radioactive contamination from the Chelyabinsk facility. Even today, standing for one hour on the shore of Lake Karachay will cause radiation poisoning and death within a week. The lake dried up in the summer of 1967, and winds blew radioactive dust with 5 million curies across the countryside, contaminating hundreds of thousands of people. This lake is considered to be the most polluted lake on earth, a legacy of the Cold War and an enormous continuing source of radioactive contamination. Recognizing the dangers of spreading contamination, Russian authorities have filled the lake with hollow concrete blocks, rocks, and soil, and plants now spread their greenery over this permanently contaminated site.

The end of the Cold War has brought the welcome dismantling of nuclear weapons by the United States and the nations of the former U.S.S.R. Thousands of nuclear weapons are being dismantled, many as a result of agreements between President Clinton and Russian President Boris Yeltsin. The agreements include a joint closure of all remaining plutonium weapons production reactors by the year 2000, funds to aid Russia in destroying missile silos and dismantling nuclear submarines and bombs, and vast cuts in nuclear arsenals of the two former enemies. The radioactive components of the weapons—at least 100 tons of weapons-grade plutonium—must be handled with great care and disposed of where they are safe from illegal access and where they will pose no danger to human health for centuries to come—a daunting task. Already, there have been several attempts to smuggle weapons-grade plutonium. A suitcase containing 363 grams of plutonium and 200 grams of lithium-6 was confiscated in Munich in August of 1994. The closing down of much of the nuclear warfare enterprise is certainly a welcome development, but the problems of security and of disposal of the wastes will remain as a long-term legacy of the Cold War.

High-level Nuclear Waste Disposal. The United States and most other countries using nuclear power have decided on geologic burial for the ultimate consignment of nuclear wastes, but no nation has developed plans to the point of actually carrying out the burial. Many nuclear nations have not even been able to find a site that may be suitable for receiving wastes. Where sites have been selected, many questions about their safety have surfaced. The basic problem is that no rock formation can be guaranteed to remain stable and dry for tens of thousands of years. Everywhere scientists look, there is evidence of volcanic or earthquake activity or groundwater leaching within the last 10,000 years or so, which is to say that it may occur again in a similar period of time. If such events did occur, the still-radioactive wastes could escape into the environment and contaminate water, air, or soil, with consequent effects on humans and wildlife.

In the United States, efforts to locate a long-term containment facility, which have been going on for about 30 years, have been hampered by the *NIMBY* syndrome (*Not in My Back Yard*—see Chapter 19). A number of states, under pressure from citizens, have passed legislation categorically prohibiting the disposal of nuclear wastes within their boundaries. In the meantime, the need to select and develop a long-term repository has become increasingly critical. With the current lack of public support for nuclear power, it is proving extremely difficult for the United States or any other nation to establish a broadly acceptable long-term nuclear waste facility. (See "Ethics" essay, p. 354.)

At the end of 1987, Congress called a halt to the debate and arbitrarily selected a remote site, Yucca Mountain in southwestern Nevada (Fig. 14–12), to be the nation's civilian nuclear waste disposal site. No other sites are being considered. Not surprisingly, Nevadans have fought this

◀ **FIGURE 14–12** *Yucca Mountain, Nevada.* A shift changes at the tunnel entrance to the facility, which has been chosen as the nation's sole high-level nuclear waste depository.

ETHICS

SHOWDOWN IN THE NEW WEST

In a scenario reminiscent of the Civil War, two western states have become the focus of a controversy over states' rights. On one side is the federal government, in the form of the Department of Energy, which is trying to solve the problem of what to do with nuclear wastes and which thinks it has a reasonable solution. On the other side are politicians in the states designated as repository sites, who are responding to the voices of citizens of those states. At stake may be the future of nuclear energy in the United States.

The Department of Energy has constructed a $1.5 billion Waste Isolation Pilot Plant (WIPP) in salt caves 2,150 feet beneath the desert in southwestern New Mexico, 26 miles from the city of Carlsbad. The plant is on federal land and has passed all necessary safety reviews. It was ready to open for business in October 1991. "Business," in this case, was for the plant to become the repository for up to a million barrels of plutonium wastes from nuclear weapons plants and laboratories around the country, as well as for the plutonium from dismantled nuclear warheads. (No commercial nuclear wastes will go to the plant.)

New Mexico sued the Department of Energy on the grounds that it was in viola-

tion of federal law prohibiting the opening of such a repository without congressional approval. The state of Texas and four environmental groups joined New Mexico in bringing suit. In October 1992, Congress gave approval to the department to begin testing the facility's ability to store the plutonium wastes—with one significant catch: The EPA was ordered to strengthen the radiation standards protecting the public from the stored wastes. In 1998, after a lengthy review, the EPA gave the WIPP certification to begin receiving radioactive materials from nuclear weapons facilities around the country, and in 1999 the site began to receive wastes from a number of facilities. To date, the WIPP has received over 60 shipments without incident.

Opposition to the long-term storage of nuclear wastes is even stronger in Nevada, which has the dubious distinction of having been selected for the nation's only repository for commercial nuclear wastes, at Yucca Mountain. The site is a barren ridge in the desert about 100 miles northwest of Las Vegas (Fig. 14–12). In 1989, the Nevada legislature passed a bill making it unlawful for any agent or agency to store high-level radioactive waste in the

state. All sorts of surveys have been carried out to test the mood of Nevadans regarding the long-term storage of wastes in their state. The responses have been uniformly negative: Three-fourths of Nevadans agreed that the state should continue to do all it can to prevent the establishment of the Yucca Mountain site. Clearly, ordinary Nevadans perceive the risks of such a site as enormous and unacceptable, in contrast to the opinions of many scientists who have performed risk calculations regarding the site. In this case, however, the state's attempts to pose a legal blockade before the Department of Energy have failed: The Supreme Court has ruled that Nevada must process applications for permits to continue work on the site.

The dilemma is plain. Public and political support for any form of nuclear energy is at an all-time low, and people in these two western states are especially indignant at their states' becoming the repository of nuclear wastes from around the nation. Yet we need to do something with the legacy of nuclear energy other than pass the problem along to the next generation. What do you think should be done?

selection, passing a law in 1989 that prohibits anyone from storing high-level radioactive waste in the state. The federal government has the power to override state prohibitions, though, and the Nevada site is now being studied intensively. To date, the studies have shown that the storerooms, 1,000 feet (300 m) above current groundwater levels, will presumably be safe from invasion by groundwater. However, the rock formations above the storerooms have proven to be riddled with cracks, allowing a much more rapid permeation of rainwater than had previously been estimated. Water would speed up corrosion of the waste containers and subsequent leakage of radioactivity into groundwater. Furthermore, earthquakes or volcanic activity could change the current conditions in a short time, and the site lies in a zone of fault lines and has a history of volcanic activity only 5,000 years ago. The Yucca Mountain facility could begin receiving wastes from commercial facilities all over the country in the year 2010, although these plans are presently on hold, pending further scientific research. In the meantime, an underground storage facility in New Mexico, the Waste Isolation Pilot Plant, has begun to receive shipments

of wastes from nuclear weapons facilities. (See "Ethics" essay, this page.) Citing health and safety concerns, President Clinton vetoed a bill proposed by Congress in 2000 to speed up the approval process for Yucca Mountain.

The Nuclear Waste Policy Act of 1982 committed the federal government to begin receiving nuclear waste from commercial power plants in 1998. This has not happened, however, and the utilities have begun trying to develop interim storage facilities on their own, on Native American lands. One site in Mescalero Apache land in New Mexico has been dropped from consideration, while a site on the Skull Valley Goshute reservation in Utah remains under consideration, pending approval from the Nuclear Regulatory Commission. The interim storage approach has been successful in other countries. For example, Sweden's facility is located in a rock cavern, receives spent-fuel rods after they have been stored on-site for a year, and will hold the rods for 30 to 40 years, until they can be transported to a deep geologic repository. With the Nuclear Waste Policy Act of 1997, the 105th Congress made an attempt to establish a similar interim facility at Yucca Mountain, but, threatened with a presidential veto, the act did not make it through the Congress. Because of this failure, and the fact that electricity customers have paid more than $12 billion into a nuclear waste fund for the purpose of spent-fuel storage and disposal, the power companies brought suit against the Department of Energy. However, a federal appeals court declined to act on their suit, and the U.S. Supreme Court also turned down an appeal from the utilities. With the continuing stalemate on federal action to establish either an interim or long-term storage facility, costly on-site storage represents the only option available to the plants short of prematurely closing them down.

The Potential for Accidents

Prior to 1986, the scenario for a worst-case nuclear power plant disaster was a matter of speculation. Then, at 1:24 A.M. local time on April 26, 1986, events at a nuclear power plant in Ukraine made such speculation irrelevant (Fig. 14–13). Since that day, Chernobyl has served as a horrible example of nuclear energy gone awry.

Chernobyl. While conducting a test of standby diesel generators, engineers disabled the power plant's safety systems, withdrew the control rods, shut off the flow of steam to the generators, and decreased the flow of coolant water in the reactor. However, they did not allow for the radioactive heat energy generated by the fuel core, and, lacking coolant, the reactor began to heat up. The extra steam produced could not escape and had the effect of rapidly boosting the energy production of the reaction. In an attempt to quell the reactor, the engineers quickly inserted the carbon-tipped control rods. The carbon tips acted as moderators, slowing down the neutrons that were produced in the reaction. The neutrons, however, were still speedy enough to trigger more fission reactions, and the result was a split-second power surge to 100 times the maximum allowed level. Steam explosions then blew the 2,000-ton top off the reactor, whereupon the reactor melted down and a fire was ignited in graphite, which burned for days. At least 50 tons of dust and debris bearing 90 million curies of radioactivity in the form of fission products were released in a plume that rained radioactive particles over thousands of square miles.

As the radioactive fallout settled, 135,000 people were evacuated and relocated. The reactor was eventually sealed in a sarcophagus of concrete and steel. A barbed-wire fence now surrounds a 1,000-square-mile exclusion zone around the reactor site. The soil remains contaminated with radioactive compounds, yet for years 2,000 Ukrainian workers were bused in daily to work at the remaining reactor in the Chernobyl complex. In December, 2000, the last reactor was permanently shut down. Only two engineers were directly killed by the explosion, but 29 of the personnel brought in to contain the reactor in the aftermath of the explosion died of radiation in a few months. Over a broad area downwind of the disaster, buildings and roadways were washed down to flush away radioactive dust. Even with these precautions, many people in or near the evacuation zone were exposed to dangerous levels of radiation, especially the short-lived radioisotope iodine-131. Because iodine collects in the thyroid gland, the radioactive iodine has been responsible for great increases in the incidence of thyroid cancer in Ukraine and neighboring Belarus. (Over 1,500 cases have occurred since 1986.) The long-term effects are estimated to range from 140,000 to 475,000 cancer deaths worldwide from the accident.

Are we in the United States in danger of such an explosion occurring at a nuclear power plant? Nuclear scientists argue that the answer is no, because U.S. power plants have a number of design features that should make a repeat of Chernobyl impossible. The Chernobyl reactor used graphite as a moderator, rather than water, as in LWRs. Further, LWRs are incapable of developing a power surge more than twice their normal power, well within the designed containment capacity of the reactor vessel. Finally, LWRs have more backup systems to prevent the core from over-

▲ **FIGURE 14–13** *Chernobyl.* An aerial view of the Chernobyl reactor on April 29, 1986, three days after the explosion. The disaster was the worst nuclear power accident, directly killing at least 31 people and putting countless numbers of people in the surrounding countryside at risk for future cancer deaths.

heating, and the reactors are housed in a thick concrete-walled containment building designed to withstand explosions such as the one that occurred in the Chernobyl reactor, which had no containment building.

However, LWRs are not immune to accidents, the most serious being a complete core meltdown as a result of a total loss of coolant. This has never happened, but there was a close call at Three Mile Island.

Three Mile Island.   On March 28, 1979, the Three Mile Island nuclear power plant near Harrisburg, Pennsylvania, suffered a partial meltdown as a result of a series of human and equipment failures and a flawed design. The steam generator (Fig. 14–8) shut down automatically because of a lack of power in its feedwater pumps, and eventually a pressure valve on top of the generator opened in response to the ensuing buildup of pressure. Unfortunately, the valve remained stuck in the open position and drained coolant water from the reactor vessel. There were no sensors to indicate that this pressure-operated relief valve was open. Operators responded poorly to the emergency, shutting down the emergency cooling system at one point and shutting down the pumps in the reactor vessel. One instrument error compounded the problem: Gauges told operators that the reactor was full of water when, actually, it needed water badly. The core was uncovered for a time and suffered a partial meltdown. Ten million curies of radioactive gas were released into the atmosphere.

The drama held the whole nation—particularly the 300,000 residents of metropolitan Harrisburg who were poised for evacuation—in suspense for several days. The situation was eventually brought under control, and no injuries or deaths occurred, but it would have been much worse if meltdown were complete. The reactor was so badly damaged, and so much radioactive contamination occurred inside the containment building, that the continuing cleanup is proving to be as costly as building a new power plant. There are no plans to restart the reactor. GPU Nuclear, operators of the plant, have since paid out $30 million to settle claims from the accident, although the company has never admitted any radiation-caused illnesses.

The real cost of the accidents at Three Mile Island and Chernobyl must be reckoned in terms of public trust. Public confidence in nuclear energy already was declining, but it plummeted after the two accidents, which pointed to human error as a highly significant factor in nuclear safety—and human error is something the public understands well. Nuclear proponents suffered a serious loss of credibility with Three Mile Island, but Chernobyl was their worst nightmare come true—a full catastrophe, just as the antinuclear movement had predicted might someday occur.

Safety and Nuclear Power

As a result of Three Mile Island and other, lesser incidents in the United States, the Nuclear Regulatory Commission has upgraded safety standards not only in the technical design of nuclear power plants, but also in maintenance procedures and in the training of operators. Thus, proponents contend, nuclear plants were designed to be safe in the beginning, and now they are safer than ever. It is estimated that as a result of new procedures instituted after the accident at Three Mile Island, nuclear plants are now six times safer than before. Some proponents of nuclear power claim that we now have the technology to build *inherently safe* nuclear reactors, designed in such ways that any accident would result in an automatic quenching of the chain reaction and suppression of heat from nuclear decay. In reality, however, there is no such thing as an inherently safe reactor, since the concept implies no release of radioactivity under any circumstances—an impossible expectation.

Instead, nuclear scientists are proposing a new generation of nuclear reactors with built-in passive safety features, rather than the *active safety* features found in current reactors. **Active safety** relies on operator-controlled actions, external power, electrical signals, and so forth. As accidents have shown, operators may override such safety factors, and electricity, valves, and meters can fail or give false information. **Passive safety**, on the other hand, involves engineering devices and structures that make it virtually impossible for the reactor to go beyond acceptable levels of power, temperature, and radioactive emissions.

In anticipation of a possible renewed interest in nuclear power, engineers are now designing LWRs that include passive safety features and much simpler, smaller power plants—the so-called advanced light-water reactors (ALWRs). For example, one passive safety feature would be to position a cold-water reservoir so that, in the event of a LOCA, the water would drain by gravity to the reactor core (Fig. 14–14). In addition, the design will make it impossible for operators to inactivate the passive safety systems. Another feature being planned is to build reactors as modular units small enough to conduct heat from nuclear decay outward into the soil; in this manner, the reactors could not incur a core meltdown. One such modular reactor, built in Germany, suffered no damage to its core when tested against a LOCA.

In East Asia, the nuclear reactor design of choice is the ALWR. In Japan, 53 nuclear plants are in operation, and there are plans for 36 more; South Korea has 16 operating plants, with 14 more in the construction or planning stage. The new plants are designed to last for 60 years, can be constructed within 5 years, and are simple enough that a single operator can control them under normal conditions. They generate electricity at a cost competitive with fossil-fuel options.

One motive behind the new designs is to restore the public's confidence in nuclear energy. Previously, nuclear proponents had emphasized the very low probabilities of accidents; as we have seen, though, improbable events can happen, and when they happen to nuclear power plants, the consequences can be dreadful. The new emphasis is on convincing the public of the fundamental safety of ALWR

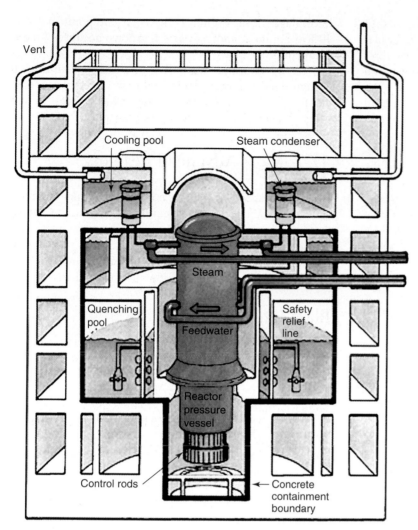

◄ FIGURE 14-14 *Advanced light-water reactor.* The core is surrounded by three concentric structures: a reactor pressure vessel, in which heat from the reactor directly boils water into steam; a concrete chamber (outlined with heavy black line) and water pool, which together contain and quench steam vented from the reactor in an emergency; and a concrete building, which acts as a secondary containment vessel and shield. Any excessive pressure in the reactor will automatically open valves that release steam into a quenching pool, reducing the pressure. Water from the quenching pool can, if necessary, flow downward to cool the core. Evaporation from a pool on top of the containment building limits the buildup of containment pressure. (From "Advanced Light-Water Reactors," by M. W. Golay and N. E. Todreas. Copyright © 1990 by Scientific American, Inc. All rights reserved.)

designs through demonstrations and explanations that non-technical people can understand.

Economic Problems with Nuclear Power

As Fig. 14-2 shows, in the United States, utilities already were turning away from nuclear power considerably before the disaster at Chernobyl. The reasons are mainly economic.

First, projected future energy demands were overly ambitious; a slower growth rate in the demand for electricity postponed orders for all types of power plants. Second, increasing safety standards for the construction and operation of nuclear power plants have caused the costs of plants to increase at least fivefold, even after inflation is considered. Adding to the rise in costs is the withdrawal of government subsidies to the nuclear industry. Third, public protests frequently have delayed the construction or start-up of a new power plant. Such delays increase costs still more, because the utility is paying interest on its investment of several billion dollars even when the plant is not producing power. As these costs are passed on, consumers become yet more disillusioned with nuclear power. Finally, safety systems may protect the public, but they do not prevent an accident from

financially ruining the utility. Since radioactivity prevents straightforward cleanup and repair, an accident can convert a multibillion-dollar asset into a multibillion-dollar liability in a matter of minutes, as Three Mile Island demonstrated. Thus, nuclear power involves a financial risk that utility executives are reluctant to take.

A major problem facing nuclear power is the rising tide of deregulation of the U.S. electric industry. (See Chapter 13.) Still undecided is what will be done about "stranded cost recovery," meaning the repayment of unpaid costs, like that connected with the construction of power plants, to the investors. This could be a real concern to nuclear power plant owners if customers choose other electricity suppliers because of cost or philosophical considerations. The Federal Energy Regulatory Commission has ruled that these unpaid costs may be recovered in order to achieve a successful transition to the competitive environment that deregulation is designed to create.

Another factor that promises to increase the cost of nuclear-generated electricity is a shorter-than-expected lifetime for nuclear power plants. Originally, it was thought that nuclear plants would have a lifetime of about 40 years. It now appears that their lifetime will be considerably less.

Worldwide, more than 60 nuclear plants have been shut down after an average lifetime of 17 years. This shorter lifetime substantially increases the cost of the power produced, because the cost of the plant must be repaid over a shorter period.

The lifetimes of nuclear power plants are shorter than originally expected because of two problems: *embrittlement* and *corrosion*. **Embrittlement** occurs as neutrons from fission bombard the reactor vessel and other hardware. Gradually, this neutron bombardment causes the metals to become brittle enough that they may crack under thermal stress—for example, when emergency coolant waters are introduced in the event of a LOCA. When the reactor vessel becomes too brittle to be considered safe, the plant must either be shut down or be repaired at great cost.

Corrosion is a normal consequence of steam generation. Very hot, pressurized water flows from the core into the steam generator through thousands of 3/4-inch-diameter pipes immersed in water (Fig. 14–8). The water inside and outside these pipes contains corrosive chemicals that, over time, cause cracks to develop in some of the pipes. If the main line conveying steam from the generator to the turbine were to rupture, the sudden increase in pressure in the generator could cause several cracked pipes to break at once. In that case, radioactive moderator–coolant water would be released and would overload safety systems, forcing the plant to vent radioactive gas to the outside. Cracked pipes are "repaired" basically by plugging them; in 1993 alone, over 10,000 worn or cracked pipes were plugged in pressurized LWRs, reducing the efficiency of the reactors and cutting their overall power. In March 1995, using a new high-tech probe, officials discovered that up to half of the steam generator pipes in the Maine Yankee plant had developed cracks, some penetrating 80% of the pipe's thickness. As a result, the plant was shut down after only 24 years of operation, and the owners of 70 other pressurized water nuclear reactors were notified by the Nuclear Regulatory Commission that they, too, could be facing similar cracking problems.

Closing down, or decommissioning, a power plant can be extremely costly. The estimated price tag for decommissioning Yankee Rowe, a small nuclear plant by current standards, is $368 million, $56 million of which is being used to construct a temporary storage facility until locations are found to receive the high-level and low-level wastes resulting from the decommissioning. Faced with these costs, some utilities are opting to repair older plants in spite of the high costs of such repairs.

Overseas, nuclear power is also facing hard economic times in some regions. The most recent nuclear reactor to be built in the United Kingdom (and possibly the last for decades), Sizewell B, cost almost $3,000 per kilowatt of capacity, almost 10 times as much as a gas-fired plant. Government subsidies keep the French and Japanese from paying what would be very high prices for electricity. Technical problems similar to those mentioned earlier have plagued the French reactors; at one point in 1991, 40% of France's reactors were shut down because of technical faults.

Yet, 31 countries are still committed to maintaining current nuclear power plants, and many are building new plants. Are there some new technologies waiting in the wings that could solve some of the many problems?

14.4 More Advanced Reactors

Uranium—especially ^{235}U—is not a highly abundant mineral on Earth. At the height of optimism about nuclear energy in the 1960s, when as many as 1,000 plants were envisioned by the turn of the century, it was forecast that shortages of ^{235}U would develop. Breeder reactors, which utilize chain reactions, were seen as the solution to this problem.

Breeder Reactors

Recall that when a ^{235}U atom fissions, two or three neutrons are ejected. Only one of these neutrons hitting another ^{235}U atom is required to sustain a chain reaction; the others are absorbed by something else. The breeder reactor is designed so that (nonfissionable) ^{238}U absorbs the extra neutrons. When this occurs, the ^{238}U is converted to plutonium (^{239}Pu), which is fissionable. Then, ^{239}Pu can be purified and used as a nuclear fuel, just like ^{235}U. Thus, the breeder converts nonfissionable ^{238}U into a useful nuclear fuel. Consequently, since there are generally two neutrons in addition to the one needed to sustain the chain reaction, the breeder may produce more fuel than it consumes. Because 99.7% of all uranium is ^{238}U, converting this to ^{239}Pu through breeder reactors effectively increases nuclear fuel reserves over a hundredfold.

Breeder reactors present all of the problems and hazards of standard fission reactors, plus a few more. If a meltdown occurred in a breeder, the consequences would be much more serious than in an ordinary fission reactor, because of the large amounts of ^{239}Pu, which has an exceedingly long half-life of 24,000 years. In addition, because plutonium can be purified and fabricated into nuclear weapons more easily than ^{235}U can, the potential for the diversion of breeder fuel to weapons production is greater. Hence, the safety and security precautions needed for breeder reactors are greater.

With its scaled-down nuclear program, the United States currently has enough uranium stockpiled. Thus, there is no urgency for the United States to develop breeders. However, in the United States and elsewhere, small breeder reactors are operated for military purposes. France and Japan are currently the only nations with commercial breeder reactors—the Superphénix and the Monju FBR, respectively. With no oil fields, these countries are determined to achieve energy independence with nuclear power. French officials point out that 1 gram of plutonium equals a ton of oil; together with a new waste fuel reprocessing facility, the French can now produce 16 tons of plutonium a year.

Fusion Reactors

The vast energy emitted by the Sun and other stars comes from fusion (Fig. 14–5b). The Sun, as well as other stars, is composed mostly of hydrogen. Solar energy is the result of fusion of this hydrogen into helium. Scientists have duplicated the process in the hydrogen bomb, but hydrogen bombs hardly constitute a useful release of energy. The aim of fusion technology is to carry out fusion in a controlled manner in order to provide a practical heat source for boiling water to power steam turbogenerators.

Since hydrogen is an abundant element on Earth (there are two atoms of it in every molecule of water) and helium is an inert, nonpolluting, nonradioactive gas, hydrogen fusion is promoted as the ultimate solution to all our energy problems—that is, pollution-free energy is produced from a virtually inexhaustible resource: water. However, most current designs do not use regular hydrogen (^1H), but rather, employ the isotopes deuterium (^2H) and tritium (^3H) in what is called the d–t reaction. Deuterium is a naturally occurring nonradioactive isotope and can be extracted in almost any desired amounts from the hydrogen in seawater. Tritium, in contrast, is an unstable gaseous radioactive

isotope that must be produced artifically. Radioactive and difficult to contain, tritium is a hazardous substance. As a result, fusion reactors could easily become a source of radioactive tritium leaking into the environment, unless effective (and costly) designs prevent leaks.

In the present state of the art, fusion power is still an energy consumer rather than a producer. The problem is that it takes an extremely high temperature—some 3 million degrees Celsius—and pressure to get hydrogen atoms to fuse. In the hydrogen bomb, the temperature and pressure are achieved by using a fission bomb as an igniter—an unthinkable way to initiate a sustained, controlled fusion reaction.

A major technical problem is how to contain the hydrogen while it is being heated to the tremendously high temperatures required for fusion into helium. No known material can withstand these temperatures without vaporizing; however, three techniques are being tested. One is the Tokamak design, in which ionized hydrogen is contained within a magnetic field while being heated to the necessary temperature. The second is laser fusion, in which a tiny pellet of frozen hydrogen is dropped into a bull's-eye, where it is hit simultaneously from all sides by powerful laser beams (Fig. 14–15). The beams simultaneously heat and pressurize the pellet to the point of fusion. A third design is a device called the Z machine, which delivers very short, but powerful, pulses of current through an array of wires. This

▼ **FIGURE 14–15** *Laser fusion.* In this experimental instrument at Lawrence Livermore Laboratory, 30 trillion watts of optical power are focused onto a tiny pellet of hydrogen smaller than a grain of sand and located in the center of a vacuum chamber. For less than a billionth of a second, the fusion fuel is heated and compressed to temperatures and densities near those found in the Sun.

concentrated electrical energy results in a blast of X-rays that recently reached 1.8 million degrees Celsius, close to the 2 to 3 million degrees it would take to initiate fusion.

Some fusion has been achieved in the first two types of devices, and in late 1994 a Tokamak facility at the Princeton Plasma Physics Laboratory in Princeton, New Jersey, achieved 10.7 megawatts of fusion power in 0.27 second of reaction. As yet, however, the break-even point has not been reached: More energy is required to run the magnets or lasers than is obtained by fusion. (It took 39.5 megawatts of energy to sustain the reaction at Princeton.) The most optimistic workers in the field believed that, with sufficient money for research—about $10 billion—the break-even point might be reached in the late 1990s. Even had this goal been achieved, however, it is still a long way from a practical commercial fusion reactor power plant. In the United States, $25 billion has already been spent on fusion energy research.

Recently, Japanese and European researchers presented a design for an International Thermonuclear Experimental Reactor (ITER), a prototype fusion reactor of the Tokamak variety that is still in the planning stage. The design is expected to cost $3 billion, and the reactor will produce 10 times as much energy as it consumes. The most optimistic view sees the ITER coming on line by 2013. As this plan indicates, developing, building, and testing a fusion power plant will require at least another 10 to 20 years and many billions of dollars. Additional plants would require additional years. Thus, fusion is, at best, a very long-term option. Many scientists believe that fusion power will always be the elusive pot of gold at the end of the rainbow. The standard joke about fusion power is that it is the energy source of the future and always will be!

14.5 The Future of Nuclear Power

With fossil fuels threatening to bring on global climate change and alternatives such as solar power (Chapter 15) only in their infancy, the long-term outlook for energy is discouraging. Using nuclear power seems to be an obvious choice; however, many are opposed to nuclear energy, and it is presently the most costly option. Indeed, worldwide, public opposition to nuclear energy is higher now than it ever has been and raises the question of whether nuclear energy has any future once the currently operating plants have exhausted their life spans. In most countries of the European Union, nuclear power has a dim future. Sweden and Germany have agreed to phase out their nuclear power plants, even though nuclear power provides 48% and 33% of their electricity, respectively. Russia and other eastern European countries are holding on to their nuclear power as a present economic necessity. Only in Asia could it be said that nuclear power is still being developed: China, Japan, South Korea, and India seem intent on pursuing a greater nuclear share of energy development.

Opposition

Opposition to nuclear power is based on several premises:

- People have a general distrust of technology they do not understand, especially when that technology carries with it the potential for catastrophic accidents or the hidden, but real, capacity to induce cancer. (See the "Earth Watch" essay, p. 361.)

- Many observers are critical of the way nuclear technology is being managed. They are aware that the same agency (the Nuclear Regulatory Commission) responsible for licensing and safety regulations is also a strong supporter of the commercial nuclear industry. Their concern is warranted: The NRC has come under fire recently in connection with the Millstone nuclear plants in Connecticut. Believing that safety rules were routinely being ignored in the facility, two employees tried to get corrective action from the plant's managers. The managers balked, and it then took three years for the employees to get action from the NRC. During that time, they were subjected to intimidation from their superiors. When the NRC finally stepped in, it became evident that the commission had been routinely waiving safety rules for years in order to let nuclear plants avoid costly shutdowns. The Millstone plant was shut down in 1996 so that it could address a host of safety violations.

- Similar problems involving lax safety, operator failures, and coverups by nuclear plants and their regulatory agencies have appeared in Canada and, as we have seen, Japan.

- The problems of high costs of construction and unexpectedly short operational lifetimes have already been mentioned. Thus, the economic argument can also be used to oppose nuclear power.

- The nuclear industry has repeatedly presented nuclear energy as extremely safe, using arguments based on the low probabilities of accidents occurring. However, when accidents do occur, probabilities become realities, and the arguments are moot.

- There remains the crucial problem of disposing of nuclear waste. All parties agree that the waste must be placed somewhere safe, but there the agreement ends, and siting a long-term nuclear waste repository has become an apparently intractable political as well as technological problem in country after country.

Completely aside from these objections, there is the basic mismatch between nuclear power and the energy problem. As we have emphasized, the main energy problem for the United States involves an eventual shortage of crude oil for transportation purposes, yet nuclear power produces electricity, which is not used for transportation. If we were moving toward a totally electric economy that included even electric cars, nuclear-generated

EARTH WATCH

RADIATION PHOBIA?

It is a familiar scene to TV viewers: Angry people with placards picket a nuclear power plant, and some are dragged off by police for obstructing access to the plant. The protesters are convinced that nuclear power is dangerous and a threat to health, even when a plant is operating normally. Often, the ranks of the protesters include medical doctors and research scientists, along with the more familiar members of environmental protest organizations. What is it about nuclear power that ignites such strong feelings? David Willis, professor emeritus of radiation biology at Oregon State University, believes that many protesters are motivated by fear of something they do not understand—that they in fact have a *radiation phobia*. This fear is made worse by the way the media portray radiation. For example, the word *radiation* is very often preceded by the adjective *deadly*. Yet only a few people suffer ill effects from radiation, and those effects are almost always due to occupational or medical accidents involving carelessness in handling sources of radiation.

Perhaps we are afraid because we are not able to perceive radiation by our normal senses of smell, taste, touch, vision, and hearing. We hear that there is radiation all around us in the form of cosmic rays—the background radiation—but there is nothing we can do about it, and life seems to go on. Willis points out that, yes, ionizing radiation can induce cancer in humans, but the kinds of cancer that are induced are identical to cancers that arise spontaneously from all sorts of causes, mostly unknown. The higher the dose of radiation, the greater is the tendency to develop cancer. As the dose declines, however—falling to levels that might be an occupational hazard for some lines of work—radiation effects simply disappear in the statistical data showing the development of cancer in a given population.

To illustrate his point, Willis asks us to consider an airplane descending into a fog bank as it comes in to land. Above the fog, the airplane's descent is quite visible to another airplane, but when the craft descends into the fog, it disappears from view even though it is still there. According to Willis, the airplane is analogous to the cancers induced from ionizing radiation, its descent

represents the lessening number of cancers induced by diminishing doses of radiation, and the fog bank represents the cancers normally occurring from a variety of causes or spontaneously. Any effects of low levels of radiation are completely masked by the large fog of cancers that occur naturally. If some cancers are induced by low-level radiation, the number is vanishingly small compared with the number of naturally occurring cancers.

Willis asserts that many opponents of nuclear power are afraid of something that actually carries a very low risk, especially when compared with other hazards in our environment. In spite of numerous studies, no linkage has ever been shown between the incidence of cancer and the presence of a nuclear facility, except for the Chernobyl accident. Willis has debated the issue with opponents of nuclear energy and finds that many of them will not listen to a reasoned explanation of the health effects of ionizing radiation and the demonstrably low risks involved with nuclear power. The result is apparent: Nuclear power is on the wane and may never recover.

electricity could be substituted for oil-based fuels. Unfortunately, electric cars have not yet proved practical, and the outlook for them in the near future remains uncertain. Consequently, nuclear power simply competes with coal-fired power plants in meeting the demands for baseload electrical power. The fact is that, given the high costs and additional financial risks of nuclear power plants, coal is cheaper, and the United States does have abundant coal reserves.

However, there are still the environmental problems of mining and burning coal, including acid precipitation and global climate change. As we saw in Chapter 13, coal is the worst choice in terms of CO_2 emissions per unit energy produced. If the long-term environmental costs were factored into the price of coal, we would find that price considerably higher than it is now. Of course, costs such as that of the long-term confinement of nuclear wastes and that of the decommissioning of power plants have yet to be factored into the cost of nuclear power.

At the bottom line, we find that one of the most unique energy options, nuclear power, is also a feared technology. The public simply does not trust the safety of the reactors

and the plans for waste disposal. Opponents believe that the technology poses long-term risks to both human health and the environment and will in no way solve our most critical energy problem: an impending shortage of fuel for transportation. Also, although the United States has relatively abundant reserves of uranium, it is not a renewable resource. Opting to meet the energy problem by all-out exploitation of this resource would assuredly lead to another energy crisis and increased imports, unless we go the route of breeder reactors.

Rebirth of Nuclear Power?

What would it take to revitalize the nuclear power option? If nuclear energy is to have a brighter future, it may be because we have found the continued use of fossil fuels to be so damaging to the atmosphere that we have placed limits on their use, but have not been able to successfully develop adequate alternative energy sources. Nuclear supporters point out that if worldwide nuclear plants were to be replaced by fossil-fuel power plants, CO_2 emissions would rise instantly by more than 8%. Observes agree that if the rebirth of nuclear power is to come, a number of changes will have to be made:

- Reactor safety concerns will have to be addressed, perhaps through the promotion of only smaller, advanced light-water reactor designs with built-in passive safety features.
- The industry's manufacturing philosophy will have to change to favor standard designs and "mass production" of the smaller reactors instead of the custom-built reactors presently in use in the United States. Recently, the Department of Energy selected two designs for future development, one of which—the Westinghouse advanced pressurized-water reactor—is a 600-megawatt unit similar to the one pictured in Fig. 14–14. The unit, designed for modular construction, is based primarily on passive safety. Westinghouse claims that it would take no more than three years to build such a unit.
- The framework for licensing and monitoring reactors must be streamlined, but without sacrificing safety concerns. This has largely been accomplished with the National Energy Policy Act of 1992.

- The political problems of siting new reactors must be resolved; perhaps the best sites will be on the grounds of present or closed reactors, taking advantage of the existing infrastructure and familiarity with (and acceptance of) a nuclear facility.
- The waste dilemma must be resolved, perhaps by lowering what some deem the unrealistic policy demands that the site must somehow be proven safe for the next 100,000 years.
- Political leadership will be required to accomplish all these developments. Currently, promoting nuclear power would be more of a liability than an asset for any politician who would forge a policy for our energy future, because of the general public mistrust of nuclear energy. Perhaps only a broad-based change in the public perception of nuclear power would bring the necessary political leadership to the surface, but it is hard to picture such a change happening, given current nuclear economics and unresolved problems. Yet, in light of the uncertainties about our energy future, many observers believe that it is only wise to keep the nuclear option alive. It remains to be seen whether this will be possible.

ENVIRONMENT ON THE WEB

COMMUNITY PERSPECTIVES IN THE SITING OF NUCLEAR FACILITIES

Cheap, clean, and plentiful—the promise of nuclear energy seems unlimited. However, in the eyes of some people, the risks also seem huge. Whether or not those dangers are real, managing public perception of risks (discussed further in Chapter 16) like these can be paramount in resolving environmental disputes.

Study after study has demonstrated that, overall, nuclear energy is cleaner and safer than fossil-fueled power generation. But although the frequency of nuclear accidents is very low, the consequences of failure can be enormous, as Three Mile Island and Chernobyl have shown. Therefore, it is not surprising that people worry about the risks of nuclear power.

Working with how the public perceives risk has now become a central problem in siting nuclear-generating and waste-disposal facilities. Although the EPA predicted that over a hundred nuclear waste facilities would be needed in the 1980s, very few were actually built. Over the past 20 years, the public has become increasingly knowledgeable about environmental risks and impacts—and more skeptical about the government's ability to make good choices. People are also more willing to be active in opposing projects, like nuclear facilities, that they believe are unacceptable. Yet, nuclear power is still a viable energy option in many cases. How can the problem of public perception of risk be overcome?

Recent research has shown that the most successful siting of nuclear facilities occurred when there was a strong atmosphere of trust between the proponent of the facility and the people in the community where the site was proposed. Such a process can also be used to resolve issues of how people perceive risk. For example,

by beginning with a community consensus that current methods of nuclear waste disposal are unacceptable, the community can then proceed to examine any positive or negative changes that would result from building the facility. In the most successful cases, the public is also involved in the development of monitoring and emergency-response programs. Thus, trust is achieved through a broad public consultation process and the development of public consensus that the proposed facility is in fact the best way to generate energy or manage the wastes.

Web Explorations

The Environment on the Web activity for this essay describes some ways that humans can anticipate and manage the negative consequences of nuclear energy. Go to the Environment on the Web activity (select Chapter 14 at **http://www.prenhall.com/wright**) and learn for yourself:

1. how new nuclear technology may improve the feasibility of the nuclear option;
2. about the forces that drive growth in energy consumption; and
3. about the ethics of compensating communities for accepting risk.

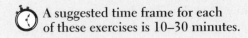 **A suggested time frame for each of these exercises is 10–30 minutes.**

REVIEW QUESTIONS

1. Compare the current outlook regarding the use of nuclear power with that of the 1960s and 1970s.
2. Describe how energy is produced in a nuclear reaction.
3. How do nuclear reactors and nuclear power plants work?
4. What environmental advantages does nuclear power offer over coal energy?
5. How are radioactive wastes produced, and what are the associated hazards?
6. Describe the two stages of nuclear waste disposal.
7. What problems are associated with the long-term containment of nuclear waste?
8. Describe what happened at Chernobyl and Three Mile Island.
9. What features might make a nuclear power plant inherently safe?
10. Discuss economic reasons that have caused many utilities to opt for coal-burning, rather than nuclear-powered, plants.
11. How do breeder and fusion reactors work? Does either one offer promise for alleviating our energy shortage?
12. Explain why nuclear power does little to address our shortfall in crude-oil production.
13. Discuss changes in nuclear power that might brighten its future.

THINKING ENVIRONMENTALLY

1. Discuss the advantages and disadvantages of nuclear power. Are we overly concerned or not concerned enough about nuclear accidents? Could an accident like Chernobyl happen in the United States?
2. Would you rather live next door to a coal-burning plant or a nuclear power plant? Defend your choice.
3. Alvin Weinberg, one of the scientists who developed the first fission reactor, believes that we have no choice other than to opt for nuclear power. He believes that in order to maintain ourselves at our present numbers and affluence, we must commit ourselves to nuclear power and the proper operation of nuclear plants. Do you agree or disagree? Why?
4. After reading the "Ethics" essay entitled "Showdown in the New West" (p. 354), what do you think should be done with our nuclear waste?

WEB REFERENCES

On-line resources for this chapter are on the World Wide Web at: **http://www.prenhall.com/wright**. (Click on Chapter 14 within the Chapter Selector.)

Renewable Energy

Key Issues and Questions

1. The total amount of solar energy reaching Earth is enormous. What is the potential for harnessing this energy?
2. Solar water-heating and space-heating systems for buildings represent well-developed technologies. What is preventing the more widespread adoption of these forms of solar heating?
3. Solar energy collected via photovoltaic cells and solar-trough collectors is used to produce electrical power. What are the current applications of these technologies, and what is their promise for the future?
4. There is a great need to develop solar-energy sources that can be coupled to fuel for transportation. What is the potential for fueling vehicles from solar hydrogen production?
5. Water, fire, and wind have provided energy for centuries. What are sustainable ways of expanding these options in the near future?
6. The options for energy into the 21st century are many. In light of global climate change, moving in the direction of renewable energy sources seems essential. What is in the way of such a move, and what is being done now to move in that direction?

One of the main roads in Belize, Central America, the Hummingbird Highway runs southeasterly from the capital of Belmopan, through second-growth rain forests and orange orchards, to Dandriga on the coast. About 15 miles down the highway from Belmopan, a small dirt road leads into Jaguar Creek, an environmental research and education center whose mission is to serve the Earth and the poor through sustainable development strategies. With 14 buildings occupying nearly 10,000 square feet, Jaguar Creek accommodates up to 40 people who might be there for short-term conferences or semester-long college courses. Belize is a developing country, and rural electrification is one facet that is still developing. Many miles from any electrical power, Jaguar Creek carries on with refrigeration, lighting, washing machines, and fans in every living space. This "off-the-grid" facility has its own solar electrical system, which covers the roof of the operations building (Fig. 15–1) and which is capable of generating 50,000 watts (50 KW) of power per day. Backup for rainy periods is provided by a propane-powered generator. The total cost for the system, including installa-

tion, was $100,000, which breaks down to about $5.00 per watt, and the system has a life expectancy of at least 25 years. Compared to the cost of a utility system with poles, wires, transformers, and the power supply, the price is a bargain. Local utility service is estimated to cost at least $25,000 per mile, and Jaguar Creek is 10 miles from the nearest power line. This sustainable energy system is consistent with Jaguar Creek's Earth-friendly approach, which also includes composting toilets and gray-water leaching fields.

Similar panels of solar cells are providing electricity in both developed and developing countries around the world. Throughout Israel and other countries in warm climates, it is now commonplace to have water heated by the Sun (Fig. 15–2). Even in temperate climates, many people have discovered that standard furnaces can be made obsolete with suitable designs to utilize the Sun's warmth. In the desert northeast of Los Angeles are "farms" with rows of trough-shaped mirrors tipped toward the Sun (Fig. 15–3). These reflectors are focusing the Sun's rays to boil water or synthetic oil and drive turbogenerators. In the

◀ **Wind power in Vermont.** Green Mountain Power Corporation installed 11 wind turbines in Searsburg, Vermont, in 1997, making Searsburg the site of the largest wind power–generating station in the eastern United States. On moderately windy days, the turbines provide enough power to satisfy the needs of 2,000 homes.

▶ **FIGURE 15–1** *Jaguar Creek.* Panels of photovoltaic cells power the Jaguar Creek facility operated by TargetEarth in Belize. The system generates 50 KW of energy a day, enough to power this facility, which houses up to 40 people.

hills east of San Francisco, regiments of "windmills" (more properly called *wind turbines*) standing in rows up the slopes and over the crests of the hills are producing electrical power equivalent to that produced by a large coal-fired power plant. Wind turbines are also becoming commonplace across northern Europe and are sprouting up as well in India, Mexico, Argentina, New Zealand, and other countries around the world.

The inertia of "business as usual" has kept the world on a track of growing dependence on fossil fuels, with nuclear power retaining its share of the energy "pie" for the present. In the meantime, however, as the preceding examples demonstrate, the use of energy from the Sun and wind has been making quiet, but steady, progress, to the point where these renewable sources of energy are becoming cost competitive with traditional energy sources—and

▲ **FIGURE 15–2** *Rooftop hot water.* In warm climates, solar hot-water heaters are becoming commonplace. Note the solar water-heating panels on each of the homes in this development in southern California.

◀ **FIGURE 15–3** *Solar thermal power in southern California.* Sunlight striking the parabolic shaped mirrors is reflected onto the central pipe, where it heats a fluid that is used, in turn, to boil water and drive turbogenerators.

in many situations are far more practical. Thus, the world has at its disposal the potential to move toward a nonpolluting, inexhaustible energy economy based on using the Sun's current energy output. In short, we have the capability of abiding by the first principle of sustainability—running on solar energy.

Our objective in this chapter is to give you a greater understanding of the potential for sustainable energy from sunlight, wind, biomass, and other sources. We will address energy policy at the end of the chapter as we compare the pros and cons of fossil fuels, nuclear power, and sustainable energy sources. Currently, renewable energy provides 7.6% of U.S. primary energy (Fig. 15–4); worldwide, renewable energy also provides 7.6% of energy use.

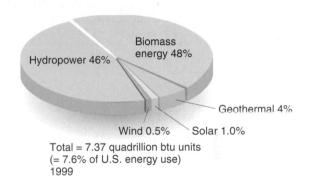

Total = 7.37 quadrillion btu units
(= 7.6% of U.S. energy use)
1999

▲ **FIGURE 15–4** *Renewable-energy use in the United States.* A mix of sources of renewable energy provided 7.6% of the nation's energy use in 1999. (*Source:* Data from Department of Energy, Energy Information Agency, *Renewable Energy Annual 1999*, March 2000.)

15.1 Principles of Solar Energy

Before turning our attention to the practical ways we capture and use solar energy, let us consider some general parameters associated with solar energy.

Solar energy originates with thermonuclear fusion reactions occurring in the Sun. (Importantly, all the chemical and radioactive polluting by-products of the reactions remain behind on the Sun.) The solar energy reaching Earth is radiant energy, ranging from ultraviolet light, which is largely screened out by ozone in the stratosphere, to visible light and infrared (heat energy) (Fig. 15–5).

The total amount of solar energy reaching Earth is vast almost beyond belief. Just 40 minutes of sunlight striking the land surface of the United States yields the equivalent energy of a year's expenditure of fossil fuel. If less than one-tenth of 1% of the Earth's surface were dedicated to collecting solar energy, that area alone could supply the electricity needs of the world.

Moreover, utilizing some of this solar energy will not change the basic energy balance of the biosphere. Solar energy absorbed by water or land surfaces is converted to heat energy and eventually lost to outer space. Even the fraction that is absorbed by vegetation and used in photosynthesis is ultimately given off again in the form of heat

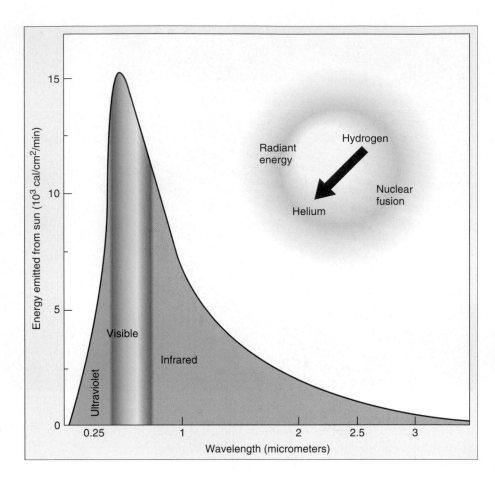

▶ FIGURE 15–5 *The solar-energy spectrum.* Equal amounts of energy are found in the visible-light and the infrared regions of the spectrum.

energy as various consumers break down food (Chapter 3). Similarly, if humans were to capture and obtain useful work from solar energy, it would still ultimately be converted to and lost as heat in accordance with the second law of thermodynamics. (See page 60.) The overall energy balance would not change.

15.2 Putting Solar Energy to Work

Solar energy is a diffuse (widely scattered) source with an intensity of about 1 KW per square meter. The main problem associated with using solar energy is one of taking a diffuse source and concentrating it into an amount and form, such as fuel or electricity, that we can use as heat and to run vehicles, appliances, and other machinery. Then there is the obvious other problem: What does one do when the Sun is not shining? These problems may be restated as ones involving the collection, conversion, and storage of solar energy. Also, in the final analysis, overcoming these hurdles must be cost effective.

Consider how natural ecosystems handle the three aspects. Leaves collect light over a wide area. Photosynthesis converts and stores light energy as chemical energy—namely, glucose and other forms of organic matter—in the plant. The plant biomass then "fuels" the rest of the ecosystem. In the sections that follow, you will see how we can overcome

or, even better, sidestep these hurdles in various ways to meet our needs from solar energy in a cost-effective manner.

Solar Heating of Water

As noted, solar hot-water heating is already popular in warm, sunny climates. A solar collector for heating water consists of a thin, broad "box" with a glass or clear plastic top and a black bottom with water tubes embedded within (Fig. 15–6). Such collectors are called **flat-plate collectors**. Faced toward the Sun, the black bottom gets hot as it absorbs sunlight—similar to how black pavement heats up in the Sun—and the clear cover prevents the heat from escaping. Water circulating through the tubes is thus heated and is conveyed to a tank where it is stored.

In an *active system,* the heated water is moved by means of a pump. In a *passive system,* natural convection currents are used. In a passive solar water-heating system, the system must be mounted so that the collector is lower than the tank. Thus, heated water from the collector rises by natural convection into the tank, while cooler water from the tank descends into the collector (Fig. 15–7). There are no pumps to buy, maintain, or remember to turn on and off. The size and number of collectors, of course, are adjusted to the need. The greatest economy is gained by carefully managing of requirements for hot water, so that the sizes of the tank and collectors can be minimized.

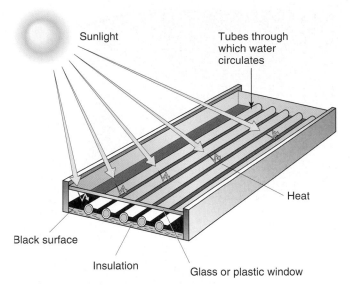

▲ FIGURE 15–6 *The principle of a flat-plate solar collector.* As it is absorbed by a black surface, sunlight is converted to heat. A clear glass or plastic window over the surface allows the sunlight to enter the collector, which traps the heat. Air or water is heated as it passes over and through tubes embedded in the black surface.

In temperate climates, where water in the system might freeze, the system may be adapted to include a heat-exchange coil within the hot-water tank. Then, antifreeze fluid is circulated between the collector and the tank. In the United States, there are an estimated 800,000 solar hot-water systems operating, but this number is still only about 0.5% of the total of all hot-water heaters.

In recent years, the most common end use of solar thermal systems has been for heating swimming pools; systems used for this purpose are very inexpensive and consist of black plastic or rubberlike sheets with plastic tubing that circulates water through the systems.

Solar Space Heating

The same concept as in solar water heating can be applied to heat spaces. Flat-plate collectors like those used in water heating can be used for space heating. Indeed, the collectors for space heating may even be less expensive, home-made devices, because it is necessary only to have air circulate through the collector box. Again, efficiency is gained if the collectors are mounted to allow natural convection to circulate the heated air into the space to be heated (Fig. 15–8).

The greatest efficiency in solar space heating, however, is gained by designing a building to act as its own collector. The basic principle is to have windows facing the sun. In the winter, because of the Sun's angle of incidence, sunlight can come in and heat the interior of the building (Fig. 15–9a). At night, insulated drapes or shades can be pulled down to trap the heat inside. The well-insulated building, with appropriately made doors and windows, would act as its own best heat-storage unit. Beyond good insulation, other systems for storing heat, such as tanks of water or masses of rocks, have not proved cost effective. Excessive heat load in the summer can be avoided by using an awning or overhang to shield the window from the high summer Sun (Fig. 15–9b).

▼ FIGURE 15–7 *Solar water heaters.* (a) In nonfreezing climates, simple water-convection systems may suffice. In freezing climates, an antifreeze fluid is circulated. (b) Solar panels (flat black collectors) and hot-water tanks on the roof of a hotel in Barbados.

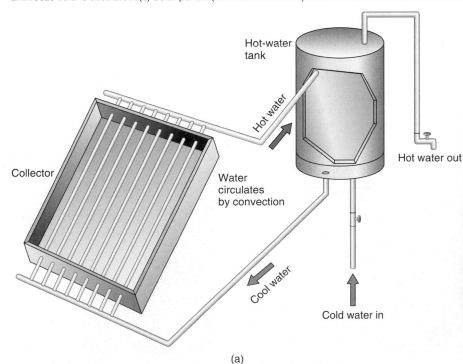

(a)

(b)

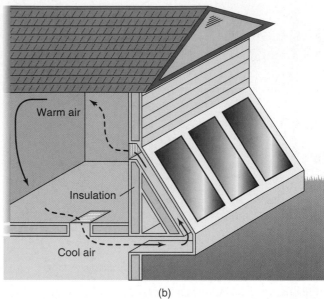

(a)   (b)

▲ **FIGURE 15–8** *Passive hot-air solar heat.* (a) Many homeowners could save on fuel bills by adding homemade solar collectors such as that shown here. (b) Air heated in the collector moves into the room by passive convection.

Along with design, positioning, and improved insulation, appropriate landscaping can contribute to the heating and cooling efficiency of both solar and nonsolar space-heating designs. In particular, deciduous trees or vines on the sunny side of a building can block much of the excessive summer heat while letting the desired winter heat pass through. An evergreen hedge on the shady side can provide protection from the cold (Fig. 15–10).

One approach that combines insulation and solar heating is **earth-sheltered housing** (Fig. 15–11). The principle is to use earth as a form of insulation and orient the building for passive solar energy. Two strategies are employed. One is to build up earth against the building's walls (earth berms) in a more conventional house design; the other is to cover the building to varying degrees with earth, exposing the interior to the outside with windows facing southward. Earth has a high capacity for heat storage, and since the walls and floors of an earth-sheltered home are usually built with poured concrete or masonry, at night the building will radi-

ate the heat stored during the day and warm the house. In the summer, the building is kept cool by its contact with earth, although moisture often will have to be controlled with dehumidifiers.

The Center for Renewable Resources estimates that in almost any climate, a well-designed passive solar home can reduce energy bills by 75% with an added construction cost of only 5% to 10%. About 25% of our primary energy is used for space and water heating. If solar heating were optimally used for this purpose, the savings on oil, natural gas, and electrical power would be considerable.

A common criticism of solar heating is that a backup heating system is still required for periods of inclement weather. Good insulation is a major part of the answer to this criticism. People with well-insulated solar homes find that they have minimal need for backup heating. When backup is needed, a small wood stove or gas heater suffices. In any case, the criticism concerning the need for backup heating misses the point. Remember that the objective of solar heat-

► **FIGURE 15–9** *Solar building siting.* In contrast to utilizing expensive and complex active solar systems, solar heating *can* be achieved by suitable architecture and orientation of the home at little or no additional cost. (a) The fundamental feature is large, Sun-facing windows that permit sunlight to enter during the winter months. Insulating drapes or shades are drawn to hold in the heat when the Sun is not shining. (b) Suitable overhangs, awnings, and deciduous plantings will prevent excessive heating in the summer.

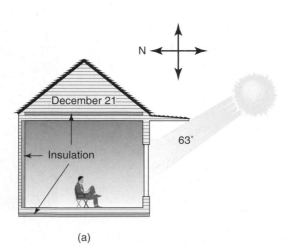

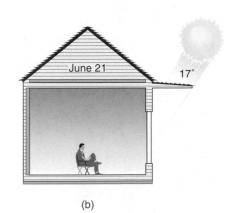

(a)   (b)

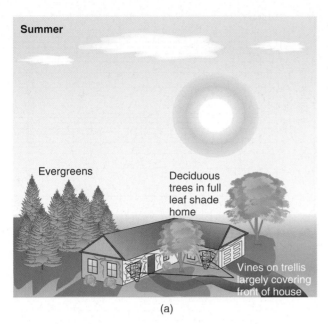

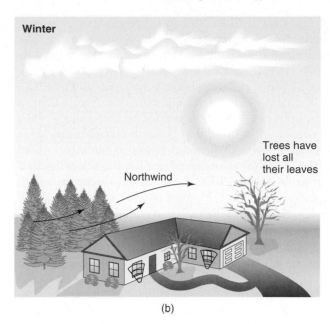

▲ **FIGURE 15–10** *Landscaping in solar heating and cooling.* (a) In summer, the house may be shaded with deciduous trees or vines. (b) In winter, leaves drop, and the bare trees allow the house to benefit from sunlight. Evergreen trees on the opposite side protect and provide insulation from cold winds.

ing is to reduce our dependency on traditional fuels. Even if solar heating and improved insulation reduced the demand for conventional fuels by a mere 20%, this would still represent sustainable savings of 20% of the traditional fuel and its economic and environmental costs. With any systematic effort, the savings could be much greater.

However, the effort to reduce the use of fossil fuels has not pleased some sectors of our society. As a result, in the 1980s, utility and oil companies launched intensive advertising campaigns purporting that solar energy is impractical and not cost effective. These campaigns did much to curtail the growth of the solar-energy industry. In fact, fossil-fuel interests successfully lobbied the government to end incentive programs for renewable energy. In the 1990s, the major factor in discouraging the use of solar heating was relatively

low energy prices, which were a disincentive to making any change. But with rising energy costs in the early 2000s, the opportunity is greater.

Solar Production of Electricity

Solar energy can also be used to produce electrical power, thus providing an alternative to coal and nuclear power. Currently, a few methods are feasible, and two in particular, *photovoltaic cells* and *solar-trough collectors*, are proving to be economically viable.

Photovoltaic Cells. A solar cell, more properly called a **photovoltaic**, or **PV cell**, looks like a simple wafer of material with one wire attached to the top and one to the bottom (Fig. 15–12). As sunlight shines on this "wafer," it puts out an amount of electrical current roughly equivalent to a

▲ **FIGURE 15–11** *Earth Hall.* Earth-sheltered building. at Au Sable Institute of Environmental Studies in Mancelona, Michigan, houses offices, classrooms, and laboratories for the institute.

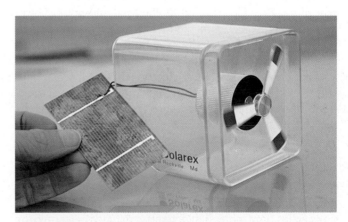

▲ **FIGURE 15–12** *Photovoltaic cell.* Converting light to electrical energy, this cell provides enough energy to run the small electric motor needed to turn these fan blades.

flashlight battery. Thus, PV cells accomplish the collection of light and its conversion to electrical power in one step. One PV cell is no more than about 2 inches (5 cm) in diameter. However, almost any amount of power can be produced by wiring cells together in panels. (Recall Fig. 15–1.)

The deceptively simple appearance of PV cells belies a very sophisticated materials science and technology. Each cell consists of two very thin layers of material. The lower layer has atoms with single electrons in the outer orbital that are easily lost. The upper layer has atoms lacking electrons in their outer orbital; such atoms readily gain electrons. (See Appendix C for some basic chemical principles.) The kinetic energy of light photons striking this "sandwich" dislodges electrons from the lower layer, creating an electrical potential between the two layers. The potential provides the energy for an electrical current to flow through the rest of the circuit. Electrons from the lower side flow through a motor or other electrical device back to the upper side. Thus, with no moving parts, solar cells convert light energy directly to electrical power, with an efficiency of about 20%.

Because they have no moving parts, solar cells do not wear out; however, deterioration due to exposure to the weather limits their life span to about 20 years. The major material used in PV cells is silicon, one of the most abundant elements on Earth, so there is little danger that the production of PV cells will ever suffer because of limited resources. The cost of these cells lies mainly in their sophisticated design and construction.

Photovoltaic cells are already in common use in pocket calculators, watches, and numerous toys. Panels of PV cells provide power for rural homes, irrigation pumps, traffic signals, radio transmitters, lighthouses, offshore oil-drilling platforms, and other installations that are distant from power lines. It is not hard to imagine a future in which every home and building has its own source of pollution-free, sustainable electrical power from an array of PV panels on the roof. Current installations in many countries and in over half of the states in the United States involve "net metering," where the rooftop electrical output is subtracted from the customer's use of power from the power grid.

The cost of PV power (cents per kilowatt-hour) is the cost of the PV cells, divided by the total amount of power they may be expected to produce over their lifetime (currently about 20 cents per kilowatt-hour). This cost must be compared with that of other power alternatives (8 cents per kilowatt-hour for residential electricity). The first PV cells had a cost factor several hundred times that of electricity from conventional power stations transmitted through the power grid, so these cells were used mainly in areas far from the grid. PV power had its first significant application in the 1950s, in the solar panels of space satellites. This application in satellites, in fact, started the development cycle rolling. As more efficient cells and less expensive production techniques evolved, costs came down. In turn, applications, sales, and potential markets expanded, creating the incentive for further development (Fig. 15–13). Recently,

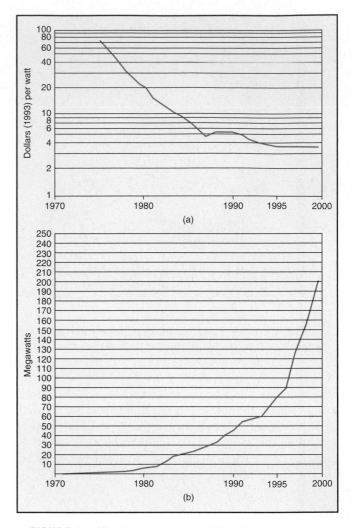

▲ **FIGURE 15–13** *The market for PV cells.* (a) Through research and development, the price of photovoltaic cells has come down about twentyfold in the past 25 years. (b) As prices have come down, sales have increased dramatically. Note the surge in sales for 1999—to 201 MW. (*Sources:* C. Flavin and N. Lenssen, *Power Surge: Guide to the Coming Energy Revolution* [New York: W. W. Norton, 1994] Figs. 8–1 and 8–2; and *Renewable Energy Annual 1999*, Energy Information Agency, Department of Energy.)

the price of PV power dropped to $3,500 per kilowatt, continuing a steady decline. In response to this trend, the industry is growing rapidly. Sales of solar panels were almost $700 million worldwide in 1999, with the bulk of the sales going to overseas customers, mostly in the developing world. The market for total PV systems has surpassed $1.6 billion and is expected to continue to grow 15–20% annually.

What will utility companies do as PV power becomes competitive? A demonstration PV power plant has been built in southern California (Fig. 15–14). However, the more promising future for PV power would appear to be in the installation of home-sized systems on rooftops, thus avoiding additional land and transmission costs. (A new technique for PV cell design–"thin-film" collectors—has the potential for constructing roof shingles that generate electrical power at a fraction of the cost of conventional PV

◀ **FIGURE 15–14 PV power plant.** The world's first photovoltaic (solar cell) power plant, located near Bakersfield, California. The array of 220 34-foot (11-m) panels produces 6.5 MW at peak, enough for 6,500 homes.

cells.) Toward this end, a local utility in Sacramento, California, has launched its PV Pioneers program. The utility installs the systems and pays homeowners $4 per month to maintain a 3- to 4-KW PV system on their roofs. These systems feed into the power grid, and their cost, including installation, is expected to continue to drop until, by 2002, the power they generate will cost much the same as peak-load power. A similar program in Japan has led to the installation of more than 30,000 systems in homes.

A new federal program, the Million Solar Roofs Initiative, aims at encouraging the installation of solar energy units on a million residential and commercial rooftops by 2010, but the initiative provides few financial incentives, focusing instead on "partnering" with other organizations. Before such systems are widespread in the United States, policy changes will be needed to provide substantial incentives through tax credits and other means.

Recently, scientists at the Oak Ridge National Laboratory (ORNL) developed an exciting technology they call hybrid lighting, in which typical PV cells are coupled with optical fibers in rooftop units. The units concentrate solar energy and transmit the visible light through the optical fibers directly to interior spaces, and the infrared light is transmitted to the PV cells to generate electricity. The system is expected to be most useful in commercial applications, a third of whose electricity is consumed by daytime lighting. Preliminary estimates suggest that such a system could triple the efficiency of use of solar energy.

Solar-Trough Collectors. The second method that is proving to be cost effective is the **solar-trough** collector sys-

tem, so named because the collectors are long, trough-shaped reflectors tilted toward the Sun (Fig. 15–15). The curvature of the "trough" is such that all of the sunlight hitting the collector is reflected onto a pipe running down the center of the system. Oil or some other heat-absorbing fluid circulating through the pipe is thus heated to very high temperatures. The heated fluid is passed through a heat exchanger to boil water and produce steam for driving a turbogenerator.

Nine solar-trough facilities in the Mojave Desert of California are now connected to the Southern California Edison utility grid. With a combined capacity of 350 MW, these facilities have about one-third the capacity of a large nuclear power plant and are converting a remarkable 22% of incoming sunlight to electrical power at a cost of 10 cents per kilowatt-hour, barely more than the cost at coal-fired facilities.

Experimental Technologies. Two additional methods of harnessing thermal energy from the Sun have been developed with government funding. In the first, commonly called a "power tower," an array of Sun-tracking mirrors focus the sunlight falling on several acres of land onto a receiver mounted on a tower in the center of the area (Fig. 15–16). The receiver transfers the heat energy collected to a molten-salt liquid, which then flows either to a heat exchanger to drive a conventional turbogenerator or to a tank at the bottom of the tower to store the heat for later use. The latest pilot plant model of this technology, Solar Two, is now connected to the region's utility grid and is generating 10 MW of electricity, enough for 10,000 homes.

▲ **FIGURE 15–15** *Solar-trough power plant.* Several solar-trough facilities are now in operation in southern California. The curved reflector focuses sunlight on, and heats oil in, the pipe. The heated oil is then used to boil water and generate steam for driving a conventional turbogenerator.

The second method of harnessing solar thermal energy is the dish–engine system (Fig. 15–17). This is a smaller system, consisting of a set of parabolic concentrator dishes that focus sunlight on a receiver. Fluid (often molten sodium) in

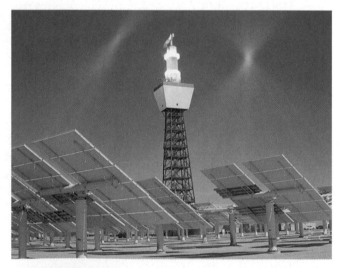

▲ **FIGURE 15–16** *Power tower Solar Two.* Sun-tracking mirrors are used to focus a broad area of sunlight onto a molten-salt receiver mounted on the tower in the center. The hot salt is stored or pumped through a steam generator that drives a conventional turbogenerator.

▲ **FIGURE 15–17** *Solar dish–engine system.* A circular mirror array focuses solar energy on a receiver, which transmits the energy to an engine that generates electric power.

the receiver is transferred to an engine (a Stirling engine) that generates electricity directly. Dish–engine systems have demonstrated efficiencies of 30%, higher than any other solar technology. They generate from 5 to 40 KW, and any number of them can be linked together to provide larger amounts of electricity. They are well suited for off-grid applications to provide power to remote sunny areas. Both new technologies require little maintenance and, of course, generate no polluting gases.

The Promise of Solar Energy

The promise of solar energy does bring with it certain disadvantages. First, the available technologies are still more expensive than conventional energy sources, although the costs are continuing to come down. Second, solar energy works only during the day, so that it requires a backup energy source or a storage battery for nighttime use. Also, the climate is not sunny enough to use solar energy in the winter in many parts of the world. Against all these drawbacks should be weighed hidden costs in the form of air pollution, strip mining, and nuclear waste disposal that are not included in the cost of power from traditional sources.

And again, the fact that the Sun provides power only during the day can be countered with the argument that 70% of electrical demand occurs during daytime hours, when industries, offices, and stores are in operation. Thus, proponents of solar energy argue that savings still can be achieved by using solar panels just for daytime needs and continuing to rely on conventional sources at night. All the solar power that can conceivably be built in the next 20 years or more could be used to supply the daytime demand, leaving traditional fuels to provide nighttime demand. In particular, the demand for air conditioning, which, after refrigeration, is the second largest power consumer, is well matched to energy from PV cells. In addition, solar-powered air conditioners could operate independently from the rest of the electrical system, thus avoiding the costs of connecting the two systems. (Such air conditioners may be expected to come onto the market within the next few years.) In the long run, we might envision the nighttime load being carried by forms of indirect solar energy, such as wind power and hydropower (discussed later in the chapter).

According to Table 13–1, about 78% of our electrical power (56% worldwide) is currently generated by coal-burning and nuclear power plants. Therefore, the development of solar electrical power can be seen as gradually reducing the need for coal and nuclear power. (See "Earth Watch" essay, this page.) Solar power has also proved suitable for meeting the needs of electrifying villages and towns in the developing world, where, because centralized power is not available, rural electrification projects based on PV cells are already beginning to spread throughout the land. The cost of the alternative—putting roughly 2 billion people who are still without electricity on the **power grid** (central power plants, high-voltage transmission lines, poles, wires, and transformers)—is unimaginable.

The promise of solar energy has not yet been realized. Although solar technology has met every expectation in terms of declining cost over time, it has not yet reached the level of market penetration that was anticipated during the years of high fossil-fuel costs, due to recent declines in the cost of conventional power generation. Right now, natural gas has the advantage over solar energy as regards the generation of electricity. But, as we saw in Chapter 13, both nat-

ural gas and oil have limited reserves, and before long in this century their cost will rise rapidly. The gap between these conventional energy sources and solar sources is narrowing, and with some encouragement from public-policy changes, solar energy will become the major source many have envisioned.

Solar Production of Hydrogen: The Fuel of the Future

So far, the methods of harnessing solar energy that have been discussed have not addressed one of our greatest concerns: the fact that our oil reserves needed to fuel transportation are dwindling. However, there may be an answer here as well. One obvious possibility is hydrogen gas. Conventional cars can be run on hydrogen gas (H_2) as a fuel in the same manner as they are now beginning to be run on natural gas (methane, or CH_4) (See Chapter 13.) This fact has been amply demonstrated. Furthermore, neither carbon dioxide nor hydrocarbon pollution is produced during the burning of hydrogen. Indeed, the only by-product of H_2 combustion is water vapor:

$$2\, H_2 + O_2 \rightarrow 2\, H_2O + energy$$

(Some nitrogen oxides will be produced, too, because the burning still uses air, which is nearly four-fifths nitrogen.)

If hydrogen gas is such a great fuel, why are we not using it? The answer is that there is virtually no hydrogen gas on Earth. Any hydrogen gas in the atmosphere has long since been ignited by lightning and burned to form water. And although many bacteria in the soil produce hydrogen in fermentation reactions, other bacteria are quick to use

EARTH WATCH

ECONOMIC PAYOFF OF SOLAR ENERGY

In addition to being a nonpolluting source of renewable energy, photovoltaic, solar trough, and wind-powered energy-generating facilities have another advantage: They can be installed quickly and added to a utility system in relatively small increments. To understand this advantage, let us draw a comparison with a nuclear power plant.

A nuclear power plant can be cost effective only if it is large—about 1,000 MW in capacity, enough to power a million average homes. Such a plant, assuming its public acceptance, would require 10 to 15 years from the time of the decision to build it until it came into operation and would cost around $5 to $10 billion. Therefore, a utili-

ty taking this route must borrow billions of dollars and keep the money tied up for 10 to 15 years before a return on the investment is realized by selling power from the new plant. Even then, if the 1,000 MW of additional power are not needed when the new plant comes on-line, the return on the investment may be meager indeed. In the late 1970s and early 1980s, a number of utility companies went bankrupt (or nearly did so) because they had huge amounts of money tied up in nuclear power plants, while demand for the power produced slackened. Clearly, building large power plants (nuclear or coal) involves considerable financial risk on the part of both utili-

ties and consumers, as consumers are the ones who ultimately pay.

On the other hand, solar or wind facilities can become operational within a few months of the decision to build them, they can be installed in small increments, and additional modules can be added as demand requires. The utility does not have to guess what power demands will be in 10 or 15 years. With solar or wind power, a utility can add capacity as it is needed, make relatively small investments at any given time, and have those investments start paying back almost immediately. Thus, this approach involves much less financial risk both for the utility and for its consumers.

the hydrogen because it is an excellent source of energy. Thus, there are abundant amounts of the *element* hydrogen, but it is all combined with oxygen in the form of water (H_2O) or other low-energy compounds.

Although we can extract hydrogen from water via electrolysis, the process *uses* energy. Recall that the first law of thermodynamics dictates that energy cannot be created ("You can't get something for nothing.") But the situation is even worse: An *input* of energy is required to convert water into hydrogen and oxygen, because of inevitable losses accounted for by the second law of thermodynamics. Therefore, the use of hydrogen as a fuel must await an energy source that is itself suitably cheap, abundant, and nonpolluting. Do these specifications sound like solar energy?

For eons, the natural world has had a method of splitting water into hydrogen and oxygen using light energy, namely, photosynthesis. (Recall that the oxygen from photosynthesis is released into the atmosphere and the hydrogen is attached to carbon dioxide to form sugars; see page 62.) Scientists, however, have not yet developed a means of mimicking the photosynthetic reaction on a scale sufficient to produce commercial amounts of hydrogen.

However, with the simple equipment used in *electrolysis*, a direct current passed through water causes the water molecules to dissociate. Hydrogen bubbles come off at the negative electrode (the cathode), while oxygen bubbles come off at the positive electrode (the anode). (You can demonstrate this process for yourself with a battery, some wire, and a dish of water. A small amount of battery acid placed in the water will facilitate the conduction of electricity and the release of hydrogen.)

An economic analysis projects that the cost of solar-produced hydrogen gas would be equivalent to gasoline costing $1.65 to $2.35 per gallon. This may not sound cheap, but the lower figure is what gas already costs in many parts of the United States, and the higher figure is less than what gasoline will cost as inevitable constraints come into play. (See Chapter 13.) Indeed, consumers in European nations and Japan are already paying much more. One important drawback to this technology is the fact that it is difficult to store enough hydrogen in a vehicle to allow it to operate over practical distances. Compressing the hydrogen into a liquid would require energy, thus reducing the energy efficiency involved. As an alternative, the hydrogen can be combined with metal hydrides, which can absorb the gas and release it when needed.

A rather elaborate plan has been suggested were we to take the concept of using solar energy-produced hydrogen for vehicular fuel to its logical conclusion. We would start by building arrays of solar-trough or photovoltaic generating facilities in the deserts of the southwestern United States, where land is cheap and sunlight plentiful. The electrical power produced by these methods would then be used to produce hydrogen gas by electrolysis. Conveniently, Texas is the hub of a network of pipelines that were originally constructed to transport natural gas from the state's oil fields throughout the nation. Many of these pipelines are now underutilized because those fields have been largely depleted. Therefore, the hydrogen gas produced in the Southwest could be transported across the land by way of these pipelines. With vehicles already being adapted to run on natural gas, hydrogen could slowly be phased in by mixing gradually increasing proportions of it with natural gas. Thus, there could be a smooth transition from our present fuel through natural gas to hydrogen.

Fuel Cells. An alternative to burning hydrogen in conventional internal combustion engines uses hydrogen in *fuel cells* to produce electricity and power the vehicle with motors—just the opposite of the technology described in the previous paragraphs. Fuel cells are devices in which hydrogen or some other fuel is recombined with oxygen chemically in a manner that produces an electrical potential rather than initiating burning (Fig. 15–18a). Because fuel cells create much less waste heat than conventional engines do, energy is transferred more efficiently from the hydrogen to the vehicle—at a percentage of 45 to 60%, versus the 15% for current combustion engine vehicles.

Both of the preceding technologies have been developed for transportation. Fuel cells are now powering buses in Vancouver and Chicago, and DaimlerChrysler plans to build 20 to 30 urban buses with fuel cell drives over the next three years. The vehicles are based on Mercedes-Benz Citaro urban buses and will be offered for sale to transportation companies in Europe. The fuel cells will develop more than 250 kilowatts of power, from hydrogen in tanks mounted on the roof of the bus. Stimulated by the **Partnership for a New Generation of Vehicles (PNGV)**, a collaboration with the federal government, DaimlerChrysler and other major automobile manufacturers have been developing "concept cars" aimed at achieving a fuel economy of 80 miles per gallon. One such car, unveiled and test-driven at the North American International Auto Show in January 2000, is the GM Precept (Fig. 15–18b), powered completely by hydrogen-based fuel cells and using metal hydrides to store the hydrogen. Major obstacles to the widespread application of fuel cell vehicles are their high cost and the lack of an infrastructure to provide hydrogen for refueling. Still, the technology is improving rapidly and is poised for commercial development.

15.3 Indirect Solar Energy

Water, fire, and wind have provided energy for humans throughout history. We are all familiar with dams, firewood, and windmills. We group these age-old sources of energy together under the rubric of "indirect solar energy" because a moment's reflection shows us that energy from the Sun is the driving force behind each. The question is, What is the

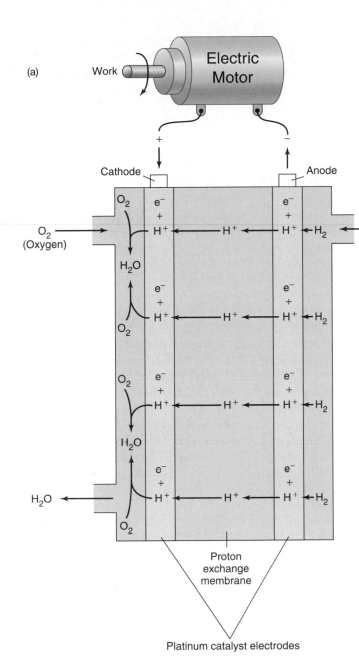

(a)

(b)

◀ **FIGURE 15–18** *Hydrogen–oxygen fuel cell.* (a) Hydrogen gas flows into the cell and meets a catalytic electrode, where electrons are stripped from the hydrogen, leaving protons (hydrogen ions) behind. The protons diffuse through a proton exchange membrane, while electrons flow from the anode to the motor. As the electrons then flow from the motor to the fuel cell cathode, they combine with the protons and oxygen to produce water. Thus, the chemical reaction pulls electrons through the circuit, producing electricity that runs the motor. Hydrogen is continuously supplied from an external tank, oxygen is drawn from the air, and the only by-product is water. *GM Precept concept car.* (b) The Precept is driven by fuel cells supplied with hydrogen gas that is stored in a tank containing a chemical hydride material to absorb the hydrogen and reduce its volume. Oxygen is supplied from a vent to the air, and only heat and water vapor are in the exhaust. The Precept can accelerate from 0 to 60 miles per hour in 9 seconds and achieves the energy equivalent of 108 miles per gallon in everyday traffic.

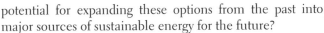

potential for expanding these options from the past into major sources of sustainable energy for the future?

Hydropower

Early in technological history, it was discovered that the force of falling water could be used to turn paddle wheels, which in turn would drive machinery to grind grain, saw logs into lumber, and do other laborious tasks. The modern culmination of this use of water power, or hydropower, is huge hydroelectric dams, where water under high pressure flows through channels, driving turbogenerators (Fig. 15–19). The amount of power generated is proportional to both the height of the water behind the dam—that is what provides the pressure—and the volume of water that flows through.

About 8.5% of the electrical power generated in the United States currently comes from hydroelectric dams, most of it from about 300 large dams concentrated in the Northwest and Southeast. Worldwide, hydroelectric dams have a generating capacity of 678,000 MW, with an additional 50,000 MW in the construction or planning stage. Hydropower generates 19% of electrical power throughout the world and is by far the most common form of renewable energy in use.

Whereas waterpower is basically a nonpolluting, renewable energy source, harnessing it by means of hydroelectric dams still involves high ecological, social, and cultural trade-offs:

1. The reservoir created behind the dam inevitably drowns farmland or wildlife habitats and perhaps towns

▲ **FIGURE 15–19** *Hoover Dam.* About 8.5% of the electrical power used in the United States comes from large hydroelectric dams such as this one.

or land of historical, archaeological, or cultural value. Glen Canyon Dam (on the border between Arizona and Utah), for example, drowned one of the world's most spectacular canyons. A number of dams have obliterated sacred lands of Native Americans.

2. Dams impede or prevent the migration of fish, even when fish ladders are provided. Federal surveys show that fish habitats are suffering in the majority of the nation's rivers because of damming. Salmon fishing, one of the great industries of the Northwest, has been heavily affected.

3. Changing from a cold-flowing river to a warm-water reservoir can have unforeseen ecological consequences. The reservoir behind the Aswan High Dam in Egypt has fostered the spread of a parasitic worm that causes a debilitating disease. The reservoir has also increased humidity over a widespread area, and the higher levels of humidity are now accelerating the deterioration of ancient monuments and artifacts that have stood virtually unchanged for many centuries.

4. Ecological consequences occur below the dam as well. Because the flow of water is regulated according to the need for power, dams play havoc downstream; water may go from near flood levels to virtual dryness and back to flood levels in a single day. Other ecological factors are also affected, because sediments with nutrients settle in the reservoir, so that smaller-than-normal amounts reach the river's mouth.

As if such trade-offs were not enough, thoughts of greatly expanding water power in the United States are nullified by the fact that few sites conducive to large dams remain. Already, 65,000 dams dot U.S. rivers; only 2% of the nation's rivers remain free flowing, and many of these are now protected by the Wild and Scenic Rivers Act of 1968, a law that effectively gives certain scenic rivers the status of national parks.

In short, proposals for new dams in the United States and elsewhere are embroiled in controversy over whether the projected benefits justify the ecological and sociological trade-offs. In 1997, the World Bank gave preliminary approval to a loan to the Lao People's Democratic Republic for construction of the Nam Theun Two Dam, a 680-MW project that is expected to help Laos begin a process of economic development. The World Bank's decision has come under fire because the project will displace 4,500 people and destroy 2,485 square miles of sensitive environments. The loan is contingent on the completion of a thorough study of the dam's impacts. Another huge dam under construction is the Three Gorges Dam on the Yangtze River in China. When completed, it will be the largest dam ever built, displacing some 1.9 million people in order to generate 18,200 MW of electricity. Offsetting the human and economic costs is the fact that it would take more than a dozen large coal-fired power plants to produce the same amount of electricity.

Wind Power

Throughout much of history, wind power—in addition to propelling sailing ships—was widely used for grinding grain—hence the term *windmill*. Then, wind-driven propellers went on to perform other tasks. Until the 1940s, most farms in the United States used windmills for pumping water and generating small amounts of electricity. In the 1930s and 1940s, windmills fell into disuse, however, as transmission lines brought abundant lower cost power from central generating plants. Not until the energy crisis and rising energy costs of the 1970s did wind begin to be seriously considered again as a potential source of sustainable energy.

Many different designs of wind machines have been proposed and tested, but the one that has proved most practical is the age-old concept of wind-driven propeller blades. The propeller shaft is geared directly to a generator. (Wind driving a generator is more properly called a **wind turbine** than a windmill.) As the reliability and efficiency of wind turbines have improved, the cost of wind-generated electricity has come down. **Wind farms**—arrays of up to several thousand such machines—are now producing pollution-free, sustainable power for about 4 cents per kilowatt-hour, competitive with traditional sources. Moreover, the amount of wind that can be tapped is immense. The American Wind Energy Association calculates that wind farms located throughout the Midwest could meet the electrical needs for the entire country, while the land beneath the turbines could still be used for farming. Hundreds of new turbines are sprouting from midwestern farms, and the farmers are paid handsome royalties for this use of their land.

For example, Hamilton, in Madison County, New York, is the site of the largest wind-power plant in the eastern United States. The facility broke ground in April 2000 and was ready to start commercial operation in six months. Seven 1.65-MW turbines provide an 11.5-MW capacity, enough electricity to power 4,000 homes. This new facility joins a rapidly rising number of wind turbines whose capacity has now surpassed 13,400 MW worldwide (Fig. 15–20) and is expanding at a rate of 25 to 30% per year. In Denmark, which is hard pressed by the lack of fossil-fuel energy, wind now supplies over 10% of electricity. California leads the United States in wind energy; the Tehachapi–Mojave area alone contains more than 5,000 turbines generating enough electricity to power 500,000 homes.

Still, wind power does have some drawbacks. First, as it is an intermittent source, problems of backup or storage must be considered. Second is an aesthetic consideration: One or two windmills can be charming, but a landscape covered with them can be visually tiresome. Third, windmills are a hazard to birds; locating wind farms on migratory routes or near critical habitats for endangered species like condors could be a problem for soaring birds in particular.

Biomass Energy

Burning firewood for heat is perhaps one of the oldest forms of energy that humans have used throughout history. However, there is nothing like a new name to put life into an old concept. Thus, "burning firewood" has become "utilizing biomass energy." "Biomass energy," or "bioconversion," means "deriving energy from present-day photosynthesis." As Fig. 15–4 indicates, biomass energy leads hydropower in renewable energy production in the United States. (Most uses of biomass energy are for heat.)

In addition to burning wood in a stove, major means of producing biomass energy include: obtaining energy from the burning of municipal wastepaper and other organic waste; generating methane from the anaerobic digestion of manure and sewage sludges, and producing alcohol from fermenting grains and other starchy materials. Let us examine the potential of each of these techniques.

Burning Firewood. Wherever forests are ample relative to the human population, firewood, or "fuelwood," can be a sustainable energy resource, and indeed, wood has been a main energy resource over much of human history. In the United States, woodstoves have enjoyed a tremendous resurgence in recent years—about 5 million homes rely entirely on wood for heating and another 20 million use wood for some heating.

The most recent development in woodstoves is the *pellet stove*, a device that burns compressed wood pellets made

◀ **FIGURE 15–20** *Wind farm at Altamont Pass, California.* Such arrays of wind turbines are proving to be an economically competitive way of generating electrical power. California now leads the United States in the use of wind energy.

from wood wastes (Fig. 15–21). The pellets are loaded into a hopper (controlled by computer chips) that feeds fuel when needed. The stove requires very little attention and burns very efficiently, relying on one fan to draw air into it and another fan to distribute heated air to the room.

The Fuelwood Crisis. In developing countries, firewood is the only source of fuel for cooking for over a billion people. In fact, 90% of the world's fuelwood is produced and used in the developing countries. Recall that two patterns of use determine how forest resources will be exploited: consumptive use and productive use (Chapter 12). Most fuelwood use in developing countries fits the *consumptive* pattern: People simply forage for their daily needs in local woodlands and forests. Unfortunately, this kind of use often results in a deforestation that spreads outward from population centers. One example is the Central American country of Honduras. PROLENA, a citizen's group, estimates that Hondurans cut 7 million cubic meters of firewood every year to use in cooking their meals, resulting in a high rate of deforestation. The country is in desperate need of a fuelwood policy that would encourage replanting and the sustainable use of wood in a land that is ecologically ideal for growing trees.

The story is similar in many other parts of the developing world: Growing numbers of people go out from villages and towns daily to gather wood. The solution to the problem is to develop a market in wood, which would encourage people with land to plant trees as a cash crop. When people have to pay for fuelwood, trees become an important resource that can be put under the stewardship of local communities or private landowners, and a sustainable use of forests can result. Then all of the other benefits of goods and services provided by forests are preserved or restored.

▲ **FIGURE 15–21** *Pellet stove.* Pellet stoves use pelletized wood waste to achieve an efficient burn that requires much less care than burns from conventional woodstoves.

Indeed, this is the trend in many parts of the developing world, but it often requires encouragement from governments or NGOs. (See Chapter 12.) Toward that end, the U.N.-sponsored Forest Trees and People Program (FTPP) works with local institutions and communities to encourage the management of forest resources for both conservation and sustainable development.

Burning Wastes. Facilities that generate electrical power from the burning of municipal wastes (waste-to-energy conversion) are discussed in Chapter 19. Many sawmills and woodworking companies are now burning wood wastes, and a number of sugar refineries are burning cane wastes to supply all or most of their power. In Spain, wastes from olive oil production are powering three electrical facilities, providing over 32 MW of power, enough to supply 100,000 households. Power from these sources may not meet more than a few percent of a country's total electrical needs, but they represent a productive and relatively inexpensive way to dispose of biological wastes.

Producing Methane. Chapter 18 examines how the anaerobic digestion of sewage sludge yields *biogas*, which is two-thirds methane, plus a nutrient-rich treated sludge that is a good organic fertilizer. Animal manure can be digested likewise. When the disposal of manure, the production of energy, and the creation of fertilizer can be combined in an efficient cycle, great economic benefit can be achieved. In China, millions of small farmers maintain a simple digester in the form of an underground brick masonry fermentation chamber with a fixed dome on top for storing gas (Fig. 15–22). Agricultural wastes, such as pig manure and cattle dung, are put into the chamber and diluted 1:1 with water, and the anaerobic digestion produces the biogas, which is used for cooking and heating. The residue slurry makes an excellent fertilizer. The concept has been expanded throughout the developing world to provide an alternative to fuelwood.

Producing Alcohol. Ethyl alcohol (ethanol) is produced by the fermentation of starches or sugars. The usual starting material is grain, sugar cane, or various sugary fruits such as grapes. Obviously, this is the process used in the production of alcoholic beverages. The only new part of the concept is that, instead of drinking the brew, we distill it and put it in our cars. However, production costs make ethanol twice as expensive as gasoline. To stimulate the industry, federal tax credits have been extended to ethanol-based fuels. The credits bring the cost down considerably. *Gasohol*, a 10% mixture of alcohol and gasoline, has been promoted and marketed in the Midwest since the late 1970s. The enthusiastic support of gasohol in the Midwest is not motivated purely by the desire to save fossil fuel; the product also provides another market for corn crops and thereby improves the economy of the region.

Recently, some midwestern states required that all gasoline be blended with ethanol. In 1999, 1.5 billion gallons of ethanol for fuel were produced, the equivalent of 35 million barrels of oil and equal to all the corn grown in

▲ **FIGURE 15–22** *Biogas power.* Animal wastes are introduced to this unit in rural India and mixed with water. The wastes then decompose under anaerobic conditions, producing biogas.

North and South Dakota. Even though it appears that ethanol as a motor fuel effectively displaces gasoline and reduces our reliance on imported oil, some studies indicate that the energy it takes to produce the ethanol is almost equal to the energy we get from its combustion, and this energy is often based on fossil fuels. Besides substituting for gasoline, ethanol is a suitable substitute for methyl tertiary butyl ether (MTBE), a petroleum derivative that has been used as a fuel additive to make the burning of the gasoline cleaner. Both MTBE and ethanol improve air quality when they are made part of reformulated gasoline, and a number of metropolitan areas depend on this type of gasoline to improve their air quality in order to meet Clean Air Act requirements. MTBE, which is a carcinogen in animals, has been showing up in water supplies and may be phased out

as an additive. If this happens, the use of ethanol could expand greatly, especially if oil prices continue to rise.

15.4 Additional Renewable-Energy Options

Geothermal Energy

In various locations in the world, such as the northwestern corner of Wyoming (now Yellowstone National Park), one finds springs that yield hot, almost boiling, water. Natural steam vents and other thermal features are also found in such areas. They occur where the hot molten rock of Earth's interior is close enough to the surface to heat groundwater, as may occur in volcanic regions. Using such naturally heated water or steam to heat buildings or drive turbogenerators is the basis of **geothermal energy**. In 1998, geothermal energy was used to generate 8,240 MW of electrical power, equivalent to the output of eight large nuclear or coal-fired power stations, in countries as diverse as Nicaragua, the Philippines, Kenya, Iceland, and New Zealand. Today, the largest single facility is in the United States, at a location known as The Geysers, 70 miles (110 km) north of San Francisco (Fig. 15–23). As impressive as this application of geothermal energy is, nearly double the amount is being used to directly heat homes and buildings, largely in Japan and China.

Some scientists feel that the potential for geothermal energy has barely been tapped; just using today's technology, they predict, an additional 35,000 to 70,000 MW of electricity could be generated. Taking into account technologies under development as well, the predicted amount rises to an impressive 65,000 to 138,000 MW, which would satisfy about 8% of total world electricity production. Recently, the U.S. Department of Energy launched a research-based initiative, called "GeoPowering the West," that aims to

◀ **FIGURE 15–23** *Geothermal energy.* One of the 11 geothermal units operated by the Pacific Gas & Electric Company at The Geysers in Sonoma and Lake counties, California. California geothermal energy produces 1,600 MW of power, roughly 7% of California's energy needs.

provide 10% of the West's energy needs by 2020. (The western U.S. was chosen because of its unique geology.)

Experience has shown that geothermal energy may not always be sustainable. The Geysers in California was steadily expanded until 1988, when it reached about 2,000 MW (equivalent to the capacity of two large nuclear power plants); it was projected that by the year 2000 the facility would be producing 3,000 MW. Actually, output from The Geysers has been declining steadily since 1988; in 1999, the facility was producing just 880 MW. The problem is that, although the source of geothermal heat may be unlimited, the amount of groundwater isn't. Tapping the hot steam is depleting the groundwater. To restore some of it, a 29-mile pipeline was constructed to conduct treated wastewater from Middletown to one of the Geysers' plants, where the water is injected into the ground to raise the level of groundwater needed to produce steam. If the strategy is successful, other communities may also be tapped for their wastewater.

Geothermal power presents some pollution problems that must be mitigated if the process is not to be environmentally ruinous. The hot steam and water brought to the surface are frequently heavily laced with salts and other contaminants, particularly sulfur compounds leached from minerals in the bedrock. These contaminants are highly corrosive to turbines and other equipment, and they cause air pollution if the steam is released into the atmosphere. Although by no means equivalent to that from a plant burning high-sulfur coal, sulfur dioxide pollution from a geothermal plant can be significant. Hot brines from geothermal sources released into streams or rivers would be ecologically disastrous.

Tidal Power

A phenomenal amount of energy is inherent in the twice-daily rise and fall of the ocean tides, brought about by the gravitational pull of the Moon and the Sun. Many imaginative schemes have been proposed for capturing this limitless, pollution-free source of energy. The most straightforward idea is to build a dam across the mouth of a bay and mount turbines in the structure. The incoming tide flowing through the turbines would generate power. As the tide shifted, the blades would be reversed, so that the outflowing water would continue to generate power.

Tantalizing as this idea sounds, it is fraught with a fundamental problem: In most regions of the world, the maximum difference between high and low tide is a matter of only 40 to 50 cm (16 to 20 inches). A head of pressure 50 cm or less is not enough to drive turbines efficiently. In about 30 locations in the world, the shoreline topography generates tides high enough—6 m (20 ft) or more—for this kind of use. Large tidal power plants already exist at two of those places—in France and Canada. The only suitable location in North America is the Bay of Fundy, where the Annapolis Tidal Generating Station has operated since 1984, at 20-MW capacity.

Thus, this application of tidal power has potential only in certain localities, and even there it would not be without adverse environmental impacts. The dams would trap sediments, impede the migration of fish and other organisms, prevent navigation, alter the circulation and mixing of saltwater and freshwater in estuaries, and perhaps have other, unforeseen, ecological effects.

A new application of tidal power that promises to avoid the problems of conventional tidal power has been proposed by Tidal Electric, Inc. With this technology, the tidal generator is located offshore and consists of a concrete impoundment enclosure that sits on the ocean bottom, together with an adjacent turbine housing. The amount of power generated is closely related to the vertical tidal range, as the output varies with the square of the range. The impoundment is filled with water as the tide rises, creating a "head" that is gradually released when the tide is low, the water running through turbines that generate electric power. Feasibility studies are under way for two locations, one in Alaska and one in India.

Ocean Thermal-Energy Conversion

Over much of the world's oceans, a thermal gradient of about 20°C (36°F) exists between surface water heated by the Sun and colder deep water. **Ocean thermal-energy conversion (OTEC)** is the name of an experimental technology that uses this temperature difference to produce power. The technology involves using the warm surface water to heat and vaporize a low-boiling-point liquid such as ammonia. The increased pressure of the vaporized liquid would drive turbogenerators. The ammonia vapor leaving the turbines would then be recondensed by cold water pumped up from as much as 300 feet (100 m) deep and returned to the start of the cycle.

Various studies indicate that OTEC power plants show little economic promise—unless, perhaps, they can be coupled with other, cost-effective operations. For example, in Hawaii, a shore-based OTEC plant uses the cold, nutrient-rich water pumped from the ocean bottom to cool buildings and supply nutrients for vegetables in an aquaculture operation, in addition to cooling the condensers in the power cycle. Even so, interest in duplicating such operations is minimal at present.

15.5 Policy for a Sustainable Energy Future

In the last three chapters, we have examined our planet's energy resources, requirements, and management options. The solar and other renewable energy alternatives we have discussed in this chapter are shown diagrammatically in Fig. 15–24. A review of the energy situation as a whole suggests that there is no reason for an immediate fear of "running out" of energy. World reserves of fossil fuels—even crude oil—are adequate for at least the next 20 to 50 years. In addition, combinations of solar and wind options can be developed and phased in over that time frame. And there is nuclear power if we should choose to go that route.

The major question confronting us is, Should we be taking any specific steps to promote or even force the transition to an economy based on solar energy? Or should we simply eliminate all subsidies for energy, to level the playing field and let free-market forces sort out the situation?

The following policies promoting the transition away from dependence on crude oil and toward reliance on solar energy were implemented in the 1970s:

- greatly expanded federal funding for research and development in the field of renewable energy
- subsidies to producers of solar energy or solar-energy products
- tax credits to people for solar installations in homes and other structures
- additional taxes on traditional fuels, making them more expensive relative to solar energy
- mandates to increase the fuel efficiency of cars
- requirement that public utilities purchase power supplied by renewable-energy facilities

Our final incentives are the moral and legal agreements we have with other nations to reduce emissions of greenhouse gases, the major one of which is carbon dioxide from the burning of fossil fuels. (See Chapter 21.) However, of those policies implemented to promote alternative sources of energy, many were phased out in the early 1980s, effectively returning the nation to the traditional policy—"business as usual"—with an economic bias toward traditional energy sources. Meanwhile, subsidies for fossil fuels and nuclear energy and incentives promoting their use have remained in force. Although we see photovoltaic cells and wind energy making headway in today's marketplace, the business-as-usual view overlooks the fact that the field of competition between solar and traditional sources of energy is steeply tilted in favor of traditional fuels. The traditional-fuel industry is heavily subsidized by

- depletion allowances (tax write-offs as a resource is depleted),
- leasing of public lands at bargain-basement prices, encouraging the exploitation of fossil-fuel reserves,
- military support to ensure access to oil in the Middle East, and
- much greater federal funding for research and development of fossil fuels and their use than for renewable energy.

Moreover, substantial costs are discounted or overlooked, including

- damage to the environment and human health from the pollution caused by burning fossil fuels,
- environmental destruction from strip mining, oil spills, and other accidents,
- the continuing and increasing economic drain as we import growing amounts of crude oil, and
- the long-term and largely irreversible impacts of rising greenhouse gases on our global climate.

On a positive note, the winds of change are sweeping through U.S. energy policy and elsewhere. The Clinton administration pursued a strategy of **energy efficiency** and **renewable-energy technology** as the path to the future. Clearly, both tactics reduce greenhouse gas emissions and air pollution and also contribute to the nation's economic strength and technological competitiveness. Consider the following developments:

- The EPA has successfully initiated two energy-efficiency programs: the **Energy Star** program and the **Green Lights** program. Both are voluntary programs in which the EPA provides companies and institutions with technical and informational support, enabling them to achieve significant savings in energy. Thus, public interest and self-interest are wedded in a way that achieves better results than could ever be obtained through the heavy hand of regulation. By mid-2000, over 3,100 entities had participated in these programs, saving more than $2.2 billion in energy costs and preventing the release of 5.8 million metric tons of carbon.

- The **Climate Change Technology Initiative (CCTI)** is the administration's package of research-and-development investments directed toward reducing greenhouse gas emissions. The initiative has been funded for several years. Congress approved over $1 billion for this important program in the fiscal-year 2000 budget. Much of the money ($310 million) is for the DOE's work on renewable-energy technologies, including some significant public–private partnerships. Some $543 million is for DOE's energy-efficiency programs, one of which is the **Partnership for a New Generation of Vehicles**, discussed earlier. (The EPA's Energy Star and Green Lights programs were funded at $109 million.)

- Basic **climate change research** (the **Global Change Research Program**) has been federally funded for several years at around $1.7 billion per year. This research by academic and independent institutions is providing much of the foundational information employed by the Intergovernmental Panel on Climate Change in its reporting.

- The ongoing **deregulation of the electric power industry** not only is saving consumers money on their electric bills, but is offering them the option of choosing electricity suppliers. In "green power pricing," now available in many localities, customers are able to buy

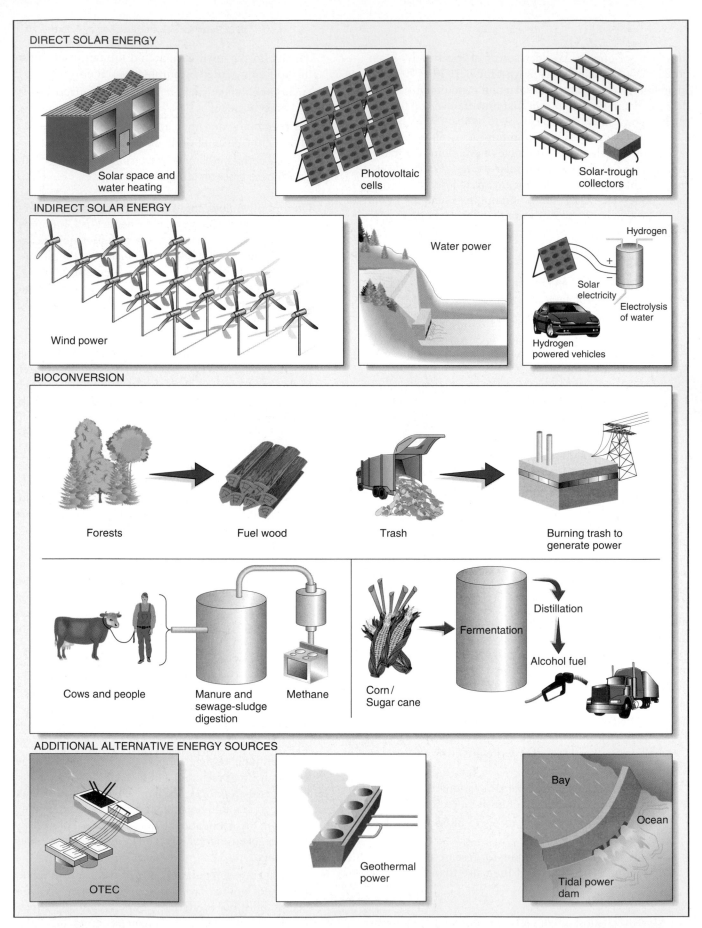

▲ FIGURE 15–24 *Renewable energy resource alternatives.* At the present time, direct solar heating of space and water, photovoltaic cells, solar-trough collectors, wind power, the production of hydrogen from solar or wind power, fuel cells, and the production of methane from animal manure and sewage sludges seem to offer the greatest potential for supplying sustainable energy with a minimum of environmental impact.

their electricity from renewable-energy sources or less polluting fossil-fuel sources, such as gas-powered plants. Many consumers prefer renewable sources, even if they have to pay a premium for them. One problem is that market competition between power sources (the promotion of which is one of the goals of deregulation) may work against the higher priced renewable-energy suppliers. Of course, removing the fossil-fuel and nuclear-energy subsidies would quickly make renewable power more price competitive.

- "Net metering" is the practice of allowing people on the power grid to subtract the cost of the energy they contribute via wind turbines or photovoltaics from their electric bills. This practice is now possible in many states and is an incentive for concerned people to take action on the energy problem by installing their own renewable-energy systems.

- Some major energy producers are looking ahead to the day when greenhouse gases will have to be cut because of very clear evidence of global climate change. Toward that end, British Petroleum is planning to cut its greenhouse gas emissions 10% below 1990 levels by 2010, and Royal Dutch/Shell will spend more than $500 million on renewable-energy technology over the next five years. Both of these firms have withdrawn from an industry lobbying group opposing the Kyoto Protocol on climate change.

- As we have seen, renewable-energy technologies are being adopted widely in the developing world. To a great extent, this development represents a "leapfrogging" over the conventional fossil fuel–based energy technology of the developed world. (See the "Ethics" essay on this page.)

It should not escape our notice that the United States stands out as the only industrialized country that has seen fit to keep gasoline prices remarkably low. Fuel in all other highly developed countries is so heavily taxed that it costs consumers $3 to $5 per gallon—unlike U.S. prices, which were below $1.50 for the last 15 years. So one future development that a number of environmental groups promote is the implementation of a **carbon tax**—a tax levied on all fuels according to the amount of carbon dioxide that is produced in their consumption. Such a tax, proponents believe, would provide both incentives to use solar sources, which would not be taxed, and disincentives to consume fossil fuels. It is hard to imagine any step that would be more effective (or, alas, more controversial) in reducing greenhouse gas emissions in the

ETHICS

TRANSFER OF ENERGY TECHNOLOGY TO THE DEVELOPING WORLD

The nations of the industrialized Northern Hemisphere have achieved their level of development using energy technologies based largely on fossil fuels (and, to a lesser extent, nuclear energy). These nations' development took place during a time when fossil fuels were inexpensive; only as those fuels became more expensive did the nations of the North begin to get serious about technologies that would make energy use more efficient. The specter of global climate change has brought a new perspective to the need to wean the industrialized North away from fossil-fuel energy in the 21st century.

The same traditional fossil-fuel-based technologies have been adopted by the developing nations of the Southern Hemisphere, except that these nations are lagging behind in their ability to implement the technology on a large scale. As we saw in Chapter 13, however, the fossil-fuel-based energy pattern of the 20th century may have to be replaced largely with renewable-energy systems in the 21st century. If the United States and other developed nations are willing to play a signif-

icant role in helping the developing nations, it is questionable whether the same developmental path the North has taken should be promoted. Instead, there is the opportunity to engage in some "leapfrogging" technology transfer. The North can put its development dollars into such technologies as the electrification of rural areas via photovoltaics, wind turbines, and efficient public transport systems for the cities.

The solar route is especially attractive for many of the climates in the developing world, where an estimated 2 billion people lack electricity. For example, some 80,000 PV systems installed in Kenya over a 10-year period now provide power to over 1% of the rural Kenyan population. The systems, with costs ranging from $300 to $1,500, are successfully marketed to people with incomes averaging less than $100 per month. Only the solar panels are imported; local companies provide the batteries and other system components. People use the energy for lighting, television, and radio. Annual sales in East Africa exceed 20,000 systems, a sign that the technology is in

great demand and is fostering appropriate development in the area. Credit and financing extended to both companies and users have been key to this success, some originating with development agencies from the North, but the majority Kenyan. Building the infrastructure to market and service the PV systems has also been an essential part of the success story, and this can be accomplished by a relatively small number of organizations and individuals.

It is entirely possible that if this strategy of development aid is followed, the nations of the South may end up pointing the way to a sustainable energy future for the rest of the world. The Rio Declaration, signed at the Earth Summit in June 1992, commits the industrial countries to providing greater levels of development aid to the poor countries. The pact also stipulates that the development should be sustainable. As the details of this commitment get fleshed out, it will be interesting to see if solar energy receives the attention it deserves. If it does, there is much the donor countries can learn from the results. What do you think?

United States. A number of European countries have already adopted such a tax.

Are the preceding developments enough to enable us to achieve a sustainable energy system and to mitigate global climate change? Quite clearly, no. Yet, they are moving us in the stewardly direction that is vital to the future of the global environment.

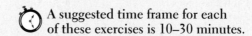

ENVIRONMENT ON THE WEB

COPING WITH "SOMETIMES" ENERGY

Renewable energy sources offer a way to supplement, or replace, much of our current reliance on traditional thermal generating technologies. But these sources also differ in several important ways from conventional energy sources. For example, wind, solar, and tidal power provide only intermittent, not continuous, generation of electricity. We cannot simply turn on the wind, the Sun, or the tides to meet a surge in power demand nor can we turn off those energy sources when demand is low. Because of this inherent uncertainty, it is often suggested that renewable energy sources are most useful as an adjunct to, rather than a replacement for, conventional thermal power generation.

Another difference is that wind, solar, and tidal generating facilities must be sited or built in places where they can best take advantage of natural conditions rather than at the convenience of the consumer. Sometimes optimal sites are far from human habitation. This means that transmission lines may be longer than for most conventional generating facilities. And when lines are longer, there is more risk of voltage loss in transmission, rather like trying to send water through a very long and leaky garden hose.

Older wind systems in particular have been plagued with problems of long transmission lines, inefficient or unreliable generating equipment, and variable voltage. If unresolved, these problems have the potential to create serious problems for the consumer. Imagine how your computer—and the work you have saved on its hard disk—could be affected by wildly fluctuating electrical current or unexpected power failure. A recent study of Hawaii Electric Light Company (HELCO), which uses some older wind-energy generating equipment, showed that modest wind power out-

put changes, and/or unscheduled changes in consumer demand, would have caused the system to lose its ability to regulate current within acceptable limits. New generating and voltage-control devices have greatly reduced the problems inherent in intermittent energy sources and will allow HELCO to expand its consumer base while maintaining its reliance on wind energy.

The future of renewable energy sources will depend partly on the ability of utilities to forecast and control intermittent sources, to store energy at peak generating periods and retrieve it when it is needed, and to maintain a consistent high-quality flow of electricity to consumers—with or without the aid of conventional generating systems.

Web Explorations

The Environment on the Web activity for this essay describes the recent trend to move to "full-cost pricing" of energy in order to pass on to the consumer the hidden costs of energy use. Go to the Environment on the Web activity (**Select chapter 15 at http://www.prenhall.com/wright**) and learn for yourself:

1. about the viability of renewable energy technologies;
2. about trends in the growth of wind and solar energy; and
3. what people are willing to pay for the choice of "green energy."

🕐 **A suggested time frame for each of these exercises is 10–30 minutes.**

REVIEW QUESTIONS

1. How does solar energy compare with fossil-fuel energy in satisfying basic energy needs? What are the three basic problems of harnessing solar energy?
2. How do active and passive solar hot-water heaters work?
3. How can a building best be designed to become a passive solar collector for heat? What are the barriers to more widespread adoption of this alternative?
4. How does a photovoltaic (PV) cell work, and what are some present applications of such cells? What is the potential for providing more energy from PV cells in the near future?
5. Describe the solar trough system, how it works, and its potential for providing power.
6. How can hydrogen gas be produced via the use of solar energy? How might hydrogen meet the need for fuel for transportation in the future?
7. What is the potential for developing more hydroelectric power in North America, and what would be the environmental impacts of such development?
8. How is wind power being harvested, and what is the future potential for wind farms?
9. What are four ways of converting biomass to useful energy, and what is the potential environmental impact of each?
10. What is geothermal energy, and how may it be harnessed? What has been California's recent experience

with geothermal power, and how has that experience changed the outlook for the use of geothermal energy?

11. What is the potential for developing tidal power in the United States?

12. What policies were moving the United States toward solar energy in the 1970s?

13. How is the traditional fossil-fuel-based system subsidized? What are the "hidden costs" involved in continuing to rely on fossil-fuel energy systems?

14. What has been the Clinton administration's energy strategy? Cite some developments in policy that might move us in a direction of a sustainable energy future.

THINKING ENVIRONMENTALLY

1. From what source(s) does a sustainable society gain all its energy needs? Consider all energy sources, both traditional and alternative, and discuss each in terms of long-term sustainability. What conclusions do you reach? Defend your conclusions.

2. State what you feel should be the energy policy of the United States, based on the total range of energy options, including fossil fuels and nuclear power as well as various solar alternatives. Which energy sources should be promoted, and which should be discouraged? Suggest laws, taxes, subsidies, and so forth that might be used to bring your policy to fruition. Give a rationale for each of your recommendations.

3. Various solar energy alternatives seem to promise a sustainable energy future. Yet, the United States seems locked into a policy of remaining dependent on fossil fuels and nuclear power. Does this mean that solar advocates are wrong? Discuss the economic and political forces that might be at work in maintaining the status quo.

4. Perform an audit of all the energy needs in your daily life. How much energy do you consume? Describe how each need might be satisfied by one or another renewable-energy option.

MAKING A DIFFERENCE

PART FOUR: Chapters 13, 14, 15

1. Examine your uses of transportation: Make a car that gets high mileage per gallon a top priority, get behind car pools, use alternative transportation, keep your car tuned up and your tires inflated and balanced properly, and choose a place to live that reduces your need to drive.

2. Take advantage of the following services from your electric company: energy conservation, home-energy audits, and subsidized compact fluorescent lighting.

3. Keep informed about any nuclear power plants in your vicinity, and lobby for their strict adherence to NRC guidelines; if you do not support nuclear power, join an organization that lobbies against it.

4. Study your own home (or plans for a future one) with regard to its thermal efficiency (insulation, appliances,

etc.) and the potential for using solar water or space heating. Improve your home's thermal efficiency, and add solar power in whatever ways you can.

5. Adopt ways to use less fuel and electricity both at home and where you work. (Turn thermostats back, turn off lights when leaving a room, wear sweaters, use energy-efficient lightbulbs, etc.)

6. Choose electricity suppliers that offer renewable energy; purchase energy-efficient appliances.

7. Using the World Wide Web, research the EPA's Green Lights and Energy Star programs, and promote their use in an institution with which you are familiar—for example, your school, college, church, or business.

WEB REFERENCES

On-line resources for this chapter are on the World Wide Web at: **http://www.prenhall.com/wright**. (Click on Chapter 15 within the Chapter Selector.)

Chapter 16
Environmental Hazards and Human Health

Chapter 17
Pests and Pest Control

Chapter 18
Water Pollution and Its Prevention

Chapter 19
Municipal Solid Waste: Disposal and Recovery

Chapter 20
Hazardous Chemicals: Pollution and Prevention

Chapter 21
The Atmosphere: Climate, Climate Change, and Ozone Depletion

Chapter 22
Atmospheric Pollution

PART 5

Pollution and Prevention

An industrial plant sends a steady stream of exhaust gases into the air, and the wind carries the pollutants away. If it didn't, life in the vicinity of the plant would be impossible. But where is "away"? The next county, or state, or country? The upper atmosphere? In reality, there is no "away" for the pollution of land, water, and air resulting from the activities of the 6 billion people inhabiting Earth. As the human population continues to grow and develop economically, we are increasingly constrained by the wastes we produce.

In this section, the first chapter (Chapter 16) starts us off with a look at the links between environmental hazards and human health. In the chapters that follow, we will examine the major forms of pollution, and we will see that whether we are talking about water pollution, hazardous wastes and household wastes, or air pollution, there are sustainable ways of dealing with the problems. We will explore the possibility of a sustainable system in which people and economic enterprises do not overburden the land, air, and water with wastes. We will also see that dealing responsibly with our wastes may sometimes call for international action, as in the case of global climate change and ozone depletion. In every chapter, we consider the public policies that have emerged to deal with each of the problems raised by pollution, and we evaluate those policies in the light of our three principles of stewardship, sustainability, and sound science. Without exaggeration, the problems of pollution pose some of the most difficult and important issues facing human society as we move into a new century.

Paper mill at sunset, Georgetown, South Carolina.

Environmental Hazards and Human Health

Key Issues and Questions

1. Life expectancy is rising worldwide, yet 10 million deaths occur yearly in children under five in the developing world. What are the major causes of death in developing and developed countries?

2. Exposure to hazards in the human environment brings about the risk of injury, disease, and death. What kinds of cultural factors, infectious diseases, physical factors, and toxic chemicals are most important in this regard?

3. Hazards take many pathways in mediating environmental risks to humans. In what ways do poverty, smoking, malarial disease, and indoor air bring harm?

4. Risk assessment is a scientific tool that the EPA is applying to its regulatory work. How is risk assessment practiced by scientists and employed in policy development?

5. The public often perceives risks differently from the experts. What is the significance of risk perception in policy development?

6. The precautionary principle is now a basis of environmental policy in the international arena. How does this principle interface with the risk-based policies of the United States?

No larger than a pinhead, with a light brown abdomen and a darker brown head and thorax, the black-legged tick, *Ixodes scapularis*, is easy to overlook (Fig. 16–1a). These eight-legged agile creatures can be picked up by way of a casual walk through the brush in many parts of the United States where deer and white-footed mice are common (Fig. 16–1b). If a tick has fed on a white-footed mouse infected with the spirochaete bacterium *Borrelia burgdorferi*, the tick can become the **vector**, or carrying agent, for Lyme disease in humans, just as it is in deer. A new bite from an infected tick usually leaves a circular red rash, which may be easy to miss. In the next few weeks, the victim develops fever, chills, fatigue, and headache, very general symptoms that may be mild at their onset and hard to diagnose. If treatment is delayed, Lyme disease can become chronic and lead to long-term arthritis, memory impairment, numbness, and pain. First discovered in Lyme, Connecticut, Lyme disease is on the increase in many parts of the United States; in 1997, 12,801 cases were reported.

A recent study by a team from the Institute of Ecosystem Studies in Connecticut, led by Clive Jones, has uncovered an intriguing relationship between oak trees, gypsy moths, and Lyme disease. The gypsy moth is an introduced pest that can so ravage a forest that it looks like winter in the middle of summer. The study indicated that when the forest thrives because of natural control of the gypsy moth by mouse predation, it also presents a much greater risk of Lyme disease to people in the vicinity. If an oak forest produces a bumper crop of acorns, which happens every few years, the white-footed mouse population will thrive on the acorns. Unfortunately, the acorns also attract deer, and as a result, all the ingredients of a Lyme disease cycle are present. However, the mice also eat the overwintering pupae left behind by the gypsy moths and thereby protect the forest from being attacked the following summer, allowing it to thrive. Paradoxically, gypsy moth defoliation, not good for the forest, leads to poor acorn crops and declines in mouse populations, therefore greatly lowering the risk of

◀ ***Smoke wars.*** Wearing a breathing mask to circumvent new laws to protect workers from second-hand smoke, a bartender in British Columbia lights a patrons cigarette.

▶ **FIGURE 16–1**
Agents in Lyme disease. (a) The deer tick, *Ixodes scapularis*, carries the spirochaete bacteria that cause Lyme disease. (b) The white-footed mouse is host to the nymph stage of the tick. The existence of mice and deer in woodlands in the eastern United States provides suitable conditions for the continued presence of Lyme disease.

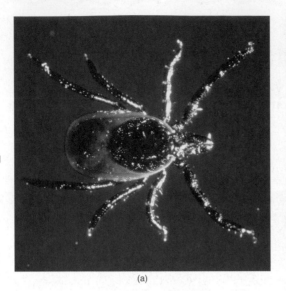

(a)

(b)

Lyme disease, which is good for the human population in the vicinity. The dilemma seems to be a choice between a healthy forest and healthy people, although as yet, no attempt has been made to intervene in this cycle.

Lyme disease is only one of a number of emerging new diseases which continually remind us that we are not always able to control the hazards posed by our inevitable contacts with the natural environment. Lassa fever, the Ebola virus, Legionnaires' disease, the hantavirus, and the West Nile virus all have taken human lives in recent years, and there is every reason to expect more new diseases to appear as we continue to manipulate the natural environment and change ecological relationships.

However, to put these diseases in perspective, it is not the new, emerging diseases that represent the greatest threat to us; rather, it is the common, familiar ones that take the greatest toll of human life and health—diseases like malaria, diarrhea, respiratory viruses, and worm infestations, which have been kept at bay in the developed countries, but continue to ravage the developing countries. In the developed countries, cancer is the killer that is most closely linked to the environment, leaving us with questions about the chemicals we are exposed to on a daily basis.

The study of the connections between hazards in the environment, on the one hand, and human disease and death, on the other, is often called **environmental health**, but it should be realized that it is *human* health that is in focus here, not the health of the natural environment. In this chapter, we will examine the nature of environmental hazards and the consequences of exposure to those hazards. Then we will consider the pathways whereby humans encounter the hazards, some interventions that can alleviate the risks to health, and how public policy can address the need for continued surveillance and improvement in our management of environmental hazards.

16.1 Links between Human Health and the Environment

Given our focus on human health, we can more precisely define how we will be using the common term "**environment**" in this chapter. By "environment," we mean the whole context of human life—the physical, chemical, and biological setting of where and how people live. Thus, the home, air, water, food, neighborhood, workplace, and even climate constitute elements of the human environment. Within human environments, there are *hazards* that can make us sick, cut our lives short, or contribute in other ways to human misfortune. In the context of environmental health, a **hazard** is defined as anything that can cause (1) injury, disease, or death to humans, (2) damage to personal or public property, or (3) deterioration or destruction of environmental components. The existence of a hazard does not mean that undesirable consequences inevitably follow. Instead, we speak of the connection between a hazard and something happening because of that hazard as a **risk**, defined here as the *probability* of suffering injury, disease, death, or some other loss as a result of exposure to a hazard. Since risks are expressed as probabilities, the analysis of the probability of suffering some harm from a hazard becomes an exercise that must be carried out by scientists or other experts, and we will eventually deal with the assessment of risk as it becomes the basis of policy decisions. For now, however, our focus is on the nature of environmental hazards.

The Picture of Health

Health has many dimensions—physical, mental, spiritual, and emotional. The World Health Organization defines *health* as "a state of complete mental, physical and social well-being, and not merely the absence of disease." This is an admirable definition, but measuring all these dimensions of health for a society is virtually impossible. For our focus on environmental health, we will admit that such dimensions exist and are very important, but, of necessity, we will focus our attention on *disease* and consider health to be simply the absence of disease. Two measures are used in studies of disease in societies: morbidity and mortality. **Morbidity** is the incidence of disease in a population and is commonly used to trace the presence of a particular type of illness, like influenza or diarrheal disease. **Mortality** refers to the incidence of death in a population; records are usually kept of the cause of death, making possible an analysis of the relative roles played by infectious diseases and factors such as cancer and heart disease. **Epidemiology** is the study of the presence, distribution, and control of disease in populations.

One universal indicator of health is human life expectancy. In 1955, average life expectancy globally was 48 years. Today, it is 66 years and rising, and it is expected to reach 73 years by 2025. This progress is the result of social, medical, and economic advances in the latter half of the 20th century that have substantially increased the well-being of large segments of the human population. In Chapter 6, we discussed the *epidemiologic transition*—the trend of decreasing death rates that is seen in countries as they modernize. We saw that, as this transition occurred in the developed countries, it involved a shift from high mortality (due to infectious diseases) toward the present low mortality rates (due primarily to diseases of aging, like cancer and cardiovascular diseases).

However, countries of the world have undergone this transition to different degrees, with very different consequences. In spite of the progress indicated by the general rise in life expectancy, one-fifth (10 million) of the annual deaths in the world are children under the age of five in the developing world. The discrepancy between the developed (high-income) and developing (low-income) countries can be seen quite clearly in the distribution of deaths by main causes, shown for 1998 in Fig. 16–2. Infectious diseases were responsible for 35% of mortality in the developing countries, compared with only 6% in the developed countries.

An entirely different perspective is brought into play as we consider some of the environmental hazards that accompany industrial growth and intensive agriculture. In addition, some of the most lethal hazards in the developed world are the outcome of purely voluntary behavior—in particular, smoking and drug use. Let us look into these different categories of environmental hazards.

Environmental Hazards

There are two fundamental ways to consider hazards to human health. One is to regard the *lack of access* to necessary resources as a hazard. No one would argue with the fact that lack of access to clean water and nourishing food is harmful to a person. Investigating hazards from this perspective means considering the social, economic, and political factors that prevent a person from having access to such basic needs. Although this is a fundamentally important perspective, much of it is outside the scope of the chapter. Instead, our focus will be primarily on the second way of considering hazards, namely, *exposure to hazards* in the environment. What is it in the environment that brings the risk of injury, disease, or death to people? Here we will consider four classes of hazards: cultural, biological, physical, and chemical.

Cultural Hazards. Many of the factors that contribute to mortality and disability are a matter of choice or at least can be influenced by choice. People engage in risky behavior and subject themselves to hazards. Thus, we may eat too much, drive too fast, use addictive and harmful drugs, consume alcoholic beverages, smoke, sunbathe, hang glide, engage in risky sexual practices, get too little exercise, or choose hazardous occupations. Why do we subject ourselves to these hazards? Basically, because we derive some pleasure or other benefit from them. Wanting the benefit, we are willing to take the risk that the hazard will not harm us. Factors such as living in inner cities, engaging in criminal activities, and so on can be added to the list of cultural sources of mortality. As Fig. 16–3 indicates, a large fraction of the mortality in the United States can be traced to such factors. It goes without saying that most of these deaths are preventable. However, trying to convince people that they should be more careful is often a losing battle.

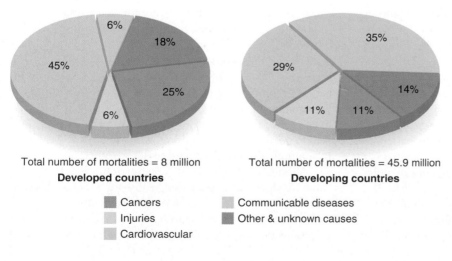

Total number of mortalities = 8 million
Developed countries

Total number of mortalities = 45.9 million
Developing countries

- Cancers
- Injuries
- Cardiovascular
- Communicable diseases
- Other & unknown causes

◀ **FIGURE 16–2** *Leading causes of mortality in developing and developed countries for 1998.* Infectious diseases are prominent in the developing countries, while cancer and diseases of the cardiovascular system predominate in the developed countries. (*Source:* Data from World Health Organization, *1999 World Health Report.*)

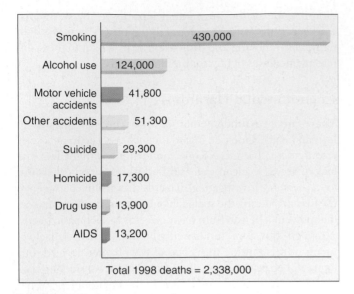

Smoking	430,000
Alcohol use	124,000
Motor vehicle accidents	41,800
Other accidents	51,300
Suicide	29,300
Homicide	17,300
Drug use	13,900
AIDS	13,200

Total 1998 deaths = 2,338,000

▲ **FIGURE 16–3   Deaths from various cultural hazards in the United States, 1998.**   Many hazards in the environment are a matter of personal choice and lifestyle. Other hazards are due to accidents and the lethal behavior of other people. (*Source*: Data from Centers for Disease Control, National Center for Health Statistics, 2000; smoking deaths = average from 1990 to 1994; alcohol-use deaths from National Council on Alcoholism and Drug Dependency.)

TABLE 16–1   Mortality from Major Infectious Diseases, 1999

Cause of Death	Estimated Yearly Deaths*
Acute respiratory infections	4,039,000
HIV/AIDS	2,673,000
Diarrheal diseases	2,213,000
Tuberculosis	1,670,000
Malaria	1,086,000
Measles	875,000
Tetanus	377,000
Pertussis (whooping cough)	295,000
Meningitis, bacterial	171,000
Syphilis	153,000
Hepatitis	124,000
Trypanosomiasis (sleeping sickness)	66,000
Leishmaniasis	57,000
Intestinal roundworms	16,000
Schistosomiasis	14,000

*Total deaths from infectious diseases in 1999 are estimated at 17.4 million by the World Health Organization.
(*Source*: Data from World Health Organization, *World Health Report 2000.*)

The connection between behavior and risk is especially lethal in the case of acquired immune deficiency syndrome (AIDS), caused by the human immunodeficiency virus (HIV). AIDS is one of a group of sexually transmitted diseases (STDs) that includes gonorrhea, syphilis, and genital herpes. As we saw in Chapter 7, AIDS is taking a terrible toll in the developing world, where over 90% of HIV-infected people live. Wherever it occurs, AIDS is largely the consequence of high-risk sexual behavior. At present, there are no cures or vaccines for the HIV virus.

Biological Hazards.   Human history can be told from the perspective of the battle with pathogenic bacteria and viruses. It is a story of epidemics like the black plague and typhus, which ravaged Europe in the Middle Ages, killing millions in every city, and of smallpox, which swept through the New World, to which explorers brought not only their weapons of warfare, but also their much more lethal diseases. The story continues in the 19th century with the first vaccinations and the "golden age" of bacteriology (a span of only 30 years), when bacteriologists discovered most of the major bacterial diseases and brought bacteria into laboratory culture. The 20th century saw the advent of virology (the study of viruses and the treatment of the diseases they cause); the great discovery of antibiotics; immunizations that led to the global eradication of smallpox, the defeat of polio, and the victory over many childhood diseases; and the growing influence of molecular biology as a powerful tool in the battle against disease.

The battle is not over, however, and never will be. Pathogenic bacteria, fungi, viruses, protozoans, and worms continue to plague every society and, indeed, every person. They are inevitable components of our environment; many are there regardless of our human presence, and others are uniquely human pathogens whose access to new, susceptible hosts is mediated by the environment. Table 16–1 shows the leading killers and the annual number of deaths they cause.

Approximately 31% of the 56 million deaths in 1999 were due to infectious and parasitic diseases. The leading causes of death in this category are the *acute respiratory infections* (e.g., pneumonia, diphtheria, tuberculosis, whooping cough, influenza, and streptococcal infections), both bacterial and viral. Pneumonia is by far the most deadly of these (Fig. 16–4). Other bacterial and viral respiratory infections represent the most common reasons for visits to the pediatrician in the developed countries, but it is primarily in the developing countries that the respiratory infections lead to death, mostly in children who are already weakened by malnourishment or other diseases.

Although AIDS has overtaken tuberculosis as the largest cause of adult deaths from a single disease, tuberculosis continues to be a major killer. Complacency about treatment and prevention in recent years has led to a resurgence of this infectious disease, even though the bacterial pathogen that causes it has been known for over 100 years. Some of the resurgence can be traced to the AIDS epidemic: HIV infection eventually leads to a compromised immune system, which allows diseases like tuberculosis to flourish.

Diarrheal diseases (e.g., cholera, dysentery, salmonellosis, giardiasis, and many viral and bacterial infections), responsible for 2.2 million deaths in 1999, have the most obvious link to the environment. Most serious cases are the consequence of ingesting food or water contaminated with pathogens, such as *Salmonella, Campylobacter,* and *E. coli,* from human wastes. Infection in small children often leads to dehydration, which can quickly be life threatening; most deaths from diarrhea occur to children under five.

Of the infectious diseases present in the tropics, malaria is by far the most serious, accounting for an estimated 300 million cases each year and over a million deaths. Caused by protozoan parasites of the genus *Plasmodium,* malaria is propagated by mosquitos of the genus *Anopheles.* The life cycle of the protozoan begins with a mosquito biting an infected person and then incubating the parasite within itself (Fig. 16–5). The infected mosquito can then transmit the disease to human hosts with every bite. Within humans, the parasite invades the liver, where it multiplies and eventually breaks out to invade red blood cells. After

▲ **FIGURE 16–4** *Jim Henson.* The famous creator of the Muppets succumbed in 1990 to a virulent pneumonia caused by the common bacterium *Streptococcus pneumoniae.*

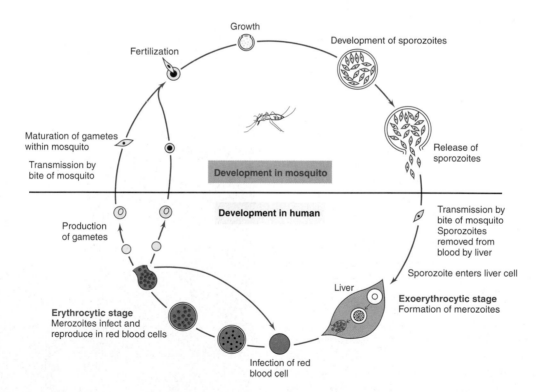

▲ **FIGURE 16–5** *Life cycle of malarial parasite.* Part of the life cycle of the protozoan parasite *Plasmodium,* which causes malaria in humans, is in the anopheles mosquito. In the mosquito, the protozoan develops from gametes (sex cells) into small, elongated cells called sporozoites, which multiply and migrate to the mosquito's salivary glands. The sporozoite stage can then be transmitted to humans through a mosquito bite. Sporozoites migrate to the liver and develop into small cells called merozoites, which multiply and then are released into the blood. There they enter red blood cells (erythrocytes) and multiply further. The parasite harms the human host by destroying red blood cells and by releasing fever-inducing substances into the bloodstream. Protozoal cells in the blood can then infect another anopheles mosquito when it bites the infected person, thus beginning the cycle again. (After Madigan, Michael T., Martinko, John M., and Parker, Jack, *Brock Biology of Microorganisms,* 8th ed., 1997. Reprinted by permission of Prentice Hall, Inc., Upper Saddle River, NJ.)

multiplying in the red blood cells, the parasites are released, destroying the cells. The parasites then invade other red blood cells. This entire chain of events occurs in synchronized cycles, leading to the periodic episodes of fever, chills, and malaise typical of malaria. Eventually, as much as 10% of the red blood cells may be destroyed in one episode, often leading to anemia. The waste products of the parasites, when they break out of the cells, produce high fever, headaches, and vomiting in the victim. If the parasites invade the brain (cerebral malaria), death occurs unless medical help is obtained quickly.

Most of the diseases responsible for mortality are also leading causes of debilitation in humans of all ages. Over 4 billion episodes of diarrhea occur to the human population in the course of a year; at any given time, more than 3.5 billion people suffer from the chronic effects of parasitic worms such as hookworms and schistosomes (Fig. 16–6).

Physical Hazards. Natural disasters, including hurricanes, tornadoes, floods, earthquakes, landslides, and volcanic eruptions (Fig. 16–7), take a toll of human life and property every year. It can be argued that much of the loss brought on by natural disasters is a consequence of poor environmental stewardship. Hillsides and mountains are deforested, leaving the soil unprotected; people build homes and towns on floodplains; villages are nestled up to volcanic mountains; and cities are constructed on known geological fault lines. A great human tendency is to assume that disasters happen only to other people in other places, and even if there is a risk, say, of a hurricane striking a coastal island, many people are willing to take that calculated risk in order to enjoy life on the water's edge.

Some disasters, however, are impossible to anticipate. Tornadoes are a good example. Each year in the United States, an average of 780 tornadoes strike, spawned from the severe weather accompanying thunderstorms. Because of the tendency for cold, dry air masses from the north to mix with warm, humid air from the Gulf of Mexico, the central United States generates more tornadoes than anywhere else on Earth. Of short duration, they develop into some of the most destructive forces known in nature, as winds reach as high as 300 miles per hour. The most intense tornadoes have killed hundreds of people; on April 3, 1974, a line of tornadoes from Canada to Georgia took more than 300 lives and caused untold property damage.

The most devastating consequences of natural hazards usually fall on those who are least capable of anticipating them and dealing with their effects. For example, in October 1998, Hurricane Mitch developed in the Caribbean Sea and slowly made its way across Central America. The combination of storm surge, winds, and rain devastated the developing countries of Honduras, El Salvador, Nicaragua, and Guatemala, killing over 10,000 people. Some locations received a year's worth of rain in one day, and the result was mud slides that obliterated whole villages. An earthquake in Izmit, Turkey, in August 1999 killed 17,000. Orissa state in

India was hit with a massive cyclone in November 1999, killing at least 3,000 and displacing no fewer than 2 million people. In December 1999, Venezuela's Caribbean coast was lashed by devastating rains, triggering massive mud slides and floods that killed more than 30,000. Mozambique and South Africa were inundated by floodwaters in early 2000, leaving over a million homeless. Every year

(a)

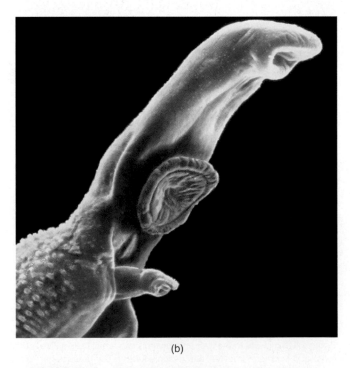

(b)

▲ **FIGURE 16–6** *Chronic biological hazards.* Parasitic worms such as (a) hookworms and (b) schistosomes infest millions and cause serious disease and debilitation.

(a)

(b)

(c)

(d)

▲ FIGURE 16–7 *Dangerous physical hazards.* Natural disasters bring death and destruction in their train. (a) Effects of Hurricane Mitch in Tegucigalpa, Honduras; (b) aftermath of an earthquake in Kobe, Japan; (c) a tornado as it swept through Cantrall, Illinois, in early May 1995; and (d) Popocatépetl, a volcano located 50 miles southeast of Mexico City, erupted again on November 27, 1998.

brings its list of disasters and the toll in deaths and the miseries of people left homeless. Most such disasters occur to people in the developing countries.

In response to this rising toll of disasters, in 2000 the World Bank launched an international partnership, the **ProVention Consortium**, to help equip the developing countries for coping with natural disasters. The consortium of agencies, governments, and universities will address in particular the link between poverty and vulnerability to disaster and will work to protect the natural environments that can shield human settlements from weather-related disasters. For instance, loans to Turkey to rebuild after the Izmit earthquake are being made contingent on the country's updating and enforcing building codes. (Poor-quality construction led to a large number of the deaths in the earthquake.)

Chemical Hazards.   Industrialization has brought with it a host of technologies that employ chemicals such as cleaning agents, pesticides, fuels, paints, medicines, and many directly used in industrial processes (Fig. 16–8). The manufacture, use, and disposal of these chemicals often bring humans into contact with them. Exposure is either through the ingestion of contaminated food and drink, breathing of contaminated air, absorption through the skin, or direct use, or else it occurs by accident (discussed at length in Chapter 20). Toxicity (the condition of being harmful, deadly, or poisonous) depends not only on exposure, but on the *dose* of a toxic substance—the amount actually absorbed by a sensitive organ. Further, it is known that for most substances, there is a threshold below which no toxicity can be detected. (See Fig. 20–1.) Making things even more difficult, different people have different thresholds of toxicity for given substances. For example, children are at greater risk than adults, because they are growing rapidly and are incorporating more of their food (and any contaminants that are in it) into new tissue. Embryos are even more sensitive: Substances transmitted across the mother's placenta can have grave impacts on the embryo's development, especially in the early stages.

(a)

(b)

▲ **FIGURE 16–8   *Industrial processes and hazards.***   Many industries produce chemicals and pollutants that contribute to the impact of toxic substances on human health. Some examples are (a) a paper mill in Port Angeles, Washington, (b) a tannery in Marrakech, Morocco, (c) an oil refinery in Philadelphia, Pennsylvania, and (d) a plastics factory.

Many chemicals, such as heavy metals, organic solvents, and pesticides, are hazardous to human health even at very low concentrations (Table 16–2). Episodes of acute poisoning are easy to understand and are clearly preventable, but the long-term exposure to low levels of many of these substances presents the greatest challenge to our understanding. Some chemical hazards are known *carcinogens* (cancer-causing agents). As Fig. 16–2 indicates, cancer takes a very large toll in the developed countries, accounting for one-fourth of mortality. Because cancer often develops over a period of 10 to 40 years, it is frequently hard to connect the cause with the effect. The 2000 *Report on Carcinogens* from the Department of Health and Human Services lists 65 chemicals now known to be human carcinogens and 153 more that are "reasonably anticipated to be human carcinogens." Many are substances used in both developed and developing countries. Other potential effects of toxic chemicals are impairment of the immune system, brain impairment, infertility, and birth defects.

Many developing countries are in double jeopardy because of both their high level of infectious disease and the rising exposure to toxic chemicals due to industrial processes and home uses of such chemicals. In both cases, the risks are largely preventable. In places where industrial growth is especially rapid, it is often the case that insufficient attention is given to the pollutants which accompany that growth. For example, 13 of the 15 cities with the worst air pollution are found in Asia, where economic growth has risen rapidly.

Cancer

The word "cancer" carries a burden of fear and uncertainty in its train. In the United States, 23% of all deaths—539,000 in 1998—are traced to cancer, and as everyone knows, even though it is largely a disease of older adults, it can strike people of all ages. How is it that a simple chemical like benzene can cause leukemia? How can aflatoxin cause liver cancer? The answer to both of these questions involves an understanding of cellular growth and metabolism. In Chapter 5, we saw that a mutation is a change in the informational content of the DNA that controls a living

---

TABLE 16–2   Hazardous Chemicals in the Environment

Chemicals found in the air in urban environments

- lead (from vehicle exhausts and industrial processes)
- hydrocarbons and other volatile organic compounds (from vehicle exhausts, and industrial processes)
- nitrogen oxides and sulfur oxides (from vehicle exhausts and fixed combustion processes like those carried out in power plants)
- particulate matter (dust, soot, and metal particles from vehicle exhausts, industrial processes, and construction)
- ozone (a secondary pollutant produced from the interaction of hydrocarbons and nitrogen oxides in sunlight)
- carbon monoxide (from vehicle exhausts and other combustion processes)

Chemicals found in food and water

- pesticides and herbicides (agricultural applications to food crops)
- heavy metals such as arsenic, aluminum, and mercury (in soil and water contaminated with industrial chemicals)
- lead (from water pipes)
- aflatoxins and other natural toxins (in foods contaminated by fungal growth)
- nitrates (in drinking water with a high nitrogen content)

Chemicals found in the home and workplace (indoors)

- lead (in household paints)
- smoke and particles (from the combustion of biofuels, coal, and kerosene)
- carbon monoxide (from the incomplete combustion of fuels)
- asbestos (in insulation)
- tobacco smoke
- household products (causing poisonings and exposure from their improper use)

Chemicals contaminating some land sites

- cadmium, chromium, and other heavy metals (in industrial wastes)
- dioxins, PCBs, and other persistent organic chemicals (in industrial wastes)

cell. During normal growth, cells develop into a variety of specific types that perform unique functions in the body: blood cells, heart muscle cells, liver cells, absorptive cells that line the small intestine, and so forth. Some cells, like the stem cells that form the cells of the blood, keep on dividing until death. Other cells, however, are programmed to stop dividing and simply perform their functions, often throughout life. A cancer is a cell line that has lost its normal control over growth. Recent research has revealed the presence of as many as three dozen genes that can bring about a malignant (out-of-control) cancer; these genes are like ticking time bombs that may or may not go off.

Carcinogenesis (the development of a cancer) is now known to be a process with several steps, often with long periods of time in between. In most cases, a sequence of five or more mutations must occur in order to initiate a cancer. Environmental carcinogens are either chemicals that are able to bind to DNA and prevent it from functioning properly or agents like radiation that can strike the DNA and disrupt it. When such an agent initiates a mutation in a cell, the cell may go for years before the next step—another mutation—occurs. Thus, it is easy to see how it could take upwards of 40 years for some cells to accumulate the sequence of mutations that leads to a malignancy. In a malignancy, the cells grow out of control and form tumors that may spread (metastasize) to other parts of the body. Early detection of cancers has improved greatly over the years, and many people survive cancer who would have succumbed to it a decade ago. The best strategy, however, is prevention, so great attention is now given to the environmental carcinogens and habits like smoking that are clearly related to cancer development. The World Health Organization estimates that 25% of cancers can be traced to environmental causes.

16.2 Pathways of Risk

The environmental hazards identified in the previous section are clearly responsible for untold human misery and death. But if many of the consequences of exposure to the different hazards are preventable, how is it that the hazards turn into mortality statistics? In other words, what are the pathways whereby risks of infection, exposure to chemicals, and vulnerability to physical hazards are mediated? If prevention is our goal—and it should be—this knowledge is crucial. We will consider a few examples of these pathways, realizing that a comprehensive examination of this question would take volumes.

The Risks of Being Poor

One major pathway for hazards is poverty. Indeed, the World Health Organization (WHO) has stated that poverty is the world's biggest killer. Environmental risks are borne more often by the poor, not only in the developing countries, but also in the advanced countries of the world. The people who are dying from infectious diseases are people who lack access to adequate health care, clean water, nutritious food, healthy air, sanitation, and shelter. Because of this, they are exposed to more environmental hazards and thus encounter greater risks.

Once again, we are reminded of the role of wealth in determining the well-being of people and of the gap between wealthy and poor countries or, what is the same thing, between developed and developing countries. Overall, the wealthier a country becomes, the healthier is its population. However, those with wealth in virtually all countries are able to protect themselves from many environmental hazards. People in developed countries tend to live longer, and when they die, the chief causes of death are diseases of old age. Sixty-seven percent of the death toll in the developed countries can be traced to the so-called degenerative diseases, such as cancers and diseases of the circulatory system (heart disease, strokes, etc.). By contrast, only 33% of the deaths in the developing countries are indicative of people who are living out their life span. Another indicator is age at death; people under the age of 50 account for two-fifths of the annual deaths in the world (more than 20 million). Most of these deaths occur in the developing world, are due to infectious disease, and are preventable.

There are many reasons for this discrepancy. One is education: Advances in socioeconomic development and advances in education are strongly correlated. As people—and especially women—become more educated, they act on their knowledge to secure relief from hazards to health. They may improve their hygiene, for example, immunize their children (Fig. 16–9), or recognize dangerous symptoms such as dehydration and seek oral rehydration therapy (the replacement of lost fluids with balanced saline solutions). Another reason for the discrepancy between health in the developed versus the developing world is nutrition. The poor in the poorest of the developing countries are most susceptible to malnutrition and hunger, as we saw in Chapter 10. Malnutrition acts in tandem with disease to make the impacts of ordinary diseases like diarrhea and respiratory infections life threatening. Indeed, malnutrition is thought to contribute to at least one-half of deaths of children in the developing countries.

Education, nutrition, and, indeed, the general level of wealth in a society do not tell the whole story, however. A nation may make a deliberate choice to put its resources into improving the health of its population above, say, militarization or the development of modern power sources. Countries like Costa Rica, China, and Sri Lanka have a much longer life expectancy and lower infant mortality than their gross domestic product would predict. They have done this by focusing public resources on such concerns as immunization, upgrading sewer and water systems, and land reform.

Another factor in the equation of environmental health is the relative distribution of wealth in a society. Where the

◀ **FIGURE 16–9** *Public health clinic.* This outreach immunization center in Bangladesh provides basic health information and services to people in the surrounding area.

gap between the rich and the poor is lower and the general well-being of the society is higher, the health status of the society is also higher. For example, both Japan and the United States have a similarly high economic status. Japan, however, now has the highest life expectancy in the world (at 81 years) and the most equitable income distribution. The United States, on the other hand, ranks 21st out of 157 countries in life expectancy and has great inequities between the wealthy and the poor.

The Cultural Risk of Smoking

Lifestyle choices such as refraining from exercise, overeating, driving fast, imbibing alcohol, climbing mountains, and so forth carry with them a significant risk of accident and death. The cultural hazard that carries the highest risk of harm, however, is **smoking**. Like ingesting alcohol and other drugs, smoking is considered to be a *personal pollutant*, because it is a cultural hazard people expose themselves to on a voluntary basis. Unfortunately, recent trends show that people are likely to start smoking at a younger age. Lured by the image of smoking as "cool" or by peer pressure, youngsters are soon addicted to the nicotine in cigarettes, an addiction that is difficult to break. Recent research indicates that nicotine and cocaine share the same neural pathways, putting the two substances on a par physiologically.

Essentially, smoking amounts to portable air pollution. The smoker is the primary recipient of the pollution, but the effects spread to all who breathe the smoke-clouded air and, in the case of pregnant women, to the unborn fetus. Worldwatch Institute's William Chandler maintains that "Tobacco causes more death and suffering among adults than any other toxic material in the environment." Cigarette smoking has been clearly and indisputably correlated with cancer

and other lung diseases. Smoking has been shown to be responsible for 30% of cancer deaths in the United States. It is the leading cause of preventable deaths in the United States, with at least 430,000 people dying each year from smoking-related causes (Fig. 16–10). The World Health Organization (WHO) puts the number of smokers worldwide at 1.1 billion, with 4 million deaths per year due to cigarette smoking. Eighty percent of smokers are in the developing world, where smoking is increasing at a rate of 2% per year. In China, where smoking is practically epidemic, 63 percent of men are smokers. Studies have shown that smokers living in polluted air experience a much higher incidence of lung disease than smokers living in clean air—demonstrating a synergistic effect. Certain diseases typically associated with occupational air pollution show the same synergistic relationship with smoking. For example,

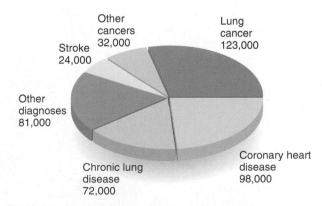

▲ **FIGURE 16–10** *Deaths caused by smoking.* Approximately 430,000 deaths in the United States are attributed to smoking. (*Source:* Data from Centers for Disease Control, *MMWR,* Mar. 3, 1999; avg. annual number of deaths, 1990–1994.)

black-lung disease is seen almost exclusively among coal miners who are also smokers, and lung disease predominates in smokers who are exposed to asbestos. Smoking also kills by impairing respiratory and cardiovascular functioning. Tobacco-related illness is estimated to cost almost $200 billion a year in lost work and health-care expenses. Huge increases in deaths will occur in the next two decades, owing to the rapid rise in smoking.

Several measures have been taken in order to regulate this cultural hazard. Surgeons general of the United States have issued repeated warnings against smoking since 1964, and public policy has taken them seriously by requiring warning labels on smoking materials, banning cigarette advertising on television, promoting smoke-free workplaces, requiring nonsmokers' areas in restaurants, and banning smoking on all domestic airline flights. Since the warnings began, the U.S. smoking population has gradually dropped from 40% to 26%.

Some recent developments in public policy in the United States are bringing major changes in the smoking arena. In January 1993, the EPA classified environmental tobacco smoke, (ETS), meaning secondhand or sidestream smoke, as a Class A (known human) carcinogen. In the words of EPA administrator Carol Browner, "Widespread exposure to secondhand smoke in the United States presents a serious and substantial public health risk." The EPA's action was unanimously supported by its science advisory board. Fallout from this action has been substantial: Smoke-free zones in public areas such as shopping malls, bars, restaurants, and workplaces are now common. In particular, the EPA has taken specific steps to protect children from ETS in all public places, and the Occupational Safety and Health Administration (OSHA) is promoting policies to protect workers from involuntary exposure to ETS in the workplace. The impact of these actions on public health should be substantial: One assessment indicated that just a 2% decrease in ETS would be the equivalent, in terms of human exposure to harmful particulates, of eliminating all the coal-fired power plants in the country. As public policy moves toward banning smoking in more and more locations, however, the tobacco industry is fighting back with advertisements emphasizing smoking as an important freedom. (See the "Ethics" essay, p. 403.)

Another significant development came from the Food and Drug Administration (FDA). In February 1994, citing the addictive properties of nicotine, then FDA administrator David Kessler proposed that the agency have the authority to regulate cigarettes as a drug. At congressional hearings following the FDA's initiative, tobacco executives denied that cigarette smoking is addictive. The hearings included allegations of a cover-up of the tobacco industry's own research, which revealed the addictive properties of nicotine. A year later, President Clinton declared nicotine an addictive drug, a move that was intended to give the FDA regulatory authority over the substance. The FDA launched efforts to curb tobacco sales to the underaged. One early casualty was "Joe Camel," a cartoon character that seemed to be targeted at the young (Fig. 16–11). However, in March 2000, the U.S. Supreme Court ruled that the FDA does not have the authority to regulate tobacco. That authority, said the Court, currently belongs to Congress. The Congress may draft its own regulations, or it may choose to give FDA regulatory authority over products containing addictive nicotine. However, the tobacco industry has many friends in Congress, and any new legislation intending to tighten regulations over tobacco products will face an uphill battle.

On the legal front, there has been more success. The attorneys general of a number of states brought suit to re-

▶ **FIGURE 16–11** *Joe Camel.* This cartoon character was featured for years in ads by R. J. Reynolds. It developed a very high recognition rate among children, and because of this, the FDA pressured the manufacturer to stop using its Joe Camel ads.

cover some of their Medicaid expenses brought about by smoking-related illnesses. In late 1998, 46 states reached a $206 billion settlement with the top four tobacco companies wherein the companies would remunerate the states over a period of years, while the states would agree to drop lawsuits asking the companies to pay for their Medicaid expenses. The money is being used not only to reimburse the states, but also to help finance programs to discourage people—especially youths—from smoking. The settlement does not preclude other suits against the tobacco companies, and these are increasingly meeting with success after decades in which juries refused to award judgments to litigants, believing that smokers should be held responsible for their actions. In July 2000, a two-year class-action case in Florida was finally decided, and the litigants (thousands of smokers and former smokers) were awarded $149 billion! The jurors were convinced by arguments that the industry had been deliberately deceiving the public about its product. The tobacco companies have appealed the judgment, of course, and the penalty will likely be reduced substantially. However, some observers believe that this suit will open a floodgate of similar suits, and they seriously question whether the tobacco industry can survive for long.

The "resolution" of this controversy, if, indeed, it is a resolution, is symptomatic of our ambivalence toward tobacco in the United States. On the one hand, we try to protect people from the serious health effects of smoking; public policy continues to move towards banning smoking in more and more locations, making it increasingly more difficult for smokers to maintain their habit. (See "Ethics"

ETHICS

THE RIGHTS OF SMOKERS?

Smoking in public is becoming socially unacceptable in many parts of the country, forcing many smokers to retreat to rest rooms, specially designated smoking areas, or sidewalks. Almost all states have laws restricting smoking in public places, and federal law now prohibits smoking on all domestic airline flights.

Although it was once considered simply a nuisance to have to breathe secondhand smoke, evidence has shown that consistent breathing of sidestream smoke can lead to some of the same consequences experienced by smokers. One study showed that nonsmoking wives of men who smoke are twice as likely to die of lung cancer as are nonsmoking wives of men who do not smoke. Some workers (bartenders, food servers, and musicians, for instance) involuntarily inhale the equivalent of 10 or more cigarettes per day. All of this new knowledge has put smokers on the defensive, as nonsmokers adopt slogans such as "Your right to smoke stops where my nose begins" and "If you smoke, don't exhale!"

Smoking, of course, has its defenders. Smokers feel that they have a right to indulge in their habit and would like nonsmokers to understand that smokers have a physical addiction that is quite difficult to break. Some smokers have become belligerent about the restrictions and are willing to take their rights into court. The newest strategy from the smoking lobby emphasizes smoking as an issue of personal freedom, like the right to use alcohol or caffeine. However, court challenges regarding the restrictions on smoking have not gone well for smokers. For example, a federal court upheld a ruling that prohibited a firefighter from smoking, whether on or off the job, because it was determined that firefighters must be in top physical condition and are subjected to hazards that smoking could aggravate. The judges ruled that smoking was not protected by the constitutional right to privacy. Civil libertarians object to such restrictions on the basis that a person has a right to smoke; others counter with arguments that individual liberty can be limited if there is a rational purpose to doing so.

In the United States, nonsmokers are in the majority, and they seem to be gaining the upper hand in the controversy over public smoking. Have we gone too far, or should we do even more to curb sidestream smoke? What do you think?

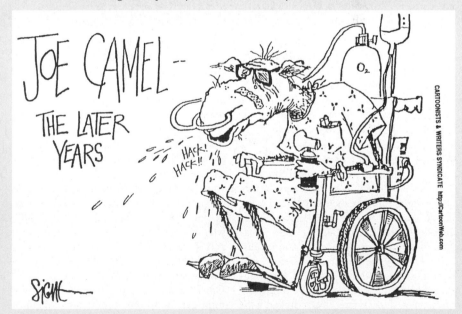

Confined to a wheelchair and breathing oxygen, Old Joe Camel displays the growing public awareness of the health hazards of many lifetime smokers, as well as of those who are exposed to sidestream smoke. The medical costs associated with this level of health care have prompted the many lawsuits mounted against the tobacco companies.

essay, p. 403.) On the other hand, we continue to subsidize the tobacco industry and tobacco exports, and we collect billions of dollars in taxes on the products. Overseas sales of American cigarettes total more than $5 billion a year and have surpassed domestic sales for the first time. The Congress and U.S. trade representatives vigorously protect this market. However, the battle against smoking is rapidly spreading worldwide. In October 2000, the World Health Organization sponsored a global meeting of health officials, with the intention of drafting an international treaty—the **Framework Convention on Tobacco Control.** It will be interesting to see what shape this agreement takes and whether the United States will support it.

Risk and Infectious Diseases

Epidemiology has been described as "medical ecology," because the epidemiologist traces a disease as it occurs in geographic locations, as well as the mode of transmission and consequences of the disease. Epidemiologists point to many reasons infectious diseases and parasites are more prevalent in the developing countries. One major pathway of risk is contamination of food and water due to inadequate hygiene and to the lack of adequate sewage treatment. If drinking water is not treated thoroughly, it becomes a vehicle for transmitting human pathogens from feces. Some of the risk

can be traced to the failure to educate people about diseases and how they may be prevented. Much, however, is simply a consequence of the general lack of resources needed to build effective public health-related infrastructure in cities, towns, and villages of the developing world. Globally, an estimated 2.4 billion people are without adequate sanitation, and 1.5 billion lack access to safe drinking water. Many diarrheal diseases are transmitted in this way, some of which, like cholera and typhoid fever, are deadly and capable of spreading to epidemic proportions. Any diarrheal infection in small children can be fatal if it is accompanied by malnutrition or dehydration.

These problems are not unique to the developing world. Inadequate surveillance in developed countries sometimes leads to major outbreaks of diarrheal diseases, with a few fatal outcomes. For instance, in 1993, the water supply of Milwaukee, Wisconsin, became contaminated with *Cryptosporidium parvum*, a protozoal parasite originating in farm animal wastes. Some 370,000 people developed diarrheal illnesses from the outbreak, and over 4,000 were hospitalized. The factorylike conditions in which food processing in the developed countries takes place often give pathogens an opportunity to contaminate food on a wide scale, resulting in diarrheal outbreaks and major recalls of food products. The occurrence of such outbreaks, the resurgence of known diseases, and the emergence of new ones

GLOBAL PERSPECTIVE

AN UNWELCOME GLOBALIZATION

We live in a time when globalization is rapidly encompassing travel, information, trade, and investment. The World Wide Web ties people together in ways unimagined a few years ago. The globalization of health, however, remains an elusive goal, similar to the globalization of economic well-being. Laurie Garrett, in her book *The Coming Plague*, describes an unwelcome form of globalization: the globalization of disease. Garrett examines the recent history of emerging diseases such as AIDS, Ebola, hantavirus, Rift Valley fever, Legionnaires' disease, and others. She also provides explanations for the resurgence of familiar diseases like tuberculosis, cholera, and pneumonia as a consequence of the widespread and unwise use of antibiotics. Many of the new diseases are clearly linked with changes in land use, which bring humans into close contact with rodents or other animals that harbor viruses previously unknown to medicine and often deadly to

humans. Resurgent diseases, on the other hand, are a creation of our medical practice. By treating people with antibiotics without restraint, we unknowingly select strains that are immune to the antibiotics and that pass on their resistant genes to unrelated bacteria by way of plasmid transfer. (Plasmids are small circular pieces of DNA that can move from one bacterium to another in a semisexual process.)

The heroes of the book are the women and men on the front lines of epidemiology, people like Joe McCormick of the CDC, who doggedly pursued lethal viruses in Africa, or Patricia Webb, who worked in maximum-containment laboratories on the deadly viruses back in the United States. Garrett makes a plea for a greater commitment from our universities, medical schools, and government agencies to train workers who will be capable of recognizing new diseases and who will be able to move about equally well in the laboratory, the

hospital, and the field in pursuit of knowledge and public health intervention around the world.

The importance of Garrett's perspective has been emphasized recently with the appearance of several new and deadly viral diseases. Wood rats in California were found to be the source of the Whitewater Arroyo virus, which caused several deaths in 2000. The Nipah virus killed 105 people in Malaysia in 1999; the virus is believed to be carried by a native fruit bat. In 1999, seven New Yorkers were killed and 60 were stricken by the West Nile virus, which is carried by birds (and is fatal to them). This virus, spread by mosquitoes, has surfaced in many Middle Eastern and African countries, and may have traveled to the United States in a bird or an infected human. A recent count indicated that three-fourths of emerging diseases are diseases of wild animals (zoonoses) and that their appearance is increasing in frequency.

like AIDS and the Ebola virus are the concern of public health agents and epidemiologists in organizations such as WHO and the Centers for Disease Control. (For a look into their work, see the "Global Perspective" essay, p. 404.)

The tropics, where most of the developing countries are found, have climates ideally suited for the year-round propagation of disease by **insects**. Mosquitoes are vectors for several deadly and debilitating diseases: yellow fever, dengue fever, elephantiasis, Japanese encephalitis, and malaria. Of these, malaria is by far the most serious. Public health attempts to control malaria have been aimed at eradicating the anopheles mosquito with the use of insecticides. Malaria was once prevalent in the southern United States, but an aggressive campaign to eliminate anopheles mosquitoes and identify and treat all human cases of malaria in the 1950s led to the complete eradication of the disease there.

Unfortunately, in the tropics, the mosquitoes have developed resistance to all of the pesticides employed, and eradication has remained an elusive goal. DDT is still used in some developing countries to spray the walls of huts and houses, but this use is highly controversial because of the pesticide's well-known harmful environmental and health impacts. Further complicating control of the disease, the *Plasmodium* protozoans have also developed resistance to one treatment drug after another. Until recently, chloro-

quine was quite effective against malaria in Africa. Now it is ineffective, and mefloquine is the drug of choice for treating cases of malaria and for prophylactic protection.

One highly promising effort funded by WHO's Tropical Disease Research program found that children provided with insecticide-treated nets over their beds (bed nets) experienced a substantial reduction in mortality from all causes. This thrust is being recommended for large-scale intervention throughout Africa as a cost-effective way of reducing the high mortality from malaria in African children. Bed nets are especially effective against anopheles mosquitoes, which emerge from hiding and feed primarily at night. In Vietnam, a combination of bed nets and more effective drugs reduced malaria deaths 98% in seven years. Elsewhere, WHO has documented increases in the incidence of malaria in association with land-use changes like deforestation, irrigation, and the creation of dams (Fig. 16–12). In semiarid areas of Africa, a seventeenfold increase in malaria resulted from development projects involving irrigation and other high-intensity agriculture. The basic problem cited by WHO is that the agencies promoting the land-use changes did not take possible health impacts into consideration.

Research continues on the development of new, more effective antimalarial drugs and on the development of an

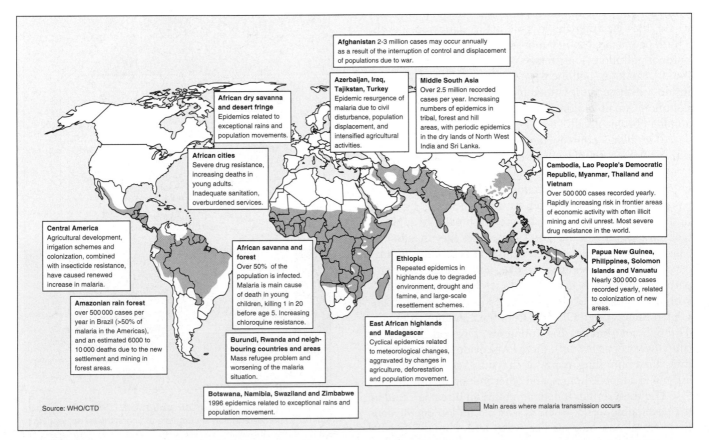

▲ FIGURE 16–12 *Malaria distribution and problem areas.* Malaria is found in the tropics and subtropics around the world. Drug resistance and environmental disruptions are contributing to the strong resurgence of malaria in the last two decades. (*Source*: World Health Organization, *Tropical Disease Research: Progress 1995–96: Thirteenth Programme Report* [Berne, Switzerland: Berteli Printers, 1997, p. 42.)

effective vaccine. The parasite's complicated life cycle and the fact that malaria does not readily induce long-term immunity have made a vaccine an elusive goal, so the current public health strategy is to inhibit access of mosquitoes to hosts (with bed nets and house spraying) and provide prompt treatment and prophylaxis to a greater proportion of the affected populations. Recently, WHO initiated a "Roll Back Malaria" campaign, designed not to eradicate the disease, but to reduce by half the malaria-caused mortality by 2010 and then to halve mortality again by 2015.

Toxic Risk Pathways

How is it that people are exposed to chemical substances that can bring them harm? Airborne pollutants represent a particularly difficult set of chemical hazards to control, as they are difficult to measure and difficult to avoid. Much attention is given to air pollutants released to the atmosphere, and rightly so. (Chapter 22 deals with these pollutants.) However, **indoor air pollution** can pose an even greater risk to health. It turns out that air inside the home and workplace often contains much higher levels of hazardous pollutants than outdoor air does! In the developed countries, the overall indoor air pollution problem is threefold. First, increasing numbers and types of products and equipment used in homes and offices give off potentially hazardous fumes. Second, buildings have become increasingly well insulated and sealed; hence, pollutants are trapped inside, where they accumulate to potentially dangerous levels. Third, people are exposed more to indoor pollution than to outdoor pollution. The average person spends 90% of his or her time indoors, and the people who spend the most time inside are those most vulnerable to the harmful effects of pollution: small children, pregnant women, the elderly, and the chronically ill.

One likely consequence of indoor air pollution is **asthma**, which is currently at epidemic proportions in the United States, afflicting 17 million people. Asthma is a chronic disease of the respiratory system in which the air passages tighten and constrict, causing wheezing, tightness in the chest, coughing, and shortness of breath. In acute attacks, the onset is sudden and sometimes life threatening. Many substances can trigger asthma attacks: dust, animal dander, mold, secondhand smoke, and various gases and particles. Asthma is a costly disease: Direct and indirect costs of the disease total more than $11 billion a year in the United States. Over 5,000 die from the condition annually. Through its Indoor Environments Division Web site (http://www.epa.gov/iaq/), the EPA provides helpful guidelines for controlling many of the substances that trigger asthma in schools, homes, and other indoor environments.

The most serious indoor air pollution threat is found in the developing world, where at least 3.5 billion people continue to rely on biofuels like wood and animal dung for cooking and heat. Fireplaces and stoves are often improperly ventilated (if at all), exposing the inhabitants to very high levels of particulates and fumes. Four major problems have been associated with indoor air pollutants in the developing world: acute respiratory infections in children, chronic lung diseases such as asthma and bronchitis, lung cancer, and birth-related problems. One study in Mexico showed that women who were continually exposed to indoor smoke had 75 times the risk of developing chronic lung disease than women who were free of that kind of exposure. Public policy solutions to this problem include ventilated stoves that burn more efficiently and the conversion from biofuels to bottled gas or liquid fuels like kerosene.

In developed countries, the sources of indoor air pollution are more varied, as is evidenced in Fig. 16–13. Among such sources are the following :

- Formaldehyde and other synthetic organic compounds emanating from plywood, particleboard, foam rubber, "plastic" upholstery, and no-iron sheets and pillowcases.
- A wide range of compounds from foods on the stove or in the oven.

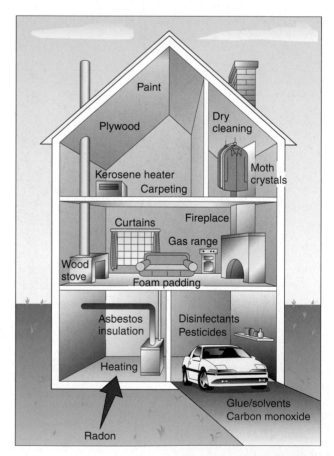

▲ **FIGURE 16–13** *Indoor air pollution.* Pollutants originate from many sources and can accumulate to unhealthy levels, leading to asthma and "sick-building syndrome" (characterized by eye, nose, and throat irritations, nausea, irritability, and fatigue), as well as other serious problems.

- Incomplete combustion and impurities from fuel-fired heating systems, such as gas or oil furnaces, kerosene heaters, and woodstoves.
- Fumes from household cleaners and other cleansing agents.
- Fumes from glues and hobby materials.
- Pesticides (addressed in Chapter 17).
- Air fresheners and disinfectants. Most air fresheners work either by dulling the sense of smell so that one dosen't notice noxious odors or by introducing "high-intensity" smells that cover up odors.
- Aerosol sprays of all sorts (e.g., oven cleaners, pesticides, hair sprays, and cooking oils).
- Radon. This radioactive gas, a by-product of the disintegration of uranium in rock, soil, and water, can accumulate in basements of homes and other buildings. As warm air escapes from the top of a house, creating a partial vacuum, radon may be drawn through the basement floor. Trapped in the house, the gas may then reach hazardous levels.
- Asbestos. This natural mineral has fiberlike crystals. It is mined from rock and was once used for heat insulation and as a fire retardation material. Pipes in steam-heating systems were wrapped in asbestos, ceilings in many public buildings were covered with it, and it was used in ironing-board covers, paints, and roofing materials. In the 1960s, researchers determined that the inhalation of asbestos fibers is associated with a unique form of lung cancer that develops as long as 20 to 30 years after exposure to the material. The EPA began regulating asbestos in the mid-1970s and initiated intensive campaigns to remove it from schools. Because of the obvious health risks, the EPA has banned practically all uses of asbestos. In the meantime, the agency has published guidelines for controlling old asbestos in public buildings, especially schools. The guidelines caution school administrators against assuming that removal is the only option; in most cases, asbestos can be sprayed with a sealant (encapsulation) or enclosed at a much lower expense than the cost of removal and with far less risk of exposure for the public.
- Smoking. Smoking carries a much higher health risk than the average exposure to any of the previously mentioned materials, and it may act synergistically to increase the risk of exposure to other pollutants. Smoking has also been shown to increase the health risks of nonsmokers subjected to ETS, which contains over 4,000 substances, at least 40 of which are known to be carcinogenic and many of which are potent respiratory irritants.

Even though the materials in this list are clearly hazardous, the link between their presence in the air and the development of a health problem is harder to establish than with the infectious diseases. As with cancer, often the best that can be done is to establish a statistical correlation between exposure levels and the development of an adverse effect. The science of **toxicology** addresses the need to study the impacts of toxic substances on human health and investigates the relationships between the presence of such substances in the environment and health problems like cancer. Chapter 20 will give us a more detailed look at toxic substances and the work of toxicology; however, one of the most important tools of the toxicologist in the developed countries is *risk assessment*, an approach to the problems of environmental health that is now a major element of public policy, but also the subject of some controversy.

16.3 Risk Assessment

All of us would like to live long and healthy lives. When confronted with the news that our days may be cut short by the effects of radiation, pesticides, secondhand smoke, or chemicals in our drinking water, we become indignant. We expect someone in authority to keep us from dying young or from spending our later years in a hospital. The high priority we place on environmental health is largely a reflection of this concern for hazards coming at us from the air we breathe, the food we eat, and the water we drink. At the grassroots level, it is this concern that is largely responsible for the billions of dollars spent on environmental protection.

Although in the developed countries our lives are much more free of hazards and risks than ever before, our society still faces hazards and the risks that they present to our health. For our own self-interest, we might like to know about and evaluate the risks to which we are subjected. Such knowledge would enable us to make informed choices as we consider the benefits and risks of the hazards around us. **Risk assessment** *is the process of evaluating the risks associated with a particular hazard before taking some action in a situation in which the hazard is present.* Table 16–3 lists some everyday risks, expressed as the probability of dying from a given hazard. For example, the table indicates that the annual risk of dying while hang gliding is 2.6 per 10,000, meaning that in the course of a year, 2.6 out of every 10,000 who engage in this sport will die as a result of an accident.

Truthfully, not very many people actually make informed choices about the hazards in their lives on the basis of such risk assessment. However, risk assessment has become an important process in the development of public policy and has been regarded as a major way of applying sound science to the hard problems of environmental regulation.

Risk Assessment by the EPA

Risk assessment began at the EPA in the mid-1970s as a way of addressing the cancer risks associated with pesticides and toxic chemicals. Since then, the process has continued to focus largely on risks to human health. As currently performed at the EPA, there are four steps in risk assessment: *hazard assessment, dose-response assessment, exposure assessment,* and *risk characterization.* The description that follows includes elements of new guidelines for risk assessment currently being evaluated by the EPA.

Hazard Assessment: Which Chemicals Cause Cancer?
Hazard assessment is the process of examining evidence linking a potential hazard to its harmful effects. In the case of accidents, the link is obvious. The use of cars, for example, involves a certain number of crashes and deaths. In these cases, *historical data*, such as the annual highway death toll, are very useful for calculating risks.

In other cases the link is not so clear, because there is a time delay between the first exposure and the final outcome. For example, establishing a link between one's exposure to certain chemicals and the development of cancer some years later is often very difficult. In cases where linkage is not obvious, the data may come from two sources: epidemiological studies and animal tests. An **epidemiological study** is a study that tracks how a sickness spreads through a community. Thus, in our example of finding a

link between cancer and exposure to some chemical, an epidemiological study would examine all the people exposed to the chemical and determine whether that population has more cancer than the general population. Data from such studies are considered to be the best data for risk assessment and have resulted in scores of chemicals being labeled as *known* human carcinogens.

The second data source, animal testing, is used when we want to find out *now* what might happen many years in the future. For example, we do not want to wait 20 years to find out that a new food additive causes cancer, so we accept evidence from **animal testing** (Fig. 16–14). A test involving several hundred animals (usually mice) takes about three years and costs more than $250,000. If a significant number of the animals develop tumors after being fed the substance being tested, the results indicate that the substance is either a *possible* or *probable* human carcinogen, depending on the strength of the results.

Three objections to animal testing have been raised: (1) Rodents and humans may have very different responses to a given chemical; (2) the doses used on the animals are often much higher than those to which humans may be exposed; and (3) some people are opposed on ethical grounds to the use of animals for such purposes. A point that sup-

TABLE 16–3 Some Commonplace Hazards, Ranked According to Their Degree of Risk

Hazardous Action	Annual Risk*
Cigarette smokers	10 per 1,000
All cancers	2.0 per 1,000
Firefighters	4.0 per 10,000
Police killed in line of duty	2.9 per 10,000
Hang gliding	2.6 per 10,000
Air pollution	2.5 per 10,000
Motor vehicle accident	1.6 per 10,000
Snowmobiling	1.3 per 10,000
Home accidents	1.1 per 10,000
Firearms	1.1 pre 10,000
Airline pilot	10 per 100,000
Mountain hiking	6.4 per 100,000
Alcohol consumption	6.4 per 100,000
Boating	5 per 100,000
Swimming	3 per 100,000
Four tablespoons peanut butter/day (cancer risk from aflatoxin)	1.3 per 100,000
Drinking water containing EPA limit of chloroform	7 per 1,000,000
Chest X-ray	3 per 1,000,000
Electrocution	1.8 per 1,000,000

*Probability of dying.
Source: R. Wilson and E. A. C. Crouch, "Risk-Benefit Analysis," 2nd edition, 2001, Harvard Center for Risk Analysis. Copyright © 2001 by Richard Wilson and Edmund Crouch. Reprinted by permission of the authors.

▲ **FIGURE 16–14** *Animal Testing.* Laboratory mice are routinely used to test the potential of a chemical to cause cancer. The mice are an important source of information used to assess the presence of a hazard in food, cosmetics, or the workplace.

ports the value of animal testing is that all chemicals shown by epidemiological studies to be human carcinogens are also carcinogenic to test animals. Besides, since it is ethically wrong to perform such tests on humans, there must be some way of testing chemicals that might be used as a food preservative or a cosmetic, in order to prevent harm to humans later on.

Another source of information is the chemical or process itself: What are its physical and chemical properties, and what mode of action might the chemical or process take in inducing cancer? Hazard assessment moves to present a "weight of evidence" narrative to provide a conclusion about the carcinogenic potential of the chemical or process. Standard descriptors are used in this conclusion, such as *"likely to be carcinogenic to humans."* Hazard assessment tells us that we *may* have a problem.

Dose-Response and Exposure Assessment: How Much for How Long?

When animal tests or human studies show a link between exposure to a chemical and an ill effect, the next step is to analyze the relationship between the concentrations of chemicals in the test (the **dose**) and both the incidence and the severity of the response. From this information, projections are made about the number of cancers that may develop in humans who are exposed to different doses of the chemical. Unless there is reason to assume otherwise, a linear response to zero dose/zero response is used, to determine an acceptable level of exposure. This process is called **dose-response assessment**.

Following dose-response assessment is **exposure assessment**. This procedure involves identifying human groups already exposed to the chemical, learning how their exposure came about, and calculating the doses and length of time of the exposure.

Risk Characterization: What Does the Science Say?

The final step, **risk characterization**, is to pull together all the information gathered in the first three steps in order to determine the risk and its accompanying uncertainties. Here is where sound science is able to inform the risk manager. The characterization summary should provide information about key supportive data and methods, their limitations, and especially, the risk estimates and uncertainties. Commonly, risk is expressed as the probability of a fatal outcome due to the hazard, as in Table 16–3 for common, everyday risks. For example, the EPA expresses cancer risk, as an "upper-bound, lifetime risk," meaning the highest value in the range of probabilities of the risk, calculated over a lifetime. The Clean Air Act of 1990 directs the EPA to regulate chemicals that have a cancer risk of greater than one in a million ($>1 \times 10^6$) for the people who are subject to the highest doses. That is the same standard employed by the FDA for regulating chemicals in food, drugs, and cosmetics. One important limitation of risk characterization is the fact that the analysts must work with the available information, which never seems to be sufficient. The final estimate of risk therefore includes a measure of the imprecision that always accompanies the use of imperfect information.

Risk Management

Regulatory Decisions. The analysis of hazards and risks is a task that falls primarily to the scientific community. The task is incomplete, however, if no further consideration is given to the information gathered. Risk management naturally follows, and this is the responsibility of lawmakers and administrators. EPA policy separates risk assessment and risk management within the agency.

Risk management involves (1) *a thorough review of the information available pertaining to the hazard in question and the risk characterization of that hazard* and (2) *a decision as to whether the weight of the evidence justifies a regulatory action.* Without doubt, public opinion can play a powerful role in both processes. In general, however, a regulatory decision will hinge on one or more of the following considerations:

1. **Cost-benefit analysis.** This type of analysis compares costs and benefits relative to a technology or proposed chemical product; if done fairly, the analysis can make a regulatory decision very clearcut.

2. **Risk-benefit analysis.** A decision can be made with regard to the benefits versus the risks of being subjected to a particular hazard, especially when those benefits cannot be easily expressed in monetary values. The use of medical X-rays is a good example. X-rays carry a calculable risk of cancer, but the benefit derived when you X-ray a broken bone is much greater than the small cancer risk involved.

3. **Public preferences.** As we will see, people have a much greater tolerance for risks that they feel are under their own control or are voluntarily accepted.

Risk management has been thoroughly incorporated into the EPA's policymaking process for at least 20 years. Its most common use has been in the design of regulations. The process has enabled the agency to target appropriate hazards for regulation and to determine where to aim the regulations—that is, at the source, at the point of use, or at the disposal of the hazardous agent. As was done for toxic chemicals and cancer, a given risk level may be adopted as a standard against which policymakers can measure new risks.

Risk Perception

In 1990, the EPA received a report from a science advisory board convened to evaluate different environmental risks in light of the most recent scientific data. One of the most important problems raised by the scientists was the significant difference between their evaluation of risks and that of the public. This discrepancy deserves a further look.

The U.S. public's increasing concern about environmental problems can be traced to the fear of hazards that pose a risk to human life and health. People *perceive* that their lives are more hazardous than ever before, but that is not true. In fact, our society is freer from hazards than it has ever been, as evidenced by increased longevity. Why, then, do people protest against nuclear power plants, waste sites, and pesticide residues in food when, according to experts, these hazards pose extremely small risks? The answer lies in people's **risk perceptions**—their intuitive judgments about risks. In short, people's perceptions are not consistent with the results coming from a scientific analysis of risk.

Table 16–4 indicates the top 11 environmental problems, based on the study presented to the EPA, compared with the top 28 concerns of the public, as indicated in a Roper poll. Clearly, the public perception of risks is quite different from that of the scientists.

Hazard vs. Outrage.   The reason for the inconsistency between public perception and actual risk calculations, according to Peter Sandman of Rutgers University, is that the public perception of risks is more a matter of outrage than hazard. Sandman holds that while the term *hazard* expresses primarily a concern for fatalities only, the term *outrage* expresses a number of additional concerns:

1. Lack of familiarity with a technology: how nuclear power is produced and how toxic chemicals are handled, for example.
2. Extent to which the risk is voluntary: Research has shown that people who have a choice in the matter will

---

TABLE 16–4   Public Concerns vs. the EPA's Top 11 Risks

The EPA's Top 11 Risks (not ranked)	Public Concerns (in rank order)*
Ecological Risks	1. Active hazardous waste sites
Global climate change	2. Abandoned hazardous waste sites
Stratospheric ozone depletion	3. Water pollution from industrial wastes
Alteration of habitat	4. *Occupational exposure to toxic chemicals*
Extinction of species and loss of biodiversity	5. *Oil spills*
	6. *Destruction of the ozone layer*
Health Risks	7. Nuclear power plant accidents
Criteria air pollutants (e.g., smog)	8. Industrial accidents releasing pollutants
Toxic air pollutants (e.g., benzene)	9. Radiation from radioactive wastes
Radon	10. *Air pollution from factories*
Indoor air pollution	11. Leaking underground storage tanks
Contamination of drinking water	12. Contamination of coastal waters
Occupational exposure to chemicals	13. Solid waste and litter
Application of pesticides	14. *Pesticide risks to farm workers*
	15. Water pollution from agricultural runoff
	16. Water pollution from sewage plants
	17. *Air pollution from vehicles*
	18. Pesticide residues in foods
	19. *Global warming*
	20. *Contamination of drinking water*
	21. Destruction of wetlands
	22. Acid rain
	23. Water pollution from city runoff
	24. Nonhazardous waste sites
	25. Biotechnology
	26. *Indoor air pollution*
	27. Radiation from X-rays
	28. *Radon in homes*

*Items in italics also appear on the EPA's list.
Source: L. Roberts, "Counting on Science at EPA," *Science* 249 [1990], 616. Copyright © 1990 by the AAAS.

accept risks roughly 1,000 times as great as when they have no choice.

3. Public impression of hazards: Accidents involving many deaths (as in Bhopal) or a failure of technology (e.g., at Three Mile Island) are thoroughly imprinted into public awareness by media coverage and are not quickly forgotten.

4. Overselling of safety: The public becomes suspicious when scientists or public relations people play up the benefits of a technology and play down the hazards.

5. Morality: Some risks have the appearance of being morally wrong. If it is wrong to foul a river, the notion that the benefits of cleaning it are not worth the costs is unacceptable. You should obey a moral imperative regardless of the costs.

6. Control: People are much more accepting of a risk if they are in control of the elements of that risk, as in automobile driving, indoor air pollution, and radon.

7. Fairness: The benefits and the risks should be connected. If the benefits go to someone else, why should you accept any risk?

The public perception of risk is strongly influenced by the media, which are far better at communicating the outrage elements of a risk than they are at communicating the hazard elements. Public concern over oil spills rose precipitously following the *Exxon Valdez* oil spill in Alaska, an accident that received extensive media coverage and had a high "outrage quotient." Cigarette smoking, however, which causes 430,000 deaths a year in the United States alone, receives minimal media attention, because it is not "news." Hence, there is no outrage factor. Indeed, it has been suggested that if all the year's smoking fatalities occurred on one day, the media would have a field day, and smoking would be banned the next day!

Risk Assessment and Public Policy. The most serious issue with the discrepancy between experts and the public is that, generally speaking, *public concern*, rather than cost–benefit analysis or risk analysis conducted by scientists, drives public policy. The EPA's funding priorities are set largely by Congress, which reflects public concern. Is this a problem?

If public outrage is the primary impetus for public policy, some serious risks may get less attention than they deserve. In particular, risks to the environment are perceived as much less important than they really are, because of the public's preoccupation with risks to human health. As Table 16–4 indicates, two major ecological risks on the EPA's list—alteration of habitat, and extinction of species and loss of biodiversity—do not even show up on the list of public concerns. Only three of the public's top 10 concerns are on the EPA's list.

This difference of opinion points to the importance of *risk communication*, a task that should not be left to the media. The value of ecosystems and their connections to human health and welfare need far greater emphasis in the public consciousness, and this responsibility falls to the scientific and educational communities, as well as to governmental agencies. Studies have shown that the most effective risk communication occurs by starting with what people already know and what they need to know and then tailoring the message so that it effectively provides people with the knowledge that helps them make an informed decision about risks.

Nonetheless, the public's concern for more than the probabilities of fatalities may have merit. Public outrage must be heard, understood, and given a reasonable response. It may not be the best source of public policy, but it reflects certain values and concerns that could easily be omitted by an "objective" risk assessment. The fact is, subjective judgments are going to play a role at every step in the risk-assessment process, from hazard assessment to risk perception to risk management. The uncertainties involved in risk assessment should remind us that the process is only a tool—and an imperfect one at best.

The use of risk assessment has come under attack in recent years, with three major concerns: (1) It does not adequately reflect the *uncertainty* inherent in much of the assessment process; (2) it has been largely employed in situations where a chemical or process is already in use; and (3) the burden of proof of harm falls largely on the regulators. In recognition of these problems, there is a strong movement toward employing the **precautionary principle** in formulating public policy to protect the environment and human health. The Wingspread Conference of 1998—a meeting of 35 academic scientists, government researchers, labor representatives, and grassroots environmentalists from the United States, Canada, and Europe—focused on providing a *consensus definition* of this principle: "Where an activity raises threats of harm to human health or the environment, precautionary measures should be taken even if some cause-and-effect relationships are not fully established scientifically. In this context, the proponent of an activity, rather than the public, should bear the burden of proof."

The precautionary principle is well established in Europe, where it originated, and has become the basis for European environmental law. For example the principle has been employed to restrict the import of genetically modified foods (Chapter 10). It is also an essential component of many U.N. environmental treaties. How it might relate to U.S. risk-based environmental policy is still under debate. The principle operates in some measure in our approach to pharmaceuticals and pesticides, according to which a substance is presumed guilty until proven innocent—that is, the substance may not be put on the market until tests show that it is safe to use.

In a very real sense, public health ministries and agencies are the primary stewards of the health and welfare of a country's people. Public policy should clearly reflect stewardship principles that put the health of people and the environment above the economic bottom line that often seems to drive our decisions. To quote the 2000 *World Health Report*, "Stewardship encompasses the tasks of defining the vision and direction of health policy, exerting influence through regulation and advocacy, and collecting and using information." Beyond the government, health care providers are also stewards whose responsibility is vitally important to environmental health. The precautionary principle represents a potent tool for implementing stewardship in the environmental health arena. Where risk assessment is useful in many, but not all, situations, it should

not be used as a blanket policy. The precautionary principle enables us to act to prevent some potential environmental health problems even at the risk of being wrong or at the risk of spending more than a problem technically deserves. The old saying, "An ounce of prevention is worth a pound of cure" captures the sense of this approach. Where uncertainty is substantial, and especially where the penalty for being wrong is great, it would seem wise to make the precautionary principle a guiding principle overseeing the entire risk assessment process.

ENVIRONMENT ON THE WEB

TO CHLORINATE OR NOT TO CHLORINATE?

With the wealth of media coverage on environmental hazards and human health issues, we are often overwhelmed with fears about environmental quality, and it can be difficult to know what risks to take most seriously. In some cases, we can make lifestyle choices that reduce our risk of accident, cancer, or infectious disease. Yet, in other cases, we are involuntarily exposed to hazards beyond our control. Examples of such involuntary risks include ingestion of trace organic compounds in food and in water, exposure to low-level radiation—for instance, from the Sun—and infection by drug-resistant pathogens.

Over the last decade, human exposure to chlorine has been a particular focus of concern in the media and in public debate. First discovered in 1774, chlorine is now widely used in manufacturing, bleaching, and disinfection. Its most common uses are in the production of safe drinking water (where it is used to kill microorganisms), in the bleaching of paper and textiles, and in the manufacture of solvents, plastics, and insecticides (DDT, for instance, is a chlorine-based insecticide). In gaseous form, chlorine is a respiratory irritant and was used as a war gas in World War I. In the modern environment, chlorinated compounds are found in air, water, soil, and biological tissue.

The use of chlorine to disinfect water, or chlorination, has been the single largest factor in controlling outbreaks of waterborne disease and has undoubtedly saved millions of lives that would otherwise have been lost to cholera, typhoid, and similar diseases. However, the debate over the use of chlorine to disinfect

water arises because chlorination can produce a range of chlorine by-products, including chloroform and bromoform. These chemicals, called trihalomethanes (THMs), are produced when chlorine reacts with organic material present in the raw water and can have serious health implications.

Does this mean that we should stop chlorinating drinking water supplies? The debate about water chlorination is a good example of the factors that must be considered when we weigh the risks we take in accepting—or avoiding—biological or chemical hazards.

Web Explorations

The Environment on the Web activity for this essay expands on the debate around drinking-water chlorination and explores both the chemical and biological hazards of both sides of the argument. Go to the Environment on the Web activity (select Chapter 16 at **http://www.prenhall.com/wright**) and learn for yourself:

1. about the arguments for and against chlorination of drinking water;
2. about a recent cholera epidemic in Latin America; and
3. about tradeoffs in selecting drinking water disinfection systems.

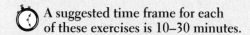 **A suggested time frame for each of these exercises is 10–30 minutes.**

REVIEW QUESTIONS

1. Differentiate between a hazard and a risk.
2. Define morbidity, mortality, and epidemiology.
3. Name the two fundamental ways of considering human health hazards.
4. What are the four categories of human environmental hazards? Give examples of each.
5. What are the pathways of risk described in this chapter? What are some components of each category?
6. Discuss the four steps of risk assessment used by the EPA.
7. What two steps are involved in risk management? What are the three considerations used to make a regulatory decision?

8. How does people's risk perception relate to the reality of a situation?

9. What factors have been known to generate public outrage?

10. Discuss the relationship between public risk perception and assessment, on the one hand, and public policy, on the other.

11. How would employing the precautionary principle better our approach to environmental public policy?

THINKING ENVIRONMENTALLY

1. Consider Table 16–4. Which ecological and health risks would you rank as the top three? Which of the seven public-outrage factors might be playing a role in your ranking?

2. Imagine that you have been appointed to a risk–benefit analysis board. Explain why you approve or disapprove the widespread use of

   • genetically modified foods
   • drugs to slow the aging process
   • nuclear power plants
   • the abortion pill

3. Suppose your town wanted to spray trees to rid them of a deadly pest. How would you use cost–benefit analysis to determine whether spraying is a good idea? Is there a better approach?

4. Design an advertisement, a song, or some other form of communication that can be used to inform the public about the risks of a hazardous activity (e.g., smoking) and what the public might do to reduce its exposure to the hazard.

WEB REFERENCES

On-line resources for this chapter are on the World Wide Web at: **http://www.prenhall.com/wright**. (Click on Chapter 16 within the Chapter Selector.)

Pests and Pest Control

Key Issues and Questions

1. For thousands of years, human enterprises have been frustrated by what we call "pests." What are pests, and why do we want to control them?
2. Different methods are used to control different agricultural pests. What are the basic philosophies behind these methods?
3. As new and more effective pesticides, such as DDT, were employed, some serious problems began to appear. What are the problems resulting from the use of chemical pesticides?
4. There are alternatives to using pesticides to control pests. How well do alternative control methods work?
5. Biotechnology has revolutionized many biological applications. How does biotechnology add to the battery of genetic methods to control pests?
6. Integrated pest management is an important way to reduce the use of pesticides. What is integrated pest management, and how has it worked?
7. Public policy for controlling pesticides has been revised. What are the objectives of the current policy, and how do they correct past problems?
8. Large quantities of pesticides are exported to developing countries. How are pesticide export and use in developing countries regulated?

The French Quarter of New Orleans is a world-famous tourist destination; in its 10 square blocks it has great jazz, colonial architecture, Bourbon Street, and outstanding restaurants. It also has serious termites, including an overseas invader called the Formosan subterranean termite (Fig. 17–1) that is eating away at its buildings. The termites are believed to have arrived in the United States on military ships returning from the Pacific theater after World War II. More robust and voracious than native species, the Formosan termites build large nests above and below ground, with millions in each nest. They infest trees as well as buildings and have damaged more than 30% of the historic live oaks of New Orleans. The termites have expanded their range from a few coastal cities to major areas of 14 states in the South and West; currently,

they cost property owners more than $1 billion a year for damage control and repair.

The Agricultural Research Service (ARS) of the U.S. Department of Agriculture has launched a program called Operation FullStop against the Formosan subterranean termite. One of the reasons the termites have become a major problem in recent years is that the pesticides once used to control them (organochlorine products like chlordane) were banned because of health and environmental concerns. Ordinary treatment with approved pesticides can hold them at bay for a time, but the termites seem very adept at exploiting the smallest contact between soil and house. Once inside, they can eat rafters, studs, door frames and flooring, often remaining undetected until something collapses.

◀ **French Quarter, New Orleans.** The French Quarter, 10 square blocks of residences, offices, restaurants, and jazz emporiums, is infested with the Formosan subterranean termite. In the foreground is a silver cap marking the location of a station for monitoring and baiting the termites with poisons.

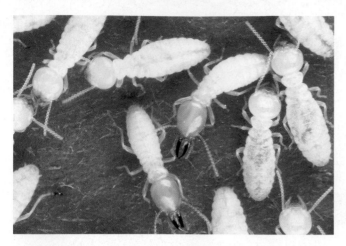

▲ **FIGURE 17–1** *Formosan subterranean termites.*
These termites are feeding on Sudan-red impregnated paper, which
allows researchers to trace their numbers and foraging range.

The ARS is trying to reduce the use of pesticides and has been employing a system of baiting the termites by putting out blocks of wood to monitor for their presence. Once termites are found in a block, a toxic baited block is switched with it, and the termites carry off the toxic bait to share with their unsuspecting fellows back in the nest. The ARS recently announced the discovery of a cottony mold that is able to kill 100% of the termites contacting it within a week; FullStop team members are now working on a method for producing the mold and employing it as part of the baiting approach. The researchers envision infected termites picking up the fungal spores and carrying them back to their colony, initiating a lethal epidemic that wipes out the millions of nestmates and eradicates the colony. If this tack works, it may be only jazz bands that bring down the house in the French Quarter of New Orleans.

17.1 The Need for Pest Control

Since earliest times, humans have suffered frustration and food losses brought on by destructive pests. To this day, farmers, pastoralists, and homeowners wage a constant battle against the insects, plant pathogens, and unwanted plants that compete with them for the biological use of crops, animals, and homes.

One dictionary defines a **pest** as "any organism that is noxious, destructive, or troublesome." This definition includes a broad variety of organisms that interfere with humans or with our social or economic endeavors: pathogens, nuisance wild animals, annoying insects, molds, etc. Earlier chapters considered nuisance wild animals (Chapter 11) and pathogens (Chapter 16); our emphasis in this chapter will be on those pests that interfere with crops, grasses, domestic animals, and structures—primarily the agricultural pests and weeds.

Defining Pests

Agricultural pests are organisms that feed on ornamental plants or agricultural crops or animals. The most notorious of these organisms are various insects, but certain worms, snails, slugs, rats, mice, and birds also fit into the category. **Weeds** are plants that compete with agricultural crops, forests, and forage grasses for light and nutrients. Some unwanted plants poison cattle or have other serious effects; others simply detract from the appearance of lawns and gardens. We try to bring these pests under control for three main purposes: to protect our food, to protect our health, and for convenience.

The Importance of Pest Control

Part of the credit for modern human prosperity can be attributed to pest control. On the credit side of the ledger, pesticides are vital elements in the prevention of the diseases that kill and incapacitate humans. In addition to having an agricultural use, pesticides have become important public health tools used to combat diseases such as malaria and sleeping sickness.

Insects, plant pathogens, and weeds destroy an estimated 37% (before and after harvest) of potential agricultural production in the United States, at a yearly loss of $122 billion to consumers and producers. Efforts to control these losses in 1997 involved the use of 975 million pounds (443 thousand metric tons) of **herbicides** (chemicals that kill plants) and **pesticides** (chemicals that kill animals and insects considered to be pests) annually (Fig. 17–2), at a direct cost of $11.9 billion. Worldwide, 2.84 million tons of pesticides were used in 1997, at a cost of $37 billion. Many of the changes in agricultural technology, such as monoculture and the widespread use of genetically identical crops, which have boosted yields, have also brought on an increase in the proportion of crops lost to pests—from 31% in the 1950s to 37% today. During this past half-century, the use of herbicides and pesticides has multiplied manyfold, leading to a disturbing and unsustainable dependency on them.

Different Philosophies of Pest Control

Medical practice employs two basic means of treating infectious diseases. One approach is to give the patient a massive dose of antibiotics, hoping to eliminate the pathogen causing the problem or to stop the pathogen before it can get established. The other approach is to stimulate the patient's immune system with a vaccine to produce long-lasting protection against any future invasion. In practice, both means are often used to keep a particular pathogen under control.

The same basic philosophies govern the control of agricultural pests. The first is **chemical treatment**. Like the use of antibiotics, chemical treatment seeks a "magic bullet" that will eradicate or greatly lessen the numbers of the pest organism. Although it has had much success, this

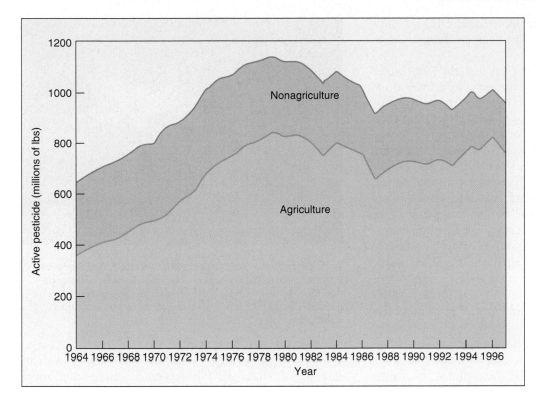

◄**FIGURE 17–2** *Pesticide use in the United States.* Nonagricultural and agricultural use is shown for the period 1964–1997. (*Source*: Data from the Environmental Protection Agency, Office of Pesticide Programs.)

approach gives only short-term protection. Furthermore, the chemical often has side effects that are highly damaging to other organisms.

The second philosophy is **ecological control**. Like stimulating the body's immune system, this approach seeks to give long-lasting protection by developing control agents on the basis of knowledge of the pest's life cycle and ecological relationships. Such agents, which may be other organisms or chemicals, work in one of two ways: Either they are highly specific for the pest species being fought, or they manipulate one or more aspects of the ecosystem. Ecological control emphasizes the protection of people and domestic plants and animals from damage from pests, rather than eradication of the pest organism. Thus, the benefits of pest control can be obtained while maintaining the integrity of the ecosystem.

These two philosophies are combined in the approach called **integrated pest management (IPM)**. IPM is *an approach to controlling pest populations by using all suitable methods—chemical and ecological—in a way that brings about long-term management of pest populations and also has minimal environmental impact.* This approach is increasing in usage, especially where pesticides are seen as undesirable because of health risks and in developing countries, where the cost of pesticides is often prohibitive.

17.2 Promises and Problems of the Chemical Approach

Pesticides are categorized according to the group of organisms they kill. There are insecticides (for insects), rodenticides (for mice and rats), fungicides (for fungi), herbicides (for plants), and so on. None of these chemicals, however, is entirely specific for the organisms it is designed to control; each poses hazards to other organisms, including humans.

Development of Chemical Pesticides and Their Successes

Finding effective materials to combat pests is an ongoing endeavor. The early substances used (frequently referred to as **first-generation pesticides**) included toxic heavy metals such as lead, arsenic, and mercury. We now recognize that these substances may accumulate in soils, inhibit plant growth, and poison animals and humans. In addition, these chemicals lost their effectiveness as pests became increasingly resistant to them. For example, in the early 1900s, citrus growers were able to kill 90% of injurious **scale insects** (minute insects that suck the juices from plant cells) by placing a tent over an infested tree and piping in deadly cyanide gas for a short time. By 1930, this same technique killed as little as 3% of the pests.

The next step had its origins in the science of organic chemistry in the early 1800s. During the 19th century, chemists synthesized thousands of organic compounds, but for the most part, these compounds sat on shelves because uses for them had not been found. By the 1930s, however, with agriculture expanding to meet the needs of a rapidly increasing population and with first-generation (inorganic) pesticides failing, farmers were begging for new ones. In time, **second-generation pesticides**, as they came to be called, were found as a result of synthetic organic chemistry.

The DDT Story. In the 1930s, a Swiss chemist, Paul Müller, began systematically testing some organic chemicals for their effect on insects. In 1938, he hit upon the chemical dichloro-diphenyltrichloroethane (DDT), a chlorinated hydrocarbon that had first been synthesized some 50 years earlier.

DDT appeared to be nothing less than the long-sought "magic bullet," a chemical that was extremely toxic to insects and yet seemed nontoxic to humans and other mammals. It was very inexpensive to produce. At the height of its use in the early 1960s, it cost no more than about 20 cents a pound. It was **broad spectrum**, meaning that it was effective against a multitude of insect pests. It was also **persistent**, meaning that it did not break down readily in the environment and hence provided lasting protection. This last attribute provided additional economy by eliminating both the material and labor costs of repeated treatments.

DDT quickly proved successful in controlling important insect disease carriers. During World War II, for example, the military used DDT to control body lice, which spread typhus fever among the men living in dirty battlefield conditions. As a result, World War II was one of the first wars in which fewer men died of typhus than of battle wounds. The World Health Organization of the United Nations used DDT throughout the tropical world to control mosquitoes and greatly reduced the number of deaths caused by malaria. (See the "Ethics" essay, p. 419.) There is little question that DDT saved millions of lives. In fact, the virtues of DDT seemed so outstanding that Müller was awarded the Nobel prize in 1948 for his discovery.

Postwar uses of DDT expanded dramatically. The chemical was sprayed on forests to control defoliating insects such as the spruce budworm. It was routinely sprayed on salt marshes to deal with nuisance insects. It was sprayed on suburbs to control the beetles that spread Dutch elm disease. And, of course, DDT proved very effective in controlling agricultural insect pests. (Fig. 17–3). Indeed, DDT was

▲ **FIGURE 17–3** *Aerial spraying.* A crop is being sprayed with pesticides to keep insects under control. Spraying is the basic technique still used in most control programs, although nowadays the pesticides in use do not persist in the environment for more than a week or two.

so effective, at least in the short run, that many crop yields increased dramatically. Growers could ignore other, more painstaking methods of pest control such as crop rotation and the destruction of old crop residues. They could grow varieties that were less resistant, but more productive. They could grow certain crops in warmer or moister regions, where formerly the damage from pests had been devastating. In short, DDT gave growers more options for growing the most economically productive crop.

It is hardly surprising that DDT ushered in a great variety of synthetic organic pesticides. Examples and some characteristics of insecticides and herbicides are listed in Table 17–1.)

TABLE 17–1 Characteristics of Pesticides

Insecticides

Type	Examples	Toxicity to Mammals	Persistence
Organophosphates	Parathion, malathion, phorate, chloropyrifos	High	Moderate (weeks)
Carbamates	Carbaryl, methomyl, aldicarb, aminocarb	Moderate	Low (days)
Chlorinated hydrocarbons	DDT, toxaphene, dieldrin, chlordane, lindane	Relatively low	High (years)
Pyrethroids	Permethrin, bifenthrin, esfenvalerate, decamethrin	Low	Low (days)

Herbicides

Type	Examples	Effects on Plants
Triazines	Atrazine, simazine, cyanizine	Interfere with photosynthesis, especially in broadleaf plants
Phenoxy	2, 4-D, 2, 4, 5, -T, methylchlorophenoxybutyrate (MCPB)	Cause hormonelike effects in actively growing tissue
Acidamine	Alachlor, Propachlor	Inhibit germination and early seedling growth
Dinitroaniline	Trifluralin, oryzalin	Inhibit cells in roots and shoots; prevent germination
Thiocarbamate	ethylpropylthiolcarbamate (EPTC), cycloate, butylate	Inhibit germination, especially in grasses

Source: U.S. Environmental Protection Agency, *Private Pesticide Applicator Training Manual*, 1993.

ETHICS

DDT FOR MALARIA CONTROL: HERO OR VILLAIN?

Malaria exacts a terrible toll in death and disease: 300 million bouts of sickness and a million deaths a year. Fifty years ago, malaria seemed to be on the way out; new synthetic drugs killed the protozoan parasites that cause the disease, and DDT wiped out mosquitoes everywhere. But resistance emerged with a vengeance, and now the parasites are highly resistant to most common malaria drugs, and the mosquitoes are resistant to DDT. Anyway, DDT has been banned in developed countries because of its human health and environmental dangers. It is still in use in many developing countries, however, largely because it is so inexpensive, but it should be banned worldwide because of its dangers. DDT is one of a group of 12 persistent organic pollutants targeted for phasing out by the UNEP, with the deadline set for 2007. This is the story from the developed world.

In malarial regions in the developing world, there is a different story. Here villagers are exposed to mosquitoes because they cannot afford screens on their homes, mosquito nets to sleep under, or repellent to spray on their children, four of whom die from malaria every minute. The anopheles mosquito, which carries malaria, is a night hunter, biting people as they rest or sleep. In 24 countries, DDT is still the main line of defense against the mosquito. The typical application involves spraying inside houses and eaves at a concentration of 2 g/m^2, once or twice a year. Even where the mosquitoes are resistant to its killing effects, the DDT acts as a repellent or irritant that keeps the mosquitoes from staying around. World Health Organization scientists have referred to indoor DDT spraying as the most cost-effective and easily applied control measure for malaria. To quote one angry malaria pro-

gram chief from a developing country, "Banning DDT is eco-imperialism. It's a model of putting the interests of the environment ahead of human lives."

Those in favor of pushing forward with the ban cite the problems of keeping DDT out of agricultural fields, where it can enter food chains; once the pesticide is made available, controlling its use is difficult. Alternative pesticides, such as synthetic pyrethroids, are already available. An integrated public-health approach is needed wherein mosquito breeding places are eliminated, bed nets are employed, and malaria cases are promptly treated to intercept the life cycle of the parasite. However, the public-health sector is unconvinced, and organizations like the Malaria Foundation International and the Malaria Project are actively campaigning against the proposed action for a worldwide ban on DDT. What do you think should be done?

Problems Stemming from Chemical Pesticide Use

The problems associated with the use of synthetic organic pesticides can be placed into three categories:

- development of resistance by pests
- resurgences and secondary-pest outbreaks
- adverse environmental and human health effects

Development of Resistance by Pests. The most fundamental problem for growers is that chemical pesticides gradually lose their effectiveness. Over the years, it becomes necessary to use larger and larger quantities, to try new and more potent pesticides, or to do both to obtain the same degree of control. Synthetic organic pesticides fared no better than first-generation pesticides in this respect. For example, in 1946, 1 kg (2.2 lb) of pesticides provided enough protection to produce about 60,000 bushels of corn. By 1971, it took 64 kg (141 lb) to produce the same amount, and losses due to pests actually *increased* during the intervening years.

Resistance builds up because pesticides destroy the sensitive individuals of a pest population, leaving behind only those few that already have some resistance to the pesticide. Resistant insect populations develop rapidly because insects have a phenomenal reproductive capacity. A single pair of houseflies, for example, can produce several

hundred offspring that may mature and reproduce themselves only two weeks later. Consequently, repeated pesticide applications result in the unwitting selection and breeding of genetic lines that are highly, if not totally, resistant to the chemicals that were designed to eliminate them. Cases have been recorded in which the resistance of a pest population has increased as much as twenty-five-thousandfold.

Over the years of pesticide use, the number of resistant species has climbed steadily. Many major pest species are resistant to all of the principal pesticides; over a thousand species of insects, plant diseases, and weeds have shown resistance to pesticides (Fig. 17–4). Interestingly, as a pest population becomes resistant to one pesticide, it also may gain resistance to other, unrelated pesticides, even though it has not been exposed to the other chemicals. (See the "Earth Watch" essay, p. 420.)

Resurgences and Secondary-Pest Outbreaks. The second problem with the use of synthetic organic pesticides is that after a pest has been virtually eliminated with a pesticide, the pest population not only recovers, but explodes to higher and more severe levels. This phenomenon is known as a **resurgence**. To make matters worse, small populations of insects that were previously of no concern because of their low numbers suddenly start to explode, creating new problems. This phenomenon is called a **secondary pest outbreak**. For example, with the use of synthetic organic pesticides, mites

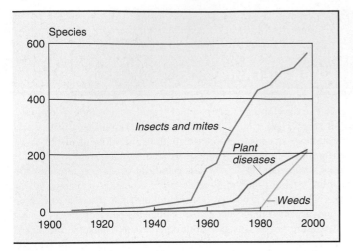

▲ FIGURE 17–4 *Number of species resistant to pesticides, 1908–98.* Using pesticides causes selection (survival of the fittest) for those individuals that are resistant. As we continue to use pesticides, we breed insects and other pests that are increasingly resistant to those pesticides. (*Source*: Lester Brown et al., *Vital Signs 1999* [Washington, DC: Worldwatch Institute, 1999].)

have become a serious pest problem, and the number of serious pests on cotton has increased from 6 to 16.

To illustrate the seriousness of resurgences and secondary-pest outbreaks, a recent study in California listed a series of 25 major pest outbreaks, each of which caused more than $1 million worth of damage. All but one involved resurgences or secondary-pest outbreaks. Of course, the species appearing in secondary-pest outbreaks

quickly became resistant to pesticides, thus compounding the problem. At first, pesticide proponents denied that resurgences and secondary-pest outbreaks had anything to do with the use of pesticides. However, careful investigations have shown otherwise: Resurgences and secondary-pest outbreaks occur because the insect world is part of a complex food web.

Thus, the chemical approach fails because it is contrary to basic ecological principles. It assumes that the ecosystem is a static entity in which one species, the pest, can simply be eliminated. In reality, the ecosystem is a dynamic system of interactions, and a chemical assault on one species will inevitably upset the system and produce other undesirable effects. Populations of plant-eating insects are frequently held in check by other insects that parasitize or prey on them (Fig. 17–5). Pesticide treatments often have a greater impact on these natural enemies than on the plant-eating insects they are meant to control. Consequently, with natural enemies suppressed, both the population of the original target pest and populations of other plant-eating insects explode. The path to sustainability demands that we understand how ecosystems work and that we adapt our interventions accordingly.

The late entomologist Robert van den Bosch coined the term **pesticide treadmill** to describe attempts to eradicate pests with synthetic organic chemicals. It is an apt term: The chemicals do not eradicate the pests; they increase resistance and secondary-pest outbreaks, which lead to the use of new and larger quantities of chemicals, which in turn lead to more resistance and more secondary-pest

EARTH WATCH

THE ULTIMATE PEST?

If we were to conjure up the ultimate insect pest, we might imagine it to have the characteristics that follow. The ultimate insect pest would

1. attack a broad variety of plants and fruits.

2. spread viruses that can devastate crop plants.

3. be resistant to a host of pesticides

4. be highly prolific and have a rapid life cycle

5. thrive on roadside weeds

6. have few natural predators

Such a pest, if unleashed on agriculture, could cause millions of dollars in crop losses and devastate many farmers.

Unfortunately, something very close to the ultimate pest is already with us. The silverleaf whitefly (*Bemisia argentifolii*) is a tiny white insect that emerged from Florida's poinsettia greenhouses in 1986 and has become established in California, Texas, Florida, and Arizona. It has all the characteristics listed and has been dubbed the "superbug" by farmers who have encountered it. The fly is known to eat at least 500 species of plants—just about everything except asparagus and onions—and it lives on roadside weeds when it can't get its favorite crop food. The insects swarm all over the plants, sucking them dry and leaving them withered and rotten. Swarms become so dense that they interfere with the breathing and vision of any-

one caught in them. Total crop losses to this insect are greater than $200 million each year, and it continues to cause extensive damage. A national campaign against the insect has been launched by the Agricultural Research Service of the USDA, complete with five-year plans. A few pesticides remain effective, but resistance is a lurking problem whenever these are used. So far, the judicious use of pesticides and cultural controls like plowing under a crop as soon as it is harvested are holding the bug at bay. For now, the entire farm community in the affected areas is involved in programs to suppress the silverleaf whitefly, and research programs are frantically looking for effective biological control agents.

(a)

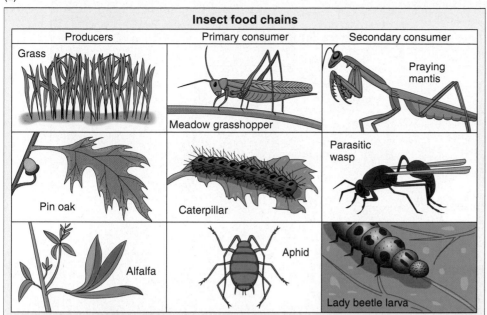

Insect food chains

Producers	Primary consumer	Secondary consumer
Grass	Meadow grasshopper	Praying mantis
Pin oak	Caterpillar	Parasitic wasp
Alfalfa	Aphid	Lady beetle larva

(b)

◀ **FIGURE 17–5** *Insect food chains.* (a) Food chains exist among insects, just as they do among higher animals. (b) Aphid lion (on right) impaling and eating a larval aphid (a small insect that sucks sap from plants).

outbreaks, and so on. The process is an unending cycle constantly increasing the risks to human and environmental health and is clearly not sustainable (Fig. 17–6).

Human Health Effects. As with all toxic substances, pesticides can be responsible for both acute and chronic health effects. According to the American Association of Poison Control Centers, over 86,000 persons suffered acute poisonings from pesticides during 1998; 16 of the victims died. Most of the victims were farmworkers or employees of pesticide companies who came in direct contact with the chemicals. The World Health Organization surveyed pesticide poisoning and estimated that there are between 3.5 and 5 million cases of acute occupational pesticide poisoning each year in developing countries, of which at least 20,000 result in death. The use of pesticides by untrained persons is considered to be the major cause of these poisonings, but in many cases children and families come in contact with the pesticides through aerial spraying, the dumping of pesticide wastes, or the use of pesticide containers for storage of drinking water.

Among the chronic effects of pesticides is the potential for causing cancer, as indicated by animal testing; evidence of carcinogenicity has been recorded for 20 pesticides. Epidemiological evidence has implicated organochlorine pesticides in various cancers, including lymphoma and breast cancer. Other chronic effects include dermatitis, neurological disorders, birth defects, and infertility; male sterility has been clearly linked to dibromochloropropane (DBCP), once used to control nematodes and now banned. Two disorders recently associated with pesticide use are suppression of the immune system and disruption of the endocrine system. Again, the evidence for an impact on humans comes from laboratory animal testing and epidemiological studies.

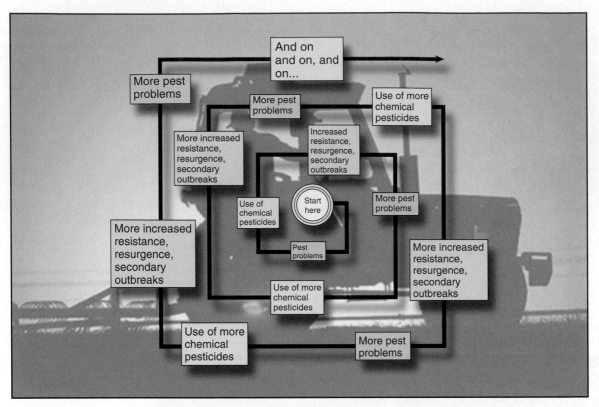

▲ **FIGURE 17–6**   *The pesticide treadmil.*   Treatment with chemical pesticides aggravates many pest problems. The continued use of these products demands ever-increasing dosages of pesticides, which further aggravate pest problems and produce more contamination of foodstuffs and ecosystems.

Persistent exposure to pesticides among factory workers in India led to abnormally low white blood cell counts (the white blood cells are crucial elements of the immune system); similar results were found for agricultural workers in the former Soviet Union. Laboratory studies show many forms of immune suppression by different types of pesticide.

Laboratory tests have shown a number of pesticides to have estrogenlike effects: atrazine (a weed killer), DDT, endosulfan (an insecticide), methoxychlor (an insecticide), and others. Both a rise in the incidence of breast cancer among humans and reports of abnormal sexual development in alligators and other animals in the wild have given credence to the possibility that very low levels of a number of chemicals are able to mimic or disrupt the effects of estrogenic hormones (sexual hormones that are very potent at low concentrations). Concern about this mimicry has reached Congress, and the EPA was directed to develop procedures for testing chemicals for endocrine disruption activity. An EPA-established advisory committee strongly recommended proceeding with a screening program, and in 1999 the EPA established the Endocrine Disruptor Screening Program. Currently, the program is evaluating processes to be used in screening the potential endocrine disruptors. Pesticides are prime suspects, and chemical manufacturers will soon be required to test many pesticides in use in the United States.

Environmental Effects.   Aerial spraying for insect control has pointed most clearly to the adverse environmental ef-

fects of pesticides. The story of DDT, used so widely during the 1940s and 1950s (80,000 metric tons in 1962 alone), illustrates the hazards.

The Conclusion of the DDT Story.   In the 1950s and 1960s, ornithologists (people who study birds) observed drastic declines in populations of many species of birds that fed at the tops of food chains. Fish-eating birds such as the bald eagle and osprey (Fig. 17–7) were so affected that their extinction was feared. Investigators at the U.S. Fish and Wildlife National Research Center near Baltimore, Maryland, showed that the problem was reproductive failure: Eggs were breaking in the nest before hatching. The investigators also showed that the fragile eggs contained high concentrations of dichloro-diphenyl-dichloroethylene (DDE), a product of the partial breakdown of DDT by the animal's body. DDE interferes with calcium metabolism, causing birds to lay thin-shelled eggs. Further study revealed that birds were acquiring high levels of DDT and DDE by bioaccumulation and biomagnification, the process of accumulating higher and higher doses through the food chain.

Because of accumulation, small, seemingly harmless amounts received over a long period of time may reach toxic levels. This phenomenon—referred to as **bioaccumulation**—can be understood as follows: Many synthetic organics are highly soluble in lipids (fats or fatty compounds), but sparingly soluble in water. As they pass through cell membranes, which are lipid, they come out of water solution and enter into the lipids of the body. Thus, traces of

▲ FIGURE 17–7   *Bald eagle.*   Populations of fish-eating birds like the osprey, brown pelican, and bald eagle were decimated in the 1950s and 1960s by the effects of widespread spraying with DDT. With the banning of the pesticide, these populations have greatly recovered. The bald eagle was taken off the endangered species list in 1994.

synthetic organics like pesticides and their breakdown products that are absorbed with food or water are trapped and held by the body's lipids, while the water and water-soluble wastes are passed in the urine. Since synthetics are unnatural compounds, the body can not fully metabolize them and has no mechanism to excrete them. Thus, trace levels consumed over time gradually accumulate in the body and may produce toxic effects sooner or later.

Bioaccumulation, which occurs in the individual organism, may be compounded through a food chain. Each organism accumulates the contamination in its food, so it accumulates a concentration of contaminant in its body that is many times higher than that in its food. In effect, the next organism in the food chain now has a more contaminated food and accumulates the contaminant to yet a higher level. Essentially all the contaminant accumulated by the large biomass at the bottom of the food pyramid is concentrated, through food chains, into the smaller and smaller biomass of organisms at the top of the food pyramid. This multiplying effect of bioaccumulation that occurs through a food chain is called **biomagnification**. One of the most distressing aspects of bioaccumulation and biomagnification is that there are no warning symptoms until concentrations of contaminant in the body are high enough to cause problems. Then it is too late to do much about it. As is often the case, bioaccumulation and biomagnification go unrecognized until serious problems bring the phenomena to light. Fig. 17–8 shows how DDT and its metabolic products worked their way up the food chain to the top predators.

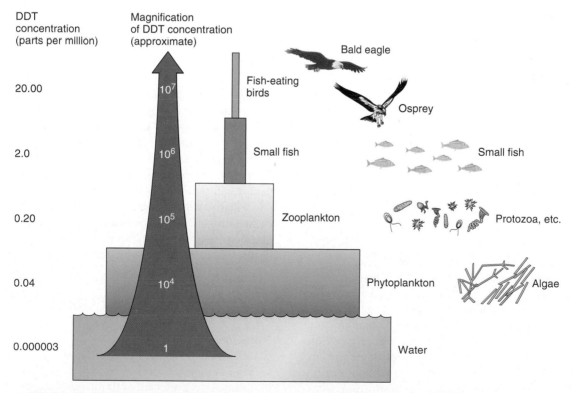

▲ FIGURE 17–8   *Biomagnification.*   Organisms on the first trophic level absorb the pesticide from the environment and accumulate it in their bodies (bioaccumulation). Then, each successive consumer in the food chain accumulates the contaminant to a yet higher level. Thus, the concentration of the pesticide is magnified manyfold throughout the food chain. Organisms at the top of the food chain are likely to accumulate toxic levels.

Silent Spring. In the 1950s, Rachel Carson, a U.S. Fish and Wildlife Service biologist and accomplished science writer, began reading the disturbing scientific accounts of the effects of DDT and other pesticides on wildlife. Carson was finally galvanized into action by a letter from a friend distressed over the large number of birds killed when the friend's private bird sanctuary was sprayed for mosquito control. By 1962, Carson finished a book-length documentation of the effects of the almost uncontrolled use of insecticides across the United States. The book, *Silent Spring*, became an instant best-seller. Its basic message was that if insecticide use continued as usual, there might someday come a spring with no birds—and with ominous impacts on humans as well.

Silent Spring triggered a national debate that continues today. Representatives of the agricultural and chemical industries claimed that the book was an unreasonable and unscientific account that, if taken seriously, would lead to a halt of human progress. At the same time, however, the book was hailed as an unparalleled breakthrough in environmental understanding. Eventually, concerns about environmental and long-term health effects led to the banning of DDT in the United States and most other industrialized countries in the early 1970s. Numerous other related chlorinated hydrocarbon pesticides (e.g., chlordane, dieldrin, endrin, and heptachlor) were also banned because of their propensity for bioaccumulation in the environment and suspected potential for causing cancer. In the years since the banning of DDT, the bird species that were adversely affected have recovered.

Forty years later, Rachel Carson is credited with stimulating the start of the modern environmental movement and the creation of the EPA. *Silent Spring* has become a classic, and the regulation of insecticides and other toxic chemicals is in a sense a monument to Rachel Carson, who died of cancer only two years after her book was published.

Nonpersistent Pesticides: Are They the Answer?

A key characteristic of chlorinated hydrocarbons is their persistence—that is, their slowness to break down. Lingering in the environment for years, they can contaminate many organisms and become biomagnified. This persistence results because these chemicals have such complex structures that soil microbes are unable to fully break them down. DDT, for example, has a half-life on the order of 20 years.

Because the persistent pesticides have been banned, the agrochemical industry has substituted *nonpersistent* pesticides for the banned compounds. For example, the synthetic organic phosphates malathion, parathion, and chlorpyrifos, as well as carbamates such as aldicarb and carbaryl, have been used extensively in place of chlorinated hydrocarbons (see Table 17–1). These compounds break down into simple nontoxic products within a few weeks after their application. Thus, there is no danger of their migrating long distances through the environment and affecting wildlife or humans long after being applied.

For several reasons, however, nonpersistent pesticides are not as environmentally sound as they might appear. First of all, the total environmental impact of a pesticide is a function not only of its persistence, but also of three other important factors: its toxicity, its dosage, and the location where it is applied. Many of the nonpersistent pesticides are far more toxic than DDT. This higher toxicity, combined with the frequent applications needed to maintain control, presents a significant hazard to agricultural workers and others exposed to these pesticides. For instance, the organophosphates are responsible for an estimated 70% of all pesticide poisonings. The Food Quality Protection Act of 1996 required the EPA to develop new health-based standards for such pesticides that address the risk of children's exposure. The fallout from the new standards has led to the banning of any use of parathion on produce (1999) and of chlorpyrifos (2000) for all over-the-counter uses, such as on flea collars and in lawn products. (Chlorpyrifos may still be used on apples, although late-season spraying may soon be banned.)

Second, nonpersistent pesticides may still have far-reaching environmental impacts. For example, to control outbreaks of grasshoppers that eat sunflowers, farmers in Argentina began spraying monocrotophos, an acutely toxic organophosphate pesticide, in 1995. Shortly afterwards, thousands of dead Swainson's hawks were seen in the sunflower fields. The hawks, which spend summers in the United States, winter in great numbers in the Argentine pampas and feed voraciously on grasshoppers there. It was estimated that 20,000 hawks died in one year, about 5% of the world's population of the species. In response, the Argentine government banned the use of monocrotophos; it had already been withdrawn from use in the United States in 1988.

Third, desirable insects may be just as sensitive as pest insects to these substances. Bees, for example, which play an essential role in pollination, are highly sensitive to nonpersistent pesticides. Thus, the use of these compounds creates an economic problem for beekeepers and jeopardizes pollination. Also, regular spraying of neighborhoods with malathion to control mosquitoes leads inevitably to great declines in butterflies and fireflies.

Finally, nonpersistent chemicals are just as likely to cause resurgences and secondary-pest outbreaks as are persistent pesticides, and pests become resistant to nonpersistent pesticides just as readily as they do to persistent pesticides.

17.3 Alternative Pest Control Methods

Numerous ecological and biological factors affect the relationship between a pest and its host. Ecological control seeks to manipulate one or more of these natural factors so that crops are protected without upsetting the rest of the ecosystem or jeopardizing environmental and human health. Since ecological control involves working with natural factors instead of synthetic chemicals, the techniques are referred to as **natural control** or **biological control** methods. This natural approach, unlike the chemical treatment approach, depends on an understanding of the pest and its relationship with its host and with its ecosystem. The more we know about the organisms involved, the greater are our opportunities for natural control.

To illustrate, the life cycle typical of moths and butterflies is shown in (Fig. 17–9). Many groups of insects have a similarly complex life cycle. The development of each stage may be influenced by numerous abiotic factors, and at each stage the insect may be vulnerable to attack by a parasite or predator. The proper completion of each stage depends on internal chemical signals provided by insect hormones. Locating mates, finding food, and other behaviors depend on external chemical signals. All these findings suggest ways in which pest populations may be controlled without resorting to synthetic chemical pesticides.

The four general categories of natural or biological pest control are (1) **cultural control** (2) **control by natural enemies** (3) **genetic control** and (4) **natural chemical control**. We will consider each of these in turn.

Cultural Control

A cultural control is a nonchemical alteration of one or more environmental factors in such a way that the pest finds the environment unsuitable or is unable to gain access to it.

Cultural Control of Pests Affecting Humans. We routinely practice many forms of cultural control against diseases and parasitic organisms. Some of these practices are so familiar and well entrenched in our culture that we no longer recognize them for what they are. For instance, disposing properly of sewage and avoiding drinking water from unsafe sources are cultural practices that protect against waterborne disease-causing organisms. Combing and brushing the hair, bathing, and wearing clean clothing are cultural practices that eliminate head and body lice, fleas, and other parasites. Regular changing of bed linens protects against bedbugs. Properly and systematically disposing of garbage and keeping a clean house with all cracks sealed and with good window screens are effective methods for keeping down populations of roaches, mice,

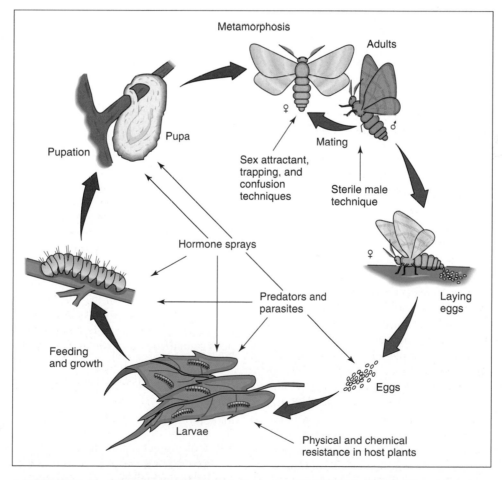

▲ **FIGURE 17–9** *Complex life cycle of insects.* Like the moth shown here, most insects have a complex life cycle that includes a larval stage and an adult stage. Biological control methods recognize the different stages and attack the insect, using knowledge of its needs and life cycle.

flies, mosquitoes, and other pests. Sanitation requirements in handling and preparing food are cultural controls designed to prevent the spread of disease. Refrigeration, freezing, canning, and drying of foods are cultural controls that inhibit the growth of organisms that cause rotting, spoilage, and food poisoning.

If such practices of personal hygiene and sanitation are compromised, as they usually are in any major disaster, there is very real danger of additional widespread mortality resulting from outbreaks of parasites and diseases. As we saw in Chapter 16, sickness and death are the consequences where these practices are not broadly pursued in a society, as in some less developed countries.

Cultural Control of Pests Affecting Lawns, Gardens, and Crops.   Homeowners are prone to use excessive amounts of herbicides to maintain a weed-free lawn. Weed problems in lawns are frequently a result of cutting the grass too short. If grass is left at least 3 inches (8 cm) high, it will usually maintain a dense enough cover to keep out crabgrass and many other noxious weeds. Thus, monitoring the height of grass is a form of cultural weed control. Also, many homeowners allow a diversity of plants in their lawns, tolerating a variety of plants some people might consider weeds.

Some plants are particularly attractive to certain pests; others are especially repugnant. In either case, the effect may spill over to adjacent plants. Thus, a gardener may control many pests by paying careful attention to eliminating plants that act as attractants (for example, roses) and growing those that act as repellents. Marigolds and chrysanthemums are justly famous for being insect repellents.

Some parasites require an alternative host and so can be controlled by eliminating this host (Fig. 17-10). Also, hedgerows, fencerows, and shelterbelts can provide refuges where natural enemies of pests (birds, amphibians, praying mantises, and so on) can be maintained.

Managing any crop residues not harvested is important. Spores of plant disease organisms and insects may overwinter or complete part of their life cycle in the dead leaves, stems, or other plant residues that remain in the fields after harvesting. Plowing under or burning the material may be quite effective in keeping pest populations to a minimum. In gardens, a clean mulch of material such as grass clippings or hay will keep down the growth of weeds and protect the soil from drying and erosion.

Growing the same crop on a plot of land year after year keeps the pest's food supply continuously available. Hence, crop rotation, the practice of changing crops from one year to the next, may provide control because pests of the first crop cannot feed on the second crop and vice versa. Crop rotation is especially effective in controlling root nematodes (roundworms that live in the soil and feed on roots) and other pests that do not have the ability to migrate appreciable distances.

For economic efficiency, agriculture has moved progressively toward monoculture—the Corn Belt, Cotton Belt, and so on. When a pest outbreak occurs, monoculture is most

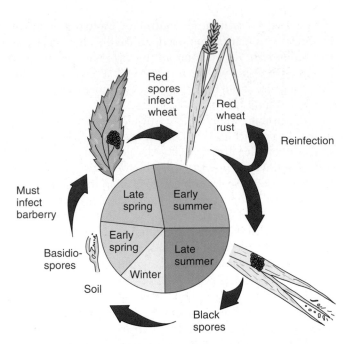

▲ **FIGURE 17-10**   *Cultural control.*   Part of the life cycle of wheat rust, a parasitic fungus that is a serious pest on wheat, requires that the rust infest barberry, an alternative host plant. The elimination of barberry in wheat-growing regions has been an important cultural control.

conducive to its rapid multiplication and spread. The spread of a pest outbreak is impeded, and other natural controls may be more effective, if there is a mixture of crop species, some of which are not vulnerable to attack. One approach that works well in the British Isles is to intersperse cultivated with uncultivated strips that are not treated with pesticides. Natural enemies of pests are maintained in the uncultivated strips.

Most of our pests that are hardest to control were unwittingly imported from other parts of the world, and we realize that many other species in other regions would be serious pests if introduced into the United States. Therefore, it is important to keep would-be pests out of the country. This is a major function of the U.S. Customs Bureau and of the agriculture departments of some states. Biological materials that may carry pest insects or pathogens either are prohibited from crossing the border or are subjected to quarantines, fumigation, or other treatments to ensure that they are free of pests. The cost of such procedures is small in comparison to the costs that could be incurred if these pests gained entry and became established in the nation.

Control by Natural Enemies

The following examples illustrate the range of possibilities for controlling pests with natural enemies:

- Scale insects, potentially devastating to citrus crops, have been successfully controlled by vedalia (ladybird) beetles, which feed on them.
- Various caterpillars have been controlled by parasitic wasps (Fig. 17-11).

- Japanese beetles are controlled in part with the bacterium *Bacillus thuringiensis* (Bt), which produces toxic crystals when soil containing the bacterium is ingested by the larvae.

- Prickly pear cactus and numerous other weeds have been controlled by plant-eating insects (Fig. 17–12). (More than 30 weed species worldwide are now limited by insects introduced into their habitats.)

- Rabbits in Australia are controlled by an infectious virus.

- Water hyacinth, which has blanketed many African lakes, is coming under control following the introduction of Brazilian weevils.

- Mealybugs in Africa are controlled by a parasitic wasp. (See the "Global Perspective" essay, p. 428.)

- In a test spraying in Niger, Africa, the swarming desert locust was completely controlled with a newly developed product called "Green Muscle," a mix of dry spores of the fungus *Metarhizium anisopliae* with oil. The product is sprayed on infested fields.

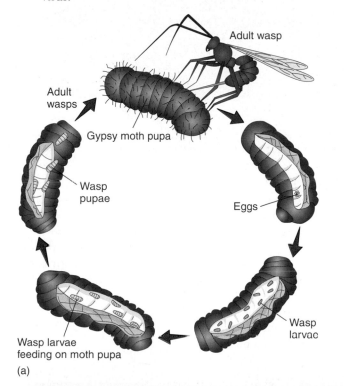

(a)

(b)

(c)

▲ **FIGURE 17–11** *Parasitic wasps.* (a) The life cycle of the parasitic wasp that uses the gypsy moth as its host. (b) A wasp depositing eggs in a gypsy moth larva. (c) Another insect parasite, a braconid wasp, lays its eggs on the pest known as the tomato hornworm, which is the larva of the sphinx moth. The wasp larvae feed on the caterpillar, and shortly before the caterpillar dies, they emerge and form the cocoons seen here.

(a)   (b)

▲ **FIGURE 17–12   *Prickly pear cactus.***   Biological control of the weed prickly pear cactus by a cactus-eating moth in Queensland, Australia. (a) The land shown here, once settled, had to be abandoned because of prickly pear infestation. (b) The same land was reoccupied following the destruction of the prickly pear by the moth.

The problem with using natural enemies is finding organisms that provide control of the target species without attacking other, desirable species. Entomologists estimate that of the 50,000 known species of plant-eating insects that have the potential for being serious pests, only about 1% actually are. The populations of the other 99% are held in check by one or more natural enemies, so they do not cause significant amounts of damage. Therefore, the first step in using natural enemies for control should be *conservation*—protecting the natural enemies that already exist. Conserva-

tion means avoiding the use of broad-spectrum chemical pesticides, which may affect natural enemies of the target species even more than they do the target pests themselves. Eliminating or placing considerable restrictions on the use of broad-spectrum chemical pesticides will, in many cases, allow natural enemies to reestablish themselves and control secondary pests, those that become serious problems only after the use of pesticides.

However, effective natural enemies are not always readily available. In many cases, the absence of a pest's natural

GLOBAL PERSPECTIVE

WASPS 1, MEALYBUGS 0

The cassava (manioc) plant originated in South America, but has been cultivated throughout the tropical world. Currently, in sub-Saharan Africa, it is the primary food for more than 200 million people, one-third of the human population in that region. It is a high-yielding crop that requires no modern technology and, accordingly, is raised by subsistence farmers everywhere.

The mealybug, an insect not previously seen in Africa, appeared in Congo in the early 1970s and spread across sub-Saharan Africa, leaving a trail of ruined harvests and hunger in its wake. Zaire and Congo, unable to afford pesticides, turned for help to the International Institute for Tropical Agriculture in Nigeria. The group

formed the Biological Control Program and began to look for natural enemies of the mealybug.

Returning to the land of origin of the cassava, a researcher found the mealybug in Paraguay and observed that the insect was kept under control by natural predators and parasites. After extensive testing, researchers identified a parasitic wasp, *Epidinocarsis lopezi*, as the prime candidate for controlling the bugs. No larger than a comma on this page, the female wasp seeks out mealybugs, paralyzes them with a sting, and deposits eggs that hatch inside the mealybug and eat their way through the insect.

Once this wasp–bug relationship was identified, the wasps were reared by the

thousands on captive mealybugs and then spread across the continent over a period of eight years. The Biological Control Program has trained 400 workers to monitor the progress of the program, and all indications are that the battle has been won. It is estimated that every dollar invested in the control program has yielded $149 in crops saved from destruction. The best news is that the control is permanent and does not require the repeated application of expensive and environmentally damaging pesticides. In the wake of the project, national biological control programs have been established and are now focusing on pests of other crops of sub-Saharan Africa.

enemies is the result of accidentally importing the pest without also importing its natural enemies. Quite often, effective natural enemies have been found by systematically combing the home region of an introduced pest and finding its various predators or parasites. To date, the U.S. Department of Agriculture (USDA) has released almost 1,000 insects to control pests, with generally favorable results. Identifying the enemy species is vital, and the species must be carefully tested before intentional releasing it, something that has not always been done well. Some 16% of 313 parasitoid wasp species introduced into North America have attacked native species. In one recent case, a weevil (a type of beetle) imported to control an exotic musk thistle (a weed), attacked five native thistle species that pose no threat to agriculture. When effective natural enemies are found, however, they can provide control indefinitely, saving millions of dollars per year.

Genetic Control

Most plant-eating insects and plant pathogens attack only one species or a few closely related species. This specificity implies a genetic incompatibility between the pest and species that are not attacked. Most genetic-control strategies are designed to develop genetic traits in the host species that provide the same incompatibility — that is, resistance to attack by the pest. The technique has been utilized extensively in connection with plant diseases caused by fungal, viral, and bacterial parasites. For example, in the years 1845–1847, the potato crop in Ireland was devastated by late blight, a fungal parasite. Nearly a million people starved, and another million people emigrated to escape the same fate. Nowadays, protection against such disasters is provided in large part by growing varieties of potato that are resistant to the blight. It is no overstatement to say that the world owes much of its production of potatoes, corn, wheat, and other cereal grains to the painstaking work of plant geneticists who selected and bred disease-resistant varieties.

The same potential exists for breeding plants that are resistant to insect pests. Traits that provide resistance may act as chemical or physical barriers.

Control with Chemical Barriers. A chemical barrier is a chemical produced by the plant we want to protect; the substance is lethal or at least repulsive to the would-be pest. Once they have identified such a barrier, plant breeders can use selection and crossbreeding to enhance the trait in the desirable plant. The relationship between wheat and the Hessian fly provides an example. The fly lays its eggs on wheat leaves, and the larvae move down the leaves and into the main stem as they feed. The weakened stem either dies or is broken in the wind. The Hessian fly was introduced into the United States in the straw bedding of Hessian soldiers during the Revolutionary War. The fly eventually spread throughout much of the Midwest, causing widespread devastation until scientists at the University of Kansas developed a variety of wheat that produces a chemical toxic

to the insect; the chemical kills the larvae when they feed on the leaves. Increasing resistance through breeding may not provide 100% protection, but even partial protection can make the difference between profit and loss for the grower. In addition, any degree of resistance conferred upon the plant lessens the need for chemical pesticides.

Control with Physical Barriers. Physical barriers are structural traits that impede the attack of a pest. For example, leafhoppers are significant worldwide pests of cotton, soybeans, alfalfa, clover, beans, and potatoes, but they can damage only plants that have relatively smooth leaves. Hooked hairs on the leaf surfaces of some plants tend to trap and hold immature leafhoppers until they die. Similarly, glandular hairs that exude a sticky substance fatally entrap alfalfa weevil larvae. Such traits can be enhanced in vulnerable plants through selective breeding.

Control with Sterile Males. Another genetic-control strategy involves flooding a natural population with sterile males that have been reared in laboratories. Combating the screwworm fly provides a prime illustration. This fly, which is closely related to the housefly and looks much like it, lays its eggs in open wounds of cattle and other animals. The larvae feed on blood and lymph, keeping the wound open and festering. Secondary infections frequently occur and often lead to the death of the animal.

Early in the 20th century, the problem became so severe that cattle ranching from Texas to Florida and northward was becoming economically impossible. In studying the situation during the 1940s, Edward Knipling, an entomologist with the USDA, observed two essential features of screwworm flies: (1) Their populations are never very large, and (2) the female fly mates just once, lays her eggs, and then dies. Knipling reasoned that if the female mated with a sterile male, no offspring would be produced. His hypothesis was correct, and today sterile males are routinely used to control this pest. After huge numbers of screwworm larvae are grown on meat in laboratories, the resulting pupae are subjected to just enough high-energy radiation to render them sterile. These sterilized pupae are then air-dropped into the infested area. Ideally, 100 sterile males are dropped for every normal female in the natural population, giving a 99% probability that wild females will mate with one of the sterile males.

This technique proved so successful that it eliminated the screwworm fly from Florida in 1958–59, and it continues to be used to control the problem in the Southwest to this day. The savings to the cattle industry are estimated at more than $300 million a year. Recently, the technique was used to eradicate the tsetse fly from the island of Zanzibar, thus eliminating the disease trypanosomosis (sleeping sickness), which the tsetse fly carries. Indeed, now herdsmen can raise their cattle without the devastating losses the disease causes elsewhere in sub-Saharan Africa, more than 10 million square miles of which are inhabited by tsetse flies.

Research efforts are under way to extend this success to other parts of the continent.

Strategies Using Biotechnology.   Biotechnology has multiplied the potential for genetic control. More complex than basic plant breeding, genetic engineering makes it possible to introduce genes into crop plants from a variety of sources: other plant species, bacteria, and viruses. The new transgenic crops are undergoing rapid development and testing; by mid-2000, 40 transgenic crops had received regulatory approval. Over 75 million acres of transgenic crops were planted in the United States that year. This approach is not without its problems, however: As we saw in Chapter 10, public acceptance of transgenic crops is far behind the exploding technology, especially in Europe.

One promising strategy is to incorporate the protein coat of a plant virus into the plant itself. When the plant "expresses" (that is, manufactures) the virus's protein coat, the plant becomes resistant to infection by the real virus. In this way, crop plants have been made resistant to more than a dozen plant viruses.

Several strategies that employ biotechnology have the potential for reducing pesticide use. Endowing plants with resistance to insects has been engineered with the incorporation of a potent protein produced by a bacterium, *Bacillus thuringiensis* (Bt). This protein kills the larvae of a number of plant-eating insects and is harmless to mammals, birds, and most other insects. Scientists have engineered the gene into a number of plants, including cotton, potatoes, and corn (Fig. 17–13). This development alone is expected to cut in half the use of pesticides on cotton—a crop that consumes more than 10% of the pesticides used throughout the world. The protection comes from a naturally occurring protein used for three decades by home gardeners, organic growers, and other farmers. Wide-scale use of the Bt cotton

▲ **FIGURE 17–13   *Bioengineered potatoes.***   The center row shows potato plants that have the Bt gene incorporated into the plant genome. These plants are protected from the Colorado potato beetle, while surrounding rows show typical devastation of nonengineered potato plants by beetles.

(Monsanto's "Bollgard" cotton) revealed that it is effective against two of three pests, but farmers had to employ pesticides to control the bollworm (a kind of moth larva) in 40% of the 2 million acres planted of the new crop. Bt corn has been developed to combat a huge problem: the European corn borer; over $1 billion in losses and control costs are traced to this one pest.

One biotech strategy has been to give a crop species resistance to a broad-spectrum herbicide. Although this might appear to encourage the use of herbicides, it is actually expected to make it possible to use less total amounts of herbicide. One such herbicide is glyphosate (Roundup®), which inhibits an amino acid pathway that occurs only in plants and not in animals, biodegrades rapidly, and can control both grasses and broadleaf plants. (Because it is nonspecific and will kill every plant, application and cleanup must be done carefully.) Genes resistant to glyphosate have been successfully introduced into cotton, canola, soybean, and corn. Farmers save money on herbicides; they need only one spraying of Roundup to control weeds in their soybean fields. (Half of the U.S. soybean crop is now planted with "Roundup Ready" soybeans.)

There is a downside, however, to these current biotech innovations. First, they are not well suited to developing countries, as they foster a dependence on a costly annual supply of seeds under current corporation practices. Also, there is the definite possibility of developing "superweeds" that are resistant to the herbicides or plant viruses, as resistance genes are shared with genetically close wild species. Furthermore, insects are likely to develop resistance to Bt plants much more readily than they would to occasional spraying of the Bt insecticide directly, because they are constantly exposed to Bt toxic effects season after season. The same potential for developing resistance applies to chemical and physical control strategies. Thus, scientists have had to develop new resistant varieties of wheat seven times in the case of wheat and the Hessian fly. Nevertheless, it is hoped that the applications of biotechnology will gradually reduce the burden of pesticides in the environment and their potential impact on human health. One sign that this strategy is working is the rapid decline in the crop-duster industry (Fig. 17–3), which is in a tailspin because of the spreading use of genetically modified crops.

Natural Chemical Control

As in humans and other animals, hormones control each stage in the development of an insect. Hormones are chemicals produced in the organism that provide "signals" to control developmental processes and metabolic functions. In addition, insects produce many pheromones—chemicals secreted by one individual that influence the behavior of another individual of the same species.

The aim of natural chemical control is to isolate, identify, synthesize, and then use an insect's own hormones or pheromones to disrupt its life cycle. Two advantages of natural chemicals are that they are nontoxic and that they are highly specific to the pest in question. (They do not affect the natural enemies of the pest to any appreciable extent.) If the affected pest is eaten by another organism, it is simply digested.

Scientists have discovered that caterpillar pupation is triggered by a decrease in the level of the chemical called **juvenile hormone**. If this chemical is sprayed on caterpillars, pupation does not occur. The larvae simply continue to feed and grow, become grossly oversized, and eventually die. One newly developed insecticide, called Mimic, is a synthetic variation of **ecdysone**, the molting hormone of insects. Mimic begins the molting process in insect larvae, but doesn't complete it. As a result, the larva is trapped in its old skin and eventually starves to death. Mimic is specific to moths and butterflies and is viewed as a potent agent in the future control of the spruce budworm, the gypsy moth, the beet army worm, and the codling moth—all highly devastating pests. Field-test results with Mimic have been promising.

Adult insects secrete pheromones that attract mates. Once identified and synthesized, these pheromones may be used in either of two ways: the trapping technique or the confusion technique. In the trapping technique, the pheromone is used to lure males or females into traps or into eating poisoned bait (Fig. 17–14). In the confusion technique, the pheromone is dispersed over the field in such quantities that males become confused, cannot find females, and thus fail to mate.

The enormous potential of natural chemicals for controlling insect pests without causing ecological damage or disruption has been recognized for at least 30 years. Research has identified a large number of pheromones, and these natural chemicals are now key tools employed in IPM for many insects: the boll weevil and the pink bollworm for cotton, the codling moth for pears and apples, the tomato pinworm for tomatoes, and several bark beetles that infest forest trees, among others.

17.4 Socioeconomic Issues in Pest Management

The increasing availability of alternative pest control methods and the accumulating evidence of the failures of the chemical approach to pest management have led to a growing movement toward avoiding the use of pesticides, particularly on foods. This movement must contend with a strong tendency on the part of farmers to employ pesticides.

Pressures to Use Pesticides

With any method of pest control, it is important to keep in mind that a species becomes a pest only when its population multiplies to the point of causing significant damage. Natural controls are generally aimed at keeping pest populations below damaging levels, not at total eradication of the

▲ **FIGURE 17–14** *Trapping technique.* Adult Japanese beetles are lured into a trap by scented bait they find attractive. Once inside the trap, baffles prevent the beetles from escaping. These traps are available in hardware stores and are quite effective. (Japanese beetles are notorious lawn grubs as larvae; as adults, they are voracious consumers of garden flowers and vegetables.)

populations. By keeping pest populations down, natural controls avert significant damage while preserving the integrity of the ecosystem.

Therefore, the question to be asked in facing any pest species is, Is the species causing significant damage? Damage should be deemed significant only when the economic losses due to the damage considerably outweigh the cost of applying a pesticide. This point is called the **economic threshold** (Fig. 17–15). If significant damage is not occurring, natural controls are already operating, and the situation is probably best left as is. Spraying with synthetic chemicals at this stage is more than likely to upset the natural balance and make the situation worse through resurgences. On the other hand, if significant damage *is* occurring, a pesticide treatment may be in order.

It makes little difference whether the threat of loss as a result of pest infestation is real or imagined, or close at hand

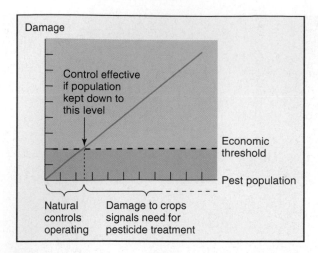

▲ **FIGURE 17–15** *The economic threshold.* The objective of pest control should not be to eradicate a pest totally. All that is needed is to keep population levels below the economic threshold.

or remote; what is important is how the grower perceives the threat. Even if there is no evidence of immediate damage from a pest, a grower who believes that his or her plantings are at risk is likely to resort to **insurance spraying**, the use of pesticides "just to be safe."

Consumers also put indirect pressure on growers to use pesticides. From customers to supermarket chains to canneries, there is a tendency to select the best-looking fruits and vegetables, leaving the remainder to be sold at lower prices or trashed. Like Snow White, we all tend to reach for the apparently perfect, unblemished apple. Growers know that blemished produce means less profit, so they indulge in **cosmetic spraying**—the use of pesticides to control pests that harm only the item's outward appearance. Cosmetic spraying accounts for a significant fraction of pesticide use, does nothing to enhance crop yield or nutritional value, and results in an increase in pesticide residues remaining on the produce.

Integrated Pest Management

Integrated pest management aims to minimize the use of synthetic organic pesticides without jeopardizing crops. This is made possible by addressing all the interacting sociological, economic, and ecological factors involved in protecting crops. With IPM, the crop and its pests are seen as part of a dynamic ecosystem; the goal is not the eradication of pests, but maintaining crop damage below the economic threshold.

Cultural and biological control practices form the core of IPM techniques. Such practices as crop rotation, polyculture (instead of monoculture), the destruction of crop residues, the maintenance of predator populations, and carefully timed planting and fertilizing are basic to IPM. "Trap crops" are often used: Early strips of a crop such as cotton are planted to lure existing pests, and then the pests are destroyed by hand or by the limited use of pesticides be-

fore they can reproduce. Spraying is not performed during the growing season, in order to encourage natural enemies of the pests to flourish. Pest populations are monitored, often by persons employed by local agricultural extension services or farm cooperatives or by persons acting as independent consultants. These **field scouts** are trained in monitoring pest populations (traps baited with pheromones are used for this purpose) and in determining whether the population exceeds the economic threshold. If this happens, remedial measures are taken to decrease the pest population. Here, pesticides are often used, in quantities and brands that do the least damage to natural enemies of the pest. In addition, **pest-loss insurance**, which pays the farmer in the event of loss due to pests, has enabled insured growers to refrain from unnecessary "insurance spraying."

Making the economic benefits of natural controls known to growers is an important aspect of IPM. Because pesticides increased yields and profits when they were first used, many farmers still cling to them, believing that they offer the only way to bring in a profitable crop. In some instances, however, the rising costs of pesticides, along with their tendency to aggravate pest problems, have eliminated their economic advantage.

Particularly in the developing world, governmental agricultural policy often determines the extent to which IPM is adopted. Governments and aid agencies usually subsidize the purchase of pesticides, thereby strongly encouraging growers to step onto the pesticide treadmill. By contrast, an Indonesian experience has provided a viable IPM model for other rice-growing countries. Faced with declining success in controlling the brown plant hopper (Fig. 17–16), Indonesian rice growers have switched from heavy pesticide spraying to a light spraying regime that preserves the natural enemies of the insect. The success of the program can be traced to close cooperation between the Indonesian government and the FAO (Fig. 17–17). FAO workers ran training sessions for farmers ("farmer field schools), weekly meetings that lasted throughout the growing season. Eventually, more than 200,000 farmers were taught about the rice agroecosystem and IPM techniques, and these farmers formed a corps that could in turn teach others. The economic and environmental benefits of the Indonesian IPM program have been remarkable. The government has saved millions of dollars annually by not purchasing pesticides, farmers have not had to invest in pesticides and spraying equipment, the environment has been spared the application of thousands of tons of pesticide, fish are once again thriving in the rice paddies, and the health benefits of reduced pesticide use have been spread from applicators to consumers and wildlife. (The FAO is sponsoring similar IPM training programs in eight other rice-growing nations and has started programs for vegetable crops in six nations.)

Recently, the FAO, the World Bank, UNDP, and UNEP have joined to cosponsor the IPM facility. Its mission is to carry forward the objectives of Agenda 21 and to establish networks among farmers, researchers, and extension services. Although IPM does rely on chemical pesticides when needed, if IPM is broadly adopted, it is a logical pathway to a sustainable future in which pesticides are not polluting groundwater, contaminating food, killing pollinating insects, and, in the end, creating new, more resistant pests.

Organically Grown Food

Public feeling is often strong against the use of chemical pesticides. Cosmetic or other unnecessary spraying persists in large part because consumers are kept ignorant about the kinds or amounts of pesticides used. Experience shows that when consumers are informed, many of them abandon the Snow White attitude. Food outlets selling **organically grown** produce (produce that is grown without synthetic chemical pesticides or fertilizers) are appearing, although the produce is not as cosmetically perfect and usually costs more than chemically treated produce (Fig. 17–18). Nonetheless, an increasing number of growers are relying on natural controls and selling through these specialized markets. Sales of organics have increased 20% a year for the past decade; organically grown food is now a $6 billion enterprise in the United States and $15.6 billion worldwide. Expanding markets for

▲ FIGURE 17–16 *Brown plant hoppers.* Immature brown plant hoppers, shown on the stem of a rice plant.

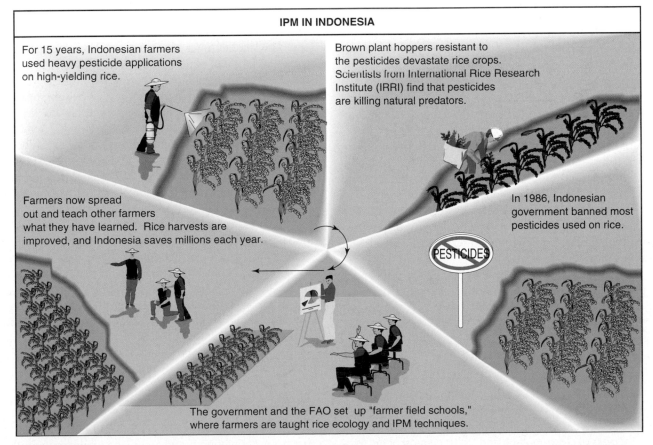

IPM IN INDONESIA

For 15 years, Indonesian farmers used heavy pesticide applications on high-yielding rice.

Brown plant hoppers resistant to the pesticides devastate rice crops. Scientists from International Rice Research Institute (IRRI) find that pesticides are killing natural predators.

In 1986, Indonesian government banned most pesticides used on rice.

PESTICIDES

The government and the FAO set up "farmer field schools," where farmers are taught rice ecology and IPM techniques.

Farmers now spread out and teach other farmers what they have learned. Rice harvests are improved, and Indonesia saves millions each year.

▲ FIGURE 17–17 *Integrated pest management.* IPM has helped Indonesian rice farmers bring the brown plant hopper under control after years of frustration on the pesticide treadmill.

▲ FIGURE 17–18   *Organic foods.*   A roadside farm stand in Concord, Massachusetts, specializes in organic produce. Organically grown foods represent an expanding market in the United States.

certified organic foods in the United States, Japan, and Europe have encouraged increasing numbers of growers to abandon the use of agricultural chemicals.

Certification itself has been a contentious process, and the industry has requested help from the government. In response, in 1990 Congress passed the Organic Foods Protection Act, which established the National Organic Standards Board (NOSB), under USDA auspices. After some years of work, proposed guidelines for certification from the NOSB were aired in early 1998; possibly in response to lobbying from the food industry, the board allowed genetically engineered foods, irradiated foods (subjected to radiation to kill bacteria), and crops fertilized with sewage sludge to be included as organic. A firestorm of protest from the organic food community ensued. Newly revised guidelines were published by the USDA's National Organic Program in 2000; no product involving genetically engineered or irradiated foods or fertilized with sewage sludge will be certified as organic. Official USDA certification of organic foods will soon reach closure.

With this episode as an introduction, let us turn to public policy as it involves the use and regulation of pesticides.

17.5 Public Policy

In regulating the use of pesticides, three concerns are obvious: (1) Pesticides must be evaluated for their intended uses and for their impacts on human health and the environment; (2) those who use the pesticides—especially agricultural workers—must be appropriately trained and protected from the risks of close contact; and (3) since most of the agricultural applications involve food, the public must be protected from the risks of pesticide residues on food products.

FIFRA

The **Federal Insecticide, Fungicide, and Rodenticide Act**, commonly known as FIFRA, addresses the first two of the preceding concerns. This law, established by Congress in 1947 and amended in 1972, is administered by the EPA, which has total jurisdiction over the manufacture, sale, use, and testing of pesticides. FIFRA requires manufacturers to register pesticides with the EPA before marketing them. The registration procedure includes testing for toxicity to animals (and, by extrapolation, to humans). From the test results, usage standards are set. For example, chlordane, which was used for 40 years as a pesticide on crops, lawns, and gardens, was tested and shown to be acutely toxic to animals in high doses; chronic exposure led to damaging effects on most vital systems and caused liver cancer in animals. Acute and chronic exposure of humans to chlordane showed similar results, except for the development of cancer. On the basis of these results, the EPA canceled all uses of chlordane for food crops in 1978, but allowed continued use against termites until 1988, when all uses of the chemical in the United States were canceled. The law also stipulates that the EPA must determine whether a product "generally causes unreasonable adverse effects on the environment," and if it so determines, the product may be restricted or canceled. This was the basis for the cancellation of DDT.

All pesticides must have an EPA-approved label that lists their active ingredients, instructs the public on their proper use, and informs users of the risks involved with their use. The large number of existing pesticides and the continuing development of new products have posed a huge challenge to the EPA's ability to provide a thorough evaluation of each substance. Since the law allows existing pesticides to remain in use unless they are proved hazardous, many products on the market still have not been subjected to adequate tests, and existing standards may allow hazardous levels of pesticide residues in some foods. However, there are enough complaints from the pesticide industry about the EPA's "overaggressive regulation" to judge that the agency is doing reasonably well in carrying out its congressional mandate. In protecting workers from risks posed by occupational exposure to pesticides, the EPA works with state environmental agencies, which have primary responsibility for enforcement. Under its Worker Protection Standard program, the EPA has issued regulations involving requirements for safety training, the use of protective equipment, the provision of emergency assistance, and permitting standards. The states must then demonstrate to the EPA that, on the basis of the federal regulations, they have adequate regulation and enforcement mechanisms.

FQPA of 1996

Protecting consumers from exposure to pesticides on food has been a contentious and confusing scene for many years. Three agencies are involved: the EPA, the FDA, and the Food Safety and Inspection Service of the USDA. Basically, the EPA sets the allowable tolerances for pesticide residues, based on its assessment of toxicity, and the FDA monitors and enforces the tolerances on all foods except meat, poultry, and egg products, which are managed by the USDA. In previous years, the **Federal Food, Drug, and Cosmetic Act of 1958 (FFDCA)** was the enabling legislation assigning these responsibilities. One clause of the FFDCA, the so-called **Delaney clause**, was highly controversial. The clause states that "no [food] additive shall be deemed to be safe if it is found to induce cancer when ingested by man or animal." Since pesticides represent one of the largest categories of toxic chemicals to which people are exposed, this clause was applied in prohibiting many pesticides from being used on foodstuffs when those pesticides had been found to cause cancer in laboratory tests with animals. In essence, the law states that if a given pesticide presents any risk of cancer, no detectable residue may remain on the food. As analytical chemical techniques became more and more sensitive, extremely low traces of a pesticide could be measured, and according to the Delaney clause, if that pesticide caused cancer in experimental animals, farmers could not use it.

The National Research Council examined this dilemma and recommended in 1987 that the anti-cancer clause be replaced with a "negligible risk" standard, based on risk-analysis procedures that had been developed subsequent to the Delaney clause. After years of debate, Congress passed the **Food Quality Protection Act (FQPA)** in 1996 and did away with the Delaney clause. The following are the major requirements of the act:

- The new safety standard is "a reasonable certainty of no harm" for substances applied to foods.
- Special consideration must be given to the exposure of young children to pesticide residues.
- Pesticides or other chemicals are prohibited if, when consumed at average levels over the course of a lifetime, they can be shown to carry a risk of more than one case of cancer per million people.
- All possible sources of exposure to a given pesticide must be evaluated, not just exposure from food.
- A special attempt must be made to assess the potential harmful effects of the so-called hormone disrupters.
- The same standards are to be applied to raw and processed foods (previously dealt with separately).

The focus on children is the outcome of a 1993 report by the National Research Council, *Pesticides in the Diets of Infants and Children*. The focus is on children both because they typically consume more fruits and vegetables per unit of body weight than do adults and because studies have shown that youths are frequently more susceptible than adults to carcinogens and neurotoxins.

The USDA Pesticide Data Program (PDP) provides annual analyses of pesticides in fresh and processed foods.

Under the FQPA, the EPA is to evaluate toxicity data on individual pesticides and then, using the USDA's PDP analyses and other information on risk assessment (see Chapter 16), establish a risk characterization for each pesticide. To date, the EPA is on target with the directives of the FQPA, having published safe tolerances for about 3,000 pesticide products (out of a total of over 9,700). As mentioned, two pesticides—methyl parathion and chlorpyrifos—recently were banned on the basis of these reevaluations, because they were found to pose unacceptable risks to children. Playing the role of public watchdog, the Consumers Union has published annual evaluations of pesticides in food and what the EPA is doing about them. Their evaluations have shown, for example, that U.S.-grown food is just as likely (and in some cases more likely) to be contaminated as foreign-grown foods, demolishing a popular stereotype of the latter. The Consumers Union concluded that the EPA needs to step up the pace of evaluating a number of key pesticides that are responsible for the lion's share of risk.

Pesticides in Developing Countries

The United States currently exports more than 250,000 tons of pesticides each year, mostly to developing countries. Some 25% of this total consists of products banned in the United States itself. FIFRA requires "informed permission" prior to the shipment of any pesticides banned in the United States. This permission comes from the purchaser, which can often be a foreign subsidiary of the exporting company. The EPA must then notify the government as to the identity of the importing country.

Fortunately, the international community has erected a more effective system, through the cooperative work of two U.N. agencies: the FAO and the United Nations Environment Program (UNEP). There is now a process of prior informed consent (PIC), whereby exporting countries inform all potential importing countries of actions they have taken to ban or restrict the use of pesticides or other toxic chemicals. Governments in the importing country respond to the notifications via the U.N. agencies, which then disseminate all the information they receive to the exporting countries and follow up by monitoring the export practices of those countries. The PIC process was officially approved and signed by 80 government representatives in 1998 and will become effective after 50 countries have ratified it.

The more serious problem of unsafe pesticide use in the developing countries has also been addressed by the FAO. In spite of early opposition from the pesticide industry and exporting countries, an international "Code of Conduct" has been drawn up whereby conditions of safe pesticide use are addressed in detail. The code makes clear the responsibilities of private companies and of countries receiving pesticides in promoting their safe use. As with PIC, the code is not yet legally binding, but it has proved useful in holding private industry and importing countries to standards of safe use. In the view of some observers, both PIC and the FAO Code of Conduct are not as strong as they should be; however, it is acknowledged that their very existence represents great progress in addressing a difficult problem in the developing countries.

New Policy Needs

A group of leading entomologists called for a 50% reduction in pesticide use in the United States, supporting their recommendation with the economic and environmental benefits of such a reduction. This goal could become public policy if it is included in legislative or regulatory law. Shortly after the Clinton administration came into power in 1992, the heads of the EPA, FDA, and USDA issued a joint announcement of their commitment to reduce pesticide use in the United States. A year later, the three federal agencies announced the Pesticide Environmental Stewardship Program, a public–private partnership committed to pesticide use and risk reduction. The program focuses on two goals: (1) developing strategies of use and risk reduction that rely on biological control rather than chemical treatment and

(2) having 75% of U.S. agricultural acreage adopt IPM programs by 2000. Partners from the private sector include some 70 agricultural organizations, most of which are pesticide users. Grants are awarded for projects consistent with the program goals, and efforts on the part of the partners and the government to reduce pesticide use and risks are shared.

Pesticide reform is in the wind. Yet, as Fig. 17–2 indicates, we have a long way to go to achieve a 50% reduction from the 1979 high point of pesticide use. As with so many of the issues we consider in this book, significant progress in pesticide control has benefited from grassroots action and pressure from public-interest groups and nongovernmental organizations like the Consumers Union. Such groups are pressing for a continued movement in the direction of ecological pest management and continued progress in keeping food free of pesticide residues. Also, with three governmental agencies involved in pesticide stewardship, there is hope that we will soon substantially move toward helping farmers jump off the pesticide treadmill and toward aiding consumers in achieving a pesticide-free food supply.

ENVIRONMENT ON THE WEB

PESTS: KILL 'EM OR EAT 'EM

Chapter 11 discusses the value of biodiversity in the natural environment. In food production, however, diversity of plant and animal species is far from welcome. Instead, species that are not intentionally cultivated are often actively removed through physical, chemical, or biological processes. Yet a pest in one situation may be highly valued in another. For example, the common dandelion is aggressively controlled in most urban lawns but is a valued foodstuff in some cuisines. Similarly, molds are unappealing in most foods but can also be an important source of pharmaceuticals, such as is the case with penicillin. Thus, the situation raises the question, "What, then, should we consider a 'pest'?"

In fact, the definition of "pest" will depend on the social and economic context of its occurrence—there are few species that are universally regarded as objectionable. For example, in many developing countries, where sources of protein such as beef are scarce for people of limited means, insects can provide an alternative protein source. "Mopani," or the emperor moth larvae, are one of many insects that are sold and eaten in South Africa; similarly, ants, worms, locusts, and cockroaches are eaten in other developing countries across Asia and Africa.

In the United States, insects and dandelions seldom appear on the dinner table, or indeed in lawns or agricultural fields. Urban dwellers strive to achieve a pest-free, unblemished lawn dotted with trees and weed-free flower beds. To maintain such a standard, however, requires intensive maintenance through fertilizer and pesticide inputs, regular mowing, aeration, and so on. It has been estimated that Americans spend $6 billion a year on lawn care. Yet in trying to solve the pest problem, Americans may be creating other environmental concerns. Lawn maintenance uses almost 70

million pounds of pesticides and 4 to 5 million tons of fertilizer every year, generates high levels of noise and air pollution (from gas-powered lawn mowers and trimmers), and contaminates urban lakes and streams with nutrient-rich runoff and high levels of residual pesticides.

Like other personal choices, how people define pests is determined by their values and the social, cultural, and economic context of their lives. However, the definition they choose can mean consequences and trade-offs for the environment.

Web Explorations

The Environment on the Web activity for this essay describes the environmental issues surrounding lawn maintenance in the urban environment. Go to the Environment on the Web activity (select Chapter 17 at **http://www.prenhall.com/wright**) and learn for yourself:

1. about insects as a source of food;
2. about the differences between traditional and organic gardening methods; and
3. about integrated pest management strategies.

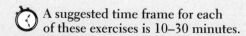 **A suggested time frame for each of these exercises is 10–30 minutes.**

REVIEW QUESTIONS

1. Define pests. Why do we control them?
2. Discuss the basic philosophies of pest control.
3. What were the apparent virtues of the synthetic organic pesticide DDT?
4. What adverse environmental and human health effects can occur as a result of pesticide use?
5. Define bioaccumulation and biomagnification.
6. Why are nonpersistent pesticides not as environmentally sound as first thought?
7. Describe the four categories of natural or biological pest control. Cite examples of each and discuss their effectiveness.
8. Define the term "economic threshold" as it relates to pest control.
9. How does integrated pest management work? Give examples.
10. How do the FIFRA and FFDCA attempt to control pesticides? What are the shortcomings of this type of legislation? What new perspective does FQPA bring to the policy scene?
11. Discuss recent policy regarding the export of pesticides to developing countries.
12. What recent amendments to public policy are promoting pest control that is environmentally safer than before, and what needs still have to be addressed?

THINKING ENVIRONMENTALLY

1. U.S. companies export pesticides that have been banned or restricted in this country. Should the practice be allowed to continue? Support your answer.
2. Almost one-third of the chemical pesticides bought in the United States are for use in houses and on gardens and lawns. What should manufacturers and users do to ensure their limited and prudent use?
3. Should the government give farmers economic incentives to switch from pesticide use to integrated pest management? Why or why not?
4. Investigate how bugs and weeds are controlled on your campus. Are IPM techniques being used? Organic fertilizers? If not, consider lobbying for their use.
5. Read or reread the "Ethics" essay entitled "DDT for Malaria Control: Hero or Villain?" (p. 419). How would you resolve this issue?

WEB REFERENCES

On-line resources for this chapter are on the World Wide Web at: **http://www.prenhall.com/wright**. (Click on Chapter 17 within the Chapter Selector.)

Water Pollution and Its Prevention

The Mississippi River watershed encompasses 40% of the land area of the United States, and as it collects water from this huge area, it eventually delivers the water to the Gulf of Mexico. Because much of the watershed is America's agricultural heartland, the Mississippi is a reflection of what happens on the farm. As U.S. farm production experienced its great growth during the latter half of the 20th century, more and more fertilizers were spread on agricultural fields, and more animals were raised on feedlots. At the same time, wetlands bordering river tributaries were drained and no longer intercepted agricultural runoff. The result was a tripling of soluble nitrogen delivered to the river and, finally, to the Gulf of Mexico.

The impact of this major flow of nitrogen to the Gulf of Mexico was first detected by marine scientists in 1974 in the form of areas where oxygen had disappeared from bottom sediments and much of the water column above them. This absence of oxygen is deadly to the bottom-dwelling animals and any oxygen-breathing creatures in the water column not able to escape. At first, the **hypoxia**, or "dead zone" as it

is called, was thought to be similar to other bodies of water, like the lower Chesapeake Bay, that received high nutrient levels—a minor disturbance that would disappear seasonally. But then the hypoxic area suddenly doubled in size following the 1993 floods in the Midwest, and it has remained large ever since. In 1999, the area extended over 7,800 square miles (20,000 km^2) off the Louisiana coast, an area larger than the state of New Jersey. The shrimpers and fishers who worked these waters knew something was happening and quickly learned to avoid the hypoxic areas.

To the scientists working on the problem, the connection between nitrogen and hypoxia was clear. Nitrogen is a limiting factor in coastal marine waters. Where it is abundant, it promotes the dense growth of phytoplankton—photosynthetic microorganisms that live freely suspended in the water. As the phytoplankton die and sink toward the bottom, they are decomposed by bacteria, a process that consumes dissolved oxygen. Eventually, the oxygen is used up, creating the dead zone of hypoxia. In the Gulf, the dead zone can extend up from the bottom to within a few meters of the

◀ **The Mississippi Delta.** The Mississippi basin consists of 40% of the landmass of the lower 48 states. Runoff from this huge area delivers enormous amounts of sediment and nutrients to the Gulf of Mexico.

surface; typically, it stays from May through September, finally being dispersed by the mixing of the water column as colder weather sets in. Because the Gulf fishery is a $2.8 billion enterprise, the problem eventually got national attention. In 1998, Congress passed the **Harmful Algal Bloom and Hypoxia Research and Control Act**, establishing an interagency task force of scientists charged with assessing the Gulf of Mexico hypoxia and another water pollution problem we shall meet later in the chapter: harmful algal blooms. The task force released its report in May 2000. The report, on the World Wide Web at http://www.nos.noaa.gov/products/pubs_hypox.html, confirmed the connection between nitrogen and the dead zone and laid out options for reducing the nitrogen load coming downriver. Two strategies were key to mitigation of the problem: (1) reducing nitrogen loads to streams and rivers in the Mississippi basin (by using less fertilizer and greatly reducing farm runoff) and (2) restoring and encouraging nitrogen retention and denitrification processes in the basin. Fertilizer and farm groups are unhappy with the findings, but the EPA is moving ahead to develop an action plan to mitigate the hypoxia in the Gulf of Mexico, as called for by the congressional act.

The Gulf dead zone turns out to be one more costly lesson about **eutrophication** (literally, enrichment). We are learning that it is not reasonable to use vital water resources to receive pollutants and then expect them to continue to provide us with their usual bounty of goods and services.

In this chapter, we will explore eutrophication and other forms of water pollution. The next four chapters focus on pollution by solid and hazardous wastes and air pollution.

18.1 Water Pollution

Before beginning, we need to clarify the definition of pollution. The EPA defines pollution as "the presence of a substance in the environment that because of its chemical composition or quantity prevents the functioning of natural processes and produces undesirable environmental and health effects." Any material that causes the pollution is called a **pollutant**.

Pollution Essentials

Pollution is not usually the result of deliberate mistreatment of the environment; the additions that cause pollution are almost always the *by-products* of otherwise worthy and essential activities—producing crops, creating comfortable homes, providing energy and transportation, and manufacturing products—and of our basic biological functions (excreting wastes). Pollution problems have become more pressing over the years because both growing population and expanding per capita use of materials and energy increase the amounts of by-products that go into the environment. Also, many materials now widely used, such as aluminum cans, plastic packaging, and synthetic organic chemicals, are **nonbiodegradable**; that is, they resist attack and breakdown by detritus feeders and decomposers and consequently accumulate in the environment. An overview of the categories of pollutants that result from various activities is shown in Fig. 18–1.

It is important to note the breadth and diversity of pollution. Any part of the environment may be affected, and almost anything may be a pollutant. The only criterion is that the addition of a pollutant result in undesirable changes. The impact of an undesirable change may be largely aesthetic—hazy air obscuring a distant view or the unsightliness of roadside litter, for instance. The impact may be on ecosystems as a whole—the die-off of fish or forests, for example. Or the impact may be on human health—toxic wastes contaminating water supplies. The impact also may range from very local—the contamination of an individual well, for instance—to global. We tend to think of pollution as the introduction of human-made materials into the environment, but undesirable changes may be caused by the introduction of too much of otherwise natural compounds; fertilizer nutrients introduced into waterways and carbon dioxide introduced into the atmosphere (Chapter 21) are examples.

Therefore, the slogan "Don't pollute" is a gross oversimplification. The very nature of our existence necessitates the production of wastes. Our job in remediating present and future pollution problems is parallel to the concept of sustainable development itself. It is to adapt the means of meeting our present needs so that by-products are managed in ways which will not cause alterations that will jeopardize present and future generations. The general strategy in each case must be to

1. identify the material or materials that are causing the pollution—the undesirable change;
2. identify the sources of the pollutants;
3. develop and implement pollution control strategies to prevent the pollutants from entering the environment; and
4. develop and implement alternative means of meeting the need that do not produce the polluting by-product—in other words, avoid the pollution altogether.

Basically, the strategy for addressing pollution may be seen as conforming to the second principle of sustainability: *For sustainability, ecosystems dispose of wastes and replenish nutrients by recycling all elements*, thereby avoiding both pollution and the depletion of resources.

Water Pollution: Sources, Types, Criteria

Water pollutants originate from a host of human activities and reach surface water or groundwater through an equally diverse host of pathways. For purposes of regulation, it is customary to

▲ **FIGURE 18–1** *Categories of pollution.* Pollution is an outcome of otherwise worthy human endeavors. Major categories of pollution and activities that cause them are shown here.

distinguish between **point sources** and **nonpoint sources** of pollutants (Fig. 18–2). Point sources involve the discharge of substances from factories, sewage systems, power plants, underground coal mines, and oil wells. These sources are relatively easy to identify and therefore are easier to monitor and regulate than nonpoint sources, which are poorly defined and scattered over broad areas. Pollution occurs as rainfall and snowmelt move over and through the ground, picking up pollutants as they go. Some of the most prominent nonpoint sources of pollution are agricultural runoff (from farm animals and croplands), storm-water drainage (from streets, parking lots, and lawns), and atmospheric deposition (from air pollutants washed to earth or deposited as dry particles).

Two basic strategies are employed in attempting to bring water pollution under control: (1) reduce the sources and (2) treat the water so as to remove pollutants or convert them to harmless forms. Water treatment is the best option for point sources; source reduction can be employed for both kinds of sources and is the best option available for nonpoint sources.

Pathogens. The most serious water pollutants are the infectious agents that cause sickness and death. (See Chapter 16.) The excrement from humans and other animals infected with certain **pathogens** (disease-causing bacteria, viruses, and other parasitic organisms) contains large numbers of

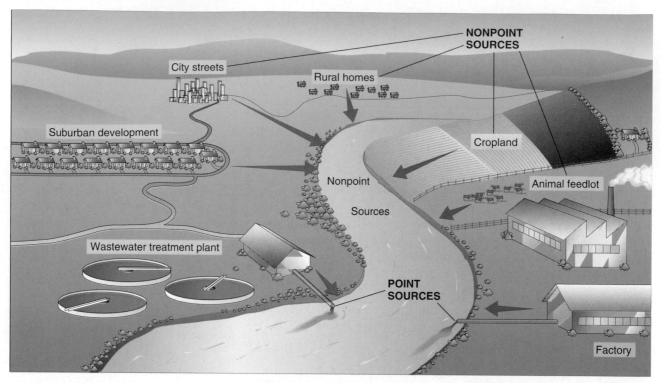

▲ **FIGURE 18–2** *Point and nonpoint sources.* Point sources are far easier to identify and correct than the diffuse nonpoint sources.

these organisms or their eggs. (See Table 18–1.) Even after symptoms of disease disappear, an infected person or animal may still harbor low populations of the pathogen, thus continuing to act as a carrier of disease. If wastes from carriers contaminate drinking water, food, or water used for swimming or bathing, the pathogens can gain access to, and infect, other individuals (Fig. 18–3).

Before the connection between disease and sewage-carried pathogens was recognized in the mid-1800s, disastrous epidemics were common in cities. Epidemics of typhoid fever and cholera killed thousands of people before the 20th century. Today, public-health measures that prevent this disease cycle have been adopted throughout the developed world and, to a considerable extent, in the developing world. These measures involve

- purification and disinfection of public water supplies with chlorine or other agents (Chapter 9);
- sanitary collection and treatment of sewage wastes;
- maintenance of sanitary standards in all facilities in which food is processed or prepared for public consumption; and
- public education in personal and domestic hygiene practices.

Standards regarding these items are set and enforced by government public health departments. A variety of other measures are enforced as well. For example, if bathing areas are contaminated with raw sewage, health departments close them to swimming. Implicit in all measures is monitoring for sewage contamination. (See the "Earth Watch" essay, p.444.) Of course, our own personal hygiene, sanita-

tion, and health precautions, such as not drinking water from untested sources and making sure that foods like pork and chicken are always well cooked, remain the last and most important line of defense against disease.

TABLE 18–1	Pathogens Carried by Sewage
Disease	**Infectious Agent**
Typhoid fever	*Salmonella typhi* (bacterium)
Cholera	*Vibrio cholerae* (bacterium)
Salmonellosis	*Salmonella* species (bacteria)
Diarrhea	*Escherichia coli*, *Campylobacter* species (bacteria) *Cryptosporidium parvum* (protozoan)
Infectious hepatitis	Hepatitis A virus
Poliomyelitis	Poliovirus
Dysentery	*Shigella* species (bacteria) *Entamoeba histolytica* (protozoan)
Giardiasis	*Giardia intestinalis* (protozoan)
Numerous parasitic diseases	(Roundworms, flatworms)

◀ **FIGURE 18–3** *The Ganges River in India.* In many places in the developing world, the same waterways are used simultaneously for drinking, washing, and disposal of sewage. In the case of the Ganges River, religious practice and spiritual cleansing are also common. A high incidence of disease, infant and childhood mortality, and parasites is the result.

Many people attribute good health in a population to modern medicine, but good health is more a result of the prevention of disease through public-health measures. An estimated 1.5 billion people do not have access to treated drinking water. Some 2.4 billion people live in areas having poor (or no) sewage collection or treatment. Largely because of poor sanitation regarding water and sewage, a significant portion of the world's population is chronically infected with various pathogens. More than 70 million new cases of diarrheal disease are reported each year, over 2 million of them resulting in death, and about half of those deaths are among children under five. Moreover, populations in areas where there is little or no sewage treatment are extremely vulnerable to deadly epidemics of any and all diseases spread via the sewage vector. Because of unsanitary conditions, an outbreak of cholera in Peru in 1990 killed several thousand people as it spread through Latin America. The WHO reported over 30 different outbreaks of cholera between 1998 and 2000; cases number in the tens of thousands annually, with over a thousand fatalities.

Organic Wastes. Along with pathogens, human and animal wastes contain organic matter that creates serious problems if it enters bodies of water untreated. Other kinds of organic matter (leaves, grass clippings, trash, etc.) can enter bodies of water as a consequence of runoff and, in the case of excessive phytoplankton, can grow within the water body. With the exception of plastics and some human-made chemicals, these wastes are biodegradable. As bacteria and detritus feeders decompose organic matter in water, they consume oxygen, which is dissolved as a gas in water. The amount of oxygen that water can hold in solution is severely limited. In cold water, dissolved oxygen (DO) can reach concentrations up to 10 ppm (parts per million); even less can be held in warm water. Compare this ratio with that of

oxygen in air, which is 200,000 ppm (20%), and you can understand why even a moderate amount of organic matter decomposing in water can deplete the water of its DO. Bacteria keep the water depleted in DO as long as there is dead organic matter to support their growth.

Thus, a common water-quality test is **biochemical oxygen demand (BOD)**, a measure of the amount of organic material in the water, in terms of how much oxygen will be required to break it down biologically, chemically, or both. The higher the BOD measure, the greater is the likelihood that dissolved oxygen will be depleted in the course of breaking it down. A high BOD causes so much oxygen depletion that animal life is severely limited or precluded, as in the bottom waters of the Gulf of Mexico. Fish and shellfish are killed when the DO drops below 2 or 3 ppm; some are less tolerant at even higher DO levels. If the system goes anaerobic (without oxygen), only bacteria can survive, using their abilities to switch to fermentation or anaerobic respiration (metabolic pathways that do not require oxygen). A typical BOD value for raw sewage would be around 250 ppm. Even a moderate amount of sewage, added to natural waters containing at most 10 ppm DO, can deplete the water of its oxygen and cause highly undesirable consequences well beyond those caused by the introduction of pathogens.

Chemical Pollutants. Because water is such an excellent solvent, it is able to hold many chemical substances in solution that have undesirable effects. Water-soluble **inorganic chemicals** constitute an important class of pollutants that include heavy metals (lead, mercury, cadmium, nickel, etc.), acids from mine drainage (sulfuric acid) and acid precipitation (sulfuric and nitric acids), and road salts employed to melt snow and ice in the colder climates (sodium and calcium chlorides). The **organic chemicals** are another group of substances that are found in polluted waters. Petroleum

EARTH WATCH

MONITORING FOR SEWAGE CONTAMINATION

An important aspect of public health is the detection of sewage in water supplies and other bodies of water with which humans come in contact. It is worth understanding how this detection is done.

It is virtually impossible to test for each specific pathogen that might be present in a body of water. Therefore, an indirect method called the **fecal coliform test** has been developed. This test is based on the fact that huge populations of a bacterium called *E. coli* (*Escherichia coli*) normally inhabit the lower intestinal tract of humans and other animals, and large numbers of the bacterium are excreted with fecal material. In temperate regions at least, *E. coli* does not last long in the outside environment. Therefore, the presumption is that when *E. coli* is found in natural waters, it is an indication of recent and probably persisting contamination with sewage wastes. In most situations, *E. coli* is not a pathogen itself, but is referred to as an **indicator organism**: Its presence indicates that water is contaminated with fecal wastes and that sewage-borne pathogens may be present. Conversely, the absence of *E. coli* is taken to mean that water is free from such pathogens.

The fecal coliform test, one technique of which is shown in the accompanying figures, detects and counts the number of coliform bacteria in a sample of water. Thus, the results indicate the relative degree of contamination of the water and the relative risk of pathogens. For example, to be safe for drinking, water should have an average *E. coli* count of no more than one *E. coli* per 100 milliliters (about 0.4 cup) of water. Water with as many as 200 *E. coli* per 100 ml is still considered safe for swimming. Beyond that level, a river may be posted as polluted, and swimming and other direct contact should be avoided. By contrast, raw sewage (99.9% water, 0.1% waste) has *E. coli* counts in the millions.

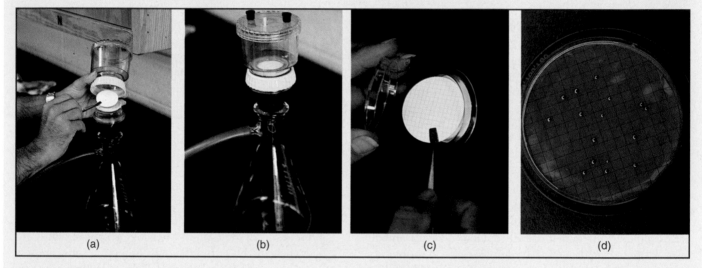

Testing water for sewage contamination by the Millipore technique. (a) A Millipore filter disk is placed in the filter apparatus. (b) A sample of the water being tested is drawn through the filter, and any bacteria present are entrapped on the disk. (c) The disk is then placed in a petri dish on a special medium that supports the growth of bacteria and that will impart a particular color to fecal *E. coli* bacteria. The dish is then incubated for 24 hours at 38°C, during which time each bacterium on the disk will multiply to form a colony visible to the naked eye. (d) *Escherichia coli* bacteria, indicating sewage contamination, are identifiable as the colonies with a metallic green sheen.

products pollute many bodies of water, from the major oil spills in the ocean (Chapter 13) to small streams receiving runoff from parking lots. Other organic substances with serious impacts are the pesticides that drift down from aerial spraying or that run off from land areas (Chapter 17) and the various industrial chemicals, like polychlorinated biphenyls (PCBs), cleaning solvents, and detergents.

Many of these pollutants are toxic, even at low concentrations (Chapter 20). Some may be concentrated through passage up the food chain in the process called biomagnification (Chapter 17). Even at very low concentrations, they can render water unpalatable for humans and dangerous for aquatic life. At higher concentrations, they can change the properties of bodies of water so as to prevent them from serving any useful purpose except perhaps for navigation. Acid mine drainage, for example, pollutes thousands of miles of streams in coal-bearing regions of the United States (Fig. 18–4). Stream bottoms are coated with orange deposits of iron, and the water is so acidic that only a few hardy bacteria and algae can tolerate it.

Sediments.   As natural landforms weather, a certain amount of sediment enters streams and rivers. However, erosion from farmlands, deforested slopes, overgrazed

▲ **FIGURE 18–4** *Acid mine drainage.* Acidic water from mines contains high concentrations of sulfides and iron, which are oxidized when they come into the open, thus producing unsightly and sterile streams.

rangelands, construction sites, mining sites, stream banks, and roads can greatly increase the load of sediment entering waterways. Sediments (sand, silt, and clay) have direct and extreme physical impacts on streams and rivers. When erosion is slight, streams and rivers of the watershed run clear and support algae and other aquatic plants that attach to rocks or take root in the bottom. These producers, plus miscellaneous detritus from fallen leaves and so on, support a complex food web of bacteria, protozoa, worms, insect larvae, snails, fish, crayfish, and other organisms, which keep themselves from being carried downstream by attaching to rocks or seeking shelter behind or under rocks. Even fish that maintain their position by active swimming occasionally need such shelter to rest (Fig. 18–5a).

Sediment entering waterways in large amounts has an array of impacts. Sand, silt, clay, and organic particles (humus) are quickly separated by the agitation of flowing water and are carried at different rates. Clay and humus are carried in suspension, making the water muddy and reducing the amount of light penetrating the water and hence photosynthesis. As the material settles, it coats everything and continues to block photosynthesis. It also kills the animals by clogging their gills and feeding structures. The eggs of fish and other aquatic organisms are particularly vulnerable to being smothered by sediment.

Equally destructive is the **bed load** of sand and silt, which is not readily carried in suspension, but is gradually washed along the bottom. As particles roll and tumble along, they scour organisms from the rocks. They also bury and smother the bottom life and fill in the hiding and resting places of fish and crayfish. Aquatic plants and other organisms are prevented from reestablishing themselves because the bottom is a constantly shifting bed of sand (Figs. 18–5b and c).

Sediments do not receive the attention the news media give to hazardous wastes and certain other pollution problems, but erosion is so widespread throughout the world that few streams and rivers escape the harsh impact of excessive sediment loads.

Nutrients. Some of the inorganic chemicals carried in solution in all bodies of water are classified as nutrients—essential elements required by plants. The two most important nutrients for aquatic plant growth are phosphorus and nitrogen, and they are often in such low supply in water that they are the limiting factors for phytoplankton or other aquatic plants. More nutrients mean more plant growth; therefore, nutrients become water pollutants when they are added from point or nonpoint sources and stimulate undesirable plant growth in bodies of water. The most obvious point sources of excessive nutrients are sewage outfalls. As we will see later, it takes a special effort to remove nutrients from sewage during the treatment process, so that is not always done. Of course, if the sewage is not treated at all, higher levels of nutrients will enter the water; this is indeed the situation in much of the developing world.

Agricultural runoff is the most notorious nonpoint source of nutrients, which are applied to agricultural crops in many forms: chemical fertilizers, manure, water for irrigation, sludge, and crop residues. When nutrients are applied in excess of plant needs, or when runoff from agricultural fields and feedlots is heavy, the nutrients are picked up by water that eventually enters streams, rivers, and lakes. Other nonpoint sources of nutrients include lawns and gardens, golf courses, and storm drains.

Water Quality Standards. When is water considered polluted? Many water pollutants, like pesticides, cleaning solvents, and detergents, are substances that are found in water only because of human activities. Others, like nutrients and sediments, are always found in natural waters, and they are a problem only under certain conditions. In both cases, "pollution" means "any quantity that is harmful to human health or the environment." The mere presence of a substance in water does not necessarily pose a problem; rather, it is the *concentration* of the pollutant that must be of primary concern.

But what concentration is worrisome? To provide standards for assessing water pollution, the EPA has established the National Recommended Water Quality Criteria. On the basis of the latest scientific knowledge, EPA has listed 157 chemicals and substances as criteria pollutants. The majority of these are toxic chemicals, but many

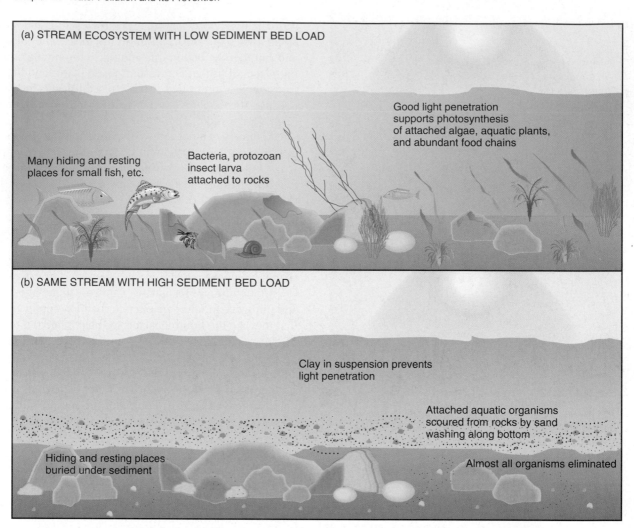

(a) STREAM ECOSYSTEM WITH LOW SEDIMENT BED LOAD

Many hiding and resting places for small fish, etc.

Bacteria, protozoan insect larva attached to rocks

Good light penetration supports photosynthesis of attached algae, aquatic plants, and abundant food chains

(b) SAME STREAM WITH HIGH SEDIMENT BED LOAD

Clay in suspension prevents light penetration

Attached aquatic organisms scoured from rocks by sand washing along bottom

Hiding and resting places buried under sediment

Almost all organisms eliminated

(c)

▲ **FIGURE 18–5** *Impact of sediment on streams and rivers.* (a) The ecosystem of a stream that is not subjected to a large sediment bed load. (b) The changes that occur with large sediment inputs. (c) Platte River at Lexington, Nebraska. The sandbars seen here constitute the bed load; they shift and move with high water, preventing the reestablishment of aquatic vegetation.

are also natural chemicals or conditions that describe the state of water, such as nutrients, hardness (a general measure of dissolved chemicals), or pH (a measure of the acidity of the water). The list identifies the pollutant and then provides a maximum permitted concentration of it. The criteria are meant to be used by the states, which are given the primary responsibility for upholding water pollution laws. Two important applications of water quality criteria are the **National Pollution Discharge Elimination System (NPDES)** and **Total Maximum Daily Load (TMDL)** programs. The NPDES program addresses point source pollution and involves issuing permits that regulate discharges from wastewater treatment plants and industrial sources. The TMDL program, by contrast, evaluates all sources of pollutants entering a body of water, especially nonpoint sources, according to the water body's ability to assimilate the pollutant. For both of these programs, water quality criteria provide an essential means of assessing the condition of the receiving body of water, as well as the various pollutant discharges.

On the basis of data supplied to the EPA by the states, over 20,000 rivers, lakes, and estuaries—40% of our waters—are not meeting water quality standards. The major pollutants are sediments, excess nutrients, pathogens, and toxic substances. At the end of the chapter we will address the specifics of public-policy responses to this serious national problem. We turn now to the most widespread water pollution problem: eutrophication.

18.2 Eutrophication

Recall from Chapter 2 that the term "trophic" refers to feeding. Literally, "eutrophic" means "well nourished." As we will see, there are some quite undesirable consequences of high levels of "nourishment" in bodies of water. Although eutrophication can be an entirely natural process, the introduction of pollutants into water bodies has greatly increased the scope and speed of eutrophication. Let us investigate the causes of this problem, as well as the means of preventing and eventually reversing it.

Different Kinds of Aquatic Plants

To understand eutrophication, we need to consider distinct kinds of aquatic plants, namely, *benthic* plants and *phytoplankton*.

Benthic plants (from the Greek *benthos*, deep) are aquatic plants that grow attached to or are rooted in the bottom of a body of water. All common aquarium plants and sea grasses are examples (Fig. 18–6a). As shown in the figure, benthic plants may be divided into two categories: **submerged aquatic vegetation (SAV)**, which generally grows totally under water, and **emergent vegetation**, which grows with the lower parts in water, but the upper parts emerging from the water.

The most important point in understanding eutrophication is this: SAV requires water that is clear enough to allow sufficient light to penetrate so as to support photosynthesis. As water becomes more turbid, light is diminished; in extreme situations, it may be reduced to penetrating to just a few centimeters beneath the water's surface. Thus, increasing turbidity decreases the depth at which SAV can survive.

A second important feature of SAV is that it absorbs its required mineral nutrients from the bottom sediments through the roots, just as land plants do. SAV is not handicapped by water that is low in nutrients. Indeed, enrichment of the water with nutrients is counterproductive for SAV, because it stimulates the growth of phytoplankton.

Phytoplankton consists of numerous species of photosynthetic algae, protists, and chlorophyll-containing bacteria (cyanobacteria, formerly referred to as blue-green algae) that grow as microscopic single cells or in small groups, or "threads," of cells. Phytoplankton lives suspended in the water and is found wherever light and nutrients are available (Fig. 18–6b). In extreme situations, water may become literally pea soup green (or tea colored, depending on the species involved), and a scum of phytoplankton may float on the surface and absorb essentially all the light. However, phytoplankton reaches such densities only in nutrient-rich water, because, not being connected to the bottom, it must absorb its nutrients from the water. A low level of nutrients in the water limits the growth of phytoplankton accordingly.

Considering the requirements of phytoplankton and SAV, you can see how the balance between them is altered when nutrient levels in the water are changed. As long as water remains low in nutrients, populations of phytoplankton are suppressed; the water is clear, and light may penetrate to support the growth of SAV. As nutrient levels increase, phytoplankton can grow more prolifically, making the water turbid and shading out the SAV.

The Impacts of Nutrient Enrichment

A lake in which light penetrates deeply—one in which the bottom is visible beyond the immediate shoreline—is **oligotrophic** (low in nutrients). Such a lake is fed by a watershed that holds its nutrients well. A forested watershed, for example, holds nitrogen and phosphorus tightly and allows very little of those elements to enter the water draining into the streams and rivers. If the streams feed into a lake, it will reflect their low nutrient content.

In an oligotrophic lake, the low nutrient levels limit the growth of phytoplankton and allow enough light to penetrate to support the growth of SAV, which draws its nutrients from the bottom sediments. In turn, the benthic plants support the rest of a diverse aquatic ecosystem by providing food, habitats, and dissolved oxygen. The oligotrophic body of water is prized for its aesthetic and recreational qualities, as well as for its production of fish and shellfish.

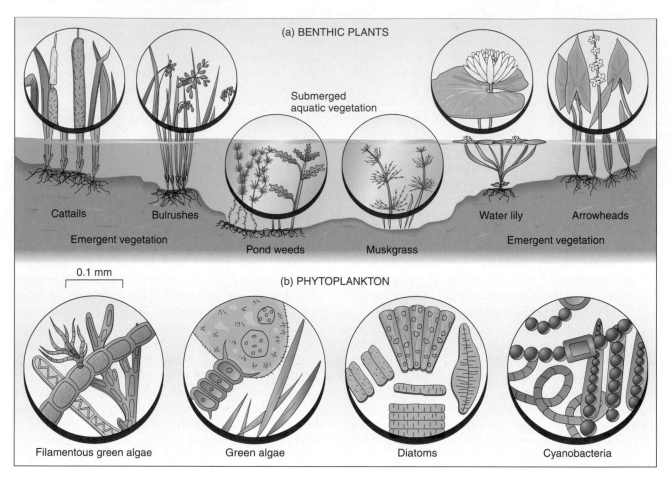

▲ **FIGURE 18–6** *Aquatic photosynthesizers.*   (a) Benthic, or bottom-rooted, plants. These are subdivided into submerged aquatic vegetation (SAV) and emergent vegetation. (b) Phytoplankton, various photosynthetic organisms that are either single cells or small groups or filaments of cells, float freely in the water.

Eutrophication.   As the water of an oligotrophic body becomes enriched with nutrients, numerous changes are set in motion. First, the nutrient enrichment allows the rapid growth and multiplication of phytoplankton, causing increasing turbidity of the water. The increasing turbidity shades out the SAV that live in the water. With the die-off of SAV, there is an obvious loss of food, habitats, and dissolved oxygen from their photosynthesis.

Phytoplankton has a remarkably high growth and reproduction rate. Under optimal conditions, phytoplankton biomass can double every 24 hours, a capacity far beyond that of benthic plants. Thus, phytoplankton soon reach a maximum population density, and continuing growth and reproduction are balanced by die-off. Dead phytoplankton settles out, resulting in heavy deposits of detritus on the lake or river bottom. In turn, the abundance of detritus supports an abundance of decomposers, mainly bacteria. The explosive growth of bacteria, consuming oxygen via respiration, creates an additional demand for dissolved oxygen. The result is the depletion of dissolved oxygen, with the consequent suffocation of fish and shellfish.

In sum, **eutrophication** refers to this whole sequence of events, starting with nutrient enrichment, and proceeding to the growth and die-off of phytoplankton, the accumulation of detritus, the growth of bacteria, and, finally, the depletion of dissolved oxygen and the suffocation of higher organisms (Fig. 18–7). From the human perspective, the eutrophic system is less than appealing for swimming, boating, and sportfishing. Also, if the lake is a source of drinking water, its value may be greatly impaired because phytoplankton rapidly clog water filters and may cause a foul taste. Even worse, some species of phytoplankton secrete various toxins into the water that may kill other aquatic life and be injurious to human health as well. (See the "Earth Watch" essay, p. 451.)

Eutrophication of Shallow Lakes and Ponds.   In lakes and ponds whose water depth is 6 feet (2 m) or less, eutrophication takes a somewhat different course. There, SAV may grow to a height of a meter or more, reaching the surface. Thus, with nutrient enrichment, the SAV is not shaded out, but grows abundantly, sprawling over and often totally covering the water surface with dense mats of vegetation that make boating, fishing, or swimming impossible (Fig. 18–8). Any vegetation beneath these mats is shaded out. As the mats of vegetation die and sink to the bottom, they create a BOD that often depletes the water of dissolved oxygen, causing the death of aquatic organisms other than bacteria.

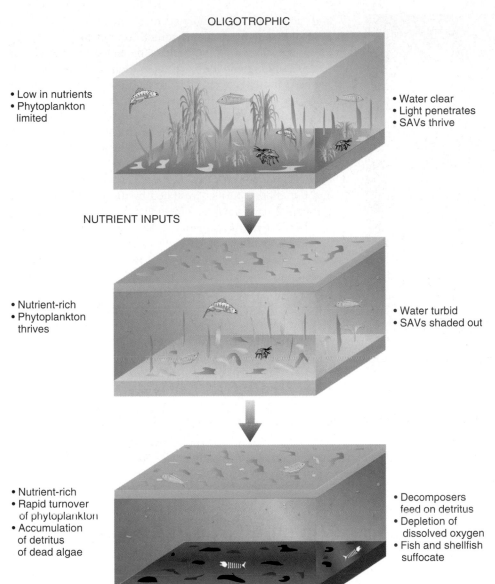

OLIGOTROPHIC

- Low in nutrients
- Phytoplankton limited

- Water clear
- Light penetrates
- SAVs thrive

NUTRIENT INPUTS

- Nutrient-rich
- Phytoplankton thrives

- Water turbid
- SAVs shaded out

- Nutrient-rich
- Rapid turnover of phytoplankton
- Accumulation of detritus of dead algae

- Decomposers feed on detritus
- Depletion of dissolved oxygen
- Fish and shellfish suffocate

EUTROPHIC

◀ **FIGURE 18–7** *Eutrophication.* As nutrients are added from sources of pollution, an oligotrophic system rapidly becomes eutrophic and undesirable.

Natural vs. Cultural Eutrophication. In nature, apart from human impacts, eutrophication is part of the process of aquatic succession, discussed in Chapter 4. Over periods of hundreds or thousands of years, bodies of water are subject to gradual enrichment with nutrients. Thus, we can speak of *natural eutrophication* as a normal process. Wherever nutrients come from sewage-treatment plants, poor farming practices, urban runoff, and certain other human activities, humans have inadvertently managed to vastly accelerate the process of nutrient enrichment. We refer to the accelerated eutrophication caused by humans as **cultural eutrophication**.

Combating Eutrophication

There are two approaches to combating the problem of eutrophication. One is to *attack the symptoms*—the growth of vegetation, the lack of dissolved oxygen or both.

The other is to *get at the root cause*—excessive inputs of nutrients and sediments.

Attacking the Symptoms. Attacking the symptoms is applicable in certain situations in which immediate remediation is the goal and costs are not prohibitive. Methods of attacking the symptoms of eutrophication include (1) chemical treatments (with herbicides), (2) aeration, (3) harvesting aquatic weeds, and (4) drawing water down.

Herbicides are often applied to ponds and lakes to control the growth of nuisance plants. To control phytoplankton growth, copper sulfate and diquat are frequently used. For controlling SAV and emergent vegetation, fluridone, glyphosate, and 2,4-D are employed. However, these compounds are toxic to fish and aquatic animals, sometimes at concentrations required to bring the vegetation under control. Often, fish are killed after herbicide is applied, because

▲ **FIGURE 18–8** *Eutrophication in shallow lakes and ponds.* In shallow water, sufficient light for photosynthesis continues to reach the submerged aquatic vegetation. The oversupply of nutrients stimulates growth, so that vegetation reaches the surface and forms mats.

the rotting vegetation depletes the water of dissolved oxygen. Also, one is required to wait a certain length of time after an herbicide is applied before using the water for swimming, irrigation, or fishing. The use of herbicides in this connection is a good example of the pesticide treadmill (Chapter 17), in which repeated applications are required indefinitely. Herbicides provide only cosmetic treatment, and as soon as they wear off, the vegetation grows back rapidly.

The depletion of dissolved oxygen by decomposers and the consequent suffocation of other aquatic life is the final and most destructive stage of eutrophication. *Artificial aeration* of the water can avert this terminal stage. An aeration technique currently gaining in popularity is to lay a network of plastic tubes with microscopic pores on the bottom of the waterway to be treated. High-pressure air pumps force microbubbles from the pores, and the bubbles dissolve directly into the water. The method is proving effective in speeding up the breakdown of accumulated detritus, improving water quality, and enabling the return of more desirable aquatic life. Despite its high cost, the technique is applicable to harbors, marinas, and some water-supply reservoirs, where the demand for better water quality justifies the cost.

In shallow lakes or ponds where the problem is bottom-rooted vegetation reaching and sprawling over the surface (see Fig. 18–8), *harvesting the aquatic weeds* may be an expedient way to improve the water's recreational potential

and aesthetics. Commercial mechanical harvesters are used, and nearby residents sometimes get together to remove the vegetation by hand. The harvested vegetation makes good organic fertilizer and mulch. But even harvesting has a limited effect: The vegetation soon grows back because roots are left in the nutrient-rich sediments.

Harvesting also does not take removing phytoplankton into consideration. To remove the microscopic cells would require filtering them from the water. Mechanically "harvesting" phytoplankton is not at all practical. The volume of water to be filtered in even a modestly sized pond is overwhelming, and the effort is further frustrated by the fact that plankton cells rapidly clog the filter, precluding the passage of water.

Because many recreational lakes are dammed, another option for shallow-water weed control is to draw the lake down for a period each year. This process kills most of the rooted aquatic plants along the shore; however, they grow back in time.

All of these approaches have in common the fact that they are only temporary fixes for the problem and will have to be repeated often, at significant cost, to keep the unwanted plant growth under control.

Getting at the Root Causes. Controlling eutrophication means employing long-term strategies for correcting the problem, which ultimately involves reducing the inputs of nutrients and sediments—two of the types of pollution discussed at the beginning of this chapter. The first step is to identify the major point and nonpoint sources of nutrients and sediments. Then it is a matter of developing and implementing strategies for correction. Which source or factor is most significant will depend on the human population and the land uses within the particular watershed. Therefore, each watershed must be analyzed as a separate entity, and appropriate measures must be taken to reduce the nutrients and sediments exiting from that watershed. In the sections that follow, we shall discuss major strategies used to control eutrophication.

First, recall the concept of limiting factors (Chapter 2): Only one nutrient need be lacking to suppress growth. In natural freshwater systems, **phosphorus** is the most common limiting factor. In marine systems (like the coastal Gulf of Mexico), the limiting nutrient is most often **nitrogen**. Both in the environment and in biological systems, phosphorus (P) is present as phosphate (PO_4^{-3}), and nitrogen is present in a variety of compounds, most commonly as nitrate (NO_3^-) or ammonium (NH^{+4}). Therefore, we will focus primarily on phosphate and nitrogen compounds.

Control Strategies for Point Sources. In heavily populated areas, discharges from sewage-treatment plants are recognized as major sources of nutrients entering waterways. The levels of phosphate in effluents from these plants are elevated even more in areas where laundry detergents containing phosphate are used. Phosphate contained in detergents for cleaning purposes goes through the system and out with the discharge. In regions where eutrophication has been recognized as a problem, a key step toward prevention has been to ban the sale of phosphate detergents or at least to regulate

EARTH WATCH

THE ALGAE FROM HELL

Surprises dot the environmental landscape of the last 50 years, but one of the most bizarre and unexpected ones emerged when, in 1988, a North Carolina State University plankton biologist, Dr. Joanne Burkholder, examined some open sores that were found on fish. She identified a new organism, a microscopic dinoflagellate (a type of algae) she named *Pfiesteria piscicida*. (Recently, a second toxic species, *Pfiesteria shumwayae*, has been identified.) Subsequent work in laboratory cultures of *Pfiesteria* revealed that the species has no fewer than 24 different lifeforms, a phenomenon unprecedented in the annals of science. Sometimes *Pfiesteria* is an amoeba, sometimes it is a flagellate, and sometimes it is in a cyst. Sometimes it secretes a toxin that stuns fish, and then it turns into a form that proceeds to feed on the fish, which have been disoriented by the toxin. Sometimes it behaves like a decent, ordinary alga and lives by photosynthesis. *Pfiesteria* has killed billions of fish in Albemarle and Pamlico Sounds, North Carolina, and has been identified in waters from Virginia, Maryland, Delaware, and Florida. In the summer of 1997, however, *Pfiesteria* struck several tributaries of Chesapeake Bay, particularly the Pocomoke River; fish kills and catches of fish with ulcerous lesions finally led the state to close the river to all fishing and catching of shellfish.

Pfiesteria is not just toxic to fish and shellfish; it has now been implicated in skin rashes and neurological disorders in people who have encountered water containing the organism. The problems first appeared in Burkholder's laboratory; later, watermen and others in close contact with infested waters began reporting similar symptoms: short-term memory loss, respiratory problems, fatigue, disorientation, and blurred vision. When news of the fish kills in Chesapeake Bay and the impacts on humans became public, there was a predictable panic response, and consumers all around the Chesapeake stopped eating seafood. In Maryland alone, the economic loss was estimated at between $15 and $20 million in sales during 1997.

The recent increase in *Pfiesteria* episodes joins a long list of what are called harmful algal blooms (HABs). Red tides, consisting of concentrations of different species of toxic dinoflagellates, have been occurring more and more frequently in coastal waters all around North America. Many scientists believe that nutrients—especially nitrogen and phosphorus—are the basic culprits in feeding the growth of the algae in estuaries and coastal waters. In the Delmarva peninsula, comprising parts of Delaware, Maryland, and Virginia, some 600 million chickens are produced each year, and runoff from the poultry farms is believed to be a major source of nutrients to Chesapeake tributaries. North Carolina State University has just opened a new *Pfiesteria* research facility to provide Dr. Burkholder with the capability of analyzing samples from suspected infestations and to continue her work with the basic biology of the organism. Recently, the National Oceanic and Atmospheric Administration (NOAA) announced a $15 million research effort to study the HABs occurring all around the continent. There is little doubt that the blooms are increasing in frequency, and there is a good possibility that the explanation will be found in the pollutants that are still entering our waterways.

---

the maximum allowable level of phosphates. Total or partial phosphate bans are now in effect in 20 states and the District of Columbia. Major detergent manufacturers are shifting to the general production of nonphosphate formulations; however, the consumer should read a product's active ingredients. Also, the bans do not cover dishwashing detergents, some of which brands are high in phosphate. Accordingly, one should check the labels on dishwashing detergents in order to find a brand with little or no phosphate.

In addition, there is an ongoing EPA program aimed at upgrading sewage-treatment plants to remove nutrients from their effluents or to handle the effluents in alternative ways so that the nutrients will not go into waterways. These processes and alternatives will be considered shortly.

Bans on detergents with phosphate and upgrades of sewage-treatment plants have brought about marked improvements in waterways that were heavily damaged by effluents from the plants. In a sense, however, these are the easy measures in that the target for correction, the effluent, is an obvious *point source*, and methods for correcting the problem are clear cut. The NPDES permitting process is the basic regulatory tool for reducing point source pollutants.

Control Strategies for Nonpoint Sources. Correction becomes more difficult, but not less important, when the source is diffuse, as in the case of farm and urban runoff. Remediation of such *nonpoint sources* will involve thousands— even millions—of individual property owners' adopting new practices regarding management and the use of fertilizer and other chemicals on their properties. Nevertheless, that is the challenge. When the Clean Water Act was reauthorized in 1987, a new section (section 319) was added, requiring states to develop management programs to address nonpoint sources of pollution. Thus, there is a legal mandate to address the issues of agricultural and urban runoff.

The EPA has opted to develop regulations via the TMDL program. The concept is straightforward:

- Identify the pollutants responsible for degrading a body of water
- Estimate the pollution coming from all sources
- Estimate the ability of the body of water to assimilate the pollutants while remaining below the threshold designating poor or inadequate water quality
- Determine the maximum allowable pollution load
- Within a margin of safety, allocate the allowable level of pollution among the different sources such that the water quality standards are achieved.

As with most attempts at solving environmental problems, the devil is in the details. The states are responsible for administering TMDL pollution abatement, but the EPA maintains oversight and approval of the results. State administrators have a number of options for managing nonpoint pollution sources, including regulatory programs, technical assistance, financial assistance, education and training, and demonstration projects. Many of these are likely to meet with some opposition, especially if their cost is significant.

Reducing or eliminating pollution from nonpoint sources will involve different strategies for different sources. For example, where a significant portion of the watershed is devoted to agricultural activities, the major source of nutrients and sediments is likely to be erosion and leaching of fertilizer from croplands and runoff of animal wastes from barns and feedlots (Fig. 18–9). All the practices that may be used to minimize such erosion, runoff, and leaching are lumped under a single term, **best management practices (BMPs)**, which includes all the methods of soil conservation discussed in Chapter 8. Table 18–2 lists examples of BMPs for a variety of nonpoint sources.

Once control measures have been put in place, the polluted body of water must be monitored to determine whether water quality standards are being attained. Several years of data collection may be necessary to detect genuine trends in quality. Once water quality standards are reached and sustained, the body of water is removed from the list of waters subject to TMDL action. If standards are not met, the TMDL process must be revisited and new pollution allowances allocated.

Recovery. The good news is that cultural eutrophication and other forms of water pollution can be controlled and often reversed. Of course, a total watershed-management approach is required. For a small lake fed from a modest-sized watershed, the task may be quite straightforward because major sources of nutrients can be identified and addressed. For large bodies of water, such as the Chesapeake Bay—the watershed of which includes some 12,000 square miles (27,000 km²) in five states and the District of Columbia—the task is significantly greater. The Chesapeake has suffered from heavy nutrient pollution, with cultural eutrophication destroying all but a tenth of the 600,000 acres of vital sea grasses that once formed the basis of the bay's rich ecosystem. The governmental jurisdictions involved have been cooperating under a Chesapeake Bay program with the common goal of achieving federal clean water standards by 2010. The watershed is divided into the smaller watersheds of each of the tributaries, and each jurisdiction is taking the responsibility of reducing the nutrients emanating from its tributaries.

One remarkable success story is Lake Washington, east of Seattle (Fig. 18–10). This 34-square-mile lake is at the center of a large metropolitan area. During the 1940s and 1950s, 11 sewage-treatment plants were sending state-of-the-art treated water into the lake at a rate of 20 million gallons per day. At the same time, phosphate-based detergents came into wide use. The lake responded to the massive input of nutrients by developing unpleasant "blooms" of noxious algae. The water lost its clarity, the desirable fish populations declined, and masses of dead algae accumulated on the shores of the lake.

Citizen concern led to the creation of a system that diverted the treatment-plant effluents into nearby Puget Sound, where tidal flushing would mix them with open-ocean water. The diversion was complete by 1968, and the lake responded quickly. The algal blooms diminished, the water regained its clarity, and by 1975 recovery was complete. Careful studies by a group of limnologists from the University of Washington showed that phosphate was the culprit. Before the diversion, the lake was receiving 220 tons of phosphate per year from all sources. Afterward, the total dropped to 40 tons per year, well

▶ **FIGURE 18–9** *A collection pond for dairy-barn washings.* When washings from animal facilities are flushed directly into natural waterways, they contribute significantly to eutrophication. This may be avoided by collecting the flushings in ponds from which both the water and the nutrients may be recycled.

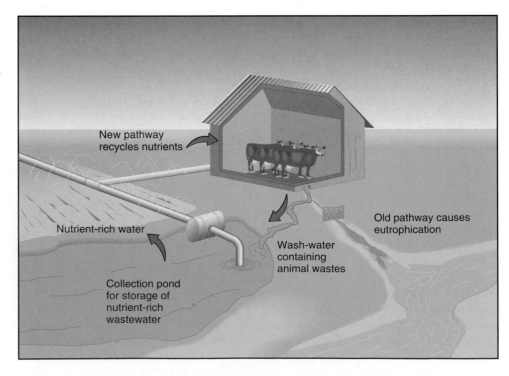

New pathway recycles nutrients

Old pathway causes eutrophication

Nutrient-rich water

Wash-water containing animal wastes

Collection pond for storage of nutrient-rich wastewater

TABLE 18-2 Best Management Practices for Reducing Pollution from Nonpoint Sources

Agriculture	Forestry
Animal waste management	Ground cover maintenance
Conservation tillage	Limiting disturbed areas
Contour farming	Log removal techniques
Strip cropping	Pesticide and herbicide
Cover crops	management
Crop rotation	Proper management of roads
Fertilizer management	Removal of debris
Integrated pest management	Riparian zone management
Range and pasture management	
Terraces	**Mining**
	Underdrains
Construction	Water diversion
Limiting disturbed areas	
Nonvegetative soil stabilization	**Multicategory**
Runoff detention and retention	Buffer strips
	Detention and
Urban	sedimentation basins
Flood storage	Devices to encourage
Porous pavements	infiltration
Runoff detention and retention	Vegetated waterways
Street cleaning	Interception and diversion
	of runoff
	Sediment traps
	Streamside management
	zones
	Vegetative stabilization and
	mulching

Source: *Guidance for Water Quality-Based Decisions: The TMDL Process* (USEPA, 1991m /EPA440-4-91-001).

▲ **FIGURE 18–10 Lake Washington.** A testimony to the fact that bodies of water can recover from cultural eutrophication, Lake Washington is now thriving because sources of nutrients enriching the lake have been removed.

within the lake's normal capacity to absorb phosphate by depositing it in deep bottom sediments. One clear lesson learned was that sewage-treatment effluent must never be allowed to enter a lake unless the removal of nutrients is part of the treatment. The major lesson, however, was that eutrophication can be reversed by paying attention to nutrient inputs and addressing the various sources. Lake management, in other words, is a matter of controlling the phosphate loading into a lake from all sources: human and animal sewage, agricultural and yard fertilizers, street runoff, and failing septic systems. Many lakes are now protected by associations of concerned citizens who see themselves as stewards of their lake.

In recent years, the United States and other developed nations have recognized the problem of eutrophication and have made substantial progress in developing or modifying sewage-treatment systems that will keep nutrients in a cycle on the land and not discharge them into waterways. But, at the other extreme, particularly in developing countries, much **raw sewage**—completely untreated—is still discharged, creating not only eutrophication of waterways, but also a significant disease hazard. We will now turn our attention to the important task of managing our organic wastes, another form of water pollution.

18.3 Sewage Management and Treatment

Before the late 1800s, the general means of disposing of human excrement was the outdoor privy. Seepage from the privy frequently contaminated drinking water and caused disease, especially in places where privies and wells were located near one another. In the late 1800s, Louis Pasteur and other scientists showed that sewage-borne bacteria were responsible for many infectious diseases. This important discovery led to intensive efforts to rid cities of human excrements as expediently as possible. Cities already had drain systems for storm water, but using these systems for human wastes had been prohibited. With the urgency of the situation, however, minds quickly changed. The flush toilet was introduced, and sewers were tapped into storm drains. Thus, Western civilization initiated the one-way flow of flushing sewage wastes into natural waterways.

The results of this practice are obvious: Receiving waters that had limited capacity for dilution became essentially open cesspools of foul odors, vermin, and filth as the overload of organic matter depleted dissolved oxygen and all aquatic life suffocated. For increasing distances around or downstream from the sewage outfall, the water became unfit to swim in because of the sewage contamination.

Development of Collection and Treatment Systems

To alleviate the problem of sewage-polluted waterways, facilities were designed and constructed to treat the outflow before it entered the receiving waterway. The first sewage-treatment plants in the United States were built around 1900. However, the combined volumes of sewage and storm water soon proved impossible to handle. During heavy rains, wastewater would overflow the treatment plant and carry raw sewage into the receiving waterway. Gradually, regulations were passed requiring developers to install separate systems: **storm drains** for collecting and draining runoff from precipitation, and **sanitary sewers** to receive all the wastewater from sinks, tubs, and toilets in homes and other buildings. (Note the distinction in the two terms; it is incorrect to speak of storm drains as sewers.)

The ideal modern system, then, is a system in which all sewage water is collected separately from storm water and is fully treated to remove all pollutants before the water finally is released. But progress toward this goal has been extremely uneven. Up through the 1970s, even in the United States and other developed countries, countless communities still discharged untreated sewage directly into waterways, and for countless more the degree of treatment was minimal. Indeed, the increasing sewage pollution of waterways and beaches was the major impetus behind the passage of the Clean Water Act of 1972. In the meantime, much of the developing world still exists in the most primitive stage with regard to sewage. Innumerable poor villages and poor areas of mushrooming cities of the developing world lack sewer systems for collecting wastewater. In such areas, it is not uncommon to find raw sewage littering the ground and overflowing gutters and streams, with children playing in the filth. Even where flush toilets and collection systems exist, the discharge of raw sewage into waterways is still common. Many of the people living in these regions must use these badly polluted waters for bathing, laundering, and even drinking (Fig. 18–3).

Before addressing methods of treating sewage, let us clarify which contaminants and pollutants we are talking about.

The Pollutants in Raw Sewage

Raw sewage is not only the flushings from toilets; it is also the collection of wastewater from all other drains in homes and other buildings. A sewer system brings all tub, sink, and toilet drains from all homes and buildings together into larger and larger sewer pipes, just as the twigs of a tree come together eventually into the trunk. This total mixture collected from all drains, which comes out at the end of the "trunk" of the collection system, is termed **raw sewage** or **raw wastewater**. Because we use such large amounts of water to flush away small amounts of dirt, especially as we stand under the shower or often just run the water with no waste at all, most of what goes down sewer drains is water. Raw sewage is about 1,000 parts water for every 1 part of waste—99.9% water to 0.1% waste.

Given our voluminous use of water, the quantity of raw-sewage output is on the order of 150–200 gallons (600–800 liters) per person per day. That is, a community of 10,000 persons will produce on the order of 1.5 to 2.0 million gallons (6–8 million liters) of wastewater each day. With the addition of storm water, raw sewage is diluted still more. Nevertheless,

the pollutants are sufficient to make the water brown and smell foul.

The pollutants generally are divided into the following four categories, which, as we shall find, correspond to techniques used for their removal:

1. Debris and grit: rags, plastic bags, coarse sand, gravel, and other objects flushed down toilets or washed through storm drains.
2. Particulate organic material: fecal matter, food wastes from garbage-disposal units, toilet paper, and other matter that tends to settle in still water.
3. Colloidal and dissolved organic material: very fine particles of particulate organic material, bacteria, urine, soaps, detergents, and other cleaning agents.
4. Dissolved inorganic material: nitrogen, phosphorus, and other nutrients from excretory wastes and detergents.

In addition to the four categories of pollutants in "standard" raw sewage, variable amounts of pesticides, heavy metals, and other toxic compounds may be found in sewage because people pour unused portions of products containing such materials down sink, tub, or toilet drains. Also, industries may discharge various toxic wastes into sewers.

Removing the Pollutants from Sewage

The challenge of treating sewage is more than installing a technology that will do the job: It is finding one that will do the job at a reasonable cost. A diagram of wastewater-treatment procedures is shown in Fig. 18–11.

Preliminary Treatment (Removal of Debris and Grit). Because debris and grit will damage or clog pumps and later treatment processes, removing them is a necessary first step and is termed **preliminary treatment**. Usually, preliminary treatment involves two steps: a screening out of debris and a settling of grit. Debris is removed by letting raw sewage flow through a **bar screen**, a row of bars mounted about 1 inch (2.5 cm) apart. Debris is mechanically raked from the screen and taken to an incinerator. After passing through the screen, the water flows through a **grit chamber**, a swimming-pool-like tank, where its velocity is slowed just enough to permit the grit to settle. The settled grit is mechanically removed from these tanks and taken to landfills.

Primary Treatment (Removal of Particulate Organic Material). After preliminary treatment, the water moves on to **primary treatment**, where it flows very slowly (about 6 ft, or 2 m, per hour) through large tanks called **primary clarifiers**. Because it flows slowly through these tanks, the water is nearly motionless for several hours. The particulate organic material, about 30% to 50% of the total organic material, settles to the bottom, where it can be removed. At the same time, fatty or oily material floats to the top, where it is skimmed from the surface. All the material that is removed, both particulate

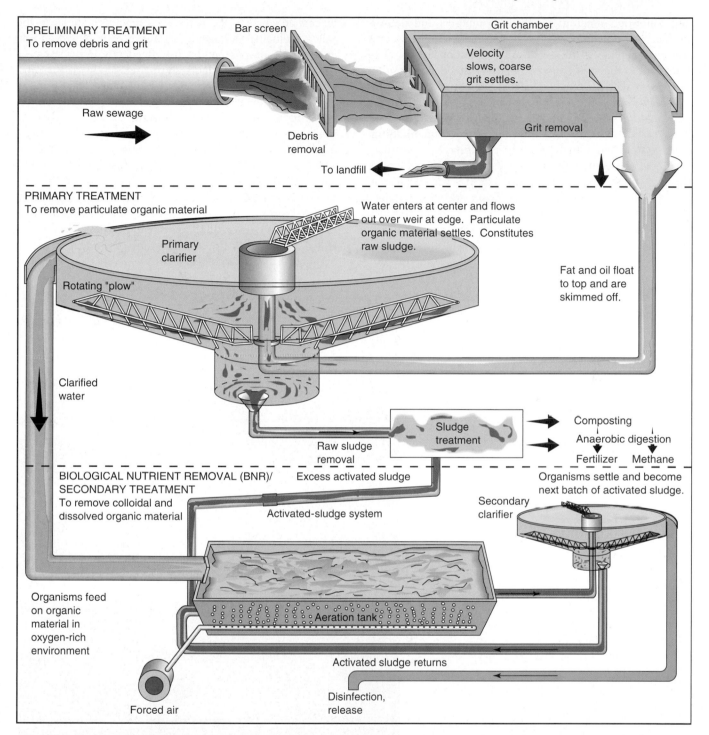

PRELIMINARY TREATMENT
To remove debris and grit

Bar screen

Grit chamber

Velocity slows, coarse grit settles.

Raw sewage

Debris removal

Grit removal

To landfill

PRIMARY TREATMENT
To remove particulate organic material

Water enters at center and flows out over weir at edge. Particulate organic material settles. Constitutes raw sludge.

Primary clarifier

Rotating "plow"

Fat and oil float to top and are skimmed off.

Clarified water

Sludge treatment

Composting

Anaerobic digestion

Fertilizer   Methane

Raw sludge removal

BIOLOGICAL NUTRIENT REMOVAL (BNR)/ SECONDARY TREATMENT
To remove colloidal and dissolved organic material

Excess activated sludge

Organisms settle and become next batch of activated sludge.

Activated-sludge system

Secondary clarifier

Organisms feed on organic material in oxygen-rich environment

Aeration tank

Activated sludge returns

Forced air

Disinfection, release

▲ **FIGURE 18–11** *A diagram of wastewater treatment.* Raw sewage moves from the grit chamber to undergo primary treatment, in which sludge is removed. The clarified water then proceeds to undergo secondary treatment (shown here as activated-sludge treatment).

organic material and fatty material, is combined into what is referred to as **raw sludge**; we shall consider its treatment shortly.

Note that primary treatment involves nothing more complicated than putting polluted water into a "bucket," letting material settle, and pouring off the water. Nevertheless, such treatment removes much of the particulate organic matter at minimal cost.

Secondary Treatment (Removal of Colloidal and Dissolved Organic Material). **Secondary treatment** is also called **biological treatment**, because it makes use of organisms—natural decomposers and detritus feeders. Basically, an environment is created that enables these organisms to feed on the colloidal and dissolved organic material and break it down to carbon dioxide and water via their cell respiration. The sewage water from primary treatment is a food-

and water-rich medium for the decomposers and detritus feeders. The only thing that needs to be added to the water is oxygen to enhance the organisms' respiration and growth. Recall that the oxygen consumed in the decomposition process is called the biochemical oxygen demand (BOD). Either of two systems may be used to add oxygen to the water: a *trickling-filter system* or an *activated-sludge system*. Historically the earlier of the two, and still widely used, is the **trickling-filter system**, in which the water exiting from primary treatment is sprinkled onto, and allowed to percolate through, a bed of fist-sized rocks 6–8 feet (2–3 m) deep (Fig. 18–12). The spaces between the rocks provide good aeration. Like a natural stream, this environment supports a complex food web of bacteria, protozoans, rotifers (organisms that consume protozoans), various small worms, and other detritus

feeders attached to the rocks. The organic material in the water, including pathogenic organisms, is absorbed and digested by decomposer and detritus feeders as it trickles by.

The **activated-sludge system** (Fig. 18–11) is the most common secondary-treatment system. Water from primary treatment enters a large tank that is equipped with an air-bubbling system or a rapidly churning system of paddles. A mixture of detritus-feeding organisms, referred to as **activated sludge**, is added to the water as it enters the tank, and the water is vigorously aerated as it moves through the tank. Organisms in this well-aerated environment reduce the biomass of organic material, including pathogens, as they feed. As the organisms feed on each other, they tend to form into clumps, termed *floc*, that settle readily when the water is stilled. Thus, from the aeration tank, the water is passed

(a)

(b)

▶ FIGURE 18–12 *Trickling filters for secondary treatment.* (a) The water from primary clarifiers is sprinkled onto, and trickles through, a bed of rocks 6–8 feet deep. (b) Various bacteria and other detritus feeders adhering to the rocks consume and digest organic matter as it trickles by in the water, which is collected at the bottom of the filters and goes on for final clarification and disinfection.

into a secondary **clarifier tank**, where the organisms settle out and the water—now with more than 90% of all the organic material removed—flows on. The settled organisms are pumped back into the entrance of the aeration tank; these organisms are the activated sludge that is added at the beginning of the process. Surplus amounts of activated sludge, which occur as populations of organisms grow, are removed and added to the raw sludge. The organic material is oxidized by microbes to form carbon dioxide, water, and mineral nutrients that remain in the water solution.

Biological Nutrient Removal (Removal of Dissolved Inorganic Material). The original secondary-treatment systems were designed and operated in a manner to simply maximize biological digestion, because the elimination of organic material and its resulting BOD was considered the prime objective. Ending up with a nutrient-rich discharge was not recognized as a problem. Today, with increased awareness of the problem of cultural eutrophication, secondary activated-sludge systems have been added and are being modified and operated in a manner that both removes nutrients and oxidizes detritus, in a process known as **biological nutrient removal (BNR)**.

Recall that in the natural nitrogen cycle (Fig. 3–18), various bacteria convert nutrient forms of nitrogen (ammonia and nitrate) back to nonnutritive nitrogen gas in the atmosphere through **denitrification**. For the biological removal of nitrogen, then, the activated-sludge system is partitioned into zones, and the environment in each zone is controlled in a manner that promotes the denitrifying process (Fig. 18–13).

With regard to phosphate, in an environment that is rich in oxygen, but relatively lacking in food—the environment of zone 3—bacteria take up phosphate from solution and store it in their bodies. Thus, phosphate is removed as the excess organisms are removed from the system. (See again Fig. 18–13). These organisms, together with the phosphate they contain, are added to, and treated with, the raw sludge, ultimately producing a more nutrient-rich treated-sludge product.

As an alternative to BNR, one may use various chemical processes. One such process is simply to pass the effluent from standard secondary treatment through a filter of lime, which causes the phosphate to precipitate out as insoluble calcium phosphate. Another is to treat the effluent with ferric chloride, which produces insoluble ferric phosphate, or with an organic polymer, which gives rise to a floc. With these methods, phosphorus is removed as the sludge is.

Final Cleansing and Disinfection. With or without BNR, the wastewater is subjected to a final clarification and disinfection. Although few pathogens survive the combined

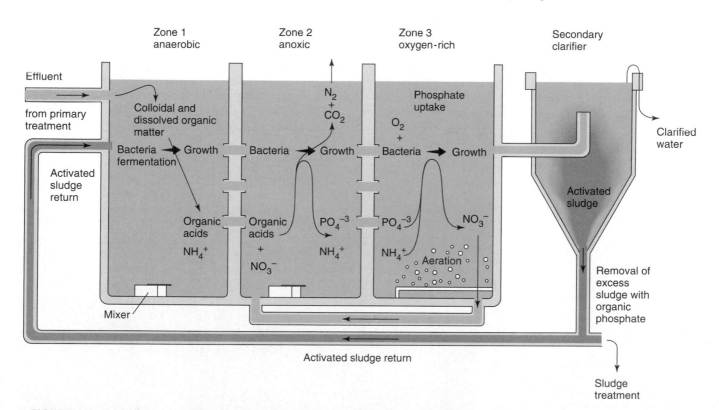

▲ **FIGURE 18–13** *Biological nutrient removal (BNR).* The secondary treatment—the activated-sludge process—may be modified to remove nitrogen and phosphate while at the same time breaking down organic matter. For this BNR process, the aeration tank is partitioned into three zones, only the third of which is aerated. As seen in the diagram, ammonium (NH_4^+) is converted to nitrate (NO_3^-) in zone 3. Recycled to zone 2, which is without oxygen gas (anoxic), the nitrate supplies the oxygen for cell respiration. In the process, the nitrate is converted to nitrogen gas and is released into the atmosphere. Phosphate is taken up by bacteria in zone 3 and removed with the excess sludge.

stages of treatment, public-health rigors still demand that the water be disinfected before being discharged into natural waterways. The most widely used disinfecting agent is chlorine gas, because it is both effective and relatively inexpensive. But this treatment also introduces chlorine into natural waterways, and even minute levels of chlorine can harm aquatic animals. Also, chlorine gas is dangerous to work with. More commonly, sodium hypochlorite (Clorox) provides a safer way of adding chlorine to achieve disinfection.

Other disinfecting techniques are coming into use as well. One alternative disinfecting agent is ozone gas, which is extremely effective in killing microorganisms and breaks down to oxygen gas in the process, which actually improves water quality. But because ozone is unstable and hence explosive, it must be generated at the point of use, a step that demands considerable capital investment and energy. Another disinfection technique is to pass the effluent through an array of ultraviolet lights mounted in the water. (Standard fluorescent lights without the white coating emit ultraviolet.) The ultraviolet radiation kills microorganisms, but does not otherwise affect the water.

After these treatment steps, the wastewater has a lower organic and nutrient content than most bodies of water into which it is discharged; BOD values are commonly 10 to 20 ppm, as opposed to 200 ppm in the incoming sewage. In other words, discharging the wastewater actually contributes toward improving water quality in the receiving body. In water-short areas, there is every reason to believe that the treated water itself, with little additional treatment, could be recycled into the municipal water-supply system.

Bear in mind that the techniques we have described here are state of the art. Many cities, even in the developed world, are still operating with antiquated systems that provide lower quality treatment. A few coastal cities still persist in discharging sewage that has received only primary treatment directly into the ocean, although in the United States this practice has almost disappeared. (See the "Ethics" essay, p. 459.)

Sludge Treatment

Recall that the particulate organic matter that settles out or floats to the surface of sewage water in primary treatment forms the bulk of *raw sludge*, although the sewage also contains excesses from activated sludge and BNR systems. Raw sludge is a gray, foul-smelling, syrupy liquid with a water content of 97% to 98%. The likelihood that pathogens are present in raw sludge is significant because it includes material directly from toilets. Indeed, raw sludge is considered a biologically hazardous material. However, as nutrient-rich organic material, it has the potential to be used as organic fertilizer if it is suitably treated to kill pathogens and if it does not contain other toxic contaminants.

Three commonly used methods for treating sludge and converting it into organic fertilizer are *anaerobic digestion*, *composting*, and *pasteurization*. Since this is a developing industry, it is unclear which method will prove most cost effective and environmentally acceptable over the long run. Also, none of these methods is capable of removing toxic substances such as heavy metals and nonbiodegradable synthetic organic compounds. The presence of such toxins can preclude the use of sludge as fertilizer.

Anaerobic Digestion.   **Anaerobic digestion** is a process of allowing bacteria to feed on the detritus *in the absence of oxygen*. The raw sludge is put into large airtight tanks called **sludge digesters** (Fig. 18–14). In the absence of oxygen, a consortium of anaerobic bacteria accomplishes the breakdown of organic matter. The end products of this decomposition are carbon dioxide, methane, and water. Thus, a major by-product of anaerobic processes is **biogas**, a gaseous mix-

▶ **FIGURE 18–14** *Anaerobic sludge digesters.*   In these tanks, the bacterial digestion of raw sludge in the absence of oxygen leads to the production of methane gas, which is tapped from the tops of the tanks, and humuslike organic matter that can be used as a soil conditioner. The egglike shape of the tanks facilitates mixing and digestion.

ETHICS

CLEANING UP THE FLOW

Coastal cities have a reputation for dragging their heels with regard to sewage treatment. With the enormous potential of an ocean for tidal mixing and flushing, coastal cities became accustomed to using the "dilution solution" for disposing of their sewage. The city of Boston, Massachusetts, is a case in point. For decades, Boston sewage—370 million gallons each day—received only primary treatment at a site on Deer Island before being discharged into Boston Harbor. Making matters worse, scum and sludge from the treatment process were also released to the harbor. Indeed, the harbor had such a well-deserved reputation for gross pollution that the Republican presidential candidate George Bush used it to demolish the environmental record of Michael Dukakis (his opponent and governor of Massachusetts) in the 1988 presidential campaign.

Under federal government pressure for years to clean up the harbor, the state established the Massachusetts Water Resources Authority (MWRA) in 1984 to oversee the project. The MWRA mapped out a plan for the cleanup that conformed with the federally mandated full implementation by 1999. The first milestone occurred in December 1991, when the flow of raw sludge was stopped. Now all sludge now goes to a pelletizing plant, where it is turned into fertilizer marketed as Bay State Fertilizer. The second milestone took place in August 1997, when the sewage began receiving secondary treatment. By December 1999, the secondary-treatment phase of the plant was in full operation. The third and

final milestone occurred in September 2000, when effluent was diverted to a 9.5-mile-long, 26-foot-diameter tunnel that carries it well out of Boston Harbor and into Massachusetts Bay, where diffusers release the effluent.

The Boston Harbor facility is one of the largest in the world, able to treat 1.3 billion gallons per day of wastewater. The project cost $3.1 billion, shared by federal, state, and local communities. The facility serves 2.5 million people from 43 communities and 5,500 businesses. Costs for a typical household have increased by 300% since 1986 and now average around $675 a year. However, as a result of significant federal funding, that amount is well below earlier projections.

Boston Harbor itself has shown its approval of the project. Harbor porpoises and seals are now seen in unprecedented numbers, even in the inner harbor, where effluents had been released for years. People can enjoy eight miles of recreational beaches and swim in their waters all but a few days of the year. A recreational fishery for bluefish, striped bass, cod, and winter flounder is flourishing. Commercial lobster and shellfish are harvested at a value of more than $10 million a year. Further improvements are planned to eliminate the combined sewer–storm-drain overflows at an additional $450 million. When this is accomplished, Boston Harbor will be one of the cleanest municipal bodies of water in the world.

The project has not been without controversy, however. The 9.5-mile outfall pipe has stirred up a firestorm of criticism from

citizens and advocacy groups on Cape Cod. Their concern centers on the impact of the effluent on water quality in Cape Cod Bay, which adjoins Massachusetts Bay to the southeast. The opponents would prefer that the outfall remain in Boston Harbor. They claim that, because of the location of the outfall pipe, coastal-water circulation may tend to carry the high-nutrient effluent into Cape Cod Bay. The Cape Codders cite numerous studies which show that nitrogen acts as a limiting factor in many coastal marine systems, and they fear that the added nitrogen from the Boston treatment plant will create conditions leading to blooms of algae, especially those that cause red tides, a recurring problem in recent years not only in Boston Harbor, but up and down the entire northeastern coast of the United States, often closing down shellfish harvesting because of the dangers of paralytic shellfish poisoning.

The MWRA claims that, besides benefiting Boston Harbor, the new treatment system will improve water quality in Massachusetts Bay and Cape Cod Bay. The MWRA points out that those bays were being polluted, since the effluent was eventually carried out of the harbor by tidal action. On the surface, this conflict appears to be another case of the "not in my backyard" (NIMBY) phenomenon: One municipality is attempting to resolve its environmental problem in an apparently effective fashion, but there are risks that the solution will create another problem many miles away. What do you think?

ture that is about two-thirds *methane*. The other one-third is made up of carbon dioxide and various foul-smelling organic compounds that give sewage its characteristic odor. Because of its methane content, biogas is flammable and can be used for fuel. In fact, it is commonly collected and burned to heat the sludge digesters, because the bacteria working on the sludge do best when maintained at about 104°F (38°C).

After four to six weeks, anaerobic digestion is more or less complete, and what remains is called **treated sludge** or **biosolids**, consisting of the remaining organic matter, which is now a relatively stable, nutrient-rich, humuslike material suspended in water. Pathogens have been largely, if not entirely, eliminated, so they no longer present any significant health hazard.

Such treated sludge makes an excellent organic fertilizer that may be applied directly to lawns and agricultural fields in the liquid state in which it comes from the digesters, providing the benefit of both the humus and the nutrient-rich water. Alternatively, the sludge may be dewatered by means of belt presses, whereby the sludge is passed between rollers that squeeze out most of the water (Fig. 18–15a) and leave the organic material as a semisolid **sludge cake** (Fig. 18–15b) that is easy to stockpile, distribute, and spread on fields with traditional manure spreaders.

Composting. Another process sometimes used to treat sewage sludge is **composting**. Raw sludge is mixed with

▲ **FIGURE 18–15** *Dewatering treated sludge.* (a) Sludge (98% water) may be dewatered by means of belt presses, as shown here. The liquid sludge is run between canvas belts going over and between rollers, so that much of the water is pressed out. (b) The resulting "sludge cake" is a semisolid humus-like material that may be used as an organic fertilizer.

wood chips or some other water-absorbing material to reduce the water content. It is then placed in **windrows**—long, narrow piles that allow air to circulate conveniently through the material and that can be turned with machinery. Bacteria and other decomposers break down the organic material to rich humuslike material that makes an excellent treatment for soil.

Pasteurization. After the raw sludge is dewatered, and the resulting sludge cake may be put through ovens that operate like oversized laundry dryers. In the dryers, the sludge is **pasteurized**—that is, heated sufficiently to kill any pathogens (exactly the same process that makes milk safe to drink). The product is dry, odorless organic pellets. Milwaukee, Wisconsin, which has a particularly rich sludge resulting from the brewing industry, has been using this process for over 60 years. The city bags and sells the pellets throughout the country as an organic fertilizer under the trade name Milorganite.

Alternative Treatment Systems

Individual Septic Systems. Despite the expansion of sewage-collection systems, in rural and suburban areas there always will be countless homes not connected to a municipal system. For these homes, individual septic systems are required. Currently, 28% of the U.S. population is served by such systems. The traditional and still most common system is the septic tank and a drain field (Fig. 18–16). Wastewater flows into the tank, where particulate organic material settles to the bottom. The tank acts like a primary clarifier in a municipal system. Water containing colloidal and dissolved organic material, as well as dissolved nutrients, flows into the drain field and gradually percolates into the soil. Organic material that settles in the tank is digested by bacteria, but accumulations still must be pumped out every three to five years. Soil bacteria decompose the colloidal and dissolved organic

material that comes through the drain field. Some people establish successful vegetable gardens over septic drain fields, thus instantiating the sound principle of recycling the nutrients. Prerequisites for this traditional system are suitable land area for the drain field and subsoil that allows sufficient percolation of water.

Using Effluents for Irrigation. The nutrient-rich water coming from the standard secondary-treatment process is beneficial for growing plants. The problem is that we don't want to put that water into waterways, where it will stimulate the growth of undesirable algae. But why not use it for irrigating plants we do want to grow? This is a way of complet-

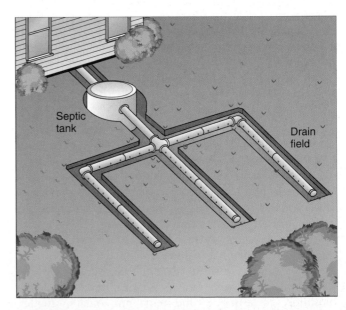

▲ **FIGURE 18–16** *Septic tank treatment.* Sewage treatment for a private home, using a septic tank and drain field. Normally, the pipes and the tank are buried underground. They are shown uncovered here only for illustration. (*Source:* USDA, Soil Conservation Service.)

ing the nutrient cycle. Indeed, the concept has been put into practice in a considerable number of locations as an alternative to upgrading the treatment to remove nutrients. Again, effluents must not be contaminated with toxic materials.

To illustrate, the nutrient-rich effluent from standard secondary treatment from St. Petersburg, Florida, was causing cultural eutrophication in Tampa Bay. Now St. Petersburg uses the effluent to irrigate 4,000 acres (1,600 hectares) of urban open space, from parks and residential lawns to a golf course. Revenues from the water sales help offset operating costs. Similarly, Bakersfield, California, receives a $30,000 annual income from a 5,000-acre (2,000-ha) farm irrigated with its treated effluent. And Clayton County, Georgia, is irrigating 2,500 acres (1,000 ha) of woodland with partially treated sewage. Hundreds of other similar projects are underway around the country.

A number of developing countries irrigate croplands with raw (untreated) sewage effluents. This practice is getting the bad with the good. The crops respond well, but parasites and disease organisms can easily be transferred to farmworkers and consumers. Therefore, it is important to emphasize that only treated effluents should be used for irrigation.

Reconstructed Wetland Systems. In treating wastewater, it is also possible to utilize the nutrient-absorbing capacity of wetlands in suitable areas and under suitable climatic conditions. The project may be part of a wetlands recovery program, or artificial wetlands may be constructed.

As one example of the former, in the 1960s and 1970s much of the land around Orlando, Florida, which was originally wetlands, was drained and converted to cattle pasture. At the time, Orlando was discharging 13 million gallons (50 million liters) per day of nutrient-rich effluent into the James River following secondary treatment. Through the Orlando Easterly Wetlands Reclamation Project, 1,200 acres (480 ha) of pastureland has been converted back to wetlands. The project involved scooping soil from, and building berms (mounds of earth) around, pastures to create a chain of shallow lakes and ponds. In addition, 1.2 million wetland plants

ranging from bulrushes and cattails to various trees were planted. The effluent entering the upper end now percolates through the wetland for about 30 days before entering the James River virtually pure. Thus, the project has re-created a wildlife habitat. Wetland systems can be designed for small as well as large areas and are becoming an increasingly popular alternative for small communities. The key to success for such systems is to ensure that they are kept in balance and not loaded beyond their ability to handle inputs.

18.4 Public Policy

The responsibility for overseeing the health of the nation's waters rests with the EPA, which, however, can develop regulations only if Congress gives it the authority to do so. Hence, the foundation for public policy must be the laws passed by Congress. Some major legislative milestones for protecting the nation's waters are shown in Table 18–3. The landmark legislation is the **Clean Water Act (CWA) of 1972**, which gave the EPA jurisdiction over, and for the first time required permits for, all point-source discharges of pollutants. This act and subsequent amendments provided $69 billion to help cities build treatment plants to meet the federal requirement for secondary treatment of all sewage. As the table shows, the 1987 amendments to the CWA established a revolving loan fund to replace the direct-grants program—the Clean Water State Revolving Fund program. To build treatment facilities, local governments borrow at low interest rates, and as they repay the loans, the funds received are used for more loans. To date, some $26 billion in loans have been made by the states for treatment facilities. However, the EPA estimates that an additional $128 billion will be needed over the next 20 years to meet funding needs for all eligible municipal wastewater treatment systems.

Reauthorization of the Clean Water Act is long overdue. Programs have been funded on a year-to-year basis, and Congress has been hung up on a debate over whether

---

TABLE 18–3 Legislative Milestones in Protecting the Nation's Waters

1899—**Rivers and Harbors Act:** First federal legislation protecting the nation's waters to promote commerce.

1948—**Water Pollution Control Act:** The federal government provides technical assistance and funds to state and local governments to promote efforts to protect water quality.

1972—**The Clean Water Act:** Legislation establishing a comprehensive federal program to achieve the goal of protecting and restoring the physical, chemical, and biological integrity of the nation's waters. Requires permits for any discharge of pollution, strengthens water quality standards, and encourages the use of the best achievable pollution control technology. Provided billions of dollars for construction of sewage-treatment plants.

1972—**Marine Protection, Research and Sanctuaries Act:** Prevents unacceptable dumping in oceans.

1977—**Clean Water Act amendments:** Strengthens controls on toxic pollutants and allows states to assume responsibility for federal programs.

1987—**Water Quality Act:** These major amendments to the Clean Water Act create a revolving loan fund to provide ongoing support for the construction of treatment plants; address regional pollution with a watershed approach; and address nonpoint source pollution in Section 319 of the act, with requirements for states to assess the problem and develop and implement plans for dealing with it. Makes available $400 million in grants to carry out the provisions of the act.

requirements should be strengthened or weakened, whether additional mandates should be subjected to a cost–benefit analysis, and whether regulatory relief should be provided to industries, states, cities, and individuals who are required to take actions to comply with the regulations. Recently, impatient with congressional inaction to reauthorize and strengthen the Clean Water Act, the Clinton administration directed federal agencies to develop a Clean Water Initiative to improve and strengthen our water pollution control efforts. An action plan was subsequently developed in 1998, and almost 100 activities were identified as part of the initiative. Most are existing activities and have been funded by way of appropriations to the federal agencies involved (EPA, USDA, Department of the Interior, etc.), but the funding is well below the president's request for $2.2 billion. New initiatives include a national response to deal with harmful algal blooms (HABs—see the "Earth Watch" essay, p. 451), more attention to the assessment of watersheds in pollution abatement, and a greater response to the problems associated with animal-feeding operations.

The EPA has identified nonpoint-source pollution as the nation's number-one water pollution problem, with the construction of new wastewater treatment facilities not far behind. Much progress has been made in the 30 years since the enactment of the Clean Water Act. The number of people in the United States served by adequate sewage-treatment plants has more than doubled, from 85 million to 173 million. Soil erosion has been reduced by 1 billion tons annually. And two-thirds of the nation's waterways are safe for fishing and swimming, double the number in 1972. Many of the nation's most heavily used rivers, lakes, and bays have been cleaned up and restored; examples are the Androscoggin River in Maine, Boston Harbor, the upper Mississippi River, the South Platte River, the upper Arkansas River, Lake Erie, the Illinois River, and the Delaware River. Fish now swim in rivers once so polluted that only bacteria and sludge worms could survive. Significantly, a 70% increase in bottom vegetation has been achieved since the mid-1980s in the Chesapeake Bay. Clearly, a national sense of stewardship has been applied to the rivers, lakes, and bays that are our heritage from a previous generation, and public policy has been enacted and billions of dollars spent to bring our waters back from a polluted condition.

ENVIRONMENT ON THE WEB

CHALLENGES IN WATER QUALITY IMPROVEMENT

Thirty years ago, eutrophication of lakes and streams was one of our central water-quality concerns. The evidence of eutrophication could be seen in the excessive aquatic plant and algal growth, nighttime oxygen depletion, and fish kills. So great was public concern about the issue that it prompted a continent-wide effort to reduce phosphorus and nitrogen concentrations in natural waters. Key elements of this strategy were the reformulation of detergents to reduce phosphorus concentrations, and improved sewage treatment at publicly owned sewage-treatment plants throughout the United States and Canada.

Today, point sources of nutrients—industrial and municipal discharges—are strictly controlled through a variety of regulatory programs. Although nutrient levels and fish kills had declined significantly by the 1980s, eutrophication problems continue to persist in many systems. In part, this phenomenon is thought to be due to the re-release of older deposits of nutrients from lake and river sediments. A simpler explanation may be that some key sources—especially nonpoint sources—remain uncontrolled.

Nonpoint sources of nutrients are numerous in our society. They include diffuse runoff of rainwater and associated contaminants from urban lawns, roofs, roads, and parking lots, and drainage from agricultural fields, feedlots, and manure storage areas. These sources have long been known to contribute to water pollution. Yet compared to the costly technologies usually needed to clean industrial or sewage wastewater, nonpoint source controls are usually relatively inexpensive and freely available.

Why, then, has progress in controlling urban and agricultural diffuse source pollution been slow? Central challenges include the problem of who "owns" the pollution and who therefore should pay for cleanup. Selection of appropriate remedial actions is also difficult, because the most effective approaches often require people to change the way they behave; for instance, by fertilizing lawns or fields less frequently or by "stooping and scooping" after pets.

Because a combination of point and nonpoint sources of pollution creates observed water quality problems such as eutrophication, water quality improvement means finding the most cost-effective combination of point- and nonpoint source controls. Where human attitudes and behavior patterns slow implementation, the "best" strategy may also involve a lengthy process of public education and citizen involvement. However, although public education programs can improve the success of nonpoint source controls, they can be time-consuming and costly in their own right.

Web Explorations

The Environment on the Web activity for this essay describes the use of computer simulation models to test alternative management strategies for water quality improvement. Go to the Environment on the Web activity (select Chapter 18 at **http://www.prenhall. com/wright**) and learn for yourself:

1. how modeling can be used to answer different key environmental management questions;

2. how computer models can be used to assess and plan watershed restoration projects;

3. about the role of watershed management in water quality improvement; and

4. about the differences between physical and computer simulation models.

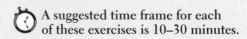 A suggested time frame for each of these exercises is 10–30 minutes.

REVIEW QUESTIONS

1. Define *pollution, pollutant, nonbiodegradable, point source,* and *nonpoint source of pollutants*.

2. Discuss each of the five categories of water pollutants and the problems they cause.

3. Describe and compare submerged aquatic vegetation (SAV) and phytoplankton. Where and how does each get nutrients and light?

4. Explain the difference between oligotrophic and eutrophic waters. Describe the entire process of eutrophication.

5. Distinguish between natural and cultural eutrophication.

6. How does the NPDES program address point-source pollution by nutrients?

7. Describe the TMDL program. How does it address nonpoint-source pollution, and what role do water quality criteria play in the program?

8. Give a brief history of how humans' handling of sewage wastes has changed as the risks and potential benefits have become more understood.

9. What is the water content of, and what are the four categories of, pollutants present in raw sewage?

10. Name and describe the facility and the process used to remove debris, grit, particulate organic matter, colloidal and dissolved organic matter, and dissolved nutrients from wastewater.

11. Why is secondary treatment also called biological treatment? What is the principle involved? What are the two alternative techniques used?

12. What are the principles involved in, and what is accomplished by, biological nutrient removal? Where do nitrogen and phosphate go in the process?

13. Raw sludge is a by-product of what treatment processes? What is the chemical and physical nature of raw sludge?

14. Name and describe three methods of treating raw sludge, and give the end product(s) that may be produced from each method.

15. Describe alternative methods for treating raw sewage.

16. How may sewage from individual homes be handled in the absence of municipal collection systems?

17. What are some of the important issues in public policy relating to sewage treatment?

THINKING ENVIRONMENTALLY

1. A large number of fish are suddenly floating dead on a lake. You are called in to investigate the problem. You find an abundance of phytoplankton and no evidence of toxic dumping. Suggest a reason for the fish kill.

2. A number of regions in the United States have banned the use of phosphate-containing detergents in recent years. What harm is caused by such detergents, and what is hoped to be achieved by the bans?

3. Describe how planting trees on an eroding hillside may protect aquatic life in an estuary many miles away.

4. Arrange a tour to the sewage-treatment plant that serves your community. Contrast it with what is described in this chapter. Is the water being purified or handled in a way that will prevent cultural eutrophication? Are sludges being converted to, and used as, fertilizer? What improvements, if any, are in order? How can you help promote such improvements?

5. Suppose a new community of several thousand people is going to be built in Arizona (with a warm, dry climate). You are called in as a consultant to design a complete sewage system, including the collection, treatment, and use or disposal of by-products. Write an essay describing the system you recommend, and give a rationale for the choices involved.

WEB REFERENCES

On-line resources for this chapter are on the World Wide Web at: **http://www.prenhall.com/wright**. (Click on Chapter 18 within the Chapter Selector.)

Municipal Solid Waste: Disposal and Recovery

Key Issues and Questions

1. Two hundred twenty million tons of municipal solid waste (MSW) are disposed of annually in the United States. What are the components of MSW, and how is this waste handled?

2. Fifty-five percent of MSW is disposed of in landfills. What are the problems with landfills, and how are these addressed in newer sites?

3. Seventeen percent of MSW is combusted, mostly in waste-to-energy (WTE) combustion facilities. What are the advantages and disadvantages of WTE combustion?

4. The best solution to solid-waste problems is to reduce waste at its source. How can the total volume of refuse be reduced?

5. More than 75% of MSW is recyclable. What role is recycling playing in waste management, and how is recycling best promoted?

6. Although most management of MSW occurs at the local level, state and federal regulations are increasing. What regulations have affected the management of MSW?

7. Much more can be done to move MSW management in a more sustainable direction. What are some recommendations to improve MSW management?

anehy Park features four soccer fields, four softball fields, three basketball courts, a baseball diamond, two playgrounds, and a two-mile jogging trail. The 55-acre park, opened in 1990, is located close to a heavily populated area of North Cambridge, Massachusetts, and is in almost constant use in good weather (Fig. 19–1). An unusual feature of the park is a big red light in the public rest room that warns users to vacate the park if the light goes on. The light is tied into an elaborate venting system that prevents methane from building up below ground. The system is necessary because Danehy Park is built on the former city dump. Aside from the presence of the light and a few minor settling problems that have interfered with water drainage on soccer fields, a newcomer would never know that this was once a blight on the neighborhood—an open, burning dump in the 1950s and then a "sanitary landfill" that was closed in 1972. The park increased Cambridge's open space by 20% when it was created.

Danehy Park is a lesson in land use that is being learned only slowly in the United States—and, for that matter, in the rest of the world. The lesson is that if we could more effectively picture what 50 years can bring in the way of change, we could solve one of the most contentious problems facing local communities: what to do with municipal waste. As old dumps and landfills were closed because of environmental concerns, they created a temporary problem identified as the "solid-waste crisis" in the 1970s and 1980s. Now many of those dumps and landfills are being converted into parks, golf courses, and nature preserves.

In some ways, the solid-waste crisis is still with us, as we will see when we examine the problem of interstate trash movement. It is commonly said that we are running out of space to put our trash and garbage. Perhaps we are, but if so, it is because of a fundamentally irresponsible feature of our modern society: We are happy to purchase the goods displayed so prominently in our malls and advertised in the media, but we are reluctant to accept the consequences of getting rid of them responsibly.

This chapter is about solid-waste issues. We examine current patterns of disposal—landfills, combustion, and

◀ *A tired landscape.* Scrap tires as far as the eye can see, accumulated by a private contractor in Smithfield, Rhode Island.

▲ **FIGURE 19–1** *Landfills to playing fields.* Thomas W. Danehy Park in Cambridge, Massachusetts, a former landfill that has been recycled into a recreational park. In addition to the playing fields, the park features a half-mile "glassphalt" pathway (built with recycled glass and asphalt), shown in the foreground.

recycling—and look for sustainable solutions to our solid-waste problems. The ideal would be to imitate natural ecosystems and reuse everything. Recall the second principle of sustainability: *Ecosystems dispose of wastes and re-*

plenish nutrients by recycling all elements. Some solutions do well in conforming to this principle, whereas others do not—and possibly cannot.

19.1 The Solid-Waste Problem

The focus of this chapter is **municipal solid waste** (MSW), defined as the total of all the materials (commonly called trash, refuse, or garbage) thrown away from homes and commercial establishments and collected by local governments. MSW is distinguished from two other categories of waste: hazardous waste (covered in Chapter 20), and nonhazardous industrial waste. The latter is no small matter: industrial facilities generate and manage 7.6 billion tons of nonhazardous industrial waste annually. Included in the category are demolition and construction wastes, agricultural and mining residues, combustion ash, sewage treatment sludge, and industrial process wastes. The states oversee these wastes, because Congress has not delegated any authority to the EPA to regulate them, as it has for MSW and hazardous waste.

Disposal of Municipal Solid Waste

Over the years, the amount of MSW generated in the United States has grown steadily, in part because of a growing population, but more so because of changing lifestyles and

the increasing use of disposable materials and excessive packaging. MSW now amounts to 4.5 pounds (2 kg) per person per day. At the current (2001) U.S. population of 283 million, that is enough waste to fill 84,000 garbage trucks each day, a total of 220 million tons (200 million metric tons) per year. The *solid-waste problem* can be stated simply: *We should make every effort to reduce our waste, but what we produce we must dispose of in the most effective and efficient way, while protecting human and environmental health.*

The refuse generated by municipalities is a mixture of materials from households and small businesses, in the proportions shown in Fig. 19–2. However, the proportions vary greatly, depending on the generator (commercial versus residential), the neighborhood (affluent versus poor), and the time of year (during certain seasons, yard wastes, such as grass clippings and raked leaves, add greatly to the solid-waste burden). It is fair to say that little attention is given to just what people throw away in the trash; even if there are restrictions and prohibitions, these can be bypassed with careful packing of the trash containers. Thus, many environmentally detrimental substances—paint, used motor oil, small batteries, and so on—are discarded with the feeling that they are gone forever.

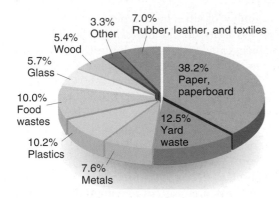

▲ **FIGURE 19–2** *U.S. municipal solid-waste composition.* The composition of municipal solid waste in 1998 in the United States. (*Source*: Data from EPA, Office of Solid Waste, *Characterization of Municipal Solid Waste in the United States*, 1999.)

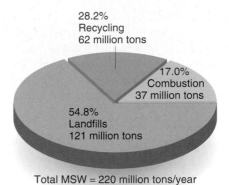

▲ **FIGURE 19–3** *U.S. municipal solid-waste disposal.* Disposal of solid waste to landfills, combustion, and recycling, 1998. (*Source*: Data from EPA, Office of Solid Waste, *Characterization of Municipal Solid Waste in the United States*, 1999.)

Customarily, local governments have assumed the responsibility for collecting and disposing of MSW. The local jurisdiction may own the trucks and employ workers, or it may contract with a private firm to provide the collection service. Traditionally, the cost of waste pickup is passed along to households via taxes; alternatively, some municipalities have opted for a "pay-as-you-throw" (PAYT, as it is called) system, in which households are charged for waste collection on the basis of the amount of trash they throw away. Some municipalities have put all trash collection and disposal in the private sector, with the collectors billing each home by volume and weight of trash. The MSW that is collected is then disposed of in a variety of ways, and it is at the point of disposal that state and federal regulations begin to apply.

Until the 1960s, most MSW was burned in open dumps. The waste was burned to reduce its volume and lengthen the life span of the dump site, but refuse does not burn well. Smoldering dumps produced clouds of smoke that could be seen from miles away, smelled bad, and created a breeding ground for flies and rats. Some cities turned to incinerators, or combustion facilities, as they are called today—huge furnaces in which high temperatures allow the waste to burn more completely than in open dumps. Without controls, however, incinerators were also prime sources of air pollution. Public objections and air pollution laws forced the phaseout of open dumps and many incinerators during the 1960s and early 1970s. Open dumps were then converted to landfills.

In the United States in 1998, 55% of MSW was disposed of in landfills, 28% was recovered for recycling and composting, and the remainder (17%) was combusted (Fig. 19–3). Over the last 10 years, the landfill component has been declining, recycling has shown a steady increase, and combustion has remained fairly constant. The pattern is different in countries where population densities are higher and there is less open space for landfills. High-density Japan, for instance, combusts 75% of its trash. Most Western European countries also deposit less than half of their MSW in landfills and combust most of the rest.

Landfills

In a **landfill**, the waste is put on or in the ground and is covered with earth. Because there is no burning, and because each day's fill is covered with at least six inches of earth, air pollution and populations of vermin are kept down. Unfortunately, aside from those concerns and the minimizing of cost, no other factors were given real consideration when the first landfills were opened. Municipal waste managers generally had no understanding of, or interest in, ecology, the water cycle, or what products decomposing wastes would generate, and they had no regulations to guide them. Therefore, in general, any cheap, conveniently located piece of land on the outskirts of town became the site for a landfill. Frequently, this site was a natural gully or ravine, an abandoned stone quarry, a section of wetlands, or a previous dump (Fig. 19–4). Once the municipality acquired the land, dumping commenced without any precautions. After the site was full, it would be covered with earth and ignored; only recently have landfills been seen as a valuable open-space resource.

▲ **FIGURE 19–4** *New Orleans dump.* This burning dump, sited on a wetlands, demonstrates the worst of the MSW disposal practices of the past.

Problems of Landfills.   Landfills are subjected to biological and physical factors in the environment and will undergo change over time as a consequence of the operation of those factors on the waste that is deposited. Several of the changes are undesirable, because, if they are not dealt with effectively, they present the following problems:

- leachate generation and groundwater contamination
- methane production
- incomplete decomposition
- settling

Leachate Generation and Groundwater Contamination.   The most serious problem by far is groundwater contamination. Recall that, as water percolates through any material, various chemicals in the material may dissolve in the water and get carried along in a process called *leaching*. The water with various pollutants in it is called *leachate*. As water percolates through MSW, a noxious leachate is generated that consists of residues of decomposing organic matter combined with iron, mercury, lead, zinc, and other metals from rusting cans, discarded batteries, and appliances—all generously "spiced" with paints, pesticides, cleaning fluids, newspaper inks, and other chemicals. The nature of the landfill site and the absence of precautionary measures noted earlier can funnel this "witches' brew" directly into groundwater aquifers.

All states have some municipal landfills that are, or soon will be, contaminating groundwater, but Florida has some unique problems. Flat and with vast areas of wetlands, much of the state is only a few feet above sea level and rests on water-saturated limestone. No matter where Florida's landfills were located, they were either in wetlands or just a few feet above the water table. As a result, more than 150 former municipal landfill sites in Florida are now on the Superfund list. (Superfund is the federal program to clean up sites that are in imminent danger of jeopardizing human health through groundwater contamination.) All landfills in Florida now have state-of-the-art landfill liners.

Methane Production.   Because it is about two-thirds organic material, MSW is subject to natural decomposition. However, buried wastes do not have access to oxygen. Therefore, their decomposition is anaerobic, and a major by-product of the process is *biogas*, which is about two-thirds methane and the rest hydrogen and carbon dioxide, a highly flammable mixture. (See Chapter 18.) Produced deep in a landfill, biogas may seep horizontally through the soil and rock, enter basements, and even cause explosions if it accumulates and is ignited. Over 20 homes at distances up to 1,000 feet from landfills have been destroyed, and some deaths have occurred as a result of such explosions. Also, gases seeping to the surface kill vegetation by poisoning the roots. Without vegetation, erosion occurs, exposing the unsightly waste.

A number of cities have exploited the problem by installing "gas wells" in old and existing landfills. The wells tap the biogas, and the methane is purified and used as fuel. There are now 102 commercial landfill gas facilities in the United States; the largest, in Sunnyvale, California, generates enough electricity to power 100,000 homes. In 1998, commercial landfill gas produced 108 trillion BTUs of energy (equivalent to 20 million barrels of oil); landfill gas production has been increasing at a rate of 11% a year for the last five years.

Incomplete Decomposition.   The commonly used plastics in MSW are resistant to natural decomposition because of their molecular structure. Chemically, they are polymers of petroleum-based compounds that microbes are unable to digest. Biodegradable plastic polymers have been developed from sources such as cornstarch, cellulose, lactic acid, soybean protein, amides, and others. For example, the German company Bayer AG recently announced plans to market a 100% biodegradable plastic structured as a polyester amide polymer. Plastic film made from the polymer degrades completely in 70 days, according to company tests. So far, however, none of the biodegradable plastics has seen common usage in consumer products.

A team of "archaeologists" from the University of Arizona, led by William Rathje, has been carrying out research on old landfills. The research has shown that even materials formerly assumed to be biodegradable—newspapers, wood, and so on—are degraded only slowly, if at all, in landfills. In one landfill, 30-year-old newspapers were recovered in a readable state, and layers of telephone directories, practically intact, were found marking each year. Since paper materials are 38% of MSW, this is a serious matter. The reason paper and other organic materials decompose so slowly is the absence of suitable amounts of moisture. The more water percolating through a landfill, the better is the biodegradation of paper materials; however, the more percolation there is, the more toxic leachate is produced!

Settling.   Finally, waste settles as it compacts and decomposes. Luckily, this eventuality was recognized from the beginning, so buildings have never been put on landfills. Settling does present a problem in landfills that have been converted to playgrounds and golf courses, because it creates shallow depressions (and sometimes deep holes) that collect and hold water. This problem can be addressed by continual monitoring of the facility and the use of fill to restore a level surface.

Improving Landfills.   Recognizing the foregoing problems, the EPA upgraded siting and construction requirements for new landfills. Under current regulations,

- New landfills are sited on high ground, well above the water table, not in a geologically unstable area, and away from airports (because of bird hazards). Often, the top of an existing hill is bulldozed off to supply a source of cover dirt and at the same time create a floor that is above the water table.

- The floor is contoured so that water will drain into a tile leachate-collection system. The floor is then covered with a plastic liner and at least two feet of compacted soil. On top of this is a layer of coarse gravel and a layer of porous earth. With such a design, any leachate percolating through the fill will encounter the gravel layer and then move through that layer into the leachate collection system. The clay layer or plastic liner prevents leachate from ever entering the groundwater. Collected leachate can be treated as necessary.

- Layer upon layer of refuse is positioned such that the fill is built up in the shape of a pyramid. Finally, it is capped with at least 18 inches of earthen material and a layer of topsoil and then seeded. The cap and the pyramidal shape help the landfill shed water. In this way, water infiltration into the fill is minimized, and less leachate is formed.

- Finally, the entire site is surrounded by a series of groundwater-monitoring wells that are checked periodically, and such checking must go on indefinitely.

These design features are summarized in Fig. 19–5. Most landfills currently in operation have the improved technologies, which protect both human health and the environment.

Although the regulations protect groundwater, the landfill pyramids themselves may well last as long as the Egyptian pyramids. (They are not likely to become tourist attractions, however!) And if they do break down, they become a threat to the groundwater; therefore, the need for monitoring remains. Nonetheless, the creative siting and construction of landfills has the potential to address some highly significant future needs, as we have seen: The abandoned landfill can become an attractive golf course, recreational facility, or wildlife preserve.

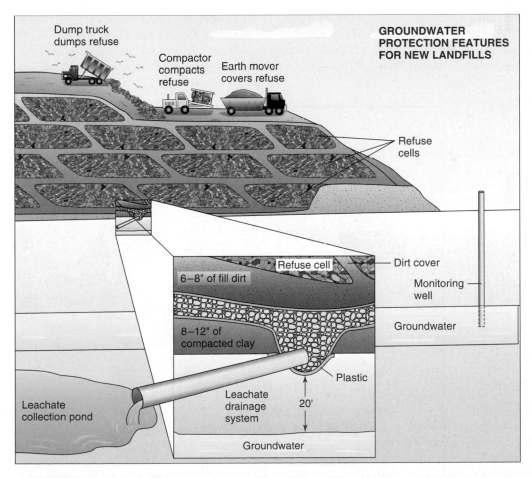

▲ **FIGURE 19–5** *Features of a modern landfill.* The landfill is sited on a high location well above the water table. The bottom is sealed with compacted soil and a plastic liner, overlain by a rock or gravel layer with pipes to drain the leachate. Refuse is built up in layers as the amount generated each day is covered with soil, so that the completed fill has a pyramidal shape that sheds water. The fill is provided with wells for monitoring groundwater.

Siting New Landfills. From 1988 to 1998, the number of municipal landfills has declined from around 8,000 to 2,314. Because the size of landfills has increased, and because recycling is on the rise, the EPA does not believe that landfill capacity is a problem. However, as the agency puts it, "regional dislocations" may be expected. This is due to the one problem associated with landfills that gets more attention than any other: **siting**. It is not that landfills take up enormous amounts of land; one landfill occupying 121 acres (Central Landfill) serves the entire state of Rhode Island, and Fresh Kill, the largest landfill in the world, has served much of New York City and its environs and is only 2,400 acres in area. But as old landfills are closed, it has become increasingly difficult to find land for the new ones needed to take their place.

People in residential communities (where MSW is generated) invariably reject proposals to site landfills anywhere near where they live. And those who already live close to existing landfills are anxious to close them down. Weary of the odor and heavy truck traffic, residents of Staten Island, New York, have pressured the city of New York to shut down the huge Fresh Kill landfill, once the recipient of 27,000 tons of garbage per day and now down to 6,000 tons. The city has promised to phase out the landfill by the end of 2001, but where the city's waste will go is still not clear.

With spreading urbanization, few suburban areas are not already dotted with residential developments. Practically any site selection, then, is met with protests and legal suits. The problem has been repeated in so many parts of the country (and globe!), that it has given rise to several inventive acronyms: **LULU**, **NIMBY**, and **NIMTOO**. "LULU" means "locally unwanted land use," "NIMBY" is "Not in my backyard," and "NIMTOO" is "Not in my term of office." The applications of these attitudes to the landfill siting problem are obvious.

The siting problem has some undesirable consequences. First, it drives up the costs of waste disposal, as alternatives to local landfills are invariably more expensive. Second, it leads to the inefficient and equally objectionable practice of long-distance transfer of trash, as waste generators look for private landfills anxious to receive trash. Very often, this transfer occurs across state and even national lines, leading to real resentment and opposition on the part of citizens of the recipient state or nation. Such opposition has resulted in numerous state efforts to restrict or prohibit landfills from receiving out-of-state trash. These efforts have invariably been struck down by the courts as unconstitutional interference with interstate commerce. Nine states import more than 1 million tons of MSW a year (Fig. 19–6). New Jersey, New York, Maryland, and Missouri lead the exporting states, with New York far in the lead at 3.8 million tons a year.

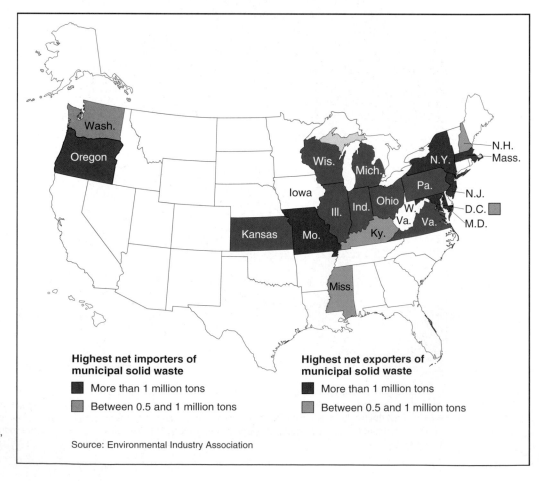

▶ **FIGURE 19–6** *Trash on the move.* Net import and export of municipal solid waste, by state, in 1997. (*Source*: Data from James E. McCarthy, Congressional Research Service, *Interstate Shipment of Municipal Solid Waste: 1998 Update.*)

Highest net importers of municipal solid waste
■ More than 1 million tons
■ Between 0.5 and 1 million tons

Highest net exporters of municipal solid waste
■ More than 1 million tons
■ Between 0.5 and 1 million tons

Source: Environmental Industry Association

One positive impact of the siting problem is to encourage residents to reduce their waste and recycle as much as possible; this alternative will be explored later. Another effect of the siting problem is to stimulate the use of combustion as an option for waste disposal.

Combustion: Waste to Energy

Because it has a high organic content, refuse (especially the plastic portion) can be burned. Currently, more than 110 combustion facilities are operating in the United States, burning about 37 million tons of waste annually—17% of the waste stream. Note, however, that this process is not really waste *disposal*; rather, it is waste *reduction*, because, after incineration, the ash must still be disposed of.

Advantages of Combustion. Combustion of MSW has some advantages:

- Combustion can reduce the weight of trash by over 70% and the volume by 90%, thus greatly extending the life of a landfill (which is still required to receive the ash).
- Toxic or hazardous substances are concentrated into two streams of ash, which are easier to handle and control than the original MSW. The *fly ash* (captured from the combustion gases by air pollution control equipment) contains most of the toxic substances and can be safely put into a landfill. The *bottom ash* (from the bottom of the boiler) can be used as fill in some construction sites and roadbeds. Some combustion facilities process the bottom ash further to recover metals and then convert the remainder into concrete blocks.
- No changes are needed in trash collection procedures or people's behavior; trash is simply hauled to a combustion facility instead of the landfill.
- Practically all modern combustion facilities are designed to generate electricity, which is sold to offset some of the costs of disposal. The newer facilities are equipped with scrubbers and filters or electrostatic precipitators, which remove most of the gases and particles to bring the emissions into compliance with Clean Air Act regulations.

Simple incineration is an outmoded method of combustion; it accomplishes a reduction in volume of the trash, but without the recovery of energy. When burned, unsorted MSW releases about 35% as much energy as coal, pound for pound. Almost all of the currently operating combustion facilities are **waste-to-energy** (WTE) facilities equipped with modern emission-control technology. To their waste processing many of these facilities add **resource recovery**, in which many materials are separated and recovered before (and sometimes after) combustion. Because the sale of the electricity offsets some of the costs of MSW disposal, the combustion facilities are usually able to compete successfully with landfills for MSW.

Drawbacks of Combustion. Combustion has some drawbacks, however:

- Health effects can result, especially from older or poorly managed facilities. Particulate matter, lead, mercury, dioxins, and furans are the most serious pollutants entering the air and contributing to local and regional air pollution. Workers at such facilities are especially at risk for health problems.
- Combustion facilities are expensive to build, and their siting has the same problem as that of landfills: No one wants to live near one.
- Combustion ash is often loaded with metals and other hazardous substances and must be disposed of in secure landfills.
- To justify the cost of its operation, the combustion facility must have a continuing supply of MSW. For that reason, the facility enters into long-term agreements with municipalities, and these agreements can lessen the flexibility of the community's solid-waste management options.
- Even if the combustion facility generates electricity, the process wastes both energy and materials, unless it is augmented with recycling and recovery. However, a number of combustion facilities compete directly with recycling for burnable materials such as newspapers and represent a major impediment to recycling in some municipalities.

An Operating Facility. Let us look at the operation of a typical modern WTE facility, which might serve a number of communities or a large metropolitan area. Servicing a population of a million or more, the plant receives about 3,000 tons of MSW per day. The waste comes in by rail and truck, and the communities pay tipping fees (the costs assessed at the disposal site) ranging from $15 to $100 per ton. Waste processing is efficient: Overall, about 80% of the MSW is burned for energy, 12% is recovered, and 8% is put in landfills. The process, pictured in Fig. 19–7, is as follows:

1. Incoming waste is inspected, and obvious recyclable and bulky materials are removed.
2. Waste is then pushed onto conveyers that feed shredders capable of reducing the width of waste particles to 6 inches or less.
3. Strong magnets remove about two-thirds of ferrous metals for recycling before combustion.
4. The waste is then blown into boilers, where light materials burn in suspension and heavier materials burn on a moving grate.
5. Water circulated through the walls of the boilers produces steam, which drives turbines for generating electricity.

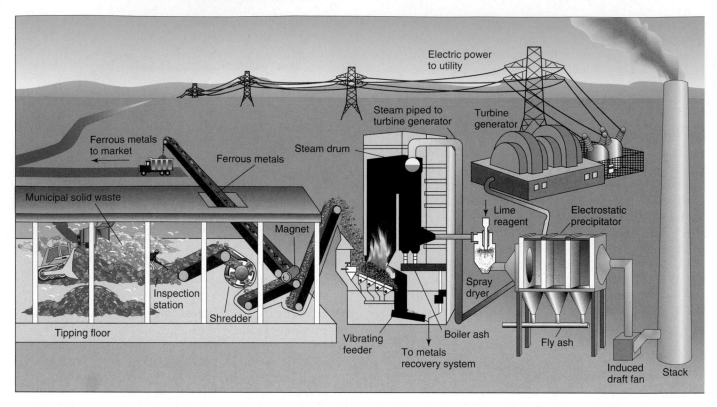

▲ FIGURE 19–7 *Waste-to-energy combustion facility.* Schematic flow for the separation of materials and combustion in a typical modern waste-to-energy combustion facility.

6. After the waste is burned, the bottom ash is conveyed to a processing facility, where further separation of metals may occur in a process that recovers brass, aluminum, gold, copper, and iron. In some facilities, this process nets $1,000 a day in coins alone!

7. Combustion gases are passed through a lime-based spray dryer–absorber to neutralize sulfur dioxide and other noxious gases. Then the gases go through electrostatic precipitators that remove particles. The resulting waste stream is significantly lower in pollutants than that which would be emitted by an energy-equivalent utility based on coal or oil combustion.

8. The fly ash and bottom ash residues are put into landfills.

An appreciation for the impact of such a facility can be gained from looking at the outcome of a year's operation. In a year, 1 million tons of MSW are processed, 40,000 tons of metal are recycled, and 570,000 megawatt-hours of electricity are generated—the equivalent of more than 60 million gallons of fuel oil and enough electricity to power 65,000 homes. All this comes from stuff that people have thrown away!

The advantages of combustion over landfills are obvious; the major disadvantage is that some recycling and composting opportunities are missed. This drawback can be addressed at the collection point, as we will see when we ex-

amine the recycling option. In any event, substantial recovery of recyclable materials—and energy—certainly can be accomplished with a modern WTE facility.

Costs of Municipal Solid Waste Disposal

The costs of disposing of MSW are becoming prohibitive. Increasing costs are not just a result of the new design features of landfills; more and more, they reflect the expenses of acquiring a site and providing transportation. Tipping fees average about $30 a ton, and at some landfills they exceed $100 a ton; the waste collector must recover this cost, as well as transportation costs to the site.

Getting rid of *all* trash is getting more expensive, and one sad consequence of this increasing expense is illegal dumping. Some towns are charging up to $5.00 a bag for MSW disposal, and it now costs $1 or more to get rid of an automobile tire and $30 or more to dispose of a refrigerator. As a result, tires, refrigerators, yard waste, car parts, and construction waste are appearing all over the landscape. Institutions and apartments that operate with dumpsters often must put padlocks on them in order to prevent their unauthorized use by people trying to avoid disposal costs. Many states have established a corps of environmental police to track down midnight dumpers and bring them to justice.

19.2 Solutions

An examination of our domestic wastes and their disposal reveals that a mammoth stream of material flows in one direction: from our resource base to disposal sites. Just as natural ecosystems depend on recycling nutrients, if we are to move in the direction of sustainability, we will also ultimately need to learn to recycle our wastes. There is strong evidence that we are moving in this direction; however, the best strategy of all is to reduce waste at its source.

Source Reduction

Waste prevention, or "source reduction" as it is often called, accomplishes two goals: It reduces the amount of waste that must be managed, and it conserves resources. It is noteworthy that, after rising rapidly during the latter decades of the 20th century, the amount of waste per capita in the United States peaked at 4.5 lb in 1990 and has since leveled off. (We still lead the world in this dubious statistic, however.) The leveling off is a signal that some changes in lifestyle may be occurring that have an impact on waste generation.

Source reduction can involve a broad range of activities on the part of homeowners, businesses, communities, manufacturers, and institutions. Consider the following examples:

- Lightening the weight of many items has reduced the amount of materials used in manufacturing. Steel cans are 60% lighter than they used to be; disposable diapers contain 50% less paper pulp, due to absorbent-gel technology; and aluminum cans contain only two-thirds as much aluminum as they did 10 years ago.
- The Information Age may be having an impact on paper use: Electronic communication, data transfer, and advertising are increasingly performed on personal computers, and the number of homes with personal computers continues to rise. For example, 1.7 trillion e-mail messages were sent in 1996, and the rate of growth is 50% per year. The growth of the World Wide Web as a medium for information transfer has been phenomenal.
- Many durable goods are reusable, and indeed, the Unites States has a long tradition of resale of furniture, appliances, and carpets. Other goods may be donated to charities and put to good use. The growing popularity of yard sales, flea markets, consignment clothing stores, and other markets for secondhand goods (Fig. 19–8) is an encouraging development.
- Lengthening product life can keep materials out of the waste stream; lengthening a product's life involves both manufacturers and consumers. If products are designed to last longer and be easier to repair, consumers will learn that they are worth the extra cost. (You get what you pay for; cheap products usually take the shortest route to the waste bin.)
- We are all on bulk-mailing lists, because these lists are sold and shared widely; thus, we are guaranteed to receive increasing volumes of advertising. To stay off such lists, simply inform mail-order companies and other organizations involved that you do not want your name and address shared. (You can write to Mail Preference Service, Direct Marketing Association, P.O. Box 9008, Farmingdale, NY 11735-9008. The service provides, at no charge, a master list of people who do not wish to receive mass advertising; it works as long as marketers choose to consult it.)

The Recycling Solution

In addition to reuse, recycling is an obvious solution to the solid-waste problem. More than 75% of MSW is recyclable material. There are two levels of recycling: **primary** and **secondary**. *Primary recycling* is a process in which the original waste material is made back into the same material—for example, newspapers recycled to make newsprint. In *secondary*

▲ **FIGURE 19–8** *Waste reduction by reuse.*

recycling, waste materials are made into different products that may or may not be recyclable—for example, cardboard from waste newspapers.

The primary items from MSW currently being heavily recycled are cans (both aluminum and steel), bottles, plastic containers, newspapers, and yard wastes. Yet, there are many alternatives for reprocessing various components of refuse, and people are coming up with new ideas and techniques all the time. A few of the major established techniques, together with current percentages of their recovery by recycling, are as follows:

- Paper and paperboard (41.6% recovery) can be remade into pulp and reprocessed into recycled paper, cardboard, and other paper products; finely ground and sold as cellulose insulation; or shredded and composted.
- Most glass (25.5% recovery) that is recycled is crushed, remelted, and made into new containers; a smaller amount is used in fiberglass or "glasphalt" for highway construction.
- Some forms of plastic (5.4% recovery) can be remelted and fabricated into carpet fiber, outdoor wearing apparel, irrigation drainage tiles, building materials, and sheet plastic (Fig. 19–9).
- Metals can be remelted and refabricated. Making aluminum (28% recovery) from scrap aluminum saves up to 90% of the energy required to make aluminum from virgin ore. In addition, aluminum ore is imported and is part of the mounting U.S. trade deficit. National recycling of aluminum saves energy, creates jobs, and reduces the trade deficit.
- Yard wastes (leaves, grass, and plant trimmings—45.3% recovery) can be composted to produce a humus soil conditioner.
- Textiles (12.8% recovery) can be shredded and used to strengthen recycled paper products.
- Old tires (22% recovery) can be remelted or shredded and incorporated into highway asphalt or burned in special combustion facilities.

Recycling is both an environmental and an economic issue. Many people are motivated to recycle because of environmental concern, but the use of recycled materials is also driven by economic factors.

Municipal Recycling. Recycling is probably the most direct and obvious way most people can become involved in environmental issues. The appeal of recycling is obvious: If you recycle, you save some natural resources from being used (trees, in the case of paper), and you prevent landfills from becoming "landfulls."

Recycling's popularity is well established: Virtually every state has specific recycling goals. EPA sources report that only 6.7% of MSW was recycled in 1960, versus 28% in 1998 (Fig. 19–10). There is a great diversity of approaches to

▲ **FIGURE 19–9** *Plastic recycling.* Bales of plastic bottles in a recycling center.

recycling in municipalities, from recycling centers requiring residents to drive miles to recycle to curbside recycling with sophisticated separation processes. The most successful programs have the following characteristics:

1. There is a strong incentive to recycle, in the form of pay-as-you-throw charges for general trash and no charge for recycled goods.
2. Recycling is not optional; mandatory regulations are in place, with warnings and sanctions for violations.
3. Residential recycling is curbside (Fig. 19–11), with free recycling bins distributed to households. (Curbside recycling has risen rapidly; over 50% of the U.S. populace is now served via 8,900 curbside programs.)
4. Recycling goals are ambitious, yet clear and feasible. Some percent of the waste stream is targeted, and progress is followed and communicated.
5. A concerted effort is made to involve local industries in recycling.
6. The municipality employs an experienced and committed recycling coordinator.

Nevertheless, municipalities experience very different recycling rates. New York City, for example, has made great progress—from recycling only 5% to 30% in four years—by mandating curbside recycling for all 3 million households.

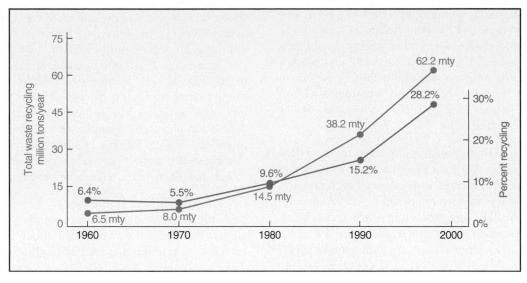

▲ FIGURE 19–10 *Waste Recycling Rates.* MSW recycled from 1960 to 1998, as total waste and percentage recycled. EPA's new goal is to reach 35% by 2005 (Source: EPA. Office of Solid Waste, Environmental Fact Sheet, April 2000).

Boston, on the other hand, with only 12% recycling, has held Massachusetts back in its goal to achieve 46% by 2000. Worcester, farther to the west in the state, has achieved an outstanding 55% recycling rate.

Recycling has its critics, who base their arguments primarily on economics. If the costs of recycling (from pickup to disposal of recycled components) are compared with the costs of combustion or placing waste in landfills, recycling frequently comes out second best. Markets for recyclable materials fluctuate wildly, and residents often end up subsidizing the recycling effort. The shortfall between recycling costs and market value range from $20 to $135 a ton for some recyclables (items such as green glass and colored plastic bottles). In some cases (notably, newspaper and glass), prices are controlled by the industry of origin. Competition between landfills and combustion facilities often lowers tipping fees, creating an even greater disincentive to recycle. Thus, critics of current recycling practices argue that, unless recycling pays for itself through the sale of recovered materials, it should not be done. However, the critics ignore the subsidies that make virgin materials less expensive to use than recycled materials. Also, garbage collection is big business, and those involved see recycling as cutting their trash business.

In spite of these economic considerations, the demonstrated support for recycling programs is strong: Experience has shown that at least two-thirds of households will recycle if presented with a curbside pickup program. The percentage

◀ FIGURE 19–11 *Curbside recycling.* Curbside pickup by a waste hauler using a special truck. This approach is reaching over 50% of the U.S. population.

goes up when recycling is combined with PAYT programs. Finally, a recent report by Franklin Associates showed that recycling is a hands-down winner when energy use and pollution are considered. For every ton of waste processed, a thorough recycling program will eliminate 620 pounds of carbon dioxide, 30 pounds of methane, 5 pounds of carbon monoxide, 2.5 pounds of particulate matter, and smaller amounts of other pollutants.

Paper Recycling.   By far the most important recycled item is newspapers, because of their predominance in the waste stream. It is a simple matter to tie up or bag household newspapers, and the amount recovered by recycling is increasing dramatically. Currently, more newspaper is being recycled than discarded (55%). Since more than 25% of the trees harvested in the United States are used to make paper, recycling paper obviously saves trees. Depending on the size and type of trees, a 1-meter stack of newspapers equals the amount of pulp from one tree.

There is often some confusion in what is meant by "recycled paper." The key is the amount of *postconsumer* recycled paper in a given product. In manufacturing processes, much paper is "wasted" and is routinely recovered and rerouted back into processing and called "recycled paper." Thus, the total recycled content of a paper can be 50%, although the actual postconsumer amount recovered by recycling might be only 10% of the total amount of paper.

As with the recycling of other materials, paper recycling is highly dependent on the existence of a market for the recycled product. Cost considerations are involved, from the point of collection of the wastepaper to the point of use by consumers. For example, there are all types of paper, and recycling mills want specific grades; gross mixtures have little market appeal. So the processor of recycled paper must perform some separation and packaging for specific uses.

After the wastepaper is incorporated into a final product, the market becomes a critical factor. Is there a demand for recycled paper? Several years ago, much recycled paper was rough and off color. In order to guarantee a market for the product, many organizations—including state and federal agencies—required purchasing and using paper with recycled content. The technology for producing high-quality paper from recycled stock has improved greatly, to the point where it is virtually impossible to distinguish the recycled product from "virgin" paper. Although the mandates for "closing the loop," as the practice of creating markets for recycled products is called, are important, they may soon disappear as paper manufacturers make the use of recovered paper a standard practice.

The market for used newspapers has fluctuated greatly over the past several decades. During the late 1980s, the market was saturated, and municipalities often had to pay to get rid of newspapers. In 1995, discarded papers were so valuable—up to $160 a ton—thieves were stealing them off the sidewalks before the recycling trucks could pick them up. A year later, as more recycling programs came on line,

the market collapsed again, and many cities once more paid as much as $25 a ton to have the newspapers hauled away—but this is still less expensive than the cost of landfills.

There is a lively international trade as well in used paper (Fig. 19–12). Forest-poor countries in Europe and Asia purchase wastepaper from the United States and other industrial countries in the Northern Hemisphere, where there continues to be a surplus of such paper. The largest importer is Taiwan (almost 2 million tons per year). Taiwan also boasts the highest percentage of reused paper content in its paper: 98%. The United States is at a much lower 33%.

Glass Recycling and Bottle Laws.   The glass in MSW is primarily in the form of containers, most of which held beverages. The average person drinks about a quart of liquid each day. Given that there are 280 million Americans, this daily consumption amounts to some 27 billion gallons of liquid per year for the nation as a whole. Most of this volume is packaged in single-serving containers that are used once and then thrown away. Nonreturnable containers constitute 6% of the solid waste stream in the United States and about 50% of the nonburnable portion; they also constitute about 90% of the nonbiodegradable portion of roadside litter. Broken bottles along roads, beaches, and parklands are responsible for innumerable cuts and other injuries, not to mention flat tires. Both the mining of the materials and the process used to manufacture the beverage containers create pollution. All of these factors produce hidden costs that do not appear on the price tag of the item. However, we pay for them with taxes to clean up litter, as well as with our injuries, flat tires, environmental degradation, and so on.

In an attempt to reverse these trends, environmental and consumer groups have promoted *bottle laws* that facilitate the recycling or reuse of beverage containers. Such laws generally call for a deposit on all beverage containers,

▲ **FIGURE 19–12** *Wastepaper exports.* Bales of wastepaper being loaded for shipment overseas.

both reusable and throwaway. Retailers are required to accept the used containers and pass them along for recycling or reuse. Bottle laws have been proposed in virtually all state legislatures over the last decade. Nevertheless, in every case, the proposals have met with fierce opposition from the beverage and container industries and certain other special-interest groups. The reason for their opposition is obvious—economic loss to their operations—but the arguments they put forth are more subtle. The container industry contends that bottle laws will result in loss of jobs and higher beverage costs for the consumer. The industry also claims that consumers will not return the bottles and that, therefore, the amount of litter will not diminish. In a 1996 media blitz, these same arguments were used in defeating an expansion of Oregon's successful bottle bill.

In most cases, the industry's well-financed lobbying efforts have defeated bottle laws. However, some states—10, as of 2000—have adopted bottle laws of varying types despite industry opposition (Table 19–1). The experience of these status has proved the beverage and bottle industry's claims false: More jobs are gained than lost, costs to the consumer have not risen, a high percentage of bottles is returned, and there is a marked reduction in the can and bottle portion of litter.

A final measure of the success of bottle laws is continued public approval. Despite industry efforts to repeal these laws, no state that has one has repealed it. Repeated attempts have been made to get a national bottle law through Congress—to date, unsuccessfully. Opponents (the same ones that oppose state-level bottle bills) argue that such a law will threaten the newly won successes in curbside recycling, with some justification: Beverage containers represent the most important source of revenue in curbside recycling. However, since curbside recycling currently reaches only half of the U.S. population, a national bottle law would undoubtedly recover a greater proportion of beverage containers. (States with bottle laws report 80% to 97% rates of return of containers.) Also, a national bottle law would be labor intensive, employing tens of thousands of workers and creating countless new jobs.

In 1998, recycling and bottle laws resulted in recovering 29% of glass containers, 44% of aluminum containers, 57% of steel containers, and about 10% of plastic containers, with an aggregate weight of 6.6 million tons.

Plastics Recycling. Plastics have a bad reputation in the environmental debate for many reasons. First, plastics have many uses that involve rapid throughput—for example, packaging, bottling, the manufacture of disposable diapers, and incorporation into a host of cheap consumer goods. Second, plastic production has increased 10% a year for the past three decades. Third, plastics are conspicuous in MSW and litter; ironically, most trash is disposed of in plastic bags manufactured just for that purpose. Finally, plastics do not decompose in the environment, because no microbes (or any other organisms) are able to digest them. (Imagine plas-

TABLE 19–1 States that Have Passed Bottle Laws			
State	Year Passed	State	Year Passed
Oregon	1972	Iowa	1978
Vermont	1973	Massachusetts	1978
Maine	1976	Delaware	1982
Michigan	1976	New York	1983
Connecticut	1972	California	1991

tics in landfills delighting archaeologists hundreds of years from now.) Therefore, the possibility of recycling at least some of the plastic in products—liquid containers, for example—interests the environmental community.

If you look on the bottom of a plastic container, you will see a number (or some letters) inside the little triangle of arrows that has come to represent recycling (Fig. 19–13). The number or the letters are used to differentiate between and sort the many kinds of plastic polymers, not all of which are recyclable. The two recyclable plastics in most common use are high-density polyethylene (HDPE), code 2, and polyethylene terephthalate (PETE), code 1. In the recycling process, plastics must be melted down and poured into molds, and unfortunately, some contaminants from the original containers may carry over. This makes it difficult to reuse the plastic for food containers; thus, uses for the recycled plastic are necessarily restricted. However, some new uses for recycled plastic are appearing, with more in the development stage: PETE, for example, is turned into carpet fiber and fill for outdoor wearing apparel, and HDPE becomes irrigation drainage tiles, sheet plastic, and, appropriately, recycling bins.

Critics of recycling of plastics point out that if the process were purely driven by the market, it wouldn't happen at all: Recovering plastics is more costly than starting from scratch—that is, beginning with petroleum derivatives—and manufacturers are getting involved mainly because

▲ **FIGURE 19–13** *Recycling symbol* This symbol representing high-density polyethylene is on the bottom of a plastic bottle.

environmentally concerned consumers are demanding their involvement. Industry-supported critics also point out that plastics in landfills create no toxic leachate or dangerous biogas; also, plastics in combustion facilities burn wonderfully hot and leave almost no ash. However, as oil reserves continue to dwindle, the petroleum corporations will find it increasingly difficult to promote a once-through waste stream for plastics. Perhaps we can conclude that even though plastic recycling has its problems, continued attention to plastics in MSW will keep us moving in a sustainable direction.

Composting

An increasingly popular way of treating yard waste (currently 13% of MSW) is composting, especially as more and more states are banning yard waste from MSW collections. Recall that composting involves the natural biological decomposition (rotting) of organic matter in the presence of air. Composting can be carried out in a backyard by homeowners, or it can be promoted by municipalities as a way of dealing with a large fraction of MSW. On a large scale, composting involves laying down the wastes in windrows up to 12 feet high and 24 feet wide. Microbes and detritus feeders (worms, grubs, etc.) will decompose the organic matter in the compost heap and greatly reduce the volume of waste. The process is speeded by aeration, which is accomplished by turning the material with large mechanical forks. The end product is a residue of humuslike material, which can be used as an organic fertilizer and soil builder. Composting is one method of treating sewage sludge, as described in Chapter 18.

Composting has clear economic and environmental benefits over putting waste in landfills and over combustion if the yard waste can be collected and processed independently of collecting trash. Some 3,800 municipal yard waste composting programs were reported in 1998.

19.3 Public Policy and Waste Management

Management of MSW used to be entirely under the control of local governments. However, in recent years, state and federal agencies have played an increasingly important role in waste management, partly through regulation and partly through encouragement and facilitation.

The Regulatory Perspective

At the federal level, the following legislation has been passed:

- The first attempt of Congress to address the problem was the Solid Waste Disposal Act of 1965. The legislation gave jurisdiction over solid waste to the Bureau of Solid Waste Management, but the agency's mandate was basically financial and technical rather than regulatory.

- With the creation of the EPA in 1970, the Resource Recovery Act of 1970 gave jurisdiction over waste management to the EPA and directed attention to recycling programs and other ways of recovering resources in MSW. The act also encouraged the states to develop some kind of waste management program.

- The passage of the Resource Conservation and Recovery Act (RCRA) of 1976 saw a more regulatory ("command and control") approach to dealing with MSW, as the EPA was given power to close local dumps and set regulations for landfills. Combustion facilities were covered by air pollution and hazardous waste regulations, again under the EPA's jurisdiction. The RCRA also required the states to develop comprehensive solid-waste management plans.

- The Superfund Act of 1980 (described in Chapter 20) addressed abandoned hazardous waste sites throughout the country, many of which (41%) are old landfills.

- The Hazardous and Solid Waste Amendments of 1984 gave the EPA greater responsibility to set solid-waste criteria for all hazardous-waste facilities. Since even household waste must be assumed to contain some hazardous substances, this meant that the EPA had to determine and regard all landfill and combustion criteria more closely.

Largely in response to these federal mandates, the states have been putting pressure on local governments to develop integrated waste management plans (see next section), with goals for recycling, source reduction, and landfill performance. In many states, particularly in the Northeast, where landfills have been closing, solid-waste management is comprehensive and ambitious. States are setting goals and policies for recycling, one significant manifestation of which is the *materials recovery facility* (see the "Earth Watch" essay, p. 480), in which processing of waste allows for improved recovery of materials for recycling. There are now more than 380 of these facilities, spread throughout the country and capable of processing 32,000 tons of waste per day.

Integrated Waste Management

It is not necessary to fasten onto just a single method of handling wastes. Source reduction, waste-to-energy combustion, recycling, recovery facilities, landfills, and composting all have roles to play in waste management. Different combinations of these options will work in different regions of the country. A system having several alternatives in operation at the same time is called *integrated waste management*. Let us examine the available MSW management options, with a view toward developing recommendations that make environmental sense—that is, push us in a more sustainable direction.

◀ **FIGURE 19–14** *The throwaway society.* A view from space.

Waste Reduction. Many observers have called the modern United States a "throwaway society." We probably deserve that label for setting the world record for per capita waste production, along with per capita energy consumption (Fig. 19–14). How can we turn this situation around? True management of MSW begins in the home (or dorm room). Materialistic lifestyles, affluence, and overconsumption can pack our homes and our trash cans with stuff we simply do not need; indeed, some observers see this as a malady called *affluenza*. (See the "Ethics" essay, p. 481.)

Several incentives have been put in place on the governmental level to encourage waste reduction. For example, the EPA sponsors the WasteWise program, targeting the reduction of MSW by establishing partnerships with not only local governments, schools, and organizations, but multinational corporations as well. This is a voluntary partnership that allows the partners to design their own solid-waste reduction programs. In the growing program, accomplishments of the partners in 1999 included eliminating more than 9 million tons of waste through targeted waste prevention and recycling activities. The EPA regularly honors partner organizations that have achieved exemplary waste reduction. For example, Target Stores achieved savings of 22 million pounds of packaging over two years by receiving goods ready for retailing from suppliers. The company also recycled over 350 million pounds of fiber, plastic, and other materials.

The EPA also brought attention to a national trend of "unit pricing," or charging households and other "customers" for the waste they dispose, with the agency's "pay-as-you-throw" (PAYT) program. Instead of paying for trash collection and disposal through local taxes, communities levy curbside charges for all unsorted MSW—say, $1–5 per container (Fig. 19–15). Between 1990 and 1998, the population served by PAYT MSW collection doubled (to 35 million). Over 4,000 cities and towns now employ the PAYT method. Following this highly successful strategy, PAYT communities reduce their waste from 14 to 27%, on average,

▲ **FIGURE 19–15** *"Pay-as-you-throw" trash pickup.* Consumers purchase the empty bags for a set price, and only these bags are picked up by the trash hauler. Next to the bags are three recycling containers.

EARTH WATCH

REGIONALIZED RECYCLING

Currently, most recycling is on a town-by-town basis. Either the towns or their waste contractors must find markets for the recycled goods: cans, bottles, newspapers, and plastics. This problem has kept many municipalities from getting into recycling. So what is the solution? It might be a regionalized materials recovery facility (MRF), referred to in the trade as a "murf." Here's how the state-owned MRF in Springfield, Massachusetts, works:

Basic sorting takes place when waste is collected, either through curbside collection or by town recycling stations (sites to which townspeople can bring wastes to be recycled). The waste is then trucked to the MRF and handled on three tracks—one for metal cans and glass containers, another for paper products, and a third for plastics. The materials are moved through the facility by escalators and conveyor belts, tended by workers who inspect and do further sorting. The objective of the process is to prepare materials for the recycled-goods market. Glass is sorted by color, cleaned, crushed into small pebbles, and then shipped to glass companies, where it replaces the raw materials that go into glass manufacture—

sand and soda ash—and saves substantially on energy costs. Cans are sorted, flattened, and sent either to detinning plants or to aluminum-processing facilities. Paper is sorted, baled, and sent to reprocessing mills. Plastics are sorted into four categories, depending on their color and type of polymer, and then sold.

The facility's clear advantages are its economy of scale and its ability to produce a high-quality end product for the recycled materials market. Towns know where to bring their waste, and they quickly become familiar with the requirements for initial sorting during the collection and transfer of the waste.

Currently, there is only one MRF in Massachusetts, built with state funds at a cost of $5 million and operated by the state Department of Environmental Protection. Several more are on the drawing board. After its first year in operation, the Springfield MRF was declared a success. It took in 42,886 tons of materials from 91 towns representing a population of 750,000. Depending on the community, the process has diverted 22% to 50% of the waste stream from landfills and incinerators and has

saved the towns millions of dollars in disposal costs. The facility has attracted recycled wastes from Connecticut and New York while those states are setting up their own regional recycling centers.

Nevertheless, some unanticipated problems appeared at the Springfield facility during its first year of operation. Several towns reduced their trash so much that they wound up not fulfilling their contractual obligations to a regional combustion facility. Also, landfill revenues declined, forcing landfill operators to lower their fees in order to attract more haulers. These appear to be temporary problems that will be worked out in time, however. In theory, landfills will operate mainly as recipients of combustion ash and materials that can't be either recycled or burned.

Very likely, the future of regional recycling will be in the private sector. In 1997, 380 MRFs were operating in the United States, some with the high technology of magnetic pulleys, optical sensors, and air sorters and many with simple manual sorting. These facilities have a daily capacity of 32,000 tons. All indications point to such facilities as increasing in the future.

---

and increase their recycling from 32 to 59%. As with the WasteWise program, EPA's role is to be a facilitator, not a regulator.

Another policy for bringing about waste reduction would be to establish a program of *extended product responsibility* (EPR), a concept that involves assigning some responsibility for reducing the environmental impact of a product at each stage of its "life cycle," especially the end. For example, Hewlett-Packard and Xerox make it easy for customers to return spent copier cartridges, and they then recycle the components of the cartridges. EPR originated in Western Europe; Germany, the Netherlands, and Sweden have well-developed regulations that require companies to take back many items, that promote the reuse of products, and that promote the manufacture of more durable products. Again, the EPA is active in providing information to manufacturers and purchasing departments to help them design and buy more environmentally sound products.

Waste Disposal.   No human society can avoid generating some waste. There will always be MSW, no matter how much we reduce, reuse, and recycle. Integrated waste man-

agement requires providing several options for waste disposal. Most experts see WTE combustion facilities holding their own, a decrease in the percentage of waste going to landfills, and a continued rise in recycling.

It will be important to break the gridlock at local and state levels on the siting of new landfills and WTE combustion facilities. Landfills will still be needed, although they should last longer than in the past. Policymakers have long known about the landfill shortage, but have opted for short-term solutions with the lowest political cost. One result of this choice is the long-distance hauling of MSW by truck and rail described earlier. If regions and municipalities are required to handle their own trash locally, as some states do, they will find places to site landfills, regardless of NIMBY considerations. Of course, new landfills must use the best technologies: proper lining, leachate collection and treatment, groundwater monitoring, biogas collecting, and final capping. If these are employed, a landfill will not be a health hazard. Also, sites should be selected with a view to some future use that is attractive.

With the imminent closing of New York City's huge Fresh Kill landfill, the city is looking to other states for fu-

ETHICS

"AFFLUENZA": DO YOU HAVE IT?

First, take this sample quiz to see if you have the bug:

   1. I'm willing to pay more for a T-shirt if it has a cool corporate logo on it.

   2. I'm willing to work at a job I hate so I can buy lots of stuff.

   3. I usually make just the minimum payment on my credit cards.

   4. When I'm feeling blue, I like to go shopping and treat myself.

   5. I spend much more time shopping each month than I do being involved in my community.

   6. I'd rather be shopping right now.

   7. I'm running out of room to store my stuff.

Give yourself two points for true and one point for false. If you scored 10 or above, you may have a full-blown case of "affluenza."

Just what is "affluenza"? Jessie H. O'Neill, who coined the term, defines it as a "dysfunctional relationship with wealth or money." Symptoms of "affluenza" include a love of shopping, a glut of stuff in your home or dorm that you really don't need or use, and a dissatisfaction with what you have. It can strike anyone, regardless of economic status. According to two PBS programs produced about it, "Affluenza" and "Escape from Affluenza," "Affluenza" is a plague of materialism and overconsumption that is so pervasive that it actually characterizes our modern society.

Recall that the United States leads the world in per capita waste generation. This is a symptom of a societywide problem, according to O'Neill. It begins at an early age, as children are bombarded by TV commercials urging them to get the latest toys. Peer pressure makes it worse, with children sometimes "forced" to wear only the approved apparel. Every fad that comes along must be accommodated. Eventually, the conditioned children grow up, carry credit cards, and drive, and then "affluenza" takes on more serious consequences. Adults caught by the disease acquire so much that they have to rent a self-storage bin to hold things they can't part with. Bankruptcy and credit card overloads are commonplace, a consequence of people's inability to control their spending.

How do you escape from "affluenza"? There are several steps you can take. First, admit that you have a problem. Then, begin to take small withdrawal steps. Before you buy something, ask yourself, Do I need it? Could I borrow it from a friend or relative? How many hours do I have to work to pay for it? Another suggested step is to avoid those recreational shopping trips to the mall. Take a walk or play ball with some kids instead. Become an advertising critic. Make a budget. These are a few of the many possible pathways for escape from "affluenza." As you do these things, you may be amazed at how challenging and rewarding it can be to live more simply—and, it might be said, more sustainably.

---

ture disposal of its waste. Yet most citizens in the other states object to being New York's dumping ground, and there is the deeper stewardship issue of local responsibility for local wastes and the inefficiency of long-distance waste hauling. It will take an act of Congress to give states or local jurisdictions the right to ban imports of waste, because of the interstate commerce clause in the Constitution. Numerous bills have been introduced and considered in recent years, but none have made it into law yet. The simplest approach would be a law giving blanket authority to state and local governments to impose restrictions on interstate commerce in waste.

Another policy goal should be to encourage more WTE combustion of MSW. Again, the technologies employed should be the best available, and once more, if these are used, there should be no significant threat to human health. This option seems to be the best way to deal with mixed waste that is nonrecyclable. Much of the energy content of the waste is converted to electricity, metals are recovered, and the waste is greatly reduced in volume.

Recycling and Reuse. Recycling is certainly the wave of the future; it should not, however, be pursued in lieu of waste reduction and reuse. The move toward more durable goods is an overlooked and underutilized option. In fact, waste reduction remains the most environmentally sound and least costly goal for MSW management: Wastes that are never generated do not need to be managed.

Many states and municipalities set ambitious recycling goals in the 1980s and early 1990s. Some set the bar quite high; California, for example, mandated a 50% recycling rate by 2000, with heavy fines for cities that failed to meet the goal. Many California cities have reached and surpassed the goal, but others are lagging behind. The fines have not been levied, however, and extensions are being granted to 2006. Massachusetts opted for a statewide goal of 46% by 2000 and had achieved a 36% rate in 1998. A new solid-waste master plan has raised the goal to 60% of MSW reduction through source reduction and recycling combined, to be achieved by 2010. One objective of the plan is to completely phase out the export of solid waste to other states. EPA's target goal for the country is 35% by 2005, which seems readily achievable.

Banning the disposal of recyclable items in landfills and combustion facilities makes very good sense. Many states have incorporated this regulation into their management program; for example, Massachusetts phased out yard waste in 1991, metals and glass in 1992, and recyclable papers and plastic in 1994. Landfill- and combustion-facility operators are supposed to conduct random truck searches and are authorized to turn back trucks with significant amounts of banned items.

Enacting a national bottle-deposit law would be a giant stride toward the reuse and recycling of beverage containers and would also greatly reduce roadside litter in states lacking bottle laws.

Finally, closing the "recycling loop" remains a significant action to be taken by governments to encourage recycling. A number of states have opted for one or more of the following approaches: (1) minimum postconsumer levels of recycled content for newsprint and glass containers; (2) requirements stating that purchases of certain goods include recycled products even if they are more expensive than "virgin" products; (3) requirements that all packaging be reusable or be made (at least partly) of recycled materials; (4) tax credits or incentives that encourage the use of recycled materials in manufacturing; (5) assistance in the development of recycling markets.

ENVIRONMENT ON THE WEB

DIFFICULTIES OF WASTE STREAM COMPOSITION

Like every other species, humans generate waste. On average, people in the United States and Canada generate 3 to 6 pounds of solid waste per person per day—an enormous difference from the 20 to 30 pounds per person *per year* that is typical of less developed countries. One of the most challenging aspects of managing all this waste is the composition of the municipal solid waste stream.

Typically, municipal waste is made up of about 40 to 50% paper, wood, and plastics (combustibles), about 10% metal and glass (non-combustibles), about 20 to 40% food wastes, yard wastes, and other compostable materials, and small proportions of inert solids, ash, sewage sludge, and bulky items such as furniture and appliances. Composition of a waste stream, however, can vary from community to community. In some places, the proportion of construction wastes—scrap wood, wallboard, plaster, concrete, and similar materials—may be high, while in other communities, the waste stream contains a high proportion of yard and food wastes, which are easily composted by homeowners. Regardless, most communities must contend with highly variable waste streams—for instance, very high compostable materials during some times of year (such as fall), or certain days of the week, but lower at other times. This uncertainty makes it difficult for municipalities to optimize treatment and disposal methods, even within the same region.

An integrated waste management strategy therefore needs to accommodate the full range of materials that can be expected in the waste stream, including provisions for waste collection, transportation, and disposal. Increasingly, municipalities are tackling problems of variable waste streams and rising waste management costs by emphasizing alternatives that encourage residents to reduce, reuse, and recycle.

Web Explorations

The Environment on the Web activity for this essay discusses the "pay-as-you-throw" (PAYT) policy that requires residents to pay a fee for each unit of garbage they generate. Go to the Environment on the Web activity (select Chapter 19 at **http://www.prenhall.com/wright**) and learn for yourself:

1. about communities that use pay-as-you-throw policies;
2. how communities make the decision to adopt user-pay waste management policies; and
3. about illegal diversion and enforcement of pay-as-you-throw systems.

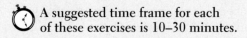 **A suggested time frame for each of these exercises is 10–30 minutes.**

REVIEW QUESTIONS

1. List the major components of municipal solid waste (MSW).
2. Trace the historical development of refuse disposal. What method of disposal is now most common in the United States?
3. What are the major costs and limitations of placing waste in landfills?
4. Outline the EPA's latest regulations for the construction of new landfills.
5. Why are landfills not a sustainable option for solid-waste disposal?
6. What are the advantages and disadvantages of combustion?
7. How can the volume of waste in the United States be reduced?
8. Define primary and secondary recycling.
9. Discuss the attributes of successful recycling programs.
10. Which waste materials may be composted?
11. What laws has the federal government adopted to control solid-waste disposal?
12. What is integrated waste management? Name some options for solid-waste management.

THINKING ENVIRONMENTALLY

1. Compile a list of all the plastic items you used and threw away this week. How can you reduce the list?

2. How and where does your school dispose of solid waste? Is a recycling program in place? How well does it work?

3. Suppose your town planned to build a combustion facility or landfill near your home. Outline your concerns, and explain your decision for or against the site.

4. Does your state have a bottle bill? If not, what has prevented the bill from being adopted?

5. Read the Ethics box on "affluenza." Take the sample quiz. If you score 10 or above, check out the "affluenza" Web site (http://www.affluenza.com/) and get help!

WEB REFERENCES

On-line resources for this chapter are on the World Wide Web at: **http://www.prenhall.com/wright**. (Click on Chapter 19 within the Chapter Selector.)

Hazardous Chemicals: Pollution and Prevention

Key Issues and Questions

1. Toxicology is the study of the harmful effects of chemicals on human health. How do toxicologists do their work? How are the results of their work expressed and made available to health practitioners?

2. Hazardous materials (HAZMATs) are chemicals that present a hazard or risk. What does the Toxics Release Inventory tell us about the extent of the release of hazardous chemicals? Why are heavy metals and synthetic organic compounds the most notorious hazardous chemicals?

3. Before 1970, chemical wastes were disposed of indiscriminately. What were the consequences of the unregulated disposal of such wastes?

4. One of the most daunting tasks our society faces is cleaning up the thousands of existing toxic waste sites. How does the Superfund Act of 1980 address this task?

5. A number of laws have been passed to protect the public and the environment from toxic wastes. What roles do the Clean Air Act, the Clean Water Act, RCRA, EPCRA, and TSCA play in accomplishing this objective?

6. A future objective is pollution avoidance rather than pollution control. How has the EPA changed its strategies in order to bring about more cooperation from states and industries to accomplish this objective?

"There are strange things done in the midnight sun by the men who moil for gold;
The arctic trails have their secret tales that would make your blood run cold.
The Northern Lights have seen queer sights, but the queerest they ever did see
Was that night on the marge of Lake LeBarge I cremated Sam McGee."

So begins the famous poem by Robert Service, "The Cremation of Sam McGee." In the poem, Lake LeBarge, in the Canadian Yukon Territory, eventually received the cremated ashes of Sam McGee. Unfortunately, the lake has been receiving a much different burden in recent years, to the point where its fish have become hazardous to eat because of their DDT content. Much to the surprise of research scientists, fish, birds, and mammals all over the Arctic are showing elevated body burdens of a number of **persistent organic pollutants** (POPs)—in particular, DDT, toxaphene, chlordane, PCBs, and dioxins.

And because the Inuit people of the far north depend on these animals for food, they, too, are carrying high loads of POPs—indeed, some of the highest in the world.

The real mystery is how these chemicals—all of which are toxic—get to such remote and pristine environments. No one is spraying pesticides there, and there are no industries to produce PCBs and dioxins. The key to the presence of such chemicals can be found in several factors that characterize POPs: persistence, bioaccumulation, and the potential for long-range transport. Added to these factors is the unique climate of the Arctic, where the extreme cold promotes a process called "cold condensation." POPs are semivolatile chemicals carried in the global air patterns to the Arctic, where they condense on the snowpack and are then washed into the water with the spring thaw. Once in the lakes and coastal waters, the chemicals are picked up by plankton and passed up the food chain through bioaccumulation and biomagnification. (See Fig. 17–8.) Dioxins in Nunavut territory are transferred to the Inuit via deposition on lichens and mosses to caribou, a major food source, as

◀ *A Nunavut hazard.* This Inuit mother is scraping the fat from a sealskin on Baffin Island, Nunavut. Seals and other animals used by the Inuit for food harbor high levels of toxic pollutants.

well as through the aquatic food chain. A recent study traced the dioxins in air masses thousands of miles back from Nunavut to a relatively small cluster of incinerators and factories in the U.S. Midwest.

The potential health effects of the chemical load carried by the Arctic natives range from immune-system disorders to disruption of their hormone systems, and, over time, to cancer. (All five of the POPs mentioned are on the National Toxicology Program's list of carcinogens.) Stories like that of the Inuit, and others that can only be called disasters, have convinced the public that significant dangers are associated with the manufacture, use, and disposal of many chemicals. But very few would advocate giving up the advantages of the innumerable products that derive from our modern chemical industry. Better to learn to handle and

dispose of chemicals in ways that minimize the risks. Thus, over the past 25 years, regulations surrounding the production, transport, use, and disposal of chemicals have mushroomed. From having almost no controls at all, the chemical industry is now among the most thoroughly regulated of all industries, and everyone using or handling chemicals that are deemed hazardous is affected by the regulations in one way or another.

This chapter focuses on hazardous chemicals—their nature and how they are investigated. An overview is presented of laws and regulations that have come into play to protect human and environmental health, and the strengths and weaknesses of those measures are examined. Finally, future directions aimed at reducing chemical hazards at less cost are examined.

20.1 Toxicology and Chemical Hazards

Toxicology is the study of the harmful effects of chemicals on human and environmental health. A toxicologist might, for example, be involved with evaluating the potential for *phenolphthalein* (a common chemical laboratory reagent that is also used in over-the-counter laxatives) to cause health problems. The toxicologist might conduct a study investigating the **acute** toxicity effects that might occur upon ingestion of the chemical or upon its contact with the skin, as well as the effects of **chronic** exposure over a period of years and, finally, the **carcinogenic** potential of phenolphthalein. Indeed, such studies have recently shown that the substance causes tumors in mice, and because of this finding, the Food and Drug Administration (FDA) has begun to cancel all laxative uses of phenolphthalein, which is now listed as "reasonably anticipated to have human carcinogenic potential."

Data on toxic chemicals are made available to health practitioners and the public via a number of sources, including the National Toxicology Program (NTP) Chemical Repository and the National Institute of Environmental Health Sciences (NIEHS), which provides an annual updated *Report on Carcinogens*, with a complete list of carcinogenic agents. The EPA also makes available data on toxic chemicals in its Integrated Risk Information System (IRIS). All of these information sources are on the Internet; Environmental Defense uses these and other data sources in its "Scorecard," an exhaustive Web-based profile of 6,800 chemicals, including information on their manufacture and uses.

Dose Response and Threshold

In our discussion of risk assessment (Chapter 16), we introduced the concepts of *dose response* and *exposure*. In investi-

gating a suspect chemical, a toxicologist would conduct animal tests, investigate human involvement with the chemical, and present information linking the **dose** (the level of exposure multiplied by the length of time over which exposure occurs) with the **response** (some acute or chronic effect or the development of tumors). For example, studies of *phenolphthalein* toxicity indicate that the LDL_0 (lowest dose at which death occurred in animal testing) was 500 mg/kg. This is a low toxicity, so concern about the chemical would center on *chronic* or *carcinogenic* issues.

Human **exposure** to a hazard is a vital part of its risk characterization, and such exposure can come through the workplace, food, water, or the surrounding environment. For example, information in the NTP Chemical Repository gives the various uses of *phenolphthalein*, followed by precautions to take in handling it and the symptoms of exposure (there are many!). Getting an accurate evaluation of human exposure is often the most difficult area of risk assessment.

In the dose-response relationship, there may or may not be a *threshold*. For many substances, organisms are able to deal with certain levels without suffering ill effects. The level below which no ill effects are observed is called the **threshold level**. Above this level, the effect of a substance depends on both its concentration and the duration of exposure to it. Higher levels may be tolerated if the exposure time is short. Thus, the threshold level is high for short exposures, but gets lower as the exposure time increases (Fig. 20–1). This reminds us that it is not the absolute amount, but rather the *dose*, that is important.

Where carcinogens are concerned, the EPA generally takes a zero-dose, zero-response approach. That is, there is no evidence of a discrete threshold level for any carcinogenic chemicals. However, the lower the dose, the more likely it is that the response cannot be distinguished from the background level of cancers in a population. In such cases, the risk becomes very low and drops below the level for which regulatory action is needed.

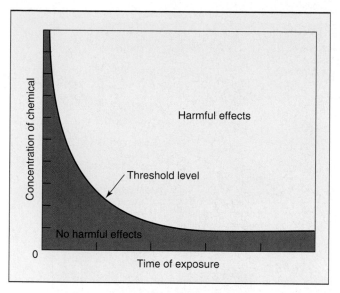

▲ FIGURE 20–1 *The threshold level.* The threshold level for harmful effects of toxic pollutants is lower as the exposure time increases. It is also different for different chemicals.

These are just some basics of toxicology. The field is well established and is the most important source of sound scientific information that leads to regulation by the FDA and EPA. The NTP was established in 1978 to provide this kind of information and has become the world's leader in assessing chemical toxicity and carcinogenicity. We turn our attention now to the chemicals themselves.

The Nature of Chemical Hazards: HAZMATs

A chemical that presents a certain hazard or risk is known as a **hazardous material (HAZMAT)**. The EPA categorizes substances on the basis of the following hazardous properties:

- **Ignitability**: substances that catch fire readily (e.g., gasoline and alcohol)
- **Corrosivity**: substances that corrode storage tanks and equipment (e.g., acids)
- **Reactivity**: substances that are chemically unstable and that may explode or create toxic fumes when mixed with water (e.g., explosives, elemental phosphorus (not phosphate), and concentrated sulfuric acid)
- **Toxicity**: substances that are injurious to health when they are ingested or inhaled (e.g., chlorine, ammonia, pesticides, and formaldehyde)

Radioactive materials, discussed in Chapter 14, are probably the most hazardous of all and are treated as an entirely separate category.

Sources of Chemicals Entering the Environment

To understand how HAZMATs enter our environment, we need to look at various aspects of how people in our society live and work. First, observe that the materials making up almost everything we use, from the shampoo and toothpaste in the morning to the TV set we watch in the evening, are products of chemical technology. Our use constitutes only one step in the **total product life cycle**, a term that considers all steps, from obtaining raw materials to final disposal of the product. Implicit in our use of hair spray, for example, is the fact that raw materials were obtained and various chemicals were produced to make both the spray and its container. Chemical wastes and by-products are inevitable in production processes. In addition, consider the risks of accidents or spills occurring in the manufacturing process and in the transportation of raw materials, the finished product, or wastes. Finally, ask yourself, What are the risks of breathing hair spray? What happens to the container you throw into the trash, which still holds some of the hair spray when the propellant is used up?

Multiply these steps by all the hundreds of thousands of products used by billions of people in homes, factories, and businesses, and you can appreciate the magnitude of the situation. More than 80,000 chemicals are registered for commercial use within the United States. At every stage—from mining raw materials through manufacturing, use, and final disposal—various chemical products and by-products enter the environment, with potential consequences for both human and environmental health (Fig. 20–2).

The use of many products—pesticides, fertilizers, and road salts, for example—involves their direct introduction into the environment. Or the intended use may leave a fraction of the material in the environment—the evaporation of solvents from paints and adhesives, for example. Then there are the product life cycles of materials that are used tangentially to the desired item. Consider lubricants, solvents, cleaning fluids, cooling fluids, and so on, with whatever contaminants they may contain. Likewise, there are the product life cycles of gasoline, coal, and other fuels that are consumed for energy. Again, in addition to the unavoidable wastes and pollutants produced, in every case there is the potential for accidental releases, ranging from minor leaks in storage tanks to supertanker wrecks such as the *Exxon Valdez*, which spilled 11 million gallons of crude oil into Prince William Sound in Alaska in 1989.

Toxics Release Inventory. Introductions of chemicals into the environment may occur in every sector, from major industrial plants to small shops and individual homes. Whereas single events involving large amounts of one chemical may constitute a disaster and make headlines, the total amounts entering the environment from millions of homes and businesses is far greater and presents much more of a health risk to society. Some idea of the quantities involved can be obtained from the *Toxics Release Inventory (TRI)*. The **Emergency Planning and Community Right-to-know Act (EPCRA) of 1986** requires industries to report the locations and quantities of toxic chemicals stored on each site to state and local governments and to report releases of toxic chemicals to the

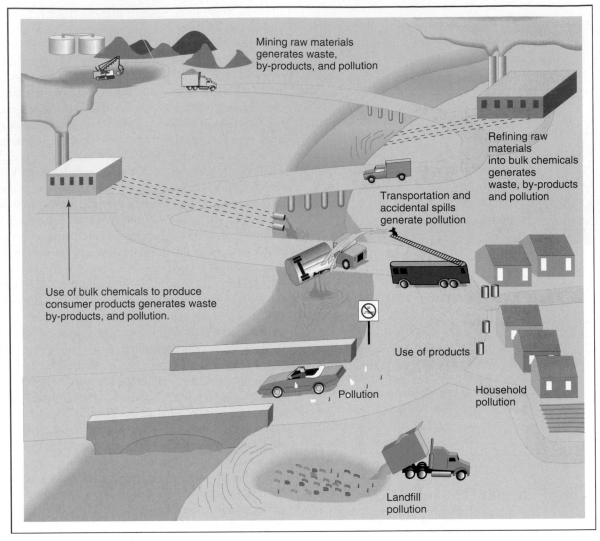

▲ FIGURE 20–2   *Total product life cycle.*   The life cycle of a product begins with the obtaining of raw materials and ends when the used-up product is discarded. At each step—or in transport between steps—wastes, by-products, or the product itself may enter the environment, causing pollution and creating various risks to human and environmental health.

environment. [The TRI does not cover small businesses, like dry-cleaning establishments and gas stations, or household hazardous wastes (HHW).] The TRI is disseminated on the Internet and provides an annual record of releases of more than 600 designated chemicals. For 1998, the TRI reveals the following information:

- Releases of toxic chemicals to the air: 628,000 tons.
- Releases of toxic chemicals to the water: 112,000 tons.
- Total environmental releases: 1,247,000 tons.

Add to this the estimated 1.6 million tons each year of HHW and an unknown quantity coming from small businesses, and you have some idea of the dimensions of the problem. The good news is that over the 10 years since the TRI has been in effect, the quantities of virtually all categories of toxic waste keep going down; the total release declined by 45% during that time.

The Threat from Toxic Chemicals

All toxic chemicals, by definition, are hazards that pose a potential risk to humans. (See Chapter 16.) Fortunately, a large portion of the chemicals introduced into the environment are gradually broken down and assimilated by natural processes. Therefore, once these chemicals are diluted sufficiently, they pose no long-term human or environmental risk, even though they may be highly toxic in acute doses (high-level, short-term exposures).

Two major classes of chemicals do not readily degrade in the environment, however: (1) *heavy metals and their compounds* and (2) *synthetic organics.* Again, if sufficiently diluted in air or water, most of these compounds do not pose a hazard. Yet there are some notable exceptions.

Heavy Metals.   The most dangerous heavy metals are lead, mercury, arsenic, cadmium, tin, chromium, zinc, and copper. These metals are used widely in industry, particularly in metalworking or metal-plating shops and in such products as batteries and electronics. They are also used in certain pesticides and medicines. In addition, because compounds of heavy metals can have brilliant colors, they are used in paint pigments, glazes, inks, and dyes. Thus, heavy metals may enter the environment wherever any of these products are produced, used, and ultimately discarded.

Heavy metals are extremely toxic because, as ions or in certain compounds, they are soluble in water and may be readily absorbed into the body, where they tend to combine with and inhibit the functioning of particular vital enzymes. Even very small amounts can have severe physiological or neurological consequences. The mental retardation caused by lead poisoning and the insanity and crippling birth defects caused by mercury are particularly well-known examples.

Organic Compounds. Petroleum-derived and *synthetic* organic compounds are the chemical basis for all plastics, synthetic fibers, synthetic rubber, modern paintlike coatings, solvents, pesticides, wood preservatives, and hundreds of other products. Because of their chemical structure, many synthetic organics are resistant to biodegradation. Ironically, this is an important part of what makes many such compounds useful: We wouldn't want fungi and bacteria attacking and rotting our tires, and paints and wood preservatives function only insofar as they are both non-biodegradable and toxic to decomposer organisms.

These compounds are toxic because they are often readily absorbed into the body. There they interact with particular enzymes, but their nonbiodegradability prevents them from being broken down or processed further. When a person ingests a sufficiently high dose, the effect may be acute poisoning and death. With low doses over extended periods, however, the effects are insidious and can be mutagenic (mutation causing), carcinogenic (cancer causing), or teratogenic (birth defect causing). Synthetic organic compounds may cause serious liver and kidney dysfunction, sterility, and numerous other physiological and neurological problems.

A particularly troublesome class of synthetic organics is the **halogenated hydrocarbons**, organic compounds in which one or more of the hydrogen atoms have been replaced by atoms of chlorine, bromine, fluorine, or iodine. These four elements are classed as *halogens*, hence the name *halogenated hydrocarbons* (Fig. 20–3). Of the halogenated hydrocarbons, the **chlorinated hydrocarbons** (also called **organic chlorides**) are by far the most common. Organic chlorides are widely used in plastics (polyvinyl chloride), pesticides (DDT, Kepone, and Mirex), solvents (carbon tetrachlorophenol and tetrachloroethylene), electrical insulation (polychlorinated biphenyls), and many other products. Most of the so-called "dirty dozen" POPs (Table 20–1) are halogenated hydrocarbons. All are toxic to varying extents, and most are known animal carcinogens. Many are also suspected endocrine disruptors at very low levels. Under the auspices of the UNEP, these 12 POPs are now the subject of consideration for a total global ban via an international treaty.

Let us consider one of these compounds, tetrachloroethylene (Fig. 20–3), in a simple case study. This substance, also called perchloroethylene, or PERC, is colorless and nonflammable and is the major substance in dry-cleaning fluid. It is an effective solvent and finds uses in all kinds of industrial cleaning operations. It can also be found in shoe polish and typewriter correction fluid. PERC evaporates readily when exposed to air, but in the soil it can enter groundwater readily because it does not bind to soil

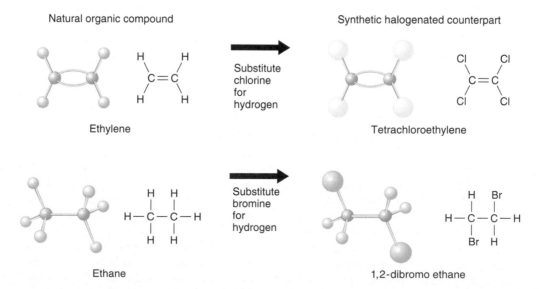

▲ FIGURE 20–3   *Halogenated hydrocarbons.* These are organic (carbon-based) compounds in which one or more hydrogen atoms has been replaced by halogen atoms (chlorine, fluorine, bromine, or iodine). Such compounds are particularly hazardous to health because they are nonbiodegradable and they tend to bioaccumulate. Shown here are tetrachloroethylene and 1,2-dibromo ethane.

TABLE 20–1   The "Dirty Dozen" Persistent Organic Pollutants (POPs)

Chemical Substance	Designed Use	Major Concerns
Aldrin	Pesticide to control soil insects and to protect wooden structures from termites	Toxic to humans, may be carcinogenic
Chlordane	Broad-spectrum insecticide to protect crops	Biomagnification in food webs
DDT	Widely used insecticide, malaria control	Biomagnification in food webs
Dieldrin	Termite control, crop-pest control	Toxic, biomagnification in food webs, high persistence
Endrin	Insecticide and rodenticide	Toxic, especially in aquatic systems
Heptachlor	General insecticide	Toxic, carcinogenic
Hexachlorobenzene	Fungicide	Toxic, carcinogenic
Mirex™	Insecticide against ants	High aquatic animal toxicity, carcinogenic
Toxaphene™	General insecticide	High aquatic animal toxicity, carcinogenic
PCBs	Variety of industrial uses, especially in transformers and capacitors	Toxic, teratogenic, carcinogenic
Dioxins	No known use; by-products of incineration and paper bleaching	Toxic, carcinogenic, reproductive system effects
Furans	No known use; by-products of incineration, PCB production	Toxic, especially in aquatic systems

Source: UNEP Persistent Organic Pollutants from United Nations Environment Programme–Chemicals Division. (http://irptc.unep.ch/pops/)

particles. Human exposure can occur in the workplace, especially in connection with dry-cleaning operations. It can also occur if people use products containing PERC, like dry-cleaned garments. PERC enters the body most readily when breathed in with contaminated air. Breathing PERC for short periods of time can bring on dizziness, fatigue, incoordination, and unconsciousness. Over longer periods of time, PERC can cause liver and kidney damage.

Laboratory studies also show PERC to be carcinogenic, and it is listed in NTP's *Report on Carcinogens* as "reasonably anticipated to be [a] human carcinogen." PERC is not known to cause any environmental harm at levels normally found in the environment, except that it does add to the burden of volatile organic chemicals causing photochemical smog. PERC is produced in large amounts in the United States, and according to the EPA's Toxic Release Inventory, over 14 million pounds of PERC were released to the environment in 1998 from industrial sources, mostly by evaporation.

An especially troublesome issue emerged as officials implemented Clean Air Act regulations to create a cleaner-burning gasoline. Methyl tertiary butyl ether (MTBE) is an oxygenate that is now added to a third of the gasoline sold in the United States. It is also a suspected carcinogen and is being found in increasing levels in wells and surface waters throughout the country. Because it has been added to gasoline, it shows up in surface waters wherever two-stroke engines are used for watercraft and in groundwater wherever underground gasoline storage tanks develop leaks. MTBE imparts a nasty odor to water at very low concentrations. The EPA has recently moved to ban the chemical (it will take several years to do so, however), which is good news for Midwestern corn growers, as ethyl alcohol is the likely substitute for MTBE as an oxygenate. (See Chapter 15.)

Involvement with Food Chains

The trait that makes heavy metals and nonbiodegradable synthetic organics particularly hazardous is their tendency to accumulate in organisms. We discussed the phenomena of bioaccumulation and biomagnification in connection with DDT in Chapter 17.

A tragic episode in the early 1970s, known as Minamata disease, revealed the potential for biomagnification of mercury and other heavy metals. The disease is named for a small fishing village in Japan where the episode occurred. In the mid-1950s, cats in Minamata began to show spastic movements, followed by partial paralysis, coma, and death. At first, this was thought to be a syndrome peculiar to felines, and little attention was paid to it. However, when the same symptoms began to occur in people, concern escalated quickly. Additional symptoms, such as mental retardation, insanity, and birth defects were also observed. Scientists and health experts eventually diagnosed the cause as acute mercury poisoning.

A chemical company near the village was discharging wastes containing mercury into a river that flowed into the bay where the Minamata villagers fished. The mercury, which settled with detritus, was first absorbed and bioaccumulated by bacteria and then biomagnified as it passed up the food chain through fish to cats or to humans. Cats had suffered first and most severely because they fed almost exclusively on the remains of fish. By the time the situation was brought under control, some 50 persons had died and 150 had suffered serious bone and nerve damage. Even now, the tragedy lives on in the crippled bodies and retarded minds of some Minamata descendants.

20.2 A History of Mismanagement

In the past, chemical wastes of all kinds were disposed of as expediently as possible. From the dawn of the Industrial Age to the relatively recent past, it was common practice to exhaust all combustion fumes up smokestacks, vent all evaporating materials and solvents into the air, and flush all waste liquids and contaminated wash water into sewer systems or directly into natural waterways. Many human health problems occurred, but they either were not recognized as being caused by the pollution or were accepted as the "price of progress." Indeed, much of our understanding regarding human health effects of hazardous materials is derived from those uncontrolled exposures. For example, the expression "mad as a hatter" comes from the fact that people who made hats in the 1800s frequently became insane. The insanity, it was later found, was caused by poisoning from the mercury used in the production process.

In the 1950s, as production expanded and synthetic organics came into widespread use in the developed countries, many streams and rivers essentially became open chemical sewers, as well as sewers for human waste. These waters were not only devoid of life; they were themselves hazardous. For example, in the 1960s, the Cuyahoga River, which flows through Cleveland, Ohio, carried so much flammable material that it actually caught fire and destroyed seven bridges before the fire burned itself out (Fig. 20–4). Worsening pollution (both chemical and sewage)

and increasing recognition of adverse health effects finally created a degree of public outrage that pushed Congress to pass the Clean Air Act of 1970 and the Clean Water Act of 1972. These acts set standards for allowable emissions into air and water and timetables for reaching those standards.

The Clean Air and Clean Water Acts and their subsequent amendments remain cornerstones of environmental legislation. However, the passage of these acts in the early 1970s left an enormous loophole. If you can't vent wastes into the atmosphere or flush them into waterways, what do you do with them? Industry turned to *land disposal*, which was essentially unregulated at the time, as an expedient alternative. Indiscriminate air and water disposal became indiscriminate land disposal. Thus, in retrospect, we see that the Clean Air and Clean Water Acts, for all their benefits in improving air and water quality, also succeeded in transferring pollutants from one part of the environment to another.

Methods of Land Disposal

In the early 1970s, there were three primary land-disposal methods: (1) deep-well injection, (2) surface impoundments, and (3) landfills. With the conscientious implementation of safeguards, each of these methods has some merit. Without adequate regulations or enforcement, however, contamination of groundwater is virtually inevitable.

Deep-Well Injection. Deep-well injection involves drilling a "well" into dry, porous material below groundwater (Fig. 20–5). In theory, hazardous waste liquids pumped into the well soak into the porous material and remain isolated indefinitely. In practice, however, it is almost impossible to

◄ FIGURE 20–4 *Cuyahoga River on fire.* Prior to laws and regulations curtailing pollution, all manner of wastes were indiscriminately discharged. In the 1960s, the Cuyahoga River actually caught fire. Incidents such as this contributed to public outrage that led to the passage of the Clean Water Act of 1972.

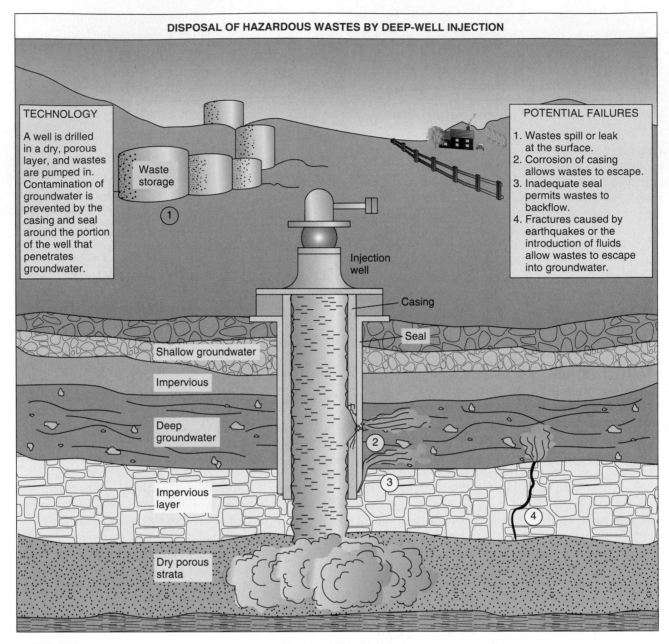

DISPOSAL OF HAZARDOUS WASTES BY DEEP-WELL INJECTION

TECHNOLOGY

A well is drilled in a dry, porous layer, and wastes are pumped in. Contamination of groundwater is prevented by the casing and seal around the portion of the well that penetrates groundwater.

Waste storage
①

POTENTIAL FAILURES

1. Wastes spill or leak at the surface.
2. Corrosion of casing allows wastes to escape.
3. Inadequate seal permits wastes to backflow.
4. Fractures caused by earthquakes or the introduction of fluids allow wastes to escape into groundwater.

Injection well

Casing

Seal

Shallow groundwater

Impervious

Deep groundwater

②

③

④

Impervious layer

Dry porous strata

▲ **FIGURE 20–5** *Deep-well injection.* This technique is used for the disposal of large amounts of liquid wastes. The concept is that toxic wastes may be drained into dry, porous strata below ground, where they may reside harmlessly "forever." However, as the figure shows, failures can occur and allow the liquid wastes to contaminate groundwater. (Adapted with permission from Environmental Action, 1525 New Hampshire Ave., N.W., Washington, D.C. 20036.)

guarantee that injected wastes will never escape and contaminate groundwater. For this reason, deep-well injection is a less used disposal source today. A number of commercial deep-well facilities accept wastes like chemical cleaning solutions, saline wastes, metals, cyanides, and corrosive materials. The total amount of deep-well injection has declined over the years, from 685 million tons in 1988 to 113 million tons in 1998.

Surface Impoundments. Surface impoundments are simple excavated depressions ("ponds") into which liquid wastes are drained and held. They were the least expensive

and hence most widely used way to dispose of large amounts of water carrying relatively small amounts of chemical wastes. As waste is discharged into the pond, solid wastes settle and accumulate while water evaporates (Fig. 20–6). If the bottom of the pond is well sealed, and if evaporation equals input, impoundments may receive wastes indefinitely. However, inadequate seals may allow wastes to percolate into groundwater, exceptional storms may cause overflows, and volatile materials can evaporate into the atmosphere, adding to air pollution problems. Where surface impoundments are used today, they are seen as strictly temporary sites for the storage of hazardous wastes.

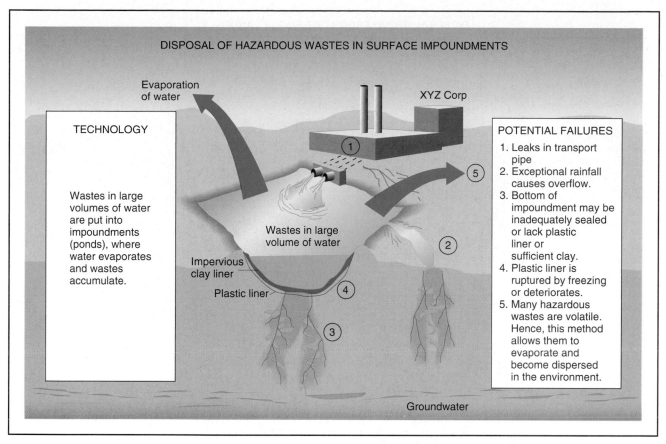

DISPOSAL OF HAZARDOUS WASTES IN SURFACE IMPOUNDMENTS

Evaporation of water

XYZ Corp

TECHNOLOGY

Wastes in large volumes of water are put into impoundments (ponds), where water evaporates and wastes accumulate.

Wastes in large volume of water

Impervious clay liner

Plastic liner

Groundwater

POTENTIAL FAILURES

1. Leaks in transport pipe
2. Exceptional rainfall causes overflow.
3. Bottom of impoundment may be inadequately sealed or lack plastic liner or sufficient clay.
4. Plastic liner is ruptured by freezing or deteriorates.
5. Many hazardous wastes are volatile. Hence, this method allows them to evaporate and become dispersed in the environment.

▲ **FIGURE 20–6** *Surface impoundment.* This is an inexpensive technique for disposing of large amounts of lightly contaminated liquid wastes. The supposition is that only water leaves the impoundment, by evaporation, while wastes remain and accumulate in the impoundment indefinitely. In reality, this method is subject to failure at a number of points and is limited today to use as a short-term measure.

Landfills. When hazardous wastes are in a concentrated liquid or solid form, they are commonly put into drums and buried in landfills. If a landfill is properly lined, supplied with a system to remove leachate, provided with monitoring wells, and properly capped, it may be reasonably safe and is referred to as a **secure landfill** (Fig. 20–7). The various barriers are subject to damage and deterioration, however, requiring adequate surveillance and monitoring systems to prevent leakage.

Because early land disposal was not regulated, in numerous instances not even the most rudimentary precautions were taken. Indeed, in many cases, deep wells were injecting wastes directly into groundwater, abandoned quarries were sometimes used as landfills with no additional precautions being taken, and surface impoundments frequently had no seals or liners whatsoever. Even worse, considerable amounts of waste failed to get to any disposal facility at all.

Midnight Dumping and Orphan Sites. The need for alternative methods to dispose of waste chemicals created an opportunity for a new enterprise: waste disposal. Many reputable businesses entered the field, but—again in the absence of regulations—there were also disreputable operators. As stacks of drums filled with hazardous wastes "mysteriously" appeared in abandoned warehouses, vacant lots, or municipal landfills, it became clear that some operators simply were pocketing the disposal fee and then unloading the wastes in any available location, frequently under cover of darkness, an activity termed **midnight dumping** (Fig. 20–8). Authorities trying to locate the individuals responsible would find that they had gone out of business and were nowhere to be found.

Some companies or individuals simply stored wastes on their own properties and then went out of business, abandoning the property and the wastes. These locations became known as **orphan sites**—hazardous-waste sites without a responsible party to clean them up. As drums containing hazardous chemicals corroded and leaked, there was great danger of reactive chemicals combining and causing explosions and fires. One of the most famous abandoned sites is the "Valley of the Drums," in Kentucky (Fig. 20–9).

Scope of the Mismanagement Problem

The mounting problem of unregulated land disposal of hazardous wastes was brought vividly to public attention by the episode at Love Canal, near Niagara Falls, New York. The area was occupied by a school and a number of houses, all of which were perched on top of a chemical waste dump that had been filled over and developed. The surface of the dump began to collapse, exposing barrels of chemical

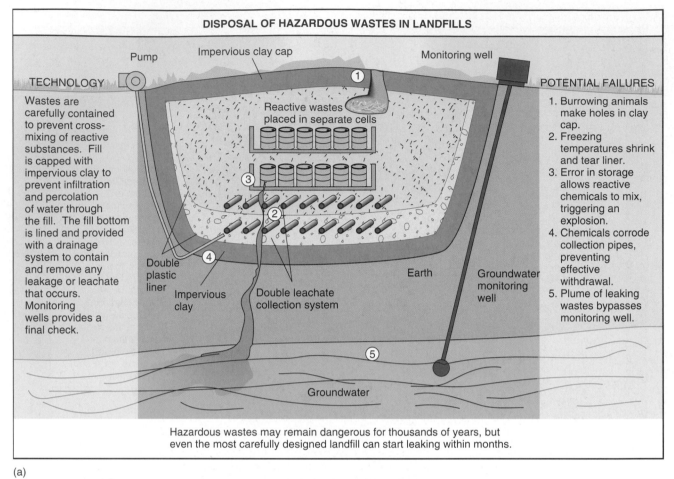

DISPOSAL OF HAZARDOUS WASTES IN LANDFILLS

Pump
Impervious clay cap
Monitoring well

TECHNOLOGY

Wastes are carefully contained to prevent cross-mixing of reactive substances. Fill is capped with impervious clay to prevent infiltration and percolation of water through the fill. The fill bottom is lined and provided with a drainage system to contain and remove any leakage or leachate that occurs. Monitoring wells provides a final check.

Reactive wastes placed in separate cells

Double plastic liner
Impervious clay
Double leachate collection system

Earth
Groundwater monitoring well

POTENTIAL FAILURES

1. Burrowing animals make holes in clay cap.
2. Freezing temperatures shrink and tear liner.
3. Error in storage allows reactive chemicals to mix, triggering an explosion.
4. Chemicals corrode collection pipes, preventing effective withdrawal.
5. Plume of leaking wastes bypasses monitoring well.

Groundwater

Hazardous wastes may remain dangerous for thousands of years, but even the most carefully designed landfill can start leaking within months.

(a)

(b)

▲ FIGURE 20–7   *Secure landfill.*   Landfills are a technique widely used for disposing of concentrated chemical wastes. (a) Precautionary measures to make this method safe are listed. Before the 1980s, these measures frequently were not taken. Although they are taken today, potentials for failure remain. (b) An operating hazardous-waste landfill in Alabama. (Adapted from an illustration by Rick Farrell. All rights reserved.)

wastes. Fumes and chemicals began seeping into cellars. Ominously, people began reporting serious health problems, including birth defects and miscarriages. An aroused neighborhood, led by activist Lois Gibbs, began demanding

that the state do something about the problem. Following confirmation of the contamination by toxic chemicals, in 1978, President Carter signed an emergency declaration to relocate hundreds of residents. The state and federal gov-

▲ **FIGURE 20–8** *Midnight dumping.* Hazardous wastes were often left on remote or unoccupied properties by unscrupulous haulers.

ernments closed the school and demolished a number of the homes nearest the site (Fig. 20–10). In all, more than 800 families moved out of the area because of the fear of damage to their health from the toxic chemicals.

The absence of a public policy for dealing with the disposal of hazardous chemicals contributed to the situation at Love Canal. Hooker Chemical and Plastics Company had purchased an abandoned canal near Niagara Falls in 1942 and proceeded to fill it with an estimated 17,000 tons of hazardous wastes. Hooker covered the canal over with a clay cap and sold it to the school board for a small sum, reportedly after warning the board that there were chemicals buried on the property. Subsequent construction penetrated the clay cap, and rain seeped in and leached chemicals

in all directions. Hooker Chemical's parent company, Occidental Petroleum, eventually spent more than $233 million on the cleanup and subsequent lawsuits.

As both government and independent researchers began surveying the extent of the problem, bad disposal practices were found to be rampant. The World Resources Institute estimated that in the United States in the early 1980s, there existed 75,000 active industrial landfill sites, along with 180,000 surface impoundments and 200 other special facilities that were or could be sources of groundwater contamination. In most cases, the contaminated area was relatively small—200 acres (80 ha) or less—but in total, the problem was immense and affected every state in the country. As studies and tests proceeded, thousands of individual wells and some major municipal wells were closed because they were contaminated with toxic chemicals.

Many cases of private-well contamination caused by careless disposal were discovered only after people experienced "unexplainable" illnesses over prolonged periods. While these incidents did not receive the media attention of Love Canal, they were nevertheless devastating to the people involved. Even more important than the damage already done was the recognition that the problem was ongoing. The problems concerning toxic chemical wastes can be divided into three areas:

- cleaning up the "messes" already created, especially where they threaten drinking-water supplies;
- regulating the handling and disposal of wastes currently being produced, so as to protect public and environmental health; and

◀ **FIGURE 20–9** *"Valley of the Drums," an orphan waste site.* Thousands of drums of waste chemicals, many of them toxic, were unloaded near Louisville, Kentucky, around 1975 and left to "rot," seriously threatening the surrounding environment, waterways, and aquifers.

▶ **FIGURE 20–10** *Love Canal.* These are a few of the nearly 100 homes that the state was forced to buy and later destroy as a result of contamination from Love Canal. The area has now been cleaned up and redeveloped with new homes.

• looking toward future solutions, such as reducing the quantity of hazardous waste produced.

We shall address each of these areas in the sections that follow.

20.3 Cleaning Up the Mess

A major public-health threat from land-disposed toxic wastes is the contamination of groundwater that is subsequently used for drinking. The first priority is to assure that people have safe water. The second priority is to clean up or isolate the source of pollution so that further contamination does not occur.

Assuring Safe Drinking Water

To protect the public from the risk of toxic chemicals contaminating drinking-water supplies, Congress passed the **Safe Drinking Water Act of 1974**. Under this act, the EPA sets national standards to protect the public health, including allowable levels of various contaminants. If any known contaminants are found to exceed maximum contaminant levels (MCLs), the water supply is closed until adequate purification procedures or other alternatives are adopted. The act was amended in 1986 to require the EPA to set MCLs for 83 contaminants and to give the EPA jurisdiction over groundwater. Public water agencies were required to monitor drinking water to be sure that it met the standards.

Although the major objectives of the Safe Drinking Water Act have been accomplished, many municipalities wanted to see the act amended because they felt that monitoring was too costly relative to the risks involved, especially for small towns supplying only a few hundred homes. These and other concerns led the Congress to amend the act in 1996. The amendments provide small municipalities with greater flexibility in water treatment and monitoring and require the EPA to assess risks, costs, and benefits before proposing regulations.

Because of concerns about public drinking water, many people have turned to drinking bottled water to avoid presumed contamination of municipal water supplies. Bottled water is considered a food and, as such, is regulated by the Food and Drug Administration (FDA). However, FDA standards for bottled water specify only that it be as safe as tap water. Ironically, consumers pay for a product for which there are no guarantees that it is any better than tap water. Indeed, numerous samples of bottled water in one study in 1991 showed higher levels of contamination than did municipal supplies.

Groundwater Remediation

If dumps, leaking storage tanks, or spills of toxic materials have contaminated groundwater and such groundwater is threatening water supplies, still, all is not lost. **Groundwater remediation** is a developing and growing technology. Techniques involve drilling wells, pumping out the contaminated groundwater, purifying it, and reinjecting the purified water back into the ground or discharging it into surface waters (Fig. 20–11). Of course, cleaning up the source of the contamination is mandatory. If, however, the contamination is extensive, remediation may not be possible, and the groundwater must be considered unfit for use as drinking water.

Superfund for Toxic Sites

Probably the most monumental task we are facing is the cleanup of the tens of thousands of toxic sites resulting from the history of mismanaged disposal of toxic materials. Where facilities were still operating, pressures were brought

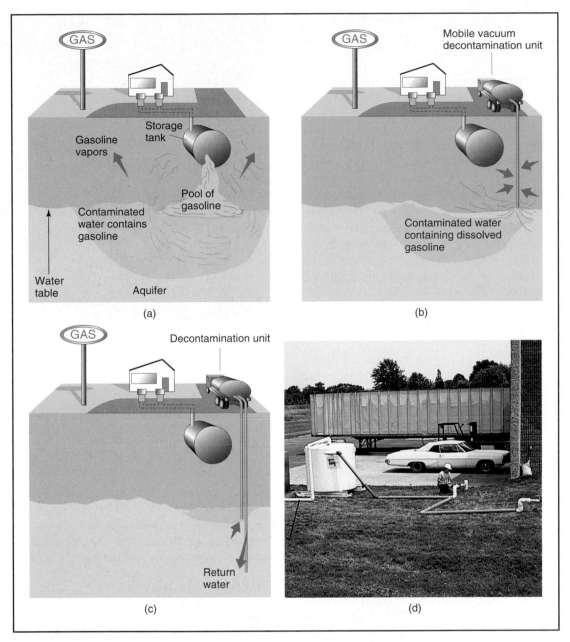

▲ **FIGURE 20–11** *Groundwater remediation.* (a) Typical subsurface contamination from a leaking fuel tank at a gas station. (b) After the leak has been repaired, the vacuum-extraction process causes gasoline and residual hydrocarbons in the soil and on the water table to evaporate and then removes the vapors, preventing further contamination of the groundwater. (c) Contaminated groundwater is pumped out, treated, and returned to the ground. (d) Photo of remediation system.

to bear on the operators to clean up the sites. Many operators who could not afford to do so, however, simply declared bankruptcy, and the sites joined those already abandoned.

The **Comprehensive Environmental Response, Compensation, and Liability Act** of 1980 (CERCLA), popularly known as Superfund, initiated a major federal program aimed at cleaning up abandoned chemical waste sites. Through a tax on chemical raw materials (the authorization for collecting more tax monies expired in 1995), this legislation provides a trust fund for the identification

of abandoned chemical waste sites, protection of groundwater near the site, remediation of groundwater if it has been contaminated, and cleanup of the site. Federal funding for the program has steadily grown from about $300 million per year in the early 1980s to $1.4 billion in 2000. But the cleanup job is nowhere near complete. It is estimated that it will go on for at least another 30 years, at a total cost of some $300 billion; such is the magnitude of the problem. Superfund, one of the EPA's largest ongoing programs, works as follows:

Setting Priorities.   It should be obvious that resources are not sufficient to clean up all sites at once. Therefore, a system for setting priorities has been developed.

- As sites are identified—note that many abandoned sites had long since been forgotten—they are first analyzed in terms of their current and potential threat to groundwater supplies by taking samples of the waste, determining its characteristics, and testing groundwater around the site for contamination. If it is determined that no immediate threat exists, nothing more may be done.

- If a threat to human health does exist, the most expedient measures are taken immediately to protect the public. These measures may include digging a deep trench, installing a concrete dike around the site, and recapping it with impervious layers of plastic and clay to prevent infiltration. Thus, the wastes are isolated, at least for the short term. If the situation has gone past the threat stage and contaminated groundwater is or will be reaching wells, the remediation procedures discussed earlier are begun immediately.

- The worst sites (those presenting the most immediate and severe threats) are put on a **National Priorities List (NPL)** and scheduled for total cleanup. A site on this list is reevaluated to determine the most cost-effective method of cleanup, and finally cleanup begins.

Cleanup Technology.   If the chemical wastes are contained in drums, they can be picked up, treated, and managed in proper fashion. The bigger problem is the soil (often millions of tons) contaminated by leakage. One procedure is to excavate contaminated soil and run it through an incinerator or kiln assembled on the site to burn off chemicals (Fig. 20–12). (Synthetic organic chemicals can be broken down by incineration, and heavy metals are converted to insoluble, stable oxides.) Another method is to drill a ring of injection wells around the site and a suction well in the center. Water containing a harmless detergent is injected into the injection wells and drawn into the suction well, cleansing the soil along the way. The withdrawn water is treated to remove the pollutants and is then reused for injection.

Another rapidly developing cleanup technology is *bioremediation*. In many cases, the soil is contaminated with toxic organic compounds that are biodegradable. The problem is that they do not degrade, because the soil lacks organisms, oxygen, or both. In **bioremediation**, oxygen and organisms are injected into contaminated zones. The organisms feed on and eliminate the pollutants (as in secondary sewage treatment) and then die when the pollutants are gone. Considerable research is being done to find and develop microbes that will break down certain kinds of wastes more readily. Bioremediation, which may be used in place of detergents to decontaminate soil, is a rapidly expanding and developing technology of waste cleanup applicable to leaking storage tanks and spills, as well as to waste-disposal sites.

▲ **FIGURE 20–12**   *Mobile incinerator for toxic waste.* The EPA is cleaning up some Superfund sites by running contaminated materials through this incinerator, which is erected at the site.

When the soil contaminants are heavy metals and nonbiodegradable organic compounds, *phytoremediation* has been employed with some success. **Phytoremediation** uses plants to accomplish a number of desirable cleanup steps: stabilizing the soil, preventing further movement of contaminants by erosion, and extracting the contaminants by direct uptake from the soil. After full growth, the plants are removed and treated as toxic waste products. However, this is a slow process and can be used only at sites where the contaminants and their concentrations are not toxic to plants.

Evaluating Superfund.   Of the literally hundreds of thousands of various waste sites, over 40,000 have been deemed serious enough to be given Superfund status. Over time, the EPA judged that about 28,000 of these sites did not pose a significant public-health or environmental threat. Such sites were accordingly assigned to the category "No Further Removal Action Planned" (NFRAP). Still, over 12,000 sites remain on the active list. It is assumed that this figure now includes most, if not all, of the sites in the United States that pose a significant risk. Interestingly, some of the worst that have come to light are on military bases, the result of what many characterize as the totally heedless and unconscionable discarding of toxic materials from military operations.

To date (2000), 1,224 sites are still on the NPL; additions are made to the list as sites from the master list are assessed, and subtractions are made when remediation on a site is complete. Since 1980, when CERCLA went into effect, about 757 sites have received all the necessary cleanup-related construction, and 226 have been crossed off the list. Cleanup of sites takes an average of 12 years, at a mean cost of $20 million. Total cleanup often takes that long because groundwater remediation is a slow process. The other NPL sites are in various stages of analysis, remediation, and construction. (An example of a Superfund site and its progress toward cleanup is given in the "Earth Watch" essay, this page.)

CERCLA is based on the principle that "the polluter pays." However, the history of a waste-disposal site may go back 50 years or more, and users may have included everything from schools, hospitals, and small businesses to large corporations, all of which mount legal defenses to disclaim responsibility. When liability is difficult to track down or the responsible parties are unable to pay, the Superfund trust fund kicks in. However, only about 25% of the cleanup costs have been paid for with public funds to date.

Over the years, the EPA has gained a great deal of experience in dealing with Superfund sites. The technology has become quite sophisticated, and many hazardous-waste remediation companies have become established. A great deal of progress has been made, and interestingly, although hazardous waste was once a top environmental concern, the American public now places it only eighth among a host of other environmental problems, according to a 1997 Roper survey.

Nevertheless, critics of the program are skeptical about its success, citing the high costs and slow progress. Industries claim that they are unfairly blamed for pollution that reaches back to activities that were legal before the enactment of CERCLA. Many feel that overly stringent standards of cleanup are costing large sums of money without providing any additional benefit to public health. (This problem of finding a suitable balance between costs and benefits will be explored in more depth in Chapter 23.) Many of the critics' concerns have been reflected in attempts to reauthorize CERCLA by the last three Congresses, but bipartisan support for the proposed legislation has been lacking, and the law continues to operate on previously established regulations.

Brownfields. One highly successful recent Superfund development is the *brownfields* program. **Brownfields** are "abandoned, idled, or underused industrial and commercial facilities where expansion or redevelopment is complicated by real or perceived environmental contamination" (EPA definition). It has been estimated that environmental hazards impair the value of $2 trillion of real estate in the United States. Federal and state initiatives are forging ahead with legislation and regulations that clear the way for the cleanup and development of such properties by limiting liability and providing incentives to developers. Many brownfield sites lie in economically disadvantaged communities, and their rehabilitation contributes jobs and exchanges a functional facility for an unsightly blight on the neighborhood. For example, the city of Chicago recently acquired and cleaned up a former bus barn in west Chicago that was

EARTH WATCH

THE CASE OF THE OBEE ROAD NPL SITE

The following is but one example of the more than 1,200 sites still on the EPA's Superfund National Priorities List:

Conditions at listing (January 1987): The Obee Road Site consisted of a plume of contaminated groundwater in the vicinity of Obee Road in the eastern section of Hutchinson, Reno County, Kansas. The Kansas Department of Health and Environment (KDHE) has been investigating the area since July 1983. At that time, the state detected volatile organic chemicals, including benzene, trans-1,2-dichloroethylene, chlorobenzene, 1,1-dichloroethylene, tetrachloroethylene, trichloroethylene, vinyl chloride, and toluene, in wells drawing on a shallow aquifer. An estimated 1,900 residents of suburban Obeeville obtained drinking water from private wells in the aquifer.

To protect public health, Hutchinson connected the homes of the Obeeville residents and a school that was drawing water from a contaminated well to the Reno County Rural Water District.

Preliminary work by the state identified the source of the contamination as the former Hutchinson City landfill, which is located at the eastern edge of what is now the Hutchinson Municipal Airport. Before it was closed in 1968, the landfill had accepted unknown quantities of liquid wastes and sludges from local industries. Later, an adjacent industrial area, the Airport Road subsite, was also identified as a source of organic solvents. The Department of Defense (DOD), which owned or maintained the airport until 1963, may also have disposed of solvents in the landfill. No records have been found for the design, use, or closure of the landfill or for the wastes that were delivered there.

Status as of 2000: Further testing and analysis showed that the original landfill had "stabilized." That is, leaching had already run its course, and there was little likelihood of further contamination from the landfill. Therefore, excavation or other cleanup of the landfill subsite has been deemed unnecessary, but monitoring of the site to detect any further leakage will continue for the foreseeable future. According to the EPA's assessment, there are no present risks to human health or the environment from the site. Remedial action that was selected includes restricting access to, and preventing future development of, the landfill, which is now covered with vegetation. Eight wells have been installed, and groundwater monitoring is to occur for a five-year period to ascertain that no further groundwater contamination is occurring. For the Airport Road subsite, three pumping wells have been installed, and groundwater is continually being pumped and treated via air stripping towers.

becoming an illegal indoor garbage dump. The Illinois EPA cleared the site for reuse, and it was acquired by Scott Peterson Meats, which erected a $5.2 million smokehouse that employs 100 workers. Frequently, the rehabilitation of brownfield sites provides industries and municipalities with centrally located, prime land for facilities that would otherwise have been carved out of suburban or "greenfield" lands (land occupied by natural ecosystems).

20.4 Management of New Wastes

The production of chemical wastes is, and will continue to be, an ongoing phenomenon as long as modern societies persist. Therefore, human and environmental health can be protected only if we have management procedures to handle and dispose of wastes safely so that we will not be creating more and more Superfund sites. We have already noted that the Clean Water Act and the Clean Air Act limit discharges into water and air, respectively. When problems regarding the disposal of wastes on land became evident in the mid-1970s, Congress passed the **Resource Conserva-** **tion and Recovery Act of 1976 (RCRA)** in order to control land disposal. Thus, a company producing hazardous wastes in the United States today is under the regulations of these three major environmental acts.

It is important to understand the acts in somewhat more detail. First, let us briefly review their enactment and administration. The general process is that the U.S. Congress writes the legislation, and after it is passed, it becomes the EPA's responsibility to administer and enforce. Typically, the legislation requires the EPA to set particular standards and write rules regarding how certain objectives should be met. This requirement often puts the EPA at the center of controversies, because almost invariably, environmentalists feel that standards and regulations are too weak and industry feels that they are too strict. The whole package of laws, standards, and regulations is passed on to respective agencies in each of the states to administer to local businesses and industries, but the EPA still oversees state activities. States may also apply their own standards and regulations if they are at least as strong as federal standards.

Checking for compliance with all the standards and regulations necessitates a tremendous amount of collecting

ETHICS

ENVIRONMENTAL JUSTICE AND HAZARDOUS WASTE

- The largest commercial hazardous-waste landfill in the United States is located in Emelle, Alabama. African-Americans make up 90% of Emelle's population. This landfill receives wastes from Superfund sites and every state in the continental United States.

- A Choctaw reservation in Philadelphia, Mississippi, was targeted to become the home of a 466-acre hazardous-waste landfill. The reservation is entirely Native American.

- A recent study found that 870,000 U.S. federally subsidized housing units are within a mile of factories that reported toxic emissions to the EPA. Most of the occupants of these apartments are minorities.

The issue is *environmental justice*, introduced in Chapter 1. The EPA defines **environmental justice** as "the fair treatment and meaningful involvement of all people regardless of race, color, national origin, or income with respect to the development, implementation, and enforcement of environmental laws, regulations, and policies. Fair treatment means that no group of people, including racial, ethnic, or socioeco- nomic group should bear a disproportionate share of the negative environmental consequences resulting from industrial, municipal, and commercial operations or the execution of federal, state, local, and tribal programs and policies."

Several recent studies have documented the fact that all across the United States, waste sites and other hazardous facilities are more likely than not to be located in towns and neighborhoods where the majority of residents is non-Caucasian. These same towns and neighborhoods are also less affluent, a further element of environmental injustice. The wastes involved are generated primarily by affluent industries and the affluent majority, but somehow the wastes tend to end up well away from where they were generated and in the backyards of people of color. It seems fair to assume that the siting of hazardous facilities is a matter of political power, and those with the power would like to have the sites well away from their own backyards.

The federal administration has taken this problem seriously. In 1994, President Clinton issued Executive Order 12898, focusing federal agency attention on environmental justice. The EPA's response has been to establish an Environmental Justice (EJ) program, put in place early in 1998, to capture the intent of Executive Order 12898 and to further a number of strategies already established by the EPA's Office of Solid Waste and Emergency Response. For example, in connection with the EPA's Brownfields Initiative, 78 communities are putting abandoned properties back into productive use. Many EJ efforts are directed toward addressing justice concerns *before* they become problems. In 1997, the EPA rejected a permit for a plastics plant in Convent, Louisiana; the plant was to be built in a predominantly minority community. The chemical company decided later to build the plant in an industrial area near Baton Rouge.

What is still uncertain about the EJ movement is the response of the Caucasian majority in the United States. Opposition to the EPA's EJ program has mounted from states, industries, and major cities, which claim that the EJ standards will impede economic progress. At times, minorities themselves have opposed the EPA on the grounds that their low-income communities need the jobs that the industrial facilities will bring. What do you think?

and analyzing of samples, as well as inspection and enforcement. Consequently, the EPA and related agencies provide an important source of jobs for people interested in environmental improvement (see "Career Link," p. 502.)

The Clean Air and Water Acts

The Clean Air Act of 1970, the Clean Water Act of 1972, and their various amendments make up the basic legislation limiting discharges into the air or water. More will be said about the Clean Air Act in Chapter 22. Under the Clean Water Act (see Chapter 18), any firm (including facilities such as sewage-treatment plants) discharging more than a certain volume into natural waterways must have a **discharge permit** (NPDES permit). Short of closing a firm down, one cannot require a firm to stop polluting instantaneously. Discharge permits are a means of seeing who is discharging what. Establishments with discharge permits are required to report all discharges of substances covered by the Toxics Release Inventory. The renewal of the permits is then made contingent on reducing pollutants to meet certain standards within certain periods of time. Standards are being made continually stricter as technologies for pollution control improve. Some manufacturing firms discharging wastewater into municipal sewer systems are required to pretreat such water to remove any pollutant that cannot be removed by sewage-treatment plants—that is, nonbiodegradable organics and heavy metals.

Nevertheless, even this restriction does not end all water pollution. Certain amounts of wastes are still legally discharged under permits, and though they may be a low percentage of total emissions, they still add up to large numbers, as we have seen. Moreover, small firms, homes, and farms are exempt from regulation and contribute an unknown quantity of toxics to air and water. Further still, a great deal of pollution comes from nonpoint sources such as urban and farm runoff.

The Resource Conservation and Recovery Act (RCRA)

The 1976 RCRA and its subsequent amendments are the cornerstone legislation designed to prevent unsafe or illegal disposal of all solid wastes on land. RCRA has three main features. First, it requires that all disposal facilities, such as landfills, be sanctioned by *permit*. The permitting process requires that they have all the safety features described in Fig. 20–7a, including monitoring wells. This demand caused most old facilities to shut down—many subsequently became Superfund sites—and new high-quality landfills with safety measures to be constructed.

Second, RCRA requires that toxic wastes destined for landfills be pretreated to convert them to forms that will not leach. Such treatment now commonly includes biodegradation or incineration in various kinds of facilities, including cement kilns (Fig. 20–13). Biodegradation involves the use of systems similar to secondary sewage treatment, as discussed in Chapter 18, and perhaps new species of bacteria capable of breaking down synthetic organics. If treatment is thorough, there may be little or nothing to put in a landfill. This is the ultimate objective.

For whatever is still going to disposal facilities, the third major feature of RCRA is to require "cradle-to-grave" tracking of all hazardous wastes. The generator (the company that originated the wastes) must fill out a form detailing the exact kinds and amounts of waste generated. Persons transporting the waste, who are also required to be permitted, and those operating the disposal facility must each sign the form, vouching that the amounts of waste transferred are accurate.

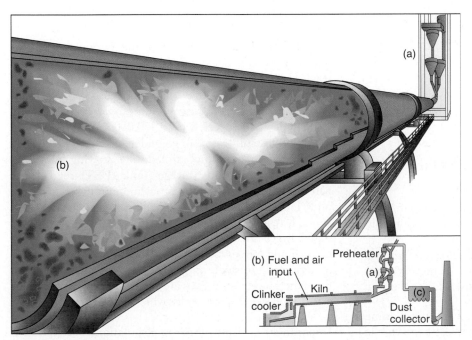

◀ **FIGURE 20–13** *Cement kiln to destroy hazardous wastes.* A cement kiln is a huge, rotating "pipe," typically 15 feet in diameter and 230 feet long, mounted on an incline. (a) Solid wastes fed in with raw materials are fully incinerated and made to react with cement compounds as they gradually tumble toward the combustion chamber. (b) Flammable, liquid wastes added with the fuel and air impart fuel value as they are burned. (c) Waste dust is trapped and recycled into the kiln. (Redrawn with permission. Southdown, Inc., Houston, TX 77002.)

CAREER LINK

DANIEL S. GRANZ, EPA ENVIRONMENTAL ENGINEER

There are times when Dan Granz, environmental engineer in EPA's Lexington, Massachusetts, Office of Environmental Measurement and Evaluation, has to suit up with a full-body protective white suit, complete with respirator, gloves, and booties. As uncomfortable as the suits are, they are essential when one is dealing with some of the unknown and potentially hazardous chemicals encountered in the field. Properly suited, Granz has dealt with some serious hazards: mislabeled drums and tanks at a hazardous waste transfer station, open lagoons on the site of a defunct tannery, and an aging factory whose floorboards were still saturated with PCBs from a manufacturing process that had ended 20 years ago. He regards the great variety of projects that come his way as a major attraction of his job.

Currently, Granz is working on a project involving lead sampling from vacant lots in Providence, Rhode Island. The project's objective is to rehabilitate the urban lots so that they can be put to good use and be brought back on the city's tax rolls. The problem is contamination by the lead-based paint used virtually everywhere a few decades ago. Lead is especially hazardous to children, who might be playing on the vacant lots. If the lead levels prove to be moderately low, EPA may recommend that uncontaminated soil be brought in to mix

Daniel S. Granz

with the urban soil and dilute its lead content. If the levels are high, the topsoil might have to be removed or covered to seal it in place. Granz and his team measure soil samples on the spot with portable equipment and then follow up with testing back in the laboratory for confirmation.

Granz brings a unique kind of preparation to his occupation. He graduated with a B.A. in biology from Gordon College in Massachusetts in 1977, but his eye was on the job market (not too promising for biologists at the time). He enrolled in a civil engineering program at Northeastern University in Boston and graduated in 1980 with a bachelor-of-science degree in civil engineering. One of the attractions of Northeastern was its COOP program, which places students as interns with different agencies or businesses. Granz elected a placement with the EPA. When his degree was granted, he was offered two jobs: a position with EPA in the same office where he now works and a job with Exxon Corpora-

tion. His COOP experience had convinced him that he would like to be on the regulatory side of the ledger, working on improving the environment.

When asked about how students might get to work in a federal agency like the EPA, Granz admitted that it was not necessarily easy, but that turnover does occur and jobs do open up regularly. His advice is to have a specialty; he believes that the combination of biology and civil engineering gave him a unique profile that was attractive to prospective employers. The EPA and other agencies often hire college students as summer interns; COOP programs are also a good idea, Granz stated. His job satisfaction comes from knowing that he can make a difference as he works with industries and the problems they have created. Quite often he is able to work directly with polluters, convincing them to do the right thing before legal action becomes necessary. Over the years, Granz has seen great improvement in the environment and is convinced that, without the work of the EPA, our air, water, and land would be far more polluted and hazardous than it is now. That is to say, there is still work to be done as the EPA addresses some of the harder problems like nonpoint source pollution, groundwater contamination, and the cleanup of brownfield sites.

Copies of their signed forms go to the EPA. All phases are subject to unannounced EPA inspections. The generator remains responsible for any waste "lost" along the way or any inaccuracies in reporting. It is easy to see how this provision of RCRA ensures that generators will deal only with responsible parties and curtails midnight dumping.

Reduction of Accidents and Accidental Exposures

A significant risk to personal and public health lies in exposures that occur as a result of leaks, accidents, and the misguided use of hazardous chemicals in the home or workplace. A considerable number of laws bear on reducing the probability of accidents and on minimizing the exposure of both workers and the public should accidents occur.

Leaking Underground Storage Tanks: UST Legislation.
One consequence of our automobile-based transportation system is millions of underground fuel-storage tanks at ser-

vice stations. Putting such tanks underground greatly diminishes the risk of explosions and fires, but it also hides leaks. Underground storage tanks have a life expectancy of about 20 years before they may spring leaks. Many thousands of tanks are crossing this threshold each year. (By 1999, states had reported over 370,000 failures.) Without monitoring, small leaks can go undetected until nearby residents begin to smell fuel-tainted water flowing from their faucets. **Underground Storage Tank (UST)** regulations, part of RCRA, now require strict monitoring of fuel supplies, tanks, and piping so that leaks may be detected early. When leaks are detected, remediation must begin within 72 hours. Rules now require all USTs to be upgraded with interior lining and cathodic protection (to retard electrolytic corrosion of the steel) and new tanks to be provided with the same protection if they are steel. Many service stations are turning to fiberglass tanks, which are noncorrodible.

Department of Transportation Regulations.   Transport is an area particularly prone to accidents. As modern society uses increasing amounts and kinds of hazardous materials, the stage is set for accidents to become widescale disasters. To reduce this risk, **Department of Transportation Regulations (DOT Regs)** specify the kinds of containers and methods of packing to be used in the transport of various hazardous materials. Such regulations are intended to reduce the risk of spills, fires, and poisonous fumes that could be generated from mixing certain chemicals in case an accident occurs.

In addition, DOT Regs require that every individual container and the outside of a truck or railcar carry a standard placard identifying the hazards (flammability, corrosiveness, the potential for poisonous fumes, and so on) of the material inside (Fig. 20–14). Such placards enable police and firefighters to identify the potential hazard and respond appropriately in case of an accident. You may also see highway HAZMAT signs restricting truckers with hazardous materials to particular routes or lanes or to using a highway only during certain hours.

Worker Protection: OSHA Act and the "Worker's Right to Know."   In the past, it was not uncommon for industries to require workers to perform jobs that entailed exposure to hazardous materials without informing the workers of the hazards involved. This situation is now addressed by certain amendments to the **Occupational Safety and Health Act (OSH Act of 1970)** known as the **hazard communication standard**, or **"worker's right to know."** Basically, the law requires businesses and industries to make available information regarding hazardous materials and suitable protective equipment. One form taken by this information is **material safety data sheets (MSDSs)**, which must accompany the shipping, storage, and handling of over 600 chemicals. These sheets contain information about the reactivity and toxicity of the chemicals, with precautions to follow when using the chemical. Notably, however, the responsibility to read the information and exercise proper precautions remains with the worker.

Community Protection and Emergency Preparedness: SARA, Title III.   In 1984, an accident at a Union Carbide pesticide plant in Bhopal, India, caused the release of some 30–40 tons of methyl isocyanate (MIC), an extremely toxic gas. An estimated 600,000 people in communities surrounding the plant were exposed to the deadly fumes. The official death toll stands at 2,500, but unofficial estimates start at 7,000 and range much higher. At least 50,000 people suffered various degrees of visual impairment, respiratory problems, and other injuries from the exposure. Ironically, most of those deaths and injuries could have been avoided if people had known a simple way to protect themselves: MIC is very soluble in water; thus, a wet towel over the head would have greatly reduced exposure, and showers would have alleviated aftereffects. Unfortunately, neither the people affected nor the medical authorities involved had any idea of the chemical they were confronted with, much less of the way to protect or treat themselves.

After the Bhopal disaster, Congress passed legislation to address the problem. For expediency, the measure was added as Title III to a bill reauthorizing Superfund: the Superfund Amendments and Reauthorization Act of 1986 (SARA). Title III of SARA is better known as the **Emergency Planning and Community Right-to-know Act (EPCRA)**, which we have already encountered in connection with the Toxics Release Inventory.

EPCRA requires companies that handle in excess of 5 tons of any hazardous material to provide a "complete accounting" of storage sites, feed hoppers, and so on. This information goes to a Local Emergency Planning Committee, which is also required in every governmental jurisdiction. The committee is made up of officials representing local fire and police departments, hospitals, and any other groups that might be involved in case of an emergency, as well as the executive officers of the companies in question.

The task of the committee is to draw up scenarios for possible accidents involving the chemicals on-site and to have a contingency plan for every case. This means everything from having firefighters trained and properly equipped to fight particular kinds of chemical fires, to having hospitals stocked with medicines to treat exposures to the particular chemicals, to implementing procedures. Thus, there can be an immediate and appropriate response to any kind of accident. Together with the information in the Toxics Release Inventory, the act enables communities to draw up a chemical profile of their local area and encourages them to initiate pollution prevention activities and risk reduction analyses.

▲ **FIGURE 20–14   *HAZMAT placards.*** These are examples of some of the placards that are mandatory on trucks and railcars carrying hazardous materials. Numbers in place of the word on the placard or on an additional orange panel will identify the specific material. Placards alert workers, police, and firefighters to the kinds of hazards they face in case an accident occurs.

The Toxic Substances Control Act (TSCA). In the past, new synthetic organic compounds were introduced for a specific purpose without any testing of their potential side effects. For example, in the 1960s, a new compound known as TRIS was found to be a very effective flame retardant and was widely used in children's sleepwear. It was only later discovered that TRIS is a potent carcinogen. Treated sleepwear was immediately withdrawn from the market. It is not known how many children (if any) developed cancer from TRIS, but the warning from this and other such cases is obvious.

Congress responded by passing the **Toxic Substances Control Act of 1976 (TSCA),** which requires that, before manufacturing a new chemical in bulk, manufacturers submit a "pre-manufacturing report" to the EPA in which the environmental impacts of the substance are assessed (including those that may derive from its ultimate disposal) and it is indicated whether the substance is a carcinogen. Depending on the results of the assessment, a product's uses may be restricted or the product may be kept off the market altogether.

Laws applying to hazardous wastes are summarized in Fig. 20–15. Note that nongovernmental consumer advocate groups have been a major force behind the passage of these laws, as well as of regulations requiring ingredients to be labeled on all products. Citizen action does make a difference!

20.5 Looking toward the Future

Essentially everyone—environmentalists as well as their critics—agrees that pollution control has become an enormously complex and costly business. Consider the costs to business of installing and operating all the pollution-control or -treatment equipment. Then reflect on the costs to federal, state, and local governments for all the inspection, monitoring, and enforcement of all the regulations. The EPA estimates that the current yearly cost of complying with federally mandated pollution-control and cleanup programs is over $100 billion. Isn't there a less costly way to accomplish the same objectives of protecting human and environmental health?

Too Many or Too Few Regulations?

Considering all the complexities and costs of pollution control, it is not hard to see why some critics feel that business is overregulated. Nor is it hard to see why some U.S. companies have moved operations to developing countries, where regulations are, in general, less strict. Yet which laws or regulations would you abandon? The environmental nightmare of toxic wastes that has come to light in the former U.S.S.R. shows what can happen in the absence of regulations.

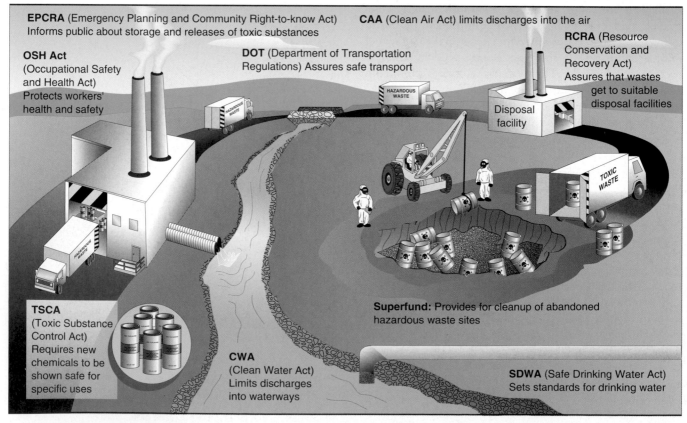

▲ **FIGURE 20–15** *Major hazardous-waste laws.* Summary of the major laws pertaining to the protection of workers, the public, and the environment from hazardous materials.

Further, regulation is far from complete. One area of concern is legally permitted discharges. The low concentrations that standards still allow in discharges add up. Although the figure is gradually going down, there still are more than 2 million tons of toxic chemicals legally discharged into the environment annually. Do the regulations need to be made even stricter?

A second area of concern is the exemption of companies that produce less than 100 kilograms (220 pounds) of hazardous waste per month, a category that homeowners and farmers fall into as well. Items such as batteries and unemptied pesticide containers go into the trash and end up in municipal landfills, most of which don't have protective measures for toxic wastes. Can we regulate such activities without creating a police state?

In short, pollution control seems to have come to a regulatory impasse. However, a new approach is developing that may provide the best of both worlds, reducing environmental pollution while at the same time cutting costs and simplifying regulatory complexity. The concept is called *pollution avoidance* or *pollution prevention*.

Pollution Avoidance for a Sustainable Society

First, let us be clear regarding the distinction between *pollution control* and *pollution prevention*. Pollution control involves adding a filter or some other device at the "end of the pipe" to prevent pollutants from entering the environment. The disposal of the captured pollutants still has to be dealt with, and this entails more regulation and control. **Pollution avoidance**, on the other hand, involves changing the production process, the materials used, or both so that harmful pollutants won't be produced in the first place. Take a simple example: Adding a catalytic converter to the exhaust pipe of your car is a case of pollution control; redesigning the engine so that less pollution is produced or switching to an electric car is a case of pollution avoidance.

Pollution prevention often translates as better product or materials management—that is, less wastage. Thus, pollution prevention frequently creates a cost savings. For example, Exxon Chemical Company added simple "floating roofs" to tanks storing its most volatile chemicals, thereby reducing evaporative emissions by 90% and gaining a savings of $200,000 per year. That paid for the cost of the roofs in six months. As another example, Carrier, Inc., a maker of air conditioners, installed equipment for more precise metal cutting. According to a spokesperson for the company, the new equipment greatly reduces wastes accumulated during cutting, eliminates the need for a toxic cutting lubricant, and leads to the production of a better product—all results that make the company more competitive. These are examples of the *minimization* or *elimination* of pollution.

A second angle on pollution avoidance is *substitution*: finding nonhazardous substitutes for hazardous materials. For example, the dry-cleaning industry, which uses large quantities of toxic organic chemicals, is exploring the use of water-based cleaning, or **wet cleaning**. Preliminary results from some entrepreneurs indicate that wet cleaning is quite comparable in cost and performance to dry cleaning; there are now more than 100 wet cleaners in North America, and the number is growing. Clairol switched from water to foam balls for flushing pipes during the manufacture of hair products, reducing wastewater by 70% and saving $250,000 per year. Water-based inks and paints are being substituted for those that contain synthetic organic solvents, and inks and dyes based on biodegradable organic compounds are being substituted for those based on heavy metals.

A third approach is *reuse*: to clean up and recycle solvents and lubricants. Some military bases, for example, have been able to distill solvents and reuse them instead of discarding them into the environment.

Perhaps the most profound feature of pollution prevention is that it is putting environmentalist groups, government regulators, and industry representatives around the same table. Pollution control, by its very nature, tends to create an adversarial relationship—"us against them"—the regulators against the regulated (industry). Over the years, this adversarial relationship has cost billions of dollars in lawyers' fees and court costs of parties suing one another. Now we are beginning to find citizens and government workers seated side by side with company executives, each looking for ways to reduce both pollution and costs through reaching consensus.

Industry is also acting unilaterally to clean up its act. The American Chemistry Council, a trade group of chemical industry executives, has initiated what it calls a **Responsible Care Program**. The aim of the program is to go beyond what is required by the law and commit members to an ongoing, long-term reduction of toxic releases and of all hazardous wastes generated by their industry.

The government's involvement in pollution prevention is crucial. The EPA has recently been changing its image from a strict "command and control" regulatory agency to one of providing leadership in ways that encourage states, industries, and consumers to cooperate to find better solutions to toxic waste problems. The EPA has initiated programs such as the following:

- Common Sense Initiative—EPA and industries sharing information about industrial processes and environmental goals;
- Project XL—EPA encouraging industries or individuals to test new pollution prevention and control technologies; and
- Design for the Environment—a voluntary program to bring environmental considerations into the design and redesign of technical processes and products.

Because of its monitoring responsibilities, the EPA has often had contentious relationships with state agencies that are implementing environmental programs. To address this problem, the EPA has established the National Environmental Performance Partnership System (NEPPS), designed to give states more flexibility and to encourage the EPA and state offices to work together to accomplish pollution-control objectives.

Finally, pollution avoidance can also be applied to the individual consumer. So far as you are able to reduce or avoid the use of products containing harmful chemicals, you are preventing those amounts of chemicals from going into the environment. You are also reducing the by-products resulting from producing those chemicals. The average American home contains as much as 100 pounds of HHW—substances like paints, stains and varnishes, batteries, pesticides, motor oil, oven cleaners, etc. These materials need to be stored safely, used responsibly, and disposed of properly if they are not to pose a risk of injury or death and a threat to environmental health. (Many communities hold regular HHW collection days; if yours doesn't, you could initiate one!)

Increasingly, companies are gradually beginning to produce and market **green products**, a term used for products that are more environmentally benign than their traditional counterparts. How fast and to what degree these green products replace traditional products will in large part depend on how we behave as consumers. In other words, consumerism can be an extremely potent force.

In conclusion, there are four ways to address the problems of chemical pollution: (1) pollution prevention, (2) recycling, (3) treatment (breaking down or converting the material to harmless products), and (4) safe disposal. The first three of these methods promote a minimum of waste, in harmony with the second principle of sustainability. There is every reason to believe that we can have the benefits of modern technology without destroying the sustainability of our environment by polluting it.

E N V I R O N M E N T   O N   T H E   W E B

DEFINING HAZARDOUS

Effective management of hazardous wastes requires that we be able to define those materials that are of most concern and to develop approaches for handling, treating, and disposing of those materials that are different from those used for ordinary solid wastes. As discussed earlier in this chapter, the EPA determines whether a substance is hazardous according to four categories: ignitability, corrosivity, reactivity, and toxicity. These categories are simple to understand and make sense to us—for instance, we can imagine that a highly ignitable substance, for instance, is of much greater concern than an inert material. However, because there is no absolute definition for any of these terms used to categorize hazardous materials, the practical limits on each category are essentially arbitrary and subject to debate.

An example of this debate can be found in the definition of toxicity. The EPA defines a toxic material as one that is injurious to health when ingested or inhaled, but the term "injurious" is a vague one. In the past, our understanding of human health injury centered on acute and visible damage such as death or obvious illness. More recently, we have become concerned about chronic effects such as cancers, miscarriages, and persistent respiratory distress. Still, it is possible to broaden the definition of "toxic" to include changes in an individual's perception—for instance, the quality of peripheral vision. Gait changes, typical of the effects of some heavy metals on human health, could also be considered a measure of toxic impact. Other possible evidence might include tiny chromosomal changes, even if no impacts on human health are apparent.

Intergenerational impacts could also be part of a definition of toxicity. Some substances, such as diethyl stilbestrol (DES, an anti-miscarriage drug given to millions of pregnant women between 1938 and 1971), for example, have few or no impacts on

the current generation but more significant or very serious impacts on the children or grandchildren of those exposed to the substance. In the case of diethyl stilbestrol, there are minor impacts on the mothers who took the drug, but more serious impacts on those women's children: rare vaginal cancer in daughters, and genital problems in sons.

There is little consensus about how toxicity can or should be defined, and different jurisdictions evaluate toxicity differently. As is the case with many other environmental issues, our working definition of this term, and indeed of the category of substance we currently consider "hazardous," reflects our society's current best understanding of the problem. As our knowledge of the nature and range of impacts grows, our definition of "hazardous wastes"—and the regulatory framework for their management—will also evolve.

Web Explorations

The Environment on the Web activity for this essay describes the challenges in managing hazardous wastes. Go to the Environment on the Web activity (select Chapter 20 at **http:// www.prenhall.com/wright**) and learn for yourself:

1. how the definition of "hazardous" and "waste" can affect regulatory approaches to managing waste;

2. about the Basel Convention on the international trade and disposal of hazardous wastes; and

3. about the strengths and weaknesses of the Basel Ban on hazardous wastes.

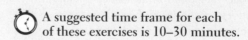 **A suggested time frame for each of these exercises is 10–30 minutes.**

REVIEW QUESTIONS

1. How do toxicologists investigate hazardous chemicals? How is this information disseminated to health practitioners and the public?

2. Into what four categories does the EPA divide hazardous chemicals?

3. Define what is meant by "total product life cycle," and describe the many stages at which pollutants may enter the environment.

4. What are the two classes of chemicals that pose the most serious long-term toxic risk, and how do they affect food chains?

5. What are the "dirty dozen" persistent organic pollutants (POPs)? Why are they on a list?

6. How were chemical wastes generally disposed of before 1970?

7. What two laws pertaining to the disposal of hazardous wastes were passed in the early 1970s? Describe how the passage of the laws shifted pollution from one part of the environment to another.

8. Describe three methods of land disposal that were used in the 1970s. What was a common denominator of all of them in terms of pollution?

9. Can contaminated groundwater be cleaned up? How?

10. What law was passed to cope with the problem of abandoned hazardous-waste sites? What are the main features of the legislation?

11. What law was passed to ensure the safe land disposal of hazardous wastes? What are the main features of the legislation?

12. What laws exist to protect the public against exposures resulting from accidents? What are the main features of the legislation?

13. What is the distinction between pollution control, pollution prevention, and pollution avoidance?

14. Discuss how environmentalists and businesspeople can work together to reduce pollution. What are some ways the EPA is using to bring this unity about?

THINKING ENVIRONMENTALLY

1. Select a chemical product or drug you suspect could be hazardous, and investigate it, using the resources made available on the Web by federal agencies and Environmental Defense.

2. Before the 1970s, it was not illegal to dispose of hazardous chemicals in unlined pits, and many companies did so. Should they be held responsible today for the contamination those wastes are causing, or should the government (taxpayers) pay for the cleanup? Give a rationale for your position.

3. Use the Toxics Release Inventory on the Web to investigate the locations and amounts released of toxic chemicals in your state or region.

4. Suppose an incineration facility is to be built near your community. It is proposed that hazardous wastes currently being landfilled at the same location will be disposed of in the new facility. Would you support or oppose the proposal? Give a rationale for your position.

5. Do you favor cutting back on regulations pertaining to hazardous chemicals in order to help balance the federal budget? What laws or regulations, if any, would you cut back? Defend your answer.

WEB REFERENCES

On-line resources for this chapter are on the World Wide Web at: **http://www.prenhall.com/wright**. (Click on Chapter 20 within the Chapter Selector.)

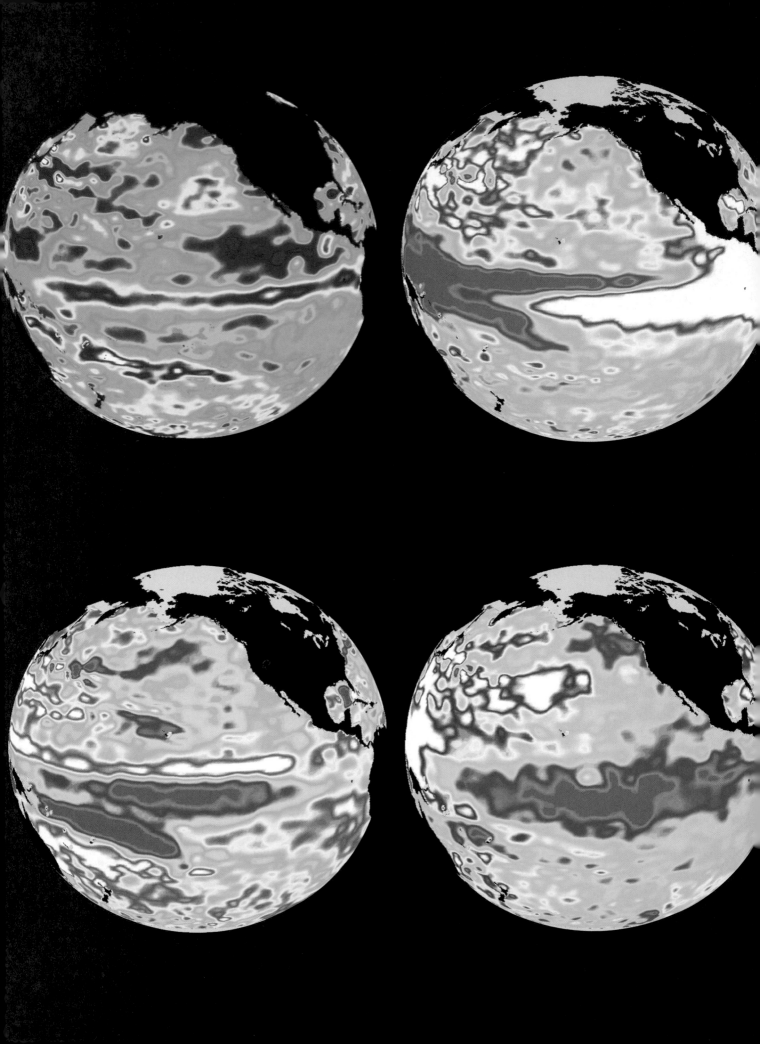

The Atmosphere: Climate, Climate Change, and Ozone Depletion

Key Issues and Questions

1. The atmosphere is the site and source of our weather. How is the atmosphere structured, and how does it function to bring us the major features of our weather?

2. There are ways of investigating the past to reveal features of earlier climates. What does the past tell us about climate change?

3. The oceans and the atmosphere are closely linked in creating climate. What is the ocean Conveyor, and how could the functioning of this system change to bring about a major climate change?

4. The interaction of solar radiation with atmospheric gases controls the balance of warming and cooling of Earth. How have atmospheric gases from human activities affected this balance?

5. Greenhouse gases are increasing in the troposphere. What is the probability that these gases will bring on global warming? If it occurs, what will be the major effects of global warming?

6. Global warming poses a serious threat to the world's climate. How have the countries of the world responded to the threat?

7. The stratospheric ozone shield is vital in protecting life from damaging ultraviolet radiation. How is ozone both formed and destroyed in the stratosphere?

8. Chlorofluorocarbons (CFCs) and other gases destroy stratospheric ozone. What evidence confirms these destructive processes?

9. Continued ozone destruction represents a threat to life on Earth. What is the global community doing about slowing or reversing ozone loss?

El Niño has become a household name. Beginning in April 1997 and extending through the spring of 1998, this climatic event linked the world together. In California and Oregon, unusually severe storms battered the coastline, causing major coastal erosion and flooding rivers (Fig. 21–1). On the eastern seaboard and the Gulf of Mexico, residents relaxed through a hurricane season that was the mildest in many years. By contrast, rainfall at least five times the normal deluged East Africa, often a region of drought. Fires blackened 1,400 square miles of drought-affected forests in Indonesia, creating huge clouds of smoke that blanketed much of Southeast Asia. In New Guinea, the lack of rain brought crop failures, necessitating a massive effort in food relief in order to prevent famine. Record crop harvests were enjoyed in India, Australia, and Argentina. Unusual rainfall in California and Florida triggered lush growth of vegetation.

Shortly after El Niño was officially declared over in May 1998, the world began to hear about La Niña, as weather conditions shifted 180°. Florida was hit by hot, dry weather, triggering wildfires that swept through woods and suburbs, fueled by the lush vegetation nurtured by El Niño. Heavy rains returned to drought-ridden areas, and Venezuela, China, and Mozambique experienced disastrous floods. By spring of 2000, La Niña had dissipated and weather conditions had returned to normal. With global damage estimated upwards of $100 billion, and with over 30,000 deaths, the 1997–2000 El Niño/La Niña has been a lesson in global climate the world will not soon forget. The atmosphere and oceans teamed up to produce a reminder that we live at the mercy of a system we do not control.

What caused these incredible changes in weather over so much of the globe? Briefly, El Niño occurs when a major

◀ *Development of the 1997–2000 El Niño/La Niña.* Satellite images of the central Pacific Ocean from May 25 and December 18, 1997, (upper left and right) showing El Niño conditions, and from June 26, 1998, and March 10, 1999, (lower left and right) showing La Niña conditions. (White is warmest, red is next, and blue is coldest.)

▲ **FIGURE 21–1** *Impacts of El Niño.* (a) Landslides on the California coast. (b) Food relief in New Guinea. (c) Smoke and fires in Indonesia. (d) Flooding in Kenya.

shift in atmospheric pressure over the central equatorial Pacific Ocean leads to a reversal of the trade winds that normally blow from an easterly direction. Warm water spreads to the east, the jet streams strengthen and shift from their normal courses, patterns in precipitation and evaporation are affected, and the system is sustained for more than a year. La Niña conditions are just the reverse: The easterly trade winds are reestablished with even greater intensity, upwelling of colder ocean water from the depths replaces the surface water blown westward, the jet streams are weakened, and weather patterns are again affected.

Meteorologists are quite good at explaining what El Niño and La Niña are and where and how they may influence the different continents and oceans, but they are still unable to explain why they happen. What we do know, however, is that there have been many El Niños in the past, but

none so intense as the latest one. In the past 12 years, there have been five El Niños, a frequency that is unprecedented. If such changes in major weather patterns persist, we will in fact be experiencing a climate change. The most recent El Niño/La Niña has revealed to people everywhere that the atmosphere, oceans, and land are linked together and that when normal patterns are disrupted, the climate on the whole Earth can be affected. Could it be that the warming trend now evident in global temperatures is responsible for triggering the latest intense El Niño? Are we in for a major climate change? Indeed, what does control the climate?

We will seek answers to these questions as we investigate the atmosphere, how it is structured, and how it brings us our weather and climate. Then we will consider the evidence for global climate change, finishing with a look at what is happening to ozone in the upper atmosphere.

21.1 Atmosphere and Weather

Atmospheric Structure

As we learned in Chapter 3, the atmosphere is a collection of gases that gravity holds in a thin envelope around Earth. The gases within the troposphere are responsible for moderating the flow of energy to Earth and are involved with the biogeochemical cycling of many elements and compounds—oxygen, nitrogen, carbon, sulfur, and water, to name the most crucial ones. The lowest layer, the **troposphere**, extends up to 10 miles (16 km) in the tropics and 5 miles (8 km) in higher latitudes. This layer contains practically all of the water vapor and clouds in the atmosphere; it is the site and source of our weather. Except for local temperature inversions, the troposphere gets colder with altitude (Fig. 21–2). Air masses in this layer are well mixed vertically; pollutants can reach the top within a few days. Substances entering the troposphere—including pollutants—may be changed chemically and washed back to Earth's surface by precipitation (Chapter 22). Capping the troposphere is the **tropopause**.

Over the tropopause is the **stratosphere**, a layer within which temperature *increases* with altitude, up to about 40 miles above the surface of Earth. The temperature increases primarily because the stratosphere contains ozone (O_3), a form of oxygen that absorbs high-energy radiation emitted by the Sun. Because there is little vertical mixing of air masses in the stratosphere and no precipitation from it, sub-stances that enter it can remain there for a long time. Beyond the stratosphere are two more layers, the *mesosphere* and the *thermosphere*, where the ozone concentration declines and only small amounts of oxygen and nitrogen are found. Because none of the reactions we are concerned with occur in the mesosphere or thermosphere, we shall not discuss those two layers. Table 21–1 summarizes the characteristics of the troposphere and stratosphere.

Weather

The day-to-day variations in temperature, air pressure, wind, humidity, and precipitation—all mediated by the atmosphere—constitute our **weather**. **Climate** is the result of long-term weather patterns in a region. The scientific study of the atmosphere—both of weather and of climate—is **meteorology**. It is fair to think of the atmosphere–ocean–land system as an enormous weather engine, fueled by the Sun and strongly affected by the rotation of Earth. Solar radiation enters the atmosphere and then takes a number of possible courses (Fig. 21–3); some is reflected by clouds and Earth's surfaces, but most is absorbed by the atmosphere, oceans, and land, which are heated in the process. The land and oceans can then radiate some of their heat back upward as infrared energy. Some of this back-radiated heat is transferred to the atmosphere. Thus, air masses will become warmer at the surface of Earth and will tend to expand, becoming lighter. The lighter air will then rise, creating vertical air currents; on a large scale, it creates the major

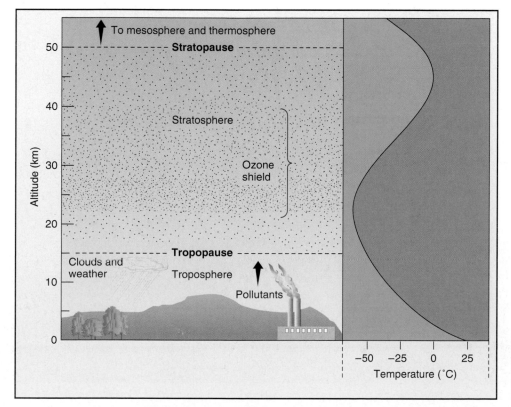

◀ **FIGURE 21–2** *Structure and temperature profile of the atmosphere.* The left-hand plot shows the layers of the atmosphere, while the plot on the right shows the vertical temperature profile.

TABLE 21–1   Characteristics of Troposphere and Stratosphere

Troposphere	Stratosphere
Extent: Ground level to 10 miles (16 km)	Extent: 10 miles to 40 miles (16 km to 65 km)
Temperature normally decreases with altitude, up to –70°F (–59°C)	Temperature increases with altitude, up to +32°F (0°C)
Much vertical mixing	Little vertical mixing, slow exchange of gases with troposphere, via diffusion
Substances entering may be washed back to Earth	Substances entering remain unless attacked by sunlight or other chemicals
All weather and climate take place here	Isolated from the troposphere by the tropopause

convection currents we encountered in Chapter 9 (Fig. 9–5). Of course, air must flow in to replace the rising warm air, and this leads to horizontal airflows, or wind. The ultimate source of the horizontal flow is cooler air that is sinking, and the combination is the Hadley cell (Fig. 9–5). As we saw in Chapter 9, these major flows of air create regions of high rainfall (equatorial), deserts (25° to 35° north and south of the equator), and horizontal winds (trade winds).

On a smaller scale, **convection currents** bring us the day-to-day changes in our weather as they move in a general pattern from west to east. Weather reports inform us of regions of high and low pressure, but where do these come from? Rising air (due to solar heating) creates high pressure up in the atmosphere, leaving behind a region of lower

pressure close to Earth. Conversely, once the moist, high-pressure air has cooled through radiating heat to space and losing heat through condensation (generating precipitation), the air then flows horizontally toward regions of sinking cool, dry air (where the pressure is lower). There, the air is warmed at the surface and creates a region of higher pressure (Fig. 21–4). The differences in pressure lead to airflows, which are the winds we experience. As the figure shows, the winds tend to flow from high-pressure regions toward low-pressure regions.

The larger scale air movements of Hadley cells are influenced by Earth's rotation from west to east. Thus, we have the trade winds over the oceans and the general flow of weather from west to east. Higher in the troposphere,

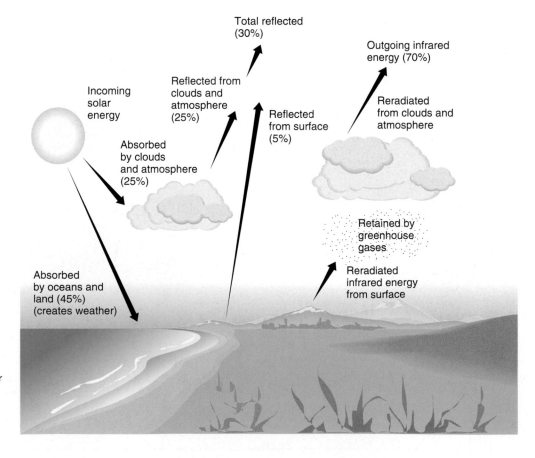

▶ **FIGURE 21–3** *Solar energy balance.* Much of the incoming radiation from the Sun is reflected back to space (30%), but the remainder is absorbed by the oceans, land, and atmosphere (70%), where it creates our weather and fuels photosynthesis. Eventually, this absorbed energy is radiated back to space as infrared energy, or heat.

Earth's rotation and air-pressure gradients generate veritable rivers of air, called **jet streams**, that flow eastward at speeds of over 300 mph and that meander considerably. Jet streams are able to steer major air masses in the lower troposphere. One example is the polar jet stream, which steers cold air masses into North America when it dips downward in latitude.

Air masses of different temperatures and pressures meet at boundaries we call **fronts**, which are regions of rapid weather change. Other movements of air masses due to differences in pressure and temperature include hurricanes and typhoons and the local, but very destructive, tornadoes. Finally, we also have major seasonal airflows, the **monsoons**, which often represent a reversal of previous wind patterns. Monsoons are created by major differences in cooling and heating between oceans and continents. The summer monsoons of the Indian subcontinent are famous for the beneficial rains they bring and notorious for the frequently devastating floods that can occur when the rains are heavy. When we put together the general atmospheric circulation patterns and the resulting precipitation, then add the wind and weather systems generating it, and finally mix this with the rotation of Earth along with the tilt of the planet on its axis, which brings us the seasons, we have the general patterns of weather that characterize different regions of the world. These patterns are referred to as the regions' **climate**.

21.2 Climate

In Chapter 2, we described climate as the average temperature and precipitation expected throughout a typical year in a given region. Recall that the different temperature and moisture regimes in different parts of the world "created" different types of ecosystem that we call *biomes*. These biomes represent the adaptations of plants, animals, and microbes to the prevailing weather patterns, or climate, of a region. Of course, the temperature and precipitation patterns are actually caused by other forces, namely, the major determinants of weather previously outlined. Humans have shown that they can adjust to practically any climate (short of the brutal conditions on high mountains or burning deserts), but this is not true of the other inhabitants of the particular regions we occupy. If other living organisms in a region are adapted to a particular climate, then a change in the climate represents a major threat to the structure and function of the existing ecosystems. The subject of climate change is such a burning issue today because we depend on these other organisms for a host of vital goods and services without which we could not survive (Chapter 3). If the climate changes, can these ecosystems change with it in such a way as not to interrupt the vital support they provide us? How rapidly can organisms and ecosystems adapt to changes in climate? How rapidly do climates change? One

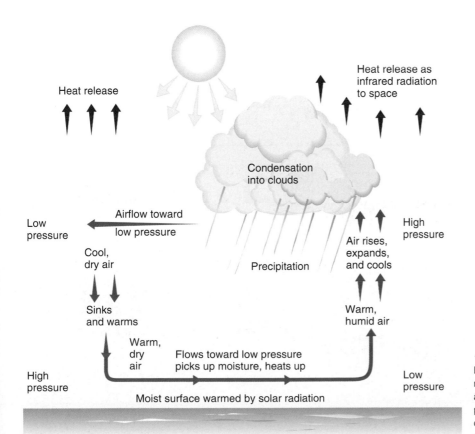

◀ **FIGURE 21–4** *A convection cell.* Driven by solar energy, these cells produce the main components of our weather as evaporation and condensation occur in rising air and precipitation results, followed by the sinking of dry air. Horizontal winds are generated in the process.

way to answer these questions is to look into the past, which may harbor "records" of climate change.

Climates in the Past

Searching the past for evidence of climate change has become a major scientific enterprise, one that becomes more difficult the further into the past we try to search. Systematic records of the factors making up weather—temperature, precipitation, storms, and so forth—have been kept for little more than a hundred years. Nevertheless, these already inform us that our climate is far from constant. The record of surface temperatures, from weather stations around the world and from literally millions of observations of temperatures at the surface of the sea, tells an interesting story (Fig. 21–5): Since 1855, global average temperature has shown periods of cooling and warming, but, in general, has increased 0.6°C (1°F).

Using indirect methods, we can extend observations on climatic changes much further back in time. By delving into historical accounts, we find that from 1100 to 1300 A.D. the Northern Hemisphere appears to have enjoyed a warming period. This was followed by the "Little Ice Age," between 1400 and 1850 A.D. Some work done quite recently on ice cores has provided a startling view of a global climate that oscillates according to several cycles, as well as some evidence that remarkable changes in the climate can occur within as little as a few decades. Ice cores in Greenland and the Antarctic have been analyzed for thickness, gas content—specifically, carbon dioxide (CO_2) and methane

(CH_4), two greenhouse gases—and **isotopes**—alternative chemical configurations of a given compound, due to different nuclear components. Isotopes of oxygen, as well as isotopes of hydrogen, behave differently at different temperatures when condensed in clouds and incorporated into ice.

The record based on these analyses indicates that Earth's climate has oscillated between ice ages and warm periods (Fig. 21–6); during major ice ages, huge amounts of water were tied up in glaciers and ice sheets, and the sea level was lower by as much as 400 feet (120 m). The most likely explanation for these major oscillations is the existence of known variations in Earth's orbit, such that in different modes of orbital configuration the distribution of solar radiation over different continents and latitudes varies substantially. These oscillations take place according to several periods, called **Milankovitch cycles** (after the Serbian scientist who first described them): 100,000, 41,000, and 23,000 years. Superimposed on the major oscillations is a record of rapid climatic fluctuations during periods of glaciation and warmer times (Fig. 21–6c). One such rapid change, called the *Younger Dryas* event (Dryas is a genus of Arctic flower), occurred toward the end of the last ice age. Earth had been warming up for 6,000 years and then plunged again into 1,500 years of cold weather. At the end of this event, 10,700 years ago, Arctic temperatures rose 7°C in 50 years! The impact on living systems must have been enormous. One possible explanation for this warming that has been discarded is variations in solar output: There is simply no evidence that the Sun has changed over the last million years, let alone undergone major changes in just a

▶ **FIGURE 21–5** *Annual mean global surface atmospheric temperatures.* The baseline, or zero point, is the 1880–1999 long-term average temperature. The warming trend since 1970 is conspicuous. (*Source*: National Climatic Data Center, NOAA.)

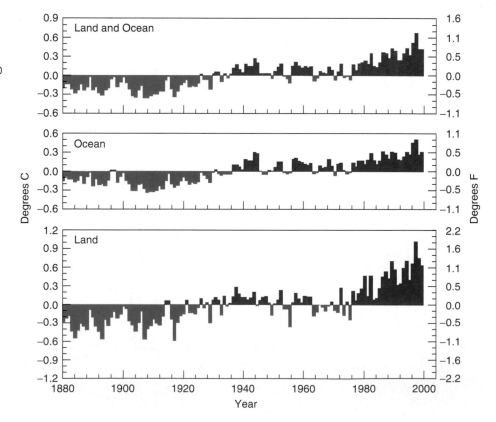

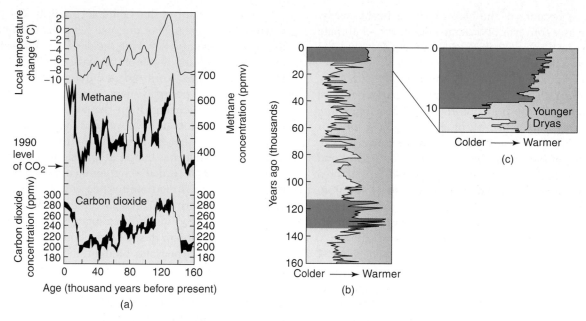

(a)

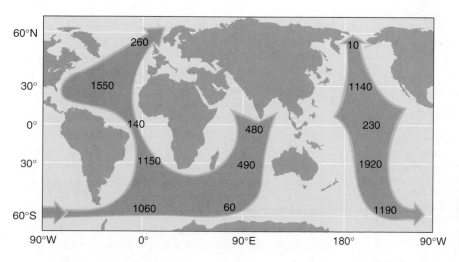

▲ FIGURE 21–6 *Past climates, as determined from ice cores.* (a) Temperature, methane, and carbon dioxide data from antarctic ice cores, covering the past 160,000 years. (b) Temperature patterns of the last 160,000 years, demonstrating climatic oscillations. (c) Higher resolution of the last 12,000 years. The "Younger Dryas" cold spell occurred at the start of this record. Note how rapidly the cold spell dissipated, at around 10,700 years before the present. (After Christopherson, Robert, *Geosystems*, 4th ed., 2000.)

few decades. In search of other explanations for these rapid changes, scientists have narrowed the field to the link between the atmosphere and the oceans. We return now to this link, having already seen the crucial role played by the oceans in El Niño events.

Ocean and Atmosphere

Earth is mostly a water planet, covered more than two-thirds by oceans. Since we all live on land, it is hard for us to imagine that the oceans play a dominant role in determining our climate. Yet the oceans are the major source of water for the hydrologic cycle and the main source of heat entering the atmosphere. As we saw in Chapter 9, the evaporation of

ocean water supplies the atmosphere with water vapor, and when water vapor condenses in the atmosphere (Fig. 9–5), it supplies the atmosphere with heat (latent heat of condensation). The oceans also play a vital role in climate because of their innate heat capacity—the ability to absorb energy when water is heated. Indeed, the entire heat capacity of the atmosphere is equal to that of just the top 3 m of ocean water! The well-known ameliorating effect of oceans on the climate of coastal land areas is a consequence of this property. Finally, through the movement of ocean currents, the oceans are highly important as conveyors of heat. Enormous quantities of heat are moved laterally through water, from hot equatorial regions to higher latitudes (Fig. 21–7); an example is the Gulf Stream, which keeps Western Europe

◄ FIGURE 21–7 *Heat transport by the oceans.* These estimates are in units of terawatts (10^{12} W). Note the connections between the oceans, indicating that some heat transported by the North Atlantic originates in the Pacific. (*Source*: Houghton, J., *Global Warming: The Complete Briefing*, 2d ed. 1997.)

warm. Thus, because of the ocean's innate heat capacity and its role as a conveyor of heat, any discussion of climate must examine the links between the atmosphere and the oceans.

A crucial concept to our concerns here is the thermohaline circulation pattern dominating oceanic currents; thermohaline refers to the contributions that temperature and salinity make to the density of seawater. This **Conveyor** system (Fig. 21–8) acts as a giant, complex conveyor belt, moving water masses from the surface to deep oceans and back again, according to the density of the mass. A key area is the high-latitude North Atlantic, where salty water from the Gulf Stream moves northward on the surface and is cooled by Arctic air currents. Cooling increases the density of the water, which then sinks to depths of up to 4,000 m—the North Atlantic Deep Water (NADW). This deep water spreads southward through the Atlantic to the southern tip of Africa, where it is joined by cold Antarctic waters, to spread northward into the Indian and Pacific oceans as deep currents. The currents gradually slow down and warm, becoming less dense and welling up to the surface, where they are further warmed and begin a movement of surface waters back again toward the North Atlantic. This movement transfers enormous quantities of heat toward Europe, providing a climate that is much warmer than the high latitudes there would suggest. This circulation pattern operates over a period of about 1,000 years for one complete cycle and is vital to the maintenance of current climatic conditions.

Recent evidence indicates that the Conveyor system has been interrupted in the past, changing climate drastically. One mechanism that could accomplish such a shutdown is the appearance of unusual quantities of fresh water in the North Atlantic, lowering the density of the water and therefore preventing much of the massive sinking that normally occurs there. North Atlantic marine sediments show evidence of the periodic invasion (six times in the last 75,000 years) of icebergs from the polar ice cap that supplied huge amounts of fresh water as they melted—called Heinrich events, after the scientist who first described them. The evidence indicates that these invasions coincided with rapid cooling, as recorded in ice cores, and suggests that the Conveyor system shifted southward, with deep water forming nearer to Bermuda than Greenland. When this occurred, a major cooling of the climate happened within a few decades. Although it is not clear how, the normal Conveyor pattern returned and brought about another abrupt change, this time warming conditions in the North Atlantic. The Younger Dryas event is believed to have involved just such a shift in the Conveyor system.

It is significant that one of the likely consequences of extended global warming is increased precipitation over the North Atlantic. If such a pattern is sustained, it could lead to a breakdown in the normal operation of the Conveyor and a rapid reversal of climatic conditions into a new ice age, a possibility that oceanographer Wallace Broecker has called "The Achilles' heel of our climate system."

21.3 Global Climate Change

The Earth as a Greenhouse

Warming Processes. You are familiar with the way the interior of a car heats up when the car is sitting in the Sun with the windows closed. This heating occurs because sunlight comes in through the windows and is absorbed by the seats and other interior objects. In being thus absorbed, the light energy is converted to heat energy, which is given off in the form of infrared radiation. Unlike sunlight, infrared radiation is blocked by glass and so cannot leave the car. The trapped heat energy causes the interior air temperature to rise. This is the same phenomenon that keeps a greenhouse warmer than the surrounding environment.

On a global scale, CO_2, water vapor, and other gases in the atmosphere play a role analogous to the glass in a greenhouse. Therefore, they are called **greenhouse gases**. Light energy comes through the atmosphere and is absorbed by Earth and converted to heat energy at the planet's surface. The infrared heat energy radiates back upward through the atmosphere and into space. The greenhouse gases that are naturally present in the troposphere absorb some of the infrared radiation and reradiate it back toward the surface; other gases (N_2 and O_2) in the troposphere do not. This greenhouse effect was first recognized in 1827 by French scientist Jean-Baptiste Fourier and is now firmly established. The greenhouse gases are like a heat blanket, insulating Earth and delaying the loss of infrared energy (heat)

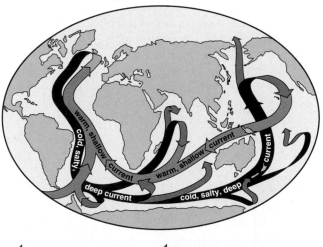

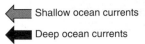 Shallow ocean currents 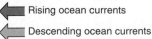 Rising ocean currents

Deep ocean currents   Descending ocean currents

▲ **FIGURE 21–8** *The oceanic Conveyor system.* Salty water flowing to the North Atlantic is cooled and sinks, forming the North Atlantic Deep Water system. This deep flow extends southward and is joined by Antarctic water, whereupon it extends into the Indian and Pacific oceans. Surface currents then proceed in the opposite direction, returning the water to the North Atlantic. (Adapted from Figure 3.15 from *Our Changing Planet* 2/e by Fred T. MacKenzie. Copyright © 1998 by Prentice Hall, Inc. Reprinted by permission of Pearson Education, Inc., Upper Saddle River, NJ 07458.)

to space (Fig. 21–9). Without this insulation, average surface temperatures on Earth would be 21°C colder, and life as we know it would be impossible. Therefore, our global climate is dependent on Earth's concentrations of greenhouse gases. If these concentrations increase or decrease significantly, our climate will change accordingly.

Global Cooling Earth's atmosphere is also subject to cooling factors. For example, on average, clouds cover 50% of Earth's surface and reflect some 21% of solar radiation away to space. This reflection of sunlight is called the **planetary albedo**, and it contributes to overall cooling. (More accurately, it prevents warming from occurring.) The effect of the albedo is especially true for *low* clouds and can be readily appreciated if you think of how you are comforted on a hot day when a large cloud passes over the Sun. The radiation that was heating you and your surroundings is suddenly intercepted higher up in the atmosphere, and you feel cooler. However, *high* clouds have a greenhouse effect, absorbing some of the infrared radiation and emitting some infrared themselves. Overall, the net impact of clouds is judged to be a slightly cooling effect, although there is debate about this.

Volcanic activity can also lead to planetary cooling. When Mount Pinatubo in the Philippines erupted in 1991, some 20 million tons of particles and aerosols entered the atmosphere and contributed to a significant drop in global temperature as radiation was reflected and scattered away. This global cooling effect lasted until the volcanic debris was finally cleansed from the atmosphere by chemical change and deposition, a process that took several years.

Finally, climatologists have found that anthropogenic sulfate aerosols (from ground-level pollution) play a significant role in canceling out some of the warming from greenhouse gases. Sulfur dioxide from industrial sources enters the atmosphere and reacts with compounds there to form a high-level aerosol—a sulfate haze. This haze reflects and scatters some sunlight and also contributes to the formation of clouds, with a concomitant increase in planetary albedo. The mean residence time of sulfates forming the aerosol is about a week; thus, the aerosol does not increase over time, as the greenhouse gases do. However, because anthropogenic sulfate is more than double that coming from natural sources, its effect is substantial and persistent. Climatologists estimate that the cooling effect of these pollutants has counteracted some 20–30% of the global warming

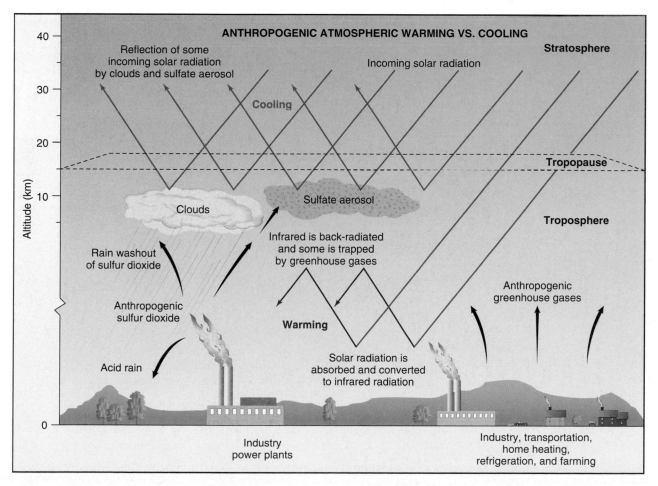

▲ **FIGURE 21–9** *Global warming and cooling.* Anthropogenic factors involved in atmospheric warming and cooling are illustrated. Some gases promote global warming; others, such as sulfur dioxide, lead to cooling.

in recent years. Newer findings indicate that sulfate aerosols and other pollutants can prevent precipitation from occurring within clouds, adding substantially to their impact on climate.

Global atmospheric temperatures are a balance of the effects of factors leading to cooling and factors leading to warming. The net result varies, depending on one's location. As we will see, this balance contributes much uncertainty to our predictions of what will happen in the future as greenhouse gases continue to increase. Figure 21–9 gives a summary of the anthropogenic factors that interact to influence the temperature at any given location.

Carbon Dioxide: Major Greenhouse Gas

More than 100 years ago, Swedish scientist Svante Arrhenius reasoned that differences in CO_2 levels in the atmosphere could greatly affect Earth's energy budget. Arrhenius suggested that, in time, the burning of fossil fuels might change the atmospheric CO_2 concentration, although he reasoned that it would take centuries before the associated warming would be noticeable. He was not concerned about the impacts of this warming, arguing that such an increase would be beneficial. (He lived in Sweden!)

In 1958, Charles Keeling began measuring CO_2 levels on Mauna Loa, in Hawaii. Measurements there have been recorded continuously, and they reveal a striking increase in atmospheric levels of the gas (Fig. 21–10). The concentrations increased exponentially until the energy crisis in the mid-1970s, rose linearly for two decades, and then resumed their exponential increase in the mid-1990s. The data also reveal an annual oscillation of 5 ppm, which reflects seasonal changes of photosynthesis and respiration in terrestrial ecosystems in the Northern Hemisphere. When respiration predominates (late fall through spring), CO_2 levels rise; when photosynthesis predominates (late spring through early fall), CO_2 levels fall. Most salient, however, is the relentless rise in CO_2 levels, ranging in recent years from 0.8 to 1.7 ppm per year. As of early 2001, atmospheric CO_2 levels were over 370 ppm, 35% higher than they were before the Industrial Revolution. Thus, our insulating blanket is thicker, and there is reason to expect that this will have a warming effect.

As Arrhenius suggested long ago, the obvious place to look for the source of increasing CO_2 levels is our use of fossil fuels. Every kilogram of fossil fuel burned results in the production of about 3 kg of CO_2. (The mass triples because each carbon atom in the fuel picks up two oxygen atoms in

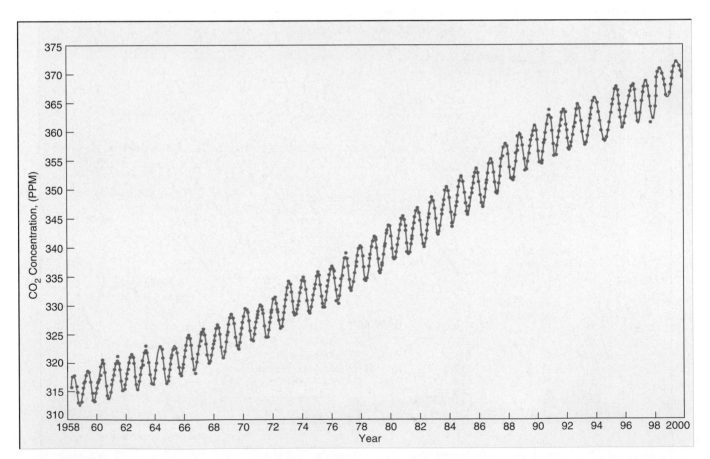

▲ **FIGURE 21–10**   *Atmospheric carbon dioxide concentrations.*   The concentration of CO_2 in the atmosphere fluctuates between winter and summer because of seasonal variation in photosynthesis. The average concentration is increasing owing to human activities—in particular, burning fossil fuels and deforestation. (All measurements made at Mauna Loa Observatory, Hawaii, by Dave Keeling and Tim Whorf, Scripps Institute of Oceanography.)

the course of burning and becoming CO_2.) Currently, 6.6 billion metric tons (gigatons) of fossil-fuel carbon are burned each year, adding about 24 billion tons of CO_2 to the atmosphere. The vast majority of this amount, as well as past fossil-fuel CO_2, has come from the industrialized countries (Fig. 21–11). It is estimated that the burning of forest trees is another anthropogenic source adding some 1.6 billion tons annually to the carbon already coming from industrial processes.

Careful calculations show that if all the CO_2 emitted from burning fossil fuels accumulated in the atmosphere, the concentration would rise by at least 3 ppm per year, not the 1.7 ppm or less charted in Fig. 21–10. In the 1990s, this meant an addition of approximately 8 gigatons per year added by anthropogenic sources, instead of the 2.8 gigatons of carbon per year actually added. This knowledge has prompted an intense search for the carbon "*sinks*" that are absorbing CO_2 and keeping it from accumulating at a more rapid rate in the atmosphere. Recent work with stable carbon and oxygen isotopes and careful measurements of atmospheric CO_2 from 120 stations around the globe have brought us much closer to quantifying annual fluxes in CO_2 and identifying the missing sinks. There is broad agreement that the oceans serve as a sink for much of the CO_2 emitted; some of this is due to uptake of CO_2 by phytoplankton and subsequent sinking, and some is a consequence of the undersaturation of CO_2 in seawater. There are limitations to the ocean's ability to absorb CO_2, however, because only the top 300 m of the ocean is in contact with the atmosphere. (As mentioned earlier, the deep ocean layers do mix with the upper layers, but the mixing time is over 1,000 years.) Calculations indicate that from 1991 to 1997, the ocean sink accounted for an uptake of 2.0 ($\pm$ 0.6) gigatons of CO_2 annually.

The seasonal swings that are obvious in Fig. 21–10 show that the biota can influence atmospheric CO_2 levels. Measurements indicate that terrestrial ecosystems can also serve as a carbon sink. Indeed, these land ecosystems apparently stored a net 1.4 ($\pm$ 0.8) gigatons of carbon annually over the 1991–1997 period. Note that this includes the losses from deforestation and thus implies an average gross annual uptake of 3 gigatons of carbon during those years. It is important to note that both ocean and land sinks fluctuate in their uptake of carbon, and indeed, during the 1980s, the land was essentially neutral. The best explanation for the observed variability in the behavior of these sinks is the climate itself, perhaps as a consequence of El Niño/La Niña phenomena, but this is as yet unproven. A simple model of the major dynamic pools and fluxes of carbon, as presently understood, is presented in Fig. 21–12.

Other Greenhouse Gases

Several other gases absorb infrared radiation and add to the insulating effect of CO_2 (Table 21–2). Some of these gases also have anthropogenic sources and are increasing in concentration, raising the concern that future warming will extend well beyond the calculated effects of CO_2 alone.

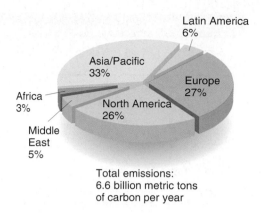

▲ **FIGURE 21–11** *Sources of carbon dioxide emissions from fossil-fuel burning.* Total emissions in 1997 were approximately 22 billion metric tons of CO_2. (*Source*: Data from World Energy Council.)

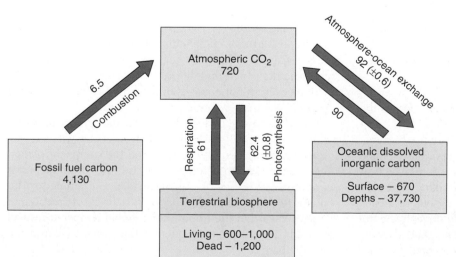

◀ **FIGURE 21–12** *Global carbon cycle of the 1990s.* Data are given in gigatons (billion metric tons) of carbon. Pools are in the boxes, and fluxes indicated by the arrows. (Data sources: Falkowski et al., 2000; Battle et al., 2000; Bosquet et al., 2000)

TABLE 21–2   Anthropogenic Greenhouse Gases in the Atmosphere

Gas	Average Concentration 100 years ago (ppb)[1]	Approximate Current Concentration (ppb)	Average Residence Time in Atmosphere (years)
Carbon dioxide (CO_2)	288,000	370,000	120
Methane (CH_4)	848	1,800	12
Nitrous oxide (N_2O)	285	312	120
Chlorofluorocarbons and halocarbons	0	1.2	50–100

[1] Parts per billion.
(*Source*: Carbon Dioxide Information Analysis Center, Oak Ridge National Laboratory.)

Water Vapor.   Water vapor absorbs infrared energy and is the most abundant greenhouse gas. Although it plays an important role in the greenhouse effect, its concentration in the troposphere is quite variable. Through evaporation and precipitation (the hydrologic cycle—see Chapter 9), water undergoes rapid turnover in the lower atmosphere, and water vapor does not tend to accumulate over time. Water vapor does, however, appear to be a major factor in what has been called the "supergreenhouse effect" in the tropical Pacific Ocean. As it traps energy that has been radiated back to the atmosphere, the high concentration of water vapor contributes significantly to the heating of the ocean surface and the lower atmosphere in the tropical Pacific. This informs us that if temperatures over the land and oceans rise, evaporation increases and water vapor concentration rises, which in turn causes even more warming.

Methane.   Methane (CH_4), the third most important greenhouse gas, is a product of microbial fermentative reactions; its main natural source is wetlands. Anthropogenic sources include livestock (methane is generated in the stomachs of ruminants), landfills, coal mines, natural gas production and transmission, rice cultivation, and manure. Although methane is gradually destroyed in reactions with other gases in the atmosphere, it is being added to the atmosphere faster than it can be broken down. The concentration of atmospheric methane has doubled since the Industrial Revolution, as revealed in core samples taken from glacial ice (Table 21–2).

Nitrous Oxide.   Nitrous oxide (N_2O) levels are also on the increase. Sources of the gas include agriculture and the burning of biomass; lesser quantities come from fossil-fuel burning. N_2O production in agriculture is a consequence of anaerobic denitrification processes (see Chapter 3) and occurs wherever nitrogen is highly available in soils. The buildup of nitrous oxide is particularly unwelcome, because its long residence time (120 years) makes the gas a problem in not only the troposphere, where it contributes to warming, but also the stratosphere, where it contributes to the destruction of ozone.

CFCs and Other Halocarbons.   Emissions of halocarbons are entirely anthropogenic. Like nitrous oxide, halocarbons are long lived and contribute both to global warming in the troposphere and to ozone destruction in the stratosphere. Used as refrigerants, solvents, and fire retardants, halocarbons have a much greater capacity (10,000 times) for absorbing infrared radiation than does CO_2. The rate of production of chlorofluorocarbons (CFCs) has declined since the Montreal Accord of 1987, and the concentration of CFCs in the troposphere leveled off in the late 1990s. However, these gases are very stable and will continue to exert their warming effects for many decades.

Together, the other anthropogenic greenhouse gases are estimated to trap about 60% as much infrared radiation as CO_2 does. Although the tropospheric concentrations of some of these gases are rising, it is hoped that they will gradually decline in importance because of steps being taken to reduce their levels in the atmosphere, leaving CO_2 as the primary greenhouse gas to cope with in the future.

Amount of Warming and Its Probable Effects

It is certain that levels of CO_2 and the other greenhouse gases are increasing in the troposphere as a result of human activities. How can we be certain, however, that the increase in greenhouse gases will bring on a significant increase in Earth's average temperature? The short answer is that we can't—at least, not yet. Many variables affect global temperatures, including the reflection of solar radiation by cloud cover, minor changes in the Sun's intensity, airborne particulate matter from volcanic activity, and sulfate aerosol. However, as Fig. 21–5 indicates, global temperatures are on the rise—a trend that coincides with the observed rise in greenhouse gases. And the strong correlation between temperature and greenhouse gases from the past is impressive (Fig. 21–6). All the evidence examined so far points to the *strong probability* that, as levels of greenhouse gases increase in the troposphere, global temperatures will indeed rise, and as a consequence, the climate will undergo major changes.

The Contributions of Modeling. Weather forecasting employs powerful computers capable of handling large amounts of atmospheric data and applying appropriate mathematical equations involved with the processes taking place in the atmosphere, oceans, and land. Forecasts of conditions for 72 hours or more have become quite accurate. Even though computing power has increased greatly in recent years, the somewhat chaotic behavior of weather parameters prevents more long-term forecasting.

Modeling climate is an essential strategy for exploring the potential future impacts of rising greenhouse gases. Climatologists employ the same powerful computers used for weather forecasting and have combined global atmospheric circulation patterns with ocean circulation and cloud–radiation feedback to produce *coupled general circulation models* (CGCMs) that are capable of simulating long-term climatic conditions. Fourteen centers around the world are now intensely engaged in running models coupling the atmosphere, oceans, and land. The latest information on the cooling effects of sulfate aerosols has been incorporated into current simulations, giving modelers more confidence in their results (Fig. 21–13). Many models have now successfully simulated past climates, based on known differences in Earth's orbit and on ocean–atmosphere coupling. The most complex models, requiring supercomputers, recently scored a major triumph in predicting the intensity and locations of the latest El Niño.

The main purpose of the models is to project the future global climate. Typically, a model will be run for a few simulated decades under standard conditions and then repeated with a doubled greenhouse gas concentration to determine the climate change "caused" by the doubling. The projections from the most recent (2000) model work agree with earlier (1995) assessments: If the concentration of greenhouse gases were to double, Earth would warm up between 1.5° and 4.5°C, with most estimates around 3.0°C. This doesn't sound like much, but consider that temperatures were only 5°C cooler during the last ice age, 18,000 years ago. At that time, ice sheets up to a mile thick stretched across North America, from New York through the Great Lakes states and all of Canada. The range is a reflection of a major limitation in the models: They are still unable to anticipate the precise impacts of clouds. However, significant improvements in the models since 1995 have greatly increased the confidence of climatologists in attributing climate changes that took place in the latter part of the 20th century to human factors and have strengthened their general conclusions about the impacts of future increases in greenhouse gases. To illustrate, let us look at the recent U.S. National Assessment.

U.S. National Assessment. In the Global Change Research Act of 1990, Congress mandated an assessment of the findings of research on global climate change and its implications for the 21st century in the United States. The assessment, titled *Climate Change Impacts on the United States: The Potential Consequences of Climate Variability and Change*, was the work of a team of climate experts from major U.S. universities and agencies and was released in November 2000, after extensive peer reviews and numerous drafts.

The assessment team employed several CGCMs in projecting changes, but based most of their work on two, from the Canadian Climate Centre, and the Hadley Centre in the United Kingdom. Figure 21–14, adapted from the assessment, shows the key parameters and processes incorporated into the models. These parameters are then programmed into powerful computers. Simulations run on the models can make any number of assumptions about future releases of greenhouse gases, but the team based its projections on

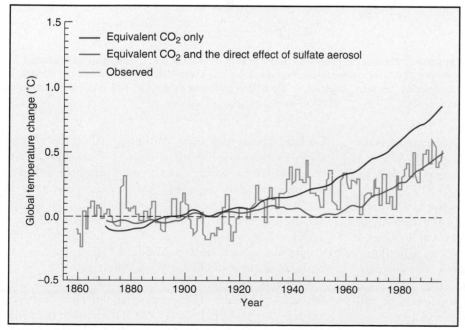

◀ **FIGURE 21–13** *Computer-simulated global warming.* Model results with and without the effects of sulfate aerosol are compared with the actual temperature record. (*Source:* Houghton, J., *Global Warming: The Complete Briefing*, 2d ed. 1997.)

MODELING THE CLIMATE SYSTEM

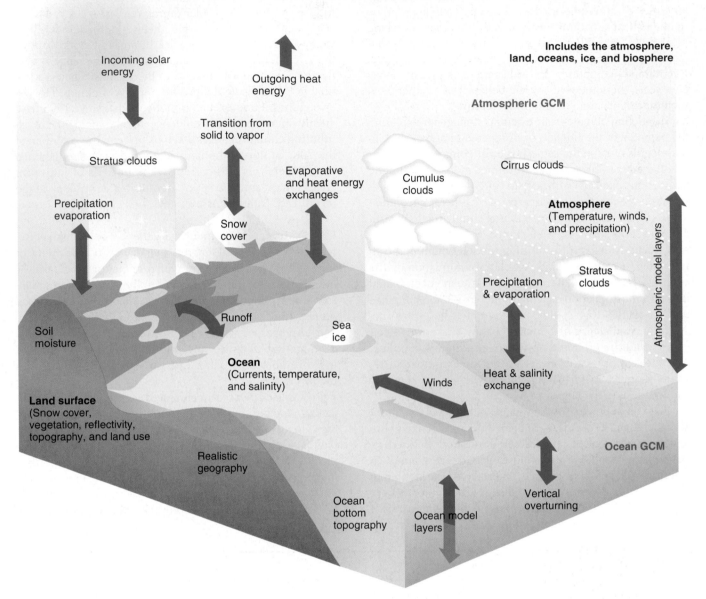

▲ **FIGURE 21–14** *Modeling the climate system.* The coupled general circulation models used in the U.S. climate change assessment employed physical and mathematical parameters involving all of the processes illustrated here. The models were used to simulate changes in these parameters over the Earth for all the seasons and for periods of decades. (*Source*: U.S. Global Change Research Program, U.S. National Assessment, 2000.)

the assumption that there would be no significant interventions to reduce the path of continued growth of world greenhouse gas emissions. Figure 21–15 shows the projected warming based on that assumption: a 5–10°F temperature rise for the United States by the end of the 21st century. Many other impacts also are projected in the assessment, which examines them for the various regions of the United States, as well as for various sectors such as agriculture, forests, human health, and water resources. The key findings of the assessment are presented in Table 21–3. It is clear that the consequences of global climate change will eventually be experienced by everyone, everywhere in the United States.

Global Impacts of Future Warming. All of the impacts shown for the United States in Table 21–3 are applicable to the global arena. Rising global temperatures are linked to two major impacts: *regional climatic changes* and a *rise in sea level*. Both of these effects show up in all of the models and are expected to become evident in a matter of a few decades. Responding to these changes will likely involve unprecedented and costly adjustments. The impact on natural ecosystems could be highly destabilizing.

Climate changes. Warming will seriously affect rainfall and agriculture. The present-day difference between temperatures at the poles and those at the equator is a major driving force for atmospheric circulation. The warming as-

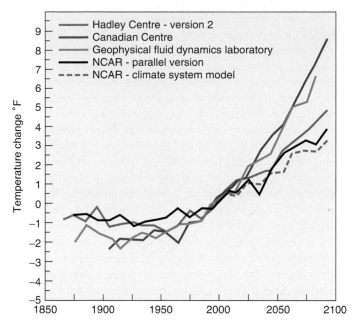

▲ **FIGURE 21–15** *Changes in temperature over the United States.* Assuming a 1%-per-year increase in greenhouse gas emissions, the U.S. assessment models forecast a temperature increase ranging from 5 to 9°F by the end of the 21st century. (*Source*: U.S. Global Change Research Program, U.S. National Assessment, 2000.)

sociated with a doubling of greenhouse gas levels is likely to be more pronounced in polar regions (as much as 10°C) and less pronounced in equatorial regions (1 to 2°C). Greater heating at the poles than at the equator will change atmospheric circulation patterns, as well as rainfall distribu-

tion patterns. Higher temperatures promote more evaporation and a greater ability of the atmosphere to hold moisture, which can enhance precipitation. Some regions will be drier, and some will be wetter. The frequency of droughts will increase, as will the frequency of rainstorms. Further changes in climate could be initiated by alterations in the global oceanic circulation, which could be brought on by greenhouse warming. Oceanographers fear that if the ocean's Conveyor goes into a stall, sudden shifts in land temperature may occur.

Another expected impact of global warming is weather *change*. The following changes have already been observed: (1) Winters in Europe have been warmer and wetter during the last decade, as a result of differences in North Atlantic circulation. (2) As noted earlier, El Niño events are becoming more frequent and more intense. (3) Annual precipitation in the United States has increased 5–10% compared with precipitation during the early 20th century, and the frequency of heavy rainstorms has increased. (4) Droughts have become more frequent in recent decades. (5) Thunderstorms, windstorms, and hurricanes have been more frequent and more severe in recent years, a fact well noted by the insurance industry, which has become interested in global-warming issues, as unprecedented sums for weather-related losses accrued in the 1990s.

The greatest difficulty for the agricultural community in coping with climatic change, however, is not knowing what to expect. Already, farmers lose an average of one in five crops because of unfavorable weather. As the climate shifts, the vagaries of weather will become more pronounced, and crop losses are likely to increase. Fortunately,

TABLE 21–3   Key Findings of the 2000 U.S. Climate Change Assessment

1. Increased Warming. Temperatures in the United States will rise 5–10°F (3–6°C) by 2100. Heat waves will become more intense and frequent in urban areas.

2. Differing Regional Impacts. Temperature and precipitation changes will vary from one region to the next. Instances of heavy and extreme precipitation will increase, and some regions will get drier.

3. Vulnerable Ecosystems. Many ecosystems are highly vulnerable to the projected climate changes. Alpine meadows may disappear, coral reefs will be threatened by the higher sea level, and some barrier islands will disappear. Many of the valuable goods and services we expect from natural ecosystems will be affected and possibly lost.

4. Widespread water concerns. Droughts, floods, and water quality will be a concern in many regions. Snowpack changes will occur in the West and Alaska.

5. Agriculture. The agriculture sector is likely to survive the changes, and U.S. crop productivity may actually increase over the first half of the 21st century, although the gains will not be uniform.

6. Forests. Forest growth may increase in the first decades, but changes in species will occur, with some significant losses, such as the sugar maple forests of the Northeast. Fire, insects, drought, and diseases will likely start impairing forest growth in the long term.

7. Coastlines. Rising sea levels and increasing frequencies and intensities of storms will inundate many beaches along the east and south coasts, damaging the coastal infrastructure.

8. Surprises and Uncertainty. The complexity of the atmosphere–land–ocean system will undoubtedly bring us some surprises, because of the uncertainty involved in the model's projections and in human responses to real climate changes.

farmers are capable of rapidly switching crops and land uses, so the impact may not be as bad as some observers think.

Rising Sea Level. With global warming, the sea level will rise because of two factors: thermal expansion as ocean waters warm and melting of glaciers and ice fields. Sea level has risen 8 cm since 1961 and is continuing to rise at rate of 2 mm per year. The warming at the poles could have an enormous effect when the ice melts, because so much water is stored in the world's remaining ice—enough to raise sea level by 75 m. The area of greatest concern is the Antarctic, which holds most of the world's ice. West Antarctica's ice sheet is not greatly elevated above sea level; in fact, most of it rests on land that is below sea level. The Ross and Weddell Seas extend outward from this ice field, and the landward fringes of these seas are covered with thick floating shelves of ice. Currently, the rate of melting and of buildup seem to be about equal, and many climate models indicate that ice could build up even more in the Antarctic with increasing greenhouse gases.

There is great uncertainty about the magnitude of the rise in sea level; model projections for the next century range from 15 to 95 cm (6 to 37 in). Even the lowest estimated rise, however, will flood many coastal areas and make them much more prone to damage from storms, forcing people to abandon properties and migrate inland. For many of the small oceanic nations, such as the Maldives, off the coast of India, or the Marshall Islands, in the Pacific, a rise in sea level would mean obliteration, not just alteration. The highest estimated rise would cause disaster for most coastal cities, which are home to half the world's population and its business and commerce. Moreover, the estimate of 15 to 95 cm extends only through the next century; the impact will certainly be greater beyond the year 2100. Are inland cities and communities ready to accommodate the billions of people that will be displaced? Are we ready to build dikes or modify all ports to accommodate the higher sea level?

Coping with Global Warming

Is Global Warming Here? There is much natural variation in weather from year to year, and local temperatures do not necessarily follow globally averaged ones. It is a fact, however, that 15 of the hottest years on record have occurred since 1980 (Fig. 21–5); indeed, 1990 was the hottest decade ever recorded. This is a revealing piece of evidence, especially since there is a likely explanation for cooler temperatures during the remaining five years: When Mount Pinatubo erupted in the Philippines in 1991, climatologists using their CGCMs predicted that it would produce a temporary cooling trend—and they were correct. After debris from Mount Pinatubo cleared from the atmosphere, we returned to the warming trend that was so evident in the 1980s. Both 1997 and 1998 set new records for global temperatures, adding further compelling evidence for global warming.

Another significant impact of global warming is the retreat of glaciers and thinning of ice caps. The World Glacier Monitoring Service has followed the melting and slow retreat of glaciers all over the world. The response time of glaciers to climatic trends is slow—10 to 50 years—but the data collected clearly show a wide-scale recession of glaciers over the past 100 years, which is consistent with an increase in greenhouse gases. Recent work has documented huge decreases in the thickness and extent of sea ice in the Arctic. The polar ice cap has lost 40% of its volume since 1958, undoubtedly a response to the rise in temperatures in the northern latitudes. (While global temperature has increased 1°F, temperatures in Alaska, Siberia, and northwest Canada have risen 5°F.)

Scientists have puzzled about the absence of a more direct correlation between CO_2 and global atmospheric temperatures over the last several decades; calculations predict significantly higher temperature increases than have actually occurred. Critics have used this discrepancy to bolster their arguments that global warming is, so to speak, a lot of "hot air." Recent work has addressed the problem. First, it is apparent that sulfate aerosol in the industrialized regions of the Northern Hemisphere appears to be canceling out much of the greenhouse warming over those regions. The cooling effects of the aerosol occur over the very regions most responsible for greenhouse gas emissions. However, scientists point out that the aerosol cooling effect is temporary. As the industrialized nations continue to reduce sulfate emissions (because of acid rain and its impacts on human health) and greenhouse gases continue to build up, the Northern Hemisphere will likely experience its own share of warming.

The second finding confirms what climatologists have thought, but, until now, have never been able to prove: The oceans are absorbing much of the heat, slowing down the rate at which the atmosphere is warming. Recently, oceanographers analyzed millions of temperature records from oceanographic data from 1950 to 1995 and found that, between 1970 and 1995, the heat content of the global oceans increased dramatically. Indeed, the heat record in the oceans follows the same general trend as that in the atmosphere, accelerating during the 1990s (Fig. 21–16).

In its 1995 second assessment report, the Intergovernmental Panel on Climate Change (IPCC), which includes over 1,000 scientists from all the member countries of the United Nations, stated cautiously that "the balance of evidence suggests a discernible human influence on global climate." (See "Ethics" essay, p. 526.) The degree of consensus reached by the IPCC is summarized in Table 21–4. In 2001, a third assessment was released in which the tone of consensus had shifted to the more definite statement that anthropogenic greenhouse gases have "contributed substantially to the observed warming over the last 50 years." The IPCC has also revised upward its assessment of the potential rise in global temperatures over the next 100 years, from a range (in 1995) of 1.5–4.0°C to 1.5–6.0°C.

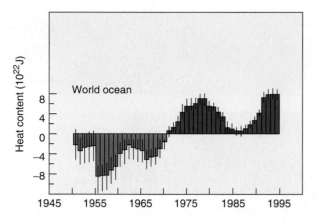

▲ **FIGURE 21–16** *Heat capture by the oceans.* The plot shows the changes in heat content of the upper 3,000 meters of the world's oceans, based on an analysis of millions of temperature data points. (*Source*: Reprinted with permisssion from *Warming of the World Ocean* by S. Levitus, et al. in *Science*, 287, March 24, 2000. Copyright © 2000 by American Association for theAdvancement of Science. Reprinted by permission.)

Political Developments. The world's industries and transportation networks are so locked into the use of fossil fuels, that massive emissions of CO_2 and other greenhouse gases seem sure to continue for the foreseeable future. It is possible, however, to lessen the rate at which emissions are added to the atmosphere and, eventually, to bring about a sustainable balance—although no one thinks that this will be easy. The means for accomplishing this goal lie in the political arena.

What Could Be Done? A variety of steps have been suggested to combat global climate change, with the goal of *stabilizing the greenhouse gas content of the atmosphere*:

- Place a worldwide cap on CO_2 emissions by limiting the use of fossil fuels in industry and transportation.

- Encourage the development of nuclear power, but only if issues concerning cost-effectiveness, reliability, spent fuel, and high-level waste are resolved.
- Stop the loss of tropical forests and encourage the planting of trees over vast areas now suffering from deforestation.
- Make energy conservation rules much more stringent. (Tighten building codes to require more insulation, use energy-efficient lighting, and so forth.)
- Reduce the amount of fuels used in transportation by raising mileage standards, encouraging car pooling, stimulating mass transit in urban areas, and imposing increasingly stiff carbon taxes on fuels.
- Invest in and deploy known renewable-energy technologies: wind power, solar collectors, solar thermal energy, photovoltaics, tidal power, and geothermal energy, among others.
- Sequester CO_2 emitted from burning fossil fuels by capturing it at the site of emission, converting it into liquid CO_2, and pumping it into the deep oceans, where the low temperatures and high pressure would preserve it as a solid mass.
- Make every effort to slow the growth of the human population; a continually growing population will inevitably produce increasing amounts of greenhouse gas emissions.

All of these steps move societies in the direction of sustainability and have many benefits beyond their impacts on global climate change. For example, investing in more efficient use and production of energy reduces acid deposition, is economically sound, lowers the harmful health effects of air pollution, lowers our dependency on foreign oil—and, of course, reduces CO_2 emissions.

TABLE 21–4 1995 IPCC Consensus on Global Climate Change

Issue	Statement	Consensus
Basic characteristics	Fundamental physics of the greenhouse effect	Virtually certain*
	Added greenhouse gases add heat energy	Virtually certain
	Greenhouse gases increasing because of human activity	Virtually certain
	Significant reduction of uncertainty will require a decade or more	Virtually certain
	Full recovery will require many centuries	Virtually certain
Projected effects by mid-21st century	Reduction of sea ice	Very probable**
	Arctic winter surface warming	Very probable
	Rise in global sea level	Very probable
	Local details of climatic change	Uncertain***
	Increases in number of tropical storms	Uncertain

*Nearly unanimous agreement among scientists; no credible alternative.
**Roughly 9 out of 10 chance.
***Roughly 2 out of 3 chance.

See Houghton, John T., ed., et al., *Climate Change 1995: The Science of Climate Change*, Working Group I; Watson, Robert T., ed., et al., *Impacts, Adaptations, and Mitigation of Climate Change*, Working Group II; Bruce, James P., ed., et al., *Economic and Social Dimensions of Climate Change*, Working Group III. All three IPCC volumes published by Cambridge University Press, New York, 1996.

ETHICS

STEWARDSHIP OF THE ATMOSPHERE

The atmosphere is a global commons—a resource used by all countries, yet not owned by any. Countries are free to draw from it to meet their needs and to add to it to get rid of pollutants. Because the air in the troposphere moves so quickly, what one country does to it can easily affect others downwind. The stratosphere is less turbulent, but inevitably it is affected by many gases that enter the lower atmosphere—as we see with the depletion of the ozone layer. Management of a global commons presents a difficult problem: Treaties must be made between countries and compliance must be assured if the global commons is to be protected. Some progress has been made in forging international agreements to curb acid emissions, which represent one form of transboundary air pollution. The Montreal Protocol and its amendments represent the most far-reaching and potentially effective international effort to protect a global resource: the stratospheric ozone layer.

Yet, taking action on ozone-depleting chemicals is child's play compared with what lies ahead in dealing with global climate change. Perhaps the most challenging aspect of the problem is that what we do about it now will do little to affect the present generation. It will likely take several generations for greenhouse gas concentrations to come to a doubling of preindustrial levels, with almost certain profound impacts. So our dilemma is this: Shall we take action *now* to restrain our use of fossil fuels in order to benefit *future* generations? Increasing numbers of people are answering this question with a definite *yes*! For example, more than 2,000 economists agreed that there are many policies to reduce greenhouse gas emissions for which the total benefits outweighed the costs. Over 1,500 scientists, including 104 Nobel laureates, signed a statement urging government leaders attending the 1997 Kyoto conference to

Sir John Houghton (left) and text coauthor Richard T. Wright (right) discuss global climate change at a conference in Cambridge, United Kingdom, in 1998.

"act immediately to prevent the potentially devastating consequences of human-induced global warming." In fact, these people are assuming the role of stewards of the atmosphere and are concerned about what they are passing on to their children and grandchildren.

One scientist whose efforts have been crucial to the work of the Intergovernmental Panel on Climate Change (IPCC) is British meteorologist Sir John Houghton. Houghton is cochair of the IPCC Working Group I and chief editor of the group's published reports (e.g., *Climate Change 1995: The Science of Climate Change*). He also has published a succinct and very readable book on climate change, *Global Warming: The Complete Briefing*, currently in its second edition. Chapter 8 of this book, entitled "Why Should We Be Concerned," addresses the question of human responsibility for caring for the planet. Houghton points out that current human exploitation of the natural world is unsustainable and that going "back to nature" or hoping for some "technical fix" to correct the damage of global warming is not a reasonable option. Houghton speaks of "a basic instinct that we wish to see our children and our grandchildren well set up in the world and wish to pass on to them some of our most treasured possessions," in particular "an

earth that is well looked after and which does not pose to them more difficult problems than those we have had to face." He states that "we have no right to act as if there is no tomorrow."

Houghton makes the point that action on environmental problems depends not only on our understanding of how the world works, but also on how we value it. He examines our responsibility to care for all living things and presents perspectives on this view from a number of the world's religions. He then speaks of the concept of stewardship—one of the continuing themes of this text—that is derived from the Judaeo–Christian Old Testament. He recommends as a model for stewardship the biblical picture of the Garden of Eden, where the first humans were placed "to work and take care of it," as Genesis puts it. He states, "At the basis of this stewardship should be a principle extending what has traditionally been considered wrong—or in religious parlance as sin—to include unwarranted pollution of the environment or lack of care for it."

Stewardship of the atmosphere suggests that the world's nations act on the Precautionary Principle and begin to curb greenhouse gas emissions now and more deeply in the future. What do you think, and what will you do?

Skeptics about global warming can be found in many sectors, from predictable sources like the fossil-fuel industry to members of the scientific community. They stress the uncertainties of the global climate models and, in particular, emphasize that there is much we do not know about the role of the oceans, the clouds, the biota, and the chemistry of the atmosphere. In the absence of "convincing evidence," many think that the threat of global warming may well be overplayed. Why take enormously costly steps, they ask, to prevent something that may have positive results or may even never happen?

Here, we once again refer to the "**Precautionary Principle**," as articulated in the 1992 Rio Declaration. The principle states, "In order to protect the environment, the precautionary approach shall be widely applied by States according to their capabilities. Where there are threats of serious or irreversible damage, lack of full scientific certainty shall not be used as a reason for postponing cost-effective measures to prevent environmental degradation." Employing this principle is like taking out insurance: We are taking measures to avoid a highly costly, but uncertain, situation (albeit one that is beginning to appear more likely).

What Has Been Done? One of the five documents signed by heads of state at the United Nations Conference on Environment and Development (UNCED) Earth Summit in Rio de Janeiro in 1992 was the **Framework Convention on Climate Change (FCCC)**. This convention agreed to aim at reducing greenhouse gas emissions to 1990 levels by the year 2000, but countries were to achieve the goal by voluntary means. Five years later, it was obvious that the voluntary approach was failing. All the developed countries except those of the European Union *increased* their greenhouse gas emissions by 7 to 9% in the ensuing five years. The developing countries increased theirs by 25%!

Prompted by a coalition of island nations (whose very existence is threatened by global climate change), the third Conference of Parties to the FCCC met in Kyoto, Japan, in December 1997 to craft a binding agreement on reducing greenhouse gas emissions. In Kyoto, 38 industrialized countries agreed to reduce emissions of six greenhouse gases below 1990 levels, to be achieved by 2012. The percentages of the reductions varied (Table 21–5). At the conference, the developing countries refused to agree to *any* reductions, arguing that the developed countries had created the prob-

lem and that it was only fair that the developing countries continue on their path to development as the developed countries did, energized by fossil fuels.

Each participating country must ratify the accord, which signifies compliance. In the United States, to achieve a 7% reduction in emissions below 1990 levels by 2010 actually means a reduction of close to 30% below what we would be generating in the absence of the Kyoto accord. Because of the obvious economic impacts of such a reduction, there is fierce opposition to such cuts from many sectors. Although the Clinton administration favored the accord, the Republican-controlled Congress was opposed; in fact, the Senate went on record, 95–0, against even considering ratification unless there was "meaningful participation" by the developing countries.

Even if the United States were to ratify the agreement, emissions would not actually be reduced until close to 2010, and by that time, the developing countries would have increased their emissions such that the total world production of greenhouse gas emissions will have *increased* by 30%. Further, the reductions crafted at Kyoto are far less than what is needed to stabilize atmospheric concentrations of greenhouse gases. The IPCC calculates that it would take emission reductions of at least 60% worldwide to stabilize greenhouse gas concentrations at today's levels. Since this has no chance of happening in the foreseeable future, the concentration of greenhouse gases in the atmosphere will continue to rise. Nevertheless, Kyoto was an important first step toward the stabilization of the atmosphere.

Subsequent negotiations have been held since Kyoto, the latest at The Hague in November 2000. Unfortunately,

TABLE 21–5 Kyoto Emission Allowances and Eight-Year Greenhouse Gas Emission Changes

Country	Kyoto allowances (%) (1990–2010)	Observed change (%) (1990–1998)	1998 Greenhouse Gas Emissions (10^6 metric tons CO_2 equivalent)
United Kingdom	−12.5	−8.9	695
Germany	−21	−16	986
France	0	−1.1	489
Italy	−6.5	+5.1	518
Australia	+8	+5.4	520
Canada	−6	+17	670
Japan	−6	+8.5	1,226
United States	−7	+22	5,954
Russian Federation	0	−58	1,122
Ukraine	0	−55	386

(*Source*: U.N. Framework Convention on Climate Change.)

these meetings have not been successful in achieving broad-scale agreement on how to accomplish the Kyoto emissions goals. Three issues continue to divide the participants: (1) To what extent should countries be allowed to claim credit for carbon "sinks" like forests and farmlands? (2) What should be the rules for emissions trading, wherein countries that exceed their emissions targets can buy credits from countries that more than meet theirs? (3) Can developed countries team up with developing countries in emissions-reducing projects in the developing countries, thus giving credits to the developed countries to be counted against their own emissions? The United States has been adamant about broad-scale emissions trading and the generous use of carbon sinks, and the European Union has opposed these moves, claiming that they will allow the United States to continue its high rate of fossil-fuel use and to take credit for steps that would probably be taken anyway. On the U.S. side, negotiators have been trying to find ways to build political support at home for eventual ratification of the Kyoto Protocol, by bring forests and farms into the picture. At present, only the United Kingdom and Germany seem to be on track to meet their target reductions, and so far, no developed country has actually ratified the agreement.

It is encouraging that some countries and corporations are taking unilateral action to reduce their greenhouse gas emissions, and others are complying with the Kyoto Protocol. At a recent meeting of the World Economic Forum, the CEOs of the world's 1,000 largest corporations voted climate change as the most urgent problem facing humanity. Yet, international negotiations to bring about enforceable cuts in greenhouse gas emissions are essential to any meaningful, sustainable response to global climate change—and

these are proving to be very difficult, as we have seen. Whatever the outcome, it is certain that we are conducting an enormous global experiment, and our children and their descendants will be living with the consequences.

Let us now turn our attention to ozone depletion, an atmospheric problem the public often confuses with global warming.

21.4 Depletion of the Ozone Layer

In another major atmospheric challenge that has been traced to human technology, the work of environmental scientists again has uncovered a serious problem for which there was no prior warning. Our knowledge of the workings of the atmosphere has been appallingly poor, and one consequence of that lack of understanding is the strong possibility that ultraviolet radiation will increase in intensity all over Earth. As with global warming, however, there are skeptics who are not convinced that the problem is serious. Let us examine the evidence.

Radiation and Importance of the Shield

Radiation from the Sun includes *ultraviolet (UV) radiation*, along with visible light. Radiation is the emission of electromagnetic waves of a whole range of energies and wavelengths (Fig. 21–17). Visible light is that part of the electromagnetic spectrum that can be detected by the eye's photoreceptors. UV radiation is like visible-light radiation, except that UV wavelengths are slightly shorter than the wavelengths of violet light, which are the shortest wavelengths visible to the human eye. It is important to distin-

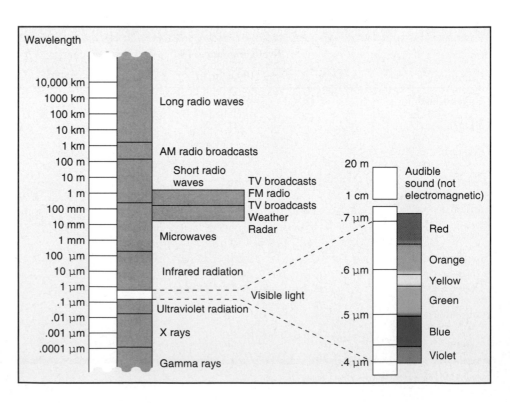

▶ **FIGURE 21–17** *The electromagnetic spectrum.* Ultraviolet, visible light, infrared, and many other forms of radiation are just different wavelengths of the electromagnetic spectrum.

guish between UVB and UVA radiation. UVB wavelengths range from 280 to 320 nanometers (0.28 to 0.32 mm), UVA from 320 to 400 nanometers. Since energy is inversely related to wavelength, UVB is more energetic and therefore more dangerous.

On penetrating the atmosphere and being absorbed by biological tissues, UV radiation damages protein and DNA molecules at the surfaces of all living things. (This is what occurs when you get a sunburn.) If the full amount of ultraviolet radiation falling on the stratosphere reached Earth's surface, it is doubtful that any life could survive. We are spared more damaging effects from ultraviolet rays because most UVB radiation (over 99%) is absorbed by ozone in the stratosphere (Fig. 21–2). For that reason, stratospheric ozone is commonly referred to as the **ozone shield**.

However, even the small amount (less than 1%) of UVB radiation that does reach us is responsible for all the sunburns and more than 700,000 cases of skin cancer and precancerous ailments per year in North America, as well as for untold damage to plant crops and other life-forms. (See "Global Perspective" essay, p. 530.)

Formation and Breakdown of the Shield

The need to maintain the ozone shield is obvious. However, certain anthropogenic pollutants are causing it to break down. Ozone in the stratosphere is a product of UV radiation acting on oxygen (O_2) molecules. The high-energy UV radiation first causes some molecular oxygen (O_2) to split apart into free oxygen (O) atoms, and these atoms combine with the molecular oxygen to form ozone via the following reactions:

$$O_2 + UVB \rightarrow O + O \qquad (1)$$
$$O + O_2 \rightarrow O_3 \qquad (2)$$

Not all of the molecular oxygen is converted to ozone, however, because free oxygen atoms may also combine with ozone molecules to form two oxygen molecules, in the reaction:

$$O + O_3 \rightarrow O_2 + O_2 \qquad (3)$$

Finally, when ozone absorbs UVB, it is converted back to free oxygen and molecular oxygen:

$$O_3 + UVB \rightarrow O + O_2 \qquad (4)$$

Thus, the amount of ozone in the stratosphere is dynamic; there is an equilibrium due to the continual cycle of reactions of formation [Eqs. (1) and (2)] and reactions of destruction [Eqs. (3) and (4)]. Because of seasonal changes in solar radiation, ozone concentration in the Northern Hemisphere is highest in summer and lowest in winter. Also, in general, ozone concentrations are highest at the equator and diminish as latitude increases—again, a function of higher overall amounts of solar radiation. However, as we have learned in recent years, the presence of other chemicals in the stratosphere can upset the normal ozone equilibrium and promote unsustainable reactions there.

Halogens in the Atmosphere. Chlorofluorocarbons (**CFCs**) are a type of halogenated hydrocarbon. (See Chapter 20.) They are nonreactive, nonflammable, nontoxic organic molecules in which both chlorine and fluorine atoms replace some of the hydrogen atoms. At room temperature, CFCs are gases under normal (atmospheric) pressure, but they liquefy under modest pressure, giving off heat in the process and becoming cold. When they revaporize, they reabsorb the heat and become hot. These attributes led to the widespread use of CFCs (over a million tons per year in the 1980s) for the following applications:

- In refrigerators, air conditioners, and heat pumps as the heat-transfer fluid.
- In the production of plastic foams.
- By the electronics industry for cleaning computer parts, which must be meticulously purified.
- As the pressurizing agent for aerosol cans.

All of these uses led to the release of CFCs into the atmosphere, where they mixed with the normal atmospheric gases and eventually reached the stratosphere. In 1974, chemists Sherwood Rowland and Mario Molina published a classic paper (for which they were awarded the Nobel prize in 1995) concluding that CFCs could be damaging the stratospheric ozone layer through the release of chlorine atoms and, as a result, would increase UV radiation and cause more skin cancer. Rowland and Molina reasoned that although CFCs would be stable in the troposphere (they have been found to last 60 to 100 years there), in the stratosphere they would be subjected to intense UV radiation, which would break them apart, releasing free chlorine atoms via the reaction:

$$CFCl_3 + UV \rightarrow Cl + CFCl_2 \qquad (5)$$

Ultimately, all of the chlorine of a CFC molecule would be released as a result of further photochemical breakdown. The free chlorine atoms would then attack stratospheric ozone to form chlorine monoxide (ClO) and molecular oxygen:

$$Cl + O_3 \rightarrow ClO + O_2 \qquad (6)$$

Furthermore, two molecules of chlorine monoxide may react to release more chlorine and an oxygen molecule:

$$ClO + ClO \rightarrow 2 Cl + O_2 \qquad (7)$$

Reactions 6 and 7 are called the **chlorine cycle**, because chlorine is continuously regenerated as it reacts with ozone. Thus, chlorine acts as a **catalyst**: a chemical that promotes a chemical reaction without itself being used up in the reaction. Since every chlorine atom in the stratosphere can last from 40 to 100 years, it has the potential to cause the breakdown of 100,000 molecules of ozone. Thus, CFCs are judged to be damaging because they act as transport agents that continuously move chlorine atoms into the stratosphere. The damage can persist, since the chlorine atoms are

GLOBAL PERSPECTIVE

COPING WITH UV RADIATION

People living in Chile, Australia, and New Zealand are no strangers to the effects of the thinning ozone shield. UV alerts are given in those parts of the Southern Hemisphere during their spring months as lobes of ozone-depleted stratospheric air move outward from the Antarctic. At times, the ozone above these regions can be less than half its normal concentration, and the consequence is, of course, greater intensities of UV radiation reaching Earth's surface. It is now predicted that one out of every three Australians will develop serious and perhaps fatal skin cancer in his or her lifetime.

A thinning ozone layer has now appeared above the Northern Hemisphere, too, and is getting serious attention. Concerned over the rising incidence of skin cancer and cataract surgery in the United States, the EPA, together with the National Oceanic and Atmospheric Administration and the Centers for Disease Control, has initiated a new "UV index." The index is in the form of daily forecasts of UV exposure, issued by the Weather Service for 58 cities. (See inset table.) Satellite measurements of stratospheric ozone are combined with other weather patterns. The goal of the index is to remind people of the dangers of UV radiation and to prompt them to take appropriate action to avoid cancer, premature aging of the skin, eye damage, cataracts, and blindness. For at least the next two decades, the dangers will be quite noticeable, ranging from a greater incidence of sunburn to a likely epidemic of malignant melanoma.

Less obvious, but no less important, are the chronic effects of exposure to the Sun.

The "normal aging" of skin is now known to be largely the result of damage from the Sun: coarse wrinkling, yellowing, and the development of irregular patches of heavily pigmented and unpigmented skin. This can happen even in the absence of episodes of sunburn. Further, a large proportion of the million cataract operations performed annually in the United States can be traced to UV exposure, another of the chronic effects of this kind of radiation.

The most serious impact of chronic exposure to the Sun is skin cancer, which occurs in some 700,000 new cases each year in the United States. Three types of skin cancer are traced to UV exposure: basal cell carcinoma (BCC), squamous cell carcinoma (SCC), and melanoma. Most of the skin cancers (75% to 90%) are BCC, slow growing and therefore treatable. SCC accounts for about 20% of skin cancers. This form can also be cured if treated early. However, it metastasizes (spreads away from the source) more readily than BCC and thus is potentially fatal. Melanoma is the most deadly form of skin cancer, as it metastasizes easily; it is also the form of cancer that causes the greatest mortality. Melanoma is often traced to occasional sunburns during childhood or adolescence or to sunburns in people who normally stay out of the Sun.

What is an appropriate response to this disturbing information? First, know when UV intensity is greatest (three hours on either side of midday) and take precautions accordingly. Protect your eyes with sunglasses and a hat, and apply sunscreen with a sunscreen protection factor (SPF) rating of 15 or higher during such times. Think twice before considering sunbathing or tanning beds, and never proceed without sunscreen that is strong enough to prevent sunburn. Always protect children with sunscreen; their years of potential exposure are many, and their skin is easily burned. If you detect a patch of skin or a mole that is changing in size or color or that is red and fails to heal, consult a dermatologist for treatment.

UV Index (EPA)

Exposure Category Index Value		Minutes to Burn for "Never Tans" (Most Susceptible)	Minutes to Burn for "Rarely Burns" (Least Susceptible)
Minimal	0–2	30 minutes	>120 minutes
Low	4	15 minutes	75 minutes
Moderate	6	10 minutes	50 minutes
High	8	7.5 minutes	35 minutes
Very high	10	6 minutes	30 minutes
"	15	<4 minutes	20 minutes

removed from the stratosphere only very slowly. Figure 21–18 shows the basic processes of ozone formation and destruction, including recent refinements of those processes that will be explained shortly.

After studying the evidence, the EPA became convinced that CFCs were a threat and, in 1978, banned their use in aerosol cans in the United States. Manufacturers quickly switched to nondamaging substitutes, such as butane, and things were quiet for several years. CFCs continued to be used in applications other than aerosols, however, and critics demanded more convincing evidence of their harmfulness.

Atmospheric scientists reason that any substance carrying reactive halogens to the stratosphere has the potential to deplete ozone. These substances include halons, methyl chloroform, carbon tetrafluoride, and methyl bromide. Because of its extensive use as a soil fumigant and pesticide, methyl bromide is thought to cause as much as 10% of current stratospheric ozone loss. (Bromine is 40 times as potent as chlorine in ozone destruction.)

The Ozone "Hole." In the fall of 1985, some British atmospheric scientists working in Antarctica reported a gaping "hole" (actually, a thinning of one area) in the stratospheric

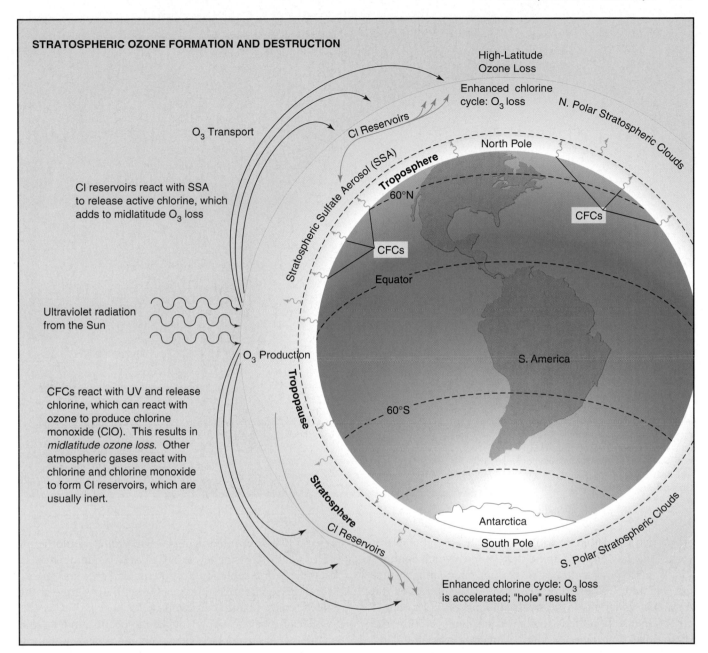

STRATOSPHERIC OZONE FORMATION AND DESTRUCTION

High-Latitude Ozone Loss

Enhanced chlorine cycle: O_3 loss

N. Polar Stratospheric Clouds

O_3 Transport

Cl Reservoirs

North Pole

Cl reservoirs react with SSA to release active chlorine, which adds to midlatitude O_3 loss

Enhanced chlorine cycle: O_3 loss

Stratospheric Sulfate Aerosol (SSA)

Troposphere

60°N

CFCs

CFCs

Equator

Ultraviolet radiation from the Sun

O_3 Production

S. America

Tropopause

CFCs react with UV and release chlorine, which can react with ozone to produce chlorine monoxide (ClO). This results in *midlatitude ozone loss*. Other atmospheric gases react with chlorine and chlorine monoxide to form Cl reservoirs, which are usually inert.

60°S

Stratosphere

Cl Reservoirs

Antarctica

South Pole

S. Polar Stratospheric Clouds

Enhanced chlorine cycle: O_3 loss is accelerated; "hole" results

▲ **FIGURE 21–18** *Stratospheric ozone formation and destruction.* UV radiation stimulates ozone production at the lower latitudes, and ozone-rich air migrates to high latitudes. At the same time, chlorofluorocarbons and other compounds carry halogens into the stratosphere, where they are broken down by UV and release chlorine and bromine. Ozone is subject to high-latitude loss during winter, as the chlorine cycle is enhanced by the polar stratospheric clouds. Midlatitude losses occur as chlorine reservoirs are stimulated to release chlorine by reacting with stratospheric sulfate aerosol.

ozone layer over the South Pole (Fig. 21–19). There, in an area the size of the United States, ozone levels were 50% lower than normal. The "hole" would have been discovered earlier by NASA satellites monitoring ozone levels, except that computers were programmed to reject data showing a drop as large as 30% as due to instrument anomalies. Scientists had assumed that the loss of ozone, if it occurred, would be slow, gradual, and uniform over the whole planet. The "hole" came as a surprise, and if it had occurred any-

where but over the South Pole, the UV damage would have been extensive.

News of the ozone "hole" stimulated an enormous scientific research effort. A unique set of conditions was found to be responsible for the ozone "hole." In the summer, gases such as nitrogen dioxide and methane react with chlorine monoxide and chlorine to trap the chlorine, forming so-called **chlorine reservoirs** (Fig. 21–18), preventing much ozone depletion.

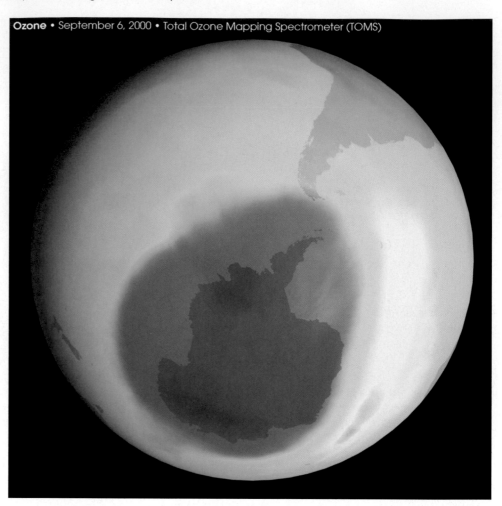

Ozone • September 6, 2000 • Total Ozone Mapping Spectrometer (TOMS)

▶ **FIGURE 21–19** *The Antarctic ozone "hole."* Total Ozone Mapping Spectrometer (TOMS) instruments aboard a satellite recorded the largest ever ozone "hole" over Antarctica during September 2000. NASA's *Earth Probe* satellite was the instrument platform. The 2000 "hole" exceeded 28.3 million km^2 (11 million square miles), an area three times as large as the entire United States land mass. The color scale represents ozone concentrations in Dobson units (D.U.), with blues and purples for amounts of less than 200. (*Source:* Earth Observatory, NASA.)

When the Antarctic winter arrives in June, it creates a vortex (like a whirlpool) in the stratosphere, which confines stratospheric gases within a ring of air circulating around the Antarctic. The extremely cold temperatures of the Antarctic winter cause the small amounts of moisture and other chemicals present in the stratosphere to form the south polar stratospheric clouds. During winter, the cloud particles provide surfaces on which chemical reactions release molecular chlorine (Cl_2) from the chlorine reservoirs.

When sunlight returns to the Antarctic in the spring, the Sun's warmth breaks up the clouds; UV light then attacks the molecular chlorine, releasing free chlorine and initiating the chlorine cycle, which rapidly destroys ozone. By November, the beginning of the Antarctic summer, the vortex breaks down and ozone-rich air returns to the area. By this time, however, ozone-poor air has spread all over the Southern Hemisphere. Shifting patches of ozone-depleted air have caused UV radiation increases of 20% above normal in Australia. Television stations there now report daily UV readings and warnings for Australians to stay out of the Sun. On the basis of current data, estimates indicate that in Queensland, where the ozone shield is thinnest, three out of four Australians are expected to develop skin cancer. The "hole" has reappeared every spring (our autumn) in Antarc-

tica and has been intensifying—exactly what would be expected with increasing levels of CFCs. In September 2000, Antarctic ozone depletion was intense and grew to a record 11 million square miles (an area three times as large as the United States) before it dissipated.

Because severe ozone depletion occurs under polar conditions, observers have been keeping a close watch on the Arctic. To date, no "hole" has developed. The higher temperatures and weaker vortex formation there have so far prevented the severe losses that have become routine in the Antarctic. However, ozone depletion does occur in the Arctic, with ozone levels about 15% lower than the pre-1976 average. Increasingly lower ozone in the Arctic is amplified by greenhouse gas effects on the stratosphere. Because CO_2 also emits radiation, increasing CO_2 concentrations in the stratosphere mean a greater loss of heat to outer space. The colder stratospheric temperatures cause (north polar stratospheric) clouds to form and persist, providing the same opportunities for ozone depletion as in the Southern Hemisphere.

Controversy over Ozone Depletion. Skeptics about the role of CFCs in ozone depletion claim, among other things, that there probably has always been an Antarctic ozone "hole," but we didn't use the right instruments to look for it. They also claim that there is little evidence of a thinning ozone layer and almost no evidence of ground-level UVB increases. One of the most persistent claims of the skeptics is that volcanoes have been ejecting enormous amounts of chlorine into the stratosphere for eons, with no apparent impact on the ozone shield. This chlorine, the claim goes, dwarfs any chlorine carried up into the stratosphere by CFCs.

Careful scrutiny reveals that this is an invalid argument. Whereas volcanoes do inject hydrochloric acid into the atmosphere, very little of it reaches the stratosphere, because of the speed with which it is washed out of the atmosphere by precipitation. Furthermore, the volcanic production was highly exaggerated because of a serious calculational error by one of the skeptics, Dixy Lee Ray, former director of the Atomic Energy Commission.

More Ozone Depletion. Ozone losses have not been confined to Earth's polar regions, although they are most spectacular there. A worldwide network of ozone-measuring stations sends data to the World Ozone Data Center in Toronto, Canada. Reports from the center reveal a rate of loss of 2–4% per decade since 1979 over midlatitudes in the Northern Hemisphere. Ozone loss is expected to peak in about the year 2001, when, hopefully, the chlorine and bromine concentrations in the stratosphere will start to decline as a consequence of the international agreements that have been forged.

Is the ozone loss significant to our future? The EPA has calculated that the ozone losses of the 1980s will eventually have caused 12 million people in the United States to develop skin cancers over their lifetime and that 93,000 of these cancers will be fatal. Americans are estimated to have developed more than 900,000 new cases of skin cancer a year in the 1990s. The ozone losses of that decade allowed more UVB radiation than ever to reach Earth. Studies have confirmed increased UVB levels, especially at higher latitudes.

Coming to Grips with Ozone Depletion

The dramatic pictures of the growing "hole" in the ozone layer (Figure 21–19) have galvanized a response around the world. In spite of the skepticism in the United States, scientists and politicians here and in other countries have achieved treaties designed to avert a UV disaster.

International Agreements. Under the auspices of its environmental program, the United Nations convened a meeting in Montreal, Canada, in 1987 to address ozone depletion. Member nations reached an agreement, known as the **Montreal Protocol**, to scale CFC production back 50% by 2000; to date, 140 countries (including the United States) have signed the agreement.

The Montreal protocol was written even before CFCs were so clearly implicated in driving the destruction of

ozone and before the threat to Arctic and temperate-zone ozone was recognized. Because ozone losses during the late 1980s were greater than expected, an amendment to the protocol was adopted in June 1990. The amendment requires participating nations to completely phase out the major chemicals destroying the ozone layer by 2000 in developed countries and by 2010 in developing countries. In the face of evidence that ozone depletion was accelerating even more, another amendment to the protocol was adopted, in November 1992, moving the target date for the complete phaseout of CFCs to 1996. Timetables for phasing out all of the suspected ozone-depleting halogens were shortened at the 1992 meeting.

Unfortunately, even with the ban, large quantities of CFCs are still being manufactured, in Russia, India, and China in particular. A lively black market has developed for CFCs, and by 1998, the U.S. Justice Department had charged 62 individuals and seven businesses for smuggling CFCs into the United States, leading to fines of $58 million, significant jail time, and recovery of tons of the banned chemicals.

Action in the United States. The United States was the leader in the production and use of CFCs and other ozone-depleting chemicals, with du Pont Chemical Company being the major producer. Following 15 years of resistance, du Pont pledged in 1988 to phase out CFC production by 2000. In late 1991, a spokesperson announced that in response to new data on ozone loss, the company would accelerate its phaseout by three to five years. Many of the large corporate users of CFCs (AT&T, IBM, and Northern Telecom, for example) completely phased out their CFC use by 1994. Also, du Pont spoke in opposition to three bills introduced in September 1995 in the 104th Congress that were designed to terminate U.S. participation in, and compliance with, the CFC-banning protocols.

The Clean Air Act of 1990 also addresses this problem, in Title VI, "Protecting Stratospheric Ozone." Title VI is a comprehensive program that restricts the production, use, emissions, and disposal of an entire family of chemicals identified as ozone depleting. For example, the program calls for a phaseout schedule for the hydrochlorofluorocarbons (HCFCs), a family of chemicals being used as less damaging substitutes for CFCs until nonchlorine substitutes are available. Halons—used in chemical fire extinguishers—were banned in 1994. The act also regulates the servicing of refrigeration and air-conditioning units.

January 1, 1996, has come and gone, and in most of the industrialized countries CFCs are no longer being produced. Substitutes for CFCs are available and, in some cases, are even less expensive than the CFCs. The most commonly used substitutes are HCFCs, which still contain some chlorine and are scheduled for a gradual phaseout. The most *promising* substitutes are HFCs—hydrofluorocarbons, which contain no chlorine and are judged to have no ozone-depleting potential. CFCs in the troposphere have peaked and are slowly declining, and scientists predict that the ozone shield will recover entirely by 2050.

Final Thoughts. The ozone story is a remarkable episode in human history. From the first warnings in 1974 that something might be amiss in the stratosphere because of a practically inert and highly useful industrial chemical, through the development of the Montreal Protocol and the final steps of CFC phaseout that are still occurring, the world has shown that it can respond collectively and effec-

tively to a clearly perceived threat. The scientific community has played a crucial role in this episode, first alerting the world and then plunging into intense research programs to ascertain the validity of the threat. Scientists continue to influence the political process that has forged a response to the threat. Although skeptics still stress the uncertainties in our understanding of ozone loss and its consequences, the strong consensus in the scientific community convinced the world's political leaders that action was clearly needed. This is a most encouraging development as we look forward to the creation of a sustainable society.

ENVIRONMENT ON THE WEB

METHANE: THE "OTHER" GREENHOUSE GAS

Over the last hundred years, the average temperature of Earth's surface has increased about 0.6°C. There is now persuasive evidence that some—perhaps most—of this increase is attributable to the actions of humans. Human activities are thought to have a significant influence on levels of carbon dioxide, methane, nitrous oxide, and, anthropogenic chemicals such as chlorofluorocarbons. Of these greenhouse gases, carbon dioxide seems to make the greatest contribution to global warming, but the role of methane (CH_4) is also significant and often overlooked. Atmospheric methane levels have almost doubled since the early 1800s and now account for about 20% of the overall global-warming effect.

Methane is produced as anaerobic bacteria break down organic wastes. Sources of methane therefore occur where abundant organic material is present but oxygen levels are low, such as in wetlands (the term "swamp gas" refers in part to methane) and where organic wastes are stored or burned.

Scientists have few accurate measurements of methane emissions to the atmosphere, so available estimates are only approximate. Of the global total of 500 million tons/year, the most important sources are thought to be wetland systems (120 million tons/year), fossil fuel extraction (100 million tons/year), human and animal waste management activities (80 million tons/year), landfill degradation (40 million tons/year), deforestation and biomass burning (40 million tons/year), and agriculture (160 million tons/year).

Agriculture's contribution of methane per year includes 60 million tons/year from rice paddies and up to 100 million tons/year from enteric (intestinal) fermentation in the guts of ruminant animals such as cows, sheep, and water buffalo. It seems incredible that processes such as cattle digestion could have an impact on the vast mass of Earth's atmosphere, currently estimated at almost 6×10^{15} tons. But it is precisely because the quantities of trace gases

like methane are so small that modest emissions—and modest controls—can alter their proportions in important ways. The 12-year lifetime of methane in the atmosphere is short compared to other greenhouse gases, so the impact of emission reductions could be apparent in only a few years.

Web Explorations

Such reductions, like those mentioned above, may be well worth consideration. Molecule for molecule, methane has 21 times more warming effect than carbon dioxide in the atmosphere. Methane levels are also rising eight times faster than those of carbon dioxide. Programs, like those described in the Environment on the Web activity for this essay, developed in partnership with agriculture and other industrial sectors, are helping methane producers increase the efficiency—and productivity—of their operations while reducing their impact on the atmospheric environment. Go to the Environment on the Web activity (select Chapter 21 at **http://www.prenhall.com/wright**) and learn for yourself:

1. why varying the quality of livestock feed can affect methane production;

2. how governments are preparing for potential methane increases expected from their expanding agriculture base; and

3. about the Kyoto Agreement and the rift separating many developing nations from their industrialized counterparts.

🕐 **A suggested time frame for each of these exercises is 10–30 minutes.**

REVIEW QUESTIONS

1. What are the important characteristics of the troposphere? Of the stratosphere?

2. Differentiate between weather and climate.

3. How does solar radiation create our weather patterns?

4. What has been discovered about global climate trends in the past? What is the contemporary trend showing?

5. What is the Conveyor system? How does it work? Why is it important?

6. How do greenhouse gases work? What is planetary albedo?

7. Which of the greenhouse gases are the most significant contributors to global warming? Describe the heat-trapping effects of CO_2.

8. What is modeling? How can it help our efforts to improve the condition of the atmosphere?

9. What are the major impacts of rising global temperature on the United States?

10. What steps could be taken to stabilize the greenhouse gas content of the atmosphere?

11. What does the Precautionary Principle of the 1992 Earth Summit state?

12. Trace the political history of the Framework Convention on Climate Change (FCCC) from 1992 to the present.

13. How is the ozone shield formed? What causes its breakdown?

14. Describe four sources of CFCs entering the stratosphere. How do CFCs affect the concentration of ozone in the stratosphere and contribute to the formation of the ozone "hole"?

15. Discuss several efforts that are currently underway to protect our ozone shield.

THINKING ENVIRONMENTALLY

1. Compile a list of ways you are producing CO_2. What steps could you take to decrease this production?

2. In your opinion, does the greenhouse effect exist? What about ozone depletion? Consider the Precautionary Principle introduced at the 1992 Earth Summit. What might be the short- and long-term economic effects of using this approach to solve environmental problems such as the greenhouse effect and ozone depletion?

3. Suggest a range of international agreements that would lead to the kind of reductions in greenhouse gas emissions called for by the Kyoto Protocol and that would satisfy the concerns of global equity between developed and developing countries.

4. How would a ban on HCFC production change the way you live? Give your reasoning for or against such a ban.

WEB REFERENCES

On-line resources for this chapter are on the World Wide Web at: **http://www.prenhall.com/wright**. (Click on Chapter 21 within the Chapter Selector.)

Atmospheric Pollution

1. Understanding air pollution begins with understanding normal atmospheric function. How is the atmosphere normally cleansed?

2. Ever since the Industrial Revolution, unhealthy smogs have plagued human cities. How are industrial and photochemical smogs generated?

3. The variety and effects of the major air pollutants have now been identified. What are the eight major air pollutants and their most serious effects?

4. Much is known now about the origins and chemistry of air pollutants. Where do primary pollutants originate, and how do they form secondary pollutants?

5. Acid deposition affects natural ecosystems as well as industrial centers. What is acid deposition, and what are its most significant effects on the environment?

6. Public policy now identifies standards for air pollutants based on their impacts on human health. What are the existing U.S. air pollution standards?

7. The most recent legislation addressing air pollution is the Clean Air Act Amendment of 1990. What are the main provisions of this act, and how are they related to earlier legislation?

8. Coping with acid deposition requires both technological and political change. What is being done in the United States to reduce acid deposition?

9. Further improvements in air quality may require rethinking how we structure our society. What further steps could be taken to improve air quality and our way of life?

On Tuesday morning, October 26, 1948, the people of Donora, Pennsylvania (population 13,000), awoke to a dense fog (Fig. 22–1). Donora lies in an industrialized valley along the Monongahela River. On the outskirts of town, a sooty sign read, "DONORA: NEXT TO YOURS, THE BEST TOWN IN THE U.S.A." The town had a large steel mill, which used high-sulfur coke, and a zinc-reduction plant that roasted ores having a high sulfur content. At the time, most homes in the area were heated with coal. At first, the fog did not seem unusual: Most of Donora's fogs lifted by noon, as the Sun warmed the upper atmosphere and then the land. This one, however, didn't lift for five days.

Through Wednesday and Thursday, the air began to smell of sulfur dioxide—an acrid, penetrating odor. By Friday morning, the town's physicians began to get calls from people in trouble. At first, the calls were from elderly citizens and those with asthmatic conditions. They were having difficulty breathing. The calls continued into Friday afternoon; people young and old were complaining of stomach pain, headaches, nausea, and choking. Work at the mills went on, however. The first deaths occurred on Saturday morning. By 10 A.M., one mortician had nine bodies; two other morticians each had one. On Sunday morning, the mills were shut down. Even so, the owners were certain that their plants had nothing to do with the trouble. Mercifully, the rain came on Sunday and cleared the air—but not before more than 6,000 townspeople were stricken and 20 elderly people had died. During the next month, 50 more people died. Eventually, it was determined that the cause of their deaths was a combination of polluting gases and particles, a thermal inversion in the lower atmosphere, and a stagnant weather system that, together, brought home the deadly potential of *air pollution*.

◀ **Smog in Mexico City.** Because of its geographic location and heavy automobile traffic, Mexico City has some of the worst air pollution anywhere.

▶ **FIGURE 22–1** *Donora, Pennsylvania.* Donora was the scene of a major air pollution disaster. This is what the town looked like at noon on October 29, 1948, in the third day of the killer smog.

With the advent of the Industrial Revolution, the mixture of gases and particles in our atmosphere began changing rapidly, and the effects on natural ecosystems and human health have proved to be dramatic and serious. In the previous chapter, we encountered the structure and function of the atmosphere and looked at its links to climate, climate change, and ozone depletion. Here we will consider what happens when we add pollutants to the air.

22.1 Air Pollution Essentials

Pollutants and Atmospheric Cleansing

For a long time, we have known that the atmosphere contains numerous **gases**: The major constituents of the atmosphere are N_2 (nitrogen), at a level of 78.08%; O_2 (oxygen), at 20.95%; Ar (argon), at 0.93%; CO_2 (carbon dioxide), at 0.03%; and water vapor, ranging from 0 to 4%. Smaller amounts of at least 40 "trace gases" are normally present as well, including ozone, helium, hydrogen, nitrogen oxides, sulfur dioxide, and neon. In addition, aerosols are present in the atmosphere—microscopic liquid and solid particles such as dust, carbon particles, pollen, sea salts, and microorganisms, originating primarily from land and water surfaces and carried up into the atmosphere.

Air pollutants are substances in the atmosphere—certain gases and aerosols—that have harmful effects. Three factors determine the level of air pollution:

- The amount of pollutants entering the air
- The amount of space into which the pollutants are dispersed
- The mechanisms that remove pollutants from the air

We must make a distinction between natural and anthropogenic air pollutants, however. For millions of years, volcanoes, fires, and dust storms have sent smoke, gases, and particles into the atmosphere. Trees and other plants emit volatile organic compounds into the air around them as they photosynthesize. However, there are mechanisms in the biosphere that remove, assimilate, and recycle these natural pollutants. First, the pollutants disperse and become diluted in the atmosphere. Then, as shown in Fig. 22–2, a naturally occurring cleanser, the **hydroxyl radical** (OH), oxidizes many pollutants to products that are harmless or that can be brought down to the ground or water by precipitation. Microorganisms in the soil further convert some of the products into harmless compounds. These three processes hold natural pollutants below toxic levels (except in the immediate area of a source, such as around an erupting volcano). Many of the pollutants oxidized by the hydroxyl radical are of concern because human activities have raised their concentrations far above normal levels.

Recent studies have shown that the hydroxyl radical plays the key role in the removal of anthropogenic pollutants from the atmosphere. Thus, highly reactive hydrocarbons are oxidized within an hour of their appearance in the atmosphere; nitrogen oxides (NO_x) are converted to nitric acid (HNO_3) within a day. It takes months, however, for less reactive substances, like carbon monoxide (CO), to be oxidized by hydroxyl. It appears that the atmospheric levels of hydroxyl are in turn determined by the levels of anthropogenic pollutant gases—thus, hydroxyl's cleansing power

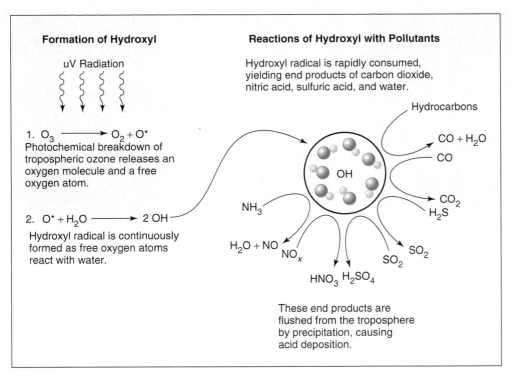

Formation of Hydroxyl

uV Radiation

1. $O_3 \longrightarrow O_2 + O^*$
Photochemical breakdown of tropospheric ozone releases an oxygen molecule and a free oxygen atom.

2. $O^* + H_2O \longrightarrow 2\ OH$
Hydroxyl radical is continuously formed as free oxygen atoms react with water.

Reactions of Hydroxyl with Pollutants

Hydroxyl radical is rapidly consumed, yielding end products of carbon dioxide, nitric acid, sulfuric acid, and water.

Hydrocarbons

$CO + H_2O$

CO

OH

CO_2
H_2S

NH_3

SO_2

$H_2O + NO$
NO_x

SO_2

HNO_3 H_2SO_4

These end products are flushed from the troposphere by precipitation, causing acid deposition.

◄ **FIGURE 22–2** *The hydroxyl radical.* This is a simplified model of atmospheric cleansing by the hydroxyl radical. The first step is the photochemical destruction of ozone, which is the major process leading to ozone breakdown in the troposphere. The second step produces hydroxyl radicals, which react rapidly with many pollutants, converting them to substances that are less harmful or that can be brought down to Earth in precipitation.

is "used up" when high concentrations of pollutants are oxidized, and the pollutants are subsequently able to build up to damaging levels. On the other hand, as Fig. 22–2 shows, the photochemical breakdown of tropospheric ozone is the major source of the hydroxyl radical, and, as we will see, higher levels of ozone result from higher concentrations of other polluting gases.

The Appearance of Smog

Down through the centuries, the practice of venting the products of combustion and other fumes into the atmosphere remained the natural way to avoid their obviously noxious effects within buildings. With the Industrial Revolution of the 1800s came crowded cities and the use of coal for heating and energy. It was then that air pollution began in earnest.

In *Hard Times*, Charles Dickens commented on a typical English scene: "Coketown lay shrouded in a haze of its own, which appeared impervious to the sun's rays. You only knew the town was there because there could be no such sulky blotch upon the prospect without a town." This shrouding haze became known as **industrial smog** (a combination of *smoke* and *fog*!), an irritating, grayish mixture of soot, sulfurous compounds, and water vapor (Fig. 22–3a). This kind of smog continues to be found wherever industries are concentrated and where coal is the primary energy source. The Donora incident is a classic example. Currently, industrial smog can be found in the cities of China, Korea, and a number of Eastern European countries.

After World War II, the booming economy in the United States led to the creation of vast suburbs and a mushrooming use of cars for commuting to and from work. As

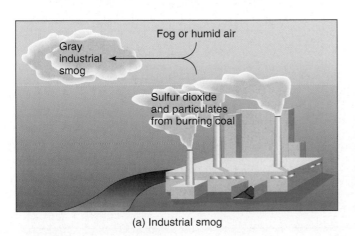

(a) Industrial smog

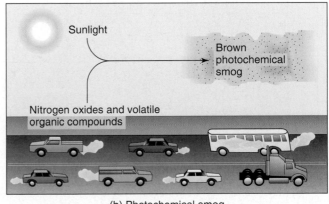

(b) Photochemical smog

▲ **FIGURE 22–3** *Industrial and photochemical smog.* (a) Industrial smog, or gray smog, occurs where coal is burned and the atmosphere is humid. (b) Photochemical smog, or brown haze, occurs where sunlight acts on vehicle pollutants.

(a)

(b)

▲ **FIGURE 22–4** *Los Angeles air.* Views of Los Angeles under (a) clear atmospheric conditions, and (b) in the smog.

other fossil fuels were used in place of coal for heating, industry, and transportation, industrial smog was largely replaced by another kind of smog. Increasingly, Los Angeles and other cities served by huge freeway systems began to be enshrouded daily in a brownish, irritating haze clearly different from the more familiar industrial smog (Fig. 22–4). Weather conditions were usually warm and sunny rather than cool and foggy, and typically the new haze would arise during the morning commute and only begin to dissipate by the time of the evening commute. This **photochemical smog**, as it came to be called, is produced when several pollutants from automobile exhausts—nitrogen oxides and volatile organic carbon compounds—are acted on by sunlight (Fig. 22–3b).

Certain weather conditions intensify levels of both industrial and photochemical smog. Under normal conditions, the daytime air temperature is highest near the ground, because sunlight strikes Earth and the absorbed heat radiates to the air near the surface. The warm air near the ground rises, carrying pollutants upward and dispersing them at higher altitudes (Fig. 22–5a). At night, when the sun is no longer heating Earth, the currents cease. This condition of cooler air below and warmer air above is called a **temperature inversion** (Fig. 22–5b). Inversions are usually very short lived, as the next morning's Sun begins the process anew and any pollutants that accumulated overnight are carried up and away. During cloudy weather, however, the Sun may not be strong enough to break up the

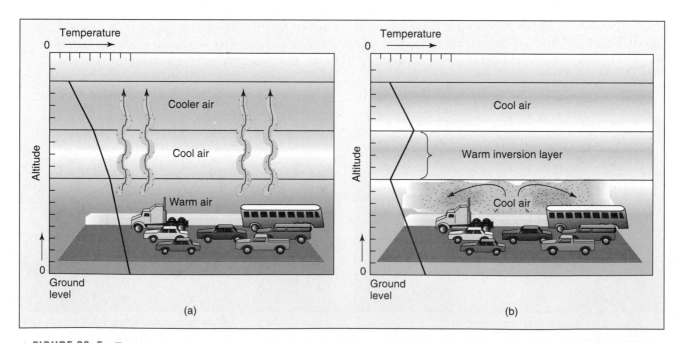

▲ **FIGURE 22–5** *Temperature inversion.* A temperature inversion may cause episodes of high concentrations of air pollutants. (a) Normally, air temperatures are highest at ground level and decrease at higher elevations. (b) A temperature inversion is a situation in which a layer of warmer air overlies cooler air at ground level.

inversion for hours or even days. Or a mass of high-pressure air may move in and sit above the cool surface air, trapping it. Topography can also intensify smogs, as the airflow is affected by nearby mountain ranges or cool ocean air. Los Angeles has both!

When such long-term temperature inversions occur, pollutants can build up to dangerous levels, prompting local health officials to urge people with breathing problems to stay indoors. For many people, smog causes headaches, nausea, and eye and throat irritation; it may aggravate preexisting respiratory conditions, such as asthma and emphysema. In some industrial cities (Donora, Pennsylvania, for example), air pollution reached lethal levels under severe temperature inversions. These cases became known as *air pollution disasters.* London experienced repeated episodes of inversion-related disasters in the mid-1900s; One episode in 1952 resulted in 4,000 pollution-related deaths.

The effects of air pollution are not limited to people. In recent years, many species of trees and other vegetation in and near cities began to die back, and farmers near cities suffered damage to, or even total destruction of, their crops because of air pollution. A conspicuous acceleration in the rate of metal corrosion and the deterioration of rubber, fabrics, and other materials was also noted.

22.2 Major Air Pollutants and Their Impact

Major Pollutants

The following eight pollutants have been identified as most widespread and serious:

1. **Suspended particulate matter.** This is a complex mixture of solid particles and aerosols (liquid particles) suspended in the air. We see these particles as dust, smoke, and haze (Fig. 22–6). Particulates may carry any or all of the other pollutants dissolved in or adsorbed to their surfaces. Because they can be inhaled, the particles impair many respiratory functions, especially in individuals with chronic respiratory problems.

2. **Volatile organic compounds (VOCs).** These include materials such as gasoline, paint solvents, and organic cleaning solutions, which evaporate and enter the air in a vapor state, as well as fragments of molecules resulting from the incomplete oxidation of fuels and wastes. VOCs are prime agents of ozone formation.

3. **Carbon monoxide (CO).** This is an invisible, odorless gas that is highly poisonous to air-breathing animals because of its ability to block the delivery of oxygen to the organs and tissues.

4. **Nitrogen oxides (NO$_x$).** These include several nitrogen–oxygen compounds, all gases. They are converted to nitric acid in the atmosphere and are a major source of acid deposition. Nitrogen dioxide is a lung irritant that can lead to acute respiratory disease in children.

5. **Sulfur oxides (SO$_x$).** The most common species, sulfur dioxide, is a gas that is poisonous to both plants and animals. Children and the elderly are especially sensitive to SO$_2$. It is converted to sulfuric acid (H$_2$SO$_4$) in the atmosphere and is also a major source of acid deposition.

6. **Lead and other heavy metals.** Lead is very dangerous at low concentrations and can lead to brain damage and death. It accumulates in the body and impairs many tissues and organs.

7. **Ozone and other photochemical oxidants.** Recall that ozone in the upper atmosphere (the stratosphere) must be preserved to shield us from ultraviolet radiation (Chapter 21). However, ozone is also highly toxic to both plants and animals; it damages lung tissue and is implicated in many lung disorders. Therefore, ground-level (tropospheric) ozone is a serious pollutant.

8. **Air toxics and radon.** Toxic chemicals in the air include carcinogenic chemicals, radioactive materials, and other chemicals (such as asbestos, vinyl chloride, and benzene) that are emitted as pollutants, but are not included in the preceding list of conventional pollutants. The Clean Air Act identifies 188 hazardous air pollutants (HAPs) in this category, many of which are known human carcinogens. Radon is a radioactive gas produced by natural processes within Earth. All radioactive substances have the potential to be damaging to any living matter they contact.

▲ **FIGURE 22–6** *Industrial pollution.* Because the stack of this wood-products mill in New Richmond, Quebec Province, Canada, lacks electrostatic precipitators, the plume contains substantial amounts of suspended particulate matter that can be seen from a great distance.

Adverse Effects of Air Pollution

It is important to recognize that air pollution is not a single entity, but an alphabet soup of the foregoing materials mixed with the normal constituents of air. Further, the amount of each pollutant present varies greatly, depending on proximity to the source and various conditions of wind and weather. As a result, we are exposed to a mixture that varies in makeup and concentration from day to day—even from hour to hour—and from place to place. Consequently, the effects we feel or observe are rarely, if ever, the effects of a single pollutant; they are the combined impact of the whole mixture of pollutants acting over the total life span, and frequently these effects are *synergistic*—that is, two or more factors combine to produce an effect greater than their simple sum.

For example, plants may be so stressed by pollution that they become more vulnerable to other environmental factors, such as drought or attack by insects. Humans may develop lung disease due to the combined effects of ozone and NO_x. Given the complexity of this situation, it is extremely difficult to determine the role of any particular pollutant in causing an observed result. Nevertheless, some significant progress has been made in linking cause and effect.

Effects on Human Health.   Humans breathe 30 lb (14 kg) of air into their lungs each day. Although some of the symp-toms of pollution that people suffer involve the moist surfaces of the eyes, nose, and throat, the major site of impact is the respiratory system (Fig. 22–7). Three categories of impact can be distinguished:

1. **Chronic:** Pollutants cause the gradual deterioration of a variety of physiological functions over a period of years.
2. **Acute:** Pollutants bring on life-threatening reactions within a period of hours or days.
3. **Carcinogenic:** Pollutants initiate changes within cells that lead to uncontrolled growth and division (cancer), which is frequently fatal.

Chronic Effects.   Almost everyone living in areas of urban air pollution suffers from chronic effects. Long-term exposure to sulfur dioxide can lead to bronchitis (inflammation of the bronchi). Chronic inhalation of ozone and particulates can cause inflammation and, ultimately, fibrosis of the lungs, a scarring that permanently impairs lung function. Carbon monoxide reduces the capacity of the blood to carry oxygen, and extended exposure to low levels of carbon monoxide can contribute to heart disease. Chronic exposure to nitrogen oxides is known to impair the immune system, leaving the lungs open to attack by bacteria and viruses.

GLOBAL PERSPECTIVE

MEXICO CITY: LIFE IN A GAS CHAMBER

"I'm getting out of here one of these days," said one resident of Mexico City. "The ecologists are right. We live in a gas chamber." For many years, Mexico City residents have had the worst air in the world. The city exceeded safety limits for ozone 300 days a year, with values sometimes almost 400% over norms set by the World Health Organization. Smog emergencies have been frequent occurrences. The city's 3.5 million vehicles (a large proportion of which are old and in disrepair) and 60,000 industries pour more than 5 million tons of pollutants into the air every day. Because it is nestled in a natural bowl, Mexico City experiences thermal inversions every day during the winter. In 1999, the World Resources Institute named Mexico City the world's most dangerous city for children in terms of air pollution. One study showed that 40% of the city's young children suffer chronic respiratory sickness during the winter months. The pollution is estimated to cost $11.5 million a day in terms of health effects and lost indus-trial production; an estimated 500,000 residents suffer from some form of smog-related health problem. Writer Carlos Fuentes has nicknamed the metropolis "Makesicko City"!

Mexico City's problems are indicative of a general dilemma throughout the developing world. Jobs and industrial activity, vital for a nation's economy, depend on the use of motor vehicles and on the industrial plants surrounding the city. Many automobile owners are absolutely dependent on their cars, but can ill afford to keep them repaired in order to reduce exhaust emissions. Some people see cars as status symbols and will acquire one as soon as possible—often a "clunker" discarded from the developed world. Industries in Mexico are not regulated as they are in the United States and lack the technology and capital to control emissions.

The Mexican government has initiated an air-monitoring system and, more recently, mandated emissions tests and catalytic converters for automobiles. The city has adopted a "Day without a Car" policy: Drivers are given decals that permit driving only on specific days. Residents, however, get around the rule by purchasing another used car for use on alternate days. New air pollution laws are beginning to be enforced: One major refinery responsible for 4% of the air-quality problem was shut down by then President Salinas in the spring of 1991, and 80 other industries were temporarily or permanently closed for violating the new laws. New subway lines are being built, and buses that are less polluting are being added to the fleet. Lead, sulfur dioxide, and carbon monoxide levels are declining. By the mid-1990s, levels of air pollution were declining by 10% a year, but a rising city population and a growing automobile fleet threaten to cancel the gains made. In the meantime, Mexico City's 20 million inhabitants are part of an unintentional experiment in demonstrating the health effects of severe air pollution.

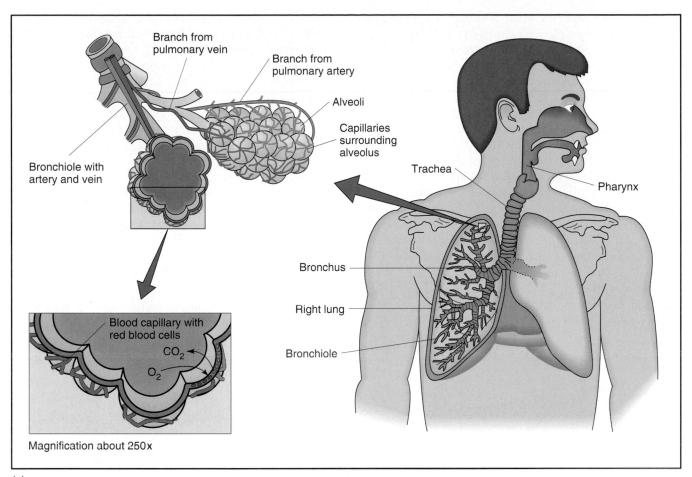

(a)

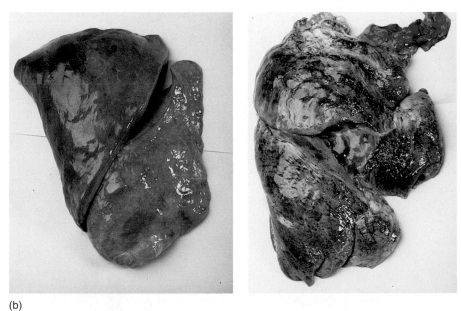

(b)

▲ **FIGURE 22–7** ***The respiratory system.*** (a) In the lungs, air passages branch and rebranch and finally end in millions of tiny sacs called alveoli. These sacs are surrounded by blood capillaries. As blood passes through the capillaries, oxygen from inhaled air diffuses into the blood from the alveoli. Carbon dioxide diffuses in the reverse direction and leaves the body in the exhaled air. (b) On the left is normal lung tissue, on the right lung tissue from a person who suffered from emphysema, a chronic lung disease in which some of the structure of the lungs has broken down. Cigarette smoking and heavy air pollution are associated with the development of emphysema and other chronic lung diseases.

Those most sensitive to air pollution are small children, asthmatics, people with chronic pulmonary or heart disease, and the elderly. Asthma, an immune system disorder characterized by impaired breathing caused by the constriction of air passageways, is brought on by contact with allergens and many of the compounds in polluted air. According to the Clean Air Task Force, ground-level ozone and the smog it generates are responsible for sending 53,000 asthmatics to the hospital each summer. Between 1980 and 1995, the prevalence of asthma in the United States rose by 75%; over 17 million people (including 4.8 million children) suffer from the condition today.

Two important studies recently analyzed by the Health Effects Institute have mounted strong evidence of the harmful effects of fine particles and sulfur pollution. These studies followed thousands of adult subjects living in 154 U.S. cities for as long as 16 years. In the studies, higher concentrations of fine particles were clearly correlated with increased mortality, especially from cardiopulmonary disease. The more polluted the city, the higher was the mortality (Fig. 22–8). The studies were used by the EPA to step up its regulatory attention to fine particles, as we will see later.

One of the heavy-metal pollutants, lead, deserves special attention. For several decades, lead poisoning has been recognized as a cause of mental retardation. Researchers once thought that eating paint chips containing lead was the main source of lead contamination in humans. In the early 1980s, however, elevated lead levels in blood were shown to be much more widespread than previously expected, and they were present in adults as well as children. Learning disabilities in children, as well as high blood pressure in adults, were correlated with high levels of lead in the blood. The major source of this widespread contamination was traced to leaded gasoline. (The lead in exhaust fumes may be inhaled directly or may settle on food, water, or any number of items that are put into the mouth.) This knowledge led the EPA to mandate the elimination of leaded gasoline, completed at the end of 1996. The result has been a dramatic reduction in lead concentrations in the environment.

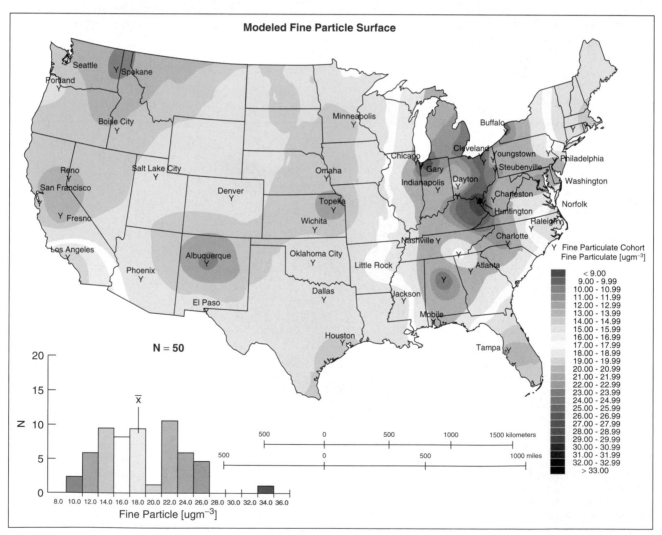

▲ FIGURE 22–8   *Distribution of fine particles.*   A map of the distribution of fine particles in the United States. [*Source*: American Cancer Society study (Pope et al., 1995).]

Acute Effects. In severe cases, air pollution reaches levels that cause death, although it should be noted that such deaths usually occur among people already suffering from severe respiratory or heart disease or both. The gases present in air pollution are known to be lethal in high concentrations, but such concentrations occur in cases of accidental poisoning only. Therefore, deaths attributed to air pollution are not the direct result of simple poisoning. However, intense air pollution puts an additional stress on the body, and if a person is already in a weakened condition (as for example, the elderly or asthmatics), this additional stress may be fatal. Most of the deaths in air pollution disasters reflect the acute effects of air pollution. However, recent research shows that even moderate air pollution can cause changes in cardiac rhythms of people with heart disease, triggering fatal heart attacks.

Carcinogenic Effects. The heavy-metal and organic constituents of air pollution include many chemicals known to be carcinogenic in high doses. According to industrial reporting required by the EPA, 4.6 million tons of hazardous air pollutants (air toxics) are released annually into the air in the United States. The presence of trace amounts of these chemicals in air may be responsible for a significant portion of the cancer observed in humans. One major source of such carcinogens is diesel exhausts; the EPA has just classified diesel exhaust as a "likely human carcinogen" and is moving ahead with plans to require drastic reductions in the pollutants from this source.

In some cases, exposure to a pollutant can be linked directly to cancer and other health problems by way of epidemiological evidence. One pollutant that clearly and indisputably is correlated with cancer and other disorders is benzene. This organic chemical is present in motor fuels. It is also used as a solvent for fatty substances and in the manufacture of detergents, explosives, and pharmaceuticals. Environmentally, benzene is found in emissions from fossil fuel combustion: motor-vehicle exhaust and burning coal and oil. Benzene is also present in tobacco smoke, which accounts for half of the public's exposure to the chemical. The EPA has classified benzene as a Group A, known human carcinogen, linked to leukemia in persons encountering the chemical through occupational exposure. Chronic exposure to benzene can also lead to numerous blood disorders and damage to the immune system.

Effects on Agriculture and Forests. Experiments show that plants are considerably more sensitive to gaseous air pollutants than are humans. Before emissions were controlled, it was common to see wide areas of totally barren land or severely damaged vegetation downwind from smelters or coal-burning power plants (Fig. 22–9). The pollutant responsible was usually sulfur dioxide.

The dying off of vegetation in large urban areas and the damage to crops, orchards, and forests downwind of urban centers are caused mainly by exposure to ozone and other photochemical oxidants. Estimates of crop damage by

▲ **FIGURE 22–9** *Air pollution damage.* The countryside around this Butte, Montana, smelter is devastated by toxic pollutants from the industrial processes that take place in the smelter.

ozone range from $2 billion to $6 billion per year, with an estimated $1 billion in California alone. It is significant that much of the world's grain production occurs in regions that receive enough ozone pollution to reduce crop yields. The same parts of the world that produce 60% of the world's food—North America, Europe, and eastern Asia—also produce 60% of the world's air pollution.

The negative impact of air pollution on wild plants and forest trees may be even greater than on agricultural crops. Significant damage to valuable ponderosa and Jeffrey pines occurred along the entire western foothills of the Sierra Nevada in California. U.S. Forest Service studies have concluded that ozone from the nearby Central Valley and San Francisco–Oakland metropolitan region was responsible for this damage. Forests under stress from pollution are more susceptible to damage by insects and other pathogens than are unstressed forests. The deaths of the ponderosa and Jeffrey pines in the Sierra Nevada were attributed to western pine beetles, which invade trees weakened by ozone.

Heavy metals, ozone, and acids carried in by clouds from the Midwest were implicated in the death of red spruce in Vermont's Green Mountains (Fig. 22–10). During the 1970s in the San Bernardino Mountains of California, an area that receives air pollution from Los Angeles, 50% of the trees died in some areas. As a result of air pollution controls, those same areas have shown significant improvement in tree growth in recent years, an encouraging sign.

Effects on Materials and Aesthetics. Walls, windows, and other exposed surfaces turn gray and dingy as particulates settle on them. Paints and fabrics deteriorate more

▲ **FIGURE 22–10** *Effect of pollution on forests.* Heavy metals and acids from distant urban areas were responsible for the deaths of red spruce trees in Vermont's Green Mountains.

rapidly, and the sidewalls of tires and other rubber products become hard and checkered with cracks because of oxidation by ozone. Metal corrosion is dramatically increased by sulfur dioxide and acids derived from sulfur and nitrogen oxides, as are weathering and the deterioration of stonework (Fig. 22–11). These and other effects of air pollutants on materials increase the costs for cleaning or replacing them by hundreds of millions of dollars a year. Many of the damaged materials are irreplaceable.

A clear blue sky and good visibility are matters of health, but they also carry significant aesthetic value and can have a deep psychological impact on people. Can a

value be put on these benefits? Many of us spend thousands of dollars and hundreds of hours commuting long distances to work so that we can live in a less polluted environment than the one in which we work. Ironically, the resulting traffic and congestion cause much of the very pollution we are trying to escape. Then we travel to national parks and wilderness areas and too often find that the visibility in *these* natural areas is impaired by what has been called "regional haze," coming from particulates and gases originating hundreds of miles away. In a surprise move, the EPA has proposed regulations to reduce regional haze, with the goal of a 10% improvement in visibility every 10 to 15 years during the next century. This may be the first time the government has attempted to regulate pollutants for aesthetic reasons!

Having considered the identity of the major air pollutants and their impacts, we now turn our attention to their origins.

22.3 Sources of Pollutants

In large measure, air pollutants are direct and indirect by-products of the burning of coal, gasoline, other liquid fuels, and refuse (wastepaper, plaster, and so on). These fuels and wastes are organic compounds. With complete combustion, the by-products of burning are carbon dioxide and water vapor, as shown by the following equation for the combustion of methane:

$$CH_4 + 2\,O_2 \rightarrow CO_2 + 2\,H_2O$$

Unfortunately, oxidation is seldom complete, and substances far more complex than methane are involved.

The first six of the major pollutants listed earlier in this chapter (**particulates**, **VOCs**, **CO**, **NO**$_x$, **SO**$_x$, and **lead**) are

▶ **FIGURE 22–11** *Effect of pollution on monuments.* The corrosive effects of acids from air pollutants are dissolving away the features of many monuments and statues, as seen in the faces of these statues in Brooklyn, New York.

called **primary pollutants**, because they are the direct products of combustion and evaporation. Some primary pollutants may undergo further reactions in the atmosphere and produce additional undesirable compounds, called **secondary pollutants**. Sources of air pollution are summarized in Fig. 22–12. Notice that while industrial processes are the major source of particulates, transportation and fuel combustion account for the lion's share of the other pollutants. Strategies for controlling air pollutants are different for the different sources.

Primary Pollutants

When fuels and wastes are burned, particles consisting mainly of carbon are emitted into the air; these are the particulates we see as soot and smoke. In addition, various unburned fragments of fuel molecules remain; these are the VOC emissions. Incompletely oxidized carbon is carbon monoxide (CO), in contrast to completely oxidized carbon, which is carbon dioxide (CO_2). Combustion takes place in the air, which is 78% nitrogen and 21% oxygen. At high combustion temperatures, some of the nitrogen gas is oxidized to form the gas nitric oxide (NO). In the air, nitric oxide immediately reacts with additional oxygen to form nitrogen dioxide (NO_2), nitrogen tetroxide (N_2O_4), or both. (These compounds are collectively referred to as the **nitrogen oxides**.) Nitrogen dioxide absorbs light and is largely responsible for the brownish color of photochemical smog.

In addition to organic matter, fuels and refuse contain impurities and additives, and these substances, too, are emitted into the air during burning. Coal, for example, contains from 0.2% to 5.5% sulfur. In combustion, this sulfur is oxidized, giving rise to the gas sulfur dioxide (SO_2). Coal may contain heavy-metal impurities, and refuse, of course, contains an endless array of "impurities."

The EPA Office of Air Quality tracks trends in national **emissions** of the major pollutants from all sources; the EPA also follows **ambient concentrations** of the pollutants at thousands of monitoring stations across the country. According to EPA data, 1998 *emissions* in the United States of the first five primary pollutants amounted to 154 million tons. By comparison, in 1970, when the first Clean Air Act became law, these same five pollutants totaled 244 million tons. The relative amounts emitted in the United States, as well as their major sources, are illustrated in Fig. 22–13. The 10-year trends in the emissions of these primary pollutants are represented in Fig. 22–14. The general picture shows gradual improvement for most emissions over the last decade, continuing a trend that reflects the effectiveness of Clean Air Act regulations. This progressively improving trend has occurred in spite of the

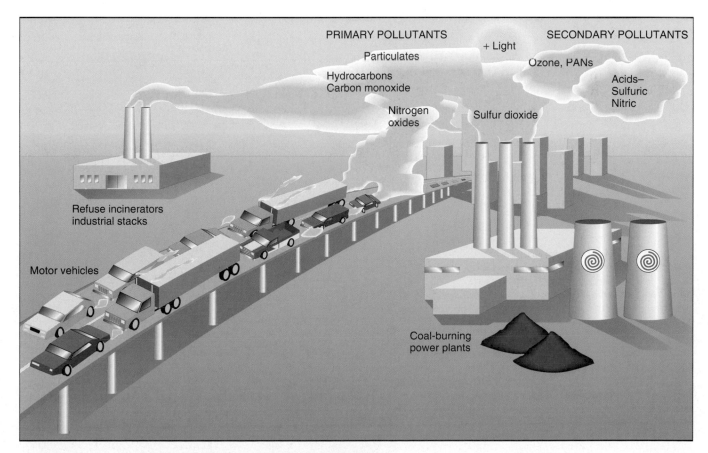

▲ FIGURE 22–12 *The prime sources of the major air pollutants.*

▶ **FIGURE 22–13** *Current emissions in the United States of five primary air pollutants, by source, for 1998.* *Fuel combustion* refers to fuels burned for electrical power generation and for space heating. Note especially the different contributions by transportation and fuel combustion, the two major sources of air pollutants. PM-10 particulate matter percentages are for nonfugitive dust sources. (*Source:* EPA Office of Air Quality, 2000.)

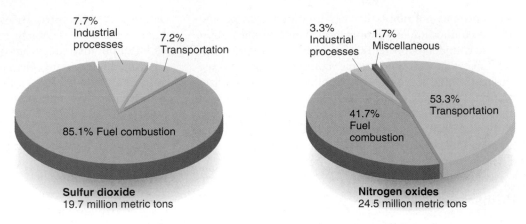

7.7%
Industrial
processes

7.2%
Transportation

85.1% Fuel combustion

Sulfur dioxide
19.7 million metric tons

3.3%
Industrial
processes

1.7%
Miscellaneous

53.3%
Transportation

41.7%
Fuel
combustion

Nitrogen oxides
24.5 million metric tons

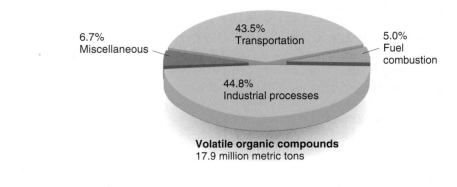

6.7%
Miscellaneous

43.5%
Transportation

5.0%
Fuel
combustion

44.8%
Industrial processes

Volatile organic compounds
17.9 million metric tons

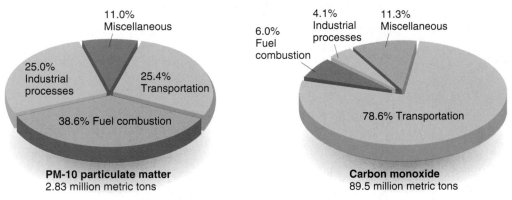

11.0%
Miscellaneous

25.0%
Industrial
processes

25.4%
Transportation

38.6% Fuel combustion

PM-10 particulate matter
2.83 million metric tons

6.0%
Fuel
combustion

4.1%
Industrial
processes

11.3%
Miscellaneous

78.6% Transportation

Carbon monoxide
89.5 million metric tons

▶ **FIGURE 22–14** *Ten-year trends in emissions of six criteria pollutants since 1988.* Improvements in carbon monoxide, sulfur dioxide, and VOC emissions reflect Clean Air Act successes. Nitrogen oxides have not improved, because little attention has been paid to them.

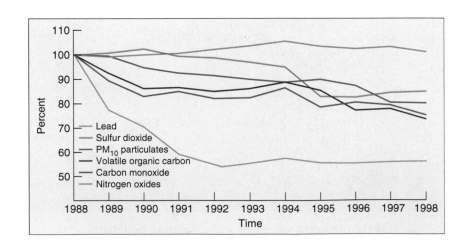

Lead
Sulfur dioxide
PM₁₀ particulates
Volatile organic carbon
Carbon monoxide
Nitrogen oxides

fact that over that time, the economy has increased by 147%, and vehicular miles traveled by 140%. Decreases in SO_2 emissions are linked to the acid rain program, but NO_x emissions have not declined at all in recent years. The good news is that ambient levels of NO_x have improved to the point that all areas in the country are now in compliance with the ambient standards.

The sixth type of primary pollutant—lead and other heavy metals—is discussed separately because the quantities emitted are far less than the levels for the first five. Before the EPA-directed phaseout, lead was added to gasoline as an inexpensive way to prevent engine knock. Emitted with the exhaust from gasoline-burning vehicles, the lead remained airborne and traveled great distances before settling as far away as the glaciers of Greenland. Since the phaseout, concentrations of lead in the air of cities in the United States have shown a remarkable decline: Between 1977 and 1998, ambient lead concentrations fell 97%. At the same time, levels of lead in children's blood have dropped greatly. Lead concentrations in Greenland ice have also decreased significantly. All these declines indicate that lead restrictions in the United States and a few other Northern Hemisphere nations have had a global impact. The data indicate that we have reached a steady state in lead emissions and that any new reductions must target the metal-processing industries, which are now by far the largest source of lead emissions.

As with lead, the concentrations of toxic chemicals and radon in the air are small compared with those of the other primary pollutants. Some of the air's toxic compounds—benzene, for example—originate with transportation fuels. Most, however, are traceable to industries and small businesses. Radon, on the other hand, is produced by the spontaneous breakdown of fissionable material in rocks and soils. Radon escapes naturally to the surface and seeps into buildings through cracks in foundations and basement floors, sometimes collecting in the structures.

Secondary Pollutants

Ozone and numerous reactive organic compounds are formed as a result of chemical reactions between nitrogen oxides and volatile organic carbons. Because sunlight provides the energy necessary to propel the reactions, these products are known collectively as **photochemical oxidants**. In preindustrial times, ozone concentrations ranged from 10 to 15 parts per billion (ppb), well below harmful levels. Summer concentrations of ozone in unpolluted air in North America now range from 20 to 50 ppb. Polluted air may contain ozone concentrations of 150 ppb or more, a level considered very unhealthy if encountered for extended periods of time. EPA data indicate that ambient ozone levels, measured at over 700 sites across the United States, have been declining over the past 20 years, but little improvement has been seen since 1994. Ozone air quality standards are far and away the leaders in noncompliance

across the country, with over 120 million living in counties still not meeting the standards.

The major reactions in the formation of ozone and other photochemical oxidants are shown in Fig. 22–15. Nitrogen dioxide absorbs light energy and splits to form nitric oxide and atomic oxygen, which rapidly combines with oxygen gas to form ozone. If other factors are not involved, ozone and nitric oxide then react to form nitrogen dioxide and oxygen gas. A steady-state concentration of ozone results, and there is no appreciable accumulation of the gas (Fig. 22–15a). When VOCs are present, however, the nitric oxide reacts with them instead of with the ozone, causing several serious problems. First, the reaction between nitric oxide and the VOCs leads to highly reactive and damaging compounds known as peroxyacetyl nitrates, or PANs (Fig. 22–15b). Second, numerous aldehyde and ketone compounds are produced as the VOCs are oxidized by atomic oxygen, and these compounds are also noxious. Finally, with the nitric oxide tied up in this way, the ozone tends to accumulate. Because of the complex air chemistry involved, ozone concentrations usually peak 30 to 100 miles (50 to 160 km) downwind of urban centers where the primary pollutants were generated. As a result, ozone concentrations above the health standard in the United States are often found in rural and wilderness areas.

Particulates take on additional potency when they adsorb other pollutants onto their surfaces. (*Ad*sorption means that the pollutants simply adhere to a surface, like flies sticking to flypaper. By contrast, *ab*sorption means "soaking in.") When inhaled, many of these contaminated particulates are fine enough to penetrate deeply into the lungs, releasing the adsorbed pollutants onto the moist lung surfaces, where they often remain trapped for life.

In a sense, **sulfuric** and **nitric acids** can also be considered secondary pollutants, since they are products of sulfur dioxide and nitrogen oxides reacting with atmospheric moisture and oxidants such as hydroxyl. Sulfuric and nitric acids are the sources of the well-known atmospheric pollution commonly termed acid rain (and technically referred to as acid deposition). We now turn our attention to this important and widespread problem.

22.4 Acid Deposition

Acid precipitation refers to any precipitation—rain, fog, mist, or snow—that is more acidic than usual. Because dry acidic particles are also found in the atmosphere, the combination of precipitation and dry-particle fallout is called **acid deposition**. In the late 1960s, Swedish scientist Svante Odén first documented the acidification of lakes in Scandinavia and traced it to air pollutants originating in other parts of Europe and Great Britain. Since then, careful monitoring has shown that broad areas of North America, as well as most of Europe and other industrialized regions of the world, are regularly experiencing precipitation that is between 10 and

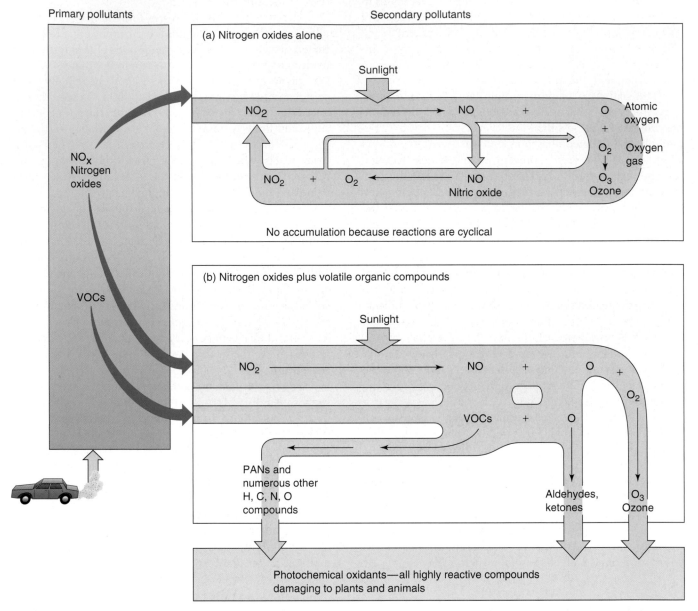

▲ FIGURE 22–15 *Formation of ozone and other photochemical oxidants.* (a) Nitrogen oxides, by themselves, would not cause ozone and other oxidants to reach damaging levels, because reactions involving nitrogen oxides are cyclic. (b) When VOCs are also present, however, reactions occur that lead to the accumulation of numerous damaging compounds—most significantly, ozone, the most injurious.

1,000 times more acidic than usual. This is affecting ecosystems in diverse ways, as illustrated in Fig. 22–16.

To understand the full extent of the problem, first we must understand some principles about acids and how we measure their concentration.

Acids and Bases

Acidic properties (e.g., a sour taste and corrosiveness) are due to the presence of hydrogen ions (H^+, a hydrogen atom without its electron), which are highly reactive. Therefore, an **acid** is any chemical that releases hydrogen ions when

dissolved in water. The chemical formulas of a few common acids are shown in Table 22–1. Note that all of them ionize—that is, their components separate—to give hydrogen ions plus a negative ion. The higher the concentration of hydrogen ions in a solution, the more acidic is the solution.

A **base** is any chemical that releases hydroxide ions (OH^-, oxygen–hydrogen groups with an extra electron) when dissolved in water. (See Table 22–1.) The bitter taste and caustic properties of all alkaline, or basic, solutions are due to the presence of hydroxide ions.

The concentration of hydrogen ions is expressed as **pH**. The pH scale goes from 0 (highly acidic) through 7 (neutral)

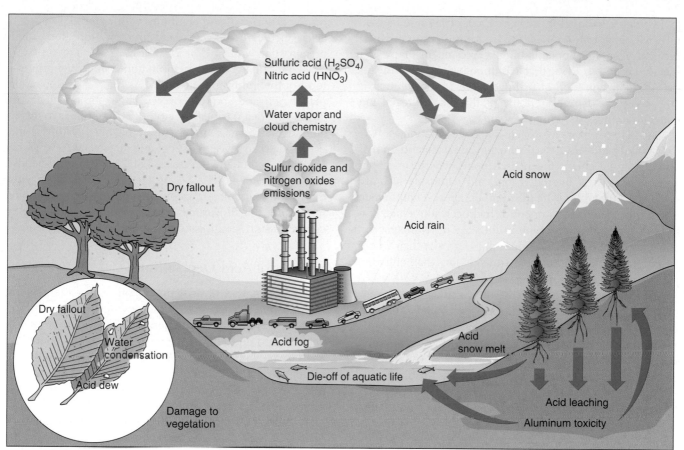

▲ **FIGURE 22–16** *Acid deposition.* Emissions of sulfur dioxide and nitrogen oxides react with hydroxyl radicals and water vapor in the atmosphere to form their respective acids, which come back down either as dry acid deposition or, mixed with water, acid precipitation. Various effects of acid deposition are noted.

TABLE 22–1 Common Acids and Bases

Acid	Formula	Yields	H^+ Ion(s)	Plus	Negative Ion	
Hydrochloric acid	HCl	→	H^+	+	Cl^-	Chloride
Sulfuric acid	H_2SO_4	→	$2 H^+$	+	SO_4^{2-}	Sulfate
Nitric acid	HNO_3	→	H^+	+	NO_3^-	Nitrate
Phosphoric acid	H_3PO_4	→	$3 H^+$	+	PO_4^{3-}	Phosphate
Acetic acid	CH_3COOH	→	H^+	+	CH_3COO^-	Acetate
Carbonic acid	H_2CO_3	→	H^+	+	HCO_3^-	Bicarbonate
Base	**Formula**	**Yields**	OH^- **Ion(s)**	**Plus**	**Positive Ion**	
Sodium hydroxide	NaOH	→	OH^-	+	Na^+	Sodium ion
Potassium hydroxide	KOH	→	OH^-	+	K^+	Potassium ion
Calcium hydroxide	$Ca(OH)_2$	→	$2 OH^-$	+	Ca^{2+}	Calcium ion
Ammonium hydroxide	NH_4OH	→	OH^-	+	NH_4^+	Ammonium

to 14 (highly basic) (Fig. 22–17). Each of the numbers on the scale represents the negative logarithm (power of 10) of the hydrogen ion concentration, expressed in grams per liter. For example, to say that a solution has a pH of 1 means that the concentration of hydrogen ions in the solution is 10^{-1} g/L (0.1 g/L); pH = 2 means that the hydrogen ion concentration is 10^{-2} g/L, and so on.

At pH = 7, the hydrogen ion concentration is 10^{-7} (0.0000001) g/L, and the hydroxide ion concentration is also 10^{-7} g/L. This is the neutral point, where small, but equal, amounts of hydrogen ions and hydroxide ions are present in pure water. The pH numbers above 7 continue to express the negative exponent of hydrogen ion concentration, but they also represent an increase in hydroxide ion concentration. Because solutions of pH greater than 7 contain higher concentrations of hydroxide ions than of hydrogen ions, they are called basic solutions.

We use the same negative logarithm arrangement to express the concentration of hydroxide ions: **pOH**. The relationship between pH and pOH is reciprocal: As the pH of a given solution goes up, its pOH goes down, and vice versa. In fact, we can state this as a rule: For any aqueous solution, pH + pOH = 14. For example, pH = 13 means that the hydrogen ion concentration is 10^{-13} (a decimal followed by 12 zeros and a 1) g/L. However, the hydroxide ion concentration of this same solution is 10^{-1} g/L, so the pOH = 1. In a neutral solution, equal amounts of hydrogen ions and hydroxide ions are present; thus, pH = 7, pOH = 7, and pH + pOH = 14.

Since numbers on the pH scale represent powers of 10, there is a *tenfold difference* between each unit and the next. For example, pH 5 is 10 times as acidic (has 10 times as many H^+ ions) as pH 6, pH 4 is 10 times as acidic as pH 5, and so on.

One easy way to measure pH is with indicator paper, which is available from any laboratory supply house. This paper contains pigments that change color when they are wetted with an acidic or a basic solution. The pH is determined by dipping a strip of indicator paper in the solution and matching the color of the wet paper with a color chart provided with the paper. When accuracy and precision are important, however, it is necessary to use electronic instruments for measuring pH.

Extent and Potency of Acid Precipitation

In the absence of any pollution, rainfall is normally slightly acidic, with a pH of 5.6, because carbon dioxide in the air readily dissolves in and combines with water to produce an acid called carbonic acid. **Acid precipitation**, then, is any precipitation with a pH of 5.5 or less.

Unfortunately, acid precipitation is now the norm over most of the industrialized world. As Fig. 22–18 shows, the pH of rain and snowfall over a large portion of eastern North America is typically below 4.5. Many areas in this region regularly receive precipitation having a pH of 4.0 and, occasionally, as low as 3.0. Fogs and dews can be even more acidic; in mountain forests east of Los Angeles, scientists found fog water of pH 2.8—almost 1,000 times more acidic than usual—dripping from pine needles. Acid precipitation has been heavy in Europe as well, from the British Isles to central Russia. Also, it is now well documented in Japan.

Sources of Acid Deposition

Chemical analysis of acid precipitation in eastern North America and Europe reveals the presence of two acids, **sulfuric acid** (H_2SO_4) and **nitric acid** (HNO_3), in a ratio of about two to one. (In the western United States and in the western provinces of Canada, nitric acid predominates, principally formed from tailpipe emissions.) As we have seen, burning fuels produce sulfur dioxide and nitrogen oxides, so the source of the acid deposition problem is evident. These oxides enter the troposphere in large quantities from both anthropogenic and natural sources. Once in the troposphere, they are oxidized by hydroxyl radicals (Fig. 22–2) to sulfuric and nitric acids, which dissolve readily in water or adsorb to particles and are brought down to Earth in acid deposition. This usually occurs within a week of the oxides' entering the atmosphere.

We must recognize that *natural sources* contribute substantial quantities of pollutants to the air: 50 to 70 million tons per year of sulfur (from volcanoes, sea spray, and microbial processes) and 30 to 40 million tons per year of nitrogen oxides (from lightning, the burning of biomass, and microbial processes). *Anthropogenic sources* are estimated at 100 to 130 million tons per year of sulfur dioxide and 60 to 70 million tons per year of nitrogen oxides. The vital difference between these two sources is that anthropogenic oxides are strongly concentrated in industrialized regions, whereas the emissions from natural sources are spread out over the globe and are a part of the global environment. Levels of the anthropogenic oxides have increased at least fourfold since 1900, while levels of the natural emissions have remained fairly constant.

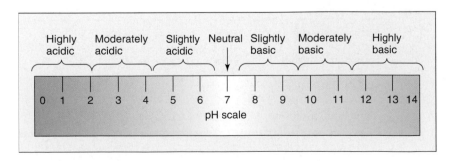

▶ **FIGURE 22–17** *The pH scale.*

As Fig. 22–13 indicates, 19.7 million tons of sulfur dioxide were released into the air in 1998 in the United States; 85% was from fuel combustion (mostly from coal-burning power plants). Some 24.5 million tons of nitrogen oxides were released, 53% traced to transportation emissions and 42% to fuel combustion at fixed sites. In the eastern United States, the source of much of the acid deposition was identified as the tall stacks of 50 huge coal-burning power plants (Fig. 22–19). The tall stacks were built to alleviate local sulfur dioxide pollution at ground level. These same plants are now reducing their emissions as a result of the Clean Air Act, along with over 200 other large coal-fired power plants targeted by the EPA, which states that 64% of SO_2 emissions and 24% of NO_x emissions still originate from fossil-fuel-burning electric utility plants.

Effects of Acid Deposition

Acid deposition has been recognized as a problem in and around industrial centers for over 100 years. Its impact on ecosystems, however, was noted only about 40 years ago, when anglers started noticing sharp declines in fish populations in many lakes in Sweden, Ontario, and the Adirondack Mountains of upper New York State. Since that time, as ecological damage has continued to spread, studies have revealed many ways in which acid deposition alters and may destroy ecosystems.

Impact on Aquatic Ecosystems. The pH of an environment is extremely critical, because it affects the function of virtually all enzymes, hormones, and other proteins in the bodies of all organisms living in that environment. Ordinarily, organisms are able to regulate their internal pH; however, consistently low environmental pH often overcomes this regulatory ability in many life-forms. Most freshwater lakes, ponds, and streams have a natural pH in the range of 6 to 8, and organisms are adapted accordingly. The eggs, sperm, and developing young of these organisms are especially sensitive to changes in pH. Most are severely stressed, and many die, if the environmental pH shifts as little as one unit from the optimum.

As aquatic ecosystems become acidified, virtually all higher organisms die off rapidly, either because the acidified water kills them or because it keeps them from reproducing. Figure 22–16 illustrates the fact that acid precipitation may leach aluminum and various heavy metals from the soil as the water percolates through it. Normally, the presence of these elements in the soil does not pose a problem, because they are bound in insoluble mineral compounds and therefore are not absorbed by organisms. As the compounds are dissolved by low-pH water, however, the metals are freed; they may then be absorbed and are highly toxic to both plants and animals. For example, mercury tends to accumulate in fish as lake waters become more acidic. Indeed, mercury levels are so high in the Great Lakes that many bordering states advise against eating fish caught in these waters.

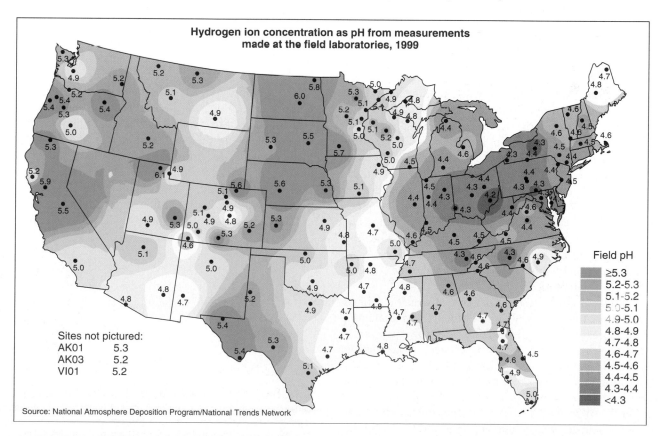

▲ FIGURE 22–18 *Acid deposition in the United States.* Data from over 200 monitoring sites across the United States indicate that acid deposition continues to fall throughout the East and much of the Midwest (AK sites are in Alaska, and VI refers to the Virgin Islands.)

(a)

(b)

▲ **FIGURE 22–19** *Midwestern coal-burning power plant.* (a) Standard smokestacks of this coal-burning power plant were replaced by new 1,000-foot (330-m) stacks to aid in the dispersion of pollutants into the atmosphere. The taller stacks alleviated local air pollution problems, but created more widespread distribution of acid-generating pollutants. (b) Locations of the 50 largest sulfur dioxide emitters, all of which are utility coal-burning power plants.

In Norway and Sweden, all the fish have died in at least 6,500 lakes and in seven rivers that once abounded in Atlantic salmon. In Ontario, Canada, approximately 1,200 lakes now harbor no fish life. In the Adirondacks, a favorite recreational region for New Yorkers, more than 200 lakes are without fish, and many are devoid of all life save resistant algae and bacteria. More than 190 New England lakes and rivers are suffering from recent acidification. The physical appearance of such lakes is deceiving. From the surface, they are clear and blue, the outward signs of a healthy condition. However, a view under the surface is eerie: In spite of ample light shimmering through the clear water, there is not a sign of life in them.

As Fig. 22–18 indicates, wide regions of the United States receive roughly equal amounts of acid precipitation, yet not all areas have acidified lakes. Apparently, many areas remain healthy, whereas others have become acidified to the point of becoming lifeless. How is this possible? The key lies in the system's *buffering capacity*: Despite the addition of acid to it, a system may be protected from pH change by a **buffer**—a substance that, when present in a solution, has a large capacity to absorb hydrogen ions and thus hold the pH relatively constant.

Limestone ($CaCO_3$) is a natural buffer (Fig. 22–20) that protects lakes from the effects of acid precipitation in many areas of the North American continent. Lakes and streams receiving their water from rain and melted snow that has percolated through soils derived from limestone will contain dissolved limestone. The regions sensitive to acid precipitation are those containing much granitic rock that does not yield good buffers. For these areas, the most critical time of year is the spring thaw, when accumulated

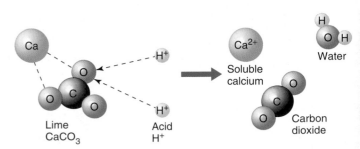

▲ **FIGURE 22–20** *Buffering.* Acids may be neutralized by certain nonbasic compounds called *buffers*. A buffer such as limestone (calcium carbonate) reacts with hydrogen ions as shown. Hence, the pH of a lake or river remains close to neutral despite the additional acid.

winter snow melts. If the thaw is sudden, streams, rivers, and lakes are hit with what has been called "acid shock," as accumulated acids send pH levels plummeting in a sudden burst of meltwater. Making matters worse, the acid shock often coincides with the times of spawning and egg laying in aquatic animals, when they are at their most vulnerable.

Any buffer has limited capacity. Limestone, for instance, is used up by the buffering reaction and so is no longer available to react with more added hydrogen ions. Ecosystems that have already acidified and collapsed are those that had very little buffering capacity. Those that remain healthy have greater buffering capacity. However, many of these still healthy lakes are gradually losing their buffering capacity as more and more acid continues to be deposited. In time, many healthy lakes will join the growing number of dead lakes if acid precipitation is not curbed.

Impact on Forests.   Along with dying lakes, the decline of forests has been conspicuous. From the Green Mountains of Vermont to the San Bernardino Mountains of California, the die-off of forest trees in the 1980s caused great concern. Red spruce forests are especially vulnerable; in New England, 1.3 million acres of high-elevation forests have been devastated. Commonly, the damaged trees lost needles and often fell prey to insects and diseases before they succumbed.

Much of the damage of acid precipitation to forests is due to chemical interactions within the forest soils. Sustained acid precipitation at first adds nitrogen and sulfur to the soils, which stimulate tree growth. In time, these chemicals affect the soils by leaching out large quantities of the buffering chemicals, usually calcium and magnesium salts. When these buffering salts no longer neutralize the acid rain, aluminum ions, which are toxic, are dissolved from minerals in the soil. The combination of aluminum and the increasing scarcity of calcium, which is essential to plant growth, leads to reduced tree growth. Research at the Hubbard Brook Experimental Forest in the White Mountains of New Hampshire has shown a marked reduction in calcium and magnesium in the forest soils from the 1960s on, which is reflected in the calcium of tree rings over the same period. The net result of these changes has been a severe decline in forest growth.

Declining forests are a serious problem in some parts of Europe, and the evidence indicates that the same kinds of soil–chemical exchanges are occurring there. Because of the variations in the buffering capacity of soils and the differing amounts of sulfur and nitrogen brought in by acid precipitation, forests are affected to varying degrees. Some forests continue to grow, whereas others are in decline. One result of the sustained acidification is a gradual shift toward more acid-tolerant species. For example, in New England, the balsam fir is moving in to replace the dead spruce.

Impact on Humans and Their Artifacts.   One of the more noticeable effects of acid deposition is the deterioration of artifacts. Limestone and marble (which is a form of limestone) are favored materials for the outsides of buildings and for monuments (collectively called **artifacts**). The reaction between acid and limestone is causing these structures to erode at a tremendously accelerated pace. Monuments and buildings that have stood for hundreds or even thousands of years with little change are now dissolving and crumbling away, as was shown in Fig. 22–11. The corrosion of buildings, monuments, and outdoor equipment by acid precipitation costs billions of dollars for replacement and repair each year in the United States. In an extreme case of corrosion, a 37-foot bronze Buddha in Kamakura, Japan, is slowly dissolving away as precipitation from Korea and China bathes the statue in rain that is more acidic than that recorded in the United States.

Whereas the decay of such artifacts is a tragic loss in itself, it should also stand as a grim reminder of how we are dissolving away the buffering capacity of ecosystems. In addition, some officials are concerned that acid precipitation's mobilization of aluminum and other toxic elements may result in the contamination of both surface water and groundwater. Increased acidity in water also mobilizes lead from the pipes used in some old plumbing systems and from the solder used in modern copper systems. As the song says, "What goes up must come down": The sulfur and nitrogen oxides pumped into the troposphere at the rate of over 40 million tons per year in the United States come down as acid deposition, generally to the east of their origin, because of the way weather systems flow. The deposits cross national boundaries: Canada receives half of its acid deposition from the United States, and Scandinavia gets it mostly from Great Britain and other Western European nations. Japan receives the windborne pollution from widespread coal burning in Korea and China. There is now a broad scientific consensus that the problem of acid deposition must be addressed at national and international levels. We turn next to the efforts being made to curb air pollution.

22.5 Bringing Air Pollution under Control

By the 1960s, it was obvious that pollutants produced by humans were overloading natural cleansing processes in the atmosphere. The unrestricted discharge of pollutants into the atmosphere could no longer be tolerated.

Under grassroots pressure from citizens, the U.S. Congress enacted the **Clean Air Act of 1970 (CAA)**. Together with amendments passed in 1977 and 1990, this law, administered by the EPA, represents the foundation of U.S. air pollution control efforts. It calls for identifying the most widespread pollutants, setting **ambient standards**—levels that need to be achieved to protect environmental and human health—and establishing control methods and timetables to meet these standards.

The CAA mandated the setting of standards for four of the primary pollutants—particulates, sulfur dioxide, carbon monoxide, and nitrogen oxides—and for the secondary pollutant ozone. At the time, these five pollutants were recognized as the most widespread and objectionable; today, with the recent addition of lead, they are known as the **criteria pollutants** and are covered by the **National Ambient Air Quality Standards** (**NAAQS**) (Table 22–2). The **primary standard** for each pollutant is based on the highest level that can be tolerated by humans without noticeable ill effects, minus a 10% to 50% margin of safety. For many of the pollutants, long-term and short-term levels are set. The short-term levels are designed to protect against acute effects, while the long-term standards are designed to protect against chronic effects.

The table reflects two standards announced by EPA in July 1997: the new PM$_{2.5}$ category and a revision of the ozone standard from 0.12 ppm for one hour to 0.08 ppm for eight hours. These changes were established in response to health studies which indicated that the modifications could substantially improve lung function and prevent early death. Cost–benefit estimates indicated that the health benefits far outweighed the costs of compliance (estimated at $9.7 billion per year). The new ozone standards are also believed to prevent a substantial amount of damage to vegetation by ozone. Both changes met with strong opposition from industry groups, which obtained a court injunction in 1999 prohibiting the EPA from enforcing the new standards. However, the U.S. Supreme Court has upheld the EPA in this dispute.

In addition, **National Emission Standards for Hazardous Air Pollutants** (**NESHAPS**) have been issued for eight toxic substances: arsenic, asbestos, benzene, beryllium, coke oven emissions, mercury, radionuclides, and vinyl chloride. The Clean Air Act of 1990 greatly extended this section of the EPA's regulatory work by specifically naming 188 toxic air pollutants for the agency to track and regulate.

Control Strategies

The basic strategy of the 1970 Clean Air Act was to regulate air pollution so that the criteria pollutants would remain below the primary standard levels. This approach is called **command and control**, because industry was given regulations to achieve a set limit on each pollutant, to be accomplished by specific control equipment. The assumption was that human and environmental health could be significantly improved by a reduction in the output of pollutants. If a particular region was in violation for a given pollutant, a local government agency would track down the source(s) and order reductions in emissions until the area came into compliance.

Unfortunately, this strategy proved difficult to implement. Most of the regulatory responsibility fell on the states and cities, which were often unable or unwilling to enforce control. Many areas violated the standards. Even today, after 30 years of legislated air pollution control, many metropolitan areas still consistently fail to meet the standards. The good news is that total air pollutants have been reduced by some 37% during a time when both population and economic activity have substantially increased. However, we are still sending huge quantities of pollutants into the atmosphere.

The Clean Air Act Amendments (CAAA) of 1990 address these failures by more directly targeting specific pollutants and more aggressively demanding compliance, through such means as the imposition of sanctions. As with the earlier law, the states do much of the work in carrying out the mandates of the 1990 act. Each state must develop a *State Implementation Plan* (*SIP*) that is required to go through a process of public comment before being submitted to the EPA for approval. The SIP is designed to reduce emissions of every NAAQS pollutant whose control standard (Table 22–2) has not been attained. One major change is a permit application process (already in place for the release of pollutants into waterways). Polluters must apply for a permit that identifies the kinds of pollutants they release, the quantities of those pollutants, and the steps they are taking to reduce pollution. Permit fees provide funds the states can use to support their air pollution control activities. The amendments also afford more flexibility than the earlier command-and-control approach, by allowing polluters to choose the most cost-effective way to accomplish the goals. In addition, the legislation uses a market system to allocate pollution among different utilities.

TABLE 22–2 National Ambient Air Quality Standards for Criteria Pollutants

Pollutant	Averaging Time*	Primary Standard
PM$_{10}$ particulates	1 year	50 μg/m^3
	24 hours	150 μg/m^3
PM$_{2.5}$ particulates¥	1 year	15 μg/m^3
	24 hours	65 μg/m^3
Sulfur dioxide	1 year	0.03 ppm
	24 hours	0.14 ppm
	3 hours	0.5 ppm
Carbon monoxide	8 hours	9 ppm
	1 hour	35 ppm
Nitrogen oxides	1 year	0.05 ppm
Ozone	8 hours	0.08 ppm
Lead	3 months	1.5 mg/m^3

*The averaging time is the period over which concentrations are measured and averaged.

¥PM$_{2.5}$ is the particulate fraction having a diameter smaller than or equal to 2.5 micrometers. It supplements the existing PM$_{10}$ standard.

Source: EPA Office of Air and Radiation, Office of Air Quality Planning and Standards, 2000.

EARTH WATCH

PORTLAND TAKES A RIGHT TURN

For many of us, the American lifestyle includes spending time in the automobile each day going to and from work. Vehicle miles per year in the United States have increased much more rapidly than population, and too often the only response of state and city governments is to build more lanes on the expressways.

Portland, Oregon, had its share of expressways 20 years ago, but took a different tack when faced with the prospect of more and more commuters on the road. In response to an Oregon land-use law, Portland threw away its plans for more expressways and instead built a light rail system, the **Metropol-** **itan Area Express** (MAX). This public transportation system now carries the equivalent of two lanes of traffic on all the roads feeding into downtown Portland. As a result, smoggy days have declined from 100 to 0 per year, the downtown area has added 30,000 jobs with no increase in automobile traffic, and Portland's economy has prospered.

This was clearly a situation in which everyone won. MAX has made a major contribution to the clean air and continued economic success of downtown Portland. The Portland solution seems so sensible that you have to ask why it is the exception and not the rule.

Reducing Particulates. Prior to the 1970s, the major sources of particulates were industrial stacks and the open burning of refuse. The CAA mandated the phaseout of open burning of refuse and required that particulates from industrial stacks be reduced to "no visible emissions."

The alternative generally taken to dispose of refuse was landfilling, a solution that has created its own set of environmental problems (discussed in Chapter 19). To reduce stack emissions, many industries were required to install filters, electrostatic precipitators, and other devices. Unfortunately, the solid wastes removed from exhaust gases frequently contain heavy metals and other toxic substances (Chapter 20). Although these measures have markedly reduced the levels of particulates since the 1970s (see Fig. 22–14), particulates continue to be released from steel mills, power plants, cement plants, smelters, construction sites, diesel engines, and so on. Wood-burning stoves and wood and grass fires also contribute to the particulate load, making regulation even more difficult.

Under the CAAA, the regions of the United States that have failed to attain the required levels must submit *attainment plans*. These plans must be based on **reasonably available control technology** (**RACT**) measures, and offending regions must convince the EPA that the standards will be reached within a certain time frame.

Also, the EPA added a new standard for particulates ($PM_{2.5}$) based on information that smaller particulate matter (less than 2.5 micrometers in diameter) has the greatest effect on health because the finer particulate matter tends to originate from combustion processes and from atmospheric chemical reactions between pollutants. The coarser material (PM_{10}) often comes from windblown dust, and although it is still regulated, it is not considered to play as significant a role in health problems as do the finer particles.

Limiting Pollutants from Motor Vehicles. Cars, trucks, and buses release nearly half of the pollutants that foul our air. Vehicle exhaust sends out the VOCs, carbon monoxide, and nitrogen oxides that lead to ground-level ozone and PANs. Additional VOCs come from the evaporation of gasoline and oil vapors from fuel tanks and engine systems.

The CAA mandated a 90% reduction in these emissions by 1975. This timing proved to be unrealistic, but enough improvements have been made over the years that a new car today emits 75% less pollution than did pre-1970 cars (Fig. 22–21). This is fortunate, because driving in the United States has been increasing much more than the population has: Between 1970 and 1999, the number of vehicle miles increased from 1 trillion to 2.4 trillion miles per year, and between 1980 and 1998, the number of vehicles on the road increased over 36%. (See inset data in Fig. 22–21.) It is hard to imagine what the air would be like without the improvements mandated by the CAA.

The reductions in automobile emissions have been achieved with a general reduction in the size of passenger vehicles, along with a considerable array of pollution control devices. Among these devices is the computerized control of fuel mixture and ignition timing, allowing more complete combustion of fuel and decreasing VOC emissions. To this day, however, the most significant control device on cars is the **catalytic converter** (Fig. 22–22). As exhaust passes through this device, a chemical catalyst made of platinum-coated beads oxidizes most of the VOCs to carbon dioxide and water. The catalytic converter also oxidizes most of the carbon monoxide to carbon dioxide. Although newer converters reduce nitrogen oxides as well, the reduction is not impressive.

Despite these efforts, the continuing failure to meet standards in many regions of the United States, as well as

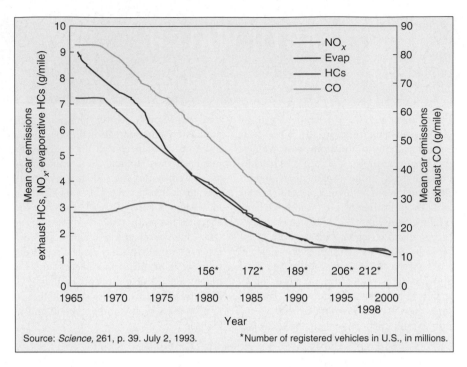

▶ **FIGURE 22–21** *Trends in automobile emissions.* Average car emissions from vehicles, in grams per vehicle mile traveled, from 1965 to (estimated) 2000. Note the numbers of vehicles on the road in the United States from 1980 to 1998.

Source: *Science*, 261, p. 39. July 2, 1993.   *Number of registered vehicles in U.S., in millions.

the aesthetically poor air quality in many cities, made it clear during the 1980s that further legislation was in order. Following are the highlights of the CAAA of 1990 on motor vehicles and fuels:

1. New cars sold in 1994 and thereafter must emit 30% less VOCs and 60% less nitrogen oxides than cars sold in 1990. Emission-control equipment must function properly (as represented in warranties) to 100,000 miles; on 1990 cars, this equipment need work only for 50,000 miles. Buses and trucks must meet more strin-

gent standards. Also, the EPA is given authority to control emissions from all nonroad engines that contribute to air pollution, including lawn and garden equipment, motorboats, off-road vehicles, and farm equipment.

2. Starting in 1992 in the 38 cities with continuing carbon monoxide problems, oxygen has been added to gasoline in the form of methyl tertiary butyl ether (MTBE) and alcohols to stimulate more complete combustion and to cut down on carbon monoxide emissions.

3. Before 1990, only a few states and cities required that vehicle inspection stations be capable of measuring emissions accurately. The new amendments require more than 40 metropolitan areas to initiate inspection and maintenance programs while other cities improve their current programs.

4. Employers in cities with high smog levels are required to take steps to increase the number of their employees who carpool or who use mass transit.

Two factors now affect fuel efficiency and consumption rates. First, the elimination of federal speed limits in early 1996 reduced fuel efficiencies because of higher speeds. Second, the sale of sport utility vehicles (SUVs) has surged since 1990. In 1999, SUVs made up 19% of all vehicle sales. In city traffic, these vehicles get 10 to 16 mpg and no higher than 22 mpg on the highway. Remember, as fuel efficiency decreases, tailpipe emissions increase.

Managing Ozone. Because ozone is a secondary pollutant, the only way to control ozone levels is to address the compounds that lead to its formation. For a long time, it was assumed that the best way to reduce ozone levels was simply to reduce emissions of VOCs. The steps pertaining to motor

▲ **FIGURE 22–22** *Cutaway of a catalytic converter.* Engine exhaust gas is routed through the converter, where catalysts promote chemical reactions that chemically change the harmful hydrocarbons, carbon monoxide and nitrogen oxides into less harmful gases.

vehicles in the Clean Air Act Amendments of 1990 address the sources of about half of VOC emissions. **Point sources** (industries) account for 30% of such emissions, and **area sources** (numerous small emitters, such as dry cleaners, print shops, and users of household products) represent the remaining 20%. RACT measures have already been mandated for many point sources, and much progress has been made through EPA, state, and local regulatory efforts to reduce emissions from these sources. In the last 20 years, VOC emissions have declined by 33%.

However, recent understanding of the complex chemical reactions involving NO_x, VOCs, and oxygen has thrown some uncertainty into this strategy of emphasizing a reduction in VOCs. The problem is that *both* NO_x and VOC concentrations are crucial to the generation of ozone. Either one or the other can become the rate-limiting species in the reaction that forms ozone. (See Fig. 22–15b.) Thus, as the ratio of VOCs to NO_x changes, the concentration of NO_x can become the controlling chemical factor because of an excess of VOCs. This happens more commonly in air pollution episodes that occur over a period of days than in the daily photochemical smog of the urban city.

In a response to petitions from states in the Northeast that were having trouble with ozone and smog because of out-of-state emissions, the EPA ordered 400 power plants in the Midwest and South to cut their NO_x emissions in half by 2003. The EPA also has issued further regulations designed to address urban and suburban smog and ozone levels, which have turned out to be more recalcitrant to reduction than expected. Under the new rules, which will be phased in gradually between 2004 and 2009, NO_x emissions for all SUVs and light trucks must be reduced to .07 gram per mile. This is a 90% reduction over current emissions regulations, and because the new technologies will affect the sticker price of the already expensive SUVs, the auto industry opposes the new regulations. Further, by 2004, gasoline manufacturers will have to cut the sulfur content of gasoline from 300 ppm to 30 ppm and of diesel fuel down to 15 ppm. The EPA believes that the combined effects of these new regulations should remove a further 2 million tons of pollutants emitted annually.

Controlling Toxic Chemicals in the Air. By EPA estimates, the total amount of toxic substances emitted into the air in the United States is around 4.6 million tons annually. Cancer and other adverse health effects, environmental contamination, and catastrophic chemical accidents are the major concerns associated with this category of pollutants.

Under the CAAA of 1990, Congress identified 188 toxic pollutants. It then directed the EPA to identify major sources of these pollutants and to develop **maximum achievable control technology** (**MACT**) standards. Besides affecting control technologies, the setting of these standards includes options for substituting nontoxic chemicals, giving industry some flexibility in meeting goals. State and local air pollution authorities will be responsible for seeing that industrial plants achieve the goals. In response to the CAAA, the EPA developed the National Toxics Inventory to track emissions of these substances. The agency has taken steps to reduce the contribution from vehicles with requirements for cleaner burning fuels in urban areas. Industrial sources are being addressed by setting emission standards for some 82 stationary-source categories (e.g., paper mills, oil refineries, etc.). The program is expected to reduce emissions by at least 1.5 million tons annually.

Coping with Acid Deposition

Ways to Reduce Acid-Forming Emissions. Scientists have calculated that a 50% reduction in current acid-causing emissions in the United States would effectively prevent further acidification of the environment. This reduction may not correct the already bad situations, but together with natural buffering processes, it is estimated to be capable of preventing further environmental deterioration. Because we know that about 50% of acid-producing emissions come from the tall stacks of coal-burning plants that generate electricity, control strategies are centered on these sources. Six main strategies have been proposed: (1) coal washing, (2) fluidized-bed combustion, (3) fuel switching, (4) scrubbers, (5) alternative power plants, and (6) reductions in the consumption of electricity.

Economic factors do not favor the first two strategies. Washing coal to remove sulfur is costly, both economically and environmentally; large amounts of polluted water would be the outcome. In fluidized-bed combustion, coal is burned in a mixture of sand and lime, churned by air forced in from underneath. The sulfur in the coal combines with the lime during combustion and is removed with the ash that is formed. When new plants are built, this may be the preferred method of emissions control, but installing the technology in existing plants would entail tearing down and rebuilding a major portion of each plant.

Oil is more expensive than coal, and switching to oil as a fuel for generating electricity would increase the dependence of the United States on foreign oil. Low-sulfur coal however, is available in central Appalachia and the western United States, and—despite the higher cost of the coal and its transportation—switching to this source represents an attractive strategy for power plants seeking to reduce their sulfur dioxide emissions without changing their technology.

Scrubbers are "liquid filters" that put exhaust fumes through a spray of water containing lime. Sulfur dioxide reacts with the lime and is precipitated as calcium sulfate ($CaSO_4$). Scrubbers have been required for all major power plants and smelters built in the United States since 1977, but the law has not required retrofitting the large, older plants in the Midwest that are the source of much of the acid rain plaguing the east and Canada. Plants in Europe, Canada, and the United States have been fitted with scrubbers relatively quickly; the scrubbers are highly effective in controlling emissions and are not prohibitively expensive.

The last two strategies in the foregoing list require major shifts in the energy policy of the United States. Currently, nuclear power represents the only alternative technology for generating electricity. It accounts for 21% of electrical power generation in the United States. However, as Chapter 14 explains, the future of nuclear energy is in serious question because of concerns about safety and the storage of nuclear waste. Reducing the consumption of electricity (conservation) makes very good sense and has already been implemented in some applications, such as appliances and insulation (Chapter 13). Conservation and energy-efficiency strategies represent a continuing potential for reductions in air pollution, although much more can be accomplished.

Political Developments. Although evidence of the link between power-plant emissions and acid deposition was well established by the early 1980s, no legislative action was taken until 1990. The problem was one of different regional interests. Western Pennsylvania and the states of the Ohio River valley, where older coal-burning power plants produce most of the electrical power of the region argued that controlling their sulfur dioxide emissions would make electricity in the region unaffordable. Throughout the 1980s, a coalition of politicians from these states, fossil-fuel corporation representatives of high-sulfur coal producers, and representatives from the electric power industry effectively blocked all attempts at passing legislation that would take action on acid deposition.

On the other side of the issue were New York and New England, as well as most of the environmental and scientific community, which argued that it was both possible and necessary to address acid deposition and that the best way to do it was to control sulfur dioxide emissions. Also, since 70% of Canada's acid-deposition problem came from the United States, diplomatic pressure toward a resolution was applied.

Title IV of the Clean Air Act Amendments of 1990. The outcome of the controversy is now history. However, we lost two decades of action because of political delays. **Title IV** of the CAAA is the first law in U.S. history to address the acid-deposition problem, by mandating reductions in both sulfur dioxide and nitrogen oxide levels. The major provisions of Title IV are as follows:

1. By the year 2010, total sulfur dioxide emissions must be reduced 10 million tons below 1980 levels. This is the 50% reduction called for by scientists, and it involves the setting of a permanent cap of 8.9 million tons on such emissions. The reduction will be implemented in phases.

2. In a major departure from the command-and-control approach, Title IV authorizes the EPA to use a free-market approach to regulation. Each plant is granted **emission allowances** based on formulas in the legislation. The penalty for exceeding these allowances is severe: a fee of $2,000 per ton and the requirement that the following year the utility must compensate for the

excess emissions. A utility may not emit more sulfur dioxide than allowances permit, but if it emits less, the difference in allowance may be sold, so that another utility may purchase the first plant's difference in place of reducing its own emissions. Judged highly successful, the sulfur dioxide emissions market is currently worth $900 million a year!

3. In the future, new utilities will not receive allowances; they must buy into the system by purchasing existing allowances. Thus, there will be a finite number of allowances in existence.

4. Nitrogen oxide emissions must be reduced by 2 million tons by the year 2000. This is to be accomplished by regulating the boilers used by the utilities and by mandating the continuous monitoring of emissions.

In the wake of the passage of the CAAA of 1990, Canada and the United States have signed a treaty stipulating that Canada will cut its sulfur dioxide emissions by half and cap them at 3.2 million tons by the year 2000. As a result of several treaties in Europe, sulfur dioxide emissions there are down 40% from 1980 levels and should decline another 18% by 2010 if the signatories fulfill their obligations. The Europeans also targeted nitrogen oxide emissions, with the goal of a 30% reduction by 1999.

The main beneficiary of these agreements is the remaining healthy aquatic and forested ecosystems, which will hopefully be protected from future acid deposition. In addition, it is hoped that ecosystems already harmed will be able to recover from current damage and that we will be back on the road to a sustainable interaction with the atmosphere.

Accomplishments of Title IV. The utilities industry has responded to the new law with three actions:

1. Many utilities are switching to low-sulfur coal. This will enable the targeted high-capacity utilities to achieve compliance rapidly.

2. Many power plants are adding scrubbers during the first few years of compliance efforts. Since the technology for high-efficiency scrubbing is well established, more and more power plants will install effective scrubbers.

3. Many utilities are trading in their emissions allowances. At current prices, the purchase of allowances often represents a less costly way to achieve compliance than by purchasing low-sulfur coal or adding a scrubber. A typical transaction might involve a trade in the rights to emit, say, 10,000 tons of sulfur dioxide at a cost of $110 a ton. The combination of approaches has created so many efficiencies that it has cut compliance costs to 10% of what was expected.

To carry out Title IV of the CAAA, the EPA initiated a two-phase approach. Phase I addressed 398 of the largest and highest-emitting coal-fired power plants, aiming at a reduction of 3.5 million tons of SO_2 by 2000. Remarkably, the

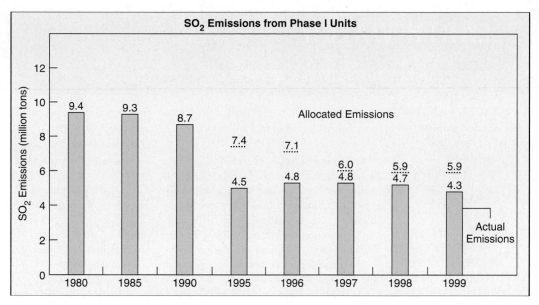

▲ **FIGURE 22–23** *SO₂ emissions from Phase I power plants.* Under Title IV of the CAAA of 1990, emissions from 263 of the highest-emitting coal-fired power plants were 1.6 million tons below the targeted amount.

goal has been met and exceeded—at a cost far below the gloomy predictions of the industries (Fig. 22–23). Phase II began in 2000 and targets the remaining sources of SO_2 in order to accomplish the 10-million-ton goal by 2010. The 2000 goal for NO_x has not been reached, however: Although the power plants have achieved a modest reduction in emissions, other sources have increased, and there has been no progress in the overall NO_x emissions picture (Fig. 22–14).

The news from the field reflects the trends. Concentrations of sulfate in rain and deposition on the land have shown a significant decline (10 to 25%) over a large part of the eastern United States in the last four years. No doubt, this decline is due to the Phase I emission reductions. However, the forests, streams, and fish of the Adirondacks and the White Mountains of New Hampshire have shown no signs of recovery, as had been hoped. The likely reason is the continued impact of nitrogen deposition; it is becoming evident that nitrogen plays a much larger role in acid deposition than was once believed. In passing the CAAA, Congress did not set curbs on nitrogen emissions, but simply opted to reduce those emissions by a minimal 2 million tons (a goal that has not been met). Most scientists believe that the recovery of affected ecosystems will be directly tied to the reductions in emissions of both sulfur and nitrogen; further cuts may well be needed, as well as a cap on nitrogen emissions. Acid rain still falls on eastern forests, as Figure 22–18 indicates.

22.6 Taking Stock

Without question, measures taken to reduce air pollution carry an economic cost. The United States sends a clear signal that human health and environmental quality rank highly in the development and implementation of public policy by investing heavily in pollution control. Some critics have charged that air pollution controls are not cost effective; that is, the benefits are not nearly as great as the costs. They see lost opportunities for economic growth and tend to disregard the *costs avoided* (from improved health), despite two recent analyses of the Clean Air Act and its amendments that found huge benefits. (See "Earth Watch" essay, p. 562.) In addition, pollution control has become a major industry providing over 2 million jobs, a significant part of the national economy.

An issue that has finally been settled is the court challenge by industries of EPA's 1997 published standards for particulates and ozone. An appeals court prevented the EPA from enforcing the rules in 1999, and the issue was brought before the U.S. Supreme Court. The industries claimed that the new rules could cost them as much as $150 billion per year. (Such claims have always proven to be inflated, however, when seen in hindsight!) The industries believed that the EPA had overstepped its power and was taking on a role that should belong to Congress, namely, to set basic public policy without considering the costs of its actions. In preliminary hearings held in late 2000, however, the Court appeared to be quite skeptical of the arguments from industry that the health gains of air pollution regulations should be subservient to cost–benefit analysis. One justice stated that the industry groups seemed to be asking the EPA to do its job "to protect the public health, provided [that] it doesn't cost too much." The U.S. Supreme Court, in February, 2001, unanimously upheld the EPA standards and rejected the industry argument that the economic costs of the regulations should be weighed against their health benefits.

Future Directions

In the early 1970s, many cities were registering more than 100 days per year of pollution in the "unhealthful" range or above; Los Angeles had about 300 such days per year. The

EARTH WATCH

THE CLEAN AIR ACT BRINGS A WINDFALL

The Clean Air Act (1970, 1977, and 1990) has been the subject of open political warfare between those who think its cost has been too high for industry, taxpayers, labor, and consumers and those who think the health and environmental benefits were justified. Compliance has affected patterns of industrial production, employment, and capital investment. Although these expenditures must be viewed as investments that have generated benefits and opportunities, the dislocation in some industries was severe and included reductions in high-sulfur coal mining and cutbacks in polluting industries such as steel. A need developed for a real cost–benefit analysis.

In 1990, Congress requested the EPA to answer the question, How do the overall health, welfare, ecological, and economic benefits of Clean Air Act programs compare with the costs of these programs? In response, the EPA performed the most exhaustive cost–benefit analysis of public policy ever attempted. Here is what the EPA reported in a 1996 study:

- The total direct cost of implementing the Clean Air Act for all federal, state, and local rules from 1970 to 1990 was $436 billion (in 1990 dollars). This cost was borne by businesses, consumers, and government entities in the form of higher prices for many goods and services and for some utilities.

- The mean estimate of direct benefits from the Clean Air Act from 1970 to 1990 was $6.8 trillion.

- Therefore, the net benefit of the Clean Air Act has been $6.4 trillion!

"The finding is overwhelming. The benefits far exceed the costs of the Clean Air Act in the first 20 years," said Richard Morgenstern, associate administrator for policy planning and evaluation at the EPA. Further, the report states that "all benefits may be significantly underestimated due to the exclusion of large numbers of benefits from the monetized benefit estimate."

The benefits to society, directly and indirectly, have been widespread across the entire population. The clean Air Act has

- reduced air pollution (described in this chapter).

- improved human health: Each year, 79,000 lives were saved, and there were 18,000 fewer heart attacks, 10,000 fewer strokes, 13,000 fewer cases of hypertension, and 15 million fewer cases of respiratory illness.

- "avoided cost": Improved health has meant less debilitating disease, less hospitalization, less need for special care, and less need for medicines.

- lowered levels of lead, which is particularly harmful to children. In 1990, 220,000 tons of lead were not burned in gasoline, because of Clean Air Act measures. Because exposure to lead impairs the cognitive development of children, the huge reductions in lead levels produced a benefit of retained

IQ and the possibility of a more productive, less dependent life.

- lowered cancer rates.

- resulted in less acid deposition.

The EPA study result should encourage us in our hopes for a more sustainable future. Society knew what to do, took action despite disruptive efforts by special-interest and political partisans, and reaped about $16 in benefits for every $1 invested to control air pollution.

In 1999, the EPA published a second analysis of costs and benefits that looked at the impacts of the CAAA of 1990 and extended expected costs and benefits to 2010. The findings are consistent with the EPA's previous analysis. According to the later analysis, the new regulations will cost an estimated $27 billion, but will generate health and ecological benefits of about $110 billion. Estimates indicate that the amendments will prevent 23,000 Americans from an early death, more than 1.7 million asthma episodes, 67,000 incidences of acute and chronic bronchitis, and 22,000 respiratory-related hospital visits. Many of the benefits, such as those to crops and ecosystems, are difficult to put in dollar terms. Thus, the benefits exceed the costs by a margin of four to one or better; this still sounds like a good bargain!

(*Source:* Adapted from R. Christopherson, *Geosystems, an Introduction to Physical Geography,* 3rd ed. Copyright © 1997 by the author. Reprinted by permission of Pearson Education Inc. Upper Saddle River, NJ 07458.)

original goal of air pollution control was to achieve "good" air quality on all days by 1975. This goal was not met; however, marked progress has occurred for most of the primary air pollutants (Fig. 22–14). Certainly, the quality of the air is much better than if controls had not been instituted, and the total cost avoided is significant.

Los Angeles's smog is not gone, but it is retreating; for the first time since pollution levels were recorded in 1970, the city went all summer in 2000 without a single smog alert. (There were 101 Stage 1 alerts in 1980!) Nevertheless, the backsliding of the last few years is troubling. From 1970 to 1990, pollution control devices and greater fuel efficiency reduced a typical car's emissions by 75%. As old, polluting cars were replaced by new, "cleaner" cars over that period, emission levels decreased despite increasing numbers of cars

and miles driven. Today, this reduction by replacement has leveled off, and the increasing number of cars and steady erosion of fuel economy are the dominant factors in preventing further progress. The Clean Air Act addresses this problem primarily by mandating increased emission controls and cleaner burning fuels. Whether these measures will be successful in further reducing air pollution remains to be seen.

A more general approach that would benefit the whole country would be to increase the fuel efficiency standard for vehicles (known as the *c*orporate *a*verage *f*uel *e*conomy, or CAFE, standard), which is now locked at 27.5 mpg for automobiles and 20.7 for SUVs and light trucks. Intense lobbying by the automobile and petroleum industries prevented more stringent standards of fuel efficiency from being included in the CAAA of 1990. In 1998, the actual average fuel economy of all passenger vehicles on U.S. highways was 24.6 mpg. (It was 26.2 in 1987.) The SUVs, classified as light trucks, feature average efficiencies of only 18 mpg. The Congress, in response to the auto lobby, has regularly prohibited the Department of Transportation from even studying the possibility of raising CAFE standards.

Available technologies could raise fuel efficiency substantially. For example, two-stroke engines (which fire on every thrust of the piston instead of on every other thrust, as in the current four-stroke engine) would cut engine weight in half, and lightweight cars with smaller engines give mileage in the range of 80 to 100 mpg. Even the SUV can go "green": Using existing technologies, Union of Concerned Scientists engineers have designed an SUV (the Exemplar) that delivers the desirable features of a standard-sized SUV and gets passenger-car mileage (28.4 mpg). All of the major auto manufacturers have pledged to improve the fuel economy of their SUVs (probably anticipating future regulations requiring them to do so).

California, which by all measures has the greatest problem with photochemical smog and vehicular pollutants, has taken a completely different approach to reducing pollution from vehicles. California law requires that by 2003 up to 10% of all vehicles sold in the state must be "emission free"—in other words, powered by electricity—or nearly so. Similar laws have been enacted in nine northeastern states and the District of Columbia. In response to these developments, automakers are reluctantly devoting increasing resources to the development of affordable electric cars (Fig. 22–24a). Entrepreneurial firms such as U.S. Electricar of California and Solectria of Massachusetts are forging ahead of the big three automakers and already have electric cars on the market. These cars are considerably lighter than conventional cars and quite limited in range (traveling 50 to 100 miles before needing recharging). They also lack such amenities as air-conditioning and other power accessories. An interesting option is the hybrid electric vehicle (HEV), which is an electric car with a small internal combustion engine equipped with an electric generator to charge the batteries. HEVs are being developed by several foreign manufacturers; in Japan, Toyota has recently begun marketing the Prius, a five-passenger sedan that gets 66 mpg, reaches speeds of 100 mph, and produces one-tenth the pollution of a comparable gasoline car (Fig. 22–24b).

Electrical cars or hybrids may well be the wave of the future, especially for short trips at low speeds, which waste energy and cause much pollution. Of course, switching from gasoline to electrical power will transfer the site of pollution emission from the moving vehicle to the power plant—a trade-off with uncertain consequences, unless we make strides in renewable energy sources.

It can be argued that all this emphasis on how vehicles burn fuel is just tinkering with the mechanics of the problem. To achieve lasting progress, we need to address the fundamental way we have organized society and industry. Consider that fact that we pour billions of dollars into expanding and improving our highways, build malls beside every major highway corridor, and erect homes in bedroom communities that are scores of miles from major employment centers, but we put relatively little money into improving mass-transit systems, so prominent in the first half of the century. It would make great sense in the coming years to fashion legislation and policies that would guide new development and redevelopment of depreciated urban areas so as to reduce average commuting distances and times and to facilitate greater use of mass-transit systems. At the same time, we can make the development of efficient mass-transit systems and fuel-efficient vehicles matters of national priority.

In its 1996 report, the President's Council on Sustainable Development said, "The federal government should redirect policies that encourage low-density sprawl to foster investment in existing communities. It should encourage shifts in transportation toward transit and expansion of transit options rather than new highway construction." The revolution in information technology, which has spawned telecommuting and greater use of the home office, is an encouraging step in the right direction. These measures will do much to move our society toward sustainability.

(a)

(b)

▲ **FIGURE 22–24** *Electric cars.* (a) GM's Impact electric car. (b) Toyota's Prius, a hybrid electric vehicle.

ENVIRONMENT ON THE WEB

THE COSTS OF GROUND LEVEL OZONE

Chapter 21 emphasized the important role of stratospheric ozone (O^3) as a shield against ultraviolet radiation. However, when ozone occurs at ground level, it is much less welcome, damaging valuable crops and affecting human health. Currently, ozone accounts for $1 to $2 billion in crop damage every year in the United States, and tens of millions of dollars a year in Canada.

Tropospheric (ground-level) ozone—a major component in photochemical smog—is formed when nitrogen compounds in the atmosphere react with volatile organic compounds (Fig. 22–15). Ozone is most likely to be formed in valleys and low-lying basins when winds are calm and warm, and sunny conditions prevail. As a result, it is usually found at highest concentrations during the summer growing season. Although the precursors of ozone arise primarily in urban areas, drifting air masses can carry ozone far into rural and wilderness areas.

Tropospheric ozone damages crops by affecting vegetation in two ways. First, as a powerful oxidant, it causes thinning and discoloration of leaf tissues. Once the leaf is weakened, it becomes more vulnerable to insect pests and disease. Ozone damage can also promote premature tissue aging, leaf loss, and reduced pollen and seed production. Available evidence suggests that ozone can reduce plant growth by up to 40%. Consequently, ozone's second impact is reduced plant growth.

Overall, a farmer with ozone-damaged crops can expect up to a 20% loss in yields from the affected area. In some regions, ozone damage now outweighs pests and disease as the most serious threat to farm income. This level of damage is not surprising since ozone concentrations of 70 to 200 parts per billion (ppb) are commonly found downwind of major cities—and concentrations as low as 40 ppb are enough to injure plants.

Through its tissue-damaging effects, ozone also endangers valuable timber stands and fragile wilderness ecosystems. As a component of urban smog, ozone impairs the aesthetics of those systems and creates secondary impacts on urban and wilderness habitats. Such damage is already apparent in urban trees and in parks downwind of major cities around the world.

However, the human health impacts of ozone are also significant. Ozone is a major cause of human respiratory distress and asthma, especially in children, causing such respiratory illnesses in as many as a million individuals each year. As a result, environmental managers need a way of tracking and reporting air quality in a way that is easily understood by the public and by decision makers. In the case of ozone, for instance, regulators need to alert the public when smog levels have reached a point that is likely to cause human health problems or endanger valuable agricultural crops. To do this, agencies use simple language and a few key "indicator" parameters to convey air quality in easily understood terms.

Web Explorations

The Environment on the Web activity for this essay discusses some of the challenges in choosing, and using, suitable indicators of air quality. Go to the Environment on the Web activity (select Chapter 22 at **http://www.prenhall.com/wright**) and learn for yourself:

1. how air quality indicators are used to monitor urban pollution;
2. how ozone levels can vary between cities within a region; and
3. how the location of monitoring stations can influence pollution monitoring results.

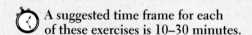 **A suggested time frame for each of these exercises is 10–30 minutes.**

REVIEW QUESTIONS

1. Give three examples each of natural and anthropogenic air pollutants.

2. What naturally occurring cleanser helps to remove pollutants from the atmosphere? What molecule (that we studied in Chapter 21) is the major source of this cleanser?

3. What is a pollutant threshold level?

4. What three factors determine the level of air pollution?

5. Describe the origin of industrial smog and photochemical smog. What are the differences in cause and appearance of each?

6. What are the major air pollutants and their most serious effects?

7. What impact does air pollution have on human health? Give the three categories of impact and distinguish among them.

8. Describe the negative effects of pollutants on crops, forests, and other materials. Which pollutants are mainly responsible for these effects?

9. Where do primary pollutants originate? How do they form secondary pollutants?

10. What is the difference between an acid and a base? What is the pH scale? The pOH scale? How are they related?

11. What two major acids are involved in acid deposition? Where does each come from?

12. How can a shift in environmental pH affect aquatic ecosystems? In what other ecosystems can acid deposition be observed? What are its effects?

13. Discuss ways in which the Clean Air Act Amendments of 1990 address the failures of former legislation.

14. What technological and political changes are taking place in the United States to reduce acid deposition?

15. What can we do in the future—as individuals and as a nation—to further improve air quality?

THINKING ENVIRONMENTALLY

1. Motor vehicles release close to half the pollutants that dirty our air. What alternatives might be introduced to encourage a decrease in our use of automobiles and other vehicles with internal combustion engines?

2. How would traditional patterns of life be different in the absence of the Clean Air Act and its amendments? Write a short essay describing the possibilities.

3. What arguments might the auto and utility industries present to delay regulatory action on particu-lates and ozone? How would health officials respond to their arguments?

4. The Swedish government has made efforts to attack the symptoms of acid deposition directly by trying to neutralize acidic lakes with lime. Discuss the pros and cons of this method.

MAKING A DIFFERENCE

PART FIVE: Chapters 16, 17, 18, 19, 20, 21, 22

1. If you are a smoker, enlist in a smoking cessation program. Insist that others do not smoke in your presence or in your home or workplace.

2. Monitor the hazards in your own life and take steps to reduce behavior ranked as risky by the experts. Be discriminating as you hear about hazards and risks. Be especially aware of the media's tendency to exploit the outrage element of risks.

3. Survey the lakes and ponds in your area and determine whether cultural eutrophication is occurring. If you find that it is, follow up by consulting a local environment group and get involved in combating the problem.

4. Use phosphate-free detergents for all washing. Information regarding the use of the product is printed on the container (often in very fine print!).

5. Explore sewage treatment in your community. To what extent is the sewage treated? Where does the effluent go? The sludge? Are there any problems? Are there alternatives that are ecologically more sound? Consult local environment groups for help.

6. Investigate how municipal solid waste is handled in your school or community and what the future plans are for waste management. Promote curbside recycling if it is not happening, and investigate the possibility of adopting a system that charges for each container of unrecycled waste.

7. Consider your consumption patterns. Live as simply as possible so as to minimize your impact on the environ-ment; buy and use durable products, and minimize your use of disposables.

8. Read labels and become informed about the potentially hazardous materials you use in the home and workplace. Wherever possible, use nontoxic substitutes. If toxic substances are used, ensure that they are disposed of properly.

9. Encourage your community to set up a hazardous waste collection center where people can take leftover hazardous materials.

10. Reduce your contribution to air pollution by amending your lifestyle to drive fewer miles: Arrange to live near your workplace, carpool, use public transportation, bicycle to work or school, avoid unnecessary trips.

11. Properly maintain all fuel-burning equipment (oil burner, gas heater, lawn mower, outboard motor, automobile, etc.) to burn efficiently.

12. When you buy a new car, consider purchasing one of the new hybrid electric vehicles or at least a car that gets good mileage.

13. Be an advocate of energy conservation on all fronts in supporting the steps we can take to combat global climate change. Insist on full U.S. participation in global efforts to curb greenhouse gas emissions, such as the Kyoto Accord of 1997.

14. Make sure anyone servicing a refrigerator or air-conditioning unit for you will recapture and recycle the CFCs if the system needs to be opened.

WEB REFERENCES

On-line resources for this chapter are on the World Wide Web at: **http://www.prenhall.com/wright**. (Click on Chapter 22 within the Chapter Selector.)

Chapter 23
Economics, Public Policy, and the Environment

Chapter 24
Sustainable Communities and Lifestyles

PART 6

Toward a Sustainable Future

How will we shape a sustainable future? In previous chapters, we uncovered many issues that need to be addressed and resolved if we are indeed to make our way to a sustainable relationship with the environment. We are convinced that a key to this mission can be found in environmental public policy and individual lifestyles.

In Chapter 23, we highlight economics and public policy as they relate to the environment. We examine how a nation can call on its wealth to address societal problems and what three factors constitute the wealth of nations: produced assets, natural capital, and human resources. In Chapter 24, the last one of the book, we explore one of the most difficult problems faced by every nation today: promoting livable and sustainable communities, particularly in the many crowded cities of the developing world. We look at examples of cities moving in a sustainable direction. We also look at some of the work being done by organizations that are grappling with the question of sustainability. Finally, we look at the vital, but difficult, task of adopting lifestyles that embrace some of the stewardship values capable of shaping a sustainable future. The outcome is uncertain, but we know enough now to be able to point in a sustainable and stewardly direction.

◀ *A street in downtown Nelson, British Columbia, Canada.*

Economics, Public Policy, and the Environment

Key Issues and Questions

1. Environmental public policy includes laws and agency-enforced regulations that deal with a society's interactions with the environment. How necessary is environmental public policy?

2. Economic systems are social and legal arrangements people make in order to satisfy their needs and wants. What are the two basic kinds of economic systems, and how do they differ?

3. The wealth of a nation is the resources a country draws on to establish and maintain an economy. What are the three components of a nation's wealth, and how do the world's nations compare in wealth as measured by these components?

4. Environmental public policy appears in response to specific problems and needs. How is environmental public policy developed in modern societies?

5. It is argued that many environmental regulations are too costly. Do the benefits of environmental public policy outweigh the costs?

6. Benefit–cost analysis is an economic tool that is applied to environmental policy. How does this tool work? What are the benefits of the last 25 years of environmental policy?

7. Politics often play an important role in the development of environmental public policy. What new developments have occurred in this arena over the last few years?

On July 18, 1997, the EPA announced a set of new air quality standards called **40 CFR Part 50** in the **Federal Register**. The new standards for particulate matter and ozone required industries and regions to take effective action to reduce the levels of these pollutants. Among the documents supporting the new rule was a regulatory impact analysis of *costs and benefits* associated with the rule. The EPA estimated that, by 2010, when the regulations are in full effect, the annual benefits of the two rules will amount to some $92 billion, whereas the costs of the regulations are estimated to be $6.4 billion. The new standards were met with strong protests from industry groups (see Chapter 22), which argued that complying with the new regulations would cost as much as $150 billion a year. The industries and the EPA became locked in a battle over the regulations that was ultimately decided by the U.S. Supreme Court. (See Chapter 22.)

The EPA was required by law to perform the benefit–cost analysis. However, that was not always the case:

In the 1970s, most environmental policies were developed with little consideration of economics. Early policies were typical command-and-control responses to air and water pollution that focused on controlling emissions from cars and factories, pesticides being sprayed over large areas, and releases from sewage outfalls and other point sources. The policies undoubtedly had a significant economic impact on businesses, consumers, and the workforce.

Reacting to the situation, and concerned that the U.S. economy in general and businesses in particular might be overregulated and thus unduly restricted, President Ronald Reagan issued Executive Order 12291 in February 1981. This order required all executive departments and agencies to support every new *major* regulation with **benefit–cost analysis**. To qualify as major, a regulation had to (1) impose annual costs of at least $100 million, *or* (2) cause a significant increase in costs or prices for some sector of the economy or geographic region, *or* (3) have a significant adverse effect on competition, investment, productivity, employment,

◄ **Capitol Building, Washington, DC.** The Senate and House of Representatives meet and establish public policy in this center of activity of the U.S. government.

innovation, or the ability of U.S. firms to compete with foreign firms. No regulations were to be issued unless the benefits outweighed the costs. It is no secret that the goal of Reagan's order was to roll back environmental regulation.

In 1993, President Bill Clinton issued Executive Order 12866, which continued most of the policies of Executive Order 12291 (with the exception of the third qualification) and added the processes of public and interagency review of proposed regulatory rules. The 1993 executive order softened the quantitative test by stating that benefits should "justify" costs and that a number of additional non-monetary factors (such as the impacts on different groups, or issues that cannot be quantified) could be considered in making a final decision on a given regulation.

Benefit–cost analysis is an economic measure applied to environmental policy. It is a reminder that all environmental decisions take place in the context of a society in which money and politics are the language of exchange and the bottom line is not always *what is right to do*, but *what is possible to do*. In this chapter, our major focus is on environmental public policy, but we will see that economics plays a dominant role in virtually every environmental issue, both domestically and internationally. This should come as no surprise, as we have seen economic issues in many previous chapters: in Chapter 3, where the value of ecosystem goods and services was discussed; in Chapter 11, where value concepts were applied to wild species; in Chapter 12, in which we addressed the question of common property resources; in Chapter 16, where benefit–cost analysis was identified as a major tool of risk management; and in Chapter 22, in which benefit–cost analysis was applied to air pollution laws. We have also seen the great disparity of wealth among nations and the profound consequences of the lack of wealth in many of the countries still in the early stages of economic development. Economic concerns are highly important, but even more important is the development of just and effective environmental public policies.

23.1 Economics and Public Policy

Environmental public policy includes all of the laws and agency-enforced regulations in a society which deal with that society's interactions with the environment. Two sets of environmental issues are comprehended in environmental public policy: the prevention or reduction of air, water, and land pollution, and the use of natural resources like forests, fisheries, oil, land, and so forth. Public policies addressing these two sets of issues are developed at all levels of government: local, state, federal, and global.

The Need for Environmental Public Policy

The purpose of environmental public policy is *to promote the common good*. Exactly what the common good consists of may be a matter of debate, but two goals stand out: the *improvement of human welfare* and the *protection of the natural world*. Both of these goals define the perspective of environmental stewardship.

What are the consequences of not having an effective environmental public policy? Human societies and their economic activities have the potential for doing great damage to the environment, and that damage has a direct impact on present and future human welfare (Table 23–1). The effects of pollution and the misuse of resources are most clearly seen in those parts of the world where environmental public policy is often not well established and implemented—the developing world. As the table shows, millions of deaths and widespread disease are directly traceable to degraded environments. The costs to human welfare are felt in the areas of health, economic productivity, and the ongoing ability of the natural environment to support human life needs. Laws to protect the environment are not luxuries to be tolerated only if they do not interfere with individual freedoms or economic development; they are an essential part of the foundation of any organized human society.

Relationships between Economic Development and the Environment

In a human society an **economy** is the system of exchanges of goods and services worked out by members of the society. Goods and services are produced, distributed, and consumed as people make **economic decisions** about what they need and want and what they will do to become players in the system—what they might provide that others would need and want. As societies develop, economic activity assumes increasingly broader dimensions, with increasingly pervasive impacts on the whole society. In previous chapters, we have examined ample instances wherein economic activity can damage the environment and human health. In addition, an unregulated economy can make intolerable inroads on natural resources. During the latter part of the 19th century in the United States, private enterprise had unrestrained access to forests, grazing lands, and mineral deposits. This exploitation of resources began to be addressed at the turn of the century, as government rules and regulations imposed necessary limits.

Economic activity in a nation also can provide the resources needed to solve the problems that very activity creates. A strong relationship exists between the level of development of a nation and the effectiveness of its envi-

TABLE 23–1	Principal Health and Productivity Consequences of Environmental Mismanagement	
Environmental problem	Effect on health	Effect on productivity
Water pollution and water scarcity	More than 2 million deaths and billions of illnesses a year are attributable to pollution; poor household hygiene and added health risks are caused by water scarcity.	Declining fisheries; rural household time (time spent fetching water) and municipal costs of providing safe water; depletion of aquifers, leading to irreversible compaction; constraint on economic activity because of water shortages.
Air pollution	Many acute and chronic health impacts: excessive levels of urban particulate matter are responsible for 300,000–700,000 premature deaths annually and for half of childhood chronic coughing; 400 million–700 million people, mainly women and children in poor rural areas, are affected by smoky indoor air.	Restrictions on vehicles and industrial activity during critical episodes; effect of acid rain on forests, bodies of water, and human artifacts.
Solid and hazardous wastes	Diseases spread by rotting garbage and blocked drains. Risks from hazardous wastes are typically local, but often acute.	Pollution of groundwater resources.
Soil degradation	Reduced nutrition for poor farmers on depleted soils; greater susceptibility to drought.	Field productivity losses in the range of 0.5–1.5% of gross national product are common on tropical soils; offsite siltation of reservoirs, river-transport channels, and other hydrologic systems.
Deforestation	Localized flooding, leading to death and disease.	Loss of sustainable logging potential and of erosion prevention, watershed stability, and carbon storage capability by forests. Loss of nontimber forest products.
Loss of biodiversity	Potential loss of new drugs.	Reduction in ecosystem adaptability and loss of genetic resources.
Atmospheric changes	Possible shifts in vectorborne diseases; risks from climatic natural disasters; diseases attributable to ozone depletion (300,000 more skin cancers per year and 1.7 million cases of cataracts per year).	Damage to coastal investments from rise in sea level; regional changes in agricultural productivity; disruption of marine food chain.

Source: The World Bank, *World Development Report, 1992* (New York: Oxford University Press, 1992).

ronmental public policies. Figure 23–1 shows a number of environmental indicators in relation to per capita income levels. (The data were combined from studies of different countries.) Several patterns emerge:

- Many problems decline (for example, sickness due to inadequate sanitation or water treatment) as income levels rise, because the society now has the resources available to address the problems with effective technologies.
- Some problems rise and then decline (urban air pollution) as a consequence of the recognition of a problem and then the development of public policy to address it.
- Increased economic activity causes some problems to rise without any clear end in sight (municipal solid waste, CO_2 emissions).

The key to solving all of these problems brought on by economic activity is the development of effective public policies and institutions. Where this does not happen, environmental degradation and human disease and death are the inevitable outcomes. To understand this relationship better, let us look at how economic systems work.

Economic Systems

Economic systems are social and legal arrangements people make in order to satisfy their needs and wants. Economic systems range from the communal barter systems of primitive societies to the current global economy, which is so complex that it defies description. In the organized societies of today's world, two kinds of economic systems have emerged: the **centrally planned economy** characteristic of

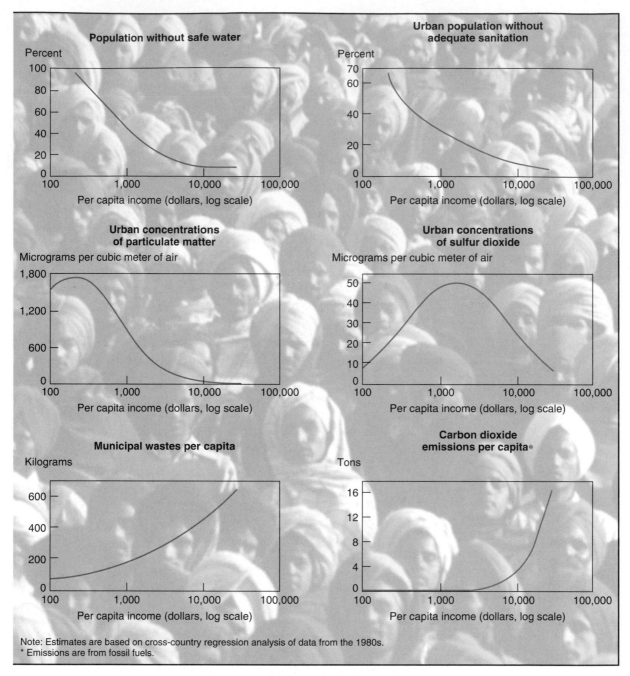

▲ FIGURE 23–1 *Environmental indicators and per capita income.* Some of the most serious environmental problems can be improved with income growth; others get worse and then improve; some problems just worsen. (*Source*: World Bank, *The World Development Report, 1992* [New York: Oxford University Press, 1992], p. 11.)

socialist countries and the **free-market economy** characteristic of the capitalist countries.

In any kind of economy, there are basic components that determine the economic flow of goods and services. Classical economic theory considers land (natural resources), labor, and capital as the three elements constituting the "factors of production." Economic activity involves the circular flow of money and products, as shown in Figure 23–2. Money flows from households to businesses as people pay for products, and from businesses to households as people are paid for their labor. Labor, land, and capital

are invested in the production of goods and services by businesses, and the products provided by businesses are "consumed" by households.

The cycles shown in the figure apply to both kinds of economic systems in existence today; they differ mostly in how economic decisions are made. In a pure centrally planned economy, the leaders make all the basic decisions regarding what and how much will be produced where and by whom. Those who adopt such a system believe that it is a more efficient system in which the real needs of the people will be met because of the wise decisions of their rulers. It is

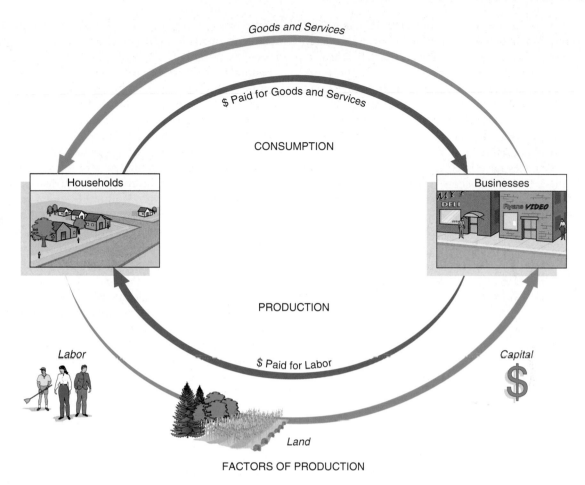

▲ FIGURE 23–2 *Classical view of economic activity.* Land (natural resources), labor, and capital are the three elements constituting the "factors of production." Economic activity involves the circular flow of money and products.

a system in which equity and efficiency are theoretically achievable ("from each according to his abilities, to each according to his needs," as the socialist saying goes). The failure of the centrally planned economy in the former Soviet Union is well known; sadly, the collapse has led to economic chaos in the countries that were once part of the Soviet Union. North Korea, Cuba, and the People's Republic of China remain as the last holdouts of this economic model.

In the pure free-market economy, the market itself determines what will be exchanged. Goods and services are offered in a market that is free from governmental interference. The system is completely open to competition and the interplay of **supply and demand**. If supply is limited relative to demand, prices rise; if there is an oversupply, prices fall, because people and corporations will economize and pay the lowest price possible. The whole system is in private hands and is driven by the desire of people and businesses to acquire goods, services, and wealth as they act in their self-interest. All "players" have free access to the market; competition spurs efficiency as inferior products and services are forced out of the market. People can make informed decisions about their purchases because there is sufficient information about the benefits and harm associated

with economic goods. The free-market economic system is thought to be at its best when left completely alone.

These are the very basics of the two economic systems. In reality, no country has a pure form of either economy. The developed countries all function with market economies, but government involvement occurs at many levels, more so in some of the "social democracies" of Western Europe than in the United States. Governments can, for example, control interest rates, determine the amount of money in circulation, and adopt policies that stimulate economic growth in times of a slowdown. Governments also maintain surveillance over financial processes like stock markets and bank operations. On the downside, powerful business interests can manipulate a market economy and unscrupulous people can exploit the freedom of the market to defraud people. Only governments can provide the policing that the society needs.

Also, many people criticize the free-market economic system as a system without a "conscience." They argue that there are many workings of the market economy that create hardships for people at the bottom of the economic ladder. If, for example, population growth is rapid and there are not enough jobs to go around, the result often is exploitation of workers, as we saw in the case of the factory workers in Nairobi, Kenya, in Chapter 6. A market economy only offers people *access* to goods and services; if people lack the means to pay, access alone will not meet their needs. Recall from Chapter 10 the importance of a "safety net" in connection with the food needs of impoverished people. Also, it is too easy for self-interest to lead businesses and individuals to exploit natural resources or to avoid costs by polluting the environment instead of producing a "clean" product. Over the years, the democracies have been learning what laws need to be put in place to control access to natural resources and to prevent damaging pollution. These two areas of concern are the substance of environmental public policy.

23.2 Resources and the Wealth of Nations

What are the resources on which a country draws to establish and maintain an economy? The classic economic paradigm would say that land which represents the environment, labor, and capital are the essential resources needed for a country to be able to mount its economy.

However, a new breed of economists, often called **ecological economists**, for want of a better term, has emerged in recent years and has taken issue with this view. These economists point out that the classic view sees the environment simply as one set of resources within the larger sphere of the human economy. The environment's vital role in supplying the goods and services on which human activities depend is largely ignored. They argue that the classical approach looks at things backwards and that the natural environment actually encompasses the economy, which is constrained by the resources found within the environment (Fig. 23–3). Without the vital raw materials provided by the environment, and without the capacity of the environment to absorb wastes (pollution), there is no economy. Thus, the ecosystems and natural resources found in a given country provide the context for that country's economy; the total global biosphere, similarly, is the context for the global economy. As described by Thomas Prugh, Robert Costanza, and Herman Daly in *The Local Politics of Global Sustainability*, **economic production** thus can be seen as it really is: the process of converting the natural world to the manufactured world. Renewable and nonrenewable resources and the ecosys-

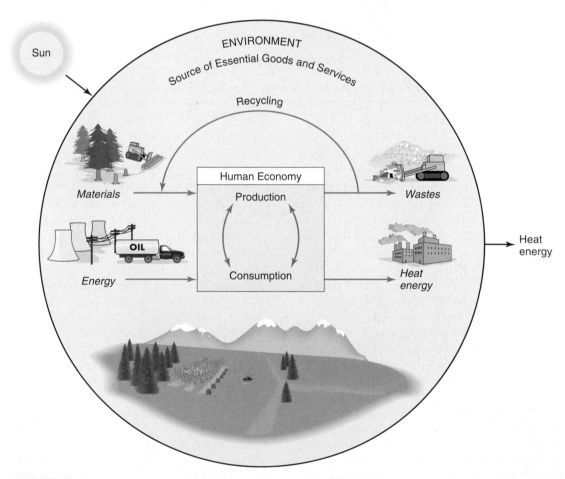

▲ **FIGURE 23–3** *Environmental economic view of economic activity.* The natural environment encompasses the economy, which is constrained by the resources found within the environment.

tems containing them are turned into cars, toys, books, food, buildings, computers, etc. Thus, to quote Prugh, Costanza, and Daly, "Resources flow into the economy from the enfolding ecosystem and are transformed by labor and capital (using energy, also a resource), and then pass out of the economy and back into the ecosystem in the form of wastes." (See Fig. 23–3.)

The ecological economists' view emphasizes the role of natural ecosystems as essential life-support elements—the *natural capital*, as we referred to it earlier. The approach brings the concepts of carrying capacity and limits into perspective, and it brings sustainability sharply into focus. If the economy continues to grow, the natural world continues to shrink; in time, economic growth must come to a sustainable steady state, and this must happen before our natural capital is consumed beyond its ability to support an economy.

Without a doubt, the ecosystems and mineral resources of a given country—its natural capital—are a major element in the wealth of that country. In recent years, the World Bank Environment Department has been working on ways of measuring the wealth of nations and has produced some insightful analyses, which we consider next.

The Wealth of Nations

The World Bank group asked the question, What are the components of a nation's wealth—its assets? In the document *Expanding the Measure of Wealth: Indicators of Environmentally Sustainable Development*, the World Bank indicates three major capital components that determine a nation's wealth: *produced assets, natural capital,* and *human resources.* **Produced assets** (Fig. 23–4a) are the stock of buildings and structures, machinery and equipment, vehicles and ships, highways and power lines, etc., that are essential to the production of economic goods and services. These are often the major focus of national eco-

nomic planning; they are the most easily measured of the three components of a nation's wealth. Produced assets have very clear income-earning potential, but they are also quite subject to obsolescence and must be renewed continually. For example, a clothing factory produces a flow of goods destined for consumers, but the machinery wears out over time, and the building itself ages. Thus, we have income—a flow of goods—and depletion, which is referred to as capital consumption.

The **natural capital** (Fig. 23–4b) refers to the goods and services supplied by natural ecosystems and the mineral resources in the ground. These represent an indispensable set of resources, some of which are *renewable* and some *nonrenewable*. Renewable natural capital is represented by forests, fisheries, agricultural soil, water resources, and the like, which can be employed in the production of a flow of goods (lumber, fish, corn). Often overlooked is the fact that this same natural capital also provides a flow of services in the form of waste breakdown, climate regulation, oxygen production, and so forth. (See Table 3–3 for an extensive list of the goods and services provided by natural capital.) As we have seen in earlier chapters, renewable natural capital is subject to depletion, but also has the capacity to yield a sustained income if it is managed responsibly. Nonrenewable natural capital, such as oil and mineral deposits, also is subject to depletion and provides no services unless it is extracted and converted into some useful form. Yet it is part of the wealth of a nation as long as it is in the ground. Some components of natural capital are easier to measure than others; for instance, the lumber of a forest can be reckoned in board feet, but how does one measure the impact of a forest on the local climate and on its capacity to sustain a high level of biodiversity?

Human resources (Fig. 23–4c) can be divided into two elements. The first is **human capital**, which refers to the population and its characteristics in terms of physical, psychological, and cultural attributes. To this is added the value imparted by education, which enables members of a

(a) Produced assets

(b) Natural capital

(c) Human resources

▲ **FIGURE 23–4** *The wealth of nations.* Three major capital components determine a nation's wealth: (a) produced assets, (b) natural capital, and (c) human resources.

population to acquire skills that are useful in an economy. This can be both formal and informal education; the point is that people acquire a productive capacity through education. Investments in health and nutrition also contribute to increases in human capital. The second element of human resources is what the World Bank group has called **social capital**—the social and political environment that people create for themselves in a society. Social capital includes more formal structures and relationships as defined by government, the rule of law, the court system, and civil liberties. It also includes the more horizontal relationships of people as they associate into religious or ethnic affiliations or join organizations in which they have a common interest. Social capital is considered a vital element of a nation's wealth, because social relationships clearly affect and are affected by economic processes. Human resources are the hardest of the three components to measure.

On the basis of these three components of the wealth of nations, the World Bank Environment Department devised various means of measuring each component and then set about evaluating the wealth of different nations. Some of the results of the bank's analysis are shown in Table 23–2 and Fig. 23–5. One finding that should come as no surprise is the notable differences in per capita wealth among nations. Interestingly, the dominant source of wealth for most nations is human resources, but natural capital ranks high in a number of countries. Although natural capital often ranks third for most countries, this does not mean that it is less important than human resources and produced assets. (The authors offer a caution in interpreting their work, as they did not assess the services component of natural capital and calculated mainly the actual resource inputs into production—that is, forests as a source of timber, mineral as-

sets, etc.) The dominance of the human resources component emphasizes the importance of investing in health, education, and nutrition in a society that needs to move further along in development. The bank has focused attention on natural capital as an essential element of economic production and has provided the basics of a tool to measure progress in sustainability. Interestingly, the World Bank did not use the commonly employed indicator of economic status known as the gross national product (GNP).

Shortcomings of the GNP

The *gross national product*, or GNP, refers to the sum of all goods and services produced in a country in a given time frame. The GNP is the most commonly used indicator of the economic health and wealth of a country. In most countries, the comparable *gross domestic product*, or GDP, is now used. The GDP is the GNP minus net income from abroad. As a per capita index, the GDP is often used to compare rich and poor countries and to assess economic progress in the developing countries. An important element in calculating the GDP is accounting for the assets that are used in the production processes. Buildings and equipment, for example, are essential to production, but they gradually wear out or depreciate. Their depreciation is charged against the value of production (called capital depreciation); that is, an accounting of capital depreciation is routinely subtracted from the production of goods and services in order to generate an accurate GDP.

The economists who invented the GNP as a measuring device 50 years ago did not take into account the depreciation of natural capital—an omission that recently has received a great deal of criticism from ecological econo-

---

TABLE 23–2 Wealth of Nations per Capita, by Region, 1994

	Dollars Per Capita				Percent Share of Total Wealth		
	Total wealth	Human resources	Produced assets	Natural capital	Human resources	Produced assets	Natural capital
North America	326,000	249,000	62,000	16,000	76	19	5
Pacific members of OECD*	302,000	205,000	90,000	8,000	68	30	2
Western Europe	237,000	177,000	55,000	6,000	74	23	2
Middle East	150,000	65,000	27,000	58,000	43	18	39
South America	95,000	70,000	16,000	9,000	74	17	9
North Africa	55,000	38,000	14,000	3,000	69	26	5
Central America	52,000	41,000	8,000	3,000	79	15	6
Caribbean	48,000	33,000	10,000	5,000	69	21	11
East Asia	47,000	36,000	7,000	4,000	77	15	8
Eastern and Southern Africa	30,000	20,000	7,000	3,000	66	25	10
West Africa	22,000	13,000	4,000	5,000	60	18	21
South Asia	22,000	14,000	4,000	4,000	65	19	16

Source: The World Bank, *Expanding the Measure of Wealth: Indicators of Environmentally Sustainable Development* (Washington, DC: The International Bank for Reconstruction and Development, 1997), p. 25. Environmentally sustainable development studies and monograph series No. 17, Table 3.3.

*Organization for Economic Cooperation and Development; Pacific members include Japan, Australia, New Zealand, and the Republic of Korea.

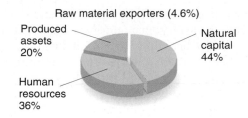

Raw material exporters (4.6%)

Produced assets 20%

Human resources 36%

Natural capital 44%

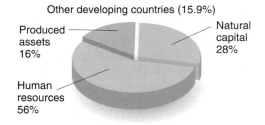

Other developing countries (15.9%)

Produced assets 16%

Human resources 56%

Natural capital 28%

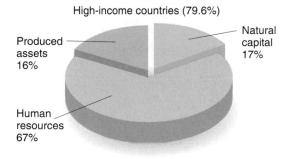

High-income countries (79.6%)

Produced assets 16%

Human resources 67%

Natural capital 17%

▲ **FIGURE 23–5** *Composition of world wealth.* Some developing countries are dependent on exporting raw materials. These countries account for 4.6% of the world's wealth. The remaining developing countries contain 15.9% of the world's wealth. The high-income developed countries possess 79.6% of the wealth. Relative percentages of produced assets, natural capital, and human resources are shown for each category. (*Source:* The World Bank, *Monitoring Environmental Progress: A Report on Work in Progress* [Washington, DC: The International Bank for Reconstruction and Development, 1995], p. 63.)

mists. Such concerns were regarded as externalities to a country's balance sheet. The natural resources and their associated natural services were considered a "gift from nature." Given this state of affairs, it is possible for a nation to cut down a million acres of forest and count the sale of the timber on the income side of the GDP ledger, whereas on the expense side, only the depreciation of chain saws and trucks would be seen. Completely hidden from accounting is the loss of all the natural services once performed by the forests and, incredibly, the disappearance of the million acres of forests as an economic asset! As long as this discrepancy remains, nations will underestimate the value of natural resources. They will be able to deplete fisheries, lose soil by intensive farming, remove forests, and degrade rangelands by overgrazing and *account for these activities as economic productivity*!

Correcting this system of accounting would require calculating the current market value of natural assets and

considering it to be part of the stock of a nation's wealth. To this value would be added the value due to the natural services performed by the ecosystems in which the resources are found. The natural assets and the services they perform are the nation's natural capital. When the natural resources are drawn down, the depreciation represented by the loss of natural capital would be entered into the ledger as the GDP is calculated.

Some economists also believe that expenditures for pollution control and environmental protection should not be included in the GDP, because they are not economically productive. It might be argued, however, that because such expenditures involve jobs and the manufacture of equipment, they *should* be counted as productive enterprises.

Agenda 21—one of the major documents that came out of the 1992 Earth Summit meeting in Rio—recognized the need for "a program to develop national systems of integrated environmental and economic accounting in all countries." The U.N. Statistical Office, which coordinates the accounting procedures of member countries, has been preparing a manual, published as a draft edition on the Internet (*Integrated Environmental and Economic Accounting*), designed to accomplish *Agenda 21*'s objective. In this manual, countries are encouraged to perform **environmental accounting** by putting their environmental assets and services in monetary units and keeping a parallel account of their *net domestic product* (similar to the GDP) as it is affected by environmental accounting. After revisions by numerous international working groups, the U.N. manual, to be named *SEEA-2000* (for "System for Integrated Environmental and Economic Accounting for the year 2000"), will be published in 2001. The manual is intended to be used as a reference document to show nations how to implement environmental accounting. It includes case studies and step-by-step instructions and strongly advocates the adoption of integrated environmental and economic accounting by U.N. member nations. In recent years, some 25 countries have experimented with different forms of environmental accounting, indicating that some form of environmental accounting may become broadly adopted before long.

Resource Distribution

None of the resources crucial to an economy are evenly distributed among nations. Table 23–2 shows clearly that great differences in wealth exist among the nations of the world, something we have seen many times in earlier chapters. It is significant that the greatest differences occur in the two categories that are most closely tied to the process of development: *human resources* and *produced assets*. With the exception of the oil-rich Middle East, the per capita differences in natural capital are not remarkable. Thus, *economic development*, which fosters growth in human resources and produced assets, is the pathway to achieving more equity in resource distribution. There are many reasons why some countries have undergone rapid development and others have not; population growth is one that has been cited in this text (Chapter 7). Perhaps the most important reasons relate to the political and legal systems prevailing in a society. For example, the definition of rights, the enforcement of those rights, and the facilitation of economic exchange are all considered essential to the successful development of human resources and produced assets. A well-developed body of law, an honest legal system, and a free press will go far in establishing rights within a society. A well-developed market economy and free entry into and exit from markets for people and businesses are also essential for development. Functioning communications and transportation networks and viable financial markets are necessary for sustaining a market economy. History tells us that if these are in place, a society will make progress in the development of human resources and produced assets. In societies where these resources are not in place, corruption, inefficiency, and human rights abuses often prevail, and continuing poverty and environmental degradation are the predictable result.

Intergenerational Equity.   What we have just discussed can be called **intragenerational equity**—the golden rule of *making possible for others what is possible for you.* Sustainable development, on the other hand, is more about **intergenerational equity**—*meeting the needs of the present without compromising the ability of future generations to meet their needs.* Thus, we can conserve resources and avoid long-term pollution partly out of a concern for future generations. Economists point to some problems with this perspective, however. Economics deals with future resources by applying what are called **discount rates**. By this, we mean the *rate* used for finding the present value of some future benefit or cost. The discount rate is often considered to approximate current interest rates. For example, a dollar is worth more to us today than it will be five years from now even if inflation were not occurring, because we can put the dollar to use now and can earn interest on that dollar if we invest it. At a 5% rate of interest, in five years our dollar would be worth $1.28. Turning this idea around, a dollar five years from now is worth only $0.78 today. The conclusion is that some resource or benefit is worth more now than it will be in the future.

This concept can be applied to a stand of trees. The owner has several options: (1) Cut all the trees and sell the timber today; (2) hold on to the trees in the hope that they will bring a better price at some time in the future; (3) harvest the trees on a sustainable basis, spreading out profits over time. If it turns out that the trees grow more slowly than current interest rates, the economically profitable decision would be to cut them now and invest the proceeds. This is clearly a problem for natural capital, since it would appear to argue in many cases for maximizing short-term profit over sustainable, long-term harvest. In other words, reap the economic value now, and put off the ecological costs to the future. Similarly, let future generations cope with global climate change; we can continue to enjoy the benefits of the unrestrained use of fossil fuels. The same can be said about other resources: Harvest those swordfish now, and when they're gone, switch to some other species of fish.

This is a problem with intergenerational equity. What is at odds here is the self-interest of present individuals versus the longer term interests of a community or a society. To make things a bit more challenging, it might also be argued that conserving resources for the future could actually be thought of as putting the interests of future generations above those of, say, the poor today; intragenerational justice is no less important than intergenerational justice. Hopefully, this topic at least gives us an insight into how difficult it is to develop appropriate public policy that deals with resource allocation.

23.3 Pollution and Public Policy

We begin now to focus more directly on environmental public policy as it is developed in order to cope with pollution. As we do, we will see important areas of intersection between public policy and economics.

Public Policy Development: The Policy Life Cycle

Environmental public policy is developed in a sociopolitical context, usually in response to a problem. Some policies are developed at local levels to solve local problems. Cities and towns, for example, establish zoning regulations in order to protect citizens from haphazard and incompatible land uses. Thus, residential and commercial zoning districts are kept separate, conservancy zoning protects sensitive watersheds, and rural zoning safeguards agricultural land.

Many problems, however, are broader in their scope and must be addressed at higher levels of government. As we saw in Chapters 21 and 22, the problems of air pollution transcend local, state, and even national boundaries. The processes that contribute to air pollution are complex, involving a myriad of activities fundamental to an industrial society—from people heating their homes and driving cars, to local dry-cleaning establishments and car-repair shops providing their services, to major power plants generating electricity for entire regions. When specific problems, such as acid deposition or the production of ground-level ozone, are addressed in a democratic society, the development of public policy often takes a predictable course, called the **policy life cycle**.

The typical policy life cycle has four stages: *recognition, formulation, implementation,* and *control* (Fig. 23–6). Each of the stages can be said to carry a certain amount of political "weight," which varies over time and is represented in the figure by the course of the life-cycle line. The thick-

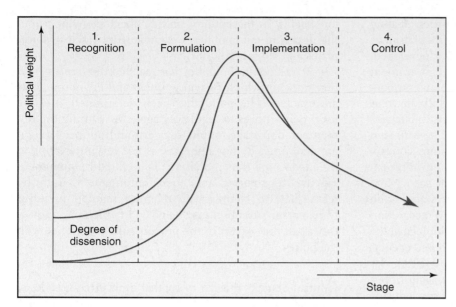

◀ **FIGURE 23–6** *The policy life cycle.* Most environmental issues pass through a policy life cycle in which an issue is accorded different degrees of political "weight" as it moves through the cycle. The final result is a policy that has been incorporated into the society and a problem that is under control.

ness of the line represents controversy and political uncertainty. We will use Rachel Carson and the formation of the Environmental Protection Agency as a case study to illustrate the policy life cycle.

Recognition Stage. Alerted in 1957 by a close friend who had experienced aerial spraying of her small bird sanctuary and the subsequent death of many birds, biologist Rachel Carson (Fig. 23–7) began to turn to technical and scientific literature to investigate what was known about the impacts of pesticides on the natural world. It took several years before Carson was ready to bring a book to publication. By that time, she was thoroughly aroused and angry at what she saw as a conspiracy to keep the public uninformed about pesticides. When her book *Silent Spring* was finally published in 1962 (and serialized in the *New Yorker*), it ignited a firestorm of criticism from the chemical and agricultural establishment. However, it also caught the public's eye, and it quickly made its way to the president's Science Advisory Committee when President John F. Kennedy read the *New Yorker* version. Kennedy charged the committee with studying the pesticide problem and recommending changes in public policy.

Illustrated by these events, the *recognition stage* is low in political weight. The stage begins with the early perceptions of an environmental problem, often coming as a result of scientific research. Scientists then publish their findings, and the media pick up the information and popularize it. When the public becomes involved, the political process is under way. During the recognition stage, dissension is high; opposing views on the problem surface as businesses or technological industries respond to the bad news that they are at the root of some new environmental problem. Eventually, the problem gets attention from some level of the government, and the possibility of addressing it with public policy is considered.

Formulation Stage. In 1963, President Kennedy's committee made recommendations that fully supported Carson's thesis in *Silent Spring*. It became clear to policymakers during the 1960s that environmental policy was weak and

▲ **FIGURE 23–7** *Rachel Carson.* Marine scientist and author Rachel Carson, who wrote the courageous book *Silent Spring* that broke the conspiracy of silence on the wide-scale use of pesticides and that jump-started the environmental movement.

often compromised by ties to the very industries creating the problems. Public debate ensued, environmentalism in the form of new organizations and legal action developed, and the executive and legislative branches of government wrestled with what to do to establish effective environmental public policy. Finally, Congress passed a bill known as the **Environmental Policy Act** in 1969, the first bill to recognize the interconnectedness of ecological systems and human enterprises. Shortly after that, a commission appointed by President Richard Nixon to study environmental policy recommended the creation of a new agency that would be responsible for dealing with air, water, solid waste, pesticide use, and radiation standards in recognition of the interrelated web of water, soil, air, and biological life so eloquently pictured by Rachel Carson. The new agency, called the **Environmental Protection Agency** (EPA), was given a mandate to protect the environment against pressures from other governmental agencies and from industry, on behalf of the public. The year was 1970, the same year that 20 million Americans celebrated the first Earth Day.

Thus, we have the *formulation stage*—a stage of rapidly increasing political weight. The public is now aroused, and debate about policy options occurs in the corridors of power. The political battles may become fierce, as questions of regulation and who will pay for the proposed changes are addressed. Media coverage is high, and politicians begin to hear from their constituencies. Lobbyists for special interests or environmental groups put pressure on legislators to soften or harden the policy under consideration. During the formulation stage, policymakers consider what may be called the *"Three E's"* of environmental public policy: **effectiveness** (whether the policy accomplishes what it intends to do in improving the environment), **efficiency** (whether the policy accomplishes its objectives at the least possible cost), and **equity** (whether the policy parcels out the financial burdens fairly among the different parties involved). Often, policymakers are prone to emphasize effectiveness over efficiency and equity at this stage of development, because they are looking for a workable solution and trying to make it into a law as soon as possible.

Implementation Stage. Given its broad jurisdiction, the newly created EPA embarked idealistically on a course that would take the agency into regulatory enforcement and environmental research needed to determine how to develop environmental protection standards. The 1970s were declared by President Nixon to be the "environmental decade." The EPA's first administrator, William Ruckelshaus, proposed that the agency's mission was, first and foremost, the "development of an environmental ethic." Within a few years, Congress enacted more than 20 major pieces of environmental legislation, giving the EPA increasing power to develop regulations. One of its most significant actions was to ban DDT and other persistent pesticides. Since then, the EPA has become the most powerful regulatory agency in the country, able to affect everything from the design of automobiles to standards for nuclear power plants and sanitary landfills.

At this point, the policy has reached the *implementation stage*, in which its real political and economic costs are exacted. The policy has been determined, and the focal point moves to regulatory agencies. During the implementation stage, public concern and political weight are declining. By now the issue is not very interesting to the media, and the emphasis shifts to the development of specific regulations and their enforcement. Industry learns how to comply with the new regulations. Over time, greater attention may be given to efficiency and equity as all the players in the process gain experience with the policy.

Control Stage. It is fair to say that in its three decades of existence, the EPA has become a mature and effective agency in carrying out its mandate to protect the environment and human health. We have seen many examples of its work in previous chapters. The final stage in the policy life cycle is the *control stage*. By this point, years have passed since the early days of the recognition stage. Problems are rarely completely resolved, but the environment is improving, with things moving in the right direction. Policies (and their derived regulations) are broadly supported and often become embedded in the society, although their vulnerability to political shifts continues. Regulations may become more simplified. The policymakers must now see that the problem is kept under control, and in due time the public often forgets that there ever was a serious problem.

The policy life cycle is, of course, a simplified view of what is frequently a highly complex and contentious political process. At any one time, different problems will be in different stages of the life cycle, as shown in Fig. 23–8, for a number of environmental problems in the industrialized countries. Also, countries in different stages of economic development will be in different stages of the policy life cycle for a given problem. For example, sickness and death due to polluted water are still serious problems in many developing countries, because public policies have not yet caught up with the need for water treatment and sewerage. On the other hand, this problem is clearly in the control stage in the industrialized societies. The reason for the discrepancy is often traceable directly to the costs that lie behind the development and implementation of public policy.

Economic Effects of Environmental Public Policy

What are the relationships between a country's economy and its environmental public policy? Are there some policies that are too costly? How should a society parcel out its limited resources to address environmental problems? Are environmental regulations a burden on the economy? Let

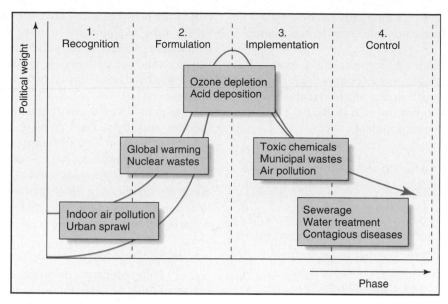

▲ **FIGURE 23–8** *Environmental problems in the policy life cycle.* Different environmental problems are in different stages of the policy life cycle in industrial societies.

us begin to sort out these questions by looking at the relationship between environmental policies and costs.

As we have seen, policies do not just appear out of thin air; they are the result of some version of the policy life cycle. In countries undergoing economic development, the choice of environmental policies can often make the difference between rapid growth and stagnation. In the already industrialized countries, the best policies are those that conform to the "Three E's."

Costs of Policies. Some policies have relatively little or no direct *monetary cost*—they do not require major investments of administration or resources. Thus, removing subsidies to special interests and denying special access to national resources can result in a more efficient and equitable operation of the economy and can protect the environment. For example, the use of national lands for cattle grazing and timber harvesting is subsidized in the United States. As a result, the real costs—and environmental consequences—of these activities are borne not just by the special-interest groups that have access to these resources, but by all taxpayers. Removing such subsidies, however, can have very real *political costs*, as powerful interests do everything they can to hold on to their privileges.

Most environmental policies involve some very real costs that must be paid by some segment of society. For instance, resources are invested in policies that control pollution because their benefits are judged to outweigh their costs—the public welfare is improved, and the environment is protected.

Who pays the costs? The equity principle implies that those benefiting from the policies should pay—the customers of businesses whose activities are regulated and the people whose health is protected by laws against polluting. (It should be recognized that when businesses are taxed or industries are forced to add new technologies, they will pass their costs along to consumers.) In short, the costs of public policies are borne by the public in one way or another.

Impact on the Economy. For years, special-interest groups have argued that environmental regulations are excessive and bad for the economy—that our concern for environmental protection costs hundreds of thousands of jobs, reduces our competitiveness in the marketplace, drives up the price of products and services, and, in general, imposes costs on the economy that are nonproductive.

How sound are these arguments? This major concern was addressed by Roger Bezdek, president of Management Information Services, Inc., in *Environment* (Sept. 1993), and we refer to his article in the following analysis.

First, we must recognize that much of the evidence used to support the charges that have been made is anecdotal. Thus, "environmental restrictions on energy extraction and production have caused the loss of 400,000 jobs," according to the American Petroleum Institute, and "Measures to protect the Everglades will cause the loss of 15,000 jobs," according to Florida sugar growers. Indeed, numerous cases can be cited wherein jobs were lost because of environmental regulations.

Anecdotal evidence can also be used in *favor* of environmental policy. "Recycling has created 14,000 jobs in California," according to the California Planning and Conservation League Foundation, and "The Clean Air Act of 1990 will generate 60,000 new jobs," according to the EPA. Many sectors of the U.S. economy most subject to environmental regulations—plastics, fabrics, etc.—have improved their efficiency and their competitiveness in the international market.

What can we conclude? Is there something more substantial to help us with this important controversy? The best evidence comes from careful studies of what is actually happening. These studies, all documented in Bezdek's article, reveal the following interesting findings:

- In general, states with the strictest environmental regulations also had the highest rates of job growth and economic performance.
- Nations with the highest environmental standards also had the most robust economies and rates of job creation.
- Only 0.1% of job layoffs were attributed by employers to environment-related causes, according to a recent study by the U.S. Bureau of Labor Statistics.

In short, the results of many careful studies of the relationship between environmental protection and economic growth refute the suggestion that environmental regulation is bad for jobs and the economy. In fact, the relationship is positive: Economic performance is highest where environmental public policy is most highly developed. (It may be no coincidence that in the former communist countries in Europe, where environmental public policy was deliberately suppressed in order to favor industry, the economies and environments are disaster areas.) The evidence indicates that concerns for energy efficiency, pollution control, and conservation of resources have encouraged businesses to modify their technologies in ways that make them *more* competitive, not less, in the national and international marketplace. And the evidence for jobs suggests that at least as many jobs have been created by environmental protection as have been lost.

Taken as a whole, environmental protection is a huge industry, and it has been growing more rapidly than the GNP. The graph in Fig. 23–9 demonstrates environmental protection as a growth industry. (Note that the EPA is in the process of revising these cost estimates, and they are turning out to be significantly *lower* than anticipated.) A study of total expenditures for the industry for one year revealed that it was responsible for

- $355 billion in total industry sales,
- $14 billion in corporate profits,
- $63 billion in federal, state, and local government revenues, and
- 4 million jobs.

In sum, we can draw several conclusions from our examination of the impact of environmental policy on the economy:

- Environmental public policy does not *diminish* the wealth of a nation; rather, it *transfers* wealth from polluters to pollution controllers and to less polluting companies.
- The "environmental protection industry" is a major job-creating, profit-making, sales-generating industry.

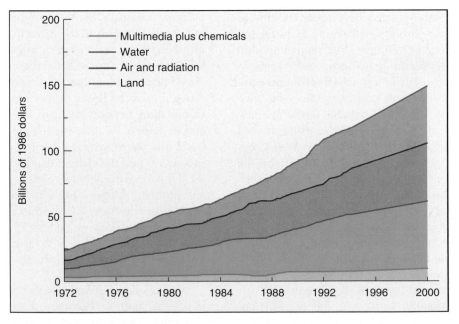

▲ FIGURE 23–9 *Costs of pollution control.*   EPA's estimate of total yearly costs of pollution control in the United States, assuming the full implementation of existing regulatory laws. (*Source*: Data from EPA, "The Cost of a Clean Environment," *Environmental Investments*, 1990.)

- The argument that environmental protection is bad for the economy is simply unsound: Not only is environmental protection good for the economy, but environmental public policy is responsible for a less hazardous, healthier, and more enjoyable environment.

Policy Options: Market or Regulatory?

Let us assume that the decision has been made to develop a policy to deal with an environmental problem. Now what? The objective of environmental public policy is to *change the behavior of polluters and resource users so as to benefit public welfare and the environment*. There are two main ways to accomplish these behavior changes: by using basic **market approaches** to set prices on pollution and the use of resources and by using a direct **regulatory approach** wherein standards are set and technologies are prescribed—the command-and-control strategy discussed in Chapter 22. Both approaches have their advantages and disadvantages. Many newer policies include elements of both.

Market-based policies have the virtues of simplicity, efficiency, and (theoretically) equity. All polluters are treated equally and will choose their responses on the basis of economic principles having to do with profitability. There are strong incentives to reduce the costs of using resources or of paying for the right to a certain amount of pollution. The trading of emission allowances under the Clean Air Act of 1990 represents a good example of the market approach to pollution control. User fees for the disposal of municipal solid waste is another example.

Many environmental problems are not readily amenable to market-based policies. For example, on the basis of knowledge of the health effects of pollutants, a society may choose to set standards that reflect the health of the most vulnerable members of the population. This is the case with the basic criteria covering air pollutants (Chapter 22). To meet these standards, regulations are established, and polluters are required by law to comply with the regulations. The regulatory approach also works well with land-use issues, in which certain values are upheld that will not necessarily be protected by a straightforward market approach. The regulatory approach has been the most commonly used approach in environmental public policy. In the interest of addressing outstanding environmental problems, policymakers are prone to turn first to this approach because it is deemed effective.

One of the shortcomings of the regulatory approach is that it practically guarantees a certain sustained level of pollution. If a polluter is told to use a particular technology or is given a cap on emissions, the polluter has no incentive to invest in technologies or a reduction in the source that would keep pollution at lower levels than allowed. Pollution prevention, as we saw in Chapter 20, is a much preferable course of action, and it is encouraging to see the EPA moving deliberately in that direction. We turn next to the details of a specific tool that is employed in developing environmental public policy: *benefit–cost analysis*. In doing so, we will address the question of how to measure the cost-effectiveness of regulating pollution.

23.4 Benefit–cost Analysis

As mentioned at the beginning of this chapter, Executive Order 12866 requires that all major regulations be supported by a **benefit–cost analysis.** Such an analysis compares the estimated costs of a project with the benefits that will be achieved. It is often used as a means of rationally deciding whether to proceed with a given project. All costs and benefits are given monetary values and compared by means of what is commonly referred to as a **benefit–cost** (or cost–benefit) **ratio.** A favorable ratio for a project means that the benefits outweigh the costs. Such a project is said to be **cost effective,** and thus there is an economic justification for proceeding with it. The analysis usually involves considering several options for accomplishing the project and selecting the option with the best benefit–cost ratio. If costs are projected to outweigh benefits, the project may be revised, dropped, or shelved for later consideration.

Benefit–cost analysis of environmental regulations is intended to build efficiency into policy so that society does not have to pay more than is necessary for a given level of environmental quality. If the analysis is done properly, it will take into consideration *all* of the costs and benefits associated with a regulatory option. In so doing, it must address the problem of *externalities*.

External and Internal Costs

In the language of economics, an **external cost** is an effect of a business process that is not included in the usual calculations of profit and loss. For example, when a business pollutes the air or water, the pollution imposes a cost on society in the form of poor health or the need to treat water before using it. This is an *external bad*. When workers improve their job performance as a result of experience and learning, the improvement is not credited on the company's ledgers, although it is considered an *external good*.

Amenities such as clean air, uncontaminated groundwater, and a healthy ozone layer are not privately owned. In the absence of regulatory controls, there are no direct costs to a business for degrading these amenities. (In other words, they are externalities.) Therefore, there are no incentives to refrain from polluting the air or water. By including *all* of the costs and benefits of a project or a regulation, benefit–cost analysis effectively brings the externalities into the economic accounting. One suggestion for

EARTH WATCH

GREEN FEES AND TAXES

Benefit–cost analysis works well only if all of the costs and benefits of a polluting activity are included in the calculations. Let us assume that this is possible. What would happen if everyone had to pay the true costs of their use of the environment as they used resources or degraded the air or water? Such charges are often called **green fees**. Oddly enough, two groups that frequently are battling—economists and environmentalists—both agree that such a policy would go a long way toward solving many of our environmental problems.

Economists point out that in a free-market economy, the market will guarantee that resources will be used in the most efficient way; people will economize. This means that most businesses and people who use, say, 100 gallons of gasoline and have to pay a levy that reflects the true costs of that gasoline and the pollution it produces will be highly motivated to keep their use of gasoline to an absolute minimum. They will adopt alternative modes of transportation

and more efficient vehicles. Including all of the external costs of the gasoline in its price would involve adding together the cost of defending Persian Gulf shipping lanes, the costs of air pollution, and even the expected costs of coping with global climate change. The green fee could be imposed as a tax on gasoline, and to ease its impact on the economy, the fee could be implemented gradually until it brings the price of gasoline to its true level. Internalizing all these costs, if it were possible, would make $5 per gallon of gasoline at the pumps seem a bargain. Many observers have suggested that a tax of, say, $100 per ton of fossil-fuel carbon would go a long way toward accomplishing the objective of paying the real costs of the fuel.

Great, say the environmentalists. This would mean that the people who do the most damage pay for their environmental impact, something environmental groups have been calling for all along. They insist that some of the revenues collected be used to mitigate the impacts of pollution and re-

source use, but if that were to be accomplished, the outcome would move our society in the direction of sustainability.

Pollution fees, gasoline taxes, and user fees all have the potential for "internalizing" the external costs of environmental degradation and the use of resources. If such green fees could become public policy, other taxes (e.g., personal income taxes) might be able to be lowered, since these are often the sources of revenue that are paying much of the external costs.

The question is, Would such a policy be politically feasible? Proponents think that it may be. If people are asked if they *like* higher taxes, the answer is a highly predictable *no*. But if asked whether they would rather be taxed on their energy use, their trash production, and their purchases of high-impact goods than be taxed on their incomes and investments, they might well choose the former, especially if they can appreciate the environmental benefits that accrue from such a policy.

accomplishing this is the use of green fees or taxes (see the "Earth Watch" essay, this page.)

The Costs of Environmental Regulations

The costs of pollution control include the price of purchasing, installing, operating, and maintaining pollution-control equipment and the price of implementing a control strategy. Even the banning of an offensive product costs money, because jobs are lost, new products must be developed, and machinery may have to be scrapped. In some instances, a pollution-control measure may result in the discovery of a less expensive way of doing something. In most cases, however, pollution control costs money. Thus, the effect of most regulations is to prevent an external bad by imposing economic costs that are ultimately shared by government, business, and consumers.

Pollution-control costs generally increase exponentially with the level of control to be achieved (Fig. 23–10). That is, a partial reduction in pollution may be achieved by a few relatively inexpensive measures, but further reductions generally require increasingly expensive measures, and 100% control is likely to be impossible at any cost. Because of this exponential relationship, regulatory control often has to

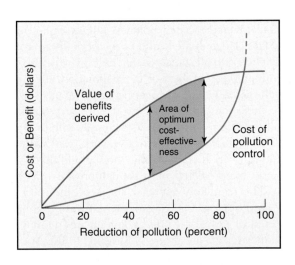

▲ **FIGURE 23–10   *The benefit–cost ratio for reducing pollution.*** The cost of pollution control increases exponentially with the degree of control to be achieved. However, benefits to be derived from pollution control tend to level off and become negligible as pollutants are reduced to near or below threshold levels. The optimum cost-effectiveness is achieved at less than 100% control.

focus on stimulating new ways of reducing pollution. Indeed, the costs of pollution control constitute a powerful incentive to make substitutions, to recycle materials, or to redesign industrial processes.

In most cases, pollution-control technologies and strategies are understood and available. Thus, equipment, labor, and maintenance costs can be estimated fairly accurately. Unanticipated problems that increase costs may occur, but as technology advances and becomes more reliable, experience is gained, lower cost alternatives frequently emerge, and such unforeseen increases are negligible. The costs of pollution control are likely to be highest at the time they are initiated; then they decrease as time passes (Fig. 23–11a). The importance of this trend will become evident when we consider the time span over which costs and benefits are compared.

How accurate are the cost estimates? A look at some historical examples reveals that industries predict higher costs of regulations than government studies do, and surprisingly, even government analysts often overestimate the costs of regulations. For example, industry sources claimed that the costs of eliminating chlorofluorocarbons from automobile air conditioners would increase the price of new cars by anywhere from $650 to $1,200. In reality, the price increase turned out to be much lower—from $40 to $400. A recent study of benefit–cost analyses revealed that overestimates of cost occurred in 12 of 14 regulations. The primary reason given for the overestimates is a failure to anticipate the impact of technological innovation, admittedly a factor that is difficult to assess.

As we saw in Figure 23–9, the costs of improving the quality of the environment represent a major economic outlay. What benefits have we received in return?

The Benefits of Environmental Regulation

In Chapter 22, we analyzed the Clean Air Act's impact on society and $6.4 trillion net benefit. Reduced air pollution, improved human health, costs avoided, and less harm to children from lead all benefited society. How does someone now argue against these regulations? The only way for special interests to weaken the Clean Air Act is if the public is not aware of these benefits in their lives.

Such benefits of regulatory policies are seldom as easy to calculate as the costs. Estimating benefits is often a matter of estimating the costs of damages that *would* occur if the regulations were not imposed (Fig. 23–11b). For example, the projected environmental damage that would be brought on by a given level of sulfur dioxide emissions from a coal-fired power plant (an external bad) becomes a benefit (an external good) when a regulatory action prevents half of those emissions. Benefits include such things as improved public health, reduced corrosion and deterioration of materials, reduced damage to natural resources, preservation of aesthetic and spiritual qualities of the environment, increased opportunities for outdoor recreation, and continued

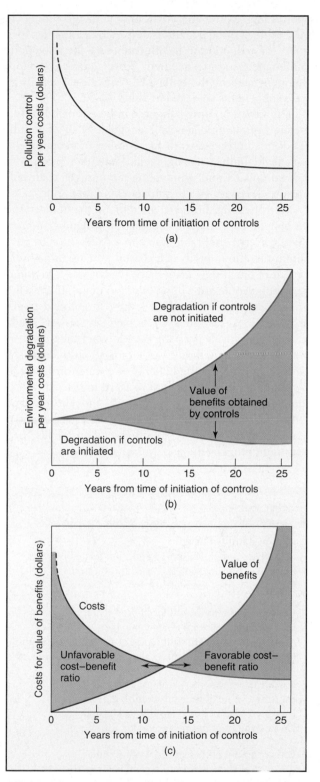

▲ FIGURE 23–11 *Cost-effectiveness of pollution control over time.* (a) Pollution-control strategies generally demand high initial costs. The costs then generally decline as those strategies are absorbed into the overall economy. (b) Benefits may be negligible in the short term, but they increase as environmental and human health recover from the impacts of pollution or are spared increasing degradation. (c) When the two curves are compared, we see that what may appear as cost-ineffective expenditures in the short term (5–10 years) may, in fact, be very cost effective in the long term.

opportunities to use the environment in the future. The dollar value of these benefits is derived by estimating, for instance, the reduction in health-care costs, the reduction in maintenance and replacement costs, and the economic value generated by the enhanced recreational activity. Examples of benefits are listed in Table 23–3.

The values of some benefits can be estimated fairly accurately. For example, it is well recognized that air pollution episodes cause increases in the number of people seeking medical attention. Since the medical attention provided has a known dollar value, eliminating the number of air pollution episodes provides a health benefit of that value.

Many benefits, however, are difficult to estimate. Accurate benefit–cost analysis depends on assigning monetary values to every benefit, but how can a dollar value be put on maintaining the integrity of a coastal wetland or on the enjoyment of breathing cleaner air? The answer is to find out how much people are willing to pay to maintain these benefits. But how can this be done if there is no free market for the benefits? Again, economists have an answer: **shadow pricing**. This involves asking people what they *might* pay for a particular benefit if it were up to them to decide. For example, to evaluate the benefits of cleaner air in the Los Angeles basin, homeowners were asked to place a value on improving their air quality from *poor* to *fair*. The average response (in 1977) was $30 a month. The total benefit was then calculated from the number of households in the basin, multiplied by the $30 average value.

TABLE 23–3 Benefits That May Be Gained by Reduction and Prevention of Pollution

1. Improved human health
   Reduction and prevention of pollution-related illnesses
   Reduction of worker stress caused by pollution
   Increased worker productivity

2. Improved agriculture and forest production
   Reduction of pollution-related damage
   More vigorous growth by removal of stress due to pollution
   Higher farm profits, benefiting all agriculture-related industries

3. Enhanced commercial or sport fishing
   Increased value of fish and shellfish harvests
   Increased sales of boats, motors, tackle, and bait
   Enhancement of businesses serving fishermen

4. Enhancement of recreational opportunities
   Direct uses such as swimming and boating
   Indirect uses such as observing wildlife
   Enhancement of businesses serving vacationers

5. Extended lifetime of materials and less necessity for cleaning
   Reduction of corrosive effects of pollution, extending the
     lifetime of metals, textiles, rubber, paint, and other coatings
   Reduction of cleaning costs
6. Enhancement of real-estate values

The Value of Human Life. Shadow pricing becomes difficult when the analysis has to place a value on human life. Many of the pollutants to which people are exposed are hazardous, exacting a toll on health and life expectancy. To estimate the benefits of regulating such pollutants, it is necessary to calculate how many lives will be saved or how many people will enjoy better health. Finding a value for these benefits is fraught with ethical difficulties. For example, the EPA calculated that the new clean-air standards for ozone and particulates would prevent 15,000 premature deaths annually. Different approaches have resulted in a range of values for human life from tens of thousands of dollars to $10 million per life saved. Any benefit–cost analysis that must factor in the risk to human life requires that human lives be valued somewhere within this very broad range. Obviously, the outcome of the analysis is determined by the value selected.

Nonhuman Environmental Components. How does shadow pricing work for the nonhuman components of natural environments—for a population of rare wildflowers, for example, or a wilderness site? These things depend entirely on how willing people are to pay for their preservation. Again, monetary values must be assigned to their existence, and again the outcome will be strongly influenced by a highly subjective element in the analysis.

Cost-Effectiveness

Cost-effectiveness analysis is an alternative option for evaluating the costs of regulations. Here, the merits of the goal are accepted, and the question is, How can that goal be achieved at the least cost? To find out, alternative strategies for reaching the goal are analyzed for costs, and the least costly method is adopted.

Cost-effectiveness can be applied to the desirable level of pollution control. As Figure 23–10 shows, a significant benefit may be achieved by modest degrees of cleanup. Note how differently the cost and benefit curves behave with increasing reduction of pollution. Little additional benefits are realized when cleanup begins to approach 100%, yet costs increase exponentially. This behavior follows from the fact that living organisms—including humans—can often tolerate a threshold level of pollution without ill effect. (See Fig. 20–1.) Therefore, reducing the level of a pollutant below threshold levels will not yield an observable improvement.

Figure 23–10 makes it clear that with a modest degree of cleanup, benefits can outweigh costs. At some point in the cleanup effort, the lines cross and the costs exceed the benefits. Consequently, while it is tempting to argue that we should strive for 100% control, demanding more than 90% control may involve enormous costs with little or no added benefit. At the point when control of a particular pollutant reaches 90%, it makes more sense to allocate dollars and effort to other projects in which greater benefits may be achieved for the money spent. Optimum cost-effectiveness that meets the efficiency criterion for public policy is achieved at the point where the benefit curve is the greatest distance above the cost curve.

However, the perspective of time should be considered in calculating cost-effectiveness (Fig. 23–11c). A situation that appears to be cost ineffective in the short term may prove extremely cost effective in the long term. This is particularly true for problems such as acid deposition or groundwater contamination from toxic wastes.

Progress

What has been the result of benefit–cost analyses to date? Pollution of air and surface water reached critical levels in many areas of the United States in the late 1960s, and since that time huge sums of money have been spent on pollution abatement. Benefit–cost analysis shows that, overall, these expenditures have more than paid for themselves in decreased health-care costs and enhanced environmental quality. Performing the analysis also forces regulators and the regulated to more carefully assess the impacts of proposed policy changes.

As we have seen, the cost of controlling pollution is high and will continue to rise. What benefits do we receive in exchange for this cost? The EPA celebrated its 25th anniversary in 1995 with a progress report and a list of the challenges ahead. The following are some of the accomplishments of environmental public policy, as reported by the EPA:

- Since 1970, total emissions of six common air pollutants have decreased by an average of 24%.
- Since 1978, average levels of lead in the blood of children have declined by nearly 75%.
- More than 1 billion pounds of toxic pollutants have been prevented from entering our waters each year, due to wastewater standards put in place over the last 25 years.
- Between 200,000 and 470,000 cases of gastrointestinal illnesses each year have been prevented by adherence to drinking-water safety standards.
- Since 1970, 73 million more people, in thousands of communities, have upgraded sewage treatment.
- Toxic air emissions have fallen by 39%.
- Since the 1970s, over 230 pesticides and 20,000 pesticide products have been banned or eliminated from use.
- More than 141,000 cleanups of underground storage tanks have been completed since 1990.
- Since 1980, 757 out of 1,450 Superfund sites on the National Priorities List have been completely cleaned up.
- Recovery of municipal solid wastes by recycling has increased from 7% in 1970 to 28% in 1998.

These improvements have occurred even as the economy grew by 90%, the population rose by 27%, and the number of motor vehicle miles driven increased by 111%. Do the benefits outweigh the costs of the improvements? To answer this question, consider the phaseout of leaded gasoline as just one example. The project has cost about $3.6 billion, according to an EPA benefit–cost report. Benefits were valued at over $50 billion, $42 billion of which were for medical costs that were avoided.

Executive Order 12866 is now in effect. A benefit–cost analysis accompanies every new regulatory rule from the EPA and other federal agencies. Even if a rule is not classified as significant, the accompanying documentation must demonstrate, by benefit–cost analysis, that the rule does not qualify as significant. And, in the Unfunded Mandates Act of 1995, Congress backed up the executive order by requiring all agencies to prepare benefit–cost analyses for any regulations imposing costs of $100 million or more in a given year. Like it or not, benefit–cost analysis is now part of public policy in the United States.

23.5 Politics, the Public, and Public Policy

Environmental public policy legislation is the responsibility of Congress. Once a piece of legislation is passed by the House of Representatives and the Senate and is signed into law by the president, it is then implemented by appropriate governmental agencies. This is where the EPA gets its authority to draw up regulations for setting new standards for air pollutants. One final role Congress must play is to pass appropriations for the various programs that have already been authorized by law. This happens as part of the budget process. The executive branch (the president) draws up a budget asking for appropriations for all of the governmental agencies and programs, and Congress decides how much of what the president asks for will actually get funded. Thus, there are two basic ways for Congress to influence a given issue: as it is formulated into law and as it receives appropriations. The appropriations must be authorized every year. Of course, new laws can always amend or abolish existing ones. Given this basic process, there are many opportunities for partisan politics to play a role in setting environmental public policy.

Politics and the Environment

The Republican party controlled the last three Congresses (104th, 105th, and 106th), operating from 1994–2000, while President Clinton, a Democrat, held the executive power. During this six-year period, there were many key struggles over environmental legislation.

The 104th Congress began with an effort from antienvironmental forces to dismantle many key laws, especially the Clean Water Act, the Clean Air Act, and Superfund.

When it became apparent what was happening, heavy grass-roots work by many environmental organizations and extensive media coverage made it clear that the public did not support this antienvironmental thrust. As a result, most of the antienvironmental legislation died in the Senate or was vetoed by the president. A new strategy was then forged by the antienvironmental elements of Congress: They attached a number of antienvironmental riders to budget legislation. (A *rider* is a piece of legislation that is attached to unrelated legislation.) President Clinton refused to sign the legislation, and the government was shut down. In the end, the White House and congressional leaders reached an agreement according to which the riders were dropped or the president was given power to cancel their implementation. Toward the end of the 104th Congress, a bipartisan coalition emerged to prevent other measures from passing—one, for example, was an attempt at a wholesale cutback of the EPA's budget, and another was to kill the Conservation Reserve Program and the Wetlands Reserve Program.

The 105th and 106th Congresses saw some of the same antienvironmental forces again attempting to get their legislation through. Again, the antienvironmentalists in Congress employed the rider strategy. For example, House and Senate versions of a 2000 appropriations bill for the Department of the Interior were loaded down with antienvironmental riders (Table 23–4). Threatened with a Clinton veto, the House and the Senate tabled the bill, and appropriations for the department were eventually included in a massive omnibus appropriations bill that the president signed into law. The latest session, like the previous ones, also failed to reauthorize the Endangered Species Act and the Superfund program.

A new dimension became evident during these struggles: the role and power of **special-interest groups**. In the 1970s and 1980s, special interests (like the coal industry or the fisheries industry) would hire lobbyists who would use their influence to try to convince congressional committee members to act in their favor on issues that were important to them. Environmental NGOs would then use the media to expose the issues to the public and call on their constituencies to telephone and send mail to their representatives in Congress. The antienvironmental interests caught on to the NGO strategy and began to do what they were doing, but more effectively, because they had deeper pockets. Soon a constant battle between special-interest groups emerged in which each would use the media and telephone calls (enhanced by phone banks, which can generate thousands of calls to legislators in a short period of time) to pressure members of Congress. Added to this picture is the massive explosion of campaign contributions. Because they have much more money, the antienvironmental forces have been able in many instances to overwhelm the environmental organizations in this battle of influence. A good example is the Global Climate Change Kyoto Accord; a huge effort was mounted by the fossil-fuel industries to discredit the scientific basis for global climate change and

TABLE 23–4 Antienvironmental Riders Attached to the 2000 Interior Appropriations Bills	
Bill	Fate of Rider
Senate Interior Appropriations Bill (S. 1292)	
• Allow grazing without environmental review	retained
• Delay national forest planning	removed
• Prevent reintroduction of grizzly bears to Idaho and Montana	removed
• Weaken 1876 mining law	revised
• Block energy efficiency programs	removed
• Allow oil industry to continue underpaying royalties	revised
House Interior Appropriations Bill (H.R. 2466)	
• Block restoration of the Kankakee River	removed
• Limit preparation for climate change protection	retained
FY2000 Omnibus Appropriations Bill (H.R. 3194)	
• Prevent National Marine Fisheries Service from taking steps to protect salmon in Alaska	removed
• Delay regulations governing flights over the Grand Canyon	removed

thereby scuttle the accord. (It failed, and the United States signed the accord, although Congress has yet to ratify it.)

What this means, of course, is that environmental concerns have become a battle of special-interest groups; environmental protection is seen as one among many interests in American politics. The issues are up for grabs, and protecting the environment depends largely on the strength of environmental-interest groups. James Skillen, executive director of the Center for Public Justice, says that this is a seriously flawed development. Ecological well-being not only must have scientific proof of its validity; it also has to intrude in the political arena in competition with opposing special interests and *win*. Skillen believes that environmental justice—meaning the protection of land, air, water, and the biota—should not have to depend on an interest group for its advocacy. Rather, it is is a matter of the *public good* and, as such, should be seen as a basic responsibility of government. To quote Skillen*, "Recognition in basic law of the necessity of ecological health as the precondition of all public and market relations should become as fundamental as the recognition that certain human rights exist as the precondition of all public rules and market regulations." The NGOs stepped into the gap because they found that doing so was necessary. One issue they have recently helped bring to light is global trade. (The World Trade Organization is highlighted in the "Global Perspective" box, p. 589.) In the end, however, the ultimate responsibility for the nation's environmental policy should reside with the public.

*Skillen, James, "How can we Do Justice to both Public and Private Trusts?" October 1996. Global Stewardship Initiative National Conference, Gloucester, MA http://cesc.montreat.edu/GSI/GSI-Conf/discussion/Skillen.html

GLOBAL PERSPECTIVE

THE WORLD TRADE ORGANIZATION

Trade is a fundamental economic activity; the exchange of goods drives the engine of the free-market economy. In today's world, trade has become increasingly globalized, and in the process, world trade has become one of the most dominant forces in society. Many, however, question whether this is good, and their concerns became headline news when the World Trade Organization (WTO) met in Seattle, Washington, in December 1999. There, tens of thousands of protesters took to the streets to block the delegates from assembling—and were the recipients of tear gas, pepper spray, and rubber bullets as things began to turn ugly. The protesters were an unusual alliance of environmental activists and organized labor, groups that are often at odds over timber harvests and fishing. What were their issues?

The WTO was created in 1993 by trade ministers from many countries meeting to build on the foundation for global trade laid by the General Agreement on Tariffs and Trade (GATT). The WTO was given the responsibility for implementing myriad trade rules and, in doing so, was empowered to enforce the rules, with the option of imposing stiff trade penalties on violators. In effect, the WTO became a force for the globalization of trade, facilitat-

ing trade by reducing or eliminating trade barriers (tariffs) between countries and subsidies to domestic industries that tended to discriminate against foreign-produced goods. Those who support the WTO claim that free trade between countries is the key to global economic growth and prosperity. This all sounds very good, but as the saying goes, "The devil is in the details."

Since its creation, the WTO has adjudicated a number of international trade issues and, in the process, has established a reputation for elevating free trade over substantial human rights and environmental resource concerns. Furthermore, many trade agreements and the dispute rulings are carried out in closed sessions. For example, the WTO ruled against a U.S. law requiring any imports of shrimp to be certified as turtle safe (i.e., the shrimp were harvested by trawlers employing "turtle exclusion devices," or TEDs). Even though TEDs have been proven to reduce turtle mortality by 97%, the WTO ruled that the import restriction was a violation of WTO rules. Other issues that have come before the WTO are: a U.S. ban on the import of wooden crates from China because they harbored the Asian long-horned beetle, a serious threat to forest trees; a ban by the European Union of hormone-treated beef

and any products involving genetically modified organisms from the United States; and bans against developing countries' products that involve child labor.

The Seattle meetings are now history; in short, the talks broke down, not because of the protests, but because of a refusal of the key players—the United States, Japan, and Europe—to move from their positions. The future of the WTO is not clear; it retains the power already delegated to it, but in the absence of any new negotiations, the controversies over secrecy, labor, and environmental issues will cloud its work. The need for reform is evident. Environmental and labor concerns must be incorporated into the workings of the WTO. In the wake of Seattle, new and interesting alliances have been forged. Several hundred labor and environmental groups have signed the "Houston Principles of the Alliance for Sustainable Jobs and the Environment," identifying concerns they share and pledging mutual support with statements like "Labor, environmental and community groups need to take action to organize as a counterbalance to abusive corporate power; The drive for short-term profits without regard for long-term sustainability hurts working people, communities and the earth."

Citizen Involvement

How is the general public involved in public policy? Another way to ask this question is, What can you do to see that environmental concerns get the attention they deserve in Congress? Here are some suggestions:

- Become involved in grassroots concern about environmental problems—often traced to the findings of environmental science—in order to initiate the recognition stage of the policy life cycle. Adopt an environmental or ecological bill of rights as part of this grassroots effort.

- Become a member of an environmental NGO that informs its constituencies of environmental concerns around the country.

- Contact your legislators to inform them of your support for the environmental public policy under consideration during the policy formulation stage.

- Take note of the viewpoint of any political candidate on matters of environmental policy. Public concern about environmental policy can play an important role in the election or defeat of political candidates.

- Comment on regulations when they are first proposed; the public can also bring a class action suit against polluters in some situations.

- The public is involved directly in public policy, as people pay for the benefits they receive through higher taxes and higher costs on the products they consume.

- Keep informed on environmental affairs; only when the public is informed is it likely that grassroots support will be maintained and public environmental policy will really reflect public opinion.

ENVIRONMENT ON THE WEB

BALANCING MULTIPLE OBJECTIVES IN PUBLIC POLICY

Economic considerations have always played a dominant role in policy, because costs affect almost every stakeholder. For industries and municipalities, the costs of an environmental action may be the key determining factor in its viability. For regulatory agencies, proposals that are too costly may not be accepted by those affected and thus may never be translated into environmental improvements. For consumers, added environmental management costs may translate into higher product costs.

Since the 1930s, cost–benefit analysis (described on p. 568) has been widely used (and misused) in the assessment of large-scale public and private projects. Another common policy tool is mathematical optimization—the use of mathematical techniques to find the policy alternative that optimizes the value of a particular variable. These techniques have become cornerstones in the development of public policy, but their principal weakness is that they generally consider only a single variable (generally cost) in a world where many considerations may be important.

Today, every potential source of pollution—industries, municipalities, institutions, and even individual households—must balance traditional costs against numerous other tangible and intangible considerations. For example, pollution reduction potential is now a central concern (or criterion) in many policy decisions. Worker safety, public health, corporate image with consumers, and aesthetics may also be of consideration.

Multicriteria decision methods, incorporating a number of decision criteria, provide a way for policymakers to consider many factors at once. Such methods rarely allow development of a "best" solution, because a solution that is "best" in terms of one criterion may be less good on another. However, multicriteria analysis does accommodate potentially conflicting criteria in a systematic way and allows the analyst to develop one or more satisfactory solutions to a complex policy problem.

The advantage of multicriteria analysis is that it reflects the multidimensional nature of environmental decision making and has the potential to address the interests of many different stakeholders. In an era where cost is no longer the only or central concern, this is a powerful attribute indeed.

Web Explorations

The Environment on the Web activity for this essay describes some of the ways in which multicriteria analysis has been used in recent environmental decisions. Go to the Environment on the Web activity (select Chapter 23 at http://www.prenhall.com/wright) and learn for yourself:

1. about tradeoffs in multicriteria analysis;

2. about the application of multicriteria analysis in evaluating ecotourism options for the Aeolian Islands, Italy; and

3. how to use multicriteria analysis in developing a policy recommendation.

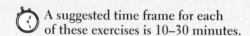 A suggested time frame for each of these exercises is 10–30 minutes.

REVIEW QUESTIONS

1. What is the overall objective of environmental public policy, and what are the objective's two most central concerns?

2. Review Table 23–1, and describe the effects on health and productivity associated with the following environmental problems: water pollution, air pollution, hazardous wastes, soil degradation, deforestation, loss of biodiversity, and atmospheric changes.

3. Three patterns of environmental indicators are associated with differences in the level of development of a nation. What problems decline, what problems rise and then decline, and what problems increase with the level of development?

4. Name the two basic kinds of economic systems and explain how they differ.

5. Describe the three components of a nation's wealth. Compare the wealth of the world's nations, as measured by these three components.

6. Why is the gross national product (GNP) not an accurate indicator of a nation's economic status?

7. What conditions are necessary for a country to make progress in the development of human resources and produced assets?

8. Explain the importance of considering both intragenerational equity and intergenerational equity in addressing resource allocation issues.

9. List the four stages of the policy life cycle, and show how *Silent Spring* and the formation of the EPA illustrate these stages.

10. What are the "Three E's" for evaluating environmental public policy?

11. What is the conclusion of careful studies regarding the relationship between environmental policies, on the one hand, and jobs and the economy, on the other?

12. What are the advantages and disadvantages of the regulatory approach vs. a market approach to policy development?

13. Define the term *benefit–cost analysis* as it relates to environmental regulation. How does this analysis method address external costs?

14. List specific costs and benefits of pollution control.

15. Discuss the cost-effectiveness of pollution control. How much progress have we made in pollution control in the last 25 years?

16. How has the political arena affected environmental public policy in the last few years? What has been the role of special-interest groups? What can your role be?

THINKING ENVIRONMENTALLY

1. Investigate the environmental public policies of a developing country, and compare the results with the information presented in Table 23–1.

2. Suppose it was discovered that bleach commonly used for laundry was carcinogenic. Referring to the policy life cycle, describe a predictable course of events until the problem is brought under control.

3. Stage a debate on the following resolution: "Environmental regulation is bad for the economy."

4. What costs of our industrial processes are now treated as externalities? How could we internalize these costs?

5. Consider the complexities and limitations of benefit–cost analysis. Should we continue to support its use to determine public policy? How great a role should it play? Support your answers.

WEB REFERENCES

On-line resources for this chapter are on the World Wide Web at: **http://www.prenhall.com/wright**. (Click on Chapter 23 within the Chapter Selector.)

Sustainable Communities and Lifestyles

1. Urban trends since World War II are marked by exurban migration, urban sprawl, urban blight, and a growing dependence on cars. Describe these trends and how they are all related.
2. Urban sprawl is at the root of many environmental problems. List and describe these problems.
3. Urban sprawl is giving way to "smart growth" in many communities. What is being done to rein in urban sprawl?
4. Exurban migration set into motion a vicious cycle of urban decline. Describe the factors involved in this cycle. What is economic exclusion?
5. Many cities around the world are attractive and "livable." What are the characteristics that make a city livable?
6. The U.N.'s Sustainable Cities Program and the work of the U.S. Council on Sustainable Development are working toward establishing sustainable communities. Describe these programs.
7. Environmental decision making is strongly influenced by values. What are some values that may be employed by people to promote stewardship and sustainability?
8. Changes in lifestyles are vital to sustainability. What are the most important changes? How can people best become involved in working toward building a sustainable society?

W aitakere City (opposite) is the sixth largest city in New Zealand, with a population of 136,000 and a land area of 39,134 hectares (96,660 acres). Waitakere City has been experiencing the typical problems of urban growth: urban sprawl, traffic congestion, and pressure on natural areas within the city. In addition, the city reflects a diverse ethnic makeup: a distinctly European ethnic majority and minorities of several cultural groups, but especially the native Maori. This city formally adopted *Agenda 21* in 1993 and was New Zealand's first ecocity.

Adopting the *Agenda 21* goal of protecting the indigenous people and cultures of countries while meeting other goals of sustainable development, Waitakere City focused on Maori principles and values first and then on New Zealand legislation and *Agenda 21* principles to develop the guidelines for the city's efforts. Using the Maori perspective of *tangata whenua*, or "people of the land," which incorporates ecological and stewardship principles into the relationship between people and land, the city has developed a

strategic plan called *Greenprint*. In this plan, the Maori are included in decision making by the city council, and Maori principles serve as guidelines in the city's ongoing growth and development. The Greenprint plan makes specific reference to the principles of *taonga* (a community's treasures, such as forests, rivers, animals, humans, and the health of the environment) and *kaitiaki* (those selected by a tribe to be stewards of the *taonga*).

With input from the Maori, the city council has developed a Green Network Plan with the objective of "protecting, restoring, and enhancing the natural environment across the whole city, while increasing human enjoyment and appreciation of that environment. This goal places human beings within the natural world, as stewards of its natural and spiritual health." The whole community is enlisted in "greening the city": restoring native plants, providing corridors for wildlife, and providing access to the parks and reserves that will maintain people's safety (from crime). Because the city's population is growing, the network plan

◀ *Waitakere City, New Zealand.* This city, New Zealand's first ecocity, covers a large area in Auckland, New Zealand. Shown here is one of the more developed areas, called Titirangi.

will be continually challenged by issues related to urban sprawl, such as loss of habitat and the impacts of exotic species and pets on native wildlife. Other Greenprint initiatives include improving housing, developing subdivisions, purchasing and retailing, and fostering local business ventures, all guided by stewardship principles. With these actions, Waitakere City has accepted the challenge of becoming an ecocity and has embarked on a course that may well make it a world leader in the development of sustainable cities.

On the other side of the world, citizens of Trenton, New Jersey, have made remarkable progress in rehabilitating their city, with the aid of Isles, Inc. (from the concept of neighborhood-scale *"islands* of redevelopment"), an urban-planning company. Trenton, with a multicultural population of 88,600, is the capital of New Jersey. Isles, Inc., was founded in 1981 by nearby Princeton University personnel, but quickly evolved into a locally owned and controlled development organization. Trenton was suffering from the all-too-common problems of a decaying urban center, with ethnic segregation and urban blight. To date, Isles has fostered the following changes: Vacant land has been transformed into community garden sites now providing $150,000 profits annually; almost 200 units of affordable housing have been created; numerous unemployed young people have been given job training in connection with the affordable housing projects; environmental education programs are provided yearly to 7,000 students; a recent capital campaign raised funding for an urban environmental education center and an endowment; and the city, with a grant from the EPA, is working on redeveloping four ugly "brownfield" sites (see Chapter 20).

Isles, with an annual budget of $1.5 million, is now an effective organization that has learned how to work with city and environmental groups to mobilize community action to accomplish the renovation of an aging city. An indication of why Isles has been so successful can be seen in one of its guiding principles: "The only sustainable solutions for urban problems are those which empower people with the knowledge and skills to help themselves." Like Waitakere City, Trenton has established a course of action and has found help in its efforts toward renewal.

This tale of two cities is tied together by a common resurgence of attention to urban living and a desire on the part of city residents to take charge of their own communities and reverse the trends that have been leading to a decline in the livability of their cities. It is significant that the direction they take is toward sustainability. These city residents recognize that things need to change, and they are becoming agents of changes that involve individual lifestyles and institutional redirection. One city is coping with sprawl, and the other is tackling urban blight.

Our objective in this chapter is first to show how these two problems—urban sprawl and urban blight—are related and how they can be remedied. Then we shall consider the global and national efforts being undertaken to forge sustainable cities and communities. Finally, we look at some ethical concerns and personal lifestyles as we consider the urgent needs of people and the environment upon entering a new millennium.

24.1 Urban Sprawl

The dominant and most environmentally salient feature of life in the United States is our near-total dependence on cars. We have developed an urban–suburban layout in which the locations where we live, work, go to school, and shop are widely separated. Thus, we have come to rely on private cars, which we drive an average of 200–300 miles (300–500 km) a week simply to meet our everyday needs. We shall refer to this lifestyle that revolves around going everywhere by car as a *car-dependent lifestyle*.

This far-flung urban–suburban network of low-density residential areas, shopping malls, industrial parks, and other facilities loosely laced together by multilane highways is referred to as **urban sprawl**. The word *sprawl* is used because the perimeters of the city have simply been extended outward into the countryside, one development after the next, with little plan as to where the expansion is going and no notion as to where it will stop. That the sprawl is continuing is all too evident. Almost everywhere we go near urban areas, we are confronted by farms and natural areas giving way to new developments, new highways being constructed, and old roadways being upgraded and expanded (Fig. 24–1).

The Origins of Urban Sprawl

Until the end of World War II, a relatively small percentage of people owned cars. Cities had developed in ways that allowed people to meet their needs by means of the transportation available—mainly walking. Every few blocks had a small grocery, a pharmacy, and other stores, as well as professional offices integrated with residences. Often, buildings had a store at the street level and residences above (Fig. 24–2). Schools were scattered throughout the city, as were parks for outdoor recreation or relaxation. Thus, walking distances were generally short, and bicycling made going somewhere even more convenient. For more specialized needs, people boarded public transportation—electric trol-

◀ **FIGURE 24–1** *New highway construction.* The hallmark of development over the past several decades has been the construction of new highways, which however, have only fed the environmentally damaging car-dependent lifestyle.

leys, cable cars, and buses—from neighborhoods to the "downtown" area, where big department stores, specialty shops, and offices were located. Public transportation did not change the compact structure of cities, because people still needed to walk to the transit line. At the outer ends of the transit lines, cities gave way abruptly to farms, which provided most of the food for the city, and open country. The small towns and villages surrounding cities—the original suburbs—were compact for the same reasons and main-

ly served farmers in the immediate area. At the end of World War II, this pattern began to change dramatically.

Despite the many advantages of urban life, many people found cities less-than-pleasant places to live. Especially in industrial cities, poor housing, inadequate sewage systems, inadequate refuse collection, pollution from home furnaces and industry, and generally congested, noisy conditions were much the norm. A decrease in services during World War II aggravated those problems. Hence, many people had the

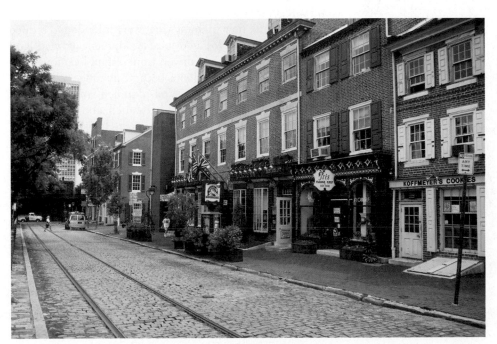

◀ **FIGURE 24–2** *The integrated city.* Before the widespread use of automobiles, cities had an integrated structure. A wide variety of small stores and offices on ground floors, with residences on upper floors, placed everyday needs within walking distance. This scene is from a historic section of Philadelphia.

desire—the "American Dream"—to live in their own house on their own piece of land away from the city.

The Ascendance of the Automobile.   Because of the massive production of war materiel, consumer goods were in short supply during World War II. As a result, when the war ended, there was a pent-up demand for consumer goods. When mass production of cars recommenced at the end of the war, people flocked to buy them. With private cars, people were no longer restricted to living within walking distance of their workplaces or of transit lines. They could move out of their cramped city apartments and into homes of their own outside the city. Their cars would allow them to drive back and forth easily to their jobs, shopping, recreation, and so on.

Returning veterans and the baby boom created a housing demand that rapidly exhausted the existing inventory. Developers responded quickly to the demand for private homes. They bought farmlands and natural areas outside the cities and put up houses. The government aided this trend by providing low-interest mortgages through the Veterans Administration and the Federal Housing Administration, and interest payments on mortgages were made tax deductible (rent was still not tax deductible). Property taxes in the suburbs were much lower than in the city. These financial factors meant that, for the first time in history, making monthly payments on one's own home in the suburbs

▲ **FIGURE 24–3   Urban sprawl.**   Prime land is being sacrificed for development around most U.S. cities and towns. The locations of developments are determined more by the availability of land than by any overall urban planning. Here suburban development is encroaching into prime farmland in Hailey, Idaho.

was cheaper than paying rent for equivalent or less living space in the city.

Thus, the mushrooming development around cities did not proceed according to any plan; rather, it happened wherever developers could acquire land (Fig. 24–3). No governing body existed to devise, much less enforce, an overall plan; cities were surrounded by a maze of more or less autonomous local jurisdictions (towns, villages, townships, and counties). Local governments were quickly thrown into a catch-up role of trying to provide schools, sewers, water systems, other public facilities, and, most of all, roads to accommodate the uncontrolled growth. Local zoning laws contributed to the sprawl by strictly separating residential and commercial uses. People could no longer expect to walk to a grocery store or an office.

Highways.   The influx of commuters into previously rural areas soon resulted in traffic congestion, creating a need for new and larger roads. To raise money to build and expand highways, Congress passed the Highway Revenue Act of 1956, which created the **Highway Trust Fund**. This legislation placed a tax on gasoline and earmarked the revenues to be used exclusively for building new roads.

It is evident how the Highway Trust Fund perpetuates development. A new highway not only alleviates existing congestion, but also encourages development at farther locations, because time, not distance, is the limiting factor for commuters. (When people describe how far they live from work, it is almost always put in terms of minutes, not distance.) The average person is willing to spend 20 to 40 minutes each way on a daily commute. Given this time limit, people who have to walk would have to live within 1 or 2 miles (1.5–3.0 km) of their work. With cars and expressways, people can live 20 to 40 miles (30–60 km) from work and still get there in the same amount of time.

Therefore, new highways that were intended to reduce congestion actually fostered the development of open land and commuting by more drivers from distant locations (Fig. 24–4). Soon, traffic conditions became as congested as ever. Average commuting *distance* has doubled since 1960, but average commuting *time* has remained about the same. The increase in commuting distance, however, requires more fuel, which generates more money for the Highway Trust Fund. Thus, the whole process is repeated in a continuing cycle (Fig. 24–5). As a result, government policy, far from curtailing urban sprawl, became a party to supporting and promoting it.

Residential developments are followed (or sometimes led) by shopping malls, industrial parks, motels, and office complexes. These commercial centers are usually situated in such a way that the only possible access is by car; this has only changed the direction of commuting. Whereas in the early days of suburban sprawl the major traffic flow was into and out of cities, it is now between suburban centers. Multilane highways connecting suburban centers are perpetually congested with traffic going in *both* directions.

(a)

(b)

◀ **FIGURE 24–4** *The new highway–traffic-congestion cycle.* (a) Growing traffic congestion creates the need for new and upgraded highways. (b) However, upgraded highways encourage development at more distant locations, thus creating more traffic congestion and hence the need for more highways.

In broad perspective, then, urban sprawl is a process of **exurban migration**—that is, a relocation of residences, shopping areas, and workplaces from their traditional spots in the city to outlying areas. Population growth, although it has occurred, has played a relatively minor part in the development of urban sprawl. The populations of many eastern and midwestern U.S. cities, excluding the suburbs, have been declining over the last 50 years as a result of exurban migration (Table 24–1). Exurban migration is continuing in

a leapfrog fashion as people from older suburbs move to **exurbs** (communities farther from cities than suburbs).

The "love affair" with cars is not just a U.S. phenomenon. Around the world, in both developed and developing countries, people aspire to own cars and adopt the car-dependent lifestyle. Consequently, urban sprawl is occurring around many developing world cities as people become affluent enough to own cars. Love of the car-dependent lifestyle, however, is not sufficient to make it sustainable. In

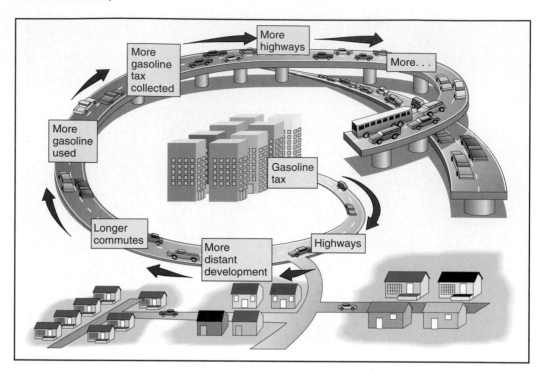

▶ **FIGURE 24–5** *The development cycle spawned by the Highway Trust Fund.*

fact, this lifestyle is at the crux of a number of the unsustainable trends described in earlier chapters.

Environmental Impacts of Urban Sprawl

The environmental impacts of urban sprawl can be put into the following categories:

- depletion of energy resources
- air pollution

- water pollution and degradation of water resources
- loss of landscapes and wildlife
- loss of agricultural land

Depletion of Energy Resources. Shifting to a car-dependent lifestyle has entailed an ever-increasing demand for petroleum. From 1950 to 1999, U.S. oil consumption nearly tripled, while population grew by just 80%, a nearly threefold increase in per capita consumption. In addition, individual suburban homes require 1.5 to 2 times more energy

| City | Population (thousands) | | | | Percent change |
	1950	1970	1990	1999	1950–1999
Baltimore, MD	950	905	736	633	−33
Boston, MA	801	641	574	555	−31
Buffalo, NY	580	463	328	296	−49
Cleveland, OH	915	751	505	502	−45
Detroit, MI	1,850	1,514	1,028	965	−48
Louisville, KY	369	362	270	253	−31
Minneapolis, MN	522	434	368	353	−32
Philadelphia, PA	2,072	1,949	1,586	1,418	−32
St. Louis, MO	857	662	396	334	−61
Washington, DC	802	757	607	519	−35

TABLE 24–1 Decline in Population of American Cities, 1950–1999

(*Source:* Data from Population Division, U.S. Bureau of the Census, 2000.)

for heating and cooling than comparable attached city dwellings. Chapter 13 described the costs and risks of becoming increasingly dependent on foreign sources of oil for fuel.

Air Pollution. Despite improvements in pollution control, many cities still fail to meet desired air-quality standards. (See Chapter 22.) Vehicles are responsible for an estimated 80% of the air pollution in metropolitan regions. Likewise, the threefold increase in per capita energy consumption exacerbates the potential for global warming as burning fossil fuels produce carbon dioxide. Automobiles also are implicated in the depletion of the stratospheric ozone shield, because a major source of CFCs entering the atmosphere is that which has escaped from vehicle air conditioners. (See Chapter 21.)

Water Pollution and Degradation of Water Resources. All the highways, parking lots, driveways, and other paved areas associated with urban sprawl lead to a substantial increase in runoff, as well as decreased infiltration over large regions. Even suburban lawns increase runoff, because they are more compacted than the original soils. Recall the consequences of increasing runoff—increased flooding, stream-bank erosion, and decreased water quality, to name a few—and the consequences of decreasing infiltration—depletion of groundwater, drying of springs and waterways, saltwater encroachment, and others. (See Chapter 9.) Also, water quality is degraded by the runoff of fertilizer, pesticides, crankcase oil (the oil that drips from engines), pet droppings, and so on.

Loss of Agricultural Land. Perhaps most serious in the long run is the loss of prime agricultural land. In the United States alone, sprawling development is eating up over 400,000 acres of prime agricultural land a year. For the time being, the United States appears to have adequate agricultural land and can tolerate this loss. However, most of the food for cities used to be locally grown on small, diversified family farms surrounding the city. With most of these farms turned into housing developments, it is estimated that food now travels an average of 1,000 miles (1,500 km) from where it is produced—mostly on huge commercial farms—to where it is eaten. The loss is not just the locally grown produce, but also the social interactions and ties with the farm community.

Loss of Landscapes and Wildlife. New developments are consuming land at an increasing pace; from 1992 to 1997, 3.2 million acres a year were developed, vs. 1.4 million acres a year from 1982 to 1992. New highways are frequently routed through parklands or along stream valleys, because such open areas provide the least expensive rights-of-way. The result is the sacrifice of aesthetic, recreational, and wildlife values in the very places where they are most important: metropolitan areas. The excess individuals of every species displaced by development are inevitably consigned to death even if similar habitat exists, because nearby ecosystems are invariably already occupied.

Furthermore, many species of wildlife need certain minimum unbroken areas to maintain viable populations, as well as unbroken "lanes" of habitat through which to migrate. Wildlife biologists are finding marked declines in countless species, ranging from birds to amphibians, as highways and new developments fragment natural areas. Also, expanding highways leads to increasing numbers of roadkills. Today, much more wildlife is killed by vehicles than by hunters.

The environmental impacts resulting from or exacerbated by urban sprawl and development are summarized in Fig. 24–6.

▼ **FIGURE 24–6** *A summary of the environmental impacts of urban sprawl.*

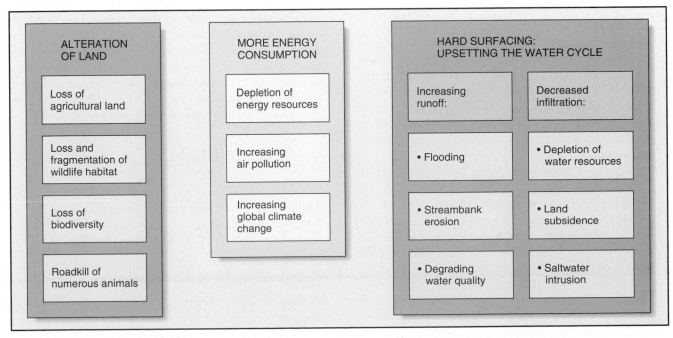

Reining In Urban Sprawl: Smart Growth

The obvious way to curtail urban sprawl might appear to be simply to pass laws that bring exurban development under control. Indeed, a number of European countries and Japan do have such laws. However, laws cannot be passed without the support of the public. In the United States, the strong sentiment toward the right of a property owner to develop property as he or she feels fit has made such restrictions, with few exceptions, politically impossible to pass.

Yet, there are signs that a sea change is occurring in attitudes of Americans toward uncontrolled growth. For example, the concept of "*smart growth*" recently has emerged. This concept involves communities and metropolitan areas addressing sprawl and purposely choosing to develop in more environmentally sustainable ways. The concept recognizes that growth will occur and focuses on economic, environmental, and community values that together will lead to more sustainable communities. Zoning laws are being changed in many localities, and a new generation of architects and developers is beginning to focus on creating integrated communities, as opposed to disassociated facilities. By directing growth to particular places, smart growth provides protection for sensitive lands. Early indications in the marketplace—the final determining factor—are that the new communities are a great success.

Smart-growth initiatives are appearing on the ballot in many states and municipalities (200 were enacted into law during 1999 alone), and the vast majority are being approved by voters who are fed up with the impacts of urban sprawl. These initiatives involve several key strategies:

1. *Setting boundaries on urban sprawl.* Oregon led the way in passing a land-use law in the 1970s requiring every city to set a boundary beyond which urban growth is prohibited. The boundary is a line permitting expected future growth—but controlled growth. Portland has had great success with this strategy, and today it is a highly desirable place to live and work.
2. *Saving open space.* The best way to preserve open space is to own it or own the development rights to the land, and states are enacting programs to acquire crucial remaining open space. New Jersey voters overwhelmingly passed a $1 billion bond issue to set aside half of the state's remaining 2 million acres of open space. In 1998 and 1999 alone, over $17 billion was allocated in numerous states for the purchase and preservation of open space.
3. *Developing existing urban space.* In 1997, Maryland's General Assembly passed the "Smart Growth and Neighborhood Conservation" initiative, which includes measures to channel new growth to community sites where key infrastructure is already in place. Uncontrolled sprawl has left behind many patches of undeveloped land in metropolitan areas, and factory relocations to suburbs and elsewhere have created "brownfield" sites that are often polluted or otherwise suffering from deterioration. These areas are already well served with roads, electricity, water, and sewerage and can be developed into new homes, workplaces, and shopping areas.
4. *Creating new towns.* A key to developing sustainable communities is changing zoning laws to allow stores, light industries, professional offices, and higher density housing to be integrated together. Affordable and attractive housing can be created around industrial parks and shopping malls for the people who work and shop in those places. Safe bicycle and pedestrian access can be provided between residential areas and workplaces. Reston, Virginia; Columbia, Maryland; and Valencia, California, were established as self-contained towns along these lines in the 1960s, and they are still highly desirable as business and residential locations. On the other hand, Fairfax County, Virginia, next to Washington, DC, is now almost 100% covered by sprawl-type development. It is estimated that if Fairfax had *allowed* the kind of development described in this chapter, the county would still be 70% open space and would be far less congested and less polluted.

In addition to these key strategies, there have been repeated attempts to break the cycle created by the Highway Trust Fund by allowing revenues to be used for purposes other than the construction of more highways. The first significant inroad was made in 1991 with the passage of the Intermodal Surface Transportation Efficiency Act (ISTEA), a bill environmentalists had been promoting for many years. Under this act, almost half of the money levied by the Highway Trust Fund is now eligible to be used for other modes of transportation, including cycling and walking as well as mass-transit facilities. The legislation also requires states to hire bicycle and pedestrian coordinators to oversee programs and to establish long-range pedestrian–bicycle plans. In addition, the act requires that states spend a total of $3 billion on "transportation enhancements," which include pedestrian and bicycle facilities. The objectives of ISTEA were significantly reinforced and extended with passage of the Transportation Equity Act for the Twenty-first Century (TEA 21) in 1998. TEA 21 authorizes a total of $218 billion for federal highway and mass-transit programs for the next five years and provides a major increase over ISTEA funding for programs that address the environmental impacts of surface transportation. The program has successfully maintained its funding into 2001.

24.2 Urban Blight

Now let us take a look at the other end of the exurban migration: the city from which people are moving. Here, we find that exurban migration is the major factor underlying an *urban decay*, or *blight*, that has occurred over the last 50 years.

Economic and Ethnic Segregation

Historically, U.S. cities have included people with a wide diversity of economic and ethnic backgrounds. However, moving to the suburbs required some degree of affluence—at least the ability to manage a down payment and mortgage on a home and the ability to buy and drive a car. Therefore, exurban migration, for the most part, excluded the poor, the elderly, and the handicapped. The poor were largely African-Americans and other minorities, because a history of discrimination had kept them from education and well-paying jobs. Discriminatory lending practices by banks and dishonest sales practices by real-estate agents kept minorities out of the suburbs even when money was not a factor. Civil rights laws passed in the 1960s made such practices illegal, but there is much evidence that they still persist in various forms. Even without the racial segregation, economic segregation continues to exist and has even intensified.

In short, exurban migration and urban sprawl have led to segregation of the population into groups sharing common economic, social, and cultural backgrounds. Moving into the new suburban and exurban developments (often labeled "white flight") were the economically advantaged, whereas many areas of the cities and older suburbs were effectively abandoned to the economically depressed, who in large part were ethnic minorities, the handicapped, and the elderly (Fig. 24–7).

(a)

(b)

◀ **FIGURE 24–7** *Segregation by exurban migration.* Exurban migration, the driving force behind urban sprawl, has also led to segregation of the population along economic and racial lines. In (a) an area of suburbia and (b) an area of inner-city Baltimore, you see the contrast.

The Vicious Cycle of Urban Blight

Affluent people moving to the suburbs set into motion a vicious cycle of exurban migration and urban blight that continues today. To understand how this downward cycle occurs, we must note several points concerning local governments. First, local governments (usually city or county agencies, although the particular entities vary from state to state) are responsible for providing public schools, maintaining local roads, furnishing police and fire protection, collecting and disposing of refuse, maintaining public water and sewers, and providing welfare services, libraries, and local parks. Second, a local government's major source of revenue to pay for these services is local property taxes, a yearly tax proportional to the market value of the property and home or other buildings on it. If property values increase or decrease, property taxes are adjusted accordingly. Third, in most cases, the central city is a governmental jurisdiction separate from the surrounding suburbs.

Because of these three characteristics of local government, exurban migration has the following consequences: By the economics of supply and demand, property values in the suburbs escalate with the influx of affluent newcomers. Thus, suburban jurisdictions enjoy an increasing tax revenue with which they can improve and expand local services. At the same time, property values in the city decline because of decreasing demand. The lower property taxes create a powerful disincentive toward maintaining property. Often, landlords allow property to deteriorate in order to reduce their taxes while keeping rents high to maximize their income. Many properties end up being abandoned by the owner for nonpayment of taxes. The city "inherits" such abandoned properties, but they are a liability rather than a source of revenue (Fig. 24–8). The declining tax revenue resulting from falling property values is referred to as an **eroding**, or **declining**, **tax base**, and it has been a serious handicap for most U.S. cities since the exurban migration started in the late 1940s. Adding to the problem is the fact that many of those remaining in the city are disadvantaged people requiring public assistance of one sort or another. Thus, cities bear a disproportionate burden of welfare obligations, as well as of the declining tax revenues.

The eroding tax base forces city governments to cut local services, increase the tax rate, or, generally, both. Hence, the property tax on a home in a city is often two to three times greater than on a comparably priced home in the suburbs, a difference that amounts to $2,000 to $3,000 per year in the tax bill on an average-priced home. At the same time, schools, refuse collection, street repair, libraries, parks, and other services and facilities deteriorate from neglect. This situation of increasing taxes and deteriorating services causes yet more people to leave the city and become part of urban sprawl, even if the car-dependent lifestyle is not their primary choice. However, it is still only the relatively more affluent individuals who can make this move, and thus the whole process of exurban migration, a declining tax base, deteriorating real estate, and worsening schools and other services and facilities is perpetuated in a vicious cycle (Fig. 24–9). This downward spiral of conditions is referred to as **urban blight** or **urban decay**.

Economic Exclusion of the Inner City

The exurban migration includes more than just individuals and families. The flight of affluent people from the city re-

▶ **FIGURE 24–8**
Abandoned buildings.
Many cities' ills are because of too *little* population, not too much. When the exurban migration causes a city's population to drop below a certain level, businesses are no longer supported. The result is abandonment and the bane of urban blight. Shown here is a section of Baltimore only a few blocks from the redeveloped Inner Harbor seen in Fig. 24–10.

moves the purchasing power necessary to support stores, professional establishments, and other enterprises. Faced with declining business, merchants and practitioners of all kinds are forced either to go out of business or to move to the new locations in the suburbs. Either way, vacant storefronts are the result, and people remaining in the city lose convenient access to goods, services, and, most important, *jobs*, because each business also represents employment opportunities.

Unemployment rates run at 50% or more in depressed areas of inner cities, and what jobs do exist are mostly at the minimum wage. By contrast, the new jobs being created are in the shopping malls and industrial parks in the outlying areas, which are largely inaccessible to inner-city residents for lack of public transportation. Therefore, not only are people who remain in the inner city poor from the outset,

but the cycle of exurban migration and urban decay has now led to their **economic exclusion** from the mainstream. As a result, drug dealing, crime, violence, and other forms of deviant social behavior are widespread and worsening in such areas.

To be sure, in the last two decades, many cities have redeveloped core areas. Shops, restaurants, hotels, convention and entertainment centers, office buildings, residences, and places for walking and relaxing are all combining to bring a new spirit of life and hope, as well as the practical assets of taxes and employment, back into cities (Fig. 24–10). This is an encouraging new trend, so far as it goes. However, walk a few blocks in any direction from the sparkling, revitalized core of most U.S. cities, and you will find urban blight continuing unabated.

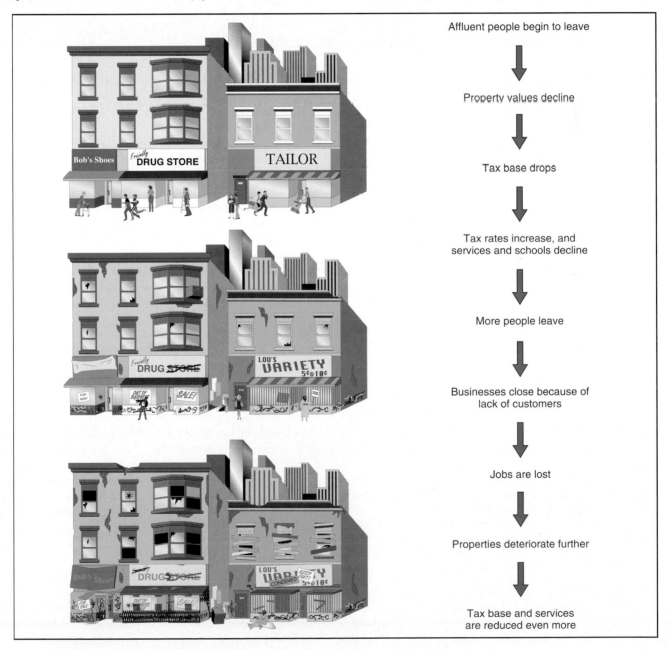

Affluent people begin to leave

↓

Property values decline

↓

Tax base drops

↓

Tax rates increase, and services and schools decline

↓

More people leave

↓

Businesses close because of lack of customers

↓

Jobs are lost

↓

Properties deteriorate further

↓

Tax base and services are reduced even more

▲ **FIGURE 24–9** *The process of urban decay.* The vicious cycle of exurban migration and urban decay.

▶ **FIGURE 24–10** *Inner Harbor, Baltimore.* High population densities help *make* a city, not destroy it. Redeveloping key areas of cities with heterogeneous mixtures of office buildings, shopping, restaurants, residences, and recreational facilities, all of which attract people, has helped to bring new life back into the cities.

What Makes Cities Livable?

It is evident that the environmental consequences of urban sprawl and the social consequences of urban blight are two sides of the same coin. There is increasing awareness that a sustainable future will depend on both reining in urban sprawl and revitalizing cities. The only possible way to sustain the global population is by having viable, resource-efficient cities, leaving the countryside for agriculture and natural ecosystems. The key word is *viable*, which means *livable*. No one wants to, or should, be required to live in the conditions that have come to typify urban blight.

Livability is a general index based on people's response to the question, "Do you like living here, or would you rather live somewhere else?" Crime, pollution, recreational, cultural, and professional opportunities, and many other social and environmental factors are summed up in the subjective answer. Although many people assume that the social ills of the city are an outcome of high population densities, crime rates and other problems in U.S. cities have climbed while populations dwindled as a result of the exurban migration. As reported in the Worldwatch Institute's *Shaping Cities: The Environmental and Human Dimensions*:

> There is no scientific evidence of a link between population density per se and social ills....Copenhagen and Vienna—two cities widely associated with urban charm and livability—each have relatively high "activity density" (the number of residents and jobs in the city), 47 and 72 people per hectare (1 hectare = 2.5 acres) respectively. By contrast, low-density cities such as Phoenix (13 people per hectare) often are dominated by unwelcoming, car-oriented commercial strips and vast expanses of concrete and asphalt.

Looking at livable cities around the world, we find that the common denominator is (1) maintaining a high population density; (2) preserving a heterogeneity of residences, businesses, stores, and shops; and (3) keeping layouts on a human dimension, so that people can incidentally meet, visit, or conduct business over coffee at a sidewalk café or stroll on a promenade through an open area. In short, the space is designed for, and devoted to, people (Fig. 24–11). In contrast, development of the past 50 years has focused on accommodating automobiles and traffic. Two-thirds of the land in cities that have grown up in the era of the automobile is devoted to moving, parking, or servicing cars, and such space is essentially alien to the human psyche (Fig. 24–12). William Whyte, a well-known city planner, remarked, "It is difficult to design space that will not attract people. What is remarkable is how often this has been accomplished."

The world's most livable cities are not those with "perfect" auto access between all points; they are cities that have taken measures to reduce outward sprawl, diminish automobile traffic, and improve access by foot and bicycle in conjunction with mass transit. For example, Geneva, Switzerland, prohibits automobile parking at workplaces in the city's center, forcing commuters to use the excellent public transportation system. Copenhagen bans all on-street parking in the downtown core. Paris has removed 200,000 parking places in the downtown area. Curitiba, Brazil, with a population of 1.5 million, is cited as the most livable city in all of Latin America. The achievement of Curitiba is due almost entirely to the efforts of Jaime Lerner, who, serving as mayor since the early 1970s, has guided development in terms of mass transit rather than cars. The space saved by not building highways and parking lots has been put into parks and shady walkways, causing the amount of green area per inhabitant to increase from 4.5 square feet ($0.5\ m^2$) in 1970 to 450 square feet ($50\ m^2$) today.

In Tokyo, millions of people ride bicycles (Fig. 24–13), either all the way to work or to stations from which they catch fast, efficient subways or the "bullet train" to their destination. By sharply restricting development outside certain city limits, Japan has maintained population densities within cities and along metropolitan corridors that ensure the vi-

ability of commuter trains. Japan's cities have maintained a heterogeneous urban structure that mixes small shops, professional offices, and residences in such a way that a large portion of the population meets its needs without cars. In maintaining an economically active city, it is probably no coincidence that street crime, vagrancy, and begging are virtually unknown in the vast expanse of Tokyo, despite the seeming congestion.

Portland, Oregon, is a pioneering U.S. city that has taken giant steps to curtail automobile use. The first step was to encircle the city with an urban growth boundary, a line outside of which new development was prohibited. Thus, compact growth, rather than sprawl, was ensured. Second, an efficient light-rail and bus system was built (see the "Earth Watch" essay, p. 557), which now carries 45% of all commuters to downtown jobs. (In most U.S. cities, only 10% to 25% of commuters ride public transit systems.) By reducing traffic, Portland was able to con-

vert a former expressway and a huge parking lot into the now-renowned Tom McCall Waterfront Park (Fig. 24–14). Portland is now ranked among the world's most livable cities.

Other U.S. cities are discovering hidden resources in their inner core areas. Expressways, which were built with federal highway funds to speed vehicles through the cities, frequently dissected neighborhoods, separating people from waterfronts and consuming existing open space. Now, with federal help from TEA 21 programs, many cities are removing or burying the intruding expressways. Boston, for example, is in the midst of a "Big Dig," an $11 billion project that will move its central freeway below ground, add parks and open space, and reconnect the city to its waterfront.

The waterfronts themselves are often the neglected relics of bygone transportation by water and rail. In recent years, cities like Cleveland, Chicago, San Francisco, and Baltimore have redeveloped their waterfronts, turning rotting

▲ FIGURE 24–13 *Bicycles in Tokyo.* Most people in Tokyo do not own automobiles, but ride bicycles or walk to stations, from which they take fast, efficient, inexpensive subways to reach their destinations.

piers and abandoned freight yards into workplaces, high-rise residences, and public and entertainment space.

It turns out that the U.S. economic boom of the 1990s contributed immeasurably to urban renewal, filling the federal coffers and at the same time boosting employment in the inner cities. Retail businesses and manufacturers are discovering that inner-city neighborhoods can provide a valuable labor pool, as well as customers for their goods and services.

Livable Equals Sustainable. The same factors that underlie livability also lead to sustainability. The reduction of auto traffic and a greater reliance on foot and public transportation reduce energy consumption and pollution. Urban heterogeneity can facilitate the recycling of materials. Housing can be retrofitted with passive solar space heating and heating for hot water. Landscaping can provide cooling, as described in Chapter 15. A number of cities are developing vacant or cleared areas into garden plots (Fig. 24–15), and rooftop hydroponic gardens are becoming popular. Such gardens will not make cities agriculturally self-sufficient, of course, but they add to urban livability, provide an avenue for recycling compost and nutrients removed from sewage, give a source of fresh vegetables, and have the potential to generate income for many unskilled workers. If urban sprawl is curbed, relatively close-in farms could provide most of the remaining food needs for the city.

This discussion of urban issues may seem to have strayed far from nature and environmental science. However, there is a close connection between the two: The decay of our cities is hastening the breakdown of our larger environment. As the growing human population spreads outward from the old cities, we are encroaching on all other life. Thus, without the creation of sustainable human communities, there is little chance for sustainability of the rest of the biosphere.

By making urban areas more appealing and economically stable, we can not only improve the lives of those who choose to remain or must remain in cities, but also spare surrounding areas. Our parks, wildernesses, and farms will

▶ FIGURE 24–14 *Portland, Oregon.* By reducing traffic, Portland, Oregon, was able to convert a former expressway and a huge parking lot into the now-renowned Tom McCall Waterfront Park. Due to such planning, Portland is now ranked among the world's most livable cities.

▲ **FIGURE 24–15** *Urban gardens.* Urban garden plots on formerly vacant lots in Philadelphia. Urban gardening is becoming recognized as having sociological, economic, environmental, and aesthetic benefits.

not be replaced by exurbs and shopping malls, but rather, will be saved for future generations. In the last decade, some remarkable developments have begun to accelerate the movement toward building sustainable communities. We turn to these developments now.

24.3 Moving toward Sustainable Communities

In Chapter 1, we noted that the 1992 U.N. Conference on Environment and Development (UNCED) created the Commission on Sustainable Development. This new body was given the responsibility for monitoring and reporting on the implementation of the agreements that were made at UNCED—in particular, those related to sustainability. According to *Agenda 21*, one of the accords signed at UNCED, countries agreed to work on sustainable development strategies and action plans that would put them on a track toward sustainability. Some 100 countries have now developed national environmental action plans or sustainable development strategies that lay out public policy priorities.

Most of these countries are currently actively engaged in transforming the policies into action plans, and we will investigate two of the outcomes of these activities: the *Sustainable Cities Program* and the work of the *U.S. Council on Sustainable Development*. We will also see that many of the objectives of these programs have already been adopted in the United States in an ongoing movement called the "Sustainable Communities Movement."

Sustainable Cities

A joint facility of the U.N. Environment Program and the U.N. Center for Human Settlements, the Sustainable Cities Program (SCP) is designed to foster the planning and management needed to move cities in the developing countries toward the objective of sustainability. The program defines a

sustainable city as a city in which "achievements in social, economic and physical development are made to last." Cities are the focus of the program because they are absorbing two-thirds of population growth in the developing countries and, in the process, are experiencing serious environmental degradation in and around the growing urban centers.

The program is viewed as a "capacity-building program," meaning that it is intended to help cities and countries mobilize resources and capabilities that are mostly available within the cities themselves by building management capacity. The approaches fostered by the program are thus "bottom up" in nature, calling for the involvement of people at all economic and social levels and reconciling their interests where conflicts are evident. Common principles seen in sustainable development theory are put into practice—principles such as social equity, economic efficiency, and environmental planning and management. Working groups specific to unique city issues are the core element of the program.

More than 15 SCP cities have been identified and are working on implementing the program. Among these cities are Madras (India), Dar es Salaam (Tanzania), Accra (Ghana), Shenyang (China), Concepción (Chile), and many others (Fig. 24–16). The process involves a preparatory phase and then moves into demonstration activities at the level of the whole city, finally extending to replications in other cities of the country. The following are the processes in focus at the outset of a city's involvement:

- clarification of environmental issues
- agreement on joint strategies
- coordination of action plans
- implementation of technical support and capital investment programs
- establishment of a continuing environmental planning and management process

Funding for the demonstration phase comes primarily from the city itself, but additional aid often originates in partnerships with other countries. Denmark, Canada, France, the United States, and Italy have provided support and expertise at this level of involvement. The World Bank and other U.N. agencies are frequently participants in SCP activities; indeed, the program's approach is quite conducive to collaboration among NGOs, other countries, and the host country and city. Although the program is in its early stages, it is a promising approach to sustainable development in exactly the places where it is most needed.

Sustainable Communities

Revitalizing urban economies and rehabilitating cities requires coordinated efforts on the part of all sectors of society. In fact, such efforts are beginning to occur. What is termed a "sustainable communities movement" is taking root in cities around the United States. Chattanooga, Tennessee, is now

▲ **FIGURE 24–16** *Dar es Salaam, Tanzania.* The first Sustainable Cities Program demonstration city, Dar es Salaam is growing at a rate of 8% per year and is struggling to cope with deteriorating environmental conditions caused by its rapid growth.

- Low-cost housing was renovated.
- A new industry to build pollution-control equipment was spawned, as was another industry to build electric buses, which now serve the city without noise or pollution.
- A recycling center employing mentally handicapped adults to separate materials was built.
- An urban greenway demonstration farm, which school-children may visit, was created.
- A zero-emissions industrial park utilizing pollution-avoidance principles was built.
- The river was cleaned up, and parks were created along the riverbanks.
- Theaters, museums, and a freshwater aquarium were renovated or built.
- All facilities and a renovated business district were made pedestrian friendly and accessible.

With these and numerous other projects, many of them still ongoing, Chattanooga has moved its reputation from one of the worst to one of the best places to live (Fig. 24–17). Chattanooga Venture developed a step-by-step guide for community groups to assist them in similar efforts to build sustainable communities; the city's experience is being modeled in cities throughout the United States as well as internationally. Recently, Chattanooga sponsored *Revision 2000*, with over 2,600 participants taking up where Vision 2000 left off. Revision 2000 identified an additional 27 goals and more than 120 recommendations for further improving Chattanooga. The city's experience demonstrates that visioning and change must occur and continue for success in creating and maintaining sustainable community development. Although Chattanooga Venture is no longer in operation, its past work has built an expectation throughout the city that public projects will involve the public and that the process will produce results. The latest manifestation of this expectation was *Recreate 2008*, in which hundreds of people engaged in a visioning process in 1998 that created a plan for revitalizing the city's park system.

regarded as a prototype of what can occur with such a movement.

Thirty years ago, Chattanooga, which straddles the Tennessee River, was a decaying industrial city with high levels of pollution. Employment was falling, and residents were fleeing to the suburbs, leaving abandoned properties and increasing crime. Then, **Chattanooga Venture**, a non-profit organization founded by community leaders, launched *Vision 2000*, the first step of which was to bring people from all walks of life together to build a consensus about what the city *could* be like. Literally thousands of ideas were gradually distilled into 223 specific projects. Then, with the cooperation of all sectors—including government, business, lending institutions, and average persons—work on the projects began, providing employment in construction for more than 7,300 people and permanent employment for 1,380 and investing more than $800 million. Among the projects were the following:

- With the support of the Clean Air and Clean Water Acts, local government clamped down hard on industries to control pollution.

President's Council on Sustainable Development

What role, if any, should the federal government play in the development of sustainable communities? In the United States, the President's Council on Sustainable Development had the responsibility of guiding the country toward greater sustainability. We have seen examples of the council's accomplishments in earlier chapters; here we give an overview of the council's work.

President Clinton created the council in 1993 to find ways to "meet the needs of the present without jeopardizing the future." Made of members from the administration, the business community, and the environmental community, the council met for six years and delivered three major reports. The first, published in 1996, was *Sustainable America: A New Consensus*. In it, the council explained how it began its work by first developing a "We Believe Statement" that helped to forge a cooperative spirit among members. One of the beliefs relates to change and the nature of sustainability: "Change is inevitable and necessary for the sake of future generations and for ourselves. We can choose a course for change that will lead to the mutually reinforcing goals of *economic growth*, *environmental protection*, and *social equity*." These three goals, which appear in Fig. 1–8 as the key to sustainable development, gave structure to the 186-page report, as it went on to present a series of 10 more specific national goals for achieving sustainable development (Table 24–2). The national goals were then given closer attention in the report in the form of policy recommendations, specific actions, and indicators of progress.

The introduction to the report is a remarkable challenge and call for change. Major global changes affecting everyone are brought into focus, such as the continuing population growth, major impacts on natural resources, the link between economic growth and increasing global pollution, and the globalization of information and economic activity. The council cited the need to move "from conflict to collaboration," meaning that businesses, the federal government, communities, and citizens must learn to work together in forging public policies. The guide for accomplishing a sustainable America, in the eyes of the council, needs to be a *stewardship ethic*: "Principles of stewardship help define appropriate human interaction with the natural world. Stewardship ... is a set of values that applies to a variety of decisions. It provides moral standards that cannot be imposed but can be taught, encouraged, and reinforced.... Principles of stewardship can illuminate complex policy choices and guide individuals toward the common good."

The report also points to the need for individual responsibility as a counter to apathy and inaction. The common good is good for individuals also, and for sustainable development to be successful, people must work together and also be held accountable for their actions.

The council produced a number of major reports, all of which are available on the Web (http://clinton3.nara.gov/PCSD/Publications/). The second, *Sustainable Communities*, was published in 1997 as the work of the Sustainable Communities Task Force. The report contains a number of case studies (including that of Chattanooga), a set of community profiles that highlight community actions and action groups from each of the 50 states, and a number of examples of states providing leadership on sustainable development. In presenting these examples, the task force captured the best of the sustainable communities movement and extended the work of the president's council in producing a set of goals and recommendations to be implemented by federal, state, and local governments, businesses, and, especially, communities and the people in them.

The final report of the council was released in May 1999 at a major event it sponsored: the National Town Meeting for a Sustainable America, in Detroit. This event featured examples of sustainability from around the country and emphasized building individual and institutional capacity so that the best sustainability practices can be replicated elsewhere. With the Detroit meeting and the last report (*Towards a Sustainable America: Advancing Prosperity, Opportunity, and a Healthy Environment for the 21st century*), the council concluded its work. This report focused on four issues of specific concern to President Clinton: (1) policies to reduce greenhouse gas emissions; (2) building a new environmental management

◀ **FIGURE 24–17**
Chattanooga, Tennessee. With numerous projects to reduce pollution and traffic and to provide attractions and amenities for people, Chattanooga, Tennessee, has changed its reputation from one of the worst to one of the best places to live in the United States.

TABLE 24–2 National Goals toward Sustainable Development
Goal 1: Health and the Environment. Ensure that every person enjoys the benefit of clean air, clean water, and a healthy environment at home, at work, and at play.
Goal 2: Economic Prosperity. Sustain a healthy U.S. economy that grows sufficiently to create meaningful jobs, reduce poverty, and provide the opportunity for a high quality of life for all in an increasingly competitive world.
Goal 3: Equity. Ensure that all Americans are afforded justice and have the opportunity to achieve economic, environmental, and social well-being.
Goal 4: Conservation of Nature. Use, conserve, protect, and restore natural resources—land, air, water, and biodiversity—in ways that help ensure long-term social, economic, and environmental benefits for ourselves and future generations.
Goal 5: Stewardship. Create a widely held ethic of stewardship that strongly encourages individuals, institutions, and corporations to take full responsibility for the economic, environmental, and social consequences of their actions.
Goal 6: Sustainable Communities. Encourage people to work together to create healthy communities in which natural and historic resources are preserved, jobs are available, sprawl is contained, neighborhoods are secure, education is lifelong, transportation and health care are accessible, and all citizens have opportunities to improve the quality of their lives.
Goal 7: Civic Engagement. Create full opportunity for citizens, businesses, and communities to participate in and influence the natural-resource, environmental, and economic decisions that affect them.
Goal 8: Population. Move toward stabilization of U.S. population.
Goal 9: International Responsibility. Take a leadership role in the development and implementation of global sustainable development policies, standards of conduct, and trade and foreign policies that further the achievement of sustainability.
Goal 10: Education. Ensure that all Americans have equal access to education and lifelong learning opportunities that will prepare them for meaningful work, a high quality of life, and an understanding of the concepts involved in sustainable development.

Source: President's Council on Sustainable Development, *Sustainable Development: A New Consensus*. (Washington, DC: U.S. Government Printing Office, 1996), pp. 12–13.

system; (3) further work for building partnerships to strengthen communities; and (4) policies to strengthen American leadership in international sustainable development issues. Like the council's other major reports, this final one contains specific recommendations (140 of them!) that address the four major issues.

The President's Council on Sustainable Development has made a valuable contribution to the country's need to think about its future. In winding up its work, the council asked the president to "find the means to continue the progress on sustainable development initiated by the Council, and to assure the existence of some forum for the thoughtful consideration of sustainable development issues by high-level leaders from all sectors." This forum has not yet been established. Policymakers, leaders in industry, and environmental organizations would do well to study the council's publications and work toward implementing its recommendations. The question is, Will they?

24.4 Epilogue

Sustainability, stewardship, and sound science—these are the themes that have kept our eyes on the basics of how we must live on our planet. As we pointed out in Chapter 1,

sustainability is the practical goal that our interactions with the natural world should be working toward, stewardship is the ethical and moral framework that informs our public and private actions, and sound science is the basis for our understanding of how the world works and how human systems interact with it. Are we making progress in incorporating these themes into our society?

Our Dilemma

Our intention in this book has been to avoid dwelling on the bad news and instead to raise the hope—indeed, the certainty—that all environmental problems *can* be addressed successfully. Throughout the text, we have pointed to policies and possibilities that can help move human affairs in a sustainable direction. There is much at stake. Is sustainable development going to work to reduce the desperate poverty and lowered life expectancy of the billions who live on less than $2 a day? Will we reduce our use of fossil fuels to halt global warming and the rising sea level? Can we hope to feed the 4 billion more who will join us by midcentury and to simultaneously reduce the malnutrition that still plagues 800 million?

As the late Nobel Laureate Henry Kendall said, environmental problems at root are human, not scientific or technical. Even though our scientific understanding is not complete and our grasp of sustainable development is still tentative, we know enough to be able to act decisively in most circumstances. Therefore, it is human decisions, at both the personal and societal level, that can bring about change, and it is these decisions that define our stewardship relationship with Earth.

However, making stewardly decisions is not a simple task. Decision making is affected by our personal values and needs, and competing values and needs exist at every turn. One group wins, and another loses. If we stop all destruction of the rain forests, the peasant who needs land to farm will not eat. If we force a halt to the harvesting of shrimp because of concerns about the by-catch, thousands of shrimpers will be out of work. Making decisions means considering competing needs and values and reaching the best conclusion in the face of the numbing complexity of demands and circumstances. It is the nature of our many dilemmas that "business as usual" resolves only a very small fraction of them. Even though public policies at the national and international level are absolutely essential elements of our success in turning things around, these in the end are the outcome of decisions by very human people. And even if thoughtful decisions are made by those in power, they will fail unless they are supported by the people who are affected by them.

What compels people to act to promote the common good? We can identify a number of values that can be the basis of stewardship and help us in our ethical decisions involving both public policies and personal lifestyles. For example, *compassion* for those less well off—those suffering from malnutrition or other forms of deprivation—can become a major element of our foreign policy, energizing our country's efforts in promoting sustainable development in the countries most in trouble. Compassion can also motivate young people to spend two years of their life in the Peace Corps or a hunger relief agency. A *concern for justice* is another crucial value. Just policies can become the norm for international economic relations, and a concern for justice can also move people to protest the placement of hazardous facilities in communities of color. Another stewardly value is *honesty*, or a concern for the truth—in the sense of keeping the laws of the land and in openly examining issues from different perspectives. Other important values for stewardship are *frugality*—using no more than is necessary of Earth's resources; *humility*—instead of demanding our rights, being willing to share and even defer to others; and *neighborliness*—being concerned for other members of our communities, such that we do not engage in activities that are harmful to them, but instead make positive contributions to their welfare.

To achieve sustainability, we will have to couple values like these with the knowledge that can come from our understanding of how the natural world works and what is happening to it as a result of human activities. For this to work, people must be willing to grapple with scientific evidence and basic concepts and develop a respect for the consensus that scientists have reached in most areas of environmental concern. It has been our objective to convey that consensus

to you in this text. Recall also that sustainable solutions have to be economically feasible, socially desirable, and ecologically viable (Fig. 1–8). Again, we have tried to emphasize all three of these components as we point to the political decisions that are shaping our future in so many of our environmental concerns.

The good news is that many people *are* engaging environmental problems, working with governments and industries, bringing relief to people in need, and demonstrating exactly the kinds of decisions that are needed for effective stewardship. (See "Ethics" essay, p. 612.) Thus, traditional environmentalist organizations, as well as people from other sectors of society, are working together on the many issues that we have highlighted in this text, bringing about policy changes and also *changes in personal lifestyles*. We conclude the text with this very important topic.

Lifestyle Changes

What are some of the lifestyle changes that are needed, and are they in fact occurring? We should be encouraged and inspired by the literally millions of people in all walks of life who are acutely aware of the problems and who are making outstanding efforts to bring about solutions. Every pathway toward solutions that we have mentioned represents the work of thousands of dedicated professionals and volunteers ranging from scientists and engineers to businesspeople, lawyers, and public servants. Indeed, we are all involved, whether we recognize it or not. Simply by our existence on the planet, everything we do—the car we drive, the products we use, the wastes we throw away, virtually every choice we make and action we take—has a certain environmental impact and a certain consequence for the future. Therefore, it is not a matter of having an effect, but of what and how great that effect will be. It is a matter of each of us asking ourselves, Will I be part of the problem or part of the solution? The outcome will depend on how each of us responds to the challenges ahead.

There are a number of levels on which we may participate to work toward a sustainable society:

- individual lifestyle changes
- political involvement
- membership and participation in nongovernmental environmental organizations
- volunteer work
- career choices

Lifestyle changes may involve such things as switching to a more fuel-efficient car or using a bicycle for short errands; recycling paper, cans, and bottles; retrofitting your home with solar energy; starting a backyard garden and composting and recycling food and garden wastes into your soil; putting in time at a soup kitchen to help the homeless; choosing low-impact recreation, such as canoeing rather than speedboating; living closer to your workplace; and any number of additional things.

ETHICS

THE TANGIER ISLAND COVENANT

Chesapeake Bay, the largest estuary in the United States, is famous for its blue crabs and the unique "watermen" who fish for the crabs and other marine animals. Watermen are independent fishers who have lived on the bay for generations and fiercely defend their way of life. In recent years, the watermen of Tangier, a small island on the Virginia part of the eastern shore of the bay, have been in constant battle with the Chesapeake Bay Foundation and the state regulators over crab and oyster laws restricting when and where they can harvest. The watermen skirted restrictions by keeping undersized crabs and dredging clams and oysters out of season. They were also accustomed to throwing their trash overboard, causing the shores of the island to be littered with debris.

In 1997, the island community of Tangier became part of an unusual, but significant, process called "values-based stewardship," a form of conflict resolution that requires working within the value system of the community to bring about sustained changes in behavior toward the environment.

Susan Drake, a doctoral candidate in anthropology at the University of Wisconsin, came to the island to carry out her thesis research on how conflicts are dealt with in homogeneous communities like Tangier. Originally attracted to the project because of the highly religious community present on the island—the two churches are the center of life and authority there—Drake knew that the watermen considered themselves part of a rich bounty in an environment created by God. After working with

watermen on their boats and spending time in their churches and with their families, Drake became concerned about the apparent lack of a connection between what the God-fearing watermen believed and what they actually did in their profession.

Convinced that she might be able to intervene in a positive way, Drake requested permission by her thesis committee to engage in "participatory action research," wherein the researcher may enter the conflict situation and attempt to bring about a resolution. Working within the value system of the community—in this case, a value system based on faith—Drake addressed a joint meeting of the two churches on the island one Sunday morning. Referring to an image of Jesus standing behind a young fisherman at the wheel of a boat in rough seas that almost all Tangier watermen displayed in the cabins of their vessels, Drake addressed the issues of dumping trash overboard and skirting the laws. She showed the picture to the congregation and then put paper blinders over the eyes of Jesus as she listed the ways in which the watermen were disobeying the law. The day of her sermon coincided with a very high tide that gathered the debris from around the shores of the island and washed it up into the streets and people's yards.

The meeting ended with a call by Drake to the watermen to adopt a covenant she had drafted earlier with the help of some of the island's women. Fifty-six watermen went to the altar and promised to abide by the following covenant: "The Watermen's Stewardship Covenant is a covenant among all watermen regardless of their pro-

fession of religious faith. As watermen we agree to: (1) be good stewards of God's Creation by setting a high standard of obedience to civil laws (fishery, boat and pollution laws), and (2) commit to brotherly accountability. If any person who has committed to this Covenant is overtaken in any trespass against this covenant, we agree to spiritually restore such a one in a spirit of gentleness. We also agree to fly a red ribbon on the antennas of our boats to signify that we are part of the Covenant."

The next day, and up to the present, scores of watermen began flying red ribbons, taking trash bags out on their boats, and fishing according to the laws. Trucks began to pick up the trash on the island early in the morning. The two churches proceeded to work with the entire island community to draft a plan for the future sustainability of the island and their way of life and to form the "Tangier Watermen's Stewardship for the Chesapeake" in order to implement the plan. The plan is far reaching and involves the women of the island lobbying the state legislature and opening up lines of communication with scientists and government regulators involved in overseeing the bay's health.

Values-based stewardship is a call to life-changing stewardship put within the value system of a community. In this case, putting the call within the context of the faith of the watermen resulted in the establishment of structures within the community that show great promise in helping to bring the entire island toward a sustainable future.

Political involvement ranges from supporting and voting for particular candidates to expressing your support for particular legislation through letters or phone calls. Membership in nongovernmental environmental organizations can enhance both lifestyle changes and political involvement. As a member of an environmental organization, you will receive, and may help disseminate, information, making you and others more aware of particular environmental problems and things you can do to help. In particular, you will be informed regarding environmentally significant legislation so that you may focus your political efforts at the most effective time and place. Also, your membership and contribution serve to support lobbying efforts of the organization. A lobbyist representing only him- or herself has relatively little impact on

legislators. On the other hand, if the lobbyist represents a million-member organization that can follow up with that many phone calls and letters (and, ultimately, votes), the impact is considerable. Finally, in cases where enforcement of the existing law has been the weak link, some organizations, such as the public-interest research groups, the Natural Resources Defense Council, and the Environmental Defense Fund, have been highly influential in bringing polluters or the government to court to see that the law is upheld. Again, this can be done only with the support of members.

Another form of involvement is joining a volunteer organization. Many effective actions that care for people and the environment are carried out by groups that depend on volunteer labor. Political organizations and virtually all

CAREER LINK

BRIAN HOPPER, ENVIRONMENTAL LAW INTERN

For Brian Hopper, it was a dream assignment. Hopper had just received a J.D. degree from Vermont Law School in May, 2000, and needed an internship to complete his studies for a Masters of Studies in Environmental Law. As part of his law degree studies, Brian had written a paper on the listing process for the Atlantic salmon under the Endangered Species Act. For the paper he had contacted Mary Colligan, Endangered Species Coordinator for the Northeast Regional Office of the National Marine Fisheries Service (NMFS), part of the National Oceanic and Atmospheric Administration (NOAA). A call to Colligan led to an invitation to do his internship with her at the Northeast Regional Office of NMFS in Gloucester, Mass.

"I have loved the sea all my life," Hopper said. "My father introduced me to SCUBA diving at an early age, and I became a certified diver at 13." Later, Hopper would take up sailing, eventually working on the America's Cup Organizing Committee and teaching sailing to all ages. But it

was a course in environmental science at Bennington College, in Vermont, that sparked his interest in environmental policy, and he went on to Connecticut College, where he designed his own interdisciplinary major in Environmental Policy. Along the way, Hopper took a number of biology and marine science courses, and on graduation, he decided that law school would give him the training and credentials to make a career in marine-related public policy.

During his internship, Hopper had an opportunity to put his interest in the Atlantic salmon to the test. "I spent a week in downeast Maine restocking rivers there with huge salmon. We released over 1,000 fish to several rivers for spawning stock." The salmon were raised in aquaculture from wild fish stocks until they were 15–30 lbs, and then released. Hopper said that later on some of the fish were seen engaging in spawning activities. He mentioned that the Maine Atlantic salmon was recently listed as an endangered species population, a decision that has been challenged by

the state of Maine and some industries. Some of Hopper's other work as an intern has been the crafting of cooperative agreements between federal agencies and eastern states, allowing the states to tap into federal funds and expertise. For example, in Maryland, the Chesapeake Bay contains crucial habitat for juveniles of the Kemp's Ridley sea turtle, the most endangered of the sea turtles. Hopper's combination of biological and legal knowledge uniquely equipped him to become involved in the development of a cooperative plan between NMFS and the state of Maryland Department of Marine Resources.

When asked about his future, Hopper said "First, I have to pass the upcoming bar exam, but when I do, I would love to return to NMFS. The people are great, and I've had an opportunity to get involved in conservation law, which has been my career objective. My internship has given me a great chance to see what I can do, and I simply can't see myself working for a law firm doing personal injury suits or corporate law."

nongovernmental organizations are highly dependent on volunteers. Many helping organizations, like those dedicated to alleviating hunger and to building homes for needy families (Fig. 24–18), are adept at mobilizing volunteers to accomplish some vital tasks that often fill in the gaps left behind by inadequate public policies.

Finally, you may choose to devote your career to implementing solutions to environmental problems. Environmental careers go far beyond the traditional occupation of wildlife or park management. There are any number of lawyers, journalists, teachers, research scientists, engineers, medical personnel, agricultural extension workers, and others focusing their talents and training on environmental issues or hazards (see "Career Link," this page.) Business and job opportunities abound in pollution control, recycling, waste management, ecological restoration, city planning, environmental monitoring and analysis, nonchemical pest control, the production and marketing of organically grown produce, and so on. Some developers concentrate on rehabilitation and the reversal of urban blight, as opposed to contributing more to urban sprawl. Some engineers are working on the development of pollution-free vehicles to help solve the photochemical smog dilemma of our cities. Indeed, it is difficult to think of a vocation that cannot be focused on promoting solutions to environmental problems.

▲ **FIGURE 24–18** *Habitat for Humanity volunteers.* By using charitable contributions and volunteer labor, Habitat for Humanity creates quality homes—over 70,000 built throughout the world by 1998—that needy families can afford to buy with no-interest mortgages, enabling the money to be recycled to build more homes. Prospective home buyers participate in the construction and may gain valuable skills in the process.

We are making our way into a new millennium, and it is going to be an era of rapid changes unprecedented in human history. We are engaged in an environmental revolution, a major shift in our worldview and practice from seeing nature as resources to be exploited to seeing nature as life's supporting structure that demands our stewardship. You are living in the early stages of this revolution, and we invite you to be one of the many who will make it happen.

For civilization as a whole, the faith that is so essential to restore the balance now missing in our relationship to the earth is the faith that we do have a future. We can believe in that future and work to achieve it and preserve it, or we can whirl blindly on, behaving as if one day there will be no children to inherit our legacy. The choice is ours; the earth is in the balance.

Al Gore, *Earth in the Balance*

ENVIRONMENT ON THE WEB

INTEGRATIVE FRAMEWORKS FOR ENVIRONMENTAL MANAGEMENT

In the mid-1980s, David Crombie, a popular former mayor of Toronto, Ontario, was appointed Head of a Royal Commission on the Future of the Toronto Waterfront. Crombie, affectionately termed "Toronto's tiny perfect mayor," was to lead the Commission in developing proposals to restructure Toronto's decaying waterfront area. The waterfront was typical of many older industrial cities, containing a mix of industrial uses, open space, and railway lands. An elevated highway blocked the view of Lake Ontario from the city and created a physical obstacle for citizens wanting to access the shoreline areas.

Crombie tackled the task with his normal wit and energy but soon found himself coming to a surprising conclusion. The waterfront, he declared, could not be managed without consideration of the entire watershed area. A study of the entire watershed would mean detailed examination of several major river basins, the entire city of Metropolitan Toronto, many outlying communities and agricultural areas, transportation corridors, and areas of parkland and green space. Because the current condition of those lands was affected by their human uses—and users—Crombie required that the Commission also examine community values relating to the waterfront, recreational, cultural, and other social issues, and the economics of water and waterfront use. In short, David Crombie and the Royal Commission had concluded that effective planning for sustainable use of one watershed component—the narrow waterfront strip—required the examination of that component in the context of the larger watershed system.

The work of the Royal Commission, now a permanent body called the Waterfront Regeneration Trust, was important not because of its scientific sophistication, or even because of the complexity of the issues it considered. The work was remarkable because it integrated biophysical, social, and economic factors in a framework that gave high regard to the values and opinions of average citizens. As the Commission proceeded with its investigations, holding workshops and open houses, listening to regulators, scientists, and citizens, it gradually educated watershed residents about watershed systems—and the implications for those systems of individual actions.

Web Explorations

The Environment on the Web activity for this essay describes the process of integrated watershed management as an example of an integrative framework for sustainable development. Go to the Environment on the Web activities (select Chapter 24 at **http:// www.prenhall.com/wright**) and learn for yourself:

1. how to develop a successful watershed management plan;
2. about funding for integrated management plans; and
3. about an example of integrative management in another field.

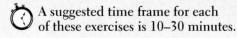 **A suggested time frame for each of these exercises is 10–30 minutes.**

REVIEW QUESTIONS

1. What problems do Trenton and Waitakere City illustrate, and how are they organizing to deal with those problems?

2. How did the structure of cities begin to change after World War II? What factors were responsible for the change?

3. Why are the terms *car dependent* and *urban sprawl* used to describe our current suburban lifestyle and urban layout?

4. What federal laws and policies tend to support urban sprawl?

5. Describe five categories of environmental consequences stemming from the car-dependent lifestyle and urban sprawl.

6. What is "smart growth"? What are four smart-growth strategies that address urban sprawl?

7. What services do local governments provide, and what is their prime source of revenue for these services?

8. What is meant by "erosion of a city's tax base"? Why does it follow from exurban migration? What are the results?

9. How do exurban migration, urban sprawl, and urban decay become a vicious cycle?

10. What is meant by "economic exclusion"? How is it related to the problems of crime and poverty in cities?

11. What are three common denominators of livable cities? What are the connections between livability and sustainability?

12. How does the U.N. Sustainable Cities Program work?

13. What is the Sustainable Communities Program? How does Chattanooga illustrate the program?

14. How did the President's Council on Sustainable Development promote sustainability in cities and suburbs?

15. What are some stewardship values, and what role could they play in fostering sustainability?

16. Give five levels on which people can participate in working toward a sustainable future.

THINKING ENVIRONMENTALLY

1. Interview an older person (65 years or older) about what his or her city or community was like before 1950 in terms of meeting people's needs for shopping, recreation, getting to school, work, and so on. Was it necessary to have a car?

2. Do a study of your region. What aspects of urban sprawl and urban blight are evident? What environmental and social problems are evident? Are they still going on? Are efforts being made to correct them?

3. Identify nongovernmental organizations in your region that are working toward reining in urban sprawl or working toward preserving or bettering the city. What specific projects are underway in these areas? What roles are local governments playing in the process? How can you become involved?

4. Suppose you are a planner or developer. Design a community that is people oriented and that integrates the principles of sustainability and self-sufficiency. Consider all the fundamental aspects: food, water, energy, sewage, solid waste, and how people get to school, shopping, work, and recreation.

MAKING A DIFFERENCE

PART SIX: Chapters 23 and 24

1. Analyze your environmental impacts as a consumer. What modes of transportation do you use, what foods do you consume or throw away, what appliances do you own, how do you heat or cool your home, and how much water do you use? Consider especially your impact on the four leading consumption-related environmental problems: air pollution, global climate change, natural-habitat changes, and water pollution.

2. Use the available Web resources to locate specific regulations developed by the EPA or other regulatory agencies, and research the details of how the agency conforms to Executive Order 12866 in developing the economic impact of the regulations.

3. Investigate land-use planning and policies in your region. Is urban sprawl or urban blight a problem? Become involved in zoning issues, and support zoning and policies that will support, rather than hinder, ecologically sound land use ("smart growth").

4. Follow up on the work of the President's Council on Sustainable Development to find out what is happening in your state or city to promote sustainable development.

5. Contact local conservation organizations or land trusts, and offer your help in protecting and preserving open land in your region.

6. Find out if there is a Habitat for Humanity project or group in your area, and if there is, volunteer some time in support of their work.

WEB REFERENCES

On-line resources for this chapter are on the World Wide Web at: **http://www.prenhall.com/wright**. (Click on Chapter 24 within the Chapter Selector.)

ABC Video Case Studies
Volume V

Each of the 10 ABC Video Case Studies that follow was selected to bring you an enhanced visual learning tool connected with the text. With the cooperation and courtesy of Prentice Hall and ABC News, we have selected video segments from such award-winning ABC news programs as "World News Tonight" (especially its feature, "A Closer Look"), "Good Morning America," and "Nightline." Presented in this section is a written overview, or abstract, for each video segment, along with some questions to encourage you to focus on the relevant issues and controversies. Each written case study has chapter or topic references that serve to emphasize the relevance of the case study to specific sections of the text. We hope that these case studies will stimulate class discussion and debate, as well as increase awareness and understanding as they remind you of the fact that the environment is often in the news, and for good reason.

Prentice Hall is making all of these ABC video segments available to your instructor for classroom use on a long-playing VCR videocassette. We encourage you to explore these cases as you use the text.

ABC CASE STUDY

Reproductive Alternatives to Abortion

One of the most controversial issues in the United States, abortion continues to generate medical and political attention. In recent years, abortions have been declining, along with the number of teen mothers. This newscast from January 25, 2000, examines some of the reasons for the downturn. Dr. Nancy Snyderman, ABC News medical correspondent, believes that many women are finding alternative ways to prevent pregnancies. Several prescription drugs are described that work as a "morning-after" pill. Eventually, the feature discusses RU-486, a drug that induces abortions within the first seven weeks of pregnancy. In September 2000, the FDA approved RU-486 for use in the United States; the drug has been widely used in Europe for 15 years. RU-486 is a hot issue. The Bush administration is examining options for withdrawing approval of the drug, which is viewed as encouraging abortions because it is an alternative to a surgical abortion. The interview ends with the parties agreeing that the real concern is to reduce the large number of unintended pregnancies. *Wright/Nebel reference: Chapter 7,* *pp. 177–178, and Environment on the Web essay, "Women as the Key to Development," p. 184.*

QUESTIONS
1. Do you think that taking a "morning-after" pill that prevents implantation of the fertilized egg is the same as abortion?
2. What do you think about the legalization of RU-486? Should Congress or the Bush administration take steps to withdraw the drug?

ABC CASE STUDY

Government Takes Down Edwards Dam in Maine

In central Maine, the Kennebec River, one of Maine's largest, runs to the sea. For 162 years, the Edwards Dam in Augusta, the state capital, blocked the free flow of the river and prevented fish from migrating upriver to the wilder regions in the north of the state. Built to provide power for a textile mill that has long since gone south, the dam continued to generate 3.5 megawatts of power, enough for a town of 2,000. The Federal Energy Regulatory Commission ruled for the first time that the environmental benefits of breaching a dam outweighed the economic value of the power generated. On July 1, 1999, the dam was breached, and the river once again flowed through Augusta. A year later, the benefits of the dam's removal were obvious: alewives, striped bass, shad, sturgeon, and Atlantic salmon were once again moving upriver, and water quality in the river above the dam improved remarkably. As a result, the Augusta riverfront was the recipient of a renewal project designed to provide public access and community development. At least 465 dams have been removed in recent years, signs of a trend toward restoration of our rivers and the fisheries they provide. *Wright/Nebel reference: Chapter 9, pp. 230–231.*

QUESTIONS
1. What river in the West is being studied for dam removal to restore historic salmon runs?
2. What criteria should be weighed in making a decision about removing a dam?

ABC CASE STUDY
Genetically Modified Food

"Nightline" host John Donvan explores the "brave new world of genetically engineered food" in this newscast from October 25, 1999. Donvan first interviews farmer Joe Hawkins, who expressed his disillusionment with the genetically modified corn he had been planting. It is not that the corn doesn't do well in the field; rather, it is being shunned by consumers, especially in Europe, where such notables as Paul McCartney and Prince Charles are opposing it. Donvan then turns to then Secretary of Agriculture Dan Glick-man, Val Giddings (representing the biotechnology industry), and Jeremy Rifkin, an outspoken critic of biotechnology. The discussion centers on the issues of testing and approval. Rifkin questions the testing procedures and is countered by Glickman and Giddings, who are supportive of the approval that has been given in the United States. In the last segment of the broadcast, the guests debate the issue of labeling genetically modified foods and disagree intensely on whether such foods should be labeled in the United States, even though U.S. consumers would prefer to see it labeled. ("It's like putting a skull and crossbones on the product," according to Giddings.) *Wright/Nebel reference: Chapter 10, pp. 241–242 and 249–251.*

QUESTIONS
1. Why is Rifkin dubious about the testing that is done on genetically modified food?
2. Would you advocate labeling genetically modified food? Or should such food be banned altogether?

ABC CASE STUDY
Back from the Brink of Extinction

Two endangered species brought back from the brink of extinction—the bald eagle and the peregrine falcon—are highlighted in the introduction to this news piece. But the focus of the report is a little-known sea turtle, the Kemp's ridley turtle, whose numbers declined dramatically during the latter part of the last century, due to poaching and the taking of eggs, but especially to the Gulf of Mexico shrimp fishery. The turtles are caught in shrimp nets and drowned before they can be released. Now both U.S. and Mexican regulations require shrimpers to employ turtle exclusion devices (TEDs) on their nets, which greatly reduces (but does not eliminate) mortality. In addition, the Mexican government has established a sanctuary on the 15 miles of Gulf beach that is sought out by almost the whole population of Kemp's ridley turtles for nesting. The newscast shows cooperative efforts being made to protect the eggs and young hatchlings to give them a chance to make it to the ocean. In recent years the nesting population has tripled, suggesting that the restoration measures have been effective. *Wright/Nebel references: Chapter 11, pp. 271–275.*

QUESTIONS
1. What are some of the hazards faced by Kemp's ridley turtles?
2. What other endangered species have been brought back by efforts to increase their reproductive success under artificial conditions?

ABC CASE STUDY
Buffalo Wars

Montana is home to the crown jewel of the U.S. National Park System—Yellowstone National Park, which is itself home to a population of bison, or buffalo, several thousand strong, descendants of the last few buffalo left in the country at the turn of the 20th century. Montana is also a cattle state, with hundreds of thousands of free-ranging cattle. As the text points out, the park is part of a much larger ecosystem: the Greater Yellowstone ecosystem. During the winter, some of the buffalo wander out of the national park in search of forage, and when they do, the buffalo wars begin. This "Nightline" segment explores an ugly battle that is reminiscent of the old cattlemen-vs.-sheep-ranchers wars. Concerned about the possible spread of brucellosis (a disease that is specific to bovine species), Montana Department of Livestock personnel have been shooting the wandering buffalo in large numbers; more than 1,000 were killed during the hard winter of 1996–97. The newscast shows members of Defenders of Wildlife and park personnel as they try to prevent the slaughter and also shows graphic footage of buffalo being shot by Montana riflemen. Chris Bury interviews cattlemen, buffalo protectors, and Interior Department personnel in attempting to bring out the deeper issues at the root of this conflict. On a positive note, the federal agencies and Montana agreed on a joint management plan in December 2000 that should end the unnecessary killing of the bison outside the park, while carefully managing the risk of transmission of brucellosis. *Wright/Nebel reference: Chapter 12, pp. 308–309.*

QUESTIONS
1. What are the real issues behind this conflict (as discussed by Keith Oney and Chris Bury)?
2. What solution has been proposed to bring a halt to the buffalo wars?

ABC CASE STUDY
Problems Associated with the U.S. Supply of Energy

ABC News takes a "Closer Look" at the energy problems plaguing California. The state is the victim of its own choices. Having opted for deregulation in 1996, California then went on a splurge of electricity-demanding growth without any new power plant construction. The form of deregulation California adopted divested the utilities from owning power plants. The utilities also were prevented by the state from signing long-term contracts with power plants and were forced to buy electricity on a day-to-day basis, a costly exercise. Making matters worse, the nation's natural gas supplies have failed to keep up with demand, forcing gas prices upwards. Customers in California are powerless (so to speak) to make the kinds of choices the proponents of deregulation promised and are facing huge increases in their electric bills as the utilities begin to pass along their own inflated costs. In other words, deregulation has not led to the wonderful free-market economizing often promised, but instead has brought on rolling blackouts and public fury. The newscast moves on to Wisconsin, where things are not much better. In the meantime, the Bush administration is citing California's woes in calling for oil exploration in the Arctic National Wildlife Refuge, a connection that has many minds boggling. *Wright/Nebel reference: Chapter 13, p. 336.*

QUESTIONS
1. Why are power companies reluctant to build new power plants?
2. What is the latest news from California and other states that have deregulated their electricity?

ABC Case Study

Mosquitoes, West Nile Virus, and Malaria

The West Nile virus attracted the nation's attention in 1999 when seven New Yorkers died of the encephalitis caused by the virus, which belong to a group of "emerging infectious diseases" such as the Ebola virus or Legionnaires' disease. The West Nile virus is carried by mosquitoes, and it is spreading. The newscast focuses first on the virus, pointing out that it can be spread by infected migratory birds. Wildlife surveys carried out in the year 2000 have revealed 70 species of birds killed by the virus in 12 eastern U.S. seaboard states. Many domestic and wild animals also have been found infected by the virus. Fortunately, the human toll in 2000 was down: only 17 cases and one death. This "Closer Look" news segment then compares the impact of the virus with that of another mosquito-borne disease, malaria—a far more devastating disease worldwide. Next, a new strategy is discussed: Scientists have learned how to genetically modify mosquitoes and are working on developing a mosquito that can no longer transmit malaria. If this works, it may be possible to apply a similar approach to mosquitoes that carry the West Nile virus. *Wright reference: Chapter 16, pp. 392, 395–396, and 404–406, and Global Perspective essay, "An Unwelcome Globalization," p. 404.*

QUESTIONS
1. Compare the impacts of the West Nile virus and malaria. Which deserves more attention and control?
2. What are the potential problems associated with releasing genetically modified mosquitoes into the environment?

ABC Case Study

EPA Calls for Reduction in Use of MTBE

In one approach to meeting Clean Air Act regulations, methyl tertiary butyl ether (MTBE) is added to a third of the gasoline sold in the United States to promote cleaner burning. However, MTBE, which has a nasty smell, has been showing up in a disturbing number of water supplies as it leaks from underground storage tanks and other sources. The main reason EPA is phasing it out, however, is that it has been shown to develop tumors in rats and is a suspected human carcinogen. The winners here might be Midwestern corn growers, because ethanol is another chemical additive that promotes clean-burning gasoline. In 1999, 1.5 billion gallons of ethanol were produced for fuel, and more ethanol plants are under construction. Corn futures, anyone? *Wright/Nebel reference: Chapter 15, pp. 380–381; Chapter 20, p. 490.*

QUESTIONS
1. What are the main sources of MTBE in drinking-water supplies?
2. From your reading in the text, what are the main arguments for and against substituting ethanol for MTBE?

ABC Case Study

Beach Erosion

This newscast comes from Belmar, on the New Jersey shore. Belmar's problem is beach erosion, and it is not unique. Worldwide, global warming is causing a rise in sea levels, and there is no end in sight. Making matters worse, coastal dunes and beaches are pounded by storms and hurricanes, both of which are becoming more intense and more frequent. Add to this the problem on the Jersey shore of little new sand being brought down its rivers, and you have a problem of billion-dollar dimensions. Dr. Norbert Psuty, a Rutgers University scientist, is interviewed in this segment. The New Jersey beaches are helped with "beach replenishment," which is simply moving sand in from offshore. Doing so, however, is costly, and, as Dr. Psuty points out, it doesn't address the basic problems. *Wright/Nebel reference: Chapter 21, pp. 521–524.*

QUESTIONS
1. What is the conclusion reached at the end of this news segment about beachfront property?
2. According to the text, how rapidly is sea level rising? What are the projections of the IPCC for the rise in sea level in the 21st century?

ABC CASE STUDY
Cleaner Air in Los Angeles

Los Angeles, the city that practically invented photochemical smog, has turned the corner and is a shining example of what rigorous air pollution regulations can do. For two years running, LA has had no smog alerts; there were 101 such alerts in 1980. The smog is retreating as a result not only of catalytic converters on cars, but also many smaller changes, as this newscast shows. Small engines now must pass emissions standards, and hair sprays, deodorants, and other household products have been reformulated. Wet–dry cleaning is replacing the organic solvent-based traditional methods. Clearly, the measures taken have greatly reduced the load of volatile organic compounds that are so essential to ozone formation. Los Angeles still has sunlight, a lot of automobile traffic, and some smog, but the city's residents are breathing freer because of the high air quality standards in place. *Wright/Nebel reference Chapter 22, pp. 539–541, pp. 561–562.*

QUESTIONS

1. What are the amounts of aerosol pollutants released in just hair sprays before 1993?
2. According to the text (p. 563), what is the next step being taken by California to achieve even cleaner air?

ABC CASE STUDY
Volumes I, II, III, and IV

Also available to instructors who adopt the eighth edition of Environmental Science are the set of videos from the fourth-, fifth-, sixth-, and seventh-edition Video Case Studies. Most of these videos are still quite relevant to current environmental concerns. The following are titles included in the Case Studies as they appear at the end of the previous editions, which refer to the videos available in Volumes I through IV.

VOLUME I (FOURTH EDITION)

Famine and Relief Efforts in Somalia
Hazardous Wastes in the Wrong Place
Environmental Concerns under Communism
Global Warming and the Scientific Community
Break in the Ozone Shield
The Place of Risk Analysis in Pollution Cleanup
Biodiversity: What Can Be Done to Protect It
Challenge on the Right: Environmental Movement as Subversive
Mismanaging the National Forests
Who Wants the Trash?
Oil and the Arctic National Wildlife Refuge
The Case of Energy Conservation
Fuel-Efficient Cars: Detroit vs. Japan
Oil on the Sound: Alaskan Oil Spill
The Nuclear Waste Dilemma
Energy Alternatives in Our Future

VOLUME II (FIFTH EDITION)

The World's Population Explosion
Assignment Rwanda: Profiles in Caring
Nature's Warning: Environmental Poisons and Sterility
Tobacco on Trial
Ozone and UV Radiation
Restoring Rivers for Wildlife
Environmental Backlash and the Scientific Community

Empty Nets: Collapse of the Fisheries
Turning Landfills into Nature Preserves
Nuclear Waste and Apaches

VOLUME III (SIXTH EDITION)

Reviving the Biosphere
World Food Prize Doctors
Drinking-Water Safety
GM's Electric Car
Global Warming
Yellowstone's Wolves
Mollie Beattie
Lobbyists and Environmental Law
Chernobyl: Ten Years Later

VOLUME IV (SEVENTH EDITION)

Yellowstone Fires: Ten Years Later
China's One-Child Population Policy
Famine Relief in Sudan
The Ndoki Rain Forest
Return to Chernobyl
Fuel-Cell Technology for Transportation
Tobacco Deals
El Niño and Global Warming
The Ad Campaign to Torpedo the Kyoto Summit

Appendix A

Environmental Organizations

This appendix presents a list of selected nongovernmental organizations that are active in environmental matters. Listed are national organizations, as well as some smaller, specialized ones. These organizations offer a variety of fact sheets, brochures, newsletters, publications, educational materials, and annual reports, many of which are available on the World Wide Web. Web addresses are provided in the list. Requests for information should be specific and are best made by e-mail; Web sites have e-mail addresses and query options. Some organizations have internship positions available for those wishing to do work for an environmental group. An exhaustive listing of over 3,000 environmental, governmental, and educational organizations can be found in the *Conservation Directory*, available from the National Wildlife Federation ($70.00).

AMERICAN FARMLAND TRUST. Focuses on the preservation of American farmland. 1200 18th Street, N.W., Suite 800, Washington, DC 20036. http://www.farmland.org

AMERICAN LUNG ASSOCIATION. Research, education, legislation, lobbying, advocacy: indoor and outdoor air pollution effects, smoking. 1740 Broadway, New York, NY 10019. http://www.lungusa.org

AMERICAN RIVERS, INC. Mission is "to preserve and restore America's rivers' systems and to foster a river stewardship ethic." Education, litigation, lobbying: wild and scenic rivers, hydropower relicensing. Information on specific rivers available. 1025 Vermont Ave., N.W., Suite 720, Washington, DC 20005. http://www.amrivers.org

BREAD FOR THE WORLD. National advocacy group that lobbies for legislation dealing with hunger. 50 F Street, N.W., Suite 500, Washington, DC 20001. http://www.bread.org

CENTER FOR SCIENCE IN THE PUBLIC INTEREST. Research and education: alcohol policies, food safety, health, nutrition, organic agriculture. 1875 Connecticut Avenue, N.W., Suite 300, Washington, DC 20009. http://www.cspinet.org

CHESAPEAKE BAY FOUNDATION. Research, education, and litigation: environmental defense and management of Chesapeake Bay and surrounding area. Philip Merrill Environmental Center, 6 Herndon Avenue, Annapolis, MD 21403. http://www.cbf.org

CLEAN WATER ACTION PROJECT. Lobbying, education, research: water quality. 4455 Connecticut Ave., N.W., Suite A300, Washington, DC 20008-2328. http://www.cleanwateraction.org

COMMITTEE FOR THE NATIONAL INSTITUTE FOR THE ENVIRONMENT. Mission is to improve the scientific basis for making decisions about environmental issues. Provides briefings on environmental issues and the legislation addressing them. 1725 K Street, N.W., Suite 212, Washington, DC 20006. http://www.cnie.org

COMMON CAUSE. Lobbying: government reform, energy reorganization, clean air. 1250 Connecticut Ave., N.W., Washington, DC 20036. http://www.commoncause.org

COMMUNITY TRANSPORTATION ASSOCIATION OF AMERICA. Provides technical assistance for rural and specialized transportation systems. 1341 G Street, N.W., Suite 600, Washington, DC 20005. http://www.ctaa.org

CONGRESS WATCH. Lobbying: consumer health and safety, pesticides. 215 Pennsylvania Avenue, S.E., Washington, DC 20003. http://www.citizen.org/congress

CONSERVATION INTERNATIONAL. Education and research to preserve and promote awareness about the world's most endangered biodiversity. 1919 M Street, N.W., Suite 600, Washington, DC 20036 http://www.conservation.org

CORAL REEF ALLIANCE. Works with the diving community and others to promote coral reef conservation around the world. 2014 Shattuck Avenue, Berkeley, CA 94704-1117 http://www.coral.org

CRITICAL MASS ENERGY PROJECT. Research and education: alternative energy and nuclear power (a function of Public Citizen). 1600 20th St., N.W., Washington, DC 20009. http://www.citizen.org/cmep

DEFENDERS OF WILDLIFE. Research, education, and lobbying: endangered species. 1101 14th Street, N.W., #1400, Washington, DC 20005. http://www.defenders.org

DUCKS UNLIMITED, INC. Hunters working to fulfill the annual life cycle needs of North American waterfowl by protecting, enhancing, restoring, and managing important wetlands and associated uplands. One Waterfowl Way, Memphis, TN 38120. http://www.ducks.org/

EARTHWATCH EXPEDITIONS, INC. Environmental research is encouraged by organizing teams to make expeditions to various locations all over the world. Team members contribute to the expenses and spend from several weeks to several months on the site. 3 Clock Tower Place, Suite 100, Box 75, Maynard, MA 01754. http://www.earthwatch.org

ENVIRONMENTAL DEFENSE FUND. Research, litigation, and lobbying: cosmetics safety, drinking water, energy, transportation, pesticides, wildlife, air pollution, cancer prevention, radiation. 1875 Connecticut Ave. N.W., Washington, DC 20009. http://www.edf.org

ENVIRONMENTAL LAW INSTITUTE. Training, educational workshops, and seminars for environmental professionals, lawyers, and judges on institutional and legal issues affecting the environment. 1616 P Street, N.W., Suite 200, Washington, DC 20036. http://www.eli.org

FOREST STEWARDSHIP COUNCIL. Promotes sustainable forest management and a system of certification of sustainable forest products. 1134 29th Street, N.W., Washington, DC 20007. http://www.fscus.org

FREEDOM FROM HUNGER. Develops programs for the elimination of hunger worldwide. 1644 DaVinci Court, Davis, CA 95617. http://www.freefromhunger.org

FRIENDS OF THE EARTH. Research, lobbying: all aspects of energy development, preservation, restoration, and rational use of the earth. 1025 Vermont Ave., N.W., Washington, DC 20005. http://www.foe.org

GLOBAL ACTION PLAN. An international program enlisting families and organizations in sustainable lifestyle changes, with special programs for children. P.O. Box 428, Woodstock, NY 12498. http://www.globalactionplan.org

GREENPEACE, USA, INC. An international environmental organization dedicated to protecting the planet through nonviolent, direct action providing public education, scientific research, and legislative lobbying. 702 H Street, N.W., Washington, DC 20001. http://www.greenpeace.org

HABITAT FOR HUMANITY INTERNATIONAL. Fosters homebuilding and ownership for the poor. Education on housing issues. 121 Habitat Street, Americus, GA 31709-3498. http://www.habitat.org

HEIFER PROJECT INTERNATIONAL. Works throughout the developing world to bring domestic animals to poor families. P.O. Box 8058, Little Rock, AR 72203. http://www.heifer.org

INSTITUTE FOR LOCAL SELF-RELIANCE. Research and education: appropriate technology for community development. 2425 18th Street, N.W., Washington, DC 20009. http://www.ilsr.org

IZAAK WALTON LEAGUE OF AMERICA, INC. Research, education, endowment grants: conservation, air and water quality, streams. 707 Conservation Lane, Gaithersburg, MD 20878. http://www.iwla.org

LAND TRUST ALLIANCE. Works with local and regional land trusts to enhance their work. 1331 H Street, N.W., Suite 400, Washington, DC 20005. http://www.lta.org

LEAGUE OF CONSERVATION VOTERS. Political arm of the environmental community. Works to elect candidates to the U.S. House and Senate who will vote to protect the nation's environment, and holds them accountable by publishing the National Environmental Scorecard each year, which can be ordered for $6 and is free to students. 1920 L Street, N.W., Suite 800, Washington, DC 20036. http://www.lcv.org

LEAGUE OF WOMEN VOTERS OF THE U.S. Education and lobbying, general environmental issues. Publications on groundwater, agriculture, and farm policy, with additional topics available. 1730 M Street, N.W., Suite 1000, Washington, DC 20036. http://www.lwv.org

NATIONAL AUDUBON SOCIETY. Research, lobbying, education, litigation, and citizen action: broad-based environmental issues. 700 Broadway, New York, NY 10003. http://www.audubon.org

NATIONAL PARK FOUNDATION. Education, land acquisition, management of endowments, grant making. National Park Foundation, 1101 17th Street, N.W., Suite 1102, Washington, DC 20036. http://www.nationalparks.org

NATIONAL PARKS CONSERVATION ASSOCIATION. Research and education: parks, wildlife, forestry, general environmental quality. 1300 19th St., N.W., Suite 300, Washington, DC 20036. http://www.npca.org

NATIONAL WILDLIFE FEDERATION. Research, education, lobbying: general environmental quality, wilderness, and wildlife. 8925 Leesburg Pike, Vienna, VA 22184. http://www.nwf.org

NATURAL RESOURCES DEFENSE COUNCIL. Research and litigation: water and air quality, land use, energy, pesticides, toxic waste. 40 West 20th Street, New York, NY 10011. http://www.nrdc.org

NATURE CONSERVANCY. "Preserve plants, animals, and natural communities that represent the diversity of life on Earth by protecting the land and the water they need to survive." 4525 North Fairfax Drive, Suite 100, Arlington, VA 22203. http://www.tnc.org/koa

OXFAM AMERICA. Funding agency for projects to benefit the "poorest of the poor" in South America, Africa, India, Central America, the Caribbean, and the Philippines. Provides whatever resources are needed. 733 15th Street, N.W., Suite 340, Washington, DC 20005. http://www.oxfamamerica.org

PLANNED PARENTHOOD FEDERATION OF AMERICA. Education, services, and research: fertility control, family planning. 810 7th Avenue, New York, NY 10019. http://www.plannedparenthood.org

THE POPULATION INSTITUTE. Education, research, and speaking engagements: population control. 107 2nd Street, N.E., Washington, DC 20002. http://www.populationinstitute.org

POPULATION REFERENCE BUREAU, INC. Organization engaged in collection and dissemination of objective population information. Excellent publications. 1875 Connecticut Avenue, N.W., Suite 520, Washington, DC 20009. http://www.prb.org

RACHEL CARSON COUNCIL, INC. Publication and distribution of information on pesticides and toxic substances, educational conferences, and seminars. 8940 Jones Mill Road, Chevy Chase, MD 20815. http://members.aol.com/rccouncil/ourpage/basic.htm

RAIN FOREST ACTION NETWORK. Information and educational resources: world's rain forests. 221 Pine St., Suite 500, San Francisco, CA 94104. http://www.ran.org

RAINFOREST ALLIANCE. Education, medicinal plants project, timber project to certify "smart wood," and news bureau in Costa Rica. 65 Bleeker St., New York, NY 10012. http://www.rainforest-alliance.org

RENEW AMERICA. "A nationwide clearinghouse for environmental solutions by seeking out and promoting successful programs. We offer positive, constructive models to help communities meet environmental challenges." Many different environmental categories are addressed. 1200 18th Street, N.W., Suite 1100, Washington, DC 20036. http://www.solstice.crest.org/environment/renew_america/index.html

RESOURCES FOR THE FUTURE. Research and education think tank: connects economics and conservation of natural resources, environmental quality. 1616 P Street, N.W., Washington, DC 20036. http://www.rff.org

SIERRA CLUB. Education and lobbying: broad-based environmental issues. 85 Second Street, 2nd Floor, San Francisco, CA 94105. http://www.sierraclub.org

TRUST FOR PUBLIC LAND. Works with citizen groups and government agencies to acquire and preserve open space. 116 New Montgomery, 4th Floor, San Francisco, CA 94105. http://www.igc.org/tpl

UNION OF CONCERNED SCIENTISTS. Connecting citizens and scientists, augments scientific analysis with innovative thinking and committed citizen advocacy to build a cleaner, healthier environment and a safer world. 2 Brattle Square, Cambridge, MA 02238. http://www.ucsusa.org/

U.S. PUBLIC INTEREST RESEARCH GROUP. Research, education, and lobbying, working through state chapters: alternative energy, consumer protection, utilities regulation, public interest. 218 D Street, S.E., Washington, DC 20003. http://www.pirg.org

WATER ENVIRONMENT FEDERATION. Research, education, and lobbying. 601 Wythe Street, Alexandria, VA 22314. http://www.wef.org

WILDERNESS SOCIETY. Research, education, and lobbying: wilderness, public lands. 1615 M St., N.W., Washington, DC. http://www.tws.org

WORLD RESOURCES INSTITUTE. Conducts research and publishes reports for educators, policymakers, and organizations on environmental issues. 10 G Street, N.E., Suite 800, Washington, DC 20002. http://www.wri.org

WORLD WILDLIFE FUND. Preservation of wildlife habitats and protection of endangered species. 1250 Twenty-Fourth Street, N.W., P.O. Box 97180, Washington, DC 20037. http://www.worldwildlife.org

WORLDWATCH INSTITUTE. Research and education: energy, food, population, health, women's issues, technology, the environment. 1776 Massachusetts Avenue, N.W., Washington, DC 20036. http://www.worldwatch.org

ZERO POPULATION GROWTH, INC. Public education, lobbying, and research: population. 1400 16th Street, N.W., Suite 320, Washington, DC 20036. http://www.zpg.org

Appendix B

Units of Measure

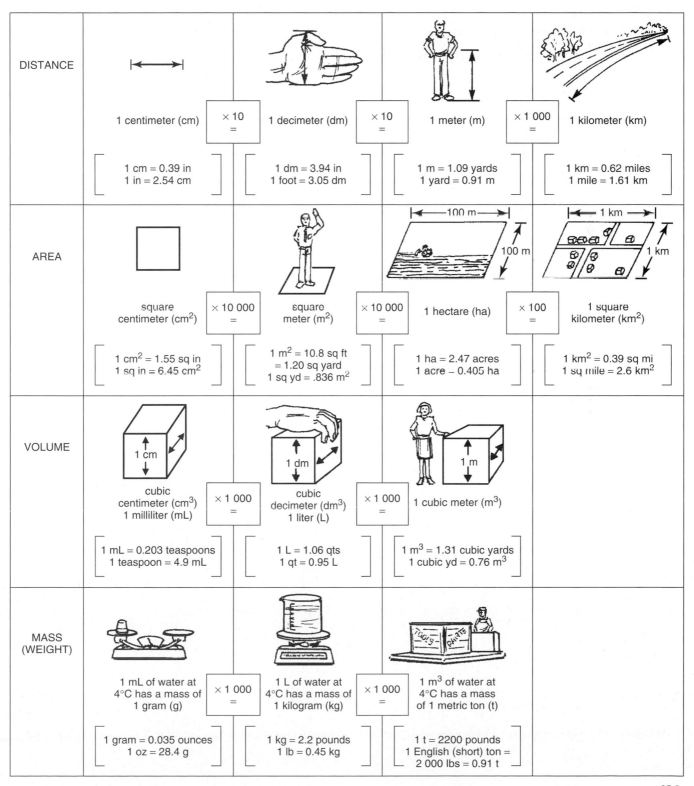

DISTANCE	1 centimeter (cm) — ×10 = — 1 decimeter (dm) — ×10 = — 1 meter (m) — ×1 000 = — 1 kilometer (km)		
	1 cm = 0.39 in 1 in = 2.54 cm	1 dm = 3.94 in 1 foot = 3.05 dm	1 m = 1.09 yards 1 yard = 0.91 m	1 km = 0.62 miles 1 mile = 1.61 km
AREA	square centimeter (cm^2) — ×10 000 = — square meter (m^2) — ×10 000 = — 1 hectare (ha) — ×100 = — 1 square kilometer (km^2)		
	1 cm^2 = 1.55 sq in 1 sq in = 6.45 cm^2	1 m^2 = 10.8 sq ft = 1.20 sq yard 1 sq yd = .836 m^2	1 ha = 2.47 acres 1 acre = 0.405 ha	1 km^2 = 0.39 sq mi 1 sq mile = 2.6 km^2
VOLUME	cubic centimeter (cm^3) 1 milliliter (mL) — ×1 000 = — cubic decimeter (dm^3) 1 liter (L) — ×1 000 = — 1 cubic meter (m^3)		
	1 mL = 0.203 teaspoons 1 teaspoon = 4.9 mL	1 L = 1.06 qts 1 qt = 0.95 L	1 m^3 = 1.31 cubic yards 1 cubic yd = 0.76 m^3	
MASS (WEIGHT)	1 mL of water at 4°C has a mass of 1 gram (g) — ×1 000 = — 1 L of water at 4°C has a mass of 1 kilogram (kg) — ×1 000 = — 1 m^3 of water at 4°C has a mass of 1 metric ton (t)		
	1 gram = 0.035 ounces 1 oz = 28.4 g	1 kg = 2.2 pounds 1 lb = 0.45 kg	1 t = 2200 pounds 1 English (short) ton = 2 000 lbs = 0.91 t	

623

Energy Units and Equivalents

1 Calorie, food calorie, or kilocalorie—The amount of heat required to raise the temperature of one kilogram of water one degree Celsius (1.8°F).

1 BTU (British Thermal Unit)—The amount of heat required to raise the temperature of one pound of water one degree Fahrenheit.

1 joule—A force of one newton applied over a distance of one meter. (A newton is the force needed to produce an acceleration of 1 m per sec per sec to a mass of one kg)

1 Calorie = 3.968 BTUs = 4,186 joules
1 BTU = 0.252 Calories = 1,055 joules

1 therm = 100,000 BTUs
1 quad = 1 quadrillion BTUs

1 watt = standard unit of electrical power

1 watt-hour (wh) = 1 watt for 1 hr. = 3.413 BTUs

1 kilowatt (kw) = 1000 watts

1 kilowatt-hour (kwh) = 1 kilowatt for 1 hr = 3413 BTUs

1 megawatt (Mw) = 1,000,000 watts

1 megawatt-hour (Mwh) = 1 Mw for 1 hr. = 34.13 therms

1 gigawatt (Gw) = 1,000,000,000 watts or 1,000 megawatts

1 gigawatt-hour (Gwh) = 1 Gw for 1 hr. = 34,130 therms

1 horsepower = 0.7457 kilowatt; 1 horsepower-hour = 2,545 BTUs

1 cubic foot of natural gas (methane) at atmospheric pressure = 1,031 BTUs

1 gallon gasoline = 125,000 BTUs

1 gallon No. 2 fuel oil = 140,000 BTUs

1 short ton coal = 25,000,000 BTUs

1 barrel (oil) = 42 gallons

Appendix C

Some Basic Chemical Concepts

Atoms, Elements, and Compounds

All matter, whether gas, liquid, or solid, living or nonliving, or organic or inorganic, is composed of fundamental units called **atoms**. Atoms are extremely tiny. If all the world's people, about 6 billion of us, were reduced to the size of atoms, there would be room for all of us to dance on the head of a pin. In fact, we would only occupy a tiny fraction (about 1/10,000) of the pin's head. Given the incredibly tiny size of atoms, even the smallest particle that can be seen with the naked eye consists of billions of atoms.

The atoms making up a substance may be all of one kind, or they may be of two or more kinds. If the atoms are all of one kind, the substance is called an **element**. If the atoms are of two or more kinds bonded together, the substance is called a **compound**.

Through countless experiments, chemists have ascertained that there are only 94 distinct kinds of atoms that occur in nature. (An additional 14 have been synthesized.) These natural and synthetic elements are listed in Table C–1, together with their chemical symbols. By scanning the table, you can see that a number of familiar substances, such as aluminum, calcium, carbon, oxygen, and iron, are elements; that is, they are a single, distinct kind of atom. However, most of the substances with which we interact in everyday life, such as water, stone, wood, protein, and sugar, are not on the list. Their absence from the list is indicative that they are not elements; rather, they are compounds, which means that they are actually composed of two or more different kinds of atoms bonded together.

Atoms, Bonds, and Chemical Reactions

In chemical reactions, atoms are neither created nor destroyed, nor is one kind of atom changed into another. What occurs in chemical reactions, whether mild or explosive, is simply a rearrangement of the ways in which the atoms involved are bonded together. An oxygen atom, for example, may be combined and recombined with different atoms to form any number of different compounds, but a given oxygen atom always has been, and always will be, an oxygen atom. The same can be said for all the other kinds of atoms. To understand how atoms may bond and undergo rearrangement to form different compounds, it is necessary to examine several concepts concerning the structure of atoms.

Structure of Atoms

Every atom consists of a central core called the **nucleus** (not to be confused with a cell nucleus). The nucleus of an atom contains one or more **protons** and, except for hydrogen, one or more **neutrons**. Surrounding the nucleus are particles called **electrons**.

Each proton has a positive (+) electric charge, and each electron has an equal and opposite negative (−) electric charge. Thus, in any atom the charge of the protons may be balanced by an equal number of electrons, making the whole atom neutral. Neutrons have no charge.

Atoms of all elements have this same basic structure, consisting of protons, electrons, and neutrons. The distinction between atoms of different elements is in the number of protons. The atoms of each element have a characteristic number of protons that is known as the **atomic number** of the element. (See Table C–1.) The number of electrons characteristic of the atoms of each element also differs, in a manner corresponding to the number of protons. The combined total of protons and neutrons in the nucleus of an element is the **mass number**, or **atomic weight**. The general structure of the atoms of several elements is shown in Figure C–1.

The number of protons and electrons in an atom of an element (i.e., the atomic number of the element) determines the chemical properties of the element. The number of neutrons may vary. For example, most carbon atoms have six neutrons in addition to the six protons, as indicated in Figure C–1. But some carbon atoms have eight neutrons. Atoms of the same element that have different numbers of neutrons are known as *isotopes* of the element. The total number of protons plus neutrons is used to define different isotopes. For example, the usual isotope of carbon is referred to as carbon-12, while the isotope with eight neutrons is referred to as carbon-14. The chemical reactivities of different isotopes of the same element are identical. (However, certain other properties may differ.) Many isotopes of various elements prove to be radioactive, such as carbon-14. The atomic weights in Table

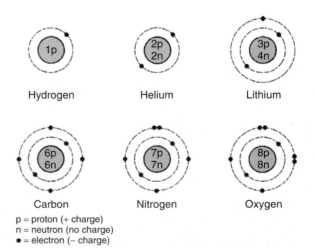

p = proton (+ charge)
n = neutron (no charge)
● = electron (− charge)

FIGURE C-1 *Structure of atoms.* All atoms consist of fundamental particles: protons (*p*), which have a positive electric charge, neutrons (*n*), which have no charge, and electrons, which have a negative charge.

Table C–1   The Elements

Element	Symbol	Atomic Number	Atomic Weight	Element	Symbol	Atomic Number	Atomic Weight
Actinium	Ac	89	(227)	*Meitnerium	Mt	109	(265)
Aluminum	Al	13	27.0	*Mendelevium	Md	101	(256)
*Americium	Am	95	(243)	Mercury	Hg	80	200.6
Antimony	Sb	51	121.8	Molybdenum	Mo	42	95.9
Argon	Ar	18	39.9	Neodymium	Nd	601	44.2
Arsenic	As	33	74.9	Neon	Ne	10	20.2
Astatine	At	85	(210)	Neptunium	Np	93	237.0
Barium	Ba	56	137.3	Nickel	Ni	28	58.7
*Berkelium	Bk	97	(245)	Niobium	Nb	41	92.9
Beryllium	Be	4	9.01	Nitrogen	N	7	14.0
Bismuth	Bi	83	209.0	*Nobelium	No	102	(254)
*Bohrium	Bh	107	(262)	Osmium	Os	76	190.2
Boron	B	5	10.8	Oxygen	O	8	16.0
Bromine	Br	35	79.9	Palladium	Pd	46	106.4
Cadmium	Cd	48	112.4	Phosphorus	P	15	31.0
Calcium	Ca	20	40.1	Platinum	Pt	78	195.1
*Californium	Cf	98	(251)	Plutonium	Pu	94	(242)
Carbon	C	6	12.0	Polonium	Po	84	(210)
Cerium	Ce	58	140.1	Potassium	K	19	39.1
Cesium	Cs	55	132.9	Praseodymium	Pr	59	140.9
Chlorine	Cl	17	35.5	Promethium	Pm	61	(145)
Chromium	Cr	24	52.0	Protoactinium	Pa	91	231.0
Cobalt	Co	27	58.9	Radium	Ra	88	226.0
Copper	Cu	29	63.5	Radon	Rn	86	(222)
*Curium	Cm	96	(245)	Rhenium	Re	75	186.2
*Dubnium	Db	105	(262)	Rhodium	Rh	45	102.9
Dysprosium	Dy	66	162.5	Ruthenium	Ru	44	101.1
*Einsteinium	Es	99	(254)	*Rutherfordium	Rf	104	(261)
Erbium	Er	68	167.3	Samarium	Sm	62	150.4
Europium	Eu	63	152.0	Scandium	Sc	21	45.0
*Fermium	Fm	100	(254)	*Seaborgium	Sg	106	(263)
Fluorine	F	9	19.0	Selenium	Se	34	79.0
Francium	Fr	87	(223)	Silicon	Si	14	28.1
Gadolinium	Gd	64	157.3	Silver	Ag	47	107.9
Gallium	Ga	31	69.7	Sodium	Na	11	23.0
Germanium	Ge	32	72.6	Strontium	Sr	38	87.6
Gold	Au	79	197.0	Sulfur	S	16	32.1
Hafnium	Hf	72	178.5	Tantalum	Ta	73	180.9
*Hassium	Hs	108	(265)	Technetium	Tc	43	98.9
Helium	He	2	4.00	Tellurium	Te	52	127.6
Holmium	Ho	67	164.9	Terbium	Tb	65	158.9
Hydrogen	H	1	1.01	Thallium	Tl	81	204.4
Indium	In	19	114.8	Thorium	Th	90	232.0
Iodine	I	53	126.9	Thulium	Tm	69	168.9
Iridium	Ir	77	192.2	Tin	Sn	50	118.7
Iron	Fe	26	55.8	Titanium	Ti	22	47.9
Krypton	Kr	36	83.8	Tungsten	W	74	183.8
Lanthanum	La	57	138.9	Uranium	U	92	238.0
*Lawrencium	Lr	103	(257)	Vanadium	V	23	50.9
Lead	Pb	82	207.2	Xenon	Xe	54	131.3
Lithium	Li	3	6.94	Ytterbium	Yb	70	173.0
Lutetium	Lu	71	175.0	Yttrium	Y	39	88.9
Magnesium	Mg	12	24.3	Zinc	Zn	30	65.4
Manganese	Mn	25	54.9	Zirconium	Zr	40	91.2

*Elements that do not occur in nature.
Parentheses indicate the most stable isotope of the element.

C–1 are expressed as the average of the isotopes of an element as they occur in nature. The weights listed in parentheses are those of the most stable known isotope.

Bonding Of Atoms

The chemical properties of an element are defined by the ways in which its atoms will react and form bonds with other atoms. By examining how atoms form bonds, we shall see how the number of electrons and protons determines these properties. There are two basic kinds of bonding: (1) **covalent bonding** and (2) **ionic bonding**.

In both kinds of bonding, it is important to recognize that electrons are not randomly distributed around the atom's nucleus. Rather, there are, in effect, specific spaces in a series of layers, or

orbitals, around the nucleus. If an orbital is occupied by one or more electrons, but is not filled, the atom is unstable; it will then tend to react and form bonds with other atoms to achieve greater stability. A stable state is achieved by having all the spaces in the orbital filled with electrons. It is important, however, to keep the charge neutral (i.e., the total number of electrons should be equal to that of the protons).

Covalent Bonding

These two requirements—filling all the spaces and keeping the charge neutral—may be satisfied by adjacent atoms sharing one or more pairs of electrons, as shown in Figure C–2. The sharing of a pair of electrons holds the atoms together in what it is called a **covalent bond**.

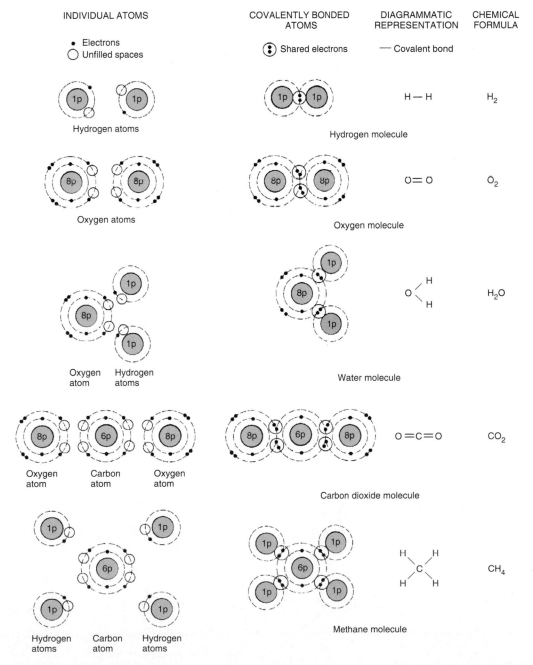

FIGURE C-2 *Atoms to molecules.* Molecules are composed of atoms covalently bonded to each other in stable configurations in which electrons are shared.

Covalent bonding, by satisfying the charge–orbital requirements, leads to discrete units of two or more atoms bonded together. Units of two or more covalently bonded atoms are called **molecules**. A few simple, but important, examples are shown in Figure C–2.

A chemical formula is simply a shorthand description of the number of each kind of atom in a given molecule. The element is given by the chemical symbol, and a subscript following the symbol gives the number of atoms of that element present in the molecule, no subscript being understood as one atom. A molecule with two or more different kinds of atoms may be called a compound, but a molecule composed of a single kind of atom—oxygen (O_2), for example—is still defined as an element.

Only a few elements—carbon, hydrogen, oxygen, nitrogen, phosphorus, and sulfur—have configurations of electrons that lend themselves readily to the formation of covalent bonds. Car-

bon in particular, with its ability to form four covalent bonds, can produce long, straight, or branched chains or rings (Fig. C–3). Thus, an infinite array of molecules can be formed by using covalently bonded carbon atoms as a "backbone" and filling in the sides with atoms of hydrogen or other elements. Accordingly, it is covalent bonding among atoms of carbon and these few other elements that produces all natural organic molecules, those molecules which make up all the tissues of living things, and also synthetic organic compounds such as plastics.

Ionic Bonding

Another way in which atoms may achieve a stable electron configuration is to gain additional electrons to complete the filling of an orbital or to lose excess electrons in completed orbital. In general, the maximum number of electrons that can be gained or lost

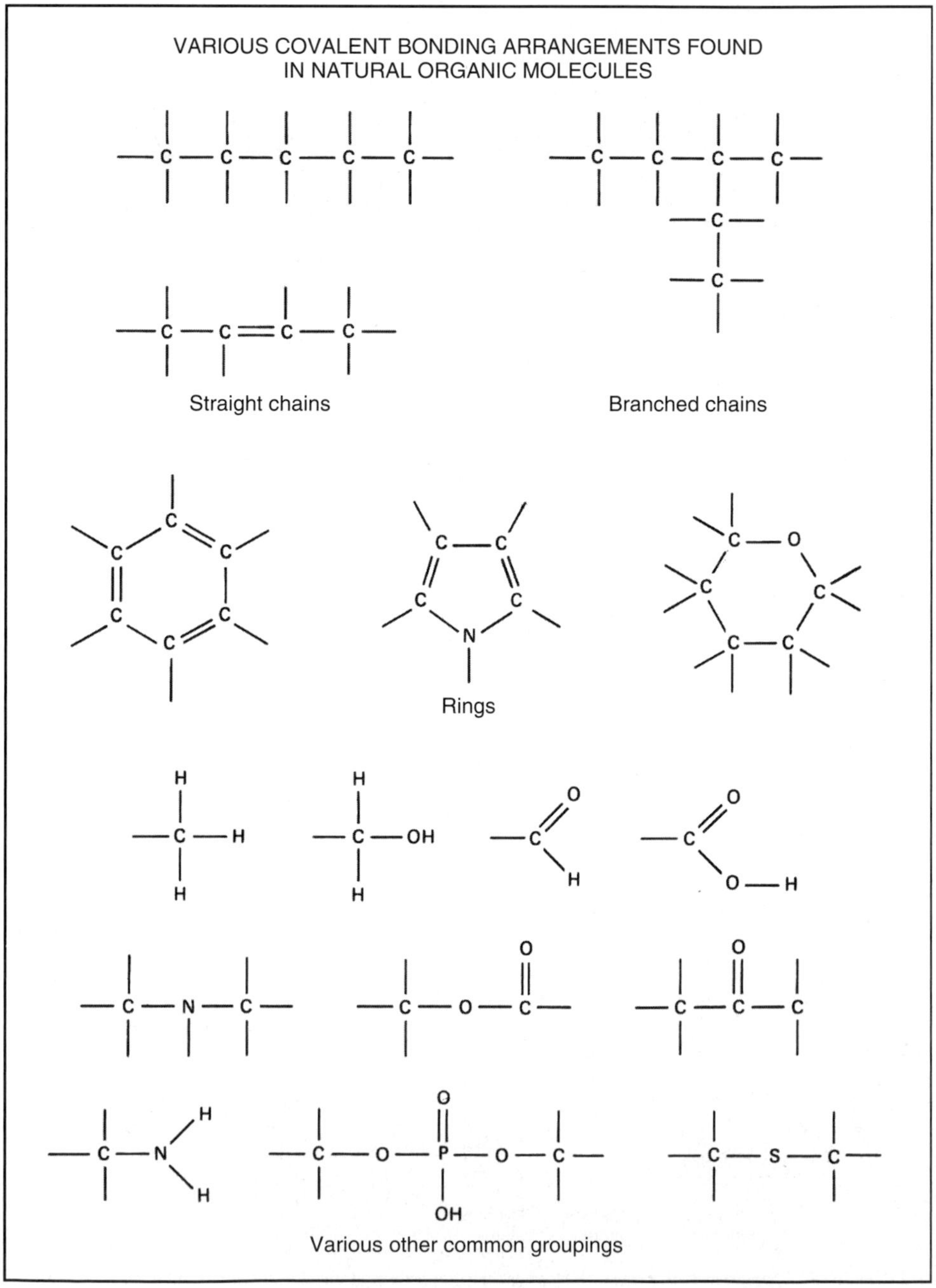

FIGURE C-3 *Covalent bonding and organic molecules.* The ability of carbon and a few other elements to readily form covalent bonds leads to an infinite array of complex organic molecules, which constitute all living things. A few major kinds of groupings are shown here. Note that each element forms a characteristic number of bonds: carbon, 4; nitrogen, 3; oxygen, 2; hydrogen, 1; sulfur, 2; and phosphorus, 5. Bonds left hanging (lines without atoms attached) indicate attachments to other atoms or groups of atoms.

by an atom is three. Therefore, an element's atomic number determines whether one or more electrons will be lost or gained. If an atom's outer orbital is one to three electrons short of being filled, it will always tend to gain additional electrons. If an atom has one to three electrons in excess of its last complete orbital, it will always tend to lose those electrons.

Of course, gaining or losing electrons results in the number of electrons in an atom being greater or less than the number of protons and the atom consequently having an electric charge. The charge will be one negative unit for each electron gained and one positive unit for each electron lost (Fig. C–4). A covalently bonded group of atoms may acquire an electric charge in the same way. An atom or group of atoms that has acquired an electric charge in this way is called an **ion**, positive or negative. Ions are designated by a superscript following the chemical symbol, which denotes the number of positive or negative charges. The absence of superscripts indicates that the atom or molecule is neutral. Some important ions are listed in Table C–2.

Since unlike charges attract, positive and negative ions tend to join and pack together in dense clusters in such a way as to neutralize an atom's overall electric charge. This joining together of ions through the attraction of their opposite charges is called

Table C–2 Ions of Particular Importance to Biological Systems			
Negative (−) Ions		**Positive (+) Ions**	
Phosphate	PO_4^{3-}	Potassium	K^+
Sulfate	SO_4^{2-}	Calcium	Ca^{2+}
Nitrate	NO_3^-	Magnesium	Mg^{2+}
Hydroxyl	OH^-	Iron	Fe^{2+}, Fe^{3+}
Chloride	Cl^-	Hydrogen	H^+
Bicarbonate	HCO_3^-	Ammonium	NH_4^+
Carbonate	CO_3^{2-}	Sodium	Na^+

ionic bonding. The result is the formation of hard, brittle, more or less crystalline substances of which all rocks and minerals are examples (Fig. C–5).

It is significant to note that whereas covalent bonding leads to discrete molecules, ionic bonding does not. Any number and combination of positive and negative ions may enter into an ionic bond to produce crystals of almost any size. The only restriction is that the overall charge of positive ions be balanced by that of neg-

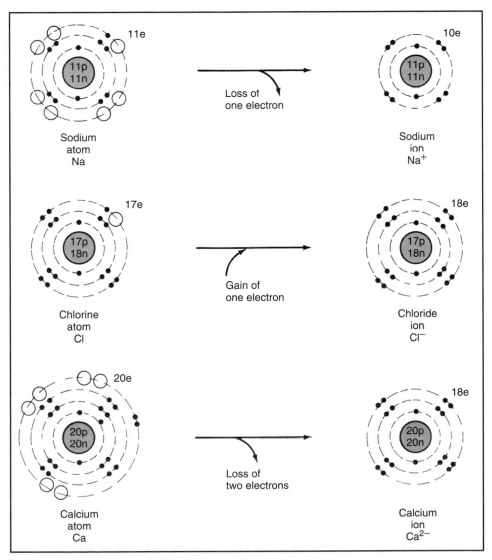

Sodium atom Na → Loss of one electron → Sodium ion Na^+

Chlorine atom Cl → Gain of one electron → Chloride ion Cl^-

Calcium atom Ca → Loss of two electrons → Calcium ion Ca^{2-}

FIGURE C-4 *Formation of ions.* Many atoms will tend to gain or lose one or more electrons in order to achieve a state of complete (electron-filled) orbitals. In doing so, these atoms become positively or negatively charged ions, as indicated.

FIGURE C-5 *Crystals.* Positive and negative ions bond together by their mutual attraction and form crystals.

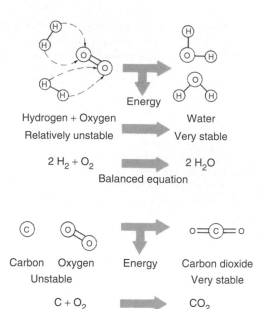

FIGURE C-6 *Stable bonding arrangements.* Some bonding arrangements are more stable than others. Chemical reactions will go spontaneously toward more stable arrangements, releasing energy in the process. However, reactions may be driven in the opposite direction with suitable energy inputs.

ative ions. Thus, substances bonded in an ionic fashion are properly called compounds, but not molecules. When chemical formulas are used to describe such compounds, they define the *ratio* of various elements involved, not specific molecules.

Chemical Reactions and Energy

While atoms themselves do not change, the bonds between atoms may be broken and re-formed, with different atoms producing different compounds or molecules. This is essentially what occurs in all chemical reactions. What determines whether a given chemical reaction will occur? Earlier, we noted that atoms form bonds because they achieve a greater stability by doing so. However, some bonding arrangements may provide greater overall stability than others. Consequently, substances with relatively unstable bonding arrangements will tend to react to form one or more different compounds that have more stable bonding arrangements. Common examples are the reaction between hydrogen and oxygen to produce water and the reaction between carbon and oxygen to produce carbon dioxide (Fig. C–6).

Energy is always released when the constituents of a reaction gain greater overall stability, as indicated in the figure. Thus, when energy is released in a chemical reaction, the atoms achieve more stable bonding arrangements. Therefore, it may be said that chemical reactions always tend to go in a direction that releases energy, as well as giving greater stability.

However, chemical reactions can be made to go in a reverse direction. With suitable energy inputs and under suitable conditions, stable bonding arrangements may be broken and less stable arrangements formed. As described in Chapter 2 of the text, this is the basis of the photosynthesis that occurs in green plants. Light energy causes the highly stable hydrogen–oxygen bonds of water to split and form less stable carbon–hydrogen bonds, thus creating high-energy organic compounds.

Obviously, much more could be said about the involvement of chemistry in environmental science. We would highly recommend a course (or several courses) in chemistry to the student who wants to gain a deeper understanding of how the subject is involved in environmental concepts and issues.

Bibliography and Additional Reading

GENERAL REFERENCES

Andrews, Richard N. L. *Managing the Environment, Managing Ourselves: A History of American Environmental Policy.* New Haven, CT: Yale University Press, 1999.

Brower, Michael, and Warren Leon. *The Consumer's Guide to Effective Environmental Choices: Practical Advice from the Union of Concerned Scientists.* New York: Three Rivers Press, 1999.

Daly, Herman E., and John B. Cobb, Jr. *For the Common Good: Redirecting the Economy toward Community, the Environment, and a Sustainable Future.* Boston: Beacon Press, 1989.

Gore, Al. *Earth in the Balance: Ecology and the Human Spirit.* Boston: Houghton Mifflin, 1992.

Heathcote, Isobel W. *Environmental Problem-Solving: A Case Study Approach.* New York: The McGraw-Hill Companies, Inc., 1997.

Keohane, Robert O., and Marc A. Levy, eds. *Institutions for Environmental Aid.* Cambridge, MA: The MIT Press. 1996.

Kirkby, John, Phil O'Keefe, and Lloyd Timberlake. *The Earthscan Reader in Sustainable Development.* London: Earthscan Publications, Ltd., 1995.

McConnell, Robert L., and Daniel C. Abel. *Environmental Issues: Measuring, Analyzing and Evaluating.* Upper Saddle River, NJ: Prentice Hall, 1999.

President's Council on Sustainable Development. *Sustainable America: A New Consensus for Prosperity, Opportunity and a Healthy Environment.* Washington, DC: U.S. Government Printing Office, 1996.

Prugh, Thomas, Robert Costanza, and Herman Daly. *The Local Politics of Global Sustainability.* Washington, DC: Island Press, 2000.

Switzer, Jacqueline Vaughn. *Environmental Politics: Domestic and Global Dimensions.* New York: St. Martin's Press, 1994.

Watson, Robert T., et al. *Protecting Our Planet, Securing Our Future: Linkages among Global Environmental Issues and Human Needs.* Washington, DC: United Nations Environment Programme, U.S. National Aeronautics and Space Administration, and the World Bank, 1998.

World Bank. *World Development Report, 2000/2001: Attacking Poverty.* New York: Oxford University Press, 2000. An annual report of the World Bank. Available as PDF files on the Web at **http://www.worldbank.org/poverty/wdrpoverty**.

World Resources Institute. *World Resources, 2000–2001.* New York: Oxford University Press, 2000. A biannual compilation of data and reports covering a wide range of environmental issues. The current one is subtitled *People and Ecosystems: The Fraying Web of Life.*

WorldWatch Institute. *State of the World: A Worldwatch Institute Report on Progress toward a Sustainable Society.* Washington, DC: Worldwatch Institute. (A collection of articles from the institute covering an array of environmental issues, published annually since 1984.)

WorldWatch Institute. *Vital Signs, The Trends That Are Shaping Our Future.* Washington, DC: WorldWatch Institute (A presentation of data showing trends that will affect our future, published annually since 1992.)

CHAPTER 1

Introduction: Sustainability, Stewardship, and Sound Science

Alpert, J., et al. "The Sustainable Biosphere Project of SCOPE." *Ambio* 24(2) (Mar 1995): 133–134.

Brown, Lester R. "Launching the Environmental Revolution." In *State of the World, 1992,* 174–190. Washington, DC: WorldWatch Institute, 1992.

Brown, Lester R. "Challenges of the New Century." In *State of the World 2000,* 3–21. Washington, DC: WorldWatch Institute, 2000.

Callicott, J. Baird. *Earth's Insights: A Multicultural Survey of Ecological Ethics from the Mediterranean Basin to the Australian Outback.* Berkeley, CA: Univ. of California Press, 1994.

Conniff, Richard. "Easter Island Unveiled." *National Geographic* 183 (Mar 1993): 54–79.

Cracraft, Joel, and Francesca T. Grifo. *The Living Planet in Crisis: Biodiversity Science and Policy.* New York: Columbia University Press, 1999.

Daily, Gretchen C. "Developing a Scientific Basis for Managing Earth's Life Support Systems. *Conservation Ecology* 3 (Dec 1999): **http://www.consecol.org/Journal/vol3/iss2/index.html** (Jan 9, 2001).

DeWitt, Calvin. *Caring for Creation: Responsible Stewardship of God's Handiwork.* Grand Rapids, MI: Baker Books, 1998.

Diamond, Jared. "Easter's End." *Discover* 16 (Aug 1995): 62–69.

Dobson, Andrew. *Justice and the Environment: Conceptions of Environmental Sustainability and Dimensions of Social Justice.* Oxford, U.K.: Oxford University Press, 1999.

Glick, Daniel. "Having Owls and Jobs Too." *National Wildlife* 33 (Aug/Sept 1995): 8–13.

Greence, George. "Caring for the Earth." *Environment* 36 (Sept 1994): 25–28.

Helvarg, David. *The War against the Green: The Wise Use Movement, the New Right and Anti-Environmental Violence.* San Francisco, CA: Sierra Club Books, 1994

Holloway, Marguerite. "Nurturing Nature." *Scientific American* 270 (Apr 1994): 98–108.

Homer-Dixon, Thomas F., et al. "Environmental Change and Violent Conflict." *Scientific American* 268 (Feb 1993): 38–45.

Kates, Robert W. "Sustaining Life on the Earth." *Scientific American* 271 (Oct 1994): 114–125.

Kenny-Gilday, Cindy, et al. "The Question of Sustainable Development." *Nature Conservancy* 45 (Jan/Feb 1995): 10–15.

Knopman, Debra S., Megan M. Susman, and Marc K. Landy. "Civic Environmentalism: Tackling Tough Land-Use Problems with Innovative Governance." *Environment* 41 (Dec 1999): 24–32.

Meadows, Donella, et al. *Beyond the Limits.* Post Mills, VT: Chelsea Green Publishing, 1992.

Muschett, F. Douglas. *Principles of Sustainable Development.* Delray Beach, FL: St. Lucie Press, 1997.

National Commission on the Environment. *Choosing a Sustainable Future.* Washington, DC: Island Press, 1993.

Oelschlaeger, Max. *Caring for Creation.* New Haven, CT: Yale University Press, 1994.

Oppenheimer, Michael. "Context, Connection, and Opportunity in Environmental Problem Solving." *Environment* 37 (June 1995): 10–15, 34–38.

Orians, Gordon H. "Ecological Concepts of Sustainability." *Environment* 32 (Nov 1990): 21–24.

Prance, Ghillean. *The Earth under Threat: A Christian Perspective.* Glasgow: Wild Goose Publications, 1996.

President's Council on Sustainable Development. *Towards a Sustainable America: Advancing Prosperity, Opportunity, and a Healthy Environment for the 21st Century.* Washington, DC: U.S. Government Printing Office, 1999.

Ruckelshaus, William D. "Toward a Sustainable World." *Scientific American* (Sept 1989): 166–175.

Sitarz, Daniel, ed. *Agenda 21: The Earth Summit Strategy to Save Our Planet.* Boulder, CO: Earth Press, 1993.

Smith, Courtland L. "Assessing the Limits to Growth." *BioScience* 45 (July/Aug 1995): 478–483.

White, Allen L. "Sustainability and the Accountable Corporation: Society's Rising Expectations of Business." *Environment* 41 (Dec 1999): 30–43.

World Commission on Environment and Development. *Our Common Future.* New York: Oxford University Press, 1987.

Young, Louise B. "Easter Island: Scary Parable." *World Monitor* (Aug 1991): 40–45.

CHAPTER 2
Ecosystems: Units of Sustainability

Acharya, Anjali. "The Fate of the Boreal Forests." *WorldWatch* 8 (May/June 1995): 20–29.

Audesirk, Teresa, and Gerald Audesirk. *Biology: Life on Earth*, 5th ed. Upper Saddle River, NJ: Prentice Hall, 1999.

Barnes, Burton V., et al. *Forest Ecology*, 4th ed. New York: John Wiley & Sons, 1998.

Begon, Michael, et al. *Ecology: Individuals, Populations and Communities*. 3d ed. Boston: Blackwell Scientific, Inc., 1998.

Bertness, Mark D. *The Ecology of Atlantic Shorelines*. Sunderland, MA: Sinauer Associates, Inc., 1999.

Chadwick, Douglas. "Ndkoki—the Last Place on Earth." *National Geographic* 188 (July 1995): 22–45.

Chadwick, Douglas H., and Jim Brandenburg. "The American Prairie." *National Geographic* 184 (Oct 1993): 90–119.

Christopherson, Robert W. *Geosystems*, 4th ed. Upper Saddle River, NJ: Prentice Hall, 2000.

Cloud, Preston. "The Biosphere." *Scientific American* 249 (Sept 1983): 176–189.

Ecology, Ecological Monographs, Ecological Applications, Conservation Ecology, and *Issues in Ecology*. Journals published regularly by the *Ecological Society of America*. Rich source of ecological information. Consult Web site **http://esa.sdsc.edu/**.

Krebs, Charles J. *Ecology: The Experimental Analysis of Distribution and Abundance*, 4th ed. New York: HarperCollins, 1994.

Leakey, Mauve, and Alan Walker. "Early Hominid Fossils from Africa." *Scientific American* 276 (June 1997): 74–79.

Pickett, S.T.A., and M. L. Cadenasso. "Landscape Ecology: Spatial Heterogeneity in Ecological Systems. *Science* 269 (July 21, 1995): 331–334.

Pringle, Heather. "The Slow Birth of Agriculture." *Science* 282 (Nov 20, 1998): 1446–1450.

Risser, Paul G. "The Status of the Science Examining Ecotones." *BioScience* 45 (May 1995): 318–325.

Root, Terry, and Stephen H. Schneider. "Ecology and Climate: Research Strategies and Implications." *Science* 269 (July 21, 1995): 334–341.

Solbrig, Otto T., and Michael D. Young. "Toward a Sustainable and Equitable Future for Savannas." *Environment* 34 (Apr 1992): 6–15.

Stiling, Peter. *Ecology: Theories and Applications*. 3d ed. Upper Saddle River, NJ: Prentice Hall, 1999.

Tattersall, Ian. "Out of Africa Again…and Again?" *Scientific American* 276 (April 1997): 60–67.

The Nature Conservancy. *Terrestrial Vegetation of the United States. Volume I: The National Vegetation Classification System: Development, Status and Applications*. 1998. **http://consci.tnc.org/library/pubs/class/index.html Volume I:** (Jan 9, 2001).

U.S. Geological Survey. *Status and Trends of the Nation's Biological Resources*. Washington, DC: U.S. Government Superintendent of Documents, 1998.

Western, David. *In the Dust of Kilimanjaro*. Washington, DC: Island Press/Shearwater Books, 1997.

Wiewandt, Thomas A. "Desert Flowers—Now You See 'em, Now You Don't." *Smithsonian* 25 (Mar 1995): 78–93.

World Wildlife Fund. *Living Planet: Preserving Edens of the Earth*. New York: Crown Publishers, Inc., 1999.

CHAPTER 3
Ecosystems: How They Work

Bazzaz, Fakhri, and Eric D. Fajer. "Plant Life in a CO_2-Rich World." *Scientific American* 266 (Jan 1992): 68–77.

Boyce, Mark S., and Alan Haney, eds. *Ecosystem Management: Applications for Sustainable Forest and Wildlife Resources*. New Haven, CT: Yale University Press, 1997.

Chapin, F. Stuart, et al. "Biotic Control over the Functioning of Ecosystems." *Science* 277 (July 25, 1997): 500–504.

Costanza, Robert, et al. "The Value of the World's Ecosystem Services and Natural Capital." *Nature* 387 (May 15, 1997): 253–260.

Croke, Vicki. "Thorns of Plenty." *International Wildlife* 29 (November/December 1999), 34–43.

Daley, Gretchen C., et al. "Ecosystem Services: Benefits Supplied to Human Societies by Natural Ecosystems. *Issues in Ecology* 2 (Spring 1997): 1–16.

Field, Christopher B., et al. "Primary Production of the Biosphere: Integrating Terrestrial and Oceanic Components." *Science* 281 (July 10, 1998), 237–240.

Kirshner, Robert P. "The Earth's Elements." *Scientific American* 271 (Oct 1994): 58–65.

Kreeyer, Karen Young, and John Gamble. "Plankton Pyramid." *Sea Frontier* 41.2 (1995): 32–35.

Morell, Virginia. "Counting Creatures of the Serengeti, Great and Small." *Science* 278 (December 19, 1997): 2058–2060.

Naeem, Shahid. "Biodiversity and Ecosystem Functioning: Maintaining Natural Life Support Processes." *Issues in Ecology* 4 (Fall 1999): 1–11. **http://esa.sdsc.edu/issues.htm** (Jan 9, 2001).

Nelson, Mark, et al. "Using a Closed Ecological System to Study Earth's Biosphere." *BioScience* 43 (Apr 1993): 225–236.

Prugh, Thomas, et al. *Natural Capital and Human Economic Survival*. Sunderland, MA: Sinauer Associates, Inc., 1995.

Sinclair, A. R. E., and Peter Arcese, eds. *Serengeti II: Dynamic Management and Conservation of an Ecosystem*. Chicago: University of Chicago Press, 1995.

Smil, Vaclav. "Global Population and the Nitrogen Cycle." *Scientific American* 277 (July 1997): 76–82.

Tilman, David. "Biodiversity: Population versus Ecosystem Stability." *Ecology* 77, no. 2 (1996): 350–363.

U.S. Global Change Research Program. *A U.S. Carbon Cycle Science Plan: A Report of the Carbon and Climate Working Group*. Washington, DC: U.S. Global Change Research Program, 1999.

Vitousek, Peter M., Harold A. Mooney, Jane Lubchenko, and Jerry M. Melillo. "Human Domination of Earth's Ecosystems." *Science* 277 (July 25, 1997): 494–499.

Vitousek, Peter M., et al. "Human Alteration of the Global Nitrogen Cycle: Causes and Consequences." *Issues in Ecology* 1 (Spring 1997): 1–15.

Vogel, Gretchen. "An Institute for Planet Earth." *Science* 280 (May 22, 1998): 1182–1185.

CHAPTER 4
Ecosystems: Population and Succession

Alcover, Josep Antoni, and Miquel McMinn. "Predators of Vertebrates on Islands." *BioScience* 44 (Jan 1994): 12–18.

Askins, Robert A. "Hostile Landscapes and the Decline of Migratory Songbirds." *Science* 267 (Mar 1995): 1956–1958.

Bader, Mike, et al. "Fires of '88." Series of articles on the Yellowstone fires of 1988. *Alliance for the Wild Rockies*, Web site **http://www.wildrockies.org/fires-of-88/** (Jan 9, 2001).

Bright, Chris. "Biological Invasions: The Spread of the World's Most Aggressive Species." *World Watch* 8 (Jul/Aug 1995): 10–19.

Carpenter, Stephen R., and Kathryn L. Cottingham. "Resilience and Restoration of Lakes." *Conservation Ecology* 1 (June 1997). **http://www.consecol.org/Journal/vol1/iss1/art2/index.html** (Jan 9, 2001).

Cochrane, Mark A., et al. "Positive Feedbacks in the Fire Dynamic of Closed Canopy Tropical Forests." *Science* 284 (June 11, 1999): 1832–1835.

Dobson, Andy P., A. D. Bradshaw, and A. J. M. Baker. "Hopes for the Future: Restoration Ecology and Conservation Biology." *Science* 277 (July 25, 1997): 515–522.

Hooper, David U., and Peter M. Vitousek. "The Effects of Plant Composition and Diversity on Ecosystem Processes." *Science* 277 (August 29, 1997): 1302–1305.

Hubbell, S. P., et al. "Light-Gap Disturbances, Recruitment Limitation, and Tree Diversity in a Neotropical Forest." *Science* 283 (January 22, 1999): 554–557.

Hudson, Peter J., Andy Dobson, and Dave Newborn. "Prevention of Population Cycles by Parasite Removal." *Science* 282 (December 18, 1998): 2256–2258.

Illius, A. W., and T. G. O'Connor. "On the Relevance of Nonequilibrium Concepts to Arid and Semiarid Grazing Systems." *Ecological Applications* 9 (3), 1999: 798–813.

Klein, David R. "The Introduction, Increase, and Crash of Reindeer on St. Matthew Island." *Journal of Wildlife Management* 32 (Apr 1968): 350–367.

Krebs, Charles J., et al. "Impact of Food and Predation on the Snowshoe Hare Cycle." *Science* 269 (Aug 25, 1995): 1112–1115.

Line, Les. "Curse of the Cowbird." *National Wildlife* 32 (Dec/Jan 1994): 40–45.

Mack, Richard, et al. "Biotic Invasions: Causes, Epidemiology, Global Consequences and Control. *Issues in Ecology* 5 (Spring 2000): 1–20. **http://esa.sdsc.edu/issues.htm** (Jan 9, 2001).

Malecki, Richard A., et al. "Biological Control of Purple Loosestrife." *BioScience* 43 (Nov 1993): 680–686.

McLaren, B. E., and Peterson, R. O. "Wolves, Moose, and Tree Rings on Isle Royale." *Science* 266 (Dec 1994): 1555–1558.

McLean, Herbert. "Fighting Fire with Fire." *American Forests* (Jul/Aug 95): 13–18, 56–63.

Mills, Edward L., et al. "Exotic Species and the Integrity of the Great Lakes." *BioScience* 44 (Nov 1994): 666–676.

Navarette, Sergio A., and Bruce A. Menge. "Keystone Predation and Interaction Strength: Interactive Effects of Predators on Their Main Prey." *Ecological Monographs* 66 (1996): 409–429.

Newhouse, Joseph R. "Chestnut Blight." *Scientific American* (July 1990): 106–111.

Noss, Reed F., and J. Michael Scott. "Ecosystem Protection and Restoration: The Core of Ecosystem Management." Ch. 12 in *Ecosystem Management: Applications for Sustainable Forest and Wildlife Resources*, Mark S. Boyce and Alan Haney, eds. New Haven, CT: Yale University Press, 1997.

Reice, Seth. "Nonequilibrium Determinants of Biological Community Structure. *American Scientist* 82 (Sept–Oct 1994): 424–435.

Shafer, Craig L. "Values and Shortcomings of Small Reserves." *BioScience* 15 (Feb 1995). 80–88.

CHAPTER 5
Ecosystems and Evolutionary Change

Allegre, Claude J., and Stephen H. Schneider. "The Evolution of the Earth." *Scientific American* 271 (Oct 1994): 66–75.

Amabile-Cuevas, Carlos F., et al. "Antibiotic Resistance." *American Scientist* 83 (July/Aug 1995): 320–329.

Benton, J. "Diversification and Extinction in the History of Life." *Science* 268 (Apr 1995): 52–58.

Broecker, Wallace S., and George Denton. "What Drives Glacial Cycles?" *Scientific American* 262 (Jan 1990): 48.

Christopherson, Robert W. *Geosystems*, 4th ed. Upper Saddle River, NJ: Prentice Hall, 2000.

Clouser, Roy A. "Genesis: On the Origin of the Human Race." *Perspectives on Science and Christian Faith* 43 (Mar 1991): 2–13.

Coppens, Yves. "East Side Story: The Origin of Humankind." *Scientific American* 270 (May 1994): 88–95.

Dalziel, Ian W. D. "Earth before Pangea." *Scientific American* 272 (Jan 1995): 58–63.

Doolittle, W. Ford. "Phylogenetic Classification and the Universal Tree." *Science* 284 (June 25, 1999): 2124–2128.

de Duve, Christian. "The Beginning of Life on Earth." *American Scientist* 83 (Sept 95): 428–437.

Eldedge, N., and S. J. Gould. "Punctuated Equilibria: An Alternative to Phyletic Gradualism," pp. 82–115 in Schopf, Thomas J., ed. *Models in Paleobiology*. San Francisco: Freeman, 1972.

Ensrink, Martin. "Life on the Edge: Rainforest Margins May Spawn Species. *Science* 276 (June 20, 1997): 1791–1792.

Freeman, Scott, and Jon C. Herron. *Evolutionary Analysis*. Upper Saddle River, NJ: Prentice Hall, 1998.

Gibbons, Ann. "A New Face for Human Ancestors." *Science* 276 (May 30, 1997): 1331–1333.

Gittleman, John L. "Are the Pandas Successful Specialists or Evolutionary Failures?" *BioScience* 44 (July/Aug 1994): 456–464.

Gore, Rick, and O. Louis Mazzatenta. "Explosion of Life: The Cambrian Period." *National Geographic* 184 (Oct 1993): 120–136.

Gould, Stephen Jay. "The Evolution of Life on the Earth." *Scientific American* 271 (Oct 1994): 84–91.

Grant, Peter R. "Natural Selection and Darwin's Finches." *Scientific American* 265 (Oct 1991): 82–87.

Jablonski, David. "The Future of the Fossil Record." *Science* 284 (June 25, 1999): 2114–2116.

Kerr, Richard A. "Huge Impact Tied to Mass Extinction." *Science* 257 (Aug 1992): 878–880.

Levinton, Jeffrey S. "The Big Bang of Animal Evolution." *Scientific American* 267 (Nov 1992): 84–93.

May, Robert M. "How Many Species Inhabit the Earth?" *Scientific American* 267 (Oct 1992): 42–49.

Mayr, Ernst. *One Long Argument*. Cambridge, MA: Harvard University Press, 1991.

Powledge, Fred. "The Food Supply's Safety Net." *BioScience* 45 (Apr 1995): 235–243.

Sereno, Paul C. "Dinosaurs and Drifting Continents." *Natural History* (Jan 1995): 40–47.

Tarbuck, Edward J., and Frederick K. Lutgens. *Earth Science*, 9th ed. Upper Saddle River, NJ: Prentice Hall, 2000.

Tuxill, John. "Losing Strands in the Web of Life: Vertebrate Declines and the Conservation of Biological Diversity. WorldWatch Paper *141*. Washington, DC: WorldWatch Institute, 1998.

Weinberg, Steven. "Life in the Universe." *Scientific American* 271 (Oct 1994): 44–51.

Wilson, Edward O., et al., eds. *Biodiversity*. Washington, DC: National Academy Press, 1988.

Wright, Richard T. *Biology through the Eyes of Faith*. San Francisco: Harper/Collins, 1989.

CHAPTER 6
The Human Population: Demographics

Bender, William, and Margaret Smith. "Population, Food, and Nutrition." *Population Bulletin* 51 (Dec 1996). Washington, DC: The Population Reference Bureau.

Bongarts, John "Demographic Consequences of Declining Fertility." *Science* 282 (Oct 16, 1998): 419–420.

Brown, Lester R., Gary Gardner, and Brian Halweil. "Beyond Malthus: Sixteen Dimensions of the Population Problem." *Worldwatch Paper 143*. Washington, DC: WorldWatch Institute, 1998.

Daily, Gretchen, et al. "Optimum Human Population Size." *Population and Environment* 15(6) (July 1994): 469–475.

Gelbard, Alene, et al. "World Population Beyond Six Billion." *Population Bulletin* 54 (Mar 1999). Washington, DC: The Population Reference Bureau.

Hinrichsen, Don. "6,000,000 Consumption Machines." *International Wildlife* 29 (Sept/Oct 1999): 22–29.

Horiuchi, Shiro. "World Population Growth Rate." *Population Today* 21 (June 1993): 6–7.

Kates, Robert W. "Population and Consumption: What We Know, What We Need to Know." *Environment* 42 (Apr 2000): 10–19.

Livernash, Robert, and Eric Rodenburg. "Population Change, Resources, and the Environment." *Population Bulletin* 53 (Jan 1998). Washington, DC: The Population Reference Bureau.

McDevitt, Thomas M., et al. *World Population Profile, 1998*. Washington, DC: U.S. Government Printing Office, 1999. **http://www.ccnsus.gov/ipc/www/world.html** (Jan 12, 2001).

McFalls, Joseph A., Jr. *Population: A Lively Introduction*. 3d ed. Washington, DC: *Population Bulletin* 53 (Sept 1998). The Population Reference Bureau.

McGranahan, Gordon, and Jacob Songsore. "Wealth, Health, and the Urban Household: Weighing Environmental Burdens in Accra, Jakarta, and São Paulo." *Environment* 36 (July/Aug 1994): 4–11.

Pimentel, David, et al. "Natural Resources and an Optimum Human Population." *Population and Environment* 15(5) (May 1994): 347–370.

Population Division, Department of Economic and Social Affairs, United Nations. *The World at Six Billion.* New York: United Nations, 1999. **http://www.undp.org/popin/** (Jan 12, 2001)

Population Reference Bureau. *2000 World Population Data Sheet.* Washington, DC: Population Reference Bureau, 2000. The Population Reference Bureau annually publishes a summary of demographic data from the countries and regions of the world.

Saltz, I. S. "Income Distribution in the Third World: Its Estimation via Proxy Data." *American Journal of Economics and Sociology* 54.1 (Jan 1995): 15–32.

U.S. Commission on Immigration Reform. *Becoming an American: Immigration and Immigrant Policy.* Washington, DC: U.S. Government Printing Office, 1997 (also available on the Internet at **http://www.utexas.edu/lbj/uscir/.**

World Bank. *World Development Report, 2000/2001: Attacking Poverty.* New York: Oxford University Press, 2000. An annual report on population and development by the World Bank. Available as PDF files on the Web at **http://www.worldbank.org/poverty/wdrpoverty/.**

CHAPTER 7
Issues in Population and Development

Ashford, Lori S. "New Perspectives on Population: Lessons from Cairo." *Population Bulletin* 50 (Mar 1995).

Bartlett, Sheridan, et al. *Cities for Children: Children's Rights, Poverty, and Urban Management.* London: Earthscan Publications, 1999.

Brockerhoff, Martin P. "An Urbanizing World." *Population Bulletin* 55 (3) (Sept 2000).

Brown, Lester R., Gary Gardner, and Brian Halweil. *Beyond Malthus: Sixteen Dimensions of the Population Problem.* Worldwatch Paper 143. Washington, DC: WorldWatch Institute, 1998.

Brown, Lester R., Gary Gardner, and Brian Halweil. "Breaking Out or Breaking Down." *WorldWatch* 12 (5) (Sept/Oct 1999): 20–29.

Caldwell, John C., and Pat Caldwell. "The African AIDS Epidemic." *Scientific American* 274 (Mar 1996): 62–67.

Cohen, Joel E. *How Many People Can the Earth Support?* New York: W.W. Norton & Co., 1995.

Dasgupta, Partha S. "Population, Poverty and the Local Environment." *Scientific American* 272 (Feb 1995): 40–45.

Ewart, Wendy R., and Beverly Winikoff. "Toward Safe and Effective Medical Abortion." *Science* 281 (July 24, 1998): 520–521.

Feldman, Linda. "Why Abortion Is the Albatross of U.S. Politics." *The Christian Science Monitor* (Apr 30 1997): 1, 10.

Horiuchi, Shiro. "Stagnation in the Decline of the World Population Growth Rate during the 1980s." *Science* 257 (August 7, 1992): 761–765.

Jacobson, Jodi L. "Gender Bias: Roadblock to Sustainable Development." WorldWatch Paper 110. Washington, DC: WorldWatch Institute, 1992.

Kahn, Peter H., Jr. *The Human Relationship with Nature: Development and Culture.* Cambridge, MA: MIT Press, 1999.

Kelber, Mim. "The Women's Environment and Development Organization." *Environment* 36 (Oct 1994): 43–45.

Martin, Philip, and Elizabeth Midgley. "Immigration to the United States." *Population Bulletin* 54 (June 1999). Washington, DC: The Population Reference Bureau.

McIntosh, Alison C., and Jason L. Finkle. "The Cairo Conference on Population and Development." *Population and Development Review* 21(2) (June 1995): 223–260.

Mortimore, Michael, and Mary Tiffen. "Population Growth and a Sustainable Environment: The Machakos Story." *Environment* 36 (Oct 1994): 10–20.

Narayan, Deepa, et al. *Voices of the Poor: Can Anyone Hear Us?* New York: Oxford University Press, 2000. Also at **http://www.worldbank.org/poverty/voices/reports.htm/cananyone** (Jan 12, 2001).

Narayan, Deepa, et al. *Voices of the Poor: Crying Out for Change.* New York: Oxford University Press, 2000. Also at **http://www.worldbank.org/poverty/voices/reports.htm/cananyone** (Jan 12, 2001).

Pearce, David, et al. "Debt and the Environment." *Scientific American* 272 (June 1995): 52–57.

Pollard, Kelvin M., and William P. O'Hare. "America's Racial and Ethnic Minorities." *Population Bulletin* 54 (3), Sept 1999.

Riche, Martha Farnsworth. "America's Diversity and Growth: Signposts for the 21st Century." *Population Bulletin* 55 (2), June 2000.

Schettler, Ted, et al. *Generations at Risk: Reproductive Health and the Environment.* Cambridge, MA: MIT Press, 1999.

Schwartländer, Bernard, et al. "AIDS in a New Millennium." *Science* 289 (July 7, 2000): 64–67.

Sen, Gita. "The World Programme of Action: A New Paradigm for Population Policy." *Environment* 37 (Jan/Feb 1995): 10–15.

Serageldin, Ismail, and Andrew Steer, eds. *Making Development Sustainable: From Concepts to Action.* Washington, DC: World Bank, 1994.

Thirunarayanapuram, Desikan. "Resource Conservation, Population and Sustainable Development." *Toward the 21st Century* (#3, 1997), Population Institute.

United Nations, et al. *A Better World for All.* 2000. **http://www.paris21.org/betterworld/home.htm** (Jan 12, 2001).

United Nations Population Fund (UNFPA). The International Conference on Population and Development—a Five-Year Update. **http://www.unfpa.org/ICPD/ICPD.HTM.**

World Bank. *World Development Report, 2000/2001: Attacking Poverty.* New York: Oxford University Press, 2000. An annual report of the World Bank. Available as PDF files on the Web at **http://www.worldbank.org/poverty/wdrpoverty/.**

World Bank. *Five Years after Rio: Innovations in Environmental Policy.* Environmentally Sustainable Development Studies and Monographs Series, No. 18. Washington, DC: World Bank, 1997.

Yunus, Muhammad. *The Grameen Bank. Scientific American* (Nov 1999), 114–119.

CHAPTER 8
Soil and the Soil Ecosystem

Ambio. *Royal Swedish Academy of Sciences.* Eight issues per year. Entire Issue of Feb 1998 (Volume 26, Number 1) devoted to soil problems.

Battersby, Simon, and Tim Forsyth. "Fighting Back: Human Adaptations in Marginal Habitats." *Environment* 41 (July/Aug 1999): 6–11, 25–30.

Belsie, Laurent. "American Farmers at the Brink." *Christian Science Monitor* (Sept 24, 1998): 1, 10.

Biocycle: Journal of Composting and Recycling. Published since 1960, a monthly journal that provides feature articles on various aspects of composting and recycling.

Bray, Francesca. "Agriculture for Developing Nations." *Scientific American* 271 (July 1994): 30–37.

Brookfield, Harold, and Christine Padoch. "Appreciating Agrodiversity: A Look at the Dynamism and Diversity of Indigenous Farming Practices." *Environment* 36 (June 1994): 6–11, 37–43.

Cody, Betsy, and Pamela Baldwin. "Grazing Fees and Rangeland Management." *CRS Issue Brief* 96006, 1998. Committee for the National Institute for the Environment **http://www.cnie.org/nle/ag-17.html** (Jan 12, 2001).

Crosson, Pierre. "Will Erosion Threaten Agricultural Productivity?" *Environment* 39 (Apr 1997): 4–9, 29–31.

Dodd, Jerrold L. "Desertification and Degradation in Sub-Saharan Africa." *BioScience* 44 (Jan 1994): 28–34.

Durning, Alan B., and Holly B. Brough. "Taking Stock: Animal Farming and the Environment." *Worldwatch Paper* 103. Washington, DC: WorldWatch Institute, 1991.

Farmer-Centered Agricultural Resource Management Programme (FARM). A United Nations Web site providing information on sustainable agriculture around the world. **http://www.unchina.org/undp/regional/html/ras92078.html** (Jan 12, 2001).

Gardner, Gary. *Shrinking Fields: Cropland Loss in a World of Eight Billion.* WorldWatch Paper 131 (July 1996). Washington, DC: WorldWatch Institute.

Hafeez, Shaheena (ed.). "Selected Technologies for Land and Soil Management in the Cold and Dry Regions of the Hindu Kush–Himalayas." *Arid Lands Newsletter* 46 (Fall/Winter 1999). World Wide

Web **http://ag.arizona.edu/OALS/ALN/aln46/aln46toc.html** (Jan 12, 2001).

Hoekstra, Thomas W., and Moshe Shachak. *Arid Lands Management: Towards Ecological Sustainability*. Urbana, IL: Univ. of Illinois Press, 1999.

Kerr, Richard. "The Sahara Is Not Marching Southward." *Science* 281 (July 31, 1998): 633–634.

Milton, Suzanne J., et al. "A Conceptual Model of Arid Rangeland Degradation." *BioScience* 44 (Feb 1994): 70–76.

Natural Resources Conservation Service. *1997 National Resources Inventory*. USDA (updated Dec 2000) **http://www.nhq.nrcs.usda.gov/ NRI/1997/** (Jan 12, 2001).

Paarlberg, Robert L. "The Politics of Agricultural Resource Abuse." *Environment* 36 (Oct 1994): 6–9.

Page, Jake. "Ranchers Form a 'Radical Center' to Protect Wide-Open Spaces." *Smithsonian* (June 1997): 50–61.

Pimentel, David, et al. "Environmental and Economic Costs of Soil Erosion and Conservation Benefits." *Science* 267 (Feb 24, 1995): 1117–1123.

Pretty, J. N., et al. "Agricultural Regeneration in Kenya: The Catchment Approach to Soil and Water Conservation." *Ambio* 24(1) (Feb 1995): 7–15.

Speth, J. G. *Towards an Effective and Operational International Convention on Desertification*. New York: International Negotiating Committee, International Convention on Desertification, United Nations, 1994.

Tanji, Kenneth, et al. "Selenium in the San Joaquin Valley." *Environment* 28 (July/Aug 1986): 6–39.

Trimble, Stanley W., and Pierre Crosson. "U.S. Soil Erosion Rates: Myth and Reality." *Science* 289 (July 14, 2000): 248–250.

UN Convention to Combat Desertification. A UN convention with ongoing work in promoting action to combat desertification. **http://www.fao.org/desertification** (Jan 12, 2001).

Wolkomir, Richard. "Enearthing Secrets Locked Deep inside Each Fistful of Soil. *Smithsonian* (March 1997): 74–82.

CHAPTER 9
Water: Hydrologic Cycle and Human Use

American Rivers. *The Nation's Most Endangered Rivers of 2000*. American Rivers, Inc., **http://amrivers.actioni.net/template2. asp?cat=2page=22id=905filter=-1** (Jan 16, 2001).

Baylery, Peter. B. "Understanding Large River–Floodplain Ecosystems." *BioScience* 45 (Mar 1995): 153–158.

Boucher, Norman. "Back to the Everglades." *Technology Review* (Aug/Sept 1995): 24–35.

Cosgrove, William J., and Frank R. Rijsberman. *Vision Report: Making Water Everybody's Business*. World Water Council, 2000. Available on the Internet at **http://watervision.org/**.

Del Moral, Leandro, and David Saurí. "Changing Course: Water Policy in Spain." *Environment* 41 (July/Aug 1999): 12–15, 31–36.

Donahue, John M., and Barbara Rose Johnston, eds. *Water, Culture, and Power*. Washington, DC: Island Press, 1998.

Gardner, Gary. "A Tale of Three Aquifers." *WorldWatch* 8 (May/June 1995): 30–36.

Gauch, Sarah. "It's No Mirage: Egyptian Oases Vanishing Fast. *Christian Science Monitor*, May 29, 1998: 8, 9.

Gleick, Peter H. "Making Every Drop Count." *Scientific American* 284 (February 2001): 40–45.

Joseph, Patrick. "The Battle of the Dams." *Smithsonian* 29 (November 1998): 49–61.

Kaiser, Jocelyn. "Battle over a Dying Sea." *Science* 284 (Apr 2, 1999): 28–30.

Knickerbocker, Brad. "U.S. Busts Up Dams in Bid to Help Fish." *Christian Science Monitor*, July 15, 1998: 3.

Kwai-cheong, Chau. "The Three Gorges Project of China: Resettlement Prospects and Problems." *Ambio* 24(2) (Mar 1995): 98–102.

Lei, Xiong. "Going against the Flow in China." *Science* 280 (Apr 3, 1998): 24–26.

Ligon, Franklin K. "Downstream Ecological Effects of Dams." *BioScience* 45 (Mar 1995): 183–192.

"Mono Lake: Saved." *National Wildlife* 33 (June/July 1995): 36–37.

Line, Les. "Bringing the Magic Back to the Platte." *National Wildlife* 38 (April/May 2000): 22–29.

Naiman, Robert J., and Monica G. Turner. "A Future Perspective on North America's Freshwater Ecosystems." *Ecological Applications* 10 (Aug 2000): 958–970.

Nakamura, Masahisa. "Preserving the Health of the World's Lakes." *Environment* 39 (Nov 1997): 16–20, 34–40.

Pimentel, David, et al. "Water Resources: Agriculture, the Environment, and Society." *Bioscience* 47 (Feb 1997): 97–106.

Platt, Rutherford H., et al. "A Full, Clean Glass? Managing New York City's Watersheds." *Environment* 42 (June 2000): 8–20.

Postel, Sandra. "Freshwater Ecosystem Services." *In Nature's Services: Societal Dependence on Natural Ecosystems*. Washington, DC: Island Press, 1997: 195–214.

Postel, Sandra. "Entering an Era of Water Scarcity: The Challenge Ahead." *Ecological Applications* 10 (Aug 2000): 941–948.

Postel, Sandra. "Growing More Food with Less Water." *Scientific American* 284 (Feb 2001): 46–51.

Reisner, Marc. Coming Undammed. *Audubon* 100 (Sept–Oct 1998): 58–65.

Rockstrom, Johan, et al. Linkages between Water Vapor Flows, Food Production and Terrestrial Ecosystem Services. *Conservation Ecology* 3(2): 5. [on-line] URL: **http://www.consecol.org/vol3/iss2/art5**.

Sampat, Payal. "Groundwater Shock." *Worldwatch* 13 (Jan/Feb 2000): 10–22.

Sparks, Richard E. "Need for Ecosystem Management of Large Rivers and Their Floodplains." *BioScience* 45 (Mar 1995): 168–182.

Stone, Richard. "Coming to Grips with the Aral Sea's Grim Legacy." *Science* 284 (Apr 2, 1999): 30–33.

Uitto, Juha I., and Asit K. Biswas, eds. *Water for Urban Areas: Challenges and Perspectives*. Tokyo: United Nations University Press, 2000.

Vorosmarty, Charles J., et al. "Global Water Resources: Vulnerability from Climate Change and Population Growth." *Science* 289 (July 14, 2000): 284–288.

White, Gilbert F. "Water Science and Technology: Some Lessons from the 20th Century." *Environment* 42 (Jan/Feb 2000): 30–38.

Williams, William David. "What Future for Saline Lakes?" *Environment* 38 (Nov 1996): 12–20, 38–39.

World Water Vision Commission. A *Water Secure World: Vision for Water, Life and the Environment*. World Water Council, 2000. Available on the Internet at **http://watervision.org/** (Jan 23, 2001).

CHAPTER 10
The Production and Distribution of Food

Aker, Jerry, ed. *Livestock for a Small Earth*. Washington, DC: Seven Locks Press, 1994.

Avery, Dennis. "Saving Nature's Legacy through Better Farming." *Issues in Science and Technology* (Fall 1997): 59–64.

Bender, William H. "How Much Food Will We Need in the 21st Century?" *Environment* 39 (Mar 1997): 7–11, 27–28.

Bread for the World. *Hunger 1999: The Changing Politics of Hunger*. Washington, DC: Bread for the World Institute, 1999.

Bread for the World. *Hunger 2000: A Program to End Hunger*. Washington, DC: Bread for the World Institute, 2000.

Brown, J. Larry. "Malnutrition, Poverty and Intellectual Development." *Scientific American* 274 (Feb 1996): 38–43.

Brown, Lester R. "Feeding Nine Billion." In *State of the World, 1999* (pp. 115–132). Washington, DC: WorldWatch Institute, 1999.

Conway, Gordon. "Food for All in the 21st Century." *Environment* 42 (Jan/Feb 2000): 8–18.

Daily, Gretchen, et al. "Food Production, Population Growth and the Environment." *Science* 281 (Aug 28, 1998): 1291–1292.

DellaPenna, Dean. "Nutritional Genomics: Manipulating Plant Micronutrients to Improve Human Health." *Science* 285 (July 16, 1999): 375–379.

Durning, Alan B., and Holly W. Brough. "Taking Stock: Animal Farming and the Environment." WorldWatch Paper 103. Washington, DC: WorldWatch Institute, 1991.

Easterbrook, Greg. "Forgotten Benefactor of Humanity." *Atlantic Monthly* 279 (Jan 1997): 75–82.

Ervin, David E., et al. "Agriculture and the Environment: A New Strategic Vision." *Environment* 40 (July–Aug 1998): 8–15, 35–40.

Faeth, Paul. "Building the Case for Sustainable Agriculture: Policy Lessons from India, Chile, and the Philippines." *Environment* 36 (Jan/Feb 1994): 16–20, 34–39.

FAO. *The State of Food Insecurity in the World, 1999*. Rome: Food and Agricultural Organization of the United Nations, 1999. Available on the Internet at **http://www.fao.org/FOCUS/E/SOFI/home-e.htm** (Jan 15, 2001).

FAO. *Agriculture: Towards 2015/30, Technical Interim Report, April 2000*. Rome: Food and Agricultural Organization of the United Nations, 1999. Available on the Internet at **http://WWW.FAO.ORG/es/ESD/at2015/toc-e.htm** (Jan 15, 2001).

Gardner, Gary, and Brian Halweil. *Underfed and Overfed: The Global Epidemic of Malnutrition*. WorldWatch Paper 150. Washington, DC: WorldWatch Institute, 2000.

Gaskell, George, et al. "Worlds Apart? The Reception of Genetically Modified Foods in Europe and the U.S." *Science* 285 (July 16, 1999): 384–387.

Halweil, Brian. "The Emperor's New Crops." *WorldWatch* 12 (July/Aug 1999): 21–29.

Hanrahan, Charles E. "Agricultural Export and Food Aid Programs." The National Council for Science and the Environment. *CRS Issue Brief* 98006, Nov 16, 2000. **http://www.cnie.org/nle/ag-41.html** (Jan 15, 2001).

Haslberger, Alexander G. "Monitoring and Labeling for Genetically Modified Products." *Science* 287 (Jan 21, 2000): 431–432.

Kurlansky, Mark. "On Haitian Soil." *Audubon* (Jan/Feb 1995): 50–57.

Levidow, Les. "Regulating BT Maize in the United States and Europe: A Scientific–Cultural Comparison. *Environment* 41 (Dec 1999): 10–22.

Mann, Charles C. "Reseeding the Green Revolution." *Science* 277 (Aug 22, 1997): 1038–1043.

Mann, Charles C. "Genetic Engineers Aim to Soup Up Crop Photosynthesis." *Science* (Jan 15, 1999): 314–316.

Mann, Charles C. "Crop Scientists Seek a New Revolution." *Science* (Jan 15, 1999): 310–314.

Meitzner, Laura S., and Martin L. Price. *Amaranth to Zai Holes: Ideas for Growing Food under Difficult Conditions*. Fort Myers, FL: Educational Concerns for Hunger Organizations, Inc., 1996.

Mortimore, Michael, and Mary Tiffen. "Population Growth and a Sustainable Environment: The Machakos Story." *Environment* 36 (Oct 1994): 10–20, 28–32.

National Research Council. *Genetically Modified Pest-Protected Plants: Science and Regulation*. Washington, DC: National Academy of Science, 2000.

Paarlberg, Robert. "Genetically Modified Crops in Developing Countries: Promise or Peril?" *Environment* 42 (Jan/Feb 2000): 19–27.

Postel, Sandra. "Redesigning Irrigated Agriculture." In *State of the World, 2000* (pp. 39–58). Washington, DC: WorldWatch Institute, 2000.

Prosterman, Roy, Tim Hanstad, and Li Ping. "Can China Feed Itself?" *Scientific American* 275 (Nov 1996): 90–96.

Prusher, Ilene R. "Inside an African Famine." *Christian Science Monitor* (Oct 9, 1998): 7–9.

Ronald, Pamela C. "Making Rice Disease-Resistant." *Scientific American* 277 (Nov 1997): 98–105.

Serageldin, Ismael. "Biotechnology and Food Security in the 21st Century." *Science* 285 (July 16, 1999): 387–389.

Smith, Miranda, ed. *The Real Dirt: Farmers Tell about Organics and Low-Input Practices in the Northeast*. Burlington, VT: Northeast Region Sustainable Agriculture Research and Education Program, 1994.

Tuxill, John. "The Biodiversity That People Made." *Worldwatch* 13 (May/June 2000): 24–35.

Ye, Xudong, et al. "Engineering the Provitamin A (ß-Carotene) Biosynthetic Pathway into (Carotenoid-Free) Rice Endosperm. *Science* 287 (Jan 14, 2000): 303–305.

CHAPTER 11
Wild Species: Biodiversity and Protection

Blaustein, Andrew R., and David B. Wake. "The Puzzle of Declining Amphibian Populations. *Scientific American* 272 (Apr 1995): 52–57.

Bright, Chris. "Crawling Out of the Pipe: The Hazardous Waste That Makes More of Itself." *WorldWatch* 12 (Jan–Feb 1999): 22–33.

Carroll, R., et al. "Strengthening the Use of Science in Achieving the Goals of the Endangered Species Act: An Assessment by the Ecological Society of America." *Ecological Applications* 6 (Jan 1996): 1–11.

Cincotta, Richard P., and Robert Engelman. *Nature's Place: Human Population and the Future of Biological Diversity*. Washington, DC: Population Action International, 2000. Also on the Internet at **http://www.populationaction.org/pubs/biodiv00/biodiv_index.htm** (Jan 16, 2001).

Cox, Paul Alan, and Michael J. Balick. "The Ethnobotanical Approach to Drug Discovery." *Scientific American* (June 1994): 82–87.

Cracraft, Joel, and Francesca T. Grifo. *The Living Planet in Crisis: Biodiversity Science and Policy*. New York: Columbia University Press, 1999.

Daily, Gretchen, et al. "The Value of Nature and the Nature of Value." *Science* 289 (July 21, 2000): 395–396.

Dobson, A. P., et al. "Geographic Distribution of Endangered Species in the United States." *Science* (January 24, 1997): 550–553.

Eisner, Thomas, Jane Lubchenco, Edward O. Wilson, David S. Wilcove, and Michael J. Bean. "Building a Scientifically Sound Policy for Protecting Endangered Species." *Science* 268 (Sept 1, 1996): 1231–1232.

Graham, Frank, Jr. "The Puffin Project Goes Global." *Audubon* 101 (July/Aug 1999): 90–95.

Graham, Frank, Jr. "The Day of the Condor." *Audubon* 102 (Jan/Feb 2000): 46–53.

Groombridge, B., and M. D. Jenkins. *Global Biodiversity: Earth's Living Resources in the 21st Century*. Cambridge, U.K.: World Conservation Monitoring Centre, 2000.

Hargrove, Eugene C. *Foundations of Environmental Ethics*. Englewood Cliffs, NJ: Prentice Hall, 1989.

Hector, A., et al. "Plant Diversity and Productivity Experiments in European Grasslands. *Science* 286 (Nov 5, 1999): 1123–1126.

Kaiser, Jocelyn. "Stemming the Tide of Invading Species." *Science* 285 (Sept 17, 1999): 1836–1841.

Kumar, Ashok, et al. "How We Busted the Tiger Gang." *International Wildlife* 24 (May/June 1994): 38–43.

Kuznik, Frank. "Eating Themselves out of House and Home." *National Wildlife* 36 (Oct–Nov 1998): 38–43.

Luoma, Jon R. "Habitat Conservation Plans: Compromise or Capitulation?" *Audubon* 100 (Jan–Feb 1998): 36–43.

Mack, Richard, et al. "Biotic Invasions: Causes, Epidemiology, Global Consequences and Control. *Issues in Ecology* 5 (Spring 2000): 1–20. **http://esa.sdsc.edu/issues.htm** (Jan 9, 2001).

May, Robert M. "How Many Species Inhabit the Earth?" *Scientific American* (Oct 1992): 42–48.

Myers, Norman. "What We Must Do to Counter the Biotic Holocaust." *International Wildlife* 29 (March/April 1999): 30–39.

Naeem, Shahid. "Biodiversity and Ecosystem Functioning: Maintaining Natural Life Support Processes." *Issues in Ecology* 4 (Fall 1999): 1–11. **http://esa.sdsc.edu/issues.htm** (Jan 9, 2001).

Nicolaou, K. C., Rodney K. Guy, and Pierre Potier. "Taxoids: New Weapons against Cancer." *Scientific American* 274 (June 1996): 94–98.

Posey, Darrell A. *Cultural and Spiritual Values of Biodiversity*. London: Intermediate Technology Publications, 1999.

Power, Thomas Michael. "The Wealth of Nature." *Issues in Science and Technology* (Spring 1996): 48–54.

Raustiala, Kal, and David G. Victor. "Biodiversity since Rio: The Future of the Convention on Biological Diversity." *Environment* 38 (May 1996): 16–20, 37–43.

Reed, Christina. "Get Along, Little Longhorn: Riding Herd on a Troublesome, Tree-Eating Beetle." *National Wildlife* 38 (Apr/May 2000): 12–13.

Reid, Walter V. "Strategies for Conserving Biodiversity." *Environment* 39 (Sept 1997): 16–20, 39–43.

Rolston, Holmes, III. *Environmental Ethics: Duties to and Values in the Natural World.* Philadelphia: Temple University Press, 1988.

Rolston, Holmes, III. "Creation: God and Endangered Species." Ch. 3 in Ke Chung Kim and Robert D. Weaver, eds., *Biodiversity and Landscapes: A Paradox of Humanity.* New York: Cambridge Univ. Press, 1994.

Sala, Osvaldo E., et al. "Global Biodiversity Scenarios for the Year 2100." *Science* 287 (Mar 10, 2000): 1770–1774.

Shilling, Fraser. 1997. "Do Habitat Conservation Plans Protect Endangered Species?" *Science* (June 13, 1997): 1662–1663.

Soulé, Michael E. "Conservation: Tactics for a Constant Crisis." *Science* 253 (1991): 744–750.

Stein, Bruce A., and Stephanie R. Flack. "Conservation Priorities: The State of U.S. Plants and Animals." *Environment* 39 (May 1997): 6–11, 34–39.

Stevens, William K. "Make Way for Moose." *National Wildlife* 37 (Aug/Sept 1999): 28–33.

Stolzenburg, William. "America the Bountiful." *Nature Conservancy* 50 (May/June 2000): 10–25.

Temple, Stanley A. "Easing the Travails of Migratory Birds." *Environment* 40 (Jan–Feb 1996): 6–9, 28–32.

Tobias, Michael. *Nature's Keepers.* New York: John Wiley & Sons, 1998.

Tuxill, John. *Losing Strands in the Web of Life.* WorldWatch Paper 141. Washington, DC: WorldWatch Institute, 1998.

Tuxill, John. *Nature's Cornucopia: Our Stake in Plant Diversity.* WorldWatch Paper 148. Washington, DC: WorldWatch Institute, 1999.

United Nations Environment Program. *Global Biodiversity Assessment.* Cambridge, U.K.: Cambridge University Press, 1995.

Wheelwright, Jeff. "Condors: Back from the Brink." *Smithsonian* (May 1997): 48–55.

Wilkinson, Todd. "Wolves Return as Kings of the American Serengeti." *Christian Science Monitor* (Dec 11, 1997): 3, 4.

Wilson, Edward O., ed. *Biodiversity.* Washington, DC: National Academy Press, 1988.

World Conservation Union. *The 2000 IUCN Red List of Threatened Species.* On the Internet at **http://www.redlist.org/** (Jan 15, 2001).

Youth, Howard. "Watching vs. Taking." *WorldWatch* 13 (May/June 2000): 12–23.

CHAPTER 12
Ecosystems as Resources

Aber, John, et al. "Applying Ecological Principles to Management of U.S. National Forests." *Issues in Ecology* 6 (Spring 2000): 1–20. **http://esa.sdsc.edu/issues.htm** (Jan 16, 2001).

Abramovitz, Janet N. *Taking a Stand: Cultivating a New Relationship with the World's Forests.* WorldWatch Paper 140. Washington, DC: WorldWatch Institute, 1998.

Allen, Scott. "Return to Georges Bank." *Boston Globe* (Apr 5, 1999): C1, C3.

Ascher, William. "Understanding Why Governments in Developing Countries Waste Natural Resources." *Environment* 42 (March 2000): 8–18.

Barber, Charles V., and Vaughan R. Pratt. "Poison and Profits: Cyanide Fishing in the Indo-Pacific." *Environment* 40 (Oct 1998): 4–9, 28–34.

Botsford, Lojis W., Juan Carlos Castilla, and Charles H. Peterson. "The Management of Fisheries and Marine Ecosystems. *Science* 277 (July 25, 1997): 509–514.

Boyd, Claude E., and Jason W. Clay. "Shrimp Aquaculture and the Environment." *Scientific American* 278 (June 1998): 58–65.

Burger, Joanna, and Michael Gochfeld. "The Tragedy of the Commons: 30 Years Later." *Environment* (Dec 1998): 4–13, 26–27.

Carey, John. "Where Have All the Animals Gone?" *International Wildlife* 29 (Nov/Dec 1999): 12–21.

Daley, Beth. "A Race to Save Maine Woods." *Boston Globe* (Sept 18, 2000): 1, B5.

Deal, Carl. *The Greenpeace Guide to Anti-Environmental Organizations.* Berkeley, CA: Odonian Press, 1992.

Denniston, Derek. *High Priorities: Conserving Mountain Ecosystems and Cultures.* WorldWatch Paper 123. Washington, DC: WorldWatch Institute, 1995.

Dewees, Christopher M. "Effects of Individual Quota Systems on New Zealand and British Columbia Fisheries." *Ecological Applications* 8 (1998): 133–138.

Fluharty, David. "Habitat Protection, Ecological Issues, and Implementation of the Sustainable Fisheries Act." *Ecological Applications* 10 (April 2000): 325–337.

Food and Agricultural Organization of the UN. *State of the World's Forests, 1999.* **http://www.fao.org/forestry/FO/SOFO/SOFO99/ sofo99-e.stm** (January 16, 2001).

Forest Service. *1997 Northwest Forest Plan.* (Updates also available.) USDA. **http://www.fs.fed.us/r6/nwfp.htm** (Jan 16, 2001).

Gustansky, Julie Ann, and Roderick H. Squires, eds. *Protecting the Land: Conservation Easements Past, Present, and Future.* Washington, DC: Island Press, 2000.

Hardin, Garrett, and John Baden, eds. *Managing the Commons.* San Francisco: W. H. Freeman, 1977. Contains Hardin's original paper on the tragedy of the commons and many other papers related to the concept of the commons.

Harvell, C. D., et al. "Emerging Marine Diseases—Climate Links and Anthropogenic Factors." *Science* 285 (Sept 3, 1999): 1505–1509.

Honey, Martha S. "Treading Lightly? Ecotourism's Impact on the Environment." *Environment* 41 (June 1999): 4–9, 28–33.

Iudicello, Suzanne, et al. *Fish, Markets and Fishermen: The Economics of Overfishing.* Washington, DC: Island Press, 1999.

Jordan, William R., III, et al. *Restoration Ecology: A Synthetic Approach to Ecological Research.* Cambridge, U.K.: Cambridge University Press, 1987.

Knickerbocker, Brad. "Activists' New Tune in the West: This Land Is Our Land." *Christian Science Monitor,* Feb 7, 1995: 1, 5.

Kremen, C., et al. "Economic Incentives for Rain Forest Conservation across Scales." *Science* 288 (June 9, 2000): 1828–1831.

Mastny, Lisa. "Coming to Terms with the Arctic." *WorldWatch* 13 (Jan/Feb 2000): 24–36.

McGinn, Anne Platt. *Rocking the Boat: Conserving Fisheries and Protecting Jobs.* WorldWatch Paper 142. Washington, DC: WorldWatch Institute, 1998.

McGinn, Anne Platt. *Safeguarding the Health of Oceans.* WorldWatch Paper 145. Washington, DC: WorldWatch Institute, 1999.

Musters, C. J. M., et al. "Can Protected Areas Be Expanded in Africa?" *Science* 287 (Mar 10, 2000): 1759–1760.

Myers, R. W., et al. "Why Do Fish Stocks Collapse? The Example of Cod in Atlantic Canada." *Ecological Applications* 7 (1997): 91–106.

National Marine Fisheries Service. *Our Living Oceans: Report on the Status of U.S. Living Marine Resources.* (NOAA Tech. Memo. NMFS-F/SPO-41.) Washington, DC: U.S. Department of Commerce, 1999.

Oelschlager, Max. *The Idea of Wilderness: From Prehistory to the Age of Ecology.* New Haven, CT: Yale University Press, 1991.

O'Malley, Robin, and Kate Wing. "Forging a New Tool for Ecosystem Reporting. *Environment* 42 (Apr 2000): 20–31.

Ostrom, Elinor, et al. "Revisiting the Commons: Local Lessons, Global Challenges." *Science* 284 (Apr 9, 1999): 278–282.

Pauly, Daniel, et al. "Fishing Down Marine Food Webs." *Science* 279 (Feb 6, 1998): 860–863.

Renner, Michael. *Working for the Environment: A Growing Source of Jobs.* WorldWatch Paper 152. Washington, DC: WorldWatch Institute, 2000.

Revkin, Andrew. *The Burning Season: The Murder of Chico Mendes and the Fight for the Amazon Rain Forest.* Boston: Houghton Mifflin, 1990.

Robinson, John G., et al. "Wildlife Harvest in Logged Tropical Forests." *Science* 284 (Apr 23, 1999): 595–596.

Rutzler, Klaus, and Ilka C. Feller. "Caribbean Mangrove Swamps." *Scientific American* 274 (Mar 1996): 94–99.

Safina, Carl. "The World's Imperiled Fish." *Scientific American* 273 (Nov 1995): 46–53.

Sellars, Richard West. *Preserving Nature in the National Parks: A History.* New Haven, CT: Yale University Press, 1999.

Sinclair, A. R. E., and Peter Arcese, eds. *Serengeti II: Dynamic Management and Conservation of an Ecosystem*. Chicago: University of Chicago Press, 1995.

Skore, Mari. "Whaling: A Sustainable Use of Natural Resources or a Violation of Animal Rights?" *Environment* 36 (Sept 1994): 12–20, 30–31.

Tenneson, Michael. "Mexico's Turtle Wars." *International Wildlife* 29 (Nov/Dec 1999): 44–51.

Terman, Max. "Natural Links: Naturalistic Golf Courses as Wildlife Habitat." *Landscape and Urban Planning* 38 (1997): 183–197.

U.S. Geological Survey. *Status and Trends of the Nation's Biological Resources*. Washington, DC: U.S. Government Superintendent of Documents, 1998.

Wallace, Aubrey. *EcoHeroes: Twelve Tales of Environmental Victory*. San Francisco: Mercury House, 1993.

Weber, Peter. *Net Loss: Fish, Jobs, and the Marine Environment*. WorldWatch Paper 120. Washington, DC: WorldWatch Institute, 1994.

Williams, Ted. "Clinton's Last Stand." *Audubon* 102 (May/June 2000): 46–49, 90–97.

Wood, Charles, and Robert Walker. "Saving the Trees by Helping the Poor: A Look at Small Producers along Brazil's TransAmazon Highway." *Resources* 136 (Summer 1999): 14–17.

World Resources Institute. *World Resources, 2000–2001. People and Ecosystems: The Fraying Web of Life*. New York: Oxford University Press, 2000.

CHAPTER 13
Energy from Fossil Fuels

Alaska Wilderness League. *Preventing the Next Valdez: Ten Years after Exxon's Spill, New Disasters Threaten Alaska's Environment*. Alaska Wilderness League, Mar 1999. **http://www.alaskawild.org/** (Jan 16, 2001).

Barnes, Douglas F., and Willem M. Floor. "Rural Energy in Developing Countries: A Challenge for Economic Development." *Annual Review of Energy and the Environment* 21 (1996): 497–530.

Birkland, Thomas A. "In the Wake of the *Exxon Valdez*." *Environment* 40 (Sept 1998): 4–9, 27–32.

Burger, Joanna. *Oil Spills*. New Brunswick, NJ: Rutgers University Press, 1997.

Campbell, Colin J., and Jean H. Laherrère. "The End of Cheap Oil." *Scientific American* 278 (Mar 1998): 78–83.

Corn, M. Lynne. *The Arctic National Wildlife Refuge: The Next Chapter*. The National Council for Science and the Environment. CRS Issue Brief 10055, June 1, 2000. **http://www.cnie.org/nle/nrgen-23.html** (Jan 16, 2001).

Cullen, Robert. "The True Cost of Coal." *Atlantic Monthly* (Dec 1993): 38–52.

DeCardis, Joseph F., et al. "Searching for Energy Efficiency on Campus." *Environment* 42 (May 2000): 8–20.

Dunn, Seth. "Power of Choice." *WorldWatch* 10 (Sept–Oct 1997): 30–35.

Dunn, Seth. "King Coal's Weakening Grip on Power." *WorldWatch* 12 (Sept/Oct 1999): 10–19.

Dunn, Seth. *Micropower: The Next Electrical Era*. WorldWatch Paper 151 (July 2000). Washington, DC: WorldWatch Institute.

Flavin, Christopher. "Energy for a New Century." *WorldWatch* 13 (Mar/Apr 2000): 8–9.

Fouda, Safea, A. "Liquid Fuels from Natural Gas." *Scientific American* 278 (March 1998): 92–95.

George, Richard L. "Mining for Oil." *Scientific American* 278 (Mar 1998): 84–85.

Hirsh, Richard F., and Adam H. Serchuk. "Power Switch: Will the Restructured Electric Utility System Help the Environment?" *Environment* 41 (Sept 1999): 4–9, 32–39.

Hirst, Eric. "A Bright Future: Energy Efficiency Programs at Electric Utilities." *Environment* 36 (Nov 1994): 10–15.

Kerr, Richard. "USGS Optimistic on World Oil Prospects." *Science* 289 (July 14, 2000): 237.

Lenssen, Nicholas. "All the Coal in China." *WorldWatch* 6 (Mar/Apr 1993): 22–29.

Rogner, H-H. "An Assessment of World Hydrocarbon Resources." *Annual Review of Energy and the Environment* 22 (1997): 217–262.

Romm, Joseph J. *The Internet Economy and Global Warming*. Center for Energy and Climate Solutions, 1999. **http://www.cool-companies.org/energy/paper1.cfm** (Jan 16, 2001).

Romm, Joseph J., and Charles B. Curtis. "Mideast Oil Forever?" *Atlantic Monthly* (Apr 1996): 57–74.

U.S. Department of Energy. *Annual Energy Review, 1999*. Review of data on all energy sources. **http://www.eia.doe.gov/emeu/aer/contents.html**.

U.S. Department of Energy. *International Energy Review, 1998*. Review of data on all international energy sources. **http://www.eia.doe.gov/emeu/iea/** (Jan 16, 2001).

USGS World Energy Assessment Team. *U.S. Geological Survey World Petroleum Assessment, 2000—Description and Results*. U.S. Geological Survey Digital Data Series-DDS-60. U.S. Department of the Interior, 2000. **http://greenwood.cr.usgs.gov/energy/WorldEnergy/DDS-60/** (January 16, 2001).

CHAPTER 14
Nuclear Power: Promise and Problems

Anspaugh, Lynn R., et al. "The Global Impact of the Chernobyl Reactor Accident." *Science* 242 (Dec 16, 1988): 1513–1519.

Balter, Michael. "Children Become the First Victims of Fallout." *Science* 272 (April 19, 1996): 357–360.

Barkenbus, Jack N., and Charles Forsberg. "Internationalizing Nuclear Safety: The Pursuit of Collective Responsiblility." *Annual Review of Energy and the Environment* 20 (1995): 83–118.

Beck, Peter W. "Nuclear Energy in the Twenty-First Century: Examination of a Contentious Subject." *Annual Review of Energy and the Environment* 24: 113–137.

Cochran, Thomas B., et al. "Radioactive Contamination at Chelyabinsk-65, Russia." *Annual Review of Energy and the Environment* 18 (1993): 507–528.

Craig, Paul P. "High-Level Nuclear Waste: The Status of Yucca Mountain." *Annual Review of Energy and the Environment* 24: 461–486.

Domenici, Pete V. "Future Perspectives on Nuclear Issues." *Issues in Science and Technology* (Winter 1997–98): 53–59.

Ewing, Rodney C. "Less Geology in the Geological Disposal of Nuclear Waste." *Science* 286 (Oct 15, 1999): 415–416.

Fain, Paul. "Remembering the Meltdown: Three Mile Island 20 Years After." *Nucleus* 21 (Summer 1999): 6–7.

Flynn, James, and Paul Slovic. "Yucca Mountain: A Crisis for Policy: Propects for America's High-Level Nuclear Waste Program. *Annual Review of Energy and the Environment* 20 (1995): 83–118.

Flynn, James, et al. "Overcoming Tunnel Vision: Redirecting the U.S. High-level Nuclear Waste Program." *Environment* 39 (Apr 1997): 6–11, 25–30.

Forsberg, C. W., and A. M. Weinberg. "Advanced Reactors, Passive Safety, and Acceptance of Nuclear Energy." *Annual Review of Energy* 15 (1990): 133–152.

French, Howard W. "Accident Makes Japan Re-examine A-Plants." *The New York Times*, Jan 13, 2000.

Furth, Harold P. "Fusion." *Scientific American* 273 (Sept 1995): 174–177.

Golay, Michael W., and Neil E. Todreas. "Advanced Light-Water Reactors." *Scientific American* 262 (Apr 1990): 82–89.

Hollister, Charles D., and Stephen Nadis. "Burial of Radioactive Waste under the Seabed." *Scientific American* 278 (Jan 1998): 60–65.

Holt, Mark, and Carl E. Behrens. *IB88090: Nuclear Energy Policy*. Congressional Research Service Issue Brief 88090, Nov 22, 2000. **http://www.cnie.org/nle/eng-5.html** (Jan 16, 2001).

Kazimi, Mujid S., and Neil E. Todreas. "Nuclear Power Economic Performance: Challenges and Opportunities." *Annual Review of Energy and the Environment* 24: 139–171.

Lichtenstein, Kenneth, and Ira Helford. "Radiation and Health: Nuclear Weapons and Nuclear Power." In Eric Chivian, et al., eds., *Critical Condition: Human Health and the Environment* (pp. 93–122). Cambridge, MA: MIT Press, 1993.

Makhijani, Arjun, Howard Hu, and Katherine Yik, eds. *Nuclear Wasteland: A Global Guide to Nuclear Weapons Production and Its Health and Environmental Effects*. Cambridge, MA: MIT Press, 1995.

Medvedev, Zhores A. *The Legacy of Chernobyl*. New York: W. W. Norton, 1990.

Morgan, M. Granger. "What Would It Take to Revitalize Nuclear Power in the United States?" *Environment* 35 (Mar 1993): 6–9, 30–33.

O'Neill, Kate. "International Nuclear Waste Transportation: Flashpoints, Controversies and Lessons." *Environment* 41 (May 1999): 12–15, 34–39.

Pooley, Eric. "Nuclear Warriors." *Time* (Mar 4, 1996): 47–54.

Pooley, Eric. "Nuclear Safety Fallout." *Time* (Mar 17, 1997): 34–36.

Probst, Katherine, and Adam Lowe. "The $200 Billion Question: Does Anyone Care about Cleaning Up the Nation's Nuclear Weapons Sites?" *Resources* 138 (Winter 2000): 11–13.

Redfern, Judy. "Europe, Japan Finalizing Reduced ITER Design." *Science* 286 (Dec 3, 1999): 1829–1830.

Rochlin, Gene I., and Alexandra von Meier. "Nuclear Power Operations: A Cross-Cultural Perspective." *Annual Review of Energy and the Environment* 19 (1994): 153–187.

Shcherbak, Yuri M. "Ten Years of the Chornobyl Era." *Scientific American* 274 (April 1996).

Vogelsang, W. F., and H. H. Barschall. "Nuclear Power." In Ruth Howes and Anthony Fainberg, *The Energy Sourcebook: A Guide to Technology, Resources, and Policy* (pp. 127–152). New York: American Institute of Physics, 1991.

Whipple, Chris G. "Can Nuclear Waste Be Stored Safely at Yucca Mountain?" *Scientific American* 274 (June 1996): 40–44.

Williams, Phil, and Paul N. Woessner. "The Real Threat of Nuclear Smuggling." *Scientific American* 274 (January 1996): 40–44.

Yonas, Gerald. "Fusion and the Z Pinch." *Scientific American* 279 (August 1998): 41–47.

Zorpette, Glenn. "Hanford's Nuclear Wasteland." *Scientific American* 274 (May 1996): 88–97.

CHAPTER 15
Renewable Energy

Aitken, Donald, and Paul Neuffer. "Low Cost, High Value Passive Solar." *Solar Today* 9 (Mar/Apr 1995): 18–20.

American Solar Energy Society. *Solar Today*. (A monthly journal with many timely articles on solar energy.) 2400 Central Ave, G-1, Boulder CO 80301.

American Wind Energy Association. "1999 Best Year Ever for Wind Energy." *1999 Global Wind Energy Market Report*. 1999. http://www.awea.org/faq/global99.html (Jan 16, 2001).

Anderson, Dennis. "Renewable Energy Technology and Policy for Development." *Annual Review of Energy and the Environment* 22 (1997): 187–215.

Appleby, A. John. "The Electrochemical Engine for Vehicles." *Scientific American* 281 (July 1999): 74–79.

Barnes, Douglas F., and Willem M. Floor. "Rural Energy in Developing Countries: A Challenge for Economic Development." *Annual Review of Energy and the Environment* 21 (1996): 497–530.

Cates, Michael R. "Hybrid Lighting: Illuminating Our Future." *Oak Ridge National Laboratory Review* 29 (3), 1996. http://www.ornl.gov/ORNLReview/rev29_3/text/contents.htm.

Coutts, Timothy J., and Mark. C. Fitzgerald. "Thermophotovoltaics." *Scientific American* 279 (Sept 1998): 90–95.

Dracker, Raymond, and Pascal De Laquil, III. "Progress Commercializing Solar–Electric Power Systems. *Annual Review of Energy and the Environment* 21 (1996): 371–402.

Dunn, Seth. "The Hydrogen Experiment." *WorldWatch* 13 (Nov/Dec 2000): 14–25.

Flavin, Christopher, and Nicholas Lenssen. *Beyond the Petroleum Age: Designing a Solar Economy*. WorldWatch Paper 100. Washington, DC: WorldWatch Institute, 1990.

Flavin, Christopher, and Molly O'Meara. "Shining Examples: Solar Financing in the Developing World. *WorldWatch* 10 (May–June 1997): 28–36.

Food and Agricultural Organization of the United Nations. *Biogas Technology: A Training Manual for Extension*. Kathmandu, Nepal: FAO/Consolidated Management Services, 1996. http://www.fao.org/sd/egdirect/EGre0021.htm (Jan 17, 2001).

Gawell, Karl, et al. "Preliminary Report: Geothermal Energy, the Potential for Clean Power from the Earth." Washington, DC: Geothermal Energy Association, 1999. http://www.geotherm.org/PotentialReport.htm (Jan 17, 2001).

Greenberg, David A. "Modeling Tidal Power." *Scientific American* 257 (Nov 1987): 128–133.

Jensen, Marc W., and Marc Ross. "The Ultimate Challenge: Developing an Infrastructure for Fuel Cell Vehicles." *Environment* 42 (Sept 2000): 10–22.

Kammen, Daniel M. "Bring Power to the People: Promoting Appropriate Energy Technologies in the Developing World." *Environment* 41 (June 1999): 10–15, 34–41.

Löfstedt, Ragnar. "Sweden's Biomass Controversy: A Case Study of Communicating Policy Issues." *Environment* 40 (June 1998): 16–20, 42–45.

McVeigh, James, et al. *Winner, Loser, or Innocent Victim? Has Renewable Energy Performed as Expected?* Discussion Paper 99–28. Washington, DC: Resources for the Future, 1999. http://www.rff.org/nat_resources/energy.htm (Jan 16, 2001).

Muller, Frank. "Mitigating Climate Change: The Case for Energy Taxes." *Environment* 38 (Mar 1996): 12–20, 36–45.

Shah, A., et al. "Photovoltaic Technology: The Case for Thin-Film Solar Cells." *Science* 285 (July 30, 1999): 692–698.

Shapiro, Andrew M. *The Homeowner's Complete Handbook for Add-On Solar Greenhouses and Sunspaces*. Emmaus, PA: Rodale Press, 1985.

Sissine, Fred. *Energy Efficiency: Budget, Climate Change, and Electricity Restructuring Issues*. CRS Issue Brief 10020, Aug 21, 2000. http://www.cnie.org/nle/eng-48.html (Jan 16, 2001).

Sissine, Fred. *Renewable Energy: Tax Credit, Budget, and Electricity Restructuring Issues*. CRS Issue Brief 10041, Aug 31, 2000. http://www.cnie.org/nle/eng-54.html (Jan 16, 2001).

Turner, John A. "A Realizable Renewable Energy Future." *Science* 285 (July 30, 1999): 687–698.

U.S. Department of Energy. *Renewable Energy Annual 1999, with Data for 1998*. Energy Information Administration (Mar 2000). http://www.eia.doe.gov/cneaf/solar.renewables/rea_data99/toc.html (Jan 16, 2001).

Yacobucci, Brent D., and Jasper Womach. *Fuel Ethanol: Background and Public Policy Issues*. CRS Issue Brief RL30369, June 27, 2000. http://www.cnie.org/nle/eng-59.html (Jan 17, 2001).

CHAPTER 16
Environmental Hazards and Human Health

Arnold, Steven F., et al. "Synergistic Activation of Estrogen Receptor with Combinations of Environmental Chemicals." *Science* 272 (7 June 1996): 1489–1491.

Arrow, Kenneth J., et al. "Is There a Role for Benefit–Cost Analysis in Environmental, Health, and Safety Regulation?" *Science* 272 (Apr 12, 1996): 221–222.

Bahrambegi, Ramine, et al. *The Modern Malaria Handbook: A PSR Guide to Sources and Strategies*. Washington, DC: Physicians for Social Responsibility, 1999. http://www.psr.org/malaria_handbook.htm (Jan 17, 2001).

Bartecchi, Carl E., et al. "The Global Tobacco Epidemic." *Scientific American* 272 (May 1995): 44–51.

Board on Natural Disasters. "Mitigation Emerges as Major Strategy for Reducing Losses Caused by Natural Disasters." *Science* 284 (June 18, 1999): 1943–1947.

Caron, Mary. "The Politics of Life and Death: Global Responses to HIV and AIDS." *WorldWatch* 12 (May/June 1999): 30–38.

Chivian, E., et al. *Critical Condition: Human Health and the Environment*. Cambridge, MA: MIT Press, 1993.

Cohen, John. "The Scientific Challenge of Hepatitis C." *Science* 285 (July 2, 1999): 26–30.

Daszak, Peter, et al. "Emerging Infectious Diseases of Wildlife—Threats to Biodiversity and Human Health." *Science* 287 (Jan 21, 2000): 443–448.

Ezzell, Carol. "AIDS Drugs for Africa." *Scientific American* 283 (Nov 2000): 98–103.

Foster, Kenneth R., et al. "Science and the Precautionary Principle." *Science* 288 (May 12, 2000): 979–980.

Freeze, R. Allan. *The Environmental Pendulum: A Quest for the Truth about Toxic Chemicals, Human Health, and Environmental Protection*. Berkeley, CA: Univ. of California Press, 2000.

Garrett, Laurie. *The Coming Plague: Newly Emerging Diseases in a World Out of Balance*. New York: Penguin Books, 1994.

Hattis, Dale. "Drawing the Line: Quantitative Criteria for Risk Management." *Environment* 38 (July–Aug 1996): 10–15, 35–39.

Jones, Clive, et al. "Chain Reactions Linking Acorns to Gypsy Moth Outbreaks and Lyme Disease Risk. *Science* 279 (Feb 13, 1998): 1023–1026.

Jukofsky, Diane. "Unnatural Disaster: Conservation Lessons from Hurricane Mitch." *Nature Conservancy* 49 (Sept/Oct 1999): 18–26.

McGinn, Anne Platt. "The Nicotine Cartel." *WorldWatch* 10 (July–Aug 1997): 18–27.

Morgan, M. Granger. "Risk Analysis and Management." *Scientific American* (July 1993): 32–41.

Nabarro, David N., and Elizabeth M. Taylor. "The 'Roll Back Malaria' Campaign." *Science* 280 (June 26, 1998): 2067–2068.

Nash, James A. "Moral Values in Risk Decisions." Ch. 14 in C. Richard Cothern, ed., *Handbook for Environmetnal Risk Decision Making: Values, Perceptions and Ethics* (pp. 195–212). Boca Raton, FL: CRC Press/Lewis Publishers, 1996.

Olshansky, S. Jay, et al. "Infectious Diseases—New and Ancient Threats to World Health." *Population Bulletin* 52, No. 3 (1997).

Ostfeld, Richard S., and Clive E. Jones. "Peril in the Understory." *Audubon* 101 (July/Aug 1999): 74–82.

Perera, Frederica P. "Environment and Cancer: Who Are Susceptible?" *Science* 278 (Nov 7, 1997): 1068–1073.

Pich, Emilio M., et al. "Common Neural Substrates for the Addictive Properties of Nicotine and Cocaine." *Science* 275 (Jan 3, 1997): 83–86.

Raffensperger, Carolyn, and Joel Tickner, eds. *Protecting Public Health and the Environment: Implementing the Precautionary Principle*. Washington, DC: Island Press, 1999.

Ratzan, Scott C., et al. "Attaining Global Health: Challenges and Opportunities." *Population Bulletin* 55 (1)(2000): 1–48.

Reich, Michael R. "The Global Drug Gap." *Science* 287 (Mar 17, 2000): 1979–1981.

Rockett, Ian R. H. "Injury and Violence: A Public Health Perspective." *Population Bulletin* 53, No. 4 (1998).

Rogers, David J., and Sarah E. Randolph. "The Global Spread of Malaria in a Future, Warmer World." *Science* 289 (Sept 8, 2000): 1763–1765.

Schettler, Ted, et al. *Generations at Risk: Reproductive Health and the Environment*. Cambridge, MA: MIT Press, 1999.

Smith, Kirk R. "Environmental Health—for the Rich or for All?" *Bulletin of the World Health Organization* 78 (2000): 1156–1161.

U.S. Environmental Protection Agency. *Reducing Risk: Setting Priorities and Strategies for Environmental Protection*. Washington, DC: Science Advisory Board, U.S. EPA, 1990.

Wilson, Richard, and E. A. C. Crouch. "Risk Assessment and Comparisons: An Introduction." *Science* 236 (Apr 17, 1987): 267–270.

World Health Organization. *The World Health Report 2000. Health Systems: Improving Performance*. Geneva, Switzerland: World Health Organization, 2000. http://www.who.int/whr/ (Jan 17, 2001).

World Resources Institute. *World Resources, 1998–99: Environmental Change and Human Health*. New York: Oxford University Press, 1998.

CHAPTER 17
Pests and Pest Control

Beckage, Nancy E. "The Parasitic Wasp's Secret Weapon." *Scientific American* 277 (Nov 1997): 82–87.

Blus, Lawrence J., and Charles J. Henny. "Field Studies on Pesticides and Birds: Unexpected and Unique Relations." *Ecological Applications* 7 (1997): 1125–1132.

Carson, Rachel. *Silent Spring*. Boston: Houghton Mifflin, 1962.

Gardner, Gary. "Preserving Agricultural Resources." *State of the World, 1996*. Washington, DC: WorldWatch Institute, 1996.

Groth, Edward, III, et al. *Update: Pesticides in Children's Foods*. Washington, DC: Consumer's Union of United States, 2000. http://www.ecologic-ipm.com/PDP/Update_Childrens_Foods.pdf (Jan 17, 2001).

Hussey, N. W., and N. Scopes. *Biological Pest Control*. Ithaca, NY: Cornell University Press, 1986.

Hynes, H. Patricia. *The Recurring Silent Spring*. New York: Pergamon Press, 1989.

Jaret, Peter. "What Pests Want in Your Home." *National Wildlife* 37 (Aug/Sept 1999): 40–49.

Kinley, III, David H. "Aerial Assault on the TseTse Fly." *Environment* 40 (September 1998): 14–18, 40–41.

Kling, James. "Could Transgenic Supercrops One Day Breed Superweeds?" *Science* 274 (Oct 11, 1996): 180–181.

Line, Les. 1996. "Lethal Migration." *Audubon* 98 (Sept–Oct 1996): 50–56, 94–95.

Louda, S. M., et al. "Ecological Effects of an Insect Introduced for the Biological Control of Weeds." *Science* 277 (Aug 22, 1997): 1088–1090.

Natural Resources Defense Council. *A Report on Intolerable Risk: Pesticides in Our Children's Food*. Washington, DC: Natural Resources Defense Council, 1989.

Paarlberg, Robert L. "Managing Pesticide Use in Developing Countries." Chapter 7 in Peter M. Haas, Robert O. Keohane, and Marc A. Levy, eds. *Institutions for the Earth*. Cambridge, MA: MIT Press, 1993.

Perring, Thomas M., et al. "Identification of a Whitefly Species by Genomic and Behavioral Studies. *Science* 259 (Jan 1, 1993): 74–77.

Pimentel, David. "Reducing Pesticide Use through Alternative Agricultural Practices: Fungicides and Herbicides." Chapter 23 in J. Altman, ed., *Pesticide Interactions in Crop Production: Beneficial and Deleterious Effects*. Boca Raton, FL: CRC Press, 1993.

Radcliffe, Edward B., and William D. Hutchison, eds. *Radcliffe's IPM World Textbook*. Electronic textbook of Integrated Pest Management. University of Minnesota. http://ipmworld.umn.edu/ipmchap.htm (Jan 17, 2001).

Schierow, Linda-Jo. "Pesticide Residue Regulation: Analysis of Food Quality Protection Act Implementation." *CRS Issue Brief* RS20043 (Aug 3, 1999). http://www.cnie.org/nle/pest-10.html (Jan 17, 2001).

Schulten, Gerard G. M. "Integrated Pest Management in Developing Countries." In *Environmentally Sound Technology for Sustainable Development*. ATAS Bulletin No. 7. New York: United Nations, 1992.

Smith, Miranda, ed. *The Real Dirt: Farmers Tell about Organic and Low-Input Practices in the Northeast*. Burlington, VT: Northeast Organic Farming Association, 1994.

Stevens, Jane Ellen. "In Africa, Imported Wasps Defeat Imported Pests." *Boston Globe*, Jan 3, 1994: 67, 70.

Stewart, Doug. "Luck Be a Ladybug." *National Wildlife* 32 (June/July 1994): 30–33.

Stolzenburg, William. "Double Agents: The Peril and Potential of Biological Control." *Nature Conservancy* 49 (July/Aug 1999): 18–24.

Strobel, Gary. "Biological Control of Weeds." *Scientific American* 265 (July 1991): 72–78.

Wargo, John. *Our Children's Toxic Legacy: How Science and Law Fail to Protect Us from Pesticides*. New Haven, CT: Yale University Press, 1996.

Winston, Mark L. *Nature Wars: People vs. Pests*. Cambridge, MA: Harvard University Press, 1997.

Wolff, M. S., et al. "Blood Levels of Organochlorine Residues and Risk of Breast Cancer." *Journal of the National Cancer Institute* 85 (1993): 648–652.

CHAPTER 18
Water Pollution and Its Prevention

Anderson, Donald M. "Red Tides." *Scientific American* 271 (Aug 1994): 62–69.

Assessment and Watershed Protection Division. *Guidance for Water Quality-Based Decision: The TMDL Process.* Washington, DC: U.S. EPA, Apr 1991. **http://www.epa.gov/OWOW/tmdl/decisions/** (Jan 17, 2001).

Boyle, Robert H. "Bringing Back the Chesapeake." *Audubon* 101 May/June 1999): 78–84.

Briscoe, John. "When the Cup Is Half Full: Improving Water and Sanitation Services in the Developing World." *Environment* 35 (May 1993): 6–15.

Carmichael, Wayne W. "The Toxins of Cyanobacteria." *Scientific American* 270 (Summer 1994): 78–86.

Carpenter, Stephen, et al. "Nonpoint Pollution of Surface Waters with Phosphorus and Nitrogen." *Issues in Ecology* 3 (Summer 1998): 1–12.

Committee on Environment and Natural Resources. *Integrated Assessment of Hypoxia in the Northern Gulf of Mexico.* Washington, DC: National Science and Technology Council and NOAA (May 2000). **http://www.nos.noaa.gov/products/pubs_hypox.html** (Jan 17, 2001).

Copeland, Claudia. *Clean Water Act Reauthorization.* CRS Issue Brief 10001. Nov 2, 2000. **http://www.cnie.org/nle/h2o-2.html** (Jan 17, 2001).

Daley, Beth. "Sewage Tunnel Set in New Era for Harbor." *The Boston Globe* (Sept 5, 2000): 1, 16.

Dugan, Eugene, ed. *Wetlands in Danger: A World Conservation Atlas.* New York: Oxford University Press, 1993.

Ervin, David E., et al. "A New Strategic Vision." *Environment* 40 (July–Aug 1998): 8–15, 35–40.

Gadgil, Ashok. "Drinking Water in Developing Countries." *Annual Review of Energy and Environment* 23 (1998): 253–286.

Gleick, Peter H. "Water in Crisis: Paths to Sustainable Water Use." *Ecological Applications* 8 (1998): 571–579.

Graff, Gordon. "The Chlorine Controversy." *Technology Review* (Jan 1995): 54–60.

Hammer, Donald A., ed. *Constructed Wetlands for Wastewater Treatment.* Chelsea, MI: Lewis Publishers, 1989.

Langenkamp, Daniel. "Signs of Life on Tisza and Danube." *Christian Science Monitor* (Mar 8, 2000).

Lisk, Ian. "Tertiary-Treated Wastewater Irrigates Monterey Peninsula's Recreational Acreage." *Water Engineering and Management* (July 1995): 18–23.

Malle, Karl-Geert. "Cleaning Up River Rhine." *Scientific American* 274 (January 1996): 70–75.

Morris, J. R., and J. Glenn. "Harmful Algal Blooms: An Emerging Public Health Problem with Possible Links to Human Stress on the Environment." *Annual Review of Energy and Environment* 24 (1999): 367–390.

Salvesen, David. *Wetlands: Mitigating and Regulating Development Impacts,* 2d ed. Washington, DC: The Urban Land Institute, 1994.

Schneider, Paul. "Clear Progress: 25 Years of the Clean Water Act." *Audubon* 99 (Sept–Oct 1997): 36–47, 106–108.

Van Deveer, Stacy D. "Protecting Europe's Seas: Lessons from the Last 25 Years." *Environment* 42 (July/Aug 2000): 10–26.

Warrick, Joby. "Death in the Gulf of Mexico." *National Wildlife* 37 (June/July 1999): 48–53.

Wittrup, Larry. "Biogas Project Advances in California." *Biocycle* (Apr 1995): 48–49.

CHAPTER 19
Municipal Solid Waste: Disposal and Recovery

Abramovitz, Janet N., and Ashley T. Mattoon. "Paper Cuts: Recovering the Paper Landscape." WorldWatch Paper 149 (Dec 1999). Washington, DC: WorldWatch Institute.

Anderson, Mikail Skou. "Assessing the Effectiveness of Denmark's Waste Tax." *Environment* 40 (May 1998): 10–15, 38–41.

Boener, Christopher, and Kenneth Chilton. "False Economy: The Folly of Demand-Side Recycling." *Environment* 36 (Jan 1994): 6–15, 32–33.

Buchholz, Rogene A., et al. "The Politics of Recycling in Rhode Island." In Rogene A. Buchholz et al., *Managing Environmental Issues: A Casebook.* Englewood Cliffs, NJ: Prentice Hall, 1992.

Davis, Gary A., et al. "Extended Product Responsibility: A Tool for a Sustainable Economy." *Environment* 39 (Sept 1997): 10–15, 36–38.

Denison, Richard T. "Environmental Life-Cycle Comparisons of Recycling, Landfilling, and Incineration." *Annual Review of Energy and the Environment* 21 (1996): 191–237.

Gaige, C. David, and Richard T. Halil, Jr. "Clearing the Air about Municipal Waste Combustors." *Solid Waste & Power* (Jan/Feb 1992): 12–17.

Hilts, Michael E. "WTE: Building a Record of Dependable Waste Disposal and Environmental Safety." *Solid Waste Technologies/ Industry Sourcebook* (1994): 12–16.

Iranpour, R., et al. "Environmental Engineering: Energy Value of Replacing Waste Disposal with Resource Recovery." *Science* 285 (July 30, 1999): 706–710.

McCarthy, James E. *Interstate Shipment of Municipal Solid Waste.* CRS Report 98–689 ENR (Aug 6, 1998). **http://www.cnie.org/nle/waste-7.html** (Jan 17, 2001).

McCarthy, James E. *Interstate Waste Transport: Legislative Issues* CRS Issue Brief RS20106 (June 16, 1999). **http://www.cnie.org/nle/waste-29.html** (Jan 17, 2001).

National Research Council. *Waste Incineration and Public Health.* Washington, DC: National Academy Press, 1999.

Platt, Brenda, et al. *Beyond 40 Percent: Record-Setting Recycling and Composting Programs.* Washington, DC: Island Press, 1991.

Preer, Robert. "Sealed and Buried, Closed Dumps Get a Life." *Boston Globe,* Oct 10, 1993: 28, 30.

Russell, Gerard F. "Trash Bag Fee Buys Recyclers: Worcester's Rules Pay Off in Compliance." *Boston Globe,* Jan 13, 1994: 25, 31.

Scherer, Ron. "Staten Island Has Had Its Fill of Nation's Largest Landfill." *Christian Science Monitor* (Apr 26, 1996): 1, 4.

U.S. Environmental Protection Agency. *Municipal Solid Waste Generation, Recycling and Disposal in the United States: Facts and Figures for 1998.* April 2000. **http://www.epa.gov/epaoswer/osw/publicat.htm** (Jan 17, 2001).

U.S. Environmental Protection Agency. "The Municipal Solid Waste Handbook." Office of Solid Waste Factbook. **http://www.epa.gov/epaoswer/non-hw/muncpl/factbook/index.htm**.

CHAPTER 20
Hazardous Chemicals: Pollution and Prevention

Bullard, Robert D., ed. *Unequal Protection: Environmental Justice and Communities of Color.* San Francisco: Sierra Club Books, 1994.

Egeland, Grace M., and John P. Middaugh. "Balancing Fish Consumption Benefits with Mercury Exposure." *Science* 278 (Dec 12 1997): 1904–1905.

Glass, David J. "Waste Management—Biological Treatment of Hazardous Wastes." *Environment* 33 (Nov 1991): 5, 43–45.

Grumbly, Thomas P. "Lessons from Superfund." *Environment* 37 (Mar 1995): 33–35.

Krimsky, Sheldon. *Hormonal Chaos: The Scientific and Social Origins of the Environmental Endocrine Hypothesis.* Baltimore: Johns Hopkins University Press, 2000.

Krueger, Jonathan. "What's to Become of Trade in Hazardous Wastes? The Basel Convention One Decade Later." *Environment* (November 1999): 11–21.

Luoma, Jon R. "Havoc in the Hormones." *Audubon* 97 (May/June 1995): 60–67.

Mastny, Lisa. "Coming to Terms with the Arctic." *WorldWatch* 13 (Jan/Feb 2000): 24–36.

McCarthy, James E., and Mary Tiemann. *MTBE in Gasoline: Clean Air and Drinking Water Issues.* CRS Issue Brief 98-290 (Oct 2, 2000). **http://www.cnie.org/nle/air-26.html** (Jan 17, 2001).

McGinn, Anne Platt. *Why Poison Ourselves? A Precautionary Approach to Synthetic Chemicals.* WorldWatch Paper 153 (Nov 2000). Washington, DC: WorldWatch Institute.

McGinn, Anne Platt. "POPS Culture." *WorldWatch* 13 (Mar/Apr 2000): 26–36.

Mukerjee, Madhusree. "Persistently Toxic." *Scientific American* (June 1995): 16–18.

Murphy, Steven. "Law—the Basel Convention on Hazardous Wastes." *Environment* 35 (Mar 1993): 42–44.

Nabholz, J. V., R. G. Clements, and M. G. Zeeman. "Information Needs for Risk Assessment in EPA's Office of Pollution Prevention and Toxics." *Ecological Applications* 7 (1997): 1094–1098.

National Research Council. *Hormonally Active Agents in the Environment.* Washington, DC: National Academy Press, 2000.

National Toxicology Program. *Report on Carcinogens,* 9th ed. Washington, DC: Department of Health and Human Services, 2000. **http://ehis.niehs.nih.gov/roc/toc9.html** (Jan 17, 2001).

Nriagu, Jerome O. "Tales Told in Lead." *Science* 281 (Sept 11, 1998): 1622–1623.

Probst, Katherine N., and Thomas C. Beierle. "Hazardous Waste Management: Lessons from Eight Countries." *Environment* 41 (Nov 1999): 22–30.

Reisch, Mark. *Superfund Reauthorization Issues in the 106th Congress.* CRS Issue Brief 10011. Oct 30, 2000. **http://www.cnie.org/nle/waste-28.html** (Jan 17, 2001).

Reisch, Mark. *Superfund and the Brownfields Issue.* CRS Report 97–731. July 24, 2000.

Scheberle, Denise. "Partners in Policymaking: Forging Effective Federal–State Relations. *Environment* 40 (Dec 1998): 14–20, 28–30.

Schierow, Linda-Jo. *Dioxin: Reassessing the Risk.* CRS Report 95–1059 ENR. Sept 11, 1998. **http://www.cnie.org/nle/pest-7.html** (Jan 17, 2001).

Tiemann, Mary. *Leaking Underground Storage Tank Cleanup Issues.* CRS Report 97–471. Feb 17, 1999. **http://www.cnie.org/nle/waste-18.html** (Jan 17, 2001).

Tiemann, Mary. *Waste Trade and the Basel Convention: Background and Update.* CRS Report 98–638 ENR. Dec 30, 1998. **http://www.cnie.org/nle/waste-26.html** (Jan 17, 2001).

U.S. EPA. *Toxics Release Inventory: 1998 Data Release.* Washington, DC: U.S. EPA. May 2000. **http://www.epa.gov/tri/** (Jan 17, 2001).

Walters, Laurel Shaper. "The Legacy of an Unnatural Disaster." *Christian Science Monitor* (Apr 23, 1997): 4.

CHAPTER 21

The Atmosphere: Climate, Climate Change, and Ozone Depletion

Alley, Richard B., and Michael L. Bender. "Greenland Ice Cores: Frozen in Time." *Scientific American* 278 (Feb 1998): 80–85.

Bard, Edouard, et al. "Hydrological Impact of Heinrich Events in the Subtropical Northeast Atlantic." *Science* 289 (Aug 25, 2000): 1321–1323.

Battle, M., et al. "Global Carbon Sinks and Their Variability Inferred from Atmospheric O_2 and ^{13}C." *Science* 287 (Mar 31, 2000): 2467–2470.

Benedick, Richard E. *Ozone Diplomacy: New Directions in Safeguarding the Planet.* Cambridge, MA: Harvard University Press, 1991.

Bindschaler, Robert, and Patricia Vornberger. "Changes in the West Antarctic Ice Sheet since 1963 from Declassified Satellite Photography. *Science* 279 (Jan 30, 1998): 689–692.

Bolin, Bert. "The Kyoto Negotiations on Climate Change: A Science Perspective." *Science* 279 (Jan 16, 1998): 330–331.

Bousquet, Philippe, et al. "Regional Changes in Carbon Dioxide Fluxes of Land and Oceans since 1980." *Science* 290 (Nov 17, 2000): 1342–1346.

Broeker, Wallace S. "Chaotic Climate." *Scientific American* 273 (Nov 1995): 62–68.

Broeker, Wallace S. "Thermohaline Circulation, the Achilles Heel of Our Climate System: Will Man-made CO_2 Upset the Current Balance?" *Science* 278 (Nov 28, 1997): 1582–1588.

Charlson, Robert J., and Tom M. L. Wigley. "Sulfate Aerosol and Climatic Change." *Scientific American* (Feb 1994): 48–57.

Coppock, Rob. "Implementing the Kyoto Protocol." *Issues in Science and Technology* (Spring 1998): 66–74.

Falkowski, P., et al. "The Global Carbon Cycle: A Test of Our Knowledge of Earth as a System." *Science* 290 (Oct 13, 2000): 291–295.

Grassl, Helmut. "Status and Improvement of Coupled General Circulation Models." *Science* 288 (June 16, 2000): 1991–1996.

Hansen, James E., et al. "Global Climate Data and Models: A Reconciliation." *Science* 281 (Aug 14, 1998): 930–932.

Hinrichsen, Don. "The Oceans Are Coming Ashore." *WorldWatch* 13 (Nov/Dec 2000): 26–35.

Houghton, J. T., et al. *Climate Change 1995–the Science of Climate Change: Contribution of Working Group I to the Second Assessment Report of the Intergovernmental Panel on Climate Change.* Cambridge, U.K.: Cambridge University Press, 1996.

Houghton, John. *Global Warming: The Complete Briefing,* 2d ed. Cambridge, U.K.: Cambridge University Press, 1997.

Intergovernmental Panel on Climate Change. *Climate Change 2001: The Scientific Basis. Summary for Policymakers.* IPCC Working Group I Third Assessment Report. Jan, 2001. **http://www.ipcc.ch/** (Jan 23, 2001).

Karl, Thomas R., Neville Nichols, and Jonathan Gregory. "The Coming Climate." *Scientific American* 276 (May 1997): 78–83.

Kasemir, Bernd, et al. "Involving the Public in Climate and Energy Decisions." *Environment* 42 (April 2000): 32–41.

Kerr, J. B., and C. T. McElroy. "Evidence for Large Upward Trends of Ultraviolet-B Radiation Linked to Ozone Depletion." *Science* 262 (Nov 12, 1993): 1032–1034.

Levitus, Sydney, et al. "Warming of the World Ocean." *Science* 287 (Mar 24, 2000): 2225–2229.

Levy, David L., and Peter Newell. "Oceans Apart? Business Responses to Global Environmental Issues in Europe and the United States." *Environment* 42 (November 2000): 8–20.

McCarthy, James J., and Malcolm C. McKenna. "How Earth's Ice Is Changing." *Environment* 42 (December 2000): 8–18.

Minnis, P., et al. "Radioactive Climate Forcing by the Mount Pinatubo Eruption." *Science* 259 (Mar 5, 1993): 1411–1415.

Montzka, Stephen, et al. "Decline in the Tropospheric Abundance of Halogen from Halocarbon: Implications for Stratospheric Ozone Depletion. *Science* 272 (May 31, 1996): 1318–1322.

Muller, Frank. "Mitigating Climate Change: The Case for Energy Taxes." *Environment* 38 (Mar 1996): 12–20, 36–43.

Ott, Hermann E. "The Kyoto Protocol: Unfinished Business." *Environment* 40 (July–Aug 1998): 16–20, 41–45.

Parson, E. A., and D. W. Keith. "Fossil Fuels without CO_2 Emissions. *Science* 282 (Nov 6, 1998): 1053–1054.

Randel, William J., et al. "Trends in the Vertical Distribution of Ozone." *Science* 285 (Sept 10, 1999): 1689–1692.

Rowland, F. Sherwood. "President's Lecture: The Need for Scientific Communication with the Public." *Science* 260 (June 11, 1993): 1571–1576.

Sarmiento, Jorge L., and Steven C. Wofsy. *A U.S. Carbon Cycle Science Plan.* Washington, DC: U.S. Global Change Research Program, 1999.

Schneider, David. "The Rising Seas." *Scientific American* 276 (Mar 1997): 112–117.

Scolnikoff, Eugene B. "The Role of Science in Policy: The Climate Change Debate in the United States." *Environment* 41 (June 1999): 16–20, 42–45.

Sellers, P. J., et al. "Modeling the Exchanges of Energy, Water and Carbon between Continents and the Atmosphere." *Science* 275 (Jan 14, 1997): 502–509.

Solomon, Barry, and Russell Lee. "Emissions Trading Systems and Environmental Justice." *Environment* 42 (October 2000): 32–45.

Suplee, Curt. "El Niño, La Niña: Nature's Vicious Cycle." *National Geographic* 195 (March 1999): 72–95.

Swift, Byron. "A Low-Cost Way to Control Climate Change." *Issues in Science and Technology* (Spring 1998): 75–81.

Tabazadeh, A., and R. P. Turco. "Stratospheric Chlorine Injection by Volcanic Eruptions: HCl Scavenging and Implications for Ozone." *Science* 260 (May 21, 1993): 1082–1086.

Toman, Michael, and Jean-Charles Hourcade. "From Bonn to the Hague, Many Questions Remain." *Resources* 138 (Winter 2000): 14–16.

United Nations Environment Program. "Environmental Effects of Ozone Depletion: 1998 Assessment" (seven articles and commentary). *Journal of Photochemistry and Photobiology: B. Biology*: 46, Oct 1998.

U.S. Country Studies Program. *Climate Change: Mitigation, Vulnerability, and Adaptation in Developing and Transition Countries.* Washington, DC: U.S. Country Studies Program, 1999.

U.S. Global Change Research Program. *Climate Change Impacts on the United States: The Potential Consequences of Climate Variability and Change.* New York: Cambridge University Press, 2000. http://www.gcrio.org/NationalAssessment/ (Jan 23, 2001).

White, Robert M. "Kyoto and Beyond." *Issues in Science and Technology* (Spring 1998): 59–65.

Young, Oran R. "Hitting the Mark: Why Are Some International Environmental Agreements More Successful than Others?" *Environment* (Oct 1999): 20–29.

CHAPTER 22
Atmospheric Pollution

Andreae, Mainrat O., and Paul J. Crutzen. Atmospheric Aerosols: Biogeochemical Sources and Role in Atmospheric Chemistry. *Science* 276 (May 16, 1997): 1052–1058.

Bamberger, Robert. *Automobile and Light Truck Fuel Economy: Is CAFE Up to Standards?* CRS Issue Brief 90122. Sept 22, 2000. http://www.cnie.org/nle/air-10.html (Jan 23, 2001).

Blackman, Allen. "Small Is Not Necessarily Beautiful: Coping with Dirty Microenterprises in Developing Countries. *Resources* 141 (Fall 2000): 9–13.

Brune, William H. "Stalking the Elusive Atmospheric Hydroxyl Radical." *Science* 256 (May 22, 1992): 1154–1155.

Carson, Nancy, and Don Munton. "Flaws in the Conventional Wisdom on Acid Deposition." *Environment* 42 (March 2000): 33–35.

Chameides, W. L., et al. "Growth of Continental-Scale Metro-Agro-Plexes, Regional Ozone Pollution, and World Food Production *Science* 264 (Apr 1, 1994): 74–77.

Coglianese, Cary. "The Constitution and the Costs of Clean Air: Is the Clean Air Act Unconstitutional?" *Environment* 42 (Nov 2000): 32–37.

Committee on Environment and Natural Resources. *National Acid Precipitation Assessment Program Biennial Report to Congress: An Integrated Assessment.* Washington, DC: National Science and Technology Council, May 1998. http://www.nnic.noaa.gov/CENR/NAPAP/NAPAP_96.htm (Jan 23, 2001).

Dillon, John. Fading Colors: Are We Losing the Sugar Maples to Acid Rain? *Audubon* 100 (Sept–Oct 1998): 46–51.

Findlayson-Pitts, Barbara J., and James N. Pitts. *Atmospheric Chemistry: Fundamentals and Experimental Techniques.* New York: John Wiley & Sons, 1986.

Findlayson-Pitts, Barbara J., and James N. Pitts. Tropospheric Air Pollution: Ozone, Airborne Toxics, Polycyclic Aromatic Hydrocarbons, and Particulates. *Science* 276 (May 16, 1997): 1045–1051.

Hedin, Lars O., and Gene E. Likens. Atmospheric Dust and Acid Rain. *Scientific American* (Dec 1996): 86–92.

Kerr, Richard. "Acid Rain Control: Success on the Cheap." *Science* 282 (Nov 6, 1998): 1024–1027.

Kley, Dieter. Tropospheric Chemistry and Transport. *Science* 276 (May 16, 1997): 1043–1044.

Lents, James M., and William J. Kelley. "Clearing the Air in Los Angeles." *Scientific American* (Oct 1993): 32–39.

Lerdau, Manuel T., and Heather L. Throop. "Isoprene Emission and Photosynthesis in a Tropical Forest Canopy: Implications for Model Development." *Ecological Applications* 9 (4) (1999): 1109–1117.

Likens, G. E., et al. "Long-term Effects of Acid Rain: Response and Recovery of a Forest Ecosystem. *Science* 272 (Apr 12, 1996): 244–246.

Mahlman, J. D. Dynamics of Transport Processes in the Upper Troposphere. *Science* 276 (May 16, 1997): 1079–1083.

McCormick, John. Acid Pollution: The International Community's Continuing Struggle. *Environment* 40 (July–Aug 1998): 16–20, 41–45.

Meng, Z., D. Dabdub, and J. H. Seinfeld. Chemical Coupling between Atmospheric Ozone and Particulate Matter. *Science* 277 (July 4, 1997): 116–119.

Morey, Candace, et al. *Pollution Lineup: An Environmental Ranking of Automakers.* Cambridge, MA: Union of Concerned Scientists, Mar 2000. http://www.ucsusa.org/ (Jan 23, 2001).

Munton, Don. Dispelling the Myth of the Acid Rain Story. *Environment* 40 (July–Aug 1998): 4–7, 27–34.

Particle Epidemiology Reanalysis Project. *Reanalysis of the Harvard Six Cities Study and the American Cancer Society Study of Particulate Air Pollution and Mortality.* Health Effects Institute, July 2000. http://www.healtheffects.org/pubs-recent.htm (Jan 23, 2001).

Pope, C. Arden, et al. "Particulate Air Pollution as a Predictor of Mortality in a Prospective Study of U.S. Adults." *American Journal of Respiratory and Critical Care Medicine* 151 (1995): 669–674.

Sillman, Sanford. "Tropospheric Ozone: The Debate over Control Strategies." *Annual Review of Energy and the Environment* 18 (1993): 31–56.

Soroos, Marvin S. Preserving the Atmosphere as a Global Commons. *Environment* 40 (Feb 1998): 6–13, 32–35.

Sperling, Daniel. "The Case for Electric Vehicles." *Scientific American* 275 (Nov 1996): 54–59.

Stoddard, J. L., et al. "Regional Trends in Aquatic Recovery from Acidification in North America and Europe." *Nature* 401 (Oct 7, 1999): 575–578.

Tuinstra, Willemijn, et al. "Using Computer Models in International Negotiation: The Case of Acidification in Europe." *Environment* 41 (Nov 1999): 32–42.

United States Environmental Protection Agency. *Latest Findings on National Air Quality: 1999 Status and Trends.* Office of Air Quality, Planning and Standards. Aug 2000. http://www.epa.gov/airtrends/ (Jan 23, 2001).

Wennberg, P. O., et al. Hydrogen Radicals, Nitrogen Radicals, and the Production of Ozone in the Upper Troposphere. *Science* 279 (Jan 2, 1998): 49–53.

Wouk, Victor. "Hybrid Electric Vehicles." *Scientific American* 277 (Oct 1997): 70–77.

CHAPTER 23
Economics, Public Policy, and the Environment

Arrow, Kenneth J., et al. "Is There a Role for Benefit–Cost Analysis in Environmental, Health, and Safety Regulation?" *Science* 272 (April 12, 1996): 221–222.

Barrett, Christopher B. "Markets, Social Norms, and Governments in the Service of Environmentally Sustainable Economic Development: The Pluralistic Stewardship Approach." *Christian Scholars Review* 29 (Spring 2000): 435–454.

Bezdek, Roger H. "Environment and Economy: What's the Bottom Line?" *Environment* 35 (Sept 1993): 6–11, 25–32.

Bhagwate, Jagdish, and Herman E. Daly. "Debate: Does Free Trade Harm the Environment?" *Scientific American* (Nov 1993): 41–57.

Buchholz, Rogene A. *Principles of Environmental Management,* 2d ed. Upper Saddle River, NJ: Prentice Hall, 1998.

Costanza, Robert, et al. *An Introduction to Ecological Economics.* Boca Raton, FL: St. Lucie Press, 1997.

Cruz, Wilfrido, et al. "Greening Development: Environmental Implications of Economic Policies." *Environment* 38 (June 1996): 6–11, 31–38.

Farrow, Scott, and Michael Toman. "Using Benefit–Cost Analysis to Improve Environmental Regulations." *Environment* 41 (Feb 1999): 12–15, 33–37.

French, Hilary. "Challenging the WTO." *WorldWatch* 12 (Nov/Dec 1999): 22–27.

Hamilton, Kirk, and Ernst Lutz. *Green National Accounts: Policy Uses and Empirical Experience.* World Bank Environment Department Paper No. 039. Washington, DC: World Bank, 1996.

Harashima, Yongi. "Effects of Economic Growth on Environmental Policies in Northeast Asia." *Environment* 42 (July/Aug 2000): 28–40.

Harrington, Winston, et al. "Predicting the Costs of Environmental Regulations: How Accurate Are Regulators' Estimates?" *Environment* 41 (Sept 1999): 10–14, 40–44.

Hawken, Paul, Amory Lovins, and L. Hunter Lovins. *Natural Capitalism: Creating the Next Industrial Revolution.* Boston: Little, Brown and Co., 1999.

Hecht, Joy E. "Environmental Accounting: Where We Are Now, Where We Are Going." *Resources* 135 (Spring 1999): 14–17.

Hoffman, Andrew J. "Integrating Environmental and Social Issues into Corporate Practice." *Environment* 42 (June 2000): 22–33.

Krupnick, Alan J., and Paul R. Portney. "Controlling Urban Air Pollution: A Benefit–Cost Assessment." *Science* 252 (Apr 26, 1991): 522–528.

Kunte, Arundhati, et al. *Estimating National Wealth: Methodology and Results.* Environment Department Paper No. 57. Washington, DC: World Bank, 1998.

Portney, Paul R. "Counting the Cost: The Growing Role of Economics in Environmental Decision-making." *Environment* 40 (Mar 1998): 14–18, 36–38.

Portney, Paul R. "Environmental Problems and Policy: 2000–2050." *Resources* 138 (Winter 2000): 6–10.

Prugh, Thomas, et al. *Natural Capital and Human Economic Survival.* Sunderland, MA: Sinauer Associates, Inc., 1995.

Renner, Michael. *Working for the Environment: A Growing Source of Jobs.* WorldWatch Paper 152. Washington, DC: WorldWatch Institute, 2000.

Repetto, Robert. "Accounting for Environmental Assets." *Scientific American* (June 1992): 94–100.

Repetto, Robert, and Roger C. Dower. "Green Fees: Charging for Environmentally Damaging Activities." *Environmental Science and Technology* 27 (1993): 214–216.

Serageldin, Ismail. *Sustainability and the Wealth of Nations: First Steps in an Ongoing Journey.* World Bank Environmentally Sustainable Development Studies and Monographs Series No. 5. Washington, DC: World Bank, 1996.

Simpson, R. David. "Economic Analysis and Ecosystems: Some Concepts and Issues." *Ecological Applications* 8 (1998): 342–349.

U.S. Department of Commerce. *Meeting the Challenge: U.S. Industry Faces the 21st Century* U.S. Environmental Industry Executive Summary. (1998). **http://www.ta.doc.gov/Reports/Environmental/Execsum.pdf**) (Jan 23, 2001).

White, Allen R. "Sustainability and the Accountable Corporation: Society's Rising Expectations of Business." *Environment* 41 (Oct 1999): 30–43.

Winsemius, Pieter, and Ulrich Guntram. "Responding to the Environmental Challenge." *Business Horizons* (Mar/Apr 1992): 12–20.

World Bank. *Monitoring Environmental Progress: A Report on Work in Progress.* Washington, DC: World Bank, 1995.

World Bank. *Expanding the Measure of Wealth: Indicators of Environmentally Sustainable Development.* World Bank Environmentally Sustainable Development Studies and Monographs Series No. 17. Washington, DC: World Bank, 1997.

CHAPTER 24
Lifestyle and Sustainability

Bacow, Adele Fleet. *Designing the City.* Washington, DC: Island Press, 1995.

Brockerhoff, Martin P. "An Urbanizing World." *Population Bulletin* 55 (September 2000). Washington, DC: Population Reference Bureau.

Buttimer, Anne. "Close to Home: Making Sustainability Work at the Local Level." *Environment* 40 (Apr 1998): 12–15, 32–40.

Chen, Donald D. T. "The Science of Smart Growth." *Scientific American* 283 (Dec 2000): 84–91.

Clinton–Gore Administration. *Building Livable Communities: Sustaining Prosperity, Improving Quality of Life, Building a Sense of Community.* (June 2000). **http://www.livablecommunities.gov/** (Jan 23, 2001).

Fuller, Millard, and Diane Scott. *No More Shacks.* Waco, TX: Word Books, 1986.

Gangloff, Deborah, et al. "Urban Forests" (special section). *American Forests* (May/June 1995): 30–37.

Gregory, Robin. "Using Stakeholder Values to Make Smarter Environmental Decisions." *Environment* 42 (June 2000): 34–44.

Johnson, Kenneth M. "The Rural Rebound." *Reports on America* 1 (Sept 1999). Washington, DC: Population Reference Bureau.

Lowe, Marcia D. *Alternatives to the Automobile: Transport for Livable Cities.* WorldWatch Paper 98. Washington, DC: WorldWatch Institute, 1990.

Lowe Marcia D. *Shaping Cities: The Environmental and Human Dimensions.* WorldWatch Paper 105. Washington, DC: WorldWatch Institute, 1991.

Morris, David. *Getting from Here to There: Building a Rational Transportation System.* Washington, DC: Institute for Local Self-Reliance, 1992.

National Research Council. *Our Common Journey: A Transition toward Sustainability.* Washington, DC: National Academy of Sciences, 1999.

O'Meara, Molly. *Reinventing Cities for People and the Planet.* WorldWatch Paper 147. Washington, DC: WorldWatch Institute, June 1999.

President's Council on Sustainable Development. *Sustainable America: A New Consensus for Prosperity, Opportunity and a Healthy Environment.* Washington, DC: U.S. Government Printing Office, 1996.

President's Council on Sustainable Development. *Sustainable Communities.* Report of Task Force on Sustainable Communities, 1998. (www.whitehouse.gov/PCSD/Publicatiions/suscomm/ind_suscom.htm).

President's Council on Sustainable Development. *Towards a Sustainable America: Advancing Prosperity, Opportunity, and a Healthy Environment for the 21st Century.* Washington, DC: U.S. Government Printing Office, 1999.

Renew America. *Environmental Success Index, 1992.* Washington, DC: Renew America, 1992. A compilation of noteworthy environmental projects throughout the country, with names and telephone numbers to contact.

Renew America. *National Awards for Environmental Sustainability.* Renew America (Jan 1995).

Rosaland, Mark, ed. *Eco-City Dimensions: Healthy Communities, Healthy Planet.* New Haven, CT: New Society Publishers, 1997.

Serageldin, Ismail, and Andrew Steer. *Making Development Sustainable: From Concepts to Action.* Environmentally Sustainable Development Occasional Paper Series No. 2. Washington, DC: World Bank, 1994.

Stoel, Thomas B., Jr. "Reining In Urban Sprawl." *Environment* 41 (May 1999): 6–11, 29–33.

The Sustainable Cities Program. *City Experiences in Improving the Urban Environment: A Snapshot of an Evaluation of Six City Initiatives in Africa.* Nairobi: United Nations Centre for Human Settlements/United Nations Environment Program, 1999. **http://www.unchs.org/scp/** (Jan 23, 2001).

United Nations Environment Program. *The Sustainable Cities Program.* Provides link to *The Environmental Planning and Management Sourcebook.* 1999. **http://www.unchs.org/scp/** (Jan 23, 2001).

U.S. Environmental Protection Agency. *Sustainability in Action: Profiles of Community Initiatives across the United States.* Fifty case studies from around the nation. **http://www.sustainable.org/casestudies/studiesindex.html** (Jan 23, 2001).

Watson, Robert T., et al. *Protecting Our Planet, Securing Our Future: Linkages among Global Environmental Issues and Human Needs.* Washington, DC: NASA, World Bank, UNEP, 1998.

Wood, Daniel B. "Backlash against Urban Sprawl Broadens." *Christian Science Monitor* (Dec 16, 1999): 1, 4.

World Bank. *Five Years After Rio: Innovations in Environmental Policy.* World Bank Environmentally Sustainable Development Studies and Monographs Series No. 18. Washington, DC: World Bank, 1997.

Photo Credits

11-8a, C. Allan Morgan/Peter Arnold, Inc.; Figure 11-8b, Mickey Gibson/Animals Animals/Earth Scenes; Figure 11-8c, Julia Sims/Peter Arnold, Inc.; Figure 11-8d, A.H. Rider/Photo Researchers, Inc.; Figure 11-8e, Jeff Lepore/Photo Researchers, Inc.; Figure 11-8f, C.C. Lockwood/Animals Animals/Earth Scenes; Figure 11-8g, Tom McHugh/Steinhart Aquarium/Photo Researchers, Inc.; Figure 11-8h, Steve Kaufman/Peter Arnold, Inc.; Figure 11-08i, Douglas Faulkner/Photo Researchers, Inc.; Figure 11-9, Darek Karp/Animals Animals/Earth Scenes; Figure 11-12, John Colwell/Grant Heilman Photography, Inc.; Figure 11-13, /Corbis; Figure 11-14, Klaus Uhlenhut/Animals Animals/Earth Scenes; Figure 11-15a, Richard Mack/Ecological Society of America; Figure 11-15b, Richard Mack/Ecological Society of America; Figure 11-16, Reproduced with permission from Richard P. Cincotta and Robert Engelman, Nature's Place: Human Population; CO-11, Richard T. Wright

Chapter 12
Figure 12-1-1, Bud Lehnhausen/Photo Researchers, Inc.; Figure 12-2-2, Ray Pfortner/Peter Arnold, Inc.; Figure 12-3, Andrew L. Young/Photo Researchers, Inc.; Figure 12-4, BIOS (A. Compost)/Peter Arnold, Inc.; Figure 12-5, Grant Heilman/Grant Heilman Photography, Inc.; Figure 12-7, Ted Levin/Animals Animals/Earth Scenes; Figure 12-8, Tom Fox/SWA Group/Houston; Figure 12-11, Michael S. Yamashita/Woodfin Camp [Associates] Figure 12-15, Jan Robert Factor/Photo Researchers, Inc.; Figure 12-16, Richard T. Wright; CO-12, Corbis; Figure 12-10UN-EW, William Hubbel/Woodfin Camp Associates

Chapter 13
Figure 13-1, Huw Evans/AP/Wide World Photos] Figure 13-3, Tony Wells/Mason Dixon Historical Society, Inc.; Figure 13-6d, C.P. Hickman/ Visuals Unlimited; Figure 13-8, Neal Palumbo/Liaison Agency, Inc.; Figure 13-14, Wesley Bocxe/Photo Researchers, Inc.; Figure 13-16, Chris Jones/Corbis/Stock Market; Figure 13-18, Bernard J. Nebel; CO-13, Mark Marten/ NASA/ Photo Researchers, Inc.; PO-04, Paul McCormick/ The Image Bank

Chapter 14
CO-14 Pete Saloutos/Corbis/Stock Market; Figure 14-1, JCO Co. Ltd./ AP/Wide World Photos; Figure 14-3, DiMaggio/Kalish/Corbis/Stock Market; Figure 14-7c, Erich Hartmann/Magnum Photos, Inc.; Figure 14-12, Laura Rauch/AP/Wide World Photos; Figure 14-13, Shone-Zoufarov/Liaison Agency, Inc.; Figure 14-15, Alexander Tsiaras/Science Source/Photo Researchers, Inc.

Chapter 15
Figure 15-1, James D. Beard; Figure 15-2, Tom McHugh/Photo Researchers, Inc.; Figure 15-3, Gabe Palmer/Corbis/Stock Market; Figure 15-7b, 15-8a, Bernard J. Nebel; Figure 15-8b, Reuters/STR/Archive Photos; Figure 15-11, /Au Sable Institute of Environmental Studies; Figure 15-12, Bernard J. Nebel; Figure 15-14, T.J. Florian/Rainbow; Figure 15-15, /Solel Solar Systems Ltd.; Figure 15-16, 15-17, Sandia National Laboratories; Figure 15-19, Lowell Georgia/Photo Researchers, Inc.; Figure 15-20, Tom McHugh/Photo Researchers, Inc.; Figure 15-21, Creative Energy Corporation/Avanti Pellet Stove; Figure 15-22, Glenn Edwards/Panos Pictures; Figure 15-23, /Pacific Gas [Electric Company] CO-15, Mark Wilson/The Boston Globe

Chapter 16
Figure 16-1a, E.R. Degginger/Animals Animals/Earth Scenes; Figure 16-1b, Zig Leszczynski/Animals Animals/Earth Scenes; Figure 16-4, AP/Wide World Photos; Figure 16-6a1, Jan Hinsch/Science Photo Library/Photo Researchers, Inc.; Figure 16-6b1, National Cancer Institute/Science Photo Library/Photo Researchers, Inc.; Figure 16-7a1, Gregory Bull/AP/Wide World Photos; Figure 16-7b1, Mike Yamashita/Woodfin Camp [Associates] Figure 16-7c1, Rich Saal/State Journal Register/AP/Wide World Photos; Figure 16-7d1, Blanca Gutierrez/AP/Wide World Photos; Figure 16-8a1, Calvin Larsen/Photo Researchers, Inc.; Figure 16-8b1, Doug Wechsler/Animals Animals/Earth Scenes; Figure 16-8c1, Gilles Mingasson/Liaison Agency, Inc.; Figure 16-8d1, Deep Light Productions/Science Photo Library/Photo Researchers, Inc.; Figure 16-9, Shehzad Noorani/Woodfin Camp [Associates] Figure 16-11, A. Ramey/Woodfin Camp [Associates] Figure 16-14, Slide Works/Jackson Laboratory; Figure CO-16, Reuters/Lyle Stafford/Archive Photos; Figure 16-11UN EB, Signe Wilkinson/Cartoonists [Writers Syndicate] PO-05, Jeremy Walker/Stone

Chapter 17
Figure 17-1, USDA/ARS/Agricultural Research Service; Figure 17-3, U.S. Department of Agriculture; Figure 17-5b, Courtesy of U.S. Department of Agriculture and David Pimentel, Cornell University; Figure 17-6 bkgd, PhotoDisc, Inc.; Figure 17-7, Johnny Johnson/Animals Animals/Earth Scenes; Figure 17-11b, Ed Degginger/Color-Pic, Inc.; Figure 17-11c, Harry Rogers/Photo Researchers, Inc.; Figure 17-12a1, Department of Lands, Queensland, Australia; Figure 17-12b1, Department of Lands, Queensland, Australia; Figure 17-13, NatureMark Potatoes, a unit of Monsanto Company; Figure 17-14, Jack Dermid/Photo Researchers, Inc.; Figure 17-16, Nigel Cattlin/Holt Studios International/Photo Researchers, Inc.; Figure 17-17, Tim Lucas/Stone; CO-17, USDA/ARS/Agricultural Research Service

Chapter 18
Figure 18-3, Tom Stoddart/Katz Pictures/Woodfin Camp [Associates] Figure 18-4, Thomas D. Brock; Figure 18-5c, Charles R. Belinky/Photo Researchers, Inc.; Figure 18-8, Bernard J. Nebel; Figure 18-10, Loren Callahan/AP/Wide World Photos; Figure 18-12a1, 18-12b1, 18-14, 18-15a, 18-15b, Bernard J. Nebel; CO-18, NASA/Mark Marten/Science Source/Photo Researchers, Inc.; Figure 18-05UN-EW1, EW2, EW3, EW4 Bob Hudson/George Waclawin/ Bernard J. Nebel

Chapter 19
Figure 19-1, Photo © Cymie Payne, courtesy of Cambridge Arts Council; Figure 19-4, Van Bucher/Photo Researchers, Inc.; Figure 19-8, Rafael Macia/Photo Researchers, Inc.; Figure 19-9, Alan L. Detrick/Photo Researchers, Inc.; Figure 19-11, Susan Greenwood/Liaison Agency, Inc.; Figure 19-12, John Griffin/The Image Works; Figure 19-13, Leonard Lessin/Peter Arnold, Inc.; Figure 19-14 © Tribune Media Services, Inc. All Rights Reserved. Reprinted with permission.; Figure 19-15, Robert J. Fiore, City of Worcester, Massachusetts; CO-19, Matt York/AP/Wide World Photos

Chapter 20
Figure 20-4, The Plain Dealer, Cleveland, Ohio; Figure 20-7b, Alon Reininger/Contact Press/Woodfin Camp [Associates] Figure 20-8, Nancy J. Pierce/Photo Researchers, Inc.; Figure 20-9, Van Bucher/Photo Researchers, Inc.; Figure 20-10, Joe Traver/Liaison Agency, Inc.; Figure 20-11d, Richard T. Wright; Figure 20-12, Tim Lynch/Liaison Agency, Inc.; CO-20, Gordon Wiltsie/Peter Arnold, Inc.; Figure 20-01UN, Richard T. Wright

Chapter 21
Figure 21-1a, NASA/John F. Kennedy Space Center/Jet Propulsion Laboratory/CNES; Figure 21-1b, David Gray/AP/Wide World Photos; Figure 21-1c, Tantyo Bangun/Still Pictures/Peter Arnold, Inc.; Figure 21-1d, Reuters/George Mulala/Archive Photos; Figure 21-19, /NASA/Goddard Space Flight Center; CO-21b, c, d, Jet Propulsion Laboratory/ NASA/ CNES; Figure 21-14UN, Russ Camp/Russ Camp

Chapter 22
Figure 22-1, Copyright Pittsburgh Post-Gazette, 1999. All rights reserved. Reprinted with permission. Photo originally from Pittsburgh Sun-Telegraph.; Figure 22-4a, Joe Sohm/Chromosohm/The Image Works; Figure 22-4b, Frank Hanna/Visuals Unlimited; Figure 22-6, Richard T. Wright; Figure 22-7b1, b2 Matt Meadows/Peter Arnold, Inc.; Figure 22-09, Joern Gerdis/Photo Researchers, Inc.; Figure 22-10, Ray Pfortner/Peter Arnold, Inc.; Figure 22-11, Ray Pfortner/Peter Arnold, Inc.; Figure 22-19a, Grapes-Michaud/Photo Researchers, Inc.; Figure 22-20-1-EW, Tri-Met, Portland, Oregon; Figure 22-22, Delphi Energy [Chassis Systems, Flint, Michigan] Figure 22-24a, General Motors/Peter Arnold, Inc.; Figure 22-24b, Toyota Motor Corporation; CO-22, Steven D. Elmore/Corbis/Stock Market;

Chapter 23
Figure 23-1-1, Explorer/Yves Gladu/Photo Researchers, Inc.; Figure 23-4a, Larry Lefever/Grant Heilman Photography, Inc.; Figure 23-4b, Noah Poritz/MacroWorld/Photo Researchers, Inc.; Figure 23-4c, David Weintraub/Photo Researchers, Inc.; Figure 23-7, UPI/Corbis; CO-23, Wes Thompson/Corbis/Stock Market; PO-06, Thomas Kitchin/Tom Stack Associates

Chapter 24
Figure 24-1, David Pollack/Corbis/Stock Market] Figure 24-2, Esbin-Anderson/The Image Works; Figure 24-3, Michael Wickes/The Image Works; Figure 24-4a, Steve Elmore/Corbis/Stock Market; Figure 24-4b, David Lawrence/Corbis/Stock Market; Figure 24-7a, b, 24-8, 24-10, Bernard J. Nebel; Figure 24-11, Geri Engberg/Corbis/Stock Market; Figure 24-12, Pete Saloutos/Corbis/Stock Market; Figure 24-13, Kim Newton/Woodfin Camp [Associates] Figure 24-14, © Mr. Janis Miglavs 1995; Figure 24-15, 18th [Glenwood Garden, photo courtesy of the Pennsylvania Horticultural Society] Figure 24-16, Bill Cardoni/Liaison Agency, Inc.; Figure 24-17, Chattanooga Area Convention and Visitors Bureau; Figure 24-18, Karl Gehring/Liaison Agency, Inc.; CO-24, G.R. 'Dick' Roberts Photo Library; Figure 24-01UN, Richard T. Wright

Glossary

abiotic. Pertaining to factors or things that are separate and independent from living things; nonliving.

abortion. The termination of a pregnancy by some form of surgical or medicinal intervention.

absolute poverty. The lack of sufficient income in cash or exchange items for meeting the most basic human needs for food, clothing, and shelter.

acid. Any compound that releases hydrogen ions when dissolved in water. Also, a water solution that contains a surplus of hydrogen ions.

acid deposition. Any form of acid precipitation and also fallout of dry acid particles. (See **acid precipitation**.)

acid precipitation. Includes acid rain, acid fog, acid snow, and any other form of precipitation that is more acidic than normal (i.e., less than pH 5.6). Excess acidity is derived from certain air pollutants, namely, sulfur dioxide and oxides of nitrogen.

activated sludge. Sludge made up of clumps of living organisms feeding on detritus that settles out and is recycled in the process of secondary wastewater treatment.

activated sludge system. A system for removing organic wastes from water. The system uses microorganisms and active aeration to decompose such wastes. The system is used most as a means of secondary sewage treatment following the primary settling of materials.

active safety features. Those safety features of nuclear reactors that rely on operator-controlled reactions, external power sources, and other features that are capable of failing. (See **passive safety features**.)

adaptation. An ecological or evolutionary change in structure or function that enables an organism to adjust better to its environment and hence enhances the organism's ability to survive and reproduce.

adiabatic cooling. The cooling that occurs when warm air rises and encounters lower atmospheric pressure. Adiabatic *warming* is the opposite process, whereby cool air descends and encounters higher pressure.

adsorption. The process whereby chemicals (ions or molecules) stick to the surface of other materials.

aeration. *Soil:* The exchange within the soil of oxygen and carbon dioxide necessary for the respiration of roots. *Water:* The bubbling of air or oxygen through water to increase the amount of oxygen dissolved in the water.

aerosols. Microscopic liquid and solid particles originating from land and water surfaces and carried up into the atmosphere.

affluenza. A term used to describe a dysfunctional relationship with wealth or money.

Agenda 21. A definitive program to promote sustainable development that was produced by the 1992 Earth Summit in Rio de Janeiro, Brazil, and adopted by the U.N. General Assembly.

age structure. Within a population, the different proportions of people who are old, middle aged, young adults, and children.

AID. The U.S. Agency for International Development, the development arm of the U.S. State Department responsible for administering international aid from the United States.

AIDS. Acquired immune deficiency syndrome, a fatal disease caused by the human immunodeficiency virus (HIV) and transmitted by sexual contact or the use of nonsterile needles (as in drug addiction).

air pollution disaster. Short-term situation in industrial cities in which intense industrial smog brings about a significant increase in human mortality.

air toxics. A category of air pollutants that includes radioactive materials and other toxic chemicals which are present at low concentrations, but are of concern because they often are carcinogenic.

alga, pl. algae. Any of numerous kinds of photosynthetic plants that live and reproduce while entirely immersed in water. Many species (the planktonic forms) exist as single cells or small groups of cells that float freely in the water. Other species (the seaweeds) may be large and attached.

algal bloom. A relatively sudden development of a heavy growth of algae, especially planktonic forms. Algal blooms generally result from additions of nutrients, whose scarcity is normally limiting.

alkaline, alkalinity. A *basic* substance; chemically, a substance that absorbs hydrogen ions or releases hydroxyl ions; in reference to natural water, a measure of the base content of the water.

alleles. The two or more variations of a gene for any particular characteristic (e.g., blue and brown are alleles of the gene for eye color).

alley cropping. Agricultural cropping in which rows of shade-producing trees are alternated with rows of food crops to promote growth in dry climates.

ambient standards. Air-quality standards (set by the EPA) determining certain levels of pollutants that should not be exceeded in order to maintain environmental and human health.

anaerobic. Oxygen free.

anaerobic digestion. The breakdown of organic material by microorganisms in the absence of oxygen. The process results in the release of methane gas as a waste product.

anaerobic respiration. Respiration carried on by certain bacteria in the absence of oxygen. Methane, which can be used as fuel gas (it is the same as natural gas), may be a by-product of the process.

animal testing. Procedures used to assess the toxicity of chemical substances using rats, mice, and guinea pigs as surrogates for humans who might be exposed to the substances.

anthropogenic. Referring to pollutants and other forms of impacts on natural environments that can be traced to human activities.

appropriate technology. Technology that seeks to increase the efficiency and productivity of hand labor without displacing workers. That is, it seeks to enable people to improve their well-being without disrupting the existing social and economic system.

aquaculture. A propagation or rearing of any aquatic (water) organism in a more or less artificial system.

aquifer. An underground layer of porous rock, sand, or other material that allows the movement of water between layers of nonporous rock or clay. Aquifers are frequently tapped for wells.

artificial selection. Plant and animal breeders' practice of selecting individuals with the greatest expression of desired traits to be the parents of the next generation.

asbestos fibers. Crystals of asbestos, a natural mineral, that have the form of minute strands; asbestos is a serious health hazard in indoor spaces.

association. A unique combination of plants on a given site; the fundamental level of classification used in the United States National Vegetation Classification system.

asthma. Chronic disease of the respiratory system in which the air passages tighten and constrict, causing breathing difficulties; acute attacks may be life threatening.

atom. The fundamental unit of all elements.

autotroph. Any organism that can synthesize all its organic substances from inorganic nutrients, using light or certain inorganic chemicals as a source of energy. Green plants are the principal autotrophs.

background radiation. Radiation that comes from natural sources apart from any human activity. We are all exposed to such radiation.

bacterium, pl. bacteria. Any of numerous kinds of prokaryotic microscopic organisms that exist as simple single cells that multiply by simple division. Along with fungi, bacteria constitute the decomposer component of ecosystems. A few species cause disease.

balanced herbivory. A diversified plant community held in balance by various herbivores specific to each plant species.

base. Any compound that releases hydroxyl (OH^-) ions when dissolved in water. A solution that contains a surplus of hydroxyl ions.

base-load power plant. A power plant, usually large and fueled by coal or nuclear energy, that is kept operating continuously to supply electricity.

bed load. The load of coarse sediment, mostly coarse silt and sand, that is gradually moved along the bottom of a riverbed by flowing water rather than being carried in suspension.

benefit–cost analysis. An analysis or a comparison of the benefits in contrast to the costs of any particular action or project. (See **cost–benefit ratio**.)

benthic plants. Plants that grow under water and that are attached to or rooted in the bottom of the body of water. For photosynthesis, these plants depend on light penetrating the water.

best management practice. Farm management practice that serves best to reduce soil and nutrient runoff and subsequent pollution.

bioaccumulation. The accumulation of higher and higher concentrations of potentially toxic chemicals in organisms. Bioaccumulation occurs in the case of substances that are ingested, but that cannot be excreted or broken down (nonbiodegradable substances).

biochemical oxygen demand (BOD). The amount of oxygen that will be absorbed or "demanded" as wastes are being digested or oxidized in both biological and chemical processes. Potential impacts of wastes are commonly measured in terms of the BOD.

biocide. Pesticide or other chemical that is toxic to many, if not all, kinds of living organisms.

bioconversion. The use of biomass as fuel. Burning materials such as wood, paper, and plant wastes directly to produce energy, or converting such materials into fuels such as alcohol and methane.

biodegradable. Able to be consumed and broken down to natural substances, such as carbon dioxide and water, by biological organisms, particularly decomposers. Opposite: nonbiodegradable.

biodiversity. The diversity of living things found in the natural world. The concept usually refers to the different species, but also includes ecosystems and the genetic diversity within a given species.

biogas. The mixture of gases—about two-thirds methane, one-third carbon dioxide, and small portions of foul-smelling compounds—resulting from the anaerobic digestion of organic matter. The methane content enables biogas to be used as a fuel.

biological control. Control of a pest population by the introduction of predatory, parasitic, or disease-causing organisms.

biological nutrient removal. A process employed in sewage treatment to remove nitrogen and phosphorus from the effluent coming from secondary treatment.

biological species concept. The concept that defines a species as a group of interbreeding natural populations that does not breed with other groups because of various barriers.

biological wealth. The life-sustaining combination of commercial, scientific, and aesthetic values imparted to a region by its biota.

biomagnification. Bioaccumulation occurring through several levels of a food chain.

biomass. Mass of biological material. Usually the total mass of a particular group or category; for example, biomass of producers.

biomass energy. Energy produced by burning plant-related materials such as firewood. (See **bioconversion**.)

biomass fuels. Fuels in the form of plant-related material (e.g., firewood). (See **bioconversion**.)

biomass pyramid. The structure that is obtained when the respective biomasses of producers, herbivores, and carnivores in an ecosystem are compared. Producers have the largest biomass, followed by herbivores and then carnivores.

biome. A group of ecosystems that are related by having a similar type of vegetation governed by similar climatic conditions. Examples include prairies, deciduous forests, arctic tundra, deserts, and tropical rain forests.

bioremediation. The use of microorganisms for the decontamination of soil or groundwater. Usually involves injecting organisms or oxygen into contaminated zones.

biosolids. Organic material removed from sewage effluents in the course of treatment. Formerly referred to as sludge.

biosphere. The overall ecosystem of Earth. The sum total of all the biomes and smaller ecosystems, which ultimately are all interconnected and interdependent through global processes such as the water cycle and the atmospheric cycle.

biosphere reserves. Natural areas in various parts of the world set aside to protect genetic biodiversity.

biota. The sum total of all living organisms. The term usually is applied to the setting of natural ecosystems.

biotechnology. The use of genetic-engineering techniques, enabling researchers to introduce new genes into food crops and domestic animals and thereby produce valuable products and more nutritious crops.

biotic. Living or derived from living things.

biotic community. All the living organisms (plants, animals, and microorganisms) that live in a particular area.

biotic potential. Reproductive capacity. The potential of a species for increasing its population and/or distribution. The biotic potential of every species is such that, given optimum conditions, its population will increase. (Contrast environmental resistance).

biotic structure. The organization of living organisms in an ecosystem into groups such as producers, consumers, detritus feeders, and decomposers.

birth control. Any means, natural or artificial, that may be used to reduce the number of live births.

blue water flow. In the hydrologic cycle, precipitation and its further movement in infiltration, runoff, surface water, and groundwater.

BOD. See **biochemical oxygen demand**.

bottle law (bottle bill).   A law that provides for the recycling or reuse of beverage containers, usually by requiring a returnable deposit upon purchase of the item.

breeder reactor.   A nuclear reactor that, in the course of producing energy, also converts nonfissionable uranium-238 into fissionable plutonium-239, which can be used as fuel. Hence, a reactor that produces at least as much nuclear fuel as it consumes.

broad-spectrum pesticides.   Chemical pesticides that kill a wide range of pests. These pesticides also kill a wide range of nonpest and beneficial species; therefore, they may lead to environmental disturbances and resurgences. The opposite of narrow-spectrum pesticides and biorational pesticides.

brownfields.   Abandoned, idled, or underused industrial and commercial facilities whose further development is inhibited because of real or perceived chemical contamination.

BTU (British thermal unit).   A fundamental unit of energy in the English system. The amount of heat required to raise the temperature of 1 pound of water 1 degree Fahrenheit.

buffer.   A substance that will maintain the pH of a solution by reacting with the excess acid in the solution. Limestone is a natural buffer that helps to maintain water and soil at a near-neutral pH.

buffering capacity.   The amount of acid that may be neutralized by a given amount of buffer.

calorie.   A fundamental unit of energy. The amount of heat required to raise the temperature of 1 gram of water 1 degree Celsius. All forms of energy can be converted to heat and measured in calories. Calories used in connection with food are kilocalories, or "big" calories, the amount of heat required to raise the temperature of 1 liter of water 1 degree Celsius.

cancer.   Any of a number of cellular changes that result in uncontrolled cellular growth.

capillary water.   Water that clings in small pores, cracks, and spaces against the pull of gravity (e.g., water held in a sponge).

carbon monoxide.   A highly poisonous gas, the molecules of which consist of a carbon atom with one oxygen attached. Not to be confused with carbon dioxide, a natural gas in the atmosphere.

carbon tax.   A tax levied on all fossil fuels in proportion to the amount of carbon dioxide that is released as they burn.

carcinogenic.   Having the property of causing cancer, at least in animals and, by implication, in humans.

carnivore.   An animal that feeds more or less exclusively on other animals.

carrying capacity.   The maximum population of a given species that an ecosystem can support without being degraded or destroyed in the long run. The carrying capacity may be exceeded, but not without lessening the system's ability to support life in the long term.

Cartagena Protocol.   An international agreement governing trade in genetically modified organisms. The Cartagena Protocol was signed in January 2000.

castings.   The humus-rich pellets resulting from earthworm activity.

catalyst.   A substance that promotes a given chemical reaction without itself being consumed or changed by the reaction. Enzymes are catalysts for biological reactions. Also, catalysts are used in some pollution control devices (e.g., the catalytic converter).

catalytic converter.   The device used by U.S. automobile manufacturers to reduce the amount of carbon monoxide and hydrocarbons in a car's exhaust. The converter contains a catalyst that oxidizes these compounds to carbon dioxide and water as the exhaust passes through.

cell.   The basic unit of life; the smallest unit that still maintains all the attributes of life. Many microscopic organisms consist of a single cell. Large organisms consist of trillions of specialized cells functioning together.

cell respiration.   The chemical process that occurs in all living cells whereby organic compounds are broken down to release energy required for life processes. For respiration, higher plants and animals require oxygen and release carbon dioxide and water as waste products, but certain microorganisms do not require oxygen. (See **anaerobic respiration**.)

cellulose.   The organic macromolecule that is the prime constituent of plant cell walls and hence the major molecule in wood, wood products, and cotton. Cellulose is composed of glucose molecules, but because it cannot be digested by humans, its dietary value is only as fiber, bulk, or roughage.

center pivot irrigation.   An irrigation system consisting of a spray arm several hundred meters long, supported by wheels pivoting around a central well from which water is pumped.

centrally planned economy.   An economic system in which a ruling class makes most of the basic decisions about how the economy will be structured; typical of communist countries.

CFCs.   See **chlorofluorocarbons**.

chain reaction.   Nuclear reaction wherein each atom that fissions (splits) causes one or more additional atoms to fission.

channelization/channelized.   The straightening and deepening of stream or river channels to speed water flow and reduce flooding. A waterway so treated is said to be channelized.

chemical barrier.   A chemical aspect of a plant that makes it resist attacks from certain pests. Used in genetic pest control.

chemical energy.   The potential energy that is contained in certain chemicals; most importantly, the energy contained in organic compounds such as food and fuels and that may be released through respiration or burning.

chemical technology.   The use of pesticides and herbicides to control or eradicate agricultural pests.

chemosynthesis.   Process whereby some microorganisms utilize the chemical energy contained in certain reduced inorganic chemicals (e.g., hydrogen sulfide) to produce organic material.

chlorinated hydrocarbons.   Synthetic organic molecules in which one or more hydrogen atoms have been replaced by chlorine atoms. Chlorinated hydrocarbons are hazardous compounds because they tend to be nonbiodegradable and therefore to bioaccumulate; many have been shown to be carcinogenic. Also called organochlorides.

chlorination.   The disinfection process of adding chlorine to drinking water or sewage water in order to kill microorganisms that may cause disease.

chlorine cycle.   In the stratosphere, a cyclical chemical process in which chlorine monoxide breaks down ozone.

chlorofluorocarbons (CFCs).   Synthetic organic molecules that contain one or more of both chlorine and fluorine atoms and that are known to cause ozone destruction.

chlorophyll.   The green pigment in plants responsible for absorbing the light energy required for photosynthesis.

chromosome.   In a living organism, one of a number of structures on which genes are arranged in a linear fashion.

CITES (Convention on Trade in Endangered Species).   An international treaty conveying some protection to endangered and threatened species by restricting trade in those species or their products.

Clean Air Act of 1970.   Amended in 1977 and 1990, the act is the foundation of U.S. air pollution control efforts.

Clean Water Act of 1972.   The cornerstone federal legislation addressing water pollution.

clear-cutting. In forestry, cutting every tree, leaving the area completely clear.

climate. A general description of the average temperature and rainfall conditions of a region over the course of a year.

climax ecosystem. The last stage in ecological succession. An ecosystem in which populations of all organisms are in balance with each other and with existing abiotic factors.

clone. A group of genetically identical individuals derived from the asexual propagation of a single individual.

cogeneration. The joint production of useful heat and electricity. For example, furnaces may be replaced with gas turbogenerators that produce electricity while the hot exhaust still serves as a heat source. An important avenue of conservation, cogeneration effectively avoids the waste of heat that normally occurs at centralized power plants.

combined-cycle natural-gas unit. A recent technology for generating electricity, employing both a natural-gas turbine and a steam turbine that uses excess heat coming from the gas turbine.

combustion. The practice of disposing of wastes by incineration in a special facility designed to handle large amounts of waste; modern combustion facilities also capture some of the energy by generating electricity on the site.

command-and-control strategy. The basic strategy behind most public policy having to do with air and water pollution. The strategy involves setting limits on pollutant levels and specifying control technologies that must be used to achieve those limits.

commons, common pool resources. Resources (usually natural ones) owned by many people in common or, as in the case of the air or the open oceans, owned by no one, but open to exploitation.

compaction. Packing down. *Soil:* Packing and pressing out air spaces present in the soil. Compaction reduces soil aeration and infiltration and thus diminishes the capacity of the soil to support plants. *Trash:* Packing down trash to reduce the space that it requires.

competitive exclusion principle. The concept that when two species compete very directly for resources, one eventually excludes the other from the area.

composting/compost. The process of letting organic wastes decompose in the presence of air. A nutrient-rich humus, or compost, is the resulting product.

composting toilet. A toilet that does not flush wastes away with water, but deposits them in a chamber where they will compost. (See **composting**.)

compound. Any substance (gas, liquid, or solid) that is made up of two or more different kinds of atoms bonded together. (Contrast **element**.)

Comprehensive Environmental Response, Compensation, and Liability Act of 1980. See **Superfund**.

condensation. The collecting of molecules from the vapor state to form the liquid state, as, for example, when water vapor condenses on a cold surface and forms droplets. Opposite: **evaporation**.

conditions. Abiotic factors (e.g., temperature) that vary in time and space, but are not used up by organisms.

confusion technique. Pest control method in which a quantity of sex attractant is applied to an area so that males become confused and are unable to locate females. The actual quantities of pheromones applied are very small because of their extreme potency.

conservation. The management of a resource in such a way as to assure that it will continue to provide maximum benefit to humans over the long run. Conservation may include various degrees of use or protection, depending on what is necessary to maintain the resource over the long run. *Energy:* Saving energy. Energy conservation entails not only cutting back on the use of heating, air-conditioning, lighting, transportation, and so on, but also increasing the efficiency of energy use — that is, developing and instigating the means of doing the same jobs (e.g., transporting people) with less energy.

Conservation Reserve Program (CRP). A federal program whereby farmers are paid to place highly erodible cropland in a reserve.

consumers. In an ecosystem, those organisms that derive their energy from feeding on other organisms or their products.

consumptive use. The harvesting of natural resources in order to provide for people's immediate needs for food, shelter, fuel, and clothing.

consumptive water use. Use of water for such things as irrigation, wherein the water does not remain available for potential purification and reuse.

containment building. Reinforced concrete building housing a nuclear reactor. Designed to contain an explosion, should one occur.

contour farming. The practice of cultivating land along the contours across, rather than up and down, slopes. In combination with strip cropping, contour farming reduces water erosion.

contraceptive. A device or method employed by couples during sexual intercourse to prevent the conception of a child.

control rods. Part of the core of a nuclear reactor; the rods of neutron-absorbing material that are inserted or removed as necessary to control the rate of nuclear fissioning.

convection. The vertical movement of air due to atmospheric heating and cooling.

Convention on Biological Diversity. The Biodiversity Treaty signed by 158 nations at the Earth Summit in Rio de Janeiro in 1992 calling for various actions and cooperative steps between nations to protect the world's biodiversity.

Conveyor system. The giant pattern of oceanic currents that moves water masses from the surface to the depths and back again, producing major effects on the climate.

cooling tower. A massive tower designed to dissipate waste heat from a power plant (or other industrial facility) into the atmosphere.

corrosion. In a nuclear power plant, the deterioration of pipes receiving hot, pressurized water in a circulation system that conveys heat from one part of the unit to another.

cosmetic spraying. Spraying of pesticides to control pests that damage only the surface appearance of fruits and vegetables.

cost–benefit analysis. See **benefit–cost analysis**.

cost–benefit ratio/benefit–cost ratio. The value of the benefits to be gained from a project, divided by the costs of the project. If the ratio is greater than unity, the project is economically justified; if the ratio is less than unity, the project is not economically justified.

cost-effective. Pertaining to a project or procedure that produces economic returns or benefits that are significantly greater than the costs incurred.

coupled general circulation model. A computer-based model for simulating long-term climatic conditions that combines global atmospheric circulation patterns with ocean circulation and cloud-radiation feedback.

covalent bond. A chemical bond between two atoms, formed by sharing a pair of electrons between the atoms. Atoms of all organic compounds are joined by covalent bonds.

credit associations. Groups of poor people who individually lack the collateral to assure loans, but collectively have the wherewithal to assure each other's loans. Associated with microlending.

criteria pollutants. Certain pollutants whose levels are used as a gauge for the determination of air (or water) quality.

critical level. The level of one or more pollutants above which severe damage begins to occur and below which few, if any, ill effects are noted.

critical number. The minimum number of individuals of a given species that is required to maintain a healthy, viable population of the species. If a population falls below its critical number, it will almost certainly become extinct.

crop rotation. The practice of alternating the crops grown on a piece of land—for example, corn one year, hay for two years, and then back to corn. (Contrast **monocropping.**)

crude birthrate. Number of births per 1,000 individuals per year.

crude death rate. Number of deaths per 1,000 individuals per year.

cultivar. A cultivated variety of a plant species. Genetically, all individuals of the cultivar are highly uniform.

cultural control. A change in the practice of growing, harvesting, storing, handling, or disposing of wastes that reduces their susceptibility or exposure to pests. For example, spraying the house with insecticides to kill flies is a chemical control; putting screens on the windows to keep flies out is a cultural control.

cultural eutrophication. The process of natural eutrophication accelerated by human activities. (See **eutrophication.**)

DDT (dichlorodiphenyltrichloroethane). The first and most widely used of the synthetic organic pesticides belonging to the chlorinated hydrocarbon class.

debt crisis. Exigency brought about by the great debt incurred by many less developed nations. These nations are so heavily in debt that they may not be able to meet their financial obligations (e.g., interest payments).

debt relief. Steps, including cancellation of debts and other means of reducing poverty, taken by the World Bank and donor nations to alleviate the heavy debt of many poor nations.

declining tax base. The loss of tax revenues that occurs when affluent taxpayers and businesses leave an area and property values subsequently decline. Also referred to as an eroding tax base.

decommissioning. The taking of nuclear power plants out of service after 25–35 years because the effects of radiation will gradually make them inoperable.

decomposers. Organisms whose feeding action results in decay or rotting of organic material. The primary decomposers are fungi and bacteria.

deep-well injection. A technique used for the disposal of liquid chemical wastes that involves putting them into deep dry wells where they permeate dry strata.

deforestation. The process of removing trees and other vegetation covering the soil, leading to erosion and loss of soil fertility.

Delaney clause. A provision of the Federal Food, Drug and Cosmetic Act of 1958 that prohibits the use of any food additive that induces cancer in people or animals.

demographic transition. The transition of a population from a condition of a high birthrate and a high death rate to a condition of a low birthrate and a low death rate. A demographic transition may result from economic or social development.

demography/demographer. The study of population trends (growth, movement, development, and so on). People who perform such studies and make projections from them.

denitrification. The process of reducing oxidized nitrogen compounds present in soil or water back to nitrogen gas in the atmosphere. Denitrification is a natural process conducted by certain bacteria (see discussion of the nitrogen cycle in the text), and is now utilized in the treatment of sewage effluents.

density-dependent. Attribute of population-balancing factors, such as predation, that increase and decrease in intensity in proportion to population density.

Department of Transportation regulations (DOT regs). Regulations intended to reduce the risk of spills, fires, and poisonous fumes by specifying the kinds of containers and methods of packing to be used in transporting hazardous materials.

deregulation. In the electric power industry, a policy that requires utility companies to divest themselves of power-generating facilities and allows the market to determine electrical costs.

desalinization. Process that purifies seawater into high-quality drinking water via distillation or microfiltration.

desertification. The formation and expansion of degraded areas of soil and vegetation cover in arid, semiarid, and seasonally dry areas, caused by climatic variations and human activities.

desertified. Said of land whose productivity has been significantly reduced (25% or more) because of human mismanagement. Erosion is the most common cause of desertified land.

desert pavement. A covering of stones and coarse sand protecting desert soils from further wind erosion. The covering results from the differential erosion of finer material.

detritus. The dead organic matter, such as fallen leaves, twigs, and other plant and animal wastes, that exists in any ecosystem.

detritus feeders. Organisms such as termites, fungi, and bacteria that obtain their nutrients and energy mainly by feeding on dead organic matter.

deuterium (^{2}H). A stable, naturally occurring isotope of hydrogen that contains one neutron in addition to the single proton normally in the nucleus.

developed countries. Industrialized countries—the United States, Canada, Western European nations, Japan, Australia, and New Zealand—in which the gross domestic product exceeds $9,360 per capita.

developing countries. All free-market countries in which the gross domestic product is less than $9,360 per capita. The category includes nations of Latin America, Africa, and Asia, except Japan.

differential reproduction. Within a population, the more successful reproduction of certain individuals compared with that of others.

dioxin. A synthetic organic chemical of the chlorinated hydrocarbon class. Dioxin is one of the most toxic compounds known to humans, having many harmful effects, including carcinogenic and teratogenic effects, even in extremely minute concentrations. Because of the use of certain herbicides that contain dioxin as a contaminant, the chemical has become a widespread environmental pollutant.

discharge permit. (Technically called an NPDES permit.) A permit that allows a company to legally discharge certain amounts or levels of pollutants into air or water.

discount rate. In economics, a rate applied to some future benefit or cost in order to calculate its present real value.

disinfection. The killing (as opposed to removal) of microorganisms in water or other media where they might otherwise pose a health threat. For example, chlorine is commonly used to disinfect water supplies.

dissolved oxygen (DO). Oxygen gas molecules (O_2) dissolved in water. Fish and other aquatic organisms depend on dissolved oxygen for respiration. Therefore, the concentration of dissolved oxygen is a measure of water quality.

distillation. A process of purifying water or some other liquid by boiling the liquid and recondensing the vapor. Contaminants remain behind in the boiler.

distributive justice.   The ethical process of making certain that all parties receive equal rights regardless of their economic status.

disturbance.   A natural or human-induced event or process (for example, a forest fire) that interrupts ecological succession and creates new conditions on a site.

DNA (deoxyribonucleic acid).   The natural organic macromolecule that carries the genetic or hereditary information for virtually all organisms.

dose.   The mathematical product of the concentration of a hazardous material and the length of exposure to it. The effects of any given material or radiation correspond to the dose received.

doubling time.   The time it will take a population to double in size, assuming that the current rate of growth continues.

drift netting.   The practice of harvesting marine fish and squid by laying down miles of gill nets across the open seas. The nets collapse around larger organisms and kill many whales, dolphins, seals, marine birds, and turtles.

drip irrigation.   Method of supplying irrigation water through tubes that literally drip water onto the soil at the base of each plant.

drought.   A local or regional lack of precipitation such that the ability to raise crops and water animals is seriously impaired.

drylands.   Ecosystems characterized by low precipitation (25 to 75 cm per year) and often subject to droughts; some 5.2 billion hectares are classified as drylands.

earth-sheltered housing.   Housing that makes use of the insulating properties of earth materials and that is also oriented for passive solar heating; often involves earth berms built up against the building walls.

easement.   In reference to land protection, an arrangement whereby a landowner gives up development rights into the future, but retains ownership of the land.

ecdysone.   The hormone that promotes molting in insects.

ecological economist.   An economist who thoroughly integrates ecological and economic concerns; part of a new breed of economist who disagrees with classical economic theory.

ecological pest management.   Natural control of pest populations through understanding various ecological factors and, so far as possible, utilizing those factors, as opposed to using synthetic chemicals.

ecological restoration.   See **restoration ecology**.

ecological succession.   Process of gradual and orderly progression from one ecological community to another.

ecologists.   Scientists who study ecology.

ecology.   The study of any and all aspects of how organisms interact with each other and with their environment.

economic exclusion.   The cutting of access of certain ethnic or economic groups to jobs, a good education, and other opportunities and thus preventing them from entering the economic mainstream of society—a condition that prevails in poor areas of cities.

economic threshold.   The level of pest damage that, to be reduced further, would require an application of pesticides that is more costly than the economic damage caused by the pests.

ecosystem.   A grouping of plants, animals, and other organisms interacting with each other and with their environment in such a way as to perpetuate the grouping more or less indefinitely. Ecosystems have characteristic forms, such as deserts, grasslands, tundra, deciduous forests, and tropical rain forests.

ecosystem management.   The management paradigm, adopted by all federal agencies managing public lands, that involves a long-term stewardship approach to maintaining the lands in their natural state.

ecotone.   A transitional region between two adjacent ecosystems that contains some of the species and characteristics of each one and also certain species of its own.

ecotourism.   The enterprises involved in promoting tourism of unusual or interesting ecological sites.

El Niño.   A major climatic phenomenon characterized by the movement of unusually warm surface water into the eastern equatorial Pacific Ocean. El Niño results in extensive disruption of weather around the world.

electrolysis.   The use of electrical energy to split water molecules into their constituent hydrogen and oxygen atoms. Hydrogen gas and oxygen gas result.

electrons.   Fundamental atomic particles that have a negative electrical charge and a very small mass, approximately 9.1×10^{-31} kg. Electrons surround the nuclei of atoms and thus balance the positive charge of protons in the nucleus. A flow of electrons in a wire is identical to an electrical current.

element.   A substance that is made up of one and only one distinct kind of atom. (Contrast **compound**.)

embrittlement.   Becoming brittle. Pertains especially to the reactor vessel of nuclear power plants gradually becoming prone to breakage or snapping as a result of continuous bombardment by radiation. Embrittlement is the prime factor forcing the decommissioning of nuclear power plants.

emergent vegetation.   Aquatic plants whose lower parts are under water, but whose upper parts emerge from the water.

emission allowance/standards.   See **discharge permit**.

endangered species.   A species whose total population is declining to relatively low levels such that if the trend continues, the species will likely become extinct.

Endangered Species Act.   The federal legislation that mandates the protection of species and their habitats which are determined to be in danger of extinction.

endocrine disrupters.   Any of a class of organic compounds, often pesticides, that are suspected of having the capacity of interfering with hormonal activities in animals.

energy.   The capacity to do work. Common forms of energy are light, heat, electricity, motion, and the chemical bond energy inherent in compounds such as sugar, gasoline, and other fuels.

energy flow.   The movement of energy through ecosystems, starting from the capture of solar energy by primary producers and ending with the loss of heat energy.

enrichment.   With reference to nuclear power, the separation and concentration of uranium-235 so that, in suitable quantities, it will sustain a chain reaction.

entomologist.   A scientist who studies insects, their life cycles, physiology, and behavior, and so on.

entropy.   Degree of disorder; increasing entropy means increasing disorder.

environment.   The combination of all things and factors external to the individual or population of organisms in question.

environmental accounting.   A process of keeping national accounts of economic activity that includes gains and losses of environmental assets.

environmental impact.   Effect on the natural environment caused by human actions. Includes indirect effects, for example, through pollution, as well as direct effects such as cutting down trees.

environmental impact statement.   A study of the probable environmental impacts of a development project. The National Environmental Policy Act of 1968 (NEPA) requires such studies prior to proceeding with any project receiving federal funding.

environmental justice.   The fair treatment and meaningful involvement of all people, regardless of race, color, national origin, or income, with respect to the development, implementation, and enforcement of environmental laws and regulations.

environmental movement. The upwelling of public awareness and citizen action regarding environmental issues that began during the 1960s.

environmental resistance. The totality of factors such as adverse weather conditions, shortages of food or water, predators, and diseases that tend to cut back populations and keep them from growing or spreading. (Contrast **biotic potential**.)

Environmental Revolution. In the view of some, a coming change in the adaptation of humans to the rising deterioration of the environment. The Environmental Revolution will purportedly bring about sustainable interactions with the environment.

environmental science. The branch of science concerned with environmental issues.

environmentalism. An attitude or movement involving concern about pollution, resource depletion, population pressures, loss of biodiversity, and other environmental issues. Usually also implies action to address those concerns.

environmentalist. Any person who is concerned about the degradation of the natural world and is willing to act on that concern.

EPA (U.S. Environmental Protection Agency). The federal agency responsible for the control of all forms of pollution and other kinds of environmental degradation.

epidemiology, epidemiological study. The study of the causes of disease (e.g., lung cancer) through an examination and comparison of large populations of people living in different locations or following different lifestyles or habits (e.g., smoking versus nonsmoking).

epidemiologic transition. The pattern of change in mortality from high death rates to low death rates; contributes to the demographic transition.

epiphytes. Air plants that are not parasitic, but that "perch" on the branches of trees, where they can get adequate light.

equilibrium theory. The theory that ecosystems are maintained over time by natural checks and balances; equilibrium theory is challenged by many ecologists.

erosion. The process of soil particles' being carried away by wind or water. Erosion moves the smaller particles first and hence degrades the soil to a coarser, sandier, stonier texture.

estimated reserves. See **reserves**.

estuary. A bay or river system open to the ocean at one end and receiving fresh water at the other. In the estuary, fresh and salt water mix, producing brackish water.

ETS (environmental tobacco smoke). "Secondhand" tobacco smoke to which nonsmokers are exposed in the presence of smokers.

euphotic zone. In aquatic systems, the layer or depth of water through which an adequate amount of light penetrates to support photosynthesis.

eutrophic. Characterized by nutrient-rich water supporting an abundant growth of algae or other aquatic plants at the surface. Deep eutrophic water has little or no dissolved oxygen.

eutrophication. The process of becoming eutrophic.

evaporation. Process whereby molecules leave the liquid state and enter the vapor or gaseous state as, for example, when water evaporates to form water vapor. Opposite: **condensation**.

evapotranspiration. The combination of evaporation and transpiration that restores water to the atmosphere.

evolution. The theory that all species now on Earth descended from ancestral species through a process of gradual change brought about by natural selection.

evolutionary succession. The succession of different species that have inhabited Earth at different geological periods, as revealed through the fossil record. The process whereby new species are formed through the process of speciation while other species pass into extinction.

experimental group. In an experiment, the group that receives the experimental treatment, in contrast to the control group, used for comparison, which does not receive the treatment. Synonym: test group.

exotic species. A species introduced into a geographical area to which it is not native.

exponential increase. The growth produced when a base population increases by a given percentage (as opposed to a given amount) each year. An exponential increase is characterized by doubling again and again, each doubling occurring in the same period of time. It produces a J-shaped curve.

externality/external cost. Any effect of a business process not included in the usual calculations of profit and loss. Pollution of air or water is an example of a *negative* externality—one that imposes a cost on society that is not paid for by the business itself.

extinction. The death of all individuals of a particular species. When this occurs, all the genes of that particular line are lost forever.

extractive reserves. As now established in Brazil, forest lands that are protected for native peoples and rubber tappers who harvest natural products of the forests, such as latex and Brazil nuts.

exurban migration. The pronounced trend since World War II of relocating homes and businesses from the central city and older suburbs to more outlying suburbs.

exurbs. New developments beyond the traditional suburbs, but from which most residents still commute to the associated city for work.

family planning. Making contraceptives and other health and reproductive services available to couples in order to enable them to achieve a desired family size.

famine. A severe shortage of food accompanied by a significant increase in the local or regional death rate.

FAO. Food and Agriculture Organization of the United Nations.

farm cooperatives. An association of consumers who jointly own and manage a farm for the production of produce specifically for their own consumption.

Farmer-centered Agricultural Resource Management (FARM) Program. A U.N. FAO-administered program in Asian countries fostering sustainable agriculture.

FDA. Federal Food and Drug Administration, with jurisdiction over foods, drugs, and all products that come in contact with the skin or are ingested.

fecal coliform test. A test for the presence of *Escherichia coli*, the bacterium that normally inhabits the gut of humans and other mammals. A positive test result indicates sewage contamination and the potential presence of disease-causing microorganisms.

Federal Agricultural Improvement and Reform (FAIR) Act of 1996. Major legislation removing many subsidies and controls from farming.

fermentation. A form of respiration carried on by yeast cells in the absence of oxygen. Fermentation involves a partial breakdown of glucose (sugar) that yields energy for the yeast and the release of alcohol as a by-product.

fertility rate. See **total fertility rate**.

fertility transition. The pattern of change in birthrates in a society from high rates to low; a major component of the demographic transition.

fertilizer. Material applied to plants or soil to supply plant nutrients, most commonly nitrogen, phosphorus, and potassium, but possibly others. *Organic* fertilizer is natural organic material, such as manure, that releases nutrients as it breaks down. *Inorganic* fertilizer, also called chemical fertilizer, is a mixture of one or more necessary nutrients in inorganic chemical form.

field scouts.   Persons trained to survey crop fields and determine whether applications of pesticides or other pest-management procedures are actually necessary to avert significant economic loss.

FIFRA (Federal Insecticide, Fungicide, and Rodenticide Act).   The key U.S. legislation passed to control pesticides.

fifth principle of ecosystem sustainability.   Biodiversity is maintained.

fire climax ecosystems.   Ecosystems that depend on the recurrence of fire to maintain the existing balance.

first principle of ecosystem sustainability.   Ecosystems use sunlight as their source of energy.

first-generation pesticides.   Toxic inorganic chemicals that were the first used to control insects, plant diseases, and other pests. These pesticides included mostly compounds of arsenic and cyanide and various heavy metals, such as mercury and copper.

first law of thermodynamics.   The empirical observation, confirmed innumerable times, that energy is never created or destroyed but may be converted from one form to another (e.g., electricity to light). Also called the **law of conservation of energy.** (See also **second law of thermodynamics.**)

fishery.   Fish species being exploited, or a limited marine area containing commercially valuable fish.

fish ladder.   A stepwise series of pools on the side of a dam where the water flows in small falls that fish can negotiate.

fission.   The splitting of a large atom into two atoms of lighter elements. When large atoms such as uranium or plutonium fission, tremendous amounts of energy are released.

fission products.   Any and all atoms and subatomic particles resulting from splitting atoms in nuclear reactors. Practically all such products are highly radioactive.

fitness.   State of an organism whereby it is adapted to survive and reproduce in the environment it inhabits.

flat-plate collector.   A solar collector that consists of a stationary, flat, black surface oriented perpendicular to the average angle of the sun. Heat absorbed by the surface is removed and transported by air or water (or some other liquid) flowing over or through the surface.

flood irrigation.   Technique of irrigation in which water is diverted from rivers by means of canals and is flooded through furrows in fields.

food aid.   Food of various forms that is donated or sold below cost to needy people for humanitarian reasons.

food chain.   The transfer of energy and material through a series of organisms as each one is fed upon by the next.

food guide pyramid.   A graphic presentation of six basic food needs arranged in a pyramid to indicate the relative proportions of each type of food needed for good nutrition.

food security.   For families, the ability to meet the food needs of everyone in the family, providing freedom from hunger and malnutrition.

food web.   The combination of all the feeding relationships that exist in an ecosystem.

fossil fuels.   Energy sources—mainly crude oil, coal, and natural gas—that are derived from the prehistoric photosynthetic production of organic matter on Earth.

fourth principle of ecosystem sustainability.   Ecosystems show resilience when subject to disturbance.

FQPA (Food Quality Protection Act of 1996).   Legislation that removed the Delaney clause and replaced many provisions of FIFRA.

fragmentation.   The division of a landscape into patches of habitat by road construction, agricultural lands, or residential areas.

Framework Convention on Climate Change.   A result of the 1992 Earth Summit, this international treaty was a start in negotiating agreements on steps to prevent future catastrophic climate change.

free-market economy.   In its purest form, an economy in which the market itself determines what and how goods will be exchanged. The system is wholly in private hands.

freshwater.   Water that has a salt content of less than 0.1% (1,000 parts per million).

front.   The boundary where different air masses meet.

fuel assembly.   In a nuclear reactor, the assembly of many rods containing the nuclear fuel, usually uranium, positioned close together. The chain reaction generated in the fuel assembly is controlled by rods of neutron-absorbing material between the fuel rods.

fuel elements.   The pellets of uranium or other fissionable material that are placed in tubes, which, together with the control rods, form the core of the nuclear reactor.

fuel rods.   See **fuel elements.**

fuelwood.   The use of wood for cooking and heating; the most common energy source for people in developing countries.

fungus, pl. fungi.   Any of numerous species of molds, mushrooms, brackens, and other forms of nonphotosynthetic plants. Fungi derive energy and nutrients by consuming other organic material. Along with bacteria, they form the decomposer component of ecosystems.

fusion.   The joining together of two atoms to form a single atom of a heavier element. When light atoms such as hydrogen are fused, tremendous amounts of energy are released.

gasohol.   A blend of 90% gasoline and 10% alcohol that can be substituted for straight gasoline. Gasohol serves to stretch gasoline supplies.

gene.   Segment of DNA that codes for one protein, which in turn determines a particular physical, physiological, or behavioral trait.

gene pool.   The sum total of all the genes that exist among all the individuals of a species.

genetic bank.   The concept that natural ecosystems with all their species serve as a repository of genes that may be drawn upon to improve domestic plants and animals and to develop new medicines, among other uses.

genetic control.   Selective breeding of a desired plant or animal to make it resistant to attack by pests. Also, attempting to introduce harmful genes—for example, those that cause sterility—into the pest populations.

genetic engineering.   The artificial transfer of specific genes from one organism to another.

genetically modified organism (GMO).   Any organism that has received genes from a different species through the techniques of genetic engineering; a **transgenic organism.**

genetics.   The study of heredity and the processes by which inherited characteristics are passed from one generation to the next.

genetic variation.   An expression of the range of genetic (DNA) differences that occur among individuals of the same species.

geothermal.   Refers to the naturally hot interior of Earth, where heat is maintained by naturally occurring nuclear reactions.

geothermal energy.   Useful energy derived from water heated by the naturally hot interior of Earth.

global climate change.   The cumulative effects of various impacts of rising levels of greenhouse gases on Earth's climate. These effects include global warming, weather changes, and a rising sea level.

glucose.   A simple sugar, the major product of photosynthesis. Glucose serves as the basic building block for cellulose and starches and as the major "fuel" for the release of energy through cell respiration in both plants and animals.

goods.   Products, such as wood and food, that are extracted from natural ecosystems.

grassroots movement.   An environmental movement that begins at the level of the populace, often in response to a perceived problem not being addressed by public policy.

gravitational water.   Water that is not held by capillary action in soil, but that percolates downward by the force of gravity.

graying.   The increasing average age in populations in developed countries and in many developing countries that is occurring because of decreasing birthrates and increasing longevity.

gray water.   Wastewater, as from sinks and tubs, that does not contain human excrement. Such water can be reused without purification for some purposes.

greenhouse effect.   An increase in the atmospheric temperature caused by increasing amounts of carbon dioxide and certain other gases that absorb and trap heat, which normally radiates away from Earth.

greenhouse gases.   Gases in the atmosphere that absorb infrared energy and contribute to the air temperature. These gases are like a heat blanket and are important in insulating Earth's surface. Among the greenhouse gases are carbon dioxide, water vapor, methane, nitrous oxide, chlorofluorocarbons, and other halocarbons.

green fee.   An added cost for a particular service or product that has the effect of internalizing the external costs and that reflects the real environmental cost of the service or product.

green manure.   A legume crop such as clover that is specifically grown to enrich the nitrogen and organic content of soil.

green revolution.   The development and introduction of new varieties of (mainly) wheat and rice that has increased yields per acre dramatically in many countries since the 1960s.

green water.   In the hydrologic cycle, water that is evaporated or transpired and returned as water vapor to the atmosphere.

green products.   Products that are more environmentally benign than their traditional counterparts.

grit chamber.   Part of preliminary treatment in wastewater-treatment plants; a swimming pool–like tank in which the velocity of the water is slowed enough to let sand and other gritty material settle.

gross domestic (national) product per capita.   The total value of all goods and services exchanged in a year in a country, divided by its population. A common indicator of the average level of development and standard of living for a country.

gross primary production.   The rate at which primary producers in ecosystems are fixing organic matter.

groundwater.   Water that has accumulated in the ground, completely filling and saturating all pores and spaces in rock or soil. Groundwater is free to move more or less readily, it is the reservoir for springs and wells, and is replenished by infiltration of surface water.

groundwater remediation.   The repurification of contaminated groundwater by any of a number of techniques.

gully erosion.   Soil erosion produced by running water and resulting in the formation of gullies.

habitat.   The specific environment (woods, desert, swamp) in which an organism lives.

habitat conservation plan.   Under the Endangered Species Act, a plan that may be drafted by the U.S. Fish and Wildlife Service in working with landowners to help mitigate conflicts resulting from an application of the act.

Hadley cell.   A system of vertical and horizontal air circulation predominating in tropical and subtropical regions and creating major weather patterns.

half-life.   The length of time it takes for half of an unstable isotope to decay. The length of time is the same regardless of the starting amount. Also refers to the amount of time it takes compounds to break down in the environment.

halogenated hydrocarbon.   Synthetic organic compound containing one or more atoms of the halogen group, which includes chlorine, fluorine, and bromine.

hard water.   Water that contains relatively large amounts of calcium or certain other minerals that cause soap to precipitate. (Contrast **soft water**.)

hazard.   Anything that can cause (1) injury, disease, or death to humans, (2) damage to property, or (3) degradation of the environment. *Cultural* hazards include factors that are often a matter of choice, such as smoking or sunbathing. *Biological* hazards are pathogens and parasites that infect humans. *Physical* hazards are natural disasters like earthquakes and tornadoes. *Chemical* hazards refer to the chemicals in use in different technologies and household products.

hazard assessment.   The process of examining evidence linking a particular hazard to its harmful effects.

hazardous material (HAZMAT).   Any material having one or more of the following attributes: ignitability, corrosivity, reactivity, and toxicity.

heavy metal.   Any of the high-atomic-weight metals, such as lead, mercury, cadmium, and zinc. All may be serious pollutants in water or soil because they are toxic in relatively low concentrations and they tend to bioaccumulate.

herbicide.   A chemical used to kill or inhibit the growth of undesired plants.

herbivore.   An organism such as a rabbit or deer that feeds primarily on green plants or plant products such as seeds or nuts. Such an organism is said to be herbivorous. (Synonym: **primary consumer**.)

herbivory.   The feeding on plants that occurs in an ecosystem. The total feeding of all plant-eating organisms.

heterotroph.   Any organism that consumes organic matter as a source of energy. Such an organism is said to be heterotrophic.

HHW (household hazardous wastes).   An important component of the hazardous-waste problem in the United States.

Highway Trust Fund.   The monies collected from the gasoline tax and designated for the construction of new highways.

hormones.   Natural chemical substances that control the development, physiology, and behavior of an organism. Hormones are produced internally and affect only the individual organism. Hormones are coming into use in pest control. (See also **pheromones**.)

host.   In feeding relationships, particularly parasitism, refers to the organism that is being fed upon (i.e., the organism that is supporting the feeder).

host–parasite relationship.   The relationship between a parasite and the organism upon which it feeds.

host-specific.   Attribute whereby insects, fungal diseases, and other parasites are unable to attack species other than their particular host.

human resources.   One component of the wealth of nations. Comprises the *human capital*, or the population and its attributes, and the *social capital*, or the social and political environment that people have created in a society.

human system.   The entire system that humans have created for their own support, consisting of agriculture, industry, transportation, communications networks, etc.

humidity. The amount of water vapor in the air. (See also **relative humidity**.)

humus. A dark brown or black, soft, spongy residue of organic matter that remains after the bulk of dead leaves, wood, or other organic matter has decomposed. Humus does oxidize, but relatively slowly. It is extremely valuable in enhancing the physical and chemical properties of soil.

hunger. Condition wherein the basic food required for meeting nutritional and energy needs is lacking and the individual is unable to lead a normal, healthy life.

hunter–gatherers. Humans surviving by hunting wild game and gathering seeds, nuts, berries, and other edible things from the natural environment.

hybrid. A plant or animal resulting from a cross between two closely related species that do not normally cross.

hybrid electric vehicle (HEV). An automobile with a small gasoline motor and batteries that is capable of getting over 60 mpg and that produces only one-tenth the pollution of a comparable gasoline-run car.

hybrid lighting. Experimental solar energy technology wherein photovoltaic cells are coupled with optical fibers in rooftop units.

hybridization. Cross-mating between two more or less closely related species.

hydrocarbon emissions. Exhaust of various hydrogen–carbon compounds due to incomplete combustion of fuel. Hydrocarbon emissions are a major contribution to photochemical smog.

hydrocarbons. *Chemistry:* Natural or synthetic organic substances that are composed mainly of carbon and hydrogen. Crude oil, fuels from crude oil, coal, animal fats, and vegetable oils are examples. *Pollution:* A wide variety of relatively small carbon–hydrogen molecules resulting from incomplete burning of fuel and emitted into the atmosphere. (See **volatile organic compounds**.)

hydroelectric dam. A dam and an associated reservoir used to produce electrical power by letting the high-pressure water behind the dam flow through and drive a turbogenerator.

hydroelectric power. Electric power that is produced from hydroelectric dams or, in some cases, natural waterfalls.

hydrogen bonding. A weak attractive force that occurs between a hydrogen atom of one molecule and, usually, an oxygen atom of another molecule. Hydrogen bonding is responsible for holding water molecules together to produce the liquid and solid states.

hydrogen ions. Hydrogen atoms that have lost their electrons (chemical symbol, H^+).

hydrologic cycle. The movement of water from points of evaporation, through the atmosphere, through precipitation, and through or over the ground, returning to points of evaporation.

hydroponics. The culture of plants without soil. The method uses water with the required nutrients in solution.

hydroxyl radical. The hydroxyl group (OH), missing the electron. The hydroxyl radical is a natural cleansing agent of the atmosphere. It is highly reactive, readily oxidizes many pollutants upon contact, and thus contributes to their removal from the air.

hypothesis. A tentative guess concerning the cause of an observed phenomenon that is then subjected to experiment to test its logical or empirical consequences.

hypoxia. The absence of oxygen in sediments and deeper levels of the water column brought about by microbial breakdown of organic matter; hypoxia has created the "dead zone" in the Gulf of Mexico.

ICPD (International Conference on Population and Development) A U.N.-sponsored conference held in 1994 in Cairo, Egypt, that produced a program of action which was drafted and later adopted by the United Nations.

incremental value. The value of some finite change in a natural service; used in calculating the value of natural services performed by ecosystems.

indicator organism. An organism, the presence or absence of which indicates certain conditions. For example, the presence of *Escherichia coli* indicates that water is contaminated with fecal wastes and that pathogens may be present; the absence of *E. coli* indicates that the water is free of pathogens.

individual quota system. A system of fishery management wherein a quota is set and individual fishers are given or sold the right to harvest some proportion of the quota.

industrialized agriculture. The use of fertilizer, irrigation, pesticides, and energy from fossil fuels to produce large quantities of crops and large numbers of livestock with minimal labor for domestic and foreign sale.

industrialized countries. (See **developed countries**.)

Industrial Revolution. During the 19th century, the development of manufacturing processes using fossil fuels and based on applications of scientific knowledge.

industrial smog. The grayish mixture of moisture, soot, and sulfurous compounds that occurs in local areas in which industries are concentrated and coal is the primary energy source.

infant mortality. The number of babies that die before one year, per 1,000 babies born.

infiltration. The process in which water soaks into soil as opposed to running off the surface of the soil.

infiltration–runoff ratio. The ratio of the amount of water soaking into the soil to that running off the surface. The ratio is obtained by dividing the first amount by the second.

infrared radiation. Radiation of somewhat longer wavelengths than red light, which comprises the longest wavelengths of the visible spectrum. Such radiation manifests itself as heat.

infrastructure. The sewer and water systems, roadways, bridges, and other facilities that underlie the functioning of a city and that are owned, operated, and maintained by the city.

inherently safe reactor. In theory, a nuclear reactor that is designed in such a way that any accident would be automatically corrected with no radioactivity released.

inorganic compounds/molecules. *Classical definition:* All things such as air, water, minerals, and metals, that are neither living organisms nor products uniquely produced by living things. *Chemical definition:* All chemical compounds or molecules that do not contain carbon atoms as an integral part of their molecular structure. (Contrast **organic compounds**.)

insecticide. Any chemical used to kill insects.

instrumental value. The value that living organisms or species have in virtue of their benefit to people; the degree to which they benefit humans. (Contrast **intrinsic value**.)

insurance spraying. Spraying of pesticides when it is not really needed, in the belief that it will ensure against loss due to pests.

integrated pest management (IPM). A program consisting of two or more methods of pest control carefully integrated together and designed to avoid economic loss from pests. The objective of an IPM is to minimize the use of environmentally hazardous synthetic chemicals. Such chemicals may be used in IPM, but only as a last resort to prevent significant economic losses.

integrated waste management. The approach to municipal solid waste that provides for several options for dealing with wastes, including recycling, composting, waste reduction, and landfilling and incineration where unavoidable.

International Whaling Commission (IWC). The international body that regulates the harvesting of whales; the IWC placed a ban on all whaling in 1986.

intrinsic value. The value that living organisms or species have in their own right; in other words, organisms and species do not have to be useful to have value. (Contrast **instrumental value**.)

inversion. See **temperature inversion**.

ion. An atom (or a group of atoms) that has lost or gained one or more electrons and, consequently, has acquired a positive or negative charge. Ions are designated by "+" or "−" superscripts following the chemical symbol for the element(s) involved.

ion-exchange capacity. See **nutrient-holding capacity**.

ionic bond. The bond formed by the attraction between a positive and a negative ion.

IPCC (Intergovernmental Panel on Climate Change). The U.N.-sponsored organization charged with continually assessing the science of global climate change, its potential impacts, and the means of responding to the threat.

irrigation. Any method of artificially adding water to crops.

isotope. A form of an element in which the atoms have more (or less) than the usual number of neutrons. Isotopes of a given element have identical chemical properties, but differ in mass (weight) as a result of the superfluity (or deficiency) of neutrons. Many isotopes are unstable and radioactive. (See **radioactive decay, radioactive emissions,** and **radioactive materials**.)

ISTEA (Intermodal Surface Transportation Efficiency Act). Legislation that provides for funding of alternative transportation (e.g., mass transit or bicycle paths) using money from the Highway Trust Fund.

Jubilee 2000. A worldwide coalition of religious groups and individuals dedicated to promoting debt relief efforts on the part of creditor nations.

junk science. Information presented as valid science, but unsupported by peer-reviewed research. Often, politically motivated and biased results are selected to promote a particular point of view.

juvenile hormone. The insect hormone that, at sufficient levels, preserves the larval state. Pupation requires diminished levels of juvenile hormone; hence, artificial applications of the hormone may block pupal development.

keystone species. A species whose role is essential for the survival of many other species in an ecosystem.

kinetic energy. The energy inherent in motion or movement, including molecular movement (heat) and the movement of waves (hence, radiation and therefore light).

Kyoto Protocol. An international agreement among the developed nations to curb greenhouse gas emissions. The Kyoto Protocol was forged in December 1997.

La Niña. A complete reversal of El Niño conditions in the tropical Pacific Ocean. Easterly trade winds are intense, and global weather patterns are often the reverse of those brought on by an El Niño event.

Lacey Act. Passed in 1900, the first national act that gave protection to wildlife by forbidding interstate commerce in illegally killed animals.

landfill. A site where municipal, industrial, or chemical wastes are disposed of by burying them in the ground or placing them on the ground and covering them with earth.

landscape. A group of interacting ecosystems occupying adjacent geographical areas.

land subsidence. The gradual sinking of land. The condition may result from the removal of groundwater or oil, which is frequently instrumental in supporting the overlying rock and soil.

land trust. Land that is purchased and held by various private organizations specifically for the purpose of protecting the region's natural environment and the biota that inhabit it.

larva, pl. larvae. A free-living immature form that occurs in the life cycle of many organisms and that is structurally distinct from the adult. For example, caterpillars are the larval stage of moths and butterflies.

law of conservation of energy. See **first law of thermodynamics**.

law of conservation of matter. Law stating that, in chemical reactions, atoms are neither created, nor changed, nor destroyed; they are only rearranged.

law of limiting factors. Law stating that a system may be limited by the absence or minimum amount (in terms of that needed) of any required factor. Also known as Liebig's law of minimums. (See **limiting factor**.)

leachate. The mixture of water and materials that are leaching.

leaching. The process in which materials in or on the soil gradually dissolve and are carried by water seeping through the soil. Leaching may eventually remove valuable nutrients from the soil, or it may carry buried wastes into groundwater, thereby contaminating it.

legumes. The group of pod-bearing land plants that is virtually alone in its ability to fix nitrogen; legumes include such common plants as peas, beans, clovers, alfalfa, and locust trees, but no major cereal grains. (See **nitrogen fixation**.)

Liebig's law of minimums. See **law of limiting factors**.

lifeboat ethic. An argument to the effect that we should limit the amount of food aid to high-population countries in order to prevent further population growth.

limiting factor. A factor primarily responsible for determining the growth or reproduction of an organism or a population. The limiting factor in a given environment may be a physical factor such as temperature or light, a chemical factor such as a particular nutrient, or a biological factor such as a competing species. The limiting factor may differ at different times and places.

limits of tolerance. The extremes of any factor (e.g., temperature) that an organism or a population can tolerate and still survive and reproduce.

lipids. A class of natural organic molecules that includes animal fats, vegetable oils, and phospholipids, the last being an integral part of cellular membranes.

litter. In an ecosystem, the natural cover of dead leaves, twigs, and other dead plant material. This natural litter is subject to rapid decomposition and recycling in the ecosystem, whereas human litter, such as bottles, cans, and plastics, is not.

livability. A subjective index of how enjoyable a city is to live in.

loam. A solid consisting of a mixture of about 40% sand, 40% silt, and 20% clay.

longevity. The average life span of individuals of a given population.

LULU ("locally unwanted land use"). Acronym expressing the difficulty in siting a facility that is necessary, but that no one wants in his or her immediate locality.

macromolecules. Very large organic molecules, such as proteins and nucleic acids, that constitute the structural and functional parts of cells.

MACT (Maximum achievable control technology). The best technologies available for reducing the output of especially toxic industrial pollutants.

Magnuson Act. An act passed in 1976 that extended the limits of jurisdiction over coastal waters and fisheries of the United States to 200 miles offshore.

malnutrition. The lack of essential nutrients such as vitamins, minerals, and amino acids. Malnutrition ranges from mild to severe and life threatening.

mass number.   The number that accompanies the chemical name or symbol of an element or isotope. The mass number represents the number of neutrons and protons in the nucleus of the atom.

material safety data sheets (MSDS).   Documents containing information on the reactivity and toxicity of chemicals and that must accompany the shipping, storage, and handling of over 600 chemicals.

materials recycling facility (MRF).   A processing plant in which regionalized recycling is carried out. Recyclable municipal solid waste, usually presorted, is prepared in bulk for the recycling market.

matter.   Any gas, liquid, or solid that occupies space and has mass. (Contrast **energy.**)

maximum sustainable yield.   The maximum amount of a renewable resource that can be taken year after year without depleting the resource. The maximum sustainable yield is the maximum rate of use or harvest that will be balanced by the regenerative capacity of the system—for example, the maximum rate of tree cutting that can be balanced by regrowth.

meltdown.   The event of a nuclear reactor's getting out of control or losing its cooling water so that it melts from its own production of heat. The melted reactor would continue to produce heat and could melt its way out of its containment vessel and eventually down into groundwater, where it would cause a violent eruption of steam that could spread radioactive materials over a wide area.

metabolism.   The sum of all the chemical reactions that occur in an organism.

methane.   A gas, CH_4. Methane is the primary constituent of natural gas and is also a product of fermentation by microbes. Methane is one of the greenhouse gases.

microbe.   Any microscopic organism, although primarily a bacterium, a virus, or a protozoan.

microclimate.   The actual conditions experienced by an organism in its particular location. Owing to numerous factors, such as shading, drainage, and sheltering, the microclimate may be quite distinct from the overall climate.

microfiltration.   A process for purifying water in which the water is forced under very high pressure through a membrane that is fine enough to filter out ions and molecules in solution; microfiltration used by small desalination plants to filter salt from seawater. Also called reverse osmosis.

microlending.   The process of providing very small loans (usually $50–$500) to poor people to facilitate their starting a small enterprise and becoming economically self-sufficient.

microorganism.   See **microbe.**

midnight dumping.   The illicit dumping of materials, particularly hazardous wastes, frequently under the cover of darkness.

Milankovitch cycle.   A cycle of major oscillations in the Earth's orbit, taking place over frequencies of thousands of years and known to influence the distribution of solar radiation and therefore global weather patterns.

Minamata disease.   A disease named for a fishing village in Japan where an "epidemic" was first observed. Symptoms, which included spastic movements, mental retardation, coma, death, and crippling birth defects in the next generation, were found to be the result of mercury poisoning.

mineral.   Any hard, brittle, stonelike material that occurs naturally in Earth's crust. All minerals consist of various combinations of positive and negative ions held together by ionic bonds. Pure minerals, or crystals, are one specific combination of elements. Common rocks are composed of mixtures of two or more minerals.

mineralization.   The process of gradual oxidation of the organic matter (humus) present in soil that leaves just the gritty mineral component of the soil.

mixture.   A combination of elements in which there is no chemical bonding between the molecules. For example, air contains (is a mixture of) oxygen, nitrogen, and carbon dioxide.

moderator.   In a nuclear reactor, any material that slows down neutrons from fission reactions so that they are traveling at the right speed to trigger another fission. Water and graphite are two types of moderators.

molecule.   The smallest unit of two or more atoms forming a compound. A molecule has all the characteristics of the compound of which it is a unit.

monoculture/monocropping.   The practice of growing the same crop year after year on the same land. (Contrast **crop rotation** and **polyculture.**)

monsoon.   The seasonal airflow created by major differences in cooling and heating between oceans and continents, usually bringing extensive rain.

Montreal Protocol.   An agreement made in 1987 by a large group of nations to cut back the production of chlorofluorocarbons by 50% by the year 2000 in order to protect the ozone shield. A 1990 amendment called for the complete phaseout of these chemicals by 2000 in developed nations and by 2010 in less developed nations.

morbidity.   The incidence of disease in a population.

mortality.   The incidence of death in a population.

municipal solid waste (MSW).   The entirety of refuse or trash generated by a residential and business community. Distinct from agricultural and industrial wastes, municipal solid waste is the refuse that a municipality is responsible for collecting and disposing of.

mutagenic.   Causing mutations.

mutation.   A random change in one or more genes of an organism. Mutations may occur spontaneously in nature, but their number and degree are vastly increased by exposure to radiation or certain chemicals. Mutations generally result in a physical deformity or metabolic malfunction.

mutualism.   A close relationship between two organisms from which both derive a benefit.

mycelia.   The threadlike feeding filaments of fungi.

mycorrhiza, pl. mycorrhizae.   The mycelia of certain fungi that grow symbiotically with the roots of some plants and provide for additional nutrient uptake.

NASA.   National Aeronautics and Space Administration.

National Ambient Air Quality Standards (NAAQS).   The allowable levels of ambient air pollutants set by EPA regulation.

National Emission Standards for Hazardous Air Pollutants (NESHAPS).   The standards for allowable emissions of certain toxic substances.

national forests.   Public forests and woodlands administered by the National Forest Service for multiple uses, such as logging, mineral exploitation, livestock grazing, and recreation.

national parks.   Lands and coastal areas of great scenic, ecological, or historical importance administered by the National Park Service with the dual goals of protection and providing public access.

National Pollution Discharge Elimination System (NPDES).   An EPA-administered program that addresses point-source water pollution through the issuance of permits that regulate pollution discharge.

national priorities list (NPL).   A list of the chemical waste sites presenting the most immediate and severe threats. Such sites are scheduled for cleanup ahead of other sites.

natural.   Produced as a normal part of nature, apart from any activity or intervention of humans. Opposite of artificial, synthetic, human made, or caused by humans.

natural capital.   The natural assets and the services they perform. One form of the wealth of a nation is its complement of natural capital.

natural chemical control.   The use of one or more natural chemicals, such as hormones or pheromones, to control a pest.

natural control methods.   Any of many techniques of controlling a pest population without resorting to the use of synthetic organic or inorganic chemicals. (See **biological control**, **cultural control**, **genetic control**, **hormones**, and **pheromones**.)

natural enemies.   All the predators and parasites that may feed on a given organism. Organisms used to control a specific pest through predation or parasitism.

natural increase.   The number of births minus the number of deaths in a given population. The natural increase does not take into account immigration and emigration and is the percent of growth (or decline) of a given population during a year. It is found by subtracting the crude death rate from the crude birthrate and changing the result to a percent.

natural laws.   Generalizations derived from our observations of matter, energy, and other phenomena. Though not absolute, natural laws have been empirically confirmed to a high degree and are often derivable from higher level theory.

natural resources.   Features of natural ecosystems and species that are of economic value and that may be exploited. Also, features of particular segments of ecosystems, such as air, water, soil, and minerals.

natural selection.   The process whereby the natural factors of environmental resistance tend to eliminate those members of a population which are least well adapted to cope with their environment and thus, in effect, tend to select those best adapted for survival and reproduction.

natural services.   Functions performed free of charge by natural ecosystems, such as control of runoff and erosion, absorption of nutrients, and assimilation of air pollutants.

Neolithic Revolution.   The development of agriculture by human societies around 10,000 years ago, leading to more permanent settlement and population increases.

net primary production.   The rate at which new organic matter is made available to consumers by primary producers (equals gross primary production minus plant respiration).

neutron.   A fundamental atomic particle found in the nuclei of atoms (except hydrogen) and having one unit of atomic mass, but no electrical charge.

new forestry.   Now part of the Forest Service's management practice, a forestry management strategy that places priority on protecting the ecological health and diversity of forests rather than maximizing the harvest of logs.

niche (ecological).   The total of all the relationships that bear on how an organism copes with the biotic and abiotic factors it faces.

NIMBY ("Not in my backyard").   A common attitude regarding undesirable facilities such as incinerators, nuclear facilities, and hazardous waste treatment plants whereby people do everything possible to prevent such facilities from being located near their residences.

NIMTOO ("Not in my term of office").   Attitude expressing the reluctance of officeholders to make unpopular decisions on matters such as siting waste facilities.

nitric acid (HNO_3).   One of the acids in acid rain. Formed by reactions between nitrogen oxides and the water vapor in the atmosphere.

nitrogen fixation.   The process of chemically converting nitrogen gas (N_2) from the air into compounds such as nitrates (NO_3^-) or ammonia (NH_3) that can be used by plants in building amino acids and other nitrogen-containing organic molecules.

nitrogen oxides (NO_x).   A group of nitrogen–oxygen compounds formed when some of the nitrogen gas in air combines with oxygen during high-temperature combustion. Nitrogen oxides are a major category of air pollutants and, along with hydrocarbons, are a primary factor in the production of ozone and other photochemical oxidants that are the most harmful components of photochemical smog. Nitrogen oxides also contribute to acid precipitation (see **nitric acid**). The major nitrogen oxides are nitric oxide (NO), nitrogen dioxide (NO_2), and nitrogen tetroxide (N_2O_2).

nitrous oxide.   A gas, N_2O. Nitrous oxide, which comes from biomass burning, fossil fuel burning, and the use of chemical fertilizers, is of concern because it is a greenhouse gas in the troposphere and it contributes to ozone destruction in the stratosphere.

NOAA.   National Oceanic and Atmospheric Administration.

nonbiodegradable.   Not able to be consumed or broken down by biological organisms. Nonbiodegradable substances include plastics, aluminum, and many chemicals used in industry and agriculture. Particularly dangerous are human-made nonbiodegradable chemicals that are also toxic and tend to accumulate in organisms (i.e., nonbiodegradable synthetic organic compounds). (See **biodegradable** and **bioaccumulation**.)

nonconsumptive water use.   Use of water for such purposes as washing and rinsing, wherein the water, albeit polluted, remains available for further uses. With suitable purification, such water may be recycled indefinitely.

nongovernmental organization (NGO).   Any of a number of private organizations involved in studying or advocating action on different environmental issues.

nonpersistent.   Said of chemicals that break down readily to harmless compounds, as, for example, natural organic compounds break down to carbon dioxide and water.

nonpoint sources.   Sources of pollution such as the general runoff of sediments, fertilizer, pesticides, and other materials from farms and urban areas, as opposed to specific points of discharge such as factories. Also called diffuse sources. (Contrast **point sources**.)

nonrenewable resources.   Resources, such as ores of various metals, oil, and coal, that exist as finite deposits in Earth's crust and that are not replenished by natural processes as they are mined. (Contrast **renewable resources**.)

no-till agriculture.   The farming practice in which weeds are killed with chemicals (or other means) and seeds are planted and grown without resorting to plowing or cultivation. The practice is highly effective in reducing soil erosion.

NRCS.   Natural Resources Conservation Service, formerly the SCS (U.S. Soil Conservation Service).

nuclear power.   Electrical power generated by using a nuclear reactor to boil water and produce steam that in turn, drives a turbogenerator.

Nuclear Regulatory Commission (NRC).   The agency within the Department of Energy that sets and enforces safety standards for the operation and maintenance of nuclear power plants.

nucleic acids.   The class of natural organic macromolecules that function in the storage and transfer of genetic information.

nucleus.   *Biology*: The large body residing in most living cells that contains the genes or hereditary material (DNA). *Physics*:

The central core of atoms, which is made up of neutrons and protons. Electrons surround the nucleus.

nutrient. *Animal*: Material such as protein, vitamins, and minerals required for growth, maintenance, and repair of the body and material such as carbohydrates required for energy. *Plant*: An essential element in a particular ion or molecule that can be absorbed and used by the plant. For example, carbon, hydrogen, nitrogen, and phosphorus are essential elements; carbon dioxide, water, nitrate (NO_3^-), and phosphate (PO_4^{3-}) are the respective ions or molecules containing the nutrient element.

nutrient cycle. The repeated pathway of particular nutrients or elements from the environment through one or more organisms and back to the environment. Nutrient cycles include the carbon cycle, the nitrogen cycle, the phosphorus cycle, and so on.

nutrient-holding capacity. The capacity of a soil to bind and hold nutrients (fertilizer) against their tendency to be leached from the soil.

observations. Things or phenomena that are perceived through one or more of the basic five senses in their normal state. In addition, to be accepted as factual, observations must be verifiable by others.

ocean thermal-energy conversion (OTEC). The concept of harnessing the difference in temperature between surface water heated by the sun and colder deep water to produce power.

oil field. The underground area in which exploitable oil is found.

oil sand. Sedimentary material containing bitumen, a tarlike hydrocarbon that is subject to exploitation under favorable economic conditions.

oil shale. A natural sedimentary rock that contains kerogen, a material that can be extracted and refined into oil and oil products.

oligotrophic. Nutrient poor and hence unable to support much phytoplankton (said of a body of water).

omnivore. An animal that feeds on both plant material and other animals.

OPEC. Organization of Petroleum Exporting Countries, a cartel of oil-producing nations that attempts to control the market price of oil.

optimal population. The population of a harvested biological resource that yields the greatest harvest for exploitation; according to maximum-sustained-yield equations, the optimal population is half the carrying capacity.

optimal range. With respect to any particular factor or combination of factors, the maximum variation that still supports optimal or near-optimal growth of the species in question.

optimum. The condition or amount of any factor or combination of factors that will produce the best result. For example, the amount of heat, light, moisture, nutrients, and so on that will produce the best plant growth.

organically grown. Grown without the use of hard chemical pesticides or inorganic fertilizer. Any produce grown according to USDA standards for organic foods is organically grown.

organic compounds/molecules. *Classical definition*: All living things and products that are uniquely produced by living things, such as wood, leather, and sugar. *Chemical definition*: All chemical compounds or molecules, natural or synthetic, that contain carbon atoms as an integral part of their molecular structure. The structure of organic compounds is based on bonded carbon atoms with hydrogen atoms attached. Organic compounds can be biodegradable or nonbiodegradable. (Contrast **inorganic compounds**.)

organic fertilizer. See **fertilizer**.

organic gardening/farming/food. Gardening or farming without the use of inorganic fertilizers, synthetic pesticides, or other human-made materials; food raised with organic farming techniques.

organic phosphate. Phosphate (PO_4^{-3}) bonded to an organic molecule.

organism. Any living thing—plant, animal, or microbe.

orphan site. The location of a hazardous waste site abandoned by former owners.

OSHA (Occupational Safety and Health Administration). Federal agency that promulgates regulations to protect workers.

osmosis. The phenomenon whereby water diffuses through a semipermeable membrane toward an area where there is more material in solution (i.e., where there is a relatively lower concentration of water). Osmosis has particular application in the salinization of those soils in which plants are unable to grow because of osmotic water loss.

outbreak. A population explosion of a particular pest. Often caused by an application of pesticides that destroys the pest's natural enemies.

overcultivation. The practice of repeated cultivation and growing of crops more rapidly than the soil can regenerate, leading to a decline in soil quality and productivity.

overgrazing. The phenomenon of animals' grazing in greater numbers than the land can support in the long term. There may be a temporary economic gain in the short term, but the grassland (or other ecosystem) is destroyed, and its ability to support life in the long term is vastly diminished.

oxidation. Chemical reaction that generally involves a breakdown of some substance through its combining with oxygen. Burning and cellular respiration are examples of oxidation. In both cases, organic matter is combined with oxygen and broken down to carbon dioxide and water.

ozone. A gas, O_3, that is a pollutant in the lower atmosphere, but that is necessary to screen out ultraviolet radiation in the upper atmosphere. May also be used for disinfecting water.

ozone hole. First discovered over the Antarctic, a region of stratospheric air that is severely depleted of its normal levels of ozone during the Antarctic spring because of CFCs from anthropogenic (human-made) sources.

ozone shield. The layer of ozone gas (O_3) in the upper atmosphere that screens out harmful ultraviolet radiation from the Sun.

PANs (peroxyacetylnitrates). A group of compounds present in photochemical smog that are extremely toxic to plants and irritating to the eyes, nose, and throat membranes of humans.

parasites. Organisms (plant, animal, or microbial) that attach themselves to another organism, the host, and feed on it over a period of time without killing it immediately, but usually doing harm to it. Parasites are commonly divided into ectoparasites—those that attach to the outside of the host—and endoparasites—those that live inside their hosts.

parent material. Rock material whose weathering and gradual breakdown is the source of the mineral portion of soil.

particulates. (See **PM-2 and PM-10** and **suspended particulate matter**.)

parts per million (ppm). A frequently used expression of concentration. The number of units of one substance present in a million units of another. Equivalent to milligrams per liter.

passive safety features. Those safety features of nuclear facilities which involve processes that are not vulnerable to operator intrusion or electrical power failures. Passive safety features enhance the degree of safety of nuclear reactors. (See **active safety features**.)

passive solar-heating system. A solar-heating system that does not use pumps or blowers to transfer heated air or water. Instead, natural convection currents are used, or the interior of the building itself acts as the solar collector.

pasteurization. The process of applying enough heat to milk or some other substance to kill pathogens and sufficient other bacteria to extend the shelf life of the product.

pastoralist. One involved in animal husbandry, usually in subsistence agriculture.

pathogen. An organism, usually a microbe, that is capable of causing disease. Such an organism is said to be pathogenic.

PCBs (polychlorinated biphenyls). A group of widely used industrial chemicals of the chlorinated hydrocarbon class. Polychlorinated biphenyls have become serious and widespread pollutants because they are extremely resistant to breakdown and are subject to bioaccumulation. They also are known to be carcinogenic.

percolation. The process of water seeping downward through cracks and pores in soil or rock.

permafrost. The ground of arctic regions that remains permanently frozen. Defines tundra, since only small herbaceous plants can be sustained on the thin layer of soil that thaws each summer.

persistent. Attribute of pesticides or other chemicals that are nonbiodegradable and highly resistant to breakdown by other means. Such chemicals therefore remain present in the environment for periods of years or longer.

persistent organic pollutants (POPs). Any members of a class of organic pollutants that are resistant to biodegradation and that are often toxic; for example, DDT, PCBs, and dioxin are persistent organic pollutants.

pest. Any organism that is noxious, destructive, or troublesome; usually an agricultural pest.

pesticide. A chemical used to kill pests. Pesticides are further categorized according to the pests they are designed to kill—for example, herbicides kill plants, insecticides kill insects, fungicides kill fungi, and so on.

pesticide treadmill. The idea that use of chemical pesticides simply creates a vicious cycle of "needing more pesticides" to overcome developing resistance and secondary outbreaks caused by the pesticide applications.

pest-loss insurance. Insurance that a grower can buy that will pay in the event of loss of a crop due to pests.

petrochemical. A chemical made from petroleum (crude oil) as a basic raw material. Petrochemicals include plastics, synthetic fibers, synthetic rubber, and most other synthetic organic chemicals.

pH. Scale used to designate the acidity or basicity (alkalinity) of solutions or soil, expressed as the logarithm of the concentration of hydrogen ions (H^+). pH 7 is neutral; values decreasing from 7 indicate increasing acidity, values increasing from 7 increasing basicity. Each unit from 7 indicates a tenfold increase over the preceding unit.

pheromones. Chemical substances secreted externally by certain members of a species that affect the behavior of other members of the same species. The most common examples are sex attractants, which female insects secrete to attract males. Pheromones are coming into use in pest control. (See also **hormones.**)

phosphate. An ion composed of a phosphorus atom with four oxygen atoms attached. Denoted PO_4^{3-}, phosphate is an important plant nutrient. In natural waters, it is frequently the limiting factor; therefore, additions of phosphate to natural water are often responsible for algal blooms.

photochemical oxidants. A major category of secondary air pollutants, including ozone, that are highly toxic and damaging, especially to plants and forests. Formed as a result of interactions between nitrogen oxides and hydrocarbons driven by sunlight.

photochemical smog. The brownish haze that frequently forms on otherwise clear, sunny days over large cities with significant amounts of automobile traffic. Photochemical smog results largely from sunlight-driven chemical reactions among nitrogen oxides and hydrocarbons, both of which come primarily from auto exhausts.

photosynthesis. The chemical process carried on by green plants through which light energy is used to produce glucose from carbon dioxide and water. Oxygen is released as a by-product.

photovoltaic cells. Devices that convert light energy into an electrical current.

physical barrier. A genetic feature on a plant, such as sticky hairs, that physically blocks attack by pests.

phytoplankton. Any of the many species of photosynthetic microorganisms that consist of single cells or small groups of cells that live and grow freely suspended near the surface in bodies of water.

phytoremediation. The use of plants to accomplish the cleanup of some hazardous chemical wastes.

planetary albedo. The reflection of solar radiation back into space due to cloud cover, contributing to cooling of the atmosphere.

plankton. Any and all living things that are found freely suspended in the water and that are carried by currents, as opposed to being able to swim against currents. Plankton includes both plant (phytoplankton) and animal (zooplankton) forms.

plant community. The array of plant species, including their numbers, ages, and distribution, that occupies a given area.

plate tectonics. The theory that postulates the movement of sections of the Earth's crust, creating earthquakes, volcanic activity, and the buildup of continental masses over long periods of time.

PM-2 and PM-10. Standard criterion pollutants for suspended particulate matter. PM-10 refers to particles smaller than 10 micrometers in diameter, PM-2 for particles smaller than 2 micrometers in diameter. The smaller particles are readily inhaled directly into the lungs.

point sources. Specific points of origin of pollutants, such as factory drains or outlets from sewage-treatment plants. (Contrast **nonpoint sources.**)

policy life cycle. The typical course of events occurring over time in the recognition, formulation, implementation, and control of an environmental problem requiring public policy action.

pollutant. A substance that contaminates air, water, or soil.

pollution. Contamination of air, water, or soil with undesirable amounts of material or heat. The material may be a natural substance, such as phosphate, in excessive quantities, or it may be very small quantities of a synthetic compound such as dioxin, which is exceedingly toxic.

polyculture. The growing of two or more species together. (Contrast **monoculture.**)

poor. Economically unable to afford adequate food or housing.

population. A group within a single species whose individuals can and do freely interbreed.

population density. The number of individuals per unit of area.

population equilibrium. A state of balance between births and deaths in a population.

population explosion. The exponential increase observed to occur in a population when conditions are such that a large percentage of the offspring are able to survive and reproduce in turn. A population explosion frequently leads to overexploitation and eventual collapse of the ecosystem.

population momentum. Property whereby a rapidly growing human population may be expected to grow for 50–60 years after replacement fertility (2.1 live births per female) is reached. Momentum is sustained because of increasing numbers entering reproductive age.

population profile.   A bar graph that shows the number of individuals at each age or in each five-year age group, starting with the youngest ages at the bottom of the profile.

population structure.   See **age structure.**

potential energy.   The ability to do work that is stored in some chemical or physical state. For example, gasoline is a form of potential energy because the ability to do work is stored in the chemical state and is released as the fuel is burned in an engine.

power grid.   The combination of central power plants, high-voltage transmission lines, poles, wires, and transformers that makes up an electricial power system.

pOH.   The negative logarithm of the concentration of hydroxyl (OH) ions. Like pH, pOH ranges from 0 to 14, with each unit representing a tenfold increase over the preceding one. The lower the pOH, the higher is the concentration of hydroxyl ions.

precautionary principle.   The principle that says that where there are threats of serious or irreversible damage, the absence of scientific certainty shall not be used as a reason for postponing cost-effective measures to prevent environmental degradation.

precipitation.   Any form of moisture condensing in the air and depositing on the ground.

predator.   An animal that feeds on another living organism, either plant or animal.

predator–prey relationship.   A feeding relationship existing between two kinds of organisms. The predator is the animal feeding on the prey. Such relationships are frequently instrumental in controlling populations of herbivores.

preliminary treatment.   The removal of debris and grit from wastewater by passing the water through a coarse screen and a grit-settling chamber.

primary consumer.   An organism, such as a rabbit or deer, that feeds more or less exclusively on green plants or their products, such as seeds and nuts. (Synonym: **herbivore.**)

primary energy sources.   Fossil fuels, radioactive material, and solar, wind, water, and other energy sources that exist as natural resources subject to exploitation.

primary pollutants.   Pollutants released directly into the atmosphere mainly as a result of burning fuels and wastes, as opposed to secondary pollutants.

primary producers/primary production.   Photosynthetic organisms. The activities of these organisms in creating new organic matter in ecosystems.

primary recovery.   In an oil well or oil field, the oil that can be removed by conventional pumping. (See **secondary recovery.**)

primary standard.   The maximum tolerable level of a pollutant. The standard is intended to protect human health.

primary succession.   See **succession.**

primary treatment.   The process that follows preliminary sewage treatment and that consists of passing the water very slowly through a large tank so that the particulate organic material can settle out. The settled material is **raw sludge.**

prior informed consent (PIC).   A U.N.-approved procedure whereby countries exporting pesticides must inform importing countries of actions the exporting countries have taken to restrict the use of the pesticides.

produced assets.   The stock of buildings, machinery, vehicles, and other elements of a country's infrastructure that are essential to the production of economic goods and services. One component of the wealth of nations.

producers.   In an ecosystem, those organisms (mostly green plants) which use light energy to construct their organic constituents from inorganic compounds.

productive use.   The exploitation of ecosystem resources for economic gain.

property taxes.   Taxes that the local government levies on privately owned properties, proportional to the value of the property. Property taxes are the major source of revenue for local governments.

protein.   Organic macromolecules made up of amino acids; the major structural component of all animal tissues, as well as enzymes in both plants and animals.

proton.   Fundamental atomic particle with a positive charge, found in the nuclei of atoms. The number of protons present equals the atomic number and is distinct for each element.

protozoon, pl. protozoa.   Any of a large group of eukaryotic microscopic organisms that consist of a single, relatively large complex cell or, in some cases, small groups of cells. All have some means of movement. Amoebae and paramecia are examples.

proven reserves.   See **reserves.**

punctuated equilibrium.   An evolutionary process proposed by Gould and Eldridge wherein many species show long stable periods with little change and short periods of rapid change that leave few fossils; the fossil record gives support to this theory.

RACT (reasonably available control technology).   Applied to the goals of the Clean Air Act, EPA-approved forms of technology that will reduce the output of industrial air pollutants. (See **MACT.**)

radioactive decay.   The reduction in radioactivity that occurs as an unstable isotope (radioactive substance) gives off radiation, ultimately to become stable.

radioactive emissions.   Any of various forms of radiation or particles that may be given off by unstable isotopes. Many such emissions have very high energy and can destroy biological tissues or cause mutations leading to cancer or birth defects.

radioactive materials.   Substances that are or that contain unstable isotopes and that consequently give off radioactive emissions. (See **isotope** and **radioactive emissions.**)

radioactive wastes.   Waste materials that contain, are contaminated with, or are radioactive substances. Many materials used in the nuclear industry become radioactive wastes because of such contamination.

radioisotope.   An isotope of an element that is unstable and that gives off radioactive emissions. (See **isotope** and **radioactive decay.**)

radon.   A radioactive gas produced by natural processes in Earth that is known to seep into buildings. Radon can be a major hazard within homes and is a known carcinogen.

rain shadow.   The low-rainfall region that exists on the leeward (downwind) side of a mountain range. This rain shadow is the result of the mountain range's causing precipitation on the windward side.

range of tolerance.   The range of conditions within which an organism or population can survive and reproduce—for example, the range from the highest to the lowest temperature that can be tolerated. Within the range of tolerance is the optimum, or best, condition.

raw sludge.   The untreated organic matter that is removed from sewage water by letting it settle. Raw sludge consists of organic particles from feces, garbage, paper, and bacteria.

raw wastewater.   (See **raw sludge.**)

reactor vessel.   Steel-walled vessel that contains a nuclear reactor.

recharge area.   The area over which groundwater will infiltrate and resupply an aquifer.

recruitment.   The maturation and successful entry of young into an adult breeding population.

recycling.   Recovery of materials that would otherwise be buried in landfills or be combusted. Primary recycling involves remaking the same material from the waste; in secondary recycling, waste materials are made into different products.

relative humidity.   The percentage of moisture in the air, compared with how much the air can hold at the given temperature.

rem.   An older unit of measurement of the ability of radioactive emissions to penetrate biological tissue. (See **sievert.**)

remediation.   A return to an original uncontaminated state. (See **bioremediation** and **groundwater remediation**.)

renewable energy.   Energy sources—namely, solar, wind, and geothermal—that will not be depleted by use.

renewable resources.   Biological resources, such as trees, that may be renewed by reproduction and regrowth. Conservation to prevent overcutting and protection of the environment are still required, however. (Contrast **nonrenewable resources**.)

replacement fertility/level.   The fertility rate (level) that will just sustain a stable population.

reproductive isolation.   One of the processes of speciation, involving anything that keeps individuals or subpopulations from interbreeding.

reserves.   The amount of a mineral resource (including oil, coal, and natural gas) remaining in Earth that can be exploited using current technologies and at current prices. Usually given as *proven* reserves—those which have been positively identified—and *estimated* reserves—those which have not yet been discovered, but are presumed to exist.

resilience.   The tendency of ecosystems to recover from disturbances through a number of processes known as resilience mechanisms (e.g., succession after a fire).

resistance.   The development of more hardy pests through the effects of the repeated use of pesticides in selecting against sensitive individuals in a pest population.

resource partitioning.   The outcome of competition by a group of species where natural selection favors the division of a resource in time or space by specialization of the different species.

resources.   Biotic and abiotic factors that are consumed by organisms.

Resources Conservation and Recovery Act of 1976 (RCRA).   The cornerstone legislation to control indiscriminate land disposal of hazardous wastes.

respiration.   See **cell respiration.**

restoration ecology.   The branch of ecology devoted to restoring degraded and altered ecosystems to their natural state.

resurgence.   The rapid comeback of a population (especially of pests after a severe dieoff, usually caused by pesticides) and the return to even higher levels than before the treatment.

reuse.   The practice of continuing to use items, as opposed to throwing them away and producing new items (e.g., bottles collected and refilled in recycling efforts).

riparian woodlands.   The strip of woody plants that grow along natural watercourses.

risk.   The probability of suffering injury, disease, death, or some other loss as a result of exposure to a hazard.

risk analysis.   The process of evaluating the risks associated with a particular hazard before taking some action. Often called risk assessment.

risk characterization.   The process of determining the level of a risk and its accompanying uncertainties after hazard assessment, dose-response assessment, and exposure assessment have been accomplished.

risk management.   The task of regulators whose job is to review risk data and make regulatory decisions based on the data. The process often is influenced by considerations of costs and benefits, as well as by public perception.

risk perception.   Nonexperts' intuitive judgments about risks, which often are not in agreement with the level of risk as judged by experts.

rooftop garden.   A form of aboveground garden on the top of a flat residential roof, using shallow soil beds and containers such as old automobile tires or plastic wading pools.

rubber tappers.   In tropical forests of South America, people who harvest the latex rubber and Brazil nuts from rain-forest trees and who are opposed to clearing the forests. Also called *serengueros.*

runoff.   The portion of precipitation that runs off the surface, as opposed to soaking into the ground.

Safe Drinking Water Act of 1974.   Legislation to protect the public from the risk that toxic chemicals will contaminate drinking-water supplies. The act mandates regular testing of municipal water supplies.

safety net.   In a society, the availability of food and other necessities extended to people who are unable to meet their own needs for a variety of reasons.

salinization.   The process whereby soil becomes saltier and saltier until, finally, the salt prevents the growth of plants. Salinization is caused by irrigation, because salts brought in with the water remain in the soil as the water evaporates.

saltwater intrusion, saltwater encroachment.   The phenomenon of seawater's moving back into aquifers or estuaries. Such intrusion occurs when the normal outflow of freshwater is diverted or removed for use.

sand.   Mineral particles 0.2 to 2.0 mm in diameter.

sanitary sewer.   Separate drainage system used to receive all the wastewater from sinks, tubs, and toilets.

SARA (Title III).   A section of the Superfund Amendments and Reauthorization Act that promulgates community Right-to-Know requirements. Also known as Emergency Planning and Community Right-to-Know Act of 1986. (See **toxics release inventory**.)

savanna.   A type of grassland typical of subtropical regions, particularly in Africa, usually dotted with trees and supported by wet and dry seasons and frequent natural fires.

secondary air pollutants.   Air pollutants resulting from reactions of primary air pollutants resident in the atmosphere. Secondary air pollutants include ozone, other reactive organic compounds, and sulfuric and nitric acids. (See **ozone, PANs,** and **photochemical oxidants**.)

secondary consumer.   An organism such as a fox or coyote that feeds more or less exclusively on other animals that feed on plants.

secondary energy source.   A form of energy such as electricity that must be produced from a primary energy source such as coal or radioactive material.

secondary pest outbreak.   The phenomenon of a small and therefore harmless population of a plant-eating insect suddenly exploding to become a serious pest problem. Often caused by the elimination of competitors through the use of pesticides.

secondary recovery.   In an oil well or oil field, the oil that can be removed by manipulating pressure in the oil reservoir by injecting brine or other substances; more costly than **primary recovery**.

secondary succession.   See **succession**.

secondary treatment. Also called biological treatment. A sewage-treatment process that follows primary treatment. Any of a variety of systems that remove most of the remaining organic matter by enabling organisms to feed on it and oxidize it through their respiration. Trickling filters and activated-sludge systems are the most commonly used secondary treatement methods.

second principle of ecosystem sustainability. Ecosystems dispose of wastes and replenish nutrients by recycling all elements.

second-generation pesticides. Synthetic organic compounds used to kill insects and other pests. Started with the use of DDT in the 1940s.

second law of thermodynamics. The empirical observation, confirmed innumerable times, that in every energy conversion (e.g., from electricity to light), some of the energy is converted to heat and some heat always escapes from the system because it always moves toward a cooler place. Therefore, in every energy conversion, a portion of energy is lost, and since, by the first law of thermodynamics, energy cannot be created, the functioning of any system requires an energy input.

secondary energy source. A source of energy—especially electricity—that depends on a primary energy source for its origin.

secure landfill. A landfill with suitable barriers, leachate drainage, and monitoring systems such that it is deemed secure against contaminating groundwater with hazardous wastes.

sediment. Soil particles—namely, sand, silt, and clay—carried by flowing water. The same material after it has been deposited. Because of different rates of settling, deposits generally are pure sand, pure silt, or pure clay.

sedimentation. The filling in of lakes, reservoirs, stream channels, and so on with soil particles, mainly sand and silt. The soil particles come from erosion, which in turn generally results from poor or inadequate soil conservation practices in connection with agriculture, mining, or development. Also called **siltation.**

sediment trap. A device for trapping sediment and holding it on a development or mining site.

seep. An area where groundwater seeps from the ground. Contrast with a spring, which is a single point from which groundwater exits.

selective breeding. The breeding of certain individuals because they bear certain traits and the exclusion from breeding of others.

selective pressure. An environmental factor which causes individuals with certain traits that are not the norm for the population to survive and reproduce more than the rest of the population. The result is a shift in the genetic makeup of the population. For example, the presence of insecticides provides a selective pressure to increase pesticide resistance in the pest population. Selective pressure is a fundamental mechanism of evolution.

septic system. An onsite method of treating sewage that is widely used in suburban and rural areas. Requires a suitable amount of land and porous soil.

sex attractant. A natural chemical substance (see **pheromones**) secreted by the female of many insect species that serves to attract males for the function of mating. Sex attractants may be used by humans in traps or to otherwise confuse insect pests in order to control them.

shadow pricing. In cost–benefit analysis, a technique used to estimate benefits when normal economic analysis is ineffective. For example, people could be asked how much they might be willing to pay monthly to achieve some improvement in their environment.

sheet erosion. The loss of a more or less even layer of soil from the land surface due to the impact of rain and runoff from a rainstorm.

shelterbelts. Rows of trees around cultivated fields for the purpose of reducing wind erosion.

sievert. A unit of measurement of the ability of radioactive emissions to penetrate biological tissues. 1 sievert = 100 rem.

silt. Soil particles between the size of sand particles and clay particles, namely, particles 0.002 to 0.2 mm in diameter.

siltation. See **sedimentation.**

sinkhole. A large hole resulting from the collapse of an underground cavern.

slash-and-burn agriculture. The practice, commonly found in tropical regions, of cutting and burning vegetation to make room for agriculture. The process destroys soil humus and may lead to rapid degradation of the soil.

sludge cake. Treated sewage sludge that has been dewatered to make a moist solid.

sludge digesters. Large tanks in which raw sludge (removed from sewage) is treated through anaerobic digestion by bacteria.

Smart Growth. A movement that addresses urban sprawl by protecting sensitive lands and directing growth to limited areas.

smog. See **industrial smog** and **photochemical smog.**

social modernization. A process of sustainable development that leads developing countries through the demographic transition by promoting education, family planning, and health improvements rather than simply economic growth.

soft water. Water with little or no calcium, magnesium, or other ions in solution that will cause soap to precipitate. (The soap would otherwise form a curd that makes a "ring" around the bathtub.) (Contrast **hard water.**)

soil. A dynamic terrestrial system involving three components: mineral particles, detritus, and soil organisms feeding on the detritus.

soil classes. Taxonomically arranged major groupings of soil that define the properties of the soils of different regions.

soil erosion. The loss of soil caused by particles' being carried away by wind or water.

soil fertility. Soil's ability to support plant growth; often refers specifically to the presence of proper amounts of nutrients. The soil's ability to fulfill all the other needs of plants is also involved.

soil horizons. Distinct layers within a soil that convey different properties to the soil and that derive from natural processes of soil formation.

soil profile. A description of the different naturally formed layers, called horizons, within a soil.

soil structure. The composition of soil in terms of particles (sand, silt, and clay) stuck together to form clumps and aggregates, generally with considerable air spaces in between. Soil structure affects infiltration and aeration and develops as organisms feed on organic matter in and on the soil.

soil texture. The relative size of the mineral particles that make up the soil. Generally defined in terms of the soil's sand, silt, and clay content.

solar cells. See **photovoltaic cells.**

solar energy. Energy derived from the Sun; includes direct solar energy (the use of sunlight directly for heating or the production of electricity) and indirect solar energy (the use of wind, which results from the solar heating of the atmosphere, and biological materials such as wood, which result from photosynthesis).

solar-trough collectors. Reflectors, in the shape of a parabolic trough, that reflect sunlight onto a tube of oil at the focal point of the trough. Thus heated, the oil is used to boil water to drive a steam turbine.

solid waste. The total of materials discarded as "trash" and handled as solids, as opposed to those that are flushed down sewers and handled as liquids.

solubility. The degree to which a substance will dissolve and enter into solution.

solution. A mixture of molecules (or ions) of one material in another, most commonly, molecules of air or ions of various minerals in water. For example, seawater contains salt in solution.

sound science. The results of scientific work based on peer-reviewed research. As one of the unifying themes of this text, sound science is the basis for our understanding of how the world works and how human systems interact with it.

special-interest group. In politics, people organized around a particular issue who lobby for their cause and exert their influence on the political process to bring about changes they favor.

specialization. In evolution, the phenomenon whereby species become increasingly adapted to exploit one particular niche, but thereby are less able to exploit other niches.

speciation. The evolutionary process whereby populations of a single species separate and, through being exposed to different forces of natural selection, gradually develop into distinct species.

species. All the organisms (plant, animal, or microbe) of a single kind. The "single kind" is determined by similarity of appearance or by the fact that members do or can mate and produce fertile offspring. Physical, chemical, or behavioral differences block breeding between species.

splash erosion. The compaction of soil that results when rainfall hits bare soil.

springs. Natural exits of groundwater to the surface.

standing crop biomass. The biomass of primary producers in an ecosystem at any given time.

starvation. The failure to get enough calories to meet energy needs over a prolonged period of time. Starvation results in a wasting away of body tissues until death occurs.

Statement on Forest Principles. One of the treaties signed at the 1992 Earth Summit whereby the signatory countries agreed to a number of nonbinding principles stressing sustainable management.

Sterile-male technique. Saturating an infested area with males of a pest species that have been artificially reared and sterilized by radiation. Matings between normal females and sterile males render the eggs infertile.

steward/stewardship. A steward is one to whom a trust has been given. Stewardship is an attitude of active care and concern for natural lands. As one of the unifying themes of this text, stewardship is the ethical and moral framework that informs our public and private actions.

stoma, pl. stomata. A microscopic pore in a leaf, mostly on the undersurface, that allows the passage of carbon dioxide and oxygen into and out of the leaf and that also permits the loss of water vapor from the leaf.

storm drains. Separate drainage systems used for collecting and draining runoff from precipitation in urban areas.

stormwater. In cities, the water that results directly from rainfall, as opposed to municipal water and sewage water piped to and from homes, offices, and so on. The extensive hard surfacing in cities creates a vast amount of stormwater runoff, which presents a significant management problem.

stormwater management. Policies and procedures for handling stormwater in acceptable ways to mitigate the problems of flooding and erosion of streambanks.

stormwater retention reservoirs. Reservoirs designed to hold stormwater temporarily and let it drain away slowly in order to mitigate the problems of flooding and streambank erosion.

stratosphere. The layer of Earth's atmosphere between 10 and 40 miles above the surface that contains the ozone shield. This layer mixes only slowly; pollutants that enter it may remain for long periods of time. (See **troposphere**.)

strip cropping. The practice of growing crops in strips alternating with grass (hay) at right angles to prevailing winds or slopes in order to reduce erosion.

strip mining. A mining procedure in which all the earth covering a desired material, such as coal, is stripped away with huge power shovels in order to facilitate removal of the material.

subduction. In plate tectonics, the process whereby an oceanic plate slides under a continental plate.

submerged aquatic vegetation (SAV). Aquatic plants rooted in bottom sediments and growing under water. Submerged aquatic vegetation depends on the penetration of light through the water for photosynthesis.

subsistence farming. Farming that meets the food needs of farmers and their families, but little more. Subsistence farming involves hand labor and is practiced extensively in the developing world.

subsoil. In a natural situation, the soil beneath topsoil. In contrast to topsoil, subsoil is compacted and has little or no humus or other organic material, living or dead. In many areas, topsoil has been lost or destroyed as a result of erosion or development, and subsoil is at the surface.

succession. The gradual or sometimes rapid change in the species that occupy a given area, with some species invading and becoming more numerous while others decline in population and disappear. Succession is caused by a change in one or more abiotic or biotic factors that benefits some species at the expense of others. *Primary succession:* The gradual establishment, through a series of stages, of a climax ecosystem in an area that has not been occupied before (e.g., a rock face). *Secondary succession:* The reestablishment, through a series of stages, of a climax ecosystem in an area from which it was previously cleared.

sulfur dioxide (SO₂). A major air pollutant and toxic gas formed as a result of burning sulfur. The major sources are burning coal (in coal-burning power plants) that contains some sulfur and refining metal ores (in smelters) that contain sulfur.

sulfuric acid (H₂SO₄). The major constituent of acid precipitation. Formed when sulfur dioxide emissions react with water vapor in the atmosphere. (See **sulfur dioxide**.)

Superfund. The popular name for the Comprehensive Environmental Response, Compensation, and Liability Act of 1980. This act is the cornerstone legislation that provides the mechanism and funding for the cleanup of potentially dangerous hazardous waste sites and the protection of groundwater.

surface impoundments. Closed ponds formerly used to collect and hold liquid chemical wastes.

surface water. All bodies of water, lakes, rivers, ponds, and so on that are on Earth's surface. Contrast with **groundwater**, which lies below the surface.

surge flow irrigation. Irrigation under the control of microprocessors that release water periodically, thus cutting its use substantially.

suspended particulate matter (SPM). A category of major air pollutants consisting of solid and liquid particles suspended in the air. (See **PM-2** and **PM-10**.)

suspension. A system in which materials are contained in or carried by water and are kept "afloat" only by the water's agitation. The materials settle as the water becomes quiet.

sustainability. Property whereby a process can be continued indefinitely without depleting the energy or material resources on which it depends. As one of the unifying themes of the text, sustainability is the practical goal toward which our interactions with the natural world should be working.

sustainable agriculture.   Agriculture that maintains the integrity of soil and water resources such that it can be continued indefinitely.

sustainable development.   Development that provides people with a better life without sacrificing or depleting resources or causing environmental impacts that will undercut the ability of future generations to meet their needs.

sustainable forest management.   Management of forests as ecosystems wherein the primary objective is to maintain the biodiversity and function of the ecosystem.

sustainable society.   A society that functions in a way so as not to deplete the energy or material resources on which it depends. Such a society interacts with the natural world in ways that sustain existing species and ecosystems.

sustainable yield.   The taking of a biological resource (e.g., fish or forests) that does not exceed the capacity of the resource to reproduce and replace itself.

symbiosis.   The intimate living together or association of two kinds of organisms.

synergism.   The phenomenon whereby two factors acting together have a greater effect than would be indicated by the sum of their effects separately—as, for example, the sometimes fatal mixture of modest doses of certain drugs in combination with modest doses of alcohol.

synfuels, synthetic fuels.   Fuels similar or identical to those that come from crude oil or natural gas. Synfuels are produced from coal, oil shale, or tar sands.

tar sands.   Sedimentary bitumen-containing material that can be "melted out" using heat and then refined in the same way as crude oil.

taxonomy.   The science of identifying and classifying organisms according to their presumed natural relationships.

TEA (total exposure assessment).   The analysis of the impact of air pollutants on people based on the amount of time they spend in different spaces, especially indoors.

tectonic plates.   Huge slabs of rock that make up Earth's crust.

temperature inversion.   The weather phenomenon in which a layer of warm air overlies cooler air near the ground and prevents the rising and dispersion of air pollutants.

teratogenic.   Causing birth defects.

terminator technology.   A transgenic technique that renders seeds incapable of germinating, in order to force farmers to purchase new seeds every year.

terracing.   The practice of grading sloping farmland into a series of steps and cultivating only the level portions in order to reduce erosion.

territoriality.   The behavioral characteristic exhibited by many animal species, especially birds and mammalian carnivores, to mark and defend a given territory against other members of the same species.

theory.   A conceptual formulation that provides a rational explanation or framework for numerous related observations.

thermal pollution.   The addition of abnormal and undesirable amounts of heat to air or water. Thermal pollution is most significant with respect to discharging waste heat from electric generating plants—especially nuclear power plants—into bodies of water.

third principle of ecosystem sustainability.   The size of consumer populations is maintained such that overgrazing and other forms of overuse do not occur.

third world.   See **developing countries.**

threatened species.   A species whose population is declining precipitously because of direct or indirect human impacts.

threshold level.   The maximum degree of exposure to a pollutant, drug, or some other factor that can be tolerated with no ill effect. The threshold level varies, depending on the species, the sensitivity of the individual, the length of exposure, and the presence of other factors that may produce synergistic effects.

tidal wetlands.   Areas of marsh grasses and reeds along coasts and estuaries where the ground is covered by high tides, but drained at low tide.

tilth.   In farming, the ability of a soil to support crop growth.

topsoil.   The surface layer of soil, which is rich in humus and other organic material, both living and dead. As a result of the activity of organisms living in the topsoil, it generally has a loose, crumbly structure, as opposed to being a compact mass. In many cases, because of erosion, development, or mining activity, the topsoil layer may be absent.

total allowable catch (TAC).   In fisheries management, a yearly quota set for the harvest of a species by managers of fisheries.

total fertility rate.   The average number of children that would be born alive to each woman during her total reproductive years if her fertility was average at each age.

Total Maximum Daily Load (TMDL) program.   An EPA-administered program to address non-point-source water pollution that sets pollution limits according to the ability of a body of water to assimilate different pollutants.

total product life cycle.   The sum of all steps in the manufacture of a product, from the obtaining of raw materials through the production, use, and, finally, disposal of the product. By-products and the pollution resulting from each step are taken into account in a consideration of the total product life cycle.

Toxic Substances Control Act (TOSCA).   A law requiring the assessment of the potential hazards of a chemical before the chemical is put on the market.

toxicology.   The study of the impacts of toxic substances on human health and the pathways by which such substances reach humans.

toxics release inventory.   An annual record of releases of toxic chemicals to the environment and the locations and quantities of toxic chemicals stored at all U.S. sites. Required by the Emergency Planning and Community Right-to-Know Act (EPCRA) of 1986.

trace elements.   Those essential elements, like copper or iron, that are needed in only very small amounts.

trade winds.   The more or less constant winds blowing in horizontal directions over the surface as part of Hadley cells.

traditional agriculture.   Farming methods as they were practiced before the advent of modern industrialized agriculture.

tragedy of the commons.   The overuse or overharvesting and consequent depletion or destruction of a renewable resource that tends to occur when the resource is treated as a commons—that is, when it is open to be used or harvested by any and all with the means to do so.

trait.   Any physical or behavioral characteristic or talent that an individual is born with.

transgenic organism.   Any organism with genes introduced from other species via biotechnology to convey new characteristics.

transpiration.   The loss of water vapor from plants. Water evaporates from cells within the leaves and exits through the stomata.

trapping technique.   The use of sex attractants to lure male insects into traps.

treated sludge.   Solid organic material that has been removed from sewage and treated so that it is nonhazardous.

trickling filter system.   System in which wastewater trickles over rocks or a framework coated with actively feeding microorganisms. The feeding action of the organisms in a well-aerated environment results in the decomposition of organic matter. Used in secondary or biological treatment of sewage.

tritium (^{3}H).   An unstable isotope of hydrogen that contains two neutrons in addition to the usual single proton in the nucleus. Tritium does not occur in significant amounts in nature.

trophic level.   Feeding level with respect to the primary source of energy. Green plants are at the first trophic level, primary consumers at the second, secondary consumers at the third, and so on.

trophic structure.   The major feeding relationships between organisms within ecosystems, organized into trophic levels.

troposphere.   The layer of Earth's atmosphere from the surface to about 10 miles in altitude. The tropopause is the boundary between the troposphere and the stratosphere above. The troposphere is well mixed and is the site and source of our weather, as well as the primary recipient of air pollutants. (See also **stratosphere.**)

turbid.   Cloudy due to particles present. (Said of water and water purity.)

turbine.   A rotary engine driven at a very high speed by steam, water, or exhaust gases from combustion.

turbogenerator.   A turbine coupled to and driving an electric generator. Virtually all commercial electricity is produced by such devices.

ultraviolet radiation.   Radiation similar to light, but with wavelengths slightly shorter and with more energy than violet light. Its greater energy causes ultraviolet light to severely burn and otherwise damage biological tissues.

underground storage tanks.   Tanks used to store petroleum products at service stations, now subject to mandated protection against leakage. (See **UST legislation.**)

undernutrition.   A form of hunger in which the individual lacks adequate food energy, as measured in calories. Starvation is the most severe form of undernutrition.

upwelling.   In oceanic systems, the upward vertical movement of water masses, caused by diverging currents and offshore winds, bringing nutrient-rich water to the surface.

urban blight/decay.   The general deterioration of structures and facilities such as buildings and roadways, in addition to the decline in quality of services such as education, that has occurred in inner-city areas.

urban sprawl.   The rapid expansion of metropolitan areas through building houses and shopping centers farther and farther from urban centers and lacing them together with more and more major highways. Widespread development that has occurred without any overall land-use plan.

UST legislation.   Amendments to the Resources Conservation and Recovery Act of 1976 passed in 1984 to address the mounting problem of leaking underground storage tanks (USTs).

value.   In ethical considerations, a property attributed to objects, species, or individuals that implies a moral duty to the one assigning the value; intrinsic and instrumental value are two types of value.

vector.   An agent, such as an insect or tick, that carries a parasite from one host to another.

vigor.   Property conferred upon crop plants or animals by virtue of their possession of traits for hardiness to disease, drought, cold, and other adverse factors or conditions.

vitamin.   A specific organic molecule that is required by the body in small amounts, but that cannot be made by the body and therefore must be present in the diet.

volatile organic compounds (VOCs).   A category of major air pollutants present in the air in the vapor state. The category includes fragments of hydrocarbon fuels from incomplete combustion and evaporated organic compounds such as paint solvents, gasoline, and cleaning solutions. Volatile organic compounds are major factors in the formation of photochemical smog.

waste-to-energy.   A process of combustion of solid wastes that also generates electrical energy.

water cycle.   See **hydrologic cycle.**

water-holding capacity.   The ability of a soil to hold water so that it will be available to plants.

waterlogging.   The total saturation of soil with water. Waterlogging results in plant roots' not being able to get air and dying as a result.

watershed.   The total land area that drains directly or indirectly into a particular stream or river. The watershed is generally named from the stream or river into which it drains.

water table.   The upper surface of groundwater, rising and falling with the amount of groundwater.

water vapor.   Water molecules in the gaseous state.

weather.   The day-to-day variations in temperature, air pressure, wind, humidity, and precipitation mediated by the atmosphere in a given region.

weathering.   The gradual breakdown of rock into smaller and smaller particles, caused by natural chemical, physical, and biological factors.

wet cleaning.   A water-based alternative to dry cleaning that avoids the use of hazardous chemicals.

wetlands.   Areas that are constantly wet and are flooded at more or less regular intervals. Especially, marshy areas along coasts that are regularly flooded by tide.

wetland system.   A biological aquatic system (usually a restored wetland) for removing nutrients from treated sewage wastewater and returning it virtually pure to a river or stream. Wetland systems are sometimes used when the application of treated wastewater for irrigation is not feasible.

Wilderness Act of 1964.   Federal legislation that provides for the permanent protection of undeveloped and unexploited areas so that natural ecological processes can operate freely in them. Most human intrusions are excluded from such areas, which now total 104 million acres in the United States.

wind farms.   Arrays of numerous, modestly sized wind turbines for the purpose of producing electrical power.

windrows.   Piles of organic material extended into long rows to facilitate turning and aeration in order to enhance composting.

wind turbines.   "Windmills" designed for the purpose of producing electrical power.

WIPP (waste isolation pilot plant).   A facility built by the Department of Energy in New Mexico to receive defense-related nuclear wastes.

wise use.   A form of environmental backlash reacting against regulations and restrictions on the use of public lands for recreation and extractive activities such as mining, forestry, and grazing.

work.   Change associated with the motion or state of matter. Any such change requires the expenditure of energy.

workability.   The relative ease with which a soil can be cultivated.

World Bank.   A branch of the United Nations that acts as a conduit for handling loans to developing countries.

worldview.   A set of assumptions that a person holds regarding the world and how it works.

xeriscaping.   Landscaping with drought-resistant plants that need no watering.

yard wastes.   Grass clippings and other organic wastes from lawn and garden maintenance.

Younger Dryas event.   A rapid change in global climate between 10,000 and 12,000 years ago that brought on 1,500 years of colder weather.

zones of stress.   Regions where a species finds conditions tolerable, but suboptimal. The species survives, but under stress.

zooplankton.   Any of a number of animal groups, including protozoa, crustaceans, worms, molluscs, and cnidarians, that live suspended in the water column and that feed on phytoplankton and other zooplankton.

Index

Note: Pages in **boldface** locate the definition of the term.

Abiotic factors, 26, 30
  conditions, **40**
  effects on biomes, 42, 44–45
  effects on species, 40–41
  fire as, 99–100
  resources, **40**
Abortion, 177–78
Absolute growth, 140, 141
Absolute poverty, **141**, 256
Absorption, **549**
Accounting, environmental, 577
Acid deposition, **549**
  acids/bases, 550–52
  effects of, 551, 553–55
  natural vs. anthropogenic, 552
  nitrogen and, 72
  pH scale, 552
  public policy and, 559–61
  reduction of, 559–61
  sources of, 552–53
  in the U.S., 553
Acid mine drainage, 444, 445
Acid precipitation, 549, 552. *See also* Acid deposition
Acids, 550–52
"Acid shock," 555
Acquired immune deficiency syndrome. *See* AIDS
Activated sludge, **456**
Activated-sludge system, 455, **456**
Active safety, **356**
Active system, **368**
Acute toxicity, **486**
Adaptation, **108**, 117
  Darwin's finches and, 107, 120–21
  giant anteater and, 118, 123
  limits to, 117–18
  for survival and reproduction, 111–12
  to the environment, 27, 108, 111, 112, 117–18
  to other species, 95
  in tundra plants, 119–20
Adiabatic cooling, **217**
Adiabatic warming, **217**
Adsorption, **549**
Adsorption, in water purification, 224
Advanced light-water reactors (ALWR), 356–57
Advocacy and scientists, 85
Aeration (soil), **196**
Aeration (water), 449, 450
Aerial spraying, 418, 430
Aerosols, **215**
Aerosol sprays, 407, 487
Affluence
  health and, 400–401
  impacts on environment and, 2, 4, 6, 143–44
  population growth and, 141–42, 150
"Affluenza," 479, **481**
Africa
  agricultural pests in, 427, 428
  AIDS in, 175–77
  debt crisis in, 172
  energy technology transfer to, 385
  famine in, 257, 258
  Kenya's "casual workers," 136–38
  Sahel region of, 257
  Serengeti, 52–53, 63, 122
  Southern Africa Development Community, 258
  Tanzania, 1, 607, 608
Agenda 21, 19, 174, 311, 577, 593, 607
Age structure, **140**, 150–52, **150**
Agriculture. *See also* Desertification; Food production
  air pollution and, 541, 545, 564

  beginnings of, 48
  degradation of land for, 191
  in developing countries, 145
  development of modern industrialized, 242–45
  ecological, 189
  extension programs, 210
  global climate change and, 523–24, 534
  intensifying cultivation in, 145
  machinery and, 243
  no-till, 202–3
  opening new lands for, 145, 243
  preserving genes for, 130
  public policy and, 208–10
  slash-and-burn, 246
  subdividing farms for, 145
  subsistence, 245–46
  sustainable, 189, 190, 202–3, 204
  traditional vs. modern, 243
  urban sprawl and, 599
  wild species as source for, 266
A horizon, **192**, 197
AIDS
  as cultural hazard, 394
  in developing countries, 175–77
Air, water, mineral interrelationship, 57
Air circulation, global, 217
Air fresheners/disinfectants, 407
Air pollution, **538**
  acid deposition, 549
    acids/bases, 549–52
    ecosystems and, 551, 553–55
    effects of, 553–55
    natural vs. anthropogenic, 552
    nitrogen and, 72
    reduction of, 559–61
    sources of, 552–53
    in the U.S., 553
  atmospheric cleansing, 538–39
  controlling, 555–63, 569
  cost to control, 582
  effects on aesthetics, 545–46
  effects on materials, 545–46, 555
  impacts of, 542–46, 564
  industrial, 541
  major pollutants in, 541
  natural vs. anthropogenic sources of, 538–39, 552
  public policy and, 555–63
  smog and, 536–37, 538, 539–41
  sources of, 546–49
  toxic chemicals and, 541, 549, 559
  urban sprawl and, 599
Air toxics, 541, 549
Alaska oil pipeline, 314–15
Alfisols, **194**
Algae
  "blooms" of, 72
  photosynthesis and, 30
Alkalinity, **196**
AllEnergy, 336
Alteration (habitat), 6, 277
Alternative Agriculture (The National Research Council), 208
Altitude effects, 45
Amazon River Basin, 91, 92
Ambient standards, **555**
American Academy of Arts and Sciences, 284
American buffalo (bison), 203, 264
American Chemistry Council, 505
American chestnut tree blight, 93
American Dream, 596
American Forest and Paper Association, 298
American Museum of Natural History, 284

American Water Works Association, 238
American Wind Energy Association, 379
Anaerobic digestion, **458**
Anaerobic sludge digesters, **458**
Animal Damage Control (ADC) program, 271
Animal farming
  consequences of, 246–47
  wild species as source for, 266
Animal-rights, 266
Animals. *See* Wild species
Animal testing, hazard assessment and, 408-9
Animal vs. plant farming, 247
Annan, Kofi, 311
Annapolis Tidal Generating Station, 382
Anopheles mosquito, 395–96, 405–6
Anthropocentric, **265**
Apfelbaum, Steven, 47
Aptitude traits, **111**, 113
Aquaculture, 266–67, 302
Aquatic ecosystems, **28–29**
  acid deposition and, 553–55
  energy movement in, 66
Aquatic photosynthesis, 448
Aquatic plants, 447
Aquatic succession, 98, 99
Aquifers, **220**
Aral Sea, 227, 228
Arctic, 484–86
Arctic fox, 118–19
Arctic National Wildlife Refuge (ANWR), 333
Area sources, **559**
Aridisols, **194**
Arora, Astha, 162
Arrhenius, Svante, 518
Arthrobotrys anchonia, 199
Artifacts, **555**
Asbestos, indoor air pollution and, 407
Assets, produced, **575**
Association, 27, **30**, 35
Asthma, 404
Aswan High Dam, Egypt, 231, 378
Atlantic puffin, 262–64
Atmosphere, **55**. *See also* Air pollution
  anthropogenic warming vs. cooling, 517–18
  clean air gases in, 55
  climate and, 513–16
  El Niño/La Niña and, 508–10
  gases in, 538
  global climate change and, 5–6, 516–28
  oceans and, 515–16
  ozone layer depletion and, 528–34
  structure of, 511
  temperature profile of, 511
  temperatures of (1880–1999), 5, 514
  weather and, 511–13
Atomic Energy Commission, 342
Atomic theory, 15, 16
Atoms, **54**
Attorneys general, 402–3
Australia
  marsupials in, 123
  ozone destruction above, 532
  rabbits in, 93, 95, 97, 427
Autos. *See* Motor vehicles
Autotrophs, **31**. *See also* Producers

"Baby boom," 150, 152
"Baby boom echo," 150
"Baby bust," 150
Bacillus thuringiensis (Bt), 249, 430

Bacteria
  as detritus feeder, 33, 34
  importance in soil, 196–97
  nitrogen fixing, 71–72
Bacteriology, "golden age" of, 394
Bakhtayeva, Alexandra, 189
Balanced herbivory, 91-92
Balance-of-trade deficit, crude oil and, 328, 329, 331
Bald eagle, 273–74, 422–23
Baltic 21, 20
Baltimore, MD, 602, 604
Banff National Park, 98–99
Bangladesh
  flooding, 221, 222
  health clinic, 401
  microlending, 178
Barrier islands, importance of, 24–26
Bar screen, **454**
Barter economy, 178
Basel Action Network, 10–11
Baseload, **321**, 344, 348
Bases, 550-52
Basicity, **196**
Bayer AG, 468
Bay State Fertilizer, 459
Baytown, TX, 295–96
Bear underpass, 270
Beaver, 131
Bed load, 445, 446
Behavioral traits, 113
Belize, Jaguar Creek, 365, 366
Bemisia argentifolii, silverleaf whitefly, 420
Benefit-cost analysis
  of 40 CFR Part 50, 569
  by executive offices/agencies, 569
  of Clean Air Act programs, 562
  multicriteria and, 590
Benefit-cost ratio, **583**
  air pollution and, 584
Benthic plants, **447**
Benzene, health effects from, 545
Bertini, Catherine, 260
Best management practices (BMPs), 452, 453
Better World for All, A (United Nations report), 171
Beyer, Peter, 241
Bezdek, Roger, 581–82
Bhopal, India, 503
B horizon, 192–93
Bicycle transportation, 600, 604, 606
"Big Dig," 605
Bioaccumulation, 422-23
Biochemical oxygen demand (BOD), 443, 456
Biodegradable plastic, 468
Biodiversity, **6**, **264**
  consequences of loss of, 280–81
  controversy over, 281
  Convention on Biological Diversity (CBD), 282
  decline causes
    exotic species, 278–79
    overuse, 279–80
    physical alteration of habitat, 277
    pollution, 278
    population factor, 277–78
  decline of, 6, 35, 276–81
  disturbance and, 101
  hot spots of, 283
  importance to ecosystems, 101, 129–31
  instrumental value of
    aesthetic, recreational, and scientific, 267–69

669

improving agricultural plants, 6, 266–67
  new cultivars from, 266
  pest control, 267
  sources of medicine, 6, 130, 267, 268
  intrinsic value of, 6, 130–31, 269
  measuring, 284
  numbers of species, 264, 275, 276
  protection of, 282–83
  stewardship concerns for, 283–84
  tropical forests and, 276
Biodiversity and Ecosystem Functioning: Maintaining Natural Life Support Processes (Tilman and Lawton), 281
Bioengineered potatoes, 430
Biogas, **380**
  production in landfills, 468
  use as fertilizer, **380**, 458–59
  use as fuel, **380**, 381, 458–59
Biogeochemical cycles
  atmosphere and, 511
  carbon cycle, 69
  human impacts on, 69, 70, 72, 78
  movement through ecosystems, 37, 38, 68–72
  nitrogen cycle, 71–72
  phosphorus cycle, 69–71
Biological control, 76, **425**
Biological Control Program, 428
Biological evolution, **109**. *See also* Evolution
Biological nutrient removal (BNR), **457**
Biological oxidation, in water purification, 224
Biological treatment, **455–56**
Biological wealth, **264**
Biomagnification, **423**
Biomass, **35**, **37**
Biomass energy, 379-81
Biomass pyramid, **37**
Biomes, 27–28, **30**
  biotic factors and, 45–46
  climate effects on, 42–45
  distribution and description of, 42–45
  pressures on, 296–307
Bioremediation, **498**
Biosolids, **459**
Biosphere 2, 77
Biosphere, **29**, **30**, **55**
Biosphere reserves, 307
Biota, **26**, **30**, **264**
Biotechnology
  new plant varieties and, 130
  pest control and, 430
  policies and, 251
  problems with, 250–51
  promise of, 249–50
Biotic community (biota), **26**, **30**, **264**
Biotic potential, **84-86**
Births
  estimating, 152
  rates of, 156–57
Bison (American buffalo), 203, 264
Blackout, **321**
Blue Ribbon Coalition, The, 308
Borlaug, Norman, 244
Bosch, Robert van den, 420
Boston Harbor, 459
Botswana, AIDS in, 175–76
Bottled water, 496
Bottle laws, 476–77, 482
Bottom trawling, 304
"Brain drain," 148
Brazilian pepper bush, 279, 280
Brazilian pit viper, 267
Bread for the World's Hunger Institute, 253
Breeder nuclear reactor, 358–59
British Petroleum, 385
Broad spectrum pesticides, **418**, 428
Broadview Water District, CA, 232

Brown, Lester, 49
Brown algae, 30
Browner, Carol, 402
Brownfields, **499**-500, 594, 600
Brown-headed cowbird, 277
Brownout, **321**
Brown plant hopper, 432, 433
Brown tree snake, 279
Brundtland, Gro Harlem, 181
Bt (*Bacillus thuringiensis*), 249, 430
Buckthorn, sea, 209
Buffer, **554-55**
Buffering capacity, **554-55**
Bulk-mailing lists, 473
Burkholder, Joanne, 451
Bush, George, 274, 282, 459
"Bush meat," 291
Business Council for Sustainable Development, 19
Butte, MT, smelter pollution, 545
Bycatch, **301**, 304

CAFE (corporate average fuel economy), **562**
Cairo conference
  ICPD Program of Action, 181–84
  importance of, 165, 181
Calcium in rocks, 194
Calorie, 58
Cambrian explosion, 127
Campbell, Colin, 332–33
Canada geese, 271
Canadian Standards Association (CSA), 311
Canadian Sustainable Forestry Certification Coalition, 311
Canaries in coal mines, 130–31
Cancer
  DNA and, 399–400
  medicines for, 267
  radioactivity and, 350, 355
  UV radiation and, 530
Cape Cod Bay, 459
Capillary water, **219**
Capitol Building, Washington, DC, 568–69
Capoten, 267
Carbon, storage by ecosystems of, 289
Carbon cycle, 69
  global amounts in, 519
  maintenance by ecosystems of, 289
Carbon dioxide
  atmospheric, 69
  concentrations of, 518–19
  emissions and climate, 5–6, 98
  emissions treaty in Kyota, Japan (1997), 5–6
  reduction of emissions, 527–28
  sources of emissions, 519
Carbon monoxide, 541, 546–49
Carbon sinks, 519
Carbon tax, 385-86
Carcinogenic, **486**, **489**
Carcinogens, 399, 545
Car-dependent lifestyle, 594, 596–98, 602, 605
"Caring for Creation," 19
Carnivores, **32**. *See also* Consumers
Carrier, Inc., 505
Carrying capacity
  of ecosystems, 293
  humans and, 103–4, 179, 259
Carson, Rachel, 8, 12, 579–80
Cars. *See* Motor vehicles
Cartagena Protocol, 251
Carter, Jimmy, 249, 251, 260, 494
Cash economy, 178
Cassava (manioc), 428
Castings, **197**
"Casual workers," 136–38
Catalyst, **529**
Catalytic converter, 557, 558
Cats, domestic, 94, 279
Cause and effect, 14, 15, 16

Cell respiration, **63-65**
Cellulose, **65**
Cement kiln, hazardous wastes and, 501
Center for Disease Control
  rabies and, 270
  UV radiation and, 530
Center for Energy and Climate Solutions, 338
Center-pivot irrigation, **206**, 229, 231
Center for Public Justice, 588
Center for Renewable Resources, 370
Central Landfill, RI, 470
Centrally planned economy, **571–72**, 573
CERCLA (Comprehensive Environmental Response, Compensation, and Liability Act of 1980). *See* Superfund
Certification of sustainable forestry, 311
Cetacean Research Unit, 306
CFCs (chlorofluorocarbons), global warming and, 520
Chaitanya savings program, 184
Chandler, William, 401
Channelization, **235**
Chapman, Kim, 47
Charity organizations in developing countries, 170
Chattanooga, TN, 12, 609
Chattanooga Venture, 608
Chelyabinsk-65 and radioactive wastes, 352–53
Chemical barriers (pest control), 429
Chemical energy, 58
Chemical pollutants, 443–44
Chemicals, hazardous. *See* Hazardous chemicals
Chemical treatment, **416**
Chemosynthesis, 31
Chernobyl nuclear accident, 343, 355–56, 362
Chesapeake Bay
  die-off of submerged vegetation in, 16
  eutrophication recovery plan for, 452
  importance of wetlands and, 289–90
  Pfiesteria spp. in, 451
  Tangier Island covenant and, 612
Chesapeake Bay Foundation, 612
Chestnut tree blight, 93
Chichen-Itza ruins, 103
Children
  education of, 166–67, 174–75
  as helping hands, 166
  poverty and, 147, 148, 149, 166–67
China
  fertility rate of, 179
  Three Gorges Dam in, 231
Chinese Exclusion Act of 1882, 148
Chlorinated hydrocarbons, 424
Chlorination, 412
Chlorine cycle, **529**
Chlorine reservoirs, **531-32**
Chlorofluorocarbons (CFCs)
  global warming and, 520
  ozone destruction and, 529–33
Chlorophyll, **30**, 61
Choctaw reservation, 500
Cholera, 146, 257, 404
C horizon, **193**
Chromosomes, 114
Chronic exposure, **486**
Cities
  car-centered, 605
  decline in, 597, 598
  livability and, 604–7
  migration to, 145–46, 147
  redeveloped core areas of, 603, 604
  sustainability and, 607–10
  urban blight and, 600–603
  urban sprawl and, 594–600

Citizen involvement, 589
Civilian Conservation Corps (CCC), 11
Civil war and famine, 257–58
Classification of terrestrial vegetation, 35
Clay, **193**
Clean Air Act
  air pollution and, 12, 381, 490, 491, 501, 555–56, 557
  amendments of 1990 (CAAA), 556, 557, 558–59
  antienvironmentalists and, 587
  ozone destruction and, 533
  Title IV of, 560–61
  Title VI of, 533–34
Clean Air Act, 1970, 1977, 1990, cost-benefit analysis of, 562
Clean Water Act of 1972, 12, 451, 461–62, 491, 501
  antienvironmentalists and, 587
Clean Water Act amendments of 1977, 461
Clean Water Initiative, 462
Clean Water State Revolving Fund program, 461
Clear-cutting, **297**, 298
Climate **511**, **513**. *See also* Atmosphere; Global climate change
Climate
  biomes and, 42–45, 513
  in the past, 98, 514–15
  regulation by ecosystems of, 76, 288
  tectonic movement and, 127
Climate Change 1995: The Science of Climate Change (Houghton), 526
Climate Change Impacts on the U.S. (climate experts), 521, 523
Climate Change Technology Initiative (CCTI), 383
Climax ecosystem, **96**
Clinton, William, 12
Clinton administration
  Alaska National Wildlife Refuge and, 333
  cost-benefit analysis and, 570
  dismantling of nuclear weapons and, 353
  energy and, 383
  environmental justice and, 500
  environmental public policy and, 587–88
  family planning funding and, 178
  international whaling and, 305
  national forests and, 310
  nuclear wastes and, 354
  protection of rivers by, 231
  sustainable development and, 609–10
  water pollution control and, 462
Clones, 113, **116**, 249
Closing the recycling loop, **476**, 482
Clown fish, 38
Coal
  acid deposition and, 553, 554, 559–60
  comparison with nuclear power, 348–49
  as fossil fuel, 334–35
  history of use, 319
  power plants in Singraali, India, 172
  reserves and use of, 334
  strip mining of, 334
Code of Conduct (World Bank, FAO), 435–36
Coexistence in plants, 90–91
Cogeneration, 336, 337
Cold War, and nuclear wastes, 353
Collection pond, for dairy pond washings, 452
Colligan, Mary, 613
Colonialism, as cause of poverty, 10, 168

Colorado River, overdraw of, 226
Columbia University, 77
Combined-cycle natural-gas unit, **336**
Coming Plague, The (Garrett), 404
Command and control, **556**, 569
Commerce, beginnings of, 48
Commission on Sustainable
    Development, 19, 607
Commode, low-flow, 232
Commoner, Barry, 325
Common pool resource (commons),
    293–95
Common Sense Initiative, 505
Community Aid Abroad (CAA)
    program, 184
Community-based natural resource
    management (CBNRM), 283
Community outreach, as key to soil
    conservation, 210
Community protection, hazardous
    chemicals and, 503
Compaction, **196**, 243
Competition
    between plant species, 90–92
    in ecosystems, 90–96
    territoriality, **92**
Competitive exclusion principle, 40, 90
Competitive relationships, 39–40
Composting
    as decay of organic wastes, **197**, 478
    as treatment of sewage sludge,
    159–60
Compound, 54
Comprehensive Environmental
    Response, Compensation, and
    Liability Act of 1980 (CERCLA).
    See Superfund
"Concept cars," 336
Concepts and sound science, **16–17**
Condensation, **215**-17
Conditions, as abiotic factors, **40**
Conference of the Parties (COP), 201
Conferences on the Law of the Sea
    (U.N.), 301
Coniferous forests biome, **42–43**, 44
Conservation, **291**
Conservation of energy, law of, 58-**59**,
    60–61
Conservation International, 283
Conservation of matter, law of, **54**
Conservation and preservation of
    ecosystems, 290–96
    terms defined, 290–91
    use patterns, 291–96
        consumptive use, 291, 292, 380
        maximum sustainable yield,
        292–93
        productive use, 291–92
        tragedy of the commons, 293–95,
        301
Conservation Reserve Program of
    1985, 208, 243, 244
Consolidated Edison, 337
Consultative Group on International
    Agriculture Research (CGIAR),
    244
Consultative Group to Assist the
    Poorest (CGAP), 173
Consumers, **31**
    categories of, 31–34
    cell respiration by, 63–65
    food pathways and, 65
    trophic relationships and, 33–38
Consumers Union, 435, 436
Consumption
    classes of, 143
    correlation with affluence and, 4,
    142–44, 150
    effects on biodiversity and, 6
Consumptive use, **291**, 292
    of water, **223**
Continental drift, 125
Continental-plate collision, 127
Contour strip cropping, 203–4

Contraceptives
    developing countries and, 167
    family planning and, 177
    fertility rates and, 167
Control rods, 347-48
Control stage (policy life cycle), 580
Controversies and science, 17–18
Convection cell, 513
Convection currents, **217**, **512**
Convention on Biological Diversity,
    276, 282
Convention to Combat Desertification
    (UNCCD), 201
Convention on Trade in Endangered
    Species of Wild Fauna and Flora
    (CITES), 282
Convergent boundaries, **126**
Conversion of grain to protein, 74
Conversion (habitat), **277**
Conveyor system, 515–16
Conway, Gordon, 249
Cooling towers, 323
Coon Creek Basin, WI, 195
Coral bleaching, **306**
Coral reefs and mangroves, 306–7
Corn
    Bt and monarch butterflies, 250
    yield, in U.S., 243
Corporate average fuel economy
    (CAFE), 562
Corridors for wildlife, 270
Corrosion, **358**
Corrosivity, **487**
Corruption and nepotism, as cause of
    poverty, 10
Cosmetic spraying, **432**
Costanza, Robert, 8, 574–75
Cost-benefit ratio, **583**
Cost effective, **583**
Cost-effectiveness, 586–87
    pollution control and, 585
Costs
    environmental regulations, 584–86
    pollution control, 582
Cougars (mountain lions), 270, 271
Country, food security and, 253, 254
Coupled general circulation models
    (CGCMs), 521, 524
Coyotes, 270–71
Crabgrass, 97–98
Credit associations, 178
Credit with Education, 178, 182
Criteria pollutants, **556**
Critical Ecosystem Partnership Fund,
    282–83
Critical number, **86**, 277
Crombie, David, 614
Croplands, 288
Crop rotation, 202
Crown fires, 100
Crude birth rate (CBR), **140**, 156-57
Crude death rate (CDR), **140**, 156-57
Crude oil
    adjustments to higher prices, 328,
    330
    adjustments to lower prices, 328,
    330
    Alaska and, 333
    consumption, domestic production,
    and imports of, 330
    crisis of the 1970s, 327–28
    exploitation of, 325–33
    history of use, 319
    oil spills, 317–18, 331
    problems with U.S. imports of
        costs of purchase, 329, 330–31
        military costs, 331
        resource limitations, 331–33
        risks of supply disruptions, 331
    reserves and use of, 332
    reserves vs. production, 325–26
    U.S. reserves/importation of, 327–33
Cry the Beloved Country (Paton),
    137–38

Cryptosporidium parvum, 404
Ctenophores, 94
Cultivar, **266**
Cultural eutrophication, **449**
Cultural pest control, 425–26
Cultural services, ecosystems and, 76
Curbside recycling, 474, 475
Curies, **349**
Cuyahoga River fire, 491
Cyanide
    "fishing" with, 306–7
    spill in Tisza River, 278
Cycle of poverty, 168, 256
Cycle of urban blight, 602–3

Dachshund and selective breeding, 110
DaimlerChrysler, 376
Daly, Herman, 8, 574–75
Dams, 171, 230–31, 377–78
Dar es Salaam, Tanzania, 608
Darwin, Charles, 107, 111, 119–20
Darwin's finches, 107, 120–21
DDT
    in the Arctic, 485
    effects on birds, 12, 273–74, 422–24
    history of, 417–19
    malaria control and, 405, 419
    as persistent organic pollutant, 489,
    490
    public policy and, 580
Deaths
    estimating, 152
    rates of, 156–57
Debt
    as cause of poverty, 11
    crisis, 172–74, 254
    relief, 173–74
Declining tax base, **602**
Decommissioning nuclear power
    plants, 358
Decomposers, **33–34**. *See also*
    Consumers; Detritivores
Deep Ecology, 284
Deep-well injection, 491–92
Deer, 270
Deer tick, 392
Defenders of Wildlife, payment for
    wolf predation, 265
Deforestation
    flooding and, 221–22
    in Iceland, 99, 100
    of mangroves, 307
    soil erosion and, 205, 444–45
    of tropical forests, 145, 146, 205,
    299–300
    in the U.S., 308–9
Delaney clause, 435
Delaware River Valley, 186
Democratic People's Republic of
    Korea, famine and, 258, 260
Demographers, **150**
Demographic terms, 140
Demographic transition, **140**
    developed/developing countries and,
    169
    factors in, 156–59
    key questions of, 164–65
    reassessment of, 164–68
Demography, **150**
Denitrification, **72**, 457
Deoxyribonucleic acid. *See* DNA
Department of Transportation
    Regulations, 502–3
Deprivation, sustainability failure and, 8
Deregulation, **336**, 357, 383
Desalination, **233**
Desert biome, **42–43**
Desertification, **201**
    causes of, 145, 202–5
    correcting, 202–4
    erosion and, 200–201
    in Russia, 189–90, 208
    salinization as type of, 206

Desert lands, 288
Desert pavement, **200**-201
Design for the Environment, 505
Detergents with phosphate, 450–51
Detritivores, **33**
    as consumers, 65
    fermentation by, 65–66
    food web, 34
    importance to soil, 191, 196–97
    nutrient cycles and, 38, 65
    trophic relationships/categories,
    33–34
Detritus, **33**
    soil and, 191, 192, 197
Detritus feeders, **33–34**. *See also*
    Detritivores
Deuterium, 347, 359
Developed countries, 1, 2, **141**
    population projections for, 152,
    153–54
Developed vs. developing countries,
    140–44
Developing countries (the South), 1,
    141. *See also* development
    cities in, 146–47
    demographic transition and, 164–68
    development in, 168–74
    gender-related work in, 166
    growth, impacts of, 144–49
    population projections for, 154–55
    water use in, 223–24
Development, 7
    debt crisis and, 172–73
    debt relief and, 173–74
    education and, 174–75
    family planning and, 177–78
    health and, 175–77
    income and, 178–80
    indicators of, 171
    international goals of, 171
    promotion of, 168–82
    resource management and, 180
    social modernization and, 174–84
    World Bank Reform and, 173–74
    World Bank successes and failures,
    169–73
Development cycle, 598
Developments, urban sprawl and,
    595–98
Devil's Hole pupfish, 272
Dewatering treated sludge, 460
Dibromochloropropane (DBCP), 421
Dichloro-diphenyltrichloroethane. *See*
    DDT
Dickens, Charles, 539
Differential reproduction, 109
Dirty dozen POPs, 490
Discharge permit, **501**
Discount rates, **578**
Disease
    acute respiratory infections, 394, 395
    black-lung, 402
    control of, 139, 144, 168
    diarrheal, 394, 395, 443
    globalization of, 404
    infectious, 394, 404–6, 453
    malaria, 395–96, 405–6
    as natural enemy, 139
    parasitic, 394, 395, 396
    resurgent, 404
    sexually transmitted diseases (STDs),
    394
    sustainability failure and, 8
    unsanitary conditions and, 443
Disinfection, in water purification, 224
Dissolved oxygen (DO), **443**, 448
Distillation, in water purification, 224
Disturbance, in ecosystems, 76, 96,
    98–102
Divergent boundaries, **126**
DNA, **113–14**
    cancer and, 399–400
    radiation effects on, 350
DNA, genes, and traits, 114

Doctors Without Borders, 258
Dodder, 32
Dogs, 110
Dolly, cloned sheep, 113
Dombeck, Mike, 310
Domestic cats, 94, 279
Donora, Pennsylvania, smog in, 537–38, 541
Dose-response assessment, **409**, 486–87
Doubling time, **140**
Doubly Green Revolution, 249
Drake, Susan, 612
Drifting continents, 124–25
Drip irrigation, **231**, 232
Drought, as cause of famine, 257
Drugs, illicit, 146
Dry-cleaning operations, 489, 490, 505
Drylands, 201–2
Ducks Unlimited, 270
Dukakis, Michael, 459
Du Pont Chemical Company, CFCs and, 533
Durning, Allen, 141, 256
Dust Bowl, 11, 203, 210
Dynamic balance, **85**

Earth in the Balance (Gore), 614
Earth Day 1970, 580
Earthquakes, 124, 126, 127, 396, 397
Earth-sheltered housing, **370**, 371
Earth's surface changes, 221–22, 233–35
Earthworm castings, **197**
Easements, 310
"Easter-egg hypothesis," 331
Easter Island, 2–4
Ecdysone, **431**
Ecocity, 592–94
E. coli (*Escherichia coli*), 444
Ecological agriculture, 189
Ecological control, **417**
Ecological economists, 574-75, 576–77
Ecological (natural) succession, 96–98
Ecological niche, **39**-40
Ecological Society of America
    Committee on Land Use guidelines, 102–3
    terrestrial vegetation classification, 35
Ecologists, 8, **27**
Ecology, **27**
Economic activity
    classical view of, 573
    environmental economic view of, 574
Economic decisions, **570**
Economic demand, 253
Economic development, 577–78
Economic production, 574–75
Economics
    disparity between nations, 140–44
    divisions of the world, 141, 142
    nuclear power and, 357–58
    public policy and, 569–74
Economic segregation, 600–601
Economic systems, 571–74
Economic threshold, **431**, 432
Economists, view of sustainable development, 8
Economy, **570**
    barter, 178
    cash, 178
    centrally planned, **571**–72, 573
    free-market, **572**-74
    public policy and, 568–74
    resources and, 574–78
    wealth of nations and, 574–78
Ecosystem management, **310**
Ecosystems, 27, **30**
    acid deposition and, 551, 553–55, 561
    biodiversity and, 101, 129–31
    biogeochemical cycles in, 68–72
    biomes and, 42–46
    changes in, 122–23

conservation and preservation of, 87, 290–96
    decline of, 4–5, 6, 35
    disturbance in, 98–102
    dryland, 201
    energy flow in, 58–68, 72, 73
    forests, 296–300
    function, principles of, 66–72
    global perspective of, 288–90
    human impact on, 47–49
    inventory of, 35
    land area, 288
    management of, **75**, 77–78, 102–3
    matter and, 54–58
    as models of sustainability, 26, 72–74
    as natural resources, 289–90
    as nature's corporations, 298
    nutrient flow in, 37, 38, 68–72
    ocean, 300–307
    overview of, 26–29
    patterns of use of, 291–96
    pressures and, 296–307
    productivity of, 66–67
    public policy and, 307–11
    resilience in, 98–102
    restoration of, 47
    services and functions, 74–76, 288–90
    structure of, 29–41
    succession in, 96–100
    sustainability, principles of, 54, 68, 72, 90, 101
    terminology of, 30
    value of, 74–76, 288–90
    waste disposal in, 33–34, 65, 68
Ecotone, 27-28
Ecotourism, **267**
Ecotron Experiment, 281
Education
    correlation with health, 400
    and fertility rate, 170
    ICPD Program of Action for, 183
    importance of, 166–68, 174–75
Effectiveness (of public policy), **580**
Efficiency (of public policy), **580**
E horizon, 192
Einstein, Albert, 344
Eldridge, Niles, 122
Electrical power, 320–24
    as clean energy, 323–24
    efficiency of, 323
    energy sources for, 322
    fluctuations in demand, 321, 322, 323
    methods of generating, 321, 322
Electric cars, 563
Electric generator, principle of, 321
Electromagnetic spectrum, 528–29
Elements, **54**-56
Elephants, 122–23
El Niño
    of 1997–98
        effects on land of, 509, 510
        sea surface temperatures and, 306
    occurrences of, 509–10
El Niño/La Niña of 1997–2000
    development of, 508, 509
    impacts of, 509–10
Eluviation, **192**
Embrittlement, **358**
Emelle, AL, 500
Emergency Planning and Community Right-to-know Act (EPCRA) of 1986, 487, 503
Emergency preparedness, hazardous chemicals and, 503
Emergent vegetation, **447**
Emigration, population growth and, 146–47
Emissions
    allowances, **560**
    sources of, 548
    trends of, 547, 548–49, 558, 561

Endangered Ecosystems Act, 87
Endangered species, 86, **272**, 273
Endangered Species Act of 1973
    critics of, 273, 274
    Habitat Conservation Plan (HCP) and, 274
    protection of wildlife by, 12, 86, 271–75
    reauthorization of, 272, 274, 588
Endocrine Disrupter Screening Program, 422
Energy, 58-68. *See also* Fossil fuel; Nuclear power; Solar energy
    changes in organisms and ecosystems, 61–66
    chemical, **58**, 59
    conservation of, 336–38
    consumption in the U.S., 320
    conversions, 60
    flow in a grazing food web, 68
    flow in natural systems, 37–38, 66–68, 73
    flow through fossil fuels, 326
    forms of, 58, 60
    geothermal, **381**-82
    history of sources, 318–20
    kinetic, **58**, 60
    laws of, 58–61
    matching sources to uses, 324–25
    matter and, 58–61
    measurement of, 58
    ocean thermal-energy conversion (OTEC), **382**
    policy for sustainability, 382–86
    potential, **58**, 60, 61, 62
    renewable, 365–87
        intermittent availability of, 386
        resource alternatives for, 384
    sources and uses, 318–25
    sustainable options for, 335–38
    technology transfer to developing world, 385
    tidal power, 382
    urban sprawl and, 598–99
Energy-efficient light bulbs, 336, 338
Energy Star Program, 383
Enrichment of uranium, **345**
Entropy, 60-61
Environmental accounting, 577
Environmental change and vulnerability, 119
Environmental Defense, 486
Environmental Defense Fund, 12, 612
Environmental Farm Plan (EFP, in Ontario), 210
Environmental hazards
    biological, 394–96
    chemical, 398–99
    cultural, 393–94
        sexually transmitted diseases, 394
        smoking, 401–4
        U.S. deaths from, 394
    health links, 392–400
    industrial processes and, 398, 399
    nuisance animals and, 270
    pesticides as, 414–37
    physical, 396–97
    picture of health and, 392–93
    risk assessment, 407–12
    risk pathways, 400–407
Environmental health, **392**
Environmental indicators, income and, 571–72
Environmentalism
    critics of, 12–13
    history of, 11–12
    modern movement of, 12–13
    stewardship and, 11–14, 18–20
Environmental justice
    hazardous waste and, 500
    movement of, 9
Environmental Justice (EJ) program, 500
Environmental law, 613

Environmental management, integrative frameworks and, 614
Environmental mismanagement, consequences of, 571
Environmental movement, **12**
Environmental Policy Act of 1969, 580
Environmental public policy, **570**. *See also* Public policy
    benefit-cost analysis and, 569–70, 583–87, 590
    development of, 578–80, 581
    economics and, 570–74, 580–83
    goals of, 570
    multicriteria analysis and, 590
    politics and, 587–90
    pollution and, 578–83
    progress in, 587
    reasons for, 571
    wealth of nations and, 574–78
Environmental Quality Incentive Program (EQUIP), 208
Environmental racism, 9, 500
Environmental regulation
    benefits of, 584–85
    costs of, 584–85
Environmental resistance, **84**-86
Environmental Revolution, **49**
Environmental tobacco smoke (ETS), as Class A carcinogen, 402
Environment (Sept. 1993), 581
Enzymatic proteins (enzymes), **114**
Epidemiological study, **408**
Epidemiologic transition, **140**, 157, 393
Epidemiology, 157, **393**, 404
Epidinocarsis lopezi, 428
Epiphytes, 91
Equilibrium, 83
Equilibrium mechanisms
    competition, 90–96
    introduced species, 93–96, 124
    parasite-host dynamics, 88–89
    plant-herbivore dynamics, 89–90
    predator-prey dynamics, 87–90
    territoriality, 92–93
Equilibrium theory, **96**
Equity, 8
    stewardship and, 9
Equity (of public policy), **580**
Eroding tax base, **602**
Erosion, **200**
    control by ecosystems of, 76, 288
    deposits, 233, 234
    desertification and, 145, 189–90, 200-**201**-205
    drylands and, 200–201
    Dust Bowl, 11, 203, 210
    gully, **200**, 234
    off-road vehicles and, 201
    pollution and, 205
    public policy and soils, 208–10
    rates by equation, 195
    rates by ground truthing, 195
    sheet, **200**
    splash, **200**
    stream-bank, 233, 234
Estimated reserves, **325**
Estrogenlike pesticide effects, 422
Estuaries, **227**
    description of, 29
    human effects on trophic structure in, 50
    overdraw of water and, 227
Ethics of stewardship
    description of, 9
    implementation of, 13–14
    importance of, 311
Ethnic segregation, 600–601
Ethnobotany, **267**
ETS (environmental tobacco smoke), as Class A carcinogen, 402
Eugenics, **113**
Euphotic zone, **64**
Eutrophication

aquatic plants and, 447
combating, 449–53
  attacking the symptoms, 449–50
  getting at root causes, 450–52
coral reefs and, 306
natural vs. cultural, 449
nutrient enrichment and, 447–49
phosphorus cycle and, 71
recovery from, 452–53
in shallow lakes/ponds, 448, 450
Evaluating science, 18
Evaporation, 215-17
Evaporative water loss, 196
Evapotranspiration, 219
Evapotranspiration loop, 220
Even-aged forest management, 297
Evolution, 106–33
  adaptation and, 108–9, 111, 112
  controversy about, 127–28
  Darwin and, 107, 111, 119–21
  ecosystems and, 122–24
  environmental selection and,
    109–11
  the fossil record and, 127
  geological time scale and, 128
  limits of change in, 117–18
  in perspective, 127–29
  plate tectonics and, 124–27
  selection and, 109–16
  speciation and, 118–22
  stewardship and, 128–31
Evolving ecosystems, 122–24
Executive Order 12291 (1981), 569–70
Executive Order 12866 (1993), 570
Executive Order 12898 (1994), 500
Exotic pets, 280
Exotic species, 84, 93–96
Expanding the Measure of Wealth:
  Indicators of Environmentally
  Sustainable Development (World
  bank group), 575
Experimentation, 15-16
Exponential increase, 84
Exposure assessment, 409, 486–87
Extended product responsibility (EPR),
  480
Extension programs, 210
External bad, 583
External cost, 583
External good, 583
Extinction
  causes of, 276, 277–80
  environmental change and, 117, 119
  human impact on rate of, 128–29,
    264
  poaching and, 146
Extractive reserves, 300
Exurban migration, 597, 601
Exurbs, 597
Exxon Chemical Company, 505
Exxon Valdez, 331, 333, 487
Eye color inheritance, 115

Family, ICPD Program of Action
  commitment to, 183
Family, food security and, 253, 254
Family planning
  abortion and, 177–78
  services of, 177–78
  unmet need, 177
Family size, factors influencing,
  165–68
Famine, 257-58
  distribution of food and, 257–58
  geography of, 258
  as natural enemy, 139
  North Korea and, 258, 260
  sustainability failure and, 8
Famine Early Warning System
  Network, 257
Faraday, Michael, 321
"Farmer field schools," 210, 432
Farming. See Agriculture
Fecal coliform test, 444

Federal Agricultural Improvement and
  Reform Act (FAIR) of 1996, 208
Federal Energy Regulatory
  Commission, 357
Federal Food, Drug, and Cosmetic Act
  of 1958, 435
Federal Housing Administration, 596
Federal lands in the U.S., 308
Fermentation, 65-66
Fertility rate, 140
  contraceptives and, 167
  decline in total, 171
  factors influencing, 165–68
  incentives for reducing, 179
  income and, 165
  literacy and, 170
  population growth dynamics and,
    150–59
  unmet need and, 177
Fertility transition, 140, 157
Fertilizer, 194, 203, 243–44
Field scouts, 432
FIFRA (Federal Insecticide, Fungicide,
  and Rodenticide Act of 1947),
  434, 435
Fifth basic principle of ecosystem
  sustainability, 101, 284
Filtration, in water purification, 224
Fire
  as abiotic factor, 99–100
  crown, 100
  ground, 100
  recovery from, 82
  succession and, 99–100, 104
Fire climax ecosystems, 99
Firewood gathering, 2, 145, 380
First basic principle of ecosystem
  sustainability, 68
First-generation pesticides, 417
First law of thermodynamics, 58-59,
  60–61
Fish
  effects by dams on, 378
  eutrophication and, 448, 449, 450
Fishery, 300
Fish harvest (global), 301
Fission, 344, 345, 346
Fission products, 345
Fission reactor, 347–48
Fitch, Maureen, 250
Fitness, 111
"Five golden rules of the humid tropics
  agriculture," 189
Flat-plate collectors, 368, 369
Flooding
  in Bangladesh, 221, 222
  consequence of changing Earth's
    surface, 221, 222, 233
  erosion and, 233
Flood irrigation, 206, 207, 231
Floristics, 35
Fluorescent light bulbs, 336, 338
Food aid, 254, 258–60
Food chain, 35-36
Food distribution
  patterns in trade, 252–53
  responsibility for, 253–54
Food Guide Pyramid, 254, 255
Food Insecurity and Vulnerability
  Information and Mapping
  Systems (FIVIMS), 256
Food pathways, 65
Food production
  animal farming, 246–47
  biotechnology and, 249–51
  by ecosystems, 76
  food aid, 258–60
  Green Revolution, 244–45
  hunger, malnutrition, famine,
    254–60
  increasing, 248–49
  modern industrialized agriculture,
    242–45
  organic, 433–44

patterns of, 242–51
  safety issues and GMOs, 240–42, 250
  subsistence agriculture, 245–46
Food Quality Protection Act (FQPA) of
  1996, 434-35
Food Safety and Inspection Service
  (USDA), 434
Food security, 253, 253-54
Food Security Act of 1985, 208
Food security responsibility, 253
Food Stamp Program, 253
Food web, 35, 36
Forage Improvement Act, 205
Forecasting technique, 152
Forests. See also Tropical rain forest
  acid deposition and, 546, 555
  air pollution effects on, 545, 546
  area of, 296
  certification of sustainable
    management, 311
  clearing and consequences, 297
  frontier, 296
  indigenous villagers controlling, 300
  management of, 297–99
  as natural heritage, 300
  pressures on, 296–300
  tropical forest deforestation, 145,
    146, 205, 299–300
  world wood consumption from, 297
Forest Stewardship Council, 300
Forests and woodlands, 288
Forest Trees and People Program
  (FTPP), 380
Formaldehyde, 406
Formosan termite, 415, 416
Formulation stage (policy life cycle),
  579–80
40 CFR Part 50, 569
Fossil fuel
  carbon emissions of, 335, 336
  climate change and, 5–6, 69, 74
  coal as, 334–35
  cost of imports, 329
  crude oil as, 325-33
  energy flow through, 326
  formation of, 325, 326
  hidden costs of, 338
  history of use, 318–20
  matching sources/uses and, 324–25
  natural gas as, 334
  oil shales/oil sands as, 335
  overview of, 315–20
  reserves and use of, 332
  sustainable energy options and,
    335–38
Fossil record, 127, 128
Fourier, Jean-Baptiste, 516
Fourth basic principle of ecosystem
  sustainability, 101
Fragmentation (habitat), 277, 599
Framework for Action, 237
Framework Convention on Climate
  Change (FCCC), 527
Frankenfood protest, 240, 241, 242
Franklin, Jerry, 77
Freedom from Hunger, 178, 182
Freedom to Farm Act of 1996, 208
Free-market economy, 572-74
French Quarter, New Orleans,
  414–15, 416
Fresh Kill Landfill, NY, 470, 480
Fresh water, 214
Frito Lay, 251
Frontier forests, 296
Fronts, 513
Fuel elements, 347
Fuel rods, 347
Fuentes, Carlos, 542
Fungi
  as detritus feeder, 33–34
  mycorrhizae, 16, 198
  predatory, 199
  use as pest control, 427
Fusion, 344, 345

G7 nations, debt cancellation and, 174
Galápagos Islands
  evolution and, 106–7
  marine iguanas, 106
  tortoises, 107, 123–24
Game animals in the U.S., 269–71
Garden pest control, 426
Garrett, Laurie, 404
Gasohol, 380-81
Gas regulation, by ecosystems, 76
Gates, Bill, 183
GDP (gross domestic product), 575
Gender-related work, 166
Gene, 114
  "marker," 131
  selection of, 111–16
Gene pool, 109
  variation in, 109–10, 114–16
General Agreement on Tariffs and
  Trade (GATT), 589
General Electric, 342, 357
General Motors, 298, 336
Genetically modified organisms
  (GMOs), 249–51
Genetically Modified Pest-Protected
  Plants: Science and Regulation
  (National Research Council), 251
Genetic bank, 267
Genetic pest control, 429–30
Genetic resources, from ecosystem, 76
Genetic variation, 109
  gene pools and, 114–16
  measuring species change, 131
Genomics, 249-51
"GeoPowering the West" Initiative,
  381–82
Georges Bank, MA, 302-3
Geothermal energy, 381-82
Gerber, 251
Geysers, The, CA, 382
Giant Australian stick insect, 108
Gibbs, Lois, 8, 494
Giraffe, 117
Glasphalt, 466, 474
Glass recycling, 476–77
Glazorsky, Nikita, 228
Glen Canyon Dam, 378
Global air circulation, 217
Global Biodiversity Assessment
  (UNEP), 264, 276, 283
Global Change Research Act of 1990,
  521
Global Change Research Program, 383
Global climate change
  amount of warming, 520–24
  biodiversity and, 278
  computer modeling of, 521–23
  cooling, 517–18
  coping with, 524–28
  Earth as greenhouse, 516–18
  greenhouse gases and, 5–6, 335,
    518–20
  impacts of, 522–24
  predictions of future, 124, 128
  public policy and, 525–28
  U.S. National Assessment of,
    521–22, 523
  warming process, 516–17
Global community, food security and,
  253
Global environment
  current picture of, 4–6
  new commitment to, 18–20
  three unifying themes of, 291, 310,
    311, 610
    sound science, 6, 14–18
    stewardship, 6, 8–14
    sustainability, 6-8
Globalization of disease, 404
Global precipitation, 217, 218–19
Global Warming: The Complete
  Briefing (Houghton), 526
Global warming. See Global climate
  change

GM Precept concept car, 376, 377
GNP (gross national product), 576–77
Goats, 89, 94
Golden rice, 241–42
Gonsalves, Dennis, 250
Gore, Al, 614
Gould, Stephen Jay, 20, 122
Gradualism, **122**
Grain
    consumption of, 73–74, 249
    production of, 248
    trade, 252
Grameen Bank, 178
Grand Banks, 303
Grand Teton National Park, 308
Granz, Daniel S., 502
Grassland biome, **42–43**
Grasslands and savannas, 288
Gravitational water, **219**
Gray fox, 118–19
Graying, 154
Gray water, **232**
Greater Yellowstone Coalition, 308
Greater Yellowstone Ecosystem, 309
Great Lakes basin, 284
Great Plains, 228
Great Texas Birding Trail, 295–96
"Green" energy, 336, 383–84
Green fees, **584**
Green History of the World, A
    (Ponting), 190–91
Greenhouse effect, **5**
Greenhouse gases, 516–17
    in the atmosphere, 520
    global climate change and, 516–20,
        524
    ozone depletion and, 520
Green Lights, 383
"Green Muscle," 427
Greenpeace, 11, 12, 13
Greenprint, 593–94
Green products, 506
Green Revolution, 244–45, 248, 252
Green taxes, **584**
"Green water flow," 217, 219
Grime, Phil, 281
Grit chamber, **454**
Gross domestic product (GDP), 576-
    77
Gross national product (GNP),
    576–77, **576-77**
Gross primary production, **63**
Ground fire, 100
Groundwater
    the hydrologic cycle and, 216, **219-
    20**
    landfills and, 468, 469
Groundwater loop, 220
Groundwater remediation, **496**, 497
Growth, absolute, 140, 141
Growth curves, 83
Growth rate, **140**, 141
Guatemala, 189, 190
Gulf of Maine, 302, 303, 304
Gulf of Mexico
    dead zone, 278, 439–40
    eutrophication of, 439–40
Gulf War of 1991, 331
Gully erosion, **200**
Gypsy moth, 83–84, 391–92

Habitat, 6, 39, 277
Habitat Conservation Plan (HCP), 274
Habitat for Humanity, 613
Hadley cells, 217, 512
Hafeez, Shaheena, 209
Haitian immigrants, 148
Haitian Refugee Immigration Fairness
    Act of 1998, 148
Half-life, 351–52
Halibut fishing, 286–88
Halogenated hydrocarbons, 489–90
Hardin, Garret, 259, 293–94
Harding, Warren G., 333

Hard Times (Dickens), 539
Harmful Algal Bloom and Hypoxia
    Research and Control Act of
    1998, 440
Harmful algal blooms (HABs), 451,
    462
Harvard University's Center for the
    Study of World Religions, 284
Harvesting aquatic weeds, 450
Hawaii Electric Light Company
    (HELCO), 386
Hazard, **392**
Hazard assessment, 408–9
Hazard communication standard, 503
Hazardous air pollutants (HAPs), 541
Hazardous chemicals
    clean up of, 496–500
    defining of, 506
    food chains and, 485–86, 490–91
    future of, 504–6
    land disposal methods of, 491–93
    laws and, 500–504, 505
    management (improved) of,
        500–504
    mismanagement of, 491–96
    nature of, 487
    pollution avoidance and, 505–6
    public policy and, 500–506
    sources of, 487–88
    threat from, 488–91
    toxicology and, 486–91
Hazardous material (HAZMAT), **487**
Hazardous and Solid Waste
    Amendments of 1984, 478
Hazards. *See* Environmental hazards
Hazard vs. outrage, 410–11
HAZMAT placards, 503
HDPE (high-density polyethylene),
    477
Health, **392–93**. *See also*
    Environmental hazards
    air pollution and, 542–46
        acute effects, 542, 545
        carcinogenic effects, 542, 545
        chronic effects, 542–43
    and environmental
        mismanagement, 571
    ICPD Program of Action for, 183
    improvements in, 175
    poverty and, 400–401
    vaccines and GMOs, 249
Heavily Indebted Poor Country
    Initiative (HIPC), 174
Heavy metals
    as air pollutants, 541, 549
    as hazardous material, 488–89
Heifer Project International, 247
Heirloom vegetables, 130
"Helping hands," 166
Henson, Jim, 395
Herbicides, **416**
Herbivores, **31**. *See also* Consumers
Hessian fly, 429, 430
Heterotrophs, **31**, 34. *See also*
    Consumers; Detritivores
Hewlett-Packard and EPR, 480
High income, highly developed,
    industrialized countries, **141**
Highway Revenue Act of 1956, 596
Highway-traffic-congestion cycle,
    596–97, 598
Highway Trust Fund, 596, 598, 600
High-yield plant varieties, 244
Hindu Kush Himalayas (HKA), 209
Hippophae spp., 209
Hodgkin's disease, 267
Hooker Chemical and Plastics
    Company, 495
Hopper, Brian, 613
Horizons, **192**
Hormonal proteins, 114
Hormones, and pest control, 430–31
Host-parasite relationship, **32**
Host specific, **91**

Houghton, Sir John, 526
Housefly, as adaptive species, 118, 119
Household hazardous wastes (HHW),
    488, 506
Houston-Galveston Bay, TX, 229
Houston Principles of the Alliance for
    Sustainable Jobs and
    Environment, 589
Hubbard Brook Experimental Forest,
    555
Hubbart, M. King, 327
"Hubbart" curve, 327
Hubbart curves of oil production, 332
Human capital, **575–76**
Humane Society of America, 270
Human immunodeficiency virus
    (HIV), 175–76, 394
Human population growth
    affluence and, 150
    consequences in developing world,
        149
    consumption and, 142–44
    controversy regarding, 138
    data for selected countries on, 143
    demographic transition and, 156–59,
        164–68
    in developing countries, 144–49
    developing/developed countries
        comparison, 144
    development and, 169–84
    dynamics of, 150–59
    economic disparities and, 140–44
    effects on biodiversity by, 6, 277–78
    environmental impacts and, 103–4,
        144–50
    explosion of, 138–40
    overview of, 135–38
    rates of, 4, 19, 138–39, 140, 141
    social impacts and, 144–50
Human resources, **575–76**
Human system, **49**
Humidity, **215**
Hummingbird Highway, 365
Humpback whales, 305, 306
Humus, **192**, **197**
    development of soil structure and, 198
    formation of, 197
    importance of, 198–99
    O horizon and, 192
Hunger, **254**-60
    root causes of, 256–57
Hunter-gatherer culture, 47–48
Hunting and trapping, 269–70
Hurricanes, 295, 396, 397
Hussein, Saddam, 331
Huston, Michael, 281
Hybrid electric vehicle (HEV), 563
Hybrids, **120**
Hydrogen, solar production of, 375–76
Hydrogen bonding, **55**-57, 215
Hydrogen-oxygen fuel cell, 377
Hydrologic cycle, **215**
    changing earth's surface and, 221–22
    human impacts on, 221–30
    maintenance by ecosystems, 76, 288
    observation of, 15
    overdrawing water and, 222, 226–30
    pollution of, 222
    summary of, 220
    water movement through, 215–20
Hydrologists, **214**
Hydropower
    ecological, social, cultural costs of
        dams, 377–78
    Hoover Dam, 378
Hydrosphere, **55**
Hydroxyl radical, **538**-39
Hypothesis, **16**
Hypoxia, **439**

Ice cores
    methane concentration changes
        and, 520
    past climate determined from, 514–15

Iceland, 99, 100
ICPD Program of Action (1994),
    181–84
Identical twins, **116**
Ignitability, **487**
Illegal dumping, 472, 493, 495
Illegal Reform Act of 1996, 148
Illicit activities, population growth and,
    147
Immigration, population growth and,
    146–47, 148
Impact (car), 563
Implementation stage (policy life
    cycle), 580
Income
    enhancement of, 178–79, 182
    environmental indicators and,
        571–72
    ICPD Program of Action for, 182
"Increasing block" pricing, **238**
Incremental value, 75
India
    Bhopal, 503
    billionth person, 162–63
    empowerment of women, 184
    Ganges River, 443
    Kaiga II nuclear power plant, 340
    Kerala, 163–64, 174
    Singraali coal-burning plants, 172
Indiana Dunes National Lakeshore, 104
Indicator organism, **444**
Individual quota (IQ), **287**, 303, 304
Indoor air pollution, 406, 407
Industrialized countries (the North), 1,
    2, **141**, 152, 153–54, 385
Industrial Revolution, 48–49, 242–43,
    252, 319
Industrial smog, **537**-39
Infant mortality, **140**
Infiltrate, **196**
Infiltration, **219**
Infiltration-runoff ratio, **219**
Information Age, 473
Infrared radiation, 516, 519–20
Inheritance of acquired characteristics,
    113
Inorganic chemicals, **443**
Inorganic fertilizer, **194**, 195
Inorganic materials, 30-31, 57
Insect
    as food, 436
    food chain of, 420–21
    life cycles of, 425
Instrumental value, **265**
Instruments in science, **17**
Insurance spraying, **432**
Integrated city, 595
*Integrated Environmental and
    Economic Accounting* (U.N.
    Statistical Office), 577
Integrated pest management (IPM),
    417, 432–33, 436
Integrated Risk Information System
    (IRIS), 486
Integrated waste management, 478–82
Intergenerational equity, **578**
Intergovernmental Panel on Climate
    Change (IPCC), 5, 524, 525, 526
Intermodal Surface Transportation
    Efficiency Act (ISTEA) of 1991,
    600
International Bank for Reconstruction
    and Development, 168. *See also*
    World Bank
International Centre for Integrated
    Mountain Development, 209
*International Conference on Population
    and Development Program of
    Action* (ICPD Program of
    Action), 181, 182–83, 247
International cooperation, ICPD
    Program of Action for, 183
International Food Policy Research
    Institute, 206

International Institute for Tropical Agriculture in Nigeria, 428
International Labor Organization, 254
International Pacific Halibut Commission, 287
International Society for Ecological Economics, 19
International Union for Conservation of Nature, *Red Data Book*, 304–5
International whaling, 304–6
International Whaling Commission (IWC), 304, 305
"Internet Economy and Global Warming, The" (Center for Energy and Climate Solutions), 338
Interspecific competition, **90**
Intragenerational equity, **578**
Intraspecific competition, **92**
Intrinsic value, **266**, 269
Introduced species, 84, 93–96
Inuit people, 484–86
Inventory, 35
Ion-exchange capacity of soil, 194
Iowa harvest, 188
Iraq, population projections for, 154–55
Irrigation, **196**
  increase in, 244
  as largest use of water, 223, 248
  new methods of, 231–32
  and salinization, 206–7
Isle Royale, 87–88
Isles, Inc., 594
ISO New England pool, 321
Isotopes, 344, **514**
Italy, population projections for, 152, 153–54
Ivory trade, 282
Ixodes scapularis, 391–92

Jack pine, 104
Jaguar Creek facility, 365, 366
Japan
  Minamata disease and, 490–91
  whaling by, 304–5
Japan Atomic Energy Institute, 341
Japanese beetles, 427, 431
J-curves, 83–84, 103
Jet streams, 513
Joe Camel, 402, 403
John D. and Catherine T. MacArthur Foundation, 283
Jubilee 2000, 174
Junk science, **14**, 18
Justice, stewardship and, 9
Juvenile hormone (caterpillar), 431

Kaiga II nuclear power plant, 340
Kalmykia, 190
Kalmyks, 190
Kansas Department of Health and Environment (KDHE), 499
Karachay, Lake, 352–53
Karner blue butterfly, 272
Keeling, Charles, 518
Kendall, Henry, 610
Kennedy, John F., 579
Kenya's "casual workers," 136–37, 138
Kerala, India, 163–64, 174
Kerogen, 335
Kessler, David, 402
Kesterson National Wildlife Refuge, CA, 207
Keystone species, **281**
Kinetic energy, **58**
Kirtland's warbler, 277
Klamath, OR, 227
Knipling, Edward, 429
Komsomolsky, Russia, 189, 208
Kress, Steven, 263–64
Kudzu, 94, 95, 279
Kuwait, 331
Kyoto agreement (1997)
  attempts to discredit, 18, 588

carbon dioxide emissions reduction, 5–6
  emission allowances, 527–28

Lacey Act of 1900, **271–72**
Laherrère, Jean, 332–33
Lakes, eutrophication in, 447–53
Lakes and ponds, 29
Lamont-Doherty Earth Observatory, 77
Lamprey, 32
Landfills, 467–71
  conversion of, 465, 466, 469
  improvement of, 468–69
  modern design of, 468–69, 480
  problems with, 468
  secure, **493**, 494
  siting of new, 470–71
Landholders, individual, 208, 209–10
Landscape, 27–28, 30
Landscape ecology, **27**
Land. *See* Public lands; Soil
"Land and Service Ethic" (U.S. Forest Service), 9
Land subsidence, **229**-30, 233
Land Trust Alliance, 310
Laser fusion nuclear reactor, 359
Latitude effects, 45
Law of limiting factors, 41
Law of mass-energy equivalence, 344
Lawn pest control, 426, 436
Lawton, John, 281
Leachate generation, **468**
Leaching, **194**, 468
  E horizon and, 192
Lead, 541, 546–47, 549
Leafhoppers, 429
Leaf katydid, 108
"Leapfrogging" technology transfer, 385
Legumes, and nitrogen fixation, 71–72
Less developed countries (LDCs), 141
Lethal mutation, **116**
Leukemia, childhood, 267
Lichens, 38–39
Liebig, Justus von, 41
Liebig's law of minimums, 41
Life
  emergence of, 128
  as hierarchy of organization of matter, 59
Lifeboat ethic, 259
Life span, 139
Lifestyles
  affluent, 144, 150
  car-dependent, 594, 596–98, 602, 605
  changes in, 611–14
Light bulbs, 336, 338
Light rail system, 557
Light-water reactors (LWRs), 347
Limestone, as buffer, 554–55
Limiting factor, 41
  biotic factors as, 45–46
  climate as, 42–46
  combating eutrophication and, 450–51
  microclimate as, 45, 46
  physical barriers as, 46
Limits of tolerance of abiotic factors, **40**
Literacy. *See* Education
Lithosphere, **55**
Little Ice Age, 514
Living modified organisms (LMOs), 251
Loam, **193**
Local Politics of Global Sustainability, The (Prugh, Costanza, and Daly), 8, 574–75
Locusts, 427
Logging
  forest management and, 297–98
  interests, 13
  roads, 310

Longevity, **139**, **150**
Long-term containment, **352**
López, Gabino, 189, 208
Los Alamos, NM fire (2000), 100, 101
Los Angeles
  air pollution and, 540, 541
  water supply and, 213–14
"Loss-of-coolant accident," **348**, 356
Louisiana Coastal Wetlands Interfaith Stewardship Group, 237
Louisiana Wetlands Conservation and Restoration Trust Fund, 237
Love Canal, 8, 493–96
Low Impact Sustainable Agriculture (LISA) program (1988), 208
Low-income countries, 141
LULU (locally unwanted land use), 470
Lumpers, 120
Lyme disease, 391–92

MacArthur Foundation, 283
McCormick, Joe, 404
McNamara, Robert, 141
Macroevolutionary changes, **127**
"Mad as a hatter," 491
Magnuson Act of 1976, 301, 302, 303
Maine Yankee nuclear plant, 358
Malaria
  control of, 405–6
  and DDT, 405, 419
  distribution and problem areas of, 405
  life cycle of parasite causing, 395–96
Malnutrition, 254-55
  disease and, 255
  protein-energy and, 255
Malthus, Thomas, 164
Management of ecosystems, 75, 77–78
Manatee, 272
Mangroves, 307
Manshardt, Richard, 250
Maori, 593–94
Marcos, Ferdinand, 10
Marine ecosystems, 64
Marine iguanas, 106–7
Marine Mammals Protection Act of 1972, 12
Marine Protection, Research and Sanctuaries Act of 1972, 461
Market approach, **583**
Marsupials in Australia, 123
Massachusetts Water Authority (MWRA), 459
Mass-energy equivalence, 344
Mass number, **344**
Mass-transit, 557, 563, 600, 605
Material safety data sheets (MSDSs), 503
Materials recovery facility (MRF), 478, 480
Mathematical optimization, 590
Matter, **54**, 58
Maximum achievable control technology (MACT), 559
Maximum contaminant levels, 496
Maximum sustainable yield (MSY), **292**, 293
May, Robert, 281
Mealybugs, 427, 428
Measuring biodiversity, 284
Meat consumption, 73–74, 247, 249, 252
Mech, David, 265
Medicine
  modern, 139, 144, 168
  prescription drugs and, 6, 130, 267, 268
Meltdown, **348**
Mercedes-Benz Citaro urban buses, 376
Mercury
  Great Lakes and, 553
  poisoning, 490–91
Metabolic traits, **111**, 114

Metals, heavy
  as air pollutants, 541, 549
  as hazardous material, 488–89
Metarhizium anisopliae, 427
Meteorology, **511**
Methane, 380
  agriculture and, 534
  global warming and, 247, 520, 534
  production in landfills, 468, 534
  ruminants' production of, 247, 520, 534
Metropolitan Area Express (MAX), 557
Mexico City, 536–37, 542
Microclimate, **45**-46
Microevolution process, **128**
Microlending, 178
Middle income, moderately developed countries, 141
Midnight dumping, 472, 493, 495
Migration
  exurban, 597–603
  ICPD Program of Action for, 183
  response to changing environment, 117
  to cities, 145–46, 147
Milankovitch cycles, 514
Millennium Ecosystem Assessment, **311**
Million Solar Roofs Initiative, 373
Mimic (pest control), 431
Minamata disease, 490–91
Mineral, **56**, 57
Mineral, water, air interrelationship, 57
Mineralization of soil, 198–99
Mineral nutrients, 30, 194
Mining coal, 334–35
Mississippi Delta, 439
Mississippi River
  channelization and, 235
  fertilizer pollutants and, 438–40
  flood of 1993, 235
Mobile incinerator, 498
Moderator, **347**
Molecule, **54**
Molina, Mario, 529
Mollisols, 194
Monarch butterflies, 250
Monoculture, **91**
  agricultural practices of, 129, 244
Mono Lake, CA, 212–14, 236
Mono Lake Committee, 213
Monsanto, 251
Monsoons, **513**
Montiel, Rodolfo, 8–9
Montreal Accord of 1987, 520, 526, 533
Moose population on Isle Royale, 87–88
"Mopani," 436
More developed countries (MDCs), 141
Morgenstern, Richard, 562
Mortality, **393**
  causes in developing/developed countries, 393
  of infants and children, 165, 174
Morton, Samuel, 20
Mosquitoes and disease, 395–96, 405–6
Moss and succession, 96
Moth life cycle, 425
Motor vehicles
  air pollution and, 540, 542, 545, 550, 553, 557–59
  fuel and, 336, 375–76, 562–63
  new designs for, 376–77, 562–63
  urban sprawl and, 594, 596–98, 604, 605
Mountain lions (cougars), 270, 271
Mount Pinatubo, 517, 524
Mount Saint Helens, 126
Muir, John, 11
Müller, Paul, 417–18
Multiple use (forests), **309**
Municipal solid waste (MSW), **466**
  amount of, 467
  combustion of, 471–72
  composition of, 467, 482

cost of, 472
disposal of, 466–67, 480–81
import/export by state, 470
integrated waste management of, 478–82
landfills for, 467–71
New Orleans burning dump, 467
problem of, 465–72
public policy and, 478–82
solid-waste crisis, 465
solutions to, 473–78
Municipal water use/treatment, 225, 226
Muriqui monkey, protection of, 291
Mutagenic, 489
Mutation, 116
Mutualism, 37-39
Mycorrhizae, 16, 198

Nam Theun Two Dam, 378
National Academy of Science, 18, 148, 251, 435
National Ambient Air Quality Standards (NAAQS), 556
National Audubon Society
egrets and, 271
as environmental caretaker, 13
formation of, 11
Seabird Restoration Project, 263–64
National Biological Service, 284
National Emission Standards for Hazardous Air Pollutants (NESHAPS), 556
National Energy Policy Act of 1992, 232, 362
National Environmental Performance Partnership System (NEPPS), 506
National Environmental Policy Act of 1969, 12
National forests (U.S.), 308–10
National Gardening Association, 130
National Heritage Network, 35, 276
National Institute of Environmental Health Sciences (NIEHS), 486
National Irrigation Water Quality Program, 207
National Marine Fisheries Service (NMFS), 302, 613
National Oceanic and Atmospheric Administration, 530, 613
National Organic Program standards, 434
National Organic Standards Board, 434
National Parks Conservation Association, 12
National parks (U.S.), 307–8
National Pollution Discharge Elimination System (NPDES), 447, 451, 501
National Priorities List (NPL), 498
National Recommended Water Quality Criteria, 445, 447
National Religious Partnership for the Environment, 19
National Research Council, 148, 208, 251, 435
National Resources Inventory, 195
National Town Meeting for a Sustainable America, 609–10
National Toxicology Program (NTP) Chemical Repository, 486, 487
National Wildlife Federation, 11, 13, 86
National wildlife refuges (U.S.), 307–8
National Wild Turkey Federation, 270
Native Seed/SEARCH, 130
Natural capital, 74-75, 264, 575, 576
Natural chemical cycles, effect on by humans, 5
Natural enemies
disease as, 139
famine as, 139
herbivore populations and, 89
pest control and, 426–29
Natural eutrophication, 449

Natural gas, 334
history of use, 320
reserves and use of, 332
Natural goods, 288
Natural Heritage Network
inventory of terrestrial vegetation by, 35
status of U.S. species, 276
Natural laws and sound science, 16
Natural organic compounds, 57
Natural pest control, 425
cultural, 425–26
genetics and, 429–30
natural chemicals and, 430–31
natural enemies and, 426–29
Natural Resources Conservation Foundation, 208
Natural Resources Defense Fund, 12, 612
Natural resources (of ecosystems), 289–90. See also Ecosystems
conservation/preservation of, 290–96
pressures on, 296–307
public policy and, 307–11
value of, 74–76, 288–90
wealth of nations and, 574–78
Natural selection, 111. See also Selection
Natural services, 288
Nature Conservancy, The
inventory of terrestrial vegetation by, 35
as land trust, 310
status of U.S. species, 276
Nazarbayev, Nursultan, 228
Nelson, British Columbia, 566
Nematodes
as human parasites, 32
in soil, 198
Neolithic Revolution, 47-48, 242
Neotropical migrants, 276
Net domestic product, 577
"Net metering," 372, 385
Net primary production, 63
Neutral mutation, 116
New England Fishery Management Council (NEFMC), 302, 303, 304
New Forestry, 309–10
NGOs (nongovernmental organizations)
agriculture and, 130, 209, 210
Basel Action Network, 10–11
Cairo Conference and, 181, 183
fuelwood crises and, 380
participating in, 612–13
public policy and, 588–89
seed savers as, 130
sustainable cities and, 607
water and, 237
Niche, 39-40
NIMBY syndrome (not in my back yard), 353, 470, 480
NIMTOO (not in my term of office), 470
Nitric acid, 552
Nitrogen, eutrophication in marine systems and, 450
Nitrogen cycle, 71–72, 289
Nitrogen fixation, 71-72
Nitrogen oxides
as air pollutants, 541, 546–49, 552–53, 559, 560
global warming and, 520
ozone destruction and, 520
Nixon, Richard, 580
Nonbiodegradable, 440
Nonconsumptive use, 223
Nondepletable, 68
Nonequilibrium systems, 100–101
Nonequilibrium theory, 96, 99, 100–101
Nongovernmental organizations. See NGOs
Nonpoint sources, 441, 442, 451, 462

Nonpolluting, 67
Nonrenewable natural capital, 575
North, 1, 385
North Korea, famine and, 258, 260
Northwest Forest Plan, 274
Norway, whaling by, 304–5
Noss, Reed, 87
No-till agriculture, 202–3, 249, 250
"Nuclear Age," 342
Nuclear power
advanced reactors, 358–60
comparison with coal power, 348–49
dream or delusion, 342–44
electrical power generated by, 344
fuel for, 344–46
future of, 360–62
hazards/costs of, 349–58
accidents, 355–56
economic problems, 357–58
radioactive emissions, 349–51
radioactive wastes, 349, 350, 351–55
safety, 356–57
mass to energy, 344–48
opposition to, 360–61
plant, 348
process of, 344–49
reactor, 347–48
rebirth of, 361–62
siting of facility, 362
in the U.S., 342–43
Nuclear reactor, 347–48, 358–60
Nuclear Regulatory Commission (NRC), 342, 354, 358, 360
Nuclear Waste Policy Act of 1982, 354
Nunavut territory, 484–86
Nutrient-holding capacity of soil, 194
Nutrients
cycling in natural systems, 37, 38, 68–72
enrichment of, 447–49
flow in human society, 75
in marine systems, 64
produced by photosynthesis, 30
water pollution and, 445
Nutrition
developing countries and, 19
improvement in, 139–40

Oakland, CA fire (1991), 100
Oak Ridge National Laboratory (ORNL), 373
Obee road, KS NPL site, 499
Occidental Petroleum, 495
Occupational Safety and Health Act (OSH) of 1970, 503
Ocean ecosystems
coral reefs and mangroves, 306–7
international whaling, 304–6
marine fisheries, 300–304
Oceans
coastal, 29
global warming and, 524, 525
heat capture by, 525
heat transported by, 515–16
open, 29
Ocean thermal-energy conversion (OTEC), 382
Oden, Svante, 549
Off-road vehicles (ORV), 308, 310
Ogallala aquifer, 228, 229
O horizon, 192, 197
Oil field, 325
Oil sand, 335
Oil. See Crude oil
Oil shale, 335
Oligotrophic, 447
Omnivores, 32. See also Consumers
O'Neill, Jessie H., 481
OPEC (Organization of Petroleum Exporting Countries), 328, 331
Open space, 600
Optimal population, 293

Optimum (range) of abiotic factors, 40
Organically grown produce, 433-34
Organic chemicals, 443
Organic compounds
as HAZMAT, 489–90
natural, 57
synthetic, 57
Organic fertilizer, 194
Organic Foods Protection Act of 1990, 434
Organic materials, 30-31
Organic molecules, 57-58, 59
Organic phosphate, 70
Organic wastes, 443
Organization for Economic Cooperation and Development, 170–71
Organization of Petroleum Exporting Countries (OPEC), 328, 331
Origin of Species by Natural Selection, The (Darwin), 107, 111
Orlando Easterly Wetlands Reclamation Project, 461
Orphan site, 493, 495
Overcultivation, 202–3
Overdrawing water
groundwater
falling water tables/depletion and, 227–28
land subsidence and, 229-30
saltwater intrusion and, 230
surface waters
diminishing of, 228–29
ecological effects and, 226–27
shortages of, 226
Overgrazing, 89
effects on natural ecosystems, 204
erosion and, 203–5
as form of overuse, 90
public lands and, 204–5
subsidies for, 581
and water pollution, 204
Overuse, loss of species and, 279–80
Oxfam, 178, 184
Community Aid Abroad (CAA) program, 184
Rural and Urban Center for Human Interest (RUCHI) program, 184
Oxidation, 64
Oxisols, 194, 205
Oxygen cycle, 289
Ozone layer
Antarctic hole in, 530–32
controversy on depletion in, 532–33
depletion of, 528–34, 599
formation/breakdown of, 528–33
public policy and, 533–34
recovery of, 533–34
Ozone shield, 529-33
Ozone (tropospheric), 541, 549, 550, 564

Paclitaxel (Taxol), 267
Panda bear, as vulnerable species, 118, 119
Pangaea, 125
PANs (peroxyacetyl nitrates), 549, 550, 557
Papaya plants, 250
Paper mill pollution (SC), 388
Paper recycling, 476
Parasite-host dynamics, 88–89
Parasites, 32
as biological hazard, 394, 396
diversity of, 32
Parasitic wasps, 426, 427, 428
Parker River National Wildlife Refuge, 25, 26
Partnership for a New Generation of Vehicles (PNGV), 376, 383
Passenger pigeon, 264–65
Passive safety, 356
Passive system, 368
Pasteur, Louis, 453

Pasteurization, of sewage sludge, 460
Pastoralists, 257
Pathogens, 32, 441
  in sewage, 442
  as water pollutant, 441–43
Paton, Alan, 137–38
"Pay-as-you-throw" (PAYT), 467, 476, 479–80
Peer review, 18
Pellet stove, 379–80
People for Ethical Treatment of Animals (PETA), 270
Percolation, 219
Peregrine falcon, 273–74
Permafrost, 44
Peroxyacetyl nitrates (PANs), 549, 550, 557
Persian Gulf War of 1991, 331
Persistent organic pollutants (POPs), 485-86, 489–90
Personal pollutant, 401
Personal Responsibility and Work Opportunity Reconciliation Act of 1996, 253
Pest, 416
  management by ecosystems of, 289
  varying meanings of, 436
Pest control
  alternative methods to, 424–31
    cultural, 425–26
    genetic, 429–30
    natural chemical, 430–31
    natural enemies, 426–29
  biotechnology and, 249, 250
  chemical approach to
    development of, 417–19
    nonpersistent pesticides and, 424
    problems from, 419–24
  importance of, 416
  need for, 416–17
  philosophies of, 416–17
  public policy and
    FIFRA, 434, 435
    FQPA of 1996, 434–35
    new needs for, 436
    pesticides in developing countries, 435–36
  socioeconomic issues and, 431–34
    integrated pest management, 432–33
    pressures to use pesticides, 431–32
Pesticide Data Program, 435
Pesticide Environmental Stewardship Program, 436
Pesticides, 416
  characteristics of, 418
  environmental effects and, 422–24
  human health effects and, 421–22
  problems with nonpersistent, 424
  resistance by pests and, 419
  resurgence and, 419-21
  secondary-pest outbreaks and, 419-21
  use in the U.S., 417
  wildlife and, 12
Pesticides in the Diets of Infants and Children (National Research Council), 435
Pesticide treadmill, 420, 422, 450
Pest-loss insurance, 432
PETE (polyethylene terephthalate), 477
Pfiesteria spp., 451
pH, 196, 550, 552
Phase I emission reductions, 561
Phases of demographic transition, 157–59
Pheasants Forever, 270
Phenolphthalein, 486–87
Pheromones, 430–31
Philadelphia, MS, 500
Phosphate in rocks, 70, 194
Phosphorus
  banning of detergents containing, 450–51
  and eutrophication, 450

Phosphorus cycle, 69–71
Photochemical oxidants, 549, 550
Photochemical smog, 540
Photosynthesis, 30, 31
  aquatic, 448
  energy flow in, 62–63
  glucose production in, 62–63
  in marine system, 64
Photovoltaic (PV) cells, 371–73
Physical barriers (limiting factor), 46
Physical barriers (pest control), 429
Physical traits, 111, 114
Physiognomy, 35
Phytoplankton
  eutrophication and, 439, 443, 447, 448
  in marine ecosystems, 64
Phytoremediation, 498
Pilot Analysis of Global Ecosystems (PAGE), 4–5, 311
Pinatubo, Mount, 517, 524
Pine barrens tree frog, 272
Planetary albedo, 517
Planned Parenthood, 178
Plant association, 27, 30
Plantations of trees, 300
Plant-herbivore dynamics, 89–90
Plants. See also Wild species
  aquatic, 447
  as autotrophs, 30–31
  benthic, 447
  competition between species, 90–92
  high-yield varieties, 244
  as producers, 30–31
  soil and, 194–99
Plant vs. animal farming, 247
Plasmodium protozoan parasites, 395, 405
Plastic
  biodegradable, 468
  recycling and, 474, 477–78
Plate tectonics theory, 124-27
Plum Island, MA, 24–26
Poaching, 146, 280
POH, 552
Point sources, 441, 442, 451, 462, 559
Policy life cycle, 578-80, 581
Political costs, 581
Political involvement, 611, 612
Politics, the environment and, 587–89
Pollination
  ecosystems and, 76
  mutualism and, 38
Pollutant, 440
Pollution. See also Air pollution; Water pollution
  abatement of, 12, 19
  agricultural
    fertilizer, 203
    nitrogen, 72
    phosphorus, 70–71
  categories of, 441
  effects on biodiversity by, 6, 278
  effects on wildlife by, 12
  erosion and, 204, 205
  from fossil fuel burning, 338
  overview of, 389
  problems following World War II with, 12
  of water cycle, 222
Pollution (air) disasters, 541
Pollution avoidance, 505–6
Pollution prevention, 505–6
Polyclimax condition, 96
Ponds and lakes, 29
Ponting, Clive, 190–91
Pools (utility companies), 321
Population, 26-27, 30. See also Human population growth
  biotic potential of, 84-86
  critical number, 86
  density dependent factors and, 85–86, 88
  environmental resistance, 84-86

equilibrium mechanisms, 87–96
  competition, 90–96
  introduced species, 93–96, 124
  parasite-host dynamics, 88–89
  plant-herbivore dynamics, 89–90
  predator-prey dynamics, 87–90
  territoriality, 92–93
  growth curves, 83–84
Population by age group, 155
Population conferences
  Bucharest, Romania (1974), 164
  International Conference on Population and Development (Cairo, 1994), 165, 181–84
  Mexico City (1984), 164–65
Population density, 85
Population equilibrium, 83
Population explosion, 84, 149
Population genetics, 131
Population growth. See Human population growth
Population Institute, and the ICPD Program of Action, 183
Population momentum, 140, 155–56
Population profile (age structure), 140, 150
  planning with, 152
  U.S., 150–52
Population projections
  comparison of, 156
  developed countries, 152, 153–54
  developing countries, 154–55
  the U.S., 154
  the world, 153
Portland, OR
  light rail system, 557, 605, 606
  Tom McCall Waterfront Park, 606
Postconsumer recycled paper, 476
Potassium in rocks, 194
Potato blight, Ireland, 429
Potential energy, 58, 60, 61, 62
Potrykus, Ingo, 241
Poverty, correlation with health risks, 400–401
Poverty cycle, 168, 256
Poverty in developing countries
  causes of, 10–11
  cycle of, 168
  growth and, 145–49
  hunger and, 256–57
  integrated approach to alleviating, 182
  women and children in, 147, 149, 165–68
"Poverty Reduction Strategy," (World Bank), 174
Power grid, 375
Precautionary approach, 293
Precautionary principle, 251, 411, 526–27
Precipitation, 217–19
Predator, 32
Predator-prey relationship, 32, 87–88
Prescription drugs, 6, 130
Preservation, 291
Presidential Commission on World Hunger, 260
President's Council on Sustainable Development
  immigration and, 148
  sprawl and transportation, 563
  stewardship of natural resources and, 13–14, 78, 236–37
  sustainable communities and, 608–10
  sustainable development goals, 610
Pressurized nuclear power plant, 348
Preston, Lewis, 181
Prey, 32
Price-Anderson Act of 1957, 342
Prickly pear cactus, 427, 428
Primary clarifiers, 454
Primary consumers, 31. See also Consumers

Primary detritus feeders, 33. See also Detritivores
Primary energy source, 320
Primary pollutants, 546–547-549
Primary producers, 63
Primary production, 63, 66
Primary recovery, 326
Primary recycling, 473
Primary standard, 556
Primary succession, 96, 97
Principles of ecosystem function, 66–72
Principles for ecosystem protection, 295
Prior informed consent, 435–36
Prius (car), 563
Private land trust, 310
Private ownership, 294
Prodigy (car), 336
Produced assets, 575
Producers, 30
  as chemical factories, 63
  in marine system, 64
  photosynthesis by, 62–63
  trophic relationships, 31, 33–38
"Production" of oil, 325
Productive use, 291–92
Productivity, environmental mismanagement and, 571
Productivity of ecosystems, 66–67
Program of Action (ICPD, 1994), 181–84
Project XL, 505
Protein-energy malnutrition, 255
Proteins, 114
Proven reserves, 325
ProVention Consortium, 397
Prudhoe Bay, 333
Prugh, Thomas, 8, 574–75
Public concerns vs. EPA risks, 410
Public-health measures, water pollution and, 442, 443
Public lands
  grazing and, 204–5
  in the U.S., 307–10
Public policy. See also Environmental public policy
  air pollution and, 555–63
  biodiversity stewardship and, 283–84
  biotechnology and, 251–52
  development of, 578–80
  ecosystems, and, 295, 307–11
  ESA and, 275
  food production/distribution and, 258–59
  Global climate change and, 525–28
  hazardous chemicals and, 500–506
  lands set aside in the U.S. by, 307–10
  multicriteria analysis and, 590
  Ozone depletion and, 533–34
  pest control and, 434–36
  population and development, 181–84
  risk assessment and, 411–12
  social modernization and, 174–84
  soil conservation and, 208–10
  sustainable energy and, 382–86
  Three E's of, 580, 581
  waste management and, 478–82
  water pollution and, 461–62
  water stewardship and, 236–38
Public Trust Doctrine, 213
Public Utility Regulatory Policies Act (PURPA) of 1978, 337
Puffin, Atlantic, 262–64
Punctuated equilibrium, 122, 127
Purification of water, 224
Purple loosestrife, 94, 95, 279

Qualitative (water), 223
Quantitative (water), 223

Rabbits in Australia, 93, 95, 96, 427
Racism, environmental, 9, 500
Radiation phobia, 361

Radioactive decay, 351-52
Radioactive emissions, 349
  biological effects of, 349–50
  relative doses from, 351
  sources of, 350–51
Radioactive wastes, 349
  disposal of, 352, 353–55
  half-live of, 352
  military, 352–53
  radioactive decay and, 351-52
  and states' rights, 354
Radioisotopes, 349
Radon, as air pollutant, 407, 541, 549
Rainfall, effects on biotic communities, 42, 44–45
Rain shadow, 218-19
Ranching, 172
Rancho Seco nuclear plant, CA, 343
Range of tolerance to abiotic factors, 40
Rapa Nui. See Easter Island
Rate of growth, 140
Rathje, William, 468
Raw materials, ecosystem provision of, 76
Raw sludge, 455, 458
Ray, Dixy Lee, 533
Reactivity, 487
Reactors, fusion reactor, 359–60
Reagan, Ronald, 12, 178, 569
Reasonably available control technology (RACT), 557
Recharge area, 220
Recognition stage (policy life cycle), 579
Recombinant DNA technology, 130
Recovery plans, 273
Recreation, ecosystem sources of, 76
Recruitment, 84
Recycling
  curbside, 474, 475
  glass, 476–77
  goals of, 474–75, 481
  increase in, 467, 474
  municipal, 474–76
  paper, 476
  plastics, 474, 477–78
  public policy and, 481–82
  rates of, 475
  symbol for plastics, 477
  and waste reduction, 473–78
Red algae, 30
Red-cockaded woodpecker, 272
Red Data Book, 304–5
Red spruce, 546, 555
Red tides, 451
Reef Check, 306
Refugia, ecosystems and, 76
Regional climatic changes, 522
Regional haze, 546
Regionalized materials recovery facility (MRF/murf), 480
Regulation. See Environmental public policy; Public policy
Regulatory approach, 583
Reindeer population on St. Matthew Island, 89
Relative acidity (pH), 196
Relative humidity, 215
Religion, intrinsic value in wild species and, 269
Remote sensing of forests, 300
Renewable energy. See also Solar energy
  examples of use, 364–67
  geothermal energy, 381–82
  intermittent availability of, 386
  ocean thermal-energy conversion (OTEC), 382
  overview of, 364–67
  public policy and, 382–86
  resource alternatives, 384
  tidal power, 382
  use in the U.S., 367
Renewable natural capital, 575

Renewable resource, 290
  introduction to, 187–88
Replacement level, 84
Replacement-level fertility, 140, 143
Report on Carcinogens (NIEHS), 486
Reproductive isolation, 118
Reproductive strategies, 84
Reserve capacity, 321
Resilience mechanisms, 101-2
Resilient ecosystem, 98–100, 101-2
Resistance, environmental, 84-86
Resource Conservation and Recovery Act (RCRA) of 1976, 478, 500, 501–2
Resource distribution, 576, 577–78
Resource management improvement, 180, 184
Resource partitioning, 40
Resource recovery, in waste-to-energy facilities, 471
Resource Recovery Act of 1970, 478
Resources, as abiotic factors, 40
Respiratory system, 543
Responsible Care Program, 505
Restoration of ecosystems, 47, 295–96
Resurgence, 419-21
Reuse, and waste reduction, 473, 481–82, 505
Revised Management Procedures (RMP), 304
Revolutions
  Environmental, 48–49
  Industrial, 48–49
  Neolithic, 47–48
Rhizobium, 71
Rider, 588
"Right-to-life" advocates, 165
Rio de Janeiro Earth Summit of 1992, 19, 282, 300, 385, 527
Riparian woodlands, 90
Risk, 392
  pathways of
    being poor, 400–401
    infectious diseases, 404–7
    smoking, 401–4
    toxic, 406–7
Risk assessment, 407
  by the EPA, 408–9
  commonplace hazards and, 408
  perception and, 409–12
  public policy and, 411–12
Risk characterization, 409, 486
Risk management, 409
Risk perception, 409–12
Rivers and Harbors Act of 1899, 461
Rivers and streams, 29
R.J. Reynolds, 402
Roadways and animal kills, 270
Rockefeller Foundation, 244, 249
"Roll Back Malaria" campaign (WHO), 406
Rome Declaration on World Food Security, 256
Roosevelt, Theodore, 11
Rosy periwinkle, 267
Rotation
  crop, 202
  forest management and, 297
"Roundup Ready" soybeans, 430
Rowland, Sherwood, 529
Royal Commission on the Future of the Toronto Waterfront, 614
Royal Dutch/Shell, 385
Ruckelshaus, William, 580
Ruminants' production of methane, 247
Runoff, 219
Rural and Urban Center for Human Interest (RUCHI) program, 184
Russia
  death of the Aral Sea, 227, 228
  Karachay, Lake and radioactive wastes, 352–53
  Komsomolsky desertification, 189–90

Safe Drinking Water Act of 1974, 12, 496
"Safe sex," 177
Safety net, 253, 254
Saint Helens, Mount, 126
St. Matthew Island, 89
Salinas, Carlos, 542
Salinization, 196, 206, 207
Salmon species, 227, 274, 613
Salt, soil uptake and, 196
Saltwater intrusion, 230
San Andreas Fault, CA, 126
Sand, 193
Sandman, Peter, 410
Sanitary sewers, 454
San Joaquin Valley, CA, 206
Saro-Wiwa, Ken, 9–10
Savan, Beth, 20
Scale insects, 417, 426
Science
  controversies, 17–18
  evaluation of, 18
  global environment and sound, 6, 14–18
  junk, 14, 18
  method used in, 14-17
  myth of objective, 20
Science Advisory Committee, 579
Science and controversies, 17–18
Science Under Siege (Savan), 20
Scientific community, 18
Scientific controversies, 17–18
Scientific method, 14-17, 20
Scott, Michael, 87
Screwworm, 429
Scrubbers, 559
S-curves, 83–84
Seabird Restoration Project, 263–64
Sea buckthorn, 209
Sea Empress oil spill, 318, 331
Sea level, rise in, 524
Seawater, desalting of, 233
Secondary clarifier tank, 457
Secondary consumers, 31, 32. See also Consumers
Secondary detritus feeders, 33. See also Consumers; Detritivores
Secondary energy source, 320
Secondary-pest outbreaks, 419–21
Secondary pollutants, 547, 549
Secondary production, 65
Secondary recycling, 473–74
Secondary succession, 96–98
Secondary or tertiary recovery, 326
Second basic principle of ecosystem sustainability, 68, 440
Second-generation pesticides, 417
Secondhand (sidestream) smoke
  as carcinogen, 402
  rights of nonsmokers/smokers, 403
Second law of thermodynamics, 60–61, 368
Secure landfill, 493, 494
"Security in one's old age," 165
Sediment, 205, 444–45, 446
SEEA-2000 (U.N. Statistical Office), 577
"Seed banks," 130
Seed Savers Exchange, 130
Seep, 220
Segregation, urban sprawl and, 600–601
Selection
  by the environment, 109–11
  by technology, 113
  natural, 111
  selective pressures and, 110–11, 117–18
  of traits and genes, 111–17
Selective breeding, 109-10, 111
Selective cutting, 297–98
Selective pressures, 110-11, 117, 118
Self-sufficiency in food, 254
Septic system, 460

Serengeti ecosystem, 52–53, 63, 122
Sese Seko, Mobutu, 10
Settling, in water purification, 224
Sewage
  alternative treatment systems
    individual septic system, 460
    reconstructed wetland systems, 461
    using effluents for irrigation, 460–61
  coastal cities and, 458, 459
  contamination
    disease and, 441–43
    monitoring for, 444
  management and treatment, 453–61
    development of collection and treatment systems, 453–54
    history of, 453–54
  pollutants in, 454
  removing pollutants from
    anaerobic digestion, 458–59
    biological nutrient removal, 457
    composting, 459–60
    final cleansing and disinfection, 457–58
    pasteurization, 460
    preliminary treatment, 454
    primary treatment, 454–55
    secondary treatment (biological treatment), 455-57
    sludge treatment, 458-60
  treatment plants, EPA program to upgrade, 451
Sewage treatment plants, EPA program to upgrade, 451
Sex chromosomes, 114
Sex-linked characteristics, 115–16
Sexually transmitted diseases (STDs), 394
Shadow pricing, 586
Shaping Cities: The Environmental and Human Dimensions, (Worldwatch Institute), 604
Sheet erosion, 200
Shelterbelts, 203–4
Shelter-wood cutting, 298
Shoreham Nuclear Power Plant, NY, 343
Short-term containment, 352
Sierra Club, 11, 13
Sievert, 350
Silent emergencies, 209
Silent Spring (Carson), 8, 424, 579
Silt, 193
Silverleaf whitefly (Bemisia argentifolii), 420
Silversword, 272
Silviculture, 297
Simplification of habitat, 277
Singraali coal-burning plants, 171–72
Sinkhole, 229-30
Sitarz, Daniel, 19
Site Working Groups (SWGs), 208
Skillen, James, 588
Slash-and-burn agriculture, 246
Sludge cake, 459, 460
Sludge treatment, 458–60
Smart growth, 600
Smart Growth and Neighborhood Conservation initiative, 600
Smog
  Donora, Pennsylvania and, 537–38, 541
  industrial, 537-39
  Mexico City and, 536–37, 542
  photochemical, 540
  temperature inversions and, 540-41
Smokey Bear campaign, 104
Smoking
  bartenders and, 390, 403
  cultural risk of, 401–4
  deaths caused by, 401
  rights of nonsmokers/smokers, 403
Snake River, and salmon, 230, 231

Snowy egret, 265, 271
Social capital, **575–76**
Social development, 181
Social Modernization, **174**-84
    education and, 174–75
    family planning and, 177–78
    importance of all five components
        of, 180, 181
    improving health and, 175–77
    income enhancement and, 178–79
    resource management improvement
        and, 180, 184
Social progress, 170
Social Security, 152
Sociologists, view of sustainable
    development, 8
Soil, 186–211. *See also* Erosion
    acid deposition and, 554, 555
    addressing degradation of, 207–10
    as an ecosystem, 196–99
    areas subject to degradation, 206
    characteristics of, 191–94
    classes of, 194
    community outreach and
        conservation of, 210
    degradation of, 191, 200–210, 248
    as a detritus-based ecosystem, 197
    effects on biotic communities by, 45
    formation by ecosystems of, 76, 288
    limiting factor and, 196
    mineralization of, 198–99
    organisms in, 196–99
    plants and, 191, 194–96
    profiles of, **192**-93
    public policy and, 208–10
    science, 191–92
    texture of, 193–94
    topsoil importance, 198
Soil aeration, 196
Soil erosion. *See* Erosion
Soil fertility, **194**
Soil particle arrangement, 107
Soil profiles, **192**-93
Soil science, 191–92
Soil separates, 193
Soil structure, **197**, 198
Soil texture, **193**-94
    soil properties and, 193
    triangle, 193
Solar energy
    biomass, 379–81
    building siting and, 370
    disadvantages with, 374
    earth-sheltered housing and, **370**, 371
    economic payoff with, 375
    in ecosystems, 67–68, 72, 73
    electricity production with, 371–74
        hybrid lighting, 373
        photovoltaic (PV) cells, 371–73
        "power tower," 373–74
        solar dish-engine system, 374
        solar-trough collectors, **373**, 374
    heating of water with, 366, 368–69
    hydrogen production with, 375–76
    hydropower, 377–78
    importance for humans, 72–74
    indirect types of, 376–81
    landscaping and, 370, 371
    mirrors used with, 367
    opposition to use of, 371
    principles of, 367–68
    promise of, 374–75
    space heating, 369–71
    spectrum, 368
    transportation fuel and, 375–76, 377
    water heaters and, 369
    water-heating panels and, 366
    wind power as, 378–79
Solectria of Massachusetts, 563
Solid Waste Disposal Act of 1965, 478
Solid waste. *See* Municipal solid waste
    (MSW)
Sound science, 6, 14–18
Sound Science Initiative, 18

Source reduction, 473
South, 1
Southern Africa Development
    Community, 258
Spanworm, 108
Special-interest group, **587**
Speciation, 118
    arctic and gray foxes, 118–19
    Darwin's finches, 107, 120–21
    process of, 121, 122
    rate of, 122
    in tundra plants, 119–20
Species, **26**, **30**, **120**. *See also* Wild
    species
    concept of, 120
    endangered, 86
    geographical distribution of, 123,
        124, 127
    introduced, 84, 93–96, 124, 278–79
    keystone, **281**
    numbers of, 264, 275, 276
    status in U.S. of, 276
    threatened, 86
    umbrella, **281**
Speth, James Gustave, 210
Splash erosion, 200
Splitters, 120
Sport utility vehicles (SUVs), 558, 559,
    562–63
Spotted knapweed, 94, 95
Spotted owl, 13, 274, 298
Spraying
    aerial, 418, 430
    cosmetic, 432
    insurance, **432**
Spring, 220
Stalin, Joseph, 190
Standing crop biomass, 66
*State of Food Insecurity in the World,
    The* (FIVIMS), 256
State Implementation Plan (SIP), 556
Statement on Forest Principles, 300
States' rights and nuclear waste sites,
    354
State of the Word 2000 (Worldwatch
    Institute), 49
*Status and Trends of the Nation's
    Biological Resources, The* (U.S.
    Geological Survey), 35
Steam driven tractors, 320
Steam engine, 319
Stellwagen Bank, 305–6
Sterile males (pest control), 429–30
Stewardship, 6
    the atmosphere and, 526
    biodiversity and, 283–84
    concept of, 8–9
    environmentalism and, 11–14
    ethics of, 9, 311
    health policy and, 412
    justice and equity, 9–11
    leaders in, 8–10
    of life, 128–31
    new commitment to, 18–21
    and public policy, 570
    sustainable development and, 609
    values of, 611
    of water, 236–38
Stomata, **194**, 195
Stop TOxic Pollution (STOP), 10
Stork, Nigel, 264
Storm drains, 454
Storm water
    improving management of, 235–36
    mismanagement of, 233–36
    sewage and, 453–54
Storm-water retention reservoir, **235**-36
Stoves, pellet, 379–80
Stratosphere, **511**, 512
Streams and rivers, 29
Structural proteins, **114**
Subduction, 126
Submerged aquatic vegetation (SAV),
    447

Subsistence agriculture, 245–46
Subsoil, 193
    B Horizon and, 192
Succession, 96–100
Sudan
    famine and, 257, 258
    food aid and, 258
Sukarno, 10
Sulfur dioxide, global cooling and,
    517–18, 524
Sulfuric acid, **552**
Sulfur oxides, 541, 546–49, 552–53,
    554, 560, 561
Superfund
    antienvironmentalists and, 587
    clean up and, 496–98
    evaluation of, 498–500
    landfills in Florida and, 468
    Obee road NPL site, 499
Superfund Act of 1980 (CERCLA),
    478
Superfund Amendments and
    Reauthorization Act (SARA) of
    1986, 503
"Supergreenhouse effect," 520
Supplemental Security Income
    program, 253
Supply and demand, 573
Surface impoundments, 492–93
Surface runoff loop, 220
Surface waters, 219
Surge flow irrigation, **231**
Surgeons general of the U.S., 402
Survival curve, 41
Suspended particulate matter, 541,
    544, 546–49, 557
Sustainability, 6
    communities (human) and, 607–14
    ecosystems as model for, 26, 54–78
    goal of, 310–11
    for the human population, 144
    lifestyle changes and, 611–14
    overview of, 6–8
    principles of ecosystem for, **54**
        first, 68
        second, 68, 440
        third, 90
        fourth, 101
        fifth, 101, 284
Sustainability, sound science and
    stewardship, 310–11
Sustainable agriculture
    ecological agriculture and, 189
    practices used in, 189, 190, 202–3,
        204
    public policy and, 208–10
Sustainable Agriculture Research and
    Education Program, 208
Sustainable America: A New Consensus
    (The President's Council on
    Sustainable Development),
    13–14, 608–10
Sustainable Cities Program (SCP), 607
Sustainable Communities Movement,
    607, 608
Sustainable Communities (Sustainable
    Communities Task Force), 609
Sustainable development, 7
    views of, 8–9
Sustainable ecosystems, 7
Sustainable Fisheries Act of 1996,
    303–4
Sustainable forest management,
    298–99
Sustainable Forestry Certification, 311
Sustainable Forestry Initiative, 298–99
Sustainable logging, **300**
Sustainable society, 7
Sustainable solutions, 8, 609, 611
Sustainable yields, **6**-7
Sustained yield (forestry), 298
Swainson's hawks, 424
Swamp gas, 534
Swamp pink, 272

Symbiosis, **39**
Synergistic, **542**
Synergistic effects (synergisms), **41**
Synthetic fuels, **335**
Synthetic organic compounds, 57
Systematists, 266

Tangier Island covenant, 612
Target Stores' waste reduction, 479
Tariff escalation, 254
Taxol (paclitaxel), 267
Taxonomists, **120**
Taylor Grazing Act of 1934, 205
Technology
    solutions and, 179, 181
    transfer to developing countries,
        183, 385
Tectonic plates, **124**-27
Temperate forests biome, 42, **42**–**43**
Temperature. *See also* Global climate
    change
    effects on biotic communities, 42,
        43–45
    latitude and altitude effects on, 45
Temperature inversions, **540**, 542
Teratogenic, **489**
Terminator technology, GMOs and,
    250–51
Termite, Formosan, 415, 416
Terrestrial-to-aquatic ecotone, 28
*Terrestrial Vegetation of the United
    States* (The Nature Conservancy
    and the Natural Heritage
    Network), 35
Territoriality, 92–93
Tetrachloroethylene (PERC), 489
Theory, 16
Thermal pollution, **323**
The World Bank, poverty in
    developing countries, 141
Third basic principle of ecosystem
    sustainability, **90**
Third World, 141
Thomas W. Danehy Park, Cambridge,
    MA, 465–66
Thomisus spider, 108
Threatened, 86
Threatened species, 86, **272**-73
Three Gorges Dam, China, 231
Three Mile Island, 356, 362
Three-strata forage system (TSFS), 209
Threshold level, **486**, 487
Throwaway society, 479
Tick, deer, 392
Tidal Electric, Inc., 382
Tidal power, 382
Tilman, David, 281
Tilth of the soil, **194**
Tipping fees, **471**, 472
Tisza River cyanide spill, 278
Tobacco Institute, 18
Tobacco. *See* Smoking
Toilet, low flow, 232
Tokaimura, Japan, nuclear facility
    accident, 341–42
Tokamak design nuclear reactors, 359,
    360
Tolerance, limits/range of, 40
Tom McCall Waterfront Park, 606
Topsoil, **192**. *See also* Erosion; Soil
    A horizon and, 192
    formation of, 191, 192
    importance of, 198
Tornado, 396, 397
Total allowable catch (TAC), **287**, 293
Total fertility rate, **140**, 143
Total Maximum Daily Load (TMDL),
    447, 451, 452
Total product life cycle, 487, 488
Towards a Sustainable America (The
    President's Council on
    Sustainable Development),
    609–10
Toxicity, **398**, 487

Toxicology, **486**
Toxics Release Inventory (TRI), 487, 488, 501, 503
Toxic Substances Control Act (TSCA) of 1976, 504
Toxic waste trade
  Basel Convention and, 11
  developing countries and, 10–11
Toyota, 563
Trade imbalance, 254
Trade restriction inequities, as cause of poverty, 10
Trade in wildlife, 280, 282
Trade winds, 217–19
Traditional medicines/healers, 267, 268
"Tragedy of the commons," **205**
  examples of, 293–95
  ocean ecosystems and, 301
Tragedy of the Commons (Hardin), 293–94
Traits, **111**
  and heredity, 113–14
  selection of, 111–16
Transform boundaries, **126**
Transgenic species, 249–51, 430
Transpiration, **194**-95
Transportation Equity Act for the Twenty-first Century (TEA 21) of 1998, 600, 605
Transportation fuel
  fuel cells, 376, 377
  hydrogen gas, 375
  natural gas, 334
Trapping and hunting, 269–70
Trash foraging, 149
Trawling of ocean bottom, 304
Treated sludge, 459
Trenton, NJ, 594
Trickling filter system, **456**
Trimble, Stanley W., 195
TRIS flame retardant, 504
Tritium, 359
Trophic categories, 30–34
  consumers, **31**–**32**, 33, 34
  detritus feeders and decomposers, **33**-34
  producers, **30**-31, 33, 34
Trophic levels (feeding levels), **35**-36
Trophic relationships, 35–37
Trophic structure, changes by humans, 50
Tropical Disease Research program, 405
Tropical forest deforestation, 299–300
Tropical rain forest
  biome, **42–43**, 44
  deforestation and, 145, 146, 205
  impressions of, 22–23
  phosphorus cycle in, 70–71
  ranching in, 172, 247
Tropics, deforestation in, 145, 146, 205, 299–300
Tropopause, **511**
Troposphere, **511**, 512
Trustees of Reservations in Massachusetts, 310
Tsetse fly, 429–30
Tuberculosis, 394–95
Tundra biome, **42–43**, 288
Turbidity, 447, 448
Turbine, **321**
Turbogenerator, **321**
Turner, Ted, 183
Turtle excluder devices (TEDs), 589
2000 Report on Carcinogens (Department on Health and Human Services), 399
2000 World health Report, 412
Typography, effects on biotic communities and, 45

Uganda, AIDS and, **177**
Ultraviolet (UV) radiation
  cancer and, 530, 532

index of, 530, 532
ozone destruction and, 528–29, 531, 532
Umbrella species, **281**
Underground Storage Tank (UST) regulations, 502
Undernourishment, **254**
Uneven-aged forest management, **297–98**
Union Carbide pesticide plant accident, 503
Union of Concerned Scientists (UCS), 18
United Nations. *See also* World Bank
  Center for Human Settlements, 607
  Children's Fund (UNICEF), 170, 178
  Commission on Sustainable Development, 19, 607
  Conference on Environment and Development (UNCED), 19–20
  Convention on Biodiversity, 251
  Convention to Combat Desertification (UNCCD), 201, 209
  Declaration on Human Rights and the Environment, 9
  Development Program, 210, 432
  Earth Summit in Rio de Janeiro, Brazil (1992), 19–20
  Educational, Scientific, and Cultural Organization (UNESCO), 170
  Environmental Program (UNEP)
    and biodiversity, 264
    pest management and, 419, 432–33
    sustainable cities and, 607
  Farmer-centered Agriculture Resource Management (FARM), 209–10
  Food and Agriculture Organization (FAO), 145, 170, 205, 254, 255, 296, 432, 435–36
  Fourth Conference on Women (Beijing 1995), 179
    Microcredit Summit campaign, 178–79
  General Assembly, 20
  international development goals and, 170–71
  money owed by U.S. to, 178
  population conferences
    Bucharest, Romania (1974), 164
    International Conference on Population and Development (Cairo, 1994), 165, 181–84
    Mexico City (1984), 164–65
  Population Division, 138–39, 140
  Population Fund (UNFPA), 178, 183
  Statistical Office, 577
  Subcommittee on Nutrition, 255
  World Commission on Environment and Development, 7
  World Food Conference (1974), 254
  World Food Summit (1996), 254, 256
  World Health Organization (WHO), 170, 223, 255, 267, 400, 401
    air pollution and, 542
    malaria prevention and, 405–6, 419
  world population projections, 152–53
United States
  acid deposition in, 553
  corn yield in, 243
  deaths from cultural hazards in, 394
  distribution of fine particles in, 544
  energy consumption in, 320
  game animals in, 269–71
  money owed to the U.N. by, 178
  nuclear power in, 343

pesticide use in, 417
petroleum products and, 330
population profiles of, 151
population projections of, 154
public and private lands in, 307–11
renewable energy use in, 367
status of species in, 276
temperature changes in, 523
United States agencies
  Agency for International Development
    empowerment of women, 184
    family planning and, 178
    food aid and, 257
    microlending and, 178
  Bureau of Labor Statistics, 582
  Bureau of Land Management (BLM), 205
  Bureau of Reclamation, 206
  Commission on Immigration Reform, 148
  Council on Sustainable Development, 607
  Customs Bureau, 426
  Department of Agriculture
    biotechnology and, 251
    Food Guide Pyramid and, 254, 255
    LISA program and, 208
    natural enemies pest control and, 429
    organic produce guidelines and, 434
    pesticides and, 434–36
  Department of Commerce, 303
  Department of Defense, 499
  Department of Energy
    energy efficiency and, 376, 383
    geothermal energy and, 381–82
    nuclear power and, 342, 352, 354, 355, 362
  Department of the Interior, 207, 284
  Department of Transportation, hazardous chemicals and, 502–3
  Environmental Protection Agency (EPA)
    air pollution and, 545, 546, 547, 556, 557, 559, 560, 561, 562
    biotechnology and, 251
    cost-benefit analyses by, 562, 569
    creation of, 12, 580
    energy and, 383
    hazardous chemicals and, 486–87, 496, 499, 500–502, 504, 505–6
    ozone destruction and, 530, 533
    pesticides and, 422, 434–36
    public policy and, 587
    risk assessment and, 408–9
    risk perception and, 409–10
    smoking regulations of, 402
    solid wastes and, 468–69, 470, 478, 479, 480
    water pollution and, 440, 445, 447, 451–52
  Fish and Wildlife Service, 86
    DDT and, 422
    recovery plans and, 273
    Special Operations, 280
    species recovery and, 263–64, 274
  Food and Drug Administration
    biotechnology and, 251
    pesticides and, 434–36
    smoking as a drug and, 402
  Forest Service
    management practices of, 309–10
    stewardship ethics, 9
  Geological Study, assessment of world oil, 333
  Geological Survey, 35
    management paradigm and, 77, 310
  National Biological Service, 282
  Natural Resource Conservation Service (NRCS), 203, 205
  Soil Conservation Service, 195, 203

United States Congress
  abortion, 178
  ESA, 274
  family planning, 178
  fuel efficiency, 563
  nuclear waste disposal, 353–55
  overgrazing, 205
  public policy, 587–89
  tobacco, 402, 403
United States National Vegetation Classification, USNVC (The Nature Conservancy and the Natural Heritage Network), 35
Unsustainable civilization, 103
Upwelling, **64**
Uranium, 344–46, 358
Urban blight, 600–603
Urban decay, 602–3
Urban gardens, 607
Urban sprawl, **594**
  controlling of, 600
  environmental impacts of, 598–99
  origins of, 594–98
  urban blight and, 600–603
U.S. Electrical of California, 563
Use of natural ecosystems, pattern of, 291–96
U.S.S.R. *See also* Russia
  toxic wastes and, 352, 353, 504

Vaccines, and GMOs, **250**
Valley of the Drums, KY, 493, 495
Value
  of ecosystems, 74–76, 288–90
  of wild species, 264–69
Values, personal, 283–84, 611–14
Value of the World's Ecosystem Services and Natural Capital (13 natural scientists and economists), 74
Vector, **391**
Vehicles. *See* Motor vehicles
Veterans Administration, 596
Virology, **394**
Vision 2000, 608
Voices of the Poor (World Bank), 255
Volatile organic compounds (VOCs), 541, 546–49, 558–59
Volcanic eruptions, 124, 126, 127, 396, 397
Volunteer work, 613

Waitakere City, New Zealand, **592–94**
Warfare, as cause of famine, 257–58
Washington, Lake, 452, 453
Wasps, parasitic, 426, 427, 428
Waste
  sewage. *See* Sewage
  solid. *See* Municipal solid waste (MSW)
  tires, 464, 465, 474
  treatment by ecosystems of, 76, 289
Waste Isolation Pilot Project (WIPP), 354
Wastepaper exports, 476
Waste prevention, 473
Waste reduction, 479, 480
Waste-to-energy (WTE) facilities, 471–72, 480, 481
WasteWise program, 480
Water. *See also* Water pollution
  balance in the hydrologic cycle, 220
  channelization of, **235**
  chlorination of, 412
  conservation of, 238
    irrigation and, 231–32
    municipal systems and, 232
  desalting seawater, 233
  differences in availability, 214–15, 217, 218
  Earth's, 214
  ground and, 219–20
  human impacts on, 221–22, 226–30
  hydrologic cycle, 215–20

obtaining more, 230–31
overdrawing of, 222, 226–30
precipitation, 217–19
purification of, 215, 217, 224
shortages, 226
sources of, 223–26
stewardship of, 236–37
storm water management, 233–36
    flooding and, 233
    improvement in, 235–36
    increased pollution and, 233, 235
    stream-bank erosion and, 233, 234
supply maintenance by ecosystems
    of, 76
terms used to describe, 221
three states of, 56, 57
usage categories of, 223
U.S. demands on, 223
uses of, 223–26
as vital resource, 214
Water, air, mineral interrelationship,
    57
Water cycle, 215. See also Hydrologic
    cycle
Waterfront Regeneration Trust, 614
Water-holding capacity, 196, 201
Water hyacinth, 94, 95, 427
Watermen, 612
Water pollution, 438–63
    controlling of, 440, 441
    essentials of, 440
    eutrophication and, 447–53
    landfill leaching and, 468
    public policy and, 461–62
    sewage management and treatment,
        453–61
    sources and types of, 440–45
        animal-based agriculture, 247
        chemical pollutants, 443–44, 445
        nutrients, 445
        organic wastes, 443
        pathogens, 441–43
        point and nonpoint, 440–42
        sediment, 444–45, 446
    urban sprawl and, 599
    water quality standards and, 445, 447
Water Pollution Control Act of 1948,
    461
Water Quality Act of 1987, 461
Water quality standards, 445, 447
Watershed, 219
Water table, 216, 219
Water vapor, 215
    global warming and, 520
Watts Bar nuclear plant, TN, 343
Wealth of nations, 574–78
Weather, 511
    global climate changes and, 523
    solar energy balance and, 511–12
Weathering, 194

Webb, Patricia, 404
Weeds, 416
Weight gain/loss, 63–64
Welfare measures, 253
Western hemisphere at night, 317
Westinghouse, 342
West Nile Virus, 404
Wet cleaning, 505
Wetlands, 288
    inland, 29
    loss of water to, 227
    reconstruction of, 461
    services of, 289–90
Wetlands Reserve Program, 588
Whale Conservation Institute, 306
Whales, remaining numbers of, 305
Whale watching, 305–6
Whaling, 304–6
Wheat
    traditional vs. high-yielding, 245
    yields, 249
Wheat rust, life cycle of, 426
White flight, 601
White-footed mouse, 391–92
White oryx, 272
Whooping crane, 272, 274
Whyte, William, 604
Wilderness, 307
Wilderness Act of 1964, 307
Wildlife. See also Wild species
    poaching of, 146, 280
    protection of, 12, 86
    urban sprawl and, 599
Wildlife refuges (U.S.), 307–8
Wildlife Services, 271
Wildlife smuggling, 280
Wildlife trafficking, 280
Wild and Scenic Rivers Act of 1968,
    231, 378
Wild species, 262–85. See also
    Biodiversity
    air pollution effects on, 545
    early conservationists, 265
    loss of, 276–81
    saving of, 269–75
    value of, 264–69
        biological wealth, 264
        instrumental, 265
        intrinsic, 266, 269
        recreation, aesthetic, and
            scientific, 267–69
        sources for agriculture, forestry,
            aquaculture, and animal
            husbandry, 266–67
        sources for medicine, 267, 268
Wild turkey restoration, 269
Willis, David, 361
Wilson, E.O., 276
Wind farm, 379
Wind power, 364–65, 378–79

Windrows, 460
Wind turbine, 378
Winged bean, 266
Wingspread Conference of 1998, 411
Wise-use movement, 13
Wolf
    as ancestor of dogs, 110
    feeding on elk, 32
    Mexican gray, 265
    population on Isle Royale, 87–88
    reintroduction of, 265
Women
    contraceptives availability and, 167
    education of, 167, 174
    empowerment of, 182–83, 184
    family size and, 165–68
    importance in development of, 174
    key to development, 184
    microlending to, 178–79
    poverty and, 147, 148, 149, 165–68
    status in developing countries, 167
    work in developing countries, 166
Wood consumption (world), 297
Workability, 193
Worker protection, 503
Worker Protection Standard program,
    434
Worker's right to know, 503
"Workfare," 253
World, population projections, 153
World Bank
    Aral Sea and, 228
    biodiversity and, 282, 283
    Consultative Group on International
        Agricultural Research, 130
    Consultative Group to Assist the
        Poorest (CGAP), 173
    economic categories of countries,
        141
    Environmental Department, 173
    FARM support and, 210
    formation of, 168
    Heavily Indebted Poor Country
        initiative (HIPC), 174
    ICPD Program of Action, 183
    integrated pest management and,
        432–33
    international development goals
        and, 170–71
    loans and, 168
    microlending and, 178
    Nam Theun Two Dam, 378
    nation's wealth and, 575, 576
    natural disasters and, 397
    "Poverty Reduction Strategy," 174
    Reform of, 173–74
    successes and failures of, 169–72
    Three Gorges Dam, 231, 378
World Commission on Water for the
    21st-Century, 237

World Conservation Union, 276
World Economic Forum, 528
World Food Program, 258
World Food Summit, 254, 256
World Food Summit Plan of Action,
    256
World Glacier Monitoring Service, 524
World Health Report 1999 (World
    Health Organization), 223
World Ozone Data Center, 533
World Resources Institute, 203, 204,
    495, 542
World Trade Organization (WTO),
    588–89
World Vision, 258
Worldwatch Institute
    and consumption classes, 141, 142
    livable cities and, 604
    our unsustainable economy and, 49
    poverty and, 256
World Water Council's Vision report,
    237
World Water Forum, 237
World wealth, composition of, 577
World Wildlife Fund, 86, 280, 299,
    304
Worms, parasitic, 32, 394, 396
Wright, Richard T., 526
WTO (World Trade Organization),
    588–89

X chromosome, 114
Xeriscaping, 232
Xerox and EPR, 480

Y chromosome, 114
"Year the earth caught fire" (1997), 299
Yellowstone National Park
    fires and, 80–82, 100
    Greater Yellowstone Coalition, 308
    wolf reintroduction in, 265
Yeltsin, Boris, 353
Yew, Pacific and English, 267
Yields, 6–7, 298
Younger Dryas event, 514, 516
Yucca Mountain, NV, nuclear waste
    depository, 353–55
Yunus, Muhammad, 178

Zebra mussels, 94
Zero-dose, zero response approach,
    486
Zero Population Growth, 12
Zion National Park, 307
Z machine nuclear reactor, 359–60
Zones of stress (abiotic factors), 40
Zonn, Igor, 228
Zooplankton, 64
Zooxanthellae, 306